Advances in Bio

**Related titles:**

*Waste to energy conversion technology*
(ISBN 978-0-85709-011-9)

*Alternative fuels and advanced vehicle technologies for improved environmental performance*
(ISBN 978-0-85709-522-0)

*Advances in hydrogen production, storage and utilization*
(ISBN 978-0-85709-768-2)

Woodhead Publishing Series in Energy: Number 53

# Advances in Biorefineries

Biomass and Waste Supply Chain Exploitation

Edited by
Keith Waldron

AMSTERDAM • BOSTON • CAMBRIDGE • HEIDELBERG • LONDON
NEW YORK • OXFORD • PARIS • SAN DIEGO
SAN FRANCISCO • SINGAPORE • SYDNEY • TOKYO
Woodhead Publishing is an imprint of Elsevier

Woodhead Publishing is an imprint of Elsevier
80 High Street, Sawston, Cambridge, CB22 3HJ, UK
225 Wyman Street, Waltham, MA 02451, USA
Langford Lane, Kidlington, OX5 1GB, UK

Copyright © 2014 Woodhead Publishing Limited. All rights reserved
Exceptions to the above:
Chapter 13: © 2014 R. J. Pearson and J. W. G. Turner. Published by Woodhead Publishing Limited.

No part of this publication may be reproduced, stored in a retrieval system or transmitted in any form or by any means electronic, mechanical, photocopying, recording or otherwise without the prior written permission of the publisher.
Permissions may be sought directly from Elsevier's Science & Technology Rights Department in Oxford, UK: phone (+44) (0) 1865 843830; fax (+44) (0) 1865 853333; email: permissions@elsevier.com. Alternatively you can submit your request online by visiting the Elsevier website at http://elsevier.com/locate/permissions, and selecting Obtaining permission to use Elsevier material.

**Notice**
No responsibility is assumed by the publisher for any injury and/or damage to persons or property as a matter of products liability, negligence or otherwise, or from any use or operation of any methods, products, instructions or ideas contained in the material herein. Because of rapid advances in the medical sciences, in particular, independent verification of diagnoses and drug dosages should be made.

**British Library Cataloguing-in-Publication Data**
A catalogue record for this book is available from the British Library

**Library of Congress Control Number:** 2014931606

ISBN 978-0-08-101381-6 (print)
ISBN 978-0-85709-738-5 (online)

For information on all Woodhead Publishing publications
visit our website at http://store.elsevier.com/

Typeset by Toppan Best-set Premedia Limited, Hong Kong

Transferred to Digital Printing in 2016

# Contents

|  |  |  |
|---|---|---|
| *Contributor contact details* | | *xv* |
| *Woodhead Publishing Series in Energy* | | *xxiii* |
| *Foreword* | | *xxix* |
| *Preface* | | *xxxiii* |

| Part I | Development and optimisation of biorefining processes | 1 |
|---|---|---|
| 1 | Green chemistry, biorefineries and second generation strategies for re-use of waste: an overview<br>L. A. PFALTZGRAFF and J. H. CLARK, University of York, UK | 3 |
| 1.1 | Introduction | 3 |
| 1.2 | Introduction to biorefineries | 14 |
| 1.3 | New renewable feedstocks | 17 |
| 1.4 | Conclusion and future trends | 28 |
| 1.5 | Sources of further information and advice | 28 |
| 1.6 | References | 29 |
| 2 | Techno-economic assessment (TEA) of advanced biochemical and thermochemical biorefineries<br>T. R. BROWN, Iowa State University, USA,<br>M. M. WRIGHT and Y. ROMÁN-LESHKOV, Massachusetts Institute of Technology, USA and<br>R. C. BROWN, Iowa State University, USA | 34 |
| 2.1 | Introduction | 34 |
| 2.2 | Biorefinery economic assessment | 36 |
| 2.3 | Trade of biomass and subsidies | 47 |
| 2.4 | Market establishment: national/regional facilities | 53 |
| 2.5 | Conclusion and future trends | 59 |
| 2.6 | References | 63 |

| 3 | Environmental and sustainability assessment of biorefineries | 67 |

L. SCHEBEK and O. MRANI, Technische Universität Darmstadt, Germany

| 3.1 | Introduction | 67 |
| 3.2 | Methodological foundations of environmental and sustainability assessment of technologies | 68 |
| 3.3 | Life cycle assessment (LCA) for biorefineries | 74 |
| 3.4 | Sustainability issues: synopsis of results from assessment of economic and social aspects | 81 |
| 3.5 | Conclusion and future trends | 83 |
| 3.6 | References | 84 |

| 4 | Biorefinery plant design, engineering and process optimisation | 89 |

J. B. HOLM-NIELSEN and E. A. EHIMEN, Aalborg University, Denmark

| 4.1 | Introduction | 89 |
| 4.2 | Microalgae biomass for biorefinery systems | 91 |
| 4.3 | Planning, design and development of biorefinery systems | 92 |
| 4.4 | Case study: a second generation lignocellulosic biorefinery (Inbicon® Biorefinery) | 101 |
| 4.5 | Upgrading biorefinery operations | 104 |
| 4.6 | Optimising biorefinery processes using process analysis | 106 |
| 4.7 | Conclusion and future trends | 107 |
| 4.8 | References | 108 |

| 5 | Current and emerging separations technologies in biorefining | 112 |

S. DATTA, Y. J. LIN and S. W. SNYDER, Argonne National Laboratory, USA

| 5.1 | Introduction | 112 |
| 5.2 | Separations technologies | 114 |
| 5.3 | Removal of impurities from lignocellulosic biomass hydrolysate liquor for production of cellulosic sugars | 121 |
| 5.4 | Glycerin desalting as a value added co-product from biodiesel production | 126 |
| 5.5 | Succinic acid production | 128 |
| 5.6 | Solvent extraction: the example of recovery of value added proteins from distiller's grains and solubles (DGS) | 130 |

| 5.7 | Biofuels recovery by solvent extraction in an ionic liquid assisted membrane contactor | 134 |
|---|---|---|
| 5.8 | Emerging trends in separations technology for advanced biofuels | 141 |
| 5.9 | Performance indices | 144 |
| 5.10 | Conclusion | 147 |
| 5.11 | Acknowledgements | 148 |
| 5.12 | References | 148 |

| 6 | **Catalytic processes and catalyst development in biorefining**<br>S. Morales-Delarosa and J. M. Campos-Martin, Instituto de Catálisis y Petroleoquímica, CSIC, Spain | **152** |
|---|---|---|
| 6.1 | Introduction | 152 |
| 6.2 | Catalysts for depolymerization of biomass | 153 |
| 6.3 | Catalysts for biomass products upgrading | 160 |
| 6.4 | Conclusion and future trends | 184 |
| 6.5 | References | 186 |

| 7 | **Enzymatic processes and enzyme development in biorefining**<br>S. A. Teter, K. Brandon Sutton and B. Emme, Novozymes, USA | **199** |
|---|---|---|
| 7.1 | Introduction | 199 |
| 7.2 | Biochemical conversion | 200 |
| 7.3 | Development of enzyme technology and techniques | 211 |
| 7.4 | Optimizing enzymes | 212 |
| 7.5 | Benchmarking enzymes and enzymatic conversion processes | 220 |
| 7.6 | Advantages and limitations of techniques | 225 |
| 7.7 | Conclusion and future trends | 225 |
| 7.8 | Sources of further information and advice | 226 |
| 7.9 | References | 226 |

| 8 | **Biomass pretreatment for consolidated bioprocessing (CBP)**<br>V. Agbor, C. Carere, N. Cicek, R. Sparling and D. Levin, University of Manitoba, Canada | **234** |
|---|---|---|
| 8.1 | Introduction | 234 |
| 8.2 | Process configurations for biofuel production | 235 |

| | | |
|---|---|---|
| 8.3 | Models for consolidated bioprocessing (CBP) | 243 |
| 8.4 | Microorganisms, enzyme systems, and bioenergetics of CBP | 245 |
| 8.5 | Organism development | 249 |
| 8.6 | Conclusion | 253 |
| 8.7 | References | 254 |

## 9 Developments in bioethanol fuel-focused biorefineries 259
S. Mutturi, B. Palmqvist and G. Lidén, Lund University, Sweden

| | | |
|---|---|---|
| 9.1 | Introduction | 259 |
| 9.2 | Ethanol biorefineries | 261 |
| 9.3 | The lignocellulose to ethanol process | 263 |
| 9.4 | Design options for biorefining processes | 279 |
| 9.5 | Process intensification: increasing the dry-matter content | 280 |
| 9.6 | Different types of ethanol biorefineries | 282 |
| 9.7 | Future trends | 288 |
| 9.8 | Conclusion | 294 |
| 9.9 | Sources of further information and advice | 294 |
| 9.10 | References | 295 |

## 10 Developments in cereal-based biorefineries 303
A. A. Koutinas, Agricultural University of Athens, Greece, C. Du, University of Nottingham, UK, C. S. K. Lin, City University of Hong Kong, Hong Kong and C. Webb, University of Manchester, UK

| | | |
|---|---|---|
| 10.1 | Introduction | 303 |
| 10.2 | Wheat-based biorefineries | 304 |
| 10.3 | Fuel ethanol production from wheat | 308 |
| 10.4 | Succinic acid production from wheat | 312 |
| 10.5 | Polyhydroxyalkanoate (PHA) production from wheat | 316 |
| 10.6 | Utilization of wheat straw | 320 |
| 10.7 | Conclusion and future trends | 322 |
| 10.8 | Sources of further information and advice | 326 |
| 10.9 | Acknowledgements | 328 |
| 10.10 | References | 328 |

| 11 | Developments in grass-/forage-based biorefineries | 335 |
|---|---|---|

J. McEniry and P. O'Kiely, Teagasc (Irish Agriculture and Food Development Authority), Ireland

| 11.1 | Introduction | 335 |
|---|---|---|
| 11.2 | Overview of grass-/forage-based biorefineries | 336 |
| 11.3 | Field to biorefinery – impact of herbage chemical composition | 340 |
| 11.4 | Green biorefinery products | 347 |
| 11.5 | Acknowledgements | 355 |
| 11.6 | References | 355 |

| 12 | Developments in glycerol byproduct-based biorefineries | 364 |
|---|---|---|

B. P. Pinto and C. J. De Araujo Mota, Universidade Federal do Rio de Janeiro, Brazil

| 12.1 | Introduction | 364 |
|---|---|---|
| 12.2 | Composition and purification of glycerol produced from biodiesel | 365 |
| 12.3 | Applications of glycerol in the fuel sector | 367 |
| 12.4 | Glycerol as raw material for the chemical industry | 372 |
| 12.5 | Conclusions and future trends | 379 |
| 12.6 | Sources of further information | 381 |
| 12.7 | References | 382 |

| Part II | Biofuels and other added value products from biorefineries | 387 |
|---|---|---|

| 13 | Improving the use of liquid biofuels in internal combustion engines | 389 |
|---|---|---|

R. J. Pearson and J. W. G. Turner, University of Bath, UK

| 13.1 | Introduction | 389 |
|---|---|---|
| 13.2 | Competing fuels and energy carriers | 390 |
| 13.3 | Market penetration of liquid biofuels | 394 |
| 13.4 | Use of liquid biofuels in internal combustion engines | 405 |
| 13.5 | Vehicle and blending technologies for alcohol fuels and gasoline | 417 |
| 13.6 | Future provision of renewable liquid fuels | 423 |
| 13.7 | Conclusion | 429 |

| | | |
|---|---|---|
| 13.8 | Acknowledgements | 429 |
| 13.9 | References and further reading | 430 |
| 13.10 | Appendix: List of abbreviations | 439 |

**14 Biodiesel and renewable diesel production methods** — 441
J. H. VAN GERPEN and B. B. HE,
University of Idaho, USA

| | | |
|---|---|---|
| 14.1 | Introduction | 441 |
| 14.2 | Overview of biodiesel and renewable diesel | 442 |
| 14.3 | Renewable diesel production routes | 442 |
| 14.4 | Biodiesel production routes | 444 |
| 14.5 | Traditional and emerging feedstocks | 454 |
| 14.6 | Feedstock quality issues | 458 |
| 14.7 | Advantages and limitations of biodiesel | 461 |
| 14.8 | Conclusion and future trends | 465 |
| 14.9 | Sources of further information and advice | 465 |
| 14.10 | References | 466 |

**15 Biomethane and biohydrogen production via anaerobic digestion/fermentation** — 476
K. STAMATELATOU, Democritus University of Thrace, Greece, G. ANTONOPOULOU, Institute of Chemical Engineering Sciences, Greece and P. MICHAILIDES, Democritus University of Thrace, Greece

| | | |
|---|---|---|
| 15.1 | Introduction | 476 |
| 15.2 | Basic principles of biogas and hydrogen production | 477 |
| 15.3 | Biogas and biohydrogen production: technological aspects | 481 |
| 15.4 | Production of biogas (methane) and biohydrogen from different feedstocks | 492 |
| 15.5 | Current status and limitations | 507 |
| 15.6 | Future trends | 513 |
| 15.7 | Sources of further information and advice | 513 |
| 15.8 | References | 514 |

**16 The production and application of biochar in soils** — 525
S. JOSEPH, University of New South Wales, Australia and P. TAYLOR, Biochar Solutions, Australia

| | | |
|---|---|---|
| 16.1 | Introduction | 525 |
| 16.2 | Effects of application of biochar to soil | 527 |

| | | |
|---|---|---|
| 16.3 | Agricultural uses of biochar | 529 |
| 16.4 | Production of biochar | 531 |
| 16.5 | Large-scale commercial production of biochar | 538 |
| 16.6 | Testing biochar properties | 541 |
| 16.7 | Markets and uses for biochar | 544 |
| 16.8 | Conclusion and future trends | 548 |
| 16.9 | References | 550 |
| 16.10 | Appendix: IBI; Standardized product definition and product testing guidelines for biochar used in soil | 554 |

| | | |
|---|---|---|
| **17** | **Development, properties and applications of high-performance biolubricants** D. R. KODALI, University of Minnesota, USA | **556** |
| 17.1 | Introduction | 556 |
| 17.2 | Markets for lubricants | 561 |
| 17.3 | Biolubricant performance requirements | 566 |
| 17.4 | Applications of biolubricants | 571 |
| 17.5 | Feedstocks for biolubricants: key properties | 574 |
| 17.6 | Chemical modifications of biolubricant feedstocks | 581 |
| 17.7 | Future trends | 590 |
| 17.8 | Conclusion | 591 |
| 17.9 | Acknowledgements | 592 |
| 17.10 | References | 592 |

| | | |
|---|---|---|
| **18** | **Bio-based nutraceuticals from biorefining** Y. LIANG and Z. WEN, Iowa State University, USA | **596** |
| 18.1 | Introduction | 596 |
| 18.2 | Lipid-based nutraceuticals | 599 |
| 18.3 | Protein and peptide-based nutraceuticals | 604 |
| 18.4 | Carbohydrate-based nutraceuticals | 606 |
| 18.5 | Other nutraceuticals | 609 |
| 18.6 | Conclusion and future trends | 614 |
| 18.7 | References | 615 |

| | | |
|---|---|---|
| **19** | **Bio-based chemicals from biorefining: carbohydrate conversion and utilisation** K. WILSON, European Bioenergy Research Institute, Aston University, UK and A. F. LEE, University of Warwick, UK and Monash University, Australia | **624** |
| 19.1 | Introduction | 624 |
| 19.2 | Sustainable carbohydrate sources | 625 |

| | | |
|---|---|---|
| 19.3 | Chemical hydrolysis of cellulose to sugars | 629 |
| 19.4 | Types and properties of carbohydrate-based chemicals | 635 |
| 19.5 | Routes to market for bio-based feedstocks | 646 |
| 19.6 | Conclusion and future trends | 650 |
| 19.7 | Sources of further information and advice | 651 |
| 19.8 | References | 651 |

## 20 Bio-based chemicals from biorefining: lignin conversion and utilisation 659
A. L. MACFARLANE, M. MAI and J. F. KADLA, University of British Columbia, Canada

| | | |
|---|---|---|
| 20.1 | Introduction | 659 |
| 20.2 | Structure and properties of lignin | 660 |
| 20.3 | Traditional processes for the production of lignin | 663 |
| 20.4 | Emerging processes for the production of lignin | 668 |
| 20.5 | Applications of lignin and lignin-based products: an overview | 672 |
| 20.6 | Future trends | 684 |
| 20.7 | Sources of further information and advice | 685 |
| 20.8 | References | 685 |

## 21 Bio-based chemicals from biorefining: lipid and wax conversion and utilization 693
Y. YANG and B. HU, University of Minnesota, USA

| | | |
|---|---|---|
| 21.1 | Introduction | 693 |
| 21.2 | Types and properties of lipids and waxes | 694 |
| 21.3 | Sources of lipids and waxes | 697 |
| 21.4 | Methods to extract and analyze lipids and waxes | 703 |
| 21.5 | Utilization of lipids and waxes | 707 |
| 21.6 | Conclusion and future trends | 714 |
| 21.7 | References | 715 |

## 22 Bio-based chemicals from biorefining: protein conversion and utilisation 721
E. L. SCOTT, M. E. BRUINS and J. P. M. SANDERS, Wageningen University, The Netherlands

| | | |
|---|---|---|
| 22.1 | Introduction | 721 |
| 22.2 | Protein and amino acid sources derived from biofuel production | 722 |
| 22.3 | Protein isolation, hydrolysis and isolation of amino acid chemical feedstocks | 724 |

| | | |
|---|---|---|
| 22.4 | (Bio)chemical conversion of amino acids to platform and speciality chemicals | 728 |
| 22.5 | Alternative and novel feedstocks and production routes | 730 |
| 22.6 | Conclusion and future trends | 731 |
| 22.7 | References | 732 |

## 23 Types, processing and properties of bioadhesives for wood and fibers  736
A. Pizzi, University of Lorraine, France and King Abdulaziz University, Saudi Arabia

| | | |
|---|---|---|
| 23.1 | Introduction | 736 |
| 23.2 | Tannin adhesives | 737 |
| 23.3 | Lignin adhesives | 745 |
| 23.4 | Mixed tannin-lignin adhesives | 749 |
| 23.5 | Protein adhesives | 750 |
| 23.6 | Carbohydrate adhesives | 751 |
| 23.7 | Unsaturated oil adhesives | 752 |
| 23.8 | Wood welding without adhesives | 755 |
| 23.9 | Conclusion and future trends | 762 |
| 23.10 | References | 765 |

## 24 Types, properties and processing of bio-based animal feed  771
E. J. Burton and D. V. Scholey, Nottingham Trent University, UK and P. E. V. Williams, AB Vista, UK

| | | |
|---|---|---|
| 24.1 | Introduction | 771 |
| 24.2 | Background | 772 |
| 24.3 | Types and properties of bio-based feed ingredients | 775 |
| 24.4 | Impact of processing technology on co-product quality | 784 |
| 24.5 | Improving feedstocks processes and yields | 786 |
| 24.6 | Regulatory issues | 789 |
| 24.7 | Future trends | 791 |
| 24.8 | Sources of further information and advice | 794 |
| 24.9 | References | 794 |

## 25 The use of biomass to produce bio-based composites and building materials  803
R. M. Rowell, University of Wisconsin, USA

| | | |
|---|---|---|
| 25.1 | Introduction | 803 |
| 25.2 | Fibrous plants | 804 |

| | | |
|---|---|---|
| 25.3 | Fiber types and isolation | 804 |
| 25.4 | Fiber properties | 807 |
| 25.5 | Types and properties of bio-based composites | 808 |
| 25.6 | Improving performance properties | 815 |
| 25.7 | Conclusion and future trends | 817 |
| 25.8 | Sources of further information and advice | 817 |
| 25.9 | References | 818 |
| **26** | **The use of biomass for packaging films and coatings** | **819** |
| | H. M. C. DE AZEREDO, Institute of Food Research, UK, M. F. ROSA and M. DE SÁ M. SOUZA FILHO, Embrapa Tropical Agroindustry, Brazil and K. W. WALDRON, Institute of Food Research, UK | |
| 26.1 | Introduction | 819 |
| 26.2 | Components of packaging films and coatings from the biomass | 822 |
| 26.3 | Processes for producing bio-based films | 822 |
| 26.4 | Processes for producing edible coatings | 825 |
| 26.5 | Products from biomass as film and/or coating matrices | 826 |
| 26.6 | Products from biomass as film plasticizers | 842 |
| 26.7 | Products from biomass as crosslinking agents for packaging materials | 844 |
| 26.8 | Products from biomass as reinforcements for packaging materials | 849 |
| 26.9 | Future trends | 853 |
| 26.10 | Conclusion | 855 |
| 26.11 | Acknowledgements | 856 |
| 26.12 | References | 856 |
| | *Index* | *875* |

# Contributor contact details

(* = main contact)

## Editor and Chapter 26

K. Waldron
Institute of Food Research
Norwich Research Park
Colney
Norwich NR4 7UA, UK

E-mail: keith.waldron@ifr.ac.uk

## Chapter 1

L. A. Pfaltzgraff and J. H. Clark*
Department of Chemistry
University of York
Heslington
York YO10 5DD, UK

E-mail: james.clark@york.ac.uk

## Chapter 2

T. R. Brown and R. C. Brown
Iowa State University
1140 Biorenewables Research
  Laboratory
Bioeconomy Institute
Ames, IA 50011, USA

M. M. Wright* and
  Y. Román-Leshkov
Massachusetts Institute of
  Technology
Department of Chemical
  Engineering
25 Ames Street
Cambridge, MA 02142, USA

E-mail: markjet7@gmail.com

## Chapter 3

L. Schebek* and O. Mrani
Technische Universität Darmstadt
Institut IWAR
Fachgebiet Industrielle
 Stoffkreisläufe
Petersenstr. 13
64287 Darmstadt, Germany

E-mail: l.schebek@iwar.
 tu-darmstadt.de; o.mrani@iwar.
 tu-darmstadt.de

## Chapter 4

J. B. Holm-Nielsen* and
 E. A. Ehimen
Department of Energy Technology
Aalborg University
Esbjerg Campus
Niels Bohrsvej 8
6700 Esbjerg, Denmark

E-mail: jhn@et.aau.dk

## Chapter 5

S. Datta, Y. J. Lin and S. W. Snyder*
Process Technology Research
Argonne National Laboratory
9700 South Cass Avenue
Argonne, IL 60439, USA

E-mail: seth@anl.gov

## Chapter 6

S. Morales-Delarosa and
 J. M. Campos-Martin*
Sustainable Energy and Chemistry
 Group (EQS)
Instituto de Catálisis y
 Petroleoquímica, CSIC
c/ Marie Curie, 2
Cantoblanco, 28049 Madrid, Spain

E-mail: j.m.campos@icp.csic.es

## Chapter 7

S. A. Teter*
Novozymes, Inc.
1445 Drew Ave
Davis, CA 95618, USA

E-mail: sate@novozymes.com

K. Brandon Sutton and B. Emme
Novozymes North America
PO Box 576
77 Perry Chapel Church Road
Franklinton, NC 27525, USA

E-mail: kteb@novozymes.com;
 bemm@novozymes.com

## Chapter 8

V. Agbor, C. Carere, N. Cicek and D. Levin*
Department of Biosystems Engineering
University of Manitoba
E2-376 EITC
Winnipeg
Manitoba R3T 5V6, Canada

E-mail: David.Levin@ad.umanitoba.ca; bisong01@gmail.com; carlo.carere@gmail.com; Nazim.Cicek@ad.umanitoba.ca

R. Sparling
Department of Microbiology
University of Manitoba
Winnipeg
Manitoba R3T 2N2, Canada

E-mail: Richard.Sparling@ad.umanitoba.ca

## Chapter 9

S. Mutturi, B. Palmqvist and G. Lidén*
Department of Chemical Engineering
Lund University
PO Box 124
221 00 Lund, Sweden

E-mail: Gunnar.liden@chemeng.lth.se

## Chapter 10

A. A. Koutinas*
Department of Food Science and Human Nutrition
Agricultural University of Athens
Iera Odos 75
11855, Athens, Greece

E-mail: akoutinas@aua.gr

C. Du
School of Biosciences
University of Nottingham
Sutton Bonington Campus
Sutton Bonington LE12 5RD, UK

E-mail: chenyu.du@nottingham.ac.uk

C. S. K. Lin
School of Energy and Environment
City University of Hong Kong
Tat Chee Avenue
Kowloon, Hong Kong

E-mail: carollin@cityu.edu.hk

Colin Webb
School of Chemical Engineering and Analytical Science
University of Manchester
Manchester M13 9PL, UK

E-mail: colin.webb@manchester.ac.uk

## Chapter 11

J. McEniry and P. O'Kiely*
Animal and Grassland Research
  and Innovation Centre
Teagasc (Irish Agriculture and
  Food Development Authority)
Grange
Dunsany, Co. Meath, Ireland

E-mail: josephmceniry@gmail.com;
  padraig.okiely@teagasc.ie

## Chapter 12

B. P. Pinto and
  C. J. de Araujo Mota*
Universidade Federal do Rio de
  Janeiro
Instituto de Química
Av Athos da Silveira Ramos 149
CT Bloco A
Cidade Universitária
Rio de Janeiro, 21941-909, Brazil

E-mail: cmota@iq.ufrj.br

## Chapter 13

R. J. Pearson* and J. W. G. Turner
Department of Mechanical
  Engineering
University of Bath
Claverton Down
Bath BA2 7AY, UK

E-mail: thepearsons7ar@tiscali.
  co.uk

## Chapter 14

J. H. Van Gerpen and B. B. He
University of Idaho
Moscow, ID 83844, USA

E-mail: jonvg@uidaho.edu

## Chapter 15

K. Stamatelatou*
Department of Environmental
  Engineering
Democritus University of Thrace
Vas. Sofias 12
67100 Xanthi, Greece

E-mail: astamat@env.duth.gr

G. Antonopoulou
Institute of Chemical Engineering
  Sciences
Stadiou Str., Platani
PO Box 1414
GR-26504 Patras, Greece

E-mail: geogant@chemeng.
  upatras.gr

## Chapter 16

S. Joseph*
School of Materials Science and
    Engineering
University of New South Wales
5 Kenneth Ave
NSW 2251, Australia

E-mail: joey.stephen@gmail.com

P. Taylor
Biochar Solutions
73 Mt Warning Rd
Mt Warning 2484, Australia

## Chapter 17

D. R. Kodali
Department of Bioproducts and
    Biosystems Engineering
University of Minnesota
2004 Folwell Avenue
St. Paul, MN 55108, USA
and
Global Agritech Inc.
710 Olive Ln. N.
Minneapolis, MN 55447, USA

E-mail: dkodali@umn.edu

## Chapter 18

Y. Liang and Z. Wen*
Department of Food Science and
    Human Nutrition
Iowa State University
Ames, IA 50011, USA

E-mail: wenz@iastate.edu

## Chapter 19

K. Wilson*
European Bioenergy Research
    Institute
School of Engineering and Applied
    Science
Aston University
Birmingham B4 7ET, UK

E-mail: k.wilson@aston.ac.uk

A. F. Lee
Department of Chemistry
University of Warwick
Coventry CV4 7AL, UK
and
School of Chemistry
Monash University
Victoria 3800, Australia

E-mail: A.F.Lee@warwick.ac.uk

## Chapter 20

A. L. Macfarlane, M. Mai,
   J. F. Kadla*
University of British Columbia
Vancouver, BC, Canada

E-mail: john.kadla@shaw.ca

## Chapter 21

Y. Yang and B. Hu*
Department of Bioproducts and
   Biosystems Engineering
University of Minnesota
316 BAE
1390 Eckles Ave
Saint Paul, MN 55108-6005, USA

E-mail: bhu@umn.edu

## Chapter 22

E. L. Scott*, M. E. Bruins and
   J. P. M. Sanders
Biobased Commodity Chemistry
Wageningen University
PO Box 17
6700 AA Wageningen, The
   Netherlands

E-mail: Elinor.Scott@wur.nl

## Chapter 23

A. Pizzi
ENSTIB-LERMAB
University of Lorraine
27 rue Philippe Seguin
BP 1041 88051
Epinal, cedex 9, France

and

Department of Physics King
Abdulaziz University
Jeddah Saudi Arabia

E-mail: antonio.pizzi@univ-lorraine.fr

## Chapter 24

E. J. Burton* and D. V. Scholey
School of Animal, Rural and
   Environmental Sciences
Nottingham Trent University
Nottingham NG25 0QF, UK

E-mail: Emily.Burton@ntu.ac.uk

P. E. V. Williams
AB Vista – A Division of AB Agri
   Ltd
64 Innovation Way
Peterborough Business Park
Lynch Wood
Peterborough PE2 6FL, UK

E-mail: DrPeter.Williams@abvista.com

## Chapter 25

R. M. Rowell
Biological Systems Engineering
University of Wisconsin
4510 Gregg Road
Madison, WI 53705, USA

E-mail: rmrowell@wisc.edu

## Chapter 26

H. M. C. de Azeredo and
    K. W. Waldron*
Institute of Food Research
Norwich Research Park
Colney
Norwich NR4 7UA, UK

E-mail: keith.waldron@ifr.ac.uk

M. F. Rosa and M. de Sá M. Souza
    Filho
Embrapa Tropical Agroindustry
Brazil

# Woodhead Publishing Series in Energy

1 **Generating power at high efficiency: Combined cycle technology for sustainable energy production**
   *Eric Jeffs*

2 **Advanced separation techniques for nuclear fuel reprocessing and radioactive waste treatment**
   *Edited by Kenneth L. Nash and Gregg J. Lumetta*

3 **Bioalcohol production: Biochemical conversion of lignocellulosic biomass**
   *Edited by Keith W. Waldron*

4 **Understanding and mitigating ageing in nuclear power plants: Materials and operational aspects of plant life management (PLiM)**
   *Edited by Philip G. Tipping*

5 **Advanced power plant materials, design and technology**
   *Edited by Dermot Roddy*

6 **Stand-alone and hybrid wind energy systems: Technology, energy storage and applications**
   *Edited by John K. Kaldellis*

7 **Biodiesel science and technology: From soil to oil**
   *Jan C. J. Bart, Natale Palmeri and Stefano Cavallaro*

8 **Developments and innovation in carbon dioxide ($CO_2$) capture and storage technology Volume 1: Carbon dioxide ($CO_2$) capture, transport and industrial applications**
   *Edited by M. Mercedes Maroto-Valer*

9 **Geological repository systems for safe disposal of spent nuclear fuels and radioactive waste**
   *Edited by Joonhong Ahn and Michael J. Apted*

10 **Wind energy systems: Optimising design and construction for safe and reliable operation**
    *Edited by John D. Sørensen and Jens N. Sørensen*

11 **Solid oxide fuel cell technology: Principles, performance and operations**
   *Kevin Huang and John Bannister Goodenough*

12 **Handbook of advanced radioactive waste conditioning technologies**
   *Edited by Michael I. Ojovan*

13 **Membranes for clean and renewable power applications**
   *Edited by Annarosa Gugliuzza and Angelo Basile*

14 **Materials for energy efficiency and thermal comfort in buildings**
   *Edited by Matthew R. Hall*

15 **Handbook of biofuels production: Processes and technologies**
   *Edited by Rafael Luque, Juan Campelo and James Clark*

16 **Developments and innovation in carbon dioxide ($CO_2$) capture and storage technology Volume 2: Carbon dioxide ($CO_2$) storage and utilisation**
   *Edited by M. Mercedes Maroto-Valer*

17 **Oxy-fuel combustion for power generation and carbon dioxide ($CO_2$) capture**
   *Edited by Ligang Zheng*

18 **Small and micro combined heat and power (CHP) systems: Advanced design, performance, materials and applications**
   *Edited by Robert Beith*

19 **Advances in clean hydrocarbon fuel processing: Science and technology**
   *Edited by M. Rashid Khan*

20 **Modern gas turbine systems: High efficiency, low emission, fuel flexible power generation**
   *Edited by Peter Jansohn*

21 **Concentrating solar power technology: Principles, developments and applications**
   *Edited by Keith Lovegrove and Wes Stein*

22 **Nuclear corrosion science and engineering**
   *Edited by Damien Féron*

23 **Power plant life management and performance improvement**
   *Edited by John E. Oakey*

24 **Electrical drives for direct drive renewable energy systems**
   *Edited by Markus Mueller and Henk Polinder*

25 **Advanced membrane science and technology for sustainable energy and environmental applications**
   *Edited by Angelo Basile and Suzana Pereira Nunes*

26 **Irradiation embrittlement of reactor pressure vessels (RPVs) in nuclear power plants**
   *Edited by Naoki Soneda*

27 **High temperature superconductors (HTS) for energy applications**
   *Edited by Ziad Melhem*

28 **Infrastructure and methodologies for the justification of nuclear power programmes**
   *Edited by Agustín Alonso*

29 **Waste to energy conversion technology**
   *Edited by Naomi B. Klinghoffer and Marco J. Castaldi*

30 **Polymer electrolyte membrane and direct methanol fuel cell technology Volume 1: Fundamentals and performance of low temperature fuel cells**
   *Edited by Christoph Hartnig and Christina Roth*

31 **Polymer electrolyte membrane and direct methanol fuel cell technology Volume 2: *In situ* characterization techniques for low temperature fuel cells**
   *Edited by Christoph Hartnig and Christina Roth*

32 **Combined cycle systems for near-zero emission power generation**
   *Edited by Ashok D. Rao*

33 **Modern earth buildings: Materials, engineering, construction and applications**
   *Edited by Matthew R. Hall, Rick Lindsay and Meror Krayenhoff*

34 **Metropolitan sustainability: Understanding and improving the urban environment**
   *Edited by Frank Zeman*

35 **Functional materials for sustainable energy applications**
   *Edited by John A. Kilner, Stephen J. Skinner, Stuart J. C. Irvine and Peter P. Edwards*

36 **Nuclear decommissioning: Planning, execution and international experience**
   *Edited by Michele Laraia*

37 **Nuclear fuel cycle science and engineering**
   *Edited by Ian Crossland*

38 **Electricity transmission, distribution and storage systems**
   *Edited by Ziad Melhem*

39 **Advances in biodiesel production: Processes and technologies**
   *Edited by Rafael Luque and Juan A. Melero*

40 **Biomass combustion science, technology and engineering**
   *Edited by Lasse Rosendahl*

41 **Ultra-supercritical coal power plants: Materials, technologies and optimisation**
   *Edited by Dongke Zhang*

42 **Radionuclide behaviour in the natural environment: Science, implications and lessons for the nuclear industry**
   *Edited by Christophe Poinssot and Horst Geckeis*

43 **Calcium and chemical looping technology for power generation and carbon dioxide ($CO_2$) capture: Solid oxygen- and $CO_2$-carriers**
   *Paul Fennell and E. J. Anthony*

44 **Materials' ageing and degradation in light water reactors: Mechanisms, and management**
   *Edited by K. L. Murty*

45 **Structural alloys for power plants: Operational challenges and high-temperature materials**
   *Edited by Amir Shirzadi and Susan Jackson*

46 **Biolubricants: Science and technology**
   *Jan C. J. Bart, Emanuele Gucciardi and Stefano Cavallaro*

47 **Advances in wind turbine blade design and materials**
   *Edited by Povl Brøndsted and Rogier P. L. Nijssen*

48 **Radioactive waste management and contaminated site clean-up: Processes, technologies and international experience**
   *Edited by William E. Lee, Michael I. Ojovan, Carol M. Jantzen*

49 **Probabilistic safety assessment for optimum nuclear power plant life management (PLiM): Theory and application of reliability analysis methods for major power plant components**
   *Gennadij V. Arkadov, Alexander F. Getman and Andrei N. Rodionov*

50  The coal handbook: Towards cleaner production Volume 1: Coal production
    *Edited by Dave Osborne*

51  The coal handbook: Towards cleaner production Volume 2: Coal utilisation
    *Edited by Dave Osborne*

52  The biogas handbook: Science, production and applications
    *Edited by Arthur Wellinger, Jerry Murphy and David Baxter*

53  Advances in biorefineries: Biomass and waste supply chain exploitation
    *Edited by Keith Waldron*

54  Geological storage of carbon dioxide ($CO_2$): Geoscience, technologies, environmental aspects and legal frameworks
    *Edited by Jon Gluyas and Simon Mathias*

55  Handbook of membrane reactors Volume 1: Fundamental materials science, design and optimisation
    *Edited by Angelo Basile*

56  Handbook of membrane reactors Volume 2: Reactor types and industrial applications
    *Edited by Angelo Basile*

57  Alternative fuels and advanced vehicle technologies for improved environmental performance: Towards zero carbon transportation
    *Edited by Richard Folkson*

58  Handbook of microalgal bioprocess engineering
    *Christopher Lan and Bei Wang*

59  Fluidized bed technologies for near-zero emission combustion and gasification
    *Edited by Fabrizio Scala*

60  Managing nuclear projects: A comprehensive management resource
    *Edited by Jas Devgun*

61  Handbook of Process Integration (PI): Minimisation of energy and water use, waste and emissions
    *Edited by Jiří J. Klemeš*

62  Coal power plant materials and life assessment
    *Edited by Ahmed Shibli*

63  **Advances in hydrogen production, storage and distribution**
*Edited by Ahmed Basile and Adolfo Iulianelli*

64  **Handbook of small modular nuclear reactors**
*Edited by Mario D. Carelli and Dan T. Ingersoll*

65  **Superconductors in the power grid: Materials and applications**
*Edited by Christopher Rey*

66  **Advances in thermal energy storage systems: Methods and applications**
*Edited by Luisa F. Cabeza*

67  **Advances in batteries for medium and large-scale energy storage: Types and applications**
*Edited by Chris Menictas, Maria Skyllas-Kazacos and Lim Tuti Mariana*

68  **Palladium membrane technology for hydrogen production, carbon capture and other applications**
*Edited by Aggelos Doukelis, Kyriakos Panopoulos, Antonios Koumanakos and Emmanouil Kakaras*

69  **Gasification for synthetic fuel production: Fundamentals, processes and applications**
*Edited by Rafael Luque and James G. Speight*

70  **Renewable heating and cooling: Technologies and applications**
*Edited by Gerhard Stryi-Hipp*

71  **Environmental remediation and restoration of contaminated nuclear and NORM sites**
*Edited by Leo van Velzen*

# Foreword

In the future, many consumer products presently derived from fossil fuel resources such as oil, coal and gas, are likely to be derived from renewable and sustainably produced biomass resources. In addition to the production of liquid and gaseous biofuels used for transport, structural composite materials, reinforced plastics using wood fibres, pharmaceuticals, health promoting products and food sweeteners, bio-products from industrial waste gases, innovative packaging and filtration materials, green biodegradable chemicals including polymers and resins, fine chemicals for paints and adhesives, and many other products are being researched and developed using rapidly advancing biotechnologies. It seems highly likely that these products will make a major contribution through both niche and mainstream markets in the bio-economy of tomorrow.

Very small markets are possible for high value specialty biopharmaceuticals up to $100,000 per kg and biochemicals with a market price up to $1,000 per kg, down to relatively low value, bulk, commodity products such as biofuels at around $1 per litre. So the aim of a biorefinery business should be to extract as much value as possible from the biomass feedstocks by achieving the optimum product mix. Focusing on high volume, low value commodities is usually not the most viable strategy, but neither is concentrating on low volume, high value products. The potential process options are being evaluated through international collaborations such as in the IEA Bioenergy's Task 42, *Co-production of Fuels, Chemicals, Fuels and Materials from Biomass* (www.ieabioenergy.com/Task.aspx?id=4) that was established in 2007.

The concept of a 'biorefinery' varies between a single feedstock converted into a single product (such as sugarcane to ethanol), single feedstock and multi-products (such as oilseed rape to biodiesel, high-protein animal feed, and heat and power generation from the straw), and multi-feedstocks to multi-products. This is analogous to an oil refinery processing a range of petroleum products and base chemicals. During the last century, the number of oil products marketed has grown from a few fuels and lubricants produced by the simple process of distillation to over 2,000 products today using complex thermal and catalytic cracking as well as reforming processes. Many of the lessons learned by process and chemical engineers can be applied to modern biorefineries and hence shorten the experience of learning-by-doing to add value to the bioenergy industry.

But biorefineries are not new. In the 1700s, the forest industry produced pitch, tar and resins, turpentine and rosin used in ship-building and the sailing of them, with some of these still being produced. The modern biorefinery concept has also existed for some decades as exemplified by the Norwegian company Borregaard, which has long used woody biomass from spruce to produce a range of biochemicals, biomaterials and bioethanol. Ethanol is only a minor, relatively low value product of their processing activities, along with the world market domination of some high value, specialist chemicals. The mix of products can be modified as markets dictate.

A strong economic business case can be made for integrating the production of conventional biomass products with new 'bio-products'. This can provide new employment opportunities, and benefit the local and global environment by the recycling of biomaterials or significantly reducing greenhouse gas emissions to reach a very low carbon footprint. Moving towards a future green bioeconomy that the world requires will enable heat, power, biofuels and materials resulting from the traditional use of biomass feedstocks to be complemented by adding value through the new and emerging bio-product technologies.

Biorefineries can be designed to create intermediate products for processing into end-products in facilities elsewhere, or derive products ready for market directly on-site. Where an available feedstock is seasonal, multi-feedstocks may be needed to keep the biorefinery operating all year round. This can add to the complexity and cost of the front-end of the plant. Minimizing the costs of collection, transport and storage of the feedstock is another challenge that cannot be ignored when determining the optimum scale of processing plant.

The biorefinery concept, even in its simplest form, provides a more complete utilisation of the biomass feedstock than bioenergy alone. This book provides an excellent overview of the numerous opportunities for commercial viability as a result of the science being applied to the engineering concepts of biorefining. New pilot-scale and demonstration plants that produce biofuels and plastic composites from woody biomass have already been established in Finland, Canada and elsewhere, with many others planned using a range of feedstocks. Development of biorefineries usually requires a supportive government policy framework, as well as research and development support in order to overcome the existing technological, social and environmental barriers. For example, recently in

New Zealand, my own country, an international forest paper pulp company has received government support to build a demonstration biorefinery with the aim to utilise plantation forest residues as the main feedstock, and formed a partnership with an oil company and the forest Crown Research organisation.

The biorefinery model enables the agricultural and forest sectors to diversify their traditional markets and products, to become more energy self-sufficient, and to displace fossil fuel-based products with low carbon and renewable alternatives. In a future carbon-constrained global economy, the use of fossil fuels will be constrained and there will be increased demand for renewable and sustainable products arising from biomass resources. Consequently, biorefineries and their bio-products will play an increasingly important economic role. The world-acclaimed editor and authors of this book have helped advance the knowledge needed to achieve that goal and provided a vision for the global green, bioeconomy of the future.

*Ralph E. H. Sims, Professor of Sustainable Energy,*
*Massey University, New Zealand*

# Preface

The concept of 'biorefining' has emerged over many decades, more recently stimulated by the drive for sustainability. The underlying aspiration is that renewable feedstock can be processed to partially replace the non-renewable fossil fuels that currently provide the bulk of our energy and chemicals. Of course, 'biorefining' is not new, as the exploitation of biomass for food, fuel and materials precedes the industrial revolution.

In the modern context there have been numerous attempts to define and describe the complex nature of biorefining. The International Energy Agency's Bioenergy Task 42 has agreed that 'Biorefinery is the sustainable processing of biomass into a spectrum of marketable products (food, feed, materials and chemicals) and energy (fuels, power and heat)'. Hence, the term 'biorefinery' is loosely defined, and can refer to a process, a plant, clusters of facilities, or a concept. Indeed, biorefining can range from the simple modification of biomass for use 'as is' in the production of materials, through to the complex extraction of molecular components followed by their bioconversion into higher value chemicals and fuels.

The biorefining industry is beset with many socio-economic and environmental challenges. At the time of writing, biorefining is dominated by the global production of ethanol from grain and sugar. The consequent exploitation of food-grade feedstock has stimulated an on-going debate regarding food vs. fuel. Additional arguments concerning the use of land for producing non-food feedstock are developing around the issue of indirect land use change. A further challenge to successful biorefining is the requirement to operate at scales and efficiencies which are both economically viable and environmentally sustainable. However, whilst biorefining approaches are often sought to produce environmentally beneficial products that replace current fossil-derived materials, they are competing with well-established products which have a number of key economic advantages. Fossil-derived products are relatively cheap, usually commodities, and serve mature global markets that have developed alongside the petrochemical industry over the last century. In contrast, the biorefinery industry is relatively new, undergoing rapid change as it develops, and requires the application and integration of a wide range of highly specialist disciplines, from the biosciences through to advanced chemical and process engineering.

The aim of *Advances in biorefineries* is to provide a comprehensive and systematic reference on the advanced processes used for biomass recovery and conversion in biorefineries. The volume comprises contributions from internationally recognised experts who have reviewed the latest developments in the area of biorefining, and is divided into two parts:

- Part I, 'Development and optimisation of biorefining processes', contains 12 substantial chapters. The first chapter introduces the concept of green chemistry with reference to exploitation of waste streams. This is followed by several chapters considering economic and environmental impacts and sustainability. The remainder of Part I assesses recent developments and optimisation strategies for key unit processes (including biomass pretreatment, catalytic conversion, enzyme development and separation technologies) and a range of biorefinery models.
- Part II, 'Biofuels and other added value products from biorefineries' comprises 14 chapters which concentrate on the creation and optimisation of products from biorefineries. Highlighting the diversity of products that can be created from biomass, these chapters provide up-to-date knowledge across a variety of outputs from the improvement of liquid biofuels in internal combustion engines, and the production of platform chemicals, proteins, lipids and carbohydrates through to the creation of high value and specialist products, including adhesives, films and coatings and bio-based nutraceuticals.

*Keith Waldron*

# Part I
Development and optimisation of biorefining processes

# 1
# Green chemistry, biorefineries and second generation strategies for re-use of waste: an overview

L. A. PFALTZGRAFF and J. H. CLARK,
University of York, UK

DOI: 10.1533/9780857097385.1.3

**Abstract**: Today fossil resources supply 86% of our energy and 96% of organic chemicals. Future petroleum production is unlikely to meet our society's growing needs. Green chemistry is an area which is attracting increasing interest as it provides unique opportunities for innovation via use of clean and green technologies, product substitution and the use of renewable feedstocks such as dedicated crops or food supply chain by-products for the production of bio-derived chemicals, materials and fuels. This chapter provides an introduction to the concepts of green chemistry and the biorefinery and, based on examples, discusses second generation re-use of waste and by-products as feedstocks for the biorefinery.

**Key words**: green chemistry, clean technologies, biorefinery, renewable and sustainable resources, food supply chain waste, resource intelligence.

## 1.1 Introduction

Through the combination of low environmental impact and safe technologies, the use of biomass can provide a renewable alternative to fossil resources. It can establish a new sustainable supply chain for the production of high value chemicals, including fuels and energy as well as materials.

### 1.1.1 Green chemistry

Green chemistry is the design of chemical products and processes that reduce or eliminate the use and generation of hazardous substances (Anastas *et al.*, 2000). The concept emerged 20 years ago with the introduction by Paul T. Anastas and J. C. Warner of the 12 principles of green chemistry (see Table 1.1). The subject continues to develop strongly around these principles (Anastas and Warner, 1998). Green chemistry aims to achieve (Clark and Macquarrie, 2002):

- maximum conversion of reactants into a determined product,
- minimum waste production through enhanced reaction design,

*Table 1.1* The 12 green chemistry principles

1. **Prevention**
   It is better to prevent waste than to treat or clean up waste after it has been created.

2. **Atom economy**
   Synthetic methods should be designed to maximize the incorporation of all materials used in the process into the final product.

3. **Less hazardous chemical syntheses**
   Wherever practicable, synthetic methods should be designed to use and generate substances that possess little or no toxicity to human health and the environment.

4. **Designing safer chemicals**
   Chemical products should be designed to effect their desired function while minimizing their toxicity.

5. **Safer solvents and auxiliaries**
   The use of auxiliary substances (e.g., solvents, separation agents, etc.) should be made unnecessary wherever possible and innocuous when used.

6. **Design for energy efficiency**
   Energy requirements of chemical processes should be recognized for their environmental and economic impacts and should be minimized. If possible, synthetic methods should be conducted at ambient temperature and pressure.

7. **Use of renewable feedstocks**
   A raw material or feedstock should be renewable rather than depleting whenever technically and economically practicable.

8. **Reduce derivatives**
   Unnecessary derivatization (use of blocking groups, protection/deprotection, temporary modification of physical/chemical processes) should be minimized or avoided if possible, because such steps require additional reagents and can generate waste.

9. **Catalysis**
   Catalytic reagents (as selective as possible) are superior to stoichiometric reagents.

10. **Design for degradation**
    Chemical products should be designed so that at the end of their function they break down into innocuous degradation products and do not persist in the environment.

11. **Real-time analysis for pollution prevention**
    Analytical methodologies need to be further developed to allow for real-time, in-process monitoring and control prior to the formation of hazardous substances.

12. **Inherently safer chemistry for accident prevention**
    Substances and the form of a substance used in a chemical process should be chosen to minimize the potential for chemical accidents, including releases, explosions and fires.

- the use and production of non-hazardous raw materials and products,
- safer and more energy efficient processes, and
- the use of renewable feedstocks.

Efficiency is the key, and green chemistry has continued developing around the principles, which guide both academia and industry in their pursuit of more sustainable processes. In an ideal case, according to these principles, a reaction would only produce useful material. Waste and pollutants would be prevented, improving the reaction yield and reducing losses, thus improving the overall economics of a process. Since our society and industries are governed by increasing efficiency and profit, green chemistry therefore theoretically fits the agendas of most manufacturing companies these days, not only appealing to chemical producers.

Today, 20 years after their publication, the 12 principles of green chemistry are as meaningful as ever in the light of the increasing interest the area attracts due to concerns over sustainability (Anastas and Kirchoff, 2002). Misunderstandings have arisen due to the attractiveness of the area to sectors dealing directly with public demands for 'greener and more environmentally friendly' products. It is therefore of vital importance that the message is not distorted by common misconceptions over what is or is not 'green', thus altering their original goal: to aim towards safer and cleaner chemistry.

The implementation of REACH (Registration, Evaluation, Authorization and Restriction of Chemicals), or Directive (EC 1907/2006), ROHS (Restriction of the Use of Certain Hazardous Substances in Electrical and Electronic Equipment) or Directive 2003/108/EC, and other initiatives highlighting the hazardous character of some chemicals used in day-to-day consumer products, such as the SIN list (n.d.), are pushing hard for their replacement to avoid further risks to human and/or environmental health. However, we should make sure the substitutes used are genuinely safer across the whole life cycle and as effective as what they are replacing. Investing in R&D focused on finding truly greener alternatives, thus eliminating rushed and weak substitutions that can even increase the number of components present in formulations when ingredients are added to compensate for a lack of performance in the 'greener' formulation, is important. The same applies to the substitution of fossil-derived chemicals with more sustainable bio-derived chemicals: when using renewable feedstocks such as biomass, we have to use clean and efficient synthetic routes, minimizing the amount of unwanted by-products and the use of scarce resources (i.e., scarce metals).

Scarce metals are increasingly used in clean alternative energy-producing technologies. Their reserves are sometimes only estimated to last another 50 years, or even less for key elements such as indium (a key component in solar panels) (Dodson *et al.*, 2012) and we must take this into account when modifying our energy and manufacturing infrastructure, taking advantage of the whole periodic table. This is especially relevant to the area of catalysis: re-usable catalytic metals are seen as better reagents than hazardous reagents such as $AlCl_3$. But many of the most interesting catalytic metals are also becoming scarce and their production process can be resource intensive and wasteful, making their recovery and reuse essential. Water is increasingly seen as a scarce resource too in certain areas of our planet, but its use as a green solvent is increasingly envisaged due to its non-toxicity compared to hydrocarbon-based solvents (Simon and Li, 2012). Nevertheless, contaminated water is difficult and expensive to treat and re-use. Another alternative to VOC solvents are involatile solvents such as ionic liquids designed to eliminate air-borne emissions. Ionic liquids are used in phase transfer catalysis for example (Welton, 2004), but their non-emissions are counteracted by their toxicity and their environmental impact when prepared, used and separated for end-use.

Biodegradability is an important sought-after characteristic for 'greener' products, but increasing the life-time of a molecule to promote its re-use could be another strategy. Heavily halogenated compounds are poorly degradable and there are some large volume halogenated compounds that need to be phased out (e.g., the solvent dichloromethane). But we must not bundle all halogenated compounds in the same 'red' basket. Nature turns over enormous quantities of organohalogen compounds and we need to learn from nature and avoid, as much as possible, those compounds that it cannot deal with (e.g., perhalogenated compounds).

Food waste is a feedstock rich in functionalized molecules, and although it is biodegradable, it should be valorized for new applications as a raw material for renewable chemicals, materials and bio-fuels, leading us towards waste minimization and waste valorization. Wasting resources should be avoided in any optimized process. However, waste can also represent an opportunity as we can no longer afford the luxury of waste.

This past paragraph shows you how tightly knit these issues are, illustrating how important it is to assess the greenness of a process through each of its steps, from the use of raw materials to end-use through manufacturing and use. One change can affect several steps and it is important to assess a process through its full life cycle even though it is time-consuming and its quality is dependent on the data used. Such a tool can help us assess the use of bio-processes versus chemo-processes, for example. Many believe bio-processes are preferable to chemo-processes as they are superior in terms of environmental impact, since they use non-toxic components to selectively yield the targeted product. But as they are time-consuming and expensive, it is unrealistic to believe that chemo-processes will be entirely replaced by natural organism catalysed processes in the foreseeable future.

## 1.1.2 Drivers for change

Our society faces a new challenge: as the current consumption model dominated by market demand is running out of breath, our society needs to adopt a more realistic and sustainable model based on the efficient and sustainable use of natural resources in order to sustain emerging economies at the standard established in the West over the last century.

Current manufacturing practices are strained by the increasing price of feedstocks such as oil and consequently of energy and petrochemicals, increasing waste cost (treatment or disposal) together with the increasing impact of legislation affecting almost all aspects of its operations (e.g., supply of raw material, manufacturing, end-use and disposal).

Legislation has had a dramatic impact on product manufacturing since human and environmental safety have attracted increased concern following publications of traces of chemicals in animal and human tissue in the 1970s and 1980s (e.g., dioxins) (Schecter, 1998). Legislation now has an influence on the type of process, process steps, emissions, end treatment of waste, illustrating how every stage of the supply chain of a chemical product has to be the least polluting possible (i.e., Integrated Pollution Prevention and Control legislation, IPPC) (Lancaster, 2010). With new regulations such as REACH and ROHS (Restriction of Hazardous Substances), an important number of chemicals will have to be replaced by less harmful substitutes, shaking to the core industrial sectors like home and personal care products, the pharmaceutical industry and the agricultural sector.

Resource is a stage in the product life cycle where green chemistry can have a major impact in the future. The use of renewable, typically biomass for carbon, instead of finite resources is becoming more economically and environmentally sound, being one of the main areas of research in green chemistry along with clean synthesis, greener solvents and renewable materials. Biomass is also a resource which can be renewed within a time interval relevant to our resource consumption (see Fig. 1.1), biomass being a 'biological material derived from living, or recently living organisms'.

The emergence of EU standards for bio-based products (Mandate M/429; see Section 1.3.1) will, in the near future, embrace life cycle considerations and introduce specifications along the whole supply chain for new and existing products on biomass content, and will further discourage the use of fossil resources in favour of renewable feedstocks such as biomass including bio-wastes.

The public and consequently the retail sector have been increasingly aware of the dangers of some unsafe practices in industry and unsafe chemicals in consumer product formulations. They are now asking manufacturers to produce bio-derived chemicals and question the environmental impact of their production, driving the market towards green and

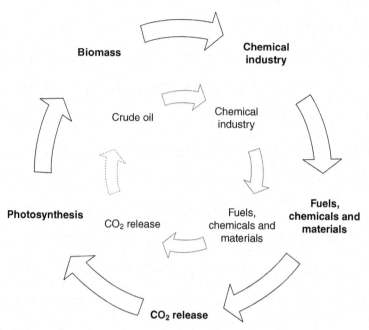

*1.1* Comparison of production cycles of chemicals derived from biomass and oil.

renewable alternatives in many sectors, especially in home and personal care products.

In line with the EU's innovation strategy and following the initiation of a new policy in 2006 aiming to support the development of high economic and societal value markets, the European Commission proposed further steps for the creation of lead markets. Bio-based products are the subject of one of the identified lead markets and fall into this category for several reasons (European Commission, 2007, 2009):

- use of renewable and expendable resources
- less dependency on limited and increasingly expensive fossil resources
- the potential to reduce greenhouse gas emissions (carbon neutral/low carbon impact)
- the potential for sustainable industrial production
- potentially improved community health
- support to rural development
- increased industrial competitiveness through innovative eco-efficient products
- potential for transfer to other regions of the world including the transfer of appropriate technologies discovered and proven in the EU.

A recent study estimates that, by 2025, over 15% of the US$3 trillion global chemical market will be derived from bio-derived sources (Vijayendran, 2010). Yet another study highlights the technical feasibility of over 90% of the annual global plastic production of 270 Mt being substituted by bioplastics. In 2005, bio-based products already accounted for 7% of global sales and around €77 billion in value in the chemical sector. EU industry accounted for approximately 30% of this value. Estimates of the ad hoc advisory group for bio-based products have identified active pharmaceutical ingredients, polymers, cosmetics, lubricants and solvents as the most important sub-segments (Commission, 2009). Active pharmaceutical ingredients in particular, with 33.7% of global chemical sales, are expected to be the chemical segment with the highest percentage sales of products produced using biotechnological processes. It is predicted that Europe will be strong in sales in the following sub-segments: active pharmaceutical ingredients, polymers and fibres, cosmetics, solvents and synthetic organic compounds.

### 1.1.3 Product substitution

The use of renewable feedstock is one of the cornerstones of modern green chemistry. Non-renewable fossil resources supply 86% of our energy and 96% of organic chemicals (Binder and Raines, 2009). But fossil resources are not renewed in a time interval relevant to our resource consumption: according to our actual consumption, the future petroleum production is unlikely to meet our society's growing needs: by 2025, our energy demands are expected to increase by 50% (Ragauskas *et al.*, 2006). Other drivers are pushing for the substitution of chemicals used daily in consumer products: safety concerns for both humans and the natural environment. Volatile chlorinated compounds used in dry cleaning, sulphonated surfactants, and polybrominated compounds in flame retardants are compounds used in formulations and processes for which replacement molecules would be preferred. Research on the production of cost-effective alternatives derived from renewable resources is an area of primary importance if we want to satisfy the requirement for green and sustainable chemicals and products. Green chemistry now embraces the whole life cycle of a product (see Fig. 1.2), rather than just focusing on the production stage. Upstream and downstream stages of the production, including the raw material employed, its use, end-use and disposal, are included, guaranteeing the true sustainability of a product (Anastas and Lankey, 2000).

Improvements in today's modern formulating-based industries at the production stage of the life cycle, are restricted (although moving towards renewable energy and zero waste is important and not trivial). The use of renewable feedstocks could offer an important margin for progress,

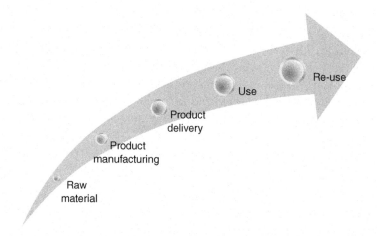

*1.2* Illustration of a product's life cycle.

especially for companies, such as consumer goods manufacturers, keen to dramatically improve the environmental performance (decrease the $CO_2$ emissions) of their products.

### 1.1.4 From petro-refineries to bio-refineries

It is important to ensure that both the resource and the process technology used as well as the products made are environmentally acceptable. The twentieth century saw the development of processes designed for the production of energy and organic chemicals based on the oil refinery. The twenty-first century must see the development of similar processes based on the biorefinery. The aim is to design an integrated process capable of generating a cost-effective source of energy and chemical feedstocks using biomass as a raw material. The key is to find alternative sources of carbon to oil, available in high quantities and process them using green chemical technologies, ensuring products obtained are truly green as well as sustainable. Technologies used should ideally be flexible enough to accommodate the natural variation of biomass associated with seasonal or variety change (Clark *et al.*, 2009). The efficiency of the process needs to be maximal: ideally every output has to have a use and a value/market. We can no longer afford the luxury of waste. Practices based on industrial symbiosis looking at re-using the waste produced by one process to feed another, or converting waste into a useful by-product with a marketable value need to be developed. The aim would be to achieve a zero waste biorefinery able to compete economically with existing systems used to produce energy and chemicals, an objective increasingly pushed by EU regulations (see Section 1.3.1).

# Green chemistry, biorefineries and second generation strategies 11

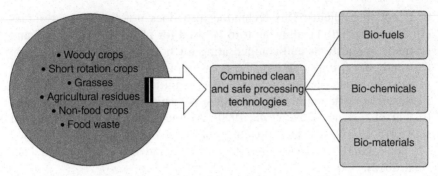

*1.3* Scheme describing an integrated biorefinery as a mixed feedstock source of chemicals, energy, fuels and materials.

Adding value to every output of the biorefinery can be achieved by combining several technologies together, using a sequential approach to extract chemicals before biomass is converted to energy. The main green extraction processes used to extract valuable compounds from biomass include liquid and supercritical $CO_2$, ultrasonic or microwave-assisted extraction and accelerated extraction. Microwave-assisted extraction is a commercial reality with Crodarom using this technique to extract purer and more degradation stable plant materials (Crodarom, n.d.). The extraction can be followed by biochemical or thermochemical processes and internal recycling of energy and waste gases. This approach ideally constitutes the basis of an economically sound starting point for the design of a biorefinery and is illustrated in Fig. 1.3. The integration of technologies for the biorefinery takes into account the complex nature of lignocellulosic biomass, in order to produce several products and render the biorefinery concept cost-effective.

Biomass contains an array of functionalized molecules, with many of them having a market value. Compounds such as natural dyes or colorants (e.g., carotenoids), polyphenols, sterols, waxes, nonacosanol or flavonoids (e.g., hesperidin), amino acids, and fatty acid derivatives can be extracted selectively using clean extraction techniques prior to the treatment of biomass by biochemical and thermochemical processes. These compounds have uses in cosmetics, as nutraceutical or semiochemicals (Clark *et al.*, 2006; Deswarte *et al.*, 2006). Often, secondary metabolites are extracted using volatile organic solvents, but clean extraction techniques such as liquid and supercritical $CO_2$ are very selective, allowing fractionation of extracted mixtures and have the advantage of being allowed for processing raw materials, foodstuffs, food components and food ingredients (together with ethanol and water) according to Directive 2009/32/EC of the European Parliament on extraction solvents used in the production of foodstuffs and food ingredients. This technique also does not leave any residues (Budarin et al., 2011), allowing it to be used for pharmaceutical, food and cosmetic applications and compensating for both high technology capital cost and energy consumption. The polarity of $CO_2$ can also be fine-tuned using co-solvents such as methanol or ethanol (Sahena et al., 2009). As a matter of comparison, the polarity of supercritical $CO_2$ can be compared to that of hexane (Deye et al., 1990). Although energy requirements of supercritical $CO_2$ are high, the technology has been commercially used for hop extraction, decaffeination of coffee and dry cleaning (Arshadi et al., 2012).

Biochemical and thermochemical processes complement each other well, the former being very selective but slow compared to the latter. Biochemical processes require low temperatures but pre-treatments are often required (e.g., ammonia fibre expansion or AFEX, dilute acid hydrolysis) to open up biomass's fibre structure and yield fuels and chemical intermediates used for further downstream processing (Eggeman and Elander, 2005; Tao et al., 2011). Processing times and space-time yields are high compared to thermochemical processes, but they are less energy intensive (Kamm and Kamm, 2004). Thermochemical processes, which include gasification, pyrolysis and direct combustion (see Table 1.2), usually operate above 500°C and are much less selective, yielding oils, gas, chars and ash (Fernández et al., 2011).

Biomass with a high acid, alkali metal and water content can be difficult to use in conventional thermal treatments: the high water content can render pyrolysis or gasification processes very difficult and the acidity of the feedstock can limit the applications of the pyrolysis oil obtained, for example.

Microwave technology has been studied for the pyrolysis of straw. This technology was proven to improve the quality of bio-oils obtained at lower temperatures (typically under 200°C), yielding oils with properties outperforming commercial fuel additives: bio-oils produced have a lower oxygen, alkali, acid and sulphur content (Budarin et al., 2009).

The properties of the oil obtained could also be modified by using additives during the heating phase, showing how microwave technology is versatile and can offer an alternative to conventional thermal processes. Microwave technology has an added advantage compared to conventional thermal heating: it activates cellulose at a temperature of 180°C, helping the conversion process (Budarin et al., 2010). It has been reported that at this precise point and under microwave heating conditions, the rate of decomposition of the amorphous part of cellulose increases due to *in-situ* pseudo acid catalysis, yielding a char and bio-oil of superior properties compared to those produced when using conventional heating methods.

Table 1.2 Examples of technological processes used as part of a biorefinery

| | Process name | Temperature (°C) | Conditions | Product(s) | Application |
|---|---|---|---|---|---|
| Thermochemical processes | Gasification | 700 | Low oxygen level | Syngas (mixture of $H_2$, CO, $CO_2$, $CH_4$) | Fuel or chemical intermediate to ethanol or dimethyl ether or isobutene |
| | Pyrolysis | 300–600 | No oxygen | Bio-oil, char and low molecular weight gases | Transportation fuel and chemicals |
| Biochemical processes | Fermentation | 5 < T°C < 30 | Presence of oxygen | Alcohol (e.g., ethanol), organic acids (e.g., succinic acid) | Transportation fuel (e.g., ethanol) |
| | Anaerobic digestion | 30–65 | No oxygen | Biogas ($CO_2$, $CH_4$) | Production of natural gas (>97% $CH_4$) |

## 1.2 Introduction to biorefineries

### 1.2.1 Defining biorefineries and bio-processing

In addition to the definition of green chemistry given previously in Section 1.1.1, two additional definitions need to be highlighted in this chapter: the term 'biorefinery' and the term 'bioprocessing'.

A biorefinery is an analogue to the current petro-refinery, in the sense it produces energy and chemicals. The major difference lies in the raw material it will use, ranging from biomass to waste. The use of clean technology is another imperative for the biorefinery, ensuring its output(s) are truly sustainable. The IEA Bioenergy Task 42 defines biorefining as 'the sustainable processing of biomass into a spectrum of bio-based products (food, feed, chemicals and/or materials) and bioenergy (biofuels, power and/or heat)' (IEA, 2009). Various biorefinery designs of varying size and output number will emerge commercially in the future (Cherubini, 2010), taking advantage of flexible technology, helping the concept of a biorefinery to process locally available biomass to its fullest in an integrated fuel-chemical-material-power cycle, improving cost-efficiency, the quality of life of the local population and lowering the environmental impact governed by the three dimensions of sustainability (environmental protection, social progress and economic development; see Fig. 1.4). Networks of biorefineries are to be considered too, for maximum resource efficiency.

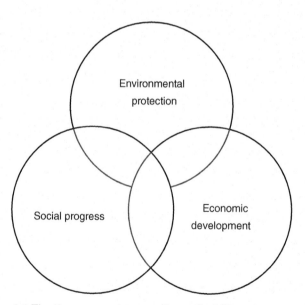

*1.4* The three cornerstones of sustainability.

Bio-refining should not be confused with bioprocessing. A bioprocess is any process which uses biological organisms (e.g., enzymes) to carry out targeted chemical or physical transformations. A bioprocess can be used as part of the conversion process in a biorefinery along with other low environmental impact technologies such as microwave chemistry or aqueous phase catalysis processing. This illustrates how important it is to interconnect different disciplines such as chemistry, chemical engineering, biotechnology and biology together with techno-economic and sustainability assessment, as they are crucial for the development of a successful fully integrated biorefinery. Biomass is a term applied which includes a great variety of different and often complex plant components embedded in a matrix that differs according to the origin of the biomass used. A multi-disciplinary approach is therefore necessary to maximize the value of the products obtained while using green chemistry technology.

### 1.2.2 Biorefinery types and product areas as defined by feedstocks and waste streams

There are three biomass feedstocks: carbohydrate (starch, cellulose and hemicellulose) and lignin from lignocellulosic biomass, triglycerides (soybean, palm, rapeseed, sunflower oil) and mixed organic residues. Lignocellulosic feedstocks can be obtained through the production of dedicated crops such as miscanthus or short rotation woody crops such as willow or poplar. Agricultural residues such as rice or wheat straw and paper pulp from the paper industry are other examples of sources of lignocellulosic material. Figure 1.5 shows the two main types of biomass feedstocks.

Biorefineries can be subdivided via over simplification into biorefineries of phase I, II and III according to the feedstock and process used, as well as product targeted (chemicals or energy) (Cherubini *et al.*, 2009; Kamm and Kamm, 2004). A table listing examples of different technological processes to be used in a biorefinery are listed in Table 1.3.

Phase I biorefineries focus on the conversion of one feedstock, using one process and targeting one product. A biodiesel production plant would be a good example of a phase I biorefinery: rapeseed or sunflower is used for oil extraction, which is subsequently transesterified to produce fatty acid methyl esters or biodiesel using methanol and a catalyst (Shahid and Jamal, 2011).

Phase II biorefineries differ from phase I biorefineries by the number of outputs they can produce. A typical example of a phase II biorefinery is the production of starch, ethanol and lactic acid together with high fructose syrup, corn syrup, corn oil and corn meal from corn wet mil operations (EPA, 2011).

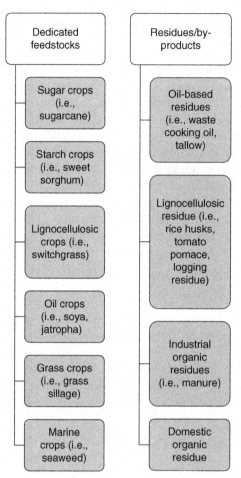

*1.5* The two main types of biomass feedstocks available (adapted from Cherubini *et al.*, 2009).

Phase III biorefineries allow for a wider range of technologies, to be combined (e.g., supercritical $CO_2$ extraction followed by biological transformation), in comparison to phase I and II biorefineries. They also allow for a higher number of valorized outputs since several constituents of the feedstock used can be treated separately. Biorefineries falling into that category can also be called 'product-driven biorefineries'. They generate two or more bio-based products and the residue is used to produce energy (either fuel, power and/or heat). Examples of phase III biorefineries include whole crop biorefineries which make use of several agricultural by-products originating from the same crop. Phase III biorefineries are typically the

*Table 1.3* Most common thermochemical and biochemical processes

| Mechanical processes | Biochemical processes | Chemical processes | Thermochemical processes |
|---|---|---|---|
| Pressing | Anaerobic digestion | Hydrolysis (basic or acidic) | Pyrolysis |
| Milling (size reduction processes) | Aerobic and anaerobic fermentation | Transesterification and esterification | Gasification |
| Pelletization | Enzymatic conversion | Hydrogenation | Combustion |
| Distillation |  | Oxidation | Steam explosion |
| Extraction |  | Methanization | Hydrothermal upgrading |
|  |  | Steam reforming | Supercritical |
|  |  | Water-gas shift |  |
|  |  | Heterogeneous and homogeneous catalysis |  |
|  |  | Water elecrolysis |  |
|  |  | Pulping |  |

ones targeting the production of chemicals and fuels. Sub-categories also exist according to the type of technology used (thermo-chemical or biochemical biorefineries).

Another classification has now been adopted by the IEA Bioenergy Task in 2010 to take into account the complexity of the biorefinery concept and its future developments around new technologies. It is based around the four cornerstones of the biorefinery concept: feedstock used (i.e., dedicated crop, process or agricultural residue, algae), platform products obtained (i.e., C5 sugars, pyrolysis oil or syngas), final products obtained (energy or chemicals) and process used (Cherubini *et al.*, 2009). This classification has the advantage of accounting for the need to apply a given technology to different feedstocks and will therefore include biorefineries developed in the future. Biorefineries should not be designed in a generic way but should be adapted to the best technology and the best feedstock available in the geographical location chosen.

## 1.3 New renewable feedstocks

### 1.3.1 Drivers for change

The EU has recognized that, in order to sustain our demands in energy, chemicals and food, while addressing environmental issues, we need to substantially reduce our dependence on oil by establishing a bio-based economy. The European Commission recently issued Mandate M/429 (European Commission, 2011) to develop a standardization programme for bio-based products, raising the general public's awareness for bio-based products (since no external, perceptible characteristics differentiate them from oil-derived products). It was developed with the contribution of industry, research organizations, sector associations and standardization bodies and is anticipated to take a life cycle approach to evaluation and be sensitive to eco-system issues which have become so evident in the bio-fuels arena. Critical issues will include moving away from first generation feedstocks, increasing use of wastes, and ensuring the use of sustainable and low environmental impact technologies throughout the supply chain alongside a consequent reduction in wastes in new feedstock industries.

Early research on renewable resources focused heavily on crops such as rapeseed, corn or sugar cane. However, the controversial competition between food and non-food uses of biomass had an negative effect on crop prices as well as on press feedback concerning biofuels (OECD, 2008). Other sources of biomass are now studied and waste is increasingly considered as another renewable feedstock for the production of bio-derived chemicals, materials and fuels.

In times which increasingly value resource efficiency, waste has become a luxury. DEFRA, the Department of Environment, Food and Rural Affairs in the UK, has estimated that businesses could save up to £23 billion by re-using resources more efficiently (DEFRA, 2012). In the EU, Council Directive 99/31/EC, better known as the Landfill Directive, will drastically reduce the amount of landfill space available as the amount of biodegradable waste sent to landfill in member countries by 2016 will have to reach 35% of the 1995 level. As a result, landfill gate fee has increased from £40–£74 to £68–£111 (including landfill tax) in the UK between 2009 and 2011 (WRAP, 2009, 2011). Policy makers support alternatives to landfill (e.g., value recovery from waste), especially in the context of achieving a zero waste economy and the vision of the European Bioeconomy 2030 (European Commission, n.d.). At the same time, our society faces a huge looming crisis of resources. Globally, '30% fewer resources [are needed] to produce one Euro or Dollar of GDP than 30 years ago; however, overall resource use is still increasing [...] as we consume growing amounts of products and services' (Giljum et al., 2009). As traditional resources such as oil and minerals become scarcer, their availability will become more politically controlled leaving them vulnerable to highly politicized negotiations and pricing.

Waste valorization represents a promising research topic from both environmental and economic points of view as 'there is a considerable emphasis on the recovery, recycling and upgrading of wastes' (Laufenberg et al., 2003). Current management practices of waste should be replaced by strategies which have a lower environmental impact and which allow the recovery of marketable products for existing or new markets, thus offering added revenues for companies. Valorizing our waste also has the potential to reduce a process's carbon footprint and dependence on fossil resources, increase its efficiency and cost-effectiveness and moving towards 'closed loop manufacturing', one of the EU's clear future strategies, highlighted in the Europe 2020 strategy document (European Commission, 2010). The use of renewables in consumer products is especially relevant at a time when public awareness of environmental issues and cradle-to-grave concerns is growing, leading to industry's increasing concern over their 'green' credentials and environmental performance.

### 1.3.2 Concept of a waste biorefinery

There is a growing recognition that the twin problems of waste management and resource depletion can be solved together through the utilization of waste as a resource. Some initiatives looking at the re-use of waste already exist, like in Spain for example, where the environmental complex of Montalban, Spain (Epremasa, Complejo Medioambiantale de Montalban), is a unique example of integrated waste management (EPREMASA, n.d.). It was built to meet the new EU directives regarding waste management; concentrating, recovering and valorizing waste in order to avoid landfilling as much as possible. The company is responsible for waste management operations in the province of Cordoba, Andalusia. It provides home collection of municipal solid waste (household waste, paper, cardboard, glass and electric appliances), transportation, processing and landfill management for 74 municipalities (approximately 475,500 inhabitants). This strategy and the scale of operations allows the facility to be cost-effective with more flexible working procedures and a rationalization of human and material resources involved in the cycle.

The complex is an integrated facility which combines high efficiency waste scanning and segregation, recycling, composting, electricity generation and landfilling activities on the same site. The complex is able to produce high quality recycled plastic by sacrificing 40% of the organic waste through the use of a more rigorous process. Its efficiency is around 90% as only 10% of the plastic arriving at the facility is landfilled (mainly plastic contained in Tetrapack® packaging). As a result, the higher quality plastic meets the specifications for being used in further plastic packaging applications which, up to now, was limited. In addition, compost is commercially produced from organic waste, as well as 1.2 MW of electricity as the composters are connected to a biogas plant.

This process illustrates how the valorization of waste can provide first generation waste-derived feedstocks (recycled plastic, compost, biogas/energy) as an alternative source of carbon. Such applications reduce the need to use virgin land and finite resources such as oil.

### 1.3.3 Opportunities offered by the use of food supply chain waste

Waste biomass from the food supply chain (i.e., agricultural residues such as wheat straw, rice husks, waste cooking oil or food manufacturing waste such as tomato peels) are an ideal renewable material as they do not compete with the food and feed industries for land. An FAO report issued in 2011, estimates that 'one-third of food produced for human consumption is lost or wasted globally, which amounts to about 1.3 billion tons per year' (Gustavsson *et al.*, 2011). It is important to note the difference between food waste and food loss, the latter being food lost due to the use of poor technological means or diseases affecting crops, for example (Parfitt *et al.*, 2010).

The agro-food supply chain includes a broad variety of manufacturing processes producing consequent cumulative quantities of different wastes, especially organic residues at every step of the supply chain (Gómez *et al.*, 2010; Laufenberg *et al.*, 2003). The increasing demand for chemicals and fuel together with other drivers are encouraging the re-use and valorization of organic waste from the food supply chain for the production of novel added-value bio-derived sustainable products. A description of a food supply chain is given in Fig. 1.6.

Food waste encompasses domestic waste produced by individuals in their homes. This represents a logistical problem as it would be difficult to collect and concentrate in one place, except in large housing complexes. On the other hand, it might be argued that if the waste produced by the agricultural and processing sectors before it reaches the consumer is generated in a more concentrate manner, it would be easier to collect and valorize. The problems associated with these wastes are:

- severe pollution problems due to high associated chemical and biological oxygen demand (COD and BOD) (Kroyer, 1995)
- varying pH (Kroyer, 1995)
- material prone to bacterial contamination (Schieber *et al.*, 2001) (e.g., fruit and vegetable by-products)
- high accumulation rate leading to disposal management problems (Zaror, 1992)
- variations in chemical content due to different varieties and seasonal variations.

Current practices for the management of food waste include:

- incineration (GHG and toxic chemical emissions)
- landfilling (polluting, GHG emissions)
- conversion to cattle feed (uneconomical process, high moisture content)

# Green chemistry, biorefineries and second generation strategies 21

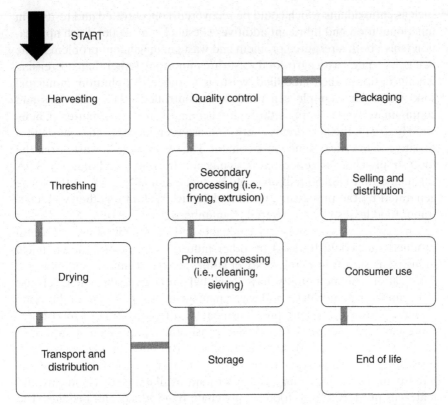

*1.6* Generic illustration of the food supply chain (adapted from Parfitt *et al.*, 2010).

- composting
- anaerobic digestion (loses much of the chemical value and low carbon efficiency).

Composting is a popular re-use practice as it lowers waste management costs, diverts waste from landfill and reduces waste disposal costs (Schaub and Leonard, 1996), but composting is also 'time consuming, location dependent and subject to contamination' (Davis, 2008).

While progress is being made in using anaerobic digestion to both treat food waste and provide some energy value, the chemical and material potential of food supply chain waste is such that we should also quickly move to realizing that potential through the use of other green chemical technologies. There are five reasons to develop the valorization of residues and by-products of food waste: they are a rich source of functionalized molecules (i.e., biopolymers, protein, carbohydrates), abundant, readily available, under-utilized and renewable. Many waste streams even contain compounds such as antioxidants which could be recovered, concentrated and re-used in functional food and lubricant additives (Peschel et al., 2006). Such applications solve both a resource problem and waste management problem as the issues associated with agro-food waste are important. Other than decreasing landfill options, when landfilled, waste is a source of pollution: municipal solid waste for example can produce uncontrolled GHG emissions and contaminate water supplies through leaching of inorganic matter (Cheng and Hu, 2010). Incineration is energy intensive, emits $CO_2$ and toxins and is sensitive to the waste's moisture content. The scale at which waste from the food supply chain is generated is significant. In the United States, USDA calculated that US$ 50 million could be saved annually if 5% of the waste generated by the processing, retail and service sectors together with consumer food losses were recovered (Laufenberg et al., 2003).

This type of waste represents a valuable and sustainable source of useful products that could be used by other industries, especially the chemical industry, as shown in Fig. 1.7. Novel strategies and technologies for waste valorization can potentially have a global impact on the chemical and biotechnological industries and waste management regulations in the years to come. However, despite the clear benefits, the utilization of food waste represents a challenge. A regular and consistent supply chain is important for the successful realization of a biorefinery. But high cumulative volumes of waste are often generated intermittently, over a period of a couple of months in a year, affecting the year-round availability of chemicals and materials produced from food supply chain by-products and residues. The large volumes of food supply chain by-products available are illustrated in Table 1.4.

*Table 1.4* Examples of food supply chain by-products and corresponding volumes available

| Nature of the food supply chain by-product | Estimated volume/year |
|---|---|
| Citrus waste produced post-juicing | 5,000,000 T in Florida, USA |
| Used cooking oil | 0.7–1 million T in Europe |
| Palm oil residues | 15,800,000 T in Indonesia |
| Olive mill residue | 30,000,000 T in the Mediterranean basin |
| Cocoa pods | 20 million T in Ivory Coast |
| Rice husks | 110 million T worldwide |
| Bagasse | 194,692,000 T in Brazil |
| Starchy wastes | 8 million T in Europe |
| Wheat straw surplus | 5.7 million T in Europe |
| Tomato pomace | 4 million T in Europe |
| Grape pomace | 15 million T in the USA |

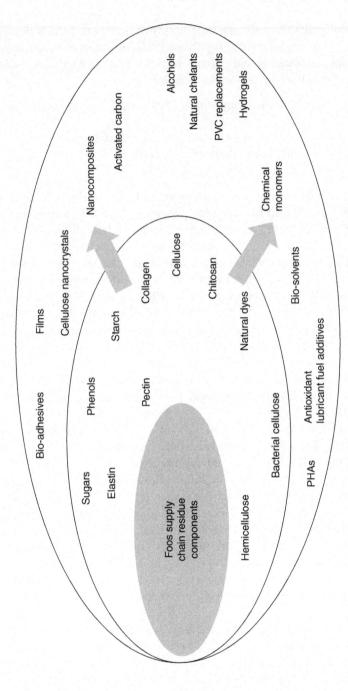

1.7 Components of food supply chain waste useful in bio-derived daily consumer products.

Although the availability of some food supply chain by-products is clearly an advantage regarding security of supply, several limitations exist and need to be taken into account as part of the logistics needed to valorize this resource. Food supply chain waste can be/can have:

- a heterogeneous variable composition (lipids, carbohydrates, proteins) (Litchfield, 1987)
- fluctuating in volumes available across the year (Litchfield, 1987)
- a high water content (Laufenberg et al., 2003) and
- a low calorific value (Laufenberg et al., 2003).

At a European level, research is being promoted via the Framework VII KBBE (Knowledge-Based Bio-economy) theme. In the UK, a number of food supply chain waste related research projects are being carried out in collaboration with industry on, for example, the use of supercritical carbon dioxide to extract chemicals from cereal straws and also the use of starch-rich wastes to make adhesives for carpet tiles and other consumer goods. In the Review of Waste Policy issued by DEFRA in June 2011, launching a zero waste economy plan, the UK Government announced it will work with industry to drive innovation in reuse and recycling for materials, such as metals, textiles and all biodegradable waste (DEFRA, 2011). The EU issued new FP7 funding calls for 2012 mentioning that 'research is needed to develop innovative concepts and practical approaches that would add value to and find markets for food waste of plant and dairy origin' (CORDIS, 2011).

In France, work on the valorization of oil crop by-products is now being supported by the French government-funded 'project PIVERT'. In Spain, a research team in Barcelona is studying the use of amino acids derived from food supply chain residues for the synthesis of amino-acid derived surfactants such as ethyl-$N$-lauroyl-L-arginate HCl or LAE, which have been successfully commercialized (Infante et al., 1992). Waste cooking oil and citrus waste produced from the juicing industry are also being studied in Spain as raw materials for the production of bio-diesel and bio-ethanol/D-limonene extraction, respectively (Kulkarni and Dalai, 2006; Sunde et al., 2011). In Greece, whey is being explored as feedstock for microbial oil production that could be used for oleochemical synthesis (Vamvakaki et al., 2010).

The topic is gaining increased attention worldwide: the NAMASTE project (EU–India) is directed at the valorization of selected by-products, such as fruit and cereal processing residues, for the global food and drink industry. In the United States, the Center for Crop Utilization Research at Iowa State University is focusing on adding value to Midwest crop (i.e., soy, corn) by-products to increase the value of the food supply chain. Scientists at the American company Cardolite have succeeded in producing

# Green chemistry, biorefineries and second generation strategies 25

thermosetting binder resins for use in the transportation and brake industries from cashew nutshell liquid (highly thermostable, impermeable and durable) (Cardolite, n.d.).

## 1.3.4 From first generation waste re-use to second generation waste re-use

The example given in Section 1.3.2 shows how one step change in a process can avoid further fossil resource deletion by recycling waste. But there are smarter ways of using food supply chain waste: this type of co-product is rich in chemical compounds and it is important to take advantage of that resource before using it for energy generation. The food supply chain generates a high amount of waste, even at a pre-consumer stage. Around 89 million tons of food waste is generated every year in the EU-27 (Bio Intelligence Service, 2010). Some 38% is generated by the manufacturing sector, 42% by the household sector (other sectors: 19%). First generation food supply chain waste re-use such as anaerobic digestion, composting, or conversion to animal feed only has marginal economic value compared to the revenue that could be generated from the production of pectin (10–12 £/kg) from citrus peels, for example.

In addition, when using waste, several criteria need to be considered in order to make sure the feedstock chosen is going to be used over the long term. Volumes available, occurrence in several geographical locations, guaranteeing a regular supply throughout the year, chemical functionalities present, extractables recoverable and their value as well as fitting the feedstock with appropriate green chemical technologies are all important parameters to consider when selecting a waste by-product for valorization.

Wheat straw is a major by-product of the agricultural sector. It is estimated that in the UK alone, 6.3 million tonnes of wheat straw was generated in 2007 (NNFCC, 2008), with a net surplus over livestock demand of 5.7 million tonnes in 2007. In the context of the UK Government's new targets on biomass generated heat and power (5% by 2020) (HM Government, 2009), wheat straw represents a good choice of feedstock for combustion for heat and power generation. However, available valuable chemical functionalities should be recovered before any thermo- or biochemical processes are applied to wheat straw for conversion to energy. Two valorization routes have been demonstrated (see Fig. 1.8): the combustion of wheat straw and subsequent valorization of the slag and fly ash produced and supercritical $CO_2$ extraction of waxes followed by char production from wheat straw by microwave pyrolysis. Both approaches are aiming for the development of a close to zero integrated wheat straw biorefinery. The first one valorizes the by-product of the combustion of wheat straw: the high

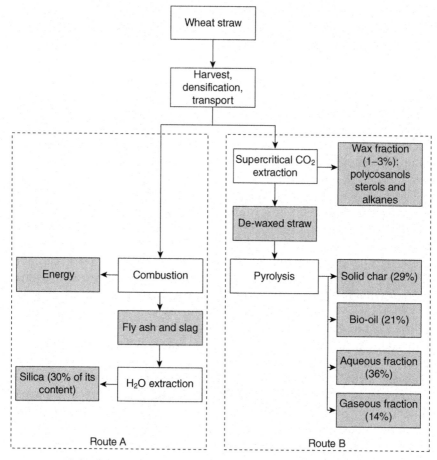

*1.8* A wheat-straw based biorefinery: comparison between two possible routes.

content of alkalis (chloride, $K_2O$ and $SiO_2$) can be extracted by water at room temperature. Up to 30% of the silica present in the ash (wheat straw ash contains 44.25% silica on a dry weight basis) can be extracted at room temperature in the form of a bio-silicate solution by using wheat straw's own alkali content. Silicates are studied as an alternative to formaldehyde-based adhesives in entirely bio-derived, fire resistant, moisture resistant construction boards and have the potential to improve the cost-effectiveness of energy producing technologies such as combustion and help the direct production of materials from agricultural biomass (Dodson *et al.*, 2011).

The second approach takes advantage of the combination of two green technologies: supercritical $CO_2$ extraction and low temperature microwave pyrolysis, benefiting from the financial return offered by the extraction of phytochemicals prior to the production of char by microwave pyrolysis at 180°C. The first step is the extraction of the wax coating the wheat straw: between 0.9 and 1.1 wt% at 32°C and 100°C, respectively, which is comparable to hexane. The added advantage associated with using supercritical $CO_2$ over hexane is that unwanted components such as pigments, free sugars and polar lipids are less soluble in supercritical $CO_2$ than in hexane. Compounds found in the extracted wax range from 6,10,14-trimethyl 2-pentadecanone used in detergents, to nonacosane, a bio-derived type of paraffin wax, and octadecanal, an aldehyde used as a flavouring additive in foods. The de-waxed wheat straw is then pyrolysed using microwaves as a heating method, producing five fractions. They are described as follows:

1. A char (29 wt%) of a calorific value of 27.2 kJ/g, which can be demineralized to avoid alkali corrosion during combustion due to the formation of alkali ash.
2. Bio-oil (21 wt%) with a reduced water (1%) and acid content (pH 7) compared to oils obtained by fast pyrolysis at temperature above 350°C, requiring less downstream processing to be used in blends with crude oil for chemical and fuel production.
3. An aqueous solution (36 wt% together with the second aqueous fraction) made of formic acid, formaldehyde, acetic acid and acetaldehyde, all of which represent interesting starting materials for further downstream chemistry. Formaldehyde has an existing market as a disinfectant.
4. An aqueous solution of sugars which can be fermented to higher volume chemicals or biofuels.
5. A gaseous fraction (14 wt%) composed of CO and $CH_4$ that could be used to fuel the process and $CO_2$ which could be used for the wax extraction.

It should be noted that both technologies are scalable and are commercially used by the food industry and yield several useful marketable products in the context of a wheat straw biorefinery. Microwave technology is less sensitive to water content than conventional convection heating. The use of biomass with a high water content can prove to be advantageous as water can dissociate at higher temperatures under microwave conditions (Vaks et al., 1994) and can generate an *in-situ* acidic pseudo catalysis process benefiting the targeted process (extraction, chemical reaction). Furthermore it is portable, tuneable (additive, temperature, pressure, power) and fast, proving to be applicable to a variety of feedstocks, or feedstock agnostic. In terms of energy consumption, the described process only requires 1.8 kJ/g of energy compared to 2.7 kJ/g when using convection heating for the pyrolysis stage (Budarin et al., 2011). Supercritical $CO_2$ may require a very high capital investment, but on a large scale, it has been proven to be more cost-competitive than using hexane (List *et al.*, 1989), as this technique is virtually residue-free, requiring less downstream separation to achieve high purity of the extracted compounds.

Straw represents 50% of the yield of a cereal crop (Clynes, 2009) and with 650,881,002 tonnes of wheat produced in the world in 2010, wheat straw represents an important agricultural by-product occurring on every continent on the planet, with Europe, Asia and North America being the largest wheat producers. In conclusion, by integrating just two green processes, several products can be obtained starting from a unique feedstock available worldwide.

## 1.4 Conclusion and future trends

Given the widespread distribution of bio-feedstocks such as dedicated non-food crops or food supply chain residues, the development of small localized biorefineries compared to traditional mega-scale refineries is attractive. This will ensure that biomass is valorized as closely as possible to its production site, avoiding high transport costs for lower value feedstocks and increasing the sustainability of the process, as well as making sure as little as possible biomass is imported to meet targets. Such an approach will also prove the feasibility and the scalability of novel clean and green technologies while requiring a lower primary investment. This will encourage further industry sectors to support biomass and food supply chain residue conversion to bio-chemicals, bio-materials and bio-fuels. Important steps in this direction include the use of continuous processing and of feedstock agnostic technologies to allow maximum biomass conversion efficiency and flexibility in operation to suit places with multiple resources (e.g., an area growing or processing fruit and vegetables). Biofuels alone are likely to become insufficient as green products as wind, solar and other clean energies develop; but the combination of bio-fuels and the higher value bio-chemicals can make biorefineries *the* sustainable production chain for the twenty-first century, just as petroleum refineries dominated the twentieth century.

## 1.5 Sources of further information and advice

- EU COST Action TD1203 'Food Waste Valorisation for Sustainable Chemicals, Materials & Fuels', http://costeubis.org/
- S. K. C. Lin *et al.*, *Energy Environ. Sci.*, 2013, 6, 426–464.
- L. Pfaltzgraff *et al.*, *Green Chem.*, 2013, 15, 307–314.
- Green Chemistry Network: http://www.greenchemistrynetwork.org/index.htm

- *Handbook of Green Chemistry and Technology*, edited by James Clark & Duncan Macquarrie, Blackwell Publishing, Oxford, 2002.
- *Renewable Raw Materials – New Feedstocks for the Chemical Industry*, edited by R. Ulber, D. Sell and T. Hirth, Wiley-VCH, Weinheim, 2011.
- *Feedstocks for the Future – Renewables for the Production of Chemicals and Materials*, edited by J. J. Bozell and M. K. Patel, American Chemical Society, Washington, DC, 2006.

## 1.6 References

Anastas, P. T. and Kirchoff, M. M. 2002. Origins, current status, and future challenges of green chemistry. *Accounts of Chemical Research*, 35, 686–694.

Anastas, P. T. and Lankey, R. L. 2000. Life cycle assessment and green chemistry: the yin and yang of industrial ecology. *Green Chemistry*, 2, 289–295.

Anastas, P. T. and Warner, J. C. 1998. *Green Chemistry: Theory and Practice*. Oxford: Oxford University Press.

Anastas, P. T., Heine, L. G. and Williamson, T. C. 2000. Green chemical syntheses and processes: introduction. In: Anastas, P. T., Heine, L. G. and Williamson, T. C. (eds) *Chemical Syntheses and Processes*. Washington, DC: American Chemical Society.

Arshadi, M., Hunt, A. J. and Clark, J. H. 2012. Supercritical fluid extraction (SFE) as an effective tool in reducing auto-oxidation of dried pine sawdust for power generation. *RSC Advances*, 2, 1806–1809.

Binder, J. B. and Raines, R. T. 2009. Simple chemical transformation of lignocellulosic biomass into furans for fuels and chemicals. *Journal of the American Chemical Society*, 31, 1979–1985.

Bio Intelligence Service 2010. Preparatory study on food waste across EU-27. European Commission (DG ENV). Available at: ec.europa.eu/environment/eussd/pdf/bio_foodwaste_report.pdf (accessed 6 October 2010).

Budarin, V. L., Clark, J. H., Lanigan, B. A., Shuttleworth, P., Breeden, S. W., Wilson, A. J., Macquarrie, D. J., Milkowski, K., Jones, J., Bridgeman, T. and Ross, A. 2009. The preparation of high-grade bio-oils through the controlled, low temperature microwave activation of wheat straw. *Bioresource Technology*, 100, 6064–6068.

Budarin, V. L., Clark, J. H., Lanigan, B. A., Shuttleworth, P. and Macquarrie, D. J. 2010. Microwave assisted decomposition of cellulose: a new thermochemical route for biomass exploitation. *Bioresource Technology*, 101, 3776–3779.

Budarin, V. L., Shuttleworth, P. S., Dodson, J. R., Hunt, A. J., Lanigan, B., Marriott, R., Milkowski, K. J., Wilson, A. J., Breeden, S. W., Fan, J., Sin, E. H. K. and Clark, J. H. 2011. Use of green chemical technologies in an integrated biorefinery. *Energy & Environmental Science*, 4, 471–479.

Cardolite. n.d. Home page. Available at: http://www.cardolite.com/ (accessed 23 March 2012).

Cheng, H. and Hu, Y. 2010. Municipal solid waste (MSW) as a renewable source of energy: current and future practices in China. *Bioresource Technology*, 101, 3816–3824.

Cherubini, F. 2010. The biorefinery concept: using biomass instead of oil for producing energy and chemicals. *Energy Conversion and Management*, 51, 1412–1421.

Cherubini, F., Jungmeier, G., Wellisch, M., Willke, T., Skiadas, I. and Van Ree, R. 2009. Toward a common classification approach for biorefinery systems. *Biofuels, Bioproducts and Biorefining*, 3, 534–546.

Clark, J. H. and Macquarrie, D. 2002. *Handbook of Green Chemistry and Technology*. Oxford: Wiley-Blackwell Publishing.

Clark, J. H., Budarin, V., Deswarte, F. E. I., Hardy, J. J. E., Kerton, F. M., Hunt, A. J., Luque, R., Macquarrie, D. J., Milkowski, K., Rodriguez, A., Samuel, O., Tavener, S. T., White, R. J. and Wilson, A. J. 2006. Green chemistry and the biorefinery: a partnership for a sustainable future. *Green Chemistry*, 8, 853–860.

Clark, J. H., Deswarte, F. E. I. and Farmer, T. J. 2009. The integration of green chemistry into future biorefineries. *Biofuels, Bioproducts and Biorefining*, 3, 72–90.

Clynes, J. 2009. *Decay Characteristics of Different Types of Straw Used in Straw Bale Building*. Available at: http://homegrownhome.co.uk/pdfs/jamesclynesmsc thesis.pdf (accessed 8 March 2012).

CORDIS. 2011. *KBBE.2012.2.3-01: Feed Production from Food Waste*. Available at: https://cordis.europa.eu/partners/web/req-2735 (accessed 5 January 2012).

CRODAROM. n.d. *Crodarom Production Facilities*. Available at: http://www. crodarom.com/home.aspx?s=110&r=124&p=896 (accessed 3 March 2012).

Davis, R. A. 2008. Parameter estimation for simultaneous saccharification and fermentation of food waste into ethanol using Matlab Simulink. *Applied Biochemistry and Biotechnology*, 147, 11–21.

DEFRA. 2011. *Government Review of Waste Policy in England 2011*. Available at: http://www.defra.gov.uk/publications/files/pb13540-waste-policy-review 110614.pdf (accessed 25 March 2012).

DEFRA. 2012. *Sustainable businesses and resource efficiency*. Available at: http://www.defra.gov.uk/environment/economy/business-efficiency/ (accessed 26 February 2012).

Deswarte, F. E. I., Clark, J. H., Hardy, J. J. E. and Rose, P. M. 2006. The fractionation of valuable wax products from wheat straw using $CO_2$. *Green Chemistry*, 8, 39–42.

Deye, J. F., Berger, T. A. and Anderson, A. G. 1990. Nile Red as a solvatochromic dye for measuring solvent strength in normal liquids and mixtures of normal liquids with supercritical and near critical fluids. *Analytical Chemistry*, 62, 615–622.

Dodson, J. R., Hunt, A. J., Budarin, V. L., Matharu, A. S. and Clark, J. H. 2011. The chemical value of wheat straw combustion residues. *RSC Advances*, 1, 523–530.

Dodson, J. R., Hunt, A. J., Parker, H. L., Yang, Y. and Clark, J. H. 2012. Elemental sustainability: towards the total recovery of scarce metals. *Chemical Engineering and Processing*, 51, 69–78.

Eggeman, T. and Elander, R. T. 2005. Process and economic analysis of pretreatment technologies. *Bioresource Technology*, 96, 2019–2025.

Environmental Protection Agency (EPA) 2011. Chapter 9: Food and Agricultural Industry, Section 9.9.7. Corn wet milling. Washington, DC: EPA.

EPREMASA. n.d. *EPREMASA, Empresa Provincial de Residuos y Medio Ambiente, S.A.* Available at: http://www.epremasa.es/index.php/quienes-somos (accessed 4 February 2012).

European Commission n.d. The European Bioeconomy in 2030. Available at: www.epsoweb.org/file/560 (accessed 6 October 2013).

European Commission 2007. *Accelerating the Development of the Market for Bio-Based Products in Europe.* Available at: http://ec.europa.eu/enterprise/policies/innovation/files/lead-market-initiative/prep_bio_en.pdf.

European Commission 2009. *Taking Bio-Based from Promise to Market.* Available at:http://ec.europa.eu/enterprise/sectors/biotechnology/files/docs/bio_based_from_promise_to_market_en.pdf.

European Commission 2010. *Europe 2020.* Available at: http://ec.europa.eu/europe2020/index_en.htm (accessed 2 February 2012).

European Commission 2011.Technical Committee TC411. M/492 Mandate addressed to CEN, CENELEC and ETSI for the development of horizontal European standards and other standardisation deliverables for bio-based products. Brussels: European Commission.

Fernández, Y., Arenillas, A. and Menéndez, J. A. 2011. Microwave heating applied to pyrolysis. In: Grundas, S. (ed.) *Advances in Induction and Microwave Heating of Mineral and Organic Materials.* Rijeka, Croatia: InTech.

Giljum, S., Hinterberger, F., Bruckner, M., Burger, E., Frühmann, J., Lutter, S., Pirgmaier, E., Polzin, C., Waxwender, H., Kernegger, L. and Warhurst, M. 2009. *Overconsumption? Our use of the world's natural resources.* Available at: http://www.foe.co.uk/resource/reports/overconsumption.pdf.

Gómez, A., Zubizarreta, J., Rodrigues, M., Dopazo, C. and Fueyo, N. 2010. An estimation of the energy potential of agro-industrial residues in Spain. *Resources, Conservation and Recycling,* 54, 972–984.

Gustavsson, J., Cederberg, C., Sonesson, U., Van Otterdijk, R. and Meybeck, A. 2011. *Global Food Losses and Food Waste.* FAO. Available at: http://www.fao.org/fileadmin/user_upload/ags/publications/GFL_web.pdf (accessed 2 February 2012).

HM GOVERNMENT 2009. *The UK Renewable Energy Strategy.* Available at: www.official-documents.gov.uk/document/cm76/7686/7686.pdf (accessed 6 October 2013).

IEA 2009. *IEA Bioenergy Annual Report 2009.* Available at: www.ieabioenergy.com/DocSet.aspx?id=6506&ret=lib (accessed 6 October 2013).

Infante, M. R., Molinero, J. and Erra, P. 1992. Lipopeptidic surfactants acid and basic N-lauroyl-L-arginine dipeptides from pure amino acids. *Journal of the American Oil Chemists' Society,* 69, 647–652.

Kamm, B. and Kamm, M. 2004. Principles of biorefineries. *Applied Microbiological Biotechnology,* 64, 137–145.

Kroyer, G. T. 1995. Impact of food processing on the environment – an overview. *Lebensmittel-Wissenschaft und Technologie – Food Science and Technology,* 28, 547–552.

Kulkarni, M. G. and Dalai, A. K. 2006. Waste cooking oils – an economical source for biodiesel: a review. *Industrial & Engineering Chemistry Research,* 45, 2901–2913.

Lancaster, M. 2010. *Green Chemistry, An Introductory Cast.* Cambridge: The Royal Society of Chemistry.

Laufenberg, G., Kunz, B. and Nystroem, M. 2003. Transformation of vegetables into value added products: (A) the upgrading concept; (B) practical implementations. *Bioresource Technology,* 87, 167–198.

List, G. R., Friedrich, J. F. and King, J. W. 1989. *Oil Mill Gazette*, 93, 28–34.

Litchfield, J. H. 1987. Microbiological and enzymatic treatments for utilizing agricultural and food processing wastes. *Food Biotechnology*, 1, 29–57.

NNFCC. 2008. *National and regional supply/demand balance for agricultural straw in Great Britain.* Available at: http://www.northwoods.org.uk/files/northwoods/StrawAvailabilityinGreatBritain.pdf (accessed 6 February 2012).

OECD. 2008. *Biofuel policies in OECD countries costly and ineffective.* Available at: http://www.oecd.org/document/28/0,3343,en_2649_37401_41013916_1_1_1_1,00.html (accessed 6 March 2012).

Parfitt, J., Barthel, M. and Macnaughton, S. 2010. Food waste within the food supply chains: quantification and potential for change to 2050. *Philosophical Transactions of the Royal Society of Biological Sciences*, 365, 3065–3081.

Peschel, W., Sanchez-Rabaneda, F., Diekmann, W., Plescher, A., Gartzia, I., Jimenez, D., Lamuela-Ravento, R., Buxaderas, S. and Codina, C. 2006. An industrial approach in the search of natural antioxidants from vegetable and fruit wastes. *Food Chemistry*, 97, 137–150.

Ragauskas, A. J., Williams, C. K., Davison, B. H., Britovsek, G., Cairney, J., Eckert, C. A., Frederick, W. J. J., Hallett, J. P., Leak, D. J., Liotta, C. L., Mielenz, J. R., Murphy, R., Templer, R. and Tschaplinski, T. 2006. The path forward for biofuels and materials. *Science*, 311, 484–489.

Sahena, F., Zaidul, I. S. M., Jinap, S., Karim, A. A., Abbas, K. A., Norulaini, N. A. N. and Omar, A. K. M. 2009. Application of supercritical $CO_2$ in lipid extraction – a review. *Journal of Food Engineering*, 95, 240–253.

Schaub, S. M. and Leonard, J. J. 1996. Composting: an alternative waste management option for food processing industries. *Trends in Food Science and Technology*, 7, 263–268.

Schecter, A. 1998. A selective historical review of congener-specific human tissue measurements as sensitive and specific biomarkers of exposure to dioxins and related compounds. *Environmental Health Perspectives*, 108, 737–742.

Schieber, A., Stintzing, F. C. and Carle, R. 2001. By-products of plant food processing as a source of functional compounds – recent developments. *Trends in Food Science and Technology*, 12, 401–413.

Shahid, E. M. and Jamal, Y. 2011. Production of biodiesel: a technical review. *Renewable and Sustainable Energy Reviews*, 15, 4732–4745.

Simon, M. and Li, C. 2012. Green chemistry oriented organic synthesis in water. *Chemical Society Reviews*, 41, 1415–1427.

SIN List n.d. Available at: http://www.sinlist.org/ (accessed 13 March 2012).

Sunde, K., Brekke, A. and Solberg, B. 2011. Environmental impacts and costs of hydrotreated vegetable oils, transesterified lipids and woody BTL – a review. *Energies*, 4, 845–877.

Tao, L., Aden, A., Elander, R. T., Ramesh Pallapolu, V., Lee, Y. Y., Garlock, R. J., Balan, V., Dale, B. E., Kim, Y., Mosier, N. S., Ladisch, M. R., Falls, M., Holtzapple, M. T., Sierra, R., Shi, J., Ebrik, M. A., Redmond, T., Yang, B., Wyman, C. E., Hames, B., Thomas, S. and Warner, R. E. 2011. Process and technoeconomic analysis of leading pretreatment technologies for lignocellulosic ethanol production using switchgrass. *Bioresource Technology*, 102, 11105–11114.

Vaks, V. L., Domrachev, G. A., Rodygin, Y. L., Selivanovskii, D. A. and Spivak, E. I. 1994. Dissociation of water by microwave radiation. *Radiophysics and Quantum Electronics*, 37, 85–88.

Vamvakaki, Z. A. N., Kandarakis, I., Kaminarides, S., Komaitis, M. and Papanikolaou, S. 2010. Cheese whey as a renewable substrate for microbial lipid and biomass production. *Engineering in Life Science*, 10, 348–360.

Vijayendran, B. 2010. Bio-based chemicals: technology, economics and markets (White Paper).

Welton, T. 2004. Ionic liquids in catalysis. *Coordination Chemistry Reviews*, 248, 2459–2477.

WRAP. 2009. *2009 Gate Fee report*. Available at: http://www.wrap.org.uk/downloads/W504GateFeesWEB.c640bae1.7613.pdf (accessed 23 March 2012).

WRAP. 2011. *2011 Gate Fee Report*. Available at: http://www.wrap.org.uk/recycling_industry/publications/wrap_gate_fees.html (accessed 23 March 2012).

Zaror, C. A. 1992. Controlling the environmental impact of the food industry: an integral approach. *Food Control*, 3, 190–199.

# 2
# Techno-economic assessment (TEA) of advanced biochemical and thermochemical biorefineries

T. R. BROWN, Iowa State University, USA,
M. M. WRIGHT and Y. ROMÁN-LESHKOV, Massachusetts Institute of Technology, USA and
R. C. BROWN, Iowa State University, USA

DOI: 10.1533/9780857097385.1.34

**Abstract**: This chapter covers techno-economic assessments (TEA) of advanced biochemical and thermochemical biorefineries. We discuss how governments, companies, and academic institutions are affecting the economic prospects of advanced biorefineries. The text describes their economic challenges and the various strategies being pursued to increase commercial adoption of advanced biorefineries: government incentives, facility scale-up, and technological innovation. Finally, we present an overall view of emerging trends in biorefinery TEAs with the intent of identifying key opportunities for improvement.

**Key words**: biorefinery techno-economic analysis (TEA), advanced biofuel incentives, biomass costs and logistics, thermochemical and biochemical conversion.

## 2.1 Introduction

The pace of biorefinery technology research and development is increasing, fueled by concerns over energy security and environmental impacts. Academic institutions and national laboratories are leading the assessment of promising biorefinery concepts. These assessments investigate concepts at various development stages – from laboratory research to plant-scale commercialization. In this chapter we summarize recent techno-economic analysis findings, discuss how policy influences biomass trade and industry subsidies, and describe the differences between national and regional biorefineries. Our concluding section contemplates the impacts of current challenges and emerging trends.

The term biorefinery encompasses different types of facilities that can convert biomass into valuable products (Brown, 2003). We define a biorefinery as an integrated facility capable of producing fuel, electricity, chemicals, and other types of bioproducts. This concept allows for the full utilization of biomass compounds and the versatility to vary the product distribution. This concept builds upon the desire to replace every product derived from a barrel of crude oil. In addition to the main industrial building blocks, researchers envision biorefineries that could produce novel types of chemicals and polymers.

There are several biorefinery concepts at various stages of development and commercialization. Their future prospects depend on technological innovation and market conditions. Given the multiple steps required to convert biomass into marketable products, research breakthroughs along any of the conversion steps could accelerate the adoption of a given pathway. Similarly, market conditions could turn formerly unprofitable schemes into commercial successes. More than likely it will be a combination of technological and economic changes that lead the way to commercially-viable biorefineries. Thus, techno-economic assessments provide a unique perspective on current and future biorefinery technologies.

The United States and European governments have established guidelines and incentives to develop renewable fuels. Biorefineries that meet specific government conditions are eligible to receive financial incentives in the form of subsidies. Biorefinery subsidies are projected to increase from US$66 billion today to almost US$250 billion by 2035 (Anon., 2011).

Some biorefinery subsidies have expired in recent years without major impacts to industry growth, signaling the maturity of the biofuel market. However, government subsidy programs have begun to set strict requirements relating to direct and indirect lifecycle greenhouse gas emissions (GHG) on advanced biorefineries as a condition of participation. In order to meet these requirements, renewable energy companies are seeking novel approaches to convert a wider range of biomass into cleaner, cheaper bioproducts. This search requires the assessment of the technical and economic prospects of novel pathways. Therefore, governments are collaborating with academic institutions, national laboratories, and commercial enterprises.

The diffuse nature of biomass availability means that biorefinery scale-up will have wide area impacts on local, regional, and national scales. Although most biorefineries today are limited to capacities of about 100 million gallons (379 million liters) per year, process development and improved biomass logistics could lead to larger biorefineries that gather biomass from hundreds of square kilometers via truck, rail, or barge transport. These biorefineries will present novel, international case scenarios. Thus, there is interest in studying the challenges and opportunities for biorefineries in the global market from a TEA perspective.

The biomass industry presents a rapidly changing landscape with challenges and opportunities. Three recent trends have emerged as the result of faltering government support, sustained high petroleum prices, and changing public opinions on biorenewables: the replacement of starch feedstocks with lignocellulosic biomass, interest in thermochemical pathways, and public and private investment in high-risk, high-reward alternatives.

Lignocellulosic biomass has historically proven difficult to convert into bioproducts with traditional biochemical approaches due to the biological recalcitrance of cellulose and antimicrobial properties of lignin. Developments in genetic and metabolic engineering have opened new pathways to convert hemicellulose into valuable products. These avenues range from enhancing biomass growth to developing bacteria strains capable of digesting formerly discarded or toxic parts of biomass crops.

Thermochemical research pathways have attracted recent attention due to their ability to inexpensively convert lignocellulosic feedstocks to energy-dense gases and liquids. Most of the recent development in the field has been on adapting conventional commercial processes (such as those employed by petroleum refineries) to handle biomass feedstocks and biobased intermediate products. However, there are growing efforts in the search for novel catalysts that can optimize the selectivity and yield of desired bioproducts. Finally, researchers have also proposed hybrid approaches that combine the strengths of the biochemical and thermochemical platforms (Brown, 2005).

There is growing commercial and political support for the development of high-risk, high-reward platforms such as microalgae-to-fuels, furan synthesis, and sugar-based hydrocarbons. Government-funded algae research was eliminated in the 1990s in favor of ethanol but has recently staged a resurgence due to concerns over land availability and GHG emissions.

## 2.2 Biorefinery economic assessment

Economic assessments are becoming important in the analysis of biorefinery concepts (Wright and Brown, 2011). This is due in part to the lack of commercial experience in establishing novel technologies that can convert alternative feedstock into products not commonly derived from renewable sources. Another factor is the recent volatility in oil prices yielding to the possibility of long-term financial viability of biorefinery projects. Finally, strong governmental support ensures that attractive processes receive financial support during development. These factors have resulted in an increase in the publication rates of biorefinery techno-economic studies.

Recent biorefinery techno-economic papers have focused on advanced biorefineries based on the thermochemical and biochemical platforms. These papers assess the technical and economic viability of technologies ranging from the development phase up to demonstration stage. Some assessments benefit from the data available for established commercial technologies employed in other industries. However, many of the advanced biorefinery technologies are first-of-a-kind facilities, which present a major engineering challenge. In this section, we discuss how insights provided by techno-economic analysis have contributed to our understanding of advanced thermochemical and biochemical biorefinery concepts.

## 2.2.1 Bioproducts from thermochemical biorefineries

Researchers have developed detailed TEAs for several biofuels including hydrogen, methanol, ethanol, mixed alcohols, Fischer–Tropsch liquids, and naphtha and diesel range blend stock fuels (Wright and Brown, 2007a). Biofuel synthesis pathways can be categorized by the feedstock intermediate, subsequent upgrading process, and type of biofuel output. Feedstock intermediates are classified here as the primary products from biomass torrefaction, pyrolysis, and gasification. Upgrading processes are associated with specific intermediate products, although torrefied biomass and bio-oil can be converted into syngas and upgraded through alternative pathways. An overview of the main thermochemical biomass-to-liquid fuel pathways is shown in Fig. 2.1.

The syngas pathway leads to several types of biofuel products depending on the upgrading process: alcohol synthesis can output mostly methanol,

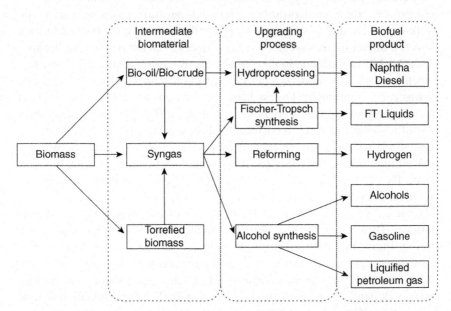

2.1 Thermochemical biomass conversion pathway intermediates, upgrading processes, and final products.

ethanol, or mixed alcohols; methanol from alcohol synthesis can be further upgraded to gasoline and liquefied petroleum gas via the methanol-to-gasoline (MTG) process; steam reforming results in high hydrogen yields, and Fischer–Tropsch synthesis generates a mixture of hydrocarbons ranging from light gases to waxes ($C_1 \rightarrow C_{120}$). Fischer–Tropsch liquids can substitute for diesel, but they have slightly different properties than conventional diesel. Therefore, some studies include a hydroprocessing unit to increase the output of naphtha- and diesel-range biofuels.

Capital costs vary significantly between different biofuel synthesis routes and plant configurations. Table 2.1 shows capital costs for biomass conversion to hydrogen, methanol, mixed alcohols, and Fischer–Tropsch liquids via syngas production and conversion. These capital costs have been re-categorized from the original analyses according to major process steps or sections. In general, pretreatment includes feedstock drying and grinding; gasification consists of the gasifier and auxiliary equipment; oxygen separation refers to the air separation island; gas cleaning includes particulate, tar, and impurity removal; syngas conditioning refers to water gas shift or reforming processes required for synthesis; synthesis/production is the main step to produce the desired fuel based on given specifications; steam and power generation, and utilities/miscellaneous group all auxiliary units required for the overall operation of the facility.

The selection of processes shown in Table 2.1 includes high and low temperature gasifiers; oxygen and air blown gasification systems; and steam and power export configurations. High temperature gasifiers tend to be more expensive due to strict metallurgy and operating constraints, but they deliver a higher quality syngas, which reduces gas cleaning and conditioning costs. Air blown gasifiers do not require air separation equipment, which is expensive, but they dilute syngas with nitrogen, which increases the size of downstream equipment. Excess heat and fuel gas are typically converted into steam and/or power depending on the quantity and quality available.

Biofuel costs include feedstock and operating costs in addition to annualized capital costs. Feedstock costs typically contribute between a quarter and more than half of the final biofuel cost because of high biomass costs. By-products, mostly heat and power, can sometimes contribute significant revenue.

Table 2.2 shows the annualized costs for selected biofuels via the syngas pathway. These costs illustrate some of the differences in assumptions found in the literature. Although the biorefinery capacities are similar, the annualized capital, operation and management, and biomass costs vary widely. Annual capital expenditures, including depreciation and capital charges, depend on the estimates described in Table 2.1 and several financial assumptions. Operation and management costs are typically based on local labor rates and factors for equipment maintenance. The differences in

Table 2.1 Capital costs for biomass to fuel conversion via the syngas pathway

| | Hydrogen (Hamelinck and Faaij, 2002) | Methanol (Hamelinck and Faaij, 2002) | Mixed alcohols (Phillips et al., 2007) | Fischer–Tropsch liquids (Tijmensen et al., 2002) | Gasoline (Phillips et al., 2011) | Naphtha and diesel via FTL (Swanson et al., 2010) |
|---|---|---|---|---|---|---|
| Cost basis (year) | 2001 | 2001 | 2007 | 2001 | 2007 | 2007 |
| Capacity (dry metric tonnes per day) | 1920 | 1920 | 2000 | 1920 | 2000 | 2000 |
| | | | Capital costs ($MM) | | | |
| Pretreatment | 38.2 | 38.2 | 23.2 | 71.6 | 25.0 | 22.7 |
| Gasification | 73.0 | 30.4 | 12.9 | 61.4 | 14.6 | 67.8 |
| Air separation | 27.7 | 0 | 0 | 51.2 | 0 | 24.3 |
| Syngas cleaning | 12.4 | 38.1 | 14.5 | 61.4 | 44.3 | 33.5 |
| Syngas conditioning | 13.3 | 62.8 | 38.4 | 3.41 | – | – |
| Synthesis/Production | 53.3 | 41.3 | 27.8 | 20.5 | 21.6 | 49.4 |
| Steam and power generation | 64.8 | 13.9 | 16.8 | 61.4 | 23.1 | 45.6 |
| Utilities/Miscellaneous | – | – | 3.6 | 10.2 | 5.9 | 33.1 |
| **Total installed equipment cost** | – | – | **137** | – | **145** | **309** |
| **Total project investment** | **282** | **224** | **191** | **341** | **200** | **606** |

Table 2.2 Operating costs for biomass to fuel conversion via the syngas pathway

| Operating costs ($MM) | Hydrogen (Hamelinck and Faaij, 2002) | Methanol (Hamelinck and Faaij, 2002) | Mixed Alcohols (Phillips et al., 2007) | Fischer–Tropsch liquids (Tijmensen et al., 2002) | Gasoline (Phillips et al., 2011) | Naphtha and diesel (Swanson et al., 2010) |
|---|---|---|---|---|---|---|
| Capital | 33.6 | 26.7 | 34.4 | 34.1 | 38.0 | 106.4 |
| Operation and management | 11.3 | 9.0 | 13.3 | 16.8 | 15.1 | 26.6 |
| Biomass | 24.7 | 24.9 | 27.0 | 34.2 | 39.1 | 51.3 |
| By-product credit | −17.4 | 0.0 | −12.8 | 0.0 | 0.0 | −5.6 |
| **Total** | **52.1** | **60.6** | **61.9** | **50.8** | **92.2** | **178.7** |
| **Biofuel MFSP[a] ($/gal)** | **$0.33** | **$0.52** | **$1.01** | **$1.89**[b] | **$1.39** | **$4.26** |
| **Biofuel MFSP ($/gge[c])** | **$1.26** | **$1.05** | **$1.54** | **$1.77** | **$1.39** | **$4.26** |

[a]MFSP: Minimum fuel selling price.
[b]Assumes 34.4 MJ/L Fischer–Tropsch liquid energy density.
[c]GGE: gallon of gasoline equivalent (32.3 MJ/L gasoline energy density).

biomass costs are due to the wide range of assumed feedstock prices ($30–$75 per metric ton). Finally, we should note that these cost assessments involve assumptions for process maturity and projections for technology improvement that significantly impact the final estimates.

There are fewer techno-economic analyses for alternatives to the syngas thermochemical pathway, such as bio-oil upgrading and hydroprocessing of lipids. This is due in part to the commercial maturity of these processes. There has been little commercial adoption in the biomass industry of alternative processes such as biomass pyrolysis to bio-oil and hydrothermal processing to bio-crude despite being under development since the 1970s. However, with rising petroleum costs, there is renewed interest in processes that replace petroleum products beyond transportation fuels.

The following processes adopt alternative routes to producing transportation fuels with valuable co-products. Biomass pyrolysis produces primarily bio-oil, which can be upgraded to fuels. Pyrolysis co-products include biochar – a soil amendment and potential carbon sequestration agent, and chemicals. Extraction and hydrolysis of bio-oil recovers sugars (pentose and hexose) that can subsequently be fermented to ethanol. Biorefineries could obtain high valued chemicals benzene, toluene, ethylene, and propylene among other hydrocarbons from bio-oil via integrated catalytic processing (ICP) using a modified HZSM-5 catalyst. Syngas fermentation could produce polyhydroxyalkonate (PHA), a biodegradable polymer, while yielding excess hydrogen. Table 2.3 shows capital and operating costs for these alternative biorefineries.

A small number of ventures have commercialized the hydroprocessing of lipids to renewable diesel and jet fuels. Commercial lipid hydroprocessing employs by-products from the food industry such as vegetable oils, waste oils, and fats. Algal biomass has the potential to become the main feedstock

Table 2.3 Capital and operating costs for alternative biorefineries producing ethanol, PHA, and aromatics and olefins

|  | Bio-oil fermentation (So and Brown, 1999) | Syngas fermentation (Choi et al., 2010) | Bio-oil integrated catalytic processing (Brown et al., 2012) |
|---|---|---|---|
| Product | Ethanol | PHA | Aromatics and olefins |
| Co-product | Sugars | Hydrogen |  |
| Capacity (per year) | 240 MM kg | 6.5 MM kg | 34.1 MM kg |
| Capital cost | $69 | $103 | $100 |
| Operating cost | $39.2 | $18.2 | $74.5 |
| Product cost | $1.59/gal | $2.80/kg | $2.18/kg |

for lipid hydroprocessing if its production costs are drastically reduced (Roesijadi et al., 2010).

### 2.2.2 Bioproducts from biochemical biorefineries

The biochemical pathway in general, and fermentation to ethanol in particular, have been employed in the US and Brazil for several decades, and its economics have been thoroughly investigated. TEAs are available for the production of ethanol from sugarcane; the production of ethanol from corn (starch); and the production of ethanol from lignocellulose. More advanced pathways such as the production of ethanol from cyanobacteria and the production of hydrocarbons from sugar fermentation are also under investigation, although TEAs for these are not yet available.

The sugarcane pathway produces ethanol, electricity, and crystallized sucrose. Sugarcane is harvested and processed to separate the plant's lignocellulose (bagasse) from the cane juice (garapa). The bagasse is combusted to provide process heat and electricity, with the latter generated in sufficient quantities to be sold onto the neighboring electricity grid. The garapa is further processed into molasses and sucrose crystals. The molasses, which are a mixture of sucrose and minerals, are sterilized and fermented with brewer's yeast (*Saccharomyces cerevisiae*) to produce a beer containing 6–10 vol% ethanol. The beer is distilled to hydrous ethanol (containing 5 vol% water) via conventional distillation. Further dehydration can occur via employment of molecular sieves or other techniques to produce anhydrous ethanol (containing less than 0.3 vol% water).

The starch ethanol pathway most commonly employs corn (maize) as feedstock, although other starch crops such as wheat and cassava are also used. It is similar to the sugarcane ethanol pathway, although a saccharification step is required to depolymerize the starch into fermentable glucose monomers. The pathway employs either dry milling or wet milling, although TEAs of wet milling are very rare and it is not covered here as a result. In dry milling the corn kernels are ground, mixed with water, and cooked to gelatinize the starch content. Enzymes are added to depolymerize the starch first to oligosaccharides and then to the monosaccharide glucose (the process is also known as saccharification). The resulting fermentation broth also contains the lipid, fiber, and protein content of the kernel, which are removed and sold following fermentation as distillers' dried grains and solubles (DDGS), a valuable livestock feed. Fermentation employs *Saccharomyces cerevisiae* and is followed by distillation and dehydration to anhydrous ethanol.

Table 2.4 reviews the capital costs for a Brazilian sugarcane ethanol biorefinery and a US corn ethanol dry mill biorefinery. Equipment components are not identical due to feedstock-specific differences between

*Table 2.4* Capital costs for first generation ethanol biorefineries

|  | Sugarcane ethanol (Efe et al., 2005) | Corn ethanol dry milling (Kwiatkowski et al., 2006) |
|---|---|---|
| Cost basis (year) | 2005 | 2005 |
| Capacity (million gallons per year) | 50 | 40 |
| | Capital costs ($MM) | |
| Milling | 3.8 | 3.4 |
| Clarification | 2.6 | – |
| Evaporation | 7.6 | – |
| Crystallization and drying | 4.0 | – |
| Saccharification | – | 5.3 |
| Fermentation | 4.2 | 10.5 |
| Distillation | 4.0 | 8.0 |
| Coproduct processing | 19.7 | 19.5 |
| **Total installed equipment cost** | **45.9** | **46.7** |
| **Total project investment** | **101.9** | **103.7**[a] |

[a] Adjusted to account for indirect costs not included in original assessment.

*Table 2.5* Operating costs for first generation ethanol biorefineries

| Operating costs ($MM) | Sugarcane ethanol (Efe et al., 2005) | Corn ethanol dry milling (Kwiatkowski et al., 2006) |
|---|---|---|
| Capital depreciation | 10.2 | 10.4* |
| Operation and management | 12.2 | 13.3 |
| Biomass | 64.2 | 35.1 |
| By-product credit | –77.4 | –11.7 |
| Total | 9.2 | 47.1 |
| Biofuel price ($/gal) | $0.82 | $1.17[a] |
| Biofuel price ($/gge) | $1.23 | $1.76[a] |

[a] Adjusted to account for indirect costs not included in original assessment.

the pathways: a sugarcane ethanol biorefinery requires equipment to process the garapa into molasses and sucrose crystals, whereas a corn ethanol biorefinery requires equipment to convert starch into dextrose via enzymatic hydrolysis. The corn ethanol facility is more expensive on an equal capacity basis as a result of the saccharification step and increased fermentation and distillation steps.

Table 2.5 reviews the operating costs for sugarcane to ethanol and starch ethanol dry milling biorefineries. While sugarcane ethanol biorefineries pay nearly twice as much in annualized costs for feedstock than do corn ethanol biorefineries of comparable ethanol output, they also derive significantly more revenue from by-product credits in the form of electricity and crystallized sucrose sales. These by-product credits cause the total annualized operating costs to be lower for a sugarcane ethanol biorefinery than a corn ethanol biorefinery despite the former's higher feedstock costs, resulting in a lower biofuel production cost for sugarcane ethanol than corn ethanol.

Lignocellulosic biomass can also be converted into ethanol via fermentation, although the recalcitrance of cellulose (a linear-chain polysaccharide) and the antimicrobial properties of lignin make it a significantly more challenging and expensive pathway than first generation ethanol pathways. The biomass is first milled to increase the surface area of the lignocellulosic material and increase hydrolysis efficiency. A pretreatment step is also commonly employed to maximize hydrolysis efficiency and dilute acid, steam explosion, and ammonia fiber explosion are considered to be the most feasible (Kazi et al., 2010). The choice of pretreatment step affects both biorefinery operating costs and ethanol yields.

Pretreatment is followed by hydrolysis. One of three hydrolysis steps is employed to convert cellulose and any hemicellulose remaining following pretreatment into fermentable monosaccharides: concentrated acid, dilute acid, or enzymatic. Concentrated acid hydrolysis is employed in multiple cellulosic ethanol biorefineries but has not been the subject of TEAs and therefore is not covered here. Dilute acid hydrolysis is faster than enzymatic hydrolysis but can generate lower yields of monosaccharides. Recent research has also called into question existing cost estimates of the enzymes employed by enzymatic hydrolysis (Klein-Marcuschamer et al., 2012), suggesting that dilute acid hydrolysis could incur lower operating costs of the two processes (Kazi et al., 2010).

Table 2.6 presents capital cost estimates from three different TEAs for lignocellulosic biorefineries employing a dilute acid pretreatment and enzymatic hydrolysis. There is some variation in the equipment costs used to calculate total project investment, although Humbird et al. (2011) and Kazi et al. (2010) are very close when adjusted for capacity. The estimate from Piccolo and Bezzo (2007) is comparatively low, although this can be attributed to low estimates for distillation and recovery equipment and exclusion of the feedstock handling area, demonstrating the importance of the assumptions used in a TEA. Total project investment from these studies is 200–300% higher than for first generation biorefineries due to the necessity of including expensive pretreatment and hydrolysis equipment.

Table 2.7 presents annual operating cost estimates from three different TEAs for lignocellulosic biorefineries employing a dilute acid pretreatment and enzymatic hydrolysis. Humbird et al. (2011) is different from the other two assessments in that it models on-site enzyme production for hydrolysis; both Kazi et al. (2010) and Piccolo and Bezzo (2009) model enzyme purchase

*Table 2.6* Capital costs for lignocellulosic ethanol biorefineries employing dilute acid pretreatment and enzymatic hydrolysis

|  | Humbird et al. (2011) (Humbird and Aden, 2009) | Kazi et al. (2010) | Piccolo and Bezzo (2007) |
|---|---|---|---|
| Cost basis (year) | 2007 | 2007 | 2007 |
| Capacity (million gallons per year) | 61 | 53 | 51 |
|  | Capital costs ($MM) | | |
| Feedstock handling | 24.2 | 10.9 | – |
| Pretreatment | 29.9 | 36.2 | 31.5 |
| Conditioning | 3.0 | – | – |
| Hydrolysis and fermentation | 31.2 | 21.8 | 12.9 |
| Enzyme production | 18.3 | – | – |
| Distillation and recovery | 22.3 | 26.1 | 4.3 |
| Wastewater | 49.4 | 3.5 | 10.4 |
| Storage | 5.0 | 3.2 | – |
| Boiler | 66.0 | 56.1 | 44.5 |
| Utilities | 6.9 | 6.3 | 11.2 |
| Other | 18.3 | – | – |
| **Total installed equipment cost** | 274.6 | 164.1 | 114.7 |
| **Total project investment** | 422.5 | 375.9 | 270.8 |

*Table 2.7* Operating costs for lignocellulosic ethanol biorefineries employing dilute acid pretreatment and enzymatic hydrolysis

| Operating costs ($MM) | Humbird et al. (2011) (Humbird and Aden, 2009) | Kazi et al. (2010) | Piccolo and Bezzo (2007) |
|---|---|---|---|
| Capital depreciation | 60.4 | 16.3 | 37.8 |
| Operation and management | 24.2 | 71.8 | 89.7 |
| Biomass | 45.2 | 57.9 | 47.6 |
| By-product credit | −6.6 | −11.7 | −2.1 |
| Total | 123.2 | 134.3 | 173.0 |
| Biofuel price ($/gal) | $2.15 | $3.40 | $2.87 |
| Biofuel price ($/gge) | $3.23 | $5.10 | $4.29 |

from external sources. On-site enzyme production generates higher capital costs (as evidenced by greater capital depreciation) and lower operation and management costs.

### 2.2.3 Power generation at biorefineries

Biorefineries can choose to generate electricity from biomass or byproducts into electricity. Although biomass electricity is typically more expensive than fossil fuel power, there are two scenarios where biomass power is an obvious choice: remote or stranded biomass supply, and production of excess byproducts.

There are large quantities of biomass in remote or stranded locations that are classified as wastes. A significant amount of this waste decomposes without yielding economic value. This loss occurs, in part, because waste biomass is difficult to gather reliably in sufficient quantities, and waste is a heterogeneous material which makes conversion difficult. Small-scale power generation is one way to capitalize on potentially low-cost feedstock.

Biomass power generation can be accomplished in several ways: biomass combustion can provide steam to drive a steam turbine; biomass gasification yields syngas that could be fed into a gas turbine; biomass pyrolysis or torrefaction yield intermediate materials that can be combusted or gasified to produce power. Representative costs for these three scenarios are given in Table 2.8.

The low capital costs ($600/kW) for biomass combustion to power are indicative of the technology's simplicity (Dornburg and Faaij, 2001; Jenkins *et al.*, 2011). However, biomass combustion is less efficient than the alternatives even at large scale. Biomass gasification for power generation requires additional capital investment for the more expensive gas turbines and auxiliary equipment to ensure that gas conditions meet strict particulate matter requirements. The higher costs are compensated by higher efficiencies and ability to scale resulting in lower operating costs. The gasification scenario allows for the use of an integrated gas combined-cycle (IGCC) design with both steam and gas turbines to further improve the process

*Table 2.8* Capital and operating costs for biomass power generation

| Technology | Capital cost ($kW^{-1}$ capacity) | Operating cost ($kWh^{-1}$) |
|---|---|---|
| Combustion to power (Dornburg and Faaij, 2001) | $600 | $0.075 |
| Gasification to power | $1600 | $0.05 |
| Pyrolysis to power (Bridgwater *et al.*, 2002) | $2400 | $0.08 |

efficiency. Finally, the power industry has shown interest in pyrolysis and torrefaction products as a means to overcome some of the challenges faced by biomass, e.g. storage, heating value, feeding. Pyrolysis and torrefaction processes yield products that can be stored with less degradation and a lower footprint, that have a higher heating value, and are easier to feed into existing equipment than raw biomass. These benefits come at a cost. Capital costs are expected to be much higher than conventional alternatives ($2400/kW) with higher operating costs as well ($0.08/kWh) (Bridgwater et al., 2002).

## 2.3 Trade of biomass and subsidies

### 2.3.1 Biomass cost estimates by feedstock type

Lignocellulosic biomass feedstocks employed by biorefineries can broadly be divided into two categories: dedicated energy crops and residues. Dedicated energy crops are crops grown specifically for use as biomass feedstocks in biorefineries. These are divided into two further categories: herbaceous energy crops and short-rotation woody crops. Herbaceous energy crops contain little to no woody material and are exemplified by grasses. Common examples include switchgrass, *Miscanthus giganteus,* and energy cane. Short-rotation woody crops are softwoods and hardwoods with short harvest rotations. Common examples include hybrid poplar and *eucalyptus*. Short-rotation woody crops have longer harvest rotations than most herbaceous crops but compensate for this by also producing higher yields by biomass weight.

Biomass residues are waste products from either urban or rural areas. Residues from urban areas include both municipal solid waste (MSW) and processing residues from factories and manufacturing centers utilizing biomass as an input. Residues from urban areas are characterized by high concentration and low costs due to the avoidance of tipping fees otherwise paid to waste haulers. The disadvantages to using urban residues as biorefinery feedstocks are their heterogeneous nature (for example, MSW frequently contains plastics, metals, and glass capable of damaging a biorefinery) and high values for nearby land, thereby increasing biorefinery costs in the form of either capital or transportation costs. Biomass residues from rural areas most commonly take the form of agricultural residues left on the field after a crop harvest, such as corn stover. These are spread out over a large area and require specialized collection equipment, resulting in higher costs as biorefinery feedstocks than urban residues. Agricultural residues have the advantages of being homogeneous and located near inexpensive land, allowing biorefineries employing them as feedstock to minimize both capital and transportation costs.

Two methods are employed for estimating biorefinery feedstock costs. The first is the use of field trials that account for detailed costs of feedstock production, collection, transportation, and mitigation of negative environmental effects (e.g., nutrient replacement necessitated by the removal of corn stover). Several studies employing field studies have calculated the cost of agricultural residues to be lower than the cost of dedicated energy crops; the delivered cost of stover is calculated to be in the range of $47/MT to $75/MT (Brechbill *et al.*, 2011; Perlack and Turhollow, 2003; Petrolia, 2008) while that of switchgrass is calculated to be in the range of $80/MT to $96/MT (Brechbill *et al.*, 2011). The disparity between the costs of agricultural residues and dedicated energy crops is due to the fact that residues do not require an accounting of production costs and opportunity costs, as they are produced during the normal course of crop production and just need to be collected and transported to the biorefinery. Dedicated energy crops must account for these costs in addition to production and opportunity costs.

The second method employed for estimating biorefinery feedstock costs is the use of economic models based on a combination of field trials, supply chain data, and macroeconomic prices. Two recent examples have been developed by researchers at North Carolina State University (Gonzalez *et al.*, 2011) and the National Research Council (Committee on Economic and Environmental Impacts of Increasing Biofuels Production, 2011). In both cases the costs estimated by the economic models have been greater than those from field trials, with the delivered cost of switchgrass ranging from $94/MT to $108/MT and stover ranging from $96/MT to $101/MT.

The higher cost estimates from the economic model methodology relative to the field trial methodology can be attributed to the highly specific and localized nature of the latter. Field trials are commonly performed at the farm- or county-scale, which are then sometimes extrapolated to the state-scale. While this entails a high degree of accuracy on smaller scales, these results are not suitable for analyses at the regional or national scale. Economic models produce results at the regional or national scale and, while they do not have the levels of detail and accuracy found in field trials, they are more suitable for large-scale analyses.

### 2.3.2 Federal subsidy programs

The United States has employed a number of biofuel subsidy and tariff programs since the 1970s that have influenced the economic feasibility of biorefineries. The majority of these programs expired at the end of 2011 (Pear, 2012) and the US government has switched the focus of biofuels policy from protectionist programs to a low-carbon mandate in the form of the Renewable Fuel Standard. Whereas past biofuels programs have focused primarily on first generation biofuels and ethanol pathways, the current mandate is broader in scope and includes biofuel pathways ranging from ethanol to butanol to biobased gasoline and diesel (so-called drop-in biofuels).

Up until their expiration at the end of 2011, the US maintained a redeemable tax credit (i.e., a credit first applied against a taxpayer's tax burden with any excess being received as a direct payment) worth $0.45 for every gallon ($0.12/liter) of pure ethanol blended with gasoline for use as transportation fuel in the form of the volumetric ethanol excise tax credit (VEETC). A concurrent tariff on ethanol imports was also employed to prevent foreign ethanol producers (particularly Brazilian, as sugarcane ethanol has historically been cheaper to produce than corn ethanol) from utilizing the subsidy. Ethanol importers were required to pay a 2.5% *ad valorem* tariff plus a fixed $0.54/gal tariff on all imported ethanol. This had the effect of making Brazilian sugarcane ethanol more expensive in the US than US corn ethanol (see Table 2.5) despite the former's smaller production costs. A number of smaller subsidy programs affected other biofuel pathways. Biodiesel producers received a $1 non-refundable tax credit (i.e., a credit applied only to a taxpayer's tax burden) for every gallon ($0.26/liter) blended with diesel or sold as fuel. Cellulosic ethanol producers received (and still receive) a $1.01 non-refundable tax credit for every gallon ($0.27/liter) of cellulosic ethanol blended with gasoline or sold as fuel in the form of the cellulosic biofuel producer tax credit (CBPTC). Non-refundable tax credits were also available for small ethanol producers and liquefied gas producers.

Popular concerns that corn ethanol production was causing starvation in the developing world (Runge and Senauer, 2007) and deforestation in the Amazon (Searchinger *et al.*, 2008) combined with a shift toward government austerity in the US to undermine political support for first generation biofuel protectionism. With the exception of the CBPTC, all of the aforementioned subsidy and tariff programs were allowed to expire by Congress at the end of 2011, leaving the Renewable Fuel Standard as the primary driver of US biofuel policy. The first iteration of the Renewable Fuel Standard (RFS1) was created by the Energy Policy Act of 2005 to serve as a simple biofuel mandate. Rapid growth in US corn ethanol production left it obsolete soon after its creation and the Energy Independence and Security Act of 2007 replaced it with a greatly expanded (both in scope and volume) Renewable Fuel Standard (RFS2). The RFS2 combines an increased biofuel mandate (36 million gallons (136 million liters) per year by 2020) with a low-carbon fuel standard (LCFS). Four separate yet nested biofuel categories exist whereas the RFS1 had only one: (1) total renewable fuels, (2) advanced biofuels, (3) biomass-based diesel, and (4) cellulosic biofuels. Each category has a particular volumetric mandate that changes over time; total renewable fuels comprise the majority of the mandate but are permanently capped in 2015, and by 2022 the cellulosic biofuel category becomes responsible for a plurality of the mandate.

The definitions of each RFS2 category encompass both biofuel type and feedstock source (Energy Independence and Security Act, 2007). To qualify for the total renewable fuels category, a biofuel must be sourced from renewable biomass (i.e., biomass meeting land-use restrictions) and achieve a 20% lifecycle greenhouse gas emission (GHG) threshold relative to gasoline. Advanced biofuels must achieve a 50% GHG reduction and cannot include corn ethanol (regardless of its lifecycle GHG analysis). Biomass-based diesel must also achieve a 50% GHG reduction and includes both biodiesel produced via transesterification and renewable diesel. Finally, cellulosic biofuels must achieve a 60% GHG reduction versus gasoline and be sourced from lignocellulosic feedstocks. Emissions from indirect land-use changes (ILUC) must be accounted for when determining whether a biofuel achieves a category's GHG reduction threshold.

The RFS2 impacts the economic feasibility of biorefineries by attaching a renewable identification number (RIN) to every gallon of biofuel blended with or sold as transportation fuel in the US. The RFS2 requires blenders to own a certain number of RINs proportionate to their market share at the end of each year to demonstrate compliance with the mandate. A blender that has met its share of the mandate can sell any excess RINs to a blender that has not, or can bank them for future use. RIN values increase when the supply of biofuels within an RFS2 category exceeds demand and can serve as an important source of income for biofuels producers, as RIN values for the biomass-based diesel category reached $1.60/gal in August 2011 (McPhail *et al.*, 2011). When demand exceeds supply (i.e., when the mandate has not been met), the core value of an RIN is the difference between the biofuel's production cost and the market price of gasoline or diesel (RIN values do not drop below 0 when this market price exceeds the biofuel's production cost). RINs are allowed to be publicly traded, however, so speculator activity can also affect RIN value.

The effect of the RINs is to ensure that biofuel producers receive the minimum value necessary to cover costs of production. When gasoline and diesel prices are greater than biofuel production costs, then the core RIN value is 0, as biofuel producers do not need additional incentive to produce up to the mandated volume. When gasoline and diesel prices are less than biofuel production costs, then the core RIN value increases to the level necessary to incentivize sufficient production to meet the mandate. As an example, assume that the three lignocellulosic ethanol TEA results presented in Table 2.7 are three different biorefineries and the cellulosic ethanol produced by each qualifies for the cellulosic biofuels category of the RFS2. Initial production will fall short of the mandated volume (the

EPA has waived the cellulosic biofuels mandate in recent years due to a complete lack of production) and, assuming a pre-tax gasoline price of $3/gal, the RIN value will be sufficiently high to incentivize production at all three biorefineries, or $2.10/gge (the difference between the highest biofuel production cost, $5.10/gge, and the pre-tax gasoline price). The biorefinery capable of achieving the lowest production cost will attain the greatest profit but all three will be profitable. This will change as total cellulosic biofuel production exceeds the mandated volume, however. Assuming the first two biorefineries produce enough to satisfy the mandate and the pre-tax gasoline price remains $3/gal, then the RIN value will decline to the difference between the pre-tax gasoline price and the second highest biofuel production cost ($4.29/gge), or $1.29/gge. In this way, the RFS2 ensures that biofuel producers remain economically feasible when gasoline and diesel prices are low while eliminating the prospect of government-subsidized windfall profits when gasoline and diesel prices are high.

### 2.3.3 State subsidy programs

A number of states have implemented subsidy programs to encourage local biofuel production. These programs range from renewable portfolio standards (RPS) to state-wide blending requirements to low- and zero-interest loans for the construction of biorefineries. Blending requirements and RPSs both indirectly affect the economic feasibility of biorefineries by creating demand for their products. Blending requirements mandate the blending of certain volumes of ethanol with gasoline and biodiesel with diesel fuel (usually 5 vol%, although this varies by state) for fuel sold within the state. RPSs mandate that a certain amount of electricity sold within the state come from renewable sources such as a cellulosic ethanol biorefinery.

States have also attempted to attract biorefinery construction by offering low- or zero-interest loans to biofuel companies for their construction within the state. For example, drop-in biofuel company KiOR has received a zero-interest $75 million loan from the state of Mississippi for the construction of a commercial-scale catalytic pyrolysis and upgrading facility within the state (Dolan, 2011). Such loans improve the economic feasibility of recipient biorefineries by eliminating interest payments on initial capital costs. Unlike blending requirements and RPSs, favorable loans are generally directed at individual companies rather than made available for all qualifying producers.

### 2.3.4 European Union subsidy programs

The European Union (EU) has implemented a number of programs incentivizing the production of biorenewable energy, both on a Eurozone scale and a national scale by its member nations. The various programs are broadly split into the categories of biorenewable electricity and biofuels. The EU's 2009 Renewables Directive (Anon., 2009a) creates two separate binding targets for member nations. First, EU members must derive 10% of their transport energy from renewable sources, including biomass, by 2020. Second, they must also derive 20% of all of their energy from renewable sources, including biomass, by 2020. Member nations are given the flexibility to determine how best to meet these targets. Additionally, the EU has established economic mechanisms to compensate participating facilities within member nations that contribute to reducing greenhouse gas emissions (GHG).

The EU has implemented an Emission Trading Scheme (ETS) to combat anthropogenic climate change resulting from GHG via a cap-and-trade mechanism. Installations located within a member nation that meet a net heat threshold are covered by the ETS. Each member nation receives annually a limited number of GHG emission allowances that are distributed to covered installations, which must in turn purchase additional allowances for any emissions that exceed this allocation.

The ETS affects biorefineries both directly and indirectly. It directly affects biorefineries by allowing them to receive offset credits in the form of emission reduction units (ERU). ERUs are awarded in exchange for activity that results in the avoidance of GHG emissions. Example projects include the production of biogas from landfills for use as fuel, the utilization of waste sawdust as electricity or biofuel feedstock, and the use of sunflower and canola oils as biodiesel fuel feedstocks (Fenhann, 2012). Each ERU represents 1 metric ton (MT) of avoided GHG emissions and can be traded with other parties. In this way, ERUs can directly contribute to the economic feasibility of a qualifying biorefinery by representing an additional value-added product.

The ETS indirectly affects biorefineries by artificially increasing the cost of fossil fuel products relative to renewable fuel products in proportion to their respective carbon footprints. Power plant operators must purchase sufficient carbon allowances to cover the plant's GHG emissions, the size of which is determined by the feedstock utilized. This increases the value of electricity derived from biorenewables by lowering its cost relative to electricity derived from fossil fuels and thereby increasing demand for it. A similar situation exists for qualifying transportation biofuels. For example, the EU's decision in 2012 to include airlines operating in Europe within the ETS (Torello *et al.*, 2012) enhanced the value of aviation biofuel, as one method by which covered airlines can avoid GHG emissions (and the need to purchase additional allowances) is by combusting biofuel instead of conventional fossil fuel-based aviation fuel during flights.

## 2.3.5 EU member nation subsidy programs

A number of EU member nations have implemented their own programs incentivizing biorenewables as part of the binding targets imposed by the 2009 Renewables Directive. A wide variety of program types are employed, ranging from feed-in tariffs to mandates to tax incentives. The United Kingdom's (UK) Renewables Obligation establishes a mandate of 20% renewable electricity generation in the country by 2020 (Swinbank, 2009). The UK's Renewable Transport Fuels Obligation also establishes a mandate for 5 vol% of UK transport fuel consumption to be derived from renewable sources by 2013. Germany initially took a different approach by levying a tax on fossil fuels that was not applied to biofuels before switching to a biofuels mandate in 2007 (Deurwaarder, 2007). The large majority of EU member nations encourage the production of biofuels via tax incentives, blending requirements, or a combination of the two (Pelkmans *et al.*, 2008).

## 2.4 Market establishment: national/regional facilities

Biomass supply chains span from local collection efforts to international networks. National and regional biorefineries can thus be classified by the extent of their supply networks and product distribution. Most biorefineries are small-scale, regional facilities that collect feedstock from within a state or region. On the other hand, large-scale, national biorefineries would transport biomass across state borders to meet demand.

A majority of US ethanol biorefineries generate less than 80 million gallons (303 million liters) per year as shown in Fig. 2.2 (Anon., 2009b). At this capacity, biorefineries can collect enough corn from surrounding counties. In Iowa, for example, corn ethanol biorefineries receive their feedstock from an average distance of 28 miles (45 kilometers) (Anon., 2008). Nearly half of corn suppliers use tractor-pulled wagons while others employ straight trucks, fifth wheels or semi-trucks. Although short transport distances characterize corn supply, corn ethanol travels much farther. Corn biorefineries employ rail, trucks, and barges to ship ethanol to demand centers both within the county and across multiple state borders.

National, and international, biorefineries are defined here as facilities that receive feedstock by multiple transportation modes and from regions hundreds of miles from the facility. The large supply networks required to feed these types of biorefineries pose key economic challenges that differ from those faced by the fossil fuel industry. The biomass industry is still trying to understand the nature of these challenges and develop ways of addressing them.

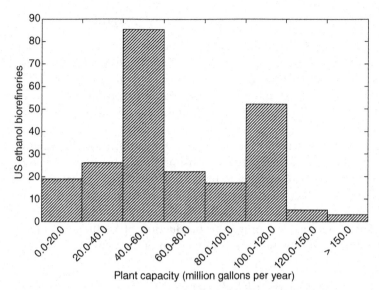

2.2 US ethanol biorefinery capacities per year.

### 2.4.1 Biomass logistics and transport infrastructure

Biomass is a diffuse resource that requires significant investment to collect and transport to a biorefinery. There are numerous studies on biomass logistics. Recent developments in geographic information systems (GIS) and operations research (OR) allow for increasing level of detail in logistic studies. In general, logistic studies attempt to estimate the costs for biomass collection, storage, and transportation.

Researchers at Iowa State developed estimates for feedstock suppliers' willingness to accept (WTA) selling price based on the following equation (Miranowski and Rosburg, 2010):

$$\text{WTA} = \left\{ \frac{C_{ES} + C_{Opp}}{Y_B} + C_{HM} + SF + C_{NR} + C_S + DFC + DVC * D \right\} - G \quad [2.1]$$

where $C_{ES}$ stands for establishment and seeding costs, $C_{Opp}$ represents land and biomass opportunity costs, $Y_B$ is the biomass yield, $C_{HM}$ are harvest and crop maintenance costs, $SF$ are stumpage fees, $C_{NR}$ are nutrient replacement costs, $C_s$ is storage, $DFC$ and $DVC$ are the fixed and variable transportation costs, respectively, $D$ is the distance to the biorefinery, and $G$ are governmental incentives.

Biomass production incurs sunk costs in the form of establishment, opportunity, and nutrient replacement costs. Establishment costs can be ignored for biomass residue, but they are important for dedicated energy crops. Biomass land and opportunity account for the loss revenue from growing alternative crops and land rental value. The removal of significant quantities of waste material would require nutrient addition in order to maintain soil quality.

Biomass collection varies between different types of feedstock. The corn, and sugarcane, industry has developed specialized machinery that collects grain at low cost: harvesting costs account for less than 15% of corn costs (Duffy, 2012). Collection costs for other types of feedstock have higher contributions to the overall cost. Harvesting cost estimates for corn stover, switchgrass, and *miscanthus* range between $14 and $84 per dry ton ($15 and $93 per dry ton) (Committee on Economic and Environmental Impacts of Increasing Biofuels Production, 2011), which could represent between 10 and over 80% of the delivered feedstock cost.

Cost estimates for biomass transport vary widely. They vary due to a lack of consistent data and because of the different methods employed. Transportation costs are typically reported as a total delivered cost or with fixed and variable components. DVC costs range between $0.09 and $0.60 per dry ton per mile ($0.06 and $0.41 per dry ton per km). DFC costs range between $4.80 and $9.80 per dry ton ($5.30 and $10.8 per dry ton).

Location is a major factor in the overall costs of delivering biomass to a facility. Land productivity, transportation networks, and storage facilities are a few of the parameters that vary significantly across various locations. GIS software provides display and analytical capabilities to investigate biomass supply chains. The US government provides a wealth of data in the form of GIS maps and dataset through their centralized portal www.data.gov. An example is shown in Fig. 2.3, which illustrates the distribution of total biomass resources in the US. There are high concentrations of biomass in the Midwest (stover), Pacific, Atlantic, and Gulf Coast (wood) regions, and around large metropolitan areas (MSW). This study estimates the biomass resources currently available in the United States by country. It includes the following feedstock categories: crop residues (5 year average; 2003–2007), forest and primary mill residues (2007), secondary mill and urban wood waste (2002), methane emissions from landfills (2008), domestic wastewater treatment (2007), and animal manure (2002). For more information on the data development, please refer to http://www.nrel.gov/docs/fy06osti/39181.pdf. Although, the document contains the methodology for the development of an older assessment, the information is applicable to this assessment as well. The difference is only in the data's time period.

The operations research field has developed mathematical formulas that evaluate biomass logistics within a geographical context. These formulas can estimate costs for a given region (city, municipality, county, state, agricultural district, nation, international) with improved relevance. Biomass supply formulations typically involve an objective function, several variables

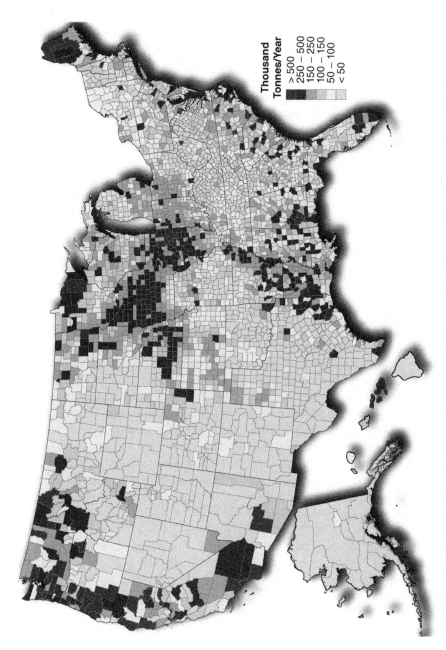

2.3 US total biomass (grain, waste, wood) county-level supply (Milbrandt 2005, produced by the National Renewable Energy Laboratory for the US Department of Energy).

or parameters, and multiple constraints. For example, we could reduce biorefinery feedstock costs by minimizing suppliers' WTA price:

$$\min \sum_{i=0, j=0}^{n,m} \left\{ \frac{C_{ES,i} + C_{Opp,i}}{Y_{B,i}} + C_{HM,i} + SF_i + C_{NR,i} + C_{S,i} + DFC + DVC * D_{ij} \right\} * F_{ij}$$
$$- G * F_{ij}$$

[2.2]

subject to

$$\sum_{i,j=0}^{m} F_{ij} \leq Biomass_i \ \forall \ i \in n$$

$$\sum_{i=0,j}^{n} F_{ij} \geq Demand_j \ \forall \ j \in m$$

variable $F_{ij}$

$\forall$: for all; $\in$: in

where $i$ and $j$ are subscripts representing biomass supply and biorefinery locations respectively, $F$ is the amount of biomass shipped from $i$ to $j$, $n$ is the set of biomass supply counties, and $m$ is the set of biorefinery locations. The *Biomass* constraint ensures that no more than the available amount of biomass gets shipped from a supply count, and the *Demand* constraint requires that the biorefinery receive enough biomass to satisfy their full capacity. This simple formulation can be expanded to include dynamic or temporal considerations, multi-level supply chains, and social and environmental parameters.

### 2.4.2 Scale-up of biorefinery operations

Biorefinery scale-up is a common approach to reducing costs by capital intensification. This approach has been proven in the fossil fuel industry where the average US coal plant generates over 227 gigawatts (in comparison, the energy production rate of a 100 million gallon (379 million liter) per year ethanol plant is equivalent to about 280 megawatts). Although biorefineries can benefit from economies of scale, they also suffer from diseconomies of scale that limits their ability to increase capacity (Wright and Brown, 2007b). Here we discuss the impact of scale on biorefinery costs and strategies to mitigate diseconomies of scale.

Product costs consist of three major categories: capital, operating, and feedstock costs. Capital costs include equipment depreciation, taxes, and the return on investment required to recoup the initial capital with a desired profit. Operating costs are expenses required to operate the facility such as labor and maintenance. Finally, feedstock costs are spent to acquire requisite raw materials. These costs are difficult to estimate, which has led to the development of sophisticated tools that depend on prior knowledge to determine costs for novel processes.

Product costs are a function of a plant's capacity. The relationship between a plant's capacity and the various cost components can be approximated with power law equations (Wright and Brown, 2007b):

$$Fuel\ Cost_M = C_0 * \left(\frac{Capacity_M}{Capacity_0}\right)^n + O_0 * \left(\frac{Capacity_M}{Capacity_0}\right)^m + F_0 * \left(\frac{Capacity_M}{Capacity_0}\right)^p$$

[2.3]

where $C$, $O$, and $F$ stand for capital, operating, and feedstock costs. Variables with subscript 0 correspond to known costs and capacities for a baseline facility. The scale factors $n$, $m$, and $p$ relate to the scaling behavior of each cost component. Scale factor values vary between technologies, but there are commonly accepted values employed in industry for major facility categories. In the thermochemical industry, $n$ is commonly assumed as 0.63 or 0.7, $m$ is approximately linear (between 0.9 and 1), and $p$ can be either smaller or greater than 1. In general, scale factors that are less than unity represent product costs that decrease with plant capacity. For example, a 0.7 scale factor suggests that every 1% increase in capacity incurs a smaller 0.7% increase in capital costs. Biorefineries are unique in the fuel industry for having large $p$ scale factors (in the order of 1.5), meaning that feedstock costs increase with plant capacity. Figure 2.4 shows unit capital costs of corn ethanol biorefineries versus plant capacity.

Economies of scale are strong incentive to build large biorefineries. However, beyond an optimal capacity, feedstock transportation costs increase at a faster rate than reductions in capital costs. Engineers are evaluating distributed processing strategies to alleviate transport costs.

Distributed processing is the notion that small-scale facilities pretreat biomass prior to shipping to a large, upgrading facility. This concept yields several economic benefits: reduced storage space requirement, slower biomass degradation rates, and improved energy densities. Pretreatment intensity varies from simple drying and grinding to torrefaction or even pyrolysis.

Biomass drying and grinding removes moisture, which can otherwise increase biomass degradation rates by encouraging microbial activity. Grinding increases the feedstock bulk density and allows for pelletization, which is the mechanical compression of loose material into dense pellets. Torrefaction increases the energy density by further lowering the moisture content, releasing low energy value compounds, and slightly modifying biomass structure. Torrefied biomass is hydrophobic, which makes it

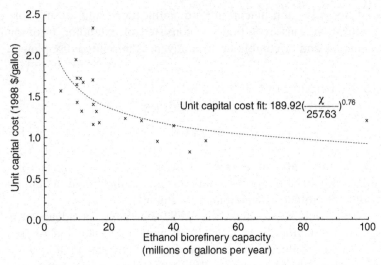

*2.4* US corn ethanol plant unit capital costs vs. capacity ($n \approx 0.84$) (Gallagher *et al.*, 2005).

ideal for long-term storage. Torrefaction allows for efficient biomass pulverization, which increases the energy density. Finally, biomass pyrolysis, or liquefaction techniques, convert biomass into a liquid form with high material density. Pyrolysis oil, combined with biomass char materials, is an energy dense material that could be shipped at low cost, but corrosion remains a challenge.

Distributed processing in specialized, small-scale facilities or depots will play an important role in a large-scale biorefinery industry (Wright and Brown, 2007b). Without pretreatment, biomass transportation will be expensive and increase the number of trucks delivering material to central facilities. Unfortunately, distributed processing faces the classical 'chicken or egg' problem: without large-scale facilities there is a limited market for pretreated biomass feedstock, and without distributed processing facilities may find it difficult to achieve large-scale capacities. Pretreatment technologies may initially feed into the existing fossil fuel infrastructure by delivering torrefied biomass to coal plants or bio-oils to oil refineries. However, some technical and economic hurdles for this alternative remain unsolved.

## 2.5 Conclusion and future trends

TEAs are evolving to address challenges faced by novel biorefinery technologies: lack of processing data, concern for environmental impacts, need for risk and uncertainty quantification, and desire for process optimization. This evolution has consisted of extending traditional TEA techniques and/or combining with advances in related disciplines.

### 2.5.1 Public support of pilot and demonstration biorefineries

Lack of industrial data to support TEA has prompted government initiatives to sponsor the construction of pilot- and demonstration-scale biorefineries such as the National Advanced Biofuels Consortium (NABC). These initiatives help gather process data that guides the direction of future public funding in biofuels research and development.

Industry has yet to meet the RFS2 mandates for advanced cellulosic biofuel production. In fact, the Environmental Protection Agency (EPA) retroactively reduced the 2011 mandate from 250 million gallons (946 million liters) to 6.5 million gallons (25 million liters). The United States Department of Agriculture attributes the lack of advanced biofuel supply to several factors including the high cost of first-of-a-kind biorefineries (Coyle, 2010). Lack of financial investment in these technologies is exacerbated by the lack of industrial data to support them. This can be alleviated with the construction of pilot- and demonstration-scale facilities.

Pilot- and demonstration-scale projects are key milestones for the commercialization of novel technologies. These facilities are not intended to generate revenue, but they sometimes require significant financial commitments. Pilot-scale biorefineries have capacities of less than 5 MT per day, and demonstration-scale facilities can operate at commercial production rates but are discontinued once operators obtain the necessary data. Industry has so far been reluctant to invest in these capital-intensive demonstration facilities. Unfortunately, there are important processing considerations that can only be tested at commercial scale.

Data from privately funded demonstration projects are almost never released to the general public. However, this data is crucial to help guide policy decisions that may affect the development of biorefinery technologies. Demonstration-plant data will drastically improve the quality of TEA studies, and conversely, TEA will improve in their ability to clarify the path from nascent technology to commercial product.

### 2.5.2 Combination of TEA and LCA

Researchers are combining TEA with life cycle analysis (LCA) to provide a comprehensive evaluation of biorefinery technologies (Hill *et al.*, 2006). These techniques share a symbiotic relationship in that they enhance findings from each discipline. TEA can quantify the economic costs associated with environmental impacts, and LCA determines the environmental effects related to TEA assumptions.

LCA researchers estimate the environmental impacts associated with biorefinery operations. Sometimes, the environmental impacts can be readily quantified in economic terms – water treatment of process effluent, for example. It is much harder to determine the economic cost of other types of environmental impacts like those associated with global climate change. TEAs can help determine the proper incentives or penalties required to encourage or mitigate these environmental impacts.

System boundaries are an important consideration for both TEA and LCA. TEAs are typically confined to the boundaries of a specific process, but can extend to include global economic activity. LCA research on the other hand encourages the expansion of system boundaries to properly account for environmental impacts. Therefore there are important tradeoffs involved in combining both techniques. In general, data availability weighs heavily on the choice of system boundaries.

There is an increasing awareness of the environmental impacts of industrial activity. The long-term implications of commissioning biorefinery projects require careful study of both economic and environmental risks. Knowledge gained from future biorefinery projects will enhance our understanding of both risks if they are investigated in a concerted fashion.

### 2.5.3 Risk and uncertainty quantification

A major challenge for TEA and LCA studies is risk and uncertainty quantification. Industry employs economic risk, measured by indicators such as rate of return and net present value, to identify investment opportunities. Policy makers rely on LCA to estimate greenhouse gas emissions and resource use. Varying degrees of uncertainty underlie these measures. Therefore, research requires additional tools to understand the implications of these uncertainties.

Researchers employ an increasing number of techniques such as case studies, sensitivity analysis, and Monte Carlo simulations to improve uncertainty quantification in TEA and LCA studies. The need for uncertainty quantification is driven by uncertainties in model parameters, their interactions, and the outputs generated by these analyses.

Case studies are the most trivial approach to quantifying uncertainty and are not always recognized as such. However, careful selection of system scenarios can provide more than enough data to understand project risks. For example, case studies based on the extreme values of historical market prices for a given commodity could be enough to rule out a potential project. The drawback of case studies is that they provide minimal insight into the interactions between different model parameters.

Sensitivity analyses improve upon case studies by evaluating several points within a range of parameter values. Their key insight is the extent to which system outputs change based on different input assumptions. Sensitivity analyses that involve a large number of randomized model evaluations are known as Monte Carlo simulations. Monte Carlo simulations benefit from inexpensive computational resources that allow rapid model evaluations. Researchers employ Monte Carlo extensively in a wide range of fields to develop model probability distributions. Increasing model complexity has limited the use of this brute-force method because it would consume significant computational time and resources. Researchers continue to adopt powerful techniques to model, collect, and assess TEA data that are beyond the scope of this chapter.

These uncertainty quantification techniques help reduce risks from assumption bias and failure to consider adverse scenarios. However, they are not a substitute for robust models with sensible built-in assumptions.

### 2.5.4 System optimization and statistical techniques

There is a growing desire to optimize TEA models and understand the implications for real systems. Modern process modeling tools include optimization functions or can couple with stand-alone optimization software like IBM ILOG CPLEX Optimizer, Gurobi™, and GAMS among others. These tools allow researchers to systematically identify optimal operating parameters that meet certain constraints.

TEA models include parameters bounded by system constraints. For example, biorefineries include both technical (reactor temperature) and economic (minimum feedstock cost) constraints that need to be considered within the model. Within the bounds of the allowable parameters there are usually one or more function maxima or minima. In this regard, TEA systems are somewhat simpler than other mathematical models – the function space is well defined. The major challenges for optimization of TEA models are large, complex models with hundreds of parameters, and models that express extremely nonlinear behavior. Techniques that address both of these challenges are the subject of much research.

Researchers employ model surrogates or reduced order models (ROMs) to optimize large models that are either too complex or computationally expensive to evaluate. ROMs can significantly reduce the time required to optimize high-fidelity models at the risk of over-simplifying the problem. Therefore, several approaches have been proposed for the identification of ROM parameters and the evaluation of ROM accuracy.

The benefits of process optimization go beyond identifying optimal values. They also identify tradeoffs between differing objectives. These tradeoffs can be illustrated by a Pareto curve. Pareto curves describe the incremental changes of a given objective value due to improving a second objective. For example, biorefineries commonly face a tradeoff between lowering process costs from the use of fossil fuels and increasing their overall environmental footprint.

These emerging trends suggest a bright future for techno-economic analysis study and its impact on the advancement of biorefineries. The study of demonstration-plant data, combination of TEA and LCA, evaluation of risk and uncertainty, and optimization of system models are fertile grounds for future research and development.

## 2.6 References

Anon. (2008) 2007 survey report on grain storage and transportation, Ames, Iowa State University Extension. Available from: http://www.soc.iastate.edu/extension/pub/farmpoll/topical/PM2049.pdf (accessed 24 June 2012).

Anon. (2009a) 'Directive 2009/28/EC On the Promotion of the Use of Energy from Renewable Sources', *Official Journal of the European Union*, L 140, 16–62.

Anon. (2009b) *2009 Ethanol Industry Outlook*. Renewable Fuels Association. Available from: http://www.cornlp.com/Adobe/outlook2009.pdf (accessed 24 June 2012).

Anon. (2011) *World Energy Outlook 2011*. Paris, International Energy Agency. Available from: http://www.iea.org/weo/ (accessed 24 June 2012).

Brechbill, S., Tyner, W. and Ileleji, K. (2011) 'The economics of biomass collection and transportation and its supply to Indiana cellulosic and electric utility facilities', *BioEnergy Research*, 4, 141–152.

Bridgwater, A.V., Toft, A.J. and Brammer, J.G. (2002) 'A techno-economic comparison of power production by biomass fast pyrolysis with gasification and combustion', *Renewable and Sustainable Energy Reviews*, 6, 181–248.

Brown, R.C. (2003) *Biorenewable Resources: Engineering New Products from Agriculture*, Ames, IA: Iowa State Press.

Brown, R.C. (2005) 'Hybrid thermochemical/biological processing – an overview' in Kamm, B. Gruber, P.R. and Kamm, M. (eds), *Biorefineries – Industrial Processes and Products*, Weinheim, Germany: Wiley-VCH.

Brown, T.R., Zhang, Y., Hu, G. and Brown, R.C. (2012) 'Techno-economic analysis of biobased chemicals production via integrated catalytic processing'. *Biofuels, Bioproducts and Biorefining*, 6, 73–87.

Choi, D.W., Chipman, D.C., Bents, S.C. and Brown, R.C. (2010) 'A techno-economic analysis of polyhydroxyalkanoate and hydrogen production from syngas fermentation of gasified biomass'. *Applied Biochemistry and Biotechnology*, 160, 1032–1046.

Committee on Economic and Environmental Impacts of Increasing Biofuels Production (2011) *Renewable Fuel Standard: Potential Economic and Environmental Effects of US Biofuel Policy*, Washington, DC: National Academies Press.

Coyle, W.T. (2010) *Next-Generation Biofuels Near-Term Challenges and Implications for Agriculture*, Washington, DC: U.S. Department of Agriculture. Available from: http://www.ers.usda.gov/publications/bio0101/ (accessed 24 June 2012).

Deurwaarder, E.P. (2007) *European Biofuel Policies*, Energy Research Centre of the Netherlands. Available from: http://www.ecn.nl/docs/library/report/2007/m07040.pdf (accessed 24 June 2012).

Dolan, K.A. (2011) 'Vinod Khosla-backed Renewable Fuel Firm KiOR Files To Go Public', *Forbes*. Available from: http://www.forbes.com/sites/kerryadolan/2011/04/12/vinod-khosla-backed-renewable-fuel-firm-kior-files-to-go-public/ (accessed 24 June 2012).

Dornburg, V. and Faaij, A. (2001) 'Efficiency and economy of wood-fired biomass energy systems in relation to scale regarding heat and power generation using combustion and gasification technologies', *Biomass and Bioenergy*, 21, 91–108.

Duffy, M. (2012) *Estimated Costs of Crop Production in Iowa – 2012*. Ames, IA: Iowa State University. Available from: http://www.extension.iastate.edu/agdm/crops/pdf/a1-20.pdf (accessed 24 June 2012).

Efe, Ç., Straathof, A.J.J. and van der Wielen, L.A.M. (2005) Technical and Economical Feasibility of Production of Ethanol from Sugarcane and Sugarcane Bagasse. Delft University of Technology. Available from: http://www.b-basic.nl/documents/Sugar_Cane_Ethanol.pdf (accessed 24 June 2012).

Energy Independence and Security Act of 2007 § 201, 42 U.S.C. § 17001 (2007).

Fenhann, J. (2012) Joint Implementation Pipeline and Database, UNEP Riso Centre. Available from: http://www.cdmpipeline.org/publications/JIpipeline.xlsx (accessed 24 June 2012).

Gallagher, P.W., Brubaker, H. and Shapouri, H. (2005) 'Plant size: capital cost relationships in the dry mill ethanol industry', *Biomass and Bioenergy*, 28, 565–571.

Gonzalez, R., Phillips, R., Saloni, D., Jameel, H., Abt, R., Pirraglia, A. and Wright, J. (2011) 'Biomass to energy in the southern United States – supply chain and delivered cost', *BioResources*, 6, 2954–2976.

Hamelinck, C.N. and Faaij, A.P.C. (2002) 'Future prospects for production of methanol and hydrogen from biomass', *J. Power Sources*, 111, 1–22.

Hill, J., Nelson, E., Tilman, D., Polasky, S. and Tiffany, D. (2006) 'Environmental, economic, and energetic costs and benefits of biodiesel and ethanol biofuels', *Proceedings of the National Academy of Sciences*, 103, 11206–11210.

Humbird, D. and Aden, A. (2009) Biochemical Production of Ethanol from Corn Stover: 2008 State of Technology Model. Golden, CO: National Renewable Energy Laboratory. Available from: http://www.nrel.gov/biomass/pdfs/46214.pdf (accessed 24 June 2012).

Humbird, D., Davis, R., Tao, L., Kinchin, C., Hsu, D., Aden, A., Schoen, P., Lukas, J., Olthof, B., Worley, M., Sextom, D. and Dudgeon, D. (2011) Process Design and Economics for Biochemical Conversion of Lignocellulosic Biomass to Ethanol. Golden, CO: National Renewable Energy Laboratory. Available from: http://www.nrel.gov/biomass/pdfs/47764.pdf (accessed 6 October 2013).

Jenkins, B.M., Baxter, L.L. and Koppejan, J. (2011) 'Biomass combustion', in Brown R.C., *Thermochemical Processing of Biomass: Conversion into Fuels, Chemicals and Power*, New York: Wiley, 13–46.

Kazi, F.K., Fortman, J.A., Anex, R.P., Hsu, D.D., Aden, A., Dutta, A and Kothandaraman, G. (2010) 'Techno-economic comparison of process technologies for biochemical ethanol production from corn stover', *Fuel*, 89(Supplement 1), S20–S28.

Klein-Marcuschamer, D., Oleskowicz-Popiel, P., Simmons, B.A. and Blanch, H.W. (2012) 'The challenge of enzyme cost in the production of lignocellulosic biofuels', *Biotechnology and Bioengineering*, 109(4), 1083–1087.

Kwiatkowski, J.R., Mcaloon, A.J., Taylor, F. and Johnston, D.B. (2006) 'Modeling the process and costs of fuel ethanol production by the corn dry-grind process', *Industrial Crops and Products*, 23, 288–296.

McPhail, L., Westcott, P. and Lutman, H. (2011) The Renewable Identification Number System and US Biofuel Mandates. Washington, DC: U.S. Department of Agriculture. Available from: http://www.ers.usda.gov/publications/bio03/bio03.pdf (accessed 24 June 2012).

Milbrandt, A. (2005) A Geographic Perspective on the Current Biomass Resource Availability in the United States. Golden, CO: National Renewable Energy Laboratory. Available from: http://www.nrel.gov/docs/fy06osti/39181.pdf (accessed 24 June 2012).

Miranowski, J. and Rosburg, A. (2010) An Economic Breakeven Model of Cellulosic Feedstock Production and Ethanol Conversion with Implied Carbon Pricing, Ames, Iowa State University. Available from: http://www.econ.iastate.edu/research/working-papers/p10920 (accessed 24 June 2012).

Pear, R. (2012) 'After three decades, federal tax credit for ethanol expires', in *The New York Times*, Washington.

Pelkmans, L., Govaerts, L. and Kessels, K. (2008) Inventory of biofuel policy measures and their impact on the market, ELOBIO Report D2.1. Available from: http://www.elobio.eu/fileadmin/elobio/user/docs/Elobio_D2_1_Policy Inventory.pdf (accessed 24 June 2012).

Perlack, R.D. and Turhollow, A.F. (2003) 'Feedstock cost analysis of corn stover residues for further processing', *Energy*, 28, 1395–1403.

Petrolia, D.R. (2008) 'The economics of harvesting and transporting corn stover for conversion to fuel ethanol: a case study for Minnesota', *Biomass and Bioenergy*, 32, 603–612.

Phillips, S., Aden, A., Jechura, J., Dayton, D. and Eggeman, T. (2007) 'Thermochemical ethanol via indirect gasification and mixed alcohol synthesis of lignocellulosic biomass', *Ind. Eng. Chem. Res.*, 46, 8887–8897.

Phillips, S.D., Tarud, J.K. and Biddy, M.J. (2011) 'Gasoline from wood via integrated gasification, synthesis, and methanol-to-gasoline technologies: a techno-economic analysis', *Ind. Eng. Chem. Res.*, 50, 11734–11745.

Piccolo, C. and Bezzo, F. (2007) 'Ethanol from lignocellulosic biomass: a comparison between conversion technologies', *Computer Aided Chemical Engineering*, 24, 1277–1282.

Piccolo, C. and Bezzo, F. (2009) 'A techno-economic comparison between two technologies for bioethanol production from lignocellulose', *Biomass and Bioenergy*, 33(3), 478–491.

Roesijadi, G., Jones, S.B., Snowden-Swan, L.J. and Zhu, Y. (2010) Macroalgae as a Biomass Feedstock: A Preliminary Analysis. Richland: Pacific Northwest National Laboratory. Available from: http://www.pnl.gov/main/publications/external/technical_reports/PNNL-19944.pdf (accessed 24 June 2012).

Runge, C.F. and Senauer, B. (2007) 'How biofuels could starve the poor', *Foreign Aff.*, 86, 41–44.

Searchinger, T., Heimlich, R., Houghton, R.A., Dong, F., Elobeid, A., Fabiosa, J., Tokgoz, S., Hayes, D. and Yu, T.-H. (2008) 'Use of US croplands for biofuels increases greenhouse gases through emissions from land-use change', *Science*, 319, 1238–1240.

So, K.S. and Brown, R.C. (1999) 'Economic analysis of selected lignocellulose-to-ethanol conversion technologies', *Applied Biochemistry and Biotechnology*, 77, 633–640.

Swanson, R., Platon, A., Satrio, J. and Brown, R.C. (2010) 'Techno-economic analysis of biomass-to-liquids production based on gasification', *Fuel*, 89(Supplement 1), S11–S19.

Swinbank, A. (2009) 'EU Policies on bioenergy and their potential to clash with the WTO', *Journal of Agricultural Economics*, 60, 485–503.

Tijmensen, M.J.A., Hamelinck, C.N. and Hardeveld, M.R.M.V. (2002) 'Exploration of the possibilities for production of Fischer Tropsch liquids and power via biomass gasification', *Biomass*, 23, 129–152.

Torello, A., Pearson, D. and Koster, B. (2012) 'Airlines, Airbus renew criticism of EU $CO_2$ scheme, *The Wall Street Journal*. Available from: http://online.wsj.com/article/BT-CO-20120524-713728.html (accessed 24 June 2012).

Wright, M.M. and Brown, R.C. (2007a) 'Comparative economics of biorefineries based on the biochemical and thermochemical platforms', *Biofuels, Bioproducts, and Biorefining*, 1, 49–56.

Wright, M.M. and Brown, R.C. (2007b) 'Establishing the optimal sizes of different kinds of biorefineries', *Biofuels, Bioproducts, and Biorefining*, 1, 191–200.

Wright, M.M. and Brown, R.C. (2011) 'Costs of thermochemical conversion of biomass to power and liquid fuels', in Brown, R.C., *Thermochemical Processing of Biomass: Conversion into Fuels, Chemicals and Power*, New York: Wiley, 307–322.

# 3
# Environmental and sustainability assessment of biorefineries

L. SCHEBEK and O. MRANI,
Technische Universität Darmstadt, Germany

DOI: 10.1533/9780857097385.1.67

**Abstract**: Given the fact that biorefineries are gaining increasing attention as a technology for mitigation of climate change and sustainable development in general, it is not surprising that sustainability assessment of biorefineries has also become an issue. The interaction of biorefineries with their environment is very complex. In general, it can be stated that biorefineries may have impacts on the natural or physical as well as on the economic and social-cultural environment. Life cycle assessment (LCA) is a systematic approach to analyze and examine the impacts of products or services on the environment. Most LCA studies on biorefineries consider or compare products or raw materials, thus such other classification features of biorefineries as platform and process appear only in the background. Challenges in the future include assessment of the expected competition between material and energetic use on one hand, and land for cultivation of food and feed production on the other hand. As a general prerequisite, the technological processes must be most efficient. In addition to that, social and economic implications of a broad implementation of biorefineries must be better understood, in order to facilitate implementation of solutions.

**Key words**: sustainability assessment, biorefineries, life cycle assessment (LCA).

## 3.1 Introduction

The concept of sustainability was introduced by the World Commission on Environment and Development (WCED, 1987) in its famous report '*Our Common Future*', delivered on behalf of the United Nations in 1987. Since then, it has been broadly discussed in scientific literature as well as in public debates at various levels of society. The WCED report defines sustainable development as 'Development that meets the needs of the present generation without compromising the ability of future generations to meet their own needs'. The popularity of this concept resides in its comprehensive and inclusive idea of fairness today and in future, with nature conservation as prerequisite to implement this idea of global equity. However, its universal applicability raised the need for further specification of goals and strategies to make it operational. A widespread idea of sustainability states that sustainable development must be based on three dimensions: environmental, economic and social (Jänicke *et al.*, 2001). From this generic view, individual strategies for stakeholders may be derived. As one example, Alles and Jenkins propose that in organizational strategies three objectives must be considered (Alles and Jenkins, 2010):

- *people* – the social consequences of its actions
- *planet* – the ecological consequences of its actions
- *profits* – the economic profitability of companies (being the source of 'prosperity').

So far, many international, national and company indicator systems to assess sustainability have been worked out. Assefa and Frostell report that more than 500 projects have been implemented to develop quantitative indicators for sustainable development (Assefa and Frostell, 2007). These may be used to assess progress of sustainability on various levels and for different applications. Research and political interest in sustainability assessment of technologies increased during the last decade. The reason is the far-reaching impact of novel technologies, which may contribute to sustainable development, but may also raise novel problems of sustainability.

Given the fact that biorefineries are gaining increasing attention as a technology for mitigation of climate change and sustainable development in general, it is not surprising that the sustainability assessment of biorefineries has also become an issue. Biorefineries are supposed to have a considerable potential to replace fossil fuels and to develop a new concept of economic production in the chemical industry. At the same time, this might pose new challenges. The demand for biomass supply and new patterns for production and workplace surroundings are an example for these. Consequently, environmental impacts are the focus of all sustainability assessments, but also other issues have to be taken into account to obtain a comprehensive picture.

## 3.2 Methodological foundations of environmental and sustainability assessment of technologies

### 3.2.1 Methods and indicators of sustainable development

Given the broad concept of sustainability outlined above, it is obvious that sustainability assessment does not mean one single method, but that different types of methodologies and assessment procedures may be applied. This calls for the structuring or a typology of approaches, and indeed such typologies can be found in the literature. Before presenting them, at the most simple, three structuring criteria shall be discussed: first, what indicators are used for the assessment, second, how the object of assessment is defined, and third, how the quantitative or qualitative values for the chosen indicators as to the defined object of assessment are generated.

As to the issue of indicators, these can be defined from the three dimensions of sustainability and on different levels; these levels have to be adequate for the object to be assessed. Many indicator systems have been defined at the country level, e.g. the United Nations Commission for Sustainable Development (UNCSD) Theme Indicator Framework (UN, 2001). Some of these indicators may be meaningful also at the company, product or technology level (e.g., the amount of greenhouse gases), but some may be not (e.g., the national debt).

The task to define the object of investigation is not as trivial as it may seem. Every object has a structure – for example, a product has its components, a technology has auxiliary processes and a demand for material and energy – so each object can be seen as a system and the definition of the objects is equal to the definition of the system boundaries. This system can be specified by technological components, but also by geographical or temporal aspects. One idea of a system boundary is specifically prominent in sustainability assessment which is the so-called life cycle of a product. 'Life cyle thinking' means looking at the full process chain from extraction of raw materials through production of a product or technology, its use by the consumer and also its end of life, where materials are transferred back to nature.

These two structuring criteria are decisive for the third one, the methodology, with which quantitative or qualitative values for the chosen indicators and the system boundary are generated. Here, usually two types of procedures are encountered: either information is gathered directly via measurement or statistics (or taken from databases which contain respective data), or a model is built in order to generate new data from a set of data fed in. The choice of methodology makes up the tools that are used for assessment.

Typologies found in the literature make use of these criteria in different ways. Singh *et al.* report on a broad literature overview of sustainability assessment methods, structured by sustainability indicators, classification and evaluation of methodologies. They address guidelines for the construction of indices; in addition, they give a comprehensive survey and description of existing sustainability indices (Singh *et al.*, 2009). Hacking and Guthrie propose a framework and a consistent terminology for approaches to sustainability assessment found in the literature, which uses three axes: 'the comprehensiveness of the SD coverage; the degree of 'integration' of the techniques and themes; and the extent to which a strategic perspective is adopted' (Hacking and Guthrie, 2008). Ness *et al.* present a proposal for assessment tools and arrange them into three main categories: indicators/indices, product-related assessment, and integrated assessment tools. There is a 'parent category' (monetary valuation tools), which acts in all categories (Ness et al., 2007).

Markevičius et al. use 35 criteria for a so-called Emerging Sustainability Assessment Framework. The majority of indicators focus on environmental issues (12 indicators), while four social indicators and one economic indicator are added (Markevičius et al., 2010). Hueting and Reijnders report on the construction of sustainability indicators. They make the general criticism that so far suggested economic and social elements for inclusion in indicators do not have plausible causal relation to nature conservation, i.e. '*sustainability defined as a production level that does not threaten the living conditions of future generations*' (Hueting and Reijnders, 2004). Böhringer and Jochem select 11 indices from 500 Sustainable Development Indicators, that are suggested to researchers and policy makers, including the Living Planet Index (LPI), Ecological Footprint (EF), City Development Index (CDI), and Human Development Index (HDI) (Böhringer and Jochem, 2007). Assefa and Frostell discuss an approach for the evaluation of indicators for social sustainability of technical systems (e.g., waste management and energy systems). Three indicators are reviewed: knowledge, perception and fear (Assefa and Frostell, 2007). Finnveden et al. mention the strategic environmental assessment (SEA), the environmental impact assessment (EIA), the environmental risk assessment (ERA), the cost-benefit analysis (CBA), the material flow analysis (MFA), the ecological footprint, and notably life cycle assesment (LCA) as most frequently used methods (Finnveden et al., 2009). Balkema et al. propose a methodology of sustainability assessment structured in three steps following the approach of life cycle assessment: (1) Goal and definition, (2) inventory analysis, and (3) optimization and results. The last step is essential for assessing sustainability (Balkema et al., 2002).

Sustainability assessment also has to be seen as a decision-making process where the interests of many stakeholders have to be taken into account. The various players have their environmental, social and economic criteria and interests for the development of a sustainable system. To support these decision-making processes and take into account different goals and interests, methods such as multi-criteria decision analysis (MCDA), multi-objective decision making (MODM), operations research and management science are proposed. These methods are used to review the assessment of various decisions and political strategies as well as to include the competing interests of various stakeholders and experts (Halog and Manik, 2011). When quantitative sustainability indicators are used, multi-objective optimization can be integrated to identify a group of favorable options for sustainable solutions (Balkema et al., 2002). Such methodology has to be included in a procedural framework of stakeholder participation (see, e.g., Stoll-Kleemann and Welp, 2006).

## 3.2.2 Technologies and their interaction with the environment

The interaction of a technology with its environment is very complex. In general it can be stated that a technology may have impacts on the natural or physical as well as on the economic and sociocultural environment. The ways in which this interaction takes place may be diverse, depending on the kind of technology. However, every technology has a reason for its application, which ultimately is to deliver either a product or a service to society. This function – if it meets the demands of the final consumer – is what makes a new technology enter the market and consequently drives its impact on the environment. This idea of the interaction of technology with its environment through a function that needs to be fulfilled is shown in Fig. 3.1 (Balkema *et al.*, 2002).

Consequently, the impact of a technology is always assessed taking into account its function. This is what makes it possible to compare a novel technology to 'old' technologies or to compare different alternatives for shaping a technology to each other. It is even possible to compare different technologies to services provided that the function for the user remains the same. Although the motivation to compare is very important, if not the most widespread one for sustainability assessment, it may also be interesting to assess one technology on its own. This motivation is encountered mostly in cases where the term technology designates not a single device, process or service, but rather a far reaching change in the sense of systems innovation which usually encompasses technology as well as organizational and social

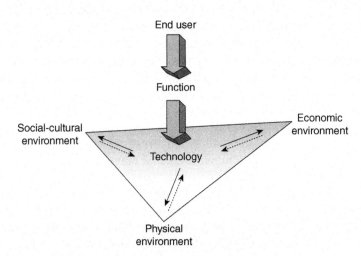

*3.1* Interaction between technology and environment (Balkema *et al.*, 2002).

innovation; a well-known example of this is the internet. Here, often the term technology assessment (TA) is used, which covers a wide procedural approach encompassing many methodological tools in order to explore and assess a multitude of possible and maybe interacting consequences on the natural or social environment (van den Ende, 1998; Bütschi et al., 2004).

In contrast, the term sustainability assessment of technologies in the literature is rather used for a more focused assessment and comparison of specific technologies and their interaction with their environment. Keeping in mind that the impact of a technology is always connected to its function, one generic approach of assessment has become most popular: the methodology of life cycle assessment (LCA). It mirrors the view of the function as driving force for the use of a technology and as a point of reference for assessment. Most approaches to sustainability assessment of technologies are based on this method, either exclusively or by incorporating it in a wider framework. This notion is also true when looking at the assessment of technologies for energetic or material use of biomass (Dewulf and van Langenhove, 2006). Consequently, in order to reduce the plethora of names and single approaches, this chapter will be restricted to life cycle assessment as the most widespread concept for sustainability assessment of technologies.

### 3.2.3 Basics of life cycle assessment (LCA)

Life cycle assessment (LCA) is a systematic approach to analyze and examine the impacts of products or services on the environment. Not only these impacts, but their whole life cycle (i.e., from resource extraction to end-of-life) is analyzed. LCA is based on the modeling of interconnections between the single processes of a product system, in order to identify the material and energy flows within the system. From this model, the so-called 'elementary flows' – i.e., material flows between the product system as a whole and the natural environment – can be derived. Finally, the impacts or damage to the environment can be assessed. The core methodology of LCA was developed during the 1990s, with the SETAC 'Code of Practice' as a first milestone (Consoli et al., 1993), and today is described in two international standards, ISO 14040 and 14044, which are part of the ISO 14000 family of standards on environmental management.

According to ISO 14040, a life cycle assessment study consists of four phases: goal and scope definition, inventory analysis, impact assessment, and finally interpretation (Fig. 3.2). The double arrows in Fig. 3.2 indicate that an LCA is seen as an iterative process; additional information gained during a study can require a backshift to a previous stage to include further aspects. The first phase of an LCA study, goal and scope definition, defines and specifies the objects and the research questions. It also defines the system

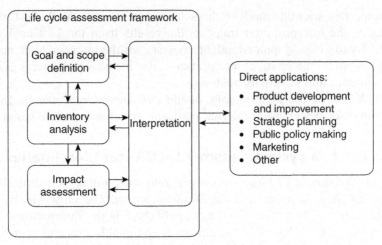

*3.2* Stages of an LCA (DIN EN ISO, 14040) cited from 'Environmental management – Life cycle assessment – Principles and framework' (ISO 14040:2006).

boundaries of the product system, e.g. as to space and time. A central task of LCA is the comparison of different products with the same function to extract particularly environmentally friendly goods or services. To fulfill this target, a clearly defined functional unit has to be determined. According to ISO 14040: 'the functional unit defines the quantification of the identified functions of the product', which means that the function has to be specified in terms of a quantity of a product or service, e.g. per kg of a product or per km of a transport service.

The second phase, life cycle inventory analysis (LCI), consists of three parts: creation of a flow model, data collection and calculation of the results. In a flow model, processes of the product system and their connections are described. The product system itself is defined by ISO 14040 as a 'collection of unit processes with elementary and product flows, performing one or more defined functions and which models the life cycle of a product.' For each unit process, data for input and output flows have to be gathered, together with other important information as to the process itself. The result of the LCI analysis is the quantification of the elementary flows resulting from the delivery of the functional unit, i.e. the product or service. The third step – the life cycle impact assessment, LCIA – transfers the results of the inventory analysis in a quantification of potential damage to the environment. To do so, the concept of impact categories is used. An impact category represents a certain environmental problem, e.g. climate change. It is quantified by a category indicator, which is based on an understanding of the underlying causes of an environmental problem as described in a respective scientific model ('characterization model'). As the last phase of LCA, the interpretation transfers the results from the LCI analysis and LCIA to a clearly understandable message for the intended audience of an LCA study. The phase of interpretation also includes procedures of quality assurance and sensitivity analysis.

Details of the methodology of life cycle assessment can be found in textbooks as well as on internet platforms (Baumann and Tillmann, 2004).

## 3.3 Life cycle assessment (LCA) for biorefineries

The assessment of biorefineries generally encounters as a difficulty, that so far there is no unified classification system. The most widely quoted definition of 'biorefinery' has been published by the International Energy Agency (IEA) Task 42. 'Biorefinery is sustainable processing of biomass into a spectrum of marketable products (food, feed, materials, and chemicals) and energy (fuels, power, heat).' In the same report, a generic classification of biorefinery systems is proposed based on four main features: platforms, products, feedstocks, and conversion processes. Most LCA studies on biorefineries consider or compare products or feedstocks, thus the other classification features such as platform and process appear only in the background. Information about platform and process only gets a closer look when systems are optimized, e.g. to use the same platform for multiple products, to optimize and reduce processing steps, or to avoid energy-intensive processes. The focus of LCA studies is on the products of biorefineries, which is not surprising, as LCA is always focused along the life cycle of its functional unit. This is an advantage on one hand, as it makes comparisons with other process routes for the same product quite easy. On the other hand, it is a shortcoming because the possible specific advantage of a biorefinery as a networking production system is difficult to account for. Still, in the following section a structure following specific products of biorefineries will be applied, while possible shortcomings by the holistic approaches of biorefineries will be discussed later.

### *LCA results for biofuels*

Given the current policy interest in biofuels, a large number of studies are available. The term biofuels denotes plant oils, biodiesel, bioethanol and biogas. Generally, LCA studies compare biofuels with the respective petrochemical fuels, which are gasoline, diesel and natural gas. An additional interest is to compare different biofuels, which are specified as so-called feedstock-technology combinations, i.e. process chains using a specific feedstock and a specific technology, where one technology may be feasible for different feedstocks and vice versa.

# Environmental and sustainability assessment of biorefineries

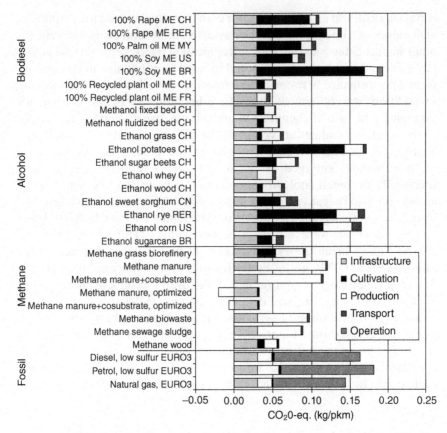

*3.3* GHG emissions for several biofuels per km for biofuels and fossil fuels (Zah *et al.*, 2007).

One comprehensive survey including the biofuels ethanol, methanol, biodiesel and biogas has been carried out in a study commissioned in 2007 by the Swiss federal administration (Zah *et al.*, 2007). It covered several technologies as well as biomass from domestic and from main production regions worldwide. The main results are presented in Fig. 3.3.

As Fig. 3.3 shows, the environmental performance as to global warming potential (GWP) is dependent on the type of fuel, the species and the regional origin of the feedstock. The lowest GHG emissions can be found for biofuels based on waste materials. Among agricultural feedstocks, those with high yield due to the species and the climatic conditions of agriculture show better performance; this is true, for example, for sugar cane from Brazil. The other way round, low yields per hectare and high use of nitrogen fertilizer along with the emissions of $N_2O$ (nitrous oxide) for certain agricultural techniques, e.g. for corn from the US, lead to comparably high values of GHG emissions. The assessment of biofuels is different for other impact categories, notably eutrophication. Here, fertilizer use generally causes a higher impact compared to fossil fuels but also shifts environmental performance between different feedstock-technology combinations. In addition, specific contributions to other impact categories exist, for example by the use of chemicals in agriculture, toxicity impacts appear.

These general findings are confirmed by the majority of studies, although controversial debates on single issues are encountered in the literature. One of these debates emerged as to the net energy ratio of starchy crops, specifically on bioethanol production from corn in the US, where some authors reported negative results and stated that more fossil fuel is consumed by production than gained as bioethanol (Pimentel and Patzek, 2005). These findings, however, were not confirmed by others (Hill *et al.*, 2006; von Blottnitz and Curran, 2007). In contrast, all studies agree that net energy ratio as well as GWP are far better for sugar cane compared to corn, and that crop yields have a major impact on the overall results, as shown in Fig. 3.4 for GHG savings per acre depending on crop yields.

The large contribution of agriculture is also confirmed for other impact categories. This is shown, for example, for sugarcane by Cavalett *et al.* (2011) (Fig. 3.5). Due to the use of agrochemicals, agriculture is generally seen as the main contributor to the impact categories of human toxicity and ecotoxicity (Bai *et al.*, 2010; Cherubini and Jungmeier, 2010).

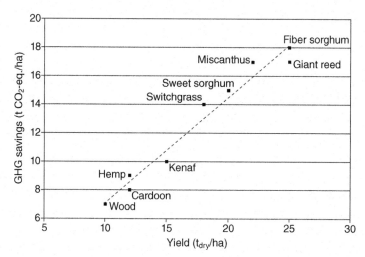

*3.4* GHG savings per hectare as a function of lignocellulosic crop yields (Cherubini *et al.*, 2009).

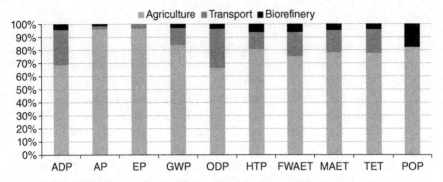

**3.5** Comparative environmental impacts breakdown for ethanol production in the E50-B. ADP, abiotic degradation; AP, acidification; EP, eutrophication; GWP, global warming; ODP, ozone depletion; HTP, human toxicity; FWAET, freshwater toxicity; MAET, marine water toxicity; TET, terrestrial toxicity; POP, photochemical oxidation (Cavalett et al., 2011).

During recent years, an additional aspect of agriculture has been recognized as a crucial issue for assessment of biofuels, which is the concern of land use changes (LUC) due to the rising demand for biomass. Land use change implies the direct or indirect change of not-cultivated land for agricultural use. Depending on former use, actual crop and agricultural techniques, carbon contained in the soil and plants can be released to the atmosphere, which in the worst case can jeopardize all GHG savings of biofuels. There are controversial statements regarding this topic since the first publication, but all studies agree that additional contributions to GHG must be expected from the conversion of very carbon-rich areas like peatland. This is confirmed, for example, for the case of palm oil diesel by a recent meta-analytic review of LCA (Manik and Halog, 2013).

The other important issue as to environmental performance of biofuels is the conversion technology itself. Here, several aspects have to be taken into account, primarily the efficiency of the technology. So-called second generation biofuels that make use of the full plant (and not only part of it, like seeds) are expected to show better performance, but this is also controversial. Von Blottnitz considered 47 publications that compare bioethanol production systems using LCA. Some of the LCA studies show a better environmental performance for second generation biofuels than first generation biofuels (Stöglehner and Narodoslawsky, 2009; Cherubini and Jungmeier, 2010). In addition to this, conversion technologies may provide by-products, e.g. glycerin from biodiesel production. In LCA, these are accounted for as to their substitution of fossil-based products. The way this is done on the methodological level may also be decisive for the outcome of an LCA study (see Cherubini and Strømman, 2011).

*LCA results for bio-based chemicals*

Bio-based basic chemicals, often called platform chemicals, are industrially produced chemicals that are used as raw material for many industrial products. Here, only few LCA studies are available (excluding ethanol, which was included above as part of the biofuels section). In 2006 a study of biotechnological production of 21 bulk chemicals from renewable resources was carried out on behalf of the EU (Patel et al., 2006). All bio-based products were compared with the respective petrochemical product using the categories of non-renewable energy use (NREU), GWP, land use in the form of land occupations and other environmental impacts. From this, significant reductions have been identified for all bio-based products. Limited availability of data and uncertainty concerning novel processes were identified as a main drawback of the assessment. Mainly based on this study, Hermann et al. presented results for the assessment of ten bio-based bulk chemicals produced by biotechnological processes (Hermann et al., 2007): 1,3-propanediol (PDO), acetic acid, acrylic acid, adipic acid, butanol, ethanol, lysine, lactic acid, polyhydroxyalkanoates (PHA), and succinic acid. In addition to that, five products produced from the aforementioned products are included: caprolactam, ethyl lactate, ethylene, polylactic acid (PLA), and polytrimethylene terephthalate (PTT). The assessment covers waste management within the system boundary and takes into account the impact categories NREU, GWP, and land use. Results show savings as to GHG and NREU for most bio-based chemicals, already for current technologies. For future technology, it is estimated that due to learning effects the savings will be 25–35% higher. This can be explained by the relatively high energy requirement for the production of petrochemical polymers.

There are some studies focusing on individual bio-based basic chemicals. Ekman and Börjesson show that propionic acid produced from by-products of agriculture leads to significant reduction of GHG emissions compared to fossil fuel alternatives. However, the contribution of propionic acid to eutrophication is higher (Ekman and Börjesson, 2011). Glutamic acid is an important component of waste from biofuel production and an interesting starting material for the synthesis of bio-based chemicals. Lammens et al. compare the environmental impacts of four bio-based chemicals from glutamic acid with their petrochemical equivalents: N-methylpyrrolidone (NMP), N-vinylpyrrolidone (NVP), acrylonitrile (ACN), and succinonitrile (SCN). The bio-based NMP and NVP show less impact on the environment, while for the ACN and SCN the petrol-based chemicals have less impact. Further optimizations indicate that the production of bio-based SCN can be improved to a level that can compete with the petrochemical process (Lammens et al., 2011).

Environmental and sustainability assessment of biorefineries 79

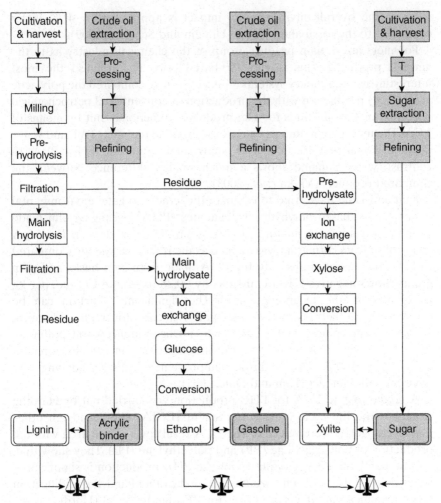

*3.6* Schematic life cycle comparison of biorefinery vs. conventional products.

Uihlein and Schebek (2009) compare the environmental impact of a lignocellulose biorefinery system with conventional production alternatives. The biorefinery delivers three products (lignin, ethanol, xylite), which are compared to their fossil counterparts (see Fig. 3.6). It was found that the biorefinery has the largest environmental impacts in the three categories fossil fuel use, respiratory effects and carcinogenics. The environmental impacts mainly arise from the provision of hydrochloric acid and to a lesser extent also from the provision of process heat. The optimal variant (acid and heat recovery) provides better results than the fossil alternatives, whereas the overall environmental impact is approximately 41% lower compared to the fossil alternatives (Uihlein and Schebek, 2009).

Polymers are a main product group of the chemical industry as to the amount produced. This is why bio-based polymers are also the most interesting for biorefinery systems. Vinka *et al.* have compared the polylactic acid (PLA) production with the production of conventional petrochemical polymers of various kinds for 1 kg product as functional unit by means of LCA. The investigated impact categories are fossil fuel use, GWP and water demand. Fossil fuel use and GWP show significant benefits for PLA. In contrast, water demand shows a much smaller difference between the compared products (Vinka *et al.*, 2003).

Kim and Dale have tried to estimate the 'cradle to gate' environmental impact of bio-based polyhydroxyalkanoates (PHA) packaging film made from crop residues (a mixture of corn grain and corn stover). The PHA production from corn grain was defined as the reference system. Compared to PHA from corn grains only, the PHA made from corn husks and corn grains shows negative GHG emissions by −0.28 to −1.9 kg $CO_2$-eq. per kg depending on the technology used. The significant reduction can be explained by the surplus energy from lignin-rich corn stover. Photochemical smog and eutrophication are related to nitrogen-induced soil pollution. PHA fermentation technology is still immature and in the development phase. The trend shows further improvements, thus reducing the environmental impact (Kim and Dale, 2005).

A cradle-to-gate LCA for PHB production was carried out by Harding *et al.* For the life cycle impact assessment (LCIA) GWP and ten other impact categories were selected. The LCA results were compared with the production of polypropylene (PP) and polyethylene (PE). They show that, on one hand, the energy required for the PHB production is significantly lower than for the polyolefin production, on the other hand, the acidification and eutrophication effects are lower for PE than for PHB (Harding *et al.*, 2007).

Roes and Patel (2007) have developed an approach, which is based on classical risk assessment methods (largely based on toxicology), as developed by the life cycle assessment (LCA) community, with statistics on technological disasters, accidents, and work-related illnesses. The approach has been applied to ethanol and four polymers from cradle to grave: polytrimethylene (PTT), polyhydroxyalkanoates (PHA), polyethylene terephthalate (PET) and polyethylene (PE). The results show lower risks for bio-based polymers compared to petrochemical equivalents. However, the uncertainties in the data need to be reduced (Roes and Patel, 2007).

Alvarenga *et al.* investigate PVC production from bioethanol as a substitute for ethylene. Two scenarios for bioethanol-based PVC for 2010 and 2018 are compared with fossil-based PVC, using several indicators for impact assessment. As to non-renewable resource use and GWP, bio-based PVC performed better; as to other impact categories, for some assumptions it performed worse for the state of 2010. As to 2018, better results turned out due to gains in efficiency and technological learning (Alvarenga *et al.*, 2013).

## 3.4 Sustainability issues: synopsis of results from assessment of economic and social aspects

### 3.4.1 Broadening the concept of LCA

Due to the widespread acceptance and use of the LCA methodology for environmental assessment of products and technologies during the last decade, efforts have been made to include the aspects of economy and social issues in LCA in order to derive a full sustainability assessment (Klöpffer, 2008). The aim is, similar to environmental LCA, to study the economic and social impacts of a product or technology, respectively, throughout its full lifetime, from manufacture through use to disposal or recycling. As a result of these efforts, methodological approaches to 'life cycle costing' (LCC) as well as 'social life cycle assessment' (SLCA) have been proposed. The currently most well-known approaches are the Code of Practice on LCC published by SETAC (Hunkeler *et al.*, 2008; Swarr *et al.*, 2011) and the Guidelines for Social Life Cycle Assessment, published the UNEP-SETAC Life Cycle Intitiative (UNEP, 2009; Benoît *et al.*, 2010). However, no standardized methodology exists, and due to the complex nature of assessment of economic as well as social issues, diverging approaches can be encountered in the literature.

Klöpffer states the conceptual 'equation' of life cycle sustainability assessment: LCSA = LCA + LCC + SLCA (Klöpffer, 2008). Prerequisite of this 'equation' is that the system boundaries and functional units of LCA, LCC and SLCA are similar. However, this, is not self-evident for approaches to economic and social assessment. The term of 'life cycle costing' was introduced during the 1970s; for a short history, see Sherif and Kolarik (1981). It originally denotes a formal analysis tool which is applied in business management. Here, the life cycle of a product is seen to cover the stages from invention, research, commercialization and end-of-life of a product, which is different from the idea of a 'physical life cycle' in LCA that covers material flows from research extraction to disposal. The idea of life cycle costing as a summary of all costs caused by a product during its entire life cycle, however, can be transferred also to the notion of a 'physical life cycle'. In this sense, LCC has also been proposed as a tool of sustainability assessment in recent years, postulating that the life cycle shall be congruent between LCC and LCA (UNEP, 2009). Even if systems boundaries are congruent, major methodological choices for LCC still exist. Notably, the question whether LCC covers monetization of externalities, i.e. environmental burdens, is crucial for any findings from this approach.

As to the method of SLCA, in the first place the question of adequate social indicators arises. Here indicators and indices are taken from different contexts mentioned in Section 3.2.1. However, given the large number of indices and indicator systems, the choice of proper indicators is often a highly controversial issue. Looking at system boundaries and matching to the phases of an environmental LCA, the methodological problem arises that indicators may not be directly related to processes, but rather to the level of companies or organizations (Jørgensen *et al.*, 2008). In addition, for some fields of interest, qualitative rather than quantitative indicators may be adequate. The emerging methodology for SLCA consequently comprises elaborate schemes for the choice of indicators which are operated by use of check-lists (Benoît-Norris *et al.*, 2011). However, it has to be emphasized again that other approaches also exist, which either use only environmental LCA, but combine it with other methods to account for economic and social aspects, or which use different methodological approaches to sustainability assessment. Examples for the variety of methods can be found in, for example, Assefa and Frostell (2007), Sugiyama *et al.* (2008) and Othman *et al.* (2010).

### 3.4.2 Results from economic and social asssessment of biorefineries

Given the far-reaching implications of the concept of biorefinery, economic and social aspects seem most important in addition to environmental assessment. However, the on-going scientific discussion on sustainability assessment methodology, combined with the open definition of the concept of 'biorefinery', results in the literature on sustainability assessments of biorefineries being scarce. Moreover, findings from the literature are often snapshots of single aspects rather than an overall sustainability assessment of the concept of 'biorefinery'.

One general finding stressed by several authors is that environmental and economic optimization go hand in hand. Several publications stress the importance of integration for efficiency improvement and consequently enhancing sustainability as well as environmental issues and costs (e.g., Mateos-Espejel *et al.*, 2011; Gassner and Maréchal, 2013). Generally speaking, the more efficient the chemistry, the lower the energy consumption during the production, and investment costs. A specific aspect is treated by Lange (2007) who points out that in fossil-based chemistry the goal of synthesis is often to add oxygen to hydrocarbons, which requires expensive

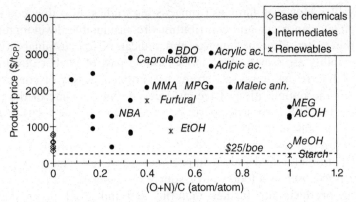

3.7 Historical average price of petro- and bio-based chemicals (1975 to 1995 average oil price of about $23/bbl) (Lange, 2007).

oxidation processes. This causes a general increase of costs along with increasing the oxygen content in the production (see Fig. 3.7) (Lange, 2007).

The generic advantage of renewable resources is that they are often rich in oxygen functionalities. This is why renewable resources can be used best when highly functional intermediates and polymer need to be obtained. Other authors point out that the selective deoxygenation of carbohydrates is more effective and therefore cheaper than the selective oxygenation of hydrocarbonates (Alles and Jenkins, 2010).

Luo *et al.* dealt with the economic and environmental analysis as well as with technical design of a lignocellulosic biorefinery (LCF), which produces ethanol, succinic acid, acetic acid and electricity. The economic analysis shows that the designed biorefinery has great potential in comparison with a sole ethanol biorefinery even if the price of succinic acid drops or the investment costs double (Luo *et al.*, 2010).

Various studies address the costs of production across the value chain. One important aspect here is transport. Several studies use mathematical models to assess the complete costs of energy use of biomass while considering the whole supply chain (biomass production, harvest, farm and road transport and conversion plant). On the basis of the assessment results, strategies have been proposed to reduce costs. In general the transport of biomass is an important aspect, so processing biomass in the vicinity of agricultural production areas has economic advantages (Yu and Tao, 2009; Akgul *et al.*, 2012; Giarola *et al.*, 2011).

## 3.5 Conclusion and future trends

Taking a long-time perspective, biorefineries are a generic part of sustainable development in the sense that they substitute fossil resources by renewables. However, findings from existing studies show than countercurrent effects can occur and contributions to sustainable development may disagree as to different impacts and indicators for sustainability. Thus, sustainability assessment is an indispensable part of further development of biorefineries, as it reveals strengths and weaknesses of concepts and supports the optimization of technologies. As an outcome of the present findings, the following aspects can be expected to be the focus of further research on sustainability of biorefineries.

A lesson learned from the present controversy on biofuels is the notion of competition for land and the importance of land use change. Agricultural products are used as a food source, source of raw materials for fuels and as starting products for further material and industrial processing. One challenge in the future is to observe the expected competition between material and energetic use on one hand, and cultivation of food and feed production on the other hand.

From this insight, a further conclusion can be drawn: as also renewables are not unlimited, efficiency of their use is the overriding challenge. In the case of biorefineries, this is notably possible by integration of processes in a biorefinery which also can be seen as a generic strength of the biorefinery concept. However, assessment of the interlinkage of processes in order to derive optimization tools up to now has not been taken into account sufficiently by sustainability assessment. Here, further methodological developments as well as close interaction with technology development in an early phase are necessary.

And last but not least, the social and economic implications of a broad implementation of biorefineries are not well understood. 'Bioeconomy' is well known as a catchphrase, but what does it denote? In a global economy, where will the feedstock for a future bioeconomy be produced? Will this be a chance or a risk for local economies, given that many of the most fertile areas are in regions of developing countries with smallholders? How will society have to change to support sustainable cultivation of biomass as well as the closing of material cycles and cascade use of carbon materials in order to enhance efficiency and lower the demand for primary resources? These questions pose scientific challenges; to address them is most important in order to provide science-based advice for policy and society as to the possibly far-reaching transformation process from a fossil-based to a future bio-based economy.

## 3.6 References

Akgul, O., Shah, N. and Papageorgiou, L.G. (2012), 'Economic optimisation of a UK advanced biofuel supply chain', *Biomass and Bioenergy*, 41, 57–72.

Alles, C.M. and Jenkins, R. (2010), Integrated corn-based biorefinery: a study in sustainable process development. In: Harmsen, J. and Powell, J.B. *Sustainable*

*Development in the Process Industries*, Hoboken, NJ: John Wiley & Sons, pp. 157–170.

Alvarenga, R.A.F., Dewulf, J., de Meester, S., Wathelet, A., Villers, J., Thommeret, R. and Hruska, Z. (2013), 'Life cycle assessment of bioethanol-based PVC', *Biofuels, Bioproducts and Biorefining*, 7(4), 386–395.

Assefa, G. and Frostell, B. (2007), 'Social sustainability and social acceptance in technology assessment: a case study of energy technologies', *Technology in Society*, 29(1), 63–78.

Bai, Y., Luo, L. and van der Voet, E. (2010), 'Life cycle assessment of switchgrass-derived ethanol as transport fuel', *Int J Life Cycle Assess*, 15(5), 468–477.

Balkema, A.J., Preisig, H.A., Otterpohl, R. and Lambert, F.J.D (2002), 'Indicators for the sustainability assessment of wastewater treatment systems', *Urban Water*, 4(2), 153–161.

Baumann, H. and Tillmann, A.-M. (2004), *The Hitch Hiker's Guide to LCA. An orientation in life cycle assessment methodology and application*. Lund: Studentlitteratur.

Benoît, C., Norris, G., Valdivia, S., Ciroth, A., Moberg, A., Bos, U., Prakash, S., Ugaya, C. and Beck, T. (2010), 'The guidelines for social life cycle assessment of products: just in time!', *International Journal of Life Cycle Assessment*, 15(2), 156–163.

Benoît-Norris, C., Vickery-Niederman, G., Valdivia, S., Franze, J., Traverso, M., Ciroth, A. and Mazijn, B. (2011), 'Introducing the UNEP/SETAC methodological sheets for subcategories of social LCA', *International Journal of Life Cycle Assessment*, 16(7), 682–690.

Böhringer, B. and Jochem, P.E.P. (2007), 'Measuring the immeasurable – a survey of sustainability indices', *Ecological Economics*, 63(1), 1–8.

Bütschi, D., Carius, R., Decker, M., Gram, S., Grunwald, A., Machleidt, P., Steyaert, S. and van Est, R. (2004), 'The practice of TA: science, interaction, and communication', *Wissenschaftsethik und Technikfolgenbeurteilung*, 22, 13–55.

Cavalett, O., Junqueira, T.L., Dias, M.O.S., Jesus, C.D.F., Mantelatto, P.E., Cunha, M.P., Franco, H.C.J., Cardoso, T.F., Filho, R.M., Rossell, C.E.V. and Bonomi, A. (2011), 'Environmental and economic assessment of sugarcane first generation biorefineries in Brazil', *Clean Technologies and Environmental Policy*, 14(3), 399–410.

Cherubini, F. and Jungmeier, G. (2010), 'LCA of a biorefinery concept producing bioethanol, bioenergy, and chemicals from switchgrass', *International Journal of Life Cycle Assessment*, 15(1), 53–66.

Cherubini, F. and Strømman, A.H. (2011), 'Life cycle assessment of bioenergy systems: state of the art and future challenges', *Bioresource Technology*, 102(2), 437–451.

Cherubini, F., Bird, N.D., Cowie, A., Jungmeier, G., Schlamadinger, B. and Woess-Gallasch, S. (2009), 'Energy- and greenhouse gas-based LCA of biofuel and bioenergy systems: key issues, ranges and recommendations', *Resources, Conservation and Recycling*, 53(8), 434–447.

Consoli, F., Allen, D., Boustead, I., Fava, J., Franklin, W., Jensen, A.A., de Oude, N., Parirish, R., Perriman, R., Postlethwaite, D., Quay, B., Seguin, J. and Vigon, B. (1993), *Guidelines for Life-Cycle Assessment: A 'Code of Practice'* Brussels: SETAC.

Dewulf, J. and van Langenhove, H. (2006), *Renewables-based Technology: Sustainability Assessment* Chichester: Wiley.

Ekman, A. and Börjesson, P. (2011), 'Environmental assessment of propionic acid produced in an agricultural biomass-based biorefinery system', *Journal of Cleaner Production*, 19(11), 1257–1265.

Finnveden, G., Hauschild, M.Z., Ekvall, T., Guinee, J., Heijungs, R., Hellweg, S., Koehler, A., Pennington, D. and Suh, S. (2009), 'Recent developments in life cycle assessment', *Journal of Environmental Management*, 91(1), 1–21.

Gassner, M. and Maréchal, F. (2013), 'Increasing efficiency of fuel ethanol production from lignocellulosic biomass by process integration', *Energy Fuels*, 27(4), 2107–2115.

Giarola, S., Zamboni, A. and Bezzo, F. (2011), 'Spatially explicit multi-objective optimisation for design and planning of hybrid first and second generation biorefineries', *Computers & Chemical Engineering*, 35(9), 1782-1797.

Hacking, T. and Guthrie, P. (2008), 'A framework for clarifying the meaning of triple bottom line, integrated, and sustainability assessment', *Environ. Impact Assess. Rev.*, 28, 73–89.

Halog, A. and Manik, Y. (2011), 'Advancing integrated systems modelling framework for life cycle sustainability assessment', *Sustainability*, 3(2), 469–499.

Harding, K.G., Dennis, J.S., Blottnitz, H. and Harrison, S.T.L. (2007), 'Environmental analysis of plastic production processes: comparing petroleum-based polypropylene and polyethylene with biologically-based poly-β-hydroxybutyric acid using life cycle analysis', *J. Biotechnol.*, 130(1), 57–66.

Hermann, B.G., Blok, K. and Patel M.K. (2007), 'Producing bio-based bulk chemicals using industrial biotechnology saves energy and combats climate change', *Environmental Science & Technology*, 41(22), 7915–7921.

Hill, J., Nelson, E., Tilman, D., Polasky, S. and Tiffany, D. (2006), 'Environmental, economic, and energetic costs and benefits of biodiesel and ethanol biofuels', *Proceedings of the National Academy of Sciences*, 103(30), 11206–11210.

Hueting, R. and Reijnders, L. (2004), 'Broad sustainability contra sustainability: the proper construction of sustainability indicators', *Ecological Economics*, 50(3–4), 249–260.

Hunkeler, D., Lichtenvort, K. and Rebotzer, G. (eds) (2008), *Environmental Life Cycle Costing*. Pensacola, FL: Society of Environmental Toxicology and Chemistry (SETAC).

ISO (2006), ISO 14040: Environmental management – Life cycle assessment – Principles and framework. Geneva: International Organization for Standardization.

Jänicke, M., Jörgens, H., Jörgensen, K., Nordbeck, R. and Busch, P.O. (2001), 'Planning for sustainable development in Germany: the former front runner is lagging behind', *Netherlands Journal of Environmental Sciences*, 16, 124–140.

Jørgensen, A., Le Bocq, A., Nazarkina, L. and Hauschild, M. (2008), 'Methodologies for social life cycle assessment', *International Journal of Life Cycle Assessment*, 13(2), 96–103.

Kim, S. and Dale, B.E. (2005), 'Life cycle assessment study of biopolymers (polyhydroxyalkanoates) derived from no-tilled corn', *International Journal of Life Cycle Assessment*, 10(3), 200–210.

Klöpffer, W. (2008), 'Life cycle sustainability assessment of products', *International Journal of Life Cycle Assessment*, 13(2), 89–95.

Lammens, T.M., Potting, J., Sanders, J.P. and de Boer, I. (2011), 'Environmental comparison of bio-based chemicals from glutamic acid with their petrochemical equivalents', *Environmental Science and Technology*, 45(19), 8521–8528.

Lange, J.-P. (2007), 'Lignocellulose conversion: an introduction to chemistry, process and economics', In Centi, G. and van Santen, R.A. (eds) *Catalysis for Renewables: From Feedstock to Energy Production*. New York: Wiley-VCH.

Luo, L., van der Voet, E. and Huppes, G. (2010), 'Biorefining of lignocellulosic feedstock – technical, economic and environmental considerations', *Bioresource Technology*, 101(13), 5023–5032.

Manik, Y. and Halog, A. (2013), 'A Meta-analytic review of life cycle assessment and flow analyses studies of palm oil biodiesel', *Integrated Environmental Assessment and Management*, 9(1), 134–141.

Markevičius, A., Katinas, V., Perednis, E. and Tamasauskiene, M. (2010), 'Trends and sustainability criteria of the production and use of liquid biofuels', *Renewable and Sustainable Energy Reviews*, 14(9), 3226–3231.

Mateos-Espejel, E., Moshkelani, M., Keshtkar, M. and Paris, J. (2011), 'Sustainability of the green integrated forest biorefinery: a question of energy', *Journal of Science & Technology for Forest Products and Processes*, 1(1), 55–61.

Ness, B., Urbel-Piirsalu, E., Anderberg, S. and Olsson, L. (2007), 'Categorising tools for sustainability assessment', *Ecological Economics*, 60(3), 498–508.

Othman, M.R., Repke, J.-U., Wozny, G. and Huang, Y. (2010), 'A modular approach to sustainability assessment and decision support in chemical process design', *Ind. Eng. Chem. Res.*, 49, 7870–7881.

Patel, M., Hermann, B. and Dornburg, V. (2006), The BREW Project: Medium and long-term opportunities and risks of the biotechnological production of bulk chemicals from renewable resources; Final Report, Utrecht. Available from: http://www.chem.uu.nl/brew/programme.html (accessed 6 October 2013).

Pimentel, D. and Patzek, T.W. (2005), 'Ethanol production using corn, switchgrass, and wood; biodiesel production using soybean and sunflower', *Natural Resources Research*, 14(1), 65–76.

Roes, A.L. and Patel, M.K. (2007), 'Life cycle risks for human health: a comparison of petroleum versus bio-based production of five bulk organic chemicals', *Risk Analysis*, 27(5), 1311–1321.

Singh, R.K., Murty, H.R., Gupta, S.K. and Dikshit, A.K. (2009), 'An overview of sustainability assessment methodologies', *Ecological Indicators*, 9(2), 189–212.

Sherif, Y.S. and Kolarik, W.J. (1981), 'Life cycle costing: concept and practice', *Omega – International Journal of Management Science*, 9(3), 287–296.

Stöglehner, G. and Narodoslawsky, M. (2009), 'How sustainable are biofuels? Answers and further questions arising from an ecological footprint perspective', *Bioresource Technology*, 100(16), 3825–3830.

Stoll-Kleemann, S. and Welp, M. (2006), *Stakeholder Dialogues in Natural Resources Management: Theory and Practice*. Berlin: Springer.

Sugiyama, H., Fischer, U., Hungerbühler, K. and Hirao, M. (2008), 'Decision framework for chemical process design including different stages of environmental, health, and safety assessment', *AICHE Journal*, 54(4), 1037–1053.

Swarr, T.E., Hunkeler, D., Klöpffer, W., Pesonen, H.-L., Ciroth, A., Brent, A.C. and Pagan, R. (2011), 'Environmental life-cycle costing: a code of practice', *International Journal of Life Cycle Assessment*, 16(5), 389–391.

Uihlein, A. and Schebek, L. (2009), 'Environmental impacts of a lignocellulose feedstock biorefinery system: an assessment', *Biomass and Bioenergy*, 33(5), 793–802.

UN (2001), Indicators of sustainable development: framework and methodologies; Background paper no. 3, 16–27, April, United Nations, New York, http://www.un.org/esa/sustdev/csd/csd9 indi bp3.pdf (last accessed 10 January 2014).

UNEP (2009), UNEP-SETAC LCI: http://www.lifecycleinitiative.org/ (accessed 15 August 2013).

Van den Ende, J., Mulder, K., Knot, M., Moors, E. and Vergragt, P. (1998), 'Traditional and modern technology assessment: toward a toolkit', *Technological Forecasting and Social Change*, 58(1–2), 5–21.

Vinka, E.T.H., Ra'bagob, K.R., Glassnerb, D.A. and Gruberb, P.R. (2003), 'Applications of life cycle assessment to NatureWorks™ polylactide (PLA) production', *Polym. Degrad. Stab.*, 80, 403–419.

von Blottnitz, H. and Curran, M.A. (2007), 'A review of assessments conducted on bio-ethanol as a transportation fuel from a net energy, greenhouse gas, and environmental life cycle perspective', *Journal of Cleaner Production*, 15(7), 607–619.

World Commission on Environment and Development (WCED) (1987), *Our Common Future*. Oxford: Oxford University Press.

Yu, S. and Tao, J. (2009), 'Economic, energy and environmental evaluations of biomass-based fuel ethanol projects based on life cycle assessment and simulation', *Applied Energy*, 86(Suppl. 1), S178–S188.

Zah, R., Böni, H., Gauch, M., Hischier, R., Lehmann, M. and Wäger, P. (2007), *Ökobilanz von Energieprodukten: Ökologische Bewertung von Biotreibstoffen*. St. Gallen: EMPa.

# 4
# Biorefinery plant design, engineering and process optimisation

J. B. HOLM-NIELSEN and E. A. EHIMEN,
Aalborg University, Denmark

DOI: 10.1533/9780857097385.1.89

**Abstract**: Before new biorefinery systems can be implemented, or the modification of existing single product biomass processing units into biorefineries can be carried out, proper planning of the intended biorefinery scheme must be performed initially. This chapter outlines design and synthesis approaches applicable for the planning and upgrading of intended biorefinery systems, and includes discussions on the operation of an existing lignocellulosic-based biorefinery platform. Furthermore, technical considerations and tools (i.e., process analytical tools) which could be applied to optimise the operations of existing and potential biorefinery plants are elucidated.

**Key words**: biorefinery, process design, process synthesis, optimisation, process up-scaling, integration.

## 4.1 Introduction

In the last 2–3 decades, biomass materials, i.e. agricultural crops, forestry products, organic fractions of household and industrial wastes and aquatic biomass (i.e., algae), have attracted increased research and commercial interest as renewable sources of fuels and high value chemicals. Increasing global energy demands and the need to reduce dependence on fossil fuel-based production systems, given the negative environmental impacts associated with their use, have been the main drivers for the use of these biologically-sourced process feedstocks. Although the advantages of using bio-derived products are well known, their production and use has been hindered by the availability and lower relative cost of fossil fuels. Advances in modern organic chemistry techniques have resulted in the increased production of high-volume low-value transportation fuels (accounting for more than 90% of the total global transportation consumption) and chemicals (Bozell, 2008). Improving the design, utilisation, energy efficiency and economics of heat, power, fuels and chemicals from biomass sources are therefore key requirements if biomass feedstocks are to become competitive with conventional fossil-based feedstocks. This

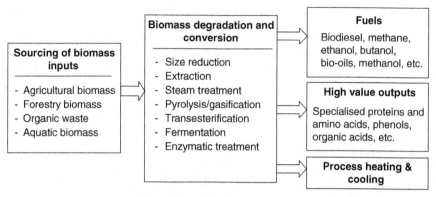

*4.1* An overview of the biorefinery concept.

would subsequently lead to the establishment of bio-materials as a potentially viable and sustainable alternative industrial raw material.

The use of the biorefinery concept has been proposed as a way to optimise the overall technical, economic and energetic efficiencies of the production processes of bio-products resulting in an array of marketable products, energy and process heating/cooling. This concept uses an extensive range of technologies to convert bio-materials into their component monomeric units which are then further reconstituted to produce high-value industrial precursors, chemicals and energy (including fuels and combined heat and power streams). An overview of factors influencing biomass inputs, process intermediates, conversion technologies and possible products and energy streams obtainable using a biorefinery set-up is shown in Fig. 4.1.

The main incentive encouraging the implementation of a biorefinery platform is the replacement of fossil-based industrial and energy feedstocks with 'green' biomass sources. The production of energy from these biomass materials is especially favoured over fossil-based feedstocks (i.e., coal combustion) since an almost neutral net carbon dioxide ($CO_2$) emission is achievable when biomass and energy production processes are managed sustainably. However, the attainment of such 'green' energy goals requires economic motivations to support the conversion of bio-materials to renewable fuels. As the fuel products are usually low-valued irrespective of the biomass feedstock, this financial incentive can be met using biorefinery systems producing high-value products at the same time. Consequently, biorefinery platforms have the potential to perform better economically and energetically than stand-alone biomass-to-fuel facilities which are currently overwhelmed by a low return on investment (Bozell, 2008).

With biorefinery research still only at a developmental stage, different definitions and categorisations for biorefinery platforms have been advanced

(IEA, 2007; Kamm and Kamm, 2004; Kamm et al., 2006). These categorisations are quite limited in scope and do not properly illustrate the full potential of biorefinery concepts for biomass processing. However, for the sake of discussion in this chapter, four main categories of potential biorefinery systems are used to demonstrate important process parameters which should be considered for integrated biomass-to-products conversion. These are:

- forestry platform biorefineries
- agricultural crop-based biorefineries
- industrial, agricultural and municipal organic waste-based biorefineries
- aquatic-based biorefineries

The emphasis of this chapter will be to illustrate various principles which should be considered in designing and implementing potential biorefinery systems. Although used to support the discussions in this chapter, an in-depth analysis of the potential economic, energetic and environmental advantages of biorefinery platforms will not be carried out since these topics are elaborated in Chapters 1–3 of this book. Furthermore, though Part II of this book specifically covers various biofuels and high-value products which could be obtained from biorefinery schemes, this chapter will draw on evidence from existing established, test and conceptual biorefinery platforms to help explain important plant design and conversion processes.

## 4.2 Microalgae biomass for biorefinery systems

Before the actual installation, implementation and operation of a proposed biorefinery system can be considered, careful planning, synthesis and design of the systems must be carried out. Section 4.3.1 looks at underlying principles and techniques which could guide biorefinery installation design and implementation. In particular, it covers the potential use of aquatic biomass inputs (specifically microalgae) in biorefinery systems since this feedstock is attracting interest as a useful 'non-food impinging, non-arable land requiring' biomass. Other factors promoting the use of this feedstock for chemicals and fuel production are its higher biomass productivity and photosynthetic efficiency compared to conventional terrestrial energy crops (Pirt, 1986; Goldman, 1980), and the ability to control the microalgae cultivation process for the production of specific macromolecular components (Illman et al., 2000). This potential to optimise the production of specific macromolecular components from microalgae biomass provides a unique platform for the production of a wide variety of products, chemicals and fuels confirming the use of this feedstock as an excellent basis for the application of a biorefinery concept.

Microalgae biomass production and conversion to fuels and chemicals is not novel. The use of microalgae biomass obtained from natural water bodies and used as human food has been well documented over time. However, most scientific work carried out on mass cultivation and use of this feedstock for fuel and chemical production is relatively recent (within the last 60 years) with the first pilot scale *Chlorella* cultivation plant reported in the 1950s (Arthur D. Little Inc., 1953) and fuel production via the production of methane ($CH_4$) using this biomass demonstrated by Meier (1955) and Golueke et al. (1957). Since then, research on algal-derived chemicals and fuels has been pursued aggressively by various applied phycology research groups worldwide leading to the present knowledge base. It is the findings of these research groups which have formed the basis of current operations of commercial large-scale microalgae plants aimed at single nutritional, mariculture and pharmaceutical applications. Coupling advances in biotechnology and conversion technologies with the vast information currently available on industrial microalgae cultivation could present an excellent foundation on which to base potential microalgae biorefinery processes. The co-production of a range of 'high- and low-valued' products and energy from microalgae biomass potentially optimises the utilisation of the various algal macromolecular components, and maximises the value which could be derived from this non-food feedstock.

Section 4.3 concentrates solely on the potential engineering and process optimisation schemes to be considered when using microalgae biomass in biorefinery schemes.

Although proposed microalgae biorefinery systems are included in the overall discussions on planning and synthesis due to their increasing interest as a model example of biomass feedstocks, which are expected to be used more and more for chemical and fuels production, other biomass inputs will also be considered. Having looked at the biorefinery planning and design aspects requirements, the technical and engineering requirements of a case study biorefinery plant is then presented in Section 4.4.

## 4.3 Planning, design and development of biorefinery systems

### 4.3.1 Initial feedstock and product considerations

A suitable starting point for the design of biorefinery systems is an awareness of the potential process biomass inputs as well as a preliminary consideration of the major product lines producible from the various applied conversion routes. The source and type of biomass raw material (i.e., purpose-grown crops, industrial or municipal organic wastes or algae) as well as the macromolecular composition of these inputs strongly influence the subsequent conversion techniques considered for the production of chemicals and fuels. A biomass input with a starch or fermentable sugar content of more than 80% (dry weight), for example, would intuitively suggest fermentation be considered for the production of bio-alcohols.

Therefore a preliminary empirical analysis of the chemical composition of the proposed biomass inputs (homogeneous or heterogeneous) must be carried out. Further research on the conversion of the identified biomass components into useful intermediates or finished product streams should then be conducted either experimentally or from previous research carried out on the selected feedstock(s). A limitation of previous research results is that most available research has focused on the production of a single product stream or the use of biomass raw materials grown under specific cultivation conditions (or specific acquisition conditions for waste biomass). This becomes an issue when proposed biorefinery systems plan to involve varied biomass inputs as process feedstocks integrating multiple treatment and conversion processes for the production of multi-product outputs.

If you were to take *Chlorella* biomass as a potential biorefinery feedstock, for example, a search of previous research for potential conversion routes and results which would form the basis for planning such a system is usually concentrated on the analysis and conversion of a particular macromolecular component. The results of research into the content of the 'high-valued' carotenoids (lutein, $\alpha$- and $\beta$-carotenes) (Iwamoto, 2006), fermentable sugars (Maršálkova *et al.*, 2010) and fatty acids and triglycerides (Milner, 1948; Ehimen *et al.*, 2010) available in *Chlorella* biomass can be found in publications using biomass inputs with different growth conditions. The same applies for the single production of different fuels and chemicals using *Chlorella* biomass as the process inputs. Care must be taken then when using such information for the planning of biorefinery systems and a comprehensive database on regional availability, optimal cultivation conditions and macromolecular compositions of various potential biomass feedstocks is essential. Where such a database is absent, additional laboratory (and possibly pilot) scale experiments must be carried out to provide information on the transformation of the proposed raw materials or to verify the adaptation of previous research findings to the specific proposed biorefinery system.

The sourcing, acquisition and transportation of the raw materials are also expected to be fundamental considerations for biorefining processes. Process logistics is a crucial parameter when considering the effective operation of such systems, determining the economic feasibility of the proposed biorefinery system. This is particularly important due to the bulky nature of most biomass raw materials and the use of relatively expensive road transportation (in relation to the product value) which could also negatively affect the $CO_2$ and energy balances of the overall production process. The decentralisation of proposed biorefinery plants by locating the conversion units as close as possible to the process raw materials is expected to be more commonly implemented to address such logistic problems.

The implications and analysis of the techno-economic costs of biomass sourcing and transportation have been detailed previously in publications and can be adapted to assess the logistics of potential biorefinery systems. The use of spatial information technologies (e.g., remote sensing and geographical information systems, GIS) to assess the practicality of establishing new decentralised biomass energy conversion plants in a selected region as presented by Ranta (2005) and Shi et al. (2008) could be applied when planning biorefinery systems. The models proposed in those studies include the influence of factors such as biomass and vegetation type, ecological retention, harvesting costs and the competing economical uses of the biomass on the eventual siting of the biomass conversion plant. Similarly, other studies such as the influence of biomass transportation costs and scale of the processing plant associated with the production of bioethanol from sugar cane and sweet sorghum, as demonstrated by Nguyen and Prince (1996), could be adapted as well as research into the optimisation of the location and capacity of a bio-processing system using a variety of lignocellulosic biomass feedstocks for ethanol and furfural production, as shown by Kaylen et al. (2000).

### 4.3.2 Design and synthesis of biorefinery systems

*Basis for biorefinery design*

A widespread approach used for the design of potential biorefinery systems is to begin with primary conversion technology, usually deemed the 'mainstay method' of the biorefinery, followed by the addition of side processing routes for the processing and upgrading of the biomass feedstock to other useful product streams. Such an approach is particularly appropriate in already established production systems, e.g., the pulp and paper industry, which could benefit from a gradual conversion of the primary milling operations to a biorefinery, reducing the overall process energy inputs, producing secondary products, and improving the overall economics of such a plant. The conversion and integration of a biorefinery scheme in existing plants could be spurred by the availability of useful waste streams. For example, the facilities at Cognis Australia Pty Ltd, Australia (the largest producer of algal beta-carotenes and carotenoids) (Ben-Amotz, 2006) could be converted to take advantage of the post-extracted residues after the carotenoid extraction and production process. The technical knowhow already accumulated on the cultivation, biochemical characteristics and conversion of the *Dunaliella salina* biomass (grown in two on-site operated saltwater lagoon farms in western and southern Australia) would help in the selection of suitable secondary co-products and conversion streams which could be integrated into the already established primary product line.

Another route for designing biorefineries is to combine and upscale laboratory-scale investigations with a revision of the process configurations based on the perceived feasibility of intended large-scale chemicals and fuel production. This route, although more time consuming, could be beneficial particularly for the implementation of novel conversion technologies avoiding any retro-fitting problems.

*Selecting best conversion biorefinery pathways*

Even having established the potential multiple conversion routes or utilisable streams in existing plant processes which could be adapted for biorefinery platforms, one of the key challenges in the design and synthesis of biorefinery systems is selecting and allocating the optimal conversion pathway for the production and separation of the product streams from the biomass input.

Various techniques have been put forward in publications to guide the design and synthesis of potential biorefinery systems (also applicable for microalgae biomass). Process synthesis approaches such as the matrix synthesis, symbol triangle, retro-synthesis and Gibbs free energy approaches were introduced in the 1970s (and still enjoy use in chemical engineering process design today). An increasing environmental awareness has led to the integration of environmental considerations in refining process designs as proposed, for example, in Crabtree and El-Halwagi (1994) and Pistikopoulos *et al.* (1994).

The use of a systematic approach to the design of optimal biorefinery pathways was reported by Bao *et al.* (2009). This is based on advancing a 'superstructure' of conversion technologies and products, then applying a tree-branching and searching technique to select the best candidate pathways and products for the intended biorefinery system. With the number of synthesised pathways being potentially extensive, such a technique provides a quick screening method to reduce the number of conversion technology alternatives to obtain the preferred pathways for the production of desired chemicals and fuels in the biorefinery setup. Other optimisation synthesis methods, such as the two-stage approach proposed by Pham and El-Halwagi (2012) for the design and synthesis of biorefinery systems, could also be easily implemented when considering biorefinery conversion.

As previously highlighted, the design considerations for microalgae biorefineries (and other biomass feedstocks) are generally complex, requiring a thorough knowledge of the component system inputs and intended conversion processes as well as an ability to develop innovative solutions to address potential problems which might be encountered. The application and integration of system tools for the initial design and synthesis of biorefinery systems is important, as it helps identify the proposed conversion routes with the minimum economic implications. Kokossis (1993) describes how such system tools could be applied in biorefinery synthesis by presenting a hierarchical cascade of information flows which should be considered. This cascade encompasses the use of forward-and-backward information flows aiding decisions on the system design, process synthesis and integration and the intended biorefinery system technologies (Fig. 4.2) (Kokossis and Yang, 2010).

Such an analysis could also provide the basis for the support of an integrated system where existing facilities already exist, with the existing processes (right of the figure) modelled and upgraded with the introduction of new process paths and products (Kokossis and Yang, 2010). This cascade analysis is thus repeated several times before selected processes are demonstrated and implemented (Kokossis and Yang, 2010).

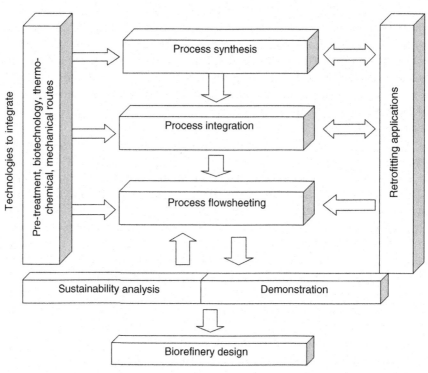

*4.2* Information flow cascades for biorefinery design (Kokossis and Yang, 2010).

With one of the credentials supporting the use of biomass feedstocks for chemicals and fuels production being the potential of the products to be $CO_2$ neutral (or optimistically negative), the optimisation design method for synthesis of biorefinery systems should establish the conversion routes requiring minimum process energy inputs. This is especially important with the potential use of microalgae biomass as the principal biorefinery raw material as is evident when its large-scale cultivation is factored into the overall framework of the production of fuels and chemicals from this feedstock. As opposed to the collection and supply of other first and second (i.e., terrestrial-based lignocellulosic biomass and waste) generation feedstocks, the industrial cultivation of single-celled microalgae biomass is a comparatively energy-intensive process mainly due to its harvesting method (i.e., centrifugation). Results of the energetic assessments carried out in Ehimen (2010) showed that the energy requirements for the biomass harvesting step could potentially account for more than 85% of the total energy requirement of the microalgae biomass production process, potentially making it the limiting step in a microalgae biorefinery. Therefore, although most biorefinery synthesis approaches concentrate on optimal conversion schemes, the biomass production (or acquisition) and harvesting stages must also be taken into account, especially when feedstocks like microalgae biomass are utilised. Also when designing the framework of potential biorefinery systems, any system analysis should include biomass production and transportation issues as well as the conversion routes as shown in Fig. 4.3.

### 4.3.3 Engineering considerations on biorefinery up-scaling and implementation

*Biomass production*

The inclusion of the biomass cultivation (or acquisition) has been included in the assessment boundaries for potential biorefinery schemes (Fig. 4.3). This factor is a particularly important engineering consideration when using microalgae biomass since the optimal choice for microalgae production and harvesting routes (as well as their energetic requirements) are still contentious and depend mainly on the specific microalgae species cultivated, type of bioreactor used, cultivation requirements and the climatic conditions available. Specific engineering solutions must be tailored to fit the particular microalgae biomass being produced. The outdoor large-scale cultivation and harvesting of *Chlorella* biomass, for example, will require technical considerations for culture stirring and harvesting techniques such as centrifugation. This is unlike the engineering requirements for microalgae species such as *Arthrospira* (*Spirulina*), where biomass harvesting could be

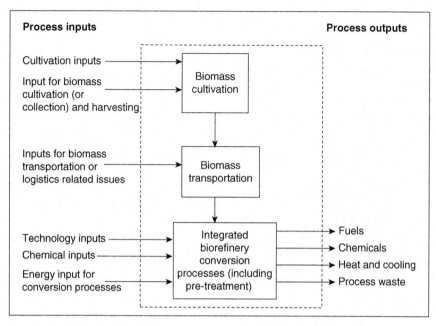

*4.3* Process boundary considerations for biorefinery systems.

accomplished using filters (Hu, 2006). A trade-off between the economics and the improved efficiency of the biomass production (and recovery) is thus essential. Although integration with waste treatment plants could potentially improve the process techno-economics of microalgae biorefineries, most high-valued products are aimed at the pharmaceutical market, where product purity and source are important parameters leading them to be sceptical about, or entirely reject, products obtained via such systems.

*Biomass conversion*

A range of conversion technologies can be applied and integrated for the production of energy carriers from microalgae biomass, depending on the intended energy application. These processes can be broadly classified into four main groups: mechanical (e.g., pressured extraction), biological (e.g., fermentation), chemical (e.g., solvent assisted extraction, transesterification), and thermal (e.g., drying, pyrolysis) (Fig. 4.4). A selected conversion route for the production of a chemical or fuel would generally involve a combination of two or more processing technologies, e.g. the combination of a mechanical oil extraction from oleaginous biomass and a thermo-chemical transesterification process for the production of the automotive fuel biodiesel from the extracted biomass oil.

# Biorefinery plant design, engineering and process optimisation 99

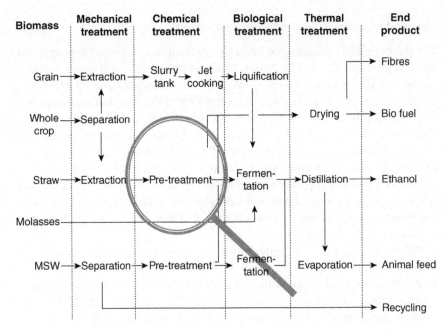

*4.4* The application of typical conversion routes using a biorefinery concept with heterogeneous feedstocks for the production of a broad range of processed products, e.g., foods/feeds/fuels/fibres and fertilisers (Nielsen, 2005).

Consequently, the engineering specifications of the conversion aspects of a proposed biorefinery systems are based on the intended product streams. Biotechnology and microbial advances have been especially important when considering the production of specialty chemicals and products. Important factors such as the sizing and control of the bioreactors, the required optimal biological conditions such as the process temperature and pH, the presence of inhibitory toxic components in the biomass and the biomass loading must be considered for such processes. However, for the production of lesser valued refining products, the application of rapid thermo-chemical conversion processes is still expected to be essential for proposed biorefinery systems, on the basis of economies of scale (Klass, 1998).

The engineering and integration of most of the constituent unit operations of the proposed biorefinery, i.e. extraction, distillation, filtration and heat utilisation, will be similar to existing systems using modern chemical engineering principles as the basis for the chemical and mechanical conversion processes and is expected to be the same for most process streams.

## Potential biorefinery integration and upscaling

For the potential up-scaling of laboratory research into biorefinery systems, process modelling tools like Aspen Plus® could be used for the initial design of the reaction and separation units of the intended biorefinery processes. Such an approach would include a description of the reaction pathways and chemical components, and selection of a suitable thermodynamic model for estimating the component's physical and thermodynamic properties. To ensure that the modelled processes reflect the observed practical results, empirical data on product yields obtained from laboratory experiments should be used in the process simulation. The user could further define the plant capacity, process input and operational conditions (e.g., flow rates, temperature and pressure) and use that as a basis to assess the technical feasibility of the biorefinery.

When considering the integration of different process streams in biorefineries, care should be taken in directly comparing refining schemes using biomass feedstocks with conventional petroleum refining systems. Unlike crude oil, which is a concentrated high-energy feedstock, most biomass raw material (especially microalgae, even after a dewatering step) usually has a high moisture content. A biorefinery scheme using such feedstocks must therefore be treated as a novel paradigm with innovative products and process streams (Subhadra, 2010).

Conceptually, the biorefinery scheme should also be treated differently from conventional refineries since in the latter, material and energy integration is usually applied to minimise the external energy inputs, with the by-products (including waste heat) of one oil refining process feeding into another complementary process stream for conversion into another product. This has led to the evolution of giant oil refineries with the associated benefit of maximum economic returns from minimum economic inputs. With the biomass inputs (especially in the case of microalgae) expected to be produced locally, it is speculated that small to medium-sized processing plants spread regionally around the biomass cultivation areas are most likely to develop as opposed to one or two giant refining plants (Willems, 2009). The scale of the proposed biorefinery system is important from an engineering perspective since smaller units could potentially limit the extent of energy recovery and even possibly hinder the re-utilisation of major stream by-products for onward conversion to other chemicals or fuels.

The integration and/or retrofitting of some stages of existing petrochemical refineries to utilise biomass feedstocks could still provide an attractive route for implementing proposed biorefinery processes. Such integration has the added advantage of potentially upgrading the conventional refinery systems to a 'greener' level. The ability to utilise waste heat streams from power generation plants could be an important criterion for the siting of a biorefinery plant, particularly when thermo-chemical processes such as biomass gasification and pyrolysis are proposed, as these processes require high operating temperatures (more than 300°C) for the production of syngas (a mixture of $CO$, $CO_2$, $CH_4$ and $H_2$) and bio-oil from a wide variety of biomass feedstocks. The syngas produced can be used directly as a fuel, starting block or intermediate for the production of chemicals (e.g., ammonia, organic acids and alcohols) and fuels (e.g., Fischer–Tropsch fuels, ethanol and di-methyl ether). Apart from the bio-oils obtained from the pyrolysis process, charcoal and light gases are also produced (Bridgwater and Peacocke, 2000). The integrated biorefinery system could also use existing platform units such as the gasification, hydro-treating and reformation refining stages of the petro-refinery to facilitate the biomass conversion or product formation processes. The availability of by-products and energy from the conventional refinery or power plant could help reduce the net material and energy inputs in the biorefinery process, improving the system's environmental impact while providing a foundation for the gradual shift to using biomass feedstocks.

To properly illustrate the operations and technological outlay of biorefinery systems and research and development considerations supporting such schemes, a case study plant using lignocellulosic biomass as the process feedstock is presented in the next section, which will also be used to highlight some of the design and integration considerations described in the earlier sections.

## 4.4 Case study: a second generation lignocellulosic biorefinery (Inbicon® Biorefinery)

Unlike the technologies associated with the production of energy and chemicals from first generation feedstocks (e.g., sugar and starch crops) which are already well established, the development of processes for the conversion of second generation biomass is still considered to be in the early developmental stages. Of the second generation feedstocks, the use of lignocellulosic biomass (e.g., residues from agriculture and forestry, as well as specially grown lignocellulosic crops) has been highlighted due to the potential availability of this raw material as a process feedstock.

The Inbicon biomass refinery (operated by Dong Energy, Denmark), selected as the case study in this chapter, uses lignocellulosic biomass for its operations. Although the conversion processes were developed and initial large-scale production carried out using wheat straw as the principal raw material, the conversion processes have been developed over time to use a variety of lignocellulosic inputs including corn stover, barley and rice straw, bagasse from sugar production, as well as garden and household waste. The biorefinery produces ethanol as its primary product, clean lignin powder and C5 molasses as a side stream, with electricity and heat co-produced via lignin combustion in a co-generation plant.

### 4.4.1 Biorefinery siting

The biorefinery is sited next to the Asnæs power station (the largest electricity and heat generation station in Denmark). As discussed in Section 4.3.1, this siting facilitates the use of waste energy (and other by-products) from the power installation. Such integration thus potentially improves the operational efficiency and environmental implications of both systems. The waste heat streams supply the thermal energy requirements of the pre-treatment and conversion steps of the biorefinery system.

### 4.4.2 Biomass transport and storage

To improve the techno-economics of the transportation of biomass feedstocks from the source point to the biorefinery, efficient baling systems have been introduced for lignocellulosic biomass. The prepared bales are then transferred to the plant using trucks or rail network where they are stored under optimal conditions.

The overall supply logistics are also closely monitored with continuous interaction between the biorefinery, feedstock suppliers and farmers, as well as government agencies and equipment manufacturers to ensure that the flow of biomass input to the refinery is kept as constant as possible (as the storage capacity of the biorefinery is limited to approximately one week's storage).

### 4.4.3 Mechanical pre-treatment

This is the first treatment step that the received biomass input undergoes. In this stage, the baled raw material (homogeneous or heterogeneous) is reduced in size. This is achieved primarily by cutting or milling processes which alter the shape, size and bulk densities of the biomass particles. The resulting biomass particles have improved particle uniformity, are easier to handle and are better suited for degradation in the next processing step.

### 4.4.4 Thermal treatment

The Inbicon biorefinery platform uses a patented hydro-thermal treatment process based on a combination of increased process pressures and temperature (i.e., pressure cooking). Here the previously mechanically treated biomass is mixed with water and the resulting mixture continuously heated to 180–200°C for between 5 and 15 minutes with process pressures of 4–8 bars. The main purpose of the heat treatment is to break down the biomass lignocellulosic structure, separating the lignin content so that the hemicelluloses and cellulose fractions can be further treated and converted to the desired ethanol product. This thermal pre-treatment step also ensures the separation of the alkali content as well as most of the inhibitors produced during the course of the partial degradation of hemicelluloses.

The Inbicon biorefinery thermal process further minimises the process energy and water consumption (and consequently the techno-energetic issues related to excess waste water treatment) by using biomass inputs with high dry matter content (30–40%, on the basis of total solid content). This biomass dry matter content requirement for the hydrothermal pre-treatment stage is higher than that of similar biomass pre-treatment processes described in publications or in practice. A solid-to-liquid separation step is then applied, allowing the separation of the solid fibre phase (containing mainly lignin and cellulose) which is pumped to the enzymatic reactors using Inbicon proprietary particle pumps.

### 4.4.5 Enzymatic hydrolysis

A cocktail of selected enzymes consisting of cellulases and liquefying enzymes are added to the fibre mass and mixed continuously for 24 hours in horizontal reactors using 'paddles'. As a result, the fibres are liquefied mainly due to the conversion of the cellulose fraction to lower sugar products (C6), leaving a solid phase largely composed of lignin. The liquid part containing fermentable sugars is then pumped to a conventional first generation sugar or starch-based fermentation system to be converted to ethanol.

### 4.4.6 Fermentation and distillation

The fermentation of the sugars is achieved by the addition of yeast over a period of 2–3 days in continuously stirred reactors. The main product from the fermentation (ethanol) is then passed through molecular sieves to purify and remove any water molecules contained in the output stream. The resulting ethanol mixture is then distilled in towers to obtain an anhydrous pure ethanol product.

### 4.4.7 Final solid/liquid separation

Further solid/liquid separation takes place integrated with other processes in the biorefinery system. The C5 sugars (i.e., molasses from the resulting liquid phase separated after the hydrothermal pre-treatment) and the solids after the enzymatic hydrolysis step are separated further into pure C5 molasses and lignin fractions. This is achieved mainly by drying these fractions using an evaporator. Some of the recovered water is separated and recycled back to form part of the water usage for earlier processing steps (thus minimising the net process water consumption). The resulting dried C5 sugars are then sold as molasses. The lignin fractions are further dried to a fine powder which is combusted in a co-generation plant to produce steam and power. The energy produced by this plant is more than the operational needs of the biorefinery plant and the excess energy is sent to the power grid.

## 4.5 Upgrading biorefinery operations

Any potential process upgrading or optimisation of integrated biomass refining systems depends on a thorough understanding of the operations of the existing plant to be improved. Before considering the addition of new process and product streams which could potentially increase the useability of the biomass inputs of the refining or processing plant, it is useful first to optimise the existing biomass acquisition, pre-treatment, conversion and product purification technologies.

### 4.5.1 Biomass feedstock production and logistics

The optimisation of existing biomass production and harvesting techniques is of vital importance to maximise overall biomass-to-product efficiency. This is especially true for proposed biorefinery systems using algae biomass where novel harvesting techniques (e.g., the application of micro-/ and ultra-filtration techniques) could potentially reduce the energy and hence cost inputs of the biomass production process.

In addition to existing biomass acquisition and transportation techniques as described in Section 4.4.2, the problem of transporting large volumes of low energy value biomass raw materials could be addressed by the application of pre-treatment technologies directly at the biomass source. The installation of mechanical pre-treatment (possibly coupled with biomass densification) on-site where the biomass is obtained, for example, could improve the overall energy content of the biomass transported to the biorefinery per volume transported as more potential chemicals and energy could be produced from the biomass using the same transportation scheme.

This scheme could be expanded to the production of the process intermediate on the biomass production site reducing the storage space requirements of the biorefinery plant. For example, with the decentralisation of the production facilities where oleaginous biomass is produced, the biomass oils can be extracted on-site, providing for an easily transportable intermediate which can be transformed to a range of fuel (e.g., biodiesel) and chemical (e.g., oleochemicals) products in a biorefinery.

## 4.5.2 Biomass pre-treatment and conversion

Optimisation routes which could improve the efficiency and overall product yield of the existing biorefinery pre-treatment and conversion processes should be implemented. The use of pre-treatment and conversion methods (including modifications to existing methods), which could potentially improve the output efficiencies per biomass input, reduce reaction times, minimise raw material handling (via a reduction of the processing steps) and optimise the economics of the biomass-to-products transformation, should be investigated continuously.

The introduction of pre-treatment methods such as the application of ultrasound technology to disrupt the biomass particles, for example, could help improve the extraction and subsequent conversion of the biomass macromolecular contents of interest. This is due to the improved availability of the macromolecules caused by the increased cellular disruption, compared to the relatively slower passive diffusion achieved with conventional mechanical stirring, leading to greater quantities available for conversion (Ehimen *et al.*, 2012; Khanal *et al.*, 2011). With the availability and ease of transformation of the cellular macromolecular biomass content, implementing such a technique as a pretreatment route could potentially improve the overall biomass utilisation without any major alteration or intensive modification of the existing conversion system.

Process improvements which can be applied to conventional techniques such as the simultaneous saccharification and fermentation of carbohydrate containing biomass as well as the *in-situ* transesterification of oleaginous biomass have been shown to potentially improve the biomass handling, reduce process energy, raw material inputs and overall time required for the conversion of these feedstocks to ethanol and biodiesel (fatty acid alkyl esters, FAME), respectively (Ehimen, 2010; Hari Krishna *et al.*, 1998). The modification and application of such *in-situ* conversion schemes facilitate the conversion of the biomass components directly to the intended products, eliminating previously required additional steps such as solvent extraction to obtain the oil feedstock, as in the conventional method. This could help simplify the chemical and fuel conversion process, potentially reducing the overall process cost and consequently the final fuel product costs.

The use of catalysis (including the use of bio-catalysts) to improve the efficiency of conversion processes should also be considered. A great deal of research, both historic and ongoing has been conducted on the application of catalysts to chemical and fuel production (e.g., Sutton *et al.*, 2001).

The application of membrane separation technologies as presented in Wang *et al.* (2009) could also be used as an energetic and cost effective alternative, replacing more energy intensive separation and purification methods (e.g., distillation) which are normally employed in biorefinery systems.

### 4.5.3 Process energy output and consumption

Although it is expected that a preliminary heat integration assessment of the biorefinery would have been conducted (i.e., as outlined by Linnhoff and Flowers (1978) and Linnhoff and Vredeveld (1984)), a continuous evaluation of available heat streams which could be applied to minimise net heating and cooling requirements of the biorefinery should be carried out.

Depending on regional availability, the integration of biorefinery systems with renewable energy sources (e.g., geothermal, wind and solar conversion systems) as described by Subhadra (2010) could be employed to improve the energy and $CO_2$ balance of the fuels and chemicals production. Furthermore, potential energy-saving methods and techniques such as heat pump technology or microwave-assisted irradiation could be investigated as a supply of process heat. This is particularly relevant where the process electricity is renewably sourced (e.g., with the use of wind power plants).

Potential energy recovery routes for the generation of additional energy streams should also be applied where possible; so, for example, $CH_4$ from the post-converted biomass residues following the anaerobic digestion process. This will be highlighted more in Section 4.7.

## 4.6 Optimising biorefinery processes using process analysis

With process analytical technologies (PAT) expected to reach maturity and become commercially viable in the pharmaceutical sector within this decade, the next step will be to introduce these advanced process control tools into the biorefinery sector (Holm-Nielsen, 2008). With such large investments being made in the biorefinery sector worldwide, there is as great a need to be pro-active in process monitoring and control regarding PAT instrumentation as in the leading food processing and pharmaceutical sectors.

In the biorefinery sectors, the competition for high quality grade fuels and products is fierce, and each percentage reduction in production prices is critical. Control tools like PAT will help monitor the large volumes of forecasted production/waste products in the coming decade and ensure the high quality of advanced biofuels as early in the process streams as possible. The ability to control and optimise the products obtainable from the various conversion schemes and to develop standardised products for the chemical and biofuels sectors in relation to conventional oil refinery products is highly relevant and presents us with extremely complex and fascinating scientific and technological challenges.

As described in Section 4.3.3, the biorefinery concept includes several mechanical, physical, chemical, and biological pre-treatment and processing steps. Integrated PAT tools are critically necessary to control and manage these very different processing steps to optimise the biorefinery plants.

## 4.7 Conclusion and future trends

The biorefinery concept (and research) has developed rapidly in the last decade. Biomass sources will increasingly become all-round competitive alternatives to crude oil as feedstocks for fuels, chemicals and materials, matching their role as raw materials for food and feed production as presented in Holm-Nielsen *et al.* (2006, 2007) and Holm-Nielsen and Kirchovas (2011). Many of these obligations will be fulfilled by *integrated* biorefinery processing facilities for optimal yield and energy efficiency. Future biomass conversion and use will be very different from biomass utilisation before the industrial revolution. It will also be different compared to the mono-processing facilities in the twentieth century (e.g., sugar factories and grain or potato starch processing plants). New synergistic multi-process flows will be developed far beyond yesterday's single-line processing plants. The resulting product portfolio for society will have the potential to replace petro-chemically synthesised products and allow new sustainable bio-products to find their way to the market, for example new biomass-based building and insulation materials made directly from processed fibres and plant materials.

Another important lesson learned from the initial stages of bio-refining is a trend towards decentralised production in strong contrast to the prevailing twentieth-century strategy of centralisation. The initial phases of these innovative steps have already begun in the design of such modern biotechnological processes and process equipment where the units will be simple and robust and *economy of scale* will be replaced by *economy of numbers* (Born, 2005). Micro fuel-cell implementation programmes running on biofuels for combined production of electricity and heat in domestic housing illustrate this well, fulfilling the idea of a bottom-up approach as a virtual power plant, compared to the top-down coal and nuclear centralised power plant concept from the last century. Biomass-based resources integrated with fuel-cells – in which the fuel itself is bio-ethanol, bio-methanol, bio-methane, or bio-hydrogen – is a fascinating concept because all the raw materials can be produced at a rural biorefinery scale. The end product, exemplified by biogas, can then be distributed through the existing natural gas grid system to houses in urban areas.

The ultimate goal for biorefineries is to be processing facilities that include centralised innovative technologies fuelled by renewable energy sources generated from wind, solar, and biomass resources. Such biorefineries will process a wide range of high- and medium-value products needed by society to replace fossil fuel-based products. The technologies involved are all known today, and are all based on flexible, more or less non-sterile, fermentation processes (Born, 2005). Biorefineries can be considered a 'bridge between agriculture and chemistry' (Braun, 2005) designed to produce basic and intermediate organic chemicals and fuels for direct or further product developments.

One of the most valuable by-product lines in the biorefinery could be the anaerobic digestion (AD) step. It is a fully integrated part of biorefinery processing due to its efficient use of by-products and waste streams from various other biological and biotechnological production processes. It is an important step producing biogas aimed for integration in the overall process energy needs but with the added advantage of producing bio-fertilisers to replace conventional chemical fertilisers in the agricultural sector. The AD process could be considered as the initial conversion step producing methane ($CH_4$) and $CO_2$ which would subsequently be used as the starting material for a variety of chemical and fuel synthesis. As highlighted in Section 4.5.1, where such gaseous intermediates are produced on decentralised sites close to the biomass source, the techno-economics and logistics of the conversion processes could potentially be improved with the use of pipelines to transport the products from these on-site facilities to the biorefinery systems. Alternatively, this step could be considered as the final conversion process applied to the biomass residues reclaiming any remaining carbon components following preceding conversion processes. The biorefinery concept provides endless new possibilities for making highly significant contributions towards the ultimate closed-system eco-solutions needed by society in the near future. Optimising the AD process is an important step towards this ecological goal. These solutions are ideologically invariant, functioning in both capitalist as well as socialist economies all over the world.

Although being a major incentive for technological advances, the use of profit to decide improving research and the implementation of biorefinery systems should be considered of secondary importance, compared to environmental concerns, pollution pressure and global warming.

## 4.8 References

Arthur D. Little Inc. (1953). Pilot Plant Studies in the Production of *Chlorella*. In *Algal Culture* (ed. J.S. Burlew). Carnegie Institution of Washington, Washington DC, pp. 235–272.

Bao, B., Ng, D.K.S., El-Halwagi, M.M. and Tay, D.H.S. (2009). Synthesis of Technology Pathways for an Integrated Biorefinery. AIChE Annual Meeting. Nashville, TN.
Ben-Amotz, A. (2006). Industrial Production of Microalgal Cell-Mass and Secondary Products – Major Industrial Species: Dunaliella. In: *Handbook of Microalgae Culture: Biotechnology and Applied Phycology* (ed. A. Richmond). Blackwell Publishing, Oxford, pp 273–280.
Born, J. (2005). *Baltic Biorefinery Symposium – From Sugar Factories to Biorefineries*. Proceedings, Aalborg University – Esbjerg Campus (ed. ACABS Research Group), Aalborg University, Esbjerg, pp. 23–32.
Bozell, J.J. (2008). Feedstocks for the Future – Biorefinery Production of Chemicals from Renewable Carbon. *CLEAN – Soil, Air, Water* 36: 641–647.
Braun, R. (2005). *Baltic Biorefinery Symposium – Biogas and Bioenergy System Developments towards Biorefineries, Trends in a Central European Context*. Proceedings, Aalborg University – Esbjerg Campus (ed. ACABS Research Group), Aalborg University, Esbjerg, pp. 93–104.
Bridgwater, A.V. and Peacocke, G.V.C. (2000). Fast Pyrolysis Processes for Biomass. *Sustainable Renewable Energy Reviews* 4: 1–73.
Crabtree, E.W. and El-Halwagi, M.M. (1994). Synthesis of Environmentally Acceptable Reactions. *AIChE Symposium Series* 90: 117–127.
Ehimen, E.A. (2010). Energy Balance of Microalgae-derived Biodiesel. *Energy Sources, A: Recovery, Utilization and Environmental Effects* 32: 1111–1120.
Ehimen, E.A., Sun, Z. and Carrington, C.G. (2010). Variables Affecting the *In Situ* Transesterification of Microalgae Lipids. *Fuel* 89: 677–684.
Ehimen, E.A., Sun, Z. and Carrington, C.G. (2012). Use of Ultrasound and Co-solvents to Improve the *In-situ* Transesterification of Microalgae Biomass. *Procedia Environmental Sciences* 15: 47–55.
Goldman, J.C. (1980). Physiological Aspects in Algal Mass Cultures. In: *Algae Biomass* (eds. G. Shelef and C.J. Soeder). Elsevier/North-Holland Biomedical Press, Amsterdam, pp. 343–359.
Golueke, C.G., Oswald, W.J. and Gotaas, H.B. (1957). Anaerobic Digestion of Algae. *Applied Microbiology* 5: 47–55.
Hari Krishna, S., Prasanthi K., Chowdaryk, G.V. and Ayyanna, C. (1998). Simultaneous Saccharification and Fermentation of Pretreated Sugar Cane Leaves to Ethanol. *Process Biochemistry* 33: 825–830.
Holm-Nielsen, J.B. (2008). Process Analytical Technologies for Anaerobic Digestion Systems – Robust biomass characterization, process analytical chemometrics and process optimization. PhD thesis, Aalborg University.
Holm-Nielsen, J.B. and Kirchovas, S. (2011). Large Scale International Bioenergy Trading – How Bioenergy trading can be realized under safe and sustainable conditions. *Proceedings of the 19th European Biomass Conference*, Berlin, June.
Holm-Nielsen, J.B., Madsen, M. and Oleskowicz-Popiel, P. (2006). Predicted Energy Crop Potentials for Bioenergy Worldwide and for EU-25. *Proceedings World Bioenergy 2006, Conference on Biomass for Energy*, Jönköping, Sweden, 30 May–1 June.
Holm-Nielsen, J.B., Oleskowicz-Popiel, P. and al Seadi, T. (2007). Energy Crop Potentials for Bioenergy in EU-27. *Proceedings 15th European Biomass Conference*, Berlin, Germany, May.
Hu, Q. (2006). Industrial Production of Microalgal Cell-Mass and Secondary Products – major Industrial Species: *Arthrospira (Spirulina) platensis*. In:

*Handbook of Microalgae Culture: Biotechnology and Applied Phycology* (ed. A. Richmond). Blackwell Publishing, Oxford, pp. 264–272.

IEA (2007). IEA Bioenergy Task on Biorefineries: Co-production, of Fuels, Chemicals, Power and Materials from Biomass. Third task meeting, Copenhagen, Denmark, 25–26 March.

Illman, A.M., Scragg, A.H. and Shales, A.H. (2000). Increase in Chlorella Strains Calorific Values when Grown in Low Nitrogen Medium. *Enzyme and Microbial Technology* 27: 631–635.

Iwamoto, H. (2006). Industrial Production of Microalgal Cell-Mass and Secondary Products – Major Industrial Species: *Chlorella*. In: *Handbook of Microalgae Culture: Biotechnology and Applied Phycology* (ed. A. Richmond). Blackwell Publishing, Oxford, pp. 255–263.

Kamm, B. and Kamm, M. (2004). Principles of Biorefineries. *Applied Microbiology and Biotechnology* 64: 134–145.

Kamm, B., Kamm, M. and Gruber, P. (2006). Biorefinery Systems – An Overview. In: *Biorefineries – Industrial Processes and Products. Status Quo and Future Directions* (eds. B. Kamm, M. Kamm and P. Gruber). Wiley-VCH, Weinheim, Vol. 1, pp. 3–40.

Kaylen, M., van Dyne, D.L., Choi, Y.S. and Blasé, M. (2000). Economic Feasibility of Producing Ethanol from Lignocellulosic Feedstocks. *Bioresource Technology* 72: 19–32.

Khanal, S.K., Takara, D., Nitayavardhana, S., Lamsal, B.P. and Shrestha, P. (2011). Ultrasound Applications in Enhanced Bioenergy and Biofuel Production. In: *Green Chemistry for Environmental Sustainability* (eds. S.K. Sharma and A. Mudho). CRC Press, Taylor & Francis Group, Boca Raton, FL, pp. 303–313.

Klass, D.L. (1998). *Biomass for Renewable Energy, Fuels and Chemicals*. Academic Press, San Diego, CA.

Kokossis, A.C. (1993). Mathematical Programming Approaches in the Context of Process Integration. 10th Annual Meeting Process Integration Research Center, UMIST, Manchester.

Kokossis, A.C. and Yang, A. (2010). On the Use of Systems Technologies and a Systematic Approach for the Synthesis and Design of Future Biorefineries. *Computers and Chemical Engineering* 34: 1397–1405.

Linnhoff, B. and Flower, J.R. (1978). Synthesis of Heat Exchanger Networks I: Systematic Generation of Energy Optimal Networks. *AIChE Journal* 24: 633–642.

Linnhoff, B. and Vredeveld, D.R. (1984). Pinch Technology has Come of Age. *Chemical Engineering Progress* 80: 33–40.

Maršálkova, B., Šimerova, M., Kuřec, M., Brányik, T., Brányiková, I., Melzoch, K. and Zachleder, V. (2010). Microalgae *Chlorella* sp. as a Source of Fermentable Sugars. *Chemical Engineering Transactions* 21: 1279–1284.

Meier, R.L. (1955). Biological Cycles in the Transformation of Solar Energy into Useful Fuels. In: *Solar Energy Research* (eds. F. Daniel and J.A. Duftie). University of Wisconsin Press, Madison WI, pp. 179–183.

Milner, H.W. (1948). The Fatty Acids of *Chlorella*. *Journal of Biological Chemistry* 176: 813–817.

Nguyen M.H. and Prince R.G.H. (1996). A Simple Rule for Bioenergy Conversion Plant Size Optimisation: Bioethanol from Sugar-cane and Sweet Sorghum. *Biomass & Bioenergy* 10: 361–365.

Nielsen, C. (2005). *Baltic Biorefinery Symposium – Integrated Biomass Utilisation Systems*. Proceedings, Aalborg University – Esbjerg Campus (ed. ACABS – Research Group), Aalborg University, Esbjerg, pp. 115–120.

Pham, V. and El-Halwagi, M. (2012). Process Synthesis and Optimization of Biorefinery Configurations. *AIChE Journal* 58: 1212–1221.

Pirt, S.J. (1986). The Thermodynamic Efficiency (Quantum Demand) and Dynamics of Photosynthetic Growth. *New Phytology* 102: 3–37.

Pistikopoulos, E.N., Stefanis, S.K. and Livingston, A.G. (1994). A Methodology for Minimum Environmental Impact Analysis. *AIChE Symposium Series* 90: 139–150.

Ranta, T. (2005). Logging Residues from Regeneration Fellings for Biofuel Production: GIS-based Availability Analysis in Finland. *Biomass and Bioenergy* 28: 171–182.

Shi, X., Elmoreb, A., Li, X., Gorenced, N.J., Jin, H., Zhang, X. and Wang, F. (2008). Using Spatial Information Technologies to Select Sites for Biomass Power Plants: A Case Study in Guangdong Province, China. *Biomass and Bioenergy* 32: 35–43.

Subhadra, B.G. (2010). Sustainability of Algal Biofuel Production Using Integrated Renewable Energy Park (IREP) and Algal Biorefinery Approach. *Energy Policy* 38: 5892–5901.

Sutton, D., Kelleher, B. and Ross, J.R.H. (2001). Review of Literature on Catalyst for Biomass Gasification. *Fuel Processing Technology* 73: 155–173.

Wang, Y., Wang, X., Liu, Y., Ou, S., Tan, Y. and Tang, S. (2009). Refining of Biodiesel by Ceramic Membrane Separation. *Fuel Processing Technology* 90: 422–427.

Willems, P.A. (2009). The Biofuels Landscape through the Lens of Industrial Chemistry. *Science* 325: 707–708.

# 5
# Current and emerging separations technologies in biorefining

S. DATTA, Y. J. LIN and S. W. SNYDER,
Argonne National Laboratory, USA

**DOI**: 10.1533/9780857097385.1.112

**Abstract**: This chapter provides an overview of separations technologies commonly employed in existing biorefineries as well as emerging separations technologies that are well poised to exhibit rapid growth in future biorefineries. Specifically, we focus on membrane-based separations, because they are becoming the separation platform of choice. In addition, we provide examples where electrochemically-driven processes could significantly reduce energy consumption during separations. We present several examples from our work, defining directions in separations technologies for integrated biorefineries.

**Key words**: separations, biorefineries, membranes, resin-wafer-eletrodeionization.

## 5.1 Introduction

Separations technologies play an important role in integrated biorefineries. Separations may account for ~50% of the total production costs in bioprocessing operations (Hestekin et al., 2002). Therefore, designing and implementing efficient separations strategies are critical factors in successful bioprocessing. In this chapter we provide an overview of separations technologies commonly employed in existing and future biorefineries. Separations are required for feedstock preparation, product recovery and purification, removal of impurities, and water management, recovery, and reuse. We focus on the chemical and biochemical aspects of separations technologies in liquid process streams.

From a thermodynamic standpoint, separations are inherently energy intensive, and therefore, strategies must consider energy demand to achieve the targeted separations. Based on the Energy Independence and Security Act, EISA 2007 (US Government Printing Office Pub L. 110–140), biofuels

---

The Publishers wish to acknowledge that this chapter is reproduced with the permission of Argonne National Laboratory, operated by UChicago Argonne, LLC, for the US Department of Energy under Contract No. DE-AC02-06CH11357.

must achieve specific targets for reduction in greenhouse gas emissions (GHGs) in the US. Conventional biofuels must achieve a 20% reduction in GHGs, advanced biofuels must achieve a 50% reduction, and cellulosic biofuels must achieve a 60% reduction. Reducing energy use in biorefining in general, and separations in specific, is critical to meeting these mandates. Europe has GHG reduction mandates that also warrant reducing energy use.

Separations are based on physical and chemical differences between species as well as the nature of the mixture. Typical physical factors include size, shape, compressibility, density, and viscosity. Chemical factors include solubility, hydrophobicity/hydrophilicity, polarity, $pK_a$, boiling and freezing points, and specific molecular interactions. Critical decisions for selection of separations strategies are based on the following questions:

- If species interact in the process stream, are the interactions based on chemical phenomena or physical trapping?
- In mixtures, are solutes dissolved or suspended? Depending on the size of the particles, suspended solid can be considered a solids/liquids separation.
- As species are concentrated, do they become reactive with themselves, other species, or the solvent?
- Are the species stable to the conditions and driving forces of the separation?

There are several critical factors that help define the most efficient separations scheme. As purity requirements increase, operation scale, energy consumption, and total costs increase exponentially. Therefore understanding purity requirements is essential in designing the separations scheme. The critical first decision to consider is the order of the separations scheme. Should a dilute solute be separated from solvent or should solvent be stripped from the dilute species or solute? Should separation schemes capture several chemically similar targets and then separate them with more specific processes? Should the separation scheme select individual species from the process stream? These decisions are based on knowledge of the process stream and targeted species, purity requirements, species stability, and available technologies. Within technologies important parameters include system footprint, substrate requirements, waste discharge requirements, and total throughput and process time.

Separations require both a method for speciation and a driving force. Speciation creates the materials needs and driving force creates the energy demand. Some common separations technologies include:

- crystallization and precipitation – based on differences in solubility;
- membranes or size exclusion chromatography – based on differences in size and shape as well as molecular interactions;

- ion exchange column chromatography – based on differences in polarity or charge;
- electrodialysis or electrodeionization – based on differences in charge and/or $pK_a$;
- distillation – based on differences in boiling points and vapor pressures;
- pervaporation or vapor permeation – based on differences in boiling points and vapor pressures as well as molecular interactions;
- adsorption – based on molecular interactions;
- solvent extraction – based on hydrophobicity/hydrophilicity or molecular interactions.

Most chemical and biological processes involve multiple reactions and separations processes. With design, integration of a specific reaction with product separations can increase overall performance of both conversion and recovery. Ultimately, choice of separations trains in integrated biorefineries is based on economics. The economics is defined by purity requirements, co-product uses, waste disposal routes, energy and substrate use, capital and operating costs, and system footprint.

There are entire textbooks written on most of these subjects (Coulson *et al.*, 1991; Green and Perry, 2007; Tarleton and Wakeman, 2008; Seader *et al.*, 2010). For the sake of brevity, we provide a few examples of the challenges in chemical and biochemical engineering associated with separations in biorefinery operations. We focus on emerging technologies that we expect to exhibit rapid growth. Specifically, we focus on membrane-based separations because they are becoming the separations platform of choice. In addition, we provide examples where electrochemical-driven processes could significantly reduce energy consumption during separations.

## 5.2 Separations technologies

### 5.2.1 Membrane separation technologies

Membrane-based separations technologies are becoming more widely deployed in integrated biorefinery operations due to their versatility, separations efficiency, energy savings, and economic benefits (Ho and Sirkar, 1992; Mulder, 1996; Cheryan, 1998; Baker, 2004; Li *et al.*, 2008). They are used in the food, pharmaceutical, biotechnological, bioprocessing, and chemical industries. A membrane is a porous, semi-permeable separation medium that fractionates different species from a solution based on size, shape, solubility, or molecular interactions. The permeate solution containing the 'smaller' species penetrates through the membrane, whereas, the retentate solution containing the 'larger' species is rejected by the membrane (Fig. 5.1a). Membranes are fabricated from many materials including inorganics such as alumina or silica or organics such as polyethersulfone,

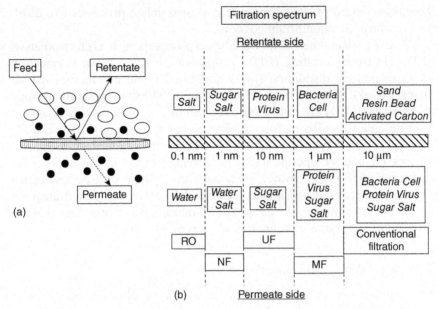

*5.1* (a) Separation scheme of a membrane process showing feed, permeate, and retentate streams. (b) Membrane filtration spectrum containing different membrane separation processes.

polyamides, or cellulose acetate. Membranes are commercially available in different module formats, including tubular, hollow fiber, flat sheet, spiral wound, etc. Membranes can be fabricated with pore diameters ranging from <1 nm (virtually non-porous) to 10 µm.

Based on pore size, membrane-based processes are classified as presented in the filtration spectrum (Fig. 5.1b). From a mixture, a particular membrane-based process rejects the species mentioned on the retentate side and allows permeation of the species mentioned on the permeate side of Fig. 5.1(b). Examples of the application of membranes based on pore size are:

- separation of activated carbon from sugar – conventional filtration
- separation of biological cells from proteins – microfiltration (MF)
- separation of proteins from salt – ultrafiltration (UF)
- separation of sugar from salt – nanofiltration (NF)
- separation of salt from water (desalination) – reverse osmosis (RO).

From the right of the filtration spectrum (conventional filtration) to the left (RO), the transmembrane pressure drop required for solvent flux increases due to decreasing membrane pore size. All of the processes from conventional filtration through RO operate under the driving force of chemical potential in the form of either concentration or pressure gradient.

Membrane-based processes are rate governed unlike processes like distillation, which are equilibrium governed.

There are some other membrane-based processes, such as electrodialysis (ED), electrodeionization (EDI), pervaporation (PV), vapor permeation (VP), membrane distillation (MD), supported liquid membranes (SLM) that are used frequently, but not included in the filtration spectrum. Among them, ED and EDI are charge-based membrane separations processes that operate under the driving force of electrochemical potential and separate charged species from uncharged species or fractionate multi-charged species. PV and VP operate under the driving force of chemical potential and fractionate organic/water mixtures with the help of a permselective (non-porous for all practical purpose) membrane. The permeate transports across the membrane in the gas phase. Membrane-based processes that are relevant in integrated biorefineries are described below.

*Size (or solubility)-based membrane separations*

MF and UF are the two widely used membrane filtration processes in biorefineries. The pore diameters of the membranes are in the range of 2 nm to 50 nm for UF and 50 nm to 5 µm for MF. Choice of membrane material and pore diameter depends on characteristics of the species present in the solution. MF is used for filtering coarse materials in biorefineries and often used as a pre-filtration step to the downstream UF. As an example, 0.45 µm polyethersulfone MF membrane (hollow fiber module) was used for removing cellular debris from the fermentation broth of a recombinant enzyme, glucose fructose oxidoreductase (GFOR). The permeate of the MF was then filtered through a 30 kDa molecular weight cut-off (MWCO) UF membrane to remove the nutrients (sugar and salts) in the permeate and collect GFOR in the retentate. The purified stream of GFOR was then used to convert sugar to biobased chemicals as described in detail in a later section. Membrane-based separations of biomolecules are susceptible to severe membrane fouling due to adsorption of biomolecules within membrane pores. A fouled membrane will exhibit dramatically declining permeate flux. A proper clean-in-place (CIP) procedure with suitable solvents and/or surfactants is necessary to restore permeate flux and reuse the fouled membrane. For example, a cleaning solution consisting of 0.2% sodium hydroxide, 25 g/L sodium chloride, and 0.2% Triton X-100 non-ionic surfactant is typically used for regenerating membranes subjected to biological fouling.

RO is typically used for desalination of process water for treatment and reuse, and, therefore, is an important part of biorefineries. In RO, a trans-membrane pressure, higher than the osmotic pressure of the solution in the feed side, is used to transfer water preferentially over the solute. This results in a purified water stream on the permeate side of the membrane. A transmembrane pressure as high as 50–80 bar is required for RO, which makes it unattractive to many users. In PV, the membrane acts as a permselective barrier between a liquid and a vapor phase. PV is commonly used to recover alcohols from very dilute aqueous solution as an alternative to energy-intensive distillation. Alcohol is preferentially solubilized in a membrane, diffused through the membrane, and then desorbed as vapor due to the applied vacuum on the permeate side. This results in an alcohol-enriched vapor phase starting from a lean solution of alcohol. An inert sweep gas is used to maintain the driving force on the permeate side.

*Charge-based membrane separations*

ED and EDI are membrane-based processes that remove ions from solution under an applied electric potential. ED is commercially used for water desalination, water treatment and reuse, and organic acid recovery. ED consists of successive alternative arrangements of cation exchange and anion exchange membranes within two electrodes to form two different types of compartments, diluate and concentrate. A schematic of the configuration of different components present in a typical ED set-up is represented in Fig. 5.2. When an electrolyte is fed through the diluate

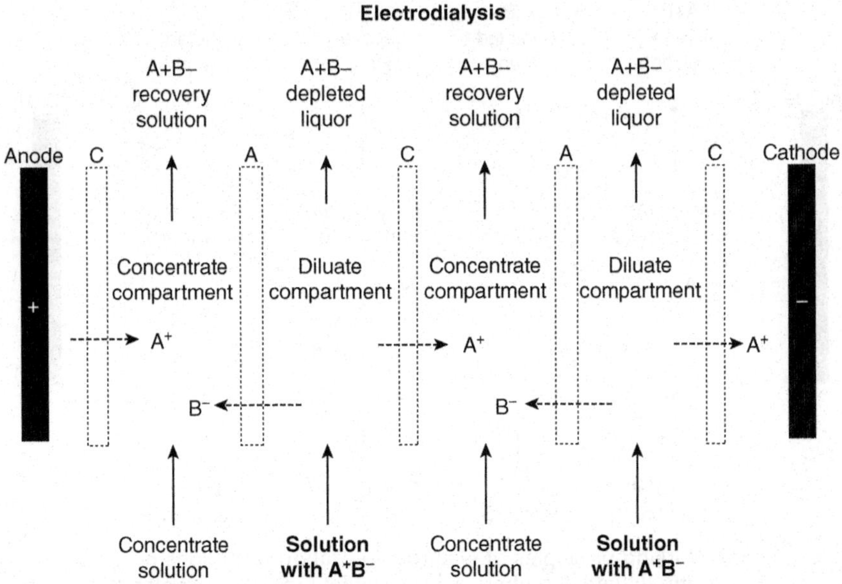

*5.2* Working principles of electrodialysis (ED). Ions move out from the diluate chamber and accumulate in the concentrate chamber due to an applied electric field.

compartment in the presence of an applied electric potential, cations move towards the cathode, transport across the cation exchange membranes, enter the concentrate compartment, and accumulate there due to the impermeable anion exchange membrane on the other boundary. Similarly, the anions move from diluate to concentrate compartment through anion exchange membranes. So, ED facilitates depletion of ions from diluate solution and accumulation of ions in concentrate solution. An electrolyte is recirculated on the two end electrode chambers to maintain the continuity in current flow.

Electrodeionization (EDI) is a modified version of ED that contains conductive ion exchange (IX) resin beads within the diluate compartment (Fig. 5.3). EDI combines the advantages of ED and IX chromatography. It utilizes *in-situ* regeneration of the IX resin beads by a phenomenon known as 'water splitting'. Water splitting on the surface of the IX resin beads regenerates the beads and ensures higher ionic conductivity within the diluate compartment. EDI outperforms ED with dilute solutions, where due to the limited ion concentration, ionic conductivity decreases and electrical energy is wasted in water splitting. In contrast, the conductive IX resin beads in EDI provide sufficient ionic conductivity, even with a dilute

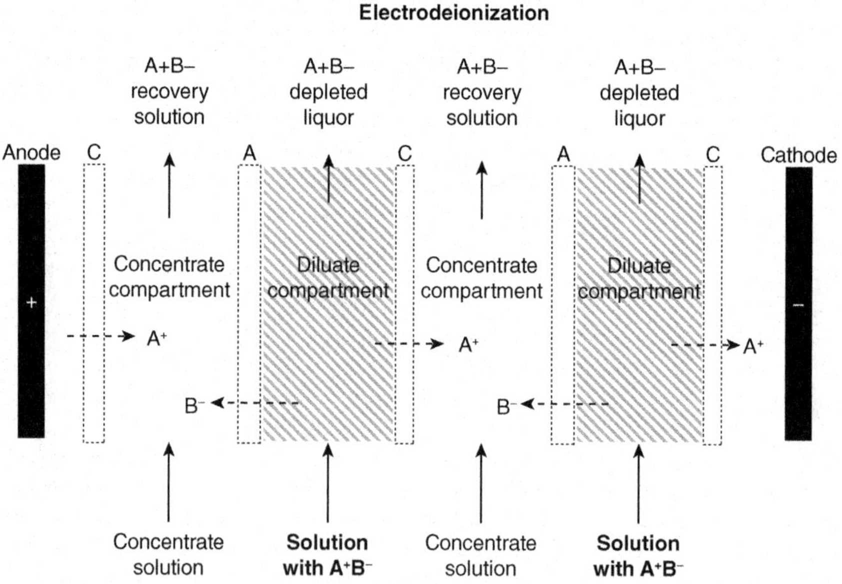

*5.3* Working principles of electrodeionization (EDI). Ions move out from the diluate chamber and accumulate in the concentrate chamber due to an applied electric field. Ion conductive resin beads in the diluate chamber provide enough conductivity to transport ions from very dilute solutions.

solution, and provide an efficient ion transport pathway through the IX resin beads. In conventional EDI, loose IX resin beads are used; however, the researchers at Argonne National Laboratory have improved the technology by using resin wafers (RW) to incorporate the loose ion exchange resin. The modified platform is called RW-EDI. Argonne patented the technology to fabricate and use the resin wafers (Lin *et al.*, 2005a, 2008). The technology offers enhanced flow distribution, higher conductivity, superior pH control, ease of materials handling and system assembly, and a porous solid support for incorporation of catalysts, biocatalysts, and other adjuvants. RW-EDI is used for production and recovery of biobased chemicals (Arora *et al.*, 2007), especially organic acids from fermentation broth (Lin *et al.*, 2005b), post-transesterification glycerin desalting (Datta *et al.*, 2009), conditioning of biomass hydrolysate liquor (Datta *et al.*, 2013), and for $CO_2$ capture from flue gas (Lin *et al.*, 2013). Argonne deploys three different ED stack sizes (Fig. 5.4) to design experiments and evaluate performance from fundamental and exploratory scale research through pilot-scale and field deployment. A small stack (14 cm$^2$ surface area) is used for proof of concept experiments, which is then scaled up to a TS2 stack (195 cm$^2$ surface area), for process optimization and more rigorous studies. At the pilot scale (1700 cm$^2$ surface area), extended campaigns are conducted to evaluate potential for commercialization.

### 5.2.2 Adsorption

Adsorption is a technique that is used frequently in biorefineries for product polishing and removal of minor impurities (Ruthven, 1984; Young, 2003; Sengupta, 2007). For example, activated carbon or resin beads packed in a cylindrical column are used as an adsorbing device; a process known as packed bed column chromatography. Resin beads, either ion exchange (charged) or adsorptive (uncharged), are used for adsorption. Ion exchange beads adsorb the charged species due to electrostatic interactions between the fixed charged groups on the resin beads with the oppositely charged counterions in solution. The adsorptive beads capture species based on physical forces, such as hydrophobic interaction. Although, there is a rate associated with it, the adsorption process is governed by the concentration of the adsorbed species in the adsorbent and solvent at equilibrium. Therefore, equilibrium adsorption capacity is an important criterion for selecting the adsorbent. Typically, for each adsorbent an adsorption isotherm plot is constructed that contains concentration of the solute within the adsorbent as a function of the concentration of the solute in the solvent at equilibrium. Chemical, mechanical, and thermal stability of the adsorbent, reusability, and ease of operation are some of the advantages of adsorptive separation techniques. Disadvantages of adsorptive separations techniques

*5.4* Photographs of three different sizes of commercially available ED stacks (Ameridia) used at Argonne National Laboratory. Photo credits: Michael Henry, Michael Henry, George Joch.

include high pressure drop, a diffusion dominated transport mechanism, and cost of regeneration.

### 5.2.3 Extraction

In biorefineries, liquid–liquid (L-L) extraction is widely implemented for recovering fuels and chemicals from biological mixtures such as fermentation broths (Treybal, 1980; Seader *et al.*, 2010). A solvent (extractant) that is immiscible with the process solution is used to extract the solute. After extraction, the extract (extracted solute + extractant) is separated from the raffinate (original solution depleted of the solute) by another unit operation, most commonly a gravity settler. The solute is recovered from the extract by evaporating the extractant. Extraction is an equilibrium-governed process which relies on the distribution of the solute between the original and extracting solvents. Important factors for selecting the extraction solvent include: partition coefficient (distribution constant), immiscibility with the original solvent, and boiling point for evaporation. We present several examples from our work defining directions in separations technologies for integrated biorefineries.

## 5.3 Removal of impurities from lignocellulosic biomass hydrolysate liquor for production of cellulosic sugars

Lignocellulosic biomass can be converted to biofuels by biochemical conversion (Fig. 5.5). Enzymatic hydrolysis is used to convert the cellulose component of the lignocellulose to soluble, monomeric sugars (primarily glucose), which are then fermented to biofuels (Aden et al., 2002). Lignocellulose, primarily composed of cellulose, hemicellulose and lignin is a rigid, crystalline structure not directly amenable to enzymatic hydrolysis. Pretreatment is required to disrupt the crystalline structure of the cellulosic moieties and expose them for the downstream enzymatic hydrolysis step. The most common pretreatment technique is dilute acid treatment with sulfuric acid at elevated temperature and pressure. The heat generates unwanted degradation products and impurities (organic acids, furans, phenolics, etc.) along with the solubilized C5 sugars (primarily xylose from hemicellulose) in the hydrolysate liquor (Tucker et al., 2003). These impurities are detrimental to downstream enzymatic hydrolysis and fermentation. The acidity of the hydrolysate liquor (e.g., by sulfuric acid and organic acids) also inhibits enzymatic hydrolysis and fermentation. Hence, removal of these byproducts and reagents is essential for further processing of the insoluble biomass component (cellulose and lignin) that will lead to an efficient biofuels production strategy (Biomass Multi Year Program Plan, 2008). The major impurities of dilute acid pretreatment are acetic acid, produced by deacetylation of hemicellulose, furfural, produced by degradation of C5 sugars, and hydroxylmethyl furfural (HMF), produced by degradation of C6 sugars. Acetic acid and sulfuric acid are ionic impurities, and therefore, could be separated based on their ionization. The non-ionic impurities, such as furfural and HMF, could be removed using adsorption techniques. Ionic impurities from corn stover hydrolysate liquor can be separated using RW-EDI followed by the removal of non-ionic impurities using adsorptive beads as described below.

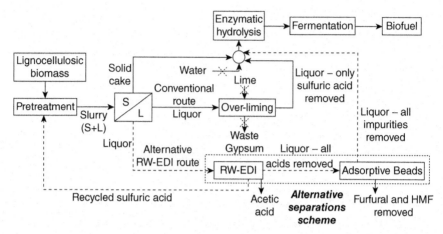

*5.5* Schematic of lignocellulosic biomass derived biofuel production route (biochemical) and the role of intermediate separation steps. The conventional route is designated by the solid line, while the proposed alternative route is designated by the dotted line. Sequential RW-EDI-based removal of ionic impurities (sulfuric acid and acetic acid) followed by adsorptive beads-based removal of non-ionic impurities (furfural and HMF) are clustered into a dotted box. Waste gypsum, lime and additional water are not involved in the alternative separations scheme (adapted from Datta *et al.*, 2013).

### 5.3.1 RW-EDI for removal of ionic impurities

Biomass hydrolysate liquor and the residual solid biomass are first separated by feeding the pretreatment slurry into a solid liquid separator (unpublished data). In conventional systems, sulfuric acid is removed by treating with lime (over-liming) and precipitating sulfate ions as $CaSO_4$ (gypsum) as shown in Fig. 5.5. Over-liming removes sulfate ions from hydrolysate liquor, albeit increasing the number of unit operations and residence time (over-liming tank + pH adjustment tank/settlement tank + separation devices) (Aden *et al.*, 2002). It also requires addition of chemicals and generation of a low-value and potentially toxic byproduct, gypsum. The flash vaporization prior to solid-liquid separation removes only 8% of the total acetic acid. Further reduction in acetic acid concentration is accomplished by aqueous dilution of the hydrolysate liquor.

This is an example of using emerging technology to remove ionic impurities, sulfuric acid and acetic acid. We treat corn stover hydrolysate liquor using resin wafer electrodeionization (RW-EDI) technology. RW-EDI-based deacidification provides a pathway for conditioning of biomass hydrolysate liquor with fewer unit operations, lower operation time, and reduced use of chemicals and water. Similar to ED, the RW-EDI stack is configured for ion transport from one compartment (diluate) to another

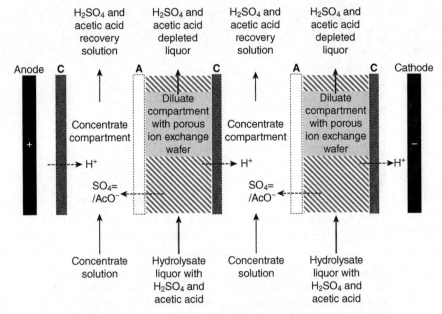

*5.6* Schematic of different components inside a resin wafer-based electrodeionization (RW-EDI) stack. Acids get transferred from the hydrolysate liquor in the diluate compartment to the recovery solution in the concentrate compartment under an applied electric field. The diluate compartment contains porous ion exchange resin wafer. C and A are cation and anion exchange membranes, respectively (adapted from Datta *et al.*, 2013).

(concentrate) by alternative arrangement of cation exchange and anion exchange membranes within two electrodes (Fig. 5.6). Diluate compartments contain ion exchange resin beads within a porous solid matrix, called a resin wafer (Lin *et al.*, 2005a, 2008) for enhanced conductivity, particularly for dilute solutions. The electrically-driven membrane process facilitates depletion of ions from one solution (feed solution within the diluate compartment) and accumulation of ions in another solution (recovery solution within the concentrate compartment) under an applied electric field. Depending on operating conditions, RW-EDI can separate the sulfate and acetate present in the corn stover hydrolysate liquor either together or sequentially. Using RW-EDI, >99% sulfuric acid and >95% of acetic acid were removed. For the neutral xylose sugar, >98% was retained (Fig. 5.7) (Lin *et al.*, 2013). By adjusting the operating conditions, selective separation of sulfuric acid and acetic acid was achieved to obtain two separate acid-enriched streams. For a typical case, the sulfuric acid-enriched stream contained around 20 g/L of sulfuric acid and 1 g/L of acetic acid. On the

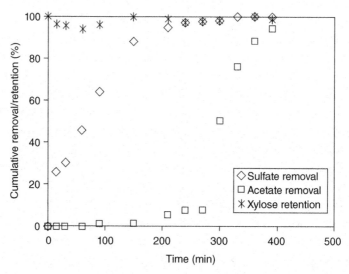

*5.7* Removal of sulfuric acid and acetic acid from corn stover hydrolysate liquor using batch mode RW-EDI. Up to 99% of sulfuric acid and >95% of acetic acid are removed, whereas, more than 98% xylose is retained from hydrolysate liquor.

other hand, the acetic acid-enriched stream contained around 0.5 g/L of sulfuric acid and 9 g/L of acetic acid. The sulfuric acid stream could be recycled back for the dilute acid pretreatment, while the acetic acid stream could be recovered as a value-added biobased co-product.

### 5.3.2 Adsorptive beads for removal of non-ionic impurities

Polymeric ion exchange resin beads (Nilvebrant *et al.*, 2001) and polymeric adsorptive beads (Weil *et al.*, 2002) have been used for the removal of fermentation inhibitory compounds (both ionic and non-ionic) from biomass hydrolysate liquor. Here, an example is presented where adsorptive beads-based removal of non-ionic impurities, such as furfural and HMF, was used as a polishing step to the RW-EDI as shown in Fig. 5.5. Furfural and HMF were removed through hydrophobic interaction with the commercially available functionalized adsorptive beads. Comparative results for removal of furfural and HMF from RW-EDI treated corn stover hydrolysate liquor using different types of commercially available resin beads are shown in Fig. 5.8. Dowex L-493 appears to be superior to the other resin beads primarily due to their very high surface area (~1100 m$^2$/g). It removed all furfural and HMF from the RW-EDI-treated corn stover hydrolysate liquor. The RW-EDI-treated and Dowex L-493-polished corn stover hydrolysate

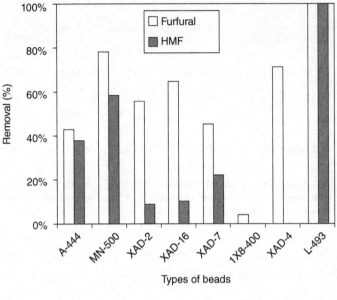
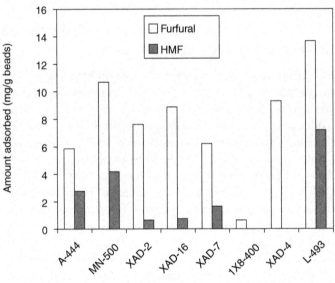

*5.8* Comparative study for removal of furfural and HMF from RW-EDI-treated corn stover hydrolysate liquor by different commercially available resin beads. Beads used are: Purolite A-444 (anion exchange) and MN-500 (cation exchange); Amberlite XAD-2, XAD-4, XAD-7 and XAD-16 (all non-ionic); Dowex 1X8-400 (anion exchange) and L-493 (non-ionic).

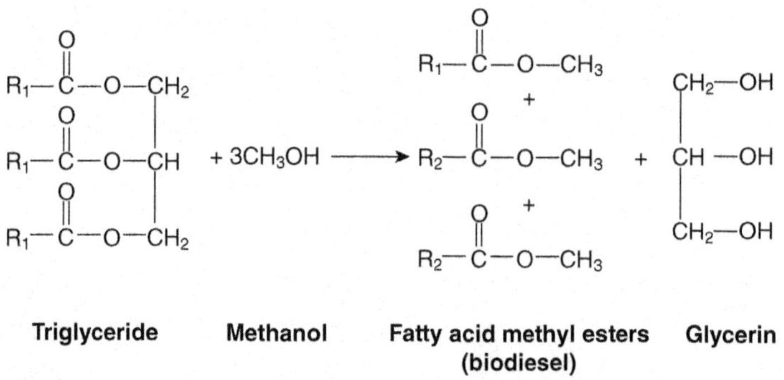

$$\begin{array}{c}
R_1-\overset{O}{\overset{\|}{C}}-O-CH_2 \\
R_1-\overset{O}{\overset{\|}{C}}-O-CH \\
R_1-\overset{O}{\overset{\|}{C}}-O-CH_2
\end{array} + 3CH_3OH \longrightarrow \begin{array}{c} R_1-\overset{O}{\overset{\|}{C}}-O-CH_3 \\ + \\ R_2-\overset{O}{\overset{\|}{C}}-O-CH_3 \\ + \\ R_2-\overset{O}{\overset{\|}{C}}-O-CH_3 \end{array} + \begin{array}{c} CH_2-OH \\ CH-OH \\ CH_2-OH \end{array}$$

Triglyceride    Methanol    Fatty acid methyl esters    Glycerin
(biodiesel)

*5.9* Reaction scheme of biodiesel. Fat (triglyceride) reacts with methanol to form biodiesel (mixture of fatty acid methyl esters) and glycerin (glycerol).

liquor is a pure sugar solution containing 40 g/L xylose. The other resin beads that were evaluated are less efficient compared to L-493, as shown by the percentage removal and amount adsorbed (mg/g of beads) values.

## 5.4 Glycerin desalting as a value added co-product from biodiesel production

The annual production of biodiesel in the United States is increasing from 75 million gallons in 2005 to an EISA 2007 mandate of 1 billion gallons (http://www.biodiesel.org/production/production-statistics). Biodiesel is primarily produced by transesterification of triglycerides (vegetable oil or fat) with an alcohol (methanol or ethanol) using inorganic catalyst (acid or base) in homogeneous phase (Fig. 5.9) (Noureddini and Zhu, 1997). The principal co-product of biodiesel production is glycerin (glycerol) (Thompson and He, 2006). Typically, 100 kg of vegetable oil is reacted with 10 kg of methanol to yield 100 kg of biodiesel and 10 kg of glycerin. The process generates inorganic salts, such as NaCl or KCl, as impurities in the glycerin phase. Removal of these impurities is critical to transform the crude glycerin into a high value, purified glycerin for use in the bioprocessing, food, pharmaceutical, and cosmetics industries. Purified glycerin from biodiesel production facilities could also be used as a starting material for useful chemicals as summarized in Fig. 5.10. The existing technologies for refining crude glycerin, such as distillation and chemical additions, are inadequate and energy intensive. Therefore, an energy efficient and cost-effective glycerin purification technique is necessary to enhance the sustainability and economics of biodiesel production.

RW-EDI could remove salt impurities from the crude glycerin layer (70–75% glycerin, 20–25% methanol, 4–5% inorganic salts and 1–2% water

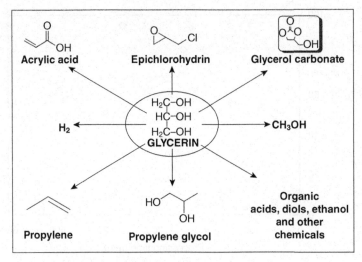

5.10 Different pathways for production of various useful chemicals using glycerin as the building block.

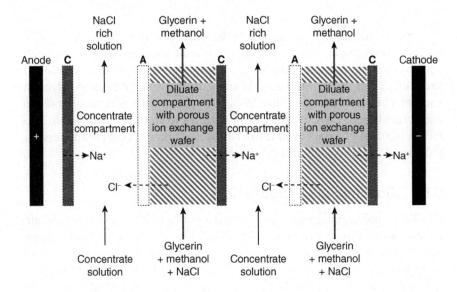

5.11 Schematic of desalting of crude glycerin stream from biodiesel production using resin-wafer-EDI (RW-EDI).

by weight). Crude glycerin solution is fed through the diluate compartment, while the salt recovery solution is fed through the concentrate compartment of the RW-EDI stack as demonstrated in Fig. 5.11. Under an applied electric potential, salt ions (e.g., $Na^+$ and $Cl^-$) from the feed crude glycerin solution transport towards the respective electrodes, cross the ion exchange

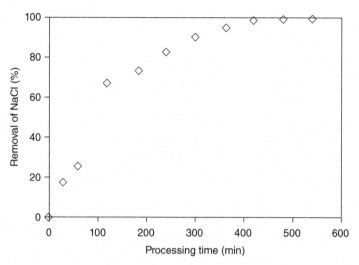

*5.12* Removal efficiency of NaCl from a simulated crude glycerin solution from biodiesel production using RW-EDI.

membrane barrier, and accumulate in the salt recovery solution. As a result, the glycerin solution becomes depleted of salt and the recovery solution becomes enriched in salt, thereby providing a purified glycerin solution at the diluate exit. A typical batch experiment with simulated crude glycerin solution (1 L) exhibits more than 99% removal of NaCl as demonstrated in Fig. 5.12. There is insignificant loss of glycerin from the feed solution. This technique could be improved further by optimizing process parameters for enhanced efficiency and cost effectiveness.

## 5.5 Succinic acid production

RW-EDI enables integration of the upstream bioreactor and downstream product separation for organic acids (Fig. 5.13). Direct recovery of organic acids from fermentations offers two immediate benefits: prevention of reactor acidification and avoidance of product inhibition. Rapid organic acid separation during production offers a new way to control reactor pH without the addition of pH buffers. We used succinic acid as a model organic acid to demonstrate the efficiency of the integrated RW-EDI for production and recovery of organic acids from fermentation broth.

Succinic acid is an important building-block chemical feedstock for polymers with wide applications (acyl halides, anhydrides, esters, amides, and nitriles for applications in drug, agriculture, and food products). Succinic acid is considered one of the most highly valued biobased products as a replacement for fossil-based maleic anhydride or butanediol (Werpy and

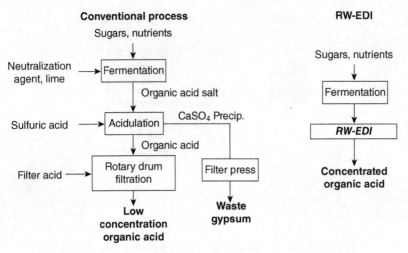

*5.13* The production ratio of succinic acid vs. acetic acid in a conventional succinic acid fermentation.

Peterson, 2004). Succinic acid can be produced by fermentation of sugars from different feedstreams by microorganisms, but due to the need for neutralization or pH control during this process, succinate salt is typically the direct product. Purification to the acid form, which is required for further conversion, is expensive and often requires further processing (Fig. 5.13) to meet purity requirements, leaving a solid gypsum waste product.

A mutant strain of *Escherichia coli*, AFP184 (pfl-, ldh-, and ptsG-), was used for the fermentation of glucose to succinic acid (Donnelly *et al.*, 2004). This strain contains mutations which enable the conversion of glucose to primarily succinic acid with minor amounts of acetic acid and ethanol. The RW-EDI stack was connected to the succinic acid fermentation tank. Fermentation broth was pumped from the fermentation tank to the RW-EDI (diluate compartment) to remove the organic acid salts from the broth. The organic acid salt was converted *in situ* to the acid in the RW-EDI using a configuration containing an alternating arrangement of bipolar membrane and anion exchange membrane (Lin *et al.*, 2011). The anion exchange membrane allows only transfer of succinate ions from diluate to concentrate compartment. The bipolar membrane splits water electrochemically (water splitting phenomenon) on its surface and produces hydroxyl ions and protons in the diluate and concentrate compartments, respectively. Protons in the concentrate compartment combine with the transferred succinate and form succinic acid. Hydroxyl ions in the diluate help maintain the basic pH required to dissolve $CO_2$ as bicarbonate ion before the broth is returned to the fermenter. The recovered succinic acid was collected from the

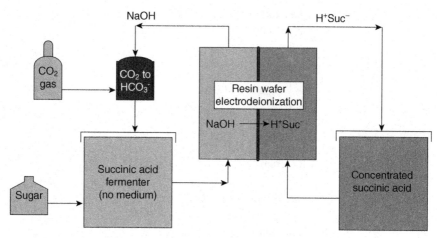

*5.14* Production of succinic acid with an integrated fermentation – RW-EDI separations system.

concentrate (product) stream in the RW-EDI. The succinic acid was biologically formed by reaction of glucose and $CO_2$.

The use of the integrated RW-EDI provides several benefits to succinic acid production. Using water splitting EDI, the succinate in the broth was simultaneously removed and converted to the acid form in the recovery stream. The alkalinity in the organic acid-depleted broth was used to dissolve the $CO_2$ before the broth returned to the fermentation tank. Figure 5.14 illustrates the configuration of the RW-EDI for succinic acid fermentation. Figure 5.15 shows succinic acid production using the RW-EDI system. Succinic acid production was increased significantly over the byproduct acetic acid. Figure 5.16 shows the concentrations of the product and byproducts in the recovery product tank. The RW-EDI provided an efficient strategy for continuous, integrated production and separation of succinic acid.

## 5.6 Solvent extraction: the example of recovery of value added proteins from distiller's grains and solubles (DGS)

The US produces ~14 billion gallons of ethanol per year, predominantly in corn dry mills, and generates distiller's grains and solubles (DGS) as the main co-product. A bushel of corn (56 lb) yields around 16 lb of DGS, or 5.4 lb of dried DGS (DDGS) per gallon of undenatured ethanol (Arora *et al.*, 2008). Compositional analysis of DGS reveals significant amounts of protein (30%), polysaccharide (18% glucan, 10% xylan and 6% arabinan)

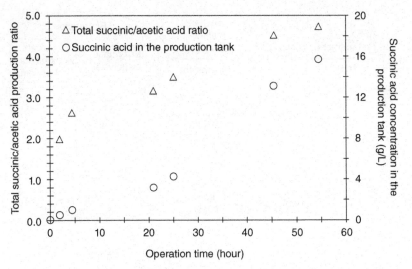

*5.15* Succinic acid fermentation using an integrated fermentation – RW-EDI separations system.

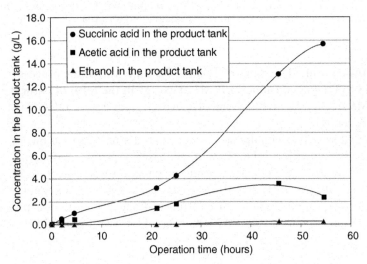

*5.16* Succinic acid, acetic acid, and ethanol in the recovery tank of the RW-EDI.

and fat (11%) (Datta *et al.*, 2010). DGS is primarily used as an animal feed for ruminants such as cattle; however, due to high fiber content, it is unsuitable for monogastrics, such as hogs and poultry. Besides that, due to high moisture content (50–60%), it is dried and transported to storage when there is a lack of immediate and local market demand. Drying DGS to produce DDGS is energy intensive and requires about one-third of the

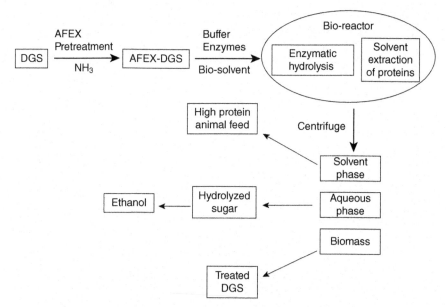

*5.17* Schematic of the experimental pathway of simultaneous bio-solvent-based extraction of proteins and enzymatic saccharification of cellulosic materials present in DGS.

energy consumed in ethanol dry mills. Typically, the whole process of drying DGS to DDGS consumes 4.77 MJ of energy in the form of natural gas and 0.5 MJ of energy in the form of electricity per kg of DDGS (McAloon *et al.*, 2000).

To address these challenges, we evaluated a new separations process to extract protein from DGS using biobased solvents and hydrolyze the unrecovered cellulosic sugars using enzymatic saccharification (Fig. 5.17) (Datta *et al.*, 2010). The goal was to extract a high value animal feed with low fiber content and produce additional sugar for ethanol production. Ammonia fiber expansion (AFEX)-treated DGS (Bals *et al.* 2006) was used because it has a more vulnerable fiber structure than the regular DGS, which enables easy access of the targeted components (protein, cellulosic materials) by the reaction mixture (biobased solvent and cellulases). The protein-rich fraction with residual biobased solvent phase could be used as a high value animal feed. The biobased solvents are food grade materials and increase the nutritional value (fatty acids) of the animal feed. The concentrated protein stream would reduce the energy required associated with drying DGS. The hydrolyzed sugar would enhance the overall ethanol yield. Overall, the process economics of the corn to ethanol production would be improved by higher ethanol yield, reduced energy inputs, and increased value of the animal feed co-products.

The solvents used were citrus-derived D-limonene (DL), soybean-derived distilled methyl esters (DME) and corn starch-derived ethyl lactate (EL). They were formulated and supplied by VertecBioSolvents (www.vertecbiosolvents.com). In the separate protein extraction study, up to 45% of protein was extracted from DGS using biobased solvents. The hydrophobic solvent DL was superior in extracting proteins in comparison to the hydrophilic EL. For the simultaneous protein extraction and enzymatic saccharification study, hydrophobic biobased solvents (DL and DME) were used along with two different enzyme recipes; cellulase (Accelerase) only, and a mixture of cellulase (Accelerase) and amylase (Stargen). Both enzyme recipes were efficient in hydrolyzing cellulose to glucose from AFEX-DGS as demonstrated in Fig. 5.18. A total of 50–80 mg glucose/gm of dry AFEX-DGS was obtained after 120 h of processing by enzymatic hydrolysis under different process conditions. Figure 5.18 also demonstrates the advantage of amylase (in the presence of cellulase) on enzymatic hydrolysis for different solvents. Approximately one-third of the glucan in DGS is starch, indicating the need for amylase in addition to cellulase (Kim *et al.*, 2008).

Further experiments with a mixture of cellulase and amylase reveal an increase in glucose production with increase in time (Fig. 5.19). In comparison to blanks, the biobased solvents did not affect glucose production. This observation implies that the biobased solvents did not adversely affect enzymatic hydrolysis and could open a new avenue for

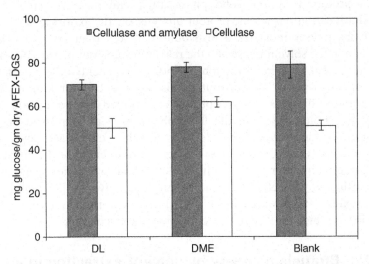

*5.18* Effect of the enzyme amylase on the enzymatic saccharification of cellulosic materials of DGS present in different solutions. The solvents used are D-limonene (DL), distilled methyl esters (DME) and buffer (blank). Time = 120 h, temperature = 50°C, AFEX-DGS:buffer:bio-solvent = 1:2:1 (adapted from Datta *et al.*, 2010).

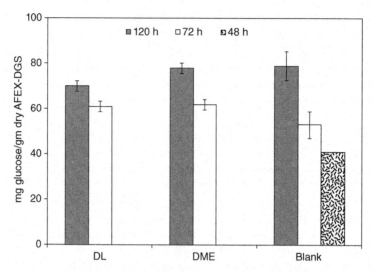

*5.19* Effect of reaction time on the enzymatic saccharification of cellulosic materials of DGS present in different solutions. The solvents used are D-limonene (DL), distilled methyl esters (DME) and buffer (blank). Temperature = 50°C, AFEX-DGS:buffer:bio-solvent = 1:2:1 (adapted from Datta *et al.*, 2010).

aqueous phase enzymatic reactions in the presence of the biobased solvents. Protein analysis on the AFEX-DGS reveals extraction of around 30% protein from the solid phase after simultaneous extraction and saccharification. However, protein analysis on the biobased solvent and aqueous phases indicates a distribution of less than 10% of extracted protein in the solvent phase and the rest in the aqueous phase. This suggests an inefficient protein extraction by biobased solvents in the presence of the aqueous phase. We attributed this to mass transfer limitations. The hydrophilic biomass was completely surrounded by the aqueous phase, thereby shielding it from the hydrophobic solvent. Improved operating conditions (size reduction, addition of other chemicals) and process engineering (mixing conditions) are necessary to enhance the efficiency of the targeted simultaneous protein extraction and enzymatic saccharification. Nevertheless, this study evaluates an innovative pathway that could potentially lead towards value added corn co-products (animal feed + additional sugar) for improved corn to ethanol economics.

## 5.7 Biofuels recovery by solvent extraction in an ionic liquid assisted membrane contactor

Development of an efficient separations technique that will lead to an integrated fermentation-separations biofuel system will reduce energy use.

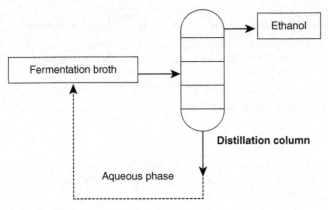

5.20 Distillation.

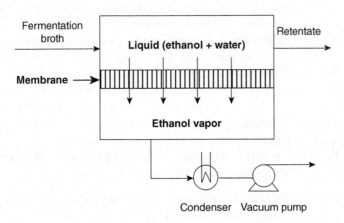

5.21 Pervaporation.

A conventional separations process, such as distillation, is used to recover ethanol from fermentation broth (Fig. 5.20) containing 2–20% ethanol. Roughly, 19,000 BTU/gal of energy is consumed for distillation treatment of a 5% ethanol fermentation broth. Most of the energy is consumed to heat/vaporize the 95% water present in the broth. An energy saving alternative strategy would be to shift to separation of the 5% ethanol rather than the 95% water as observed in processes such as pervaporation (PV) or liquid–liquid extraction (LLE). PV consists of a permselective membrane that concentrates the minor component from a solution in the feed side to the vapor phase on the permeate side (under vacuum) as shown in Fig. 5.21. Because of its simplicity and lower energy consumption compared to distillation, PV is frequently used for dehydration of organic solvents, recovery of organics from aqueous solution and separation of azeotropic

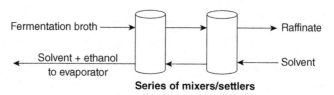

5.22 Liquid–liquid extraction.

organic mixtures. In spite of the energy benefits, the low partition coefficient and slow permeation rate of alcohols across the PV membrane limit the application of PV.

Another well-known alternative to distillation for biofuels recovery is LLE by organic solvents (Wittenberg and Arana 2010; Aravani et al., 2010; van der Wielen and Heijnen, 2010). Different organic solvents, such as hexane, acetone, methanol, ethanol, higher chain alcohols, have been evaluated; however, hexane is the preferred solvent for industries. Other specialty solvents, such as supercritical $CO_2$, have also been reported for biofuels extraction (Bothun et al., 2003), but they have not been implemented at commercial scale. For LLE, typically a series of mixers/gravity settlers are employed (Fig. 5.22), where the solvent phase is dispersed within the aqueous phase, the alcohol is transferred from the aqueous phase to the solvent phase, followed by the separation of the two phases by gravity settling. There are several challenges in using LLE for alcohol recovery (Gawronski and Wrzesinska, 2000; Bothun et al., 2003):

1. Lower interfacial contact area due to poor dispersion and mixing (it is an oil-in-water emulsion) requires large-scale reactors.
2. Phase separation using density difference is a slow process, increasing processing time and equipment volume requirements.
3. There is a chance of contamination of the fermentation broth by the solvent due to invasiveness of the method.
4. Solvent loss due to substantial vapor pressure.
5. Product loss in the portion of liquid–solvent interface that is hard to be separated.

To address the challenges associated with PV and LLE, we developed a method for LLE using a membrane contactor (Snyder et al., 2006). It contains a solvent on the permeate side (and within the membrane pores) that enhances the alcohol selectivity factor between the membrane and the fermentation broth and also improves the permeation rate of extracted alcohol through the membrane (Shukla et al., 1989; Stanojevic et al., 2003). We describe an emerging technology for alcohol recovery by liquid–liquid extraction in an ionic liquid assisted membrane contactor.

## 5.7.1 Ionic liquid (IL) assisted membrane contactor

Ionic liquids are solvents comprised entirely of cations and anions and are molten salts under atmospheric conditions (Fadeev and Meagher, 2001; McFarlane et al., 2005; Zhao et al., 2005). They are defined as salts whose melting point is below 100°C. Due to their thermal and mechanical stability, lower vapor pressure and electrochemical properties, they are gradually gaining popularity as industrial solvents, electrolytes, and other specialty applications, such as $CO_2$ capture. ILs are available commercially. Their high costs have limited deployment in commodity applications. However, the IL-assisted membrane contactor offers the opportunity to minimize IL loss, and therefore, reduce IL operating costs and make them a viable option for commercial applications (Lin and Snyder, 2012). Membrane contactors are established separations systems for small footprint, energy-efficient recovery of target species in continuous mode operation. They consist of a porous membrane with an extracting solvent (IL in this case) on the permeate side and the process solution (fermentation broth) on the feed side as demonstrated in Fig. 5.23. A small pressure differential is applied to wet the membrane with the extracting solvent. Typically, a membrane module with high pore surface area, such as hollow fiber or honeycomb, is used as membrane contactor. Both the membrane material and extracting solvent provide chemical discrimination for the target. Recovery of biofuels by liquid–liquid extraction within an IL-assisted membrane contactor has several advantages:

1. High interfacial surface area in the hollow fiber or honeycomb membrane module significantly increases the liquid–liquid contact area and thus improves mass transfer.
2. The membrane acts as a permselective barrier to isolate the two phases and selectively transfer alcohol through the membrane pores to the solvent phase. It does not rely on density difference.
3. The in-line slipstream evaporation in the IL loop removes the extracted alcohol continuously and ensures a higher driving force for extraction.

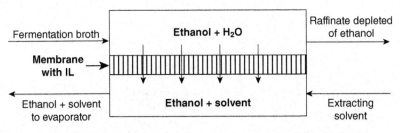

5.23 Ionic liquid assisted membrane contactor for liquid–liquid extraction of ethanol.

4. Overall, the system has a smaller footprint, reduced processing time, and reduced risk of contamination of the fermentation.
5. Near zero vapor pressure of ILs minimizes solvent loss, and therefore significantly reduces make-up solvent requirements and minimizes thermal energy needed for alcohol recovery from the solvent.
6. The reduced solvent volume enables operation of the permeate side at elevated temperatures with reduced energy consumption. Keeping the extracting at elevated temperatures enhances both extraction of the alcohol in the membrane contactor and subsequent stripping from the solvent.

The IL-assisted membrane contactor is used for liquid–liquid extraction (recovery) of ethanol or butanol from a fermentation broth. The chemical characteristics of an IL can be tuned by proper choice of the cation and anion from a vast pool of candidates. We evaluated both hydrophilic and hydrophobic ILs for ethanol recovery. However, it was observed that the hydrophobic IL has higher solubility for ethanol. The IL is used on the permeate side to extract ethanol from the fermentation broth on the feed side (Fig. 5.24). A small positive pressure head is applied on the feed side to avoid contamination of the IL into the fermentation broth. The process relies on three steps: solubility (partitioning) of ethanol in the membrane, diffusion (transfer) of ethanol within the IL-assisted membrane pores, and dissolution (desorption) of ethanol from the permeate IL phase due to a concentration gradient. The raffinate (treated fermentation broth), depleted of ethanol, is recycled back to the storage tank or fermenter. The IL phase

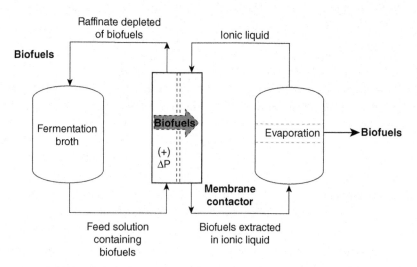

5.24 Schematic for continuous ethanol recovery using a membrane contactor.

containing the extracted ethanol is fed to an evaporator to strip the ethanol from the IL phase. The IL is recycled back to the membrane contactor and the operation is conducted in a continuous mode to maximize the recovery. Countercurrent flow (aqueous and IL phases) is employed to maximize the concentration gradient throughout the length of the membrane module.

Experimental results for the recovery of ethanol from a simulated fermentation broth are presented. Hydrophobic 1-hexyl-3-methylimidazolium bis-trifluoromethylsulfonylimide, ([hmim][Tf2N]), was used for all experiments. The experiments were conducted in a continuous batch mode operation without the evaporator, hence an accumulation of ethanol in the IL phase is observed. Figure 5.25 represents the separation factor for three different cases. It is evident from Fig. 5.25, that the IL-assisted membrane contactor has enhanced the separation factor (for both hydrophilic and hydrophobic membranes) compared to the IL without any membrane. The separation factor is defined as the ratio of the concentration of ethanol to water in IL phase divided by the ratio of the concentration of ethanol to water in aqueous phase. The separation factor has increased from 37 without the membrane contactor to more than 115 with the membrane contactor. This could be attributed to the membrane, which acts as a barrier between the two phases, thereby enhancing the selectivity of ethanol over water. The separation factor is a key process performance index to determine the energy consumption for ethanol recovery. Figure 5.26 illustrates

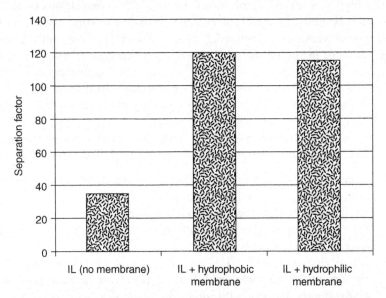

5.25 Separation factor for ethanol separation from water using ionic liquids. Direct L-L extraction, and in a membrane contactor with hydrophilic and hydrophobic membranes.

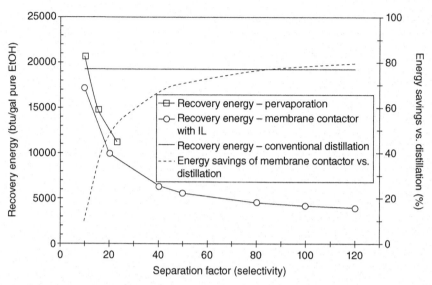

*5.26* Thermodynamic calculation of theoretical energy consumption of membrane-based recovery of ethanol at different membrane separation factors.

the calculated thermodynamic energy consumption for ethanol recovery using the IL and membrane contactor with different separation factors. The high separation factor shown in Fig. 5.25 demonstrates clear energy savings benefits for the IL/membrane contactor in comparison to distillation or pervaporation. Figure 5.27 represents a typical concentration profile for ethanol in the extracted IL phase as a function of time. As observed in Fig. 5.27, starting from a pure IL, the cumulative concentration of ethanol increases and after around 6 h the concentration reaches 8%. This experiment shows the potential for recovery of ethanol from a fermentation broth.

Potential challenges associated with the alcohol recovery using the IL-assisted membrane contactor and potential solutions are outlined below.

1. Fouling of the membrane module – selection of proper membrane material and proper CIP procedure.
2. Shell side vs. lumen side – need to determine experimentally, but solvent flow in the lumen side and broth in the shell side might be beneficial from operational point of view.
3. Identifying a suitable IL – experiments to determine the partition coefficient of alcohols in different ILs from a vast pool of ILs. Hydrophobic ILs likely the superior candidates.
4. Price of ILs – ILs are expensive, but reduced volumes and avoiding solvent losses and makeup could make them economically feasible.

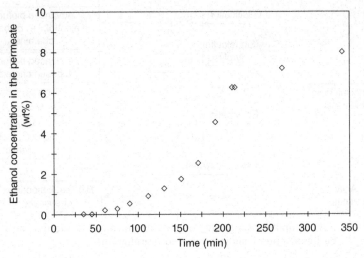

*5.27* Concentration profile of recovered ethanol in the ionic liquid phase.

5. Portability/ease of operation – IL-based extraction in the membrane contactor should require a small footprint.

## 5.8 Emerging trends in separations technology for advanced biofuels

Advanced biofuels are defined by EISA 2007 as biofuels that are derived from non-food feedstocks and result in at least a 50% reduction in life cycle GHG emissions. To achieve market penetration, advanced biofuels should be compatible with existing fuel infrastructure and 'drop-in' to existing fuel production, distribution, and utilization pathways. A simplified schematic containing an overview of advanced biofuels production routes is given in Fig. 5.28.

In general, conversion technologies involve multiple steps that require different separations techniques. Separations technologies tend to be specific to the feedstocks, products, process streams and conversion technologies. Separations are classified based on the types and concentrations of species, solvents, and reaction conditions. There are significant opportunities to improve crosscutting separations technology that will enable deployment in integrated biorefineries. Critical crosscutting separations challenges include:

- *Feedstock variability*: Feedstocks are produced from a range of agricultural materials. Even with a specific feedstock, biomass

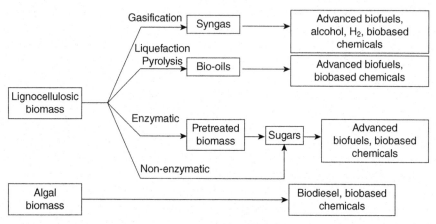

*5.28* Simplified schematic of production routes of biofuels and biobased chemicals in advanced biorefineries.

composition, as well as water and ash content, are dependent on the conditions for growth, harvesting, processing, transport, and storage. Choice of an appropriate separations technology is driven by composition and variability of the feedstock.

- *Product purity requirements*: Intermediate and products may have significantly different purity requirements depending on subsequent processing requirements. As a general rule of thumb, separations costs exhibit a logarithmic relationship to the purity requirement. Therefore targeted purification requirements must be well defined.
- *Product heterogeneity*: Most conversion systems target producing single products. To meet fuel specifications, advanced biofuels will likely require a distribution of intermediates and products. Separations systems must be designed that retain the targeted distributions.
- *Unknown contaminants*: With a limited understanding of reaction mechanisms during development phases, byproduct, contaminant, and inhibitor concentrations or even their existence are uncertain. During scale-up, general separations schemes will need to be adapted to specific feedstocks, biorefinery operations, and product portfolios.
- *Distinct conversion processes*: Advanced biofuels production could use mixtures of biochemical, thermal, and catalytic processes. Each conversion platform has distinct separations demands and limitations, including operating conditions, inhibitors, and product concentrations.
- *Low concentration of targets*: Biomass feedstocks typically require large water volumes during pre-processing and conversion. Intermediates, byproducts, and products can be present at dilute concentrations. Depending on the nature of the species, sometimes it is more efficient

to remove solutes from the solvent and sometimes vice versa. Effective separations at low concentrations are essential. Low concentrations increase energy consumption, system footprint, and capital equipment costs.
- *Water management*: A significant fraction of both the energy demand and waste discharge associated with biorefinery operations can be attributed to water management. For example, distillation of an 85–95% water fraction is a significant energy consumer. Biorefinery facilities may have significant restrictions on release of wastewater. Separations technologies that improve water treatment and reuse can reduce both water input and wastewater discharge.
- *Conversion route divergence*: Conversions frequently involve transformations through different physical forms of matter (i.e., solid, liquid, and gas) of the components for the separation. Change in physical form (e.g., precipitation or evaporation) can facilitate or complicate separations and must be considered in process design.
- *Compatibility at operating conditions*: Conversion routes typically consist of multiple process steps with different operating parameters (temperature, pressure, etc.). Therefore, intermediate and product stability, as well as materials compatibility, must be considered at all potential operating conditions. Separations must be designed to avoid incompatibility, instability, and undesired reactivity.
- *Coordination of multiple separations steps*: Design of a conversion/separations train must consider the difficulty of separating species or classes of species in the process stream. Separations systems are typically designed based on an increasing degree of difficulty or complexity, i.e. species that are very similar chemically or physically are separated last.
- *General separations platforms*: Crosscutting R&D investment in general separations platforms will enable more rapid deployment of specific applications in integrated biorefineries and facilitate commercialization of advanced biofuels.

Some of the critical separations challenges in conversion technologies for advanced biofuels are:

1. Production and upgrading of sugar intermediates
    - Sugar intermediates in non-enzymatic routes to carbohydrate derivatives: (i) developing separations capabilities for removing lignin from sugar streams; (ii) removing co-solvents from biomass hydrolysate streams.
    - Sugar intermediates in enzymatic hydrolysis: (i) solid/liquid separations and *in-situ* phase separations; (ii) inhibitor removal (low-fouling membranes and mesoporous structures); (iii) C5/C6 sugar separations/concentration.

- Catalytic sugar upgrading to hydrocarbons: (i) sugar stream contaminant removal; (ii) sugar and organic acids concentration; (iii) separations processes that operate at temperature to clean intermediate, byproduct, and product streams; (iv) recycle and recovery of reagents used in processes.
- Biological sugar upgrading to hydrocarbons: (i) increase knowledge in fundamental separations science and membrane development, flocculation, and coagulation chemistry; (ii) process integration and collaboration with upstream processes such as organism or pathway design.

2. Production (liquefaction) and upgrading of pyrolysis oil/bio-oil
   - Bio-oil production: investigate novel separations technologies to minimize the impact of destabilizing components. Examples are (i) remove biochar, ash, catalyst fines from vapors; (ii) enhance vapor condensation and maximize aerosol collection; (iii) remove biochar and catalyst fines from bio-oil intermediates to minimize upgrading catalyst fouling; (iv) separate organic hydrocarbons from aqueous streams.
   - Bio-oil upgrading: (i) separations of destabilizing components from stable components; (ii) determine effect of bio-oil chemical properties on membranes (particle fouling, acidic nature); (iii) evaluate staged condensation of bio-oil fractions; (iv) biochar removal and filtering; (v) internal hydrogen production and reuse; (vi) removal of bio-oil corrosive contaminants.

## 5.9 Performance indices

### 5.9.1 Reverse osmosis (RO)

For processes with non-porous membranes, such as RO, there are three major performance indices that are commonly used (Ho and Sirkar, 1992):

- flux
- rejection
- recovery

Flux is defined as the transfer rate of a species per unit surface area. Mass flux is generally expressed as mass/area/time in the form of $g/cm^2/s$ or $mole/cm^2/s$. Similarly, volumetric flux is expressed as $cm^3/cm^2/s$.

For RO, volumetric flux of water ($J_w$) and solute ($J_s$) are defined as:

$$J_w = A(\Delta P - \Delta \pi)$$

where $A$ = water permeability = $\dfrac{P_w}{L}$ and $P_w = \dfrac{D_w C_w \bar{V}_w}{RT}$

$$J_s = B(C_{Fm} - C_P)$$

where $B$ = solute transport number = $\dfrac{P_s}{L}$ and $P_s = D_s K_s$

$\Delta P$ = transmembrane pressure drop
$\Delta \pi$ = osmotic pressure gradient between the feed and permeate
$L$ = thickness of the membrane
$C_{Fm}$ = Concentration of the solute in the feed side at the membrane surface
$C_P$ = concentration of the solute in the permeate side
$D_w$ and $D_s$ = diffusivitities of water and solute, respectively
$C_w$ = concentration of water in the membrane
$\bar{V}_w$ = partial molar volume of water
$R$ = universal gas constant
$T$ = absolute temperature
$K_s$ = solute partition coeffient

The solute rejection, $R$, is expressed as:

$$R = \left(1 - \dfrac{C_P}{C_F}\right)$$

where $C_P$ and $C_F$ are concentration of the solute in permeate and feed, respectively.

Recovery of water ($r$) is expressed as:

$$r = \dfrac{J_w \cdot \text{Area}}{\text{Feed volumetric flow rate}}$$

## 5.9.2 Ultrafiltration (UF) and microfiltration (MF)

For processes with microporous membranes, such as UF and MF, the expression of flux is different than for RO (Ho and Sirkar, 1992). Note that for processes with microporous membranes, the osmotic pressure gradient between the feed and permeate solutions is negligible.

*Flux*

Hagen-Poiseuille's equation for laminar, incompressible flow through a uniform cylindrical channel is used to express pressure drop ($\Delta P$) as:

$$\Delta P = \dfrac{32 \mu L v}{d_p^2}$$

The above equation is used to derive an expression for permeate flux ($J$) through a microporous membrane as given below. $J$ is expressed as volumetric flow rate per unit external surface area of membrane.

$$J = \frac{qN_P}{A_m} = v\frac{\pi d_P^2}{4}\frac{N_P}{A_m} = \frac{\pi r_P^4 \Delta P}{8\mu L}\frac{N_P}{A_m} = \frac{\varepsilon r_P^2 \Delta P}{8\mu L} = A(\Delta P)$$

where $\mu$ = viscosity of solution, $L$ = thickness of membrane, $v$ = velocity, $d_P$ = diameter of membrane pore, $q$ = volumetric flow rate through a single pore, $N_P$ = number of pores, $A_m$ = membrane external surface area, $\Delta P$ = transmembrane pressure drop, $A$ = permeability, and

$$\varepsilon = \text{porosity of membrane} = \frac{N_P \pi r_P^2}{A_m}.$$

From Darcy's Law the flux, $J$, could also be expressed as

$$J = \frac{\Delta P}{\mu R_m}$$

where $R_m$ = resistance of the membrane.

At higher rejection, the rejected species tend to form a layer on the surface of the membrane and this phenomenon is known as fouling. The expression of $J$ for a fouled membrane changes to:

$$J = \frac{\Delta P}{\mu(R_m + R_F)}$$

where $R_F$ = resistance of the layer of the rejected species.

*Selectivity (S)*

Selectivity of species $a$ over $b$ for membrane-based separation is defined as:

$$S = \frac{\text{ratio of the concentration of } a \text{ to that of } b \text{ in permeate}}{\text{ratio of the concentration of } a \text{ to that of } b \text{ in feed}} = \frac{(C_a/C_b)_P}{(C_a/C_b)_F}$$

### 5.9.3 Electrodialysis (ED) and electrodeionization (EDI)

For ED/EDI, the critical performance measurement indices are briefly described below.

*Separation efficiency*

Separation efficiency ($\alpha$) in an ED/EDI process is defined as:

$$\alpha = 1 - \frac{C_o}{C_i}$$

where $C_i$ and $C_o$ are the mole concentration of species at the inlet and outlet of the diluate side of ED/EDI stack, respectively.

*Current efficiency*

Current efficiency (η) is defined as:

$$\eta = \frac{(C_i - C_o)zQF}{I}$$

where $Q$ = volumetric flow rate of the solution, $F$ = Faraday's number = 96,485 amp-s/mole, $I$ = current, and $z$ = ionic charge/mole.

*Productivity*

Productivity ($P$) is the same as the mass flux of the species and is expressed as:

$$P = \frac{(C_i - C_o)Q}{A_m}$$

where $A_m$ = the cross-sectional area of the membrane.

### 5.9.4 Adsorption

*Adsorption capacity (q)*

$$q = \frac{\text{Amount of solute adsorbed}}{\text{Mass of adsorbent used}} \text{ or } \frac{\text{Amount of solute adsorbed}}{\text{Porous surface area of adsorbent used}}$$

### 5.9.5 Liquid–liquid extraction

*Partition coefficient (K)*

$$K = \frac{\text{Concentration in solvent 1}}{\text{Concentration in solvent 2}} = \frac{(C)_1}{(C)_2}$$

## 5.10 Conclusion

Separations can account for ~50% of the total production costs in bioprocessing operations. Designing and implementing efficient separations strategies are critical factors in successful bioprocessing. We provided an overview of separations technologies commonly employed in existing biorefineries as well as emerging separations technologies that are well poised to exhibit rapid growth in future biorefineries. Success requires consideration of feedstock composition, potential conversion technologies, separations scheme, product composition, and purity requirements. Design of the separations systems will impact energy use, system footprint, and capital costs. Several examples were presented from our work.

## 5.11 Acknowledgements

Funding for the work is gratefully acknowledged from the following sponsors: US Department of Energy, Office of Energy Efficiency and Renewable Energy, Bioenergy Technologies Office, US Department of Energy, Office of Energy Efficiency and Renewable Energy, Technology Commercialization Fund, US Department of Agriculture – CSREES Grant # 68-3A75-6-505, and the Illinois Corn Marketing Board.

## 5.12 References

Aden, A., Ruth, M., Ibsen, K., Jechura, J., Neeves, K., Sheehan, J. and Wallace, B., Lignocellulosic biomass to ethanol process design and economics utilizing co-current dilute acid prehydrolysis and enzymatic hydrolysis for corn stover, Technical Report of National Renewable Energy Laboratory (NREL/TP-510-32438), June 2002.

Aravani, A.M., Goodall, B.L., Mendez, M., Pyle, J.L. and Morenoet, J.E., Methods and Synthesis for Biofuel Production, US Patent US 2010/0297749 A1, 2010.

Arora, M.B., Hestekin, J.A., Snyder, S.W., Martin, E. St., Donnelly, M., Millard, C.S. and Lin, Y.P., The separative bioreactor: a continuous separation process for the simultaneous production and direct capture of organic acids, *Separations Science and Technology*, 2 (2007) 2519–2538.

Arora, S., Wu, M. and Wang, M., Update of Distillers Grains Displacement Ratios for Corn Ethanol Life-Cycle Analysis, Argonne National Laboratory, 2008 (http://www.transportation.anl.gov/pdfs/AF/527.pdf).

Baker, R., *Membrane Technology and Applications*, Wiley, Chichester, 2004.

Bals, B., Dale, B. and Balan, V., Enzymatic hydrolysis of distiller's grain and soluble (DDGS) using ammonia fiber expansion treatment, *Energy Fuels* 20 (2006) 2732–2736.

Biomass Multi Year Program Plan, Office of the Biomass Program, Energy Efficiency and Renewable Energy, US Department of Energy, March 2008.

Bothun, G.D., Knutson, B.L., Strobel, H.J. and Nokes, S.E., Mass transfer in hollow fiber membrane contactor extraction using compressed solvents, *Journal of Membrane Science* 227 (2003) 183–196.

Cheryan, M., *Ultrafiltration and Microfiltration Handbook*, CRC Press, Boca Raton, FL, 1998.

Coulson, J.M., Richardson, J.F., Backhurst, J.R. and Harker, J.H., *Chemical Engineering Volume 2*, Fourth edition, Pergamon, Oxford, 1991.

Datta, S., Henry, M.P., Ahmad, S.F., Snyder, S.W. and Lin, Y.J., Removal of Salt Impurities from Glycerin using Electrodeionization Technique. AIChE National Meeting, 2009.

Datta, S., Bals, B.D., Lin, Y.J., Negri, M.C., Datta, R., Pasieta, L., Ahmad, S.F., Moradia, A., Dale, B.E. and Snyder, S.W., An attempt towards simultaneous biobased solvent based extraction of proteins and enzymatic saccharification of cellulosic materials from distiller's grains and solubles, *Bioresource Technology* 101 (2010) 5444–5448.

Datta, S., Lin, Y.J., Schell, D.J., Millard, C.S., Ahmad, S.F., Henry, M.P., Gillenwater, P., Fracaro, A.T., Moradia, A., Gwarnicki, Z.P. and Snyder, S.W., Removal of acidic impurities from corn stover hydrolysate liquor by resin wafer based electrodeionization, *Industrial and Engineering Chemistry Research* 52 (2013) 13777–13784.

Donnelly, M.I., Sanville-Millard, C.Y. and Nghiem, N.P., Method to Produce Succinic Acid from Raw Hydrolysates, US Patent 6,743,610, 2004.

Fadeev, A.G. and Meagher, M.M., Opportunities for ionic liquids in recovery of biofuels, *Chemical Communications* (2001) 295–296.

Gawronski, R. and Wrzesinska, B., Kinetics of solvent extraction in hollow-fiber contactors, *Journal of Membrane Science* 168 (2000) 213–222.

Green, D. and Perry, R., *Perry's Chemical Engineers' Handbook*, Eighth edition, McGraw-Hill, New York, 2007.

Hestekin, J., Snyder, S. and Davison, B., Direct Capture of Products from Biotransformations, Chemical Vision 2020, 2002. Available at: http://www.chemicalvision2020.org/pdfs/direct_capture.pdf

Ho, W. and Sirkar, K.K., *Membrane Handbook*, Springer, New York, 1992.

Kim, Y., Mosier, N.S., Hendrickson, R., Ezeji, T., Blaschek, H., Dien, B., Cotta, M., Dale, B. and Ladisch, M., Composition of corn dry-grind ethanol by-production: DDGS, wet cake, and thin stillage. *Bioresource Technology* 99 (2008) 5165–5176.

Li, N.N., Fane, A.G., Ho, W.S. and Matsuura, T., *Advanced Membrane Technology and Applications*, Wiley-AIChE, Hoboken, NJ, 2008.

Lin, Y.J. and Snyder, S.W., Membrane Contactor Assisted Extraction/Reaction System Employing Ionic Liquids and Method Thereof, US Patent 8,110,111, 2012.

Lin, Y.J., Henry, M.P. and Snyder, S.W., Electronically and Ionically Conductive Porous Material and Method for Manufacture of Resin Wafers Therefrom, US Patent 7,452,920 (Utility) and PCT, 2005a.

Lin, Y.J., Henry, M.P., Snyder, S.W., Martin, E.St., Arora, M.B. and de la Garza, L., Devices Using Resin Wafers and Applications Thereof, US Patent 7,507,318, 2005b.

Lin, Y.J., Henry, M.P. and Snyder, S.W., Electronically and Ionically Conductive Porous Material and Method for Manufacture of Resin Wafers Therefrom, US Patent 7,977,395, 2008.

Lin, Y.J., Snyder, S.W. and St. Martin, E.J., Retention of Counterions in the Separative Bioreactor, US Patent 8,007,647, 2011.

Lin, Y.J., Snyder, S.W., Trachtenberg, M.C., Cowan, R.M., Datta, S., Carbon Dioxide Capture Using Resin-Wafer Electrodeionization, US Patent 8,506,784, 2013.

McAloon, A., Taylor, F., Yee, W., Ibsen, K. and Wooley, R., Determining the Cost of Producing Ethanol from Corn Starch and Lignocellulosic Feedstocks, National Renewable Energy Laboratory (NREL/TP-580-28893), Golden, CO, 2000.

McFarlane, J., Ridenour, W.B., Luo, H., Hunt, R.D., DePaoli, D.W. and Ren, R.X., Room temperature ionic liquids for separating organics from produced water, *Separations Science and Technology* 40 (2005) 1245–1265.

Mulder, J., *Basic Principles of Membrane Technology*, Springer, New York, 1996.

Nilvebrant, N-O., Anders, R., Larsson, S. and Jonsson, L.J., Detoxification of lignocellulose hydrolysates with ion-exchange resins, *Applied Biochemistry and Biotechnology* 91–93 (2001) 35–49.

Noureddini, H. and Zhu, D., Kinetics of transesterification of soybean oil, *Journal of the American Oil Chemists' Society* 74 (1997) 1457–1463.

Ruthven, D.M., *Principles of Adsorption and Adsorption Processes*, Wiley-Interscience, Chichester, 1984.

Seader, J.D., Henley, E.J. and Roper, D.K., *Separation Process Principles*, Third edition, Wiley, Hoboken, NJ, 2010.

Sengupta, A.K., *Ion Exchange and Solvent Extraction: A Series of Advances, Volume 18*, CRC Press, Boca Raton, FL, 2007.

Shukla, R., Kang, W. and Sirkar, K.K., Acetone-butanol-ethanol (ABE) production in a novel hollow fiber fermenter-extractor, *Biotechnology and Bioengineering*, 34 (1989) 1158–1166.

Snyder, S.W., Lin, Y.J., Hestekin, J.A., Henry, M.P., Pujado, P., Oroskar, A., Kulprathipanja, A.S. and Randhava, S., Membrane Contactor Assisted Water Extraction System for Separating Hydrogen Peroxide from a Working Solution, and Method Thereof, US Patent 7,799,225, 2006.

Stanojevic M., Lazarevic B. and Radic D., Review of membrane contactors designs and applications of different modules in industry, *FME Transactions*, 31 (2003) 91–98.

Tarleton, E.S. and Wakeman, R.J., *Dictionary of Filtration and Separation*, Filtration Solutions, Exeter, 2008.

Thompson, J.C. and He, B.B., Characteristics of crude glycerin from biodiesel production from multiple feedstocks, *Applied Engineering in Agriculture* 22 (2006) 261–265.

Treybal, R.E., *Mass-Transfer Operations*, McGraw-Hill, New York, 1980.

Tucker, M.P., Kim, K.H., Newman, M.M. and Nguyen, Q.A., Effects of temperature and moisture on dilute-acid steam explosion pretreatment of corn stover and cellulose enzyme digestibility, *Applied Biochemistry and Biotechnology* 105–108 (2003) 165–177.

US Government Printing Office Pub.L. 110–140: http://www.gpo.gov/fdsys/pkg/PLAW-110publ140/content-detail.html

van der Wielen, L.A.M. and Heijnen, J.J., Process for the Continuous Biological Production of Lipids, Hydrocarbons or Mixtures Thereof, European Patent EP 2 196 539 A1, 2010.

Weil, J.R., Dien, B., Bothast, R., Hendrickson, R., Mosier, N.S. and Ladisch, M.R., Removal of fermentation inhibitors formed during pretreatment of biomass by polymeric adsorbents, *Industrial and Engineering Chemistry Research* 41 (2002) 6132–6138.

Werpy, T. and Petersen, G. *Top Value Added Chemicals From Biomass Volume I: Results of Screening for Potential Candidates from Sugars and Synthesis Gas*, US Department of Energy, 2004. Available at: http://www1.eere.energy.gov/biomass/pdfs/35523.pdf

Wittenberg, J. and Arana, F., Methods of Microbial Oil Extraction and Separation, International Patent WO 2010/120939 A2, 2010.
Young, R.T., *Adsorbents: Fundamentals and Applications*, Wiley-Interscience, Chichester, 2003.
Zhao, H., Xia, S. and Ma, P., Use of ionic liquids as 'green' solvents for extractions, *Journal of Chemical Technology and Biotechnology* 80 (2005) 1089–1096.

# 6
# Catalytic processes and catalyst development in biorefining

S. MORALES-DELAROSA and
J. M. CAMPOS-MARTIN,
Instituto de Catálisis y Petroleoquímica, CSIC, Spain

DOI: 10.1533/9780857097385.1.152

**Abstract**: In this chapter, we will focus on the catalysts used in a biorefinery for the production of fuels and base chemicals from biomass. Catalysts and catalytic processes are involved in several steps of a biorefinery, but in general, these processes can be divided in two main groups: (a) processes of biomass deconstruction to produce upgradeable gaseous or liquid platforms; and (b) processes to upgrade deconstructed biomass to useful fuels or chemicals.

**Key words**: catalyst, biomass, depolymerization, products upgrading, thermochemical processes.

## 6.1 Introduction

Biomass is a renewable carbon source that can be processed in an integrated biorefinery, in a manner similar to petroleum in conventional refineries, to produce fuels and chemicals. While commercial-scale biofuel production has been established with bioethanol (corn, sugar cane) and biodiesel (canola, soybeans), these first generation processes utilize only the edible fraction of certain food crops, thereby decreasing their widespread applicability. The development of second and third generation biofuels that utilize lignocellulosic biomass and algae could allow for the large-scale production of sustainable fuels and chemicals.

The process of refining biomass feedstocks to hydrocarbon biofuels can be subdivided into two general portions. First, whole biomass is deconstructed to produce upgradeable gaseous or liquid platforms. This step is typically carried out through thermochemical pathways to produce synthesis gas (by gasification) or bio-oils (by pyrolysis or liquefaction), or by hydrolysis pathways to produce upgradeable intermediates. In all of these steps catalysts play a crucial role, because without them the processes are uneconomical and will produce an excessive amount of waste. The use of catalysts in a biorefinery scheme has been widely reviewed (Kamm and Kamm, 2004; Lichtenthaler and Peters, 2004; Huber et al., 2006; Corma et al., 2007; Gallezot, 2007a, 2007b, 2012; Huber and Corma, 2007; Lange,

2007; Behr *et al.*, 2008; Claus and Vogel, 2008; Goyal *et al.*, 2008; Haveren *et al.*, 2008; Clark *et al.*, 2009; Simonetti and Dumesic, 2009; FitzPatrick *et al.*, 2010; Perego and Bianchi, 2010; Shanks, 2010; Langan *et al.*, 2011; Lovett *et al.*, 2011; Murzin and Simakova, 2011; Phillips *et al.*, 2011a; Stark, 2011; Zhou *et al.*, 2011; Murat Sen *et al.*, 2012; Ruppert *et al.*, 2012; Sanders *et al.*, 2012). We will focus on the catalysts used in the production of fuels and base chemicals from biomass.

## 6.2 Catalysts for depolymerization of biomass

### 6.2.1 Depolymerization of lignocellulosic materials

*Hydrolysis*

Lignocellulosic materials are complex mixtures of natural polymers – cellulose (35–50%), hemicelluloses (25–30%), and lignin (15–30%) – tightly bonded by physical and chemical interactions (Rubin, 2008). Since cellulose is the major component of lignocellulosic materials, efficient processes for hydrolysis or depolymerization of cellulose seem to be interesting entry points for the production of biofuels and biochemicals (Fig. 6.1).

However, its chemical transformation presents a serious challenge due to the problems associated with its mechanisms for structural protection, which make this polymer recalcitrant towards efficient chemical and biological transformations. One of the most important bottlenecks of commercializing lignocellulosic bioethanol is the discovery of a cost-effective hydrolysis of cellulose (Banerjee *et al.*, 2010; Kumar *et al.*, 2009). The β-glycosidic linkages of the sugar molecules contained in cellulose or lignocellulose are strongly protected by the tight packing of cellulose chains in microfibers, making hydrolysis challenging. Accordingly, the hydrolysis of cellulose requires harsh conditions such as the use of dilute strong acids at high temperatures. The hydrolysis rate of cellulose and cellobiose shows a clear dependence on the $pK_a$ of the acid used in the reaction (Morales-Delarosa *et al.*, 2012; Vanoye *et al.*, 2009). For this reason, the hydrolysis of lignocellulosic materials requires the use of strong acids as catalysts ($pK_a < 0$). Typically the catalysts are based on sulfuric, hydrochloric and p-toluenesulfonic acids (Murzin and Simakova, 2011; Rinaldi and Schüth, 2009a). The use of homogeneous catalysts has some problems of separation, reuse, and neutralization.

The application of solid acid catalysts to biomass transformation into transportation biofuels and chemicals has been receiving much attention (Rinaldi and Schüth, 2009a). The ease of catalyst separation after the reaction, which enables their reuse, is an advantage of heterogeneous catalysis for biorefineries. Furthermore, solid acid catalysts are typically less aggressive to the industrial plants than liquid mineral acids. However, the

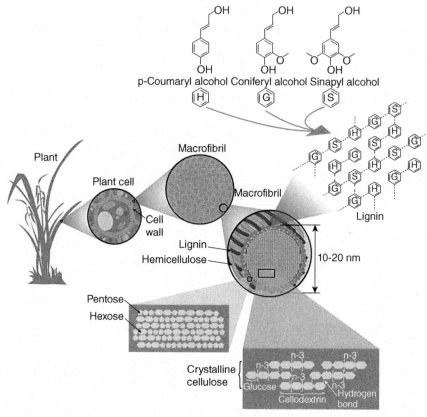

*6.1* Structure of lignocelluloses biomass (reprinted from Rubin (2008) with permission from Nature Publishing Group).

hydrolysis of solid biomass with heterogeneous catalysts must occur in two steps. Firstly, a partial hydrolysis of biomass proceeds, involving Brønsted acidic species either released by the solid material or formed in the reaction medium (Rinaldi and Schüth, 2009b). Subsequently, the reactions involving the solid catalyst take place when the oligomers are small enough to access the pore system. Thus, the porosity of the solid catalyst can play a fundamental role in the catalytic activity and selectivity, because it determines to what extent the initial hydrolysis needs to proceed in homogeneous phase before starting the heterogeneously catalysed reactions. There is also the possibility that the reaction starts on the external active sites of the catalysts. However, due to the fact that a dispersion system with two solid phases exists, such contributions are probably small. Transformation of lignocellulose into biofuels and chemicals involves water either as a reagent, a product, or even as a solvent. Only a number of materials are suitable for these applications (in terms of acidity, stability, and insolubility). Some catalysts that fulfill these characteristics are functionalized activated carbons, functionalized resins, functionalized silica, zeolites, some solid heteropoly compounds, niobic acid, $MoO_3$-$ZrO_2$, zirconium tungstates, zirconium phosphates, lanthanum phosphates, niobium phosphates, and some other materials (Okuhara, 2002).

Another alternative is the use of enzymes (Murzin and Simakova, 2011), through which the hydrolysis is produced with high selectivity. But the use of enzymes has the drawbacks of high cost and low activity. Enzyme catalysis demands the use of highly specific cellulases and is actually a heterogeneous process that depends on such physicochemical properties as crystallinity, the degree of polymerization, surface conditions, and the presence of lignin and hemicellulose, if cellulose is not used as the raw material. All this, as was mentioned above, results in a low reaction rate.

A breakthrough in lignocellulose chemistry was the finding that alkylmethylimidazolium ionic liquids (IL) can dissolve cellulose or even wood (Pinkert et al., 2009; Zhu et al., 2006; Sun et al., 2011; Olivier-Bourbigou et al., 2010). The dissolution process disrupts the fibers of cellulose, leaving the hydroxyl groups and β-glycosidic bonds accessible for the hydrolysis. It appears that the 'physical' barrier can be overcome through the formation of a cellulose solution that facilitates the acid-catalyzed hydrolysis under mild reaction conditions and a lower catalyst loading. Chemical deconstruction of cellulose to produce glucose in ionic liquids has resulted in only moderate yields (Rinaldi et al., 2008, 2010; Li et al., 2008), which contrasts with the high yields obtained from cellulose in concentrated acids and other cellulose solvents (Kumar et al., 2009). This low yield is due to the small amount of water added in the reaction in the presence of IL; a large amount of water leads to the precipitation of unreacted cellulose. However, a large excess of water is essential to reach higher sugar yields dissolved in ionic liquids because it favors the hydrolysis of cellulose and retards the dehydration of glucose. This effect is described by the 'Le Chatelier's Principle'. By adding water successively during the reaction, this issue was elegantly overcome (Binder and Raines, 2010; Morales-Delarosa et al., 2012), producing a very high yield of sugars. The recovery and the reuse of ionic liquids – in a continuous process – is a big challenge to industrialization of new technologies. Given the high costs of ionic liquids, efficient recycling is mandatory for the development of commercially viable processes.

*Hydrogenolysis*

Because the rates of hydrolysis of cellulose and degradation of glucose are close (Vanoye et al., 2009; Ruppert et al., 2012), one method to reduce the contribution from degradation could be combining hydrolysis with other processes, e.g., hydrogenation, leading to the formation of sugar alcohols, which are more stable than sugars. This concept was studied some time ago with the hydrolytic hydrogenation of cellulose, hemicellulose, and wood with Ru, Pd, and Pt catalysts in the presence of phosphoric and sulfuric acids (Murzin and Simakova, 2011), but recently there has been a renaissance of reactions that use only water, without the use of diluted acid solutions (Fukuoka and Dhepe, 2006; Yan et al., 2006; Luo et al., 2007; Ji et al., 2008; Jollet et al., 2009; Zheng et al., 2010; Zhu et al., 2010; Ruppert et al., 2012). In these studies the acidic function of the solution was substituted with the acid function of the carrier that accelerates hydrolysis and is associated with the metallic function that produces hydrogenation. In addition to cellulose, hydrolytic hydrogenation can take place for the mixtures of cellulose and hemicellulose (Käldström et al., 2011). Sorbitol is obtained and, correspondingly, the product of xylose hydrogenation, namely, xylitol.

### 6.2.2 Gasification of biomass

Gasification is a thermochemical conversion process of solid biomass into a gas-phase mixture of carbon monoxide (CO), hydrogen ($H_2$), carbon dioxide ($CO_2$), methane ($CH_4$), organic vapors, tars (benzene and other aromatic hydrocarbons), water vapor, hydrogen sulfide ($H_2S$), residual solids, and other trace species (HCN, $NH_3$, and HCl) (Bulushev and Ross, 2011). The specific fractions of the various species obtained may depend on process conditions and on the environment (inert, steam) prevailing during gasification. Catalytic biomass gasification is a complex process that includes numerous chemical reaction steps such as pyrolysis, steam gasification, and water gas shift reaction (Florin and Harris, 2008). Biomass gasification produces very low levels of particulates, as well as very small amounts of $NO_x$ and $SO_x$ when compared with fossil fuels (de Lasa et al., 2011). Moreover, biomass can be used as a source to produce various chemical species (Fig. 6.2).

However, a serious issue for the broad implementation of the biomass gasification technology is the generation of unwanted products (e.g., char, tar, coke-on-catalyst, particles, nitrogen compounds, and alkali metals) (Banowetz et al., 2008). Char or biochar is a solid carbonaceous residue, while tar is a complex mixture of condensable hydrocarbons, which includes single-ring to five-ring aromatic compounds along with other oxygen-containing hydrocarbons species. The continual build-up of tars present in the produce gas can cause blockages and corrosion and also reduce overall efficiency. Tars can be converted thermally when the gasifier works at high temperatures, but even at temperatures in excess of 1,000°C, tar cannot be removed completely. High gasification temperature reduces

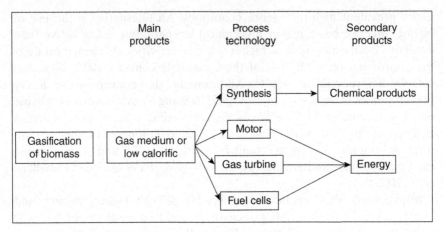

*6.2* Uses of syngas produced from biomass.

the formation of tar; but high energy consumption (i.e., high production cost for syngas) makes the process economically unviable (Asadullah *et al.*, 2001).

In consequence, there is a need to develop stable and highly active catalysts for biomass gasification with the goal of producing high-quality synthesis gas and/or hydrogen free of tars. Catalysts for use in biomass gasification conversion may be divided into two distinct groups which depend on the position of the catalytic reactor relative to that of the gasifier. In the first group, catalyst, 'primary catalysts', is present in the gasifier; in the second group, catalyst, 'secondary catalysts', is placed in a reactor downstream from the gasifier.

As primary catalysts, several compounds have been proposed such as dolomite, olivine, alkali metal, and nickel catalysts. All of them are capable of promoting several important chemical reactions such as water gas shift and steam reforming. Thus, primary catalysts can minimize tars and increase both hydrogen and $CO_2$, avoiding altogether complex downstream tar removal operations. Nickel catalysts work very well for a short time on stream but are affected by deactivation as a result of carbon deposition and sintering of the nickel metal particles in the catalyst (de Lasa *et al.*, 2011; Bulushev and Ross, 2011; Sutton *et al.*, 2001).

The use of dolomite, a magnesium ore with the general formula $MgCO_3$ $CaCO_3$, as a primary catalyst in biomass gasification has attracted much attention because it is a cheap disposable catalyst that can significantly reduce the tar content of the product gas from a gasifier (de Lasa *et al.*, 2011; Sutton *et al.*, 2001; Bulushev and Ross, 2011). The main issue with dolomite is its fragility, as it is soft and quickly attrites in fluidized beds under prevalent high-turbulence conditions. An alternative is the use of olivine; this ore has a higher mechanical strength, but is less active than dolomite. Both catalysts show higher catalytic activities of calcined catalysts than untreated ones. The use of these catalysts converts 100% of water-soluble heterocyclic at 900°C. Additionally, the conversion of heavy polyaromatics increased from 48% to 71% using 17 wt% untreated olivine mixed with sand at 900°C, whereas the conversion of heavy polyaromatics reached up to 90% with 17 wt% of calcined dolomite. Furthermore, a total tar amount of 4.0 g/m$^3$ could be reduced to 1.5 and 2.2 g/m$^3$ using calcined dolomite and olivine, respectively (Corella *et al.*, 1999; Caballero *et al.*, 2000).

Monovalent alkali metals of group 1A are all highly reactive and electropositive. Alkali metals, principally K and to a lesser extent Na, exist naturally in biomass and accumulate in the gasifier ashes. These alkali metals can have a significant impact during pyrolysis, forming a reactive char that enhances gasification (Lv *et al.*, 2010; Quyn *et al.*, 2002; Wu *et al.*, 2002). Furthermore, the use of ash itself as a catalyst solves the problem of ash waste handling and gives an added value to the gasification by increasing the gasification rate and reducing the tar content in the gas produced. However, the major disadvantage of these ash-based catalysts is their potential activity losses due to particle agglomeration. The direct addition of alkali metals has several disadvantages such as the difficult and expensive recovery of the catalyst, increased char content after gasification, and ash disposal problems.

Another possible alternative is to have a catalytic process in a reactor placed downstream from the gasifier. In this reactor, product gases are further processed using secondary catalysts. Typical materials that are used as secondary catalysts are dolomite and nickel-based formulations. These catalysts decrease the tar content of the product gas in the 750–900°C range (Bulushev and Ross, 2011; de Lasa *et al.*, 2011; Sutton *et al.*, 2001). Ni catalysts serve not only for the removal of tars and methane but also for the adjustment of the synthesis gas composition by means of water-gas shift reaction. Unfortunately, Ni catalysts often suffer from deactivation by sintering and/or coke deposition. Combined application of nickel catalysts and dolomite guard beds looks promising (Bulushev and Ross, 2011; Sutton *et al.*, 2001). Even more interesting is the use of dolomite as a catalyst support for Ni where deactivation of such a catalyst was found after 60 h of operation (Wang *et al.*, 2004). Similarly, Ni/olivine catalysts have previously been studied in different reactions related to biomass conversion (Świerczyński *et al.*, 2006). Coke deposition on these catalysts was found to be negligible (Zhao *et al.*, 2008). These catalysts showed excellent results in pilot-scale biomass gasification units and showed no deactivation during a 45 h test (Pfeifer *et al.*, 2004).

## 6.2.3 Biomass pyrolysis

Pyrolysis is an appropriate process for the conversion of large amounts of biomass into bio-oil, from which biofuels and chemicals can be produced. However, some important bio-oil characteristics are disadvantageous, such as high water and oxygen content, corrosiveness, lower stability, immiscibility with crude oil-based fuels, high acidity, high viscosity, and low calorific value, all of which make the direct use of bio-oil as motor fuel impossible. Additionally, bio-oil is unstable, subject to transformation during storage.

Pyrolysis can be performed in a fluidized bed reactor with circulation, and the composition of the products depends on the time they spend in the reactor. Prolonged contact results mainly in gas products, while so-called rapid pyrolysis with contact times of one to two seconds makes it possible to obtain up to 75% of bio-oil (Murzin and Simakova, 2011). There have been several attempts to improve the quality of bio-oil by adding catalysts to the pyrolysis reactor (Stöcker, 2008; Zhou et al., 2011). These catalysts are based on acid centers, liquid acids ($H_2SO_4$, hydrochloric acid, phosphoric acid) and solid Lewis acids (zeolites and mesoporous materials with uniform pore size distribution (MCM-41, MSU, SBA-15)).

The use of H-ZSM-5 as catalysts was studied (Stöcker, 2008; Zhou et al., 2011). The deoxygenation, decarboxylation, and decarbonylation reactions of the bio-oil components, cracking, alkylation, isomerization, cyclization, oligomerization, and aromatization are catalyzed by acidic sites of the zeolite by a carbonium ion mechanism. However, tar and coke were also formed as undesirable byproducts. The need for regeneration of catalysts reduces notably the performance of the process. Some improvements have been obtained by adding a metallic function (Ni) to the zeolite (Stöcker, 2008). However, the process is far from a possible application at industrial scale. Some improvements have been obtained with mesoporous catalysts (Fig. 6.3) (MCM-41, MSU, and SBA-15). MCM-41 (with Al, Fe, Cu or Zn) materials significantly affected the product yield and quality of the obtained bio-oil. This behavior was attributed mainly to the one-dimensional mesopores (pore diameter ca. 2–3 nm) in combination with the large surface area of the MCM-41 materials (about $1,000\,m^2g^{-1}$), and their mild acidity. All these factors provide the desired environment for a controlled conversion of the high molecular weight lignocellulosic molecules (Antonakou et al., 2006). Bio-oil with enhanced stability was produced applying these mesoporous materials by transfer of oxygen, which is known as the main cause of the instability of bio-oil, into water, carbon monoxide, and carbon dioxide. Furthermore, the MCM-41 catalysts produced larger amounts of phenolics in the bio-oil obtained. The catalytic activity of Al-MCM-41 for bio-oil upgrading was higher than that of siliceous MCM-41 because of the larger number of acid sites. Finally, improved reforming

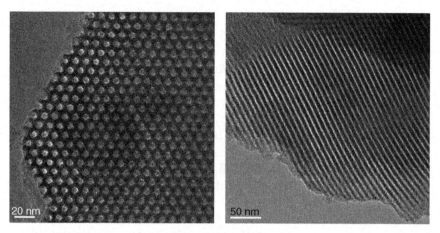

*6.3* Micrograph of an SBA-15 ordered mesoporous material.

results were obtained when the pyrolytic bio-oil vapor passed through a catalytic layer rather than if wood from Japanese larch was mixed with the catalyst directly (Stöcker, 2008).

## 6.3 Catalysts for biomass products upgrading

### 6.3.1 Triglyceride transformations

*Transesterification*

Lipids can be transformed in biodiesel by transesterification. Biodiesel, a clean renewable fuel, has recently been considered as the best candidate for a diesel fuel substitute because it can be used in any compression ignition engine without the need for modification. Chemically, biodiesel is a mixture of methyl esters with long-chain fatty acids and is typically made from nontoxic, biological resources such as vegetable oils, animal fats, or even used cooking oil. The plant oils usually contain free fatty acids, phospholipids, sterols, water, odorants and other impurities. They have high viscosities and boiling points so the oil cannot be used as fuel directly because it produces serious problems in the diesel engine (especially in direct injection engine). The most common way to produce biodiesel is through catalyzed transesterification (Fig. 6.4).

Transesterification or alcoholysis is the displacement of alcohol from an ester by another in a process similar to hydrolysis, except that alcohol is used instead of water. Methanol is mainly used in commercial biodiesel production, although some longer alcohols have also been applied, such as ethanol, butanol, propanol, and amyl alcohol. Methanol is used most often

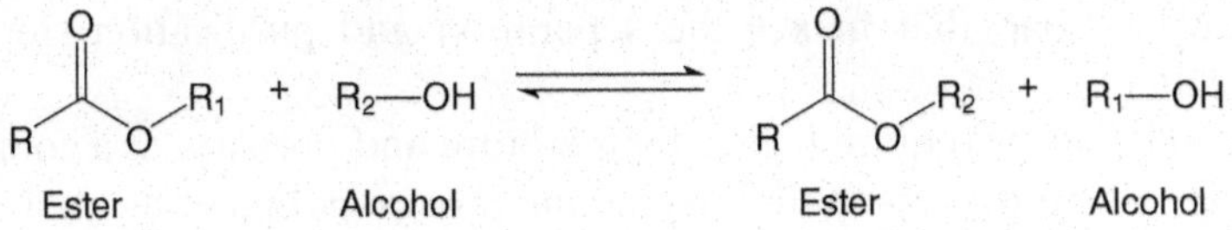

*6.4 Transesterification reaction.*

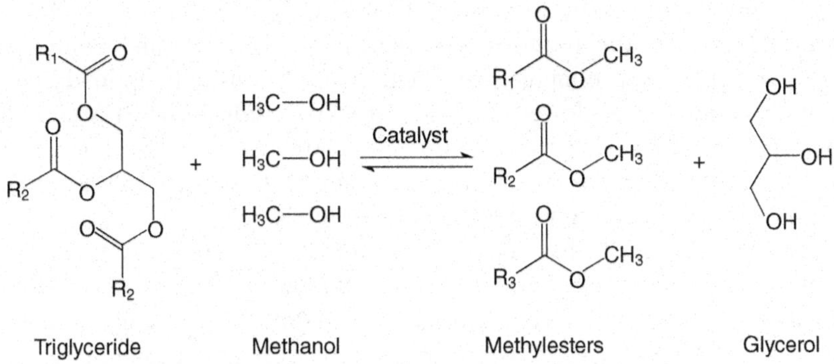

*6.5 Reaction of triglyceride transesterification with methanol.*

because of its low cost and the fact that it is readily available. This process has been widely used to reduce the high viscosity of triglycerides (Fig. 6.5). Transesterification is one of the reversible reactions and proceeds essentially by mixing the reactants. However, the presence of a catalyst (a strong acid or base) accelerates the conversion.

Catalysts used for the transesterification of triglycerides are classified as alkali, acid, or enzyme, among which alkali catalysts such as sodium hydroxide, sodium methoxide, potassium hydroxide, and potassium methoxide are more effective. Alkali-catalyzed transesterification occurs at a reaction temperature above alcohol boiling point (70°C) at atmospheric pressure, and an alcohol and triglyceride ratio 6:1 to drive the equilibrium to maximum ester yield (Lestari et al., 2009a; Leung et al., 2010). Chemical transesterification using an alkaline catalysis process gives high conversion levels of triglycerides to their corresponding methyl esters in short reaction times. However, the reaction has several drawbacks: it is energy intensive, recovery of glycerol is difficult, the alkaline catalyst has to be removed from the product, and alkaline waste water requires treatment. Additionally, the quality of the feedstock must be very high, the oil has to be free fatty acid (FFA) and the alcohol anhydrous. Because of the presence of these impurities, the yield to esters is reduced, leading to partial saponification, producing soap that makes the separation and purification steps more difficult.

Acid catalysts require higher temperature and pressure at a comparable amount of catalyst compared to alkaline processes, higher alcohol/oil ratio, and yields slower reaction rate. But, these catalysts may be applied for feedstock with high FFA content and are less sensitive to water content in alcohol.

Utilization of a lower-grade feedstock such as grease, beef tallow, and used frying oils from the food industry may reduce production costs, but on the other hand this requires processing of feeds with a high content of FFAs. In this case, a homogeneous acid catalyst is usually applied at the pre-treatment stage of lower grade feedstocks to reduce the FFA content. Therefore, an integrated process comprising acid-catalyzed pre-esterification of FFAs followed by base-catalyzed transesterification of triglyceride is the preferred way of dealing with lower-quality feedstocks.

Enzymatic catalysts like lipases are able to effectively catalyze the transesterification of triglycerides in either aqueous or non-aqueous systems, which can overcome the problems mentioned above. In particular, glycerol can be easily removed without any complex process, and also the free fatty acids contained in waste oils and fats can be completely converted to alkyl esters. On the other hand, in general the production cost of a lipase catalyst is significantly greater than that of an alkaline one.

An efficient heterogeneous catalyst would provide economic benefits as, unlike homogeneous catalysts, it does not require a separation procedure. In most cases, catalysts can be recycled and re-used for long periods, providing an opportunity for a large-scale continuous process. Many heterogeneous catalysts that are based on both acid and alkaline solid catalysts have recently been proposed in the literature (Meher *et al.*, 2006; Lopez Granados *et al.*, 2007, 2009a, 2009b, 2010, 2011; Martin Alonso *et al.*, 2007, 2009a, 2009b, 2010b; Kotwal *et al.*, 2009; Lestari *et al.*, 2009a; Leung *et al.*, 2010; Fu *et al.*, 2011; Toda *et al.*, 2005; Marchetti *et al.*, 2007). Activity comparison with a homogeneous catalyst showed that the homogeneous catalyst gave higher ester yield, using the same reaction conditions, compared to the heterogeneous catalysts. This effect is due to a double effect on the mass transfer, the presence of alkali helps the mixture of oil and alcohol (Lestari *et al.*, 2009a), and the presence of a solid introduces a third phase transfer limitations. Basic catalysts include Na/NaOH/$\gamma$-Al$_2$O$_3$, KOH/$\gamma$-Al$_2$O$_3$, hydrotalcites, CaO, Li/CaO, and basic resins, while acid solids employed include C-SO$_3$H, WO$_3$/ZrO$_2$, acid resins, sulfonic functionalized silica, and heterogenized acidic polymers. The main part of these heterogeneous catalysts suffers leaching and deactivation during its use in reaction (Alba-Rubio *et al.*, 2010; Lopez Granados *et al.*, 2009b; Martin Alonso *et al.*, 2007, 2009b); only some examples seem to be reasonably stable in reaction: poly(styrenesulfonic) acid (Lopez Granados *et al.*, 2011), sulfonic functionalized carbon (Toda *et al.*, 2005), and basic resins (Marchetti *et al.*, 2007).

*Hydrogenation*

Hydrotreating of triglycerides and fatty acids to hydrocarbon middle distillates are suitable as alternative diesel fuels (Fig. 6.6). Hydroprocessing has advantages in terms of the flexibility of feedstock, because the process can utilize low-quality feeds such as grease, used frying oil, animal fats, or tall oils from the Kraft pulping industry. The separation stage of the byproducts is not as complex as in the transesterification reaction. The main product, which consists mostly of paraffinic hydrocarbons, is superior for application in combustion engines (Table 6.1).

Hydrotreating is usually used in a petroleum refinery to remove sulfur, nitrogen, and metals from petroleum-derived feedstocks. Conventional hydrotreating catalysts containing sulfided mixed oxides such as NiMo,

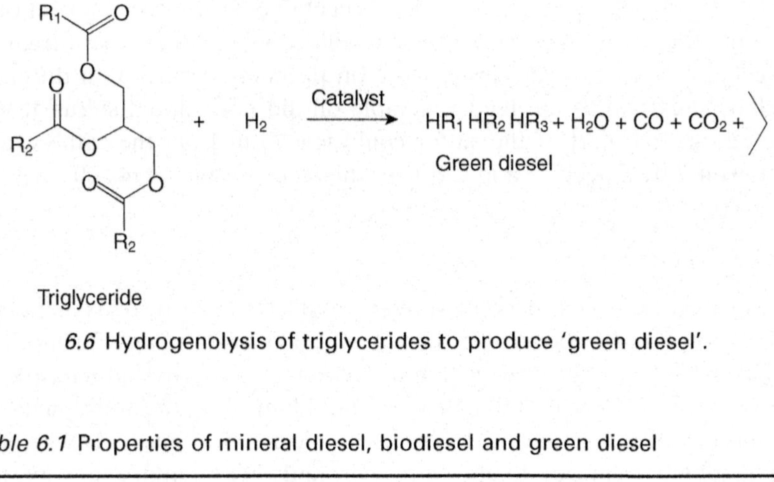

*6.6* Hydrogenolysis of triglycerides to produce 'green diesel'.

*Table 6.1* Properties of mineral diesel, biodiesel and green diesel

|  | Mineral ULSD | Biodiesel FAME | Green diesel |
|---|---|---|---|
| O, % | 0 | 11 | 0 |
| Specific gravity | 0.84 | 0.88 | 0.78 |
| Sulfur content, ppm | <10 | <1 | <1 |
| Heating value, MJ/kg | 43 | 38 | 44 |
| Cloud point, °C | −5 | −5–+15 | −10–+20 |
| Distillation, °C | 200–350 | 340–355 | 265–320 |
| Cetane number | 40 | 50–65 | 70–90 |
| Stability | Good | Marginal | Good |

NiW, and CoMo, can be used in the process. Application of this process for upgrading vegetable oils involves three main pathways: hydrogenation of double bonds in the alkyl chain of fatty acids; decarboxylation and decarbonylation; and hydrodeoxygenation (or dehydration/hydrogenation) to produce alkanes (Lestari *et al.*, 2009a). The main products are *n*-aliphatic hydrocarbons of the corresponding fatty acids, and propane from the triglyceride molecules. The oxygen content in the triglyceride is released either as carbon monoxide or carbon dioxide, together with the formation of water.

Coprocessing with the existing refinery reactor was initially preferred in the hydroprocessing of vegetable oil. However, further studies have indicated some problems due to the presence of trace metal contaminants in the vegetable oil, such as phosphorous, sodium, potassium, and calcium. Furthermore, the presence of water and carbon oxide affected the catalyst lifetime and caused problems in the separation of carbon oxides from the recycle gas. Considering all these potential issues, the refinery seems to favor the construction of a dedicated or stand-alone unit that is optimized for vegetable oil processing because of the unique nature of the feedstock (Lestari *et al.*, 2009a).

Despite the hydrotreating of pure vegetable oil in the conventional hydrotreating catalyst, $NiMo/Al_2O_3$, presulfided with $H_2S/H_2$, hydrotreating of vegetable oil and hydrodesulfurization probably occurred on different catalytic sites, and as the feed vegetable oil did not contain sulfur, it was pointed out that part of the sulfur could leach out from the catalyst and there would be a need to add sulfur to the feed (Lestari *et al.*, 2009a).

*Deoxygenation*

An interesting option is to convert bioderived feedstocks by deoxygenation (Mäki-Arvela *et al.*, 2006, 2011; Snåre *et al.*, 2006; Lestari *et al.*, 2009a). The advantage of this technology is that no hydrogen is required in the process compared to the hydrotreating process, thus eliminating the additional cost of hydrogen. This reaction is catalyzed by noble metals supported on carbon. The reaction occurs at temperatures around 300°C, and a low partial pressure of hydrogen was beneficial in increasing the turnover frequency (TOF) and final conversion, compared to pure hydrogen or an inert reaction atmosphere (Kubičková *et al.*, 2005). The presence of hydrogen also diminished the consecutive aromatization, which is undesirable because aromatics are unsuitable as diesel constituents and they accelerate catalyst deactivation. It should be noted that catalytic deoxygenation gives a higher selectivity than hydrotreating. The results of metal catalyst testing on similar supports by normalizing the results with metal content revealed the following order: Pd > Pt > Ni > Rh > Ir > Ru > Os (Snåre *et al.*, 2006). Some

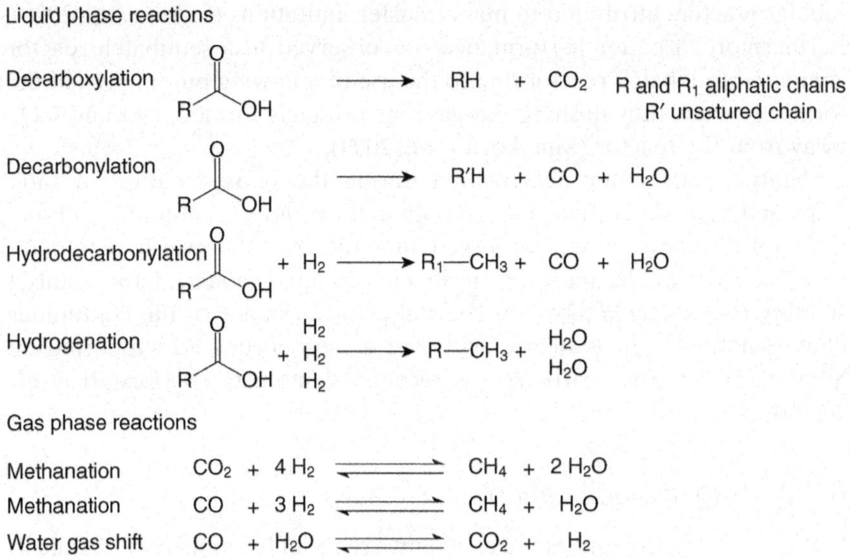

## Liquid phase reactions

Decarboxylation: R-C(=O)-OH → RH + $CO_2$    R and $R_1$ aliphatic chains
R' unsatured chain

Decarbonylation: R-C(=O)-OH → R'H + CO + $H_2O$

Hydrodecarbonylation: R-C(=O)-OH + $H_2$ → $R_1$—$CH_3$ + CO + $H_2O$

Hydrogenation: R-C(=O)-OH + $H_2$ + $H_2$ + $H_2$ → R—$CH_3$ + $H_2O$ + $H_2O$

## Gas phase reactions

Methanation    $CO_2$ + 4 $H_2$ ⇌ $CH_4$ + 2 $H_2O$
Methanation    CO + 3 $H_2$ ⇌ $CH_4$ + $H_2O$
Water gas shift    CO + $H_2O$ ⇌ $CO_2$ + $H_2$

*6.7* Reactions involved in the deoxygenation of triglycerides.

authors propose that activated carbons possess catalytic activity for the hydrothermal decarboxylation of fatty acids (Fu *et al.*, 2011).

A plausible reaction path for the production of linear hydrocarbons from fatty acids via direct deoxygenation combines liquid phase reactions and gas-phase reactions (Fig. 6.7). Fatty acids of vegetable oils can be directly decarboxylated or decarbonylated. Direct decarboxylation removes the carboxyl group by releasing carbon dioxide and paraffinic hydrocarbon, while direct decarbonylation produces an olefinic hydrocarbon via removal of the carboxyl group by forming carbon monoxide and water, as illustrated by the first two reactions in Fig. 6.7, respectively. Additionally, fatty acids can be deoxygenated by adding hydrogen; in this case the production of linear hydrocarbon can occur via direct hydrogenation or indirect decarboxylation. The gas-phase reactions are those involved in the production of CO, $CO_2$, hydrogen, and water during decarboxylation/decarbonylation. Deoxygenation of unsaturated renewables, such as oleic acid, methyl oleate, and linoleic acids also leads to saturated diesel-fuel-range hydrocarbons. The reaction occurs initially via hydrogenation of double bonds and subsequent deoxygenation of the corresponding saturated feed (Lestari *et al.*, 2009a). The reaction rates of different reactants were independent of the carbon chain length of its fatty acids (Simakova *et al.*, 2009).

Comparison of the deoxygenation activity was performed in batch, semibatch and tubular reactors. The lowest productivity was found in tubular reactors attributed to mass transfer limitations (Snåre et al., 2008). Furthermore, a better performance was observed in a semibatch reactor compared to a batch reactor due to the use of a flowing purging gas in the former reactor, thus flushing the gaseous products formed, CO and $CO_2$, away from the reactor (Simakova et al., 2009).

Catalysts can suffer deactivation during the deoxygenation of fatty acids and their derivatives due to both coking and poisoning by CO and $CO_2$. Furthermore, it was observed that the use of lower-boiling-point solvents, such as decane and mesitylene, slightly enhanced the catalyst stability (Lestari et al., 2009a). The stable performance of the continuous deoxygenation of neat stearic acid over a mesoporous Pd supported on Sibunit (mesoporous carbon) was recently demonstrated (Lestari et al., 2009b).

### 6.3.2 Sugar modification

In the present petroleum-based economy, a range of fuels and base chemicals can be produced out of a single feedstock (petroleum). If future technology is to be biomass, the possibility of converting lignocellulose materials into a series of chemicals and fuels would greatly facilitate the transition. Given the chemical disparity that exists between feedstock and end product, the preparation of fuels or chemicals from biomass will typically occur through partially deoxygenated biomass derivatives. These intermediates, usually referred to as platform chemicals, will afford a greater degree of flexibility in downstream processes. Conversion of biomass into functionalized, targeted platform molecules is unique to hydrolysis-based methods and allows for the production of a wide range of fuels and chemicals. This topic has been reviewed recently by several authors, so we will present a general overview here. For a more complete view, see these reviews by Davda et al. (2005), Huber and Dumesic (2006), Chheda et al. (2007a), Simonetti and Dumesic (2009), Martin Alonso et al. (2010a), Tong et al. (2010), Zakrzewska et al. (2010), Geboers et al. (2011), Huang and Percival Zhang (2011), and Kazi et al. (2011). We will focus on the methods for production and processing of three important platform molecules, furfural and 5-hydroxymethylfurfural (HMF), levulinic acid (LA) and γ-valerolactone (GVL).

*Furans platform*

The conversion of cellulose into furans, 5-hydroxymethylfurfural (HMF) and 2-furaldehyde (furfural) and its derivatives, a versatile feedstock not only for the production of polyesters and other plastics, but also diesel-like fuels and pharmaceuticals (Fig. 6.8) is a very interesting process. Polysaccharides can be hydrolyzed to their constituent monomers, which

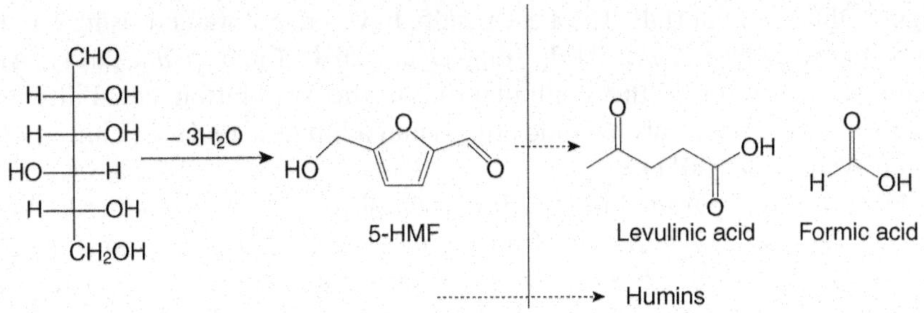

6.8 Reactions involved in the formation of HMF.

can subsequently be dehydrated by protonic acid as well as by Lewis acid catalysts (Zhao et al., 2007; West et al., 2008; Tong et al., 2010; Lam et al., 2011) to furan compounds with a carbonyl group such as HMF, 5-methylfurfural, or furfural. Furthermore, furans can be produced from both cellulose and hemicellulose fractions of biomass; thus, furan platforms utilize a large fraction of the available lignocellulosic feedstock.

Furfural and HMF can be produced by dehydration with good selectivity (e.g., 90%) from xylose and fructose, respectively, in biphasic reactors (Chheda et al., 2007b), whereas yields are lower for glucose (42% at low concentrations of 3 wt%) (Martin Alonso et al., 2010a). The addition of aprotic solvents, such as dimethyl sulfoxide (DMSO) or N,N-dimethylacetamide, improves the selectivity to HMF from fructose, with final yields of over 90% (Binder and Raines, 2009; Martin Alonso et al., 2010a). Reducing the water concentration is critical to the selective preparation of HMF, because it is readily hydrated in water to form levulinic acid and formic acid.

To minimize the incidence of side reactions, such as condensation, furfural compounds can be extracted from the aqueous layer using organic solvents (Roman-Leshkov et al., 2006; Chheda et al., 2007a; West et al., 2008), and by addition of salts to the aqueous phase (Roman-Leshkov et al., 2007), which decreases the solubility of organic species in water. The use of DMSO and an extracting solvent increases HMF selectivity to 55% when glucose is used as the feed, compared to 11% in water (Martin Alonso et al., 2010a). A potential drawback of this approach is the use of solvents; the presence of solvents requires a downstream separation step, which increases the total cost of the process; this point is very important when high boiling point solvents are employed, like DMSO. The energy requirements of downstream purification can be reduced by using other solvents like 2-butanol and methyl isobutyl ketone (MIBK) (Martin Alonso et al., 2010a). The HMF yield can be increased by the combination of metal chlorides and strong acids (HCl), especially in the transformation of glucose, cellulose, and lignocellulosic materials. The best results have been obtained using $CrCl_2$ as catalyst (Zhao et al., 2007; Tong et al., 2010; Lima et al., 2011). An interesting option is the combination of the dissolution capability of lignocellulosic materials by ionic liquids and a biphasic reactor (Lima et al., 2011; Wang et al., 2011).

HMF can be upgraded to liquid fuels employing different methods (Fig. 6.9) (Kazi et al., 2011; Jin and Enomoto, 2011). HMF can be transformed by hydrogenolysis to 2,5-dimethyl furan (DMF) with copper-based catalysts (Cu-Ru/C or $CuCrO_4$) (Roman-Leshkov et al., 2007). DMF is not soluble in water and can be used as blender in transportation fuels. A second alternative is the formation of hydrocarbons, as it has been shown that various carbohydrate-derived carbonyl compounds such as furfural, HMF, dihydroxyacetone, acetone, and tetrahydrofurfural can be condensed in aqueous and organic solvents to form larger molecules ($C_7$–$C_{15}$) that can subsequently be converted into components of diesel fuel (Barrett et al., 2006; Chheda and Dumesic, 2007). To form larger hydrocarbons, HMF and other furfural products can be upgraded by aldol condensation with ketones, such as acetone, over basic catalysts (NaOH, $MgO/ZrO_2$), or acid catalysts (West et al., 2008; Sádaba et al., 2011a, 2011b; Murzin and Simakova, 2011). Single condensation of HMF and acetone produces a C9 intermediate, which can react with a second molecule of HMF to produce a C15 intermediate. Aldol condensation can be coupled with hydrogenation steps using a bifunctional catalyst like $Pd/MgO–ZrO_2$, leading to high yields of condensation products (Chheda and Dumesic, 2007; Chheda et al., 2007a; Geboers et al., 2011; Martin Alonso et al., 2010a). By selective hydrogenation, HMF and furfural can be converted to 5-hydroxymethyltetrahydrofurfural (HMTHDA) and tetrahydrofurfural (THF2A) that, after self-condensation and hydrogenation/dehydration steps, produce C12 or C10 alkanes, respectively.

*Levulinic acid platform*

Levulinic acid is produced during the hydrolysis of sugars to furans, as a decomposition of unstable HMF (Fig. 6.8). High yield to levulinic acid can be obtained by adjusting the reaction conditions using acid minerals as catalyst (Bozell et al., 2000; Martin Alonso et al., 2010a). Levulinic acid can be upgraded into different liquid fuels (Fig. 6.10).

Levulinic acid (LA) can be converted to methyltetrahydrofuran (MTHF) by hydrogenation using $Ru(acac)_3$ and $NH_4PF_6$, or Pd-Re/C in liquid phase or nickel-promoted copper/silica in gas phase (Bozell et al., 2000; Mehdi et al., 2008; Upare et al., 2011). MTHF can be blended up to 70% with gasoline without modification of current internal combustion engines. The lower heating value of MTHF compared with gasoline is compensated by

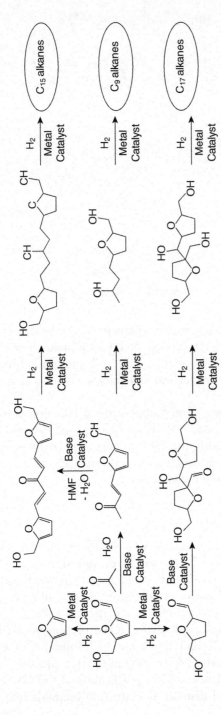

6.9 Schematic upgrade of HMF to liquid fuels.

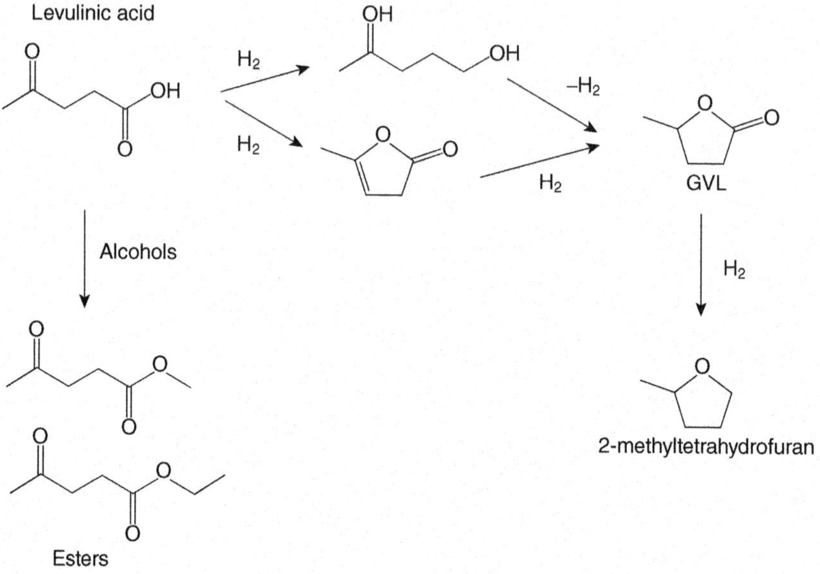

*6.10* Production of fuels from levulinic acid.

its higher specific gravity, which results in similar mileage to that achieved with gasoline. Direct conversion of levulinic acid to MTHF is possible; however, improved yields can be achieved through indirect routes, which proceed through the production of γ-valerolactone as an intermediate (see next section).

Another processing option for LA is the production of methyl and ethyl esters that can be blended with diesel fuel. This esterification can be carried out at room temperature during LA storage in the presence of methanol or ethanol. Various acid catalysts have been studied to increase yields and reaction rates (Timokhin *et al.*, 1999).

*γ-Valerolactone platform*

Derivatives of γ-valerolactone (GVL) (Lange *et al.*, 2010) are an alternative class of promising biofuels that can be produced from cellulose (Fig. 6.11). γ-Valerolactone can be used in a number of applications, ranging from direct use as a fuel additive or solvent to diverse upgrading strategies for the production of fuels and chemicals. There are limitations to its direct application as a transportation fuel in the present infrastructure, such as low energy density, blending limits, and high solubility in water.

Because GVL is derived from levulinic acid in aqueous media, the direct use of GVL as a fuel necessitates purification of the GVL, by separation/purification steps and distillation/extraction methods that remove water,

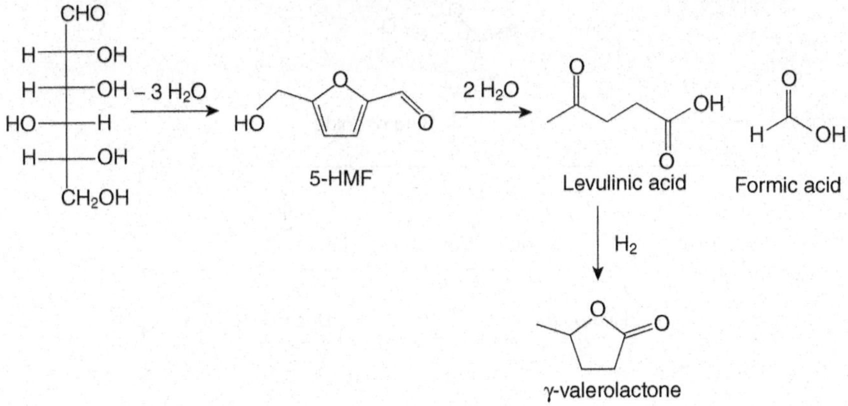

6.11 Schematic obtaining of γ-valerolactone.

but increase the overall cost of the process. Another alternative is to directly process the aqueous solutions of GVL to produce hydrophobic liquid alkanes with the appropriate molecular weight to be used as liquid fuels (Fig. 6.12).

GVL can be upgraded by liquid fuels by two strategies: (a) ring opening and hydrogenation and (b) ring opening and decarboxylation.

GVL can be converted by ring opening to pentenoic acids, and this mixture of pentenoic acids can subsequently be hydrogenated to produce pentanoic acid. These two reactions can be performed using $Pd/Nb_2O_5$ or Pt/HZSM-5 (Serrano-Ruiz et al., 2010; Lange et al., 2010). Pentanoic acid can be upgraded to 5-nonanone by ketonization over $CeZrO_x$ (Martin Alonso et al., 2010a) or to alkyl esters by esterification with alcohols with acid catalyst (Lange et al., 2010). This 5-nonanone can be hydrogenated/dehydrated to nonane over $Pt/Nb_2O_5$ (West et al., 2008). Another alternative is to hydrogenate the ketones over Ru/C at 423 K and 50 bar to produce alcohols which can subsequently be dehydrated over an acid catalyst, such as Amberlyst 70 (423 K), to produce nonene, which can be coupled by acid catalyzed oligomerization (Martin Alonso et al., 2010c). In this case, smaller ketones would also be converted to alkenes that would undergo oligomerization along with nonene to produce C6–C27 alkenes that can be hydrogenated over $Pt/Nb_2O_5$ to liquid alkanes to be used as a jet fuel or diesel blenders. The molecular weight range for the final alkanes can be modified by varying reaction conditions of temperature, pressure, or WHSV (Martin Alonso et al., 2010c).

Alternatively, GVL can be upgraded by ring opening and decarboxylation to produce an equimolar mixture of butenes and $CO_2$ (Bond et al., 2010). Both reactions take place over a solid acid catalyst, HZSM-5 or $SiO_2/Al_2O_3$,

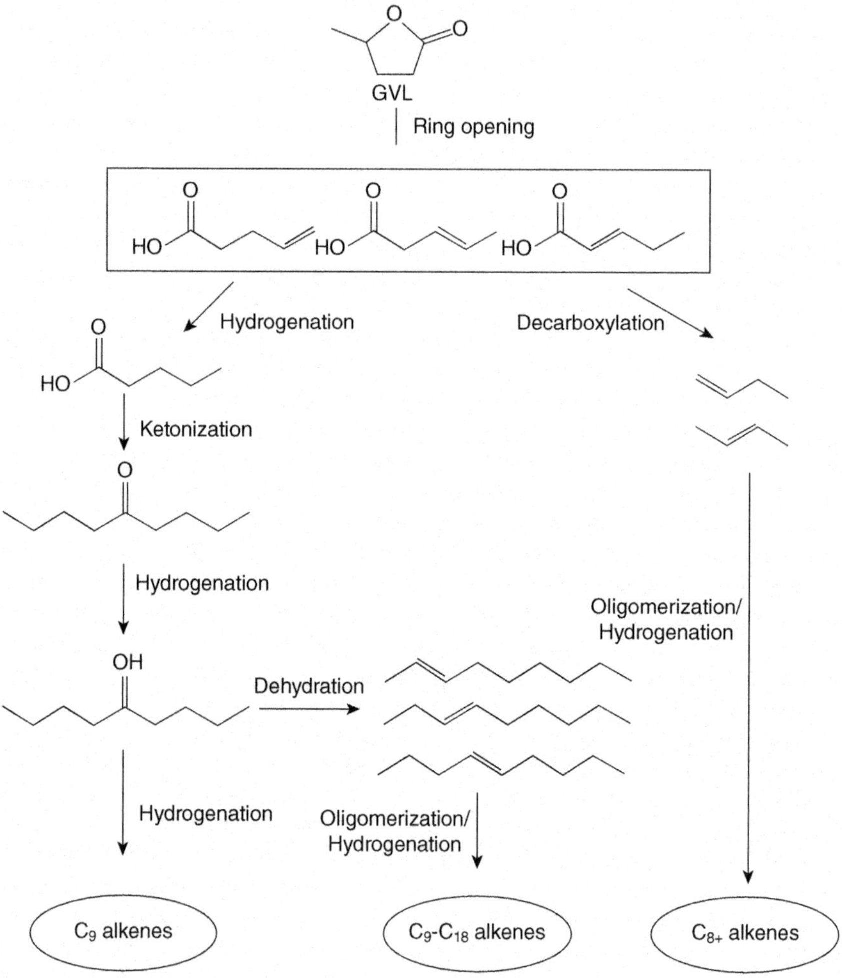

*6.12* Upgrading of γ-valerolactone to liquid fuels.

with good butene yields. These butene monomers are coupled by oligomerization over an acid catalyst (Amberlyst 70 or ZSM-5) to form C8+ alkenes that can be used as jet fuel upon hydrogenation. This process requires separated reactors because the butene oligomerization is favored at elevated pressures, while the GVL decarboxylation is favored by low pressures.

### 6.3.3 Bio-oil upgrading

Bio-oil contains 10–40 wt% oxygen with about 25 wt% water that cannot readily be separated, making it very different from standard crude oil, and it has a very low heating value (17 MJ/kg) (Bridgwater, 2012). Additionally, a significant part of the oxygenated compounds present are organic acids, like acetic or formic acid, that make the bio-oil an acid mixture. The acidic nature of the oil constitutes a problem, as it will entail harsh conditions for equipment used for both storage, transport, and processing. Common construction materials such as carbon steel and aluminum have proven unsuitable when operating with bio-oil, due to corrosion. Another important problem of bio-oil is the instability during storage, where viscosity, HV, and density are all affected. This is due to the presence of highly reactive organic compounds.

For all of these reasons it is important to upgrade bio-fuels. Catalytic upgrading of bio-oil is a complex reaction network due to the high diversity of compounds in the feed. One option not included in this block is the gasification (see Section 6.2.2), but usually bio-oil upgrading is related with the cracking, decarbonylation, decarboxylation, hydrocracking, hydrodeoxygenation, hydrogenation, and polymerization, but the most effort has been devoted to both zeolite cracking and HDO (Czernik and Bridgwater, 2004; Zhang *et al.*, 2007; Mortensen *et al.*, 2011; Bridgwater, 2012).

*Deoxygenation by cracking*

Deoxygenation cracking rejects oxygen as $CO_2$, as summarized in the conceptual overall reaction:

$$C_1H_{1.33}O_{0.43} + 0.26O_2 \rightarrow 0.65CH_{1.2} + 0.34CO_2 + 0.27H_2O$$

The deoxygenation upgrading can operate on the liquid or vapors within or close coupled to the pyrolysis process, or they can be decoupled to upgrade either the liquids or re-vaporized liquids. Deoxygenation cracking does not require co-feeding of hydrogen and can therefore be operated at atmospheric pressure. The bio-oil is converted into at least three phases in the process: oil, aqueous, and gas. Typically, reaction temperatures in the range from 300 to 600°C are used for the process. An increased temperature resulted in a decrease in the oil yield and an increase in the gas yield. This is due to an increased rate of cracking reactions at higher temperatures, resulting in the production of smaller volatile compounds. However, in order to decrease the oxygen content to a significant degree, the high temperatures were required. In conclusion, it is crucial to control the degree of cracking. A certain amount of cracking is needed to remove oxygen, but if the rate of cracking becomes too high, at increased temperatures, degradation of the bio-oil to light gases and carbon will occur instead.

The catalysts employed are based on zeolites and mesoporous materials: ZSM-5, H-Y, MCM-41, SBA-15, FCC catalysts, etc. (Czernik and Bridgwater,

2004; Zhang et al., 2007; Mortensen et al., 2011; Bridgwater, 2012). A report on hydrocarbon processing for the future of FCC and hydroprocessing in modern refineries states that 'Biomass-derived oils are generally best upgraded by HZSM-5 or ZSM-5, as these zeolitic catalysts promote high yields of liquid products and propylene. Unfortunately, these feeds tend to coke easily, and high TANs (Total Acid Number) and undesirable by-products such as water and $CO_2$ are additional challenges' (Bridgwater, 2012).

*Hydrodeoxygenation*

Hydrodeoxygenation (HDO) rejects oxygen as water by catalytic reaction with hydrogen. The process can be depicted by the following conceptual reaction:

$$C_1H_{1.33}O_{0.43} + 0.77H_2 \rightarrow CH_2 + 0.43H_2O$$

where '$CH_2$' represents an unspecified hydrocarbon product.

HDO is closely related to the hydrodesulfurization (HDS) process from the refinery industry, used in the elimination of sulfur from organic compounds (Mortensen et al., 2011). Both HDO and HDS use hydrogen for the exclusion of the heteroatom, forming respectively $H_2O$ and $H_2S$. Water is formed in the conceptual reaction, so (at least) two liquid phases will be observed as product: one organic and one aqueous. The appearance of two organic phases has also been reported, which is due to the production of organic compounds with densities less than water. In this case a light oil phase will separate on top of the water and a heavy one below. The formation of two organic phases is usually observed in instances with high degrees of deoxygenation, which will result in a high degree of fractionation in the feed.

Regarding operating conditions, a high pressure is generally used, which has been reported in the range from 75 to 300 bar in the literature (Czernik and Bridgwater, 2004; Zhang et al., 2007; Mortensen et al., 2011; Bridgwater, 2012). The high pressure has been described as ensuring a higher solubility of hydrogen in the oil and thereby a higher availability of hydrogen in the vicinity of the catalyst. This increases the reaction rate and further decreases coking in the reactor. High degrees of deoxygenation are favored by high residence times.

The catalysts originally tested were based on sulfided CoMo or NiMo supported on alumina or aluminosilicate and the process conditions are similar to those used in the desulfurization of petroleum fractions (Mortensen et al., 2011). However a number of fundamental problems arose, including that the catalyst supports of typically alumina or aluminosilicates were found to be unstable in the high water content environment of bio-oil and the sulfur was stripped from the catalysts requiring constant re-sulfurization (Bridgwater, 2012). More recently, attention has turned to precious metal catalysts on less susceptible supports, and considerable academic and industrial research has been initiated in the last few years, for instance: Pd on C (Wildschut et al., 2009; Zhao et al., 2009; Crossley et al., 2010), Ru on C (Zhao et al., 2009), and Pt (Fisk et al., 2009).

### 6.3.4 Glycerol transformation

Glycerol can be obtained from biomass (including rapeseed and sunflower oil) via hydrolysis or methanolysis of triglycerides. The reactions for the direct transformation of vegetable oils and animal fats into methyl esters and glycerol have been known for over a century. However, it is only recently, following more than 10 years of research and development, that the transesterification of triglycerides, using rapeseed, soybean and sunflower oils, has gained significance for its role in the manufacture of high quality biodiesel fuel. Glycerol can also be commercially produced by the fermentation of sugars such as glucose and fructose, either directly or as a byproduct of the industrial conversion of lignocellulose into ethanol (Zhou et al., 2008). Glycerol can be transformed into different chemicals of interest (Fig. 6.13) depending on the reaction conditions and the catalyst employed.

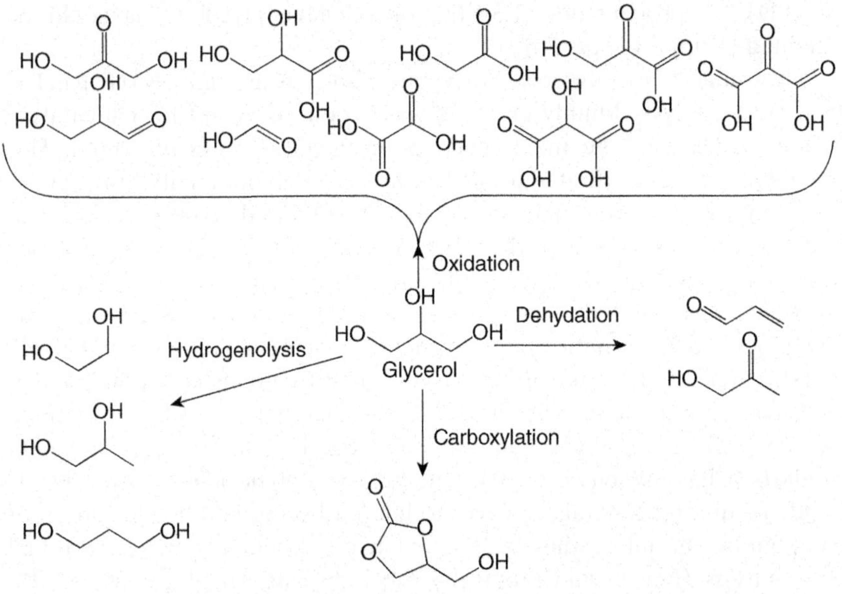

*6.13* Main reactions involved in the upgrading of glycerol.

## Oxidation

The oxidation of glycerol is conducted mainly using supported noble metal nanoparticles such as Pd, Pt, and Au as catalysts (Zhou et al., 2008). Noble metal catalysts prepared using sol–gel immobilization techniques performed better than catalysts prepared by impregnation or incipient wetness methods (Porta and Prati, 2004). Several supports have been employed: different carbons (i.e., carbon black, activated carbon, and graphite) and oxides ($TiO_2$, MgO, $CeO_2$, and $Al_2O_3$), all of which were active for the heterogeneously catalyzed liquid-phase oxidation of glycerol under atmospheric pressure conditions. However, for the same reaction conditions and using comparable metal particle size, the carbon supported catalysts showed high activity for the liquid phase oxidation of glycerol. The pH of the solution directs the selectivity to the different products, when a basic reaction solution is used, the oxidation of the primary alcohol function is promoted, whereas acidic conditions promoted the oxidation of the secondary alcohol function (Smits et al., 1987; Mallat and Baiker, 1995; Abad et al., 2006).

## Hydrogenolysis

Hydrogenolysis is a catalytic chemical reaction that breaks a chemical bond in an organic molecule with the simultaneous addition of a hydrogen atom to the resulting molecular fragments. Through the selective hydrogenolysis of glycerol in the presence of metallic catalysts and hydrogen, 1,2-propanediol (1,2-PD), 1,3-propanediol (1,3-PD), or ethylene glycol (EG) could be obtained (Ruppert et al., 2012).

First studies of the glycerol hydrogenolysis use Raney metals (Ni, Rh, Ru, Ir, Cu) as catalysts. Mainly methane was produced except for Cu catalyst, where 1,2-PD was the main reaction product (Zhou et al., 2008). This phenomenon was attributed to the low hydrogenolytic activity toward C–C bonds of copper, but high activity for C–O bond hydrogenation and dehydrogenation. However, the reaction conditions employed for glycerol hydrogenolysis using Cu Raney catalysts are very hard ($P > 10\,MPa$). This drawback was avoided using supported metal catalysts (Miyazawa et al., 2007). Several combination support metals have been studied, Cu, Ru, Pd, and Rh catalysts supported on ZnO, C, and alumina, for instance (Chaminand et al., 2004; Dasari et al., 2005; Vila et al., 2012), reaching a 100% selectivity to 1,2-PD during the hydrogenolysis of glycerol in water using CuO/ZnO catalysts (Chaminand et al., 2004). The hydrogenolytic activity toward C–C bonds of different metals can be modulated by the addition of different compounds, and interesting results have been obtained by sulfur-modified Ru catalysts (Casale and Gomez, 1994; Lahr and Shanks, 2005), Ni/Re catalysts (Werpy et al., 2003), Ni/Ce catalysts (Jiménez-Morales et al., 2012) or a combination of cobalt, copper, manganese, molybdenum, and an inorganic polyacid (Schuster and Eggersdorfer, 1996). The addition of acid to metal catalysts enhances the glycerol conversion (Zhou *et al.*, 2008). During the hydrogenolysis reaction, the acid dehydrates the glycerol to 1-hydroxyacetone, and then the metallic function hydrogenates more easily to 1-hydroxyacetone (Miyazawa *et al.*, 2006). An excellent combination is Ru/C with an acidic resin, that yields better results than other metals (Rh/C, Pd/C and Pt/C) and acids (zeolites, sulfated zirconia, $H_2WO_4$, and liquid $H_2SO_4$) (Kusunoki *et al.*, 2005; Miyazawa *et al.*, 2006, 2007).

Despite much research effort, the potential importance of the glycerol hydrogenolysis reaction is limited to the laboratory scale as the common drawbacks of high temperature and pressure, dilute solutions and the low selectivity toward propylene glycol still require further investigation.

*Dehydration*

Acid dehydration of glycerol to acrolein could offer an alternative for the currently commercial catalytic petrochemical process based on propylene (Fig. 6.14).

This reaction was studied using reactive distillation in the presence of organic or mineral acids (Waldmann and Petrū, 1950), and in gas phase over solid acids (Schwenk *et al.*, 1933; Haas *et al.*, 1994). However, this process is energy intensive because it is necessary to evaporate glycerol and water present in crude glycerol. An alternative is the use of liquid phase dehydration. Sub- (HCW) and supercritical water (SCW) reactions can be used for the production of acrolein from glycerol (Ramayya *et al.*, 1987; Bühler *et al.*, 2002; Ott *et al.*, 2006). The yield to acrolein in SCW is very low, but can be increased by the presence of acids (Watanabe *et al.*, 2007). The rate constant of acrolein decomposition was always higher than that of acrolein formation in the absence of an acid catalyst, but the rate constant of acrolein formation could overcome that of acrolein decomposition by the addition of an acid in supercritical condition. The use of SCW is only

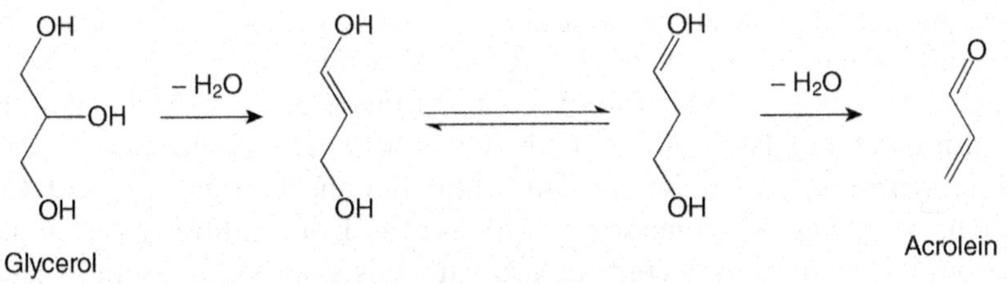

*6.14* Steps in dehydration of glycerol to acrolein.

promising if acid is added, but this produces tremendous corrosion problems, because SCW itself induces corrosion, and the presence of an acid compound intensifies the corrosive effect. Several authors confirmed that the formation of acrolein from glycerol was controlled by ionic species (such as proton) and can be increased by the presence of an acid and HCW conditions (Bühler et al., 2002; Watanabe et al., 2007). A catalytic alternative is the use of zinc sulfate, which is an effective catalyst for the acrolein synthesis from glycerol in HCW (573–663 K, 25–34 MPa, 10–60 s), achieving an acrolein selectivity of 75% at 50% of glycerol conversion (Ott et al., 2006). These results indicate that an improvement in the technology is necessary to find any practical application in the formation of acrolein from glycerol.

*Carboxylation*

Glycerol carbonate is a new and interesting material in the chemical industry. Inexpensive glycerol carbonate could serve as a source of new polymeric materials for the production of polycarbonates and polyurethanes. Several procedures of glycerol carbonate have been described in the literature, but the most interesting is the direct reaction with carbon dioxide (Vieville et al., 1998; Aresta et al., 2006; Zhou et al., 2008). First studies focused on the use of basic zeolites and resin catalysts (Vieville et al., 1998). Under supercritical $CO_2$ and in the presence of ethylene carbonate, the glycerol carbonate can be formed by direct reaction with carbon dioxide. Another alternative is the use of metal alcoxides, especially Sn catalysts. The carbonate was formed with an appreciable rate until a 1.14:1 molar ratio of carbonate to catalyst was reached.

### 6.3.5 Syngas transformations

The mixture of gases ($CO$, $CO_2$, and $H_2$) produced in the gasification of biomass (syngas or synthesis gas) can be converted in methanol or hydrocarbons by the traditional technologies that are well known and have been developed for a long time (Chinchen et al., 1988; Iglesia et al., 1993; Khodakov et al., 2007; Mokrani and Scurrell, 2009; Klerk, 2011).

The composition of syngas from biomass gasification is generally different from that from natural gas reforming and coal gasification. Syngas from natural gas mainly consists of $H_2$ or $CO$, with a small amount of $CO_2$, while syngas from biomass gasification contains much more $CO_2$ and less $H_2$, resulting in a low H/C ratio and a high $CO_2$/CO ratio. Conditioning of the crude syngas can increase the H/C ratio but the concentration of $CO_2$ will be very high in comparison with syngas from natural gas. For this reason, the catalysts selected for use with this syngas have some special characteristics.

## Methanol synthesis

Commercially, methanol is produced from natural gas or coal via syngas, mainly containing CO and $H_2$ along with a small amount of $CO_2$, which has been developed over the last century (Klier, 1982; Chinchen et al., 1988; Lange, 2001). Methanol synthesis catalyst can be produced from a mixture of copper, zinc and aluminum nitrates by coprecipitation with sodium carbonate followed by filtration, washing, drying, and calcination of the purified precipitate (Klier, 1982; Chinchen et al., 1988). Ternary Cu–Zn–Al oxide catalyst produces methanol at 5.0–10.0 MPa and 473–523 K. However, the ternary catalyst that was active for CO-rich feedstock was not so active for the $CO_2$-rich sources (syngas from biomass). In fact, a small quantity of $CO_2$ (3–5%) is present in the conventional synthesis of methanol (Chinchen et al., 1988; Liu et al., 2003). For this reason, only modifications of traditional Cu/ZnO catalysts have been proposed to improve the catalytic activity. Cu/ZnO-based catalysts have been modified with $Al_2O_3$, $Ga_2O_3$, $ZrO_2$, and $Cr_2O_3$ (Saito and Murata, 2004; Yang et al., 2006); they showed a better stability and a slight increase in methanol productivity. But the most interesting results have been obtained with multicomponent catalysts (Cu/ZnO/$ZrO_2$/$Al_2O_3$ and Cu/ZnO/$ZrO_2$/$Al_2O_3$/$Ga_2O_3$). Multicomponent catalysts were more active than the ternary or the binary catalyst. In addition, the multicomponent catalysts were found to be highly active even after the treatment of the catalysts in flowing $H_2$ at 723 K, indicating that their thermal stabilities were extremely high (Yang et al., 2006). Methanol synthesis from biomass syngas is a technology ready to be implanted at industrial scale (Phillips et al., 2011b).

## Fischer–Tropsch (FT) synthesis

In 1922, Hans Fischer and Franz Tropsch proposed the synthol process, which gave, under high pressure (>100 bar), a mixture of aliphatic oxygenated compounds via reaction of carbon monoxide with hydrogen over alkalized iron chips at 673 K. This product was transformed after heating under pressure into 'Synthine', a mixture of hydrocarbons (Khodakov et al., 2007). Since its discovery, the significance of the process has been amply demonstrated by the enormous amount of research and development effort achieved concomitantly to the oil price increase, that is, more than 4,000 papers on FT were published between 2000 and 2011, and a great number of patents dealing with FT synthesis can be found in the literature, and have been thoroughly reviewed (Iglesia et al., 1993; Dry, 2002a, 2002b; Khodakov et al., 2007; de Smit and Weckhuysen, 2008; Abelló and Montané, 2011; Klerk, 2011; De la Peña O'Shea et al., 2003, 2005; González Carballo et al., 2011; Ojeda et al., 2004; Pérez-Alonso et al., 2007).

*Table 6.2* Comparison of cobalt and iron Fischer–Tropsch catalysts

| Parameter | Cobalt catalyst | Iron catalyst |
|---|---|---|
| Cost | More expensive | Less expensive |
| Lifetime | Resistant to deactivation | Less resistant to deactivation |
| Activity at low conversion | Comparable | |
| Productivity at high conversion | Higher | Lower |
| Maximal chain growth probability | 0.94 | 0.95 |
| Water gas shift reaction | Not very significant | Significant |
| Maximal sulfur content | <0.1 ppm | <0.2 ppm |
| Flexibility (pressure and temperature) | Less flexible | Flexible |
| $H_2$/CO ratio | ≈2 | 0.5–2.5 |
| Attrition resistance | Good | Not very resistant |

All group VIII metals have noticeable activity in the hydrogenation of carbon monoxide to hydrocarbons. But only ruthenium, iron, cobalt, and nickel have catalytic characteristics which allow them to be considered for commercial production. Nickel catalysts under practical conditions produce too much methane. Ruthenium is too expensive; moreover, its worldwide reserves are insufficient for large-scale industry. Thus, cobalt and iron are the metals which were proposed by Fischer and Tropsch as the first catalysts for syngas conversion. Both cobalt and iron catalysts have been used in the industry for hydrocarbon synthesis. A brief comparison of cobalt and iron catalysts is given in Table 6.2.

Iron-based catalysts are especially suited for the production of liquid hydrocarbon products from syngas derived from sources, such as coal (CTL) and biomass (BTL), which typically have too low a $H_2$ to CO ratio to stoichiometrically produce longer chain hydrocarbon products. Because they have unique water–gas shift (WGS) capabilities, they catalyze the reaction between carbon monoxide and water to form hydrogen and carbon dioxide. Iron-based FTS catalyst precursors consist of nanometer-sized $Fe_2O_3$ crystallites to which often promoters are added to improve the catalyst performance. A typical catalyst contains promoters like copper to enhance catalyst reducibility, potassium to improve CO dissociation, along with some silica or zinc oxide to improve the amount of iron atoms interacting with the gas phase (i.e., catalyst dispersion). The catalyst is treated in $H_2$, CO, or syngas to convert it to its active form. During FTS, a complex mixture of iron phases is formed. The nature of active sites in iron-based catalysts is still a subject of debate. Several forms of iron oxides and iron carbides ($Fe_xC_y$) may co-exist during the FT reaction: α-Fe, γ-Fe, $Fe_3O_4$, coexisting with ε-$Fe_2C$, ε′-$Fe_{2.2}C$, $Fe_5C_2$ (Hägg carbide), $Fe_7C_3$, and θ-Fe₃C (Pérez-Alonso *et al.*, 2007; de Smit and Weckhuysen, 2008; González-Carballo and Fierro, 2010; Abelló and Montané, 2011).

The BTL process is one of the most important and sustainable paths to produce liquid fuels, such as gasoline and diesel, and chemicals from renewable resources. However, a still unresolved major problem of FTS is to achieve good selectivity control toward certain products of interest. Such control depends mainly on the nature of active ingredients in the catalyst, the presence of promoters, and the choice of an adequate support. Accordingly, tailoring the catalyst to achieve specific product distributions is a major objective for current investigations.

Biomass syngas feeds exhibit low conversion efficiency owing to their $H_2$-deficient or $CO_2$-rich nature. Many studies on the carbon monoxide hydrogenation with regard to hydrocarbon production have been performed by using iron-based catalysts independent of the $H_2/CO$ ratio in the syngas feed. The flexibility of the WGS reaction in such cases makes it possible to achieve the production of hydrocarbons by $H_2/CO$ ratios < 2 in the syngas feed.

### 6.3.6 Lignin use

Following the biomass pre-treatment, the lignin polymer is susceptible to a wide range of chemical transformations to form valuable chemicals (Zakzeski *et al.*, 2010). The fragmentation reactions can be principally divided into lignin cracking or hydrolysis reactions, catalytic reduction reactions, and catalytic oxidation reactions. It should be noted that selecting effective catalytic processes of lignin transformation into valuable chemical source remains a problem (Zakzeski *et al.*, 2010).

*Catalytic cracking*

The cracking or hydrocracking of lignin can be performed on the same catalysts used at oil refineries. The catalysts used in hydrocracking are predominantly bifunctional, combining a support active in cracking with a (noble) metal for the hydrogenation reaction. The hydrogenation catalyst is typically composed of noble metal, cobalt, tungsten, or nickel, and the cracking component typically consists of zeolites or amorphous silica-alumina of various compositions. Lignin can also be treated with hydrocracking catalysts, which leads to cleavage of the β-O-4 bond and relatively unstable carbon–carbon bonds. The resulting low molecular weight aromatic compounds are then susceptible to further conversion to valuable products (Thring and Breau, 1996). The acidic function of the catalysts is proportioned by zeolites or silica alumina. H-ZSM-5 and H-mordonite produced more aromatic than aliphatic hydrocarbons from fast pyrolysis bio-oil, whereas H-Y, silicalite, and silica-alumina produced more aliphatic than aromatic hydrocarbons.

*Hydrolysis*

Hydrolysis of lignin is catalyzed by bases. This process has been studied in detail, because delignification is one of the main processes in the manufacture of cellulose and paper. The reaction was favored by strong bases, and combinations of bases gave either positive synergistic effects, such as with NaOH and $Ca(OH)_2$, or negative synergistic effects, such as with LiOH or CsOH with $Ca(OH)_2$ (Miller *et al.*, 1999; Zakzeski *et al.*, 2010). The obtained products consist in a mixture of products derived from phenol (Miller *et al.*, 1999; Nenkova *et al.*, 2008; Zakzeski *et al.*, 2010). Several solvents have been used for the hydrolysis of lignin, mainly polar solvents under standard conditions (Nenkova *et al.*, 2008) or supercritical conditions (Miller *et al.*, 1999; Wahyudiono *et al.*, 2007, 2009). Among them, some attention has been focused on supercritical water, which is able to hydrolyze lignin in the absence of catalysts (Wahyudiono *et al.*, 2007, 2009). However, the high temperature necessary to obtain supercritical water (374°C) can yield coke formation (Zakzeski *et al.*, 2010).

*Reduction*

For lignin reductions, typical reactions involve the removal of the extensive functionality of the lignin subunits to form simpler monomeric compounds such as phenols, benzene, toluene, or xylene. These simple aromatic compounds can then be hydrogenated to alkanes or used as platform chemicals for use in the synthesis of fine chemicals using technology already developed in the petroleum industry.

The first catalysts studied in the hydrogenation of lignin are based on metal-supported Raney Ni, Pd/C, Rh/C, $Rh/Al_2O_3$, Ru/C, $Ru/Al_2O_3$. Employing these catalysts, a significant amount of the original lignin was converted into the monomeric products 4-propylguaiacol and dihydroconiferyl alcohol under mild conditions (3.4 MPa, 468 K) (Pepper and Hibbert, 1948; Pepper and Lee, 1969; Pepper and Fleming, 1978; Pepper and Supathna, 1978). However, the majority of studies have been focused on the use of catalysts based on industrial hydrotreatment catalysts (cobalt- and nickel-promoted molybdenum catalysts) and their modification; sulfided Co-Mo catalyst provided the best results (Zakzeski *et al.*, 2010). This phenomenon was explained by the systematic study of C–O bond hydrogenolysis of diphenyl ether of a series of sulfided $M-Mo/Al_2O_3$ catalysts (M = Cr, Fe, Co, Ni, Ru, Rh, Pd, Re, Ir, or Pt, at 623 K, 13.8 MPa $H_2$), where Co-Mo, Rh-Mo, and Ru-Mo catalysts showed the highest hydrogenolysis activity in this order.

Some disadvantages that are associated with conventional hydrodeoxygenation catalysts are possible contamination of products by incorporation of sulfur, rapid deactivation by coke formation, and potential poisoning by water. These issues arise especially with biomass feedstocks and thus have prompted efforts to explore alternative hydrogenation catalysts (Zhao et al., 2009). Ni-W/$SiO_2$-$Al_2O_3$, Ni-Cu/$ZrO_2$, Ni-Cu/$CeO_2$, transition metal carbides, and noble metals on carbon have been proposed as alternative catalysts. The supported platinum group catalysts are known to be more active than the sulfided Mo-based ones, and can therefore be used at lower temperatures, and non-alumina supports such as carbon avoid water instability associated with $Al_2O_3$ (Elliott and Hart, 2008).

*Oxidation*

For lignin oxidation, lignin is converted to more complicated platform chemicals with extensive functionality or converted directly to target fine chemicals. Oxidative catalysts have played an important role in the pulp and paper industry as a means to remove lignin and other compounds from wood pulps in order to increase the quality of the final paper product.

Some studies have focused on the use of heterogeneous catalysts, using different catalysts and oxidants. Some systems are based on the use of titanium oxide as photocatalyst, or its modification with Pt or Fe in the lignin oxidation at room temperature (Portjanskaja and Preis, 2007; Ma et al., 2008; Portjanskaja et al., 2009). Other authors propose the use of copper-based catalysts operating at higher temperature (373 K); these catalysts have only Cu, Cu-Ni, Cu-Mn, Cu-Fe, etc. (Bhargava et al., 2007; Zhang et al., 2009). Some authors have studied the use of alternative oxidants like hydrogen peroxide. In this case, they propose the use of hydrogen peroxide with methylrhenium trioxide catalysts immobilized on poly(4-vinyl pyridine) or polystyrene (Crestini et al., 2006; Herrmann et al., 2000).

Several studies have focused on the use of homogeneous catalysts; these catalysts are based in coordination compounds of transition metals. The oxidation of lignin by homogeneous catalysts represents one of the most promising approaches toward the production of fine chemicals from lignin and lignin pulp streams. Several homogeneous catalysts that are capable of performing selective oxidation of lignin have been reported in the literature. Homogeneous catalysts offer several advantageous properties that make them particularly suitable for lignin oxidation, especially the ability to use a wide range of ligands, the electronic and steric properties which drastically influence the activity, stability, and solubility of the catalyst. It thus becomes possible to tune the reactivity and selectivity of the homogeneous catalyst to the oxidation of specific lignin linkages or functionalities with appropriate choice of ligands.

Generally, the homogeneous catalysts used for lignin oxidation can be subdivided into several categories depending on the ligand set employed: metalloporphyrins, Schiff-base catalysts, polyoxometalates (POM), simple metal salts, and other kind of catalysts. A significant disadvantage of using the porphyrin complexes is the susceptibility to degradation in the presence of excess oxidant, particularly $H_2O_2$, or through the formation of catalytically inactive μ-oxo species (Zakzeski *et al.*, 2010). In contrast, the Schiff-base catalysts have several advantages over the metalloporphyrin complexes discussed above in that they are often cheaper, easier to synthesize, and relatively stable (Zakzeski *et al.*, 2010). Similarly to the case of metalloporphyrins, the original objective for the design of POMs focused on the ability to selectively degrade lignin rather than cellulose and other materials in the paper industry. That is, active catalysts rapidly oxidized lignin to carbon dioxide and water with minimal degradation of the polysaccharides, leaving a lignin-free white pulp suitable for paper production. For this reason, a modification of POM composition is necessary for the selective oxidation of Kraft pulps to chemicals (Voitl and von Rohr, 2008). In general, the order of activity for the homogeneous catalysts was $Cu^{2+} > Fe^{2+} > Mn^{2+} > Ce^{2+} > Bi^{2+} > Co^{2+} > Zn^{2+} > Mg^{2+} > Ni^{2+}$ (Bhargava *et al.*, 2007) when the same category of catalysts was studied.

## 6.4 Conclusion and future trends

The biobased economy is expected to grow significantly in the coming years. A pillar of this, both now and in the future, is biorefining, the sustainable processing of biomass into a spectrum of marketable products and energy. Biorefineries will use a wider range of feedstocks and will produce a greater variety of end-products than today. This characteristic introduces a big challenge in the application of catalytic processes to a biorefinery scheme, because very flexible catalysts and processes are needed.

Currently, over 90% of petrochemicals are produced via catalytic processes. The petrochemical industry is based on only a few hydrocarbons (ethylene, propylene, C4-olefins, benzene, toluene, and xylenes), from which all other chemicals and materials are derived. Specific chemical functionality (often derived from functional groups including heteroatoms – elements other than carbon or hydrogen – such as oxygen) is added in subsequent catalytic processes. In contrast, biomass-derived molecules already contain large numbers of oxygen-containing functional groups and are in effect 'over-functionalized'. Because of this basic difference in chemistry, one of the biggest challenges over the next few years will be to retrofit existing chemical technology to start with more oxidized carbon and go to less oxidized carbon, whereas the current chemical industry takes things in the opposite direction. The challenge will be to adapt hydrocarbon-based 'petrochemical thinking' to oxygen-rich, biomass-derived feedstocks. A specific challenge in processing these highly functionalized, biomass-derived molecules (with several different types of functional groups) is the selectivity issue (the need to selectively hydrogenate different groups).

Another challenge is the design of robust catalysts, which is made difficult for two main reasons, both specific to biomass. Firstly, biomass-derived molecules are highly functionalized and therefore very reactive. A drawback of this high reactivity is the rapid deactivation of catalysts by accumulation of carbonaceous compounds on their surface. The second reason is the large percentage of water in the reaction media, either mixed with the substrate or generated by the reaction. Therefore, catalysts must be resistant to water (with no leaching, no destruction of the active phase and/or support), but the main issue is the surface properties of the catalyst in an aqueous environment. Understanding the detailed structure of the active sites is a challenge due to the presence of water, which considerably modifies the way catalysts behave under real conditions. Acidity or alkalinity, for example, are extremely difficult to control during processing, because conditions are modified by the presence of water, while catalysts are usually characterized under laboratory conditions. The conventional view of catalysis – structure controls morphology, which determines function – has to be totally rethought because of this discrepancy between laboratory work and operating conditions.

*Trends in biomass depolymerization*

New, flexible biomass pre-treatment processes should be developed and tailored to suit improved biomass feedstocks in order to obtain fully functional fractions (e.g., lignin and carbohydrates from lignocellulose). Current research on lignocellulose breakdown must be accelerated, by improving existing technologies to develop efficient and cost-effective processes.

For gasification and pyrolysis processes, research should focus on scaling up and integrating them into existing production units, together with end-product quality improvement (e.g., syngas purification for catalytic conversion and pyrolysis oil upgrading and fractionation).

*Trends in product upgrading*

Biomass products contain a large number of oxygenated functional groups. The number of these groups must be reduced by different reactions. There is therefore a need to perform research in order to develop novel catalysts that are able to perform these reductions of oxygenated group reactions, in contrast to the oxidation reactions typical of the conventional chemical industry. Processing these highly functionalized, biomass-derived molecules leads to selectivity issues (with selective hydrogenation being needed to modify some groups but not others). The catalytic processes need to remove this functionality selectively, often to generate 'bi-functional' molecules that can be used as building blocks for bulk chemicals and bulk bio-based polymers. At this point the development of bi- or even multi-functional catalysts is a critical issue.

Further research needs to be focused on the development of catalysts that are capable of selectively transforming biomass-derived monomers (sugars, fatty acids, etc.) to platform molecules, or catalyzing the reaction from these intermediates to final products. At the same time, the work should also focus on the possible modification of catalyst surface properties to improve their functionality in the presence of water. R&D is needed to understand catalyst functionality and to revise the structure/morphology/function approach to catalyst development using new tools (such as process spectroscopy, etc.). The high reactivity of biomass-derived molecules leads to catalyst coking issues (rapid deactivation of catalysts by accumulation of carbonaceous compounds on their surface). Solutions to these issues can be found via the development of more robust catalytic formulations and also in process design, by designing new reactors and introducing small doses of oxygen in the reaction medium, optionally with the addition of an oxygen-splitting function on the catalyst surface to facilitate the process.

*Trends in syngas transformations*

Syngas made by biomass gasification can be used to produce hydrogen (via the catalytic water–gas shift reaction to $H_2$ and $CO_2$), biofuels (e.g., synthetic diesel by Fischer–Tropsch synthesis), or chemicals (mainly short-chain alcohols). The main problems arise from contamination of the syngas by impurities that 'poison' or inactivate the catalyst. Future catalysts for syngas conversion have to be developed which have a greater resistance to poisoning, allowing syngas purification costs to be reduced.

## 6.5 References

Abad, A., Almela, C., Corma, A. and García, H. 2006. Efficient chemoselective alcohol oxidation using oxygen as oxidant. Superior performance of gold over palladium catalysts. *Tetrahedron*, 62, 6666–6672.

Abelló, S. and Montané, D. 2011. Exploring iron-based multifunctional catalysts for Fischer–Tropsch synthesis: a review. *ChemSusChem*, 4, 1538–1556.

Alba-Rubio, A. C., Vila, F., Martin Alonso, D., Ojeda, M., Mariscal, R. and Lopez Granados, M. 2010. Deactivation of organosulfonic acid functionalized silica catalysts during biodiesel synthesis. *Applied Catalysis B – Environmental*, 95, 279–287.

Antonakou, E., Lappas, A., Nilsen, M. H., Bouzga, A. and Stöcker, M. 2006. Evaluation of various types of Al-MCM-41 materials as catalysts in biomass pyrolysis for the production of bio-fuels and chemicals. *Fuel*, 85, 2202–2212.

Aresta, M., Dibenedetto, A., Nocito, F. and Pastore, C. 2006. A study on the carboxylation of glycerol to glycerol carbonate with carbon dioxide: the role of the catalyst, solvent and reaction conditions. *Journal of Molecular Catalysis A: Chemical*, 257, 149–153.

Asadullah, M., Fujimoto, K. and Tomishige, K. 2001. Catalytic performance of Rh/$CeO_2$ in the gasification of cellulose to synthesis gas at low temperature. *Industrial & Engineering Chemistry Research*, 40, 5894–5900.

Banerjee, S., Mudliar, S., Sen, R., Giri, B., Satpute, D., Chakrabarti, T. and Pandey, R. A. 2010. Commercializing lignocellulosic bioethanol: technology bottlenecks and possible remedies. *Biofuels, Bioproducts and Biorefining*, 4, 77–93.

Banowetz, G. M., Griffith, S. M. and El-Nashaar, H. M. 2008. Mineral content of grasses grown for seed in low rainfall areas of the Pacific Northwest and analysis of ash from gasification of bluegrass (*Poa pratensis* L.) straw. *Energy & Fuels*, 23, 502–506.

Barrett, C. J., Chheda, J. N., Huber, G. W. and Dumesic, J. A. 2006. Single-reactor process for sequential aldol-condensation and hydrogenation of biomass-derived compounds in water. *Applied Catalysis B: Environmental*, 66, 111–118.

Behr, A., Westfechtel, A. and Pérez Gomes, J. 2008. Catalytic processes for the technical use of natural fats and oils. *Chemical Engineering & Technology*, 31, 700–714.

Bhargava, S., Jani, H., Tardio, J., Akolekar, D. and Hoang, M. 2007. Catalytic wet oxidation of ferulic acid (a model lignin compound) using heterogeneous copper catalysts. *Industrial & Engineering Chemistry Research*, 46, 8652–8656.

Binder, J. B. and Raines, R. T. 2009. Simple chemical transformation of lignocellulosic biomass into furans for fuels and chemicals. *Journal of the American Chemical Society*, 131, 1979–1985.

Binder, J. B. and Raines, R. T. 2010. Fermentable sugars by chemical hydrolysis of biomass. *Proceedings of the National Academy of Sciences*, 107, 4516–4521.

Bond, J. Q., Alonso, D. M., Wang, D., West, R. M. and Dumesic, J. A. 2010. Integrated catalytic conversion of γ-valerolactone to liquid alkenes for transportation fuels. *Science*, 327, 1110–1114.

Bozell, J. J., Moens, L., Elliott, D. C., Wang, Y., Neuenscwander, G. G., Fitzpatrick, S. W., Bilski, R. J. and Jarnefeld, J. L. 2000. Production of levulinic acid and use as a platform chemical for derived products. *Resources, Conservation and Recycling*, 28, 227–239.

Bridgwater, A. V. 2012. Review of fast pyrolysis of biomass and product upgrading. *Biomass and Bioenergy*, 38, 68–94.

Bühler, W., Dinjus, E., Ederer, H. J., Kruse, A. and Mas, C. 2002. Ionic reactions and pyrolysis of glycerol as competing reaction pathways in near- and supercritical water. *The Journal of Supercritical Fluids*, 22, 37–53.

Bulushev, D. A. and Ross, J. R. H. 2011. Catalysis for conversion of biomass to fuels via pyrolysis and gasification: a review. *Catalysis Today*, 171, 1–13.

Caballero, M. A., Corella, J., Aznar, M.-P. and Gil, J. 2000. Biomass gasification with air in fluidized bed: hot gas cleanup with selected commercial and full-size nickel-based catalysts. *Industrial & Engineering Chemistry Research*, 39, 1143–1154.

Casale, B. and Gomez, A. M. 1994. Catalytic method of hydrogenating glycerol. EP patent application EP19920830357.

Chaminand, J., Djakovitch, L. A., Gallezot, P., Marion, P., Pinel, C. and Rosier, C. 2004. Glycerol hydrogenolysis on heterogeneous catalysts. *Green Chemistry*, 6, 359–361.

Chheda, J. N. and Dumesic, J. A. 2007. An overview of dehydration, aldol-condensation and hydrogenation processes for production of liquid alkanes from biomass-derived carbohydrates. *Catalysis Today*, 123, 59–70.

Chheda, J. N., Huber, G. W. and Dumesic, J. A. 2007a. Liquid-phase catalytic processing of biomass-derived oxygenated hydrocarbons to fuels and chemicals. *Angewandte Chemie International Edition*, 46, 7164–7183.

Chheda, J. N., Roman-Leshkov, Y. and Dumesic, J. A. 2007b. Production of 5-hydroxymethylfurfural and furfural by dehydration of biomass-derived mono- and poly-saccharides. *Green Chemistry*, 9, 342–350.

Chinchen, G. C., Denny, P. J., Jennings, J. R., Spencer, M. S. and Waugh, K. C. 1988. Synthesis of methanol: Part 1. Catalysts and kinetics. *Applied Catalysis*, 36, 1–65.

Clark, J. H., Deswarte, F. E. I. and Farmer, T. J. 2009. The integration of green chemistry into future biorefineries. *Biofuels, Bioproducts and Biorefining*, 3, 72–90.

Claus, P. and Vogel, H. 2008. The roll of chemocatalysis in the establishment of the technology platform 'renewable resources'. *Chemical Engineering and Technology*, 31, 678–699.

Corella, J., Aznar, M.-P., Gil, J. and Caballero, M. A. 1999. Biomass gasification in fluidized bed: where to locate the dolomite to improve gasification? *Energy & Fuels*, 13, 1122–1127.

Corma, A., Iborra, S. and Velty, A. 2007. Chemical routes for the transformation of biomass into chemicals. *Chemical Reviews (Washington, D. C.)*, 107, 2411–2502.

Crestini, C., Caponi, M. C., Argyropoulos, D. S. and Saladino, R. 2006. Immobilized methyltrioxo rhenium (MTO)/$H_2O_2$ systems for the oxidation of lignin and lignin model compounds. *Bioorganic & Medicinal Chemistry*, 14, 5292–5302.

Crossley, S., Faria, J., Shen, M. and Resasco, D. E. 2010. Solid nanoparticles that catalyze biofuel upgrade reactions at the water/oil interface. *Science*, 327, 68–72.

Czernik, S. and Bridgwater, A. V. 2004. Overview of applications of biomass fast pyrolysis oil. *Energy & Fuels*, 18, 590–598.

Dasari, M. A., Kiatsimkul, P.-P., Sutterlin, W. R. and Suppes, G. J. 2005. Low-pressure hydrogenolysis of glycerol to propylene glycol. *Applied Catalysis A: General*, 281, 225–231.

Davda, R. R., Shabaker, J. W., Huber, G. W., Cortright, R. D. and Dumesic, J. A. 2005. A review of catalytic issues and process conditions for renewable hydrogen and alkanes by aqueous-phase reforming of oxygenated hydrocarbons over supported metal catalysts. *Applied Catalysis B: Environmental*, 56, 171–186.

De la Peña O'Shea, V. A., Menéndez, N. N., Tornero, J. D. and Fierro, J. L. G. 2003. Unusually high selectivity to $C_{2+}$ alcohols on bimetallic CoFe catalysts during CO hydrogenation. *Catalysis Letters*, 88, 123–128.

De la Peña O'Shea, V. A., Alvarez-Galvan, M. C., Campos-Martin, J. M. and Fierro, J. L. G. 2005. Strong dependence on pressure of the performance of a Co/$SiO_2$

catalyst in Fischer–Tropsch slurry reactor synthesis. *Catalysis Letters*, 100, 105–110.
De Lasa, H., Salaices, E., Mazumder, J. and Lucky, R. 2011. Catalytic steam gasification of biomass: catalysts, thermodynamics and kinetics. *Chemical Reviews (Washington, DC)*, 111, 5404–5433.
De Smit, E. and Weckhuysen, B. M. 2008. The renaissance of iron-based Fischer–Tropsch synthesis: on the multifaceted catalyst deactivation behaviour. *Chemical Society Reviews*, 37, 2758–2781.
Dry, M. E. 2002a. The Fischer–Tropsch process: 1950–2000. *Catalysis Today*, 71, 227–241.
Dry, M. E. 2002b. High quality diesel via the Fischer–Tropsch process – a review. *Journal of Chemical Technology & Biotechnology*, 77, 43–50.
Elliott, D. C. and Hart, T. R. 2008. Catalytic hydroprocessing of chemical models for bio-oil. *Energy & Fuels*, 23, 631–637.
Fisk, C. A., Morgan, T., Ji, Y., Crocker, M., Crofcheck, C. and Lewis, S. A. 2009. Bio-oil upgrading over platinum catalysts using *in situ* generated hydrogen. *Applied Catalysis A: General*, 358, 150–156.
FitzPatrick, M., Champagne, P., Cunningham, M. F. and Whitney, R. A. 2010. A biorefinery processing perspective: treatment of lignocellulosic materials for the production of value-added products. *Bioresource Technology*, 101, 8915–8922.
Florin, N. H. and Harris, A. T. 2008. Enhanced hydrogen production from biomass with *in situ* carbon dioxide capture using calcium oxide sorbents. *Chemical Engineering Science*, 63, 287–316.
Fu, J., Shi, F., Thompson, L. T., Lu, X. and Savage, P. E. 2011. Activated carbons for hydrothermal decarboxylation of fatty acids. *ACS Catalysis*, 1, 227–231.
Fukuoka, A. and Dhepe, P. L. 2006. Catalytic conversion of cellulose into sugar alcohols. *Angewandte Chemie International Edition*, 45, 5161–5163.
Gallezot, P. 2007a. Catalytic routes from renewables to fine chemicals. *Catalysis Today*, 121, 76–91.
Gallezot, P. 2007b. Process options for the catalytic conversion of renewables into bioproducts. In: Centi, G. and Van Santen, R. A. (eds) *Catalysis for Renewables: From Feedstock to Energy Production*. Weinheim: WILEY-VCH.
Gallezot, P. 2012. Conversion of biomass to selected chemical products. *Chemical Society Reviews*, 41, 1538–1558.
Geboers, J. A., Van De Vyver, S., Ooms, R., Op De Beeck, B., Jacobs, P. A. and Sels, B. F. 2011. Chemocatalytic conversion of cellulose: opportunities, advances and pitfalls. *Catalysis Science & Technology*, 1, 714–726.
González-Carballo, J. M. and Fierro, J. L. G. 2010. Fundamentals of syngas production and Fischer–Tropsch synthesis. In: Ojeda, M. and Rojas, S. (eds) *Biofuel from Fischer–Tropsch Synthesis*. New York: Nova Science Publishers, pp. 1–32.
González Carballo, J. M., Finocchio, E., García, S., Rojas, S., Ojeda, M., Busca, G. and Fierro, J. L. G. 2011. Support effects on the structure and performance of ruthenium catalysts for the Fischer–Tropsch synthesis. *Catalysis Science and Technology*, 1, 1013–1023.
Goyal, H. B., Seal, D. and Saxena, R. C. 2008. Bio-fuels from thermochemical conversion of renewable resources: a review. *Renewable and Sustainable Energy Reviews*, 12, 504–517.
Haas, T., Neher, A., Arntz, D., Klenk, D. and Girke, W. 1994. Method of preparation of 1,2- and 1,3- propanediol. European patent application EP0598228 A1.

Haveren, J. V., Scott, E. L. and Sanders, J. 2008. Bulk chemicals from biomass. *Biofuels, Bioproducts and Biorefining*, 2, 41–57.

Herrmann, W. A., Weskamp, T., Zoller, J. P. and Fischer, R. W. 2000. Methyltrioxorhenium: oxidative cleavage of CC-double bonds and its application in a highly efficient synthesis of vanillin from biological waste. *Journal of Molecular Catalysis A: Chemical*, 153, 49–52.

Huang, W.-D. and Percival Zhang, Y. H. 2011. Analysis of biofuels production from sugar based on three criteria: thermodynamics, bioenergetics, and product separation. *Energy & Environmental Science*, 4, 784–792.

Huber, G. W. and Corma, A. 2007. Synergies between bio- and oil refineries for the production of fuels from biomass. *Angewandte Chemie International Edition*, 46, 7184–7201.

Huber, G. W. and Dumesic, J. A. 2006. An overview of aqueous-phase catalytic processes for production of hydrogen and alkanes in a biorefinery. *Catalysis Today*, 111, 119–132.

Huber, G. W., Iborra, S. and Corma, A. 2006. Synthesis of transportation fuels from biomass: chemistry, catalysts, and engineering. *Chemical Reviews (Washington, DC)*, 106, 4044–4098.

Iglesia, E., Reyes, S. C., Madon, R. J. and Soled, S. L. 1993. Selectivity control and catalyst design in the Fischer–Tropsch synthesis: sites, pellets, and reactors. In: Eley, D. D., Pines, H. and Weisz, P. B., (eds) *Advances in Catalysis, Vol. 39*. Academic Press, San Diego, CA, pp. 221–302.

Ji, N., Zhang, T., Zheng, M., Wang, A., Wang, H., Wang, X. and Chen, J. G. 2008. Direct catalytic conversion of cellulose into ethylene glycol using nickel-promoted tungsten carbide catalysts. *Angewandte Chemie International Edition*, 47, 8510–8513.

Jiménez-Morales, I., Vila, F., Mariscal, R. and Jiménez-López, A. 2012. Hydrogenolysis of glycerol to obtain 1,2-propanediol on Ce-promoted Ni/SBA-15 catalysts. *Applied Catalysis B: Environmental*, 117–118, 253–259.

Jin, F. and Enomoto, H. 2011. Rapid and highly selective conversion of biomass into value-added products in hydrothermal conditions: chemistry of acid/base-catalysed and oxidation reactions. *Energy & Environmental Science*, 4, 382–397.

Jollet, V., Chambon, F., Rataboul, F., Cabiac, A., Pinel, C., Guillon, E. and Essayem, N. 2009. Non-catalyzed and Pt/[gamma]-$Al_2O_3$-catalyzed hydrothermal cellulose dissolution-conversion: influence of the reaction parameters and analysis of the unreacted cellulose. *Green Chemistry*, 11, 2052–2060.

Käldström, M., Kumar, N. and Murzin, D. Y. 2011. Valorization of cellulose over metal supported mesoporous materials. *Catalysis Today*, 167, 91–95.

Kamm, B. and Kamm, M. 2004. Principles of biorefineries. *Applied Microbiology and Biotechnology*, 64, 137–145.

Kazi, F. K., Patel, A. D., Serrano-Ruiz, J. C., Dumesic, J. A. and Anex, R. P. 2011. Techno-economic analysis of dimethylfuran (DMF) and hydroxymethylfurfural (HMF) production from pure fructose in catalytic processes. *Chemical Engineering Journal*, 169, 329–338.

Khodakov, A. Y., Chu, W. and Fongarland, P. 2007. Advances in the development of novel cobalt Fischer–Tropsch catalysts for synthesis of long-chain hydrocarbons and clean fuels. *Chemical Reviews*, 107, 1692–1744.

Klerk, A. de 2011. Fischer–Tropsch fuels refinery design. *Energy & Environmental Science*, 4, 1177–1205.

Klier, K. 1982. Methanol synthesis. In: Eley, D. D., Pines, H. and Weisz, P. B., (eds) *Advances in Catalysis*, Vol. 31. Academic Press, San Diego, CA, pp. 243–313.

Kotwal, M. S., Niphadkar, P. S., Deshpande, S. S., Bokade, V. V. and Joshi, P. N. 2009. Transesterification of sunflower oil catalyzed by flyash-based solid catalysts. *Fuel*, 88, 1773–1778.

Kubičková, I., Snåre, M., Eränen, K., Mäki-Arvela, P. and Murzin, D. Y. 2005. Hydrocarbons for diesel fuel via decarboxylation of vegetable oils. *Catalysis Today*, 106, 197–200.

Kumar, S., Singh, S. P., Mishra, I. M. and Adhikari, D. K. 2009. Recent advances in production of bioethanol from lignocellulosic biomass. *Chemical Engineering and Technology*, 32, 517–526.

Kusunoki, Y., Miyazawa, T., Kunimori, K. and Tomishige, K. 2005. Highly active metal–acid bifunctional catalyst system for hydrogenolysis of glycerol under mild reaction conditions. *Catalysis Communications*, 6, 645–649.

Lahr, D. G. and Shanks, B. H. 2005. Effect of sulfur and temperature on ruthenium-catalyzed glycerol hydrogenolysis to glycols. *Journal of Catalysis*, 232, 386–394.

Lam, E., Majid, E., Leung, A. C., Chong, J. H., Mahmoud, K. A. and Luong, J. H. 2011. Synthesis of furfural from xylose by heterogeneous and reusable nafion catalysts. *ChemSusChem*, 4, 535–541.

Langan, P., Gnanakaran, S., Rector, K. D., Pawley, N., Fox, D. T., Cho, D. W. and Hammel, K. E. 2011. Exploring new strategies for cellulosic biofuels production. *Energy & Environmental Science*, 4, 3820–3833.

Lange, J.-P. 2001. Methanol synthesis: a short review of technology improvements. *Catalysis Today*, 64, 3–8.

Lange, J.-P. 2007. Lignocellulose conversion: an introduction to chemistry, process and economics. *Biofuels, Bioproducts and Biorefining*, 1, 39–48.

Lange, J.-P., Price, R., Ayoub, P. M., Louis, J., Petrus, L., Clarke, L. and Gosselink, H. 2010. Valeric biofuels: a platform of cellulosic transportation fuels. *Angewandte Chemie International Edition*, 49, 4479–4483.

Lestari, S., Mäki-Arvela, P., Beltramini, J., Lu, G. Q. M. and Murzin, D. Y. 2009a. Transforming triglycerides and fatty acids into biofuels. *ChemSusChem*, 2, 1109–1119.

Lestari, S., Mäki-Arvela, P. I., Bernas, H., Simakova, O., Sjöholm, R., Beltramini, J., Lu, G. Q. M., Myllyoja, J., Simakova, I. and Murzin, D. Y. 2009b. Catalytic deoxygenation of stearic acid in a continuous reactor over a mesoporous carbon-supported Pd catalyst. *Energy & Fuels*, 23, 3842–3845.

Leung, D. Y. C., Wu, X. and Leung, M. K. H. 2010. A review on biodiesel production using catalyzed transesterification. *Applied Energy*, 87, 1083–1095.

Li, C., Wang, Q. and Zhao, Z. K. 2008. Acid in ionic liquid: an efficient system for hydrolysis of lignocellulose. *Green Chemistry*, 10, 177–182.

Lichtenthaler, F. W. and Peters, S. 2004. Carbohydrates as green raw materials for the chemical industry. *Comptes Rendus Chimie*, 7, 65–90.

Lima, S., Antunes, M. M., Pillinger, M. and Valente, A. A. 2011. Ionic liquids as tools for the acid-catalyzed hydrolysis/dehydration of saccharides to furanic aldehydes. *ChemCatChem*, 3, 1686–1706.

Liu, X.-M., Lu, G. Q., Yan, Z.-F. and Beltramini, J. 2003. Recent advances in catalysts for methanol synthesis via hydrogenation of CO and $CO_2$. *Industrial & Engineering Chemistry Research*, 42, 6518–6530.

Lopez Granados, M., Zafra Poves, M. D., Martin Alonso, D., Mariscal, R., Cabello Galisteo, F., Moreno-Tost, R., Santamaria, J. and Fierro, J. L. G. 2007. Biodiesel from sunflower oil by using activated calcium oxide. *Applied Catalysis B: Environmental*, 73, 317–326.

Lopez Granados, M., Martin Alonso, D., Alba-Rubio, A. C., Mariscal, R., Ojeda, M. and Brettes, P. 2009a. Transesterification of triglycerides by CaO: increase of the reaction rate by biodiesel addition. *Energy & Fuels*, 23, 2259–2263.

Lopez Granados, M., Martin Alonso, D., Sadaba, I., Mariscal, R. and Ocon, P. 2009b. Leaching and homogeneous contribution in liquid phase reaction catalysed by solids: the case of triglycerides methanolysis using CaO. *Applied Catalysis B: Environmental*, 89, 265–272.

Lopez Granados, M., Alba-Rubio, A. C., Vila, F., Martin Alonso, D. and Mariscal, R. 2010. Surface chemical promotion of Ca oxide catalysts in biodiesel production reaction by the addition of monoglycerides, diglycerides and glycerol. *Journal of Catalysis*, 276, 229–236.

Lopez Granados, M., Alba-Rubio, A. C., Sadaba, I., Mariscal, R., Mateos-Aparicio, I. and Heras, A. 2011. Poly(styrenesulphonic) acid: an active and reusable acid catalyst soluble in polar solvents. *Green Chemistry*, 13, 3203–3212.

Lovett, J. C., Hards, S., Clancy, J. and Snell, C. 2011. Multiple objectives in biofuels sustainability policy. *Energy & Environmental Science*, 4, 261–268.

Luo, C., Wang, S. and Liu, H. 2007. Cellulose conversion into polyols catalyzed by reversibly formed acids and supported ruthenium clusters in hot water. *Angewandte Chemie International Edition*, 46, 7636–7639.

Lv, D., Xu, M., Liu, X., Zhan, Z., Li, Z. and Yao, H. 2010. Effect of cellulose, lignin, alkali and alkaline earth metallic species on biomass pyrolysis and gasification. *Fuel Processing Technology*, 91, 903–909.

Ma, Y.-S., Chang, C.-N., Chiang, Y.-P., Sung, H.-F. and Chao, A. C. 2008. Photocatalytic degradation of lignin using $Pt/TiO_2$ as the catalyst. *Chemosphere*, 71, 998–1004.

Mäki-Arvela, P., Kubickova, I., Snåre, M., Eränen, K. and Murzin, D. Y. 2006. Catalytic deoxygenation of fatty acids and their derivatives. *Energy & Fuels*, 21, 30–41.

Mäki-Arvela, P. I., Rozmysłowicz, B., Lestari, S., Simakova, O., Eränen, K., Salmi, T. and Murzin, D. Y. 2011. Catalytic deoxygenation of tall oil fatty acid over palladium supported on mesoporous carbon. *Energy & Fuels*, 25, 2815–2825.

Mallat, T. and Baiker, A. 1995. Catalyst potential: a key for controlling alcohol oxidation in multiphase reactors. *Catalysis Today*, 24, 143–150.

Marchetti, J. M., Miguel, V. U. and Errazu, A. F. 2007. Heterogeneous esterification of oil with high amount of free fatty acids. *Fuel*, 86, 906–910.

Martin Alonso, D., Mariscal, R., Moreno-Tost, R., Zafra Poves, M. D. and Lopez Granados, M. 2007. Potassium leaching during triglyceride transesterification using $K/gamma-Al_2O_3$ catalysts. *Catalysis Communications*, 8, 2074–2080.

Martin Alonso, D., Lopez Granados, M., Mariscal, R. and Douhal, A. 2009a. Polarity of the acid chain of esters and transesterification activity of acid catalysts. *Journal of Catalysis*, 262, 18–26.

Martin Alonso, D., Mariscal, R., Lopez Granados, M. and Maireles-Torres, P. 2009b. Biodiesel preparation using Li/CaO catalysts: activation process and homogeneous contribution. *Catalysis Today*, 143, 167–171.

Martin Alonso, D., Bond, J. Q. and Dumesic, J. A. 2010a. Catalytic conversion of biomass to biofuels. *Green Chemistry*, 12, 1493–1513.

Martin Alonso, D., Vila, F., Mariscal, R., Ojeda, M., Lopez Granados, M. and Santamaria-Gonzalez, J. 2010b. Relevance of the physicochemical properties of CaO catalysts for the methanolysis of triglycerides to obtain biodiesel. *Catalysis Today*, 158, 114–120.

Martin Alonso, D., Bond, J. Q., Serrano-Ruiz, J. C. and Dumesic, J. A. 2010c. Production of liquid hydrocarbon transportation fuels by oligomerization of biomass-derived C9 alkenes. *Green Chemistry*, 12, 992–999.

Mehdi, H., Fábos, V., Tuba, R., Bodor, A., Mika, L. T. and Horváth, I. T. 2008. Integration of homogeneous and heterogeneous catalytic processes for a multi-step conversion of biomass: from sucrose to levulinic acid, γ-valerolactone, 1,4-pentanediol, 2-methyl-tetrahydrofuran, and alkanes. *Topics in Catalysis*, 48, 49–54.

Meher, L., Vidyasagar, D. and Naik, S. 2006. Technical aspects of biodiesel production by transesterification – a review. *Renewable and Sustainable Energy Reviews*, 10, 248–268.

Miller, J. E., Evans, L., Littlewolf, A. and Trudell, D. E. 1999. Batch microreactor studies of lignin and lignin model compound depolymerization by bases in alcohol solvents. *Fuel*, 78, 1363–1366.

Miyazawa, T., Kusunoki, Y., Kunimori, K. and Tomishige, K. 2006. Glycerol conversion in the aqueous solution under hydrogen over Ru/C + an ion-exchange resin and its reaction mechanism. *Journal of Catalysis*, 240, 213–221.

Miyazawa, T., Koso, S., Kunimori, K. and Tomishige, K. 2007. Development of a Ru/C catalyst for glycerol hydrogenolysis in combination with an ion-exchange resin. *Applied Catalysis A: General*, 318, 244–251.

Mokrani, T. and Scurrell, M. 2009. Gas conversion to liquid fuels and chemicals: the methanol route – catalysis and processes development. *Catalysis Reviews*, 51, 1–145.

Morales-Delarosa, S., Campos-Martin, J. M. and Fierro, J. L. G. 2012. High glucose yields from the hydrolysis of cellulose dissolved in ionic liquids. *Chemical Engineering Journal*, 181–182, 538–541.

Mortensen, P. M., Grunwaldt, J. D., Jensen, P. A., Knudsen, K. G. and Jensen, A. D. 2011. A review of catalytic upgrading of bio-oil to engine fuels. *Applied Catalysis A: General*, 407, 1–19.

Murat Sen, S., Henao, C. A., Braden, D. J., Dumesic, J. A. and Maravelias, C. T. 2012. Catalytic conversion of lignocellulosic biomass to fuels: process development and technoeconomic evaluation. *Chemical Engineering Science*, 67, 57–67.

Murzin, D. and Simakova, I. 2011. Catalysis in biomass processing. *Catalysis in Industry*, 3, 218–249.

Nenkova, S., Vasileva, T. and Stanulov, K. 2008. Production of phenol compounds by alkaline treatment of technical hydrolysis lignin and wood biomass. *Chemistry of Natural Compounds*, 44, 182–185.

Ojeda, M., Granados, M. L., Rojas, S., Terreros, P., García-García, F. J. and Fierro, J. L. G. 2004. Manganese-promoted Rh/Al$_2$O$_3$ for C2-oxygenates synthesis from syngas: effect of manganese loading. *Applied Catalysis A: General*, 261, 47–55.

Okuhara, T. 2002. Water-tolerant solid acid catalysts. *Chemical Reviews*, 102, 3641–3666.

Olivier-Bourbigou, H., Magna, L. and Morvan, D. 2010. Ionic liquids and catalysis: recent progress from knowledge to applications. *Applied Catalysis A: General*, 373, 1–56.

Ott, L., Bicker, M. and Vogel, H. 2006. Catalytic dehydration of glycerol in sub- and supercritical water: a new chemical process for acrolein production. *Green Chemistry*, 8, 214–220.

Pepper, J. M. and Fleming, R. W. 1978. Lignin and related compounds. V. The hydrogenolysis of aspen wood lignin using rhodium-on-charcoal as catalyst. *Canadian Journal of Chemistry*, 56, 896–898.

Pepper, J. M. and Hibbert, H. 1948. Studies on lignin and related compounds. LXXXVII. High pressure hydrogenation of maple wood. *Journal of the American Chemical Society*, 70, 67–71.

Pepper, J. M. and Lee, Y. W. 1969. Lignin and related compounds. I. A comparative study of catalysts for lignin hydrogenolysis. *Canadian Journal of Chemistry*, 47, 723–727.

Pepper, J. M. and Supathna, P. 1978. Lignin and related compounds. VI. A study of variables affecting the hydrogenolysis of spruce wood lignin using a rhodium-on-charcoal catalyst. *Canadian Journal of Chemistry*, 56, 899–902.

Perego, C. and Bianchi, D. 2010. Biomass upgrading through acid–base catalysis. *Chemical Engineering Journal*, 161, 314–322.

Pérez-Alonso, F. J., Herranz, T., Rojas, S., Ojeda, M., López Granados, M., Terreros, P., Fierro, J. L. G., Gracia, M. and Gancedo, J. R. 2007. Evolution of the bulk structure and surface species on Fe-Ce catalysts during the Fischer–Tropsch synthesis. *Green Chemistry*, 9, 663–670.

Pfeifer, C., Rauch, R. and Hofbauer, H. 2004. In-bed catalytic tar reduction in a dual fluidized bed biomass steam gasifier. *Industrial & Engineering Chemistry Research*, 43, 1634–1640.

Phillips, S. D., Tarud, J. K., Biddy, M. J. and Dutta, A. 2011a. Gasoline from woody biomass via thermochemical gasification, methanol synthesis, and methanol-to-gasoline technologies: a technoeconomic analysis. *Industrial & Engineering Chemistry Research*, 50, 11734–11745.

Phillips, S. D., Tarud, J. K., Biddy, M. J. and Dutta, A. 2011b. Gasoline from woody biomass via thermochemical gasification, methanol synthesis, and methanol-to-gasoline technologies: a technoeconomic analysis (Addition/Correction). *Industrial & Engineering Chemistry Research*, 50, 14226.

Pinkert, A., Marsh, K. N., Pang, S. and Staiger, M. P. 2009. Ionic liquids and their interaction with cellulose. *Chemical Reviews*, 109, 6712–6728.

Porta, F. and Prati, L. 2004. Selective oxidation of glycerol to sodium glycerate with gold-on-carbon catalyst: an insight into reaction selectivity. *Journal of Catalysis*, 224, 397–403.

Portjanskaja, E. and Preis, S. 2007. Aqueous photocatalytic oxidation of lignin: the influence of mineral admixtures. *International Journal of Photoenergy*, 2007, Article 76730.

Portjanskaja, E., Stepanova, K., Klauson, D. and Preis, S. 2009. The influence of titanium dioxide modifications on photocatalytic oxidation of lignin and humic acids. *Catalysis Today*, 144, 26–30.

Quyn, D. M., Wu, H. and Li, C.-Z. 2002. Volatilisation and catalytic effects of alkali and alkaline earth metallic species during the pyrolysis and gasification of

Victorian brown coal. Part I. Volatilisation of Na and Cl from a set of NaCl-loaded samples. *Fuel*, 81, 143–149.

Ramayya, S., Brittain, A., Dealmeida, C., Mok, W. and Antal Jr., M. J. 1987. Acid-catalysed dehydration of alcohols in supercritical water. *Fuel*, 66, 1364–1371.

Rinaldi, R. and Schüth, F. 2009a. Acid hydrolysis of cellulose as the entry point into biorefinery schemes. *ChemSusChem*, 2, 1096–1107.

Rinaldi, R. and Schüth, F. 2009b. Design of solid catalysts for the conversion of biomass. *Energy & Environmental Science*, 2, 610–626.

Rinaldi, R., Palkovits, R. and Schüth, F. 2008. Depolymerization of cellulose using solid catalysts in ionic liquids. *Angewandte Chemie International Edition*, 47, 8047–8050.

Rinaldi, R., Engel, P., Büchs, J., Spiess, A. C. and Schüth, F. 2010. An integrated catalytic approach to fermentable sugars from cellulose. *ChemSusChem*, 3, 1151–1153.

Roman-Leshkov, Y., Chheda, J. N. and Dumesic, J. A. 2006. Phase modifiers promote efficient production of hydroxymethylfurfural from fructose. *Science*, 312, 1933–1937.

Roman-Leshkov, Y., Barrett, C. J., Liu, Z. Y. and Dumesic, J. A. 2007. Production of dimethylfuran for liquid fuels from biomass-derived carbohydrates. *Nature*, 447, 982–985.

Rubin, E. M. 2008. Genomics of cellulosic biofuels. *Nature*, 454, 841–845.

Ruppert, A. M., Weinberg, K. and Palkovits, R. 2012. Hydrogenolysis goes bio: from carbohydrates and sugar alcohols to platform chemicals. *Angewandte Chemie International Edition*, 51, 2564–2601.

Sádaba, I., Ojeda, M., Mariscal, R., Fierro, J. L. G. and Granados, M. L. 2011a. Catalytic and structural properties of co-precipitated Mg-Zr mixed oxides for furfural valorization via aqueous aldol condensation with acetone. *Applied Catalysis B: Environmental*, 101, 638–648.

Sádaba, I., Ojeda, M., Mariscal, R., Richards, R. and Granados, M. L. 2011b. Mg-Zr mixed oxides for aqueous aldol condensation of furfural with acetone: effect of preparation method and activation temperature. *Catalysis Today*, 167, 77–83.

Saito, M. and Murata, K. 2004. Development of high performance Cu/ZnO-based catalysts for methanol synthesis and the water-gas shift reaction. *Catalysis Surveys from Asia*, 8, 285–294.

Sanders, J. P. M., Clark, J. H., Harmsen, G. J., Heeres, H. J., Heijnen, J. J., Kersten, S. R. A., Van Swaaij, W. P. M. and Moulijn, J. A. 2012. Process intensification in the future production of base chemicals from biomass. *Chemical Engineering and Processing: Process Intensification*, 51, 117–136.

Schuster, L. and Eggersdorfer, M. 1996. Process for the preparation of 1,2-propanediol. US Patent US5616817 A.

Schwenk, E., Gehrke, M. and Aichner, F. 1933. Production of acrolein. US Patent US1916743 A.

Serrano-Ruiz, J. C., Wang, D. and Dumesic, J. A. 2010. Catalytic upgrading of levulinic acid to 5-nonanone. *Green Chemistry*, 12, 574–577.

Shanks, B. H. 2010. Conversion of biorenewable feedstocks: new challenges in heterogeneous catalysis. *Industrial and Engineering Chemistry Research*, 49, 10212–10217.

Simakova, I., Simakova, O., Mäki-Arvela, P., Simakov, A., Estrada, M. and Murzin, D. Y. 2009. Deoxygenation of palmitic and stearic acid over supported Pd

catalysts: effect of metal dispersion. *Applied Catalysis A: General*, 355, 100–108.
Simonetti, D. A. and Dumesic, J. A. 2009. Catalytic production of liquid fuels from biomass-derived oxygenated hydrocarbons: catalytic coupling at multiple length scales. *Catalysis Reviews*, 51, 441–484.
Smits, P. C. C., Kuster, B. F. M., Van Der Wiele, K. and Van Der Baan, S. 1987. Lead modified platinum on carbon catalyst for the selective oxidation of (2–) hydroxycarbonic acids, and especially polyhydroxycarbonic acids to their 2-keto derivatives. *Applied Catalysis*, 33, 83–96.
Snåre, M., Kubičková, I., Mäki-Arvela, P., Eränen, K. and Murzin, D. Y. 2006. Heterogeneous catalytic deoxygenation of stearic acid for production of biodiesel. *Industrial & Engineering Chemistry Research*, 45, 5708–5715.
Snåre, M., Kubičková, I., Mäki-Arvela, P., Chichova, D., Eränen, K. and Murzin, D. Y. 2008. Catalytic deoxygenation of unsaturated renewable feedstocks for production of diesel fuel hydrocarbons. *Fuel*, 87, 933–945.
Stark, A. 2011. Ionic liquids in the biorefinery: a critical assessment of their potential. *Energy & Environmental Science*, 4, 19–32.
Stöcker, M. 2008. Biofuels and biomass-to-liquid fuels in the biorefinery: catalytic conversion of lignocellulosic biomass using porous materials. *Angewandte Chemie International Edition*, 47, 9200–9211.
Sun, N., Rodríguez, H., Rahman, M. and Rogers, R. D. 2011. Where are ionic liquid strategies most suited in the pursuit of chemicals and energy from lignocellulosic biomass? *Chemical Communications*, 47, 1405–1421.
Sutton, D., Kelleher, B. and Ross, J. R. H. 2001. Review of literature on catalysts for biomass gasification. *Fuel Processing Technology*, 73, 155–173.
Świerczyński, D., Courson, C., Bedel, L., Kiennemann, A. and Guille, J. 2006. Characterization of Ni–Fe/MgO/Olivine catalyst for fluidized bed steam gasification of biomass. *Chemistry of Materials*, 18, 4025–4032.
Thring, R. W. and Breau, J. 1996. Hydrocracking of solvolysis lignin in a batch reactor. *Fuel*, 75, 795–800.
Timokhin, B. V., Baransky, V. A. and Eliseeva, G. D. 1999. Levulinic acid in organic synthesis. *Russian Chemical Reviews*, 68, 73–84.
Toda, M., Takagaki, A., Okamura, M., Kondo, J. N., Hayashi, S., Domen, K. and Hara, M. 2005. Green chemistry: biodiesel made with sugar catalyst. *Nature*, 438, 178.
Tong, X., Ma, Y. and Li, Y. 2010. Biomass into chemicals: conversion of sugars to furan derivatives by catalytic processes. *Applied Catalysis A: General*, 385, 1–13.
Upare, P. P., Lee, J. M., Hwang, Y. K., Hwang, D. W., Lee, J. H., Halligudi, S. B., Hwang, J. S. and Chang, J. S. 2011. Direct hydrocyclization of biomass-derived levulinic acid to 2-methyltetrahydrofuran over nanocomposite copper/silica catalysts. *ChemSusChem*, 4, 1749–1752.
Vanoye, L., Fanselow, M., Holbrey, J. D., Atkins, M. P. and Seddon, K. R. 2009. Kinetic model for the hydrolysis of lignocellulosic biomass in the ionic liquid, 1-ethyl-3-methyl-imidazolium chloride. *Green Chemistry*, 11, 390–396.
Vieville, C., Yoo, J. W., Pelet, S. and Mouloungui, Z. 1998. Synthesis of glycerol carbonate by direct carbonation of glycerol in supercritical $CO_2$ in the presence of zeolites and ion exchange resins. *Catalysis Letters*, 56, 245–247.
Vila, F., López Granados, M., Ojeda, M., Fierro, J. L. G. and Mariscal, R. 2012. Glycerol hydrogenolysis to 1,2-propanediol with $Cu/\gamma\text{-}Al_2O_3$: effect of the activation process. *Catalysis Today*, 187, 122–128.

Voitl, T. and Von Rohr, P. R. 2008. Oxidation of lignin using aqueous polyoxometalates in the presence of alcohols. *ChemSusChem*, 1, 763–769.

Wahyudiono, Kanetake, T., Sasaki, M. and Goto, M. 2007. Decomposition of a lignin model compound under hydrothermal conditions. *Chemical Engineering & Technology*, 30, 1113–1122.

Wahyudiono, Sasaki, M. and Goto, M. 2009. Conversion of biomass model compound under hydrothermal conditions using batch reactor. *Fuel*, 88, 1656–1664.

Waldmann, H. and Petrū, F. 1950. Über die Dehydratisierung von Alkoholen mittels Phthalsäureanhydrids (1. Mitteil.). *Chemische Berichte*, 83, 287–291.

Wang, P., Yu, H., Zhan, S. and Wang, S. 2011. Catalytic hydrolysis of lignocellulosic biomass into 5-hydroxymethylfurfural in ionic liquid. *Bioresource Technology*, 102, 4179–4183.

Wang, T., Chang, J., Lv, P. and Zhu, J. 2004. Novel catalyst for cracking of biomass tar. *Energy & Fuels*, 19, 22–27.

Watanabe, M., Iida, T., Aizawa, Y., Aida, T. M. and Inomata, H. 2007. Acrolein synthesis from glycerol in hot-compressed water. *Bioresource Technology*, 98, 1285–1290.

Werpy, T. A., Frye, J. G. J., Zacher, A. H. and Miller, D. J. 2003. Hydrogenolysis of 6-carbon sugars and other organic compounds. WO patent application, WO 2003035582.

West, R. M., Liu, Z. Y., Peter, M. and Dumesic, J. A. 2008. Liquid alkanes with targeted molecular weights from biomass-derived carbohydrates. *ChemSusChem*, 1, 417–424.

Wildschut, J., Arentz, J., Rasrendra, C. B., Venderbosch, R. H. and Heeres, H. J. 2009. Catalytic hydrotreatment of fast pyrolysis oil: model studies on reaction pathways for the carbohydrate fraction. *Environmental Progress & Sustainable Energy*, 28, 450–460.

Wu, H., Quyn, D. M. and Li, C.-Z. 2002. Volatilisation and catalytic effects of alkali and alkaline earth metallic species during the pyrolysis and gasification of Victorian brown coal. Part III. The importance of the interactions between volatiles and char at high temperature. *Fuel*, 81, 1033–1039.

Yan, N., Zhao, C., Luo, C., Dyson, P. J., Liu, H. and Kou, Y. 2006. One-step conversion of cellobiose to C6-alcohols using a ruthenium nanocluster catalyst. *Journal of the American Chemical Society*, 128, 8714–8715.

Yang, C., Ma, Z., Zhao, N., Wei, W., Hu, T. and Sun, Y. 2006. Methanol synthesis from $CO_2$-rich syngas over a $ZrO_2$ doped CuZnO catalyst. *Catalysis Today*, 115, 222–227.

Zakrzewska, M. E., Bogel-Łukasik, E. and Bogel-Łukasik, R. 2010. Ionic liquid-mediated formation of 5-hydroxymethylfurfural – a promising biomass-derived building block. *Chemical Reviews*, 111, 397–417.

Zakzeski, J., Bruijnincx, P. C. A., Jongerius, A. L. and Weckhuysen, B. M. 2010. The catalytic valorization of lignin for the production of renewable chemicals. *Chemical Reviews*, 110, 3552–3599.

Zhang, J., Deng, H. and Lin, L. 2009. Wet aerobic oxidation of lignin into aromatic aldehydes catalysed by a perovskite-type oxide: $LaFe_1\text{-}xCuxO_3$ (x = 0, 0.1, 0.2). *Molecules*, 14, 2747–2757.

Zhang, Q., Chang, J., Wang, T. and Xu, Y. 2007. Review of biomass pyrolysis oil properties and upgrading research. *Energy Conversion and Management*, 48, 87–92.

Zhao, C., Kou, Y., Lemonidou, A. A., Li, X. and Lercher, J. A. 2009. Highly selective catalytic conversion of phenolic bio-oil to alkanes. *Angewandte Chemie International Edition*, 48, 3987–3990.

Zhao, H., Holladay, J. E., Brown, H. and Zhang, Z. C. 2007. Metal chlorides in ionic liquid solvents convert sugars to 5-hydroxymethylfurfural. *Science*, 316, 1597–1600.

Zhao, Z., Kuhn, J. N., Felix, L. G., Slimane, R. B., Choi, C. W. and Ozkan, U. S. 2008. Thermally impregnated Ni–olivine catalysts for tar removal by steam reforming in biomass gasifiers. *Industrial & Engineering Chemistry Research*, 47, 717–723.

Zheng, M.-Y., Wang, A.-Q., Ji, N., Pang, J.-F., Wang, X.-D. and Zhang, T. 2010. Transition metal–tungsten bimetallic catalysts for the conversion of cellulose into ethylene glycol. *ChemSusChem*, 3, 63–66.

Zhou, C. H., Beltramini, J. N., Fan, Y. X. and Lu, G. Q. 2008. Chemoselective catalytic conversion of glycerol as a biorenewable source to valuable commodity chemicals. *Chemical Society Reviews*, 37, 527–549.

Zhou, C.-H., Xia, X., Lin, C.-X., Tong, D.-S. and Beltramini, J. 2011. Catalytic conversion of lignocellulosic biomass to fine chemicals and fuels. *Chemical Society Reviews*, 40, 5588–5617.

Zhu, S., Wu, Y., Chen, Q., Yu, Z., Wang, C., Jin, S., Ding, Y. and Wu, G. 2006. Dissolution of cellulose with ionic liquids and its application: a mini-review. *Green Chemistry*, 8, 325–327.

Zhu, Y., Kong, Z. N., Stubbs, L. P., Lin, H., Shen, S., Anslyn, E. V. and Maguire, J. A. 2010. Conversion of cellulose to hexitols catalyzed by ionic liquid-stabilized ruthenium nanoparticles and a reversible binding agent. *ChemSusChem*, 3, 67–70.

# 7
Enzymatic processes and enzyme development in biorefining

S. A. TETER, K. BRANDON SUTTON and B. EMME, Novozymes, USA

DOI: 10.1533/9780857097385.1.199

**Abstract**: Improvements in enzyme cocktails for converting biomass have driven down costs for production of bioproducts and have allowed commercialization of the first cellulosic biorefineries. Innovations in biochemical engineering have resulted in cocktails that require dramatically less enzyme for a given level of biomass conversion than were required just a few years ago. Developments in the enzymes themselves must be accompanied by optimization of the biochemical conversion process to allow for the lowest possible product costs to be achieved. This chapter describes both enzyme innovations and critical themes in process optimization.

**Key words**: cellulase, hemicellulase, process optimization, biomass biorefinery.

## 7.1 Introduction

Over the past decade, there has been immense interest in alleviating our dependence on oil. Biobased products, produced from abundant and renewable plant resources, are favored by governments worldwide as promising alternatives to current petroleum-derived liquid fuels and numerous chemicals. It is projected that biofuels in total have the potential to meet 30% or more of gasoline demand (Somerville *et al.*, 2010), reducing dependency on imported oil in many countries worldwide, helping to keep oil prices under control and substantially decreasing greenhouse gas (GHG) emissions. As cellulosic biofuels and chemicals permeate through the marketplace, significant positive benefits to rural economies and career creation are forecast as well (Boyle and Labastida, 2012).

In a relatively short period of time, technologies for production of a broad range of biobased products have been developed. Liquid fuel production has long been a focus due to the large market potential for petroleum replacement in the transportation sector. Ethanol has the longest history and is the first commercially produced liquid fuel derived from biomass (PRNewswire/, 2010). Other fuels which are currently being commercialized include butanol, as well as fuels based on long-chain hydrocarbons and isoprenoid derived fuels (Fortman *et al.*, 2008; Wackett, 2008). Given that cost-competitiveness for low-margin products like liquid fuels is a critical issue, some have argued that higher value products may allow for earlier commercialization. Catalogues of potential 'building block' chemicals which could be produced from biomass have been published (see, e.g., Werpy and Petersen, 2004) and a number of microbially produced replacements for commodity chemicals that are currently derived from petroleum have been successfully produced (Fortman *et al.*, 2008).

Transforming how fuels and chemicals are produced is an immense challenge. Creation of a new industry which converts lignocellulosic plant matter into sugars, then transforms this biological currency into fuels and chemicals, will take time. While the focus of this chapter is the underlying technical accomplishments behind biomass to sugar conversion, other hurdles remain. Progress has been made in addressing other milestones, including expansion of infrastructure for delivery of fuels, ensuring that fuels can easily be accommodated in engines of existing cars and trucks, providing means for financing of biorefineries in a constrained global economy, and securing a stable market for the first advanced biofuel and chemical producers. Attention to these issues will ensure a more level playing field between biobased products and the entrenched petroleum-derived products which have enjoyed a century of subsidies and political support.

For biorefineries to be viable, the costs of production must be cost-competitive with petroleum-based products both in processing and capital costs. Production of low cost sugars is critical along with microbial transformation of those sugars to product at high yield in simple, low cost fermentation schemes. Advances in biomass conversion technology will generate more positive economic balance sheets across a range of process scenarios. However, maximizing the benefit that can come from bioconversion technologies requires holistic evaluation of process technology, including unit steps for transformation of sugars to product. This integrated process development, toward reduction in total costs to find the 'sweet spot' for total production costs, will be a theme of this chapter. The focus will be on understanding how one technical achievement, the biochemical improvement in enzyme performance, can best be integrated into a designed process.

## 7.2 Biochemical conversion

The essential steps for biochemical conversion of lignocelluloses to sugars are shown in Fig. 7.1. Sources of biomass are diverse (Somerville *et al.*, 2010), with the ideal feedstock depending on the geographical location. Categories of lignocellulosic material include agricultural residues, wood, industrial waste, energy crops, and municipal wastes. Following biomass harvest and transport to the biorefinery, the process begins with particle size reduction to enable the second step, thermochemical pretreatment, to be more efficient. Pretreatment is required to expose and disrupt the recalcitrant, crystalline cell wall matrix. A wide variety of pretreatment methodologies have been described, and impact of the selected process on downstream unit steps will be discussed. Following pretreatment, biomass is prepared for enzymatic hydrolysis. Hydrolysis (saccharification) is accomplished by enzymatic deconstruction of cellulose and hemicellulose to sugars. Sugars are then fermented or transformed through chemical conversion to product. Finally the product is recovered from spent hydrolysate streams, entailing separation from aqueous solutions containing a range of biomass-derived chemicals.

The highly simplified schematic of unit process steps in Fig. 7.1 forms the basis of refining complex plant biomass to bioproducts, but as will be discussed in this overview, the range of process options are quite broad, and there are many permutations of the basic process which are being explored and commercialized.

Plant cell wall polysaccharides are the most abundant of all biological materials on earth, which makes these energy-rich sugar polymers an attractive renewable feedstock for biorefineries. Plant cell walls form the bulk of lignocellulosic biomass. These serve structural and protective roles for plants and have evolved over hundreds of millions years to be highly recalcitrant and resistant to degradation. It is worth considering that microbes have co-evolved enzymatic systems that allow them to access

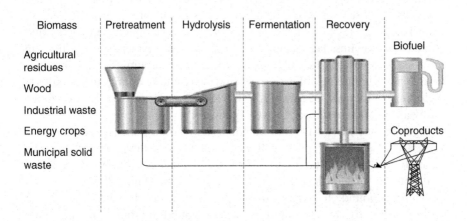

*7.1* Biochemical conversion of biomass to product. The basic unit process steps for conversion of biomass materials to products such as biofuel and electricity (coproduct) are shown.

plant nutrients. The complete breakdown of most lignocellulose occurs over weeks to years in a typical natural environment through microbial action, which attests to the biostability of plant cell walls. This recalcitrance poses a challenge to the biorefinery where material must be completely converted to monomeric sugars in hours to days.

Somerville *et al.* (2004) provide an excellent overview of plant cell wall structure and functional principles. In particular, this review gives a structural overview of polysaccharides in a model plant cell wall, as well as presenting 'average' structures for cell wall polysaccharides in the supplementary online materials which are useful for understanding the diversity of hemicellulose compositions in different types of plants. Of interest to biorefiners are the types of monomeric carbohydrates that make up the major polysaccharides. The cellulose fraction comprises cellobiose units (two glucose units) repeated in beta-1,4-linked chains. These chains are packed together into microfibrils, which are compact and highly stable due to extensive hydrogen bonding and van der Waals interactions. They are generally 3–6 nm across, each containing 36 glucan chains with regions of high crystallinity. The compactness and stability of the fibers makes them resistant to microbial deconstruction and lend strength to the cell wall. Surrounding the cellulose microfibrils are hemicellulose polymers, which are structurally and compositionally diverse in the plant kingdom. Unlike cellulose, which is made up of only one hexose sugar (glucose), hemicelluloses are branched, containing pentoses (xylose, arabinose), a range of hexose sugars (mannose, glucose, galactose), as well as other components such as uronic acids. An exemplary hemicellulose is glucuronoarabinoxylan, a common structure in grasses, which is abundant in many agricultural residues. Glucuronoarabinoxylan consisting of beta-1,4-linked xylose as a 'xylan backbone' may be decorated with secondary branches such as arabinose, ferulic acid, acetyl, glucoronic acid, and galactosyl moieties. Hemicellulose fibers are not typically crystalline in nature. However, the heterogeneity of chemical building blocks demands equally heterogeneous collections of enzymes for their complete breakdown including numerous transporters for uptake and entry into metabolic pathways for use by fermentative organisms.

Lignin is the next most abundant material in biomass and is predominately present in 'secondary' plant cell walls. Lignin is a highly heterogeneous aromatic polymer, built of various syringyl, guaiacyl, or other hydroxyphenyl units linked by a wide array of chemical bonds. Hemicellulose is often covalently bound to lignin, partially preventing breakdown of the carbohydrate to sugars and it is also a potent inhibitor of enzymatic breakdown due to its 'sticky' nature; protein components adsorb non-specifically to lignin. While oxidoreductase enzymes have evolved to catalyze the decomposition of lignin, the use of lignin building blocks in the biorefinery has lagged behind the harvest of sugars from cellulose and hemicelluloses.

Generally speaking, it is the crystalline nature of the cellulose and the inhibitory properties of lignin which make plant biomass resistant to biological attack. Other components that make up biomass are less abundant, and are often not specifically targeted as sources for biomaterials. These include lipid, proteins, and other carbohydrate polymers such as mannans and pectin.

### 7.2.1 Process integration: pretreatment/hydrolysis interface

An effective pretreatment alters or removes impediments to hydrolysis, both structural and compositional. The result is an improved rate of enzyme hydrolysis and increased yield of fermentable sugars. The key technical challenges in pretreatment optimization are to significantly disrupt cell wall structure to allow for enzyme access, while minimizing degradation and inhibitor formation, without driving up capital and process costs. At the interface of pretreatment and enzyme science, researchers have aimed to understand the fundamental properties of biomass that is 'well-pretreated.' A rigorous understanding of the physical and chemical characteristics of optimally pretreated biomass may allow rational design of both improved pretreatment technologies and improved enzymes. Researchers are not in agreement about what factor(s) correlate most with increased enzyme accessibility. Some consider that the increase in surface area and 'pore' size of biomass is the property which best correlates with enzyme digestibility (Kumar and Wyman, 2009a, 2009b; Chandra et al., 2007). Others cite reduced cellulose crystallinity as the controlling factor. Since pretreatment serves as a type of fractionation by removing and/or relocating lignin as well as disrupting the hemicellulose (Kim and Holtzapple, 2006; Liu and Wyman, 2005), a number of authors have noted a direct link between enzymatic accessibility and degree of hemicellulose and lignin removal during pretreatment. However, it has also been observed that a near complete removal of hemicellulose and lignin can have a detrimental impact on enzymatic accessibility (Ishizawa et al., 2009). Other frequently addressed parameters include fiber size or degree of polymerization (DP) of the cellulose.

As the goal of the combined pretreatment and hydrolysis steps is to convert as much of the polymeric sugar to fermentable sugar as possible, care must be taken during these steps not to degrade or irreversibly transform the sugars, as they will then be lost, representing a costly reduction in product per ton of starting material. The impact of their loss on process cost depends on the feedstock; for less expensive feedstocks, a process may be optimized wherein some of the sugars may be sacrificed due to inaccessibility following a less severe pretreatment, or at the other extreme, due to degradation during a particularly harsh pretreatment. Much consideration is taken in optimizing the pretreatment conditions. Commonly, a composite design of experiments is used to explore variables such as catalyst concentration, temperature, and duration of process, with the aim of balancing resultant accessibility to enzyme, while minimizing loss of sugar potential due to degradation.

Pretreated slurries have physical and chemical characteristics that can prevent the enzyme proteins from catalyzing the depolymerization of the cellulose and hemicellulose into monomeric sugars for further fermentation. Pretreatment process design also impacts the performance of downstream fermentation processes. A critical area of pretreatment research seeks to minimize inhibitors of enzymatic hydrolysis and fermentation.

Many leading pretreatment technologies take advantage of acid or base catalysts to accelerate biomass deconstruction. The extremes of the pH scale are not well tolerated by the majority of microorganisms and enzymes, and thus pH modification is often required following pretreatment in order to bring biomass slurries to an appropriate pH for optimal enzymatic conversion and sugar fermentation. Commercial fungal cellulases currently on the market have pH optima in the range of pH 4–6. When adding an alkaline chemical to an acid pretreated slurry, or acidic chemical to an alkaline pretreatment, formation of salts is unavoidable. Direct costs for neutralization of chemical inputs, and also for salt disposal that result from the need to manage waste streams at the end of the process flow are key cost drivers in pretreatment. Salts that end up in the soluble fraction after fermentation must be dealt with in wastewater treatment steps, and salts that remain in the insoluble fraction of residual material after fermentation may produce contaminant emissions which are regulated and pose a treatment cost. One example is sulfur oxide, as in the case of a sulfuric acid enriched process.

In addition to these process costs, accumulation of salts in process streams can be inhibitory to enzymes during hydrolysis and to the fermenting organism due to osmotic stress. Salt formation from pH adjustments is a burden to the point that some biorefineries have been designed to avoid using acid as a pretreatment technology as they require greater neutralization chemical inputs.

In addition to the salts generated during pretreatment, there are other compounds that are well known as inhibitors in downstream processes, particularly as biological inhibitors of commonly used fermentation organisms, such as furans, weak acids, and phenolic compounds.

The pretreatment is one of most expensive processing steps in the biomass conversion pathway, and thus has a great potential for improvement. Some of the operating expenses have been touched upon in the above discussion, but the capital costs associated with the specialized equipment needed to achieve high temperatures and pressures and resist corrosion in the presence of aggressive catalysts comprise a major part of the combined biorefinery capital expense. Sulfuric acid catalyzed pretreatments in particular have been highlighted as being associated with expensive capital inputs due to the need to build reactors of exotic metals that withstand degradation by strong acid. For other processes, the need to recycle expensive catalysts provides a driver for high pretreatment costs.

Reducing inhibitor generation and reducing the chemical loading in pretreatment can provide a multitude of benefits. Maintaining the effectiveness of a pretreatment in opening up the cell wall to enzymatic attack while decreasing catalyst load such that cheaper metallurgy can be used has been a recent focus of effort. Displacing chemical inputs may require more complex and robust enzymes, but the overall savings in chemical consumption and waste treatment and salt disposal are positive. Perhaps the most substantial benefit is seen in the fermentation process; the impact of lowered salts and decreased inhibitors improves the rates and yields of conversion of sugars to product. For this reason, convergent efforts in the industry have recently pushed to decrease the need for externally added catalysts in pretreatment (Chen *et al.*, 2012).

In addition to reducing the negative impact of inhibitors through improved pretreatments, a range of biological innovations has helped reduce the magnitude of the problem. On the enzyme front, screening for enzymes that are highly active in the presence of biomass-derived inhibitors such as soluble lignin has led to improved cocktails for use of 'real life' pretreatment slurries. On the fermentation front, organisms have been developed that are more tolerant to specific inhibitors like furfural and HMF (Geddes *et al.*, 2011). Multiple studies utilizing model substrates have shown that some inhibitors are synergistically detrimental in their action. Due to the synergism between different classes of inhibitors, evolutionary engineering and adaptation to biomass pretreatment slurries has been an attractive path for improving stress tolerance of organisms (Geddes *et al.*, 2011).

Different types of pretreatment lead to different enzyme requirements downstream. While dilute acid pretreatments effectively convert a majority of hemicellulose to monomeric sugars, alkali pretreatments leave a substantial amount of insoluble and/or soluble oligomeric xylan, with or without side chain branching. Many fermentation processes are unable to utilize soluble oligomeric sugars; for this reason, the degree of polymerization of soluble sugars is important, as enzymes may need to convert soluble oligomers to monomers.

In the case of dilute acid pretreated corn stover, insoluble hemicelluloses may be reduced to less than 5% of total insoluble solids in pretreated material. Interestingly, hydrolysis of this material still benefits from the presence of enzymes that can attack this insoluble hemicellulose.

Beta-xylosidases, which can convert soluble hemicellulose oligomers to monomers, are required at relatively low levels due to the low content of oligomerized hemicelluloses.

### 7.2.2 Enzymatic hydrolysis and product fermentation: process design

There are several viable hydrolysis and fermentation options available, each with benefits and drawbacks. The most economically viable process options select for configurations that maximize enzyme performance. Research is dedicated to understanding how feedstock, pretreatment, and fermentation methods holistically change the process economy with respect to hydrolytic conversion of biomass. Ultimately, each feedstock and pretreatment combination should be evaluated on an individual basis to determine the best process configuration to enable the industry. Fermentation of biomass hydrolysates to ethanol has been explored in much more detail than any other biobased product, and thus most of the process design discussion in this chapter focuses on ethanol as the end product.

A central consideration in deriving an optimal process is the impact of total solids (TS) loadings on overall process economy. A constant challenge to biomass-based biorefiners is the need for high ethanol titers at the end of fermentation; ethanol titers less than 5–6% in general can be considered cost prohibitive due to the power input needed for distillation to drive off the additional water. The simplest way to obtain high ethanol concentrations is to ensure high potential sugar content in upstream processes by maintaining the total solids concentration between 20 and 30% TS, or even higher if possible. However, at these high total and insoluble solids concentrations, efficient mixing is difficult. Unlike starch-based fermentation processes where relatively high solids consistencies can be achieved, lignocellulose slurries at high total solids have much higher viscosity due to their intrinsic high water retention properties. The first hours of hydrolysis require comparatively high energy input until the bulk viscosity 'breaks.' Sugar concentrations increase as hydrolysis progresses, introducing product inhibition to the existing enzyme inhibitors in the liquid phase; this often results in lower overall hydrolysis yields. To avoid these inhibitor sources, a dilute or lower total solids process could be coupled with an evaporation step or the addition of a concentrated syrup stream (e.g., from a first generation co-located starch biorefinery) to allow for sufficient sugar concentrations to reach ethanol titer targets. However, the capital and operational expenses must be weighed against the distillation cost savings.

Relative to the basic process configuration like the one in Fig. 7.1, process designs with greater and lesser complexity can be explored with respect to the impact of these designs on the enzymes and fermentation organisms. In general, increasing complexity, as illustrated here by considering the impact of splitting process streams, may open possibilities to allow for providing a more optimal environment to hydrolytic enzymes and fermentation organisms.

Solid–liquid separations are often considered at various stages in the process, as this can allow for detoxification, recycling of catalysts, and integration of other water streams. We will consider the impact of solid–liquid separation at two different stages of the process, though this procedure could be employed at several additional points in the process.

As was explained above, pretreatments that hydrolyze hemicellulose to a high degree often produce side products that are inhibitory both to enzymes and to organisms. With this in mind, it may be desirable to introduce a solid–liquid separation step following pretreatment to overcome problems with soluble inhibitors. Detoxification of soluble streams (liquor) generated from pretreatment can be performed, such as over-liming in the case of acid pretreatment, to remove inhibitory components such as furfural (Mohagheghi *et al.*, 2006). Separating the process stream at this stage also offers the potential to further reduce soluble inhibitors in the solid fraction by washing solids to varying levels. The advantage is a dramatic improvement in enzyme performance that is often observed following removal of soluble inhibitors, and an even more pronounced improvement in fermentability of resulting sugars from hydrolysis of the insoluble fraction. Streams can be recombined following hydrolysis of the insoluble fraction, or alternatively, splitting the process streams allows for an option to ferment pentose and hexose streams separately. This translates into shorter fermentation times and faster uptake rates for the individual sugars. The significant drawback to either of these strategies is the addition of water or salts into the process, diluting sugar titers. In all cases, the benefits gleaned from splitting the process stream must outweigh the significant costs of the solid–liquid separation combined with additional costs for downstream evaporation or similar concentration strategy.

Another point where solid–liquid separation is often considered is after the hydrolysis. Separation of lignin-enriched residues also allows for their potential use in high value co-products (Inyang *et al.*, 2010; Kim and Kadla, 2010), and may be necessary for effective recovery of soluble fermentation products. Unfortunately, separation of the very fine particles remaining after near complete enzyme hydrolysis is quite difficult, requiring the use of flocculants to aid separation, which increases operational costs. Loss of sugar potential is also an issue.

While splitting process streams increases the process complexity, consolidation has been an area of intense interest in the past few years. Enzymatic hydrolysis and fermentation are areas which may be consolidated to varying extents.

*Separate hydrolysis and fermentation (SHF)*

Fungal enzymes which efficiently convert pretreated lignocellulose to sugars require temperatures in the range of 45–60°C for process relevant hydrolysis time frame (3 days or longer); current industrially relevant fermentation strains cannot tolerate these temperatures. A separate hydrolysis and fermentation (SHF) can be run to allow each of the processes to take place at the optimal temperature. The dedicated hydrolysis in this process configuration also allows for pH to be adjusted following conversion to sugars in cases where there is a mismatch between the pH optima for the two processes. Furthermore, separation of the hydrolysis and fermentation phases allows process flexibility in the fermentation such as enabling batch and fed-batch processes. Batch fermentation of mixed sugar streams typically takes two to three days (depending on the amount of yeast inoculation, or 'pitch' used) because of the organism's response to depletion of the preferred sugar (glucose) and requirement to switch gears metabolically to utilize other sugars like xylose (diauxic growth). The second lag phase in the organism that is observed as it switches carbon source can be diminished by using a fed-batch fermentation to limit the effective concentration of the preferentially consumed sugar, and forcing co-fermentation of both sugars concurrently. The result is often a dramatic decrease in fermentation time.

*Simultaneous saccharification and fermentation (SSF)*

There are several arguments concerning running a consolidated hydrolysis and fermentation step (called SSF for simultaneous saccharification and fermentation). First, biomass degrading enzyme cocktails may be inhibited by high sugar concentrations; glucose, in particular, is a potent end-product inhibitor of some cellulase enzyme cocktails. SSF prevents product inhibition of the enzyme system, as sugar monomers are consumed as soon as they are produced. In some cases where enzymes are especially prone to end-product inhibition (such as when beta-glucosidase is limiting or ineffective, see Section 7.4.1), the benefits of preventing glucose from accumulating may outweigh the drawbacks associated with reduced enzyme performance resulting from the suboptimal temperature for hydrolysis (SSF is often run at a temperature that is optimal for the fermentation organism). Also, a depletion of monomeric sugars as they are produced may reduce the risk of contamination. Osmotic stress to the fermenting organism is minimized as the initial high dose of sugars experienced in a batch SHF is avoided. Counter to these arguments are the considerations that ethanol is an inhibitor to many biomass-degrading enzymes (Holtzapple et al., 1990; Wu and Lee, 1997), and also that the lower temperatures used for SSF are more friendly to contaminating organisms than the high temperatures used for a dedicated hydrolysis step. Improving beta-glucosidase performance in enzyme cocktails has been a focus area for enzyme developers (see Section 7.4.1).

*Hybrid hydrolysis and fermentation (HHF)*

A combination of the SHF and SSF process aims to take advantage of the benefits of both these systems through use of a 'hybrid hydrolysis and fermentation' or HHF configuration. In this case, hydrolysis is performed under conditions that are optimal for enzyme performance, but the fermentation is initiated before the target hydrolysis conversion level is reached under conditions that are optimal for fermentation; thus, enzymatic conversion and fermentation occur simultaneously in the later part of an HHF. Enzymatic hydrolysis rates will drop over the course of hydrolysis due to product inhibition; reducing the temperature of hydrolysis to accommodate concurrent fermentation after the hydrolysis rate reaches a certain threshold does not compromise hydrolysis yields. A schematic of enzymatic hydrolysis rates versus time is shown as a function of process time in Fig. 7.2. At the optimal enzyme hydrolysis temperature of 50°C, the hydrolysis rate gradually slows due in part to feedback inhibition since

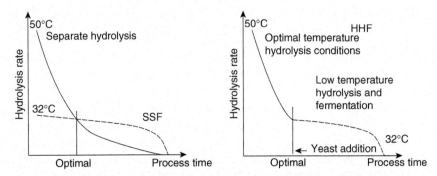

*7.2* Hybrid hydrolysis and fermentation (HHF) incorporates the best features of a separate hydrolysis and a simultaneous saccharification and fermentation (SSF). Hydrolysis rates versus process time for a typical fungal enzymatic process are shown at 50°C and 32°C.

there is no organism to take up glucose and cellobiose. At 32°C, the optimal yeast fermentation temperature, the cellulase enzymes work at a suboptimal rate due to the lower temperature but no feedback inhibition is experienced due to the simultaneous fermentation of the released sugars. The optimal duration of hydrolysis versus fermentation in the HHF configuration is substrate and pretreatment dependent; examining the rates of product formation in a separate hydrolysis process and in an SSF process will allow for prediction of the optimal timing for switching from hydrolysis to SSF mode in an HHF. When the feedback inhibition at the optimal temperature gets too great, yeast is added and the reaction conditions are altered to match the optimum conditions for the fermentation organism.

Clearly, the choice between an SHF, SSF, or HHF format will depend on the characteristics of the enzyme cocktail employed in a process. For an enzyme preparation that has a very narrow high temperature optimum, and high tolerance to end-product inhibition, a process with extensive conversion at high temperature is favored. A recent trend in enzyme development has been to take advantage of higher rates of reaction accomplished by thermostable cellulases (Viikari et al., 2007), and to engineer beta-glucosidase enzymes that can effectively catalyze cellobiose to glucose conversion even in the presence of high levels of glucose. A result is that the relative performance of enzymes at hydrolysis temperature optima is much higher than at the fermentation temperature, favoring extensive high temperature hydrolysis, even at high solids loadings.

*Consolidation bioprocessing (CBP/DMC)*

Consolidated bioprocessing (CBP, also known as direct microbial conversion, or DMC) is a more extreme consolidation of unit steps than the scenario in Fig. 7.1. In CBP, a single organism produces the enzymes necessary to hydrolyze biomass in addition to fermenting the resulting sugars to product. The resulting process is an SSF with no exogenous enzyme added. While elegant in the simplicity of application design, the process places high demands on the microorganism. Proponents of the CBP concept have argued this configuration has the potential to significantly reduce costs (Lynd et al., 2008).

Two strategies have been discussed for development of a suitable CBP organism. In one strategy, recruitment of native cellulolytic organisms is coupled with genetic modification to introduce metabolic pathways to bioproducts such as ethanol. In a second strategy, enzyme expression is engineered into an organism where bioproduct formation has been well established, and in particular the case of *S. cerevisiae* for ethanol production of mixed sugars is pursued. Much progress in developing a CBP organism has been made by both these strategies over the past five years, but outstanding questions remain. For the case of *S. cerevisiae* expression of cellulases, the challenge will be whether sufficient enzyme expression titers can be achieved under industrially relevant conditions. Two publications suggest that the necessary titers may be reached, though some extrapolation to industrial conditions is required to make these assumptions (Ilmén et al., 2011; Agbogbo et al., 2011). The challenge for transforming novel cellulolytic organisms into domesticated bioproduct production strains is mainly in assuring that they can tolerate biomass inhibitors and high product titers (ethanol, in particular, is not well tolerated by many of these organisms). For both classes of production organisms, it remains to be seen whether heterologous protein expression poses a significant metabolic burden such that bioproduct formation suffers. Negative impacts on microbial functions such as reduced gene transcription and translation have been documented and the impact may be reduced growth rate and reduced cell biomass yields.

Essentially, the CBP configuration removes the complexity from the realm of process design, relocating the same pathways to a single organism. Whether or not process and capital expense savings can be realized from employment of a full-fledged CBP configuration will depend on whether the competing pathways for enzyme production, product formation, and withstanding a range of exogenous stressors can be balanced. A 'partial' CBP configuration may also be explored, where a fraction of enzymes required for biomass hydrolysis is added exogenously, and another portion of the total dose is produced concomitantly with the bioproduct during SSF.

## 7.3 Development of enzyme technology and techniques

Commercial enzyme technology for biorefineries has become available in the past five years. In 2000, the US Department of Energy (DOE) funded enzyme development at Novozymes, Inc. and Genencor International, for dedicated use in biorefineries. Both companies started work with cocktails produced by *Trichoderma reesei*, which had been developed for use in a range of textile and other applications (Bhat, 2000). An example of an application for cellulases is their use in treatment of cotton fabrics, creating a softer, 'acid-washed' look and texture (Miettinen-Oinonen and Suominen, 2002). The properties of cellulases for this application are clearly different than for enzymes employed in a biorefinery. While incomplete degradation of cellulose in cotton is desirable in textile applications, near-complete destruction of cellulose materials to soluble components is desired in biorefineries. *T. reesei* is, however, an excellent 'chassis' for production of biorefinery enzymes; decades of research in academic and industrial labs have domesticated this fungus and transformed it into a powerful enzyme production host. Yields exceeding 100g/L soluble protein have been reported for strains of *T. reesei* (Martinez *et al.*, 2008). In addition, complete genome sequencing and development of molecular tools for genetic engineering of the organism, as well as fundamental studies of the biomass gene regulation, limitations and regulators or protein expression machinery, and optimization of fermentation conditions allow biotechnologists to manipulate this host organism successfully (Schuster *et al.*, 2012; Le Crom *et al.*, 2009; Uzbas *et al.*, 2012; Portnoy *et al.*, 2011; Martinez *et al.*, 2008). Finally, *T. reesei* cellulases are well characterized; a relatively simple, yet highly effective repertoire of glycoside hydrolases is produced by the organism (Le Crom *et al.*, 2009). Addition and replacement strategies for improving upon the consortium of biomass-degrading enzymes have proved an excellent strategy (Merino and Cherry, 2007). In addition to Novozymes, AB Enzymes GmbH, Iogen Corporation, and Dupont utilize *T. reesei* isolates to produce enzymes for the biomass industry.

Within filamentous fungi, other platform systems have been exploited as production hosts. Dyadic International has improved the production yield and suitability of *Myceliopthora thermophila* (originally mischaracterized as *Chrysosporium lucknowense*) (Gusakov, 2011).

Meiji Seika Co. produces a consortium of enzymes from *Acremonium cellulolyticus* (Gusakov, 2011). DSM has described biomass-degrading enzyme production in *Talaromyces emersonii* (Los *et al.*, 2011). Several companies have developed strains for CBP, including Mascoma, who has worked to develop both recombinant *S. cerevisiae*, as well as GM strains of the cellulolytic bacteria *Clostridium thermocellum* (Mascoma, 2009). TMO Renewables has developed a strain of *Geobacillus thermoglucosidasius* with introduced metabolic pathways, taking advantage of endogenous biomass-degrading enzymes (Atkinson *et al.*, 2010). Likewise, other companies such as Aemetis, Direvo, and Deinove have aimed to exploit the ability of celluolytic bacteria to degrade biomass.

## 7.4 Optimizing enzymes

Developing enzymes for conversion of biomass begins with in-depth and systematic study of microbial diversity among biomass utilizing organisms. Despite decades of work, we have only started to appreciate the diversity of strategies that are employed by microbes to convert the cellulose and hemicellulose in biomass to sugars. Recent advances in DNA sequencing technologies have allowed an explosion of sequence diversity from cultured and uncultured organisms, and curation of these genetic resources has alerted researchers to the existence of diversity within known biomass active families, and also suggested the existence of novel enzymes which have yet to be characterized (Mba Medie *et al.*, 2012). Classification of biomass active enzymes may emphasize the specificity of the catalyzed reaction, such as the system used by the International Union of Biochemistry and Molecular Biology's Enzyme Nomenclature and Classification (http://www.chem.qmul.ac.uk/iubmb/enzyme/) (Webb, 1992), or may emphasize the evolutionary relationships between enzyme classes (based on sequence identity), such as the scheme developed by Bernard Henrissat and colleagues: Carbohydrate-Active EnZYmes (http://www.cazy.org/) (Cantarel et al., 2009). When describing enzyme innovations, we will list the relevant EC classification and glycoside hydrolase family (GH) according to these two schemes.

Biomass active enzymes are often 'modular,' a term used to describe the occurrence of more than one discrete domain in a single protein, with distinct function. A common arrangement among secreted fungal enzymes is a combination of a catalytic core and a carbohydrate binding module (CBM). CBMs may be specific for various types of carbohydrate polymer found in biomass and are classified by sequence identity (Guillen et al., 2010). The presence of CBMs directs enzymes to their site of action, and increases the affinity of their interaction to the substrate. In some cases CBMs have been reported to disrupt crystallinity of cellulose microfibrils, but this effect has also been disputed (Wang et al., 2008; Hildén and Johansson, 2004).

Here, we limit the discussion to a few recent examples of enzyme-based improvements made within an industrial context, with focus on recent innovations to improve fungal cocktails. Fungal enzyme types that are abundant in well-characterized fungal secreted enzyme systems are outlined in Fig. 7.3. Cellobiohydrolases (CBHs) processively hydrolyze cellulose to soluble cellobiose, and endoglucaneses (EGs) cleave cellulose internally, creating new reducing ends. CBH Is cleave from the reducing end, while CBH IIs are specific for non-reducing ends. GH61s are lytic polysaccharide monoxygenases. Beta-glucosidases (bGs) convert soluble glucooligomers such as cellobiose to glucose, and are important for alleviating end-product inhibition of cellobiose to other cellulases. This oversimplified description of key cellulases aptly describes the biomass-degrading machinery that are secreted by many cellulolytic fungi; however, recently genome sequencing has revealed that fungal systems are more diverse, and include, for example, white rot systems where recognizable genes encoding cellobiohydrolases are absent (Martinez et al., 2009). Bacterial systems are either cellulosomal (cell associated), or are found as free, soluble secreted enzymes, and frequently include multi-domain proteins where a range of activities are combined in one polypeptide (Mba Medie et al., 2012). As the bacterial systems have not been exploited commercially to the degree that fungal cellulases have, they will not be included in this section, but numerous reviews are available (e.g., Doi et al., 2003; Demain et al., 2005; Fontes and Gilbert, 2010).

214 Advances in Biorefineries

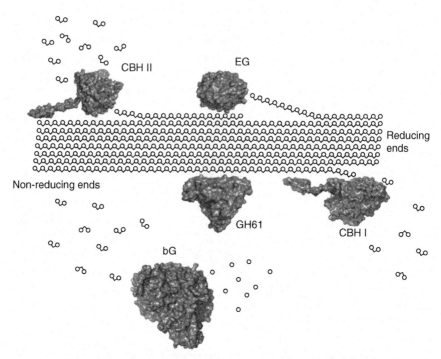

*7.3* Fungal cellulases synergistically deconstruct cellulose microfibrils to monomeric glucose. A schematic of the major enzyme types involved in cellulose hydrolysis includes cellobiohydrolases I and II (CBH I and CBH II respectively), endoglucanases (EG), beta-glucosidases (bG), and glycoside hydrolase family 61 proteins (GH61).

### 7.4.1 Beta-glucosidase

Beta-glucosidase (bG) (or cellobiase, cellobiose hydrolases) (EC 3.2.1.21) catalyzes conversion of cellobiose and other soluble glucooligomers to glucose. Enzymes with bG activity are found in glycoside hydrolyase families GH1, 3 and 9; they cleave soluble oligosaccharides (Langston *et al.*, 2006; Eyzaguirre *et al.*, 2005). Lower quantities of bG are required for an SSF reaction relative to HHF or SHF reactions, as in SSF, the enzyme cleaves soluble glucooligomers to glucose, which is rapidly taken up by the fermentation organism. Many cellulolytic fungi produce one or more bGs, but often the total bG protein is not commonly abundant, making up less than 1% of total secreted protein in most wild type secretomes (Chundawat *et al.*, 2011). Interestingly, some bacteria do not depend on exogenous beta-glucosidase, instead using cellobiose phosphorylase to drive metabolism from celluolytic degradation, a strategy that was recently engineered into yeast (Sadie *et al.*, 2011). Engineering of fermentation strains with an effective cellobiose transporter (Ha *et al.*, 2011) may allow for improved sugar uptake when enzymes are optimized for SSF.

To support SHF and HHF, biotechnological improvement of bG levels and biochemical properties in fungal enzyme cocktails is required. Beta-glucosidase activity must be maintained in the presence of high concentrations of its product, glucose, to prevent cellobiose from accumulating. Cellobiose is a potent inhibitor of CBH and EG. For industrial biomass conversion targeting high feedstock loads, supplementing bG to non-engineered microbial cellulolytic enzyme preparations can be imperative, because of high cellobiose level during the enzymatic conversion.

With regard to developing enzyme cocktails with improved bG activity, technology for improving bG expression in filamentous fungi is critical. To support optimal performance of bGs in high temperature hydrolysis, at process-relevant pH, improvement of bG stability has been accomplished (Wogulis *et al.*, 2011; Postlethwaite and Clark, 2010; Lamsa *et al.*, 2004; Hill *et al.*, 2012). Characterization of bGs that are resilient to glucose inhibition may also be relevant to their industrial application (Fang *et al.*, 2010).

### 7.4.2 Cellobiohydrolases

Cellobiohydrolases (CBHs) make up the bulk of many fungal cellulolytic cocktails. The enzymes have tunnel-like active sites, and they processively cleave cellobiose units from the ends of cellulose chains. CBH I enzymes (EC 3.2.1-) cleave sugars from the reducing ends of cellulose, while CBH IIs (EC 3.2.1.91) are specific towards non-reducing ends. In fungi, CBH I and CBH II activities are associated with GH7 and GH6 families, respectively. Many fungal CBHs are associated with type 1 carbohydrate binding modules (CBM1s), and these may assist the catalytic core with processivity (Beckham *et al.*, 2010; Tavagnacco *et al.*, 2011).

In the past few years, insights about molecular limitations of cellobiohydrolases have highlighted potential for their improvement by rational protein design. Processive CBH movement can be obstructed by kinks or other impediments on the cellulose surface (Igarashi *et al.*, 2011) and it has been suggested that k(off) values may be a major factor in CBH efficiency (Kurasin and Valjamae, 2011; Praestgaard *et al.*, 2011). CBH variants with potential for improved activity due to decreased binding of product (product expulsion) have been constructed (Bu *et al.*, 2011). Further, engineering of CBH I to reduce end-product inhibition has been performed (Healey *et al.*, 2012). Stability of CBH I and II is known to be limiting to high temperature hydrolysis performance in a range of cellulase cocktails, and recent developments in stabilization of these enzymes are promising (Heinzelman *et al.*, 2009, 2010; Voutilainen *et al.*, 2009, 2010).

### 7.4.3 endo-1,4-β-Glucanases

Endoglucanases (EGs, EC 3.2.1.4) hydrolyze internal glycosidic bonds in cellulose. EGs generally have 'cleft'-like active sites, and they are described as hydrolyzing and then dissociating, though some bacterial EGs are known to act 'processively' on crystalline cellulose (Li et al., 2007). As endoglucanases create new 'ends' for CBHs to bind to, it is not surprising that significant synergism is observed between EGs and CBHs. EGs are found among numerous GH families, with fungal examples predominantly in GH5, 7, 12, and 45. Fungal EGs are often associated with CBM1s.

### 7.4.4 GH61s

GH61 proteins are abundant in diverse fungal systems. The presence of this type of protein in fungal secretomes and genomes was described 15 years ago (Saloheimo et al., 1997), but their industrial relevance was first appreciated by scientists who uncovered GH61s as potent enhancers of cellulolytic cocktails. Addition of recombinant GH61s to *Trichoderma* cellulase cocktails was shown to greatly enhance the activity of these cellulases in lignocellulose degradation, lowering the required enzyme concentration for substrate breakdown by a factor of two (Harris et al., 2010). Originally, *T. reesei* and *Aspergillus kawachii* GH61s were classified as weak endoglucanases (Saloheimo et al., 1997; Karlsson et al., 2001; Koseki et al., 2008). The first published crystal structures of a GH61 indicated that these are novel enzymes and not EGs (Karkehabadi et al., 2008; Harris et al., 2010). No obvious cleft or hydrolytic active site could be identified in the structure; instead, a divalent metal-binding site was observed on the surface of the protein. Structural similarities between GH61 and a chitin binding protein 21 (CBP21, belonging to CBM33) were reported. Further clues to the mechanism of GH61 were obtained when it was shown that the protein could not stimulate hydrolysis in the presence of cellulases on relatively pure cellulose substrates, which suggested that a chemical present in lignocellulose but absent in cellulose might be a requirement for GH61 activity (Harris et al., 2010).

Elucidation of the function of GH61 (and the related chitin cleaving CBM33 enzymes from bacteria) has recently been accomplished. Cleavage of cellulose (or chitin, for CBM33) has been demonstrated, and a clear requirement for a redox-active factor is shown (Langston et al., 2011; Phillips et al., 2011; Beeson et al., 2012; Forsberg et al., 2011; Quinlan et al., 2011; Westereng et al., 2011). Quinlan et al. (2011) showed that pretreated biomass contains soluble redox cofactors that potentiate GH61 activity, which explains the previous results where no activity was found on pure cellulose in the absence of these soluble cofactors. GH61s use reducing equivalents (provided by small molecule reductants or by cellobiose dehydrogenase) and oxidatively cleave cellulose in a copper-dependent fashion. The mechanistic details of this novel reaction mechanism are still being explored. In some studies, cleavage products are oxidized on the reducing end (Langston et al., 2011; Phillips et al., 2011; Westereng et al., 2011), while others show non-reducing end oxidation (Langston et al., 2011; Phillips et al., 2011) or oxidation at both positions (Quinlan et al., 2011). Recently, the structures of two more members of the GH61 family were solved, and the observation of trapped oxygen species in the crystal supports a recently framed mechanism of action wherein oxygen is inserted into cellulose at C–H bonds adjacent to the glycosidic bond, rendering the cellulose chain unstable, which ultimately causes the bond to break (Beeson et al., 2012; Li et al., 2012).

### 7.4.5  endo-β-Xylanases and β-xylosidases

As was mentioned in Section 7.2.1, the nature of the biomass pretreatment greatly impacts the degree of hemicellulose conversion, and thus has a direct effect on the relative requirement for xylanases and beta-xylosidases. The total xylanase activity secreted by wild type fungi is often insufficient for effective conversion of pretreated biomass. Even when near-complete conversion of xylan to monomeric sugars is accomplished during a pretreatment, as is the case with dilute acid catalyzed pretreatments, a xylanase may benefit the conversion of cellulose by removing the small amounts of remaining insoluble xylan material, thus allowing greater access to cellulose. For biomass feedstocks that have high insoluble xylan content, xylanase and beta-xylosidase enzymes are critical to effective hydrolysis.

While the structure of hemicellulose is diverse and depends on the source of the feedstock, endo-xylanases (EC 3.2.1.8, xylanases) are known to be active on materials with different O-substitutions by acetyl, glucuronoyl, arabinosyl, and other modifications, although the ability to cleave within the xylan backbone close to these substitutions is variable. Xylanases are found widely among different GH families, including GH8, 10, 11, 30, and 43, but the ones from fungi that are most commonly used in biomass applications belong to GH10 and GH11, and these differ in specificity. GH10 xylanases are more active on substituted xylan relative to GH11s and produce shorter oligosaccharides (Ustinov et al., 2008). Xylanases often contain carbohydrate binding modules (CBMs).

Many pretreatments can solubilize hemicellulose, but do not completely convert oligomeric xylan to xylose. Inclusion of beta-xylosidases (BX, EC 3.2.1.37) in enzyme cocktails is beneficial in many cases to complete conversion of soluble xylooligosaccharides. BXs have catalytic cores belonging to the GH3, 30, 39, 43, 52, and 54 families.

As for cellulases, the hemicellulases show synergy between different classes of activity, so it is often of significant benefit to include a complete array of xylanase, beta-xylosidase, and debranching activities. When working with hardwood or feedstocks (such as many agricultural residues and energy crops) that are classified as grasses, the enrichment in arabinoxylan may require careful optimization of hemicellulase components.

### 7.4.6 Hemicellulase 'debranching' enzymes

Among 'debranching' enzymes which can catalyze the removal of arabinosyl substitutions on xylan, $\alpha$-L-arabinofuranosidases (EC 3.2.1.55) may be beneficial for industrial enzyme applications where the hemicellulose structure has not been greatly disrupted during pretreatment. These enzymes are found in GH3, 43, 51, 54, and 62, and often in addition contain CBMs (Saha, 2000). Depending on GH origin, they prefer singly substituted xylose sites (O2, 3, or 5), or rather disubstituted xylose sugars (Ara esterifying O2 and 3). As for arabinose substitutions, it is sometimes also beneficial to include enzymes that remove glucuronoyl or glucuronoyl methyl esters from the xylan backbone, through use of alpha-glucuronidases (EC 3.2.1.139), which can be found in GH67 and GH115. Likewise, removal of galactose substitutions, which are found in galactomannan, pectin, and other hemicelluloses, can sometimes improve performance, and this is accomplished by alpha-galactosidases (EC 3.2.1.22), a group of enzymes whose catalytic cores belong to GH4, 27, 36, 57, and 110 families. Softwoods and their abundant arabinoglucuronoxylan, and grasses which contain arabinoxylans are substrates which may require these debranching activities, depending on the pretreatment type and severity.

Several types of carbohydrate esterases are relevant to biomass hydrolytic enzyme cocktails. Esterases that can cleave acetyl, feruloyl, and glucoronyl moieties within hemicellulose are advantageous in order to allow for complete conversion of some highly decorated hemicellulose backbones. Acetyl xylan esterases (EC 3.1.1.72), feruloyl esterases (EC 3.1.1.73), and/or glucuronoyl esterases (EC 3.1.1.-) may enhance the activity of other hemicellulase components, especially when the substrate is acetylated hardwood xylan or ferulated grass arabinoxylan, and when a lower severity pretreatment is used. Different feruloyl esterases have different specificity towards different hydroxycinnamoyl ester bonds, which are involved in linking hemicellulose to lignin (Benoit et al., 2008). Glucuronoyl esterases can assist $\alpha$-glucuronidases by hydrolyzing glucuronyl ester linkages.

### 7.4.7 Other activities

A range of other substrates may also be beneficial to biomass hydrolytic cocktails, depending on the feedstock that is used. These include mannanases, which may help with softwood degradation. For substrates that contain pectin, such as sugar beet pulps and orange peels, a wide range of pectinolytic enzymes are required. Xyloglucanases and beta-glucanases may be of use, and are often present in complex secretomes of naturally occurring cellulose-degrading microbes.

In addition to components where the biochemical activity is known, various other proteins have been reported to confer positive benefit to hydrolysis when included in combination with cellulases and hemicellulases. It should be noted that even non-catalytic proteins such as bovine serum albumin (BSA) have an effect if the protein to biomass ratio is high enough. Acting as a 'blocking agent,' non-catalytic protein can non-specifically bind to components such as lignin and thereby prevent inhibitory adsorption by cellulases and hemicellulases. Including BSA or other 'negative control' proteins may help to discriminate proteins with biomass specific, and potentially catalytically, active candidates.

One class of proteins which may have potential in a biorefinery setting are the expansins; these proteins play a role in cell wall 'loosening' in plants (Sampedro and Cosgrove, 2005). Fungi and bacteria have been noted to secrete proteins with sequence homology to plant expansins; fungal candidates have been dubbed 'swollenins' or 'loosenins.' Swollenin from *T. reesei* was noted to disrupt the structure of cotton and filter paper cellulose without increasing reducing ends (Saloheimo *et al.*, 2002). Reports in the literature describing cellulose enhancement by fungal swollenins and loosenin (Wang *et al.*, 2010; Wang *et al.*, 2011; Jager *et al.*, 2011; Chen *et al.*, 2010; Quiroz-Castaneda *et al.*, 2011) and also by bacterial expansin-like proteins (Lee *et al.*, 2010; Kim *et al.*, 2009) have emerged in the past few years. Further study will be required in order to determine the mechanism by which these molecules increase lignocellulose conversion and allow industrial exploitation of this class of protein.

### 7.4.8 Thermostabilization and development of 'thermally active' enzyme cocktails

Including appropriate biochemical activities and designing appropriate mixtures to optimize synergy is a fundamental activity in enzyme development. In addition it is helpful to obtain highly active representatives from each enzyme type. Enzyme kinetics are very much temperature driven; in general, the higher the temperature, the faster the reaction. However, as temperature increases, so does protein denaturation/inactivation. Discovery of wild type enzymes from thermotolerant fungi is a popular strategy for obtaining highly efficient enzymes (Viikari *et al.*, 2007). In addition, engineering of proteins so that they retain activity during high temperature is effective, through use of rational design and/or protein evolution campaigns. The multicomponent nature of effective cocktails often requires that a number of protein components be stabilized. By iteratively introducing more and more stable components, the temperature profiles of a complete mixture are often shifted to a higher temperature. The goal is to achieve higher performance in conversion per unit protein by increasing the rates of reaction of the biological catalysts through improvement of 'thermal activity.'

## 7.5 Benchmarking enzymes and enzymatic conversion processes

### 7.5.1 Benchmarking the state of enzyme technology

Iterative improvements to enzyme cocktails by corrections of individual enzymes, addition of new components, and optimizing of mixtures for improved synergy and high temperature performance all contribute positively to biochemical conversion economy. Arguably, monitoring enzyme performance improvements is best done by comparing the enzyme dose required for a given degree of carbohydrate to monomer conversion of a given, pretreated feedstock. Comparative observations of enzyme initial rates of reaction, while important toward understanding the fundamental properties and mechanisms of individual enzymes, may not predict performance in an industrially relevant setting, where high extents of conversion of biomass to sugars are required. Looking at enzyme dose to achieve an economically relevant conversion target allows for monitoring of dose requirement, and the reduction in enzyme dose that is achieved through engineering of a complex cocktail is a measure of technological improvement. In benchmarking performance, the ideal conditions should be ones that mimic industrially relevant conversion process flows. As was described in Section 7.2.2, this may necessitate that hydrolysis is performed in the presence of soluble inhibitors, as the costs of solid–liquid separations may be prohibitive. Also, the benchmarking of enzyme performance should be done at high total solids loadings, reflecting the importance of maintaining high sugar concentrations for production of high product titers.

As an example of published fold enzyme reductions, Novozymes in 2005 reported progress on a Department of Energy (DOE) funded subcontract to the National Renewable Energy Laboratory (NREL) (Teter and Cherry, 2005a, 2005b). During the course of a four-year project, Novozymes and NREL reduced the enzyme requirement for converting 80% of insoluble cellulose to glucose in dilute acid pretreated corn stover (PCS) by six-fold (Fig. 7.4). The lab-scale assay (50 g) was run under conditions that are relevant to an industrial process (pH 5.0 and 50°C for 7 days), with the caveat that the substrate used was washed NREL dilute acid PCS.

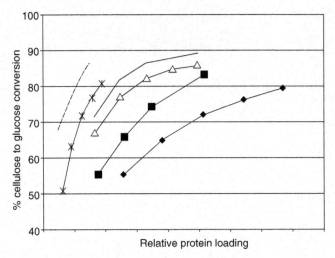

*7.4* Novozymes achieved a six-fold enzyme dose reduction during a DOE-funded subcontract to NREL. The plot shows enzyme dose response curves for various enzyme cocktails produced during the Bioenergy project, performed at Novozymes, Inc. in Davis California from 2000 to 2005.

A second enzyme improvement project has been concluded at Novozymes (Project DECREASE, September 2008–October 2011). Researchers have further reduced the enzyme dose required for 80% conversion of total glucan and xylan-based carbohydrate in NREL dilute acid PCS by 1.9-fold (Fig. 7.5). In the most recent benchmarking, a larger lab-scale reaction was used (500 g), with 21.5% TS whole slurry (unwashed) PCS as the substrate for 7 days of hydrolysis. For 80% conversion, a 1.9-fold dose reduction is observed for CTec3 relative to the reference enzyme cocktail.

### 7.5.2 Techno-economic modeling and optimizing biochemical conversion processes

While measuring the relative improvements in enzyme performance through benchmarking dose reductions is valuable, the industrially relevant question is whether their use enables a process which allows for competitive production costs. Techno-economic modeling is performed to quantify costs that the current technology would allow in a mature industry (so-called '$n^{th}$ generation' plant assumptions). As the process is composed of discrete unit steps, the modeling seeks to quantify the contribution of individual steps to capital and operating costs. In this context, various models have been used to quantify the enzyme cost component.

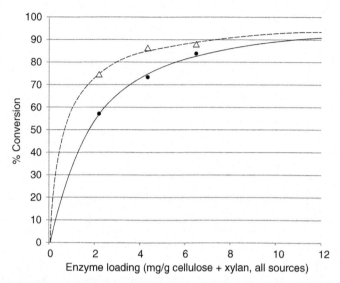

*7.5* Novozymes achieved a further 1.9-fold dose reduction during a second DOE-funded project. The plot shows enzyme dose response curves for a starting advanced enzyme cocktail (solid line, filled circles) that was available in 2008, and Cellic® CTec3 (dotted line, open triangles), which was launched after the conclusion of project DECREASE in February 2012.

Examples of techno-economic modeling include NREL's design reports, formulated to model the economics of producing ethanol from cellulose. Published over the last two decades, these ASPEN PLUS-based models show the assumptions and crude data used to build the models in exhaustive detail (Aden *et al.*, 2002; Wooley *et al.*, 1999; Humbird *et al.*, 2011). These models aimed to quantify the impact of technology that the researchers deemed reasonable to achieve in 2012. The key parameter calculated was the minimum ethanol selling price (MESP, $/gal), defined as the lowest price required to generate a 10% internal rate of return (IRR). Important design considerations can be found in the reports themselves, and summarized in Foust *et al.* (2009). Underscoring interest in producing a range of fuels and commodity chemicals from biomass sugars, NREL's 2011 design report included a section that 'backed-out' the cost of producing sugar at a cellulosic ethanol plant, toward capturing reference costs for producing mixed pentose and hexose sugars from biomass in this context. The model postulated a 2012 minimum ethanol selling price of $2.15/gal (2007 dollar basis) for a 2,205 ton/day plant.

Numerous other examples of techno-economic models for production of cellulosic ethanol exist, with published methodology presented in less detail than in the NREL design cases. An example includes a model for costs associated with conversion of AFEX-pretreated corn stover where various co-products in addition to combined heat and power (CHP) were assumed, and where enzymes were produced on site (Lau et al., 2012). A model simulating a mature CBP process with 'aggressive' performance parameters using NREL-based/ASPEN-based models was published by Lynd et al. (2005).

In the studies cited, the enzyme costs are quite variable, as is also noted by Olson et al. (2012). While this suggests that there is disagreement regarding economy of enzyme use, in fact the differences are not surprising given that the processes themselves are quite variable. As an illustrative example, the quantity of requirements for conversion of well-washed pretreated corn stover is lower than for unwashed slurry, which is responsible in part for the differences in estimated enzyme use costs. Optimizing a process for total cost reduction is not the same as optimizing for lowest enzyme cost. As a new generation of enzymes is developed, producers may elect to keep total enzyme quantity per gallon at the same level as for a previous less effective cocktail, but alter the process to allow for more dramatic cost reductions. Improved enzymes may allow for a higher total solids loading, or for a shorter process time, or both (Simms-Borre, 2012).

Toward understanding how Cellic® enzymes impact total ethanol production costs, Novozymes has developed a techno-economic model. The cost model is an Excel-based model derived from industrial partners' input and the NREL 2002 Lignocellulosic Biomass to Ethanol Process Design and Economics report describing a dilute sulfuric acid pretreatment followed by hydrolysis and fermentation. The model accounts for the mass flows through the plant and scales equipment using vendor quotes and scalability factors from the NREL report or from Novozymes' partner input. Table 7.1 describes some of the key assumptions used in the model. Lab-scale data reflecting mass balance flows of the pretreatment, hydrolysis, and fermentation processes are input into the model; thus, the model reflects the current state of technology.

While enzyme hydrolysis in a dedicated hydrolysis was shown in Fig. 7.5 as an example of benchmarking, for total process cost optimization, the best format for enzyme use is a hybrid hydrolysis and fermentation (HHF), and performance of Cellic®CTec3 with an advanced pentose/hexose co-fermenting strain is shown in Fig. 7.6. Cellic CTec3 was added at a commercially relevant dose at day 0. The proprietary C5/C6 co-fermenting yeast strain was added at a 1 g/L initial pitch at days 5, 0, and 3, respectively. As shown, the SHF and HHF have very similar and superior performance to the SSF configuration, both reaching high ethanol titers by day 7 or 8. Final ethanol titers show a clear benefit for several days of dedicated hydrolysis prior to fermentation, but there is significant flexibility in process time distribution. Using the base case model, sensitivity analysis reveals an

*Table 7.1* Key assumptions: Novozymes' techno-economic cost model

| Parameter | Input | Unit |
|---|---|---|
| Feedstock cost (corn stover) | $65 | dry MT |
| Scale of plant | 2000 | MT/day |
|  | 52 | MGal/yr |
| Ethanol yield | 74 | gal/dry ton |
| Target cellulose conversion (pretreatment and hydrolysis) | 68 | % |
| Total HHF residence time | 7 | days |
| Enzyme use cost | 0.50 | Gal |
| Electricity export value | $35 | MWh |
| Depreciation method | 10 yr MACRS |  |
| IRR | 10% |  |

*Note*: Prices are indicated in 2008 US dollars; MACRS = modified accelerated cost recovery system.

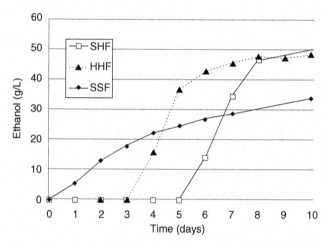

*7.6* Comparison of CTec3 performance in SHF, SSF, and HHF modes. The plot shows ethanol titers from simultaneous saccharification and fermentation (SSF), separate hydrolysis and fermentation (SHF), and hybrid hydrolysis and fermentation (HHF) of 20% TS NREL dilute acid steam exploded pretreated corn stover.

ethanol selling price of $2.50/gal, before subsidies. The assumption is made that first generation plants that will be constructed will likely not achieve the economies predicted, but it is expected that a mature plant using the modeled technology should achieve the predicted economic parameters ($n^{th}$ generation plant design).

The model is particularly useful for providing a basis for understanding where the total of process and capital costs can be reduced. Predicting a process design whereby lowered costs can be achieved is performed on a case-by-case basis, using input from partners, and data with variously pretreated feedstocks. 'Sweet spots' in the process indicate how a globally optimized process can best be designed. In the hydrolysis step, the most sensitive inputs are the solids and enzyme loading and the glucan conversion to glucose.

## 7.6 Advantages and limitations of techniques

Biochemical conversion pathways and the enzyme catalysts that enable their success have significant advantages. Biotechnology is a relatively new toolkit, and innovations are being rapidly developed. Exemplary areas include synthetic biology, as well as the genomics realm. Genetic manipulation of organisms for industrial purposes has a long history, but engineers have recently pushed the ease of manipulating microbial genomes with the aim of synthesizing entirely new biological organisms. Biofuel and bioproduct synthesis has been a popular 'test bed' for emerging tools of synthetic biology. In the area of genomics and bioinformatics, available microbial sequence information that drives discovery of novel biocatalysts has exploded, driven by dramatic reduction in DNA sequencing costs. With the ability to sequence DNA from uncultured organisms, a vast unexplored territory of microbe diversity is accessed. While techno-economic modeling by NREL currently predicts that the MESP for ethanol produced by biochemical conversion and by thermochemical conversion are quite similar (Foust *et al.*, 2009), thermochemical approaches utilize relatively more mature technologies, and arguably may demonstrate with slower rates of progress relative to the 'younger' field of biotechnology.

A limitation associated with use of biochemical catalysts for biomass conversion is their intrinsic stability. While the use of biological agents as catalysts for converting recalcitrant lignocellulose to sugars is attractive from a capital cost viewpoint due to their compatibility with mild reaction conditions, this also means that care must be taken to maintain appropriate conditions such as temperature and pH during hydrolysis. Since processes cannot be undertaken under sterile conditions, industrial-scale reactions must utilize enzymes and organisms that are robust enough to withstand contamination by naturally occurring organisms.

## 7.7 Conclusion and future trends

From a technical viewpoint, challenges to the biochemical conversion process should continue to be identified through careful techno-economic modelling. Sensitivity analysis can help to identify areas where innovations have the most impact in reducing total process costs. An emerging trend that has great promise is the tendency to seek developments that have impact across unit process steps. An example is the use of lower severity pretreatment conditions, which allow for lower wastewater treatment costs, reduction in capital expenditures, and increased xylan yields (Chen *et al.*, 2012). The feasibility of the change in pretreatment configuration can be enabled by advances in enzyme cocktails that allow for efficient polymeric xylan conversion (Blake, 2012).

Another challenge and emerging opportunity is the expansion in use of biomass-derived sugars to produce a wider array of bioproducts, in addition to advanced ethanol. Interest in diversifying biofuel types has been driven from both the supply and demand sides; technologies have advanced for producing new fuels (Fortman *et al.*, 2008); and parties such as the US Navy and Department of Defense have proposed funding and off-take agreement structures to enable production of non-ethanol fuels for use in jets and marine vessels where ethanol is not an option (Office of the Press Secretary, 2011). While progress has been made toward developing technologies for advanced biofuel production, cost reductions will likely be possible through holistic integration of process steps in the complete value chain.

## 7.8 Sources of further information and advice

Due to the very fast pace of progress in this area, web-based synopses of technical and market issues are an excellent source of information. Federally funded national laboratories in the US are very active in setting the direction for technological progress, and in particular NREL research publications demonstrate the rapid rate of progress and the directive of research as prioritized by the DOE (http://www.nrel.gov/biomass/publications.html).

At the time of writing (early 2012), a number of biorefineries around the world are under construction, with plans to start producing ethanol and other products at commercial scale. New players announce their near-term production plans quite frequently, and keeping track of where various companies are situated on the path to production can be quite complicated. Biofuels Digest (http://www.biofuelsdigest.com/) provides a daily account of new developments and commentary on progress.

## 7.9 References

Aden, A., Ruth, M., Ibsen, K., Jechura, J., Neeves, K., Sheehan, J., Wallace, B., Montague, L., Slayton, A. and Lukas, J. (2002). Lignocellulosic biomass to ethanol process design and economics utilizing co-current dilute acid prehydrolysis and enzymatic hydrolysis for corn stover. Golden, CO: NREL.

Agbogbo, F., Argyos, A., Bardsley, J. S., Barrett, T., Blecher, A., Brevnova, E., Caiazza, N., Ying, C. Y., Deleault, K. M., Den Haan, R., Foster, A. S., Froehlich, A. C., Gandhi, C. V., Gosselin, J., Hau, H. H., McBride, J. E., Mellon, M., Rajgarhia,

V. B., Rice, C. F., Shikhare, I., Skinner, R., Stonehouse, E., Tripathi, S. A., Warner, A. K., Wenger, K. S., Wiswall, E. and Xu, H. (2011). *Yeast Expressing Saccharolytic Enzymes for Consolidated Bioprocessing Using Starch and Cellulose.* World patent application PCT/US2011/039192. Filing date: 3 June 2011.

Atkinson, A., Cripps, A., Rudd, B., Eley, K., Martin, S., Milner, P. and Mercier, C. (2010). *Thermophilic micro-organisms for ethanol production.* US patent application 12/376,826. Filing date: 28 September 2007.

Beckham, G. T., Matthews, J. F., Bomble, Y. J., Bu, L., Adney, W. S., Himmel, M. E., Nimlos, M. R. and Crowley, M. F. (2010). Identification of amino acids responsible for processivity in a Family 1 carbohydrate-binding module from a fungal cellulase. *J Phys Chem B*, 114, 1447–53.

Beeson, W. T., Phillips, C. M., Cate, J. H. and Marletta, M. A. (2012). Oxidative cleavage of cellulose by fungal copper-dependent polysaccharide monooxygenases. *J Am Chem Soc*, 134, 890–2.

Benoit, I., Danchin, E. G., Bleichrodt, R. J. and De Vries, R. P. (2008). Biotechnological applications and potential of fungal feruloyl esterases based on prevalence, classification and biochemical diversity. *Biotechnol Lett*, 30, 387–96.

Bhat, M. K. (2000). Cellulases and related enzymes in biotechnology. *Biotechnology Advances*, 18, 355–83.

Blake, J. (2012). On the right path: Novozymes Cellic® HTec3-Unlocking Synergies for a Stronger Cellulosic Future. In: BIO World Congress, 2012 Orlando, FL.

Boyle, H. and Labastida, R. R. (2012). Moving towards a next-generation ethanol economy. Bloomberg New Energy Finance.

Bu, L., Beckham, G. T., Shirts, M. R., Nimlos, M. R., Adney, W. S., Himmel, M. E. and Crowley, M. F. (2011). Probing carbohydrate product expulsion from a processive cellulase with multiple absolute binding free energy methods. *J Biol Chem*, 286, 18161–9.

Cantarel, B. L., Coutinho, P. M., Rancurel, C., Bernard, T., Lombard, V. and Henrissat, B. (2009). The Carbohydrate-Active EnZymes database (CAZy): an expert resource for glycogenomics. *Nucleic Acids Res.*, 37, D233–8.

Chandra, R. P., Bura, R., Mabee, W. E., Berlin, A., Pan, X. and Saddler, J. N. (2007). Substrate pretreatment: the key to effective enzymatic hydrolysis of lignocellulosics? *Adv Biochem Eng Biotechnol*, 108, 67–93.

Chen, X. A., Ishida, N., Todaka, N., Nakamura, R., Maruyama, J., Takahashi, H. and Kitamoto, K. (2010). Promotion of efficient saccharification of crystalline cellulose by *Aspergillus fumigatus* Swo1. *Appl Environ Microbiol*, 76, 2556–61.

Chen, Y., Stevens, M., Zhu, Y., Holmes, J., Moxley, G. and Xu, H. (2012). Reducing acid in dilute acid pretreatment and the impact on enzymatic saccharification. *Journal of Industrial Microbiology & Biotechnology*, 39, 691–700.

Chundawat, S. P. S., Beckham, G. T., Himmel, M. E. and Dale, B. E. (2011). Deconstruction of lignocellulosic biomass to fuels and chemicals. *Annual Review of Chemical and Biomolecular Engineering*, 2, 121–45.

Demain, A. L., Newcomb, M. and Wu, J. H. (2005). Cellulase, clostridia, and ethanol. *Microbiol Mol Biol Rev*, 69, 124–54.

Doi, R. H., Kosugi, A., Murashima, K., Tamaru, Y. and Han, S. O. (2003). Cellulosomes from mesophilic bacteria. *J Bacteriol*, 185, 5907–14.

Eyzaguirre, J., Hidalgo, M. and Leschot, A. (2005). Beta-glucosidases from filamentous fungi: properties, structure, and applications. In Yarema, K. J. (ed.) *Handbook of Carbohydrate Engineering.* Boca Raton, FL: CRC Press.

Fang, Z., Fang, W., Liu, J., Hong, Y., Peng, H., Zhang, X., Sun, B. and Xiao, Y. (2010). Cloning and characterization of a beta-glucosidase from marine microbial metagenome with excellent glucose tolerance. *J Microbiol Biotechnol*, 9, 1351–8.

Fontes, C. M. and Gilbert, H. J. (2010). Cellulosomes: highly efficient nanomachines designed to deconstruct plant cell wall complex carbohydrates. *Annu Rev Biochem*, 79, 655–81.

Forsberg, Z., Vaaje-Kolstad, G., Westereng, B., Bunaes, A. C., Stenstrom, Y., Mackenzie, A., Sorlie, M., Horn, S. J. and Eijsink, V. G. (2011). Cleavage of cellulose by a CBM33 protein. *Protein Sci*, 20, 1479–83.

Fortman, J. L., Chhabra, S., Mukhopadhyay, A., Chou, H., Lee, T. S., Steen, E. and Keasling, J. D. (2008). Biofuel alternatives to ethanol: pumping the microbial well. *Trends Biotechnol*, 26, 375–81.

Foust, T. D., Aden, A., Dutta, A. and Phillips, S. (2009). An economic and environmental comparison of a biochemical and a thermochemical lignocellulosic ethanol conversion processes. *Cellulose*, 16, 547–65.

Geddes, C. C., Nieves, I. U. and Ingram, L. O. (2011). Advances in ethanol production. *Curr Opin Biotechnol*, 22, 312–19.

Guillen, D., Sanchez, S. and Rodriguez-Sanoja, R. (2010). Carbohydrate-binding domains: multiplicity of biological roles. *Appl Microbiol Biotechnol*, 85, 1241–9.

Gusakov, A. V. (2011). Alternatives to *Trichoderma reesei* in biofuel production. *Trends Biotechnol*, 29, 419–25.

Ha, S. J., Galazka, J. M., Kim, S. R., Choi, J. H., Yang, X. M., Seo, J. H., Glass, N. L., Cate, J. H. D. and Jin, Y. S. (2011). Engineered *Saccharomyces cerevisiae* capable of simultaneous cellobiose and xylose fermentation. *Proc. Nat. Acad. Sci. USA*, 108, 504–9.

Harris, P. V., Welner, D., Mcfarland, K. C., Re, E., Navarro Poulsen, J. C., Brown, K., Salbo, R., Ding, H., Vlasenko, E., Merino, S., Xu, F., Cherry, J., Larsen, S. and Lo Leggio, L. (2010). Stimulation of lignocellulosic biomass hydrolysis by proteins of glycoside hydrolase family 61: structure and function of a large, enigmatic family. *Biochemistry*, 49, 3305–16.

Healey, S., Luginbuhl, P., Lyon, C. S., Poland, J., Stege, J. T. and Varvak, A. (2012). New polypeptide comprising specific amino acid substitution of *Trichoderma reesei* useful for producing ethanol, used in detergent compositions and for saccharifying biomass, where the biomass is corn stover, bagasses, sorghum. BP Corp North America, INC, International publication no. WO212048171 A2.

Heinzelman, P., Snow, C. D., Smith, M. A., Yu, X., Kannan, A., Boulware, K., Villalobos, A., Govindarajan, S., Minshull, J. and Arnold, F. H. (2009). SCHEMA recombination of a fungal cellulase uncovers a single mutation that contributes markedly to stability. *J Biol Chem*, 284, 26229–33.

Heinzelman, P., Komor, R., Kanaan, A., Romero, P., Yu, X., Mohler, S., Snow, C. and Arnold, F. (2010). Efficient screening of fungal cellobiohydrolase class I enzymes for thermostabilizing sequence blocks by SCHEMA structure-guided recombination. *Protein Eng Des Sel*, 23, 871–80.

Hildén, L. and Johansson, G. (2004). Recent developments on cellulases and carbohydrate-binding modules with cellulose affinity. *Biotechnology Letters*, 26, 1683–93.

Hill, C., Lavigne, J., Whissel, M. and Tomashek, J. (2012). *Modified beta-glucosidases with improved stability.* World patent application WO2010022518 A1. 27 March 2012.

Holtzapple, M., Cognata, M., Shu, Y. and Hendrickson, C. (1990). Inhibition of *Trichoderma reesei* cellulase by sugars and solvents. *Biotechnology and Bioengineering*, 36, 275–87.

Humbird, D., Davis, R., Tao, L., Kinchin, C., Hsu, D., Aden, A., Schoen, P., Lukas, J., Olthof, B., Worley, M., Sexton, D. and Dudgeon, D. (2011). Process design and economics for biochemical conversion of lignocellulosic biomass to ethanol dilute-acid pretreatment and enzymatic hydrolysis of corn stover. Golden, CO: NREL (National Renewable Energy Laboratory).

Igarashi, K., Uchihashi, T., Koivula, A., Wada, M., Kimura, S., Okamoto, T., Penttila, M., Ando, T. and Samejima, M. (2011). Traffic jams reduce hydrolytic efficiency of cellulase on cellulose surface. *Science*, 333, 1279–82.

Ilmén, M., Den Haan, R., Brevnova, E., McBride, J., Wiswall, E., Froehlich, A., Koivula, A., Voutilainen, S. P., Siika-Aho, M., La Grange, D. C., Thorngren, N., Ahlgren, S., Mellon, M., Deleault, K., Rajgarhia, V., Van Zyl, W. H. and Penttila, M. (2011). High level secretion of cellobiohydrolases by *Saccharomyces cerevisiae*. *Biotechnology for Biofuels*, 4, 30.

Inyang, M., Gao, B., Pullammanappallil, P., Ding, W. C. and Zimmerman, A. R. (2010). Biochar from anaerobically digested sugarcane bagasse. *Bioresource Technology*, 101, 8868–72.

Ishizawa, C., Jeoh, T., Adney, W., Himmel, M., Johnson, D. and Davis, M. (2009). Can delignification decrease cellulose digestibility in acid pretreated corn stover? *Cellulose*, 16, 677–86.

Jager, G., Girfoglio, M., Dollo, F., Rinaldi, R., Bongard, H., Commandeur, U., Fischer, R., Spiess, A. C. and Buchs, J. (2011). How recombinant swollenin from *Kluyveromyces lactis* affects cellulosic substrates and accelerates their hydrolysis. *Biotechnol Biofuels*, 4, 33.

Karkehabadi, S., Hansson, H., Kim, S., Piens, K., Mitchinson, C. and Sandgren, M. (2008). The first structure of a glycoside hydrolase family 61 member, Cel61B from *Hypocrea jecorina*, at 1.6 A resolution. *J Mol Biol*, 383, 144–54.

Karlsson, J., Saloheimo, M., Siika-Aho, M., Tenkanen, M., Penttila, M. and Tjerneld, F. (2001). Homologous expression and characterization of Cel61A (EG IV) of *Trichoderma reesei*. *European Journal of Biochemistry*, 268, 6498–507.

Kim, E. S., Lee, H. J., Bang, W. G., Choi, I. G. and Kim, K. H. (2009). Functional characterization of a bacterial expansin from *Bacillus subtilis* for enhanced enzymatic hydrolysis of cellulose. *Biotechnol Bioeng*, 102, 1342–53.

Kim, S. and Holtzapple, M. T. (2006). Effect of structural features on enzyme digestibility of corn stover. *Bioresource Technology*, 97, 583–91.

Kim, Y. S. and Kadla, J. F. (2010). Preparation of a thermoresponsive lignin-based biomaterial through atom transfer radical polymerization. *Biomacromolecules*, 11, 981–8.

Koseki, T., Mese, Y., Fushinobu, S., Masaki, K., Fujii, T., Ito, K., Shiono, Y., Murayama, T. and Iefuji, H. (2008). Biochemical characterization of a glycoside hydrolase family 61 endoglucanase from *Aspergillus kawachii*. *Appl Microbiol Biotechnol*, 77, 1279–85.

Kumar, R. and Wyman, C. E. (2009a). Access of cellulase to cellulose and lignin for poplar solids produced by leading pretreatment technologies. *Biotechnol Prog*, 25, 807–19.

Kumar, R. and Wyman, C. E. (2009b). Does change in accessibility with conversion depend on both the substrate and pretreatment technology? *Bioresour Technol*, 100, 4193–202.

Kurasin, M. and Valjamae, P. (2011). Processivity of cellobiohydrolases is limited by the substrate. *J Biol Chem*, 286, 169–77.

Lamsa, M., Fidantsef, A. and Gorre-Clancy, B. (2004). *Variants of Beta-glucosidases*. World patent application PCT/US2004/013401. Filing date: 30 April 2004.

Langston, J., Sheehy, N. and Xu, F. (2006). Substrate specificity of Aspergillus oryzae family 3 beta-glucosidase. *Biochimica et Biophysica Acta – Proteins and Proteomics*, 1764, 972–8.

Langston, J. A., Shaghasi, T., Abbate, E., Xu, F., Vlasenko, E. and Sweeney, M. D. (2011). Oxidoreductive cellulose depolymerization by the enzymes cellobiose dehydrogenase and glycoside hydrolase 61. *Appl Environ Microbiol*, 77, 7007–15.

Lau, M. W., Bals, B. D., Chundawat, S. P. S., Jin, M., Gunawan, C., Balan, V., Jones, A. D. and Dale, B. E. (2012). An integrated paradigm for cellulosic biorefineries: utilization of lignocellulosic biomass as self-sufficient feedstocks for fuel, food precursors and saccharolytic enzyme production. *Energy & Environmental Science*, 5, 7100–10.

Le Crom, S., Schackwitz, W., Pennacchio, L., Magnuson, J. K., Culley, D. E., Collett, J. R., Martin, J., Druzhinina, I. S., Mathis, H., Monot, F., Seiboth, B., Cherry, B., Rey, M., Berka, R., Kubicek, C. P., Baker, S. E. and Margeot, A. (2009). Tracking the roots of cellulase hyperproduction by the fungus *Trichoderma reesei* using massively parallel DNA sequencing. *Proc Natl Acad Sci U S A*, 106, 16151–6.

Lee, H. J., Lee, S., Ko, H. J., Kim, K. H. and Choi, I. G. (2010). An expansin-like protein from *Hahella chejuensis* binds cellulose and enhances cellulase activity. *Mol Cells*, 29, 379–85.

Li, X., Beeson, W. T., Phillips, C. M., Marletta, M. A. and Cate, J. H. (2012). Structural basis for substrate targeting and catalysis by fungal polysaccharide monooxygenases. *Structure*, 20, 1051–61.

Li, Y., Irwin, D. C. and Wilson, D. B. (2007). Processivity, substrate binding, and mechanism of cellulose hydrolysis by *Thermobifida fusca* Cel9A. *Appl Environ Microbiol*, 73, 3165–72.

Liu, C. G. and Wyman, C. E. (2005). Partial flow of compressed-hot water through corn stover to enhance hemicellulose sugar recovery and enzymatic digestibility of cellulose. *Bioresource Technology*, 96, 1978–85.

Los, A., Vonk, B., Van Den Berg, M., Damveld, R. A., Sagt, C., Vollebregt, A. and Schooneveld-Bergmans, M. (2011). *Talaromyces Transformants*. World patent application WO11054899 A1. Filing date: 4 November 2010.

Lynd, L. R., Van Zyl, W. H., McBride, J. E. and Laser, M. (2005). Consolidated bioprocessing of cellulosic biomass: an update. *Curr Opin Biotechnol*, 16, 577–83.

Lynd, L. R., Laser, M. S., Bransby, D., Dale, B. E., Davison, B., Hamilton, R., Himmel, M., Keller, M., McMillan, J. D., Sheehan, J. and Wyman, C. E. (2008). How biotech can transform biofuels. *Nat Biotech*, 26, 169–72.

Martinez, D., Berka, R. M., Henrissat, B., Saloheimo, M., Arvas, M., Baker, S. E., Chapman, J., Chertkov, O., Coutinho, P. M., Cullen, D., Danchin, E. G., Grigoriev, I. V., Harris, P., Jackson, M., Kubicek, C. P., Han, C. S., Ho, I., Larrondo, L. F., De

Leon, A. L., Magnuson, J. K., Merino, S., Misra, M., Nelson, B., Putnam, N., Robbertse, B., Salamov, A. A., Schmoll, M., Terry, A., Thayer, N., Westerholm-Parvinen, A., Schoch, C. L., Yao, J., Barabote, R., Nelson, M. A., Detter, C., Bruce, D., Kuske, C. R., Xie, G., Richardson, P., Rokhsar, D. S., Lucas, S. M., Rubin, E. M., Dunn-Coleman, N., Ward, M. and Brettin, T. S. (2008). Genome sequencing and analysis of the biomass-degrading fungus (syn. *Hypocrea jecorina*). *Nat Biotechnol*, 26, 553–60.

Martinez, D., Challacombe, J., Morgenstern, I., Hibbett, D., Schmoll, M., Kubicek, C. P., Ferreira, P., Ruiz-Duenas, F. J., Martinez, A. T., Kersten, P., Hammel, K. E., Vanden Wymelenberg, A., Gaskell, J., Lindquist, E., Sabat, G., Bondurant, S. S., Larrondo, L. F., Canessa, P., Vicuna, R., Yadav, J., Doddapaneni, H., Subramanian, V., Pisabarro, A. G., Lavin, J. L., Oguiza, J. A., Master, E., Henrissat, B., Coutinho, P. M., Harris, P., Magnuson, J. K., Baker, S. E., Bruno, K., Kenealy, W., Hoegger, P. J., Kues, U., Ramaiya, P., Lucas, S., Salamov, A., Shapiro, H., Tu, H., Chee, C. L., Misra, M., Xie, G., Teter, S., Yaver, D., James, T., Mokrejs, M., Pospisek, M., Grigoriev, I. V., Brettin, T., Rokhsar, D., Berka, R. and Cullen, D. (2009). Genome, transcriptome, and secretome analysis of wood decay fungus *Postia placenta* supports unique mechanisms of lignocellulose conversion. *Proc Natl Acad Sci U S A*, 106, 1954–9.

Mascoma (2009). *Mascoma Announces Major Cellulosic Biofuel Technology Breakthrough*. Lebanon, NH. Available at: http://www.mascoma.com/download/Technology%20AdvancesRelease%20-%20050709%20FINAL.pdf (accessed 2 November 2013).

Mba Medie, F., Davies, G. J., Drancourt, M. and Henrissat, B. (2012). Genome analyses highlight the different biological roles of cellulases. *Nat Rev Microbiol*, 10, 227–34.

Merino, S. T. and Cherry, J. (2007). Progress and challenges in enzyme development for biomass utilization. *Biofuels*, 108, 95–120.

Miettinen-Oinonen, A. and Suominen, P. (2002). Enhanced production of *Trichoderma reesei* endoglucanases and use of the new cellulase preparations in producing the stonewashed effect on denim fabric. *Applied and Environmental Microbiology*, 68, 3956–64.

Mohagheghi, A., Ruth, M. and Schell, D. J. (2006). Conditioning hemicellulose hydrolysates for fermentation: effects of overliming pH on sugar and ethanol yields. *Process Biochemistry*, 41, 1806–11.

Office of the Press Secretary, The White House (2011). *President Obama Announces Major Initiative to Spur Biofuels Industry and Enhance America's Energy Security. USDA, Department of Energy and Navy Partner to Advance Biofuels to Fuel Military and Commercial Transportation, Displace Need for Foreign Oil, and Strengthen Rural America*. Washington DC.

Olson, D. G., McBride, J. E., Shaw, J. A. and Lynd, L. R. (2012). Recent progress in consolidated bioprocessing. *Curr Opin Biotechnol*, 23, 396–405.

Phillips, C. M., Beeson, W. T., Cate, J. H. and Marletta, M. A. (2011). Cellobiose dehydrogenase and a copper-dependent polysaccharide monooxygenase potentiate cellulose degradation by *Neurospora crassa*. *ACS Chem Biol*, 6, 1399–406.

Portnoy, T., Margeot, A., Seidl-Seiboth, V., Le Crom, S., Ben Chaabane, F., Linke, R., Seiboth, B. and Kubicek, C. P. (2011). Differential regulation of the cellulase transcription factors XYR1, ACE2, and ACE1 in *Trichoderma reesei* strains producing high and low levels of cellulase. *Eukaryot Cell*, 10, 262–71.

Postlethwaite, S. and Clark, L. (2010). *Beta-glucosidase variant enzymes and related polynucleotides* World patent application PCT/US2010/038879. Filing date: 16 June 2010.

Praestgaard, E., Elmerdahl, J., Murphy, L., Nymand, S., Mcfarland, K. C., Borch, K. and Westh, P. (2011). A kinetic model for the burst phase of processive cellulases. *FEBS J*, 278, 1547–60.

PRNewswire/ (2010). Statoil Now Blending Inbicon's Cellulosic Ethanol for Danish Drivers. *PR Newswire*, 4 November.

Quinlan, R. J., Sweeney, M. D., Lo Leggio, L., Otten, H., Poulsen, J. C., Johansen, K. S., Krogh, K. B., Jorgensen, C. I., Tovborg, M., Anthonsen, A., Tryfona, T., Walter, C. P., Dupree, P., Xu, F., Davies, G. J. and Walton, P. H. (2011). Insights into the oxidative degradation of cellulose by a copper metalloenzyme that exploits biomass components. *Proc Natl Acad Sci U S A*, 108, 15079–84.

Quiroz-Castaneda, R. E., Martinez-Anaya, C., Cuervo-Soto, L. I., Segovia, L. and Folch-Mallol, J. L. (2011). Loosenin, a novel protein with cellulose-disrupting activity from *Bjerkandera adusta*. *Microb Cell Fact*, 10, 8.

Sadie, C. J., Rose, S. H., Den Haan, R. and Van Zyl, W. H. (2011). Co-expression of a cellobiose phosphorylase and lactose permease enables intracellular cellobiose utilisation by *Saccharomyces cerevisiae*. *Appl Microbiol Biotechnol*, 90, 1373–80.

Saha, B. C. (2000). Alpha-L-arabinofuranosidases: biochemistry, molecular biology and application in biotechnology. *Biotechnol Adv*, 18, 403–23.

Saloheimo, M., Nakari-Setala, T., Tenkanen, M. and Penttila, M. (1997). cDNA cloning of a *Trichoderma reesei* cellulase and demonstration of endoglucanase activity by expression in yeast. *Eur J Biochem*, 249, 584–91.

Saloheimo, M., Paloheimo, M., Hakola, S., Pere, J., Swanson, B., Nyyssonen, E., Bhatia, A., Ward, M. and Penttila, M. (2002). Swollenin, a *Trichoderma reesei* protein with sequence similarity to the plant expansins, exhibits disruption activity on cellulosic materials. *Eur J Biochem*, 269, 4202–11.

Sampedro, J. and Cosgrove, D. J. (2005). The expansin superfamily. *Genome Biol*, 6, 242.

Schuster, A., Bruno, K. S., Collett, J. R., Baker, S. E., Seiboth, B., Kubicek, C. P. and Schmoll, M. (2012). A versatile toolkit for high throughput functional genomics with *Trichoderma reesei*. *Biotechnol Biofuels*, 5, 1.

Simms-Borre, P. (2012). *On the right path to cellulosic ethanol*. Bagsvaerd, DK: Novozymes. Available at: http://www.biotimes.com/en/articles/2012/March/Pages/On-the-right-path-to-cellulosic-ethanol.aspx (accessed 2 November 2013).

Somerville, C., Bauer, S., Brininstool, G., Facette, M., Hamann, T., Milne, J., Osborne, E., Paredez, A., Persson, S., Raab, T., Vorwerk, S. and Youngs, H. (2004). Toward a systems approach to understanding plant cell walls. *Science*, 306, 2206–11.

Somerville, C., Youngs, H., Taylor, C., Davis, S. C. and Long, S. P. (2010). Feedstocks for lignocellulosic biofuels. *Science*, 329, 790–2.

Tavagnacco, L., Mason, P. E., Schnupf, U., Pitici, F., Zhong, L., Himmel, M. E., Crowley, M., Cesaro, A. and Brady, J. W. (2011). Sugar-binding sites on the surface of the carbohydrate-binding module of CBH I from *Trichoderma reesei*. *Carbohydr Res*, 346, 839–46.

Teter, S. and Cherry, J. (2005a). Improved enzymes for biomass utilization. In: 14th Eur. Biomass Conference Proceedings, Paris, France.

Teter, S. and Cherry, J. (2005b). Improving cellulose hydrolysis with new enzyme compositions. In: AICHE Annual Meeting Proceedings, Salt Lake City, UT.

Ustinov, B. B., Gusakov, A. V., Antonov, A. I. and Sinitsyn, A. P. (2008). Comparison of properties and mode of action of six secreted xylanases from *Chrysosporium lucknowense*. *Enzyme and Microbial Technology*, 43, 56–65.

Uzbas, F., Sezerman, U., Hartl, L., Kubicek, C. P. and Seiboth, B. (2012). A homologous production system for *Trichoderma reesei* secreted proteins in a cellulase-free background. *Appl Microbiol Biotechnol*, 93, 1601–8.

Viikari, L., Alapuranen, M., Puranen, T., Vehmaanpera, J. and Siika-Aho, M. (2007). Thermostable enzymes in lignocellulose hydrolysis. *Adv Biochem Eng Biotechnol*, 108, 121–45.

Voutilainen, S. P., Boer, H., Alapuranen, M., Janis, J., Vehmaanpera, J. and Koivula, A. (2009). Improving the thermostability and activity of *Melanocarpus albomyces* cellobiohydrolase Cel7B. *Appl Microbiol Biotechnol*, 83, 261–72.

Voutilainen, S. P., Murray, P. G., Tuohy, M. G. and Koivula, A. (2010). Expression of *Talaromyces emersonii* cellobiohydrolase Cel7A in *Saccharomyces cerevisiae* and rational mutagenesis to improve its thermostability and activity. *Protein Eng Des Sel*, 23, 69–79.

Wackett, L. P. (2008). Biomass to fuels via microbial transformations. *Curr Opin Chem Biol*, 12, 187–93.

Wang, L., Zhang, Y. and Gao, P. (2008). A novel function for the cellulose binding module of cellobiohydrolase I. *Science in China Series C: Life Sciences*, 51, 620–9.

Wang, M., Cai, J., Huang, L., Lv, Z., Zhang, Y. and Xu, Z. (2010). High-level expression and efficient purification of bioactive swollenin in *Aspergillus oryzae*. *Appl Biochem Biotechnol*, 162, 2027–36.

Wang, Y., Tang, R., Tao, J., Gao, G., Wang, X., Mu, Y. and Feng, Y. (2011). Quantitative investigation of non-hydrolytic disruptive activity on crystalline cellulose and application to recombinant swollenin. *Appl Microbiol Biotechnol*, 91, 1353–63.

Webb, E. C. (ed.) (1992). *Enzyme Nomenclature*, San Diego, CA: Academic Press.

Werpy, T. and Petersen, G. (2004). Top value added chemicals from biomass. volume I – Results of screening for potential candidates from sugars and synthesis gas. Golden, CO: National Renewable Energy Laboratory.

Westereng, B., Ishida, T., Vaaje-Kolstad, G., Wu, M., Eijsink, V. G., Igarashi, K., Samejima, M., Stahlberg, J., Horn, S. J. and Sandgren, M. (2011). The putative endoglucanase PcGH61D from *Phanerochaete chrysosporium* is a metal-dependent oxidative enzyme that cleaves cellulose. *PLoS One*, 6, e27807.

Wogulis, M., Harris, P. and Osborn, D. (2011). *Beta-glucosidase variants and polynucleotides encoding same*. World patent application PCT/US2011/054185. Filing date: 30 Septembr 2011.

Wooley, R., Ruth, M., Sheehan, J., Ibsen, K., Majdesk, I. H. and Galvez, A. (1999). Lignocellulosic biomass to ethanol process design and economics utilizing co-current dilute acid prehydrolysis and enzymatic hydrolysis: current and futuristic scenarios. Golden, CO: National Renewable Energy Laboratory.

Wu, Z. and Lee, Y. Y. (1997). Inhibition of the enzymatic hydrolysis of cellulose by ethanol. *Biotechnology Letters*, 19, 977–9.

# 8
# Biomass pretreatment for consolidated bioprocessing (CBP)

V. AGBOR, C. CARERE, N. CICEK, R. SPARLING and D. LEVIN, University of Manitoba, Canada

DOI: 10.1533/9780857097385.1.234

**Abstract**: Biomass pretreatment and subsequent downstream processing contribute to the final cost of biocommodities produced from lignocellulosic feedstocks. Strategies that employ fewer steps for the processing of biomass can reduce costs and produce fuels and value-added products more cost effectively. This chapter reviews the various types of physico-chemical pretreatments used for lignocellulosic biomass. It describes the methods and process conditions used, as well as the physical and chemical effects of the treatment on biomass structure. Different configurations of biomass processing are presented, highlighting the increasing trend towards consolidated bioprocessing as a path to low cost biorefining for biomass-based fuels and chemicals.

**Key words**: biomass, pretreatment, consolidated bioprocessing, cellulolytic bacteria, biofuels.

## 8.1 Introduction

The depletion of 'sweet crude' oil reserves around the world and the increasing global effort to reduce dependence on petroleum-based fuels have intensified the development of biofuels as transportation and industrial fuels. The costs of production, transportation, preconditioning and pretreatment, and subsequent conversion of these feedstocks via microbial fermentation ultimately determine the costs of biomass-based ('cellulosic') fuels. The cost of biomass production drives the final cost of energy production from any given feedstock (Lynd *et al.*, 2002; Chandra *et al.*, 2007), but pretreatment of the biomass is the second most expensive unit cost in the conversion of lignocellulose to ethanol and other chemicals (Merino and Cherry, 2007).

Industrial bioethanol production is advancing beyond grain-based ethanol production because of the energy limitations and economic/environmental concerns associated with sugar/starch-based ethanol production (Brown, 2006; Groom *et al.*, 2008; Searchinger *et al.*, 2008; Simpson *et al.*, 2008). With the annual global energy demand predicted to increase to 17 billion tonnes of oil by 2035, it is evident that fuels derived from lignocellulosic biomass are an attractive and less expensive alternative for local biofuel production, compared to sugars-, starch-, or oil-based feedstocks with higher economic value.

Lignocellulosic biomass is the most abundant organic material in nature with 10–50 billion tonnes annual worldwide production (Claassen et al., 1999). Lignocellulosic biomass consists of tightly knit polymers (cellulose, hemicelluloses, lignin, waxes, and pectin) synthesized by plants as they grow. Figure 8.1 shows the inter-relationship between the three major plant polymers targeted by the biorefinery.

As a result of its complex and highly ordered structure, lignocellulosic biomass is inherently recalcitrant to bioprocessing and requires deconstruction by various pretreatments to release sugars within the biopolymers for fermentation. The term pretreatment was coined to describe any process that converts lignocellulosic biomass from its native form (which is recalcitrant to hydrolysis with cellulases) to a form that is more digestible by hydrolytic enzymes (Lynd et al., 2002). Pretreatments range from simple size reduction to more advanced biological or physico-chemical processes designed to improve the digestibility of the biomass. Table 8.1 summarizes the various physical pretreatments and their effects on biomass structure, while Table 8.2 presents a summary of the major physico-chemical pretreatments. A combination of physical and physico-chemical pretreatments is often used to improved the digestibility of lignocellulosic biomass (Agbor et al., 2011).

Pretreatments that modify the biomass composition to make it more accessible vary from neutral, to acidic, to quite alkaline (Table 8.2). Dilute acidic pretreatments will hydrolyze mostly the hemicelluloses, leaving the cellulose and lignin intact. Alkaline pretreatments will solubilize less hemicellulose and lignin than acidic pretreatments, but will alter the structural/chemical nature of the lignin, producing a hydrated cellulose product, mixed with hemicelluloses and lignin. Solvent-based pretreatment such as Organosolv will solubilize almost all of the hemicellulose, precipitate the lignin, and leave behind a purer cellulose mesh (Mosier et al., 2005; Merino and Cherry, 2007; Zhao et al., 2009).

## 8.2 Process configurations for biofuel production

Industrial-scale cellulosic biofuels production requires efficient, low cost processes that will ensure economic viability. The current paradigm for bioprocessing of lignocellulosic biomass into bioethanol involves a four-step process: (i) cellulase production, (ii) hydrolysis of polysaccharides, (iii) fermentation of soluble cellulose hydrolysis products, (iv) fermentation of soluble hemicellulose hydrolysis products (Lynd et al., 2002). This process has been segmented in different combinations over time to design different

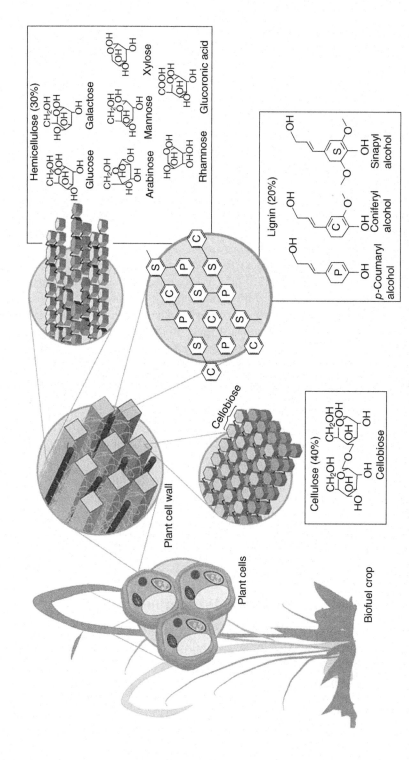

8.1 Plant biomass polymers (cellulose, hemicelluloses and lignin). The source of polymers, the inter-relatedness in the heteromatrix and the monomeric units of the polymers from a woody plant showing approximate percentages.

Table 8.1 Various physical pretreatments and their effects on biomass structure

| Method | Particle size (mm) | Main effect on biomass | References |
|---|---|---|---|
| Coarse size reduction | 10–50 | Increase in available surface area | (Cadoche and López, 1989; Palmowski and Muller, 1999) |
| Chipping | 10–30 | Decrease heat and mass transfer limitations | (Palmowski and Muller, 1999) |
| Grinding | 0.2–2 | Shearing, reduce particle size, degree of polymerization and cellulose crystallinity | (Sun and Cheng, 2002; Agbor et al., 2011) |
| Milling (disk, hammer and ball milling) | 0.2–2 | Shearing, reduce particle size, degree of polymerization and cellulose crystallinity | (J. Y. Zhu et al., 2009) |
| Use of microwaves | No change | Effects compaction of biomass and specific energy required for compression when used with chemicals, e.g. NaOH and water | (Kashaninejad and Tabil, 2011) |
| Use of gamma rays | No change | Cleave β-1,4-glycosidic bonds, thus increasing surface area and decrease cellulose crystallinity | (Takacs et al., 2000) |

Table 8.2 Summary of physico-chemical pretreatment methods

| Method | Process conditions | Mode of action | References |
| --- | --- | --- | --- |
| Steam pretreatment or steam explosion | Involves rapidly heating biomass with steam at elevated temperatures (190–240°C) and pressures between 0.7 and 4.8 MPa with residence times of 3–8 min followed by explosive decompression as the pressure is released. | Hemicellulose hydrolysis is thought to be mediated by the acetic acid generated from acetyl groups associated with hemicellulose and other acids released during pretreatment. The pressure is held for several seconds to a few minutes to promote hemicellulose hydrolysis, and then released. | (McMillan, 1994) (Mosier et al., 2005) (Weil et al., 1997) |
| Liquid hot water pretreatment | Optimally operated between 180 and 190°C, (e.g., for corn stover) and at low dry matter (~1–8%) content, leading to more poly- and oligosaccharide production. Temperatures of 160–190°C are used for pH controlled LHW pretreatment and 170–230°C have been reported depending on the severity of the pretreatment. | Hemiacetal linkages are cleaved by hot water, liberating acids during biomass hydrolysis, which facilitates the breakage of ether linkages in the biomass. | (Bobleter, 1994) (Wyman et al., 2005) |
| Dilute acid pretreatment | Dilute sulfuric acid is mixed with biomass to increase the accessibility to the cellulose in the biomass by solubilizing hemicellulose. The mixture is heated directly with the use of steam as in steam pretreatment, or indirectly via the vessel walls of the reactor. | The substrate is heated to the desired temperature in an aqueous solution and pretreated using preheated sulfuric acid (concentrations of < 4 wt%) in a stainless steel reactor. In this pretreatment the dilute acid releases oligomers and monomeric sugars by affecting the reactivity of the biomass carbohydrate polymers. | (Esteghalian et al., 1997) (Torget et al., 1990) |

| | | | |
|---|---|---|---|
| Ammonia fiber/ freeze explosion, ammonia recycle percolation and soaking aqueous ammonia | By bringing biomass in contact with anhydrous liquid ammonia at a loading ratio of 1:1 to 1:2 (1–2 kg of ammonia/kg of dry biomass) for 10–60 min at 60–90°C and pressures above 3 MPa, or 150–190°C for a few minutes. | The chemical effect of ammonia or ammonia under pressure causes the cellulosic biomass to swell, thus increasing the accessible surface area while decrystallizing cellulose as the ammonia penetrates the crystal lattice to yield a cellulose-ammonia complex. | (Alizadeh et al., 2005) (Kim and Lee, 2005) (Mittal et al., 2011) |
| Organosolv pretreatment | Organosolv pretreatments are conducted at high temperatures (100–250°C) using low boiling point organic solvents (methanol and ethanol) or high boiling point alcohols (ethylene glycol, glycerol, tetrahydrofurfuryl alcohol) and other classes of organic compounds. | Pretreatments with organic solvents extract lignin and solubilized hemicelluloses by hydrolyzing the internal lignin bonds, as well as the ether and 4-O-methylglucuronic acid ester bonds between lignin and hemicellulose and also by hydrolyzing glycosidic bonds in hemicellulose, and partially in cellulose depending on process conditions. | (Thring et al., 1990) (Zhao et al., 2009) |

*Continued*

Table 8.2 Continued

| Method | Process conditions | Mode of action | References |
| --- | --- | --- | --- |
| Lime pretreatment | Conducted over a wide temperature range 25–130°C using 0.1g Ca(OH)$_2$/g biomass at low pressures. | Solubilize hemicelluloses and lignin by deactylation and partial delignification. | (Chang et al., 1997) |
| Wet oxidative pretreatment | Treatment of biomass with air, water, and oxygen at temperatures above 120°C with or without a catalyst such as an alkali. | Oxidative factors come into play when oxygen is introduced at high pressures. | (Chang et al., 2001) (Galbe and Zacchi, 2007) |
| Carbon dioxide explosion pretreatment | Involves the use of supercritical carbon dioxide at high pressures (1,000–4,000 psi) at a given temperature up to 200°C for a few minutes. | Carbonic acid formed from the penetration of carbon dioxide into wet biomass at high pressure helps in hemicellulose hydrolysis while the release of the pressure results in disruption of biomass. | (Kim and Hong, 2001) (Zheng et al., 1995) |
| Ionic-liquid pretreatment | Using ionic liquids at temperatures < 100°C as non-derivatizing solvents to effect dissolution of cellulose. | Ionic liquids disrupt the three-dimensional network of lignocellulosic components by competing with them for hydrogen bonding. | (Moultrop et al., 2005) (Zavrel et al., 2009) |
| Fractionation solvents | Cellulose and organic solvents lignocellulose fractionation. | Solvents such as phosphoric acids, sulfite or ionic liquids enable disruption to fibrillar structure of biomass and effecting cellulose crystallinty. | (Z. G. Zhu et al., 2009) |

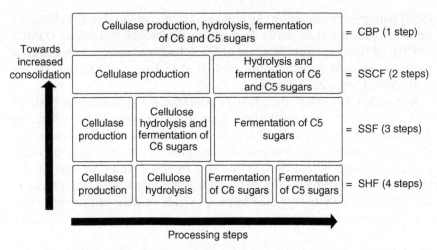

8.2 Evolution of biomass processing strategies featuring enzymatic hydrolysis. The horizontal arrow indicates the four primary treatment processes involved with cellulosic ethanol production. The vertical arrow indicates steps taken toward increased consolidation. SHF: separate hydrolysis and fermentation, SSF: simultaneous saccharification and fermentation, SSCF: simultaneous saccharification and co-fermentation, CBP: consolidated bioprocessing (adapted from Lynd et al., 2002).

configurations or processing strategies to reduce the cost of biofuel production, as shown in Fig. 8.2.

Separate hydrolysis and fermentation (SHF) is a four-stage process with a separate biocatalyst for each. Simultaneous saccharification and fermentation (SSF) combines hydrolysis and fermentation of hexose (C6) sugars, without the pentose (C5) sugars, while simultaneous saccharification and co-fermentation (SSCF) combines cellulose hydrolysis and fermentation of both hexose and pentose sugars in one step. Currently, the majority of pilot studies and industrial processes have proceeded with separate saccharification and fermentation options for bioprocessing, featuring enzymatic hydrolysis, e.g. bioethanol production.

Consolidated bioprocessing (CBP), on the other hand, combines cellulase production and substrate hydrolysis and fermentation of the hydrolysate (both hexose and pentose sugars) in one step, thus saving the cost of investing in a multi-step process (Lynd, 1996; Lynd et al., 2002; Xu et al., 2009). Of all the reported technological advances to reduce processing costs for cellulosic ethanol, CBP has been estimated to reduce production costs by as much as 41% (Lynd et al., 2008). A techno-economic evaluation for bioethanol production from softwood (spruce), hardwood (salix), and an agricultural residue (corn stover), concluded that the process configuration

(SSF) had greater impact on the cost reduction compared to the choice of substrate (Sassner *et al.*, 2008). Hence, direct microbial conversion (DMC) or CBP of lignocellulosic biomass utilizing bacteria, may be considered a preferred method for ethanol or biological hydrogen production (Levin *et al.*, 2006; Carere *et al.*, 2008).

Although CBP offers the greatest potential for reducing the costs of biofuel and co-product production, a great deal remains to be realized in the development of microbial biocatalysts that can utilize cellulose and other fermentable sugars to produce the products of interest with yields that are industrially relevant. In the rest of this chapter, we discuss the rationale for, and models of, CBP as well as strategies for development of microbial biocatalysts that may improve yields of desired products from CBP-based biorefineries.

### 8.2.1 The rationale for CBP

Bioenergy use is propagated by incentives and successful policy interventions that have been enacted to select and encourage industrial development of renewable energy sources. Bioethanol continues to be the dominant biofuel with increasing use as a transportation fuel in Brazil, USA, Canada, and the European Union (EU). The routine ways (SHF, SSF, and SSCF) for industrial ethanol production all involve a dedicated cellulase production step. Avoiding this step has been noted to be the largest potential energy-saving step. For a given amount of pretreated biomass, it has been shown that CBP offers cost savings that are not associated with the routine methods, even at the lowest amount of cellulase loading required to produce fermentation products. However, typical CBP yields of ethanol and other end-products are much lower than obtained via traditional processes (Lynd *et al.*, 2005, 2008).

Hydrogen ($H_2$), on the other hand, is considered a clean energy, having a gravimetric energy density of $122\,KJ\,g^{-1}$ and water as the only by-product of combustion. Currently $H_2$ is produced via energy intensive, environmentally harmful processes such as catalytic steam reformation of methane, nuclear or fossil fuel-mediated electrolysis of water, or by coal gasification (Levin *et al.*, 2004; Carere *et al.*, 2008). Biological $H_2$ production via anaerobic fermentation using cellulosic substrates is an attractive process for the following reasons: (i) it is less energy intensive, (ii) it utilizes a simple process design and feedstock processing, and (iii) it has the potential to utilize agricultural or agri-industrial by-product streams (Sparling *et al.*, 1997; Valdez-Vazquez *et al.*, 2005; Levin *et al.*, 2006). However, rates and yields of $H_2$ by direct microbial conversion are low, and increasing the rates and yields remains a challenge for biological $H_2$ production by CBP (Levin *et al.*, 2009).

## 8.3 Models for consolidated bioprocessing (CBP)

The trend in industrial bioprocessing for biofuels and other industrial products, such as lactic acid, glutamic acid, $n$-butanol, and pinene (Hasunuma et al., 2013) is toward increased consolidation of the different process steps (as described in Fig. 8.2). Lessons drawn from nature and industry can help further research and development of CBP for biofuels and co-products, given that CBP seeks to mimic natural microbial cellulose utilization for industrial applications.

### 8.3.1 Ruminant or natural CBP

Many animals and insects have evolved to feed on and digest raw biomass. Some well-known cellulolytic bacteria have been isolated from these organisms as their natural habitat, e.g. *Ruminococcusalbus* and *Clostridium termitidis* inhabit the gut of ruminants and termites, respectively. By taking a close look at the highly developed ruminal fermentation of cattle (i.e., ruminant CBP = rCBP), Weimer et al. (2009) proposed that breakthroughs developed by ruminants and other already existing anaerobic systems with cellulosic biomass conversion can guide future improvements in engineered CBP (eCBP) systems. Comparing the journey of the feed through the bovine digestive tract to the transformation process of a cellulosic feedstock in a biorefinery, Weimer et al. (2009) suggest that the sliding, longitudinal movement of bovine rumination is a better physical pretreatment than conventional grinding. This is because it results in substantial increase in surface area of the plant material available for microbial attack resulting in 'effective fiber' properties similar to burr mills. Burr mills consume two-thirds of the energy required by hammer mills, and although they have been considered less efficient in the grinding of grain, they could be efficient in the milling of lignocellulosic biomass for CBP as a result of the 'effective fiber' properties generated (Weimer et al., 2009).

A great amount of effort is invested chewing the feed into a fine physically pretreated state. The feed is masticated while it is moist, another strategy supported by recent studies which show that milling after chemical treatment will significantly reduce energy consumption, reduce cost of solid–liquid separation requirements, and reduce the energy required for mixing pretreated slurries (J. Y. Zhu et al., 2009; Zhu and Pan, 2010).

Another similarity is the fact that ruminal microflora (*R. albus, R. flavefaciens,* and *fibrobactersuccinogens*) found in high numbers in the rumen are capable of rapid growth on cellulose using cellulosomal complexes similar to the well-characterized cellulosomes of *Clostridium thermocellum* or *C. phytofermentans* that are being investigated for eCBP (Lynd et al., 2002; Weimer et al., 2009). Thus, the limitation of rCBP could be explored to develop better operating parameters for eCBP. In summary, it appears that ruminants have developed an efficient and elegant physical pretreatment process that could provide insight in developing a physical pretreatment process tailored for industrial CBP, as well as serve as a model for eCBP.

### 8.3.2 Engineered CBP

Unlike rCBP, eCBP seeks to utilize pure cultures of specialist native cellulolytic bacteria, or recombinant cellulolytic bacteria, to convert cellulose via direct fermentation to value-added end-products. Aerobic or anaerobic microorganisms could be used for eCBP; however, the use of a separate aerobic step for cell growth is not envisioned because it is not a characteristic feature of CBP. In eCBP, much attention is dedicated to the strategic development of cellulolytic and hemicellulolytic microorganisms both for substrate utilization and end-product formation. Among the many bacteria that have been considered as CBP-enabling microorganisms, anaerobic bacteria such as *C. thermocellulum*, *C. phytofermentans*, and the aerobic yeast *Saccharomyces cerevisiae* have been the most investigated as potential eCBP-enabling microorganisms. Figure 8.3 compares natural and engineered bioprocessing by differentiating the different unit operations.

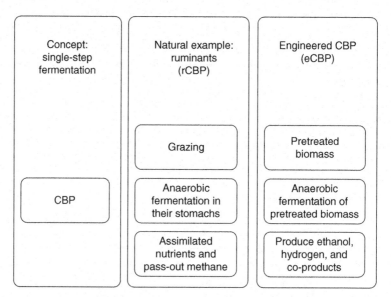

*8.3* A comparison of natural and engineered consolidated bioprocessing by differentiating the different unit operations (adapted from Weimer *et al.*, 2009).

## 8.4 Microorganisms, enzyme systems, and bioenergetics of CBP

### 8.4.1 CBP microorganisms

Based on substrate utilization, carbohydrate hydrolyzing species represent a wide range of specialist and non-specialist microbes, and specialized microbes capable of utilizing cellulose or hemicellulose-derived sugars are preferentially selected as CBP-enabling microorganisms. Cellulolytic bacteria belong to the phyla Actinobacteria, Proteobacteria, Spirochates, Thermotogae, Fibrobacteres, Bacteriodes, and Firmicutes, but approximately 80% of the cellulolytic bacteria are found within the Firmicutes and Actinobacteria (Bergquist et al., 1999). Many of these bacteria isolated from soil, insects, ruminants, compost, and sewage have the natural ability to hydrolyze cellulose and/or hemicellulose with the majority of the reported bacteria belonging to the phylum Firmicutes, and are within the class Clostridia and the genus *Clostridium*. For example, *Clostridium thermocellum* is a Gram positive, acetogenic, obligate anaerobe with the highest known growth rate on crystalline cellulose and the most investigated as a potential CBP-enabling bacteria (Lynd et al., 2002; Xu et al., 2009).

Although cellulolytic bacteria belong to aerobic or anaerobic groups of bacteria, for large-scale CBP, anaerobiosis is advantageous because of oxygen transfer limitations that are avoided when using anaerobic bacteria (Demain et al., 2005). However, because of the increasing tendency to consolidate steps for bioethanol production, *Saccharomyces cerevisiae* and the fungus *Tricoderma reesei* are being investigated as CBP-enabling candidates for bioethanol production (van Zyl et al., 2007; Xu et al., 2009). Other candidates into which saccharolytic systems have been engineered for CBP include *Zymomonas mobilis*, *Escherichia coli*, and *Klebsiella oxytoca* (van Zyl et al., 2007).

### 8.4.2 Carbohydrate active enzyme systems

To utilize plant biomass for growth, microorganisms produce multiple enzymes that hydrolyze the cellulose, hemicellulose, and pectin polymers found in plant cell walls (Warren, 1996). As a class, these carbohydrate active enzymes are referred to as glycoside hydrolases (GHs). Extracellular GHs can be secreted freely into the environment surrounding the cell (non-complex GH systems) or they can be cell-associated in large enzyme complexes (cellulosomes). Gycoside hydrolases that specifically target cellulose include:

- endoglucanases (1,4 ß-D-glucan-4-glucanohydrolases), which cleave random internal amorphous sites of a cellulose chain producing cellulo-dextrins of various lengths and thus new chain ends;

- exoglucanases (including 1,4-ß-D-glucanohydrolases or cellodextrinases and 1,4-ß-D glucancellobiohydrolases, or simply cellobiohydrolase), which act in a processive manner on either the reducing and non-reducing ends of cellulose chains liberating either D-glucose (glucanohydrolase) or D-cellobiose (cellobiohydrolase) or shorter cellodextrins; and
- ß-glucosidase (ß-glucoside glucohydrolases) which hydrolyze soluble cellodextrins and cellobiose to glucose.

The ability of cellulases to hydrolyse ß-1,4-glycosidic bonds between glucosyl residues distinguishes cellulase from other glycoside hydrolases (Lynd et al., 2002).

### 8.4.3 Non-complex glycoside hydrolase systems

Non-complex systems consist of secreted glycoside hydrolases and generally involve fewer enzymes. Aerobic fungi of the genera *Trichoderma* and *Aspergillus* have been the focus of research for non-complex cellulase systems. *Trichoderma reesei*, which is the most researched non-complex cellulase system, produces at least two exoglucanases (CBH I and CBH II), five endoglucanases (EG I, EG II, EG III, EG IV, and EG V) and two ß-glucosidases (BGL I and BGL II) that act synergistically in the hydrolysis of polysaccharides. However, CBH I and CBH II are the principal components of the *T. reesei* cellulase system representing 60% and 20%, respectively, on a mass basis of the total protein produced (Lynd et al., 2002). The ability to produce and secrete over 100g of cellulase per liter of culture has established *T. reesei* as a commercial source of cellulase enzymes (Xu et al., 2009).

### 8.4.4 Complex glycoside hydrolase systems

Some anaerobic cellulolytic microorganisms possess a specialized macromolecular complex of carbohydrate active enzymes known as the cellulosome. First described for *C. thermocellum* by Lamed et al. (1983), the cellulosome is an exocellular, multicomponent complex of GHs that mediates binding to lignocellulosic biomass and subsequent hydrolysis of the cellulose and hemicellulose polymers (Lamed et al., 1983; Carere et al., 2008). Functionally, cellulosomes are assembled on the cell walls of bacteria and enable concerted enzyme activity by minimizing distances of enzyme substrate interactions and optimizing synergies among the catalytic components, thus enabling efficient hydrolysis of the polymers and uptake of the hydrolysis products (Lynd et al., 2002).

Although cellulosome compositions can differ in the number and variety of GHs from one species to another, they generally consist of catalytic components attached to a glycosylated, non-catalytic scaffold protein that is anchored to the cell wall. In *C. thermocellum*, the anchor protein, known as the cellulose integrating protein (CipA), or 'scalffoldin', is a large (1,850 amino acid long and 2–16 MDa) polypeptide, which is anchored to the cell wall via type II cohensin domains. The *C. thermocellum* cellulosome contains nine GHs with endoglucanase activity (CelA, CelB, CelD, CelE, CelF, CelG, CelH, CelN, and CelP), four GHs which exhibit exoglucanase activity (CbhA, CelK, CelO, CelS), five or six GHs which exhibit xylanase activity (XynA, XynB, XynV, XynY, XynZ), one enzyme with chitinase activity (ManA), and one or two with lichenase activity (LicB). CelS, the major exoglucanase, and CelA, the major endoglucanase associated with the *C. thermocellum* cellulosome generate oligocellulodextrins containing two (cellobiose) to five (cellopentose) glucose residues (Lynd *et al.*, 2002; Demain *et al.*, 2005). The cellulosomes of some strains of *C. thermocellum* have been shown to degrade pectin probably via pectin lyase, polygalacturonate hydrolase, or pectin methylesterase activities. Other minor activities include ß-xylosidase, ß-galactosidase, and ß-mannosidase (Lynd *et al.*, 2002; Demain *et al.*, 2005). These modules have dockerin moieties that can associate with the cohesins of the scaffoldin to form the cellulosome (Fig. 8.4).

Cellobiosephosphorylase, which hydrolyzes cellobiose and longer chain oligocellulodextrins to glucose and glucose-1-phosphate via substrate level phosphorylation, and cellodextrinphosphorylase, which phosphorylates ß-1,4-oligoglucans via phosphorolytic cleavage, have also been associated with the *C. thermocellum* cellulase system. Unlike the fungal cellulases, the celllosome of *C. thermocellum* is able to completely solubilize crystalline cellulose such as Avicel, a characteristic referred to as Avicelase or 'true cellulase' activity (Demain *et al.*, 2005). Moreover, the *C. thermocellum* cellulase system results in the oligosaccharide hydrolysis products that are different from those of aerobic cellulolytic fungi like *T. reesei*, which generates cellobiose as the primary hydrolysis product (Zhang and Lynd, 2005).

### 8.4.5 Mode of action

Polysaccharide hydrolyzing enzymes such as cellulases and xylanases are modular proteins consisting of at least two domains: the catalytic module, and the carbohydrate binding module (Gilkes *et al.*, 1991; Horn *et al.*, 2012). Common features of most GH systems that effect binding to cellulose surface and facilitate hydrolysis are the carbohydrate binding modules (CBMs), which are known to have the following functions:

- CBMs play a non-catalytic role by 'sloughing-off' cellulose fragments from the surface of cellulosic biomass by disrupting the non-hydrolytic crystalline substrate (Lynd *et al.*, 2002);

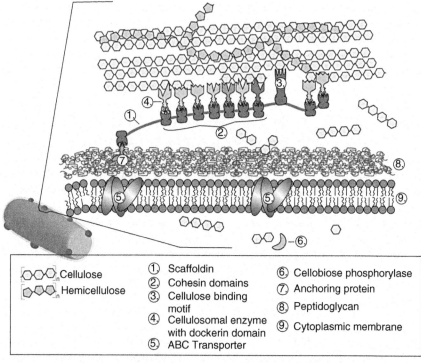

*8.4* Structural representation of the cellulosome as a macromolecular enzyme complex on the surface of a cellulolytic bacteria, displaying the various components of a complex cellulase system.

- they help concentrate the enzymes on the surface of the substrate (i.e., the proximity effect or phase transfer); and
- they help in substrate targeting/selectivity.

CBMs specific for insoluble cellulose are categorized as Type A CBMs, which interact with crystalline cellulose and Type B, which interact with non-crystalline cellulose (Arantes and Saddler, 2010a).

Recent studies that have focused on bacterial and fungal GHs have identified two GH families that have flat substrate-binding surfaces with the capability of cleaving crystalline polysaccharides via an oxidative reaction mechanism that depends on the presence of divalent metal ions and an electron donor. These two families are the Family 33 carbohydrate binding module (CBM33) proteins identified in bacteria and Family 61 glycoside hydrolases (GH61) from fungi (Vaaje-Kolstad *et al.*, 2010; Horn *et al.*, 2012). CBM33 and GH61 GHs bind to cellulose via their flat CBM substrate binding sites, which disrupt the orderly packing of the crystalline cellulose chains, creating accessible points by both introducing cuts in the polymer chains and by generating charged groups at the cut sites.

Endoglucanases and exoglucanases act synergistically. The endoglucanases generate new reducing and non-reducing ends for exoglucanases, which in turn release soluble cellodextrins and cellobiose that are converted to glucose by ß-glucosidase (Wood and McCrae, 1979; Horn et al., 2012; Kostylev and Wilson, 2012). However, for GHs to efficiently hydrolyze cellulosic biomass, they must first be able to access the cellulose chains that are tightly packed in microfibrils trapped within a heteropolymer matrix (Arantes and Saddler, 2010a; Horn et al., 2012). Factors that increase accessibility have been identified and intensely investigated (Reese, 1956; Jeoh et al., 2007; Arantes and Saddler, 2010a, 2010b; Horn et al., 2012).

### 8.4.6 Bioenergetics of CBP

The bioenergetics of CBP will differ with the type (aerobic, aerotolerant, or anaerobic) and number (pure, co-, or mixed-cultures) of microorganisms being considered as CBP-enabling agents. This was thought to be even more challenging for anaerobes from a bioenergetic standpoint given that ATP available from catabolism is used to support both cell growth and cellulase production (Lynd et al., 2002). An assessment of the bioenergetic benefits associated with growth on cellulosic substrates in terms of net cellular energy currency (ATP, ADP, or AMP) available for growth and cellulase production is vital for eCBP-enabling microorganisms.

A comprehensive bioenergtic model validating the bioenergetic feasibility of employing *C. thermocellum* on crystalline cellulose was reported by Zhang and Lynd (2005), who determined that *C. thermocellum* assimilates oligo-cellodextrins (G2–G6) of mean chain length of $n \approx 4$ (where $n$ = degree of polymerization of glucose (G) moieties). The oligo-cellulodextrins are imported into the cell and then cleaved by substrate level phosphorylation by cellodextrin- and cellobiose-phosphorylases. Phosphorylation results in cleavage of ß-glucosidic bonds releasing glucose and glucose-6-phosphate, that undergo glycolysis via the Emden–Meyerhoff pathway to generate ATP. Assimilation of oligo-cellodextrins with an average of 4.2 glucose units more than compensates for higher ATP expended on cellulase synthesis when *C. thermocellum* is grown on cellulose compared to cellobiose (Zhang and Lynd, 2005). Thus, the anaerobic fermentation of cellulose using *C. thermocellum* as a CBP-enabling microorganism is bioenergetically feasible, without the need for added saccharolytic enzymes.

## 8.5 Organism development

Although CBP of cellulosic biomass offers great potential for lower cost biofuels and fermentation products, robust, industrial microorganisms capable of both high rates of substrate conversion and high yields of the desired fermentation end-products are not available. Desirable characteristics of CBP-enabling microorganisms include production of highly active GH enzymes for rapid substrate hydrolysis, transport and utilization of the resulting hydrolytic products, high product selectivity and yield. Considerable efforts are underway to identify natural isolates with the desired characteristics and/or to develop stains with the desired characteristics via genetic engineering. The former strategy involves using naturally occurring cellulolytic microorganisms to improve end-product properties related to product yield, tolerance and titre. A classical approach is to metabolically influence end-product yield and solvent tolerance in anaerobic cellulolytic Clostridia. The latter strategy involves the use of genetic engineering of non-cellulolytic microorganisms. The best example of this is the engineering of *S. cerevisiae*, which naturally exhibits high product yields and solvent tolerance, to express a heterologous GH system that enables it to hydrolyze cellulose/hemicellulose or utilize sugars derived from hemicellulose hydrolysis (Lynd *et al.*, 2002, 2005). The vast majority of R&D towards organism development is focused on either bacteria or yeast as primary candidates for CBP-enabling microbes. However, the use of cellulolytic non-unicellular fungi as CBP has also been proposed (Xu *et al.*, 2009). This strategy can be classified under the native cellulolytic strategy from the proponents of CBP. Figure 8.5 shows the organism development strategies and commonly employed CBP-enabling microorganisms.

### 8.5.1 Metabolic engineering

Bailey (1991) defined metabolic engineering (ME) as the improvement of cellular activities by manipulation of enzymatic, transport, and regulatory functions of the cell with recombinant DNA technology. Metabolic engineering tools being used in the quest for the development of ethanologenic and currently hydrogenic microorganisms include the following:

- Mutagenesis via homologous recombination involving the mutation of a target gene that encodes a native protein to downregulate the expression of another protein or results in the synthesis of an undesired or inactive protein.
- Heterologous gene expression as a means of manipulating the metabolic fluxes toward the synthesis of a desired end product. This is the most likely metabolic engineering strategy amenable to biofuels and overexpression of enzyme catalysis flow past forks to the desired end product is a common strategy (Carere *et al.*, 2008). Pyruvate overflow in *Clostridium cellulolyticum* was established to be as a result of the inability of pyruvate-ferredoxinoxido-reductase to metabolize pyruvate

# Biomass pretreatment for consolidated bioprocessing (CBP) 251

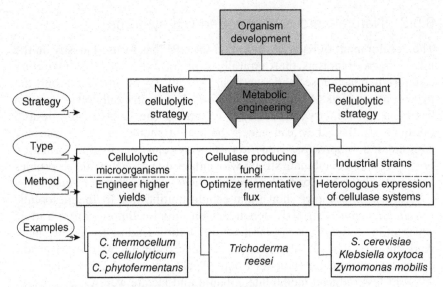

*8.5* Organism development strategies employed in research development for a CBP-enabling microorganism.

to acetyl-CoA, resulting in reduced cell growth and increased lactate. However, heterologous expression of pyruvate decarboxylase and alcohol dehydrogenase from *Zymomonas mobilis* into *Clostridium cellulolyticum* resulted in 93% acetate, 53% ethanol and hydrogen yield increased by more than 75% thus showing that cellulose fermentation can be improved by using genetically engineered strains of cellulolytic Clostridia (Guedon *et al.*, 1999).

- Antisense RNA (asRNA) attempts in redirecting metabolic flow by targeting the same genes as in mutagenesis but instead of completely abolishing protein activity as in mutagenesis, asRNA aims explicitly at downregulating the expression of a native protein by inhibiting translation due to duplex RNA structure blocking the ribosome binding site or rapid degradation of mRNA by RNases specific for RNA duplex, or by the inhibition of mRNA transcription due to premature termination. By so doing, asRNA avoids potentially lethal mutations and can be used to inducibly repress expression of proteins by using inducible promoters for asRNA. This strategy was used to reduce levels of enzymes responsible for butyrate formation in *Clostridium acetobutylicum*, demonstrating that asRNA can be used to downregulate specific protein, thus redirecting metabolic flux (Desai and Papoustakis, 1999; Carere *et al.*, 2008).

## 8.5.2 Natural versus engineered GH systems

The development of bacteria and fungi for CBP has focused mostly on the use of microorganisms that naturally express GH systems to hydrolyze cellulose/hemicellulose and synthesize products of interest from the hydrolysis products. Metabolic engineering of anaerobic cellulolytic bacteria has been the primary approach for enhancing the yields of the desired products so that they can meet the requirements of an industrially consolidated bioprocess. Gene transfers systems, electrotransformation protocols, and recombinant strains with enhanced product synthesis profiles have been described for both *C. cellulolyticum* and *C. thermocellum*.

Previous studies have shown that cellulose utilization by the mesophilic *C. cellulolyticum* is strongly dependent on initial cellulose concentration, which ultimately affects carbon flow distribution leading to end products. And the cessation of early growth was as a result of pyruvate overflow during high carbon flux (Guedon et al., 1999; Desvaux et al., 2000). Increased levels of less reduced metabolite, ethanol and lactate were observed with high levels of carbon flux, whereas at a low carbon flux, pyruvate is oxidized preferentially to acetate and lactate, thus showing an innate capability to balance carbon and electron flow or generation of reducing equivalents (Guedon et al., 1999). However, a decrease in the accumulation of pyruvate at high carbon flux was achieved by heterologous expression of pyruvate decarboxylase (PDC) and alcohol dehydrogenase (ADH) from *Zymomonas mobilis* in a shuttle vector pMG8. Growth of recombinant strain resulted in a 150% increase in cellulose utilization, 180% increase in dry cell weight, 48% decrease in lactate production, 93% increase in acetate and 53% increase in ethanol over the wild *C. cellulolyticum*, proving that genetically engineered strains could be used to greatly increase yields of cellulose fermentation by *C. cellulolyticum* (Guedon et al., 2002).

To show the potential of using *C. thermocellum* as robust platform organism for CBP, Argyros et al. (2011) constructed a mutant with novel genetic engineering tools that allow for the creation of unmarked mutations while using a replicating plasmid. A counter selection strategy was used to delete genes for lactate dehydrogenase (Ldh) and phosphotransacetylase (Pta) resulting in a stable strain with 40:1 ethanol selectivity and a 4.2-fold increase in ethanol yield over the wild-type strain (Argyros et al., 2011).

Expression of heterologous cellulases in non-cellulolytic microorganisms that are known to possess desired product formation characteristics, such as faster sugar consumption, higher ethanol yield and high resistance to ethanol and fermentation inhibitors, has also been accomplished (Hasunuma and Kondo, 2012). For example, genes encoding endoglucanse II (EG II) cellobiohydrolase II (CB II) from *T. reesei* and beta-glucosidase BGL 1 from *Aspergilus aculeatus* were integrated into the chromosome of wine yeast strain using a single vector conferring resistance to antibiotics G418. The mutant strain was able to hydrolyze corn stover cellulose and produced ethanol without the addition of exogenous saccharolytic enzymes (Khramtsov et al., 2011). Significant advances related to recombinant enzyme expression support the potential of *S. cerevisiae* as CBP host, and the number of genes expressed is not probably as important as the metabolic burden and stress responses associated with such high-level expression (van Zyl et al., 2007).

Heterologous expression of cellulolytic enzymes for the development of a cell surface which provides display of cellulolytic enzymes or cellulases that are secreted is currently being investigated in other non-cellulolytic, ethanologenic bacteria such as *E. coli*, *Zymomonas mobilis* and *Klebsiella oxytoca* to enable growth and fermentation of pretreated lignocellulosic biomass (Jarboe et al., 2007; van Zyl et al., 2007).

## 8.6 Conclusion

CBP is a less energy intensive method and a potential low cost route for production of cellulosic ethanol, as well as other industrially important products, because of the avoided cost of exogenous enzymes required for cellulose hydrolysis in SHF, SSF and SSCF (Lynd et al., 2008; Weimer et al., 2009; Xu et al., 2009). The saccharification and fermentation steps in SHF and SSF have large differences in operating temperatures which complicate development of pilot- and industrial-scale processes compared to CBP, which is conducted in a single vessel at a single optimized temperature. CBP offers simplification of the total operation process for ethanol production from cellulosic biomass compared to SHF and SSF (Hasunuma and Kondo, 2012; Hasunuma et al., 2013). CBP also has the added benefit of requiring minimal pretreatment of lignocellulosic biomass, because pretreated feedstocks for CBP do not need to be completely saccharified with costly, huge volumes of exogenous enzymes, thus the cost of pretreatments, which is a key bottleneck in lowering the net cost of production of cellulosic bioethanol, is kept very low (Xu et al., 2009; Agbor et al., 2011).

Although suitable for the production of high value products and low cost fuels, the quest for a suitable industrial CBP-enabling microorganism limits the impact of the process technology design for industrial purposes compared to sequential step processes. While the use of industrial yeast and bacterial strains used in conventional SHF and SSF processes is well established, the use of natural and/or engineered microorganisms in CBP is not yet mature and hence industrial uptake has been slow. However, research interest in CBP is growing and production of different products via CBP is under investigation (Lynd et al., 2008; Xu et al., 2009; Hasunuma and Kondo, 2012).

## 8.7 References

Agbor, V.B., Sparling, R., Cicek, N., Berlin, A. and Levin, D.B., 2011. Biomass pretreatments: fundamentals toward application. *Biotechnology Advances*, 5, 1–11.
Alizadeh, H., Teymouri, F., Gilbert, T.I. and Dale, B.E., 2005. Pretreatment of switchgrass by ammonia fibre explosion (AFEX). *Applied Biochemistry and Biotechnology*, 121–124, 1133–41.
Arantes, V. and Saddler, J.N., 2010a. Access to cellulose limits the efficiency of enzymatic hydrolysis: the role of amorphogenesis. *Biotechnology for Biofuels*, 3, 4.
Arantes, V. and Saddler, J.N., 2010b. Cellulose accessibility limits the effectiveness of minimum cellulase loading on the efficient hydrolysis of pretreated lignocellulosic substrates. *Biotechnology for Biofuels*, 4, 3.
Argyros, D.A., Tripathi, S.A., Barrett, T.F., Rogers, S.R., Feinberg, L.F., Olson, D.G., Foden, J.M., Miller, B.B., Lynd, L.R. and Hogsett, D.A., 2011. High ethanol titers from cellulose by using metabolically engineered thermophilic, anaerobic microbes. *Applied and Environmental Microbiology*, 77, 8288–94.
Bailey, J.E., 1991. Toward a science of metabolic engineering. *Science*, 252, 1668–74.
Bergquist, P.L., Gibbs, M.D., Morris, D.D, Te'o, V.S., Saul, D.J. and Morgan, H.W., 1999. Molecular diversity of thermophillic cellulolytic and hemicellulolytic bacteria. *FEMS Microbiology Ecology*, 28, 99–147.
Bobleter, O., 1994. Hydrothermal degradation of polymers derived from plants. *Progress in Polymer Science*, 19, 797–841.
Brown, L., 2006. *Exploding US grain demand for automotive fuel threatens world food security and political stability*. Earth Policy Institute, Washington, DC.
Cadoche, L. and López, G.D., 1989. Assesment of size reduction as a preliminary step in the production of ethanol from lignocellulosic wastes. *Biological Wastes*, 30, 153–7.
Carere, C.R., Sparling, R., Cicek, N. and Levin, D.B., 2008. Third generation biofuels via direct cellulose fermentation. *International Journal of Molecular Sciences*, 9, 1342–60.
Chandra, R.P., Bura, R., Mabee, W.E., Berlin, A., Pan, X. and Saddler, J.N., 2007. Substrate pretreatment: the key to effective enzymatic hydrolysis of lignocellulosics? *Advances in Biochemical Engineering/Biotechnology*, 108, 67–93.
Chang, V.S., Burr, B. and Holtzapple, M.T., 1997. Lime pretreatment of switchgrass. *Applied Biochemistry and Biotechnology*, 63–65, 3–19.
Chang, V.S., Nagwani, M., Kim, C.H. and Holtzapple, M.T., 2001. Oxidative lime pretreatment of high-lignin biomass – poplar wood and newspaper. *Applied Biochemistry and Biotechnology*, 94, 1–28.
Claassen, P., Van Lier, J., Lopez Contreras, A., Van Niel, E., Sijtsma, L., Stams, A., De Vries, S. and Weusthuis, R., 1999. Utilisation of biomass for the supply of energy carriers. *Applied Microbiology and Biotechnology*, 52, 741–55.
Demain, A.L., Newcomb, M. and Wu, J.H.D., 2005. Cellulase, clostridia, and ethanol. *Microbiology and Molecular Biology Reviews*, 69, 124–54.
Desai, R.P. and Papoustakis, E.T., 1999. Antisense RNA strategies for metabolic engineering of *Clostridium acetobutylicum*. *Applied and Environmental Microbiology*, 65, 936–45.

Desvaux, M., Guedon, E. and Petitdemange, H., 2000. Cellulose catabolism by *Clostridium cellulolyticum* growing in batch culture on defined medium. *Applied and Environmental Microbiology*, 66, 2461–70.

Esteghalian, A., Hashimoto, A.G., Fenske, J.J. and Penner, M.H., 1997. Modelling and optimization of dilute-sulfuric-acid pretreatment of corn stove, poplar and switchgrass. *Bioresource Technology*, 59, 129–36.

Galbe, M. and Zacchi, G., 2007. Pretreatment of lignocellulosic materials for efficient bioethanol production. *Advances in Biochemical Engineering/Biotechnology*, 108, 41–65.

Gilkes, N.R., Henrissat, B., Kilburn, D.G., Miller, R.C.J. and Warren, R.A.J., 1991. Domains in microbial β-1,4-glycanases: sequence conservation, function, and enzyme families. *Microbiological Reviews*, 55, 303–15.

Groom, M.J., Gray, E.M. and Townsend, P.A., 2008. Biofuels and biodiversity: principles for creating better policies for biofuel production. *Conservation Biology*, 22, 602–9.

Guedon, E., Payot, S., Desvaux, M. and Petitdemange, H., 1999. Carbon and electron flow in *Clostridium cellulolyticum* grown in chemostat culture on synthetic medium. *Journal of Bacteriology*, 181, 3262–9.

Guedon, E., Desvaux, M. and Petitdemange, H., 2002. Improvement of cellulolytic properties of *Clostridium cellulolyticum* by metabolic engineering. *Applied and Environmental Microbiology*, 68, 53–8.

Hasunuma, T. and Kondo, A., 2012. Consolidated bioprocessing and simultaneous saccharification and fermentation of lignocellulose to ethanol with thermotolerant yeast strains. *Process Biochemistry*, 47, 1287–94.

Hasunuma, T., Okazaki, F., Okai, N., Hara, K.Y., Ishii, J. and Kondo, A., 2013. A review of enzymes and microbes for lignocellulosic biorefinery and the possibility of their application to consolidated bioprocessing technology. *Bioresource Technology*, 135, 513–22.

Horn, S.J., Vaaje-Kolstad, G., Westereng, B. and Eijsink, V.G.H., 2012. Novel enzymes for the degradation of cellulose. *Biotechnology for Biofuels*, 5, 45.

Jarboe, L., Grabar, T., Yomano, L., Shanmugan, K. and Ingram, L., 2007. Development of ethanologenic bacteria. *Advances in Biochemical Engineering/Biotechnology*, 108, 237–61.

Jeoh, T., Ishizawa, C.I., Davis, M.F., Himmel, M.E., Adney, W.S. and Johnson, D.K., 2007. Cellulase digestibility of pretreated biomass is limited by cellulose accessibility. *Biotechnology and Bioengineering*, 98, 112–22.

Kashaninejad, M. and Tabil, L.G., 2011. Effect of microwave–chemical pre-treatment on compression characteristics of biomass grinds. *Biosystems Engineering*, pp.36–45.

Khramtsov, N., McDade, L., Amerik, A., Yu, E., Divatia, K., Tikhonov, A., Minto, M., Kabongo-Mubalamate, G., Markovic, Z. and Ruiz-Martinez, M., 2011. Industrial yeast strain engineered to ferment ethanol from lignocellulosic biomass. *Bioresource Technology*, 102, 8310–13.

Kim, H.K. and Hong, J., 2001. Supercritical $CO_2$ pretreatment of lignocellulose enhances enzymatic cellulose hydrolysis. *Bioresource Technology*, 77, 139–44.

Kim, T.H. and Lee, Y.Y., 2005. Pretreatment of corn stover by ammonia recycle percolation process. *Bioresource Technology*, 96, 2007–13.

Kostylev, M. and Wilson, D.B., 2012. Synergistic interactions in cellulose hydrolysis. *Biofuels*, 3, 61–70.

Lamed, R., Setter, E. and Bayer, E.A., 1983. Characterisation of a cellulose-binding, cellulose-containing complex in *Clostridium thermocellum*. *Journal of Bacteriology*, 156, 828–36.

Levin, D.B., Pitt, L. and Love, M., 2004. Biohydrogen production: prospects and limitations to practical application. *International Journal of Hydrogen Energy*, 29, 173–85.

Levin, D.B., Islam, R., Cicek, N. and Cicek, R., 2006. Hydrogen production by *Clostridium thermocellum* 27405 from cellulosic biomass substrates. *International Journal of Hydrogen Energy*, 31, 1496–503.

Levin, D.B., Carere, C.R., Cicek, N. and Sparling, R., 2009. Challenges for biohydrogen production via direct lignocellulose fermentation. *International Journal of Hydrogen Energy*, 34, 7390–403.

Lynd, L.R., 1996. Overview and evaluation of fuel ethanol from cellulosic biomass: technology, economics, the environment, and policy. *Annual Review of Energy and the Environment*, 21, 403–65.

Lynd, L.R., Weimer, P.J., Zyl, W.H.V. and Pretorius, I.S., 2002. Microbial cellulose utilization: fundamentals and biotechnology. *Microbiology and Molecular Biology Reviews*, 66, 506–77.

Lynd, L.R., Van Zyl, W.H., McBride, J.E. and Laser, M., 2005. Consolidated bioprocessing of cellulosic biomass: an update. *Current Opinion in Biotechnology*, 16, 577–83.

Lynd, L.R., Laser, M.S., Bransby, D., Dale, B.E., Davison, B., Hamilton, R., Himmel, M., Keller, M., McMillan, J.D., Sheehan, J. and Wyman, C.E., 2008. How biotech can transform biofuels. *Nature Biotechnology*, 26, 169–72.

McMillan, J.D., 1994. Pretreatment of lignocellulosic biomass. *Enzymatic Conversion of Biomass for Fuels Production*, 566, 292–324.

Merino, S.T. and Cherry, J., 2007. Progress and challenges in enzyme development for biomass utilization. *Advances in Biochemical Engineering/Biotechnology*, 108, 95–120.

Mittal, A., Katahira, R., Himmel, M.E. and Johnson, D.K., 2011. Effects of alkaline pretreatment or liquid ammonia treatment on crystalline cellulose: changes in crystalline structures and effects. *Biotechnology for Biofuels*, 4, 41.

Mosier, N., Wyman, C.E., Dale, B.E., Elander, R., Lee, Y.Y., Holtzapple, M.T. and Ladischa, M., 2005. Features of promising technologies for pretreatment of lignocellulosic biomass. *Bioresource Technology*, 96, 673–86.

Moultrop, J.S., Swatloski, R.P., Moyna, G. and Rogers, R.D., 2005. High resolution $^{13}$C NMR studies of cellulose and cellulose oligomers in ionic liquid solutions. *Chemical Communications*, 12, 1557–19.

Palmowski, L. and Muller, J., 1999. Influence of the size reduction of organic waste on their anaerobic digestion. In *II International Symposium on Anaerobic Digestion of Solid Waste*. Barcelona, Spain 15–17 June 1999, 137–44.

Reese, E. T., 1956. Enzymatic hydrolysis of cellulose. *Applied Microbiology*, 4, 39–45.

Sassner, P., Galbe, M. and Zacchi, G., 2008. Techno-economic evaluation of bioethanol production from three different lignocellulosic materials. *Biomass and Bioenergy*, 32, 422–30.

Searchinger, T., Heimlich, R., Houghton, R.A., Dong, F., Elobeid, A., Fabiosa, J., Tokgoz, S., Hayes, D. and Yu, T.H., 2008. Use of US croplands for biofuels increases greenhouse gases through emissions from land-use change. *Science*, 319, 1238–40.

Simpson, T.W., Sharpley, A.N., Howarth, R.W., Paerl, H.W. and Mankin, K.R., 2008. The new gold rush: fueling ethanol production while protecting water quality. *Environmental Quality*, 37, 318–24.

Sparling, R., Risbey, D. and Poggi-Varaldo, H.M., 1997. Hydrogen production from inhibited anaerobic composters. *International Journal of Hydrogen Energy*, 22, 563–6.

Sun, Y. and Cheng, J., 2002. Hydrolysis of lignocellulosic materials for ethanol production: a review. *Bioresource Technology*, 83, 1–11.

Takacs, E., Wojnarovits, L., Foldavary, C., Hargagittai, P., Borsa, J. and Sajo, I., 2000. Effect of combined gamma-radiation and alkali treatment on cotton-cellulose. *Radiation Physics and Chemistry*, 57, 399–403.

Thring, R.W., Chornet, E. and Overend, R., 1990. Recovery of a solvolytic lignin: effects of spent liquor/acid volume ratio, acid concentrated and temperature. *Biomass*, 23, 289–305.

Torget, R.W., Werdene, P. and Grohmann, K., 1990. Dilute acid pretreatment of two short-rotation herbaceous crops. *Applied Biochemistry and Biotechnology*, 24/25, 115–26.

Vaaje-Kolstad, G., Westereng, B., Horn, S.J., Lui, Z.L., Zhai, H., Sorlie, M. and Eijsink, V.G.H., 2010. An oxidative enzyme boostying enzymatic conversion of recalcitrant polysaccharides. *Science*, 330, 219–22.

Valdez-Vazquez, I., Sparling, R., Risbey, D., Rinderknecht-Seijas, N. and Poggi-Varaldo, H.M., 2005. Hydrogen production via anaerobic fermentation of paper mill waste. *Bioresource Technology*, 96, 1907–13.

Van Zyl, W.H., Lynd, L.R., Den Haan, R. and McBride, J.E., 2007. Consolidated bioprocessing for bioethanol production using *Saccharomyces cerevisiae*. *Biofuels*, 108, 205–35.

Warren, R.A.J., 1996. Microbial hydrolysis of polysaccharides. *Annual Reviews Microbiology*, 50, 183–212.

Weil, J.R., Sariyaka, A., Rau, S.L., Goetz, J., Ladisch, C.M., Brewer, M., Hendrickson, R. and Ladisch, M.R., 1997. Pretreatment of yellow poplar wood sawdust by pressure cooking in water. *Applied Biochemistry and Biotechnology*, 68, 21–40.

Weimer, P.J., Russell, J.B. and Muck, R.E., 2009. Lessons from the cow: what the ruminant animal can teach us about consolidated bioprocessing of cellulosic biomass. *Bioresource Technology*, 100, 5323–31.

Wood, T.M. and McCrae, S. I., 1979. Synergism between enzymes involved in the solubilization of native cellulose. *Advances in Chemistry*, 181, 181–209.

Wyman, C.E., Dale, B.E., Elander, R.T., Holtzapple, M.T., Ladisch, M.R. and Lee, Y.Y., 2005. Coordinated development of leading biomass pretreatment technologies. *Bioresource Technology*, 96, 1959–66.

Xu, Q., Singh, A. and Himmel, M.E., 2009. Perspectives and new directions for the production of bioethanol using consolidated bioprocessing of lignocellulose. *Current Opinion in Biotechnology*, 20, 364–71.

Zavrel, M., Bross, D., Funke, M., Buchs, J. and Spiess, A.C., 2009. High-throughput screening for ionic liquids dissolving (ligno-)cellulose. *Bioresource Technology*, 100, 2580–7.

Zhang, Y.H.P. and Lynd, L.R., 2005. Cellulose utilization by *Clostridium thermocellum*: bioenergetics and hydrolysis product assimilation. *National Academy of Science*, 102, 7321–5.

Zhao, X., Cheng, K. and Liu, D.H., 2009. Organosolv pretreatment of lignocellulosic biomass for enzymatic hydrolysis. *Applied Microbiology and Biotechnology*, 82, 815–27.

Zheng, Y.Z., Lin, H.M. and Tsao, G.T., 1995. Supercritical carbon-dioxide explosion as a pretreatment for cellulose hydrolysis. *Biotechnology Letters*, 17, 845–50.

Zhu, J.Y. and Pan, X.J., 2010. Woody biomass pretreatment for cellulosic ethanol: technology and energy consumption evaluation. *Bioresource Technology*, 100, 4992–5002.

Zhu, J.Y., Wang, G.S., Pan, X.J. and Gleisner, R., 2009. Specific surface to evaluate the efficiencies of milling and pretreatment of wood for enzymatic saccharification. *Chemical Engineering Science*, 64, 474–85.

Zhu, Z.G., Sathitsuksanoh, N., Vinzant, T., Shell, D.J., McMillan, J.D. and Zhang, Y.H., 2009. Comparaitive study of corn stover preteated by dilute acid and cellulose solvent-based lignocellulose fractionation: enzymatic hydrolysis, supramolecular structure, and substrate accessibility. *Biotechnology and Bioengineering*, 103, 715–24.

# 9
## Developments in bioethanol fuel-focused biorefineries

S. MUTTURI, B. PALMQVIST and G. LIDÉN,
Lund University, Sweden

DOI: 10.1533/9780857097385.1.259

**Abstract**: The production of ethanol from lignocellulose is on its way to industrial realization as evidenced by the current completion of several commercial scale plants. In these plants, biomass can be utilized for a range of different products apart from ethanol, such as electricity, biogas and heat. In this chapter, we describe the technology for production of ethanol within these biorefineries and discuss the challenges and development trends. The focus of this chapter is on the biochemical conversion routes.

**Key words**: pretreatment, hydrolysis, fermentation, lignocellulose, co-products.

## 9.1 Introduction

Ethanol is clearly a significant product in the world of biorefineries. Alternatively, you could also say that biorefineries are increasingly important in the production of ethanol. In the former case, ethanol is one of a range of products in the processing of biomass. In the latter case, biorefineries are used to increase the overall profitability (or feasibility) of ethanol production from biomass (Pham and El-Halwagi, 2012). In this review we will start from the latter perspective, i.e. we already know (or have decided) that ethanol is a desired product and we want to enable the feasibility, improve the sustainability or maximize the profitability of biorefineries. All product streams and substrate streams are clearly of interest, just as in a petrochemical refinery, from which we borrow 'refinery' in the term 'biorefinery'.

Ethanol is today mainly produced from sugars (obtained from sugar cane primarily) or starch (obtained from corn primarily). The growth in ethanol production worldwide has been impressive since the mid-1970s, driven almost exclusively by the two dominating producers, the US and Brazil, and with the purpose of producing ethanol as a fuel (Fig. 9.1). The first phase of the increase in ethanol production was a result of the Brazilian Pro-Alcohol program launched in 1975, which caused an impressive growth in ethanol production between 1975 and 1985 (Goldemberg, 2006). The starting point of the program was the oil embargo in 1973, which was

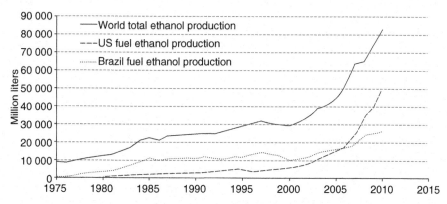

*9.1* World total ethanol production since 1975 together with the two major fuel ethanol producers, US and Brazil. (Compiled from: Dr Berg, F. O. Litchs, the Renewable Fuels Association (http://www.ethanolrfa.org/) and Goldemberg, 2006).

reinforced by the realization that Brazil had a huge unused capacity as a sugar cane producer. The first decade of rapid ethanol production growth was followed by a period of somewhat slower growth – and even a short period of decreased world production – until about the year 2000. At this point in time, the expansion of US ethanol production started. Ethanol production in the US surpassed that of Brazil in 2005 and the production gap has widened since then.

The history of ethanol as a fuel, however, dates much further back than the 1970s. The first use of ethanol was likely as a mixture with pine-derived turpentine in oil lamps in the 1860s (Songstad *et al.*, 2009). Whale oil was running out and an alternative fuel was needed. At about this time, Nicholas Otto was working on his combustion engine using a mixture containing ethanol. Decades later, Henry Ford had a vision of using ethanol derived from biomass as a motor fuel. Henry Ford allegedly shared the views of the so-called 'Chemurgy' movement[1] in the 1930s, which promoted the production of chemicals from agriculture (Giebelhaus, 1980). The people in the Chemurgy movement were, in a sense, early proponents of biorefineries. However, petroleum-derived fuels could be produced more cheaply and ethanol was soon outcompeted as a fuel. So with the exception of the two world wars, the interest in ethanol as a vehicle fuel stayed low until the oil crisis in 1973. Apart from previous periodic shortages of petroleum on the market (and the risk of a more permanent future shortage), there are two other major drivers behind production of ethanol and other chemicals from renewable resources, namely environmental concerns and agricultural policies. These driving forces are interconnected and difficult to resolve.

[1]The latter part of the word is derived from 'ergon' meaning 'work' (Finlay, 2004).

It is probably fair to conclude that environmental concerns for climate changes have had a significant impact in the introduction of renewable fuel targets in both Europe and the US. These targets have been formulated in the Renewable Energy Directive (RED) in the European Union (Commission Directive 2009/28/EC), and the Energy Independence and Security Act (EISA, 2007) in the US. The former puts forward the so-called 20-20-20 targets, i.e. a 20% reduction in EU greenhouse gas emissions (compared to 1990 levels); an increased share of EU energy consumption produced from renewable resources to 20%, and finally a 20% improvement in energy efficiency within the EU – all to be reached by 2020. Specifically, a target of 10% renewable fuels is stated. The EISA has instead a very specific absolute production target, calling for 36 billion gallons (136 billion liters) per year of renewable biofuels by 2022, out of which a maximum of 15 billion gallons can be corn-based ethanol.

The EU targets on greenhouse gas (GHG) emissions have led to much activity on standardization of calculation of 'field-to-exhaust pipe' GHG emissions using life cycle analysis (LCA), and there are now ISO standards for LCA (14040 and 14044). The requirements for a net reduction of GHG emissions are gradually increased in the coming decade and suppliers of renewable fuels must make calculations of GHG emissions using the standards. A net GHG reduction of 60% in comparison to fossil fuels should be reached by 2020, which cannot typically be met by today's starch-based ethanol production, and will favor production from lignocellulose. There is thus presently a strong regulatory incentive to push ethanol production from starch-based into lignocellulose-based production. Such a production will necessarily require efficient use of all parts of the raw material, i.e. a biorefinery approach is called for.

## 9.2 Ethanol biorefineries

A biorefinery can be depicted as in Fig. 9.2, i.e. it is a processing facility for producing a multitude of product based on a renewable carbon source. The International Energy Agency (IEA) states that 'Biorefinery is the sustainable processing of biomass into a spectrum of marketable products (food, feed, materials, chemicals) and energy (fuels, power, heat)'. The span of products and processes in a biorefinery is therefore very large and a sub-classification is often made (Kamm and Kamm, 2007). This classification can be based on:

- the type of substrate used (e.g., agricultural material, forest materials, waste and residues, or algae)
- the type of product obtained (e.g., fuel, heat, or electricity)
- the type of conversion processes used (thermochemical, biochemical or pyrolysis).

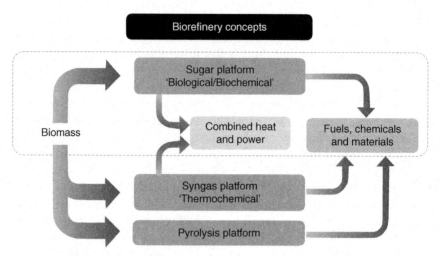

*9.2* Conceptual picture of a biorefinery, showing the main technology choices.

The IEA through its Working Group 42 introduced a fourth ground for classification, namely the so-called platforms, which are based on core intermediates in the processing, e.g. C5/C6 carbohydrates, syngas, lignin or pyrolysis oil (http://www.iea-bioenergy.task42-biorefineries.com).

The variation in technology can be quite large, but there are still enough common features to justify the use of the term biorefinery. Similar to the oil refinery, fuels and heat are prime products and out of the fuels, ethanol is a major target product. Ethanol is of course also a chemical *per se*, and a potential platform chemical, which can be used for production of, for example, ethylene to be further used in the polymer industry. The Brazilian company, Braskem has launched the production of 'green polyethylene' using this approach (http://www.braskem.com.br/plasticoverde/eng/default.html).

By 'ethanol biorefinery' we mean here a biomass-based refinery, in which ethanol is a major product. There are several options in terms of feedstocks, technology, products and platforms. Ethanol can be a product of the syngas route, where the syngas is obtained through thermochemical processing, followed by either Fischer–Tropsch synthesis using the syngas or fermentative production from syngas using various kinds of *Clostridia*, such as *Clostridium ljungdahlii*, *Clostridium carboxidivorans*, or other proprietary strains, e.g. *Clostridium* P11 (Kundiyana *et al.*, 2010; Wilkins and Atiyeh, 2011; Köpke *et al.*, 2011). This route is currently pursued by companies such as Coskata and Ineos Bio. A principal advantage of the syngas route is the potential to use also the lignin fraction for ethanol production (of course at the same time withdrawing lignin from other potential uses). A drawback is that ethanol is often formed together with acetate, and scale-up hinges on efficient gas–liquid mass transfer (Köpke et al., 2011).

In this chapter we will focus on the biochemical conversion route via the so-called sugar platform (i.e., the option which is within the dashed box in Fig. 9.2). The choice of feedstock is obviously critical. The availability of the feedstock will define and limit the production capacity and will also be geographically specific (Table 9.1). Furthermore, the net GHG emissions are to a very large extent determined by the feedstock itself – or rather the agricultural practices associated with it – and to a lesser degree by the conversion process.

## 9.3 The lignocellulose to ethanol process

The main process steps in any biochemical ethanol refinery are (Fig. 9.3):

- pretreatment
- hydrolysis
- fermentation
- product recovery
- wastewater purification.

Co-products will vary a lot depending on feedstock and market conditions, but in an energy focused biorefinery, co-products will be heat, electricity, pellets and biogas, as shown in Fig. 9.3. The relative proportion between these co-products can be changed – within certain limits – and this flexibility is an important feature of a biorefinery. We will now go through the basic process steps in more detail.

Although the same main process steps in a bioethanol biorefinery are always needed, the way to operate each individual step is highly dependent on desired products and to a large extent on the selected feedstock. Biomass is composed of three major macromolecules, namely cellulose, hemicelluloses and lignin, with smaller fractions of proteins, extractives and ash. However, the relative ratios and the composition of, in particular, hemicellulose and lignin differ greatly between different types of biomass, as indicated in Table 9.2. The main sugar in the branched hemicellulose polymer is, for example, xylose in most hardwoods and agricultural crops, whereas mannose is the dominant sugar in softwood hemicellulose. The cellulose polymer, in contrast, is always comprised of repeating cellobiose units (i.e., glucose dimers) regardless of biomass type.

### 9.3.1 Pretreatment

Pretreatment is necessary to overcome the recalcitrant nature of native lignocellulose. In contrast to the polysaccharides starch and glycogen, the

Table 9.1 Classification and availability of various lignocellulose sources

| Biomass type | Species | Typical growth region | Reported productivities |
|---|---|---|---|
| **Hardwood** | Birch (*Betula* spp.) | Northern hemisphere | |
| | Eucalyptus (*Eucalyptus* spp.) | Australia, New Guinea, Indonesia, Europe | Spain: 13.9–14.6 T/ha/year (*E. globulus*) 20.4–21.5 T/ha/year (*E. nitens*) (Pérez-Cruzado et al., 2011) |
| | Willow (*Salix* spp.) | Northern hemisphere | Worldwide average chip production: 10 T/ha/year (González-García et al., 2012b) Denmark: 11–22 T/ha/year (Callesen et al., 2010) Europe and North Central USA: 2–11 T/ha/year (Amichev et al., 2010) |
| | Poplar (*Populus* spp.) | | |
| | Aspen (*Populus tremula* L.) | | Scandinavia average: 7.9–9.5 T/ha/year (Tullus et al., 2009) |
| **Softwood** | Douglas Fir (*Pseudotsuga menziesii*) | North America | |
| | Norway Spruce (*Picea abies*) | Northern Europe | Sweden: 5–9 T/ha/year (Bergh et al., 2005) |
| | Pine (*Pinus* spp.) | Northern hemisphere, Chile, Australia, New Zealand | Australia: 17–39 T/ha/year (Snowdon and Benson, 1992) |

| | | | |
|---|---|---|---|
| **Dedicated crops** | Switch grass (*Panicum virgatum*) | North America | |
| | Miscanthus (*Miscanthus giganteus*) | Europe, Africa, South Asia | |
| | Giant Reed (*Arundo donax L.*) | Mediterranean, South Asia | Italy: 37.7 T/ha/year (Angelini et al., 2009) |
| | Cassava pulp (*Manihot esculenta*) | Africa (Nigeria), South East Asia | |
| | Hemp (*Cannabis sativa L*) | Canada, Europe | |
| | Bamboo (*Phyllostachys spp.*, *Bambusa bambos*, *Thyrsostachys siamensis*) | South Asia (India, Thailand) East Asia (China, Japan) | |
| **Crop residues** | Wheat straw | Asia, Europe, USA | |
| | Rice straw | Asia | |
| | Corn stover | USA, China | |
| | Corn cobs | USA, China | |
| | Oat straw | North-western Europe, Mid-eastern Africa | |
| | Sugarcane bagasse | Brazil, India, China | |
| | Barley straw | Europe | |
| | Cotton stalk | Asia | |
| | Sweet sorghum bagasse | South Asia Central America, Africa | |

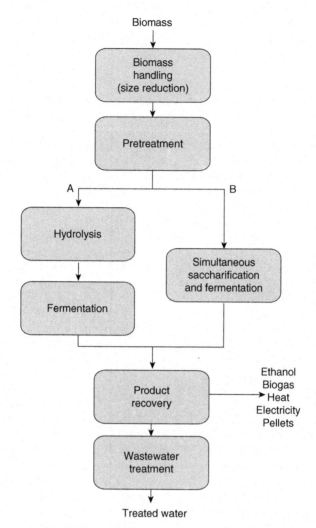

*9.3* Schematic process overview of a bioethanol-focused biorefinery. After pretreatment, the two conversion steps take place: first, degradation of cellulose into monomeric sugars, and second, fermentation of the sugars to ethanol. Traditionally this is done in either a separate hydrolysis and fermentation (A) or a simultaneous saccharification and fermentation (B) set-up.

Table 9.2 Composition of various lignocellulose feedstocks

| Type | Plant | Glucan | Xylan | Arabinan | Mannan | Galactan | Acetyl | Lignin[a] | Extractives[b] | Reference |
|---|---|---|---|---|---|---|---|---|---|---|
| **Hardwood** | Birch | 38.2 | 18.5 | NR | 1.2 | NR | NR | 22.8 | 2.3 | Hayn et al., 1993 |
| | Willow | 43.0 | 14.9 | 1.2 | 3.2 | 2.0 | 2.9 | 24.2 | NR | Sassner et al., 2006 |
| | Poplar | 49.9 | 17.4 | 1.8 | 4.7 | 1.2 | NR | 18.1 | NR | Wiselogel et al., 1996 |
| | Red Maple | 41.9 | 19.3 | 0.8 | NR | NR | NR | 24.9 | NR | Jae et al., 2010 |
| | Eucalyptus | 42.9[c] | 12.7[c] | 2.3[c] | 0.9[c] | 2.2[c] | NR | 16.7 | 19.2 | Vázquez et al., 2007 |
| | | 46.1 | 17.1 | 0.8 | 0.4 | 1.5 | NR | 19.8 | 0.6 | Rencoret et al., 2010 |
| | Aspen | 45.9 | 16.7 | 0.0 | 1.2 | 0.0 | NR | 23.0 | NR | Youngblood et al., 2010 |
| **Softwood** | Douglas Fir | 43.0 | 3.0 | 1.0 | 13.0 | 2.0 | NR | 28.0 | NR | Mabee et al., 2006 |
| | | 45.5 | 3.1 | 0.7 | 12.7 | 4.3 | NR | 30.6 | 3.8 | Johansson, 2010 |
| | Spruce | 43.4 | 4.9 | 1.1 | 12.0 | 1.8 | NR | 28.1 | 1.0 | Tengborg et al., 1998 |
| | | 41.9 | 6.1 | 1.2 | 14.3 | NR | NR | 27.1 | 3.8 | Hayn et al., 1993 |
| | Pine | 46.4 | 8.8 | 2.4 | 11.7 | NR | NR | 29.4 | NR | Wiselogel et al., 1996 |
| | | 37.7 | 4.6 | 0.0 | 7.0 | NR | NR | 27.5 | 5.4 | Hayn et al., 1993 |
| | | 41.7 | 6.3 | 1.8 | 10.8 | 3.9 | NR | 26.9 | NR | Youngblood et al., 2010 |
| | Western Hemlock | 41.4 | 3.3 | 1.0 | 12.0 | 1.8 | NR | 31.4 | 0.8 | Johansson, 2010 |
| **Crop residues** | Wheat straw | 38.2 | 21.2 | 2.5 | 0.3 | 0.7 | NR | 23.4 | 13.0 | Wiselogel et al., 1996 |
| | | 35.2 | 30.5 | 4.4 | 0.0 | 0.0 | NR | 18.5 | NR | Foyle et al., 2007 |
| | | 36.5 | 18.4 | 2.2 | 0.0 | NR | NR | 17.6 | 3.6 | Hayn et al., 1993 |
| | Rice straw | 34.2 | 24.5 | NR | NR | NR | NR | 11.9 | 17.9 | Wiselogel et al., 1996 |
| | | 38.9 | 20.4 | 3.4 | 0.0 | 0.5 | NR | 13.5 | 5.3 | Kadam et al., 2000 |
| | Corn stover | 35.6 | 18.9 | 2.9 | 0.3 | NR | NR | 12.3 | 5.5 | Hayn et al., 1993 |
| | | 38.9 | 23.0 | 3.4 | 0.4 | 1.8 | 2.6 | 16.2 | NR | Templeton et al., 2009 |
| | | 36.4 | 18.0 | 3.0 | 0.6 | 1.0 | NR | 16.6 | 7.3 | Wiselogel et al., 1996 |
| | Corn cobs | 37.0 | 27.8 | 2.2 | NR | 0.6 | NR | 13.9 | NR | Wang et al., 2011 |
| | Sugarcane bagasse | 39.0 | 22.1 | 2.1 | 0.4 | 0.5 | NR | 23.1 | NR | DOE, USA |

*Continued*

Table 9.2 Continued

| Type | Plant | Glucan | Xylan | Arabinan | Mannan | Galactan | Acetyl | Lignin[a] | Extractives[b] | Reference |
|---|---|---|---|---|---|---|---|---|---|---|
| | Barley straw | 38.1 | 18.7 | 3.9 | 0.0 | 0.0 | NR | 20.5 | NR | Kim et al., 2011 |
| | Cotton stalk | 35.6 | 21.4 | 0.0 | 0.0 | 0.0 | NR | 27.8 | NR | Akpinar et al., 2007 |
| | Sweet sorghum bagasse | 41.3 | 18.0 | 1.94 | 0.85 | 1.26 | NR | 16.5 | NR | Goshadrou et al., 2011 |
| Dedicated crops | Switch grass | 31.0 | 20.4 | 2.8 | 0.3 | 0.9 | NR | 17.6 | 17.0 | Wiselogel et al., 1996 DOE, USA |
| | | 34.2 | 22.8 | 3.1 | 0.3 | 1.4 | NR | 19.1 | NR | |
| | Miscanthus | 39.5[c] | 19.0[c] | 1.8[c] | NR | 0.4 | NR | 24.1 | 4.2 | Vrije et al., 2002 |
| | Arundo donax L. | 39.3 | 18.4 | 1.2 | 0.2 | 0.4 | NR | 26.2 | NR | Bura et al., 2012 |
| | Cassava pulp | 19.1 | 4.2 | 1.4 | 0.7 | 0.5 | NR | 2.2 | NR | Kosugi et al., 2009 |
| | Bamboo | 42.6 | 15.0 | 0.0 | 0.0 | 0.0 | NR | 26.2 | NR | Sathitsuksanoh et al., 2010 |
| | | 40.7 | 23.6 | 1.1 | 0.6 | 1.2 | NR | 27.1 | NR | Tippayawong and Chanhom, 2011 |
| | Hemp (Cannabis sativa L) | 37.4 | 21.1 | 2.9 | NR | NR | 2.9 | 18.0 | NR | González-García et al., 2012a |
| Secondary and tertiary | Sorghum fiber | 28.7 | 15.8 | 2.03 | 0.4 | 0.4 | NR | NR | NR | Godin et al., 2011 |
| | Newspaper | 35.1[c] | 5.0[c] | 3.9[c] | 10.7[c] | 2.3[c] | NR | 39.1[c] | NR | Foyle et al., 2007 |
| | White office paper | 65.4[c] | 14.4[c] | 0.0 | 0.0 | 0.0 | NR | 9.5[c] | NR | Foyle et al., 2007 |

NR: Not reported.
[a]The values denote acid insoluble or klason lignin content.
[b]The values denote organic solvent extractives (see corresponding reference for more specific details).
[c]The values denote the monomeric form of the respective carbohydrate.
DOE: Department of Energy, USA (http://www.afdc.energy.gov/biomass/progs/search1.cgi).

function of cellulose is not to serve as an energy store, but rather to serve as a construction material. Protection of the cellulose by lignin and hemicellulose reinforces that structure, and the degradation rate in nature is therefore lower than desirable for technical application (but well suited for a functioning eco-system). A pretreatment of the material makes the fibers more accessible to enzymatic attacks. The treatment aims to open up the structure of fibers, solubilize parts of the material (i.e., lignin or hemicelluloses), reduce particle size and degree of polymerization (DP), and increase the surface area of the material.

The pretreatment in many ways determines the bioethanol process, since the properties of the material after pretreatment will have an impact on all the subsequent steps in the production. In general, it is important to design the pretreatment so that not too much degradation products are formed, since these will impact primarily the fermentation, but also the hydrolysis (Almeida *et al.*, 2007).

Pretreatment methods can be divided into physical or chemical methods, although a combination of the two is often used. One can say that there are two principal strategies of the pretreatment; either you aim to remove mainly the lignin or you aim to remove mainly the hemicellulose fraction from the material. No method will completely succeed with any of these two aims, but hemicellulose is mostly removed with dilute acid catalyzed steam pretreatment, and lignin is to a larger extent removed by alkaline pretreatment, and oxidation improves the efficiency of lignin removal (Alvira *et al.*, 2010; Galbe and Zacchi, 2007). Two common pretreatment abbreviations used are STEX for steam explosion and AFEX for ammonia fiber explosion. A compilation of some of the most common technologies is given in Table 9.3.

### 9.3.2 Hydrolysis

The pretreatment will leave most of the cellulose in polymeric form, and, depending on the pretreatment, some hemicelluloses in polymeric or oligomeric form. Enzymatic hydrolysis of cellulose (and hemicelluloses) into monomeric sugars is thus needed since most microorganisms utilize only monomeric (or possibly dimeric) sugars during fermentation. The enzymatic hydrolysis was long regarded as the principal bottleneck in bioethanol production from lignocellulose due to the rather slow action of cellulases and the need for large amounts of expensive enzymes. However, impressive progress in the past decade has resulted from major R&D efforts by enzyme developers, such as Novozymes, Genencor/DuPont, DSM and others, to reduce both enzyme loadings and production costs. The US Department of Energy (DOE) set a target that enzyme costs should not exceed US$0.12/gallon ethanol by 2012 in their grants to some major

Table 9.3 Summary of different methods for lignocellulose biomass pretreatment

| Pretreatment method | Main action | Advantage | Disadvantage |
| --- | --- | --- | --- |
| Biological | Lignin and hemicellulose degradation | Low energy consumption | Slow process |
| Milling | Particle size reduction | Reduces cellulose crystallinity | High energy consumption |
| Steam explosion (STEX) | Hemicellulose removal | High glucose yields and cost effective | Partial hemicelluloses removal Some inhibitor generation |
| Ammonia fiber explosion (AFEX) | Lignin removal | Increased accessible surface area Low formation of inhibitors | Not efficient for lignin rich materials Large amounts of ammonia |
| Wet oxidation | Lignin removal | Minimizes energy demand (exotermic) Low formation of inhibitors | High cost of oxygen and alkaline catalyst |
| Organosolv | Lignin and hemicelluloses hydrolysis | Targets both hemicellulose and lignin | High cost Solvents need to be drained and recycled |
| Dilute acid | Hemicellulose removal | Less corrosion problems and less formation of inhibitors compared to concentrated acid | More inhibitor generation than STEX |
| Concentrated acid[a] | Hemicellulose and cellulose degradation | High glucose yield Ambient temperatures | Formation of inhibitors High cost of acid and acid recirculation Reactor corrosion |

[a]Can also be considered as a method for complete hydrolysis to monomeric sugars, as opposed to a pretreatment step.

enzyme companies. A benchmarking of enzymes from these companies was recently conducted by the National Renewable Energy Laboratory (NREL) (McMillan *et al.*, 2011).

Traditionally, enzymatic hydrolysis has been regarded as a synergetic reaction between three major classes of enzymes, i.e. endo-1,4-β-glucanases, exo-1,4-β-glucanases and β-glucosidases (Van Dyk and Pletschke, 2012). Endo-1,4-β-glucanases randomly cleave internal bonds in the cellulose polymer which results in the formation of two new chain ends. Exo-1,4-β-glucanases (mainly cellobiohydrolases, CBH) are the most abundant component in both natural and commercial enzyme mixtures and they mainly work in a processive manner, cutting (mainly) cellobiose units from either the reducing or non-reducing ends of the glucan chain. β-Glucosidases catalyze the hydrolytic splitting of cellobiose into glucose. The enzymes are end-product inhibited by glucose and especially cellobiose. Therefore it is important to have an enzyme blend with sufficient β-glucosidase activity in order to avoid cellobiose inhibition of the cellobiohydrolases and to be able to cope with the activity loss due to the accumulated high amount of glucose.

The synergistic effects between the different cellulase components have been extensively studied throughout the years and more recently also the synergism between cellulases and other auxiliary enzymes, e.g. xylanases, xylosidases, mannanases and esterases have been studied (Van Dyk and Pletschke, 2012). In particular it has been shown that xylobiose, and larger xylo-oligomers, exhibit a very strong inhibition on the cellulases, which can be reduced/minimized by adequate supplementation of hemicellulases (Qing and Wyman, 2011). This xylo-oligomer inhibition can be quite significant when working with xylan-rich lignocellulosic substrates (e.g., agricultural residues and hardwood; Table 9.2), especially when a rather mild pretreatment method has been used, which gives solubilization of xylan into oligomeric rather than monomeric xylose.

Another complication when hydrolyzing lignocellulosic materials is the presence of large amounts of lignin in the material (if a lignin dissolving pretreatment method has not been used). Lignin could act as a physical barrier on the cellulose surface, limiting the accessibility of the enzymes, and it has been shown to unproductively bind enzymes which lower the free amount of enzymes able to hydrolytically degrade the cellulose/cellobiose (Palonen *et al.*, 2004).

Recently a new type of enzyme and enzyme action was identified as an important component for efficient enzymatic hydrolysis of cellulosic material (Horn *et al.*, 2012). The enzyme belongs to the GH61 family (the abbreviation GH stands for glycoside hydrolases) but the enzyme is in fact an oxidizing enzyme that oxidizes the C1/C4 carbon on the glucan chain. Hence it has a completely different way of cleaving intermolecular glucose bonds compared to the hydrolytic cellulases mentioned earlier. Unlike the cellobiohydrolases, the GH61 enzyme does not work in a processive manner, and does not appear to need a specific binding site on the polymer chains. Therefore it can attack and cut the cellulose chain at any location, creating two new free ends for the cellobiohydrolases to act on. As such, it has the potential to break crystalline cellulose structures and enhance the hydrolysis rate by creating more reactive sites for the CBH enzymes.

### 9.3.3 Fermentation

Once the fermentable sugars have been obtained in the hydrolysis, the next issue is the conversion of these sugars (all sugars, or only a selected stream) by fermentation. The ideal reaction for conversion of glucose (the main hexose sugar) to ethanol is:

$$C_6H_{12}O_6 \rightarrow 2C_2H_5OH + 2CO_2$$

i.e., one mole of glucose gives a maximum of two moles of ethanol, which corresponds to 0.51 g ethanol/g glucose. In practice, one should not count on more than about 90% efficiency in the fermentation, i.e. about 0.46 g/g (Öhgren *et al.*, 2007). The other main hexose sugars are mannose and galactose (see Table 9.2). Out of these sugars, mannose is normally easily fermented, whereas the fermentation of galactose will depend on the microorganism used. The fermentation does not need to be done after the hydrolysis, but can be done concomitant with the enzymatic hydrolysis – a process option called SSF (simultaneous saccharification and fermentation; see further discussion later on). The microorganism used for the fermentation must be able to work in the process streams with as little conditioning of the stream as possible. The microorganism should therefore be:

- inhibitory tolerant (to compounds present in the hydrolysate)
- able to utilize multiple sugars (hexoses and pentoses)
- ethanol tolerant
- able to ferment at low pH to minimize contamination.

There are many microorganisms that are able to ferment sugars, including yeasts, e.g. *Saccharomyces cerevisiae*, *Scheffersomyces stipitis*,[2] *Kluyveromyces marxianus*, *Dekkera bruxellensis* (Blomqvist *et al.*, 2010), bacteria such as *Zymomonas mobilis* (Rogers *et al.*, 1982), *Escherichia coli* (reviewed by, e.g., Jarboe *et al.*, 2007) and filamentous fungi such as *Mucor indicus*, and *Rhizopus oryzae* (Abedinifar *et al.*, 2009). The organisms have different pros and cons in terms of ethanol tolerance, fermentation rate, sugar utilization range, by-product formation pattern, and tolerance to inhibitors.

[2]This yeast was previously called *Pichia stipitis* and most publications will be found under this name.

Genetic engineering of the organism is often needed to improve properties. For instance *E. coli* produces only little ethanol in its native state, but can be transformed into an ethanologenic organism (Gonzalez *et al.*, 2003).

The process design, or the entire refinery concept, will influence the choice of organism. The organism will need to be able to use all sugars if a maximum ethanol production is desired (which is not the case in all biorefinery concepts). Alternatively, if the SSF configuration is chosen, a high thermo-tolerance of the microorganism will be an advantage (apart from the above four) as the temperatures optima for enzymatic hydrolysis (typically 45–50°C) and fermentation (typically 30–35°C) differ significantly. The thermo-tolerant yeast *Kluyveromyces marxianus* may be of interest in these processes (Pessani *et al.*, 2011). However, the prime choice in the current sugar and starch-based fermentation industry has been the common Baker's yeast, *Saccharomyces cerevisiae* and there is good cause to believe that this will be the case also in lignocellulose-based biorefinery approaches. There is well-established process technology for large-scale production of the yeast (Østergaard *et al.*, 2000), and it is fully accepted in the fermentation industry today and enjoys GRAS status (i.e., it is generally regarded as safe). *S. cerevisiae* has an excellent ethanol tolerance, some strains tolerate up to 20% (v/v) (Verduyn *et al.*, 1990; Casey and Ingledew, 1986). Not least important is that the organism is relatively robust to inhibitors from the pretreatment step of lignocellulose (Hahn-Hägerdal *et al.*, 1994; Olsson and Hahn-Hägerdal, 1993).

### 9.3.4 Pentose utilization

In biorefinery concepts aiming at a maximum ethanol yield, all sugar streams should be fermented to ethanol, which requires the conversion also of pentoses. However, wild-type *S. cerevisiae* is not able to ferment pentoses, and genetic engineering of the organism is needed to enable pentose conversion. The main pentose sugar in most biomass is xylose (D-xylose) (Table 9.2) and metabolic engineering of the yeast has therefore aimed at enabling xylose fermentation in *S. cerevisiae* (reviewed by, e.g., Almeida *et al.*, 2011; Van Vleet and Jeffries, 2009; Matsushika *et al.*, 2009; Hahn-Hägerdal *et al.*, 2007). There are two main options to introduce the enzymatic pathway which will enable conversion of the pentose xylose. One option is the two-step conversion via the enzymes xylose reductase (XR) and xylitol dehydrogenase (XDH). The other option is to use the one-step isomerization catalyzed by xylose isomerase (XI). In both cases the sugar xylulose is formed, which later is phosphorylated by the native enzyme xylulokinase (XK) to xylulose-5-phosphate, and enters into the main glycolysis via the pentose phosphate pathway (PPP) (see Fig. 9.4). The mere introduction of the pathway is, however, not sufficient for obtaining efficient yeast strains

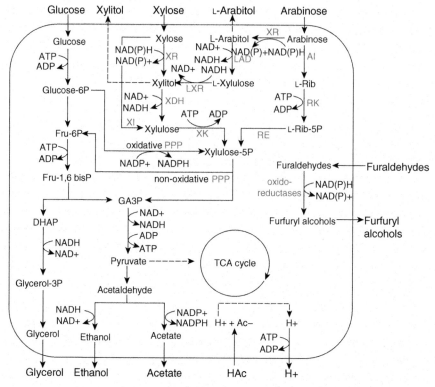

*9.4* Xylose and arabinose pathways expressed in recombinant *Saccharomyces cerevisiae*. The utilization of co-factors and ATP in the central carbon metabolism is depicted schematically. Abbreviations: XR, xylose reductase; XDH, xylitol dehydrogenase; XI, xylose isomerase; XK, xylulokinase; AI, arabinose isomerase; RK, ribulokinase; RE, ribulose-5-P 4-epimerase; LAD, L-arabitol dehydrogenase; LXR, L-xylulosereductase; PPP, pentose phosphate pathway; Fru-6P, fructose 6-phosphate; Fru-1,6 bisP, fructose 1,6 bisphosphate; GA3P: glyceraldehyde-3 phosphate; DHAP, dihydroxyacetone phosphate; L-Rib, L-ribulose; L-Rib-5P, L-ribulose 5-phosphate; HAc, undissociated weak acid; Ac-, dissociated weak acid. (Reprinted with permission from J. Almeida, D. Runquist, V. Sànchez i Nogué, G. Lidén and M. F. Gorwa-Grauslund. 'Stress-related challenges in pentose fermentation to ethanol by the yeast *Saccharomyces cerevisiae*', *Biotechnol. J*, 6, 286–299, 2011.)

and additional improvements are necessary (as discussed in, e.g., Almeida *et al.*, 2011). In the XR/XDH pathway, a main issue has been the co-factor dependencies of the two reactions, which may result in excretion of the by-product xylitol. This co-factor problem is avoided when using the XI pathway, but a main challenge has been to find an isomerase showing sufficient activity at allowable temperature for the yeast. Successful results were not really obtained until a XI from the anaerobic fungus *Piromyces* was expressed in *S. cerevisiae* (Kuyper *et al.*, 2003, 2005). More recently also other useful XIs, from the anaerobic bacterium *Clostridium phytofermentans* (Brat *et al.*, 2009), the fungus *Orpinomyces* (Madhavan *et al.*, 2009), and from *Burkholderia cenocepacia* (de Figueiredo Vilela *et al.*, 2013), have been found and expressed in *S. cerevisiae*.

In addition to the direct genetic engineering, i.e. introduction of the 'missing' heterologous genes, it has been shown to be very important to evolve the strains on the xylose-rich substrate. In particular, this was crucial in obtaining the first successful XI strain (Kuyper *et al.*, 2003). During the evolution, the relative expression levels are 'tuned' for optimal performance by the yeast itself through spontaneous mutation and selection mechanisms (Zhou *et al.*, 2012; Lee *et al.*, 2012). Some of the currently available *S. cerevisiae* strains designed for utilization of xylose are given in Table 9.4.

A second pentose found at relatively high levels in some materials is arabinose (L-arabinose). The problem of introducing arabinose utilization

*Table 9.4* Some of the xylose fermenting engineered strains of *S. cerevisiae*

| Strain | Xylose utilization[a] | Description | Reference |
|---|---|---|---|
| TMB3400 | XR/XDH/XK + RM | Xyl1, Xyl2-*P. stipitis* XK-*S. cerevisiae* | Wahlbom *et al.*, 2003 |
| MA-R5 | XR/XDH/XK | Xyl1, Xyl2-*P. stipitis* XK-*S. cerevisiae* (ScXK) | Matsushika *et al.*, 2009 |
| 424A (LNH-ST) | XR/XDH/XK | US patent application #08/148, 581; patent no. 5789210 | Sedlak and Ho, 2004 |
| – | XI | XylA-*Piromyces* sp. strain E2 | Kuyper *et al.*, 2003 |
| RWB218 | XI + EE | XylA-*Piromyces* sp. strain E2 | Kuyper *et al.*, 2005 |
| INVSc1/pRS406XKS/ pILSUT1/pWOXYLA | XI | XylA-*Orpinomyces* SUT1-*P. stipitis* XKS-*S. cerevisiae* | Madhavan *et al.*, 2009 |
| BWY10Xyl/ YEp-opt.*Xl*-Clos-K | XI + EE | XylA-*Clostridium phytofermentans* | Brat *et al.*, 2009 |
| BY4741-S1 | XI + RM + EE | XylA-*Piromyces* sp. | Lee *et al.*, 2012 |
| H131-A3-AL[cs] | XI/XK/NOPPP + EE | XylA-*Piromyces* sp. Xyl3- *P. stipitis* | Zhou *et al.*, 2012 |

[a] XR- xylose reductase; XDH, xylose dehydrogenase; XI, xylose isomerase; XK, xylose kinase; RM, random mutagenesis; EE, evolutionary engineering; NOPPP, overexpression of non-oxidative pentose phosphate pathway genes.

into yeast is similar to that for xylose utilization as both pathways end with the formation of xylulose (Fig. 9.4; Almeida *et al.*, 2011). Also similarly there is an isomerase pathway and a reduction/oxidation pathway. Arabinose utilization on its own is less interesting, and work has therefore focused on combining utilization of L-arabinose and D-xylose (Bettiga *et al.*, 2009; Sanchez *et al.*, 2010). Whereas there are currently several xylose fermenting yeast strains available, this is not yet quite true for arabinose co-utilizing yeasts.

### 9.3.5 Inhibitor tolerance

The challenge in lignocellulose fermentation is not only to develop organisms which use more sugars, but to have them do that in a complex medium containing several inhibitors (Fig. 9.5). The pretreatment process generates several compounds, which affect the performance of the fermenting microorganism (Klinke *et al.*, 2004; Palmqvist and Hahn-Hägerdal, 2000). These inhibitors can be divided into:

- furans – most important 2-furaldehyde (or furfural) and 5-hydroxymethyl-2-furaldehyde (or HMF);

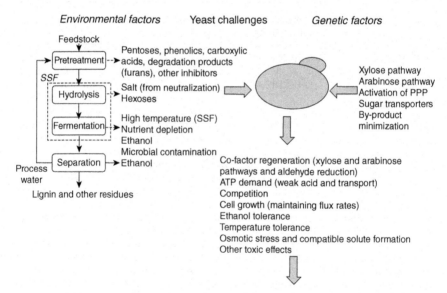

*9.5* Challenges faced by the yeast *Saccharomyces cerevisiae* during the production of ethanol from lignocellulosic feedstocks. (Reprinted with permission from J. Almeida, D. Runquist, V. Sànchez i Nogué, G. Lidén and M. F. Gorwa-Grauslund. 'Stress-related challenges in pentose fermentation to ethanol by the yeast *Saccharomyces cerevisiae*', *Biotechnol. J*, 6, 286–299, 2011.)

- weak acids – acetic acid, levulinic and formic acid formed in degradation of furans); and
- phenolics – derived from the lignin fraction.

The furans are formed in the degradation of monosaccharides, and levulinic and formic acid come from further degradation of the furans. Acetic acid is different in the sense that it is already present in the material itself in terms of acetyl groups on hemicellulose.

One can handle the inhibition problem in different ways (Fig. 9.6). Obviously, the best situation would be not to have any inhibitors in the medium. That would require either a process that does not generate any inhibitors, or the removal of inhibitors. Alternatively, one can screen for strains that are tolerant, or develop more tolerant strains. Also the fermentation process in itself may be important in avoiding problems with inhibition. The reason is that several of the inhibitors, notably several aldehyde compounds such as furfural and HMF, are converted by the yeast itself into less toxic alcohols (Almeida *et al.*, 2007, 2009) and an *in-situ* detoxification can be obtained during the process if suitably tuned.

By understanding this particular mechanism of inhibition, several genetic targets have been identified. Overexpression of the gene encoding for

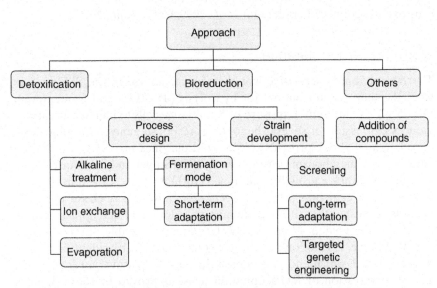

*9.6* A schematic representation on how to deal with inhibition. (Reprinted with permission from J. R. M. Almeida, M. Bertilsson, M. F. Gorwa-Grauslund, S. Gorsich and G. Lidén. 'Metabolic effects of furaldehyde and impacts on biotechnological processes', *J Appl Microbiol Biotechnol*, 82, 625–638, 2009.)

alcohol dehydrogenase (ADH6) was, for instance, found to increase the ability of *S. cerevisiae* to convert HMF (Petersson *et al.*, 2006) and also increased the ethanol formation from a hydrolysate. Alriksson *et al.* (2010) observed that overexpression of three native genes (ATR1, FLR1, YAP1) involving multidrug resistance and stress response in *S. cerevisiae* resulted in enhanced resistance to coniferyl aldehyde, HMF, and spruce hydrolysate. More targets for improved resistance are continuously reported. Very promising results have also recently been reported from adding sulfite or dithionite to the medium, which gave detoxifying effects on spruce hydrolysates (Alriksson *et al.*, 2011). The mechanism of detoxification may partly be reduction, but also sulfonation of inhibitors (Cavka *et al.*, 2011).

Acetic acid is a different story in comparison to many other inhibitors in that it is linked to the material composition *per se* since the hemicellulose is acetylated. Thus acetic acid is very difficult to avoid if the hemicellulose is to be hydrolyzed. The effects of weak acids are pH dependent, since it is the undissociated form of the acids which diffuses across the plasma membrane. There it dissociates and releases a proton due to a higher intracellular pH. To maintain the intracellular pH, the cell has to export the proton which costs ATP. The pKa value of the acid will determine at what pH the undissociated form of the acid dominates. Since many carboxylic acids have pKa in the range 4.5–5, large changes in toxicity are likely to occur for pH changes in this range. Xylose utilization appears to be extra sensitive to acetic acid (Bellissimi *et al.*, 2009; Casey *et al.*, 2010).

### 9.3.6 Thermo-tolerance

Thermo-tolerant *S. cerevisiae* would be an advantage in the SSF process since the temperature optima for hydrolysis (45–50°C) and fermentation (30–35°C) are different. Exposure of *S. cerevisiae* cells to high temperatures induces expression of several heat stress–response genes, including those encoding heat shock proteins (HSPs) and enzymes involved in trehalose and glycogen metabolism (Morano *et al.*, 1998; Singer and Lindquist, 1998). A high-temperature growth phenotype (Htg+) was categorized in thermo-tolerance of *S. cerevisiae* (Steinmetz *et al.*, 2002). Results of classical genetic analysis suggested that the Htg phenotype is dominant and approximately six genes, designated *HTG1* to *HTG6*, are responsible for conferring this phenotype. *HTG6*, one of the six genes was recently identified to be the gene *RSP5*, which encodes a ubiquitin ligase (Shahsavarani *et al.*, 2012) and overexpression of *RSP5* ubiquitin ligase improved thermo-tolerance in *S. cerevisiae*. Zhang *et al.* (2012) isolated thermo-tolerant *S. cerevisiae* with a high-energy pulse electron (HEPE) beam, and obtained a strain which could produce more than 80 g/L of ethanol at as high a temperature as 43°C.

## 9.4 Design options for biorefining processes

In designing a process, there is a choice between completing the hydrolysis before starting the fermentation – separate hydrolysis and fermentation (SHF) – or running the hydrolysis together with the fermentation – simultaneous saccharification and fermentation (SSF).

### 9.4.1 Separate hydrolysis and fermentation (SHF) or simultaneous saccharification and fermentation (SSF)?

The two basic choices have different pros and cons (Table 9.5). The fact that hydrolysis and fermentation are carried out in separate vessels in SHF introduces a degree of freedom to operate each process at its respective optimal conditions, however, at the cost of one more vessel. The extra capital cost may be quite substantial as indicated by Wingren *et al.* (2003). If the solid fraction (mainly lignin) is removed after the enzymatic hydrolysis, it is possibly to recycle the yeast in an SHF process. This is not feasible in SSF, since the yeast cannot readily be separated from the remaining lignin. However, in the removal of the lignin fraction in SHF, there may be non-negligible sugar losses, especially if a high solids loading is used in the enzymatic hydrolysis. In contrast, if the entire slurry is sent to fermentation, all monosaccharides liberated in the pretreatment can potentially be fermented. The removal of sugars as they are released in the hydrolysis in

*Table 9.5* Comparison of SHF and SSF

| SHF | | SSF | |
|---|---|---|---|
| Advantages | Disadvantages | Advantages | Disadvantages |
| Different temperature and pH for enzymatic hydrolysis and fermentation possible | Capital cost – one more reactor | Decreased sugar end-product inhibition | Same temperature and pH for both hydrolysis and fermentation |
| Yeast recirculation possible | Loss of sugars during solid/liquid separation | *In situ* detoxification | No yeast recirculation |
| | End-product inhibition | Decreased capital cost | |
| | Inhibition by other inhibitors | Improvement of xylose fermentation by increased ratio of xylose to glucose | |

SSF is furthermore an important advantage since end-product inhibition of cellulases can be minimized. An increased overall yield for these two reasons was the main benefit stated by the original inventors of SSF (Gauss et al., 1976). The enzymes are to some extent inhibited by the ethanol produced, but this inhibition is relatively low compared to that of cellobiose and glucose (Wu and Lee, 1997). Decreased end-product inhibition of cellulases, in particular cellobiose inhibition on CBH, is, however, a major target in development of more efficient enzymes, which may tilt the balance in favor of SHF processes.

Co-fermentation of pentoses and hexoses in SSF is sometimes dignified with the abbreviation SSCF, standing for simultaneous saccharification and co-fermentation. SSCF may offer an advantage over SHF in these cases since the ratio between xylose and glucose can be higher than in SHF (for a more thorough discussion, see Olofsson et al., 2008). This is true for the cases in which the pretreatment results in the release of monosaccharides from hemicellulose, e.g. in acid catalyzed steam pretreatment, but not in cases where the pretreatment does not release monosaccharides.

Since the yeast cannot be reused in the SSF process, it has to be produced for each batch. Efficient aerobic cultivation of *S. cerevisiae* has to be performed in a fed-batch mode to avoid glucose repression which would give a low biomass yield. The fed-batch cultivation of the yeast should use lignocellulose hydrolysate (derived from the pretreatment stage) to adapt the yeast since this improves the performance significantly (Alkasrawi et al., 2006). The cultivation must be designed not only to avoid glucose repression but also inhibition by the medium. RQ (respiratory quotient), DOT (dissolved oxygen tension), or ethanol concentration can be used as measured variables for feedback control during the propagation of the yeast (as discussed by Rudolf, 2007).

The choice between SHF and SSF will depend on both feedstock and the desired product spectrum in the biorefinery. There are many different variants of both these concepts (Fig. 9.7), in particular with respect to feeding of substrates and enzymes. Feeding will be a means to minimize effects of inhibition, increase xylose conversion, or handle high viscosities (Olofsson et al., 2008).

## 9.5 Process intensification: increasing the dry-matter content

A key aspect to attain a good economy in the bioethanol process is to reach a sufficiently high ethanol concentration after the fermentation step, since this reduces the cost (specific energy demand) of the subsequent distillation in a non-linear fashion (Galbe et al., 2007). The cost curve gradually flattens out and a concentration above 5 wt% is relatively satisfactory. High ethanol

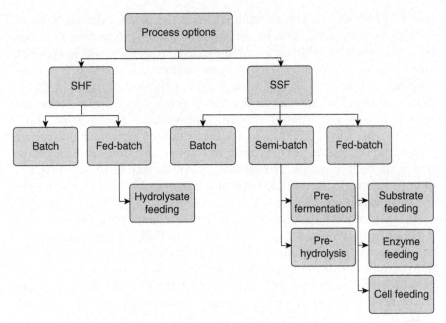

*9.7* A schematic representation of different fermentation strategies for production of bioethanol using lignocelluloses.

titers will require that the process operate at high biomass loadings to give high concentrations throughout the production. This also decreases the use of process water and capacity needed for water treatment. Increasing the biomass concentration in the hydrolysis and fermentation step is, however, not without problems since enzymatic conversion yields tend to decrease dramatically at elevated biomass loadings even at the same specific enzyme loading (Kristensen *et al.*, 2009; Mohagheghi *et al.*, 1992). There will therefore be an economic optimum at some intermediate biomass loading, which will vary with the type of biomass used, the price of enzymes, and the kind of by/co-products obtained.

## 9.5.1 Hybrid processes and novel concepts

The desire to increase the biomass content in the process has led to different 'hybrid' process configurations which employ a combination of the SHF and SSF approach – aiming at getting the best of both worlds. One such option is the inclusion of a so-called viscosity reduction step, which in essence is an enzymatic hydrolysis step operated at optimal hydrolysis temperature for a short time, but not to complete hydrolysis. A high initial hydrolysis rate can be achieved, and the process can continue as an SSF process, much like in corn-based ethanol production. One can also combine this with various fed-batch strategies in which material, enzymes and/or yeast are fed to the fermentation vessel during the process. Another option is to apply a changing temperature during the process. This is of particular interest in the SSF process, where a non-isothermal operation may give overall process improvements (Kang et al., 2012; Mutturi and Lidén, 2013). The relative rates of enzymatic hydrolysis, glucose consumption, yeast growth and decay are affected during such non-isothermal operation. Process benefits will depend on the recalcitrance of the pretreated material, and optimal and permissible temperature ranges of enzymes and yeast.

Another process concept is the so-called consolidated bioprocessing (CBP). In a CBP process, four steps (i.e., production of hydrolytic enzymes (cellulases and hemicellulases), hydrolysis of polymeric carbohydrates in the pretreated substrate, fermentation of hexoses, and fermentation of pentoses) are carried out in a single reactor (Lynd et al., 2005). One can either use a wild-type organism, such as *Clostridium phytofermentans* (Jin et al., 2011) able to both produce cellulases and ferment, or use genetic engineering to introduce genes encoding one or several cellulases into a fermenting microorganism, or alternatively engineer a good cellulase producing organism into an ethanol producer. The main approach has been to introduce cellulase encoding genes into a fermenting organism, e.g. *S. cerevisiae*. CBP is slowly gaining recognition as a potential low cost biomass processing methodology (van Zyl et al., 2007). Also here one can foresee various hybrid concepts, in which perhaps a 'base mixture' of enzymes is added, complemented by some specific enzyme activities that are expressed by tailored yeasts.

## 9.6 Different types of ethanol biorefineries

The ethanol biorefinery can be set up with focus on different combinations of co-products. To some extent there is flexibility in the operation, but the basic layout will set limits to this flexibility. With ethanol as the main product from the cellulosic part of the biomass, we can classify the biorefinery as either energy-, lignin- or C5-driven with respect to how the parts of the biomass not giving ethanol are treated (see Figs 9.8–9.10). In a recent study by Ekman et al. (2013), three different energy product scenarios were compared for a tentative straw-based ethanol biorefinery in Sweden. Ethanol was produced from either only the C6 fraction (C5 was taken for biogas production) or the C6 + C5 fractions. Surplus lignin (not needed for process heat) was used for electricity production. For the overall process economy it was essential to include a heat sink, i.e. a district

# Developments in bioethanol fuel-focused biorefineries 283

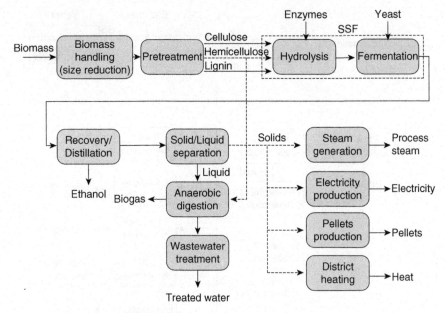

9.8 Conceptual figure for an 'energy-driven' bioethanol-based biorefinery. The focus is to produce several different energy carriers to be able to make as much use as possible of the total energy content in the raw material.

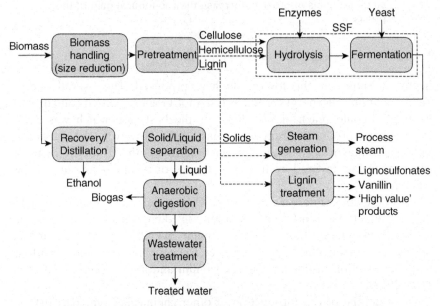

9.9 Conceptual figure for a 'lignin-driven' bioethanol-based biorefinery. The focus is to separate lignin early on in the process and use all or some of it to produce high value co/by-products in order to increase the economical gain of the process.

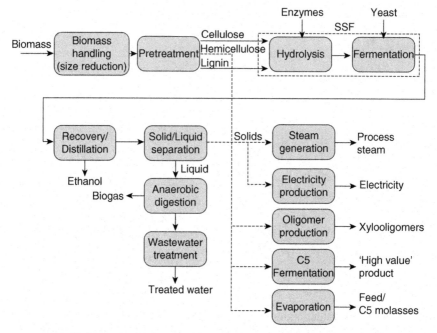

*9.10* Conceptual figure for a 'C5-driven' bioethanol-based biorefinery. The focus is to solubilize and separate the hemicelluloses early on in the process in order to have a separate C5-fermentation to high value co/by-products in order to increase the economical gain of the process.

heating system, for the low grade heat produced. The overall energy yield – counting all 'energy' products (ethanol, biogas, electricity and district heating) – could reach almost 70% for the best case which was when ethanol was produced from C6 sugars only. However, the economic optimization favored an ethanol production from both C6 and C5 fractions. This was in fact favored even in the absence of revenues for sale of low value heat.

The energy biorefinery is the concept which dominates the lignocellulose ethanol plants currently in operation, in demonstration scale or in construction phase (Table 9.6). The plant operated by Borregaard in Sarpsborg, Norway, can be said, however, to be a lignin biorefinery, with its major production of lignosulfonates, vanillin and also cellulose (Rødsrud *et al.*, 2012).

The potential of (co-)production of other chemicals from (part of) the sugar streams is large and many tentatively interesting compounds have been suggested. As pointed out by Bozell and Petersen (2010), this is a

Table 9.6 Overview of operational demonstration plants and commercial plants currently under construction

| Company | Location | Process | Feedstock | Capacity* |
|---|---|---|---|---|
| **Operational demonstration plants** | | | | |
| Borregaard Industries Ltd[a] | Sarpsborg, Norway | Biochemical | Spruce pulp | 15,800 t/a |
| Abengoa Bioenergy | Salamanca, Spain | Biochemical | Barley and wheat straw | 4,000 t/a |
| Inbicon (DONG Energy) | Kalundborg, Denmark | Biochemical | Wheat straw | 4,300 t/a |
| SEKAB/EPAP | Örnsköldsvik, Sweden | Biochemical | Primary wood chips | 160 t/a |
| DuPont | Tennessee, USA | Biochemical | Corn stover, cobs and fibre; switchgrass | 750 t/a |
| Mascoma Corporation | Rome, NY, USA | Biochemical | Wood chips, switchgrass | 500 t/a |
| Iogen Corporation | Saskatoon, Canada | Biochemical | Wheat, barley and oat straw; corn stover, sugar cane bagasse | 1,600 t/a |
| Coskata | Pennsylvania, USA | Syngas fermentation | Wood chips, natural gas | 120 t/a |
| Verenium | Jennings, LA, USA | Biochemical | Sugarcane bagasse, dedicated energy crops, wood products | 4,200 t/a |
| Sued-Chemie AG | Straubing, Germany | Biochemical | Wheat straw | 1,000 t/a |
| **Commercial plants under construction** | | | | |
| Beta Renewables (Chemtex)[b] | Crescentino, Italy | Biochemical | Arundo Donax, wheat straw | 60,000 t/a |
| Abengoa Bioenergy | Kansas, USA | Biochemical | Corn stover, wheat straw, switch grass | 75,000 t/a |
| POET-DSM Advanced Biofuels | Iowa, USA | Biochemical | Agricultural residues | 75,000 t/a |
| INEOS Bio | Florida, USA | Syngas fermentation | Waste (vegetative, wood and garden) | 24,000 t/a |

[a] The Borregaard plant is actually a full-scale biorefinery with ethanol as one of their smaller co/by-products.
[b] The Beta Renewables plant has finished construction and is under start-up during summer 2013.
*The capacities are those stated by the respective companies for the IEA database.
Data taken from IEA Task 39 database (http://demoplants.bioenergy2020.eu/, 2012-11-20) and cross-checked with the Advanced Biofuels & Biobased Materials Project Database (Released Q3 2012 (07/27/12 – revision 1.1)).

diverging problem since the choices of products are so many. The US Department of Energy, in a highly cited study, produced a list of a number of particularly interesting bio-based platform chemicals (Werpy and Petersen, 2004). These were selected based on criteria such as available biomass precursors (carbohydrates, lignin, fats, and proteins), process platforms, useful building blocks for further processing, secondary chemicals obtainable, intermediates, final products and applications. The 12 most interesting sugar-based building blocks were 1,4-diacids (succinic, fumaric, and malic), furan-2,5-dicarboxylic acid, 3-hydroxypropionic acid, aspartic acid, glucaric acid, glutamic acid, itaconic acid, levulinic acid, 3-hydroxybutyrolactone, glycerol, sorbitol, and xylitol/arabitol. A spectrum of high-value and commodity chemicals can be obtained from these platform chemicals as shown in Table 9.7. The original list was revised by Bozell and Petersen (2010) and some compounds, such as glutamic acid and glucaric acid, were taken off the list of most interesting compounds. However, the majority of the compounds originally identified stayed on the list and many are targets for current development projects. A comprehensive overview of

*Table 9.7* Commodity chemicals potentially derived from lignocellulose feedstocks

| Source (carbon) | Platform compound | Biochemicals |
|---|---|---|
| Cellulose and hemicellulose (C3) | *Glycerol* | Fermentation products, propylene glycol, malonic, 1,3-PDO, diacids, propyl alcohol, dialdehyde, epoxides |
| | Lactic acid | Acrylates, L-propylene glycol, dioxanes, polyesters, lactide |
| | 3-Hydroxypropionate | Acrylates, acrylamides, esters, 1,3-propanediol, malonic acid |
| | Propionic acid | Reagent, propionol, acrylate |
| | Malonic acid | Pharma, intermediates |
| | Serine | 2-amino-1,3-PDO, 2-aminomalonic acid |
| Cellulose and hemicellulose (C4) | *Succinic acid* | THF, 1,4-butanediol, γ-butyrolactone, pyrrolidones, esters, diamines, 4,4-bionelle, hydroxybutyric acid |
| | *Fumaric acid* | Unsaturated succinate derivatives |
| | *Malic acid* | Hydroxy succinate derivatives |
| | *Aspartic acid* | Amino scuccinate derivatives |
| | 3-Hydroxybutyrolactone | Hydroxybutyrates, epoxy-γ-butyrolactone, butenoic acid |
| | Acetoin | Butanediols, butenols |
| | Threonine | Diols, ketone derivatives |

*Table 9.7 Continued*

| Source (carbon) | Platform compound | Biochemicals |
| --- | --- | --- |
| Cellulose and hemicellulose (C5) | *Itaconic acid* | Methyl succinate derivatives, unsaturated esters |
| | Furfural | Many furan derivatives |
| | *Levulinic acid* | δ-aminolevulinate, 2-methyl THF, 1,4-diols, esters, succinate |
| | *Glutamic acid* | Amino diols, glutaric acid, substituted pyrrolidones |
| | Xylonic acid | Lactones, esters |
| | *Xylitol/Arabitol* | EG, PG, glycerol, lactate, hydroxyl furans, sugar acids |
| | Citric/Aconitic acid | 1,5-Pentanediol, itaconic derivatives, pyrrolidones, esters |
| | 5-hydroxymethylfurfural | Numerous furan derivatives, succinate, esters, levulinic acid |
| | Lysine | Caprolactam, diamino alcohols, 1,5-diaminopentane |
| | Gluconic acid | Gluconolactones, esters |
| | *Glucaric acid* | Dilactones, monolactones, other products |
| | *Sorbitol* | Glycols (EG, PG), glycerol, lactate, isosorbide |
| Lignin (C6) | Syngas products | Methanol/dimethyl ether, ethanol, mixed liquid fuels |
| | Hydrocarbons | Cyclohexanes, higher alkylates |
| | Phenols | Cresols, eugenol, coniferols, syringols |
| | Oxidized products | Vanillin, vanillic acid, DMSO, aldehydes, quinones, aromatic and aliphatic acids |
| | Macromolecules | Carbon fibers, activated carbon, polymer alloys, polyelectrolites, substituted lignins, wood preservatives, neutraceuticals/drugs, adhesives and resins |

The 12 platform chemicals are represented in italics. The table was partially adapted from Menon and Rao (2012).

recent developments on the fermentation routes and genetic engineering strategies toward many platform chemicals can be found in Jang *et al.* (2012).

Zhang *et al.* (2011) specifically looked at products from the lignin and hemicellulose streams. The main products from lignin today (when not used as a fuel) are lignosulfonates, which are used as adhesives, binders or plasticizers in, for example, concrete. However, lignin is also a potential source for various aromatic compounds. A few specific products produced from lignin, like vanillin, are already on the market. The challenge is to find economic methods for depolymerizing the lignin, and subsequently separating the various monomeric compounds. The C5-stream contains predominantly xylose. Products to be made in the short term from xylose, proposed by Zhang *et al.* (2011), include the sweetener xylitol, obtainable by reduction of xylose, and furfural which is obtained through (acid catalyzed) dehydration of xylose. Fermentation to lactic acid is another option of interest in the short term.

In the context of chemical production, it deserves to be repeated that ethanol in itself can be used not only as a fuel but also as a platform chemical. About 75% of organic chemicals were produced from only a few base chemicals; ethylene, propylene, benzene, toluene and xylene in 1987 (Coombs, 1987). Of these, ethylene can be obtained by dehydration of ethanol, but also acetaldehyde and acetic acid are relatively easily obtained from ethanol. Furthermore, it has been reported that production of chemicals, such as acetic acid, from bioethanol can actually lead to higher $CO_2$ savings than by using it as a fuel (Rass-Hansen *et al.*, 2007).

## 9.7 Future trends

### 9.7.1 Industrialization and process development

The development of lignocellulose-based ethanol biorefineries has gone through several ups and downs in the past decades. Steadily, however, research has come closer to industrial realization and there are today at least 10 demonstration-scale facilities in Europe and the US alone (see Table 9.6). A very significant step forward is the current completion of the first commercial (or semi-commercial) scale ethanol biorefinery plants. The plant in Crescentino, Italy, built by Beta Renewables leads the way, closely followed by plants built by Abengoa and POET in the US. The completion of commercial plants shows that the technology has reached a sufficient level of maturity for implementation. This, however, does not mean that there are no development needs, but rather that these now shift into more directly process-related issues and optimization. Some of the main issues worked on are summarized in Table 9.8. The number of biofuel-related patent filings is rising. Some recent innovations (Table 9.9) give a flavor of the current development efforts on process efficiency, pretreatment improvement, novel enzymes and fermenting strains as well as the 'biorefinery' issues on valorization of by-products.

*Table 9.8* List of critical process development issues at various stages of biorefinery operation

|  | Process development issues |
|---|---|
| Biomass handling | Harvest–transportation–storage |
| Pretreatment | Should be mild yet efficient so as to achieve higher sugar release along with lower degradation products which inhibit fermentation |
| Enzymatic hydrolysis | Cheap and efficient enzymes |
|  | Thermostable stable enzymes in case of SHF process |
|  | Discovery of novel enzymes |
| Fermentation | Strains capable of converting all sugars released from lignocellulosics in the presence of inhibitors |
| Solids handling | Higher solids loadings throughout the process so as to minimize the water usage |
| Co-products | Development of viable technologies to produce co-products and by-products thereby reducing the residual waste |
| Energy | Strategic regulation of power requirements during the operation of process plants (steam generation and recycle) |

### 9.7.2 A future transition of paper industries?

One may argue that paper industries are in fact full-scale biorefineries. This is true in a sense, but the product focus has been very strong on pulp and paper only. The development of integrated forest biorefineries with also other products may open up for a new kind of business (Marinova et al., 2009). The major advantage of such integrated biorefineries would be to use the existing infrastructure for transport and material handling. Some options for the transition of a traditional pulp and paper industry into an integrated forest biorefinery are:

- extraction of hemicellulose prior to pulping and to be used for conversion into ethanol, organic acids, furfural, xylitol, polymers, or chemical intermediates
- recovery of lignin from black liquor stream for production of lignin-based commodity chemicals
- usage of lignin and wood residues in gasification processes for combined heat and power, chemicals, and fuel-production.

The black liquor in traditional pulp mills is used in recovery boilers for steam generation. However, alternative technologies are being developed for black liquor gasification, wherein the lignin present in the black liquor is converted to synthesis gas (Pettersson and Harvey, 2012). The synthesis

Table 9.9 Recent patent applications related to the biorefinery concept

| Patent Number | Description | Assignee | Inventor | Publication date |
|---|---|---|---|---|
| WO 2012126099 | A solvent extraction of the lignaceous residue of a biorefining process. | Lignol Innovations Ltd, (British Columbia, CA) | Robert SC, Yurevich BM, Ewellyn C | 27/09/2012 |
| WO 2012125925 | A method for hydrolyzing biomass and viscosity reduction of biomass mixture using a composition comprising a polypeptide having glycosyl hydrolase family 61/endoglucanase. | Danisco US Inc, (Palo Alto, CA, USA) | Colin M, Mian LI, Bradley KR, Suzanne LE | 20/09/2012 |
| WO 2012085860 | Delivery of steam produced by an electric power generation plant to a lignocellulosic biomass refinery. | Inbicon A/S (Skærbæk, Denmark) | Boye LH, Henning A | 28/06/2012 |
| US 20120165582 | A method for petrochemical products from direct liquefaction of the dried biomass. | Nexxoil AG (Zurich, CH) | Willner T | 28/06/2012 |
| WO 2012088429 | A method for treating biomass by allowing gaseous ammonia to condense on biomass and react with water present in the biomass thereby increasing the reactivity of polysaccharides in the biomass. | Board of Trustees of Michigan State University (MI, USA) | Balan V, Bruce DE, Shishir C, Leonardo S | 28/06/2012 |

| Patent No. | Title | Assignee | Inventors | Date |
|---|---|---|---|---|
| WO 2012059105 | A method for producing fermentation product such as ethanol using novel anaerobic, extreme thermophilic, ethanol high-yielding bacterium 'DTU01'. | Technical University of Denmark, (Lyngby, DK) | Irini A, Ana FT, Borisov KD | 10/05/2012 |
| US 20120071591 | Method to manufacture plastic materials comprising lignin and polybutylene succinate from renewable sources. | NA | Mohanty AK, Misra M, Sahoo S | 22/04/2012 |
| WO 2012049054 | A method for producing bioethanol, by pretreating the lignocellulosic vegetable raw materials in order to separate the cellulose. | Compagnie Industrielle (CIMV), (Rue Danton, France) | Michel D, Bouchra BM | 19/04/2012 |
| US 20120052543 | A method for pretreating biomass using combination of physical and chemical methods and thereby removing the detoxification and acid reconcentration steps. | Keimyung University (Daegu, KR) | Yoon KP | 01/03/2012 |
| US 20120023000 | A method for accounting carbon flows and determining a regulatory value for a biofuel. | NA | Rhodes JS | 26/01/2012 |
| WO 2012012306 | A methodology for reducing cost of enzymes in biorefinery. | NA | Andrew D, Michael E, Vince Y | 26/01/2012 |
| US 007973199 | A method for producing acetone from hydrated ethanol derived from biomass using Zr-Fe catalyst. | Metawater Co., Ltd (Tokyo, Japan) | Masuda T, Tago T, Yanase T, Tsuboi, H | 05/07/2011 |

Source: www.freepatentsonline.com.

292    Advances in Biorefineries

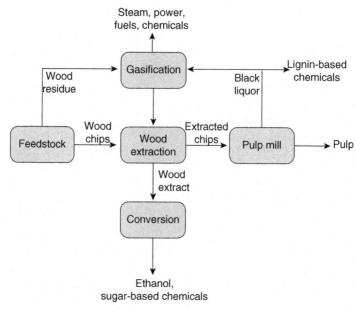

*9.11* An integrated forest biorefinery based on a Kraft pulp mill (adapted from Marinova *et al.*, 2009).

gas (or syngas) is in turn used for production of dimethyl ether, methanol or other fuel compounds. A schematic representation of an integrated forest biorefinery based on a Kraft mill is shown in Fig. 9.11.

### 9.7.3  Lessons from LCA studies

The future success of bioethanol refineries will be determined not only by the technical and economic performance of the plants, but also the environmental performance of the plants. Lowering of net greenhouse gas emissions (GHG) is, as mentioned in the introduction, particularly critical. Life cycle assessment (LCA) is necessary to monitor the above mentioned benefits when compared to that of fossil fuels, and LCA is now an accepted tool for guiding decision-making towards sustainability. LCA involves a holistic cradle-to-grave assessment approach to evaluate environmental performance by considering the potential impacts from all stages of manufacture, product use (including maintenance and recycling), and end-of-life management (von Blottnitz and Curran, 2007). Moreover, the International Organization for Standardization (ISO), a worldwide federation of national standards bodies, has standardized this framework within the ISO 14040 series on LCA.

The results from recent studies on LCA of biofuel production plants have established that global warming emissions and fossil energy consumption can be reduced significantly when conventional diesel and gasoline are substituted with biofuels (Punter *et al.*, 2004; Kim and Dale, 2005; von Blottnitz and Curran, 2007; Liska *et al.*, 2009; Cherubini and Jungmeier, 2010). However, in terms of other environmental impacts related to, for example, land usage, eutrophication of surface water, acidification and water pollution by pesticides, negative effects can result and must be carefully considered (Cherubini and Jungmeier, 2010). The environmental impact of a biorefinery depends on factors such as biomass feedstocks used, conversion routes, process configurations, and end-use applications. The LCA analysis can be used to identify further steps to be taken to, for example, reduce GHG emissions and improve overall energy efficiency of biorefineries. Examples include use of thermo-compressors for steam condensation, waste heat recovery and reuse of heat, and combustion of volatile organics using thermal oxidizers (Liska *et al.*, 2009).

### 9.7.4 Future crops

The availability of feedstocks is a strategic question, and development on future dedicated feedstocks for biofuel production has also taken a prominent place in the agenda of bioenergy research. It is not possible to point out one single new feedstock which should be used for bioethanol production, as local factors related to regio-specific agricultural practices, market forces, political directives, social and biological issues will define the suitability (Sticklen, 2008). However, factors to consider include:

- identification of indigenous woody and grass species best suited to local conditions
- yields of such lignocellulose sources and scope for improvement
- production costs in terms of fertilizer input and comparison to traditional crops
- applicability of genetic engineering tools for drought tolerance and higher biomass productivity
- farm-to-refinery logistics in terms of storage and handling
- feedstock value from by-products
- environmental sustainability.

Technologies for the conversion of lignocellulosic feedstocks, such as agriculture residues from corn, rice, sugarcane, perennial grasses (switchgrass and giant miscanthus) or short rotation woody crops (fast-growing poplar and shrub willow), which contain higher cellulose fractions, should be given more prominence (Ravindranath *et al.*, 2009; Sticklen, 2008). Newer and emerging tools in plant genetic engineering offer great potential in reducing the processing costs of these feedstocks to bioethanol. Techniques such as production of cellulases and hemicellulases within the crop biomass, genetic manipulation to modify lignin content, upregulation of cellulose and hemicellulose biosynthesis, and expression of drought tolerant traits are some of the potential research areas for improving biomass feedstocks for bioethanol production (Sticklen, 2008).

## 9.8 Conclusion

The worldwide growth in ethanol production has been spectacular in the past three decades, but so far has been based solely on starch or sugar. Continued expansion will have to be based on a broader raw material base, such as lignocellulose, and for both economic and environmental reasons a biorefinery approach will have to be used. The conceptual idea of a biorefinery is not new, but dates back to 'pre-petroleum' times. Today, however, the concept of biorefineries is about to be tested by industrial implementation, bringing together actors in agriculture, biotechnology and chemical process technology in a novel close collaboration. The technical and economic performance of the industrial pioneering efforts currently made will have a huge impact on the pace of development of the bioeconomy in the coming decade.

## 9.9 Sources of further information and advice

There is a wealth of information available on both biorefineries and bioethanol. The review paper by Kamm and Kamm (2007) is a good introduction to the concept of biorefineries, whereas the extensive report on 'Top value added chemicals from biomass' (Werpy and Petersen, 2004) gives a flavor of the chemical potential of biomass. The development of the bioethanol industry is well covered in several papers by Goldemberg (e.g., Goldemberg, 2006) which describe the development of the Brazilian ethanol industry.

Several organizations provide useful information on the internet. The International Energy Agency (IEA) has a working task force 42 entitled 'Biorefineries: Co-production of Fuels, Chemicals, Power and Materials from Biomass' (http://www.ieabioenergy.com). Useful information is also supplied by the US Department of Energy, in particular under its biomass programs (http://www.eere.energy.gov/biomass/). Links to maps of on-going biorefinery projects can be reached from this site. The renewable fuels association (RFA) (http://www.ethanolrfa.org/) is another organization which provides a wide range of information on ethanol.

## 9.10 References

Abedinifar, S., Karimi, K., Khanahmadi, M. and Taherzadeh, M. J. (2009) 'Ethanol production by *Mucor indicus* and *Rhizopus oryzae* from rice straw by separate hydrolysis and fermentation', *Biomass Bioenerg*, 33, 828–833.

Akpinar, O., Ak, O., Kavas, A., Bakir, U. and Yilmaz, L. (2007) 'Enzymatic production of xylooligosaccharides from cotton stalks', *J Agric Food Chem*, 55, 5544–5551.

Alkasrawi, M., Rudolf, A., Lidén, G. and Zacchi, G. (2006) 'Influence of strain and cultivation procedure on the performance of simultaneous saccharification and fermentation of steam pretreated spruce', *Enzyme Microb Technol*, 38, 279–287.

Almeida, J. R. M., Modig, T., Petersson, A., Hähn-Hägerdal, B., Lidén, G. and Gorwa-Grauslund, M. F. (2007) 'Increased tolerance and conversion of inhibitors in lignocellulosic hydrolysates by *Saccharomyces cerevisiae*', *J Chem Technol Biotechnol*, 82, 340–349.

Almeida, J. R. M., Bertilsson, M., Gorwa-Grauslund, M. F., Gorsich, S. and Lidén, G. (2009) 'Metabolic effects of furaldehyde and impacts on biotechnological processes', *Appl Microbiol Biotechnol*, 82, 625–638.

Almeida, J. R., Runquist, D., Sanchez i Nogué, V., Lidén, G. and Gorwa-Grauslund, M. F. (2011) 'Stress-related challenges in pentose fermentation to ethanol by the yeast *Saccharomyces cerevisiae*', *Biotechnol J*, 6, 286–299.

Alriksson, B., Horváth, I. S. and Jönsson, L. J. (2010) 'Overexpression of *Saccharomyces cerevisiae* transcription factor and multidrug resistance genes conveys enhanced resistance to lignocellulose derived fermentation inhibitors', *Process Biochem*, 45, 264–271.

Alriksson, B., Cavka, A. and Jönsson, L. J. (2011) 'Improving the fermentability of enzymatic hydrolysates of lignocellulose through chemical *in-situ* detoxification with reducing agents', *Bioresour Technol*, 102, 1254–1263.

Alvira, P., Tomás-Pejó, E., Ballesteros, M. and Negro, M. J. (2010) 'Pretreatment technologies for an efficient bioethanol production process based on enzymatic hydrolysis: a review', *Bioresour Technol*, 101, 4851–4861.

Amichev, B. Y., Johnston, M. and van Rees, K. C. J. (2010) 'Hybrid poplar growth in bioenergy production systems: biomass prediction with a simple process-based model (3PG)', *Biomass Bioenerg*, 34, 687–702.

Angelini, L. G., Ceccarini, L., Nassi, N. and Bonari, E. (2009) 'Comparison of *Arundo donax* L. and *Miscanthus x giganteus* in a long-term field experiment in Central Italy: analysis of productive characteristics and energy balance', *Biomass Bioenerg*, 33, 635–643.

Bellissimi, E., van Dijken J. P., Pronk, J. T. and van Maris, A. J. A. (2009) 'Effects of acetic acid on the kinetics of xylose fermentation by an engineered, xylose-isomerase-based *Saccharomyces cerevisiae* strain', *FEMS Yeast Res*, 9, 358–364.

Bergh, J., Linder, S. and Bergstrom, J. (2005) 'Potential production of Norway spruce in Sweden', *Forest Ecology and Management*, 204, 1–10.

Bettiga, M., Gorwa-Grauslund, M. F. and Hahn-Hägerdal, B. (2009) 'Metabolic engineering in yeasts'. In *The Metabolic Pathway Engineering Handbook*; edited by Smolke, C. London: Taylor and Francis/CRC Press, 22.21–22.46.

Blomqvist, J., Eberhard, T., Schnürer, J. and Passoth, V. (2010) 'Fermentation characteristics of *Dekkera bruxellensis* strains', *Appl Microbiol Biotechnol*, 87, 1487–1497.

Bozell, J. J. and Petersen, G. R. (2010) 'Technology development for the production of biobased products from biorefinery carbohydrates; the US Department of Energy's Top 10 revisited', *Green Chem*, 12, 539–554.

Brat, D., Boles, E. and Wiedemann, B. (2009) 'Functional expression of a bacterial xylose isomerase in *Saccharomyces cerevisiae*', *Appl Environ Microbiol*, 75, 2304–2311.

Bura, R., Ewanick, S. and Gustafson, R. (2012) 'Assessment of *Arundo donax* (giant reed) as feedstock for conversion to ethanol', *TAPPI Journal*, 11, 59–66.

Callesen, I., Grohnheit, P. E. and Østergård, H. (2010) 'Optimization of bioenergy yield from cultivated land in Denmark', *Biomass Bioenerg*, 34, 1348–1362.

Casey, E., Sedlak, M., Ho, N. W. Y. and Mosier, N. S. (2010) 'Effect of acetic acid and pH on the cofermentation of glucose and xylose to ethanol by a genetically engineered strain of *Saccharomyces cerevisiae*', *FEMS Yeast Res*, 10, 385–393.

Casey, G. P. and Ingledew, W. M. (1986) 'Ethanol tolerance in yeasts', *Crit Rev Microbiol*, 13, 219–280.

Cavka, A., Alriksson, B., Ahnlund, M. and Jönsson, L. J. (2011) 'Effect of sulfur oxyanions on lignocellulose-derived fermentation inhibitors', *Biotechnol Bioeng*, 108, 2592–2599.

Cherubini, F. and Jungmeier, G. (2010) 'LCA of a biorefinery concept producing bioethanol, bioenergy, and chemicals from switchgrass', *Int J LCA*, 15, 53–66.

Commission Directive 2009/28/EC. On the promotion of the use of energy from renewable sources and amending and subsequently repealing Directives 2001/77/EC and 2003/30/EC.2009.04.23, *OJ L*, 140, 16–62.

Coombs, J. (1987) 'EEC resources and strategies', *Phil Trans R Soc Lond Ser A*, 321, 405–422.

de Figueiredo Vilela, L., de Mello, V. M., Reis, V. C. B., da Silva Bon, E. P., Torres, F. A. G., Neves, B. C. and Eleutherio, E. C. V. (2013) 'Functional expression of *Burkholderia cenocepacia* xylose isomerase in yeast increases ethanol production from a glucose–xylose blend', *Bioresour Technol*, 128, 792–796.

Ekman, A., Wallberg, O., Joelsson, E. and Börjesson, P. (2013) 'Possibilities for sustainable biorefineries based on agricultural residues – a case study of potential straw-based ethanol production in Sweden', *Appl Energy*, 102, 299–308.

Energy Independence and Security Act of 2007. Public Law 110–140, 2007; http://frwebgate.access.gpo.gov/cgi-bin/getdoc.cgi?dbname)110_cong_public_laws&docid)f:publ140.110.pdf.

Finlay, M. R. (2004) 'Old efforts at new uses: a brief history of chemurgy and the American search for biobased materials.' *J Ind Ecol*, 7, 33–46.

Foyle, T., Jennings, L. and Mulcahy, P. (2007) 'Compositional analysis of lignocellulosic materials: evaluation of methods used for sugar analysis of waste paper and straw', *Biores Technol*, 98, 3026–3036.

Galbe, M. and Zacchi, G. (2007) 'Pretreatment of lignocellulosic materials for efficient bioethanol production', *Biofuels*, 108, 41–65.

Galbe, M., Sassner, P., Wingren, A. and Zacchi, G. (2007) 'Process engineering economics of bioethanol production', *Adv Biochem Eng Biotechnol*, 108, 303–327.

Gauss, W. F., Suzuki, S. and Takagi, M. (1976) 'Manufacture of alcohol from cellulosic materials using plural ferments', US Patent US3990944.

Giebelhaus, A. W. (1980) 'Farming for fuel: the alcohol motor fuel movement of the 1930s', *Agricultural History*, 54, 173–184.

Godin, B., Agneessens, R., Gerin, P. and Delcarte, J. (2011) 'Composition of structural carbohydrates in biomass: precision of a liquid chromatography method using a neutral detergent extraction and a charged aerosol detector', *Talanta*, 85, 2014–2026.

Goldemberg, J. (2006) 'The ethanol program in Brazil', *Environ Res Lett*, 1, 1–5.

Gonzalez, R., Tao, H., Purvis, J. E., York, S. W., Shanmugam, K. T. and Ingram, L. O. (2003) 'Gene array-based identification of changes that contribute to ethanol tolerance in ethanologenic *Escherichia coli*: comparison of KO11 (Parent) to LY01 (resistant mutant)', *Biotechnol Prog*, 19, 612–623.

González-García, S., Luo, L., Moreira, M. T., Feijoo, G. and Huppes, G. (2012a) 'Life cycle assessment of hemp hurds use in second generation ethanol production', *Biomass Bioenerg*, 36, 268–279.

González-García, S., Iribarren, D., Susmozas, A., Dufour, J., Murphy, R. J. (2012b) 'Life cycle assessment of two alternative bioenergy systems involving Salix spp. biomass: Bioethanol production and power generation', *Appl Energ*, 95, 111–122.

Goshadrou, A., Karimi, K. and Taherzadeh, M. J. (2011) 'Improvement of sweet sorghum bagasse hydrolysis by alkali and acidic pretreatments', World Renewable Energy Congress, Bioenergy Technology, Linköping, Sweden, 8–13 May, 374–380.

Hahn-Hägerdal, B., Jeppsson, H., Olsson, L. and Mohagheghi, A. (1994) 'An inter laboratory comparison of the performance of ethanol-producing microorganisms in a xylose-rich acid hydrolysate', *Appl Microbiol Biotechnol*, 41, 62–72.

Hahn-Hägerdahl, B., Karhumaa, K., Fonseca, C., Spencer-Martins, I. and Gorwa-Grauslund, M. F. (2007) 'Towards industrial pentose fermenting yeast strains', *Appl Microbiol Biotechnol*, 74, 937–953.

Hayn, M., Steiner, W., Klinger, R., Steinmüller, H., Sinner, M. and Esterbauer, H. (1993) 'Basic research and pilot studies on the enzymatic conversion of lignocellulosics', in Saddler, J. N., *Bioconversion of Forest and Agricultural Plant Residues*, CAB International, Wallingford, 33–72.

Horn, S. J., Vaaje-Kolstad, G., Westereng, B. and Eijsink, V. G. (2012) 'Novel enzymes for the degradation of cellulose', *Biotechnol Biofuels* 5, 45.

Jae, J. H., Tompsett, G. A., Lin, Y. C., Carlson, T. R., Shen, J. C., Zhang, T. Y., Yang, B., Wyman, C. E., Conner, W. C. and Huber, G. W. (2010) 'Depolymerization of lignocellulosic biomass to fuel precursors: maximizing carbon efficiency by combining hydrolysis with pyrolysis', *Energy Environ Sci*, 3, 358–365.

Jang, Y.-S., Kim, B., Shin, J. H., Choi, Y. J., Choi, S., Song, C. W., Lee, J., Park, H. G. and Lee, S. Y. (2012) 'Bio-based production of C2–C6 platform chemicals', *Biotechnol Bioeng*, 109, 2437–2459.

Jarboe, L. R., Grabar, T. B., Yomano, L. P., Shanmugan, K. T. and Ingram, L. O. (2007) 'Development of ethanologenic bacteria', *Adv Biochem Eng Biotechnol*, 108, 237–261.

Jin, M., Balan, V., Gunawan, C. and Dale, B. E. (2011) 'Consolidated bioprocessing (CBP) performance of *Clostridium phytofermentans* on AFEX-treated corn stover for ethanol production', *Biotechnol Bioeng*, 108, 1290–1297.

Johansson, J. (2010) 'Pretreatment of prevalent Canadian west coast softwoods using the ethanol organosolv process: assessing robustness of the ethanol organosolv process', Masters thesis, Lund University, Sweden.

Kadam, K. L., Forrest, L. H. and Jacobson, W. A. (2000) 'Rice straw as a lignocellulosic resource: collection, processing, transportation, and environmental aspects', *Biomass Bioenerg*, 18, 369–389.

Kamm, B. and Kamm, M. (2007) 'Biorefineries – multi-product processes', *Adv Biochem Eng Biotechnol*, 105, 175–204.

Kang, H.-W., Kim, Y., Kim, S.-W. and Choi, G.-W. (2012) 'Cellulosic ethanol production on temperature-shift simultaneous saccharification and fermentation using the thermostable yeast *Kluyveromyces marxianus* CHY1612', *Bioprocess Biosyst Eng*, 35, 115–122.

Kim, S. and Dale, B. E. (2005) 'Life cycle assessment of various cropping systems utilized for producing biofuels: bioethanol and biodiesel', *Biomass Bioenerg*, 29, 426–439.

Kim, Y., Yu, A., Han, M., Choi, G. and Chung, B. (2011) 'Enhanced enzymatic saccharification of barley straw pretreated by ethanosolv technology', *Appl Microbiol Biotechnol*, 163, 143–152.

Klinke, H. B., Thomsen, A. B. and Ahring, B. K. (2004) 'Inhibition of ethanol-producing yeast and bacteria by degradation products produced during pre-treatment of biomass', *Appl Microbiol Biotechnol*, 66, 10–26.

Köpke, M., Mihalcea, C., Bromley, J. C. and Simpson, S. D. (2011) 'Fermentative production of ethanol from carbon monoxide', *Curr Opin Biotechnol*, 22, 320–325.

Kosugi, A., Kondo, A., Ueda, M., Murata, Y., Vaithanomsat, P., Thanapase, W., Arai, T. and Mori, Y. (2009) 'Production of ethanol from cassava pulp via fermentation with a surface engineered yeast strain displaying glucoamylase', *Renew Energy*, 34, 1354–1358.

Kristensen, J. B., Felby, C. and Jorgensen, H. (2009) 'Yield-determining factors in high-solids enzymatic hydrolysis of lignocellulose', *Biotechnol Biofuels*, 2, 11.

Kundiyana, D. K., Bellmer, D. D., Huhnke, R. L., Wilkins, M. R. and Claypool, P. L. (2010) 'Influence of temperature, pH and yeast on in-field production of ethanol from unsterilized sweet sorghum juice', *Biomass Bioenerg*, 34, 1481–1486.

Kuyper, M., Harhangi, H. R., Stave, A. K., Winkler, A. A., Jetten, M. S. M., de Laat, W. T. A. M., den Ridder, J. J. J., Op den Camp, H. J. M., van Dijken, J. P. and Pronk, J. T. (2003) 'High level functional expression of a fungal xylose isomerase: the key to efficient ethanolic fermentation of xylose by *Saccharomyces cerevisiae*', *FEMS Yeast Res*, 4, 69–78.

Kuyper, M., Toirkens, M. J., Diderich, J. A., Winkler, A. A., van Dijken, J. P. and Pronk, J. T. (2005) 'Evolutionary engineering of mixed-sugar utilization by a xylose-fermenting *Saccharomyces cerevisiae* strain', *FEMS Yeast Res*, 5, 925–934.

Lee, S-M., Jellison, T. and Alper, H. S. (2012) 'Directed evolution of xylose isomerase for improved xylose catabolism and fermentation in the yeast *Saccharomyces cerevisiae*', *Appl Environ Microbiol*, 78, 5708–5716.

Liska, A. J., Yang, H. S., Bremer, V. R., Klopfenstein, T. J., Walters, D. T., Erickson, G. E. and Cassman, K. G. (2009) 'Improvements in life cycle energy efficiency and greenhouse gas emissions of corn-ethanol', *J Ind Ecol*, 13, 58–74.

Lynd, L. R., van Zyl, W. H., McBride, J. E. and Laser, M. (2005) 'Consolidated bioprocessing of cellulosic biomass: an update', *Curr Opin Biotechnol*, 16, 577–583.

Mabee, W. E., Gregg, D. J., Arato, C., Berlin, A., Bura, R., Gilkes, N., Mirochnik, O., Pan, X., Pye, E. K. and Saddler, J. N. (2006) 'Updates on softwood-to-ethanol process development', *Appl Biochem Biotechnol* 129–132, 55–70.

Madhavan, A., Tamalampudi, S., Ushida, K., Kanai, D., Katahira, S., Srivastava, A., Fukuda, H., Bisaria, V. S. and Kondo, A. (2009) 'Xylose isomerase from polycentric fungus *Orpinomyces*: gene sequencing, cloning, and expression in *Saccharomyces cerevisiae* for bioconversion of xylose to ethanol', *Appl Microbiol Biotechnol*, 82, 1067–1078.

Marinova, M., Mateos-Espejel, E., Jemaa, N. and Paris, J. (2009) 'Addressing the increased energy demand of a Kraft mill biorefinery: the hemicellulose extraction case', *Chem Eng Res Des*, 87, 1269–1275.

Matsushika, A., Inoue, H., Kodaki, T. and Sawayama, S. (2009) 'Ethanol production from xylose in engineered *Saccharomyces cerevisiae* strains: current state and perspectives', *Appl Microbiol Biotechnol*, 84, 37–53.

McMillan, J., Jennings, E. W., Mohagheghi, A. and Zuccarello, M. (2011) 'Comparative performance of precommercial cellulases hydrolyzing pretreated corn stover', *Biotechnol Biofuels*, 4, 29.

Menon, V. and Rao, M. (2012) 'Trends in bioconversion of lignocelluloses: biofuels, platform chemicals and biorefinery concept', *Prog Energ Combust*, 38, 522–550.

Mohagheghi, A., Tucker, M., Grohmann, K. and Wyman, C. (1992) 'High solids simultaneous saccharification and fermentation of pretreated wheat straw to ethanol', *Appl Biochem Biotechnol*, 33, 67–81.

Morano, K. A., Liu, P. C. and Thiele, D. J. (1998) 'Protein chaperones and the heat shock response in *Saccharomyces cerevisiae*', *Curr Opin Microbiol*, 1, 197–203.

Mutturi, S. and Lidén, G. (2013) 'Effect of temperature on simultaneous saccharification and fermentation of pretreated spruce and arundo', *Ind Eng Chem Res*, 52, 1244–1251.

Öhgren, K., Bura, R., Lesnicki, G., Saddler, J. and Zacchi, G. (2007) 'A comparison between simultaneous saccharification and fermentation and separate hydrolysis and fermentation using steam-pretreated corn stover', *Process Biochem*, 42, 834–839.

Olofsson, K., Bertilsson, M. and Lidén, G. (2008) 'A short review on SSF – an interesting process option for ethanol production from lignocellulosic feedstocks', *Biotechnol Biofuels*, 1, 7.

Olsson, L. and Hahn-Hägerdal, B. (1993) 'Fermentative performance of bacteria and yeasts in lignocellulose hydrolysate', *Proc Biochem*, 28, 249–257.

Østergaard, S., Olsson, L. and Nielsen, J. (2000) 'Metabolic engineering of *Saccharomyces cerevisiae*', *Microbiol Mol Biol Rev*, 64, 34–50.

Palmqvist, E. and Hahn-Hägerdal, B. (2000) 'Fermentation of lignocellulosic hydrolysates. II: Inhibitors and mechanisms of inhibition', *Bioresour Technol*, 74, 25–33.

Palonen, H., Tjerneld, F., Zacchi, G. and Tenkanen, M. (2004) 'Adsorption of *Trichoderma reesei* CBH I and EG II and their catalytic domains on steam pretreated softwood and isolated lignin', *J Biotechnol*, 107, 65–72

Pérez-Cruzado, C., Merino, A. and Rodríguez-Soalleiro, R., (2011) 'A management tool for estimating bioenergy production and carbon sequestration in *Eucalyptus globulus* and *Eucalyptus nitens* grown as short rotation woody crops in north-west Spain', *Biomass Bioenerg*, 35, 2839–2851.

Pessani, N. K., Atiyeh, H. K., Wilkins, M. R., Bellmer, D. D. and Banat, I. M. (2011) 'Simultaneous saccharification and fermentation of Kanlow switchgrass by thermotolerant *Kluyveromyces marxianus* IMB3: the effect of enzyme loading, temperature and higher solid loadings', *Bioresour Technol*, 102, 10618–10624.

Petersson, A., Almeida, J. R., Modig, T., Karhumma, K., Hahn-Hägerdal, B. and Gorwa-Grauslund, M. F. (2006) 'A 5-hydroxymethylfurfural reducing enzyme encoded by the *Saccharomyces cerevisiae* ADH6 gene conveys HMF tolerance', *Yeast*, 23, 455–464.

Pettersson, K. and Harvey, S. (2012) 'Comparison of black liquor gasification with other pulping biorefinery concepts – systems analysis of economic performance and $CO_2$ emissions', *Energy*, 37, 136–153.

Pham, V. and El-Halwagi, M. M. (2012) 'Process synthesis and optimization of biorefinery configurations', *AIChE J*, 58, 1212–1221.

Punter, G., Rickeard, D., Larivé, J. F., Edwards, R., Mortimer, N., Horne, R., Bauen, A. and Woods, J. (2004) 'Well-to-wheel evaluation for production of ethanol from wheat. A report by the Low CVP fuels working group', WTW sub-group, FWG-P-04-024, October.

Qing, Q. and Wyman, C. (2011) 'Supplementation with xylanase and β-xylosidase to reduce xylo-oligomer and xylan inhibition of enzymatic hydrolysis of cellulose and pretreated corn stover', *Biotechnol Biofuels*, 4, 18.

Rass-Hansen, J., Falsig, H., Jorgensen, B. and Christensen, C. H. (2007) 'Bioethanol: fuel or feedstock?', *J Chem Technol Biotechnol*, 82, 329–333.

Ravindranath, N. H., Manuvie, M., Fargione, J., Canadell, J., Berndes, G., Woods, J., Watson, H. and Sathaye, J. (2009) 'Greenhouse gas implications of land use and land conversion to biofuel crops', in Howarth, R. and Bringezu, S., *Biofuels: Environmental Consequences and Interactions with Changing Land Use*, Proceedings of the Scientific Committee on Problems of the Environment (SCOPE) International Biofuels Project Rapid Assessment, 22–25 September 2008, Gummersbach, Germany. Cornell University, NY, 11–125.

Rencoret, J., Gutiérrez, A., Nieto, L., Jiménez-Barbero, J., Faulds, C. B., Kim, H., Ralph, J., Martínez, A. T. and del Río, J. C. (2010) 'Lignin composition and structure in young versus adult *Eucalyptus globulus* plants', *Plant Physiol*, 155, 667–682.

Rødsrud, G., Lersch, M. and Sjöde, A. (2012) 'History and future of world's most advanced biorefinery in operation', *Biomass Bioenerg*, 46, 46–59.

Rogers, P., Lee, K., Skotnicki, M. and Tribe, D. (1982) '*Ethanol production* by Zymomonas mobilis', in *Microbial Reactions*, New York, Spinger-Verlag, pp. 37–84.

Rudolf, A. (2007) 'Fermentation and cultivation technology for improved ethanol production from lignocellulose', PhD thesis, Lund University.

Sanchez, R. G., Karhumaa, K., Fonseca, C., Nogué, V. S., Almeida, J. R. M., Larsson, C. U., Bengtsson, O., Bettiga, M., Hahn-Hägerdal, B. and Gorwa-Grauslund, M. F. (2010) 'Improved xylose and arabinose utilization by industrial recombinant *Saccharomyces cerevisiae* strain using evolutionary engineering', *Biotechnol Biofuels*, 3, 13.

Sassner, P., Galbe, M. and Zacchi, G. (2006) 'Bioethanol production based on simultaneous saccharification and fermentation of steam-pretreated Salix at high dry-matter content', *Enzyme Microb Technol*, 39, 756–762.
Sathitsuksanoh, N., Zhu, Z., Ho, T-J., Bai, M-D. and Zhang, Y.-H. P. (2010) 'Bamboo saccharification through cellulose solvent-based biomass pretreatment followed by enzymatic hydrolysis at ultra-low cellulase loadings', *Bioresour Technol*, 101, 4926–4929.
Sedlak, M. and Ho, N. W. (2004) 'Production of ethanol from cellulosic biomass hydrolysates using genetically engineered *Saccharomyces* yeast capable of cofermenting glucose and xylose', *Appl Biochem Biotechnol*, 113–116, 403–416.
Shahsavarani, H., Sugiyama, M., Kaneko, Y., Chuenchit, B. and Harashima, S. (2012) 'Superior thermotolerance of *Saccharomyces cerevisiae* for efficient bioethanol fermentation can be achieved by overexpression of RSP5 ubiquitin ligase', *Biotechnol Adv*, 30, 1289–1300.
Singer, M. A. and Lindquist, S. (1998) 'Thermotolerance in *Saccharomyces cerevisiae*: the yin and yang of trehalose', *Trends Biotechnol*, 16, 460–468.
Snowdon, P. and Benson, M. L. (1992) 'Effects of combinations of irrigation and fertilization on the growth and above-ground biomass production of *Pinus radiata*', *For Ecol Manag*, 52, 87–116.
Songstad, D. D., Lakshmanan P., Chen, J., Gibbons, W., Hughes, S. and Nelson, R. (2009) 'Historical perspective of biofuels: learning from the past to rediscover the future', *In Vitro Cell Dev Biol – Plant*, 45, 189–192.
Steinmetz, L. M., Sinha, H., Richards, D. R., Spiegelman, J. I., Oefner, P. J., McCusker, J. H. and Davis, R. W. (2002) 'Dissecting the architecture of a quantitative trait locus in yeast', *Nature*, 416, 326–330.
Sticklen, M. B. (2008) 'Plant genetic engineering for biofuel production: towards affordable cellulosic ethanol', *Nat Rev Genet*, 9, 433–443.
Templeton, D. W., Sluiter, A. D., Hayward, T. K., Hames, B. R. and Thomas, S. R. (2009) 'Assessing corn stover composition and sources of variability via NIRS', *Cellulose*, 16, 621–639.
Tengborg, C., Stenberg, K., Galbe, M., Zacchi, G., Larsson, S., Palmqvist, E., Hahn-Hägerdal, B. (1998) 'Comparison of $SO_2$ and $H_2SO_4$ impregnation of softwood prior to steam pretreatment on ethanol production', *Appl Biochem Biotechnol* 70–72, 3–15.
Tippayawong, N. and Chanhom, N. (2011) 'Conversion of bamboo to sugars by dilute acid and enzymatic hydrolysis', *Int J Renewable Energy Res*, 1, 240–244.
Tullus, A., Tullus, H., Soo, T. and Pärn, L. (2009) 'Above-ground biomass characteristics of young hybrid aspen (*Populus tremula* L. x *P. tremuloides* Michx.) plantations on former agricultural land in Estonia', *Biomass Bioenerg*, 33, 1617–1625.
van Dyk, J. S. and Pletschke, B. I. (2012) 'A review of lignocellulose bioconversion using enzymatic hydrolysis and synergistic cooperation between enzymes – factors affecting enzymes, conversion and synergy', *Biotechnol Adv*, 30, 1458–1480.
Van Vleet, J. H. and Jeffries, T. W. (2009) 'Yeast metabolic engineering for hemicellulosic ethanol production', *Curr Opin Biotechnol*, 20, 300–306.
van Zyl, W. H., Lynd, L. R., Den Haan, R. and McBride, J. E. (2007) 'Consolidated bioprocessing for bioethanol production using *Saccharomyces cerevisiae*', *Adv Biochem Eng Biotechnol*, 108, 205–235.

Vázquez, G., Freire, M. S. and Antorrena, G. (2007) Valorisation of lignocellulosic waste materials: tannins as a source of new products. Proceedings of European Congress of Chemical Engineering (ECCE-6), Copenhagen, 16–20 September.

Verduyn, C., Postma, E., Scheffers, W. A. and van Dijken, J. P. (1990) 'Physiology of *Saccharomyces cerevisiae* in anaerobic glucose-limited chemostate cultures', *J Gen Microbiol*, 136, 395–403.

von Blottnitz, H. and Curran, M. A. (2007) 'A review of assessments conducted on bio-ethanol as a transportation fuel from a net energy, greenhouse gas, and environmental life cycle perspective', *J Clean Prod*, 15, 607–619.

Vrije, T. D., Haas, G. G. D., Tan, G. B., Keijsers, E. R. P. and Claassen, P. A. M. (2002) 'Pretreatment of *Miscanthus* for hydrogen production by *Thermotoga elfii*', *Int J Hydrogen Energ*, 27, 1381–1390.

Wahlbom, C. F., Cordero Otero, R. R., van Zyl, W. H., Hahn-Hägerdal, B. and Jonsson, L. J. (2003) 'Molecular analysis of a *Saccharomyces cerevisiae* mutant with improved ability to utilize xylose shows enhanced expression of proteins involved in transport, initial xylose metabolism, and the pentose phosphate pathway', *Appl Environ Microbiol*, 69, 740–746.

Wang, G. S., Lee, J-W., Zhu, J. Y. and Jeffries, T. W. (2011) 'Dilute acid pretreatment of corncob for efficient sugar production', *Appl Microbiol Biotechnol*, 163, 658–668.

Werpy, T. and Petersen, G. (eds) (2004) 'Top value added chemicals from biomass', Pacific Northwest National Laboratory and National Renewable Energy Laboratory, Report 8674.

Wilkins, M. R. and Atiyeh, H. K. (2011) 'Microbial production of ethanol from carbon monoxide', *Curr Opin Biotechnol*, 22, 326–330.

Wingren, A., Galbe, M. and Zacchi, G. (2003) 'Techno-economic evaluation of producing ethanol from softwood: comparison of SSF and SHF and identification of bottlenecks', *Biotechnol Prog*, 19, 1109–1117.

Wiselogel, A., Tyson, S. and Johnson, D. (1996) 'Biomass feedstock resources and composition', in Wyman, C. E., *Handbook on Bioethanol: Production and Utilization*, Taylor & Francis, Washington, DC, 105–118.

Wu, Z. and Lee, Y. Y. (1997) 'Inhibition of the enzymatic hydrolysis of cellulose by ethanol', *Biotechnol Lett*, 19, 977–979.

Youngblood, A., Zhu, J. and Scott, C. T. (2010) 'Ethanol production from woody biomass: silvicultural opportunities for surpressed western conifers', USDA Forest Service Proceedings RMRS-P-61.

Zhang, Q., Fu, Y., Wang, Y., Han, J., Lv, J. and Wang, S. (2012) 'Improved ethanol production of a newly isolated thermotolerant *Saccharomyces cerevisiae* strain after high-energy-pulse electron beam', *J Appl Microbiol*, 112, 280–288.

Zhang, X., Tu, M. and Paice, M. G. (2011) 'Routes to potential bioproducts from lignocellulosic biomass lignin and hemicelluloses', *Bioenerg Res*, 4, 246–257.

Zhou, H., Cheng, J. S., Wang, B. L., Fink, G. R. and Stephanopoulos, G. (2012) 'Xylose isomerase overexpression along with engineering of the pentose phosphate pathway and evolutionary engineering enable rapid xylose utilization and ethanol production by *Saccharomyces cerevisiae*', *Metab Eng*, 14, 611–622.

# 10
# Developments in cereal-based biorefineries

A. A. KOUTINAS, Agricultural University of Athens, Greece,
C. DU, University of Nottingham, UK,
C. S. K. LIN, City University of Hong Kong, Hong Kong
and C. WEBB, University of Manchester, UK

DOI: 10.1533/9780857097385.1.303

**Abstract**: Restructuring conventional cereal-based processes is essential in order to create viable biorefineries for the production of fuels, chemicals and materials. Advanced biorefinery schemes should exploit the full potential of cereal grains by exploiting every component and residue to produce a wide spectrum of commodity and speciality products. This chapter presents generic biorefinery approaches that utilize the wheat grain for the production of bioethanol, polyhydroxybutyrate, succinic acid and various added-value products (e.g., arabinoxylans). Cereal-based food by-product or waste streams generated from primary processing of cereals, households, restaurants and catering services could be used for the development of second-generation biorefineries.

**Key words**: cereal-based biorefineries, bioethanol, polyhydroxybutyrate, succinic acid, food waste.

## 10.1 Introduction

Conventional cereal processes are mature industrial technologies that have been developed for the production of food or non-food products. Cereal processes are based on different fractionation schemes exploiting mainly dry and wet milling operations to break down the grain into various fractions with many end-uses (e.g., food, feed, chemicals, textiles, cosmetics, fermentation). Dry milling generally produces incomplete fractionation of grain components through physical processing involving mainly grinding and sieving unit operations. For instance, in traditional wheat milling processes, the main aim is to maximize the separation of bran from endosperm. Wet milling processes can be divided according to the type of solvent (i.e., aqueous and non-aqueous) used for selective separation of cereal grain components. The corn wet milling process constitutes an industrially mature technology that has been developed mainly in the United States for the production of a spectrum of products.

Corn refining started in 1848 by Thomas Kingsford who introduced the art of starch manufacture in a small firm of only 70 employees that by 1880 grew to become the largest company of its kind worldwide (Peckham, 2001). The success of this company led to the creation of new companies that benefited from the increasing demand for starch, as laundry aid and food ingredient, and corn syrups that were used in the candy, baking, brewing and vinegar industries (Peckham, 2001). During the twentieth century, corn refining evolved into a mature industrial process (Johnson, 2006). The wet milling process of corn fractionates the original grain into its components, namely starch, fibre, protein and oil. Starch is subsequently converted into a spectrum of products including bioethanol, dextrose, glucose syrups of 20–70 dextrose equivalent and high fructose corn syrups. NatureWorks LLC operates the only large-scale industrial facilities for the production of lactic acid from corn (180,000 t per year) and Ingeo polylactide resins (140,000 t per year) in Blair, Nebraska, USA (Vink et al., 2010).

Cereal-based biorefineries were among the first biorefining schemes – together with lignocellulosic, whole-crop, thermochemical-based and green biorefineries – proposed at the beginning of the current biorefinery era as alternative sustainable technologies to petroleum refineries (Kamm et al., 2006). The utilization of cereals as renewable resources for biorefinery development is dependent on the combination of physical, chemical, thermal and biological processing for the fractionation, extraction or (bio)conversion of the original grain and associated residues for the production of a spectrum of products including biofuels, chemical, biodegradable plastics, biomaterials, functional proteins, oils, antioxidants, polysaccharides, food and feed.

This chapter presents research carried out in the Satake Centre for Grain Process Engineering (SCGPE) at the University of Manchester (UK) on the development of cereal-based biorefineries for the production of fuel ethanol, polyhydroxybutyrate as a biodegradable polymer and succinic acid as a platform chemical for the future sustainable chemical industry. Future trends will also be presented based on the utilization of cereal-based food waste generated from cereal processors, bakeries, confectionary industries, households and restaurants that do not compete with food production.

## 10.2 Wheat-based biorefineries

Understanding the structure of wheat grain is important in order to comprehend the development of modern wheat milling operations for flour production and to design biorefinery strategies. The main parts of the wheat grain are the bran, the endosperm and the germ. Bran and germ are the terms used in traditional wheat milling operations. Bran is a generic term that includes all outer layers of the wheat grain that do not contain starch and gluten, the main components of endosperm. Germ is also a generic term to describe the embryo, the main parts of which are the embryonic axis and the scutellum. The bran is constituted by several layers including the outer

Table 10.1 Composition of major wheat grain fractions (on a dry basis, db)

| Fraction[a] | Content (%) | Ash (%) | Protein (%) | Lipids (%) | Crude fibre (%) | Carbohydrates[b] (%) |
|---|---|---|---|---|---|---|
| Whole wheat | 100 | 1.5 | 12 | 2 | 2 | 82 |
| Bran | 17 | 9 | 11 | 5 | 14 | 61 |
| Pericarp | 9 | 3 | 5 | 1 | 21 | 70 |
| Aleurone layer | 8 | 16 | 18 | 9 | 7 | 50 |
| Endosperm | 80 | 0.5 | 10 | 1 | > 0.5 | 88 |
| Germ | 3 | 5 | 26 | 10 | 3 | 56 |

[a] Bran, endosperm and germ are the major fractions of whole wheat and pericarp with the aleurone layer are the major fractions of bran.
[b] Estimated from other data.
*Source*: Pomeranz (1987).

and inner pericarp, the seed coat, the nucellar tissue and the aleurone layer. The aleurone layer surrounds the entire endosperm and part of the embryo. The bran fraction, including the aleurone layer, is approximately 17% of the whole grain weight on a dry basis. Detailed description of the structure of wheat grain is given by Evers and Bechtel (1988). Pomeranz (1987, 1988) and MacMasters *et al.* (1971) presented in detail the chemical composition of a typical wheat grain. Table 10.1 presents the composition of each major wheat grain fraction.

The viability of a whole-crop biorefinery based on wheat is dependent on the efficient fractionation of the grain and the production of added-value products with diversifying market outlets and improved production economics. Research at the SCGPE has focused on the production of, or the extraction of, added-value products from wheat bran, germ and part of the gluten, while the majority of the endosperm fraction has been used for the production of a generic fermentation feedstock. Koutinas *et al.* (2007a) presented a biorefining strategy that integrates the utilization of wheat as the sole raw material for production of both speciality and commodity products. The main target is the fractionation of wheat in order to maximize the products that could be produced either via extraction if they are present in one of the wheat layers or via microbial bioconversion employed for the production of bioethanol, biodegradable polymers or platform chemicals. In this way, minimization of waste production can be achieved.

The generic wheat-based biorefinery concept is presented in Fig. 10.1. The main focus of this biorefining concept is the fermentative production of a major commodity product from the major component of the wheat grain,

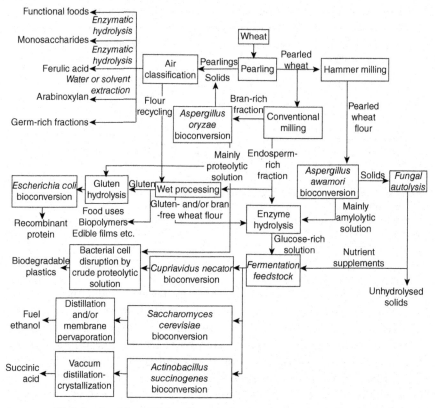

*10.1* A wheat-based biorefinery concept. (taken from Koutinas *et al.*, 2007a)

namely starch, and part of the gluten depending on the requirements for nitrogen sources by the microorganism used in each case. Thus, the process presented in Fig. 10.1 has been modified accordingly for the production of fuel ethanol (Arifeen *et al.*, 2007a,b), polyhydroxybutyrate (Xu *et al.*, 2010) or succinic acid (Du *et al.*, 2008). The other wheat components can be used for the extraction or production of commodity or speciality products with diversified market outlets.

The first unit operation employed in the process presented in Fig. 10.1 is pearling, in which bran layers are sequentially removed through friction and abrasion stages depending upon duration (Dexter and Wood, 1996; Gills and McGee, 1999; Koutinas *et al.*, 2006). Pearling eventually produces two fractions, one rich in bran and the other rich in endosperm. Bran-rich pearlings can be used for the extraction of various components including functional foods, monosaccharides, ferulic acid, arabinoxylan, and germ-rich fractions (Koutinas *et al.*, 2006). Pearled wheat grains that are enriched in starch and gluten are macerated with a hammer mill into flour which is subsequently used for the production of case-specific fermentation feedstocks and for gluten extraction. Gluten is a valuable co-product as it can be used in conventional applications (e.g., food industry) or as a raw material for the production of novel products, such as biodegradable plastics, edible films, adhesives, biomedical materials, composites, binders and raw material for the separation of amino acids as precursors for chemical production (Bietz and Lookhart, 1996; Kim, 2008; Koutinas et al., 2008; Zhang and Mittal, 2010; Lammens et al., 2012). The quantity of gluten that can be extracted from each alternative processing scheme is dependent on the requirement for nitrogen sources by the microorganism that is employed in the fermentation stage. For instance, succinic acid production is notorious for the high concentrations of yeast extract required for microbial growth and thus product formation. This means that in the wheat-based biorefinery concept, a significant quantity of gluten should be used as nitrogen supplement in the fermentation in order to eliminate the need for yeast extract.

A small fraction of the pearled wheat flour is used as the sole fermentation medium in submerged (preferably operated in continuous mode) or solid state fungal fermentations using a fungal strain of *Aspergillus awamori* for the production of crude enzyme consortia, predominantly amylolytic enzymes but with significant others such as proteases and phytase. In this way, the macromolecules (e.g., starch, gluten, phytic acid) contained in wheat can be hydrolysed into directly assimilable nutrients (e.g., glucose, amino acids, peptides). The crude filtrate from fungal fermentation is mixed with pearled wheat flour from which gluten has been entirely or partially extracted. Koutinas et al. (2007b) reported that sufficient starch hydrolysis to glucose can be achieved by following a temperature ramp process at temperatures lower than 70°C. In this process operated in batch mode, simultaneous gelatinization, liquefaction and saccharification can be achieved when enzyme consortia produced by *A. awamori* are employed.

The biorefinery concept presented in Fig. 10.1 exploits fungal autolysis as a natural process to generate nutrient supplements as substitute for commercial nutrients such as yeast extract. The fungal biomass that is produced during enzyme production can be autolysed under oxygen limiting conditions at 55°C and natural pH (Koutinas et al., 2004, 2005). Re-generating nutrients via fungal autolysis could replenish the nutrients consumed for fungal growth and also replace the nutrients lost during bran separation and gluten extraction. The glucose-rich solution produced via wheat flour hydrolysis could be mixed with fungal autolysate to produce fermentation media to suit the needs of many microorganisms. In the following sections, alternative processing schemes utilizing wheat grains or associated residues as raw material for biorefinery development will be presented.

## 10.3 Fuel ethanol production from wheat

The current energy supply relies mainly on fossil fuels, with oil, coal and gas contributing over 80% of the world energy consumption. The sustainable development of the society requires continuous supply of affordable renewable energy. One of the commercialized alternative biofuels is fuel ethanol production from plant-based biomass, e.g. corn, wheat or sugar cane. Although the debate of food versus fuel never settles, the worldwide bioethanol production has continued to increase over recent years (Fig. 10.2). In Europe, bioethanol is predominately produced from wheat. The wheat used in such first generation biorefineries is not generally of food grade. In fact, most first generation biorefineries use feed grade wheat, which is originally produced or imported for animal feed. In conventional wheat-based fuel ethanol production processes, starch is converted into bioethanol, while the protein and fibre fractions end up in the 'dried distillers grains with solubles' (DDGS). DDGS is sold as a valuable animal feed due to its high protein content. A conventional wheat-based biorefinery production process is shown schematically in Fig. 10.3.

The first step in bioethanol production is milling. The grain is milled either by wet milling or dry milling. In wet milling, wheat is soaked in a weak sulphurous acid solution at 48–52°C for about 2 days prior to milling. In comparison to dry milling, wet milling leads to relatively higher starch utilization efficiency. Then, the milled grains are cooked and starch hydrolysis enzymes, e.g. α-amylase, are added to break down the starch polymer to soluble dextrins (a process known as liquefaction). Cooking is also a way to sterilize the medium. Then, in the saccharification stage, glucoamylase is used to further hydrolyse dextrins into glucose. Saccharification could also be integated with fermentation, enabling 'simultaneous saccharification and fermentation' (SSF). In most bioethanol plants, the yeast strain *Saccharomyces cerevisiae* is most often utilized. The bacterial strain *Zymomonas mobilis* is also reported to be used in industrial-scale fermentations. After around 2–4

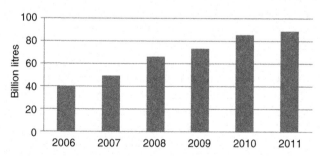

*10.2* Worldwide fuel ethanol production from 2006 to 2011 (http://ethanolrfa.org, accessed 17 May 2012).

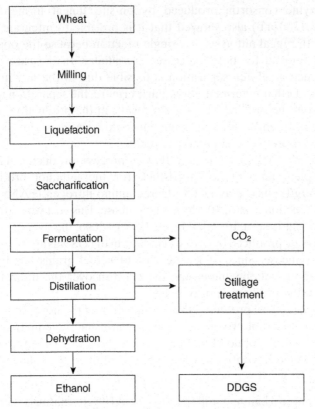

*10.3* A typical first generation wheat-based bioethanol production process.

days of fermentation, the ethanol concentration could reach around 8–10 wt%. The fermentation step is followed by distillation, which separates the stillage for DDGS production. Distillation produces an ethanol stream that contains up to 95.6% ethanol. Fuel grade ethanol is nowadays most often produced via molecular sieve dehydration.

Research carried out at the SCGPE focused on restructuring the conventional wheat-based industrial process for fuel ethanol production through the development of a novel biorefinery concept (Fig. 10.1). Arifeen *et al.* (2007a) presented the development of a continuous process converting wheat into a nutrient-complete feedstock suitable for fuel ethanol production. As presented in Fig. 10.1, starch and fungal autolysates are mainly employed as fermentation media for ethanol production, while other major wheat components such as bran and gluten are extracted and processed for different end-uses. Starch liquefaction and saccharification have been integrated in a continuous process through the utilization of complex enzyme consortia produced by on-site fungal fermentations. Arifeen *et al.* (2007a,b) also showed that it is possible to integrate starch hydrolysis with fungal autolysis in a single reaction because the operating conditions employed for both processes are similar. In the case of fuel ethanol production, gluten separation is feasible due to the low nitrogen requirements of ethanol fermentation. Furthermore, the separation of bran and gluten from the original wheat grain results in the production of pure yeast cells after the end of fermentation that could be used as a much higher market value co-product compared to DDGS.

Arifeen *et al.* (2007a) estimated that a process producing 120 $m^3$/h nutrient-complete fermentation feedstock for fuel ethanol production containing 250 g/L glucose and 0.85 g/L free amino nitrogen (FAN) would result in a production cost of $0.126 per kg glucose. This cost was estimated assuming a wheat cost of $0.16/kg. From 1997 to 2006, the average cost of starch hydrolysate produced by corn wet millers in the US was approximately $0.14 per kg glucose. Since 2006, the cost of cereal grains has become significantly more volatile, increasing the concern over the utilization of food crops for the production of biofuels.

Arifeen *et al.* (2007b) presented the optimization of the novel biorefining concept for fuel ethanol production from wheat using the equation-based software General Algebraic Modelling System (GAMS). The process flow sheet was designed in continuous mode and contained the following unit operations:

- wheat milling in a hammer mill (pearling could be alternatively employed to take advantage of the bran fraction as described in Fig. 10.1)
- gluten extraction as co-product
- fungal submerged fermentation for enzyme production
- simultaneous starch hydrolysis and fungal autolysis
- yeast fermentation with recycling integrated with a pervaporation membrane for ethanol concentration (up to 40 mol%)
- fuel-grade ethanol purification by pressure swing distillation (PSD) consisting of a low-pressure and a high-pressure column that could lead to reduced operating cost (up to 44%) for fuel ethanol production through the application of heat integration between the high- and low-pressure columns.

At production capacities in the range of 10–33.5 million gal per year (37.85–126.8 million L per year), the proposed wheat-based biorefinery could lead to a fuel ethanol production cost of $0.96–0.50 per gal ethanol ($0.25–0.13 per L ethanol). It should be stressed that the production cost of fuel ethanol is strongly dependent on the commercial value of the generated co-products (e.g., gluten, yeast and bran-rich pearlings). Bran-rich pearlings have been evaluated for the extraction of arabinoxylans

(Du et al., 2009; Misailidis et al., 2009) and gluten has been enzymatically converted into a fermentation medium for the production of recombinant proteins (Satakarni et al., 2009). Wheat bran has also been employed for the extraction of ferulic acid, which can be converted into vanillin (Di Gioia et al., 2007).

Du et al. (2009) and Misailidis et al. (2009) developed a process for the extraction of arabinoxylans from wheat bran as a potential co-product in the wheat-based fuel ethanol plant (Fig. 10.4). Arabinoxylan (AX) is a component of wheat bran that could be used as a food ingredient for viscosity enhancement and gel formation. More promisingly, the extraction of AX uses only ethanol as the main extractant. This immediately suggests scope for economical AX recovery within a fuel ethanol production plant. Experiments showed that high arabinoxylans content was presented in the outer layer of the wheat bran (Du et al., 2009). A wheat bran sample which was produced by pearling to a level of 4% (w/w) contained around 27%

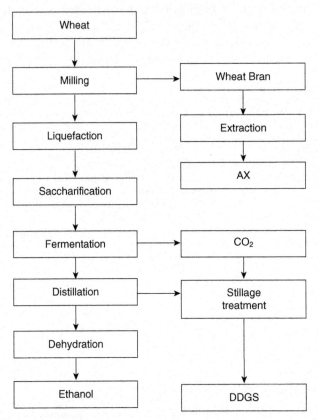

*10.4* An integrated wheat-based fuel ethanol and arabinoxylan (AX) production process.

AX (w/w). That was 50% more than the bran recovered by conventional roller milling of whole wheat. The economic assessment indicated that the cost of AX product (80% purity) was around £3.7–4.5 per kg in an integrated fuel ethanol and AX plant (Misailidis *et al.*, 2009). If the selling price of AX could reach £6/kg, the overall return on investment (ROI) would increase from 17 to 26%. Alternatively, if the ROI was kept constant at 17%, the ethanol could be sold 14% cheaper. These results suggested that the development of AX co-production in a fuel ethanol biorefinery would increase the economic competitiveness and commercialization feasibility.

## 10.4 Succinic acid production from wheat

Succinic acid (SA) is a 1,4-dicarboxylic acid which has been demonstrated to be one of the key platform molecules to be transformed into a variety of useful chemicals (Beauprez *et al.*, 2010; Bozell and Petersen, 2010; Cukalovic and Stevens, 2008; Lin *et al.*, 2012). From esters to amides through pyrrolidones, alcohols and/or biopolymers, the rich chemistry of the two carboxylic groups within the molecule (Fig. 10.5) as well as its partial solubility in water are two key relevant assets from the list of building blocks. Increasing petroleum prices favour the fermentative production of

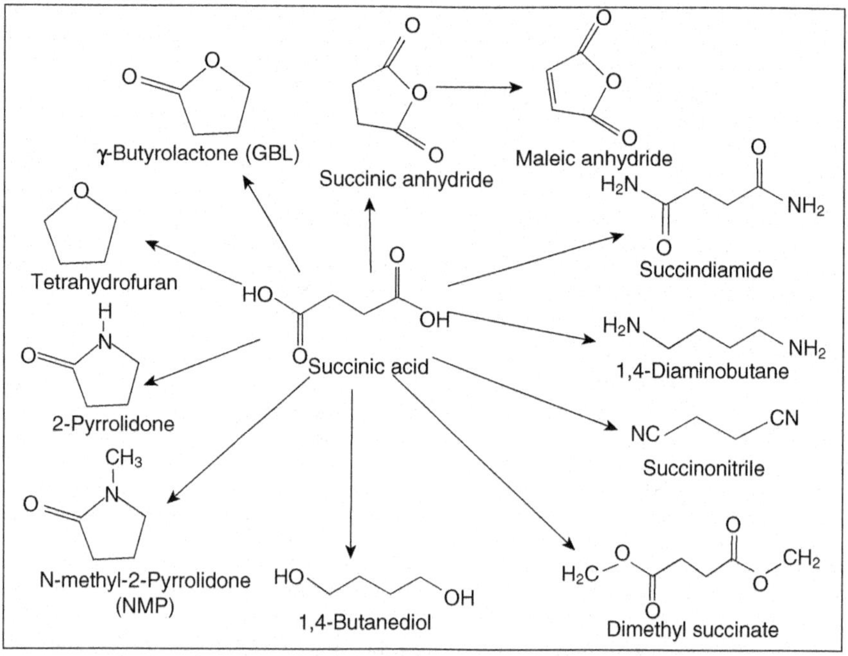

*10.5* Transformations of succinic acid to added-value chemicals.

SA as a replacement for petrochemical production from maleic anhydride. The current production of maleic anhydride is based on n-butane. The process requires high energy consumption due to demanding harsh operating conditions: high temperature (250°C) and pressure (200 bar). Therefore, low temperature and pressure bioproduction of SA could be economically favourable.

Research carried out at the SCGPE has focused on the development of wheat-based bioprocessing schemes for the production of succinic acid that integrate upstream processing of whole wheat grains or wheat milling by-products, fermentative production of succinic acid and chemical transformations of succinic acid either directly from fermentation broths or after purification in the form of crystals (Luque *et al.*, 2009; Lin *et al.*, 2010). Figure 10.6 presents a schematic diagram of a representative wheat-based bioprocess employed for the production of succinic acid followed by its conversion into various value-added products. As mentioned earlier, wheat contains all essential nutrients required for the formulation of a nutrient-complete fermentation medium. Thus, research focused on the production of fermentation media from wheat that can enhance succinic acid production. Four wheat-based upstream processing strategies (Fig. 10.7) have been developed at the SCGPE to produce carbon-rich and nitrogen-rich streams (Dorado *et al.*, 2009; Du *et al.*, 2008; Koutinas *et al.*, 2007a; Webb *et al.*, 2004). Although Strategy I has fewer processing steps, it has been demonstrated that it was not suitable for generic feedstock production due to the low glucose and free amino nitrogen (FAN) concentrations in the fungal filtrate. On the other hand, both Strategies II and III could fulfil the nutrient requirement of typical SA fermentations. In Strategy III, two fungal strains (*Aspergillus awamori* and *Aspergillus oryzae*) have been utilized to produce amylolytic and proteolytic enzymes, respectively. These enzymes

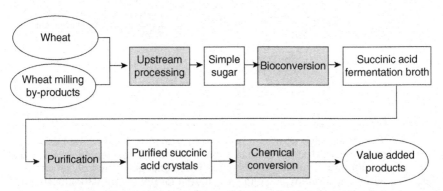

10.6 A wheat-based process using wheat and wheat milling by-products for succinic acid production as a precursor for chemical synthesis.

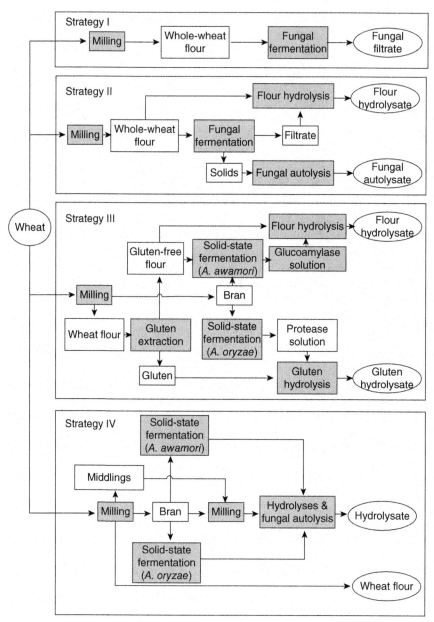

*10.7* Alternative upstream processing strategies for a wheat-based biorefinery. Strategy I: direct submerged fungal fermentation; Strategy II: submerged fungal fermentation with subsequent hydrolysis and autolysis; Strategy III: solid-state fungal fermentation with subsequent hydrolyses; Strategy IV: solid-state fungal fermentation with subsequent hydrolysis of wheat-flour milling by-products and fungal autolysis.

were subsequently employed to hydrolyse gluten-free flour and gluten, respectively. In terms of nutrient concentrations, the highest glucose concentration that could be obtained by solid state fermentation was $170\,g\,L^{-1}$.

As mentioned in Section 10.3, Arifeen *et al.* (2007a) presented an optimized upstream process for the production of a nutrient-complete wheat-based medium involving the combined hydrolytic/autolytic reaction using *A. awamori* submerged fungal fermentation solids. The simulations suggested that this process could produce $120\,m^3\,h^{-1}$ generic fermentation feedstock containing $250\,g\,L^{-1}$ glucose and $0.85\,g\,L^{-1}$ FAN, resulting in a production cost of $0.126/kg glucose. Therefore, based on a succinic acid yield of 0.81 g/g wheat (Du *et al.*, 2008), the substrate cost is $0.156/kg. The substrate cost is reduced by 5.6- and 1.7-fold as compared to the processes using maleic anhydride and glucose, respectively (Table 10.2).

The search for alternative and more sustainable water sources have been the focus of many studies in recent decades due to the shortage of fresh water in many areas. A recent report from the United Nations Office has summarized that overcoming the crisis in water and sanitation is one of the greatest human development challenges of the early twenty-first century (Harlem, 2012). The development of a seawater-based biorefinery strategy could have a potentially strong impact in this area with a holistic utilization of seawater, aiming at more efficient, low cost and low carbon footprint processes. Lin *et al.* (2011) presented the first report regarding the novel usage of seawater instead of plain water in fermentative biochemical production. Interestingly, there were no significant differences in terms of SA production in fermentations using seawater/water mixtures containing between 65% and 100% synthetic seawater. Indeed, a complete replacement

*Table 10.2* Comparison of substrate cost and conversion yield in succinic acid production

| Substrate | Average selling price ($/kg substrate) | Conversion yield (g succinic acid/g substrate) | Substrate cost ($/kg succinic acid) |
|---|---|---|---|
| Maleic anhydride[a] | 0.977 | 0.95 | 1.027 |
| Glucose[b] | 0.390 | 0.91 | 0.428 |
| Wheat-based feedstock[c] | 0.126 | 0.81 | 0.156 |

[a] Selling price and conversion yield of maleic anhydride from Song and Lee (2006).
[b] Conversion yield of succinic acid fermentation using glucose is based on Lee *et al.* (1999).
[c] Selling price of wheat-based feedstock is based on Arifeen *et al.* (2007a) and the conversion yield is based on results from Du *et al.* (2008).

of distilled/tap water by seawater could be achieved which will facilitate media preparation procedures, as well as make the process more economically and environmentally sound.

In bench-top fermentations with 50% seawater and wheat-derived media, around 45 g/L SA was produced with a yield of 1.02 g/g (consumed glucose) and a productivity of 0.84 g $L^{-1} h^{-1}$. These values were similar to those obtained in fermentations using semi-defined media and tap water. Most promisingly, 49 g/L SA was produced with a yield of 0.94 g/g and a productivity of 1.12 g $L^{-1} h^{-1}$ in a fermentation using only wheat-derived medium and natural seawater (Lin et al., 2011). In summary, mineral compounds and salts in seawater, together with wheat-derived media, were found to meet well the mineral requirements for *A. succinogenes* fermentations, without significant inhibition of cell growth. Interestingly, compounds present in seawater had a major effect on rates of reactions of a range of downstream transformations of SA including esterifications and amidations in comparison with reactions run under similar conditions using distilled water (Lin et al., 2011).

Table 10.3 shows the SA production using various raw complex materials. It indicates that almost any type of raw biomass could be used for the bioproduction of SA. The SA concentration, yield and productivity depend on the sugar, nutrient and inhibitor concentrations in the biomass-derived media.

## 10.5 Polyhydroxyalkanoate (PHA) production from wheat

Polyhydroxyalkanoates (PHAs) are biodegradable polyesters that contain 3-, 4-, 5- and 6-hydroxy-alkanoic acids as monomers. One of the main advantages of PHA production is the wide spectrum of applications, including agricultural uses (e.g., controlled release of insecticides, mulch films), food packaging, medical uses, manufacture of articles such as combs, pens and bullets, and production of flushables, scaffolds for tissue engineering applications, binders, biocomposites, adhesives, flexible packaging, thermoformed articles, synthetic paper and medical devices, among others (Du et al., 2012; Wolf et al., 2005; Philip et al., 2007). They are accumulated for energy and carbon storage in the form of intracellular granules by many microorganisms. The bacterial strain *Cupriavidus necator* (previously designated as *Ralstonia eutropha*) is the most widely studied microorganism for the production of PHAs that are accumulated by this microorganism as secondary metabolite in the presence of an abundant source of carbon and the limitation of another nutrient such as N, P, Mg, K, O or S. The homopolymer polyhydroxybutyrate (PHB) made of 3-hydroxybutyric acid (3HB) units and the

*Table 10.3* Comparison of succinic acid concentration, yield and productivity in batch fermentations using natural raw resources by *Actinobacillus succinogenes*

| Substrate | SA concentration (g/L) | Yield (g/g) | Productivity (g/L/h) | References |
|---|---|---|---|---|
| Cane molasses | 55.2 | 0.80 | 1.15 | Liu et al. (2008) |
| Corncob hydrolysate | 23.6 | 0.58 | 0.49 | Yu et al. (2010) |
| Corn core | 32.1 | 0.89 | 0.67 | Zheng et al. (2009) |
| Corn stover hydrolysate | 66.2 | 0.66 | 1.38 | Li et al. (2011) |
| Corn straw | 33.7 | 0.81 | 0.70 | Zheng et al. (2009) |
| Crop stalk (corn stalk and cotton stalk) | 15.8 | 1.23 | 0.62 | Li et al. (2010) |
| Glucose with spent Brewer's yeast hydrolysate | 46.8 | 0.69 | 0.98 | Jiang et al. (2010) |
| Pastry hydrolysate | 31.7 | 0.35 | 0.87 | Zhang et al. (2013) |
| Rapeseed meal | 23.4 | 0.115 | 0.33 | Chen et al. (2010) |
| Rice straw | 17.7 | 0.63 | 0.37 | Zheng et al. (2009) |
| Straw hydrolysate | 45.5 | 0.81 | 0.95 | Zheng et al. (2009) |
| Waste bread | 47.3 | 1.16 | 1.12 | Leung et al. (2012) |
| Wheat flour hydrolysate and fungal autolysate | 64.0 | 0.81 | 1.19 | Du et al. (2008) |
| Wheat hydrolysate with seawater | 49.0 | 0.94 | 1.12 | Lin et al. (2011) |
| Wheat milling by-products hydrolysate | 62.1 | 1.02 | 0.91 | Dorado et al. (2009) |
| Wheat straw | 19.0 | 0.74 | 0.40 | Zheng et al. (2009) |

copolymer poly(3-hydroxybutyrate-*co*-3-hydroxyvalerate) made of 3HB and 3-hydroxyvaleric acid (3HV) units are the most widely studied members of the PHA family that are produced by various strains of *C. necator* usually by consuming two different carbon sources as precursors.

PHA production can be achieved by many carbon sources including monosaccharides (e.g., glucose, fructose), disaccharides (e.g., sucrose, lactose), alcohols (e.g., methanol, ethanol), triacylglycerols derived from vegetable oils and animal fats, alkanes (e.g., hexane to dodecane) and organic acids (e.g., butyrate upwords) (Wolf *et al.*, 2005; Castilho *et al.*, 2009; Koller *et al.*, 2010). Glucose consumption by *C. necator* strains leads to the production of PHB. The biosynthetic pathway (Steinbüchel and Schlegel, 1991) of 3HB monomers leading to PHB production in *C. necator* begins with the condensation of two acetyl-CoA molecules, catalysed by the enzyme β-ketothiolase, leading to the formation of acetoacetyl-CoA. The intracellular accumulation of acetyl-CoA molecules occurs only under nutrient limiting conditions that prevent their utilization in the tricarboxylic acid (TCA) cycle. Subsequently, acetoacetyl-CoA is converted into 3-hydroxybutyryl-CoA via acetoacetyl-CoA reductase. PHB is finally produced through esterification of 3-hydroxybutyryl-CoA monomers by PHA synthase. The main driving force for the production of PHB is the availability of reducing equivalents in the form of NADPH (Madison and Huisman, 1999).

Nowadays, it is widely accepted that commercial production of PHAs can only be achieved within biorefinery strategies and through cost reduction in raw material selection and processing, fermentation in industrial-scale bioreactors and downstream separation. In the case of the raw material, research has focused on the evaluation of various agricultural products and agro-industrial waste and by-product streams as renewable resources for the production of PHAs, including cereals, whey, green grass, pulp fibre sludge, silage, molasses, and meat and bone meal (Rusendi and Sheppard, 1995; Zhang *et al.*, 2004; Nikel *et al.*, 2006; Solaiman *et al.*, 2006; Koutinas *et al.*, 2007c; Koller *et al.*, 2010). Pure starch and commercial protein supplements (e.g., yeast extract, casein hydrolysate, casamino acids) have been tested for PHB production (Bormann *et al.*, 1998; Lapointe *et al.*, 2002; Yu *et al.*, 2003; Quillaguaman *et al.*, 2005; Huang *et al.*, 2006).

Research at the SCGPE focused on the development of a wheat-based biorefining strategy for the production of PHB as the core commodity product and bran-rich pearlings and gluten as added-value co-products (Xu *et al.*, 2010). The proposed biorefinery concept is included in Fig. 10.1. The main challenge of PHB production from wheat was to formulate two types of fermentation feedstocks from wheat; one rich in all nutrients necessary for microbial growth and a second one containing high glucose concentration and optimum concentration of nitrogen sources. The nutrient-complete fermentation medium was produced by combining fungal autolysates with flour hydrolysates that were produced from pearled wheat flour after partial extraction of gluten. The second fermentation medium can be used as feeding medium during PHB accumulation where nutrient limitation is necessary. It was observed that supplying low amount of nitrogen sources is necessary in order to maintain microbial cell viability and achieve high PHB concentrations and yields.

Fermentations carried out in a 1 L bioreactor and operated in fed-batch mode using a nutrient-complete wheat-based medium at the beginning of the fermentation and a concentrated (pure) glucose feeding solution during the PHB accumulation phase (Fig. 10.8(a)), showed that PHB accumulation is enhanced with increasing microbial biomass concentration achieved

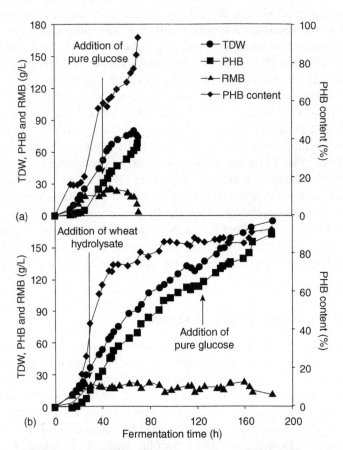

*10.8* Total dry weight (TDW), polyhydroxybutyrate (PHB), residual microbial biomass (RMB) and PHB content in microbial cells during fed-batch fermentations using pure glucose (a) or wheat hydrolysate (b) as feeding media.

during the initial growth phase (Xu *et al.*, 2010). Microbial biomass concentration was enhanced with increasing initial FAN concentration at the beginning of the fermentation up to an initial FAN concentration of less than 1000 mg/L. A maximum PHB concentration of 68.2 g/L was achieved when an initial FAN concentration of 960 mg/L was used at the beginning of the fermentation and concentrated pure glucose was used as feeding medium (Fig. 10.8(a)). Microbial cell autolysis that occurred at the end of the fermentation increased the PHB content up to 93% (w/w). Fermentations with FAN concentration higher than 1000 mg/L resulted in enhanced microbial cell growth but no PHB accumulation.

Figure 10.8(b) shows that sequential feeding with a wheat hydrolysate followed by a pure glucose solution during PHB accumulation leads to the production of PHB concentrations up to 162.8 g/L at a productivity of 0.89 g/L/h. Xu *et al.* (2010) showed that running fed-batch fermentations with addition of wheat hydrolysates at different flow rates results in varying PHB production. It is, therefore, evident that there is a critical concentration of organic nitrogen that leads to substantial increase of PHB production. In addition, the consumption of amino acids and peptides for bacterial growth and glucose mainly for PHB accumulation leads to increased glucose to PHB conversion yields. Furthermore, bacterial cell autolysis was also observed in fed-batch fermentation using wheat hydrolysate and pure glucose as feeding media leading to a final PHB content of 93% (w/w).

Considering that PHA concentrations of 60 to preferably 80 g/L should be achieved to facilitate commercial viability (Wolf *et al.*, 2005), the PHB production achieved from wheat-based media surpasses this limiting threshold. In addition, the glucose to PHB conversion yields (higher than 0.37 g/g) achieved when wheat-based media are used are among the highest reported in the literature. However, the productivity achieved (0.89 g/L/h) should be improved further to enhance the commercial potential of this process.

## 10.6 Utilization of wheat straw

The sustainability of wheat-based biorefineries could be improved by integrating wheat straw and other residue utilization through the optimization of their application in the agricultural field to maintain soil quality and avoid erosion losses, and as renewable resources for the production of chemicals, materials, biofuels and energy. The amount of crop residue that should remain in the field could be calculated through simple carbon models (Hettenhaus, 2006). Surplus amounts of wheat residues could be used for energy production via combustion or chemical production through thermochemical conversion technologies either alone or integrated, in many cases, with enzymatic hydrolysis and microbial bioconversions.

Pyrolysis and gasification are potential technologies for fuel and chemical production from wheat straw (Demirbas, 2006). Hydrogen and methanol production via gasification could be commercially viable, but production of simple alcohols, aldehydes, mixed alcohols and Fischer–Tropsch liquids are not cost-competitive yet and require further research and development activities (Werpy and Petersen, 2004).

With the pressure of continuous increases in food prices, it is increasingly desirable to produce fuel ethanol from lignocellulosic raw materials (e.g., wheat straw) rather than food-based materials. Thus, the concept of the second generation of bioethanol production process was developed (Fig. 10.9). Wheat straw, which is an abundant lignocellulosic materials in Europe, has attracted great interest as a raw material for fuel ethanol production. In comparison with the first generation wheat-based bioethanol production process, the utilization of wheat straw in ethanol fermentation is relatively complex and energy intensive. Usually, the wheat straw needs

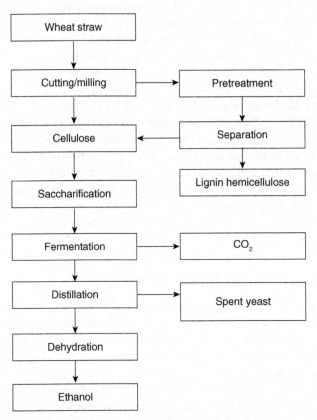

*10.9* Scheme for a second generation bioethanol production process using wheat straw as a starting material.

to be physically pre-treated (e.g., cutting, milling) to reduce the particle size. Then, wheat straw powder will be subjected to one or more chemical pretreatments, such as dilute acid pretreatment, to separate cellulose from lignin and hemicellulose. Next, cellulases are used for enzymatic hydrolysis of cellulose to obtain glucose for subsequent yeast fermentation. One of the main challenges in the wheat straw to biofuel process is the efficient conversion of lignocellulosic raw materials into simple sugars with minimum or even negligible production of inhibitors to the bioethanol fermentations. Various pre-treatment methods have been investigated, including dilute acid pretreatment, alkali pretreatment, steam explosion, ammonia fibre explosion, thermochemical pretreatment, organosolv process and biological pretreatment. These processes were recently reviewed by Talebnia *et al.* (2010). It is reported that a wheat straw-based fuel ethanol pilot plant has been constructed, but commercial-scale production has not yet been confirmed.

Although not yet economically viable, the hemicellulose fraction of wheat straw could be converted into directly assimilable sugars by many microorganisms through integrated chemical pre-treatment and enzymatic hydrolysis methods (Saha, 2003). Enzymatic hydrolysis of the hemicellulose fraction of wheat straw would result in the release of xylose, fermentative conversion of which could lead to the production of xylitol (Canilha *et al.*, 2006), which has been listed by the US Department of Energy among the top 12 value-added platform molecules for the production of a range of chemicals, including xylaric acid, propylene glycol, ethylene glycol, mixture of hydroxyl-furans, and polyesters (Werpy and Petersen, 2004). Wheat straw hydrolysates could also be used for fermentative production of 2,3-butanediol and lactic acid (Saha, 2003; Saito *et al.*, 2012) or chemical conversion to levulinic acid (Chang *et al.*, 2007). Agricultural residues, such as wheat straw, could be used for the production of biofibres with potential applications as composites, textiles, food, enzymes, fuels, chemicals, pulp and paper (Reddy and Yang, 2005).

## 10.7 Conclusion and future trends

Cereal grains were obviously among the initial renewable raw materials that were considered for the production of biofuels, chemicals and materials via biorefinery development. The existence of industrial cereal processors facilitated these initiatives because the successful implementation of a new industrial concept is highly dependent on the availability of raw material, the existence of skilled workforce and logistics (e.g., transportation). Although cereals fulfilled most of the above prerequisites, their main utilization as food and feed has created a major barrier for widespread industrial implementation.

Cherubini (2010) pointed out that most of the existing biofuels and biochemicals are currently produced in single production chains but not within a biorefinery concept, and usually require materials in competition with the food and feed industry. Food waste and by-product streams constitute renewable resources that could be utilized for chemical and material production. In this way, biorefinery concepts could be developed that do not compete but, on the contrary, coincide with food production. Dorado et al. (2009) and Leung et al. (2012) demonstrated the future potential of integrating succinic acid production in existing cereal-based industries (e.g., wheat milling industry, bakeries) through the utilization of by-products or waste streams. Two types of bioprocesses or biorefineries could be developed:

- integration of novel bioprocesses or green chemical technologies to valorize cereal-based waste or by-product streams in existing industrial plants
- collect cereal-derived food waste streams from households, restaurants and catering services for fractionation and bioconversion into chemicals, materials and biofuels.

### 10.7.1 Valorization of industrial cereal-based waste and by-product streams

Cereal-based industries, including starch production, wheat milling, confectionery industries, pasta production and bakeries among others, generate significant quantities of starch- or flour-based waste and by-product streams. For instance, wheat middlings and bran are by-products from the wheat dry milling industry. Koutinas et al. (2006) reported that out of the $5,624 \times 10^3$ t of wheat processed by UK flour millers in 1999–2000, approximately $1,148 \times 10^3$ t of by-products were generated, the predominant fraction of which is middlings that contain significant quantities of starch (20–35%), protein (15–19%) and phosphorus (approximately 1%). The poor nutritional value of wheat milling by-products reduces its versatility as animal feed only to pigs and cattle (Koutinas et al., 2006). For this reason, in many countries, surplus quantities of wheat milling by-products that cannot be used as animal feed could be treated as renewable resources for the production of chemicals and materials.

Dorado et al. (2009) devised a strategy to convert wheat milling by-product streams into a generic fermentation feedstock (Strategy IV in Fig. 10.7). Similar to Strategy III presented also in Fig. 10.7, wheat bran has been utilized as substrate for the production of amylolytic and proteolytic enzymes using the fungal strains A. awamori and A. oryzae, respectively. Simultaneous starch and protein hydrolysis together with fungal autolysis

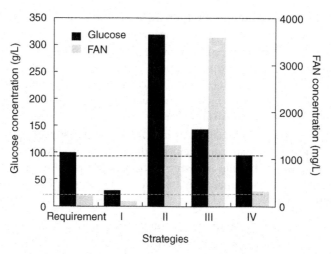

*10.10* Glucose and FAN concentrations in wheat-derived media produced from different upstream strategies compared with the requirement of typical succinic acid fermentation.

then occurred when the *A. awamori* and *A. oryzae* solid state fermentation solids were mixed with suspensions of wheat middlings and bran at 55°C. Hydrolysis of wheat middlings and bran in a paddle reactor resulted in the highest glucose and FAN concentrations of around 100 g/L and 300 mg/L, respectively. The carbon and nitrogen concentrations of the hydrolysate mixture met well the nutrient requirement of an industrial SA fermentation (Fig. 10.10). Figure 10.10 presents the glucose and FAN concentrations produced in each one of the fermentation feedstock production strategies presented in Fig. 10.7 along with the respective requirements of a typical succinic acid fermentation process. Using wheat milling by-products as the sole raw material for microbial feedstock production resulted in the production of 50.6 g/L succinic acid at a productivity of 1.04 g/L/h and a glucose to succinic acid conversion yield of 0.73 g/g (Dorado *et al.*, 2009). Therefore, it was demonstrated that wheat milling industry by-products could be effectively utilized in fermentative succinic acid production. In this way, succinic acid production could be integrated into existing wheat milling plants to provide an alternative market for surplus wheat milling by-products that are not used as animal feed.

Research at the Agricultural University of Athens in Greece focuses on the utilization of flour- or starch-based waste streams from confectionery industries as renewable resources for the production of microbial oil as raw material for biodiesel or oleochemical production. Microbial oil is produced by oleaginous microorganisms and has similar fatty acid content to vegetable oils. Koutinas and Papanikolaou (2011) presented a detailed review of microbial oil production from various renewable resources and discussed its prospect for biodiesel production.

## 10.7.2 Valorization of generic cereal-based waste streams

Food waste is considered as one of the major problem of the twenty-first century. According to a study by the Food and Agriculture Organisation of the United Nations (FAO), there are 1.3 billion metric t of food production. Around one-third of the total is lost or wasted every year worldwide (FAO, 2012). In industrial countries, 95–115 kg of perfectly good food is wasted per person per annum. It is estimated that one third of the food sold in the UK ends up as waste and half of this is still edible. Food waste is a major fraction of biodegradable municipal waste, most of which is disposed of in landfill sites where it is converted into methane, a potent greenhouse gas, by anaerobic digestion. According to the Landfill Directive, biodegradable municipal waste disposed into landfills should be decreased to 35% of 1995 levels, by 2016 in the UK. However, there is as yet no well-established solution to this problem.

Amongst the many forms of food waste, bread is a major contributor to the problem. The bakery industry is one of the world's major food industries and varies widely in terms of production scale and applications. The Western European bread industry produces 25 million t of bread per annum (BREAD4PLA, 2012). Germany and the UK are the main operators with 60% of plant sector production. France, the Netherlands and Spain produce another 20% of total production capacity. The bread market in the UK is worth over £3.5 billion and the total volume is approximately 2.9 million t annually (Melikoglu, 2008). Bakery wastes consist primarily of stale bread, bread rolls and cookies. The solid waste from baked bread (about 175 thousand t of bakery waste in Europe) (BREAD4PLA, 2012) cannot be reprocessed and generally it is used as animal feed, disposed to landfill sites or incinerated. Landfilling eventually leads to production of methane (a greenhouse gas) and groundwater pollution (organic compounds). Incineration of bakery waste can also release nitrogen oxide gases. Although utilization of waste bread as animal feed represents an environmentally friendly recycling route for this waste, economically it represents a very low added-value option.

Waste bread possesses ideal characteristics as a substrate for solid state or submerged fermentation and since it is a waste, it possesses little or no commercial value. Selection of a substrate for fermentation depends upon several factors related mainly to cost, availability and nutrient composition. Bread contains approximately 50% starch, 40% water and 7% protein; whereas bread crust contains 70% starch. If waste bread could be collected properly then it could be used to produce glucose for subsequent fermentations to produce biofuels, biodegradable polymers and chemicals. Moreover, valorization of waste bread through microbial bioconversions could be a viable way of reducing environmental burdens, meeting the Landfill Directive and producing high added-value from a waste in a sustainable manner.

Research at the SCGPE focused on the utilization of waste bread as raw material for the production of crude enzyme consortia (mainly amylolytic enzymes) via solid state fermentation using *A. awamori* (Wang et al., 2009). Enzyme-rich aqueous extracts from solid state fermentation were used for hydrolysis of untreated waste bread to produce fermentation feedstock. The results showed that solid state fermentation of waste bread pieces can be successfully used to process waste bread for the production of nutrient-rich hydrolysates. Leung et al. (2012) demonstrated the feasibility of utilizing waste bread as a generic feedstock for the production of succinic acid. The resultant succinic acid concentration was 47.3 g/L with an overall yield of 0.55 g SA per g bread, which is the highest reported, amongst food waste-derived media. Although succinic acid was selected as the test case, many other fermentation products could be produced from waste bread via fermentation or green chemical conversion. This could reduce the dependence on petroleum for chemical production. The utilization of waste bread for the production of value-added products should be seriously considered by local governments as part of their strategy for tackling the municipal solid waste problem and for environmentally friendly production of chemicals, materials and fuels.

## 10.8 Sources of further information and advice

While the use of biological raw materials for the production of value-added products through fermentation has been practised for many thousands of years, and even the production of industrial chemicals through biotechnology is almost a century old (starting in 1917 with the development, in Manchester, of the acetone-butanol fermentation by Chaim Weizmann), the biorefinery concept is rather new. It was not until the mid-1990s that the term 'biorefinery' came to mean what it does today. In Europe, the Danish Bioraf Foundation project was exploring links between agriculture and industry and published its report on 'The whole crop biorefinery' (Gyling, 1995). Meanwhile, in the US, Wyman and Goodman (1993) had recently published their seminal paper on the production of fuels, chemicals and materials from biomass. It was at this time that Colin Webb began work on the cereal-based biorefinery and established the SCGPE in Manchester (Webb, 1994).

Composition and conventional processing technologies for cereal grains is presented in various books including Pomeranz (1987, 1988), White and

Johnson (2003), MacGregor and Rattan (1993), and Webster (1989). Cereal-based biorefineries as well as lignocellulosic-based biorefineries including utilization of cereal straws have been presented in a number of books, for example those by Campbell *et al.* (1997) and Kamm *et al.* (2006). Johnson and May (2003) present the corn-based biorefinery development established by exploiting the wet milling process of corn. Other excellent texts to consult on the biorefinery concept include those by Demirbas (2010) and Stuart and El-Hawagi (2012). The latter addresses design aspects related to the integrated biorefinery concept. Several journals are now dedicated to the area. Amongst these are Wiley's *Biofuels, Bioproducts and Biorefining* (Biofpr) and Springer's *Biomass Conversion and Biorefinery*.

For general information and latest developments, the reader should consult the National Non-Food Crops Centre (http://www.nnfcc.co.uk/biorefinery), the International Forum on Industrial Biprocesses (http://www.ifibiop.org/) and the National Biorefineries Database (http://en.openei.org/datasets/node/50). For biofuels specific biorefinery information, an excellent information source can be found at the Biofuels Platform (http://www.plateforme-biocarburants.ch/en/home/).

Recent research projects focus on the utilization of cereal-based food waste for the production of biofuels, chemicals and biomaterials. BREAD4PLA (2012) is a European project (http://www.bread4pla-life.eu/) aiming to demonstrate the potential production of polylactic acid (PLA) for bio-based plastic formulation using bakery waste as raw material for fermentative production of lactic acid. The PLA produced will be used in the production of new biodegradable packaging for bakery products. This means that the wastes generated from a cereal-based food industry will be utilized as raw material for the production of a product that will be used by the same industry as a substitute for petroleum-derived plastics.

Pilot-plant facilities have been created in order to evaluate the conversion of renewable raw materials including cereal-based food waste for the production of a spectrum of products. At the Leibniz Institute for Agricultural Engineering Potsdam-Bornim (ATB) in Germany, waste bread is used as substrate for the production of lactic acid. A production capacity of 10 t of lactic acid in 200 days per year was achieved (Venus and Richter, 2007). The Biorenewables Development Centre (BDC) in York (UK) was constructed in 2012. It was built upon the research and development expertise of the Centre for Novel Agricultural Products (CNAP) and the Green Chemistry Centre of Excellence (GCCE) at the University of York, with support from the Science City of York. It contains pilot-scale processing equipment with the aim to develop a broad range of products and processes based on the use of biorenewable resources (BDC, 2012).

The School of Energy and Environment at the City University of Hong Kong has recently started collaboration with the coffee retail giant

'Starbucks Hong Kong' (Lin et al., 2013; Zhang et al., 2013). The partnership, which was facilitated by the NGO The Climate Group, focuses on the valorization of spent coffee grounds and unconsumed bakery wastes via bioprocessing. The collaboration is based on a support scheme and part of the 'Care For Our Planet' campaign: for every set of Care For Our Planet Cookies Charity Set sold, Starbucks will donate HK$8 to the School of Energy and Environment of City University of Hong Kong to support research on valorization of food waste for sustainable production of chemicals and materials. The aim of the research is to valorize the disposed coffee grounds and unconsumed baked goods to bio-plastics and detergent ingredients, facilitating the development of biomass use in Hong Kong and reducing the release of greenhouse gases and other air pollutants into the atmosphere.

## 10.9 Acknowledgements

This research was supported by the Engineering and Physical Sciences Research Council (EP/C530993/1) and gr/s24909/01 in association with the Crystal Faraday Partnership in the United Kingdom. We are also grateful for the provision of the Overseas Research Students Awards Scheme to Dr. Lin by Universities UK. The authors gratefully acknowledge the generous contribution of the Satake Corporation of Japan in providing financial support for much of the research carried out in the Satake Centre for Grain Process Engineering (SCGPE, University of Manchester, UK). Dr Carol Lin additionally acknowledges the Industrial Technology Funding from the Innovation and Technology Commission (ITS/323/11) in Hong Kong, the donation from the Coffee Concept (Hong Kong) Ltd. for the 'Care For Our Planet' campaign, as well as a grant from the City University of Hong Kong (Project No. 7200248). Dr Apostolis Koutinas acknowledges the financial support from 'NUTRI-FUEL' (09SYN-32-621) research project implemented within the National Strategic Reference Framework (NSRF) 2007–2013 and co-financed by national (Greek Ministry – General Secretariat of Research and Technology) and community funds (EU – European Social Fund).

## 10.10 References

Arifeen, N., Wang, R., Kookos, I., Webb, C. and Koutinas, A. A. (2007a) Optimization and cost estimation of novel wheat biorefining for continuous production of fermentation feedstock. *Biotechnol Progress*, 23, 872–880.

Arifeen, N., Wang, R., Kookos, I. K., Webb, C. and Koutinas, A. A. (2007b) Process design and optimization of novel wheat-based continuous bioethanol production system. *Biotechnol Progress*, 23, 1394–1403.

BDC (2012) *Biorenewables Development Centre* [Online]. Available: http://www.biorenewables.org/

Beauprez, J. J., De Mey, M. and Soetaert, W. K. (2010) 'Microbial succinic acid production: natural versus metabolic engineered producers', *Proc Biochem*, 45, 1103–1114.

Bietz, J. A. and Lookhart, G. L. (1996) 'Properties and non-food potential of gluten', *Cereal Foods World*, 41, 376–382.

Bormann, E. J., Leihner, M., Roth, M., Beer, B. and Metzner, K. (1998) 'Production of polyhydroxybutyrate by *Ralstonia eutropha* from protein hydrolysates', *Appl Microbiol Biotechnol*, 50, 604–607.

Bozell, J. J. and Petersen, G. R. (2010) 'Technology development for the production of biobased products from biorefinery carbohydrates – the US Department of Energy's "Top 10" revisited', *Green Chem*, 12, 539–554.

Bread4pla. (2012) *Welcome to BREAD4PLA* [Online]. Available: http://www.bread4pla-life.eu/index.php

Campbell, G. M., Webb, C. and McKee, S. L. (eds) (1997) *Cereals: Novel Uses and Processes*. Plenum Press, New York.

Canilha, L., Carvalho, W., Almeida, E. and Silva, J. B. (2006) 'Xylitol bioproduction from wheat straw: hemicellulose hydrolysis and hydrolyzate fermentation', *J Sci Food Agric*, 86, 1371–1376.

Castilho, L. R., Mitchell, D. A. and Freire, D. M. G. (2009) 'Production of polyhydroxyalkanoates (PHAs) from waste materials and by-products by submerged and solid-state fermentation', *Biores Technol*, 100, 5996–6009.

Chang, C., Cen, P. and Ma, X. (2007) 'Levulinic acid production from wheat straw', *Bioresource Technol*, 98, 1448–1453.

Chen, K., Zhang, H., Miao, Y., Wei, P. and Chen, J. (2010) 'Simultaneous saccharification and fermentation of acid-pretreated rapeseed meal for succinic acid production using *Actinobacillus succinogenes*', *Enz Microb Technol*, 48, 339–344.

Cherubini, F. (2010) 'The biorefinery concept: using biomass instead of oil for producing energy and chemicals', *En Conver Manag*, 51, 1412–1421.

Cukalovic, A. and Stevens, C. V. (2008) 'Feasibility of production methods for succinic acid derivatives: a marriage of renewable resources and chemical technology', *Biofuels, Bioproducts and Biorefining*, 2, 505–529.

Demirbas, A. (2010) *Biorefineries for Biomass Upgrading Facilities*. Springer-Verlag, London.

Demirbas, M. F. (2006) 'Hydrogen from various biomass species via pyrolysis and steam gasification processes', *Energy Sources, Part A: Recov Util Environ Effects*, 28, 245–252.

Dexter, J. E. and Wood, P. J. (1996) 'Recent applications of debranning of wheat before milling', *Trends Food Sci Technol*, 7, 35–41.

Di Gioia, D., Sciubba, L., Setti, L., Luziatelli, F., Ruzzi, M., Zanichelli, D. and Fava F. (2007) 'Production of biovanillin from wheat bran', *Enz Microb Technol*, 41, 498–505.

Dorado, M. P., Lin, S. K. C., Koutinas, A., Du, C., Wang, R. and Webb, C. (2009) 'Cereal-based biorefinery development: utilisation of wheat milling by-products for the production of succinic acid', *J Biotechnol*, 143, 51–59.

Du, C., Lin, S. K. C., Koutinas, A., Wang, R., Dorado, P. and Webb, C. (2008) 'A wheat biorefining strategy based on solid-state fermentation for fermentative production of succinic acid', *Biores Technol*, 99, 8310–8315.

Du, C., Campbell, G. M., Misailidis, N., Mateos-Salvador, F., Sadhukhan, J., Mustafa, M. and Weightman, R. M. (2009) 'Evaluating the feasibility of commercial arabinoxylan production in the context of a wheat biorefinery principally producing ethanol. Part 1. Experimental studies of arabinoxylan extraction from wheat bran', *Chem Eng Res Des*, 87, 1232–1238.

Du, C., Sabirova, J., Soetaert, W. and Lin, C. S. K. (2012) 'Polyhydroxyalkanoates production from low-cost sustainable raw materials', *Current Chem Biol*, 6, 14–25.

Evers, A. D. and Bechtel, D. B. (1988) 'Microscopic structure of the wheat grain', in: Pomeranz, Y., *Wheat: Chemistry and Technology*, Vol. I. AACC, St. Paul, MN, 47–95.

FAO (2012) *Statistical Yearbook 2012*, Food and Agriculture Organisation, Rome.

Gills, J. M. and McGee, B. C. (1999) Advances in durum semolina milling. British Pasta Products Association, Technical Seminar, London. Available at: http://www.satake-usa.com/brochures/Durum_Milling_Advances.pdf

Gyling, M. (1995) *The Whole Crop Biorefinery Project: Main Report*. Bioraf Denmark Foundation, Aakirkeby.

Harlem, B. G. (2012) 'The global water crisis: Addressing an urgent security issue', in: Harriet, B., Morris, T., Sandford, B. and Adeel, Z., *Papers for the Interaction Council*, 2011–2012. Available at: http://www.inweh.unu.edu/WaterSecurity/documents/WaterSecurity_FINAL_Aug2012.pdf

Hettenhaus, J. (2006) 'Achieving sustainable production of agricultural biomass for biorefinery feedstock', *Ind Biotechnol*, 2, 257–274.

Huang, T.-Y., Duan, K.-J., Huang, S.-Y. and Chen, C. W. (2006) 'Production of polyhydroxyalkanoates from inexpensive extruded rice bran and starch by *Haloferax mediterranei*', *J Ind Microbiol Biotechnol*, 33, 701–706.

Jiang, M., Chen, K., Liu, Z., Wei, P., Ying, H. and Chang, H. (2010) 'Succinic acid production by *Actinobacillus succinogenes* using spent brewer's yeast hydrolysate as a nitrogen source', *Appl Biochem Biotechnol*, 160, 244–254.

Johnson, D. L. (2006) 'The corn wet milling and corn dry milling industry – a base for biorefinery technology development', in: Kamm, B., Gruber, P. R. and Kamm, M., *Biorefineries – Industrial Processes and Products*, Weinheim, Wiley-VCH, 345–353.

Johnson, L. A. and May, J. B. (2003) 'Wet milling: the basis for corn biorefineries', in: White, P. J. and Johnson, L. A., *Corn: Chemistry and Technology*, 2nd edn, American Association of Cereal Chemists, St. Paul, MN, 449–494.

Kamm, B., Gruber, P. R. and Kamm, M. (2006) *Biorefineries – Industrial Processes and Products: Status Quo and Future Directions*. Weinheim, Wiley–VCH.

Kim, S. (2008) 'Processing and properties of gluten/zein composite', *Biores Technol*, 99, 2032–2036.

Koller, M., Atlic, A., Dias, M., Reiterer, A. and Braunegg, G. (2010) 'Microbial PHA production from waste raw materials', in: Chen, D.G-Q. *Plastics from Bacteria. Natural Functions and Applications*, Vol. 14. Springer-Verlag, Berlin, 85–119.

Koutinas, A. A. and Papanikolaou, S. (2011) 'Biodiesel production from microbial oil', in: Luque, R., Campelo, J. and Clark, J. H., *Handbook of Biofuels Production – Processes and Technologies*, Woodhead Publishing Limited, Cambridge 177–198.

Koutinas, A. A., Wang, R.-H. and Webb, C. (2004) 'Restructuring upstream bioprocessing: technological and economical aspects for production of a generic microbial feedstock from wheat', *Biotechnol Bioeng*, 85, 524–538.

Koutinas, A. A., Wang, R.-H. and Webb, C. (2005) 'Development of a process for the production of nutrient supplements for fermentations based on fungal autolysis', *Enz Microb Technol*, 36, 629–638.

Koutinas, A. A., Wang, R., Campbell, G. and Webb, C. (2006) 'A whole crop biorefinery system – a closed system for the manufacture of non-food products from cereals', in: Kamm, B., Gruber, P. R. and Kamm, M., *Biorefineries – Industrial Processes and Products*. Wiley-VCH, Weinheim, 165–191.

Koutinas, A. A., Xu, Y., Wang, R.-H. and Webb, C. (2007a) 'Polyhydroxybutyrate production from a novel feedstock derived from a wheat-based biorefinery', *Enz Microb Technol*, 40, 1035–1044.

Koutinas, A. A., Arifeen, N., Wang, R.-H. and Webb, C. (2007b) 'Cereal-based biorefinery development: integrated enzyme production for cereal flour hydrolysis', *Biotechnol Bioeng*, 97, 61–72.

Koutinas, A. A., Xu, Y., Wang, R.-H. and Webb, C. (2007c) 'Polyhydroxybutyrate production from a novel feedstock derived from a wheat-based biorefinery', *Enz Microb Technol*, 40, 1035–1044.

Koutinas, A. A., Du, C., Wang, R.-H. and Webb, C. (2008) 'Production of chemicals from biomass', in: Clark, J. H. and Fabien, E. I. D. (eds), *Introduction to Chemicals from Biomass*. Wiley-VCH, Weinheim, 77–101.

Lammens, T. M., Franssen, M. C. R., Scott, E. L. and Sanders, J. P. M. (2012) 'Availability of protein-derived amino acids as feedstock for the production of bio-based chemicals', *Biom Bioen*, 44, 168–181.

Lapointe, R., Lambert, A. and Savard, L. (2002) 'Process for production of biopolymer', World Intellectual Property Organization, Publication number WO0222841.

Lee, P. C., Lee, W. G., Kwon, S., Lee, S. Y. and Chang, H. N. (1999) 'Succinic acid production by *Anaerobiospirillum succiniciproducens*: effects of the $H_2/CO_2$ supply and glucose concentration', *Enz Microb Technol*, 24, 549–554.

Leung, C. C. J., Cheung, A. S. Y., Zhang, A. Y.-Z., Lam, K. F. and Lin, C. S. K. (2012) 'Utilisation of waste bread for fermentative succinic acid production', *Biochem Eng J*, 65, 10–15.

Li, J., Zheng, X.-Y., Fang, X.-J., Liu, S.-W., Chen, K.-Q., Jiang, M., Wei, P. and Ouyang, P.-K. (2011) 'A complete industrial system for economical succinic acid production by *Actinobacillus succinogenes*', *Biores Technol*, 102, 6147–6152.

Li, Q., Yang, M., Wang, D., Li, W., Wu, Y., Zhang, Y., Xing, J. and Su, Z. (2010) 'Efficient conversion of crop stalk wastes into succinic acid production by *Actinobacillus succinogenes*', *Biores Technol*, 101, 3292–3294.

Lin, S. K. C., Du, C., Blaga, A. C., Camarut, M., Webb, C., Stevens, C. V. and Soetaert, W. (2010) 'Novel resin-based vacuum distillation-crystallisation method for recovery of succinic acid crystals from fermentation broths', *Green Chem*, 12, 666–671.

Lin, C. S. K., Luque, R., Clark, J. H., Webb, C. and Du, C. (2011) 'A seawater-based biorefining strategy for fermentative production and chemical transformations of succinic acid', *En Environ Sci*, 4, 1471–1479.

Lin, C. S. K., Luque, R., Clark, J. H., Webb, C. and Du, C. (2012) 'Wheat-based biorefining strategy for fermentative production and chemical transformations of succinic acid', *Biofuels, Bioproducts and Biorefining*, 6, 88–104.

Lin, C. S. K., Pfaltzgraff, L. A., Herrero-Davila, L., Mubofu, E. B., Solhy, A., Clark, P. J., Koutinas, A., Kopsahelis, N., Stamatelatou, K., Dickson, F., Thankappan, S.,

Zahouily, M., Brocklesby, R. and Luque, R. (2013) 'Food waste as a valuable resource for the production of chemicals, materials and fuels: current situation and global perspective', *En Environ Sci*, 6, 426–464.

Liu, Y.-P., Zheng, P., Sun, Z.-H., Ni, Y., Dong, J.-J. and Zhu, L.-L. (2008) 'Economical succinic acid production from cane molasses by *Actinobacillus succinogenes*', *Biores Technol*, 99, 1736–1742.

Luque, R., Lin, C. S. K., Du, C., Macquarrie, D. J., Koutinas, A., Wang, R., Webb, C. and Clark, J. H. (2009) 'Chemical transformations of succinic acid recovered from fermentation broths by a novel direct vacuum distillation-crystallisation method', *Green Chem*, 11, 193–200.

MacGregor, A. W. and Rattan, S. B. (1993) *Barley: Chemistry and Technology*. American Association of Cereal Chemists, St. Paul, MN.

MacMasters, M. M., Hinton, J. J. C. and Bradbury, D. (1971) 'Microscopic structure and composition of the wheat kernel', in: Pomeranz, Y., *Wheat: Chemistry and Technology*, 2nd edn, AACC, St. Paul, MN, 51–113.

Madison, L. L. and Huisman, G. W. (1999) 'Metabolic engineering of poly(3-hydroxyalkanoates): from DNA to plastic', *Microbiol Mol Biol Rev*, 63, 21–53.

Melikoglu, M. (2008) Production of Sustainable Alternatives to Petrochemicals and Fuels Using Waste Bread as a Raw Material. PhD thesis, The University of Manchester.

Misailidis, N., Campbell, G. M., Du, C., Sadhukhan, J., Mustafa, M., Mateos-Salvador, F. and Weightman, R. M. (2009) 'Evaluating the feasibility of commercial arabinoxylan production in the context of a wheat biorefinery principally producing ethanol: Part 2. Process simulation and economic analysis', *Chem Eng Res Des*, 87, 1239–1250.

Nikel, P. I., de Almeida, A., Melillo, E. C., Galvagno, M. A. and Pettinari, M. J. (2006) 'New recombinant *Escherichia coli* strain tailored for the production of poly(3-hydroxybutyrate) from agroindustrial by-products', *Appl Environ Microbiol*, 72, 3949–3954.

Peckham, B. W. (2001) 'The first hundred years of corn refining in the United States', *Starch/Stärke*, 53, 257–260.

Philip, S., Keshavarz, T. and Roy, I. (2007) 'Polyhydroxyalkanoates: biodegradable polymers with a range of applications', *J Chem Technol Biotechnol*, 82, 233–247.

Pomeranz, Y. (1987) *Modern Cereal Science and Technology*. VCH Publishers, New York.

Pomeranz, Y. (1988) 'Chemical composition of kernel structures', in: Pomeranz, Y., *Wheat: Chemistry and Technology, Vol. I*. AACC, St. Paul, MN, 97–158.

Quillaguaman, J., Hashim, S., Bento, F., Mattiasson, B. and Hatti-Kaul, R. (2005) 'Poly($\beta$-hydroxybutyrate) production by a moderate halophile, *Halomonas boliviensis* LC1 using starch hydrolysate as substrate', *J Appl Microbiol*, 99, 151–157.

Reddy, N. and Yang, Y. (2005) 'Biofibres from agricultural byproducts for industrial applications', *Trends Biotechnol*, 23, 22–27.

Rusendi, D. and Sheppard, J. D. (1995) 'Hydrolysis of potato processing waste for the production of poly-$\beta$-hydroxybutyrate', *Biores Technol*, 54, 191–196.

Saha, B. C. (2003) 'Hemicellulose bioconversion', *J Ind Microbiol Biotechnol*, 30, 279–291.

Saito, K., Hasa, Y. and Abe, H. J. (2012) 'Production of lactic acid from xylose and wheat straw by *Rhizopus oryzae*', *Biosci Bioeng*, 114, 166–169.

Satakarni, M., Koutinas, A. A., Webb, C. and Curtis, R. (2009) 'Enrichment of fermentation media and optimization of expression conditions for the production of EAK16 peptide as fusions with SUMO', *Biotechnol Bioeng*, 102, 725–735.
Solaiman, D. K. Y., Ashby, R. D., Foglia, T. A. and Marmer, W. N. (2006) 'Conversion of agricultural feedstock and coproducts into poly(hydroxyalkanoates)', *Appl Microbiol Biotechnol*, 71, 783–789.
Song, H. and Lee, S. Y. (2006) 'Production of succinic acid by bacterial fermentation', *Enz Microb Technol*, 39, 352–361.
Steinbüchel, A. and Schlegel, H. G. (1991) 'Physiology and molecular genetics of poly(β-hydroxy-alkanoic acid) synthesis in *Alcaligenes eutrophus*', *Mol Microbiol*, 5, 535–542.
Stuart, P. R. and El-Hawagi, M. (2012) *Integrated Biorefineries: Design, Analysis, and Optimization*. CRC Press, Boca Raton, FL.
Talebnia, F., Karakashev, D. and Angelidaki, I. (2010) 'Production of bioethanol from wheat straw: an overview on pretreatment, hydrolysis and fermentation', *Biores Technol*, 101, 4744–4753.
Venus, J. and Richter, K. (2007) Development of a pilot plant facility for the conversion of renewables in biotechnological processes. *Engineering in Life Sciences*, 7, 395–402.
Vink, E. T. H., Davies, S. and Kolstad, J. J. (2010) 'The eco-profile for current Ingeo® polylactide production', *Ind Biotechnol*, 6, 212–224.
Wang, R., Godoy, L. C., Shaarani, S. M., Melikoglu, M., Koutinas, A. and Webb, C. (2009) 'Improving wheat flour hydrolysis by an enzyme mixture from solid state fungal fermentation', *Enz Microb Technol*, 44, 223–228.
Webb, C. (1994) 'Grain processing: a challenge for chemical engineers', *Chemical Technology Europe*, I (5), 21–23.
Webb, C., Koutinas, A. and Wang, R. (2004) 'Developing a sustainable bioprocessing strategy based on a generic feedstock', *Adv Biochem Eng Biotechnol*, 86, 195–268.
Webster, F. H. (1989) *Oats: Chemistry and Technology*. American Association of Cereal Chemists, St. Paul, MN.
Werpy, T. and Petersen, G. (2004) 'Top value added chemicals from biomass. Volume I – Results of screening for potential candidates from sugars and synthesis gas', Pacific Northwest National Laboratory (PNNL) and National Renewable Energy Laboratory (NREL). Available at: http://www1.eere.energy.gov/biomass/pdfs/35523.pdf
White, P. J. and Johnson, L. A. (2003) *Corn: Chemistry and Technology*, 2nd edn. American Association of Cereal Chemists, St. Paul, MN.
Wolf, O., Crank, M., Patel, M., Marscheider-Weidemann, F., Schleich, J., Husing, B. and Angerer, G. (2005) 'Techno-economic feasibility of large scale production of bio-based polymers in Europe', Technical Report EUR 22103 EN.
Wyman, C. E. and Goodman, B. J. (1993) 'Biotechnology for production of fuels, chemicals, and materials from biomass', *Appl Biochem Biotechnol*, 39–40, 41–59.
Xu, Y., Wang, R.-H., Koutinas, A. A. and Webb, C. (2010) 'Microbial biodegradable plastic production from a wheat-based biorefining strategy. *Proc Biochem*, 45, 153–163.
Yu, H., Shi, Y., Yin, J., Shen, Z. and Yang, S. (2003) 'Genetic strategy for solving chemical engineering problems in biochemical engineering', *J Chem Technol Biotechnol*, 78, 283–286.

Yu, J., Li, Z., Ye, Q., Yang, Y. and Chen, S. (2010) 'Development of succinic acid production from corncob hydrolysate by *Actinobacillus succinogenes*', *J Ind Microbiol Biotechnol*, 37, 1033–1040.

Zhang, A. Y. Z., Sun, Z., Leung, C. C. J., Wei, H., Lau, K. Y., Li, M. and Lin, C. S. K. (2013) 'Valorisation of bakery waste for succinic acid production', *Green Chem*, 15, 690–695.

Zhang, H. and Mittal, G. (2010) 'Biodegradable protein-based films from plant resources: a review', *Environ Progr Sust En*, 29, 203–220.

Zhang, S., Norrlow, O., Wawrzynczyk, J. and Dey, E. S. (2004) 'Poly(3-hydroxybutyrate) biosynthesis in the biofilm of *Alcaligenes eutrophus*, using glucose enzymatically released from pulp fiber sludge', *Appl Environ Microbiol*, 70, 6776–6782.

Zheng, P., Dong, J.-J., Sun, Z.-H., Ni, Y. and Fang, L. (2009) 'Fermentative production of succinic acid from straw hydrolysate by *Actinobacillus succinogenes*', *Biores Technol*, 100, 2425–2429.

# 11
# Developments in grass-/forage-based biorefineries

J. McENIRY and P. O'KIELY,
Teagasc (Irish Agriculture and
Food Development Authority), Ireland

DOI: 10.1533/9780857097385.1.335

**Abstract**: This chapter provides a brief overview of the potential of grassland biomass as a feedstock for a green biorefinery. It begins by discussing the current role and importance of grasslands in Europe and the impact of grassland management and field-to-industrial facility process steps on herbage chemical composition. The chapter subsequently describes some of the green biorefinery activities ongoing in Europe and outlines the potential products from the press-juice and press-cake fractions of grassland biomass.

**Key words**: green biorefinery, grassland biomass, fractionation, press-juice, press-cake.

## 11.1 Introduction

Grasslands play a major role in global agriculture, accounting for approximately 70% of the world agricultural land area and 26% of total land area (FAO, 2012). Grasslands are predominantly used for animal production, particularly as a principal source of food for ruminants. More recently, grassland biomass has been considered for the production of renewable energy, chemicals and materials. A 'green biorefinery' represents the sustainable processing of green biomass into a spectrum of marketable products and energy (Cherubini *et al.*, 2009). The chemical composition of a biomass feedstock presented to a biorefinery will determine the potential range of products produced. Furthermore, the viability of such an industrial facility will depend on the range of suitable applications identified for the separated fractions (Kromus *et al.*, 2004). In general, the recovery of plant protein is the main focus of green biorefinery concepts for processing of fresh biomass (Kamm *et al.*, 2010), while lactic acid and amino acids are the main products of green biorefinery concepts for processing ensiled biomass (Mandl, 2010). Cherubini *et al.* (2009) suggested that the main driver for the development of biorefinery concepts is the efficient and cost effective production of transportation biofuels, whereas additional economic and environmental benefits may be created from the coproduced biomaterials and biochemicals. This chapter gives an update on the green biorefinery activities ongoing in Europe and outlines some of the potential products from the press-juice and press-cake fractions of grassland biomass.

## 11.2 Overview of grass-/forage-based biorefineries

### 11.2.1 Green biorefinery

A 'green biorefinery' represents the sustainable processing of green biomass into a spectrum of marketable products and energy (Cherubini et al., 2009). In the context of a green biorefinery, green biomass can include any naturally occurring wet biomass such as agricultural crops (e.g., grass, lucerne, clover and immature cereal) and agricultural residues (e.g., sugar beet leaves). These plants represent a natural chemical factory and can be rich in basic products such as carbohydrates, proteins, lignin and lipids, as well as various other substances such as vitamins, dyes and minerals (Kamm and Kamm, 2004). Green biorefineries are often described as multiple-product systems, with the potential range of products depending on the composition of the feedstock presented to the biorefinery. This multiple-product approach was born out of economic necessity, with single-product approaches often struggling to create sufficient revenue to cover the feedstock and subsequent processing costs (Mandl, 2010). As such, the viability of an industrial facility processing green biomass will depend on the range of suitable applications identified for the separated fractions (Kromus et al., 2004). Furthermore, this multiple-product approach can facilitate the efficient utilisation of the whole plant and any process residues.

### 11.2.2 Grassland composition

Poaceae (i.e., grass family) and Fabaceae (i.e., legume family) represent two of the major plant families used as forages. The Poaceae are a large and ubiquitous family of monocotyledonous flowering plants and include three of the most important crops in the world in wheat (*Triticum* spp.), maize (*Zea mays*) and rice (*Oryza* spp.), as well as many lawn and pasture grasses (e.g. *Lolium* spp.). The family Fabaceae are a large family of dicotyledonous flowering plants and also represent a large number of important agricultural and food plants including soybean (*Glycine max*), clover (*Trifolium* spp.) and lucerne (*Medicago* spp.).

Grasses and legumes can be further classified into two physiological groups, $C_3$ (i.e., cool season or temperate) and $C_4$ (i.e., warm season or tropical) plants, based on their photosynthetic pathway. There are distinct differences between the photosynthetic pathways of $C_3$ and $C_4$ species which affect light, water and nitrogen use efficiencies (Lattanzi, 2010).

In warm regions, $C_4$ grasses can outyield $C_3$ grasses due to their more efficient photosynthetic pathway. However, the lower temperatures and shorter growing seasons in Northern Europe limit the growth of $C_4$ plants (Lewandowski *et al.*, 2003). Furthermore, $C_3$ and $C_4$ plants may be classified as either annual (i.e., life cycle completed in one growth season) or perennial (i.e., life cycle completed in two or more years).

The term 'grassland' in this chapter refers to a plant community in which true grasses are usually the dominant species, with forbs (i.e., herbaceous dicotyledon species, including legumes) present in variable amounts (Hopkins, 2000). Grasslands are a major agricultural resource, supporting both intensive and extensive systems of ruminant livestock production. The main focus of this chapter is on $C_3$ or temperate grassland biomass that is currently or has recently been used as forage for animal production.

Intensively managed temperate grasslands are dominated by a small number of grass species including perennial ryegrass (*Lolium perenne*), Italian ryegrass (*Lolium multiflorum*), timothy (*Phleum pratense*), cocksfoot (*Dactylis glomerata*) and tall fescue (*Festuca arundinacea*) (Hopkins and Holz, 2006; Plantureux *et al.*, 2005). These grasses may be present as indigenous species in permanent pastures (i.e., land used to grow grasses or other herbaceous forage naturally (self-seeded) or through cultivation (sown) and that is not included in the crop rotation of the holding for five years or longer; Commission Regulation E.U. No. 796/2004), be part of swards reseeded many years previously, or may be resown every few years as part of a crop rotation (Buxton and O'Kiely, 2003). The majority of reseeded temperate grassland in Europe is now dominated by perennial ryegrass due to its high digestibility when harvested at the appropriate growth stage, high yield in response to nitrogen fertiliser application, and ease of preservation as silage due to its relatively high water soluble carbohydrate content (Whitehead, 1995). However, in situations where ryegrasses are limited by winter survival, there is a continuing role for timothy, while in drier temperate regions there is a role for more drought-tolerant species such as cocksfoot and tall fescue (Hopkins and Wilkins, 2006). Legumes such as white clover (*Trifolium repens*), lucerne (*Medicago sativa*) and red clover (*Trifolium pratense*) can also represent a prominent component of temperate grasslands (Peeters *et al.*, 2006). These legumes can be grown alone or in combination with compatible grasses and they can reduce the requirement for fertiliser N through fixation of atmospheric N, while also increasing herbage yield and quality (Peyraud *et al.*, 2009).

## 11.2.3 Grassland in Europe – role and importance

Grasslands play a major role in global agriculture, accounting for approximately 70% of the world agricultural land area and 26% of total land area (FAO, 2012). In the EU-27, permanent and temporary grasslands represent more than 30 and 5% of the utilised agricultural area, respectively. However, this varies considerably between countries and regions with, for example, permanent grassland constituting 75% and 1.5% of the utilised agricultural area in Ireland and Finland, respectively (Eurostat, 2008). Furthermore, grasslands vary greatly in their degree and intensity of management, with grassland systems in Europe ranging from extreme Taiga vegetation in the far north to dry Mediterranean grassland in the south (Smit et al., 2008).

Grasslands are predominantly used for animal production, particularly as a principal source of feed for ruminants. However, animal production systems differ across Europe. In Northwest Europe, for example, grasslands can meet up to 100% of the nutrient requirements of the ruminant livestock population, while in other parts of Europe other forage crops and temporary grassland are more important. In many European countries grassland utilisation is decreasing due to a trend towards more controlled animal production systems or to decreasing livestock numbers. Grassland-based production is being increasingly challenged by the use of other forage and concentrate feeds (Wilkins et al., 2003). For example, the relatively low proportion of grassland in some specialised dairy farms can be explained through higher stocking rates and higher proportions of green maize and temporary grassland used as forage crops (Dillon, 2010). For grassland that is no longer needed for intensive dairy or beef production, intensive management is often unlikely to be continued (Stoate et al., 2009; Isselstein et al., 2005).

Grasslands are also becoming increasingly recognised for their contribution to environmental functions, and the protection of grassland area is integrated into the Common Agriculture Policy (CAP) cross-compliance system. This requirement for the protection of permanent grasslands is in recognition of the positive impacts of grassland compared with cropping/arable land on a range of ecosystem services (Dillon, 2010). Grasslands play an important multifunctional role in maintaining floral and faunal diversity (Isselstein et al., 2005), providing water catchment protection and soil erosion control (Cerdan et al., 2010), reducing the impact of global warming through carbon sequestration (Peeters and Hopkins, 2010) and providing landscape and amenity value (Gibon, 2005). For example, semi-natural grasslands are among the most species-rich habitats in Europe. Generally, natural and semi-natural grasslands are found in extensive, low stocking rate production systems in Europe (Dillon, 2010). The maintenance of semi-natural grassland habitats through traditional agricultural practices is vital for the protection of biodiversity. Periodic defoliation of extensive grassland, by mechanical cutting or grazing, is vital for controlling succession of plants (Rook et al., 2004) and is essential for maintenance of grassland habitats.

## 11.2.4 Grassland productivity and potential availability

Data on actual grassland productivity and its spatial distribution in Europe are scarce. Grassland productivity is affected by soil characteristics, botanical composition, climatic conditions (e.g., rainfall and temperature) and specific management practices (Peeters and Kopec, 1996). Smit et al. (2008) presented data on grassland productivity across Europe. The highest productivity, about 10 t DM ha$^{-1}$, was reported to be achieved in temperate Atlantic zones (e.g., Netherlands, Britain and Ireland, North Germany), while regions with the lowest productivity (1.5 t DM ha$^{-1}$) are located close to the Mediterranean. Central European countries (e.g., Germany, Austria and Switzerland) were reported to have intermediate yields of 6 t DM ha$^{-1}$ and higher. Aside from favourable climatic conditions, high fertiliser N inputs are also a major determinant of DM yields. In general, however, information on the biomass yields of different grassland species in Europe is difficult to compile since grass yields are rarely directly measured in farm practice. In an extensive study comparing the productivity of perennial ryegrass and timothy across 32 European sites, yields of perennial ryegrass varied from almost 2 t (Vila Real, Portugal) to 20 t DM ha$^{-1}$ (Kiel, Germany) and were higher than timothy in most cases (Peeters and Kopec, 1996).

Some EU countries are facing difficulties in ensuring regular utilisation of extensive grasslands by ruminants to meet conservation objectives (Wachendorf et al., 2009). In other regions, more controlled ruminant production systems or decreasing livestock numbers are reducing the requirement for intensively managed grassland. However, little or no information is available on the grassland biomass resource available in Europe for alternative applications. In Ireland, for example, grass is a biomass resource that is readily available (O'Keeffe et al., 2011). In a recent study, McEniry et al. (2013a) calculated the annual grassland resource available in Ireland as the difference between current estimated grass supply and the grass requirement of the national cattle herd and sheep flock. They reported that under current grassland management practices in Ireland, there is an estimated annual grassland resource of ca. 1.7 million t DM available in excess of livestock requirements. Furthermore, increasing N fertiliser input combined with increasing the grass utilisation rate by cattle has the potential to significantly increase this resource to 12.2 million t DM per annum. Thus, there is potential for grassland biomass to be a readily available resource in Europe. Consequently, numerous policy, research and commercial groups have considered alternative options for grassland biomass use, with the production of renewable energy, chemicals and materials from grassland biomass receiving considerable interest.

## 11.3 Field to biorefinery – impact of herbage chemical composition

The chemical composition of a biomass feedstock presented to a biorefinery will determine the potential range of products produced. The soil characteristics, botanical composition of the sward, environmental factors (e.g., rainfall, temperature) and specific management factors (e.g., harvest date, nutrient management) will have a significant impact on the yield and chemical composition of a biomass feedstock and are important factors for consideration. Similarly, management or preservation of the feedstock so as to ensure year-round availability, and its subsequent fractionation, represent two major process steps in the utilisation of green biomass in an industrial facility. These two process steps can also have a significant impact on the composition of the separated fractions in a biorefinery.

### 11.3.1 Green plant chemical composition

Temperate forages can be divided chemically into two main fractions: (a) cell walls composed mainly of structural carbohydrates and (b) cell contents which include the most readily and highly digestible components (Moore and Hatfield, 1994).

*Cell walls*

The structural carbohydrates cellulose and hemicellulose are the major components of the plant cell wall and provide structural integrity to the plant (Esau, 1977). Cellulose is the carbohydrate polymer of greatest abundance in plant cell walls and is composed of $\beta$-1,4-linked glucose units. Hemicellulose is composed of multiple polysaccharide polymers and consists of sugars including glucose, xylose, galactose, arabinose and mannose. The main hemicellulose in plants is xyloglucan; however, in most grasses the main hemicellulose component is arabinoxylan (Reiter, 1998). The remainder of the cell wall is composed of lignin, protein and pectin, with lipids and minerals in smaller amounts (Theander and Westerlund, 1993). Lignin is a phenolic polymer and can be formed from the polymerisation of three main monomeric units: *P*-coumaryl, coniferyl and sinapyl alcohols, with lignin in grass being formed from all three monomers (Evert, 2006).

The content of cell wall material is greater in the stem than in leaves of forages, with the difference between these parts being greater in legumes than grasses (Wilson, 1993). With advancing plant maturity, the proportion of cell wall increases in relation to cell contents (Fig. 11.1). This reflects the

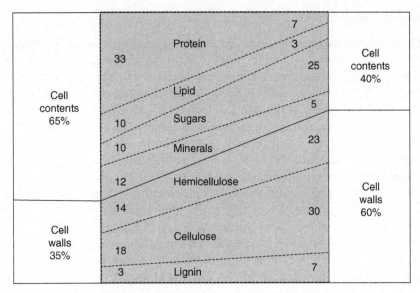

*11.1* Schematic representation of changes in the chemical composition of grass with advancing maturity (adapted from Holmes, 1980).

general decrease in leaf to stem ratio and the increasing cell wall content within the stems (Buxton, 1996). This process is accompanied by an increasing content of indigestible lignin within the cell wall fraction (Ugherughe, 1986), and it is the presence of lignin in the cell wall that limits polysaccharide breakdown (Buxton and Russell, 1988). For legumes, the cell wall concentration of leaves changes little with advancing maturity, whereas stems progressively undergo cambial growth which adds thick-walled xylem tissue during development increasing the diameter of the stem (Wilson, 1993).

*Cell contents*

The cell contents of forages are comprised primarily of sugars, proteins, lipids and minerals (Kromus *et al.*, 2006). In the early stages of plant growth, cell contents may represent up to 65% of forage DM, with protein being a major component (Fig. 11.1). With advancing plant maturity, cell contents, as a proportion of plant DM, decrease as cell wall material increases and this is accompanied by a pronounced decline in protein concentration. However, the sugar fraction of forages is highly labile, and the amounts present in the plant at any growth stage also depends on prevailing environmental conditions, especially light and temperature.

*Non-structural carbohydrates*

The main non-structural carbohydrates in temperate grasses are cold water-soluble carbohydrates and they are composed of the monosaccharides glucose and fructose, the disaccharide sucrose and the polysaccharide fructan. Pentoses are also present in limited quantities as a result of hydrolysis of hemicellulose (McDonald *et al.*, 1991). Water-soluble carbohydrate concentration in grasses and legumes may range from 50 to 300 g kg$^{-1}$ DM (Frame and Laidlaw, 2011). The relative amounts of these sugars can be influenced by herbage species, the leaf-to-stem ratio, the time of day, light intensity and temperature (Smith, 1973). For example, ryegrasses have a higher water-soluble carbohydrate concentration than many other common grass species. The monosaccharides function primarily as key intermediates of metabolic pathways (Moore and Hatfield, 1994). Sucrose, composed of glucose and fructose, plays an important role in carbohydrate transport, is the primary donor for starch and fructan synthesis, and is a storage molecule in some plants (Kandler and Hopf, 1980). Fructans are the predominant carbohydrate reserve in temperate $C_3$ grasses, while sucrose and starch are dominant in tropical $C_4$ grasses (McGrath, 1988; White, 1973).

*Protein*

Plant proteins may be categorised as seed or leaf proteins, with the latter representing the metabolic proteins concerned with the growth and biochemical functioning of the cells. Proteins are usually present in greater concentrations in the leaves of grasses and forage legumes than in stems, with soluble leaf protein occurring mainly as enzymes in the chloroplasts, mitochondria and nucleus of young dividing cells (Buxton and O'Kiely, 2003).

Herbage proteins are often referred to as fraction I or fraction II proteins based primarily on their size (Singer *et al.*, 1952). Fraction I protein is found in the chloroplast and is composed almost entirely of ribulose-1,5-bisphosphate carboxylase/oxygenase (Rubisco), the primary enzyme involved in photosynthesis (Jensen and Bahr, 1977). Fraction I protein is the single most abundant protein in forages and accounts for approximately 50% of total leaf protein. Fraction II protein is composed of a more complex mixture of proteins originating from the chloroplasts (e.g., plastocyanin and ferrodoxin) and cytoplasm (e.g., actin and tubulin; Buxton and O'Kiely, 2003).

The major amino acids in grass are asparagine, glutamine, alanine and leucine, while concentrations of cysteine, methionine, histidine and arginine are relatively low (Van Vuuren, 1993). As Rubisco makes up such a large proportion of total plant protein, it is not surprising that amino acid composition is similar among most herbaceous plants. For example, Mela and Rand (1979) reported no significant differences in amino acid composition between timothy, cocksfoot, meadow fescue and perennial ryegrass.

Crude protein content is used as an indication of the contribution of nitrogenous compounds in plant DM and is a product of total measured N concentration multiplied by 6.25 (i.e., assumes a crude protein fraction composed of organic compounds of 16% N on a molecular weight basis). True protein generally accounts for 75–85% of crude protein in grass (Buxton and O'Kiely, 2003). Typical crude protein concentrations in leguminous forages are between 150 and 250 g kg$^{-1}$ DM, with corresponding values for temperate $C_3$ grasses and tropical $C_4$ grasses of 100–200 and 50–100 g kg$^{-1}$ DM, respectively (Lyttelton, 1973). Total crude protein concentration is inversely related to the growth stage of the plant, but can also be influenced by environmental and management conditions under which the plant grows.

*Lipids*

The majority of plant lipids act as energy stores (e.g., triacylglycerols in seeds) or as plant membrane components (e.g., galactosylglycerides in chloroplast membranes and phosphoglycerides in nonchloroplast membranes). Furthermore, surface lipids such as waxes and cutin provide an impervious barrier on the plant surface to reduce water loss and to provide protection against plant pathogens and toxins (Hatfield *et al.*, 2007). Lipids are found in relatively small amounts in temperate grasses and generally decrease as the plant matures (Frame and Laidlaw, 2011; Fig. 11.1).

*Organic acids*

Of the nonvolatile organic acids generally present in grass and legume species, the most commonly found are malic, citric, succinic and quinic, with lesser amounts of fumaric and shikimic acids (Muck *et al.* 1991; Jones and Barnes, 1967). Organic acids and their salts together with phosphates, form the buffer systems in many plants. As these acids buffer within the pH range 6.0 to 4.0, their role in silage preservation can be significant, where they function as biological buffers resisting the acidification of the ensiled mass (Rooke and Hatfield, 2003).

## 11.3.2 Grassland management

The genome of each grass species imposes differences in adaptability, productivity and chemical composition. Although the majority of reseeded temperate grassland in Western Europe is dominated by perennial ryegrass, other grass species (e.g., cocksfoot, tall fescue and timothy) have different physical and chemical characteristics which may offer benefits for non-agricultural uses or be more suited to specific management (e.g., response to fertiliser) or environmental (e.g., temperature, water deficit) conditions. For example, in general, ryegrass and cocksfoot are reported to produce a higher DM yield response to high rates of N fertiliser input compared with other common grassland species (Reid, 1985). In addition, Davies and Morgan (1982) reported that perennial ryegrass had a higher crude protein concentration than other common grass species. Similarly, King *et al.* (2012a) reported that the high water soluble carbohydrate concentration and the lower buffering capacity observed for perennial and Italian ryegrasses, compared with other common grassland species, would make them more suitable for preservation as silage. Furthermore, legumes generally have a lower concentration of water soluble carbohydrate and a higher buffering capacity than grasses, making them more difficult to preserve as silage (Buxton and O'Kiely, 2003).

The most important factor influencing the chemical composition of a specific herbage is the growth stage at harvest (Buxton, 1996). Advancing plant maturity from the vegetative to the inflorescence growth stage is characterised by an increase in fibre components and a decrease in digestibility, buffering capacity, and crude protein and water soluble carbohydrate concentrations (Fig. 11.1; Buxton and O'Kiely, 2003). This decrease in buffering capacity, crude protein and water-soluble carbohydrate concentration can be attributed to changes in chemical composition resulting from the declining cell content to cell wall ratio (Buxton, 1996). A much faster rate of decline in herbage quality is observed during the primary reproductive spring growth than during subsequent vegetative regrowths (Balasko and Nelson, 2003).

As well as influencing herbage quality, harvest date also has a significant effect on herbage DM yield, with a delayed harvest generally increasing DM yield. Low DM yields may reduce the biomass availability within the catchment area of a green biorefinery (O'Keeffe *et al.*, 2011). Furthermore, the unit cost of producing a grass feedstock for a biorefinery will decrease as the biomass yield increases, but there must be an economic equilibrium between yield and herbage quality. The optimal stage for harvesting will be determined by the particular grassland biomass use. For example, grass harvested at earlier growth stages may be more suitable for biogas production through anaerobic digestion (McEniry and O'Kiely, 2013a), while grass harvested at later growth stages may be more suitable for technical fibre or combustion (McEniry *et al.*, 2012).

Fertilisation of grassland, in particular the application of N, is employed primarily to ensure that economically viable yields are available for harvesting at a time of adequate herbage quality (Keating and O'Kiely, 2000a). While the yield response is greatest for N application, P, K and regular applications of lime are also required to maintain grassland productivity (Balasko and Nelson, 2003). However, N fertiliser can also impact on herbage quality and ensilability. Increasing the rate of N fertiliser application generally increases herbage crude protein concentration (Keady and O'Kiely, 1998) and buffering capacity (O'Kiely et al., 1997), and reduces herbage DM (Whitehead, 1995) and water-soluble carbohydrate concentrations (Keating and O'Kiely, 2000b). This increase in herbage buffering capacity will make the grass more difficult to preserve as silage than indicated by the decrease in water-soluble carbohydrate concentration alone (Keating and O'Kiely, 2000b). Although inorganic N fertiliser can increase herbage yields, there are financial and environmental issues associated with very high rates of application. For example, the use of N fertiliser in agricultural systems is one of the biggest contributors to greenhouse gas emissions through fertiliser manufacture and $N_2O$ emissions from soils (Dillon, 2010). The valuable nutrient content of animal manures and/or digestate (from biogas production), and the incorporation of N-fixing legumes into grassland, all have the potential to reduce the requirement for inorganic N fertiliser resulting in positive financial and environmental gains (McEniry et al., 2011; Peyraud et al., 2009; Gerin et al., 2008).

### 11.3.3 Preservation

Under normal practical conditions, grassland biomass can be harvested and utilised immediately in a biorefinery, or it can be conserved under anaerobic, acidic conditions (i.e., silage) or following drying (i.e., hay). Variable weather conditions can make efficient conservation of hay impractical. Therefore, in order to ensure a predictable quality and a constant year-round supply of feedstock to a biorefinery facility, grass usually needs to be harvested and stored as silage. Preservation is achieved by the combination of an anaerobic environment and the bacterial fermentation of sugar, the lactic acid produced from the latter process lowering the pH and preventing the proliferation of spoilage microorganisms (Muck, 1988). The main objective of ensilage is the efficient preservation of the crop at an optimum growth stage for later use during seasons when the fresh crop is unavailable (McDonald et al., 1991). Ideally crops for preservation as silage should have an adequate content of fermentable substrate in the form of water-soluble carbohydrates, a relatively low buffering capacity and a DM content above $200\,g\,kg^{-1}$ (Buxton and O'Kiely, 2003).

A reduction in water soluble carbohydrate concentration during ensiling is generally the most evident change in herbage chemical composition and indicates its substantial use as a substrate for fermentation (King et al.,

2013a). Water-soluble carbohydrates are available for fermentation by a variety of microorganisms, of which lactic acid bacteria are the most important. Under good ensiling conditions, lactic acid bacteria become the dominant microbial population, producing mainly lactic acid as a fermentation product with a consequent decrease in pH (McEniry et al., 2008). In general, silage which has undergone a desirable fermentation is characterised by a low pH, high lactic acid content and low concentrations of butyric acid and ammonia-N (Haigh and Parker, 1985). Under suboptimal ensiling conditions, a secondary clostridial fermentation may lead to considerable DM and energy losses due to extensive production of $CO_2$ and $H_2$ from the fermentation of lactate and hexose sugars, and also extensive degradation of proteinaceous compounds to ammonia-N (McDonald et al., 1991).

Harvesting of a forage crop is followed by rapid and extensive proteolysis (Muck, 1988). Degradation of plant protein during wilting and ensiling is inevitable and results in changes in the N-constituents of the ensiled herbage. Two stages of protein degradation can be considered. Firstly, peptide bond hydrolysis takes place releasing free amino acids and peptides, and these amino acids can subsequently be degraded to a variety of end products including $NH_3$, organic acids and amines. Ammonia and amines are largely the end products of microbial activity rather than plant enzyme activity (Rooke and Hatfield, 2003). During wilting and in well-fermented silages dominated by lactic acid bacteria, proteolysis is largely a result of plant enzymes (Woolford, 1972). Proteolysis can be minimised with a rapid wilt under dry conditions (Carpintero et al., 1979). In general, the extent of proteolysis during ensiling depends largely on the herbage DM concentration and the rapidity with which acid conditions are established (McDonald et al., 1991).

Furthermore, small increases in the relative proportions of neutral detergent fibre, acid detergent fibre, ash and nitrogen can also occur during ensiling as a result of a loss in organic matter during fermentation and through effluent production (McEniry et al., 2013b). For some silages, a decrease in the hemicellulose fraction is also observed, suggesting the hydrolysis of hemicellulose by plant enzymes at the early stages of ensiling or acid hydrolysis during longer term storage, and this can provide additional substrate for lactic acid fermentation (McDonald et al., 1991).

### 11.3.4 Wet fractionation

The majority of the technological options for a grass-based biorefinery involve the essential primary process of fractionating green plant biomass into a fibre-rich press-cake and a nutrient-rich press-juice. The press-cake fraction represents mainly the cell wall fraction of the herbage, which is rich in cellulose, hemicellulose and lignin. The press-juice fraction represents mainly the cell contents and therefore contains proteins, water-soluble carbohydrates, organic acids, minerals and other substances (Kamm and Kamm, 2004).

Mandl et al. (2005, 2006) stated that the overall goal of fractionation is to transfer as much of the valuable components (e.g., lactic acid, protein) as possible into the press-juice fraction, and reported recovery rates of 85–95% for lactic acid and 55–65% for crude protein from clover-grass and lucerne silages. The process of fractionation involves the mechanical dehydration or pressing of plant biomass to remove water from the plant structural framework, with or without an additional hydrothermal conditioning step to enhance the removal of plant solubles (King et al., 2012b). Prior to pressing, the plant material is generally chopped to reduce particle size, to increase flow rates through the press and reduce blockages, and to aid in the release of plant cell contents. Screw presses, which provide a high degree of maceration of cell walls as a result of axial movement and abrasion of the tissue under high pressure, have been used with some success. The use of less effective piston and roller presses to dehydrate plant material has also been described (Carlsson, 1982).

Wachendorf et al. (2009) and Richter et al. (2009) described the process of hydrothermal conditioning of silages, in which water and silage were mixed (4:1 ratio) and heated to 60°C under continuous stirring for 15 minutes. This treatment aimed to macerate cell walls and produce a 'mash' which could be mechanically pressed to allow the transfer of soluble minerals (e.g., K, Mg, P and Cl) and organic compounds (e.g., carbohydrates, proteins and lipids) into the press-juice fraction. Consequently, there was a reduction in water and mineral content in the press-cake fraction, together with an overall enrichment of the fibrous constituents. McEniry et al. (2012) reported that fractionation resulted in an average reduction in Cl (95%) and K (98%) and greater than 55% of the N and ash in the press-cake fraction relative to the parent material. Mahmoud et al. (2011) and Arlabosse et al. (2011) have also described the use of thermally assisted mechanical dewatering processes to separate alfalfa and spinach into solid and liquid fractions without the addition of water.

## 11.4 Green biorefinery products

### 11.4.1 Press-juice fraction

The composition of the press-juice fraction will vary considerably between fresh and ensiled herbage. Silage has a different composition to fresh herbage with its N fraction dominated by amino acids instead of proteins, and can contain lactic acid and a range of other products (e.g., acetic acid,

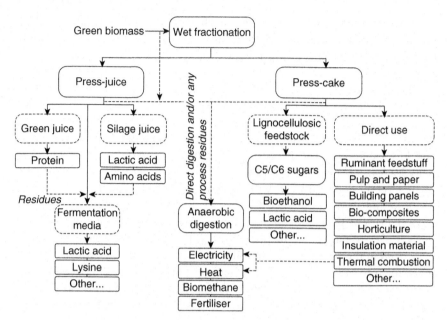

*11.2* Schematic representation of the 'green biorefinery' concept.

propionic acid, butyric acid and ethanol) from the fermentation of sugar during ensiling.

*Green juice*

The recovery of plant protein is the main focus of green biorefinery concepts based on the processing of fresh biomass (Fig. 11.2; Kamm *et al.*, 2010). Green forage plants such as grass and legumes can have a high crude protein content of up to $250 \, \text{g} \, \text{kg}^{-1}$ DM (Lyttelton, 1973). Green crop fractionation has been investigated for the last two centuries as a means to mechanically extract protein from green forage crops in a form that could be used efficiently by non-ruminants. The extracted crude protein can be used for feeding pigs and poultry, thus reducing the dependence on imported protein rich feeds. The remaining press-cake fraction can be fed to ruminants on the premise that green crops contain a much higher concentration of protein than ruminants actually require (Jones, 1977). Thus, the main objective of this partial extraction was often the provision of a more nutritionally balanced crop containing the correct amount of protein for ruminants, with the primary product of economic value being the pressed-crop.

Furthermore, during times of crisis (e.g., Second World War) the high crude protein content of green forage crops stimulated interest in the use of leaf protein concentrate for human nutrition (Carlsson, 1994; McDougall, 1980; Pirie, 1971). The removal of surplus protein prior to feeding green forage crops to ruminants was proposed as a mechanism to alleviate food-supply shortages in Europe. In this case the primary product of economic value was the green juice fraction. More recently, novel protein foods from plants have been proposed as a protein-rich replacement for meat in human diets so as to reduce the strain that intensive animal husbandry practices pose to the environment (Dijkstra *et al.*, 2003; Linnemann and Dijkstra, 2002).

The first industrial process for leaf protein extraction, the Rothamsted process, was developed by Pirie. The procedure was based on heat coagulation (70°C) of protein in green juice (Pirie, 1966). Procedures based on two-step heating of green juice subsequently enabled extraction of protein fractions of different composition (Edwards *et al.*, 1975; Defremery *et al.*, 1973). Fraction II or green fraction leaf protein concentrate consists of a mixture of proteins originating from the chloroplasts and cytoplasm and can be separated from the green press-juice fraction by thermal coagulation (60–70°C) and centrifugation. After removal of the green fraction leaf protein concentrate, the Fraction I or water-soluble white fraction leaf protein concentrate can be separated by a number of methods including thermal precipitation (~ 85°C), acid precipitation, solvent extraction or membrane filtration (Dijkstra *et al.*, 2003; Koschuh *et al.*, 2004; Edwards *et al.*, 1975; Pirie, 1971).

Kamm *et al.* (2010) proposed two protein product streams in the Havelland (Germany) green biorefinery demonstration plant (Table 11.1; Fig. 11.3). Firstly, green fraction leaf protein concentrate is produced by thermal coagulation for use as an animal feed. A second protein product stream focuses on the extraction of white fraction leaf protein concentrate, via thermal coagulation and ultrafiltration, with potential for innovative food processing and cosmetic applications. Similarly, the goal of the green biorefinery research project GRASSA in the Netherlands (www.grassnederland.nl) is the recovery of a protein-rich feed product from fresh grass, in addition to the utilisation of the press-cake fraction in the pulp and paper industries.

Deproteinated brown juice is an inevitable co-product of leaf protein concentrate extraction and will contain protein, water-soluble carbohydrates and minerals. This brown juice has been proposed as a fertiliser, a ruminant feedstuff, a feedstock for biogas production and as a fermentation medium (Worgan and Wilkins, 1977). Anderson and Kiel (2000) and Thomsen *et al.* (2004) described the use of deproteinated brown juice from the green crop drying industry as a fermentation medium for the production of amino acids and organic acids. For example, the production of lysine would be attractive as it is an essential amino acid used as a feed additive for pig and poultry

Table 11.1 Overview of some green biorefinery activities in Europe

| Location | Company/partners | Feedstock | Biorefinery products | Reference(s) |
|---|---|---|---|---|
| Schaffhausen, Switzerland | 2B AG | Fresh grass | • Blow-in insulation (2B Gratec®)<br>• Biogas | Grass (2004) |
| Obre, Switzerland | Biomass Process Solution | Grass silage | • Insulation boards (Gramitherm®)<br>• Biogas | www.bpsag.ch<br>www.gramitech.ch |
| Utzenaich, Austria | Okoenergie Utzenaich GmbH | Grass silage | • Lactic acid<br>• Amino acids<br>• Biogas | Mandl (2010) |
| Groningen, The Netherlands | Avebe | Fresh grass | • Protein-rich animal feedstuff<br>• Technical fibre | Hulst, 2002<br>www.grassanederland.nl |
| Esbjerg, Denmark | Agroferm | Fresh lucerne | • Lysine additive for pigs and poultry<br>• Lactic acid | Thomsen et al. (2004)<br>www.vitalys.dk |
| Potsdam, Germany | Leibniz Institute for Agricultural Engineering Postdam-Bornim (ATB) | Grass, cereals (e.g. rye wholemeal) | | Venus and Richter (2007) |
| Brensbach, Germany | Biowert | Grass silage | • Protein-rich feedstuff for pigs and poultry<br>• Blow-in insulation<br>• Bio-composite materials<br>• Biogas | www.biowert.de |
| Havelland, Germany | Research Institute Biopos e.V., Agro-Farm GmbH, FMS-Futtermittel GmbH, biorefinery.de GmbH, LINDE KCA, GmbH and Markischer Hof GmbH. | Fresh lucerne and grass | • Green pellets for animal feeding<br>• Protein-rich animal feedstuff<br>• White protein concentrates for R&D<br>• Biogas | Kamm et al. (2010) |

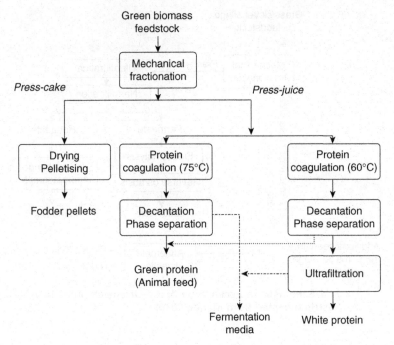

11.3 Schematic representation of the Havelland green biorefinery, Germany (adapted from Kamm et al., 2010).

feed mixtures. This technology is currently being implemented by the Danish company VitaLys (www.vitalys.dk; Agroferm).

*Silage juice*

Preservation of grassland biomass as silage ensures a more predictable quality and a constant supply of feedstock to a biorefinery facility. The microbial fermentation of water-soluble carbohydrates during ensiling results in the production of lactic acid among a range of other fermentation products. In addition, much plant protein is hydrolysed to peptides and free amino acids during ensiling. Lactic acid and amino acids are currently seen as the key compounds of green biorefinery concepts processing grass silage (Fig. 11.4; Mandl, 2010). Lactic acid is the major fermentation product produced during ensilage, with values ranging from 20 to 40 g kg$^{-1}$ in well-preserved, extensively fermented grass silage. Lactic acid is widely used in food, pharmaceutical and cosmetic applications, and as a building block for biodegradable plastic (Datta and Henry, 2006). Free amino acid values can range from 2 to 24 g kg$^{-1}$ in well-preserved grass silage. Amino acids are considered as valuable building blocks and can potentially be used for a

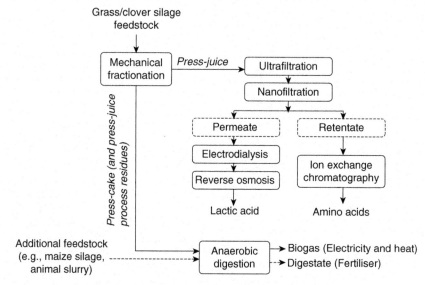

*11.4* Schematic representation of the Utzenaich green biorefinery, Austria (adapted from Mandl, 2010).

wide variety of applications in industry including the synthesis of drugs, cosmetics and food additives (Scott *et al.*, 2007).

The 'Austrian' green biorefinery concept is based on a decentralised system with small-scale production of grass silage by local farmers and the subsequent separation of lactic acid and various amino acids from the grass silage press-juice fraction (Kromus *et al.*, 2004). Results from Utzenaich highlight the importance of well-preserved grass silage as a starting material with lactic acid as the main fermentation product (Ecker and Harasek, 2010). Downstream processing of the press-juice fraction involves a series of separation technologies to purify high-grade amino acids and food-grade lactic acid including ultrafiltration, nanofiltration, electrodialysis, reverse osmosis and ion exchange chromatography (Fig. 11.4; Ecker *et al.*, 2012; Mandl, 2010; Novalin and Zweckmair, 2009; Thang and Novalin, 2008; Koschuh *et al.*, 2005; Thang *et al.*, 2005; Thang *et al.*, 2004).

### 11.4.2 Press-cake fraction

The press-cake fraction is rich in cellulose, hemicellulose and lignin and can potentially be used as a raw material for a wide range of applications (Table 11.2). Furthermore, the press-cake fraction is a potential feedstock for thermal combustion (McEniry *et al.*, 2012; Richter *et al.*, 2011; Wachendorf *et al.*, 2009), biogas (Murphy *et al.*, 2011) and bioethanol (Sieker *et al.*, 2011;

*Table 11.2* Potential applications for the direct use of separated press-cake fraction

| Application | Overview | Reference(s) |
|---|---|---|
| Ruminant feedstuff | • Can be a valuable forage source for ruminants<br>• Nutritive value will depend on the nutritive quality of the parent material and the efficiency of the wet fractionation process | McEniry and O'Kiely (2013b)<br>Nishino et al. (1997)<br>Bryant et al. (1983) |
| Pulp and paper | • An increase in paper demand together with a shortage of wood fibre has stimulated the search for alternative non-wood fibre sources<br>• Wet fractionation may alleviate the problems associated with high ash and silica content in grassland biomass | Pahkala et al. (1997)<br>Ilvessalo-Pfäffli (1995) |
| Building panels | • Grass fibre could act as a partial substitute for wood fragments in particleboard manufacture | Nemli et al. (2009)<br>Zheng et al. (2009) |
| Bio-composites | • Nano-fibrillated cellulose is a potential replacement for conventional fillers (e.g., glass, carbon fibre) in the manufacture of high strength polymer composites | Sharma et al. (2012)<br>Eichhorn et al. (2010) |
| Horticulture | • Interest in non-peat growth media for horticulture is increasing due to environmental issues related to the destruction of peatland habitats<br>• Growing pots and mulch boards | Mandl et al. (2006) |
| Insulation material | • Biomass Process Solution operate a pilot plant in Obre, Switzerland that successfully manufactures and markets thermal and acoustic insulation boards from grass fibre (Gramitherm®)<br>• Blow-in insulation | Grass (2004)<br>www.gramitech.ch<br>www.biowert.de |
| Cementitious reinforcement | • Internal and external restraints can inhibit free drying shrinkage of cementitious products generating tensile stresses in the material<br>• Replacement for polypropylene fibres in mitigating the risk of cracking due to early-age restrained shrinkage | King et al. (2013b) |

Neureiter et al., 2004; Koegel et al., 1999) production. In addition, the separated press-cake fraction is a potential feedstock for a lignocellulosic biorefinery and if this lignocellulosic matrix can be successfully depolymerised to obtain smaller molecules (e.g., sugars), a wide range of potential products (e.g., platform chemicals, biofuels) could be realised.

### 11.4.3 Energy products

*Thermal combustion*

Bioenergy generation by combustion has been proposed as an alterative use for permanent and semi-natural grassland biomass no longer required for forage production (Tonn et al., 2010; Prochnow et al., 2009b). However, high concentrations of ash, and more specifically N, S, Cl and K, may limit the suitability of this biomass for combustion (Obernberger et al., 2006). A solution to this limitation may be that wet fractionation results in a reduction in the concentration of these minerals in the press-cake fraction compared to the parent material, thereby improving the suitability of the press-cake fraction for combustion (Richter et al., 2011; Wachendorf et al., 2009). McEniry et al. (2012) recently reported that mineral concentrations in the press-cake fraction from a range of herbages were similar to other biomass fuels such as miscanthus and willow, but with the average gross calorific values being slightly lower (18.3, 19.0 and 19.9 MJ kg$^{-1}$ for the press-cake fraction, miscanthus and short-rotation coppice willow, respectively).

*Biogas*

Methane-rich biogas can be produced from a wide range of feedstocks (e.g., agricultural crops, animal manure and organic wastes from food industries) through anaerobic digestion, and can be used as a replacement for fossil fuels in both heat and power generation, and as a vehicle fuel (Weiland, 2010). The production of biogas in central Europe is closely linked to the agricultural sector and farm-based biogas plants are widespread (Murphy et al., 2011). Most plants operate on manure-based substrates, fortified with a range of dedicated energy-rich crops including maize, cereals, sugar beet and grass. Grass and grass silage can be an excellent feedstock for biogas production with a wide range of values (198–467 L $CH_4$ kg$^{-1}$ volatile solids) for specific $CH_4$ yield being reported (Nizami et al., 2012; Murphy et al., 2011; Prochnow et al., 2009a). Fresh grassland biomass, silage and/or any of the process residues from the refining process could potentially be used for biogas production. For example, the press-juice fraction represents an excellent feedstock for biogas production, with Richter et al. (2009) reporting specific $CH_4$ yields of 397–426 L $CH_4$ kg$^{-1}$ volatile solids for the press-juice fraction from semi-natural grasslands. The specific $CH_4$ yield of the press-cake fraction would be assumed to be lower than the pre-fractionated parent material due to the curtailing impact of increasing fibre concentration on specific $CH_4$ yield (McEniry and O'Kiely, 2013a).

*Bioethanol*

The separated press-cake fraction may also serve as a potential lignocellulosic feedstock for the production of second generation biofuels. Neureiter *et al.* (2004) and Koegel *et al.* (1999) investigated the hydrolysis of the press-cake fraction from grass silage and alfalfa, respectively, to produce a feedstock for bioethanol. Sieker *et al.* (2011) recently reported the simultaneous pretreatment, saccharification and fermentation of the press-cake fraction from grass silage for ethanol production, but ethanol yields were low.

The key to exploiting the chemical value of this lignocellulosic feedstock is to depolymerise the lignocellulosic matrix in order to obtain smaller molecules that can be utilised, or further converted into platform chemicals and biofuels (Hayes, 2009). The conversion of lignocellulosic biomass to ethanol generally requires three process steps including pretreatment of the biomass, acid or enzymatic hydrolysis and fermentation/distillation (Naik *et al.*, 2010). Most pre-treatments aim to break apart the lignocellulosic matrix and to hydrolyse hemicellulose, and can range from physical pretreatments such as particle size reduction, steam explosion and liquid hot water, to chemical pretreatments such as acid or alkaline catalysed treatments. Cellulose is subsequently hydrolysed through attack by the electrophilic hydrogen atoms of the $H_2O$ molecule on the glycosidic oxygen. This slow reaction can be speeded up using elevated temperature and pressures and is catalysed by acids and highly selective enzymes (Hayes, 2009). After pretreatment and hydrolysis, the glucose monomers can be converted into ethanol via microbial fermentation.

## 11.5 Acknowledgements

The authors acknowledge funding provided under the National Development Plan, through the Research Stimulus Fund (#RSF 07 557), administered by the Department of Agriculture, Food & Marine, Ireland.

## 11.6 References

Andersen, M. and Kiel, P. (2000) 'Integrated utilisation of green biomass in the green biorefinery', *Ind Crop Prod*, 11, 129–137.

Arlabosse, P., Blanc, M., Kerfai, S. and Fernandez, A. (2011) 'Production of green juice with an intensive thermo-mechanical fractionation process. Part I: Effects of processing conditions on the dewatering kinetics', *Chem Eng J*, 168, 586–592.

Balasko, J. A. and Nelson, C. J. (2003) 'Grasses for northern areas', in Barnes, R. F., Nelson, C. J., Collins, M. and Moore, K. J., *Forages – an Introduction to Grassland Agriculture*, Oxford: Blackwell Publishing, 125–147.

Bryant, A. M., Carruthers, V. R. and Trigg, T. E. (1983) 'Nutritive value of pressed herbage residues for lactating dairy cows', *New Zealand J Agr Res*, 26, 79–84.

Buxton, D. R. (1996) 'Quality-related characteristics of forages as influenced by plant environment and agronomic factors', *Anim Feed Sci Tech*, 59, 37–49.

Buxton, D. R. and O'Kiely, P. (2003) 'Preharvest plant factors affecting ensiling', in Buxton, D. R. Muck, R. E. and Harrison, J. H., *Silage Science and Technology*, Madison, WI: American Society of Agronomy, Crop Science Society of America, Soil Science Society of America, 199–250.

Buxton, D. R. and Russell, J. R. (1988) 'Lignin constituents and cell-wall digestibility of grass and legume stems', *Crop Sci*, 28, 553–558.

Carlsson, R. (1982) 'Trends for future applications of wet-fractionation of green crops', in Griffiths, T. W. and Maquire, M. F., *Forage Protein Conservation and Utilization, Proceedings of a seminar in the EEC programme of coordination of research on plant protein*, Devon: Grassland Research Institute, 241–247.

Carlsson, R. (1994) 'Sustainable primary production – green crop fractionation: effects of species, growth conditions, and physiological development', in Pessarakli, M., *Handbook of Plant and Crop Physiology*, New York: Marcel Dekker, 941–963.

Carpintero, C. M., Henderson, A. R. and McDonald, P. (1979) 'The effect of some pre-treatments on proteolysis during the ensiling of herbage', *Grass Forage Sci*, 34, 311–315.

Cerdan, O., Govers, G., Le Bissonnais, Y., Van Oost, K., Poesen, J., Saby, N., Gobin, A., Vacca, A., Quinton, J. and Auerswald, K. (2010) 'Rates and spatial variations of soil erosion in Europe: a study based on erosion plot data', *Geomorphology*, 122, 167–177.

Cherubini, F., Jungmeier, G., Wellisch, M., Willke, T., Skiadas, I., Van Ree, R., de Jong, E. (2009) 'Toward a common classification approach for biorefinery systems', *Biofuel Bioprod Bior*, 3, 534–546.

Commission Regulation (EC) No. 796/2004 of 21 April 2004 relating to laying down detailed rules for the implementation of cross-compliance, modulation and the integrated administration and control system provided for in Council Regulation (EC) No. 1782/2003 establishing common rules for direct support schemes under the common agricultural policy and establishing certain support schemes for farmers. Corrigendum *OJ* L141, 30.4.2004, p. 18.

Datta, R. and Henry, M. (2006) 'Lactic acid: recent advances in products, processes and technologies – a review', *J Chem Technol Biotechnol*, 81, 1119–1129.

Davies, D. A. and Morgan, T. E. H. (1982) 'Herbage characteristics of perennial ryegrass, cocksfoot, tall fescue and timothy pastures and their relationship with animal performance under upland conditions', *J Agr Sci*, 99, 153–161.

Defremery, D., Miller, R. E., Edwards, R. H., Knuckles, B. E., Bickoff, E. M. and Kohler, G. O. (1973) 'Centrifugal separation of white and green protein fractions from alfalfa juice following controlled heating', *J Agr Food Chem*, 21, 886–889.

Dijkstra, D. S., Linnemann, A. R. and van Boekel, T. A. J. S. (2003) 'Towards sustainable production of protein-rich foods: appraisal of eight crops for Western

Europe. Part II: Analysis of the technological aspects of the production chain', *Crit Rev Food Sci*, 43, 481–506.
Dillon, P. (2010) Managing European grasslands to increase the sustainability and competitiveness of livestock production systems', MULTISWARD. Available from: http://www.multisward.eu/multisward_eng/Output-deliverables (accessed 26 March 2013).
Ecker, J. and Harasek, M. (2010) 'Membrane based production of lactic acid and amino acids – green biorefinery Austria', in *Proceedings of the 19th International Congress of Chemical and Process Engineering and the 7th European Congress of Chemical Engineering*, Prague, Czech Republic.
Ecker, J., Schaffenberger, M., Koschuh, W., Mandl, M., Böchzelt, H. G., Schnitzer, H., Harasek, M. and Steinmüller, H. (2012) 'Green biorefinery Upper Austria – pilot plant operation', *Sep Purif Techno*, 96, 237–247.
Edwards, R. H., Miller, R. E., De Fremery, D., Knuckles, B. E., Bickoff, E. M. and Kohler, G. O. (1975) 'Pilot plant production of an edible white fraction leaf protein concentrate from alfalfa', *J Agr Food Chem*, 23, 620–626.
Eichhorn, S. J., Dufresne, A., Aranguren, M., Marcovich, N. E., Capadona, J. R., Rowan, S. J., Weder, C., Thielemans, W., Roman, M. and Renneckar, S. (2010) 'Review: current international research into cellulose nanofibres and nanocomposites', *J Materials Sci*, 45, 1–33.
Esau, K. (1977) *Anatomy of Seed Plants*, New York: Wiley.
EUROSTAT (2008) htpp://www.eurostat.ec.europa.
Evert, R. F. (2006) *Esau's Plant Anatomy: Meristems, Cells, and Tissues of the Plant Body: Their Structure, Function, and Development*, New York: Wiley.
FAO (2012) Available from: http://faostat3.fao.org/home/index.html (accessed 26 March 2013).
Frame, J. and Laidlaw, A. S. (2011) *Improved Grassland Management*, Marlborough: The Crowood Press.
Gerin, P. A., Vliegen, F. and Jossart, J. M. (2008) 'Energy and $CO_2$ balance of maize and grass as energy crops for anaerobic digestion', *Bioresource Technol*, 99, 2620–2627.
Gibon, A. (2005) 'Managing grassland for production, the environment and the landscape: challenges at the farm and the landscape level', *Livest Prod Sci*, 96, 11–31.
Grass, S. (2004) Utilisation of grass for production of fibres, protein and energy, OECD, Available from: http://www.gramitech.ch/download/OECD_Publication_Utilisation_of_Grass.pdf (accessed 27 March 2013).
Haigh, P. M. and Parker, J. W. G. (1985) 'Effect of silage additives and wilting on silage fermentation, digestibility and intake, and on liveweight change of young cattle, *Grass Forage Sci*, 40, 429–436.
Hatfield, R. D., Jung, H. G., Broderick, G. and Jenkins, T. C. (2007) 'Nutritional chemistry of forages', in Barnes, R. F., Nelson, J. F., Moore, K. J. and Collins, M., *Forages – the Science of Grassland Agriculture*, Oxford: Blackwell Publishing, 467–486.
Hayes, D. J. (2009) 'An examination of biorefining processes, catalysts and challenges', *Catal Today*, 145, 138–151.
Holmes, W. (1980) *Grass: Its Production and Utilization*, Oxford: Blackwell Scientific Publications.

Hopkins, A. (2000) *Grass: Its Production and Utilization*, Oxford: Blackwell Science Ltd.
Hopkins, A. and Holz, B. (2006) 'Grassland for agriculture and nature conservation: production, quality and multi-functionality', *Agron Res*, 4, 3–20.
Hopkins, A. and Wilkins, R. J. (2006) 'Temperate grassland: key developments in the last century and future perspectives', *J Agr Sci*, 144, 503–523.
Hulst, A. C. (2002) 'Bio-refinery of grass and other raw materials from vegetable sources and product applications', in *Proceedings of the 5th European Symposium on Industrial Crops and Products*, IENICA: Amsterdam, 37.
Ilvessalo-Pfäffli, M. S. (1995) *Fiber Atlas: Identification of Papermaking Fibers*, Berlin: Springer.
Isselstein, J., Jeangros, B. and Pavlu, V. (2005) 'Agronomic aspects of biodiversity targeted management of temperate grasslands in Europe – a review', *Agron Res*, 3, 139–151.
Jensen, R. G. and Bahr, J. T. (1977) 'Ribulose 1,5-bisphosphate carboxylase-oxygenase', *Ann Rev Plant Physio*, 28, 379–400.
Jones, A. S. (1977) 'The principles of green crop fractionation', in Wilkins, R. J., *Green Crop Fractionation – Occasional Symposium No. 9 (British Grassland Society)*, Luton: Inprint Ltd, 1–7.
Jones, E. C. and Barnes, R. J. (1967) 'Non-volatile organic acids of grasses', *J Sci Food Agri*, 18, 321–324.
Kamm, B. and Kamm, M. (2004) 'Principles of biorefineries', *Appl Microbiol Biot*, 64, 137–145.
Kamm, B., Hille, C. and Schonicke, P. (2010) 'Green biorefinery demonstration plant in Havelland (Germany)', *Biofuel Bioprod Bior*, 4, 253–262.
Kandler, O. and Hopf, H. (1980) 'Occurrence, metabolism, and function of oligosaccharides', *Biochem Plants*, 3, 221–270.
Keady, T. W. and O'Kiely, P. (1998) 'An evaluation of potassium and nitrogen fertilization of grassland, and date of harvest, on fermentation, effluent production, dry-matter recovery and predicted feeding value of silage', *Grass Forage Sci*, 53, 326–337.
Keating, T. and O'Kiely, P. (2000a) 'Comparison of old permanent grassland, *Lolium perenne* and *Lolium multiflorum* swards grown for silage: 3. Effects of varying fertiliser nitrogen application rate', *Irish J Agr Food Res*, 39, 35–53.
Keating, T. and O'Kiely, P. (2000b) 'Comparison of old permanent grassland, *Lolium perenne* and *Lolium multiflorum* swards grown for silage: 4. Effects of varying harvesting date', *Irish J Agr Food Res*, 39, 55–71.
King, C., McEniry, J., Richardson, M. and O'Kiely, P. (2012a) 'Yield and chemical composition of five common grassland species in response to nitrogen fertiliser application and phenological growth stage', *Acta Agr Scan B-SP*, 62, 644–658.
King, C., McEniry, J., O'Kiely, P. and Richardson, M. (2012b) 'The effects of hydrothermal conditioning, detergent and mechanical pressing on the isolation of the fibre-rich press-cake fraction from a range of grass silages', *Biomass Bioenerg*, 42, 179–188.
King, C., McEniry, J., Richardson, M. and O'Kiely, P. (2013a) 'Silage fermentation characteristics of grass species grown under two nitrogen fertilizer inputs and harvested at advancing maturity in the spring growth', *Grassland Sci*, 59, 30–43.

King, C., Richardson, M., McEniry, J. and O'Kiely, P. (2013b) 'Potential use of the separated press-cake fraction from grass silages as a fibre reinforcement to control the shrinkage properties of clay and cementitious building materials', *Biosys Eng*, 115, 203–210.

Koegel, R. G., Sreenath, H. K. and Straub, R. J. (1999) 'Alfalfa fiber as a feedstock for ethanol and organic acids', *Appl Biochem Biotech*, 77, 105–115.

Koschuh, W., Povoden, G., Thang, V. H., Kromus, S., Kulbe, K. D., Novalin, S. and Krotscheck, C. (2004) 'Production of leaf protein concentrate from ryegrass (*Lolium perenne x multiflorum*) and alfalfa (*Medicago sauva* subsp. *sativa*). Comparison between heat coagulation/centrifugation and ultrafiltration', *Desalination*, 163, 253–259.

Koschuh, W., Thang, V. H., Krasteva, S., Novalin, S. and Kulbe, K. D. (2005) 'Flux and retention behaviour of nanofiltration and fine ultrafiltration membranes in filtrating juice from a green biorefinery: a membrane screening', *J Membrane Sci*, 261, 121–128.

Kromus, S., Wachter, B., Koschuh, W., Mandl, M., Krotscheck, C. and Narodoslawsky, M. (2004) 'The green biorefinery Austria – development of an integrated system for green biomass utilization', *Chem Biochem Eng Q*, 18, 7–12.

Kromus, S., Kamm, B., Kamm, M., Fowler, P. and Narodoslawsky M. (2006) 'The green biorefinery concept – fundamentals and potential', in Kamm, B., Gruber, P. R. and Kamm, M., *Biorefineries – Industrial Processes and Products*, Weinheim: Wiley, 253–285.

Lattanzi, F. A. (2010) '$C_3/C_4$ grasslands and climate change', *Grassland Sci Eur*, 15, 3–13.

Lewandowski, I., Scurlock, J. M. O., Lindvall, E. and Christou, M. (2003) 'The development and current status of perennial rhizomatous grasses as energy crops in the US and Europe', *Biomass Bioenerg*, 25, 335–361.

Linnemann, A. R. and Dijkstra, D. S. (2002) 'Toward sustainable production of protein-rich foods: appraisal of eight crops for Western Europe. Part I. Analysis of the primary links of the production chain', *Crit Rev Food Sci*, 42, 377–401.

Lyttelton, J. M. (1973) 'Proteins and nucleic acids', in Butler G. W. and Bailey, R., *Chemistry and Biochemistry of Herbage*, London: Academic Press, 63–103.

Mahmoud, A., Arlabosse, P. and Fernandez, A. (2011) 'Application of a thermally assisted mechanical dewatering process to biomass', *Biomass Bioenerg*, 35, 288–297.

Mandl, M. G. (2010) 'Status of green biorefining in Europe', *Biofuel Bioprod Bior*, 4, 268–274.

Mandl, M., Graf, N., Ringhofer, J., Kromus, S., Bochzelt, H. and Schnitzer, H. (2005) 'The Austrian Green Biorefinery (A-GBR) concept overview: results of lactic acid and amino acid yields in silage juice', in *Proceedings of the 1st International Conference on Renewable Resources and Biorefineries*, Available from: http://www.joanneum.at/uploads/tx_publicationlibrary/img3037.pdf (accessed 27 March 2013).

Mandl, M., Graf, N., Thaller, A., Bochzelt, H. and Schnitzer, H. (2006) 'Green biorefinery – primary processing and utilization of fibers from green biomass', Nachhaltig Wirtschaften Project number 806111. Project report phase II. Available from: http://www.fabrikderzukunft.at/results.html/id3028 (accessed 27 March 2013).

McDonald, P., Henderson, N. and Heron, S. (1991) *The Biochemistry of Silage*, Marlow: Chalchombe Publications.

McDougall, V. D. (1980) 'Support energy and green crop fractionation in the United Kingdom', *Agr Syst*, 5, 251–266.

McEniry, J. and O'Kiely, P. (2013a) 'Anaerobic methane production from five common grassland species at sequential stages of maturity', *Bioresource Technol*, 127, 143–150.

McEniry, J. and O'Kiely, P. (2013b) 'The estimated nutritive value of three common grassland species at three primary growth harvest dates following ensiling and fractionation of press-cake', *Agr Food Res*, 22, 194–200.

McEniry, J., O'Kiely, P., Clipson, N. J. W., Forristal, P. D. and Doyle, E. M. (2008) 'The microbiological and chemical composition of silage over the course of fermentation in round bales relative to that of silage made from unchopped and precision-chopped herbage in laboratory silos', *Grass Forage Sci*, 63, 407–420.

McEniry, J., O'Kiely, P., Crosson, P., Groom, E. and Murphy, J. D. (2011) 'The effect of feedstock cost on biofuel cost as exemplified by biomethane production from grass silage', *Biofuel Bioprod Bior*, 5, 670–682.

McEniry, J., Finnan, J., King, C. and O'Kiely, P. (2012) 'The effect of ensiling and fractionation on the suitability for combustion of three common grassland species at sequential harvest dates', *Grass Forage Sci*, 67, 559–568.

McEniry, J., Crosson, P., Finneran, E., McGee, M., Keady, T. W. J. and O'Kiely, P. (2013a) 'How much grassland biomass is available in Ireland in excess of livestock requirements?' *Irish J Agr Food Res*, 52, 67–80.

McEniry, J., King, C. and O'Kiely, P. (2013b) 'Silage fermentation characteristics of three common grassland species in response to advancing stage of maturity and additive application', *Grass Forage Sci*, DOI: 10.1111/gfs.12038.

McGrath, D. (1988) 'Seasonal variation in the water-soluble carbohydrates of perennial and Italian ryegrass under cutting conditions', *Irish J Agr Res*, 27, 131–139.

Mela, T. and Rand, H. (1979) 'Amino acid composition of timothy, meadow fescue, cocksfoot and perennial ryegrass at two levels of nitrogen fertilization and at successive cuttings', *Annales Agriculturae Fenniae*, 18, 246–251.

Moore, K. J. and Hatfield, R. D. (1994) 'Carbohydrates and forage quality', in Fahey, G. C., Collins, M. C., Mertens, D. R., Moser, L. E., *Forage Quality, Evaluation, and Utilization*, Madison, WI: American Society of Agronomy, Crop Science Society of America, Soil Science Society of America, 229–280.

Muck, R. E. (1988) 'Factors influencing silage quality and their implications for management', *J Dairy Sci*, 71, 2992–3002.

Muck, R. E., Wilson, R. K. and O'Kiely, P. (1991) 'Organic acid content of permanent pasture grasses', *Irish J Agr Res*, 30, 143–152.

Murphy, J. D., Braun, R., Weiland, P. and Wellinger, A. (2011) Biogas from crop digestion, IEA Bioenergy. Available from: www.iea-biogas.net/_download/publi-task37/Update_Energy_crop_2011.pdf (accessed 27 March 2013).

Naik, S. N., Goud, V. V., Rout, P. K. and Dalai, A. K. (2010) 'Production of first and second generation biofuels: a comprehensive review', *Renew Sust Energ Rev*, 14, 578–597.

Nemli, G., Demirel, S., Gümüskaya, E., Aslan, M. and Acar, C. (2009) 'Feasibility of incorporating waste grass clippings (*Lolium perenne* L.) in particleboard composites', *Waste Manage*, 29, 1129–1131.

Neureiter, M., Danner, H., Frühauf, S., Kromus, S., Thomasser, C., Braun, R. and Narodoslawsky, M. (2004) 'Dilute acid hydrolysis of presscakes from silage and grass to recover hemicellulose-derived sugars', *Bioresource Technol*, 92, 21–29.

Nishino, N., Miyase, K., Ohshima, M. and Yokota, H. (1997) 'Effects of extraction and reconstitution of ryegrass juice on fermentation, digestion and *in situ* degradation of pressed cake silage', *J Sci Food and Agr*, 75, 161–166.

Nizami, A. S., Orozco, A., Groom, E., Dieterich, B. and Murphy, J. D. (2012) 'How much gas can we get from grass?', *Appl Energ*, 92, 783–790.

Novalin, S. and Zweckmair, T. (2009) 'Renewable resources – green biorefinery: separation of valuable substances from fluid-fractions by means of membrane technology', *Biofuel Bioprod Bior*, 3, 20–27.

Obernberger, I., Brunner, T. and Bärnthaler, G. (2006) 'Chemical properties of solid biofuels – significance and impact', *Biomass Bioenerg*, 30, 973–982.

O'Keeffe, S., Schulte, R. P. O., Lalor, S. T. J., O'Kiely, P. and Struik, P. C. (2011) 'Green biorefinery (GBR) scenarios for a two-cut silage system: investigating the impacts of sward botanical composition, N fertilisation rate and biomass availability on GBR profitability and price offered to farmers', *Biomass and Bioenerg*, 35, 4699–4711.

O'Kiely, P., O'Riordan, E. G. and Maloney, A. P. (1997) 'Grass ensilability indices as affected by the form and rate of inorganic nitrogen fertiliser and the duration to harvesting', *Irish J Agr Food Res*, 36, 93.

Pahkala, K., Paavilainen, L. and Mela, T. (1997) 'Grass species as raw material for pulp and paper', in *Proceedings of the XVIII International Grassland Congress, Winnepeg, Manitoba*, 55–60.

Peeters, A. and Hopkins, A. (2010) 'Climate change in European grasslands', *Grassland Sci Eur*, 15, 72–74.

Peeters, A. and Kopec, S. (1996) 'Production and productivity of cutting grasslands in temperate climates of Europe', *Grassland Sci Eur*, 1, 59–73.

Peeters, A., Parente, G. and LeGall, A. (2006) 'Temperate legumes: key-species for sustainable temperate mixtures', *Grassland Sci Eur*, 11, 205–220.

Peyraud, J. L., Le Gall, A. and Lüscher, A. (2009) 'Potential food production from forage legume-based systems in Europe: an overview', *Irish J Agric Food Res*, 48, 115–135.

Pirie, N. W. (1966) 'Leaf protein as a human food', *Science*, 152, 1701–1705.

Pirie, N. W. (1971) *Leaf Protein: Its Agronomy, Preparation, Quality and Use*, Oxford: Blackwell Scientific Publications.

Plantureux, S., Peeters, A. and McCracken, D. (2005) 'Biodiversity in intensive grasslands: effect of management, improvement and challenges', *Agron Res*, 3, 153–164.

Prochnow, A., Heiermann, M., Plöchl, M., Linke, B., Idler, C., Amon, T. and Hobbs, P. J. (2009a) 'Bioenergy from permanent grassland – a review: 1. Biogas', *Bioresource Technol*, 100, 4931–4944.

Prochnow, A., Heiermann, M., Plöchl, M., Amon, T. and Hobbs, P. J. (2009b) 'Bioenergy from permanent grassland – a review: 2. Combustion', *Bioresource Technol*, 100, 4945–4954.

Reid, D. (1985) 'A comparison of the yield responses of four grasses to a wide range of nitrogen application rates', *J Agr Sci*, 105, 381–387.

Reiter, W. D. (1998) '*Arabidopsis thaliana* as a model system to study synthesis, structure, and function of the plant cell wall', *Plant Physiol Bioch*, 36, 167–176.

Richter, F., Graß, R., Fricke, T., Zerr, W. and Wachendorf, M. (2009) 'Utilization of semi-natural grassland through integrated generation of solid fuel and biogas from biomass. II. Effects of hydrothermal conditioning and mechanical dehydration on anaerobic digestion of press fluids', *Grass Forage Sci*, 64, 354–363.

Richter, F., Fricke, T. and Wachendorf, M. (2011) 'Influence of sward maturity and pre-conditioning temperature on the energy production from grass silage through the integrated generation of solid fuel and biogas from biomass (IFBB): 1. The fate of mineral compounds', *Bioresource Technol*, 102, 4855–4865.

Rook, A. J., Dumont, B., Isselstein, J., Osoro, K., WallisDeVries, M. F., Parente, G. and Mills, J. (2004) 'Matching type of livestock to desired biodiversity outcomes in pastures – a review', *Biol Conserv*, 119, 137–150.

Rooke, J. A. and Hatfield, R. D. (2003) 'Biochemistry of ensiling', in Buxton, D. R., Muck, R. E. and Harrison, J. H., *Silage Science and Technology*, Madison, WI: American Society of Agronomy, Crop Science Society of America, Soil Science Society of America, 95–139.

Scott, E., Peter, F. and Sanders, J. (2007) 'Biomass in the manufacture of industrial products – the use of proteins and amino acids', *Appl Microbiol Biot*, 75, 751–762.

Sharma, H. S. S., Carmichael, E., Muhamad, M., McCall, D., Andrews, F., Lyons, G., McRoberts, C. and Hornsby, P. R. (2012) 'Biorefining of perennial ryegrass for the production of nanofibrillated cellulose', *RSC Advances*, 2, 6424–6437.

Sieker, T., Neuner, A., Dimitrova, D., Tippkötter, N., Muffler, K., Bart, H. J., Heinzle, E. and Ulber, R. (2011) 'Ethanol production from grass silage by simultaneous pretreatment, saccharification and fermentation: first steps in the process development', *Eng Life Sci*, 11, 436–442.

Singer, S. J., Eggman, L., Campbell, J. M. and Wildman, S. G. (1952) 'The proteins of green leaves. IV. A high molecular weight protein comprising a large part of the cytoplasmic proteins', *J Biol Chem*, 197, 233–239.

Smit, H. J., Metzger, M. J. and Ewert, F. (2008) 'Spatial distribution of grassland productivity and land use in Europe', *Agr Syst*, 98, 208–219.

Smith, D. (1973) 'Influence of drying and storage conditions on nonstructural carbohydrate analysis of herbage tissue – a review', *Grass Forage Sci*, 28, 129–134.

Stoate, C., Báldi, A., Beja, P., Boatman, N. D., Herzon, I., Van Doorn, A., De Snoo, G. R., Rakosy, L. and Ramwell, C. (2009) 'Ecological impacts of early 21st century agricultural change in Europe – a review', *J Environ Manage*, 91, 22–46.

Thang, V. H. and Novalin, S. (2008) 'Green biorefinery: separation of lactic acid from grass silage juice by chromatography using neutral polymeric resin', *Bioresource Technol*, 99, 4368–4379.

Thang, V. H., Koschuh, W., Kulbe, K. D., Kromus, S., Krotscheck, C. and Novalin, S. (2004) 'Desalination of high salt content mixture by two-stage electrodialysis as the first step of separating valuable substances from grass silage', *Desalination*, 162, 343–353.

Thang, V. H., Koschuh, W., Kulbe, K. D. and Novalin, S. (2005) 'Detailed investigation of an electrodialytic process during the separation of lactic acid from a complex mixture', *J Membrane Sci*, 249, 173–182.

Theander, O. and Westerlund, E. (1993) 'Quantitative analysis of cell wall components', in Jung, H. G., Buxton, D. R., Hatfield, R. D. and Ralph, J., *Forage*

*Cell Wall Structure and Digestibility*, Madison, WI: American Society of Agronomy, Crop Science Society of America, Soil Science Society of America, 83–104.

Thomsen, M. H., Bech, D. and Kiel, P. (2004) 'Manufacturing of stabilised brown juice for L-lysine production from university lab scale over pilot scale to industrial production', *Chem Biochem Eng Q*, 18, 37–46.

Tonn, B., Thumm, U. and Claupein, W. (2010) 'Semi-natural grassland biomass for combustion: influence of botanical composition, harvest date and site conditions on fuel composition', *Grass Forage Sci*, 65, 383–397.

Ugherughe, P. O. (1986) 'Relationship between digestibility of *Bromus inermis* plant parts', *J Agron Crop Sci*, 157, 136–143.

Van Vuuren, A. M. (1993) 'Digestion and nitrogen metabolism of grass fed dairy cows', PhD dissertation, Wageningen, Landbouwuniversiteit te Wageningen.

Venus, J. and Richter, K. (2007) 'Development of a pilot plant facility for the conversion of renewables in biotechnological processes', *Eng Life Sci*, 7, 395–402.

Wachendorf, M., Richter, F., Fricke, T., Grab, R. and Neff, R. (2009) 'Utilization of semi-natural grassland through integrated generation of solid fuel and biogas from biomass. I. Effects of hydrothermal conditioning and mechanical dehydration on mass flows of organic and mineral plant compounds, and nutrient balances', *Grass Forage Sci*, 64, 132–143.

Weiland, P. (2010) 'Biogas production: current state and perspectives', *Appl Microbiol Biotechnol*, 85, 849–860.

White, L. M. (1973) 'Carbohydrate reserves of grasses: a review', *J Range Manage*, 26, 13–18.

Whitehead, D. C. (1995) *Grassland Nitrogen*, Wallingford: CAB International.

Wilkins, R. J., Hopkins, A. and Hatch, D. J. (2003) 'Grassland in Europe', *Grassland Sci*, 49, 258–266.

Wilson, J. R. (1993) 'Organization of forage plant tissues', in Jung, H. G., Buxton, D. R., Hatfield, R. D. and Ralph, J., *Forage Cell Wall Structure and Digestibility*, Madison, WI: American Society of Agronomy, Crop Science Society of America, Soil Science Society of America, 1–32.

Woolford, M. K. (1972) 'Some aspects of the microbiology and biochemistry of silage making', *Herbage Abstracts*, 42, 105–111.

Worgan, J. T. and Wilkins, R. J. (1977) 'The utilisation of deproteinised forage juice', in Wilkins, R. J., *Green Crop Fractionation – Occasional Symposium No. 9 (British Grassland Society)*, Luton: Inprint Ltd, 119–129.

Zheng, Y., Pan, Z., Zhang, R., El-Mashad, H. M., Pan, J. and Jenkins, B. M. (2009) 'Anaerobic digestion of saline creeping wild ryegrass for biogas production and pretreatment of particleboard material', *Bioresource Technol*, 100, 1582–1588.

# 12
## Developments in glycerol byproduct-based biorefineries

B. P. PINTO and C. J. DE ARAUJO MOTA,
Universidade Federal do Rio de Janeiro,
Brazil

DOI: 10.1533/9780857097385.1.364

**Abstract**: Biodiesel is presently one of the most important biofuels used worldwide. Glycerol or glycerin is a byproduct of biodiesel production, with still few economical applications and can be used as a raw material for the fuel sector. Biotechnological pathways may be used for the production of ethanol, used as biofuel or in the biodiesel transesterification process. Production of syngas from glycerol opens the possibility of obtaining hydrocarbons in the diesel and gasoline range. Many glycerol derivatives can be used as additives for gasoline, diesel and biodiesel, showing the great versatility of this substance.

**Key words**: biodiesel, biorefineries, fuels, glycerol, green chemistry.

## 12.1 Introduction

The development of fuels and chemical products based on renewable resources is the aim of biorefineries. The world is still dependent on oil, but this dependence has led to critical changes in the climate of the planet. Global warming is a reality and may lead to great environmental, economic and social impacts in the coming years, if nothing is done to stop or slow down this process. Greenhouse gases, particularly $CO_2$ from fossil fuels, are mainly responsible for the global warming process. Thus, it is imperative to develop new processes for the production of fuels and chemicals, without impacting the environment, especially in terms of greenhouse gases.

The use of biofuels is spreading all over the world. Bioethanol and biodiesel will share a significant part of the fuel market, contributing to the control of global warming in the future. Biodiesel is produced mainly through the transesterification of vegetable oils or animal fat, the triglycerides. In this process, methanol reacts with the triglyceride in the presence of a basic or acidic catalyst to afford fatty acid methyl esters, the biodiesel themselves, and glycerol (Fig. 12.1). Roughly, for each 100 m³ of vegetable oil processed, about 10 m³ of glycerol is produced. In recent years, the surplus of glycerol coming from biodiesel fabrication has enormously increased, representing, today, about 65% of the world's glycerol production.

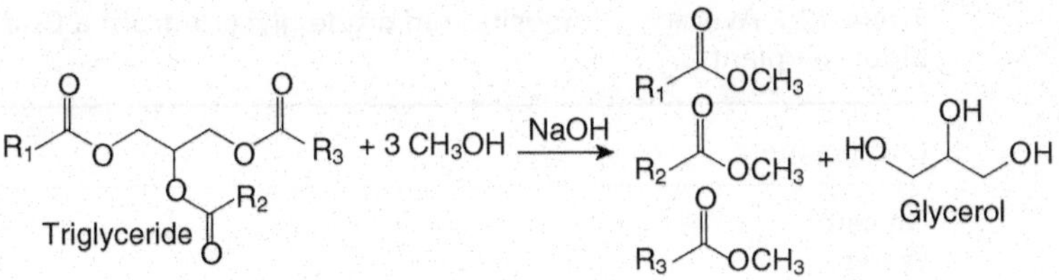

*12.1* Transesterification of triglycerides to produce fatty acid methyl esters (biodiesel) and glycerol.

Glycerol can also be obtained from algae (Muscatine, 1967). This is another potential renewable source of glycerol, alternative to the production from biodiesel that could be used in biorefineries processes. Some species of the unicellular algae *Dunaliella* possess outstanding adaptability and tolerance toward a wide range of salinities from seawater. The capability of the cell to thrive in high salt concentrations depends on its unique ability to produce intracellular glycerol (Chitlaru and Pick, 1991). *Dunaliella salina* and *Dunaliella viridis* grow in media containing different salt concentrations. Algae growth, as well as glycerol production, increased as the salinity of the medium increased (Hadi *et al.*, 2008)

Glycerol or glycerin is traditionally used in cosmetics, soaps and pharmaceuticals. However, these sectors cannot drain the enormous amount of glycerol that comes from biodiesel production. Thus, the glycerol byproduct of biodiesel production must find new applications, capable of making use of the increasing output of this chemical and adding value to the biodiesel chain. The purpose of this chapter is to highlight some potential applications of glycerol in the fuel sector and in the chemical industry.

## 12.2 Composition and purification of glycerol produced from biodiesel

Glycerol is 1,2,3-propanetriol. It was first identified by Carl Scheele in 1779, upon heating olive oil with litharge (PbO). The viscous, transparent liquid that separated from the oil phase was named glycerol due to its sweet taste (from the Greek: glykos = sweet). The term glycerin applies to commercial products, which are rich (95%) in glycerol. However, with the increasing production of biodiesel, there are many commercial glycerin products with different glycerol contents and other impurities, such as water, salts and organic compounds.

Glycerol has a boiling point of 290°C and a viscosity of 1.5 Pa.s at 20°C. Therefore, purification procedures are normally time-consuming and costly. The most traditional process involves distillation at reduced pressure. A

*Table 12.1* Average composition of crude glycerin from a Brazilian biodiesel plant

| Composition | wt% |
|---|---|
| Glycerol | 80.0 min |
| Water | 10.0 max |
| Methanol | 1.0 max |
| NaCl | 10.0 max |
| Ashes | 10.0 max |

product with a glycerol content of at least 99.5 wt% can be obtained by thin film distillation and meets the requirements of the United States Pharmacopeia (USP). Other purification processes such as extraction, ion-exchange, adsorption, crystallization and dialysis can also be applied to the glycerin phase. The commercial utilization of the glycerin will be impacted by the costs and purity of the product.

During biodiesel production, two phases are produced at the end of the transesterification process. The upper ester phase contains the main product (biodiesel). The lower phase consists of glycerol and many other substances. The exact composition of the raw glycerol phase depends on the method of transesterification and the separation conditions of biodiesel production (Hájek and Skopal, 2010). When the lower glycerol layer is removed, the ester phase is washed to remove the residual glycerol, base catalyst and soaps formed during the reaction. Then, methanol and water present in the ester phase can be removed by distillation (Van Gerpen, 2005).

The raw glycerin phase has different compositions: glycerol, soaps, inorganic salts, water, methanol and esters. Table 12.1 shows the typical composition of the glycerin phase obtained in a Brazilian biodiesel plant. The minimum glycerol content is 80 wt% and there is about 10 wt% water. The remaining 10 wt% is mainly methanol and dissolved salts, such as NaCl, formed upon acid neutralization of the homogenous basic catalyst.

Refining glycerol from biodiesel production begins with an acid treatment to split the soaps into free fatty acids and salts. Fatty acids are not soluble in glycerol and are separated from the top and recycled to the process. Excess methanol can be recovered by distillation. The salts remain in the glycerol phase and may be one of the most deleterious impurities that limit the use of this raw material in chemical processes (da Silva and Mota, 2011). Purification is required to achieve a product with the necessary purity.

The crude glycerol phase can be purified by ion exchange resins to remove the sodium from glycerol/water solutions with a high salt concentration (Carmona *et al.*, 2008, 2009). Adsorption and membrane technologies can also be used. Many purification methods are based on the distillation of the glycerol phase to strip alcohol contaminants from glycerol (Potthast et al., 2010).

Ion exchange purification is not considered economically viable when high concentrations of salts are present in the crude glycerol. Distillation and membrane technologies are commonly used to obtain ultra-pure glycerol, but membrane technologies are more cost-effective than distillation, provided that some form of prior purification that reduces salts and organic matter has taken place (Manosak et al., 2011).

Purification of crude glycerol from biodiesel production with phosphoric acid, obtaining a product with a final purity of over 86%, has been reported (Hájek and Skopal 2010; Javani et al., 2012). Potassium phosphate obtained as byproduct could potentially be used as fertilizer. Glycerol of high purity can be obtained upon the extraction of crude glycerol with ethanol (Kongjao et al., 2010). Salts and fatty acids were separated through filtration and decantation, respectively.

Thus, there are many physical and chemical methods of glycerol purification. It is important that dissolved salts are reduced to the lowest possible level, because they may affect the catalysts used in further glycerol processing.

## 12.3 Applications of glycerol in the fuel sector

The complete combustion of glycerol (Fig. 12.2) generates 4195 kcal per kg, but there are many difficulties associated with this process. Incomplete burning may generate acrolein, which is highly toxic to humans. The salts present in the crude glycerol from biodiesel production may deteriorate the equipment, leading to corrosion and other problems. All these drawbacks make the direct combustion of glycerol economically less attractive than the chemical or biochemical transformation.

Glycerol can also be used in the production of ethanol, an important biofuel used worldwide. Historically, ethanol has been produced mainly from sugars and carbohydrates via microbial fermentation. Speers and co-workers (2012) developed a microbial co-culture for the conversion of glycerol into ethanol and electricity in bioelectrochemical systems. Ethanol can be used as a feedstock for the transesterification of vegetable oil for biodiesel production and the electricity can be used to partially

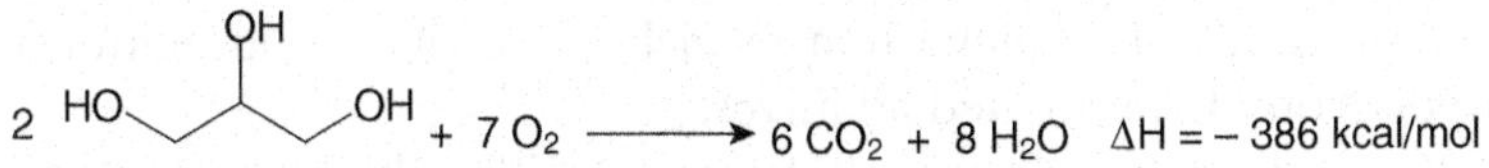

$$2\ HOCH_2CH(OH)CH_2OH + 7\ O_2 \longrightarrow 6\ CO_2 + 8\ H_2O \quad \Delta H = -386\ kcal/mol$$

12.2 Complete combustion of glycerol.

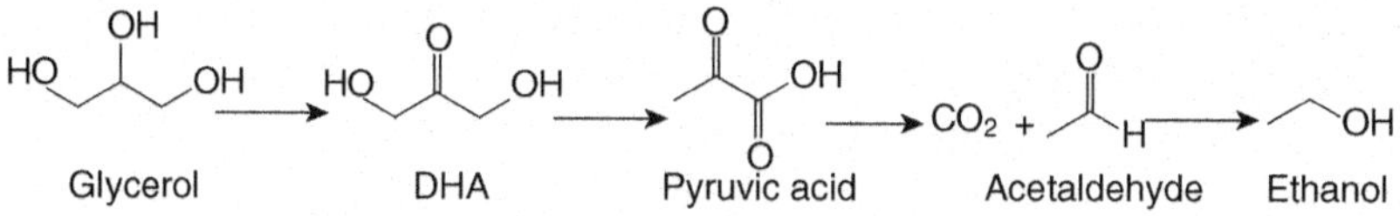

*12.3* Biothechnological pathway of ethanol production from glycerol.

offset the energy needs of the biodiesel plant. The platform includes a glycerol-fermenting bacterium, *Clostridium cellobioparum*, which produces ethanol and other fermentative byproducts (lactate, acetate, formate, and $H_2$), and *Geobacter sulfurreducens*, which can convert the fermentative byproducts into electricity. Both organisms were adaptively evolved for tolerance to industrially relevant glycerol concentrations and co-cultivation of these strains stimulated microbial growth and resulted in ethanol and complete conversion of fermentative byproducts into electricity.

The yeast *Saccharomyces cerevisiae* utilizes the general glycolytic pathway for the majority of its energy production. The carbohydrates are converted to pyruvic acid, which is decarboxylated to acetaldehyde, and then to ethanol. Starting from glycerol, the pathway involves formation of di-hydroxy-acetone (DHA) and pyruvic acid (Fig. 12.3). To further increase ethanol production and evaluate fermentative performance, the genes involved in the conversion of pyruvate to ethanol were overexpressed (Yu *et al.*, 2012). These genes included pyruvate decarboxylase, which is involved in the decarboxylation of pyruvate and thus controls the first step in the production of ethanol from pyruvate, and alcohol dehydrogenase, which is the enzyme involved in the ethanol production pathway from acetaldehyde.

There have been intensive efforts to describe methods for the efficient conversion of glycerol to ethanol via metabolic pathway engineering of *E. coli*, to minimize byproducts. The engineered *E. coli* strain produced 21 g/L of ethanol from 60 g/L of pure glycerol, with a volumetric productivity of 0.216 g/L/h under anaerobic conditions (Yazdania and Gonzalez, 2008).

Glycerol gasification to synthesis gas, a mixture of CO and $H_2$, has also been studied (Soares *et al.*, 2006). The reaction is endothermic by 83 kcal/mol, but can be carried out at temperatures around 350°C over Pt and Pd catalysts. Synthesis gas is used in many industrial processes, like methanol production and Fischer–Tropsch synthesis of hydrocarbons.

Another approach to use glycerol from biodiesel production in the fuel sector is the development of glycerol ethers, acetals/ketals and esters with potential use as fuel additives. The glycerol molecule has about 50% of its mass in terms of oxygen atoms, which makes it a good platform for the production of oxygenated additives.

The acid-catalyzed reaction of glycerol with isobutene affords tert-butyl-glyceryl ethers (Klepacova *et al.*, 2005), which are considered as an octane

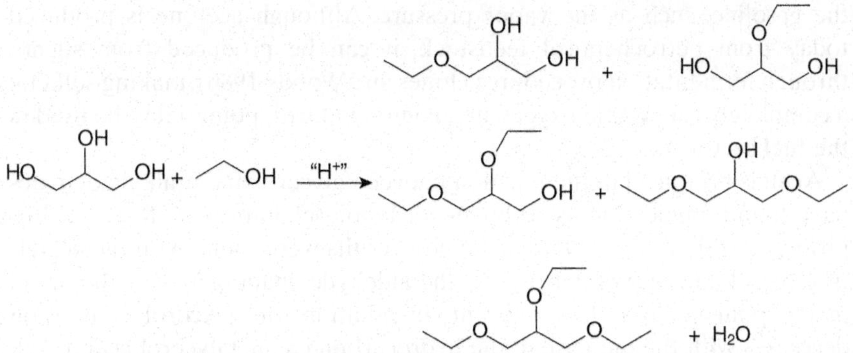

12.4 Reaction of glycerol with ethanol in the presence of acid catalysts.

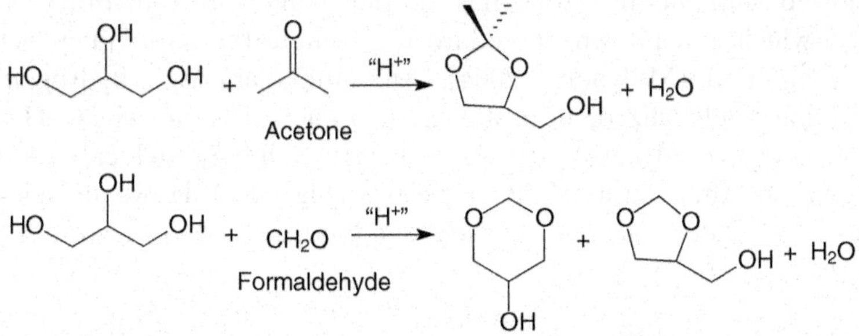

12.5 Reaction of glycerol with acetone and formaldehyde in the presence of acid catalysts.

booster for gasoline (Wessendorf, 1995). Ethyl glyceryl ethers (Fig. 12.4) can be produced through the acid-catalyzed reaction between glycerol and ethanol (Pariente et al., 2009), being a completely renewable molecule. These ethers are potential additives for biodiesel, improving the cold flow properties (Pinto, 2009). For instance, addition of 0.5 vol% of glyceryl ethyl ethers in the soybean and palm biodiesel led to a reduction of up to 5°C in the pour point, indicating that these ethers can be used in blends with biodiesel.

Glycerol acetals and ketals are another class of derivatives with potential use as fuel additives. They are produced through the acid-catalyzed reaction of glycerol with aldehydes and ketones, respectively. The reaction of acetone with glycerol produces one ketal, known as solketal, whereas reaction with formaldehyde solution affords two acetal isomers (da Silva et al., 2009) (Fig. 12.5). Solketal is a potential additive for gasoline (Mota et al., 2010). Within 5 vol% addition, it improved the octane number and significantly reduced gum formation, without affecting other important properties of the gasoline, such as the vapor pressure. Although acetone is produced today from petrochemical feedstock, it can be produced from sugars, through fermentation procedures (Jones and Woods, 1986), making solketal a completely renewable oxygenated compound with potential to be used in the fuel sector.

Acetals produced in the reaction between glycerol and n-alkylaldehydes have found application as additives for biodiesel, improving the cold flow properties (Silva et al., 2010b). The best results were found with the acetals of glycerol and butyraldehyde. As the aldehyde chain increases, the effect in the pour point is less relevant. In addition, the glycerol conversion decreases with the increase in the hydrocarbon chain. Glycerol acetals can also be used as antioxidants. The acid-catalyzed reaction of glycerol with aromatic aldehydes, such as benzaldehyde, anisaldehyde and furfural, affords acetals with a benzylic C–H bond (Fig. 12.6). These molecules showed antioxidant properties in the diphenylpicrylhydrazyl (DPPH) test, which is a known procedure to estimate the antioxidant activity of a compound (Molyneux, 2004). An antioxidant is a hydrogen donor, forming a delocalized, more stable, free radical. The benzylic C–H bonds of the aromatic glycerol acetals can afford highly delocalized radicals, explaining their antioxidant properties (Fig. 12.7). However, tests of the aromatic glycerol acetals with soybean biodiesel did not lead to significant

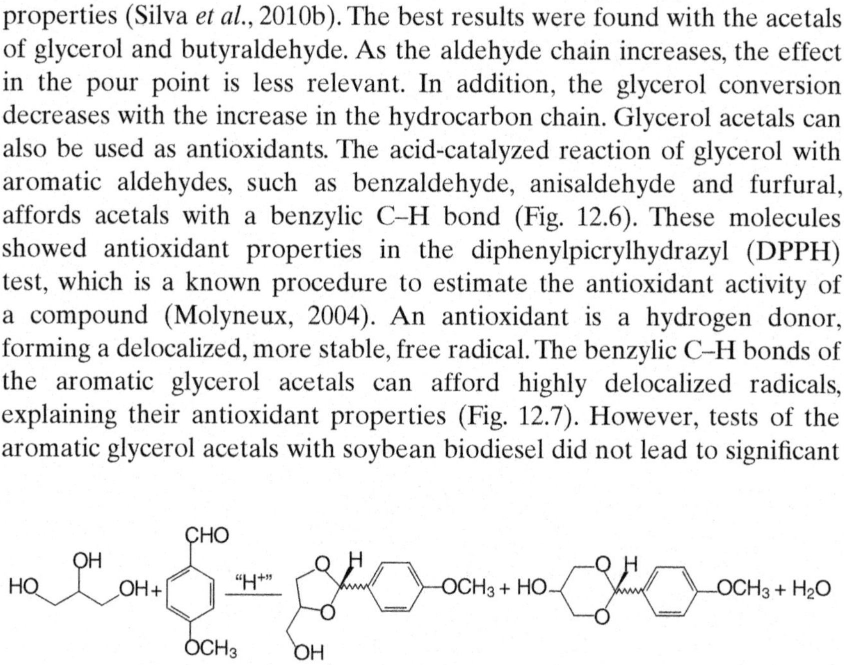

*12.6* Reaction of glycerol with anisaldehyde. Formation of aromatic glycerol acetals with antioxidant properties.

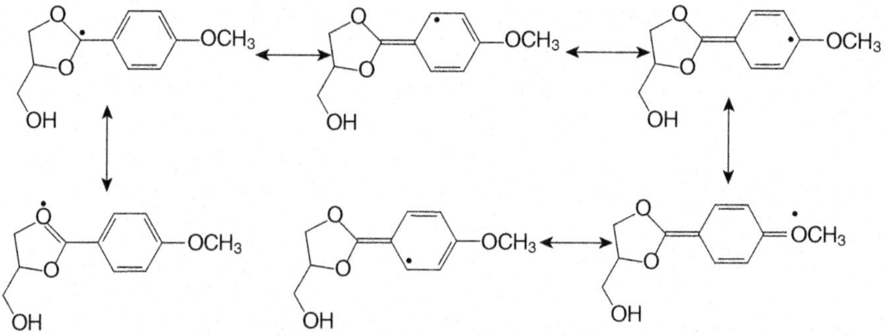

*12.7* Free radical resonance structures showing the electron delocalization in the aromatic ring of the glycerol/anisaldehyde acetal.

*Table 12.2* Oxidation stability of soybean biodiesel with furfural/glycerol acetals, according to EN 14112 standards

| Sample | Additive (%) | Induction period (h) |
|---|---|---|
| B 100[a] | 0 | 2.73 |
| B100 + furfural/glycerol acetals | 0.1 | 3.07 |
| B100 + furfural/glycerol acetals | 0.25 | 3.06 |
| B100 (BHT)[b] | 0 | 6.53 |
| B100 (BHT) + furfural/glycerol acetals | 0.1 | 8.10 |
| B100 (BHT) + furfural/glycerol acetals | 0.25 | 8.89 |

[a] Soybean biodiesel prepared in the lab (no additive).
[b] Commercial soybean biodiesel with butyl-hydroxy-toluene (BHT) to improve oxidation stability.

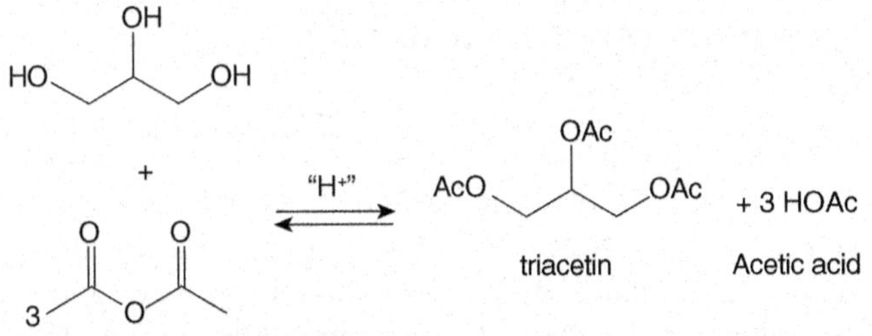

12.8 Esterification of glycerol with acetic anhydride. Selective formation of triacetin.

improvement in the oxidation resistance, measured according to the EN 14112 method, but the concomitant use of a commercial antioxidant, such as butyl-hydroxy-toluene (BHT) and the acetals gives much better results (Table 12.2), indicating a synergistic effect (Soares, 2011).

The acetins or glycerol acetates are useful compounds. Triacetin or glycerol triacetate is important in the tobacco industry and, more recently, has been tested as a fuel additive, especially for biodiesel, improving the viscosity and the pour point (Melero et al., 2007). The most traditional method of preparation of the acetins is the direct esterification of glycerol with acetic acid in the presence of an acidic catalyst (Gonçalves et al., 2008), yielding a mixture of the acetins. To increase the selectivity to triacetin, a large excess of acetic acid should be used. Another approach is to use acetic anhydride (Fig. 12.8) (Liao et al., 2009). Use of zeolite beta or K-10 montmorillonite as catalysts for the acetylation of glycerol with acetic anhydride leads to 100% selectivity to triacetin within 20 minutes of reaction time (Silva et al., 2010a).

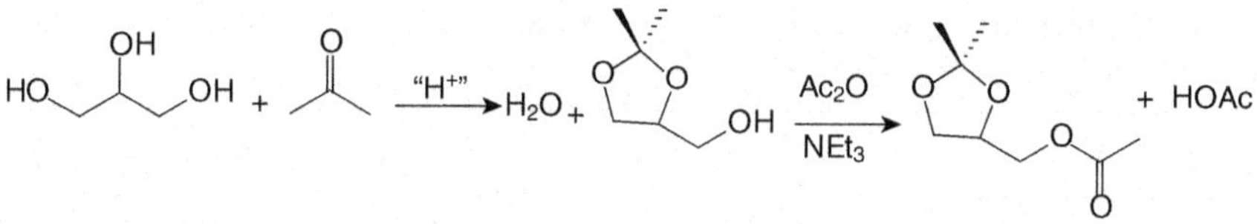

*12.9 Production of solketal acetate.*

Esterification of the free hydroxyl group of glycerol ketals and acetals has also been reported in the literature (Garcia *et al.*, 2008). The reaction of solketal with acetic anhydride in the presence of triethylamine produces solketal acetate in 90% yield (Fig. 12.9). This product may be used to improve the viscosity of biodiesel, without affecting the flash point. It may also reduce the formation of particulates in diesel.

## 12.4 Glycerol as raw material for the chemical industry

The chemical industry is still based on oil and gas, with sales of approximately 2 trillion euros, indicative of its huge and powerful economic situation. Naphtha is the main feedstock for the chemical industry. It is initially transformed into light olefins, such as ethene and propene, and aromatics, like benzene, toluene and xylenes. These compounds are then transformed into polymers and other chemicals through complex chemical processes, before being used in everyday life as plastic components, dyes, textiles, paints and other materials. The shortage and price fluctuations of oil in the near future will force the chemical industry to diversify its processes, giving more importance to renewable materials. Bioethanol produced from sugarcane is opening this new era. Recently, the major Brazilian chemical company, Braskem, has started up a plant to dehydrate ethanol to ethylene, which is subsequently polymerized to polyethylene (Braskem, 2010). Glycerol from biodiesel production may follow the same path, being a substitute for propylene-based chemicals.

### 12.4.1 Glycerol to propanediols

The hydrogenolysis of glycerol over supported metal catalyst can afford 1,2 and 1,3-propanediol (Chaminand *et al.*, 2004) (Fig. 12.10). The 1,2 isomer, also known as propylene glycol, has many uses, including as an anti-freezing agent and in the production of polyurethane foams and other polymers. The present worldwide production of propylene glycol is about 1 million tonnes per year. The traditional process involves the hydrolysis of propylene oxide, which in turn is produced from propene. The reaction can be carried out in

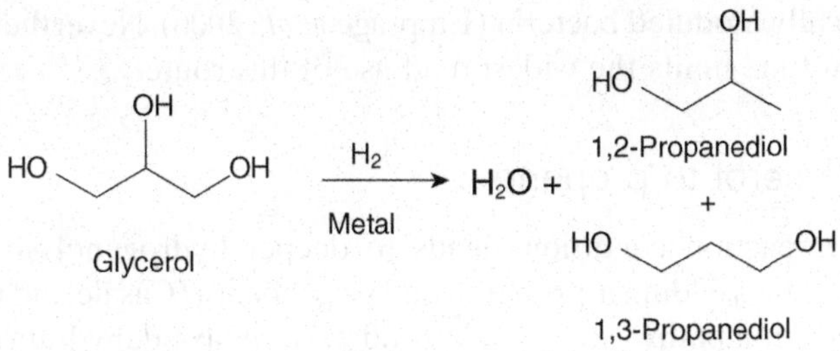

**12.10** Hydrogenolysis of glycerol over metal catalysts to afford 1,2 and 1,3-propanediols.

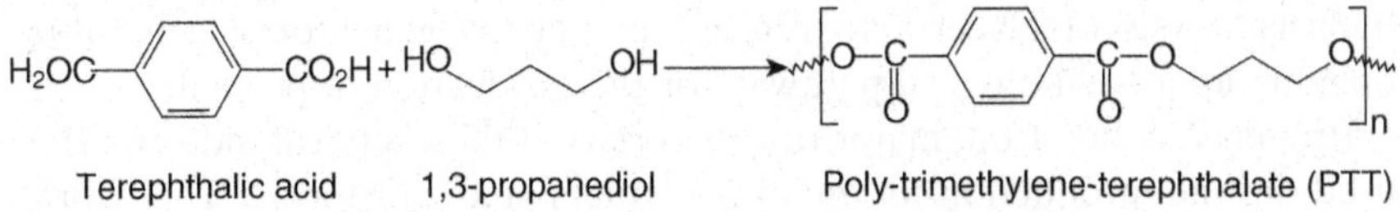

**12.11** Production of PTT from the reaction of terephthalic acid and glycerol.

batch or continuous flow conditions, in temperatures normally ranging from 180 to 250°C. Copper, palladium and ruthenium supported catalysts have normally been used (Dasari et al., 2005), but other metals like iron, nickel and rhodium can be used as well. The concomitant use of an acidic catalyst, such as sulfonic acid resins, improves the selectivity of the catalyst toward hydrogenolysis, allowing working at lower temperatures and reduced pressures in batch conditions (Miyazawa et al., 2006).

The hydrogenolysis of glycerol to propylene glycol has been industrially implemented by ADM (2012). Dow Chemical also has a technology named Propylene Glycol Renewable (PGR), that is based on glycerol hydrogenolysis. These facts show that the use of glycerol as a renewable feedstock for the chemical industry is now a reality.

In contrast to its 1,2 isomer, 1,3-propanediol has considerably fewer uses, with a global market of approximately 360,000 tonnes per year. The main utilization is in the reaction with terephthalic acid to produce a polyester fiber named poly-trimethylene-terephthalate (PTT) (Fig. 12.11), commercially known as CORTERRA. The traditional route used by Shell involves the reaction of ethylene oxide with $CO/H_2$ at high pressures. Nevertheless, the product can also be obtained by glycerol hydrogenolysis. With the choice of proper reaction conditions and modifications on the metal catalyst, the ratio between 1,3 and 1,2-propanediol can go up to 5, but at moderate glycerol conversion (Gong et al., 2009). The 1,3-propanediol can also be produced from glycerol through biotechnological processes, using genetically modified bacteria (Emptage *et al.*, 2006). Nevertheless, the long reaction time limits the widespread use of this route.

## 12.4.2 Glycerol to propene

More severe reaction conditions leads to deeper hydrogenolysis of the glycerol molecule, yielding n-propanol and isopropanol (Casale and Gomez, 1994). The reaction pathway is complex and may involve dehydration steps. This may explain why the concomitant use of acidic catalysts in the medium or as support improves the selectivity. It has been shown that CO and $CO_2$ can also be formed through decomposition reactions over supported Pt catalyst (Wawrzetz *et al.*, 2010). These results indicate that a complete hydrogenolysis of glycerol to remove all the oxygen atoms is feasible, opening up a technological pathway for the production of propene.

Propene is one of the major raw materials of the chemical industry. It is used in the production of many polymers and chemicals. The world production of propene is about 40 million tonnes per year, and it is expected to sharply increase in the coming years. Propene is normally produced from the steam cracking of naphtha. More recently, the catalytic cracking of vacuum gas oil has been an alternative source of propene. In contrast to ethene, which can be produced from dehydration of bioethanol, there are few technological routes to propene from renewable feedstock, most of them involving multistep reactions.

Mota and collaborators (2009) showed that the use of supported iron-molybdenum catalysts can be used in the selective hydrogenolysis of glycerol to propene (Fig. 12.12). The selectivity to propene can reach up to 90% (excluding water) over this catalyst at 300°C and continuous flow conditions. It is not clear why this catalyst is so selective to propene under the reaction conditions. X-ray diffraction analysis and temperature programmed reduction have indicated a strong interaction between the metals, which may affect the reducibility of the iron. The mechanism of the reaction seems to involve the participation of metal or metal oxide centers and acid sites of the support. Glycerol is initially dehydrated to acetol over the acid sites, which is then hydrogenated to 1,2-propanediol. Interaction of the diol with the acid sites leads to acetone, which is

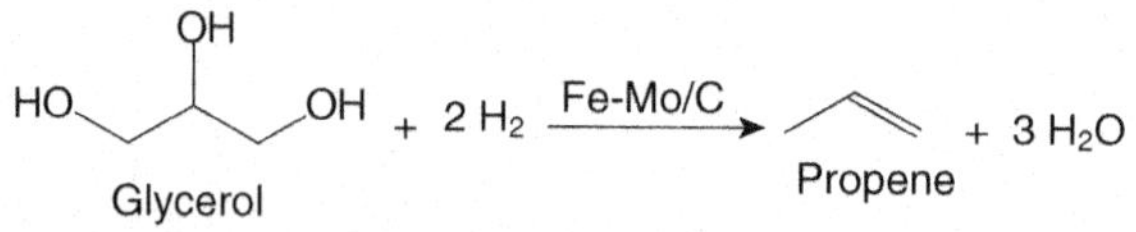

Glycerol + 2 $H_2$ $\xrightarrow{\text{Fe-Mo/C}}$ Propene + 3 $H_2O$

*12.12* Selective hydrogenolysis of glycerol to propene over Fe-Mo catalysts supported over activated carbon.

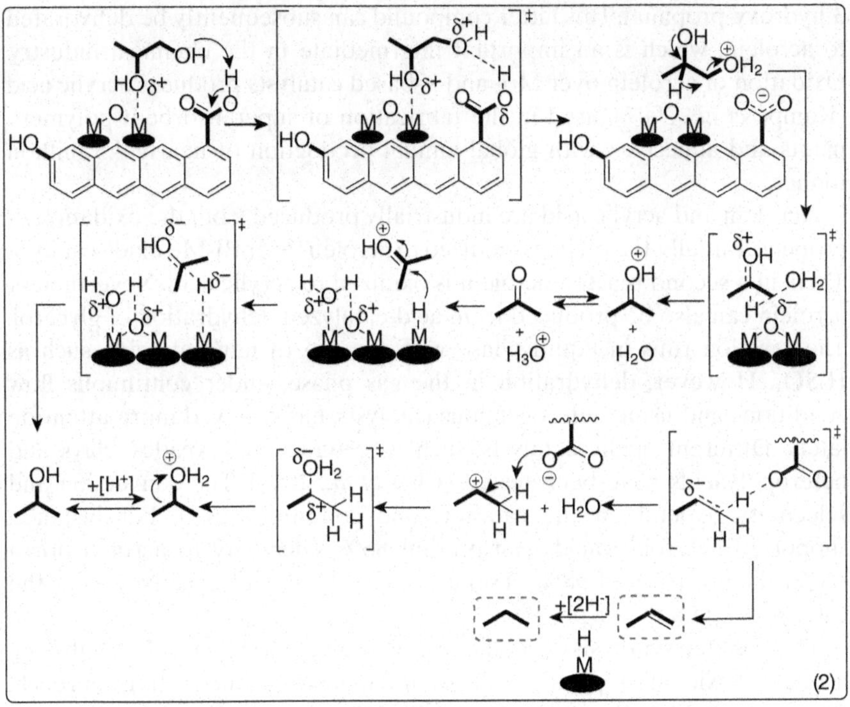

*12.13* Possible mechanistic pathway for hydrogenolysis of glycerol to propene over Fe-Mo supported catalyst (pathway from 1,2-propanediol).

hydrogenated to produce isopropanol. The final step involves dehydration of the alcohol to propene (Fig. 12.13). Further hydrogenation of propene to propane is slowed down due to the incomplete reduction of the metal catalyst. It is worth mentioning that acetol, propanediol, acetone and isopropanol can all be observed upon varying the reaction temperature, space velocity and hydrogen partial pressure, supporting the proposed reaction pathway.

The search for a 'green' propene, produced from renewable materials, is a major goal of many chemical companies. The use of glycerol from biodiesel production may be an alternative for the 'green' propene, opening up the possibility of producing plastics from this renewable feedstock.

## 12.4.3 Glycerol dehydration to acrolein and acrylic acid

Acid-catalyzed glycerol dehydration can follow two pathways: dehydration of the primary hydroxyl group affording hydroxy-acetone, also known as acetol, or dehydration of the secondary hydroxyl group yielding

3-hydroxy-propanal. This latter compound can subsequently be dehydrated to acrolein, which is an important intermediate in the chemical industry. Oxidation of acrolein over Mo- and V-based catalysts produces acrylic acid (Kampe *et al.*, 2007), used in the fabrication of superabsorbent polymers, paints and adhesives, with global annual production of nearly 4.5 million tonnes.

Acrolein and acrylic acid are industrially produced from the oxidation of propene. Initially, the olefin is oxidized to acrolein over Bi-Mo oxide catalyst. Then, in a second reactor, acrolein is oxidized to acrylic acid. Nevertheless, acrolein can also be produced from acid-catalyzed dehydration of glycerol. The reaction runs in liquid phase with the use of mineral acids, such as $H_2SO_4$. However, dehydration in the gas phase, under continuous flow conditions and using heterogeneous catalysts, has received more attention lately. Different acidic catalysts, such as zeolites, metaloxides, clays and heteropolyacids have been tested (Chai *et al.*, 2007). The conversion and selectivity depend on the catalysts and conditions used. For instance, supported heteropolyacids can present 86% selectivity to acrolein for a glycerol conversion of 98% (Tsukuda *et al.*, 2007), but deactivation of the catalyst is still a major problem.

The one-step synthesis of acrylic acid from glycerol (Fig. 12.14), combining acidic and oxidant properties in the same catalyst, is an interesting approach. The oxidative dehydration of glycerol has been studied over mixed oxide catalysts in the presence of air (Deleplanque *et al.*, 2010). Mixed molybdenum and vanadium oxides, as well as vanadium/tungsten oxides, were active in this reaction. They show 100% glycerol conversion with selectivity to acrylic acid between 24 and 28%. Acetic acid, probably coming from oxidation of acetaldehyde formed upon cracking of the 3-hydroxy-propanal, was also observed as byproduct. Supported tungsten oxide catalyzes the oxidative dehydration of glycerol to acrylic acid (Ulgen and Hoelderich, 2011), but the selectivity to acrylic acid is rather low, being within 5%.

Vanadium-impregnated zeolite beta can also be used in the oxidative dehydration of glycerol to acrylic acid (Pestana, 2009). The catalysts were prepared by wet impregnation ammonium metavanadate on the ammonium-exchanged zeolite beta, followed by air calcination. The selectivity to acrylic acid was around 20% at 70% glycerol conversion. The catalytic activity was associated to the dispersion of vanadium species inside the zeolite pores.

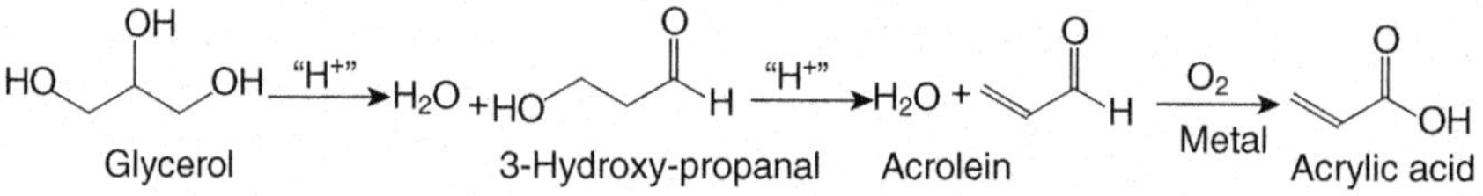

*12.14* Oxidative dehydration of glycerol to acrylic acid.

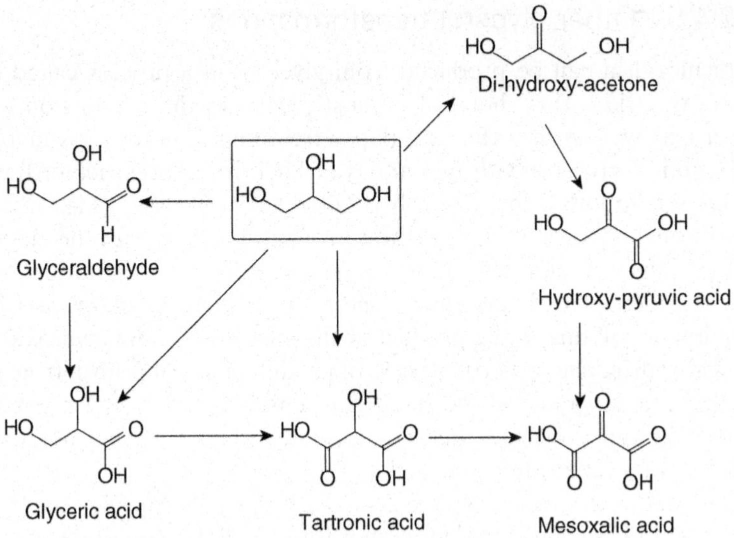

**12.15** Products formed in the oxidation of glycerol.

## 12.4.4 Glycerol oxidation

Glycerol oxidation can produce several important compounds (Fig. 12.15). The selective oxidation of the secondary hydroxyl group leads to 1,3-dihydroxy-acetone (DHA), which is an artificial tanning agent and has global production of more than two thousand tonnes per year. DHA is normally produced from glycerol fermentation (Hekmat et al., 2007), but can also be prepared by electrochemical methods (Ciriminna et al., 2006).

Glyceraldehyde is an intermediate in the carbohydrate metabolism. A good method of preparation involves the oxidation of glycerol over Pt-supported catalysts. Selectivity of 55% in glyceraldehyde with glycerol conversion of 90% can be observed with the use of Pt/C catalyst (Garcia et al., 1995).

Glyceric acid is selectively produced by the oxidation of glycerol over Au/C catalysts in the presence of oxygen (Carrettin et al., 2004). Bimetallic gold-platinum or gold-palladium catalysts show higher turnover frequencies, but are also less selective, yielding C-C cleavage products, such as oxalic and glycolic acids (Bianchi et al., 2005).

Glycerol oxidation with $H_2O_2$ in the presence of metal-containing silicate catalysts gives formic acid as a major compound, indicating the high activity of the system, which leads to C–C bond cleavage (McMorn et al., 1999). Formic acid is the major observed product.

## 12.4.5 Other glycerol transformations

Epichloridrin can be produced from glycerol in a process called epicerol (Solvay, 2011). This chemical is mostly used in the production of epoxy resins, as well as in water and paper treatment. The process involves the reaction of glycerol with two moles of HCl in the presence of Lewis acid catalysts, followed by controlled alkaline hydrolysis (Fig. 12.16). The traditional process of epichloridrin production starts with the chlorination of propene at elevated temperatures, forming allyl chloride, which is then reacted with hypochlorous acid to afford 1,3-dichloro-isopropanol. Treatment of this latter product with aqueous sodium hydroxide yields epichloridrin. The epicerol process of producing epichloridrin from glycerol has several advantages. The water consumption is 90% lower compared to the traditional process, as well as chlorinated waste materials, not to mention the use of a renewable feedstock.

Glycerol carbonate has gained increasing applications in recent years. It can be used as solvent and monomer for the production of polycarbonates, polyesters and polyamides. There are many routes for its production, such as the reaction of glycerol with cyclic organic carbonates (Vieville et al., 1998), and through the reaction of glycerol with urea (Hammond et al., 2011). This latter procedure has become popular in the last few years, but involves long reaction times (6–8 hours) and high temperatures (160°C). A simple procedure to produce glycerol carbonate involves the reaction of glycerol with $N,N'$-carbonyl-diimidazol (CDI) (Mota et al., 2007) (Fig. 12.17). The product can be obtained in quantitative yields in 15 minutes at room temperature. The use of crude glycerin, coming from biodiesel production, does not affect the yield, as the alkaline catalyst dissolved in the glycerin phase accelerates the reaction. The major drawback is the cost of CDI and its production route, which still employs phosgene ($COCl_2$), a highly toxic gas.

The direct carbonation of glycerol with $CO_2$ is a promising route for the production of glycerol carbonate, because it uses a greenhouse gas (Fig. 12.18). Organotin compounds (Aresta et al., 2006) have been employed as catalysts, but the yields are still modest, in the range of 5%, and the reaction time is over 12 hours. Metal-impregnated zeolites may be promising catalysts

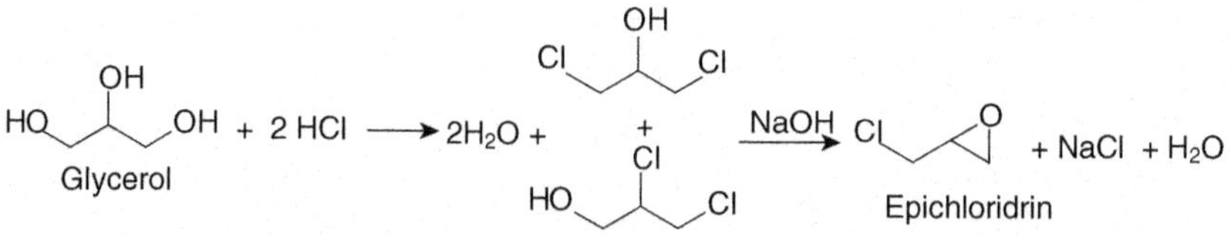

*12.16* The epicerol process: production of epichloridrin from glycerol.

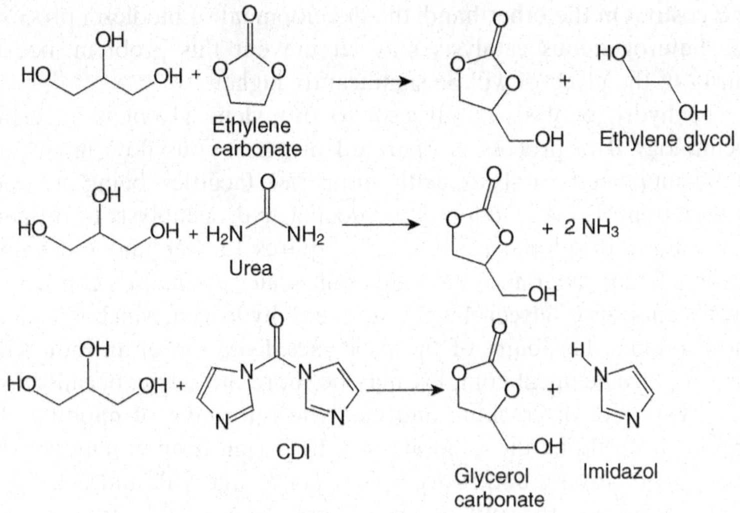

*12.17* Three different procedures for the synthesis of glycerol carbonate.

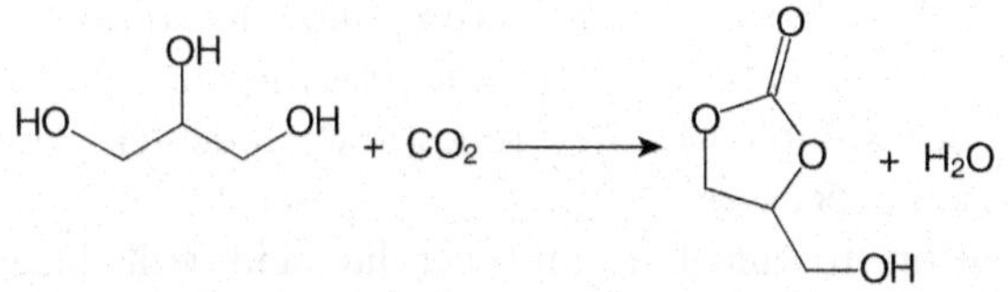

*12.18* Production of glycerol carbonate from $CO_2$.

for the direct carbonation of glycerol. Preliminary results indicate that glycerol carbonate can be produced in up to 5.8% yield within 2 hours of reaction time (Ozório, 2012). The use of metal-impregnated zeolites opens the possibility of using heterogeneous catalysts in this reaction, contributing to lowering the costs associated with catalyst recovery, which are always high when homogeneous catalysts are involved.

## 12.5 Conclusions and future trends

In the few years, glycerol will certainly occupy an important position as a renewable feedstock for the chemical industry. As the worldwide utilization of biodiesel increases, the surplus of bioglycerol will become attractive and new applications will emerge. Purity is still a major concern if chemical transformation is forecasted. The presence of salts dissolved in the glycerol phase of biodiesel production may affect the catalysts used for further chemical transformation. Purification procedures are still time-consuming and costly. On the other hand, the development of biodiesel processes that use heterogeneous catalysts may circumvent this problem, because the purity of the glycerol will be significantly higher.

The hydrogenolysis of glycerol to propylene glycol is an established technology. The process is operated in continuous flow mode and uses metal-supported catalysts, with industrial facilities being in operation. A variation of this reaction is the deeper hydrogenolysis to propene. This is a unique development, but still requires further major developments before being operated at industrial scale. A major concern in the hydrogenolysis of glycerol is the source of hydrogen, which normally limits the economic feasibility of the processes. Location of an industrial plant near a petrochemical complex may be more attractive, because hydrogen is a byproduct of cracking and catalytic reforming of naphtha. Another approach would be the generation of hydrogen from biomass gasification, but such processes are still not employed on a large scale and lack economical feasibility. Thus, the utilization of non-renewable hydrogen would be necessary for the rapid industrial implementation of glycerol hydrogenolysis to propanediols and propene.

The epicerol process, developed by Solvay, offers an opportunity to produce epichloridrin from glycerol. The company reports several advantages of the process in relation to the traditional one, based on propene. This is a good example of how a renewable feedstock process can compete with oil-based processes.

Glycerol dehydration to acrolein and acrylic acid still faces some challenges. Most of the acid catalysts tested deactivate during work conditions, making the development of a competitive industrial process still far away. The oxidative dehydration of glycerol to acrylic acid may be more interesting from a technical and economic point of view. Since the reaction is carried out in air flow, catalyst deactivation is usually a minor problem. However, selectivity is still a major concern. The best catalysts can produce up to 28% of acrylic acid, with acrolein being the main product formed. This may suggest that catalyst activity must be adjusted to perform the two reactions under the same experimental conditions. Apparently, the acidic function is working properly, but the oxidant function requires further improvement. Once the selectivity problem is resolved, the oxidative dehydration of glycerol to acrylic acid may economically compete with the traditional process, which involves two steps starting with propene. The possibility of developing a one-step process from glycerol would be a major achievement, reducing operational and capital costs.

Glycerol may also be used in the fuel sector. There are many studies for the biotechnological transformation of glycerol in ethanol, a major biofuel used worldwide. This achievement may integrate the biodiesel and the bioethanol industries. Another possibility would be the use of ethanol in the transesterification process, producing a completely renewable biodiesel. Glycerol can also be employed in the production of syngas, which can be transformed into hydrocarbons through the Fischer–Tropsch process yielding diesel, gasoline and kerosene.

Glycerol ethers, acetals/ketals and esters show potential of use as additives for gasoline, diesel and biodiesel. Solketal, produced in the reaction of glycerol with acetone, is an interesting additive for gasoline, improving the octane number and reducing gum formation. Glycerol acetals of aromatic aldehydes show antioxidant properties and may be useful in many sector, including fuels, plastics and food.

In summary, glycerol will be one of the most important renewable feedstock in the near future. Together with bioethanol, glycerol will be an important raw material for the production of plastics. Bioethanol is a good option for ethylene-based polymers, and glycerol will be the best option for propylene-based polymers. Some of the developments are already a reality, whereas others still require further developments. As fuel additives, glycerol derivatives can occupy increasing market share, especially in the biodiesel industry. Many biodiesel additives can be made from glycerol, integrating the whole industrial chain and draining huge amounts of the glycerol produced in transesterification.

## 12.6 Sources of further information

In recent years, there have been many reviews and books regarding the chemical transformation of glycerol. The reader may find additional information in the following published material.

Beatriz A, Araújo Y J K and de Lima D P (2011), 'Glycerol: a brief historic and their application in stereoselective syntheses', *Quim Nova*, 34, 306–319.

Behr A, Eilting J, Irawadi K, Leschinski J and Lindner F (2008), 'Improved utilization of renewable resources: new important derivatives of glycerol', *Green Chem*, 10, 13–30.

Huber G W, Iborra S and Corma A (2006), 'Synthesis of transportation fuels from biomass: chemistry, catalysts and engineering', *Chem Rev*, 106, 4044–4098.

Jérôme F, Pouilloux Y and Barrault J (2008), 'Rational design of solid catalysts for the selective use of glycerol as a natural organic building block', *Chem Sus Chem*, 1, 586–613.

Mota C J A, da Silva C X A and Gonçalves V L (2009), 'Glycerochemistry: new products and processes from glycerin of biodiesel production', *Quim Nova*, 32, 639–648.

Pagliaro M and Rossi M (2008), *The Future of Glycerol. New Usages for a Versatile Raw Material*, Cambridge, RSC.

Pagliaro M, Ciriminna R, Kimura H, Rossi M and Pina C D (2007), 'From glycerol to value-added products', *Angew Chem Int Ed*, 46, 4434–4440.

Zeng Y, Chen X and Shen Y (2008), 'Commodity chemicals derived from glycerol, an important biorefinery feedstock', *Chem Rev*, 108, 5253–5277.

Zhou C, Beltramini J N, Fan Y X and Lu G Q (2008), 'Chemoselective catalytic conversion of glycerol as a biorenewable source to valuable commodity chemicals', *Chem Soc Rev*, 37, 527–549.

## 12.7 References

ADM (2012), Evolution Chemicals. Available from: http://www.adm.com/en-US/products/evolution/Propylene-Glycol/Pages/default.aspx (accessed 1 April 2012).

Aresta M, Dibenedetto A, Nocito F and Patore C (2006), 'A study on the carboxylation of glycerol to glycerol carbonate with carbon dioxide: the role of the catalyst, solvent and reaction conditions', *J Molec Catal A*, 257, 149–152.

Bianchi C L, Canton P, Dimitratos N, Porta F and Prati L (2005), 'Selective oxidation of glycerol with oxygen using mono and bimetallic catalysts based on Au, Pd and Pt metals', *Catal Today*, 203, 102–103.

Braskem (2010), Braskem inaugurates green ethylene plant in Triunfo (RS), becoming the global biopolymer leader, São Paulo. Available from: http://www.braskem.com.br/site/portal_braskem/en/sala_de_imprensa/sala_de_imprensa_detalhes_10338.aspx (accessed 20 March 2012).

Carmona M, Valverde J and Pérez A (2008), 'Purification of glycerol/water solutions from biodiesel synthesis by ion exchange: sodium removal Part I', *J Chem Tech Biotechnol*, 84, 738–744.

Carmona M, Lech A, de Lucas A, Pérez A and Rodrigues J F (2009), 'Purification of glycerol/water solutions from biodiesel synthesis by ion exchange: sodium and chloride removal Part II', *J Chem Tech Biotech*, 84, 1130–1135.

Carrettin S, McMorn P, Johnston P, Griffin K, Kiely C J, Attard G A and Hutchings G J (2004), 'Oxidation of glycerol using supported gold catalysts', *Top Catal*, 27, 131–136.

Casale B and Gomez A M (1994), *Catalytic method of hydrogenating glycerol*. US patent application 5276181, 4 January.

Chai S H, Wang H P, Liang Y and Xu B Q (2007), 'Sustainable production of acrolein: investigation of solid acid–base catalysts for gas-phase dehydration of glycerol', *Green Chem*, 9, 1130–1136.

Chaminand J, Dajakovitch L, Gallezot P, Marion P, Pinel C and Rosier C (2004), 'Glycerol hydrogenolysis on heterogeneous catalysts', *Green Chem*, 6, 359–361.

Chitlaru E and Pick U (1991), 'Regulation of glycerol synthesis in response to osmotic changes in *Dunaliella*', *Plant Physiol*, 96, 50–60.

Ciriminna R, Palmisano G, Della Pina C, Rossi M and Pagliaro M (2006), 'One-pot electrocatalytic oxidation of glycerol to DHA', *Tetrahedron Lett*, 47, 6993–6995.

da Silva C X A and Mota C J A (2011), 'The influence of impurities on the acid-catalyzed reaction of glycerol with acetone', *Biomass Bioenergy*, 35, 3547–3551.

da Silva C X A, Gonçalves V L C and Mota C J A (2009), 'Water-tolerant zeolite catalyst for the acetalisation of glycerol', *Green Chem*, 11, 38–41.

Dasari M A, Kiatsimkul P P, Sutterlin W R and Suppes J (2005), 'Low-pressure hydrogenolysis of glycerol to propylene glycol', *Appl Catal A*, 281, 225–231.

Deleplanque J, Dubois J L, Devaux J F and Ueda W (2010), 'Production of acrolein and acrylic acid through dehydration and oxydehydration of glycerol with mixed oxide catalysts', *Catal Today*, 157, 351–358.

Emptage M, Haynie S L, Laffend L A, Pucci, J P and Whited G M (Du Pont) (2006), *Process for the biological production of 1,3-propanediol with high titer*. United States patent application US7067300.

Garcia R, Bessom M and Gallezot P (1995), 'Chemoselective catalytic-oxidation of glycerol with air on platinum metals', *Appl Catal A*, 127, 165–176.

Garcia E, Laca M, Perez E, Garrido A and Peinado J. (2008), 'New class of acetal derived from glycerin as a biodiesel fuelcomponent', *Energy Fuels*, 22, 4274–4280.

Gonçalves V L C, Pinto B P, Silva J C and Mota C J A (2008), 'Acetylation of glycerol catalyzed by different solid acids', *Catal Today*, 673, 133–135.

Gong L, Lu Y, Ding Y, Lin R, Li J, Dong W, Wang T and Chen W (2009), 'Solvent effect on selective dehydroxylation of glycerol to 1,3-propanediol over a Pt/WO$_3$/ZrO$_2$ catalyst', *Chin J Catal*, 30, 1189–1191.

Hadi M H, Shariati M and Afsharzadeh S (2008), 'Microalgal biotechnology: carotenoid and glycerol production by the green algae *Dunaliella* isolated from the Gave-Khooni Salt Marsh, Iran', *Biotechnol Bioprocess Eng*, 13, 540–544.

Hájek M and Skopal F (2010), 'Treatment of glycerol phase formed by biodiesel production', *Bioresour Technol*, 101, 3242–3245.

Hammond C, Lopez-Sanchez J A, Ab Rahim M H, Dimitratos N, Jenkins R L, Carley A F, He Q, Kiely C J, Knight D W and Hutchings G J (2011), 'Synthesis of glycerol carbonate from glycerol and urea with gold-based catalysts', *Dalton Trans*, 40, 3927–3937.

Hekmat D, Bauer R and Neff V (2007), 'Optimization of the microbial synthesis of dihydroxyacetone in a semi-continuous repeated-fed-batch process by *in situ* immobilization of gluconobacteroxydans', *Appl Chem Papers*, 42, 71–76.

Javani A, Hasheminejad M, Tahvildari K and Tabatabaei (2012), 'High quality potassium phosphate production through step-by-step glycerol purification: a strategy to economize biodiesel production', *Bioresour Technol*, 104, 788–790.

Jones D T and Woods D R (1986), 'Acetone-butanol fermentation revisited', *Microbiol Rev*, 50, 484–524.

Kampe P, Giebelder L, Smuelis D, Kunert J, Drochner A, Haass F, Adams A H, Ott J, Endres S, Shimanke G, Buhrmester T, Martin, M, Fuess H and Vogel H (2007), 'Heterogeneously catalysed partial oxidation of acrolein to acrylic acid – structure, function and dynamics of the V-Mo-W mixed oxides', *Phys Chem Chem Phys*, 9, 3577–3589.

Klepacova K, Mravec D and Bajus M. (2005), 'tert-Butylation of glycerol catalysed by ion-exchange resins', *Appl Catal A*, 294, 141–147.

Kongjao S, Damronglerd S and Hunsom M (2010), 'Purification of crude glycerol derived from waste used-oil methyl ester plant', *Korean J Chem Eng*, 27, 944–949.

Liao X, Zhu Y, Wang S G and Li Y (2009), 'Producing triacetylglycerol with glycerol by two steps: esterification and acetylation', *Fuel Proc Tech*, 90, 988–993.

Manosak R, Limpattayanate S and Hunsom M (2011), 'Sequential-refining of crude glycerol derived from waste used-oil methyl ester plant via a combined process of chemical and adsorption', *Fuel Process Technol*, 92, 92–99.

McMorn P, Roberts G and Hutchings, G (1999), 'Oxidation of glycerol with hydrogen peroxide using silicalite and aluminophosphate catalysts', *Catal Lett*, 63, 193–197.

Melero J A, van Grieken R, Morales G and Paniagua M (2007), 'Acidic mesoporous silica for the acetylation of glycerol: synthesis of bioadditives to petrol fuel', *Energy Fuel*, 21, 1782–1791.

Miyazawa T, Kusunoki Y, Kunimori K and Tomishige K (2006), 'Glycerol conversion in the aqueous solution under hydrogen over Ru/C plus an ion-exchange resin and its reaction mechanism', *J Catal*, 240, 213–221.

Molyneux P (2004), 'The use of the stable free radical diphenylpicrylhydrazyl (DPPH) for estimating antioxidant activity', *Songklanakarin J Sci Technol*, 26, 211–219.

Mota C J A, Gonçalves V L C and Rodrigues R C (UFRJ) (2007), *Process for the production of glycerin carbonate*. Brazilian patent application 0706121-8, 23 July.

Mota C J A, Gonçalves V L C, Gambetta R and Fadigas J C (Quattor Petrochemicals) (2009), *Preparation of heterogeneous catalysts used in selective hydrogenation of glycerin to propene, and a process for the selective hydrogenation of glycerin to propene.* WO patent application 155674 A2, 30 December.

Mota C J A, Silva C X A, Rosenbach N, Costa J and Silva F (2010), 'Glycerin derivatives as fuel additives: the addition of glycerol/acetone ketal (solketal) in gasolines', *Energy Fuels*, 24, 2733–2736.

Muscatine L (1967), 'Glycerol excretion by symbiotic algae from corals and tridacna and its control by the host', *Science*, 156, 516–519.

Ozório L P (2012), Production of glycerol carbonate from $CO_2$ and glycerol in the presence of metal-impregnated zeolite catalysts. Masters thesis, Federal University of Rio de Janeiro, Rio de Janeiro.

Pariente S, Tanchoux N and Fajula F (2009), 'Etherification of glycerol with ethanol over solid acid catalysts', *Green Chem*, 11, 1256–1261.

Pestana C F M (2009), Dehydration of glycerol with heterogeneous acid catalysts: formation of acrolein and acrylic acid. Masters thesis, Federal University of Rio de Janeiro, Rio de Janeiro.

Pinto B P (2009), Etherification of glycerol with alcohols catalyzed by solid acids. Masters thesis, Federal University of Rio de Janeiro, Rio de Janeiro.

Potthast R, Chung C and Mathur I (Johann Haltermann Ltd) (2010), *Purification of glycerin obtained as a bioproduct from the transesterification of triglycerides in the synthesis of biofuel*. United States patent application 12290728, 18 May.

Silva L N, Gonçalves V L C and Mota C J A (2010a), 'Acetylation of glycerol with acetic anhydride', *Catal Commun*, 11, 1036–1039.

Silva P H R, Gonçalves V L C and Mota, C J A (2010b), 'Glycerol acetals as antifreezing additives for biodiesel', *Bioresour Technol*, 101, 6225–6229.

Soares R R, Simonetti D A and Dumesic J A (2006), 'Glycerol as a source for fuels and chemicals by low-temperature catalytic processing', *Angew Chem Int Ed*, 45, 3982–3985.

Soares T A (2011), Glycerol-derived antioxidants for blending with biodiesel. Masters thesis, Federal University of Rio de Janeiro: Rio de Janeiro.

Solvay (2011), Epicerol – a biodiesel by-product as raw material for polyester resins. Available from: http://www.solvaychemicals.com/EN/Sustainability/Issues_Challenges/EPICEROL.aspx (accessed 14 March 2012).

Speers A, Young J and Reguera G (2012), 'Improving the economic viability of the biodiesel industry through wastewater treatment', Abstracts of Papers, 243rd ACS National Meeting and Exposition, San Diego, CA, 25 March–29 March.

Tsukuda E, Sato S, Takahashi R and Sodesawa T (2007), 'Production of acrolein from glycerol over silica-supported heteropoly acids', *Catal Commun*, 8, 1349–1353.

Ulgen A and Hoelderich W (2011), 'Conversion of glycerol to acrolein in the presence of $WO_3/TiO_2$ catalysts', *Appl Catal A*, 400, 34–38.

Van Gerpen J (2005), 'Biodiesel processing and production', *Fuel Process Technol*, 86, 1097–1107.

Vieville C, Yoo J W, Pelet S and Mouloungui Z (1998), 'Synthesis of glycerol carbonate by direct carbonatation of glycerol in supercritical $CO_2$ in the presence of zeolites and ion exchange resins', *Catal Lett*, 56, 245–247.

Wawrzetz A, Peng B, Hrabar A, Jentys A, Lemonidou A A and Lercher J A (2010), 'Towards understanding the bifunctional hydrodeoxygenation and aqueous phase reforming of glycerol', *J Catal*, 269, 411–420.

Wessendorf R (1995), 'Glycerinderivate als Kraftstoffkomponenten', *Erdoel & Kohle, Erdgas, Petrochemie*, 48, 138–143.

Yazdania S S and Gonzalez R (2008), 'Engineering *Escherichia coli* for the efficient conversion of glycerol to ethanol and co-products', *Metabolic Eng*, 10, 340–351.

Yu K O, Jung J, Ramzi A B, Kim W K, Park C and Han S O (2012), 'Improvement of ethanol yield from glycerol via conversion of pyruvate to ethanol in metabolically engineered *Saccharomyces cerevisiae*'. *Appl Biochem Biotechnol*, 166, 856–865.

# Part II
Biofuels and other added value products from biorefineries

# 13
# Improving the use of liquid biofuels in internal combustion engines

R. J. PEARSON and J. W. G. TURNER,
University of Bath, UK

DOI: 10.1533/9780857097385.2.389

**Abstract**: Liquid biofuels offer a range of attractive qualities, including the potential for increased energy independency and reduced greenhouse gas emissions. However, the market penetration and supply of biofuels may be constrained by a number of factors. In this chapter, the use of alcohols and biodiesel in internal combustion engines is discussed and the technology of flexible-fuel vehicles is described. It is shown how iso-stoichiometric ternary blends of gasoline, ethanol and methanol can serve as drop-in fuels for E85 flex-fuel vehicles. The chapter goes on to further explore ways in which biofuels can become part of a wider sustainable energy system based on carbon-neutral liquid fuels for the transport sector, enabling a gradual evolution of both vehicle and fuel technology.

**Key words**: alcohol, biodiesel, ethanol, flexible-fuel, fatty acid methyl ester (FAME), hydrated vegetable oil, hydrous ethanol, methanol, ternary blends.

## 13.1 Introduction

Liquid biofuels provide an opportunity for nations to increase energy independency or reduce greenhouse gas emissions by supplying energy-dense fuels which are miscible with petroleum gasoline and diesel. Both bioethanol and fatty acid methyl ester (FAME)-based biodiesel can be used in low concentration blends in vehicles with no modifications. Only minor changes in fuel system material specifications together with a low-cost alcohol sensor are necessary for Vehicle Operating compatibility with high concentrations of ethanol. The low level alcohol blends, particularly in the form of 10 vol% blends of ethanol in gasoline (E10), are already displacing significant quantities of gasoline in countries such as the United States. Ethanol provides beneficial properties, including high resistance to auto-ignition, which can be exploited at high-load operating conditions in modern downsized pressure-charged spark-ignition engines. Its main drawback is its low volumetric energy density which, in the absence of significant changes to the fuel taxation system, is likely to limit its market penetration in the form of high concentration blends such as E85 due to increased fuel consumption under light-load operating conditions. Biodiesel penetration, in the form of FAME, is limited by the reluctance of manufacturers to warranty their vehicles for use with this fuel above very low concentration levels (5–7 vol%). This reluctance to warranty vehicles for operation on high levels of biodiesel could lead to gradual increases in the level of ethanol in gasoline or hydrated vegetable oil in diesel in order to meet legislation which effectively mandates the use of renewable fuels.

Ultimately the availability of sustainable feedstocks constrains the supply of biofuels and this limits the level at which they are able to displace fossil fuels. As a route to going beyond the biomass supply limit, the concept of Sustainable Organic Fuels for Transport (SOFT) has been developed, based on the synthesis of methanol from recycled $CO_2$, water and renewable energy. In this way biofuels can be part of a wider energy system based on carbon-neutral liquid fuels for the transport sector, enabling a gradual evolution of both vehicle and fuel technology.

## 13.2 Competing fuels and energy carriers

The drive to reduce dependency on fossil fuels over recent years has focused attention on the use of alternative fuels for transport, particularly in the United States and Brazil, where the use of biofuels blended into fossil fuels has been adopted, in the latter case, since the mid-1970s. When they are made from feedstocks which satisfy appropriate sustainability criteria and do not give rise to the emission of significant levels of greenhouse gases in their cultivation, biofuels can help to alleviate concerns regarding energy security and climate change within the road transport sector which, globally, is over 90% dependent on oil (IEA, 2010). Through their miscibility with conventional gasoline and diesel fuels, bioethanol and biodiesel have been introduced into the fuel pool in significant quantities. The use of these alternative fuels has been possible without a quantum change in either the transport energy distribution infrastructure or the technology, and therefore cost, of the vehicles in which they are used. In the respect that they offer an *evolutionary* rather than a *revolutionary* transition, the adoption of liquid alternative fuels could serve as a more stable pathway than electrification or the use of molecular hydrogen in order to address issues of climate change security of energy supply for transport. Indeed, it is the opportunity for evolutionary transition which has led to the significant current presence of biofuels in the marketplace, as shown in Section 13.3.

### 13.2.1 On-board energy density and technology costs

Renewable liquid fuels provide, at low additional vehicle cost, the on-board energy storage levels required by vehicles for personal transport which are

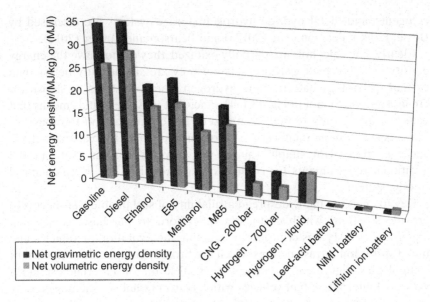

*13.1* Net system volumetric and gravimetric energy densities for various on-board energy carriers (based on lower heating values).

not dedicated solely to regular short-range routes. This is not the case for some other competing technologies. The very low net gravimetric and volumetric energy densities (including the mass/volume of the energy containment system) of current technology batteries are shown for lead-acid, nickel-metal hydride (NiMH) and lithium ion chemistries in Fig. 13.1. To match the range of a conventional gasoline-fuelled vehicle with a 50-litre fuel tank would require a useable battery capacity of approximately 100 kWh, accounting for the greater TTW efficiency of an electric vehicle. The mass of a fuel tank containing 50 litres of gasoline would be about 46 kg; that of a 100 kWh lithium ion battery would be in the range of 700–900 kg, depending on the technology and the permissible depth of discharge.

Hydrogen is also fundamentally limited as an energy carrier in that it is the least dense element in the periodic table. Figure 13.1, which includes the system package volumes and masses, shows that, while the net on-board energy density of hydrogen comfortably exceeds that of current batteries, it is still very low compared with liquid fuels. Because of the extreme physical conditions required to package hydrogen, the bulky system volume and high storage tank mass become high fractions of the net volumetric and gravimetric energy levels of the vehicle on-board energy storage system (Amaseder and Krainz, 2006; Eberle, 2006). Details of high-pressure, cryogenic, and metal hydride hydrogen storage systems are discussed by Bossel (2006), Pearson *et al.* (2012a) and Pearson and Turner (2012).

Clearly, since alcohols are partially oxidized, they do not have the energy density of liquid hydrocarbon fuels but they are significantly better than current technology batteries and hydrogen storage systems, as shown for ethanol and methanol in Fig. 13.1. The fact that they are liquids means that vehicles can be fully re-fuelled in two or three minutes and the shape of the tanks containing them can be more easily adapted to available vehicle package space with simple vapour recovery systems and without the additional requirements for cooling systems required by electrochemical storage systems.

Key to enabling the widespread availability of sustainable transport is to provide solutions which customers can afford to purchase. Renewable liquid fuels satisfy this criterion by enabling the evolution of vehicles and fuel distribution infrastructures which are broadly compatible with current technologies. Figure 13.2 compares the vehicle bill of material costs for a variety of alternative fuel vehicles with a conventional vehicle powered by an internal combustion engine (ICE). A fixed 'glider' (vehicle rolling chassis, including the body) cost is assumed for all options (Pearson *et al.*, 2012a). For the battery electric vehicle (BEV) options, a minimum state of charge (SoC) of 15% has been assumed; for the extended range electric vehicles

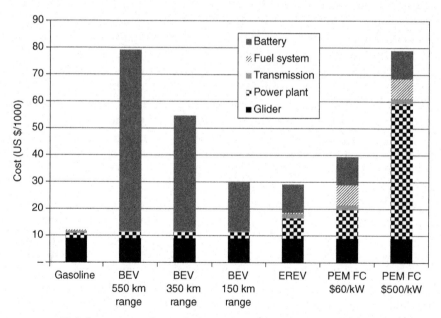

*13.2* Cost comparison of alternative energy vehicles. Assumed vehicle range (unless stated) = 550 km; battery cost = $750/kWh.

(EREV) and proton exchange membrane fuel cell (PEM FC) options, a minimum SoC of 35% has been assumed. The battery cost assumed for volume production levels is $750/kWh (slightly better than current prices). An EREV can be thought of as a plug-in hybrid electric vehicle (PHEV) with a significant electric-only range and with the engine taking the role of an electrical generator operating largely independently of vehicle speed. For EREVs, the strategy is often to try to size the battery so that a large portion of the distance travelled by the vehicle can be done in EV mode.

It is clear that, for a range-equivalent vehicle, the cost of the battery makes the BEV unaffordable to most customers. Reducing the vehicle range to 150 km from 550 km brings the costs down to a more accessible level, but this significantly range-compromised vehicle is still about 2.5 times more expensive than a conventional vehicle with a much higher utility level. This presents the customer with a very large negative price-performance differential. The EREV option, which enables lower storage capacity batteries to be used but requires both an electric motor and a fuel converter/generator (ICE assumed in this case), has a similar cost premium but is not encumbered by range compromise.

A significant portion of the high cost premium of a hydrogen-fuelled vehicle, whether using fuel cells or ICEs (the latter vehicle powertrain variant is not shown in Fig. 13.2), is the cost of the hydrogen storage system which, because the vehicle is also hybridized to manage the operating locus of the fuel cell in order to exploit its theoretically high efficiency levels, is additional to those of a 14 kWh battery (the same as that assumed for the EREV). It can be seen in Fig. 13.2 that even at the lowest fuel cell cost ($/kW), the bill of materials for hydrogen fuel cell electric vehicles is of such a level that widespread adoption is unlikely in the foreseeable future. This high vehicle price exacerbates the difficulty of justifying the expense of the necessary fuel-distribution infrastructure. Mintz *et al.* (2002) estimated the cost of providing a hydrogen infrastructure in the United States capable of re-fuelling 100 million fuel cell vehicles (40% of the US light duty vehicle fleet) as up to $650 \times 10^9$.

### 13.2.2 Environmental benefits

The debate regarding the environmental benefits of renewable fuels is complex and controversial. For a given feedstock and fuel, the specific production process, fertilizer used, transport and distribution, and, importantly, land-use change (direct and indirect) must be considered. In addition to well-to-wheel, or life cycle GHG emissions, parameters such as 'carbon pay-back' times have been calculated to quantify the time period over which a biofuel must be produced in order to offset the negative GHG impact of cultivating land which was formerly a carbon sink in the natural ecosystem. Whilst many studies have discussed these effects (Fargione et al., 2008; RFA, 2008; Searchinger et al., 2008; Bringezu et al., 2009; Zinoviev et al., 2010) they are not within the scope of the present work. However, the biofuels introduced into the EU as a result of the Renewable Energy Directive (EC, 2009a) are governed by sustainability criteria with a view to transport energy suppliers reducing the life cycle GHG emissions of their fuel by at least 6% by the end of 2020; there are also interim 2% and 4% targets to be met by the end of 2014 and 2017, respectively (EC, 2009c). Article 7b of the Fuel Quality Directive states that the GHG reductions for biofuels sold in the EU must be at least 35% (currently), rising to 50% in 2017, and 60% in 2018 for biofuels produced in installations in which production started after 1 January 2017 (EC, 2009c). Criteria for calculating these GHG benefits have been developed and default values for various fuels and pathways are defined in the Directive (EC, 2009c).

In addition to meeting supply targets discussed in Section 13.3.2, the US Renewable Fuel Standard (EPA, 2010a) also has requirements that fuels meet GHG emissions thresholds for compliance with each of four types of renewable fuel categories. California has its own initiative, the Low Carbon Fuel Standard (ARB, 2012), calling for a reduction in the carbon intensity of the transportation fuel pool used in the state of 10% by 2020. Figure 13.3 has been developed in order to indicate the well-to-wheel $CO_2$ emissions of vehicles with a range of tank-to-wheel $CO_2$ emissions as a function of the carbon intensity (g $CO_2$/MJ) of the fuel being used. The vertical lines in figures (a) and (b) represent ethanol at E20 and E85 levels, respectively, which meets the EU GHG reduction targets for biofuels mentioned above. It can be seen that using low-carbon-intensity renewable fuels at high concentration levels enables low well-to-wheel $CO_2$ emission levels to be achieved for a range of cars covering a range of operating efficiencies. These GHG emissions are similar to (E20), or substantially lower than (E85), the well-to-wheel emissions of an electric vehicle operating on electricity generated at the EU average carbon intensity (Pearson and Turner, 2012).

## 13.3 Market penetration of liquid biofuels

In the previous section it has been established that renewable liquid fuels are a potentially pragmatic route to de-carbonizing transport because they provide evolutionary transition mechanisms for both the vehicle technology and fuel distribution infrastructure. These fuels are already making a significant contribution to transport energy supply. Figure 13.4 shows that, globally, ethanol was the largest contributor to alternative road transport fuel in 2009 with consumption of 38.7 million tonnes of oil equivalent per annum (Mtoe/a) representing 29% of the alternative road transport energy supply but only 2.3% of the total global fuel consumption for this sector of

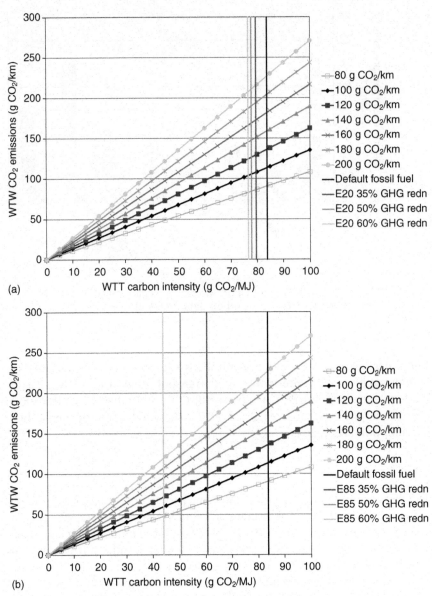

*13.3* Well-to-wheel $CO_2$ emissions (g $CO_2$/km) as a function of fuel well-to-tank carbon intensity (g $CO_2$/MJ) and vehicle tank-to-wheel $CO_2$ emissions. (a) using E20; (b) using E85. Carbon intensity of default fossil fuel is that given in EC (2009c) of 83.8 g $CO_2$/MJ. No TTW $CO_2$ benefits due to the use of ethanol are assumed.

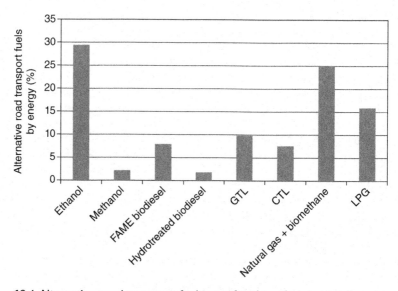

*13.4* Alternative road transport fuels as a fraction of global total alternative fuel supply. Based on data in IEA (2010).

1701 Mtoe/a (IEA, 2010). Only natural gas is close to ethanol in terms of energy supply as an alternative road transport fuel. The 13 Mtoe/a of biodiesel production gives it almost 10% of alternative road transport energy supply, but represents only 0.7% of total road transport fuel consumption (IEA, 2010).

### 13.3.1 Biodiesel

The European Union (EU) dominated the production of biodiesel in 2009 with 54% of production, followed by the United States and Brazil with 13% and 9%, respectively (IEA, 2010). About 75% of EU biodiesel is made from rapeseed feedstock, with 13% coming from soybean, and 8% from palm oil (Haer, 2011). China and India produced 2% and 1% of the world's biodiesel, respectively. The main market for biodiesel consumption is Europe, in particular Germany, which introduced the use of 'pure' biodiesel (B100) from rapeseed feedstock grown on fallow land in the 1990s. Tax relief on this fuel meant that initially the retail price was 5–15% higher than fossil diesel on an energy equivalent basis and some OEMs decreed that their existing light duty vehicles were compatible with B100 (Kramer and Anderson, 2012). These factors gave rise to 1,900 fuel stations offering biodiesel by 2006, with about 70% of sales being of B100. The imposition of taxes by the German government in 2008 in order to recoup lost revenue transformed the market to one where most biodiesel sales originated from blending with fossil diesel.

Simultaneously, vehicle manufacturers raised compatibility concerns regarding the use of B100 with modern common rail fuel injection systems and particulate filters and their regeneration strategies. Additionally, concerns regarding oil dilution and degradation, deposit formation, and materials compatibility limited the blend concentration of biodiesel for use in all vehicles in the EU to 7 vol% by volume (B7) (EC, 2010). Cooper (2011) has shown that the world cereal and grain production is far greater than that of vegetable oil. This ameliorates the threat of production of bioethanol on food prices relative to that of biodiesel. In this chapter the focus will henceforth be primarily on the production and use of alcohol fuels from biological and synthetic techniques, although some of the technology discussed regarding the onward synthesis of hydrocarbons has relevance to alternative diesel fuels.

### 13.3.2 Alcohol fuels

The potential of ethanol as a fuel in the internal combustion engine has been recognized for over 100 years (White, 1907) and its high octane index meant that it was initially used as a knock inhibitor in gasoline until tetraethyl lead came to dominate the market. The US produced 56% of the ethanol consumed by the road transport sector in 2009, with Brazil producing 33% and the EU 4% (IEA, 2010).

*Brazil*

Historically, Brazil has the world's most mature market for bioethanol in road transport fuels. Although ethanol-gasoline blending was taking place in Brazil on a significant scale in the 1930s, it was the OPEC oil crisis of 1973 which prompted the large-scale introduction of ethanol made from sugar cane as part of a national alcohol programme ('ProAlcool') in 1975 (Soccol *et al.*, 2005). The evolution of alcohol-fuelled vehicles in Brazil is well summarized by Kramer and Anderson (2012). The extent of the penetration of ethanol in the Brazilian fuel pool is such that it is not possible to purchase gasoline which does not contain bioethanol. The level of ethanol in gasoline (to form 'gasohol') is currently allowed to vary between 18 and 25 vol%, depending on the state of the global sugar market. Hydrous ethanol is also sold, consisting of at least 94.5 vol% ethanol, with the balance being a permitted mix of components consisting of water (mainly), hydrocarbons and other alcohols.

After initially blending ethanol with gasoline at around 20 vol%, dedicated E100 vehicles were introduced in 1979 and, by 1985, these vehicles represented 80% of light-duty vehicle production, assisted by ethanol prices which were sufficiently lower than the gasoline price to easily offset the volumetric energy density differential. Subsequent fluctuations in the global sugar market led to gasoline being cheaper than ethanol and the demand for dedicated E100 vehicles virtually disappeared by the mid-1990s. The introduction of flexible-fuel vehicles (FFVs), capable of using anything from gasohol to E100 (actually hydrous ethanol about E94 with 6 vol% water), around 2002 rejuvenated ethanol sales and enabled customers to exploit the price and tax advantages of E100 over gasoline whilst protecting them from the volatility of the sugar price by enabling operation on lower ethanol blends when desirable.

*United States*

In the US, the oil crisis of the 1970s and the drive to improve air quality in states such as California gave rise to an interest in methanol. Abundant availability of indigenous feedstocks (principally coal and natural gas) and low production costs drove the introduction of gaoline-M85 FFVs (the first 'modern' FFVs), capable of operating on any mixture ranging from conventional gasoline to a mixture of 85 vol% methanol/15% gasoline. A national M85 standard (ASTM D5797 – covering mixtures containing between 70 and 85% methanol in gasoline) was put in place. Political factors, combined with the drop in the oil price and a methanol shortage brought about by the sale of methanol stocks to make methyl-tert-butyl ether (MTBE) for use as an oxygenate in reformulated gasoline, required by the Clean Air Act Amendments (EPA, 2012), led to the reduction of interest in methanol in the late 1990s and the rise of bioethanol production. The use of reformulated gasoline is mandated in some areas of the US in order to reduce emissions of unburned hydrocarbons from older vehicles and to help reduce smog (ground-level ozone) formation. MTBE had been used in gasoline at low levels since 1979 as an octane enhancer when tetra-ethyl lead was banned. MTBE itself was subsequently banned in many states and its role as an oxygenate component in gasoline has now been completely replaced by ethanol.

The role of ethanol as an oxygenate source in reformulated gasoline began to rise in the early 2000s when the use of MTBE as an oxygenate additive was phased out, as shown in Fig. 13.5(a). Ethanol consumption has risen strongly since 2005 due to the federal policies which encouraged its use, reaching over 7.3 billion gallons gasoline-equivalent (gge) in 2009 and representing over 90% of US *alternative* fuel energy demand in 2009. The rapid growth in consumption has been matched by the rapid growth in production, as shown in Fig. 13.6 which reveals that the US is now by far the largest producer of ethanol, its output having risen to 62% of the total

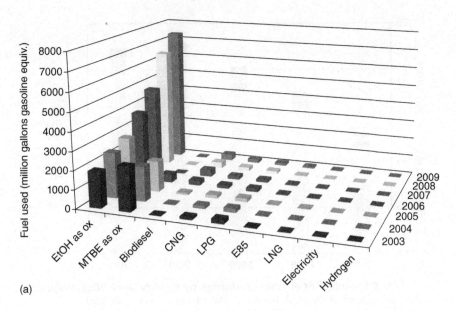

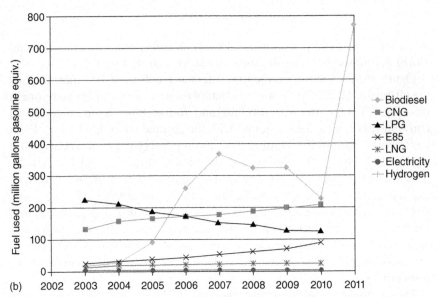

*13.5* Alternative fuel consumption (normalized to gasoline-equivalent US gallons where 1 US gallon = 3.785 litres) in the US road transport sector (a) 2003–2009 and (b) 2003–2010/11. Based on data in Davis et al., (2011) and US DoE (2012a). ('EtOH as ox' and 'MTBE as ox' indicate the use of ethanol and MTBE as oxygenates in gasoline fuel.).

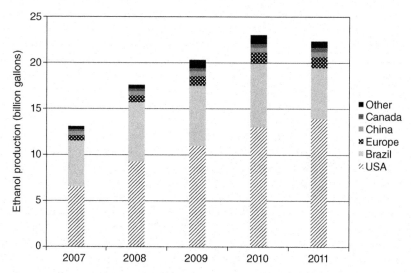

*13.6* Evolution of ethanol production by country since 2007. Volume in gallons of ethanol. Based on data from (US DoE, 2012a).

of 22.4 billion gallons produced globally in 2011. Together, the US and Brazil accounted for 87% of global production in that year.

Figure 13.5(b) focuses on the lower demand fuels in the US market; they are all dwarfed by the quantities of ethanol used as an oxygenate component (see Fig. 13.5(a)). The rise, decline, and recent huge jump in biodiesel consumption can be seen, together with the gradual fall of liquid petroleum gas (LPG) consumption. Natural gas consumption (compressed (C)NG and liquefied (L)NG in the figure) has shown a steady rise, and it might be postulated that the current low cost of natural gas resulting from the exploitation of shale gas reserves could lead to further significant increases. An alternative scenario is the conversion of natural gas to methanol for transport use (Turner *et al.*, 2012a). It is also clear that electricity and hydrogen have no significant demand up to 2009, comprising 0.06% and less than 0.002% of US alternative fuel energy demand in 2009, respectively.

The original Renewable Fuels Standard (RFS) (EPA, 2007) was created under the Energy Policy Act of 2005 and required 7.5 billion US gallons of renewable fuel to be blended into gasoline by 2012. As a result of the Energy Independence and Security Act of 2007 (US Congress, 2007) the Renewable Fuel Standard was expanded (RFS2) (EPA, 2010a) to include diesel as well as gasoline, mandated an increase of renewable fuel from 9 billion gallons in 2008 to 36 billion gallons in 2022, and established new categories of renewable fuels, setting separate targets for volume and

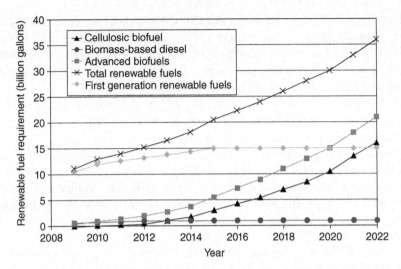

*13.7* Renewable fuel volume requirements for RFS2 (based on data in EPA, 2010a). Note: any fuel which meets the requirement for cellulosic biofuel or biomass-based diesel is also valid for meeting the advanced biofuel requirement.

greenhouse gas (GHG) reduction for each fuel. Figure 13.7 shows the ramp up in fuel volumes required to meet the RSF2 stipulation. The target for 'biomass-based diesel' is a minimum of 1 billion gallons from 2012 to 2022 (with the exact target to be set by future rulemaking) whilst that for 'cellulosic biofuel' was set to rise from 0.5 billion gallons in 2012 to 16 billion gallons in 2022, most of which is expected to be in the form of cellulosic ethanol (EPA, 2010a). In total, 21 billion gallons of 'advanced biofuels' (including cellulosic biofuels) are required in 2022, leaving 15 billion gallons to be supplied by first generation fuels, mostly in the form of corn ethanol. Many problems have been encountered meeting the targets for cellulosic biofuels and the 2012 target has recently been revised downward to 8.65 *million* gallons – this will represent less than 0.006% of US fuel usage that year (more precisely, the figure of 0.006% represents cellulosic biofuels as a fraction of non-renewable gasoline and diesel use), compared with 9.23% overall for renewable fuels, comprised mostly of 'corn ethanol' (EPA, 2011a).

Anderson *et al.* (2012a) provide an excellent account of the introduction of ethanol into the gasoline fuel pool in the US, including details of the changes in the gasoline blend stock, namely the blend stock for oxygenate blending (BOB), into which the ethanol is blended. Between 2000 and 2010, US ethanol consumption grew from 1.6 billion gallons/year to 13 billion gallons/year; the latter figure represents a hypothetical nationwide uniform ethanol-gasoline blend level of almost 10 vol% (Anderson *et al.*, 2012a). In fact, virtually all the ethanol used in US transport is used in the form of E10 (a mixture of 10 vol% of ethanol in 'gasoline') which has been available in the US since the 1980s but is now widespread due to the political and environmental developments described above.

A 'blend wall' has arisen due to all the low-level ethanol-gasoline blends, often referred to as 'gasohol', being close to the maximum ethanol content (10 vol%) which vehicle manufacturers will allow for a user to remain within their warranty provision. In order to address this issue, the EPA has granted two partial waivers (EPA, 2010b, 2011b) which allow, but do not mandate, the introduction of E15 for use in light-duty vehicles of model year 2001 and later. The approval process is inherently slow since it involves extensive testing and is open to challenge by vehicle manufacturers.

Anderson *et al.* (2012) examine various scenarios for ethanol introduction, including the most optimistic, where the RFS2 targets are met. This latter scenario would lead to notional uniform ethanol blend levels of E24 by 2022 and of E29 by 2035 (assuming a 2% annual growth post-2022). In the absence of significant growth in E85 sales, this could not be implemented with new waiver approvals since these levels of ethanol are well beyond the tolerance capability level of current conventional vehicles. New vehicle and engine specifications would be required, together with the maintenance of a protection grade fuel (E10 or E15) for existing vehicles. A discussion of octane targets for these blends is also included.

Flexible-fuel (or flex-fuel) vehicles (FFVs) are capable of using ethanol in concentration levels of up to 85 vol% (E85). However, despite the registration of over 9 million such cars at the end of 2011, incentivized by the rating of FFVs under CAFE legislation (EPA, 2010c), representing 4% of the US LDV fleet (Anderson *et al.*, 2012a), only 1% of the total ethanol use has been in the form of E85 sales, as shown in Fig. 13.5(a) and (b). This has led to a reappraisal of the Alternative Motor Fuels Act (AMFA) credits given to FFVs in future CAFE and EPA fuel economy and emissions legislation (EPA, 2011c). The fuel economy rating of an FFV is calculated by taking the harmonic mean of the fuel economy on gasoline (or diesel) and the fuel economy of the alternative fuel divided by 0.15. Thus, a vehicle which achieves 25 miles/US gallon using gasoline and 15 miles/US gallon using E85 would be rated having a fuel economy of $(2/((1/25) + (0.15/15)))$ = 40 miles/US gallon. This assumes that the FFV operates on the alternative fuel 50% of the time. US Congress has extended the FFV incentive (called the 'dual-fuelled' vehicle incentive) to model year (MY) 2019 but has provided for its phase-out between MY 2015 and MY 2019 by gradually reducing the allowed limit of the maximum fleet fuel economy increase for a manufacturer due to this credit from 1.2 miles/gallon. (MYs 1993–2014) to 0 miles/gallon after MY 2019 in 0.2 miles/gallon decrements over that period (EPA, 2011c). FFV credits for MY 2020 and beyond will reflect the 'real-world' percentage of usage of the alternative fuels (EPA, 2011c).

Ethanol in high-concentration form, E85 (now defined by ASTM D5798 as 51–83 vol% ethanol in gasoline), has suffered both from limited availability and uncompetitive pricing on an energy basis in the US. There are fewer E85 pumps in the US than there are EV charging stations (about 2,500 versus about 6,750 in 2012). The requirement for FFVs to run on any concentration of ethanol in gasoline, from 0% to 85%, also means that the vehicles are generally not configured to be capable of exploiting the high octane numbers of the higher level ethanol blends – this prevents such vehicles being able to offset the reduction in volumetric energy content of the fuel by increasing the thermal efficiency of the engine. Conversely, if FFVs were to be optimized to have high compression ratios in order to reduce their volumetric fuel consumption on E85 (as discussed in Section 13.4.2), they would give an unattractive increase in fuel consumption when operating on fuel with lower ethanol concentration such as E10 if the octane level of the BOB is maintained (Anderson *et al.*, 2012a). Although this may be viewed as a mechanism to incentivize the FFV customer to use E85, the proposition is only reasonable if there are sufficient fuel stations offering the fuel.

*European Union*

Whilst the European gasoline specification EN228 allows up to 3% methanol in gasoline, there has never been any specification for high concentration of methanol in Europe analogous to the ASTM 5797 standard. In the EU, by the end of 2020, the Renewable Energy Directive (RED) and the Fuel Quality Directive (FQD) (EC, 2009a, 2009c) together require that 10% of transport energy be supplied in renewable form and that the overall GHG intensity of fuels should be reduced by 6%. With diesel penetration at approximately 50% across the EU, it is possible that, due to lack of sufficient supplies of sustainable vegetable oils for biodiesel manufacture and some issues of achieving emission compliance of modern vehicles using more than 7 vol% of biodiesel, the RED and FQD targets may need to be met by supplying base fuel in the form of E20, when the lower volumetric energy density of ethanol is considered (Cooper, 2011). However, recent attempts to introduce E10 into the German market did not go well due to some customer confusion (Kramer and Anderson, 2012).

In contrast to the CAFE regulations which make it attractive to manufacture FFVs in the US, the fiscal penalties for GHG emissions from cars sold in the EU is based only on tailpipe $CO_2$ emissions. This gives vehicle manufacturers no incentive to spend even the small extra amount required in order to produce an FFV (ca. €100/vehicle). It has recently been suggested by a major transport energy supplier (Cooper, 2011) that attributing some $CO_2$ benefit to manufacturers will provide a more compelling reason for OEMs to make FFVs and thus produce a greater outlet for ethanol as an automotive fuel.

Despite the lack of apparent incentive, manufacturers such as Saab, Ford, Volvo, Renault, and VW have introduced FFVs into their vehicle range. The EU vehicle tailpipe $CO_2$ penalty system does, however, presently allow a 5% reduction in tailpipe $CO_2$ to be claimed for any FFV that an OEM sells, provided one-third of the fuel stations in the country in which the vehicle is sold has at least one E85 refuelling pump (EC, 2009b). In Sweden there has been co-ordinated activity to install E85 pumps so that by 2009 50% of the network was covered (Bergström et al., 2007a), rising to 59% in 2011 (Kramer and Anderson, 2012), and in 2008, 22% of all new car sales were FFVs (Kramer and Anderson, 2012), driven by government fuel tax relief, which made E85 cheaper than gasoline on an equal energy basis, and vehicle use initiatives. Kramer and Anderson (2012) show that this fuel tax benefit has been variable since 2005. They also show that the drop in FFV sales which occurred in 2009 coincided with a period when E85 had a cost disadvantage of 30% relative to gasoline.

The benefits of liquid fuels compared with their gaseous counterparts are again highlighted by the 2006 Swedish legislation requiring that all fuel stations above a certain size offer at least one alternative fuel: most stations covered by the law installed E85 pumps and storage tanks which, at €40,000–45,000, offered a ten-fold lower installation cost compared with biogas storage and dispensing equipment (Kramer and Anderson, 2012).

In the rest of Europe, legislation and incentives for high concentration ethanol use are largely absent and thus fuel pump availability for E85 and FFV sales remain sparse. For Germany in 2010 FFVs represented 0.05% of the 2.9 million new vehicle sales (Kramer and Anderson, 2012). It is, however, perhaps worth noting that for a vehicle at the 2011 EU average of $135.7\,\mathrm{g\,CO_2/km}$, and at the highest proposed fine rate in 2015 of €95/$(gCO_2/km)$, this represents a saving to the manufacturer of €541 per car (with the benefit limited to $5.7\,\mathrm{g\,CO_2/km}$ in this instance since the target would be achieved), which the authors contend is significantly greater than the additional costs of producing a vehicle which is flex-fuel capable E85 (Turner et al., 2012a).

*China*

Ethanol blends in gasoline have been used in five Chinese provinces since 2004; however, the use of methanol is favoured in order to avoid conflicts with food demand. Whilst China produced only 3% of the ethanol used in road transport globally in 2009 (US DoE, 2012a), it has dominated the production and use of methanol in this sector. The consumption of methanol in China, mainly in the form of M15, is around 3 million tonnes per annum (Niu and Shi, 2011) and is motivated by China's large coal reserve which offers the potential of greater energy independence. Processes to convert coal to ethanol are also being investigated (Pang, 2011).

## 13.4 Use of liquid biofuels in internal combustion engines

### 13.4.1 Biodiesel

Whilst alcohol fuels present the fuel blender, additive supplier, and vehicle manufacturer with a tightly defined blend component having consistent properties, the chemical composition of biodiesel formed by transesterification of seed oils or animal fats to form fatty acid methyl esters (FAMEs) is dependent on the original feedstock source and the esterification process. This results in a wide variation in FAME compostition.

The cetane number, which measures the auto-ignitability of fuels for compression-ignition engines, varies over wide ranges for FAME components from the same feedstock origin. The cetane number of FAME is dependent on the distribution of fatty acids in the original oil or fat from which it was produced. Higher cetane numbers are caused by higher saturation levels in the fatty acid molecules and longer carbon chains (Geller and Goodrum, 2004). Bamgboyne and Hansen (2008) report cetane numbers for biodiesel fuels from a wide range of feedstocks. The cetane number of biodiesel derived from soyabean oil (soyabean methyl ester, or SME) has been found to vary between 45 (lower auto-ignitability) and 60 (higher auto-ignitability), whilst that of rapeseed methyl ester (RME) can vary between 48 and 61.2. Palm oil methyl ester (POME) has been measured to have cetane numbers between 59 and 70. This compares with a typical cetane number of a premium European diesel of around 60. The minimum cetane number required by conventional diesel fuel specifications in the EU (EN 590) and the US (ASTM D975) is 51 and 40, respectively.

The fact that FAME molecules are esters and are therefore oxygenated means that they will have a lower gravimetric energy density than petroleum diesel fuels. However, the large size of the molecules (in the range $C_{12}$ to $C_{22}$) means that the impact of the two oxygen atoms which comprise the ester functional group is much lower than that of the oxygen atom contained within the methanol ($C_1$) or ethanol ($C_2$) molecules. The degree of this energetic deficit is also ameliorated by the fact that the density of FAME is higher than that of petroleum diesel due to increased chain length and the presence of carbon–carbon double bonds. Thus, whilst the gravimetric energy density of FAME can be perhaps 12 per cent lower than a mineral diesel such as US No. 2 diesel fuel, its increased density can reduce the difference in volumetric energy content to about 7%. In B10 blends the impact on fuel economy of FAME has been found to be less than 1% (Gardiner *et al.*, 2011).

The wide variations in the FAME composition and its consequent variable interaction with the base diesel in a blend can have markedly different effects on low temperature vehicle operability, with the fuel pour point and cold filter plugging point changing significantly (Saito *et al.*, 2008). The fuel's oxidation stability (McCormick *et al.*, 2006; Miyata *et al.*, 2004), its compatibility with the vehicle fuel injection equipment, its propensity to form deposits (Caprotti *et al.*, 2007), and the effects of fuel dilution on the engine lubricant (Thornton, 2009) also vary significantly with the composition of the FAME. Bespoke additives are required for specific blend compositions, making the task of ensuring fuel compliance with the vehicle fleet a complex task. The issues are well summarized by Richards *et al.* (2007). In contrast, hydrogenated vegetable oils (HVOs) and biomass to liquid (BTL) fuels have compositions which are much closer to their petroleum diesel counterparts; however, both types of fuel are significantly more energetically intensive to produce than FAME-based fuels and therefore more expensive. The ability of FAME-based biodiesel to lubricate fuel pumps and fuel injectors (its 'lubricity') is superior to that of petroleum diesel; however, the lubricity of HVO is lower, requiring lubricity additives.

As with all fuels, adherence to rigorous quality standards is necessary for increased penetration of biodiesel into the market. Standards exist in the EU and US for biodiesel quality. The EU standard, EN 14214, controls the quality of FAME used either as a fuel itself or as a blending component in diesel fuel. This biodiesel standard specifies the minimum ester content (96.5 mass%) and maximum methanol (used in the production process), glyceride, and glycerol content. Density and viscosity ranges are specified, and the minimum cetane number, set at 51, is identical to that of petroleum diesel fuel (a discussion of the relevance of the specification parameters is given by Ferrari *et al.*, 2011).

The ASTM D975 standard for conventional diesel fuel allows biodiesel concentration of up to 5 vol%. Such blends are approved for safe operation in any compression-ignition engine designed to be operated on petroleum diesel (US DOE, 2012b). Blends of up to 20% biodiesel in 80% petroleum diesel (B20), controlled by ASTM D7467 (B6–B20), are the largest outlet for biodiesel in the United States and provide a good compromise in balancing cost, emissions, cold-weather performance, and materials compatibility. Operation on B20 and lower-level blends does not in principle require engine modification. However, not all diesel engine manufacturers warrant their products for use with such blends. ASTM D6751 regulates the specification of B100 for use as both a fuel and a blending component in petroleum diesel.

Whilst there can be significant emissions benefits from the use of biodiesel, as discussed by Gardiner *et al.* (2011), it seems that, in the form of FAME, there is an industry-wide desire to limit its use to relatively low-level blends. The quality of conventional diesel fuel in the EU is controlled by the EN 590 standard which currently limits FAME content to 5 vol% (B5). An extension of the FAME limit to 7% (B7) is proposed, with no limit to levels of HVO or 'diesel-like hydrocarbons made from biomass using the Fischer–Tropsch process' (EC, 2009c). The varying properties of FAME as a function of its feedstock gives rise to conservatism on the part of the vehicle manufacturers to warrant their vehicles for use with higher blends and may be an obstacle to achieving compliance with the RED and FQD, requiring either additional bio-content to come from HVO or BTL fuels or from a disproportionate increase in the use of ethanol in gasoline blends.

### 13.4.2 Alcohol fuels for spark-ignition engines

The principal alcohols which have been used or are being considered for use as either a component in fuel blends with gasoline or as the dominant fuel blend constituent themselves are methanol, ethanol and butanol. Methanol ($CH_3OH$) and ethanol ($C_2H_5OH$) have one and two carbon atoms respectively, and no isomers, whilst butanol ($C_4H_9OH$) has four carbon atoms and four structural isomers. The higher-order alcohols, that is, those with more than two carbon atoms (primarily propanol, butanol and pentanol (amyl alcohol)), formed in the fermentation process are sometimes known as fusel alcohols.

As the number of carbon atoms increases, the influence of the functional hydroxyl (OH) group on the physico-chemical properties of the molecule diminishes. The presence of the OH group in place of one of the hydrogen atoms of an alkane induces significant polarity in the molecule due to the two lone pairs of electrons present on the oxygen atom (Pearson and Turner, 2012). The concentration of negative charge around the oxygen atom produces a net positive charge on the rest of the alcohol molecule, which, in particular, is focused around the hydrogen atom attached to the oxygen atom of the hydroxyl group. This *intra-molecular* polarity generates strong *inter-molecular* forces, known as hydrogen bonds. These forces give the low-carbon-number alcohols higher boiling points and enthalpies of vaporization than would be expected for non-polar compounds of similar molecular mass. As in FAME-based biodiesel, however, the presence of the oxygen atom reduces the energy density of the fuel. Its impact is more pronounced the smaller the molecule, so that methanol has less than half

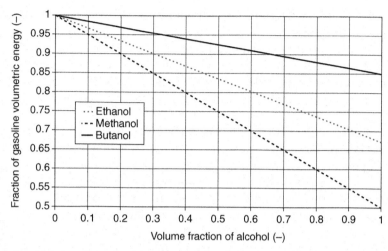

*13.8* Volumetric energy density variation of alcohol-gasoline blends.

the gravimetric energy density of gasoline (generally a mixture of $C_4$ to $C_8$ hydrocarbons) and about 40% of that of methane, the corresponding $C_1$ alkane. More detail on the properties of alcohols as fuels can be found in Pearson and Turner (2012).

*Volumetric energy density and stoichiometry*

The impact of alcohol concentration on the volumetric energy density of gasoline blends with ethanol, methanol, and butanol is shown in Fig. 13.8. For alcohol concentration of 10 vol% in gasoline, the reduction in volumetric energy density relative to the gasoline is 1.5%, 3.3% and 5%, for butanol, ethanol and methanol, respectively, whereas for 85% alcohol concentration the respective reductions in energy density are 12.8%, 28% and 42.5% for these alcohols. With vehicles which use engines not optimized for use with alcohol fuels (which is the current situation, even with FFVs), the consequence of this reduced volumetric energy content is an approximately proportional increase in volumetric fuel consumption for operating cycles which are not sufficiently aggressive to exploit the higher octane numbers of the resulting fuel blends. A consequence of this is that it is difficult to achieve significant market penetration of fuels with high alcohol concentrations with a volumetric-based fuel taxation system, rather than one based on the energetic content of the fuel, or the non-renewable carbon component (Turner and Pearson, 2008).

As a corollary to the presence of the oxygen atom in alcohol fuels having an impact on reducing the volumetric energy content which reduces as the

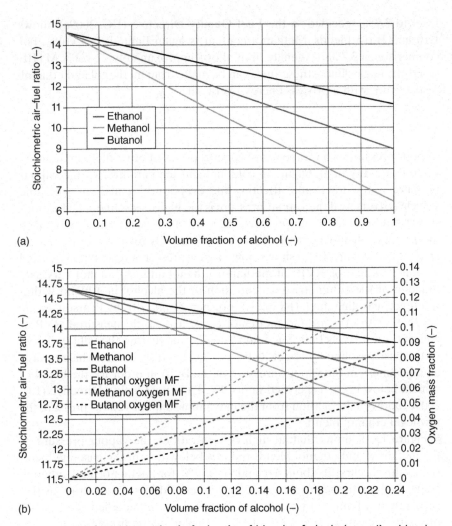

*13.9* Stoichiometric air–fuel ratio of blends of alcohol-gasoline blends.

length of the carbon chain increases, Fig. 13.9 shows how the stoichiometric air–fuel ratios of gasoline-butanol mixtures are closer to that of gasoline than those of gasoline-ethanol and gasoline-methanol mixtures. This is one of the reasons that gasoline standards have oxygenate limits. The impact of the alcohol concentration on the oxygen mass fraction is shown in Fig. 13.9. The European EN228 standard for gasoline currently limits the mass fraction of oxygenates in the fuel to 2.7% – this limit corresponds to about 7 vol% of ethanol in the blend (E7) and 11 vol% of butanol (Bu16). Methanol is limited as a specific component to 3 vol%. The increase in this level to 3.7% specified in the Fuel Quality Directive (EC, 2009c) allows E10 and Bu15 blends. Methanol content is again limited to 3 vol%. For comparison, a 3.7% oxygenate limit allows approximately 22 vol% MTBE or ETBE in gasoline – these fuels can be made from methanol and ethanol, respectively, with a process energy overhead.

*Vapour pressure*

Vapour pressure is an important property of automotive fuels, influencing the start quality of an engine in cold ambient temperatures (if the vapour pressure is too low), and affecting the evaporative emissions from the vehicle (adversely if the vapour pressure is too high). Gasoline specifications stipulate allowable vapour pressures dependent on the season, geographical location and alcohol content – these are extensively reviewed by Andersen *et al.* (2010). Figure 13.10 shows how vapour pressures (calculated as Reid vapour pressure, or RVP) of methanol, ethanol and iso-butanol vary as a function of the volumetric concentration of the alcohol in a blend with a reference gasoline fuel (RF-02-03 (Turner *et al.*, 2012b; Pearson *et al.*, 2012b)). Both ethanol and methanol cause a peak in measured RVP at concentration levels between 5 and 10 vol%, increasing the value relative to the base gasoline by about 8 kPa and 23 kPa, respectively. This requires an adjustment in the vapour pressure of the BOB in order for the blend to remain below normal specified limits. The EU Fuel Quality Directive (EC, 2009c) allows a derogation from the maximum summer vapour pressure for low-level ethanol blends.

Figure 13.10 shows that, at high concentrations, the vapour pressure of the methanol and ethanol blends reduces below that of the gasoline blend stock: this occurs at about 80 and 45 vol% for methanol and ethanol, respectively, with the gasoline used in this case. At high concentrations this can cause low temperature cold-start difficulties in unmodified engines, but this issue is easily overcome in FFVs, particularly with direct fuel injection technology where injection into the hot compressed air is found close to top-dead-centre. Siewart and Groff (1987), Kapus *et al.* (2007), Marriott *et al.* (2008) and Hadler *et al.* (2011) discuss the successful use of fuel injection strategies to augment the quality of the engine start. For port-fuel-injected engines, measures such as heating the fuel rail can enable acceptable cold-start performance down to −25°C. Bergström *et al.* (2007a) report acceptable cold starts down to −25°C in the absence of additional technology on an engine with port-injection (PI) of fuel using a Swedish winter grade bioethanol (E75 with a Reid vapour pressure of 50). In the US, the ASTM D5798 standard for E85 can allow the ethanol concentration to be as low as 51% by volume ethanol – this provides a significantly easier blend for low-temperature starting.

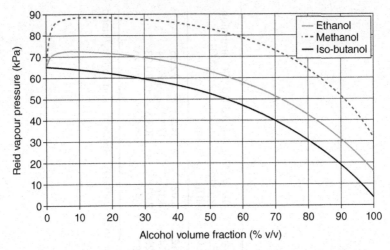

*13.10* Calculated variation of vapour pressure of methanol, ethanol and iso-butanol with alcohol volume fraction. For validation data, see Turner *et al.* (2012b) and Pearson *et al.* (2012b). The base gasoline has an RVP of 65 kPa. Courtesy of M. Davy, Loughborough University.

*Octane numbers*

Table 13.1 shows the low-carbon number alcohols have a number of physico-chemical characteristics which are synergistic with modern engine designs, and in particular highly pressure-charge downsized engines. In particular, the high research octane numbers (RONs) of methanol, ethanol and iso-butanol are facilitators for improved combustion phasing and lower component protection over-fuelling (to control exhaust gas temperatures). The motor octane numbers (MONs) of these fuels are proportionally less high and thus their sensitivity (RON-MON) is very high relative to gasoline; nevertheless, these fuels would not be considered as having a low MON relative to most gasoline sold throughout the world.

Moran and Taylor (1995) posited that, because the intake manifold temperature is not controlled during the RON test, the high heat of vaporization of the low-carbon number alcohols would cause a physical cooling effect which would have a direct bearing on the knock-limited compression ratio. In contrast, the MON test uses a heater to try to control the intake manifold temperature to 149°C, which is well above the respective boiling points of these alcohols and is thus not affected by this physical cooling effect.

Stein *et al.* (2012) showed that both denatured ethanol (97% ethanol, 1% water, 2% gasoline) and hydrous ethanol (94.2% ethanol, 5.8% water) are extremely knock resistant, allowing very high engine loads to be achieved at low speeds in an engine with a 14.0:1 compression ratio.

Table 13.1 Properties of 95 RON gasoline, methanol, ethanol, and iso-butanol

| Fuel | QLHV [MJ/(kg fuel)] | $h_{fg}$ [kJ/(kg fuel)] | Density at NTP [kg/l] | RON/MON | Boiling point at 1 bar [deg. C] | Stoich. AFR (mass) [-] | Stoich. AFR (mole) [-] | QLHV [MJ/(kg stoich. air)] | $h_{fg}$ [kJ/(kg stoich. air)] | $CO_2$ emission [g/MJ] |
|---|---|---|---|---|---|---|---|---|---|---|
| Gasoline 95 | 43.2 | 350 | 0.74 | 95.3/85.9 | 37–167 | 14.31 | 47.12 | 3.02 | 24.5 | 72.85 |
| Methanol | 20.09 | 1100 | 0.79 | 108.7/88.6 | 65 | 6.44 | 7.14 | 3.12 | 170.9 | 68.44 |
| Ethanol | 26.95 | 925 | 0.79 | 108.6/89.7 | 79 | 8.96 | 14.29 | 3.01 | 103.3 | 70.98 |
| Iso-butanol | 33.08 | 585 | 0.81 | 106.3/90.4 | 118 | 11.14 | 28.57 | 2.97 | 52.5 | 71.90 |

Note that the properties of '95 RON gasoline' vary considerably depending on local refinery streams and other blend components.

*Performance*

Nakata *et al.* (2006) used a high compression ratio (13:1) naturally aspirated port-fuel injected spark-ignition engine and found that torque increased by 5% and 20% using E100 compared with the operation on 100 RON and 92 RON gasoline, respectively. The full improvements in torque due to being able to run MBT ignition timing were apparent for E50.

Marriott *et al.* (2008) show significant performance benefits for E85 (RON measured as 107.7) compared with a 104 RON gasoline when used in a naturally aspirated direct-injection gasoline engine. Peak torque generated by the engine increased by 5% and peak power by about 4% at the same enriched air–fuel ratio. An increase in volumetric efficiency of about 3% was measured at the peak torque operating point. Smaller but still significant performance benefits were available from operation at stoichiometric conditions when using E85 fuel. The majority of the combustion-related benefit in performance using E85 was determined to come from a reduction in the cumulative heat energy rejected to the engine coolant.

Korte *et al.* (2011) and OudeNijeweme *et al.* (2011) show significant synergies between the high octane properties of alcohol fuels and pressure-charged down-sized engines. Alcohol blends were made up using a 95 RON blend stock so that the E10 blend had a RON of 99 and the 16 vol% iso-butanol blend (Bu16) had a RON of 98. Note that this value is higher than that of a typical commercial E10 blend, even within Europe, due to the use of a 95 RON gasoline blend stock. Octane numbers of alcohol fuels with various blend stocks are discussed by Anderson *et al.* (2012a, 2012b).

Other blends used were E22, which matched the 102 RON of the high-octane forecourt specification gasoline included in the study, Bu68 (RON 104) and E85 (RON 106). Korte *et al.* (2011) conclude that all the high octane fuels used show a synergy with down-sizing under high-load conditions and that such fuels enable either higher levels of downsizing or increased compression ratios to be used, leading to additional $CO_2$ improvements. In related work using substantially the same blends, Stansfield *et al.* (2012) showed that it is possible to use high-alcohol blends in an unmodified production vehicle which has not been sold as having flex-fuel capability. Here E85 gave appreciably the most power (approximately 20% more than 95 RON gasoline), with the other blends forming a fairly tight band lying between the two.

*Efficiency*

Using E100, a full-load thermal efficiency at 2,800 rev/min. of 39.6% was reported by Nakata *et al.* (2006), compared with 37.9% and 31.7% using the high- and low-octane gasoline, respectively. A thermal efficiency improvement of 3% was achieved using E100 over the 100 RON gasoline at a typical part-load operating point, where the engine was far from the area where knock becomes a limiting factor. The efficiency benefit using ethanol in the part-load region was attributed to the lower heat losses due to the reduction in combustion temperatures. At high loads, when the engine is knock-limited, spark timing must be retarded and this leads to higher exhaust gas temperatures. Typically a pre-turbine temperature limit is imposed. High-octane fuels enable more advanced combustion phasing which in turn requires lower levels of over-fuelling in order to remain within the temperature constraints.

Bergström et al. (2007a, 2007b) found that, using a production turbocharged ethanol-gasoline flex-fuel engine with port-fuel injection, the high knock resistance and concomitant lower exhaust gas temperatures experienced when running on E85 allowed the fuel enrichment level at full-load to be reduced to the extent that, for the same limiting peak pressures as those tolerated using gasoline fuel, stoichiometric operation across the engine speed range is possible. Kapus et al. (2007) found that for identical engine performance, the more favourable combustion phasing when operating on E85 at full load leads to less requirement for fuel enrichment giving a 24% improvement in efficiency compared with operation on 95 RON gasoline. Thermal efficiency improvements at full load of over 35% relative to 95 RON gasoline have been found using E100 in a direct-injection, turbocharged spark-ignition engine (Brewster, 2007) operating at high BMEP levels.

Korte et al. (2011) show detailed fuel flow rate maps for an engine with a peak BMEP of over 30 bar using the fuels described in Section 13.3.2. It was found that whilst an increase in fuel flow was required with the alcohol fuels in order to obtain the same load for part-load operation, when the load was above 20 bar BMEP, operation on E22 (rated at 102 RON) required a lower fuel flow rate than when 95 RON gasoline was used. In order to protect engine components from over-heating, 'over-fuelling' is applied. Around the peak power point, operation on E85 required approximately the same fuel flow rate as the 95 RON gasoline whilst the fuel consumption of E22 was about 20% lower. At the peak power point these fuel flow rates correspond to increases of thermal efficiency, relative to the 95 RON gasoline, of 35% and 38% for E22 and E85, respectively. The full-load torque curve of the engine could be achieved at stoichiometric air–fuel ratio using E85.

*Dedicated alcohol engines*

The greater dilution tolerance of methanol and ethanol was exploited by Brusstar et al. (2002) who converted a base 1.9-litre direct-injection turbocharged diesel engine to run on M100 and E100 by replacing the diesel injectors with spark plugs and fitting a low-pressure alcohol fuel injection system in the intake manifold. Running at the 19.5:1 compression ratio of the base diesel engine, the PI methanol variant increased the peak brake thermal efficiency from 40% to 42%, while parity with the diesel was achieved using ethanol. Cooled exhaust gas recirculation (EGR) enabled the engine to achieve close-to-optimum ignition timing at high loads, while high levels of EGR dilution were used to spread the high efficiency regions to extensive areas of the part-load operating map. This was possible due to the lower molar air–fuel ratios and higher flame speeds of the fuels providing higher combustion tolerance to cooled EGR.

Vancoille et al. (2012) have confirmed the effectiveness of this approach for methanol, investigating the viability of throttle-less load control using EGR or excess air. Experiments performed on a single-cylinder engine showed that the EGR and excess air dilution tolerances of methanol are significantly higher than for gasoline. Using a turbocharged multi-cylinder engine with a 19:1 compression ratio, similar to that used by Brusstar et al. (2002), they found it was possible to use EGR levels of up to 30 mass% at a range of engine speeds and loads with limited cycle-to-cycle combustion variation. A peak thermal efficiency of 42% was obtained. The engine operating maps from this work were used by Naganuma et al. (2012) to model the potential benefits of a dedicated methanol-fuelled vehicle. The predictions indicated that a vehicle using an engine optimized for operation on M100 would increase the average engine efficiency on the NEDC from 22.8% for operation on baseline gasoline to 26% for M100, reducing $CO_2$ emissions from 167g/km to 132g/km. Still larger benefits are predicted for other vehicles (Naganuma et al., 2012).

*Pollutant emissions, deposits, and lubricant dilution*

Bergström et al. (2007b) show that, using a vehicle with a port-fuel-injected LEV2 emissions-capable engine on the US FTP 75 cycle, $NO_x$ emissions levels are reduced for all the ethanol blends tested without secondary air injection (SAI). CO emissions levels were increased by up to 50% for the highest concentration ethanol blends tested (E64 and E85) and unburned hydrocarbon emissions (uHC) were increased by about 75% using the E85 blend. They also show that when SAI is used, the uHC and CO emissions are significantly reduced. For uHC emissions, the levels for E85 were reduced by more than a factor of three and became independent of ethanol concentration. Bergström et al. (2007a) show that a similar technology engine calibrated to EU IV emissions compliance using gasoline would also be compliant using E85. Turner et al. (2012c) report emissions results for an EU V vehicle with a boosted DI engine, using a bespoke E85 blend and the iso-stoichiometric ternary blends of gasoline, ethanol and methanol described in Section 13.5.3. While emissions analysers are known to have a lower response to oxygenated species than to pure hydrocarbons, the level of reduction in uHC shown by Turner *et al.* (2012c) for their blends is of the order that, correcting using a ratio of 1.154 from the earlier work of OudeNijeweme *et al.* (2011) (measured by them for E85), still resulted in uHC emissions no worse than for gasoline.

Of the various types of uHC emissions, those of the aldehydes in particular are a potential concern, and especially those of formaldehyde. This is because in the metabolism of formaldehyde the crucial step is that from formic acid to carbon dioxide and water, which depends on folic acid and gives rise to a wide variation in fatal dose depending on the victim's age, body mass and, in the case of women, whether they are pregnant or not. With alcohol combustion, aldehydes are intermediate species of the oxidation of the fuel: methanol primarily yields formaldehyde and ethanol acetaldehyde, together with lower amounts of formaldehyde. Early work, presumably with older engine and emissions control technology than is commonly used today, suggested that generally some changes to catalyst formulation may be necessary to ensure long-term catalyst durability with methanol (Nichols *et al.*, 1988), but this is not expected to be an issue with present-day technology: flex-fuel vehicles have been shown to be capable of meeting limits for formaldehyde when operated on E85 (West *et al.*, 2007) and are expected to be able to do so for other alcohols such as n-butanol (Gingrich *et al.*, 2009). It is known that aldehyde emissions can be successfully neutralized by the type of three-way catalysts typically used to control emissions of modern spark-ignition engines (Menrad *et al.*, 1988; Wagner and Wyszyński, 1996; Shenghua *et al.*, 2007). Gasoline may actually yield greater challenges with respect to aldehydes on legislated drive cycles in the future: Benson *et al.* (1995) reported acetaldehyde emissions being three times higher in a flex-fuel vehicle when operated on gasoline than E85, and in another study West *et al.* (2007) reported that gasoline had higher aldehyde emissions than E85 in a vehicle operated on the US06 drive cycle (which requires higher driving loads than is typically the case for other drive cycles). Furthermore, some potential technologies to improve the knock limit of down-sized engines such as cooled EGR have been found to increase aldehyde emissions from gasoline (Gingrich *et al.*, 2009). Pearson and Turner (2012) discuss the potential health effects arising from aldehyde emissions from the combustion of the alcohols as well as giving an overview of other safety aspects of the fuels.

Cairns *et al.* (2009) found, using a boosted DI engine, that E22 blends produced the highest levels of smoke emissions in their study of ethanol and butanol blends with gasoline. Negligible smoke emissions were found using E85 fuel. Additionally they found that fuel injector deposits were highest using an E10 blend whereas the use of E85 at part load produced almost no deposits. Bergström *et al.* (2007a) found significant deposits accumulating on a port injector using E85 due to the presence of polyisobutylene components in the gasoline.

Price *et al.* (2007) measured particulate matter (PM) number concentrations for alcohol-gasoline blends in a direct injection engine. They found that PM number was lowest for E85, with 95 RON gasoline, E30 and M30 blends giving similar PM number concentrations, and M85 giving the highest level. They suggested that with the high-alcohol blends, the potential for PM to form at high load is due to the fact that the local AFR is above the saturation point of the mixture, leading to droplet burning which is only partially offset by the oxygen bound in the fuel. Later work by Chen *et al.* (2011) showed that under stoichiometric conditions high-ethanol-blend fuels gave higher PM emissions than low-blend ones when operated in a DI engine under homogeneous stoichiometric conditions. Under rich conditions, the PM output from an E10 ethanol blend was much lower, however, and they suggested that this is what gives much lower PM from vehicles operating on alcohol blends when they are operated on transient drive cycles. This suggests that any stratification in the combustion chamber will have some effect on PM formation with high-ethanol-blend fuels. Conversely, Stansfield *et al.* (2012) reported data from vehicle tests showing that the two high-alcohol blends that they tested, E85 and Bu68, both failed the EU V particulate mass limit, although it should be noted that the production vehicle that they were using had not been sold as being flex-fuel capable.

Bergström *et al.* (2007a, 2007b) and Hadler *et al.* (2011) describe the potential for high levels of oil dilution in the engine lubricant using ethanol, particularly if the engine is repeatedly cold-started in a cold climate and run for a short time. Bergström *et al.* (2007a) state that the build-up of fuel is not harmful as long as the engine is given time to reach normal operating conditions, preferably within ten cold start events.

## 13.5 Vehicle and blending technologies for alcohol fuels and gasoline

### 13.5.1 Flexible-fuel vehicles

Ethanol/gasoline flex fuel vehicles have been in existence for many years, the first practical example being the Ford Model T which could operate on gasoline or ethanol (thought at that time to be an attractive feature for potential customers in rural areas). Kramer and Anderson (2012) discuss the market penetration of many forms of alternative fuels and the related vehicles on a worldwide basis, including ethanol. They discuss and define the different types of vehicle and show that some flexibility in the use of different fuels is important to vehicle purchasers for market penetration of both fuels and vehicles. The fact that when ethanol was developed as a transport fuel for Brazil, and that dedicated fuel vehicles were first used which subsequently restricted the uptake of ethanol, versus the acceleration of the market when true flex-fuel vehicles were offered, is cited by them as a clear example of the desirability of fuel flexibility onboard the vehicle. It is interesting to note that in Brazil hydrous ethanol is offered as a fuel and can be used in flex-fuel vehicles which may also have a proportion of gasoline in the tank, easing some of the concern about phase separation (because the vehicles can be made sufficiently flexible in operation to overcome this, and agitation of the tank when the vehicle moves will also tend to keep the various components mixed together). While gasoline starting systems have been used to date on Brazilian 'total flex' vehicles, there is a move towards replacing these with heated injectors to permit cold starting even on hydrous E100. On this subject, it is interesting to note that in Brazil, even gasoline (or 'gasohol') contains 22–25 vol% ethanol, meaning that the minimum level of alcohol in fuel is equivalent to that which could be realized in the entire fuel pool in the US (or Europe) were all available ethanol eventually blended into it (Anderson et al., 2012a).

In order to ascertain the exact proportion of gasoline and ethanol in the fuel tank so that the engine management parameters can be set accordingly, generally Brazilian FFVs have used a so-called 'virtual' sensor to assist in adjusting the fuelling rate. Knowledge of the ratio between the two fuels is necessary because of the different volumetric heating values and stoichiometries of the various fuels supplied to the marketplace. More recently, physical alcohol sensors have become commonplace, especially in markets with challenging emissions and onboard diagnostics requirements. The following section discusses the different fuel sensing technologies and various background reasons for their deployment.

Virtual sensors are so called because during a re-fuelling event, they utilize the fuel tank level sensor to obtain an approximate value for the volume of fuel added, and subsequently employ an algorithm to allow the engine management system (EMS) to calculate two approximate 'new' values of the stoichiometric AFR, assuming that the fuel added was either gasoline or E85 only. Because prior to key-off the EMS knew the value of stoichiometry that the vehicle was operating at that moment, since it has calculated two possible new values for stoichiometric AFR, on restart, it monitors the $O_2$ sensors and looks for a perturbation in their signal. Once it has ascertained whether the AFR is swinging in the rich or lean direction, it adjusts the operating parameters gradually until it locks on to the new value, and the vehicle has been 'conditioned'. Figure 13.11 shows a representation of a virtual sensor detection system. The initial swings to the new AFR can be quite rapid, but an extended period of conditioning is

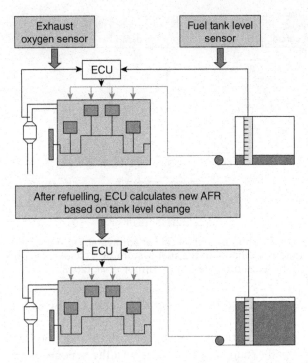

*13.11* Representation of a virtual sensor system to detect the concentration of ethanol in the fuel system of a gasoline/alcohol flex-fuel vehicle. Reproduced from Turner *et al.* (2013).

necessary because commercial E85 fuels can feature a reduced ethanol content of 51% and still be within the ASTM D5798 specification, and low ethanol contents have habitually been used in winter to facilitate easier starting and better low-temperature behaviour. Vehicles certified to EU IV level emissions standards in Europe employed such a virtual sensor, an obvious attraction of which is that they are very cost-effective, as they require no additional hardware.

For diagnosis of emissions system performance and reliability under EU V regulations, there is a requirement to diagnose all sensors used by the EMS, meaning that there is a requirement for all FFVs certified to this emissions level to employ an additional sensor capable of directly measuring the concentration of alcohol in the fuel system. This is termed a 'physical' sensor. Physical sensors generally measure the electric permittivity (or the resistance) of the fuel. Since these values vary widely between non-polar pure hydrocarbons of the type generally comprising gasoline, and compounds with high polarity (such as the alcohols), they can be used to indicate the alcohol concentration directly. Figure 13.12 shows a representation of a fuel system employing a physical alcohol concentration sensor.

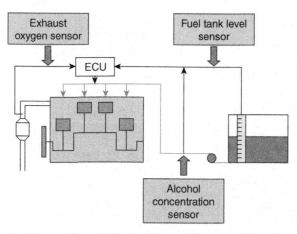

*13.12* Representation of a physical sensor system to detect the concentration of ethanol in the fuel system of a gasoline/alcohol flex-fuel vehicle. Reproduced from Turner *et al.* (2013).

Kramer and Anderson (2012) point out that in the US, FFVs have been widely introduced at no on-cost to the consumer, and from this it is reasonable to assume that the on-cost to the vehicle manufacturer, while not negligible, is small enough to be offset against the benefits of producing the vehicle from a corporate average fuel economy (CAFE) standards point of view. In other markets, such as Sweden, the vehicles carry only a modest price penalty, which can quickly be recouped given favourable fuel prices. Bergström *et al.* (2007a) and Hadler *et al.* (2011) give a detailed account of the engine and fuel system design modifications required in order to produce an FFV from a dedicated gasoline vehicle platform. In Brazil, the vehicle production cost is slightly more because of the need for a small gasoline-fuelled starting system (itself mainly necessary since hydrous ethanol can be put into the primary tank of these vehicles). Future vehicles could adopt heated components in the fuel system to vaporize the fuel and, through the consequent ability to delete the gasoline-based starting system, reduce the price to closer to that of a simple gasoline vehicle. This would mean that the payback time for the customer would be concomitantly shorter (Kramer and Anderson, 2012). Note that beyond the use of a gasoline start system, Brazilian market vehicles still use flex-fuel technology; the next section will discuss vehicles which do not have the feature of flexible fuelling from a single tank.

### 13.5.2 Tri-flex-fuel vehicles

As mentioned previously, it is possible to create vehicles capable of operating on any volumetric proportion of methanol, ethanol and gasoline. Pearson *et al.* (2009a) report one such vehicle which used a physical alcohol sensor to infer the approximate bulk alcohol concentration to set the ignition timing and a wide-range oxygen sensor to provide arbitrary control of the injector pulse-width using the oxygen sensor. Emissions capability of this vehicle was demonstrated by operating it on different ratios of the three fuels.

If a vehicle has originally been developed to operate on high-blend methanol fuels and gasoline, it is relatively straightforward to modify it to accept ethanol as well, as discussed by Nichols (2003). However, if the vehicle has not been developed to function satisfactorily on high proportions of methanol in an alcohol-gasoline system, it is still possible to introduce it into the fuel pool as a major fuel component. This can be achieved through the adoption of targeted blends of gasoline, ethanol and methanol with specific ratios targeting the stoichiometric AFRs of any ethanol-gasoline binary mixture that a flex-fuel vehicle has been developed to operate on. The following section discusses this possibility.

### 13.5.3 Iso-stoichiometric ternary blends

Gasoline, ethanol and methanol (GEM) are all miscible together (with or without cosolvents to avoid phase separation, which varies with temperature). Ternary, or three-component, blends of them can be configured to have the same target stoichiometric air–fuel ratios (AFRs) as any binary gasoline-ethanol blend. Work that has been conducted to date has concentrated on such 'GEM' ternary blends with a target stoichiometric AFR equivalent to that of E85, i.e. 9.7, but equally GEM blends equivalent to E10, E22, etc., could be arranged. For a fixed stoichiometric AFR, the relationship between the three components is defined by linear volumetric relationships as shown for E85-equivalent stoichiometry in Fig. 13.13. Importantly, when configured in such a manner, all such iso-stoichiometric blends not only have near-identical volumetric lower heating value (LHV), it has also been found that they have practically the same octane numbers and extremely close enthalpies of vaporization (to ±2%, based on the mass ratios in the blend). In Fig. 13.13 one can see that as the volume percentage of ethanol is reduced, so the rate of increase of the methanol proportion is faster than that of the gasoline proportion, because as one volume unit of ethanol is removed, a volume unit of the binary gasoline-methanol mixture with the same stoichiometric AFR as ethanol has to be used to replace it. The necessary volume ratio of gasoline:methanol to give the required 9:1 stoichiometic AFR is 32.7:67.3, as discussed by Turner *et al.* (2011).

Note that Fig. 13.13 clearly shows that E85 contains no methanol and that its binary equivalent for a gasoline and methanol mixture (where no ethanol is present) occurs at 44 and 56 vol%, respectively. This is the left-hand limit

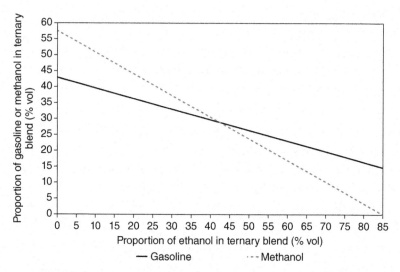

*13.13* Relationship between blend proportions of gasoline, ethanol and methanol in iso-stoichiometric GEM ternary blends configured with a stoichiometric AFR of 9.7. Reproduced from Turner *et al.* (2013).

for this stoichiometry. It is interesting to note that the ratio where the proportion of gasoline and methanol are equal occurs at approximately 42.5 vol% ethanol, which is coincidentally half the volume which would be present in E85.

Test results showing that these iso-stoichiometric blends are effectively drop-in fuels, on a vehicle operational level, to E85/gasoline FFVs with both virtual and physical sensor systems are presented by Turner *et al.* (2011, 2012b). Preliminary investigations into material compatibility have been reported by Turner *et al.* (2012c), together with results showing that the primary exhaust pollutants can be readily controlled, all E85-equivalent ternary blends essentially out-performing commercial 95 RON gasoline in this respect, although particulate and aldehyde emissions were not measured in these tests. In the US, the use of iso-stoichiometric GEM blends may be especially attractive since there are already 9 million E85/gasoline FFVs in use, a result of their improved performance in terms of gasoline usage under US CAFE regulations. In any location, a progressive subsequent rollout by methanol percentage and region could allow introduction of methanol as a major transport fuel component contributing to meaningfully reduced greenhouse gas emissions, improved energy security and better air quality before a vehicle specification change to M100 has to be adopted (e.g., as proposed by the US Open Fuel Standard). This approach was discussed in Turner *et al.* (2012c).

Note that, should their use be beneficial with regard to the utilization of all available feedstocks, it is possible to produce ternary blends of other alcohols with gasoline, or even quaternary (and higher) blends. Examples of these may be mixtures of gasoline, methanol and butanol with or without ethanol respectively. A more detailed description of the physico-chemical properties of some of these mixtures, with additional remarks on iso-stoichiometric blends of gasoline, butanol and methanol (GBM), and hydrous GEM and GBM blends, is given by Pearson et al. (2012b).

## 13.6 Future provision of renewable liquid fuels

The rise in global oil demand over the next 25 years is likely to originate entirely from the transport sectors of emerging economies as increasing prosperity drives greater levels of personal mobility and freight. The development of renewable liquid transport fuels which are not feedstock constrained could enable the continued provision of full-range affordable vehicles and will mitigate, and potentially eliminate, the wealth transfer and exposure to price volatility associated with high dependency on feedstock which have a finite supply. Here the provision of liquid and gaseous fuels which are synthesized from $CO_2$, water and renewable energy are discussed. Without the provision of synthetic carbon-neutral liquid fuels, biofuels are likely to appear as merely palliatives in the processes of alleviating issues regarding security of energy supply and reducing transport $CO_2$ emissions.

### 13.6.1 The biomass limit

The worldwide sustainable potential of biogenic wastes and residues has been estimated at approximately 50 EJ/year (1 EJ = 1 × 10$^{18}$ J). Estimates of the global sustainable potential of energy crops have a huge spread, ranging from 30 EJ to 120 EJ/year, depending mainly on the assumptions made regarding food security and retaining biodiversity. These combined values put the total potential sustainable bioenergy supply in 2050 between 80 and 170 EJ/year (Pearson and Turner, 2012). Current global energy use is about 500 EJ/year. Thus the mid-point value of the total sustainable bioenergy supply is around one quarter of this and less than one tenth of the projected global energy use in 2050 (WBGU, 2008). The global transport energy demand in 2007 was about 100 EJ (EIA, 2010) and, extrapolating the EIA value of 143 EJ for liquid fuels in 2035 (EC, 2011), is projected to grow to about 170 EJ in 2050. Assuming that half the available sustainable biomass feedstock was available for biofuel production at a conversion efficiency of 50% (Bandi et al., 1995) limits the substitution potential of biofuels to about 20% of the 2050 energy demand. The IEA make the slightly more optimistic prediction that 32 EJ of biofuels will be used globally in 2050, providing

27% of transport fuel (IEA, 2011b). For individual countries, the biomass potential could be significantly higher or lower, depending on their population densities and sustainable agricultural potential.

A further limiting factor is the requirement to feed the increasing global population, with its shift toward westernized diets demanding much greater amounts of land and water (UN, 2006). These issues may constrain biofuel production to the use of the wastes and residues quantified above.

### 13.6.2 Sustainable Organic Fuels for Transport (SOFT)

In this section, the concept of Sustainable Organic (meaning carbon-containing) Fuels for Transport (SOFT) is briefly introduced as a means of circumventing the biomass limit to the penetration of biofuels in transport. The concept is proposed as a long-term solution to supplying carbon-neutral liquid fuels which could eventually supply the bulk of transport energy demand beyond the 20–30% which can be sustainably supplied by biofuels. These fuels, like biofuels, would be miscible with current petroleum-based fuels so that an evolutionary transition from one organic liquid fuel to another could occur. Whilst Section 13.2 described the problems of using molecular hydrogen as a transport fuel, here it is proposed to 'package' the hydrogen in a more convenient manner by combining it with re-cycled $CO_2$ to synthesize energy-dense liquid fuel. In this way hydrogen is used *in* the fuel rather than *as* the fuel. Since the concept is based on the reduction of $CO_2$ and water to synthesize the fuel and its subsequent use results in oxidation returning the fuel to these components, the process is not feedstock-limited. Importantly, if all processes are powered with carbon-free energy and the $CO_2$ used to make the fuel is captured directly from the atmosphere, then the combustion of this fuel would result in zero net increase in the atmospheric $CO_2$ concentration. The concept of synthesizing fuel from water and recycled $CO_2$ was first proposed in the 1970s by Steinberg (1977) and there have been many other proposals in the meantime (Bandi *et al.*, 1995; Stucki *et al.*, 1995; Weimer *et al.*, 1996; Olah *et al.*, 2009, 2011; Jensen *et al.*, 2007; Pearson and Turner, 2007; Pearson *et al.*, 2009b, 2012a; Littau, 2008; Jiang *et al.*, 2010; Graves *et al.*, 2011).

*Recycling $CO_2$*

Synthetic carbon-neutral liquid fuels mimic the overall function of photosynthesis by reducing water and $CO_2$ to make hydrocarbon-based products. The energy input to the process can either be provided directly in the form of sunlight (to produce 'solar fuels') or may be off-peak renewable energy (see Section 13.6.3). By combining hydrogen with $CO_2$ it can be chemically liquefied into a high energy density hydrocarbon fuel.

Clearly, if the captured $CO_2$ stems from the combustion of fossil energy resources, this approach is not renewable and will still result in an increase in atmospheric $CO_2$ concentration. Rather than a recycling process, it amounts to $CO_2$ re-use and offers the potential of a notional reduction in emissions of approximately 50% (Graves et al., 2011). In a closed-cycle fuel production process, ideally there is no net release of non-renewable $CO_2$. If the hydrogen generation process is via the electrolysis of water, this represents by far the greatest energy input to the process. For this reason the fuels produced in this way may be referred to as 'electrofuels', as they are essentially vectors for the storage and distribution of electricity generated from renewable energy. When the feedstocks are water and $CO_2$ from the atmosphere, the fuel production and use cycle is materially closed and therefore sustainable. Such a cycle offers security of feedstock supply on a par with that of the 'hydrogen economy', as the timescale for mixing of $CO_2$ in the atmosphere is sufficiently short to ensure a homogeneous distribution. With access to sufficient water and renewable energy, the process has the potential to provide fuel from indigenous resources for most nations.

$CO_2$ and water are the end products of any combustion process involving materials containing carbon and hydrogen. Further reactions to form carbonates are exothermic processes. The capture of $CO_2$ in inorganic carbonates and other media is a burgeoning area of research and some of the literature is described in detail in Graves et al. (2011) and Pearson et al. (2012a).

Whereas the adoption of battery electric or hydrogen fuel cell vehicles requires paradigm shifts in the costs of the vehicles themselves or their fuel distribution infrastructure, or both (Pearson et al., 2012a), the development of carbon-neutral liquid fuels enables a contiguous transition to sustainable transport. Drop-in fuels such as gasoline, diesel and kerosene can be produced from $CO_2$ (via CO) and $H_2$ via Fischer–Tropsch (FT) synthesis, but the simplest and most efficient liquid fuel to make is methanol (Pearson and Turner, 2012). Indeed the option to make gasoline is retained even if methanol is produced initially since the former can be made via the Exxon-Mobil methanol-to-gasoline process. In addition to being the simplest fuel to synthesize from $CO_2$ and water feedstocks, methanol provides much greater biomass feedstock diversity, as it can be made from anything which is (or ever was) a plant.

*Fuel synthesis*

Once hydrogen and $CO_2$ are available, the simplest and most direct route to producing a high-quality liquid fuel is the catalytic hydrogenation of $CO_2$ to methanol via the reaction:

$$CO_2 + 3H_2 \rightarrow CH_3OH + H_2O \quad \Delta H^0_{298} = -49.9 \text{ kJ/(mol. methanol)}$$

During the production of methanol via direct hydrogenation of $CO_2$, by far the largest component of the process energy requirement is the hydrogen production (Pearson and Tuner, 2012; Pearson et al., 2012a). This is true of any electrofuel using hydrogen as an intermediate or final energy carrier. Assuming an electrolyser efficiency of 80% and a $CO_2$ extraction energy of 250 kJ/mol.$CO_2$ (representing about a 10% rational thermodynamic efficiency relative to the minimum thermodynamic work requirement of 20 kJ/mol.$CO_2$) gives a higher heating value (HHV) 'electricity-to-liquid' efficiency of about 45%, including multi-pass synthesis of the methanol and re-compression of the unconverted reactants (Pearson and Turner, 2012; Pearson et al., 2012a). In the late 1990s, Specht et al. (1998a, 1998b) measured total process $CO_2$ capture energy levels of 430 kJ/mol, in a demonstration plant using an electrodialysis process to recover the absorbed $CO_2$, representing a rational efficiency of less than 5%. Despite this low $CO_2$ capture and concentration efficiency, the measured overall fuel production efficiency was 38%.

Without policy intervention, the intermittent use of alkaline electrolysers, due to their limited current densities, is likely to be too expensive (Graves et al., 2011) to produce fuel under present market economics. Improvements to this technology are at an advanced stage of development (Graves et al., 2011; Ganley, 2009) and other promising technologies are emerging. Graves et al. (2011) describe the use of high temperature co-electrolysis of $CO_2$ and $H_2O$ giving close to 100% electricity-to-syngas efficiency for use in conventional FT reactors. This ultra-efficient high temperature electrolysis process using solid oxide cells combined with a claimed $CO_2$ capture energy (from atmospheric air) as low as 50 kJ/mol (Lackner, 2009) leads to a prediction of an electricity-to-liquid efficiency of 70% (HHV basis). With a constant power supply this high overall efficiency enables the production of synthetic gasoline at $2/gallon ($0.53/litre) using electricity available at around $0.03/kWh (Graves et al., 2011). Doty et al. (2010) state that off-peak wind energy in areas of high wind penetration in the US averaged $0.0164/kWh in 2009 and the lowest 6 hours of the day averaged $0.0071/kWh. With more pessimistic values for the cost of $CO_2$ capture such as the $1,000/tonne quoted by House et al. (2011), the gasoline cost component due to the supply of the carbon feedstock alone might be as high as $7.5/gallon (about €1.30/litre). With 20% electrolyser capacity the cost of fuel synthesis could be as high as $4/gallon at $0.03/kWh electricity (higher current density electrolysers could reduce this to $2.2/gallon) (Graves et al., 2011). For perspective, a cost of $11.5/gallon is around €2.05/litre. Currently gasoline retail costs in the EU range from €1.14/litre to €1.67/litre including duties and taxes (which can be as high as €0.6/litre).

In a reconfigured system which bases fuel duty and taxation on non-renewable life cycle carbon intensity, a fuel made from air-extracted $CO_2$ and water might be commercially attractive in the medium term, i.e. in advance of the point where sequestration of air-captured $CO_2$ becomes economically feasible. Recycling of the $CO_2$, rather than sequestrating it after it has been removed from the atmosphere, cannot result in any net greenhouse gas (GHG) reduction. Its inclusion in a closed carbon cycle to make transport fuels, however, can potentially have the effect of rendering carbon-neutral the fastest growing GHG emissions sector as well as providing a spur to the development of air capture technology for $CO_2$ which may subsequently be used for sequestration purposes.

### 13.6.3 Renewable fuels within an integrated renewable energy system

Renewable energy sources, such as those based on wind and solar power, are limited in their ability to meet current and future energy demands not by the resource potential, which for wind and solar is many times current demand, but by the intermittent nature of supply. To escape this conundrum, large-scale storage systems are required. Biomass is a form of large-scale storage of solar energy but, whilst it may be part of a sustainable system, it cannot underpin it. One possibility for large-scale energy storage is to use off-peak renewable energy to synthesize chemical energy carriers. Chemical energy storage systems, based on the conversion of renewable energy into a gaseous or liquid energy carrier, enable the stored energy to be either re-used for power generation or transferred to other energy sectors such as transport, where the de-carbonization issue is more problematic, and there is an ever-present demand to supply a high-value energy carrier.

In addition to the ready synthesis of methanol from $CO_2$ and hydrogen, methane can also be made from the same feedstocks using the Sabatier process. It then has the advantage that it can be stored in the gas grid, which, for most developed countries, far exceeds the capacity of existing renewable energy storage media (e.g., pumped hydro) or proposed systems such as large flywheels or redux fuel cells. Sterner (2009), Specht *et al.* (2009), and Breyer *et al.* (2011) describe such a concept, where the synthesized methane is stored and readily retrievable to smooth out the supply of renewable energy through conversion back to electrical energy via combustion in conventional power stations. This process is given the name 'renewable power methane' (RPM), and its operation within a renewable energy system based on wind, solar and biomass has been modelled over a period of one week on a one hour resolution based on a winter load demand. The renewable-power-to-methane efficiency is predicted to be 48% (Sterner, 2009; Specht *et al.*, 2009) using measured energy values for capture and

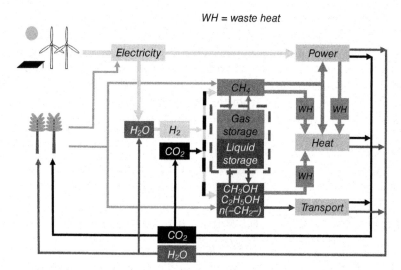

*13.14* Integrated power, heat and transport system featuring large-scale energy storage capacity and combining renewable methane and liquid fuels and using renewable energy and biomass.

concentration of $CO_2$ from air of 430 kJ/mol, which, as discussed in Section 13.6.2, is a realistic value achievable with current cost-effective 'off-the-shelf' technology.

Because the limit on the amount of renewable energy that can be stored would now be significantly beyond any expected day-to-day variation in energy utilization, the upper limit on the level of renewables in which it is economically attractive to invest is effectively raised. While the 'round-trip' losses in converting renewable electricity to methane and back to electricity are significant and would amount to an expected process efficiency of about 20% (electricity-to-electricity), the entire process is practical, achievable and, most importantly, represents an evolutionary path to a fully sustainable energy economy. The production of renewable electricity and RPM for power generation reserve (and its related use in the heat sector, where it would displace fossil natural gas) can also be integrated with the production of liquid fuels, in the form of both biofuels and electrofuels, for use directly in transport (Pearson *et al.*, 2012a). A schematic representation of such a system combining the power, heat and transport sectors is shown in Fig. 13.14, where the renewable liquid fuels are represented by methanol ($CH_3OH$), ethanol ($C_2H_5OH$) and hydrocarbon fuels ($n(-CH_2-)$). The latter can be synthesized from methanol using such processes as the methanol-to-gasoline (MTG) process. In this system, where it can be sustainably produced, biomass is able to contribute to the transport sector via the production of ethanol (or biodiesel), or the energy sector via direct combustion for power generation, as shown in Fig. 13.14.

## 13.7 Conclusion

The production of low-carbon and, ultimately, carbon-neutral liquid fuels is the most pragmatic way in which to either de-carbonize transport or increase its sustainability and thus increase the security of energy supply to the transport sector. This route also allows the continued provision of globally compatible, affordable transport via the retention of low-cost internal combustion engines supplied both on and off the vehicle by low-cost liquid fuel systems.

Low-level biofuel blends, particularly in the form of E10, are already displacing significant quantities of gasoline in countries such as the United States. Ethanol can provide fuels with high resistance to auto-ignition which are synergistic with the trend towards downsized pressure-charged spark-ignition engines. Without significant changes to the fuel taxation system, the low volumetric energy density of ethanol is likely to limit the market penetration of high concentration blends such as E85. Biodiesel penetration is limited by the reluctance of manufacturers to warrant their vehicles for use with higher levels of FAME. These factors, together with the limitations on vegetable oil feedstocks, may lead to higher levels of ethanol in gasoline in order to satisfy energy security, renewability or GHG reduction targets. Butanol and methanol may also have a part to play.

Low-carbon-number alcohols can be used for personal mobility and light-duty applications, and synthetic hydrocarbons for applications where maximum energy density is crucial (such as aviation). A range of technologies is described to enable the transition from the current vehicle fleet to equivalent-cost vehicles capable of using sustainable methanol. All transport energy can be supplied, using biofuels up to the biomass limit, and beyond it using carbon-neutral liquid fuels made using renewable energy, water and re-cycled $CO_2$ from the atmosphere – this approach uses hydrogen in the fuel rather than using it as the fuel. The role of biofuels in this transitional route and end-game would prevent them being regarded as a dead-end by vehicle manufacturers and politicians alike, thus ensuring their continued production and development.

## 13.8 Acknowledgements

The authors wish to thank the following people for the benefit of discussions with them during the compilation of this work: Arthur Bell (SASOL), Martin Davy (Loughborough University), Eelco Dekker (BioMCN), Peter Edwards (University of Oxford), Matt Eisaman (Brookhaven National

Laboratories), Stefan de Goede (SASOL), Ben Iosefa (Methanex), Leon diMarco, Richard Stone (University of Oxford), Andre Swartz (SASOL), Gordon Taylor (GT-Systems), Sebastian Verhelst (University of Ghent), Chris Woolard (University of Cape Town), and Paul Wuebben (CRI). The authors would also like to acknowledge the funding of HGCA and DEFRA during the 'HOOCH' project during which some of the concepts discussed above were developed.

## 13.9 References and further reading

Amaseder, F. and Krainz, G. (2006), 'Liquid hydrogen storage systems developed and manufactured for the first time for customer cars', SAE paper number 2006-01-0432, SAE 2006 World Congress, Detroit, MI, 3–6 April.

Andersen, V.F., Anderson, J.E., Wallington, T.J., Mueller, S.A. and Nielsen, O.J. (2010), 'Vapor pressures of alcohol–gasoline blends', *Energy Fuels*, 24, 3647–3654.

Anderson, J.E, DiCicco, D.M., Ginder, J.M., Kramer, U., Leone, T.G., Raney-Pablo, H.E. and Wallington, T.J. (2012a), 'High octane number ethanol–gasoline blends: quantifying the potential benefits in the United States', *Fuel*, 97, 585–594.

Anderson, J.E, Leone, T.G., Shelby, M.H., Wallington, T.J., Bizub, J., Foster, M., Lynskey, M.G. and Polovina, D. (2012b), 'Octane numbers of ethanol-gasoline blends: Measurements and novel estimation method from molar composition', SAE paper no. 2012-01-1274.

ARB (2012), 'Low carbon fuel standard program', California Environmental Protection Agency – Air Resources Board. Available from: http://www.arb.ca.gov/fuels/lcfs/lcfs.htm (accessed 19 February 2012).

Bamgboyne, A.J. and Hansen, A.C. (2008), 'Prediction of cetane number of biodiesel fuel from the fatty acid methyl ester (FAME) composition,' *Int. Agrophysics*, 22, 21–29.

Bandi, A., Specht, M., Weimer, T. and Schaber, K. (1995), '$CO_2$ recycling for hydrogen storage and transportation – electrochemical $CO_2$ removal and fixation', *Energy Conversion and Management*, 36(6–9), 899–902.

Benson, J.D., Koehl, W.J., Burns, V.R., Hochhauser, A.M., Knepper, J.C., Leppard, W.R., Painter, L.J., Rapp, L.A., Reuter, R.M., J.D., Rippon, B. and Rutherford, J.A. (1995), 'Emissions with E85 and gasolines in flexible/variable fuel vehicles – The auto/oil air quality improvement research program', SAE paper no. 952508 1995.

Bergström, K., Melin, S.-A. and Jones, C.C. (2007a), 'The new ECOTEC turbo BioPower engine from GM Powertrain – utilizing the power of nature's resources', *28th International Vienna Motor Symposium*, Vienna, 26–27 April.

Bergström, K., Nordin, H., Konigstein, A., Marriott, C.D. and Wiles, M.A. (2007b), 'ABC – Alcohol based combustion engines – challenges and opportunities', 16th Aachener Kolloquium Fahrzeug- und Motorentechnik, Aachen, 8–10 October.

Best-europe (2008), 'World's first ethanol powered diesel car'. Available from: www.best-europe.org/Pages/ContentPage.aspx?id=488 (accessed 17 December 2008).

Blumberg, P.N., Bromberg, L., Kang, H. and Tai, C. (2009), 'Simulation of high efficiency heavy duty SI engines using direct injection of alcohol for knock avoidance', *SAE Int. J. Engines*, 1(1), 1186–1195.

Bossel, U. (2006), 'Does a hydrogen economy make sense?', *Proc. IEEE*, 84(10), 1826–1837.

Brewster, S. (2007), 'Initial development of a turbo-charged direct injection E100 combustion system', SAE paper no. 2007-01-3625.

Breyer, Ch., Rieke, S., Sterner, M. and Schmid, J. (2011), 'Hybrid PV-wind-renewable methane power plants – a potential cornerstone of global energy supply', *26th European Photovoltaic Solar Energy Conference*, Hamburg, Germany, 5–9 September.

Bringezu, S., Schutz, H., O'Brien, M., Kauppi, L., Howarth, R.W. and McNeely, J. (2009), 'Towards sustainable production and use of resources: assessing Biofuels', International Panel for Sustainable Resource Management, United Nations Environment Programme. Available from: http://www.unep.org/pdf/Assessing_Biofuels-full_report-Web.pdf (accessed 18 February 2012).

Bruetsch, R.I. and Hellman, K.H. (1992), 'Evaluation of a passenger car equipped with a direct injection neat methanol engine', SAE paper no. 920196.

Brusstar, M.J. and Gray, C.L. (2007), 'High efficiency with future alcohol fuels in a stoichiometric medium duty spark ignition engine', SAE paper no. 2007-01-3993.

Brusstar, M., Stuhldreher, M., Swain, D. and Pidgeon, W. (2002), 'High efficiency and low emissions from a port-injected engine with neat alcohol fuels', SAE paper no. 2002-01-2743.

Brusstar, M., Haugen, D. and Gray, C. (2008), 'Environmental and human health considerations for methanol as a transportation fuel', 17th Int. Symp. on Alt. Fuels, Taiyuan, China, 14 October.

Cairns, A., Stansfield, P., Fraser, N., Blaxhill, H., Gold, M., Rogerson, J. and Goodfellow, C. (2009), 'A study of gasoline-alcohol blended fuels in an advanced turbocharged DISI engine', SAE paper no. 2009-01-0138.

Caprotti, R., Breakspear, A., Klaua, T., Weiland, P., Graupner, O. and Bittner, M. (2007), 'RME behaviour in current and future diesel fuel FIEs', SAE paper no. 2007-01-3982.

Casten, S., Teagan, P. and Stobart, R. (2000), 'Fuels for fuel cell-powered vehicles', SAE paper no. 2000-01-0001.

Chen, L., Xu, F., Stone, R. and Richardson, D. (2011), 'Spray imaging, mixture preparation and particulate matter emissions using a GDI engine fuelled with stoichiometric gasoline/ethanol blends', Paper Number C1328/002, Internal Combustion Engines: Improving Performance, Fuel Economy, and Emissions, Institution of Mechanical Engineers Conference, London, 29–30 November, pp. 43–52.

Cooper, J. (2011), 'Future fuels for transport: regulatory and supply drivers'. Internal Combustion Engines: Improving Performance, Fuel Economy, and Emissions, Institution of Mechanical Engineers Conference, London, 29–30 November.

Davis, S.C., Diegel, S.W., and Boundy, R.G. (2011), '*Transportation Energy Data Book*', Edition 30. ORNL-6986. Center for Transportation Analysis, Energy and Transportation Science Division, Oak Ridge National Laboratory, Oak Ridge, TN. Prepared for the Vehicle Technologies Program, Office of Energy Efficiency and Renewable Energy, US Department of Energy.

Doty, G.N., McCree, D.L., Doty, J.M. and Doty, F.D. (2010), 'Deployment prospects for proposed sustainable energy alternatives in 2020', Paper ES 2010-90376, Proceedings of ES2010 Energy Sustainability 2010, Phoenix, AZ, 17–22 May.

Eberle, U. (2006), 'GM's research strategy: Towards a hydrogen-based transportation system', FuncHy Workshop, Hamburg, September.

Eberle, U., Arnold, G. and von Helmolt, R. (2006), 'Hydrogen storage in metal-hydrogen systems and their derivatives', *J. Power Sources*, 154, 456–460.

EC (2009a), 'On the promotion of the use of energy from renewable sources and amending and subsequently repealing Directives 2001/77/EC and 2003/30/EC', Directive 2009/28/EC of the European Parliament and of the Council, 23 April.

EC (2009b), 'Setting emission performance standards for new passenger cars as part of the Community's integrated approach to reduce $CO_2$ emissions from light-duty vehicles'. Regulation (EC) 443/2009 of the European Parliament and of the Council, 23 April.

EC (2009c), 'Amending Directive 98/70/EC as regards the specification of petrol, diesel, and gas-oil and introducing a mechanism to monitor and reduce greenhouse gas emissions and amending Council Directive 1999/32/EC as regards the specification of fuel used by inland waterway vessels and repealing Directive 93/12/EEC'. Directive 2009/30/EC of the European Parliament and of the Council of 23 April 2009.

EC (2010), Comité Européen de Normalisation, 'Automotive fuels – diesel. Requirements and test methods', E 590. February.

EC (2011), 'Roadmap to a single European transport area – Towards a competitive and resource efficient transport system', White Paper COM(2011) 144 final, European Commission, 28 March.

Edwards, P.P. (2012), Private communication.

EIA (2010), 'International energy outlook 2010: Transport sector energy consumption', US Energy Information Administration. Available from: http://www.eia.doe.gov/oiaf/ieo/transportation.html (accessed 20 April 2011).

EPA (2007), 'Regulation of fuels and fuel additives: Renewable Fuel Standard program: final rule'. Environmental Protection Agency, 40 CFR Part 80. Federal Register vol. 72, no. 83, pp. 23900–24014, 01/05. Available from: http://www.epa.gov/otaq/renewablefuels/rfs-finalrule.pdf (accessed 19 February 2012).

EPA (2010a), 'Regulation of fuels and fuel additives: changes to Renewable Fuel Standard program: final rule'. Environmental Protection Agency, 40 CFR Part 80. Federal Register vol. 75, no. 58, pp. 14670–14904, 26/03. Available from: http://www.epa.gov/otaq/renewablefuels/rfs-finalrule.pdf (accessed 19 February 2012).

EPA (2010b), 'Partial grant and partial denial of Clean Air Act waiver application submitted by Growth Energy to increase the allowable ethanol content of gasoline to 15 percent; Decision of the administrator'. Environmental Protection Agency. Federal Register vol. 75, no. 213, pp. 68094–68150, 04/11. Available from: http://www.gpo.gov/fdsys/pkg/FR-2010-11-04/pdf/2010-27432.pdf (accessed 19 February 2012).

EPA (2010c), 'Light-duty vehicle greenhouse gas emissions standards and corporate average fuel economy standards: final rule'. Environmental Protection Agency/Department of Transportation/National Highway Traffic Safety Administration. 40 CFR Parts 85, 86, and 600; 49 CFR Parts 531, 533, 536. Federal Register vol. 75, no. 88, pp. 25324–25728, 07/05.

EPA (2011a), 'EPA finalizes 2012 Renewable Fuel Standards', EPA Regulatory Announcement, Office of Transportation and Air Quality, EPA-420-F-11-044,

December. Available from: http://www.epa.gov/otaq/fuels/renewablefuels/documents/420f11044.pdf (accessed 19 February 2012).

EPA (2011b), 'Partial grant of Clean Air Act waiver application submitted by Growth Energy to increase the allowable ethanol content of gasoline to 15 percent; Decision of the administrator'. Environmental Protection Agency. Federal Register vol. 76, no. 17, pp. 4662–4683, 26/01. Available from: http://www.gpo.gov/fdsys/pkg/FR-2011-12-01/pdf/2011-30358.pdf (accessed 19 February 2012).

EPA (2011c), '2017 and later model year light-duty vehicle greenhouse gas emissions and corporate average fuel economy standards'. Environmental Protection Agency/Department of Transportation/National Highway Traffic Safety Administration. 40 CFR Parts 85, 86, and 600; 49 CFR Parts 523, 531, 533, 536, and 537. Federal Register vol. 76, no. 231, pp. 74854–25728, 01/12.

EPA (2012), 'MTBE in fuels'. Available from: http://www.epa.gov/mtbe/gas.htm (accessed 19 February 2012).

Fargione, J., Hill, J., Tilman, D., Polasky, S. and Hawthorne, P. (2008), 'Land clearing and the biofuel carbon debt', *Science Express*, 319, 1235–1238.

Ferrari, R.A., Turtelli Pighinelli, A.L.M. and Park, K.J. (2011), 'Biodiesel production and quality', in dos Santos Bernardes, M.A. (ed.) *Biofuel's Engineering Process Technology*, Rijeka, Croatia: InTech. Available from: http://www.intechopen.com/books/biofuel-s-engineering-process-technology/biodiesel-production-and-quality

Furey, R.L. (1985), 'Volatility characteristics of gasoline-alcohol and gasoline-ether blends', SAE paper number 852116.

Ganley, J.C. (2009), 'High temperature and pressure alkaline electrolysis', *Int. J. Hydrogen Energy*, 34(9), 3604–3611.

Gardiner, T., Gaade, J., Head, B., Hygate, C., Xu, H.M. and Abdullah, N.R. (2011), 'Research into requirements for worldwide sustainable biodiesel capability', Internal Combustion Engines: Improving Performance, Fuel Economy and Emissions, Institution of Mechanical Engineers Conference, London, 29–30 November.

Geller, D.P. and Goodrum, J.W. (2004), 'Effects of specific fatty acid methyl esters on diesel fuel lubricity', *Fuel*, 83, 2351–2356.

Gingrich, J., Khalek, I., Alger, T. and Mangold, B. (2009), 'Consideration of emissions standards for a dilute spark-ignited engine operating on gasoline, butanol, and E85', SIA International Conference: The Spark-Ignition Engine of the Future, Strasbourg, France, 3–4 December.

Graves, C., Ebbesen, S.D., Mogensen, M. and Lackner, K. (2011), 'Sustainable hydrocarbon fuels by recycling $CO_2$ and $H_2O$ with renewable or nuclear energy', *Renewable and Sustainable Energy Reviews*, 15, 1–23.

Hadler, J., Szengel, R., Middendorf, H., Sperling, H., Groer, H.-G. and Tilchner, L. (2011), 'The 1.4l 118 kW TSI for E85 mode – expansion of the most economical line of petrol engine from Volkswagen', 32nd International Vienna Motor Symposium, Vienna, 5–6 May.

Haer, G. (2011), 'Delivering now: US biodiesel market update and outlook 2012', IHS World Methanol Conference, San Diego, CA, 6–7 December.

Hikino, K. and Suzuki, T. (1989), 'Development of methanol engine with autoignition for low NOx and better fuel economy', SAE paper number 891842, SAE

International Off-Highway & Powerplant Congress and Exposition, Milwaukee, WI.

House, K.Z., Baclig, A., Ranjan, M., van Nierop, E., Wilcox, J. and Herzhog, H. (2011), 'Economic and energetic analysis of capturing $CO_2$ from ambient air', *PNAS Early Edition*, www.pnas.org/cgi/doi/10.1073/pnas.1012253108

Hunwartzen, I. (1982), 'Modification of CFR test engine unit to determine octane numbers of pure alcohols and gasoline-alcohol blends', SAE paper no. 820002. Society of Automotive Engineers International Congress and Exposition, Detroit, MI, 22–26 February.

IEA (2010), 'Advanced Motor Fuels Annual Report', International Energy Agency, Paris. Available from: http://www.iea-amf.vtt.fi/pdf/annual_report_2010.pdf (accessed 02 January 2012).

IEA (2011a), *World Energy Outlook*. International Energy Agency, Paris. Available from: http://www.iea.org/Textbase/npsum/weo2011sum.pdf (accessed 05 December 2011).

IEA (2011b), 'Technology road map: biofuels for transport', International Energy Agency. Available from: http://www.iea.org/papers/2011/biofuels_roadmap.pdf (accessed 01 February 2012).

Jackson, M.D., Unnasch, S., Sullivan, C. and Renner, R.A. (1985), 'Transit bus operation with methanol fuel', SAE paper no. 850216.

Jackson, N. (2006), 'Low carbon vehicle strategies – options and potential benefits', Cost-Effective Low Carbon Engines Conference, I.Mech.E., London, November.

Jensen, S.H., Larsen, P.H. and Mogensen, M. (2007), 'Hydrogen and synthetic fuel production from renewable energy sources', *Int. J. Hydrogen Energy*, 32, 3253–3257.

Jiang, Z., Xiao, T., Kuznetsov V.L. and Edwards, P.P. (2010), 'Turning carbon dioxide into fuel', *Phil. Trans. R. Soc. A*, 368, 3343–3364.

Kapus, P.E., Fuerhapter, A., Fuchs, H. and Fraidl, G.K. (2007), 'Ethanol direct injection on turbocharged SI engines – potential and challenges', SAE paper number 2007-01-1408.

Korte, V., OudeNijeweme, D., Bisordi, A., Stansfield, P., Bassett, M., Mahr, B., Williams, J., Ali, R., Gold, M. and Rogerson, J. (2011), 'Downsizing and biofuels: synergies for significant $CO_2$ reductions', 20th Aachen Colloquium Automobile and Engine Technology, Aachen, Germany, 10–12 October.

Kramer, U. and Anderson, J.E. (2012), 'Prospects for flexible- and bi-fuel light duty vehicles: consumer choice and public attitudes', 2012 MIT Energy Initiative Symposium: Prospects for Flexible- and Bi-Fuel Light Duty Vehicles, Cambridge, MA, 19 April.

Lackner, K.S. (2009), 'Capture of carbon dioxide from ambient air', *Eurpoean Physical Journal – Special Topics*, 176(1), 93–106.

Li, W., Zhong, L. and Xie, K. (2008), 'The development of methanol industry and methanol fuel in China', 6th Annual Methanol Forum, Dubai, 3–5 November.

Littau, K. (2008), 'An "atmospherically healthy" recipe for carbon-neutral fuels: a synthetic fuel made from sunlight, $CO_2$, and water', CTSI Clean Technology & Sustainable Industries Conference & Trade Show, Boston, MA, 1–5 June.

Marriott, C.D., Wiles, M.A., Gwidt, J.M. and Parrish, S.E. (2008), 'Development of a naturally aspirated spark ignition direct-injection flex-fuel engine', SAE paper no. 2008-01-0319.

McCormick, R.L., Alleman, T.L., Waynick, J.A., Westbrook, S.R. and Porter, S. (2006), 'Stability of biodiesel and biodiesel blends: interim report', Technical Report NREL/TP-540-39721, National Renewable Energy Laboratory, April.

Menrad, H., Bernhardt, W. and Decker, G. (1988), 'Methanol vehicles of Volkswagen – a contribution to better air quality', SAE paper number 881196.

Mintz, M., Folga, S., Molburg, J. and Gillette, J. (2002), 'Cost of some hydrogen fuel infrastructure options', Argonne National Laboratory Transportation Technology R&D Center, Transportation Research Board, 6 January.

Miyata, I., Takei, Y., Tsurutani, K. and Okada, M. (2004), 'Effects of bio-fuels on vehicle performance – degradation mechanism analysis of bio-fuels', SAE paper no. 2004-01-3031.

Moran, D.P. and Taylor, A.B. (1995), 'An evaporative and engine-cycle model for fuel octane sensitivity prediction', SAE paper no. 952524.

Moreira, J.R, Coelho, S.T., Velazquez, S.M.S.G., Apolinario, S.M., Melo, E.H. and Elmadjian, P.H.B. (2008), BEST project – Contribution of ethanol usage in public urban transport. Available from: www.aea.org.br/twiki/pub/AEA/PAPERS/PAP0018-17.09-13h30-AnditrioIPE.pdf (accessed 1 November 2011).

Naganuma, K., Vancoillie, J., Verhelst, S., Sileghem, L., Turner, J.W.G., Pearson, R.J. and Martens, K. (2012), 'Drive cycle analysis of load control strategies for methanol fuelled ICE vehicle', SAE paper number 2012-01-1606, SAE Powertrains, Fuels and Lubricants Meeting, Malmo, Sweden, 18–20 September.

Nakata, K., Utsumi, S., Ota, A., Kawatake, K., Kawai, T. and Tsunooka, T. (2006), 'The effect of ethanol fuel on a spark ignition engine', SAE paper no. 2006-01-3380.

National Standard of the People's Republic of China (2009), 'Methanol gasoline (M85) for motor vehicles'. Standardization Administration of the PRC, GB/T23799-2009, ICS 75.160.20. E31. Issued 18 May; Implemented 1 December.

Nichols, R.J. (2003), 'The methanol story: a sustainable fuel for the future', *J. Sci. Ind. Research*, 62, 97–105.

Nichols, R.J., Clinton, E.L., King, E.T., Smith, C.S. and Wineland, R.J. (1988), 'A view of flexible fuel vehicle aldehyde emissions', SAE paper number 881200.

Niu, J. and Shi, L. (2011), 'Methanol used as vehicle fuel will become a main alternative fuel in China', XIX ISAF International Symposium on Alcohol Fuels, Verona, Italy, 10–14 October.

Ogawa, T., Kajiya, S., Kosaka, S., Tajima, I. and Yamamoto, M. (2008), 'Analysis of oxidative deterioration of biodiesel fuel', SAE paper 2008-01-2502.

Olah, G.A., Goeppert, A. and Prakash, G.K.S. (2009), *Beyond Oil and Gas: The Methanol Economy*, 2 edn, Wiley-VCH, Weinheim.

Olah, G.A., Prakash, G.K. and Goeppert, A. (2011), 'Anthropogenic chemical carbon cycle for a sustainable future', *J. Am. Chem. Soc.*, 133(33), 12881–12898.

OudeNijeweme, D., Stansfield, P., Bisordi, A., Bassett, M., Williams, J., Gold, M., Ali, R. and Rogerson, J. (2011), 'Significant $CO_2$ reductions by utilizing the synergies between a downsized SI engine and biofuels', Internal Combustion Engines: Improving Performance, Fuel Economy and Emissions, Institution of Mechanical Engineers Conference, London, 29–30 November.

Owen, K. and Coley, T. (1995), *Automotive Fuels Reference Book*, 2nd edn, Society of Automotive Engineers, Warrendale, PA.

Pang, P. (2011), 'China coal chemical development and outlook', IHS World Methanol Conference, San Diego, CA, 6–7 December.

Pearson, R.J. and Turner, J.W.G. (2007), 'Exploitation of energy resources and future automotive fuels', SAE paper number 2007-01-0034, SAE Fuels and Emissions Conference, Cape Town, South Africa, January.

Pearson, R.J. and Turner, J.W.G. (2012), 'Renewable fuels: an automotive perspective', in Sayigh, A. (ed.) *Comprehensive Renewable Energy*, vol. 5, pp. 305–342, Elsevier, Oxford.

Pearson, R.J., Turner, J.W.G. and Peck, A.J. (2009a), 'Gasoline-ethanol-methanol tri-fuel vehicle development and its role in expediting sustainable organic fuels for transport', 2009 I.Mech.E. Low Carbon Vehicles Conference, London, 20–21 May, pp. 89–110.

Pearson, R.J., Turner, J.W.G., Eisaman, M.D. and Littau, K.A. (2009b), 'Extending the supply of alcohol fuels for energy security and carbon reduction', SAE paper no. 2009-01-2764. SAE Powertrains, Fuels and Lubricants meeting, San Antonio, TX, 2–4 November.

Pearson, R.J., Eisaman, M.D., Turner, J.W.G., Edwards, P.P., Jiang, Z., Kuznetsov, V.L., Littau, K.A., diMarco, L. and Taylor, S.R.G. (2012a), 'Energy storage via carbon-neutral fuels made from $CO_2$, water, and renewable energy'. *Special Issue of Proc. IEEE: 'Addressing the intermittency challenge: Massive energy storage in a sustainable future'*, 100(2), 440–460.

Pearson, R.J., Turner, J.W.G., Bell, A., de Goede, S., Woolard, C. and Davy, M. (2012b), 'Iso-stoichiometric fuel blends: characterization of physico-chemical properties for mixtures of gasoline, ethanol, methanol, and water', *Journal of Automotive Engineering, Part D*, in preparation.

Pickard, W., Shen, A. and Hansing, J. (2009), 'Parking the power: strategies and physical limitations for bulk energy storage in supply–demand matching on a grid whose input power is provided by intermittent sources', *Renewable and Sustainable Sustainable Energy Reviews*, 13, 1934–1945.

Price, P., Twiney, B., Stone, R., Kar, K. and Walmsley, H. (2007), 'Particulate and hydrocarbon emissions measurements from a spray guided direct injection spark ignition engine with oxygenate fuel blends', SAE paper no. 2007-01-0472.

RFA (2008), *The Gallagher Review of the Indirect Effects of Biofuels Production*, Renewable Fuels Agency, St-Leonards-on-Sea, UK, July.

Richards, P., Ried, J., Tok, L.-H. and MacMillan, I. (2007), 'The emerging market for biodiesel and the role of fuel additives', SAE paper no. 2007-01-2033 (JSAE paper no. 20077232).

Rodgers, R.D. and Voth, G.A. (2007), 'Ionic liquids', *Acc. Chem. Res.*, 40, 1077–1078.

Saito, K., Kobayashi, S. and Tanaka, S. (2008), 'Storage stability of FAME blended diesel fuels', SAE paper no. 2008-01-2505.

Searchinger, T., Heimlich, R., Houghton, R.A., Dong, F., Elobeid, A., Fabiosa, J., Tokgoz, S., Hayes, D. and Yu, T.-H. (2008), 'Use of US croplands for biofuels increases greenhouse gases through emissions from land-use change', *Science Express*, 319, 1238–1240.

SEKAB (2008) 'Ethanol also for lorries', press information, 17 March. Available from: www.sekab.com

Shenghua, L., Clemente, E.R.C., Tiegang, H. and Yanjv, W. (2007), 'Study of spark ignition engine fueled with methanol/gasoline blends', *Applied Thermal Engineering*, 27, 1904–1910.

Siewart, R.M. and Groff, E.G. (1987), 'Unassisted cold starts to −29°C and steady-state tests of a direct-injection stratified-charge (DISC) engine operated on neat alcohols', SAE paper no. 872066.

Soccol, R., Vandenberghe, L.P.S., Costa, B., Woiciechowski, A.L., de Carvalho, J.C., Medeiros, A.B.P., Francisco, A.M. and Bonomi, L.J. (2005), 'Brazilian biofuel programme: an overview', *J. of Sci. Ind. Res.*, 64, 897–904.

Specht, M., Bandi, A., Elser, M. and Staiss, F. (1998a), 'Comparison of $CO_2$ sources for the synthesis of renewable methanol,' in *Advances in Chemical Conversion for Mitigating Carbon Dioxide*, Inui, T., Anpo, M., Izui, K., Yanagida, S. and Yamaguchi, T. (eds), Studies in Surface Science, vol. 114, Elsevier Science, Amsterdam, pp. 363–367.

Specht, M., Staiss., F., Bandi, A. and Weimer, T. (1998b), 'Comparison of the renewable transport fuels, liquid hydrogen and methanol, with gasoline – energetic and economic aspects', *Int. J. Hydrogen Energy*, 23(5), 387–396.

Specht, M., Baumgart, F., Feigl, B., Frick, V., Stürmer, B., Zuberbühler, U., Sterner, M. and Waldstein, G. (2009), 'Storing bioenergy and renewable electricity in the natural gas grid', *FVEE – AEE Topics 2009*, pp. 69–78. Available from: http://www.solar-fuel.net/fileadmin/user_upload/Publikationen/Wind2SNG_ZSW_IWES_SolarFuel_FVEE.pdf, (accessed 28 February 2012).

Stansfield, P., Bisordi, A., OudeNijeweme, D., Williams, J., Gold, M. and Ali, R. (2012), 'The performance of a modern vehicle on a variety of alcohol-gasoline fuel blends', *SAE Int. J. Fuels Lubr.*, 5(2); 813–822.

Stein, R.A., Polovina, D., Roth, K., Foster, M., Lynskey, M.G., Anderson, J.E., Shelby, M.H., Leone, T.G. and VanderGriend, S. (2012), 'Effect of heat of vaporization, chemical octane, and sensitivity on knock limit for ethanol–gasoline blends'. *SAE Int. J. Fuels Lubr.*, 5(2), 823–843.

Steinberg, M. (1977), 'Production of synthetic methanol from air and water using controlled thermonuclear reactor power – I. Technology and energy requirement', *Energy Conversion*, 17, 97–112.

Sterner, M. (2009), 'Bioenergy and renewable power methane in integrated 100% renewable energy systems', Dr.-Ing Thesis, University of Kassel, September.

Stucki, S., Schuler, A. and Constantinescu, M. (1995), 'Coupled $CO_2$ recovery from the atmosphere and water electrolysis: feasibility of a new process for hydrogen storage', *Int. J. Hydrogen Energy*, 20(8), 653–663.

Thornton, M.J. (2009), 'Impacts of biodiesel fuel blends oil dilution on light-duty diesel engine operation', SAE paper no. 2009-01-1790.

Turner, J.W.G. and Pearson, R.J. (2008), 'The application of energy-based fuel formulae to increase the efficiency relevance and reduce the $CO_2$ emissions of motor sport'. SAE paper no. 2008-01-2953. SAE Motorsports Engineering Conference and Exposition, Concord, NC, December.

Turner, J.W.G., Pearson, R.J., Holland, B. and Peck, R. (2007a), 'Alcohol-based fuels in high performance engines'. SAE paper number 2007-01-0056, SAE Fuels and Emissions Conference, Cape Town, South Africa, January.

Turner, J.W.G., Peck, A., and Pearson, R.J. (2007b), 'Flex-fuel vehicle development to promote synthetic alcohols as the basis for a potential negative-$CO_2$ energy economy', SAE paper no. 2007-01-3618.

Turner, J.W.G., Pearson, R.J., Purvis, R., Dekker, E., Johansson, K. and ac Bergström, K. (2011), 'GEM ternary blends: removing the biomass limit by using iso-stoichiometric mixtures of gasoline, ethanol and methanol', SAE paper number

2011-24-0113, The 10th International Conference on Engines and Vehicles, Capri, Naples, Italy, 11–16 September.

Turner, J.W.G., Pearson, R.J., Dekker, E., Iosefa, B., Dolan, G.A., Johansson, K. and ac Bergström, K. (2012a), 'Evolution of alcohol fuel blends towards a sustainable transport energy economy'. 2012 MIT Energy Initiative Symposium: Prospects for Flexible- and Bi-Fuel Light Duty Vehicles, Cambridge, MA, 19 April.

Turner, J.W.G., Pearson, R.J., McGregor, M.A., Ramsay, J.M., Dekker, E., Iosefa, B., Dolan, G.A., Johansson, K. and ac Bergström, K. (2012b), 'GEM ternary blends: testing iso-stoichiometric mixtures of gasoline, ethanol and methanol in a production flex-fuel vehicle fitted with a physical alcohol sensor', SAE paper number 2012-01-1279, SAE 2012 World Congress, Detroit, MI, 24–26 April.

Turner, J.W.G., Pearson, R.J., Bell, A., De Goede, S. and Woolard, C. (2012c), 'GEM ternary blends of gasoline, ethanol and methanol: investigations into exhaust emissions, blend properties and octane numbers', SAE paper number 2012-01-1586, SAE Powertrains, Fuels and Lubricants Meeting, Malmo, Sweden, 18–20 September.

Turner, J.W.G., Pearson, R.J., Dekker, E., Iosefa, B., Dolan, G., Johansson, K. and ac Bergström, K. (2013), 'Extending the role of alcohols as transport fuels using iso-stoichiometric ternary blends of gasoline, ethanol and methanol', *Applied Energy*, 102, 72–86.

UN (2006), 'World agriculture: towards 2030/2050', FAO, United Nations, Rome. Available from: http://www.fao.org/fileadmin/user_upload/esag/docs/ (accessed 20 December 2011).

Urban, C.M., Timbario, T.J. and Bechtold, R.L. (1989), 'Performance and emissions of a DDC 8V-71 engine fueled with cetane improved methanol', SAE paper no. 892064.

US Congress (2007), Energy Independence and Security Act of 2007, Public Law 110-140, 110th Congress, DOCID: f:publ140.110.

US DoE (2012a), 'Energy efficiency and renewable energy', Alternative Fuels and Advanced Vehicles Data Center. Available from: http://www.afdc.energy.gov/afdc/data/fuels.html (accessed 20 June 2012).

US DoE (2012b), 'Biodiesel blends', Alternative Fuels and Advanced Vehicles Data Center. Available from: http://www.afdc.energy.gov/fuels/biodiesel_blends.html (accessed 20 June 2012).

Vancoille, J., Verhelst, S., Demuynck, J., Galle, J., Sileghem, L. and Van De Ginste, M. (2012), 'Experimental evaluation of lean-burn and EGR as load control strategies for methanol engines', SAE paper no. 2012-01-1283.

von Helmolt, R. and Eberle, U. (2007), 'Fuel cell vehicles: status 2007', *J. Power Sources*, 165, 833–843.

Wagner, T. and Wyszyński, M.L. (1996), 'Aldehydes and ketones in engine exhaust emissions – a review', *Proc. Instn. Mech. Engrs Journal of Automotive Engineering, Part D*, 210, 109–122.

WBGU (2008), 'World in transition: future bioenergy and sustainable land use: summary for policy makers', October. Available from: http://www.cbd.int/doc/biofuel/wbgu-bioenergy-SDM-en-20090603.pdf (accessed 15 December 2011).

Weimer, T., Schaber, K., Specht, M. and Bandi, A. (1996), 'Methanol from atmospheric carbon dioxide: a liquid zero emission fuel for the future', *Energy Conversion and Management*, 37(6–8), 1351–1356.

West, B.H., López, A.J., Theiss, T.J., Graves, R.L., Storey, J.M. and Lewis, S.A. (2007), 'Fuel economy and emissions of the ethanol-optimized Saab 9-5 Biopower', SAE paper number 2007-01-3994, SAE Powertrain & Fluid Systems Conference and Exhibition, Chicago, IL, October.

White, T.L. (1907), 'Alcohol as a fuel for the automotive motor', SAE paper no. 070002.

Wuebben, P., Unnasch, S., Pellegrin, V., Quigg, D. and Urban, B. (1990), 'Transit bus operation with a DDC 6V-92TAC engine operating on ignition-improved methanol', SAE paper no. 902161.

Zinoviev, S., Muller-Langer, F., Das, P., Bertero, N., Fornasiero, P., Kaltschmitt, M., Centi, G. and Miertus, S. (2010), 'Next-generation biofuels: survey of emerging technologies and sustainability issues', *ChemSusChem*, 3, 1106–1133.

## 13.10 Appendix: List of abbreviations

| | |
|---|---|
| AFR | air:fuel ratio |
| ATDC | after top-dead-centre |
| BXX | blend of XX% by volume of biodiesel (FAME) in diesel |
| BEV | battery electric vehicle |
| BMEP | brake mean effective pressure |
| BOB | blend-stock for oxygenate blending |
| BTL | biomass-to-liquids |
| BuXX | blend of XX% by volume of butanol in gasoline |
| CAFE | corporate average fuel economy |
| CFR | Co-operative Fuels Research |
| CNG | compressed natural gas |
| CoV | coefficient of variation |
| CTL | coal-to-liquids |
| DI | direct (fuel) injection |
| DME | dimethyl ether |
| ECU | electronic control unit |
| EGR | exhaust gas recirculation |
| EMS | engine management system |
| EREV | extended-range electric vehicle |
| ETBE | ethyl-tert butyl ether |
| EXX | blend of XX% by volume of ethanol in gasoline |
| EU | European Union |
| FAME | fatty acid methyl ester |
| FFV | flexible-fuel vehicle |
| GEM | gasoline-ethanol-methanol |
| GHG | greenhouse gas |
| GTL | gas-to-liquids |
| HC | hydrocarbon |
| HFCEV | hydrogen fuel cell electric vehicle |

| | |
|---|---|
| HHV | higher heating value |
| ICE | internal combustion engine |
| ICEV | internal combustion engine vehicle |
| IMEP | indicated mean effective pressure |
| LCOE | levelized cost of energy |
| LHV | lower heating value |
| LNG | liquid natural gas |
| LPG | liquid petroleum gas |
| MBT | minimum advance for best torque |
| MFB | mass fraction burned |
| MON | motor octane number |
| MTBE | methyl-tert butyl ether |
| MTG | methanol-to-gasoline |
| Mtoe | million tonnes of oil equivalent |
| MXX | blend of XX% by volume of methanol in gasoline |
| MY | model year |
| NEDC | New European Drive Cycle |
| NiMH | nickel metal hydride |
| NMEP | net mean effective pressure |
| $NO_x$ | oxides of nitrogen |
| OPEC | Organization of Petroleum Exporting Countries |
| PEM FC | proton exchange membrane fuel cell |
| PI | port (fuel) injection |
| PHEV | plug-in hybrid electric vehicle |
| POME | palm oil methyl ester |
| RFS | renewable fuel standard |
| RME | rapeseed methyl ester |
| RON | research octane number |
| SI | spark-ignition |
| SME | soya bean methyl |
| TTW | tank-to-wheel |
| US | United States |
| WTT | well-to-tank |
| WTW | well-to-wheel |

# 14
# Biodiesel and renewable diesel production methods

J. H. VAN GERPEN and B. B. HE, University of Idaho, USA

DOI: 10.1533/9780857097385.2.441

**Abstract**: Vegetable oils and animal fats have come to be recognized as important sources of renewable fuels. Fatty acid methyl esters are known as biodiesel, and are the leading source for non-petroleum diesel fuel. With hydrogenation and isomerization, oils and fats can be converted to renewable diesel and jet fuel that are drop-in replacements for petroleum diesel fuel and jet fuel. This chapter reviews the processes used to produce these fuels and the feedstocks that will provide the supplies of oils and fats needed to meet a growing world demand.

**Key words**: biodiesel, renewable diesel, biofuels, bioenergy, renewable energy.

## 14.1 Introduction

Concern about the limited availability of fossil fuels and their impact on climate change is driving research into alternative energy sources. Replacing transportation fuels is particularly challenging because of the need for an energy dense fuel that will provide equivalent or better exhaust emissions and not contribute to corrosion or deposit formation in engines. Lignocellulosic biomass is seen as a plentiful and low cost feedstock for alternative transportation fuels (US DOE, 2011). However, starch, sugar and oil crops and animal fats are currently the most widely used feedstocks for alternative transportation fuels because the conversion processes are much simpler and more cost-effective.

Ethanol production from corn and other cereal grains is approaching 15 billion gallons per year, or about 10% of the current consumption of gasoline (EIA, 2012). Production of biomass-based diesel fuel in the US surpassed 1 billion gallons in 2011, which was about 3% of the on-highway diesel fuel consumption at the time (US EPA, 2012a). This biomass-based diesel fuel was composed primarily of fatty acid methyl esters (FAME), or *biodiesel*, along with a small quantity of non-ester-based fuel, which will be referred to here as *renewable diesel* fuel. Processes to produce diesel fuel from lignocellulosic biomass have been proposed (e.g., Huber *et al.*, 2005), but there is no commercial production at this time. This chapter describes the characteristics, production, feedstock sources and current status of diesel fuels produced from vegetable oils and animal fats.

## 14.2 Overview of biodiesel and renewable diesel

Biodiesel is defined to be the monoalkyl esters of fatty acids from vegetable oils and animal fats (ASTM D 6751). It is produced via transesterification of the triacylglycerides in the oils and fats. The same triacylglyceride feedstock used for biodiesel can be hydrogenated to a hydrocarbon product called *renewable diesel* (Furimsky, 2000). Hydrogenation can also be used to upgrade a wide variety of different low-grade oils produced by thermochemical processes such as pyrolysis and liquefaction. While the paraffinic products of these processes bear some similarity to the hydrogenated products of triacylglycerides, they are not currently commercially viable fuels and are therefore not included in this discussion.

Carefully defining the meanings of the terms used for various fuels is important as the feedstock and processing options proliferate. This has led to the development of detailed fuel specifications that describe the performance properties of the fuel along with some limited description of its composition. For example, ASTM D 6751, the biodiesel standard used in the US (ASTM, 2011), is the source of the definition statement given above. The standard used in Europe, EN 14214, uses a similar definition. These standards do not include non-ester products such as the paraffinic product produced by hydrogenating the oils and fats. The wide range of possible feedstocks and processes that have been proposed for the production of alternative diesel fuels presents a major challenge to standard development, although a specification for non-petroleum jet fuel (ASTM D 7566) has been approved.

## 14.3 Renewable diesel production routes

The basic process to produce renewable diesel starts with hydrogenation which saturates the double bonds and removes the oxygen, either as $H_2O$ or $CO_2$ depending on the availability of hydrogen, from the fatty acid chains of the triacylglyceride. Hydrogenation and decarboxylation are two of the basic reactions that occur during the production of renewable diesel and are shown in Fig. 14.1.

In this case, a representative triacylglyceride molecule, triolein, is converted to the n-paraffin molecules heptadecane or octadecane by separate pathways that depend on the availability of hydrogen. Hydrogenation preserves more of the original carbon in the fuel but requires 2.5 times more hydrogen than decarboxylation. In either case, propane is produced as a byproduct, which can be reformed and used as a hydrogen source or sold directly as fuel. The ideal liquid fuel mass yields for the reactions are 0.815 kg $C_{17}H_{36}$/kg triolein and 0.862 kg $C_{18}H_{38}$/kg triolein for decarboxylation and hydrogenation, respectively. Most of the mass loss is as $CO_2$, water,

Decarboxylation     $C_{57}H_{104}O_6 + 6H_2 \rightarrow 3C_{17}H_{36} + 3CO_2 + C_3H_8$

Hydrogenation     $C_{57}H_{104}O_6 + 15H_2 \rightarrow 3C_{18}H_{38} + 6H_2O + C_3H_8$

*14.1* Decarboxylation and hydrogenation reactions for triolein.

and propane but only the propane has a market value. Because the final product is typically sold on a volume basis, the volume yield is also of interest. Since the densities of n-paraffins are typically very low (0.777 kg/L for both $C_{17}H_{36}$ and $C_{18}H_{38}$), the volume yields are much higher, at 0.954 L $C_{17}H_{36}$/L triolein and 1.010 L $C_{18}H_{38}$/L triolein for decarboxylation and hydrogenation, respectively.

In the case shown in Fig. 14.1, the renewable diesel will consist of C17 and C18 n-paraffins. A real vegetable oil or animal fat will contain a diverse mixture of fatty acid chains in the triacylglycerides resulting in a mixture of n-paraffins suitable for blending with petroleum-based diesel fuel. One drawback of using real vegetable oils and animal fat is that the cloud point is usually found to be unacceptably high (Knothe, 2010). In this form, the renewable diesel is only suitable for use as a low-level blending component for No. 2, or heavier, diesel fuel. In a refinery, this paraffinic stream can be isomerized, cracked, and distilled to produce fuels in the boiling range of jet fuel or even gasoline. This is the product that is most commonly referred to as bio-jet fuel. The economics of bio-jet fuel are challenging because the price of diesel fuel and jet fuel are traditionally very similar, so there is no financial incentive for refiners to do the extra processing needed to upgrade the paraffinic renewable diesel stream to jet fuel specifications.

Hydrogenation of triacylglycerides may be done with dedicated processes or by co-feeding the triacylglycerides with petroleum-based products (Sebos *et al.*, 2009). In the latter case, the n-paraffins produced from the triacylglycerides enhance some properties, such as the cetane number, but may also impact cold flow properties (Sebos *et al.*, 2009). Common hydrotreatment catalysts such as NiMo or CoMo on $\gamma$-$Al_2O_3$ are used as opposed to zeolites, which tend to promote hydrocracking and lower the yield of fuel in the diesel boiling range. Although cracking and pyrolysis may provide molecules with better cold flow properties, these processes also produce substantial amounts of low-value materials with little value as transportation fuels. Therefore, isomerization is more commonly used to enhance the cold flow properties of renewable diesel (Knothe, 2010).

Although hydrogenation pathways consume more hydrogen than decarboxylation, water-gas-shift reactions with $CO_2$ were indicated as significant sinks for hydrogen (Donnis *et al.*, 2009) and may be more significant sources of CO in the product gases than decarbonylation reactions.

## 14.4 Biodiesel production routes

Biodiesel is most commonly produced via a transesterification reaction between a vegetable oil or animal fat and a simple alcohol such as methanol, as shown in Fig. 14.2 (Van Gerpen, 2005). The reaction is usually catalyzed with a strong base such as sodium methoxide although sodium or potassium hydroxides are also used.

Another option is to use solid phase, or heterogeneous, catalysts currently under development. These will be discussed in more detail in a later section. The oil or fat is dried before the reaction so that moisture does not have a chance to enhance the saponification side reaction which consumes the catalyst and decreases yield. After drying takes place, the oil, alcohol, and catalyst are mixed and then agitated for a period of 10 minutes to an hour depending on the level and type of agitation. Simple stirred reactors are frequently used, although high shear mixers, ultrasonic mixers, and even co-solvents are used to accelerate the reaction. It is very common to conduct the reaction in two or more steps with a partial reaction allowed to occur in a first stage reactor, then removal of the glycerin that has been formed (which is accompanied by a significant fraction of the catalyst) and then adding additional methanol and catalyst for a second reaction to the final equilibrium.

Lower cost feedstocks frequently contain elevated levels of free fatty acids (FFAs). These can be removed by various techniques but can also be converted to methyl esters using an acid catalyzed pretreatment (Canakci and Van Gerpen, 2001). This pretreatment will be discussed later, but basically involves adding sulfuric acid and methanol to convert the FFAs to methyl esters so that the standard alkali-catalyzed process can be used to convert the triacylglyceride portion of the feedstock into biodiesel.

After the reaction is complete, the remaining glycerin is removed by either settling in a decanter, centrifugation, or possibly with a coalescer,

$$\begin{array}{c} \text{O} \\ \parallel \\ CH_2\text{-}O\text{-}C\text{-}R_1 \\ | \\ \text{O} \\ \parallel \\ CH\text{-}O\text{-}C\text{-}R_2 \\ | \\ \text{O} \\ \parallel \\ CH_2\text{-}O\text{-}C\text{-}R_3 \end{array} + 3\,CH_3OH \xrightarrow{(NaOCH_3)} \begin{array}{c} \text{O} \\ \parallel \\ CH_3\text{-}O\text{-}C\text{-}R_1 \\ \\ \text{O} \\ \parallel \\ CH_3\text{-}O\text{-}C\text{-}R_2 \\ \\ \text{O} \\ \parallel \\ CH_3\text{-}O\text{-}C\text{-}R_3 \end{array} + \begin{array}{c} CH_2\text{-}OH \\ | \\ CH\text{-}OH \\ | \\ CH_2\text{-}OH \end{array}$$

Triacylglyceride      Methanol      Mixture of fatty esters      Glycerin

*14.2* Transesterification of triacylglyceride.

although this approach is less common. Then the methanol is removed by flash evaporation and the small amounts of residual free glycerin, soaps, and methanol are removed by washing with deionized water, with an ion exchange resin, or with a solid adsorbent such as magnesium silicate. The latter two options are sometimes referred to as 'dry washing.' The final product may be further subjected to a cold filtration process whereby the fuel is cooled to near its freezing point, held at that temperature for sufficient time to allow the crystallization of minor impurities such as sterol glucosides and saturated monoglycerides, and then filtered. The filtration usually requires the addition of a filter aid such as diatomaceous earth and may follow a period of warming back to close to ambient temperature. Additives such as antioxidants and pour point depressants may be added before the fuel is sold. Finally, the fuel is analyzed to verify compliance with the ASTM specification and to ensure fuel quality (Van Gerpen, 2005).

The following discussion will focus in greater depth on recent developments in catalyst technology as well as ultrasonic and supercritical reactors and waterless purification techniques.

### 14.4.1 Heterogeneous catalysts for biodiesel production

A major concern with using a homogeneous alkali catalyst, such as sodium methylate, for biodiesel production is the large quantity of water used for washing the residual catalyst and soap from the biodiesel. The ratio of water used is typically 1:1, i.e., for every gallon of biodiesel produced, one gallon of washing water is needed. This ratio can be reduced to 10% of the volume of biodiesel produced if the soap is split by acidulation prior to or during washing (Van Gerpen, 2005). The consumption of wash water can be virtually eliminated if facilities are in place to recycle and reuse the wash water.

Another drawback of using a homogeneous base catalyst is soap formation due to the high FFA content (e.g., greater than 5 wt%) of the feedstocks, such as waste vegetable oils or microalgal oil. Soap formation in biodiesel production not only adversely affects the biodiesel yield but also makes biodiesel separation from the byproduct glycerin very difficult. When high concentrations of FFAs are present, a two-step process, i.e., a strong acid-catalyzed esterification followed by a base-catalyzed process, is needed to avoid the problem of significant soap formation (Canakci and Van Gerpen, 1999, 2001).

The advantages of using heterogeneous catalysts in biodiesel production include the elimination of water-washing or dry-washing of the post-reaction mixture for catalyst and soap removal. Other advantages include decreased production of waste water, reuse of the heterogeneous catalysts, high productivity per unit of reactor capacity, and easier scale-up for continuous-flow processes. It is also claimed that heterogeneous catalysts can catalyze both the esterification and transesterification reactions so that the two-step process is not required to deal with high FFA feedstocks (Furuta *et al.*, 2004). A major disadvantage of using heterogeneous catalysts, however, is the high catalyst cost and higher operating cost due to the elevated temperatures usually required (Kiss *et al.*, 2010; Sakai *et al.*, 2009). To date, no cost-effective and highly efficient heterogeneous catalyst has been identified for practical commercial use.

Heterogeneous catalysts for biodiesel production are mainly metal hydroxides and oxides and other non-metal-based compounds (Borges and Diaz, 2012; Semwal *et al.*, 2011; Chouhan and Sarma, 2011). The catalyst supports are typically aluminum, activated carbon, and organic resin. Table 14.1 provides an overview of the current efforts in exploring suitable heterogeneous catalysts for biodiesel production. Generally, all of the reported catalysts lack activity at low operating temperatures (65°C or lower). Even with extended periods of reaction time, the overall biodiesel yields or vegetable oil conversion rates catalyzed by heterogeneous catalysts are low (less than 98%) and are not comparable with those by homogeneous catalysts. This incomplete transesterification usually translates to poor quality biodiesel and it is likely that the fuel will not meet the standards specified by the ASTM.

There are various methods to increase the catalyst activity, one of which is to raise the operating temperature to 120°C, or even to 200°C and higher, so that the metal catalysts start to show their advantageous catalytic activity (Jitputti *et al.*, 2006; Garcia *et al.*, 2008; Carmo *et al.*, 2009: Peng *et al.*, 2008; Park *et al.*, 2010). However, this will obviously increase operating costs. Other cost considerations must be made when using heterogeneous catalysts in biodiesel production. This process requires a much higher methanol-to-vegetable oil molar ratio, as high as 60:1, in order to achieve a reasonable reactivity (Benjapornkulaphong *et al.*, 2009; Garcia *et al.*, 2008; Carmo *et al.*, 2009; Nakagaki *et al.*, 2008). Such a high methanol-to-oil molar ratio leads to additional effort in methanol recovery, re-purification, and reuse, and thus adds to the operating cost (Sakai *et al.*, 2009; Kiss *et al.*, 2010). Some catalyst producers recommend that the fuel be vacuum distilled to remove the products of incomplete reaction, which is an indication of low catalyst activity.

Most of the heterogeneous catalysts described in the literature were tested in a powder form or a slurry suspended in a vegetable oil and methanol mixture with high mixing intensity (up to 1500 rpm) in a batch mode. Keeping these catalysts suspended in large-scale, continuous-flow reactors would be a challenge. Pelletized or other shaped heterogeneous catalysts are typically not developed unless a catalyst has proved to be effective and worth the effort. At that stage, some engineering challenges must be overcome, such as the strength and pore sizes of the structure and

Table 14.1 Examples of commonly researched heterogeneous catalysts and their effectiveness

| No. | Active ingredients/support | Conditions of use | Examples of effectiveness | References |
|---|---|---|---|---|
| 1 | CaO, activated powder | Molar ratio of methanol to oil 13:1, 60°C | Yield: 94 wt% of oil after 90 min | Granados et al., 2007 |
| 2 | MgO, nano particles | Molar ratio of methanol to oil 6–36:1, 250°C and 24 MPa | Yield: approx. 100 wt% of oil after 20 min with 2–5 wt% catalyst | Wang and Yang, 2007 |
| 3 | CaO, activated powder | Molar ratio of methanol to oil 12:1, 65°C, catalyst application 8 wt% | Yield: 95 wt% after 3 h Catalyst activity decreases with repeated uses | Liu et al., 2007 |
| 4 | CaO, powder | Ratio of methanol to oil 3.9:15 g, 60°C, catalyst application 0.1 g per 15 g of oil | Yield: approx. 90 wt% after 3 h | Kawashima et al., 2009 |
| 5 | CaO, powder | Ratio of methanol to oil 50:100 ml, ~64°C (reflux), catalyst application 0.78 g per 100 ml of oil | Yield: approx. 90 wt% after 1 h | Kouzu et al., 2008 |
| 6 | $CaZrO_3$ and $CaO-CeO_2$, powder | Molar ratio of methanol to oil 6:1, 60°C, catalyst application 10 wt% of oil | Yield: approx. 90 wt% after 10 h | Kawashima et al., 2008 |
| 7 | $Ca_xMg_{2-x}O_2$ from MgO and $Ca(NO_3)_2$ | Molar ratio of methanol to oil 12:1, catalyst application 6 wt%, approx. 65°C (reflux) | Oil conversion of 91.3% achieved after 5 h | Xie et al., 2012 |

Continued

Table 14.1 Continued

| No. | Active ingredients/support | Conditions of use | Examples of effectiveness | References |
|---|---|---|---|---|
| 8 | Ca-Si oxides | Ratio of 24 mL methanol to 1.0 g oil, catalyst application 20 wt% of oil, 65°C | Yield: approx. 100% after 4 h | Hsin et al., 2010 |
| 9 | SrO, activated powder | Molar ratio of methanol to oil 15:1–18:1, 65–70°C, catalyst application 2.5–3.0 wt% | Yield: approx. 95 wt% after 30 min<br>Catalyst activity decreases with repeated uses | Liu et al., 2008 |
| 10 | $SrO/SiO_2$ | Molar ratio of methanol to oil: 6:1, catalyst application 5 wt% of oil, 65°C | Yield: approx. 80 wt%<br>Catalyst activity decreases with repeated uses | Chen et al., 2012 |
| 11 | $K_2CO_3$ supported on MgO | Ratio of methanol to oil 1.12:5 g, catalyst 50 mg/5 g of oil, 70°C | Yield: approx. 99 wt% after 2 h | Liang et al., 2009 |
| 12 | $KAlSiO_4$ (kalsilite) | Ratio of methanol to oil: 150:300 g, catalyst application 5 wt% of oil, 120–180°C | Yield: approx. 100% | Wen et al., 2010 |
| 13 | KI/mesoporous silica | Molar ratio of methanol to oil 16:1, 70°C, a catalyst application 5.0 wt% of oil | Approx. 90% of conversion achieved after 8 h | Samart et al., 2009 |
| 14 | Tetramethylguanidine on silica gel | Ratio of methanol to oil 1.5:10.0 g, catalyst application 0.7 g of 10 g oil, 80°C | Yield: approx. 86% after 3 h<br>Catalyst activity decreases with repeated uses | Faria et al., 2008 |

| | | | | |
|---|---|---|---|---|
| 15 | Sr(NO$_3$)$_2$ on ZnO | Molar ratio of methanol to oil 12:1, approx. 65°C (reflux), catalyst application 5 wt% | Conversion: 94.7% after 4 h | Yang and Xie, 2007 |
| 16 | LiNO$_3$/Al$_2$O$_3$, NaNO$_3$/Al$_2$O$_3$, and KNO$_3$/Al$_2$O$_3$ | Molar ratio of methanol to oil 65:1, catalyst application 10 wt% or more at 60°C | Methyl esters content: approx. 94% achieved after 3 h | Benjapornkulaphong et al., 2009 |
| 17 | S-ZrO$_2$ (sulfated zirconia) | Molar ratio of methanol to oil 20:1, catalyst application 5 wt% at 120°C | Yield: approx. 99% after 2 h | Garcia et al., 2008 |
| 18 | Al-MCM-41 mesoporous molecular sieves | Molar ratio of methanol to oil 60:1, catalyst application 0.6 wt% at 130°C | Conversion: approx. 79% after 2 h | Carmo et al., 2009 |
| 19 | La/zeolite from La(NO$_3$)$_3$ | Molar ratio of methanol to oil 14.5:1, catalyst application 1.1 wt% at 60°C | Conversion: approx. 49% after 4 h | Shu et al., 2007 |
| 20 | Na$_2$MoO$_4$ (sodium molybdate) | Molar ratio of methanol to oil 54:1, catalyst application 5 wt% at 60°C | Yield: approx. 96% after 3 h | Nakagaki et al., 2008 |
| 21 | SO$_4^{2-}$/TiO$_2$–SiO$_2$[a] | Molar ratio of methanol to oil 9:1 to 12:1, catalyst application 3 wt% at 200–220°C | Yield: approx. 95% after 3–4 h | Peng et al., 2008 |
| 22 | SO$_4^{2-}$/SnO$_2$ powder, and SO$_4^{2-}$/ZrO$_2$ powder | Molar ratio of methanol to oil 6:1, catalyst application 0.5–3 wt%, 250°C and 24 MPa | Yield: 90.3 wt% after 4 h | Jitputti et al., 2006 |
| 23 | SO$_4^{2-}$/ZrO$^2$ | Molar ratio of methanol to oil: 9:1, catalyst application 3.5 wt% of oil, 120°C | Yield: approx. 98 wt% after 1 h | Niu et al., 2012 |

*Continued*

Table 14.1 Continued

| No. | Active ingredients/support | Conditions of use | Examples of effectiveness | References |
|---|---|---|---|---|
| 24 | Tungstated zirconia (W-Zr) | Molar ratio of methanol to oil 6:1, catalyst application 2 wt% at 60°C | Yield: approx. 95% after 3–4 h | Borges et al., 2011 |
| 25 | KAc (potassium acetate)-NaX zeolite interchanged | Molar ratio of methanol to oil: 12:1 to 72:1, catalyst application 3–6 wt% of oil, 100–155°C, and 6–88 bar pressure | Yield: approx. 96% at 48:1 methanol to oil ration, 6 wt% catalyst, 155°C for 3 h | Gomes et al., 2011 |
| 26 | Mg-Al hydrotalcites | Molar ratio of methanol to oil: 9:1 to 12:1, catalyst application 2.5 wt% of oil, 60–65°C | Yield: approx. 97% after 4 h | Kondamudi et al., 2011 |
| 27 | Na- and $NH_4$-quntinites, bi-functional | Molar ratio of methanol to oil: 15:1, catalyst application 10 wt% of oil, 75°C | Yield: approx. 98% after 2 h | Li et al., 2011 |
| 28 | $Nd_2O_3$-K (neodymium oxide with potassium hydroxide) | Molar ratio of methanol to oil: 14:1, catalyst application 6 wt% of oil, 60°C | Yield: approx. 92% after 1.5 h | Park et al., 2010 |
| 29 | $WO_3/ZrO_2$ | Molar ratio of methanol to oil: 9:1, catalyst application 10–20 wt% of oil, 75–200°C | Yield: approx. 93–98% | Wang et al., 2011 |
| 30 | $Li_2SiO_3$ | Molar ratio of methanol to oil: 12:1 to 36:1, catalyst application 10 wt% of oil, 60°C | Yield: approx. 92–96% after 3 h | |

[a] It is claimed that a 10,000 tonnes/year biodiesel production demonstration plant using this catalyst has been built.

ensuring adequate capability for mass transfer. Catalyst poisoning is also a serious consideration for commercialization. Another problematic phenomenon is the reusability of the heterogeneous catalysts. Some catalysts appear to lose their reactivity quickly due to leaching after being used for only a few batches (Liu et al., 2007, 2008; Faria et al., 2008; Di Serio et al., 2010; Chen et al., 2012).

### 14.4.2 Ultrasonic processing

Application of ultrasonication in chemical reaction systems has been extensively researched due to its effective production of radicals for initiating chemical reactions and its localized micro-scale cavitation and mixing (Shol, 1988; Flint and Suslick, 1991). The study of chemical changes induced by ultrasonication is now recognized as the science of sonochemistry and the chemical reactors that incorporate ultrasound are referred to as sonochemical reactors (Mason, 2000; Thompson and Doraiswamy, 1999).

The insolubility of methanol in vegetable oils is a limiting factor for the transesterification of triacylglycerides for biodiesel production (Van Gerpen, 2005). To overcome this problem, mechanical mixing is typically applied in conventional processes to improve the reaction rate. Recognizing its effectiveness in creating micro-scale cavitation and intensified local mixing, researchers have attempted to apply ultrasonication in transesterification of vegetable oils and animal fats for biodiesel production (e.g., Ji et al., 2006; Wu et al., 2007; Hanh et al., 2009).

The level of positive effect of ultrasonication on transesterification varies largely among reports. One consistent advantage is the shorter reaction time, which ranges from one minute (Teixeira et al., 2009) to a few minutes (Stavarache et al., 2005; Singh et al., 2007; Kumar et al., 2010a). Although most work has been done with conventional homogeneous catalysts, similar phenomena were observed in systems where heterogeneous catalysts were used (Kumar et al., 2010b; Yu et al., 2010).

Currently, most reports published on ultrasound-assisted transesterification are laboratory evaluations. The exact mechanism behind the enhancement by ultrasonication is not yet clearly understood. Based on experience with other ultrasound-assisted chemical systems, the ultrasonic cavitation combined with the radial motion of the ultrasonic cavitation bubbles are proposed to be the reason for the accelerated reaction rates (Mootabadi et al., 2010). The localized high intensity energy generates microbubbles that create increased interfacial areas to overcome the mass transfer limit. Such high energy density may also overcome the activation energy to initiate the transesterification (Singh et al., 2007). However, such a phenomenon is significant only in the close proximity to the ultrasound transducer (Monnier et al., 1999a, 1999b; Cintas et al., 2010). Another important finding under specific experimental conditions is that chemical radicals are not the reason for the enhanced transesterification under ultrasonication as claimed in other sonochemical systems (Kalva *et al.*, 2009). This may be due to the fact that the moderate intensity of ultrasonication used for biodiesel production does not generate enough energy to produce radical formation.

Some use of ultrasound-assisted systems in smaller biodiesel production plants has been reported but no technical details are provided. According to the nature of ultrasonication and the experience obtained from other sonochemical systems, chemically effective and economically viable sonochemical reactors for biodiesel production still require additional research and development, especially regarding the engineering aspects of the sonochemical reactors (Thompson and Doraiswamy, 1999). The widespread commercial application of ultrasonication biodiesel production may take another decade. A robust yet highly productive transesterification system may be created through a combination of sonochemical reactors with heterogeneous catalysts for biodiesel production, which may also be more suitable for small producers (He and Van Gerpen, 2012a).

### 14.4.3 Supercritical processing

Conventional processes for biodiesel production use homogeneous catalysts, such as sodium methoxide or hydroxide which require water washing or 'dry washing' with absorbent materials to remove soaps from the crude biodiesel. This washing step requires extra equipment and consumes the homogeneous catalyst, which increases the overall operating cost. One non-catalyzed approach for biodiesel production is the processing of vegetable oils in supercritical methanol. Saka and his team first reported non-catalytic supercritical biodiesel preparation starting in the late 1990s and have published extensively in this field (e.g., Saka and Kusdiana, 1999, 2001a, 2001b; Kusdiana and Saka, 2001a, 2004a; Warabi *et al.*, 2004a, 2004b; Saka *et al.*, 2006).

Supercritical processing of vegetable oils can achieve the necessary transesterification without the need for a catalyst. Once in the supercritical stage, methanol has a much stronger solvent effect that dissolves the vegetable oils and thus overcomes the insolubility between the vegetable oils and methanol, allowing the reaction to proceed quickly in a homogeneous phase at a high reaction temperature. The enhanced solvent capability of supercritical methanol is attributed to its much weakened hydrogen bonds and highly reduced polarity (Yamaguchi *et al.*, 2000). Most reported research uses supercritical methanol as the solvent. The operating conditions are close to or above the supercritical properties of methanol (i.e., 240°C and 8.1 MPa) and in the range of 270–430°C. Although a higher temperature provides a more effective supercritical fluid, considering the thermal stability of the unsaturated fatty acids and the increased operating cost due to high temperature and high pressures, a temperature range of 270–300°C is recommended (He *et al.*, 2007b; Imahara *et al.*, 2008).

In addition to the elimination of the catalyst, another advantage of supercritical methanol processing is its tolerance of a high FFA concentration in the feedstock. This is due to the fact that FFA can be esterified directly to methyl esters (Kusdiana and Saka, 2001b).

Generally water in the vegetable oil adversely affects the process (e.g., soap formation which requires extra catalyst) and the biodiesel quality (e.g., hydrolyte formation) in the conventional alkali-catalyzed transesterification process (He and Van Gerpen, 2012b). However, in supercritical methanol processing, the presence of water was found to have no negative effect on biodiesel (i.e., methyl esters) production (Kusdiana and Saka, 2004a), because hydrolysis of triacylglycerides caused by the presence of water, if it occurs, leads to the formation of free fatty acids, which are then esterified into methyl esters. Therefore, a small amount of water in the vegetable oil or animal fat is not a major concern in supercritical biodiesel production.

Short chain, primary alcohols are the logical choice for transesterifying vegetable oils and animal fats (Warabi *et al.*, 2004a). However, other solvents can also be used in supercritical processing for biodiesel production, such as dimethyl carbonate (Ilham and Saka, 2010), methyl acetate and other carboxylate esters (Saka and Isayama, 2009; Goembira *et al.*, 2012; Niza *et al.*, 2012).

Recognizing the potential advantage of triacylglyceride hydrolysis in supercritical methanol to free fatty acids and subsequent esterification to methyl esters, a two-step process has also been studied by researchers. In this two-step process, vegetable oil was hydrolyzed in sub-critical water for 20 min at 270°C into free fatty acids and glycerin byproduct. After the mixture of glycerin and excess water was separated out, the free fatty acids were fed into a supercritical methanol reactor and esterified into methyl esters at a comparable operating temperature and for a comparable reaction time (Kusdiana and Saka, 2004b; Minami and Saka, 2006). Using acetic acid to replace water for triacylglyceride hydrolysis under subcritical conditions was also studied by the same group and claimed to have better effectiveness (Saka *et al.*, 2010).

### 14.4.4 Purification by adsorbents and resins

Magnesium silicate is widely used to remove free fatty acids and other polar compounds from used cooking oils to extend their life. It can also be used to purify biodiesel by adsorbing free glycerin, soaps, methanol as well as monoglycerides and sterol glucosides. Approximately 1% magnesium silicate powder is added to the biodiesel at 60–65°C. However, the exact treatment level depends on the amount of contaminants to be removed. The mixture is agitated for 20 minutes and then the adsorbent is removed by filtration. Typically, the residual methanol is removed from the crude biodiesel before the adsorbent is added so the active sites on the adsorbent particles are not overwhelmed by the alcohol (Berrios and Skelton, 2008).

An alternative process uses ion exchange resins to remove free glycerin and soaps. The resins usually consist of small (~0.5 mm) styrene beads with treated surfaces. The beads are placed in fixed beds and the biodiesel is pumped through the beads. When soap molecules in the biodiesel contact the resin beads, hydrogen ions from the beads are exchanged with the sodium (or potassium) ions from the soap (Wall *et al.*, 2011). Thus, removal of the soap causes an increase in the acid value of the biodiesel. Soap levels above about 2500 ppm may cause the resulting fuel to exceed the ASTM specified acid value. Glycerin is removed by adsorption to the resin bead surface, a connection that can be overcome by washing the beads with methanol, this allows the resin beads to regenerate. It is also claimed that magnesium silicate can also be utilized in fixed beds with regeneration by methanol wash.

## 14.5 Traditional and emerging feedstocks

### 14.5.1 Traditional feedstocks

Currently, soybean oil is the dominant feedstock in the US while canola and corn oils, animal fats, and yellow greases, are the supplemental feedstocks. Soybean oil is a co-product of the soybean meal industry. Conventional soybean seeds contain about 18–20% oil. Despite the low oil content, soybean oil produced in the US is still available in large quantities due to the vast amount of soybean production. For example, 2.97 billion bushels or approximately 80 million metric tons of soybeans were produced in 2012 (USDA, 2012a), which is approximately equivalent to $17.6 \times 10^6 \text{m}^3$ (or approx. 4.6 billion gallons) of oil. In 2012, 1.12 million metric tons (2.48 billion pounds) of canola seed was produced, which is equivalent to approximately 146 million gallons of oil (USDA, 2012b). However, most of the traditional feedstocks have established markets and cannot easily migrate to the biodiesel industry.

The 'double-zero' rapeseed or canola oil is the major feedstock for biodiesel in Europe. Rapeseed adapts well to a wide variety of climate and soil conditions and is cultivated widely around the world. The Canadian modified cultivar of rapeseed or canola is very low in the undesirable erucic acid and glucosinolates, and is now the main cultivar planted for its oil in

Europe and North America. Total biodiesel production by the EU-27 was 9.57 million tons (approx. 2.31 billion gallons) in 2010 (European Biodiesel Board, 2012). Although there were imported feedstocks such as palm oils, this production was mainly from rapeseed/canola oils.

It is predicted that rapeseed/canola production in 2012 will be 20 million metric tons in the EU-27 and 14.5–15.5 million metric tons in Canada (Mielke, 2012). If 45% oil content (CGC, 2012) is assumed, the canola oil produced by the EU-27 and Canada in 2012 will be approximately 9.0 million tons (or 2.6 billion gallons) and 6.75 million tons (or 1.95 billion gallons), respectively. It is expected that the biodiesel production by the EU-27 will increase in 2012. Therefore, in order to keep growing, biodiesel producers in Europe must explore other feedstock opportunities while competing with the food market for canola oil.

Vegetable oils are composed mainly of triacylglycerides, which are the glycerin esters of different fatty acids. When processed for food use, these oils contain very low free fatty acids, and are very low in sulfur and other heterogeneous chemicals. This makes them high quality feedstocks for biodiesel production, especially when using homogeneous alkali-catalyzed conversion processes. When using food-grade vegetable oil as the feedstock, soap formation is typically not a concern in biodiesel production.

Different oils and fats are characterized by the fatty acid chains that are present in the triacylglycerides. The nomenclature for identifying fatty acids is commonly $C_{X:Y}$, where X is the number of carbon atoms in the fatty acid chain and Y is the number of double bonds. The fatty acid profiles for common vegetable oils and animal fats are shown in Table 14.2.

Chemically, animal fats are composed of the same compounds as those in seed oils but with fatty acid profiles that contain larger fractions of saturated fatty acids ($C_{16:0}$ and $C_{18:0}$). Animal fats are the byproducts of the livestock and poultry industries, and are relatively inexpensive feedstocks. Although large quantities of animal fats are produced annually, these fats have established markets as food and food additives (such as butter, lard, and shortenings), animal feed additives, and industrial products such as fatty acids, soap, paints, and lubricants.

Greases are generally used vegetable oils and/or animal fats. Yellow grease, or waste vegetable oil (WVO), consists mostly of used cooking oils, and brown grease is typically recovered from grease traps of restaurants and food processing plants. Greases usually contain large amounts of free fatty acids and other polymerized and/or hydrolyzed compounds after repeated uses at elevated temperatures. Due to the manner of its collection, greases are commonly high in water content which complicates the conversion process. Brown grease is even more problematic due to its contamination by cleaning agents. Therefore, greases are low quality feedstocks for biodiesel production, and special processing and/or

Table 14.2 Fatty acid compositions of common seed oils and animal fats

| Oils/fats | Fatty acid profiles | | | | | | | | |
|---|---|---|---|---|---|---|---|---|---|
| | $C_{12:0}$ | $C_{14:0}$ | $C_{16:0}$ | $C_{18:0}$ | $C_{18:1}$ | $C_{18:2}$ | $C_{18:3}$ | $C_{20:0}$ | $C_{22:1}$ |
| Coconut | 45–53 | 16–21 | 7–10 | 2–4 | 5–10 | 1–2.5 | | | |
| Corn | | 1–2 | 8–16 | 1–3 | 20–45 | 34–65 | 1–2 | | |
| Cottonseed | | 0–2 | 20–25 | 1–2 | 23–35 | 40–50 | | | |
| Palm | | 0.5–2 | 39–48 | 3–6 | 36–44 | 9–12 | | | |
| Rapeseed, low erucic or canola | | | 1–3 | 2–3 | 50–60 | 15–25 | 8–12 | | |
| Rapeseed, high erucic | | | 1–3 | 0–1 | 10–15 | 12–15 | 8–12 | 7–10 | 45–60 |
| Soybean, high linoleic | | | 6–10 | 2–5 | 20–30 | 50–60 | 5–11 | | |
| Soybean, high oleic | | | 2–3 | 2–3 | 80–85 | 3–4 | 3–5 | | |
| Lard | | 1–2 | 25–30 | 10–20 | 40–50 | 6–12 | 0–1 | | |
| Tallow | | 3–6 | 22–32 | 10–25 | 35–45 | 1–3 | | | |

pretreatment are required to avoid problems caused by impurities in order to have a quality biodiesel product.

### 14.5.2 Emerging feedstocks

Emerging feedstocks, including camelina, jatropha, pennycress, and microalgae, have also been researched and utilized for biodiesel production. Camelina, a member of the *Brassica* family, has many species but the most important for biodiesel is *Camelina sativa* (Vollmann et al., 1996). It has recently attracted the attention of farmers in the northern US (McVay and Lamb, 2008). Camelina adapts well to marginal lands having dry and cool climate conditions with a short growing season that, with adequate rainfall, allows for double cropping. The yield of camelina in North America is about 890–1,350 kg/ha (800–1,200 lb/acre) and the oil content of camelina seeds varies from 30 to 40% (McVay and Lamb, 2008). Camelina oil consists of up to 88% unsaturated fatty acids (Putnam et al., 1993), which contribute to the desirable cold flow properties of the biodiesel produced from it. However, camelina seeds are very small (only up to 2 g per 1,000 seeds) and the plant is very strong when ripe, which makes harvesting camelina very difficult.

Jatropha is a perennial shrub that grows in various harsh climates and poor soil conditions. The jatropha plant and its fruits and seeds contain poisonous toxins so it is considered to be a non-food crop (Dias et al., 2012).

Jatropha seeds contain up to 35% non-edible oil (Kumar and Sharma, 2008) and its oil yield can be as high as 1,900 kg/ha (1,690 lb/acre), making it a good candidate for biodiesel production. Jatropha fruits ripen year round and harvesting is very labor-intensive. Unless automated harvesting machines are invented, jatropha fruit harvesting is unlikely to be cost-effective. Jatropha is being actively explored by countries that have limited agricultural land but are still interested in seeking oil sources for biofuels such as India, China, and other developing countries (Kumar et al., 2012; Yang et al., 2012; Mofijur et al., 2012). However, the suitability of jatropha as a large-scale domesticated crop has been questioned and discussions are underway to find sustainable ways to produce it (Contran et al., 2013; Kumar et al., 2012).

Pennycress is a collective name applied to a group of species belonging to the *Brassicaceae* family. Field pennycress, *Thlaspi arvense*, is a variety that has been identified as having industrial potential, especially for use as a feedstock for biodiesel. The seeds of pennycress contain up to 36% oil and are high in erucic and linoleic acids (Moser et al., 2009b). Pennycress seeds are small. Seeds harvested in Peoria, Illinois, have average dimensions of 1.87×1.35×0.74 mm and 1000 seeds weighed only 0.97 g at 9.5% moisture (Evangelista et al., 2012). However, the seed yield can approach 1,420 kg/ha (1,265 pounds/acre). Field pennycress is basically considered to be an annual weed, growing in open spaces and roadsides (Mitich, 1996). This species is tough and its seeds can have a long life, lasting approximately 20–30 years (Koundinya and Hansen, 2012). Evaluation of biodiesel produced from pennycress oil has reported desirable properties due to the high contents of erucic (32.8 wt%) and linoleic (22.4 wt%) acids. The biodiesel produced from pennycress oil has a cetane number close to 60 and exhibits favorable properties at low temperatures (Moser et al., 2009a). To be established as a feasible feedstock for biodiesel production, more studies are needed on pennycress to establish its agronomic, harvesting, and processing requirements as well as its environmental and social impacts.

One of the most promising feedstocks for biodiesel production is believed to be microalgae oil (Wu et al., 2012; Ahmad et al., 2011; Stephens et al., 2010; Chisti, 2008). As a potential source for biofuel production, microalgae have been exploited and researched extensively (e.g., Sheehan et al., 1998). Microalgae can be cultivated in open ponds or in bioreactors (Chen et al., 2011). If strains of microalgae contain 30% lipids, the productivity of algal oil can be as high as 58,700 L/ha (6,270 gal/acre); as a comparison to traditional oil seeds, canola productivity is only 1,190 L/ha (127 gal/acre) (Chisti, 2007). Practical production of microalgal lipids for biodiesel production still faces many challenges. Among them, cost effectiveness is the most critical (Stephens et al., 2010; Harun et al., 2011). Difficulties exist

*Table 14.3* Fatty acid compositions of oils from emerging feedstocks

| Oils | Fatty acid profiles | | | | | | | | | |
|---|---|---|---|---|---|---|---|---|---|---|
| | $C_{14:0}$ | $C_{16:0}$ | $C_{18:0}$ | $C_{18:1}$ | $C_{18:2}$ | $C_{18:3}$ | $C_{20:0}$ | $C_{20:1}$ | $C_{20:5}$ | $C_{22:1}$ |
| Algal oil | 12–15 | 10–20 | | 4–19 | 1–2 | 5–8 | | | 35–48 | |
| Camelina | | 7–8 | 2–3 | 12–16 | 15–25 | 30–40 | | 12–15 | | 2–3 |
| Jatropha | | 11–16 | 6–15 | 34–45 | 30–50 | | 3–5 | | | |
| Pennycress[a] | 0.1 | 3.1 | 0.5 | 12.6 | 22.4 | 11.8 | 0.3 | 8.6 | | 32.8 |

[a] Moser et al., 2009a.

not only in producing oil efficiently from high lipid strains but also in cost-effective cultivation and harvesting systems. Of particular significance is the high energy requirement for the de-watering and drying processes. Currently, research and development on producing biodiesel from microalgal oils is very active, and operations at pilot and demonstration scales are reported, but no reports are available yet on commercial microalgal biodiesel production. The fatty acid profiles of these emerging plant oils are provided in Table 14.3.

## 14.6 Feedstock quality issues

### 14.6.1 Feedstocks with high content of free fatty acids

Most biodiesel is currently produced using high quality soybean oil catalyzed by homogeneous alkali catalysts. However, other feedstocks such as waste vegetable oils and/or microalgal lipids can be high in FFAs requiring a different approach. The FFA content ranges between 2 and 7% for used vegetable oils, 5 and 30% for animals fats, and 0 to 100% for some low quality trap grease (Van Gerpen et al., 2004). If a feedstock contains high levels of FFA, typically 5% or higher, alkaline catalysts are not suitable for use in triacylglyceride transesterification. This is because the alkaline catalysts, such as sodium methoxide, will react with the FFA directly to produce fatty acid salts or 'soaps'. Soap formation not only consumes the catalysts but also causes the reacting mixture to emulsify. This can lead to difficulties in processing and to a low quality biodiesel product. Early research by Freedman et al. (1984) demonstrated the significant FFA effect on the transesterification process catalyzed by alkaline catalysts. At the same operating conditions, the product yield of methyl esters was reduced from 93–98% to 67–86% due to the presence of a significant amount (approx. 6.7%) of free fatty acids in the oil. Soap generally partitions into the byproduct glycerin and is considered waste if not recovered with additional processing. Generally speaking, if a feedstock contains no more than 1–2% FFA in alkaline-catalyzed transesterification, the FFA can be ignored; if 2–5% FFA, additional alkaline catalyst is needed to account for the consumption of catalyst due to soap formation and special attention needs to be paid to the procedures. An acid-catalyzed pretreatment, or distillation to remove the FFAs, is required if the oil or fat contains more than 5% FFAs.

It has been shown that strong acid-catalyzed transesterification is too slow and requires much higher alcohol levels; however, acid catalysts are very effective in catalyzing the esterification of FFA to methyl esters (Canakci and Van Gerpen, 1999). Therefore, a two-step process was proposed and is now considered to be the standard process for converting high FFA feedstocks, e.g., waste vegetable oils, into biodiesel (Canakci and Van Gerpen, 2001; Van Gerpen et al., 2004). In this two-step process, a strong acid is used to first convert the FFA into alkyl esters through an esterification process, and then the triacylglycerides are converted to alkyl esters via alkaline-catalyzed transesterification. The byproduct of the FFA esterification reaction is water. It has to be removed before the mixture proceeds to the transesterification process, which follows the same steps as the transesterification of high quality oil described previously.

### 14.6.2 Impurities that affect product quality

*Moisture*

Moisture in biodiesel is problematic, especially during long-term storage. High moisture in biodiesel may cause the fuel to deteriorate. Hydrolytic, oxidative, and other chemical reactions may occur, especially when minerals are present, as may be the case when using metal storage tanks (Waynick, 2005). Water promotes hydrolysis of the methyl esters, which leads to an increase in acidity; high acidity accelerates the decomposition of biodiesel when oxygen is available. This destructive loop effect will eventually lead to rancidity of the fuel. Another negative consequence of high moisture content in biodiesel is the potential for microbial growth. Some airborne microorganisms can grow on biodiesel and produce biomass. This biomass will quickly plug fuel filters and cause operational problems (Zhang et al., 2011).

Biodiesel contains oxygen in the form of carboxyl groups and tends to be polar, and thus will be hygroscopic. Thorough investigation into biodiesel moisture absorption and retention has shown that biodiesel can hold 1,000–1,700 ppm of moisture at 4–35°C. This is much higher than that in petroleum diesel, in which moisture retention of 40–114 ppm was observed within the same temperature range (He et al., 2007a).

The high moisture content in biodiesel can be attributed to multiple causes. High moisture is expected if water washing is used to purify the fuel. Despite careful drying after the water wash, moisture can still exist in biodiesel due to its hygroscopic nature. Vacuum drying can reduce the moisture in biodiesel to a level of 200–300 ppm. Biodiesel can also absorb moisture from the air during long-term storage without a nitrogen blanket. Therefore, adequate drying, if a water wash is used during processing, and careful sealing of the biodiesel storage containers, including the use of nitrogen blanketing, are necessary to prevent high moisture content in the biodiesel.

*Sterol glucosides*

Sterol glucosides are compounds of plant sterols, also known as phytosterols, which are present in all plants as important structural components to stabilize the phospholipid bilayers in plant membranes (Piironen *et al.*, 2000). The major plant sterols in higher plants are β-sitosterol and its glycoside β-sitosterolin. Acylated sterol glucosides are the major form and present in all parts of vegetables including fruit, tuber, root, stem, leaf, and cereals (Sugawara and Miyazawa, 1999).

Due to the small quantity in vegetable oils, measurement of plant sterols is difficult. Biodiesel producers have observed that insoluble particles are seen in biodiesel after the fuel has been stored for a few days. According to Lee *et al.* (2007), these fine insoluble particles consist of the non-acylated form of sterol glucosides. These particles then act as nuclei where other impurities, such as saturated monoacylglycerides, crystallize and consequently plug the fuel filter. The non-acylated form appears to be produced from the acylated form, which is soluble, during transesterification.

To resolve this problem, many producers perform a large version of 'cold soak filtration', one of the standard tests in the ASTM Standard D6751 (ASTM, 2011), during bulk production of the fuel. The biodiesel is cooled down to 3–5°C overnight to allow sterol glucosides to precipitate and be removed by filtration. This process is simple and effective, but is very costly. There are some adsorbents that suppliers claim will remove sterol glucosides but third-party validation is not available.

*Phosphorus*

High levels of phosphorous in biodiesel will cause several negative consequences. It will poison the vehicle's catalytic converter and decrease its efficiency (NREL, 2009). This will negatively affect the exhaust emissions. Therefore, phosphorus is strictly regulated. The biodiesel specification for maximum phosphorous content is 0.001% or 10 ppm by both ASTM 6751 and EN 14214.

Crude vegetable oils always contain small amounts of phospholipids, commonly referred to as gums. Degumming is typically performed to remove the impurity before an oil is used for biodiesel production. Acidic solutions of phosphoric or sulfuric acid are used to hydrate the gums to an insoluble form that can be separated from the oil. However, residual phosphorous may still be present in degummed oils at the level of a few hundreds of ppm.

High levels of phosphorous in the biodiesel are mainly caused by inadequate pretreatment of the feedstock. A study by Van Gerpen and Dvorak (2002) showed that phosphorous has considerable effect on biodiesel yield. If the oil contains 50 ppm or more of phosphorus, the biodiesel yield was reduced by 3–5%. The phosphorous is removed in soap form, which stays at the interface of the biodiesel and crude glycerin layers and thus makes the product separation difficult. The phosphorous is not carried into the biodiesel during processing, which is good news to biodiesel producers. This advantageous effect is attributed to the miscibility of the saponified phospholipids with polar compounds like methanol and glycerin. Reduction of methanol in the biodiesel layer helps bring the soapy materials into the interface or glycerin layer (Mendow et al., 2011). Due to the quality of feedstock used, i.e., mainly soybean oil, and rigorous process control, biodiesel produced in the US generally has phosphorus levels of about 1 ppm or less (NREL, 2009).

## 14.7 Advantages and limitations of biodiesel

Biodiesel is made from renewable plant and animal sources such as soybean oil, animal fats, and waste vegetable oils. Biodiesel burns in diesel engines with the same efficiency as petroleum-based diesel fuel. Its high flash point makes it safer to use and store. Biodiesel generally has a higher cetane number than petroleum diesel fuel, in the range of 45–55 for soybean biodiesel and 49–62 for rapeseed biodiesel (Mittelbach and Remschmidt, 2005). Environmentally, biodiesel is non-toxic and biodegradable. It contains very low levels of sulfur and nitrogen, and emits much fewer pollutants, particularly smoke, than petroleum diesel. It can be blended with petroleum diesel in any ratio, utilize existing distribution infrastructure, and requires no engine modifications. Biodiesel possesses excellent lubricity. Even low level blending of biodiesel, e.g., 2%, with petroleum diesel fuel will improve the lubricity of the blend to a satisfactory level and thus extend the engine's life (Van Gerpen et al., 2006).

However, biodiesel also has limitations. Biodiesel contains less energy than petroleum diesel, approximately 8% less per unit volume or 12% less per unit mass. Biodiesel generally has higher cloud and pour points, which make it less favorable for use in low temperature environments. Biodiesel has good biodegradability but it is also less stable chemically, thermally, and oxidatively, compared with petroleum diesel (Jain and Sharma, 2011).

Biodiesel is a stronger solvent than petroleum diesel. It may not be compatible with some fuel lines and/or gasket materials in engines and it may cause degradation or other compatibility issues. Biodiesel may also be incompatible with some metals and cause corrosion problems (Diaz-Ballote *et al.*, 2009; Hu *et al.*, 2012; Singh *et al.*, 2012). Another drawback of biodiesel is the possibility of elevated nitrogen oxide emissions, however this is controversial. The discussions in the following two sections focus mainly on biodiesel's feedstock availability, cold flow properties, and oxidative stability.

### 14.7.1 Feedstock availability

Despite its advantages, biodiesel production is constrained by the availability of feedstock. Biodiesel production in the US was approximately 1.05 billion gallons in 2011 (US EPA, 2012a), and approximately 830 million gallons in the first nine months of 2012 (US EPA, 2012b). The National Biodiesel Board expects the total biodiesel production to be up to 2 billion gallons (or $7.3 \times 10^6 \, m^3$) in the next few years (Jobe, 2012). The estimate requires a considerable increase in feedstock supply, and most likely will require major new sources beyond soybean oil.

As the world's largest biodiesel producers, the EU-27 and North America are facing a challenge of limited feedstock supply. Biodiesel production in the US will grow as the Renewable Fuel Standard mandates expand and world demand for diesel fuel increases. Biodiesel production in the US will be limited to approximately 2 billion gallons/year because of feedstock availability, as mentioned above. Before economically viable microalgal and jatropha oils are available for biodiesel production, biodiesel producers will find no easy solution to the issue of feedstock shortage. Feedstock suppliers will balance their profitability between selling to existing markets, especially the food market, and to the biodiesel industry.

Another limitation preventing the biodiesel industry from expanding is the cost relative to petroleum diesel fuel. It is generally agreed that biodiesel is more expensive than petroleum diesel in the market, and the major cause of this is the feedstock cost. An estimate for a plant of 17,400 tons/year (5 million gallons/year) has revealed that oil is the dominant contributor, at 80.6%, of the total production cost (Van Gerpen *et al.*, 2006). This observation is consistent with an early study by Haas *et al.* (2006) on a $37,850 \, m^3$/year (or 10 million gallon/year) plant, in which the feedstock (soybean oil) cost was 88% of the total cost. Although the byproduct, glycerin, can be recovered and marketed to offset some of the cost, the dominance of the oil cost will not change much.

Varying production costs are expected when different business models are adopted; however, the production cost is still dependent on the cost of the feedstock. It is logical, therefore, to use low cost feedstocks, such as animal fats and waste vegetable oils or greases. In today's market, low cost feedstock usually means low quality feedstock. Additional cost is needed to upgrade and handle the low quality feedstocks.

## 14.7.2 Cold flow properties and oxidative stability

Biodiesel is prone to the issue of poor low temperature operability due to its content of long-chain, saturated fatty acids. The cold flow properties of biodiesel, as characterized by its cloud point (CP), pour point (PP), and cold filter plugging point (CFPP), are less satisfactory than petroleum diesel. CP is the temperature at which a fuel starts to show observable crystals upon cooling under defined conditions (ASTM D2500, 2005). PP is the lowest temperature at which a fuel can maintain its flowability or ability to be pumped. The CFPP for biodiesel is defined as the highest temperature at which a given volume of fuel fails to pass through a standardized filtration device under specified testing conditions.

All cold flow properties are related to the melting points (m.p.) of the fuel constituents and their solubility in the fuel. A high m.p. component will crystallize and precipitate out when its concentration in the fuel is beyond its solubility. The components of biodiesel generally have higher melting temperatures than those of petroleum diesel, especially the long-chain, saturated fatty acid esters. The longer the fatty acid chain, the higher the melting point of the component. When unsaturated bonds are present in the alkyl chain, the m.p. of the component decreases considerably. For example, the m.p. of methyl oleate ($C_{18:1}$) is $-20°C$, while the m.p. of linoleate ($C_{18:2}$) is $-35°C$. Another factor affecting the cold flow properties is the solubility of a component in the fuel. A high m.p. component will not crystallize unless its concentration is higher than the quantity the fuel can dissolve. The final CP or PP of a biodiesel is the collective outcome of the properties of the individual methyl esters and their liquid–solid equilibria in the fuel. The CP and PP of biodiesel largely depend on the fatty acid profiles of the feedstock, as shown in Table 14.4. Traceable impurities, such as plant sterol glucosides, can also affect the biodiesel cold flow properties. Plant sterol glucosides have higher melting points and limited solubility in biodiesel. Once in biodiesel, they may crystallize and serve as the nuclei for other high m.p. components to agglomerate.

Biodiesel is relatively unstable compared to petroleum diesel. This is mainly due to the chemical characteristics of the biodiesel constituents. The fatty acid chains contain variable numbers of unsaturated carbon to carbon (C–C) bonds, depending on the feedstock, with the carboxyl groups linked with alkyl groups at one end. These unsaturated C–C double bonds are non-conjugated in structure and are thermochemically vulnerable to

*Table 14.4* Examples of cloud points of biodiesel from different feedstocks (Imahara *et al.*, 2006)

| No. | Oil/fat | Methyl ester composition (wt%) | | | | | | Cloud point | |
|---|---|---|---|---|---|---|---|---|---|
| | | $C_{16:0}$ | $C_{18:0}$ | $C_{18:1}$ | $C_{18:2}$ | $C_{18:3}$ | Others | (K) | (°C) |
| 1 | Beef tallow | 23.9 | 17.5 | 43.9 | 2.3 | 0.1 | 12.3 | 286 | 13 |
| 2 | Palm | 39.5 | 4.1 | 43.2 | 10.6 | 0.2 | 2.4 | 283 | 10 |
| 3 | Sunflower | 6.1 | 4.2 | 24 | 63.5 | 0.4 | 1.8 | 274 | 1 |
| 4 | Soybean | 10.7 | 3.2 | 25 | 53.3 | 5.4 | 2.5 | 272 | −1 |
| 5 | Linseed | 6.7 | 3.7 | 21.7 | 15.8 | 52.1 | 0 | 268 | −5 |
| 6 | Olive | 10.7 | 2.6 | 78.7 | 5.8 | 0.7 | 1.5 | 268 | −5 |
| 7 | Safflower | 6.4 | 2.2 | 13.9 | 76 | 0.2 | 1.3 | 267 | −6 |
| 8 | Rapeseed | 4.3 | 1.9 | 61.5 | 20.6 | 8.3 | 3.1 | 267 | −6 |

*Source*: Imahara *et al.*, 2006.

oxidation. The mechanism is that alkyl radicals are formed first at the positions of unsaturation following an attack by oxidants; then hydroperoxides are formed, followed by a series of degradation and polymerization reactions. The presence of free mineral acids catalyzes the formation of alkyl free radicals (Shahidi, 2005; Frankel, 2005). The relative rate of oxidation depends on the level of unsaturation. The more double-bonds an oil or a fat has, the more it is prone to oxidation. The relative vulnerability of the oleic ($C_{18:1}$), linoleic ($C_{18:2}$), and linolenic ($C_{18:3}$) chains is 1:12:26 after 30 days of storage (Chapman *et al.*, 2009). Other literature reports even higher oxidation rates of 1:40:80 for the same fatty acids (Witting, 1965). This difference is largely attributed to the environment to which the oil or fat is exposed. A large number of chemicals can be produced from such oxidation reactions, including aliphatic alcohols, aldehydes, short-chain organic acids, and polymers (Loury, 1972; Neff *et al.*, 1993; Andersson and Lingnert, 1998; Bondioli *et al.*, 2002). The adverse consequences of such chemicals in biodiesel are reduced flash point, rancidity, increased acidity, increased viscosity, and accelerated fuel degradation. Biodiesel oxidative instability may be accelerated due to the impurities present in the fuel. Minerals such as rust from the containers serve as catalysts for decomposition reactions. Incompletely reacted products, indicated by the presence of excess monoglycerides and free glycerin, likely provide a reactive environment. The potential for high moisture retention in biodiesel (15–25 times more than that in petroleum diesel; He *et al.*, 2007a) also provides favorable conditions for biodiesel oxidation.

To ensure its quality in long-term storage, biodiesel needs to strictly meet the ASTM D6751 or EN21414 specifications. Rusty tanks need to be thoroughly cleaned before being filled with biodiesel. Direct sunlight and high temperatures should be avoided. Nitrogen blanketing should be used to reduce air contact and moisture absorption in the biodiesel. For the long-term storage of biodiesel, a biocide application is recommended to prevent biological contamination.

## 14.8 Conclusion and future trends

Current production of both biodiesel and renewable diesel are driven by the requirements of the Renewable Fuel Standard (RFS) which is mandated by the Energy Independence and Security Act of 2007 (EIA, 2012). The RFS required that petroleum refiners in the US utilize 1 billion gallons of biomass-based diesel fuel in 2011. These fuels are now designated as Advanced Biofuels, with mandated levels that will ramp up to even higher levels in future years. Although alternative diesel requirements have been met every year, the level of lignocellulosic ethanol required by the RFS has not met expected levels and this may require adjustment of the target production levels. The uncertainty regarding the government's continued commitment to the RFS is one of the greatest challenges in financing alternative fuel products in the US.

## 14.9 Sources of further information and advice

*Books and proceedings*

*The Biodiesel Handbook*, 2nd edn. 2010. Eds. G. Knothe and J. Van Gerpen. AOCS Publishing. ISBN-10:1893997626, ISBN-13: 978-1893997622.

*Building a Successful Biodiesel Business: Technology Considerations, Developing the Business, Analytical Methodologies*, 2nd edn. 2006. By J. Van Gerpen, R. Pruszko, D. Clements, and B. Shanks. Biodiesel Basics. ISBN-10: 097863490X, ISBN-13: 978-0978634902.

*Biodiesel: Blends, Properties and Applications* (Energy Science, Engineering and Technology). 2011. Eds. J. Mario Marchetti and Z. Fang. Nova Science. ISBN-10: 1613246609, ISBN-13: 978-1613246603.

*Biodiesel: Blends, Properties and Applications* (Energy Science, Engineering and Technology). 2010. Ed. J. Mario Marchetti. Nova Science. ISBN-10: 1616689633, ISBN-13: 978-1616689636.

*Biodiesel Science and Technology: From Soil to Oil*. 2010. By J. Bart, S. Cavallaro, and N. Palmeri. Woodhead Publishing. ISBN-10: 1845695917, ISBN-13: 978-1845695910.

*Advances in Biodiesel Production: Processes and Technologies*. 2012. Eds. R. Luque and J. Antonio Melero. Woodhead Publishing. ISBN-10: 0857091174, ISBN-13: 978-0857091178.

*Online documents*

*Biodiesel Handling and Use Guide*, 4th edn. 2009. NREL/TP-540-43672. National Renewable Energy Laboratory, US Department of Energy. Available at: http://www.nrel.gov/vehiclesandfuels/npbf/feature_guidelines.html.

Technical publications on biodiesel. National Renewable Energy Laboratory, US Department of Energy. Available at: http://www.nrel.gov/vehiclesandfuels/npbf/pubs_biodiesel.html.

Renewable and Alternative Fuels. US Environmental Protection Agency (EPA). Available at: http://www.epa.gov/otaq/fuels/alternative-renewablefuels/index.htm.

Online documents and information. National Biodiesel Board. Available at: http://www.biodiesel.org/.

Studies and Reports. European Biodiesel Board. Available at: http://www.ebb-eu.org/studies.php.

Online documents and information. IEA Bioenergy Task 39 – Commercializing Liquid Biofuels. Available at: http://www.task39.org/.

## 14.10 References

Ahmad, A., N. Mat Yasin, C. Derek, and J. Lim. 2011. Microalgae as a sustainable energy source for biodiesel production: a review. *Renewable and Sustainable Energy Reviews* 15(1): 584–593. DOI: 10.1016/j.rser.2010.09.018.

Andersson, K. and H. Lingnert. 1998. Influence of oxygen and copper concentration on lipid oxidation in rapeseed oil. *JAOCS* 75(8): 1041–1046.

ASTM Standard D2500-05. 2005. *Standard Test Method for Cloud Point of Petroleum Products*. West Conshohocken, PA: ASTM International. DOI: 10.1520/D2500-05.

ASTM (American Society for Testing and Materials). 2011. *D 6751 Standard Specification for Biodiesel Fuel (B100) Blend Stock for Distillate Fuels*. West Conshohocken, PA: ASTM International. DOI: 10.1520/D6751-11B.

Benjapornkulaphong, S., C. Ngamcharussrivichai, and K. Bunyakiat. 2009. $Al_2O_3$-supported alkali earth metal oxides for transesterification of palm kernel oil and coconut oil. *Chem. Eng. J.* 145: 468–474. DOI: 10.1016/j.cej.2008.04.036.

Berrios, M. and R.L. Skelton. 2008. Comparison of purification methods for biodiesel. *Chem. Eng. J.* 144: 459–465. DOI: 10.1016/j.cej.2008.07.019.

Bondioli, P., A. Gasparoli, L. Bella, and S. Tagliabue. 2002. Evaluation of biodiesel storage stability using reference methods. *Eur. J. Lipid Sci. Technol.* 104: 777–784. DOI: 10.1002/1438-9312(200212)104:12<777::AID-EJLT777>3.0.CO;2-#.

Borges M. and L. Diaz. 2012. Recent developments on heterogeneous catalysts for biodiesel production by oil esterification and transesterification reactions: a review. *Renewable and Sustainable Energy Reviews* 16(5) (2012). DOI: 10.1016/j.rser.2012.01.071.

Borges M., A. Brito, A. Hernandez, and L. Diaz. 2011. Alkali metal exchanged zeolite as heterogeneous catalyst for biodiesel production from sunflower oil and waste oil: studies in a batch/continuous slurry reactor system. *International Journal of Chemical Reactor Engineering* 9: 1–20.

Canakci, M. and J. Van Gerpen. 1999. Biodiesel production via acid catalysis. *Trans. ASAE* 42(5): 1203–1210.

Canakci, M. and J. Van Gerpen. 2001. Biodiesel production from oils and fats with high free fatty acids. *Trans. ASAE* 44(6): 1429–1436.

Carmo Jr., A. L. de Souza, C. da Costa, E. Longo, J. Zamian, and G. da Rocha Filho. 2009. Production of biodiesel by esterification of palmitic acid over mesoporous aluminosilicate Al-MCM-41. *Fuel* 88: 461–468. DOI: 10.1016/j.fuel.2008.10.007.

CGC (Canadian Grain Commission). 2012. Oil content – quality of western Canadian canola 2011. Available at: http://www.grainscanada.gc.ca/canola/harvest-recolte/2011/hqc11-qrc11-06-eng.htm, (accessed Nov. 20, 2012).

Chapman, T.M., H.J. Kim, and D.B. Min. 2009. Prooxidant activity of oxidized α-tocopherol in vegetable oils. *J. of Food Sci.* 74(7): C536–C542. DOI: 10.1111/j.1750-3841.2009.01262.x.

Chen, C., K. Yeh, R. Aisyah, D. Lee, and J. Chang. 2011. Cultivation, photobioreactor design and harvesting of microalgae for biodiesel production: a critical review. *Bioresource Technol.* 102(1): S171–S181. DOI: 10.1016/j.biortech.2010.06.159.

Chen, C., C. Huang, D. Tran, and J. Chang. 2012. Biodiesel synthesis via heterogeneous catalysis using modified strontium oxides as the catalysts. *Bioresour. Technol.* 113: 8–13. DOI: 10.1016/j.biortech.2011.12.142.

Chisti, Y. 2007. Biodiesel from microalgae. *Biotechnol. Adv.* 25: 294–306. DOI: 10.1016/j.biotechadv.2007.02.001.

Chisti, Y. 2008. Biodiesel from microalgae beats bioethanol. *Trends in Biotechnol.* 26(3): 126–131. DOI: 10.1016/j.tibtech.2007.12.002.

Chouhan, A. and A. Sarma. 2011. Modern heterogeneous catalysts for biodiesel production: a comprehensive review. *Renewable and Sustainable Energy Reviews* 15(9): 4378–4399. DOI: 10.1016/j.rser.2011.07.112.

Cintas, P., S. Mantegna, E. Gaudino, and G. Cravotto. 2010. A new pilot flow reactor for high-intensity ultrasound irradiation: application to the synthesis of biodiesel. *Ultrasonics Sonochemistry* 17(6): 985–989. DOI: 10.1016/j.ultsonch.2009.12.003.

Contran, N., L. Chessa, M. Lubino, D. Bellavitea, P. Roggero, and G. Enne. 2013. State-of-the-art of the *Jatropha curcas* productive chain: from sowing to biodiesel and by-products. *Industrial Crops and Products* 42: 202–215. DOI: 10.1016/j.indcrop.2012.05.037.

Dias L., R. Missio, and D. Dias. 2012. Antiquity, botany, origin and domestication of *Jatropha curcas* (Euphorbiaceae), a plant species with potential for biodiesel production. *Genet. Mol. Res.* 11(3): 2719–2728.

Diaz-Ballote, L., J. Lopez-Sansores, L. Maldonado-Lopez, and L. Garfias-Mesias. 2009. Corrosion behavior of aluminum exposed to a biodiesel. *Electrochemistry Communications* 11(1): 41–44. DOI: 10.1016/j.elecom.2008.10.027.

Di Serio, M., R. Tesser, L. Casale, A. D'Angelo, M. Trifuoggi, and E. Santacesaria. 2010. Heterogeneous catalysis in biodiesel production: the influence of leaching. *Topics in Catalysis* 53(11–12): 811–819. DOI: 10.1007/s11244-010-9467-y.

Donnis, B., R. G. Egeberg, P. Blom, and K.G. Knudsen. 2009. Hydroprocessing of bio-oils and oxygenates to hydrocarbons: understanding the reaction routes. *Topics in Catalysis* 52: 229–240. DOI: 10.1007/s11244-008-9159-z.

EIA 2012. Energy Information Agency. Available at: www.eia.gov, (accessed August 30, 2012).

European Biodiesel Board. 2012. Statistics – 2010 production by country. Available at: http://www.ebb-eu.org/stats.php (accessed Nov. 26, 2012).

Evangelista, R., T. Isbell, and S. Cermak. 2012. Extraction of pennycress (*Thlaspi arvense* L.) seed oil by full pressing. *Industrial Crops and Products* 37(1): 76–81. DOI: 10.1016/j.indcrop.2011.12.003.

Faria, E., H. Ramalho, J. Marques, P. Suarez, and A. Prado. 2008. Tetramethylguanidine covalently bonded onto silica gel surface as an efficient and reusable catalyst for transesterification of vegetable oil. *Appl. Catal. A: Gen.* 338: 72–78. DOI: 10.1016/j.apcata.2007.12.021.

Flint, E., and K. Suslick. 1991. The temperature of cavitation. *Science* 253(5026): 1397–1399. DOI: 10.1126/science.253.5026.1397.

Frankel, E.N. 2005. *Lipid Oxidation*. The Oily Press. Bridgwater, UK.

Freedman, B., E. Pryde, and T. Mounts. 1984. Variables affecting the yields of fatty esters from transesterified vegetable oils. *JAOCS* 61: 1638–1643. DOI: 10.1007/BF02541649.

Furimsky, E. 2000. Catalytic hydrodeoxygenation. *Appl. Catal. A: Gen.* 199: 147–190. DOI: 10.1016/S0926-860X(99)00555-4.

Furuta, S., H. Matsuhashi, and K. Arata. 2004. Biodiesel fuel production with solid superacid catalysis in fixed bed reactor under atmospheric pressure. *Catal. Commun.* 5(12): 721–723. DOI: 10.1016/j.catcom.2004.09.001.

Garcia, C., S. Teixeira, L. Marciniuk, and U. Schuchardt. 2008. Transesterification of soybean oil catalyzed by sulfated zirconia. *Bioresour. Technol.* 99: 6608–6613. DOI: 10.1016/j.biortech.2007.09.092.

Goembira, F., K. Matsuura, and S. Saka. 2012. Biodiesel production from rapeseed oil by various supercritical carboxylate esters. *Fuel* 97: 373–378. DOI: 10.1016/j.fuel.2012.02.051.

Gomes, J., J. Puna, L. Goncalves, and J. Bordado. 2011. Study on the use of MgAl hydrotalcites as solid heterogeneous catalysts for biodiesel production. *Energy* 36(12): 6770–6778. DOI: 10.1016/j.energy.2011.10.024.

Granados, M., M. Poves, D. Alonso, R. Mariscal, F. Galisteo, R. Moreno-Tost, J. Santamaría, and J. Fierro. 2007. Biodiesel from sunflower oil by using activated calcium oxide. *Appl. Catal. B: Environ.* 73: 317–326. DOI: 10.1016/j.apcatb.2006.12.017.

Haas, M., A. McAloon, W. Yee, and T. Foglia. 2006. A process model to estimate biodiesel production costs. *Bioresour. Technol.* 97: 671–678. DOI: 10.1016/j.biortech.2005.03.039.

Hanh, H., D. Nguyen, K. Okitsu, R. Nishimura, and Y. Maeda. 2009. Biodiesel production through transesterification of triolein with various alcohols in an ultrasonic field. *Renewable Energy* 34(3): 766–768. DOI: 10.1016/j.renene.2008.04.007.

Harun, R., M. Davidson, M. Doyle, R. Gopiraja, M. Danquaha, and G. Forde. 2011. Technoeconomic analysis of an integrated microalgae photobioreactor, biodiesel and biogas production facility. *Biomass Bioenergy* 35:741–747. DOI: 10.1016/j.biombioe.2010.10.007.

He, B. and J. H. Van Gerpen. 2012a. Application of ultrasonication in transesterification processes for biodiesel production. *Biofuels* 3(4): 479–488. DOI: 10.4155/bfs.12.35.

He, B. and J. H. van Gerpen. 2012b. Analyzing biodiesel for contaminants and moisture retention. *Biofuels* 3(3): 351–360. DOI:10.4155/bfs.12.19.

He, B., J. Thompson, D. Routt, and J. Van Gerpen. 2007a. Moisture absorption in biodiesel and its petro-diesel blends. *Appl. Eng. Agri.* 23(1): 71–76.

He, H., T. Wang, and S. Zhu. 2007b. Continuous production of biodiesel fuel from vegetable oil using supercritical methanol process. *Fuel* 86: 442–447. DOI: 10.1016/j.fuel.2006.07.035.

Hsin, T., S. Chen, E. Guo, C. Tsai, M. Pruski, and V. Lin. 2010. Calcium containing silicate mixed oxide-based heterogeneous catalysts for biodiesel production. *Topics in Catalysis* 53(11–12): 746–754. DOI: 10.1007/s11244-010-9462-3.

Hu, E., Y. Xu, X. Hu, L. Pan, and S. Jiang. 2012. Corrosion behaviors of metals in biodiesel from rapeseed oil and methanol. *Renewable Energy* 37(1): 371–378. DOI: 10.1016/j.renene.2011.07.010.

Huber, G.W., J.N. Chheda, C.J. Barrett, and J.A. Dumesic. 2005. Production of liquid alkanes by aqueous-phase processing of biomass-derived carbohydrates. *Science* 308: 1446–1450. DOI: 10.1126/science.1111166.

Ilham, Z. and S. Saka. 2010. Two-step supercritical dimethyl carbonate method for biodiesel production from *Jatropha curcas* oil. *Bioresour. Technol.* 101: 2735–2740. DOI: 10.1016/j.biortech.2009.10.053.

Imahara, H., E. Minami, and S. Saka. 2006. Thermodynamic study on cloud point of biodiesel with its fatty acid composition. *Fuel* 85(12–13): 1666–1670. DOI: 10.1016/j.fuel.2006.03.003.

Imahara, H., E. Minami, S. Hari, and S. Saka. 2008. Thermal stability of biodiesel in supercritical methanol. *Fuel* 8: 1–6. DOI: 10.1016/j.fuel.2007.04.003.

Jain, S. and M. Sharma. 2011. Thermal stability of biodiesel and its blends: a review. *Renewable and Sustainable Energy Reviews* 15(1): 438–448. DOI: 10.1016/j.rser.2010.08.022.

Ji, J., J. Wang, Y. Li, Y. Yu, and Z. Xu. 2006. Preparation of biodiesel with the help of ultrasonic and hydrodynamic cavitation. *Ultrasonics* 44: E411–E414. DOI: 10.1016/j.ultras.2006.05.020.

Jitputti, J., B. Kitiyanan, P. Rangsunvigit, K. Bunyakiat, L. Attanatho, and P. Jenvanitpanjakul. 2006. Transesterification of crude palm kernel oil and crude coconut oil by different solid catalysts. *Chem. Eng. J.* 116: 61–66. DOI: 10.1016/j.cej.2005.09.025.

Jobe, J. 2012. Opening address to the Annual Technical Workshop of the National Biodiesel Board, Kansas City, MO, Oct. 29, 2012.

Kalva, A., T. Sivasankar, and V. Moholkar. 2009. Physical mechanism of ultrasound-assisted synthesis of biodiesel. *Ind. & Eng. Chem. Res.* 48(1): 534–544. DOI: 10.1021/ie800269g.

Kawashima, A., K. Matsubara, and K. Honda 2008. Development of heterogeneous base catalysts for biodiesel production. *Bioresour. Technol.* 99(9): 3439–3443. DOI: 10.1016/j.biortech.2007.08.009.

Kawashima, A., K. Matsubara, and K. Honda. 2009. Acceleration of catalytic activity of calcium oxide for biodiesel production. *Bioresour. Technol.* 100: 696–700. DOI: 10.1016/j.biortech.2008.06.049.

Kiss, F., M. Jovanovic, and G. Boskovic. 2010. Economic and ecological aspects of biodiesel production over homogeneous and heterogeneous catalysts. *Fuel Processing Technology* 91(10): 1316–1320. DOI: 10.1016/j.fuproc.2010.05.001.

Knothe, G. 2010. Biodiesel and renewable diesel: a comparison. *Progress in Energy and Combustion Science*, 36: 364–373. DOI: 10.1016/j.pecs.2009.11.004

Kondamudi, N., S. Mohapatra, and M. Misra. 2011. Quintinite as a bifunctional heterogeneous catalyst for biodiesel synthesis. *Appli. Cat. A: Gen.* 393(1–2): 36–43. DOI: 10.1016/j.apcata.2010.11.025

Koundinya, V. and R. Hansen. 2012. Pennycress. Agricultural Marketing Resource Center, Iowa State University. Available at: http://www.agmrc.org/commodities__products/grains__oilseeds/pennycress/ (accessed Nov. 21, 2012).

Kouzu, M., T. Kasuno, M. Tajika, S. Yamanaka, and J. Hidaka. 2008. Active phase of calcium oxide used as solid base catalyst for transesterification of soybean oil with refluxing methanol. *Appl. Catal. A: Gen.* 334: 357–365. DOI: 10.1016/j.apcata.2007.10.023

Kumar, A. and S. Sharma. 2008. An evaluation of multipurpose oil seed crop for industrial uses (*Jatropha curcas* L.): a review. *Industrial Crops and Products* 28: 1–10. DOI: 10.1016/j.indcrop.2008.01.001.

Kumar, D., G. Kumar, J. Poonam, and C. Singh. 2010a. Fast, easy ethanolysis of coconut oil for biodiesel production assisted by ultrasonication. *Ultrasonics Sonochemistry* 17(3): 555–559. DOI: 10.1016/j.ultsonch.2009.10.018

Kumar, D., J. Poonam, and C. Singh. 2010b. Ultrasonic-assisted transesterification of *Jatropha curcus* oil using solid catalyst, Na/SiO$_2$. *Ultrasonics Sonochemistry* 17(5): 839–844. DOI: 10.1016/j.ultsonch.2010.03.001

Kumar, S., A. Chaube, and S. Jain. 2012. Sustainability issues for promotion of Jatropha biodiesel in Indian scenario: a review. *Renewable and Sustainable Energy Reviews* 16(2): 1089–1098. DOI: 10.1016/j.rser.2011.11.014

Kusdiana, D. and S. Saka. 2001a. Kinetics of transesterification in rapeseed oil to biodiesel fuel as treated in supercritical methanol. *Fuel* 80(5): 693–698. DOI: 10.1016/S0016-2361(00)00140-X

Kusdiana, D. and S. Saka. 2001b. Methyl esterification of free fatty acids of rapeseed oil as treated in supercritical methanol. *J. Chem. Eng. Japan* 34(3): 383–387. DOI: 10.1252/jcej.34.383

Kusdiana, D. and S. Saka. 2004a. Effects of water on biodiesel fuel production by supercritical methanol treatment. *Bioresour. Technol.* 91(3): 289–295. DOI: 10.1016/S0960-8524(03)00201-3

Kusdiana, D. and S. Saka. 2004b. Two-step preparation for catalyst-free biodiesel fuel production – hydrolysis and methyl esterification. *Appl. Biochem. Biotechnol.* 113: 781–791.

Lee, I., L.M. Pfalzgraf, G.B Poppe, E. Powers, and T. Haines. 2007. The role of sterol glucosides on filter plugging. *Biodiesel Magazine*, April 6. Available at: http://www.biodieselmagazine.com/articles/1566/the-role-of-sterol-glucosides-on-filter-plugging/.

Li, Y., F. Qiu, D. Yang, X. Li, and P. Sun. 2011. Preparation, characterization and application of heterogeneous solid base catalyst for biodiesel production from soybean oil. *Biomass & Bioenergy* 35(7): 2787–2795. DOI: 10.1016/j.biombioe.2011.03.009

Liang, X., S. Gao, H. Wu, and J. Yang. 2009. Highly efficient procedure for the synthesis of biodiesel from soybean oil. *Fuel Process. Technol.* 90: 701–704. DOI: 10.1016/j.fuproc.2008.12.012

Liu, X., H. He, Y. Wang, and S. Zhu. 2007. Transesterification of soybean oil to biodiesel using SrO as a solid base catalyst. *Catal. Commun.* 8: 1107–1111. DOI: 10.1016/j.catcom.2006.10.026

Liu, X., H. He, Y. Wang, and S. Zhu. 2008. Transesterification of soybean oil to biodiesel using CaO as a solid base catalyst. *Fuel* 87: 216–221. DOI: 10.1016/j.fuel.2007.04.013

Loury, M. 1972. Possible mechanisms of autoxidative rancidity. *Lipids* 7: 671–675. DOI: 10.1007/BF02533075

Mason, T. 2000. Large scale sonochemical processing: aspiration and actuality. *Ultrasonics Sonochemistry* 7(4): 145–149. DOI: 10.1016/S1350-4177(99)00041-3

McVay, K. and P. Lamb. 2008. Camelina production in Montana. Available at: http://msuextension.org/publications/AgandNaturalResources/MT200701AG.pdf (accessed Nov. 20, 2012).

Mendow, G., F. Monella, M. Pisarello, and C. Querini. 2011. Biodiesel production from non-degummed vegetable oils: phosphorus balance throughout the process. *Fuel Processing Technol.* 92(5): 864–870. DOI: 10.1016/j.fuproc.2010.11.029

Mielke, T. 2012. World supplies of rapeseed and canola likely to remain tight in the 2012/13 season. *Inform* 24(3): 136–138.

Minami, E. and S. Saka. 2006. Kinetics of hydrolysis and methyl esterification for biodiesel production in two-step supercritical methanol process. *Fuel* 85: 2479–2483. DOI: 10.1016/j.fuel.2006.04.017

Mitich, L. 1996. Field pennycress (*Thlaspi arvense* L.): the stinkweed. *Weed Technology* 10(3): 675–678.

Mittelbach, M. and C. Remschmidt. 2005. *Biodiesel, the Comprehensive Handbook*. Boersedruck, Vienna.

Mofijur, M., H. Masjuki, M. Kalam, M. Hazrat, A. Liaquat, M. Shahabuddin, and M. Varman. 2012. Prospects of biodiesel from Jatropha in Malaysia. *Renewable and Sustainable Energy Reviews* 16(7): 5007–5020. DOI: 10.1016/j.rser.2012.05.010

Monnier, H., A. Wilhelm, and H. Delmas. 1999a. Influence of ultrasound on mixing on the molecular scale for water and viscous liquids. *Ultrasonics Sonochemistry* 6(1–2): 67–74. DOI: 10.1016/S1350-4177(98)00034-0

Monnier, H., A. Wilhelm, and H. Delmas. 1999b. The influence of ultrasound on micromixing in a semi-batch reactor. *Chem. Eng. Sci.* 54(13–14): 2953–2961. DOI: 10.1016/S0009-2509(98)00335-2

Mootabadi, H., B. Salamatinia, S. Bhatia, and A. Abdullah. 2010. Ultrasonic-assisted biodiesel production process from palm oil using alkaline earth metal oxides as the heterogeneous catalysts. *Fuel* 89(8): 1818–1825. DOI: 10.1016/j.fuel.2009.12.023

Moser, B., S. Shah, J. Winkler-Moser, S. Vaughn, and R. Evangelista. 2009a. Composition and physical properties of cress (*Lepidium sativum* L.) and field pennycress (*Thlaspi arvense* L.) oils. *Industrial Crops and Products* 30(2): 199–205. DOI: 10.1016/j.indcrop.2009.03.007

Moser, B., G. Knothe, S. Vaughn, and T. Isbell. 2009b. Production and evaluation of biodiesel from field pennycress (*Thlaspi arvense* L.) oil. *Energies and Fuels* 23(8): 4149–4155. DOI: 10.1021/ef900337g

Nakagaki, S., A. Bail, V. dos Santos, V. de Souza, H. Vrubel, F. Nunes, and P. Ramos. 2008. Use of anhydrous sodium molybdate as an efficient heterogeneous catalyst for soybean oil methanolysis. *Appl. Catal. A: Gen.* 351(2): 267–274. DOI: 10.1016/j.apcata.2008.09.026

Neff, W. E., T. L. Mounts, W. M. Rinsch, and H. Konishi. 1993. Photooxidation of soybean oils as affected by triacylglycerol composition and structures. *JAOCS* 70(2): 163–168. DOI: 10.1007/BF02542620

Niu, L., L. Gao, G. Xiao, and B. Fu. 2012. Study on biodiesel from cotton seed oil by using heterogeneous super acid catalyst $SO_4^{2-}/ZrO^2$. *Asia-Pacific Journal of Chemical Engineering* 7: S222–S228. DOI: 10.1002/apj.532

Niza, N., K. Tan, K. Lee, and Z. Ahmad. 2012. Biodiesel production by non-catalytic supercritical methyl acetate: thermal stability study. *Appl. Energy* 101(special Issue): 198–202. DOI: 10.1016/j.apenergs.2012.03.033

NREL. 2009. *Biodiesel Handling and Use Guide*. NREL/TP-540-43672. Golden, CO: National Renewable Energy Laboratory, US DOE.

Park, Y.M., J.Y. Lee, S.H. Chung, I.S. Park, S.Y. Lee, D.K. Kim, J.S. Lee, and K.Y. Lee. 2010. Esterification of used vegetable oils using the heterogeneous $WO_3/ZrO_2$ catalyst for production of biodiesel. *Bioresour. Technol.* 101: S59–S61. DOI: 10.1016/j.biortech.2009.04.025

Peng, B, Q. Shu, J. Wang, G. Wang, D. Wang, and M. Han. 2008. Biodiesel production from waste oil feedstock by solid acid catalysis. *Process Saf. Environ. Prot.* 86(B6): 441–447. DOI: 10.1016/j.psep.2008.05.003

Piironen V., D.G. Lindsay, T.A. Miettinen, J. Toivo, and A.M. Lampi. 2000. Plant sterols: biosynthesis, biological function and their importance to human nutrition. *J. Sci. Food Agri.* 80: 939–966. DOI: 10.1002/(SICI)1097-0010 (20000515)80:7<939::AID-JSFA644>3.3.CO;2-3

Putnam, D., J. Budin, L. Field, and W. Breene. 1993. Camelina: a promising low-input oilseed. In: *New Crops* by J. Janick and J.E. Simon (eds), pp. 314–322. New York: Wiley.

Saka, S. and Y. Isayama. 2009. A new process for catalyst-free production of biodiesel using supercritical methyl acetate. *Fuel* 88: 1307–1313. DOI:10.1016/j.fuel.2008.12.028

Saka, S. and D. Kusdiana. 1999. Transesterification of rapeseed oils in supercritical methanol to biodiesel fuels. In: *Proceedings of the Fourth Biomass Conference of the Americas*. Vol. 1. Eds. R. Overend, and E. Chornet, p. 747. Oakland, CA. Qxford: Pergamon.

Saka, S. and D. Kusdiana. 2001a. Biodiesel fuel from rapeseed oil as prepared in supercritical methanol. *Fuel* 80: 225–231. DOI: 10.1016/S0016-2361(00)00083-1

Saka, S. and D. Kusdiana. 2001b. Kinetics of transesterification in rapeseed oil to biodiesel fuel as treated in supercritical methanol. *Fuel* 80: 693–698. DOI: 10.1016/S0016-2361(00)00140-X

Saka, S., D. Kusdiana, and E. Minami. 2006. Non-catalytic biodiesel fuel production with supercritical methanol technologies. *J. Scientif. Ind. Res.* 65(5): 420–425.

Saka, S., Y. Isayama, Z. Ilham, and X. Jiayu. 2010. New process for catalyst-free biodiesel production using subcritical acetic acid and supercritical methanol. *Fuel* 89(7): 1442–1446. DOI:10.1016/j.fuel.2009.10.018

Sakai, T., A. Kawashima, and T. Koshikawa. 2009. Economic assessment of batch biodiesel production processes using homogeneous and heterogeneous alkali catalysts. *Bioresour. Technol.* 100(13): 3268–3276. DOI: 10.1016/j.biortech.2009.02.010

Samart, C., P. Sreetongkittikul, and C. Sookman. 2009. Heterogeneous catalysis of transesterification of soybean oil using KI/mesoporous silica. *Fuel Process. Technol.* 90: 922–925. DOI: 10.1016/j.fuproc.2009.03.017

Sebos, I., A. Matsoukas, V. Apostolopoulos, and N. Papayannakos. 2009. Catalytic hydroprocessing of cottonseed oil in petroleum diesel mixtures for production of renewable diesel. *Fuel* 88: 145–149. DOI:10.1016/j.fuel.2008.07.032

Semwal, S., A. Arora, R. Badoni, and D. Tuli. 2011. Biodiesel production using heterogeneous catalysts. *Bioresour. Technol.* 102(3): 2151–2161. DOI: 10.1016/j.biortech.2010.10.080

Shahidi, F. (ed.). 2005. *Bailey's Industrial Oil and Fat Products.* Vol. 1, Section 1.7 *Lipid Oxidation: Theoretical Aspects.* 6th edn. John Wiley & Sons, New York.

Sheehan, J., T. Dunahay, J. Benemann, and P. Roessler. 1998. *A Look Back at the US Department of Energy's Aquatic Species Program–Biodiesel from Algae.* Report NREL/TP-580-24190, Golden, CO: National Renewable Energy Laboratory, US DOE.

Shol, A. 1988. Industrial applications of ultrasound. In: *Ultrasound: Its Chemical, Physical, and Biological Effects.* Ed. K. Suslick, pp.97–122. New York: VCH Publishers.

Shu, Q., B. Yang, H. Yuan, S. Qing, and G. Zhu. 2007. Synthesis of biodiesel from soybean oil and methanol catalyzed by zeolite beta modified with $La^{3+}$. *Catal. Commun.* 8: 2159–2165. DOI: 10.1016/j.catcom.2007.04.028

Singh, A., S. Fernando, and R. Hernandez. 2007. Base-catalyzed fast transesterification of soybean oil using ultrasonication. *Energy and Fuels* 21(2): 1161–1164. DOI:10.1021/ef060507g

Singh, B., J. Korstad, and Y. Sharma. 2012. A critical review on corrosion of compression ignition (CI) engine parts by biodiesel and biodiesel blends and its inhibition. *Renewable and Sustainable Energy Reviews* 16(5): 3401–3408. DOI: 10.1016/j.rser.2012.02.042

Stavarache, C., M. Vinatoru, R. Nishimura, and Y. Maeda. 2005. Fatty acids methyl esters from vegetable oil by means of ultrasonic energy. *Ultrasonics Sonochemistry* 12(5): 367–372. DOI: 10.1016/j.ultsonch.2004.04.001

Stephens, E., I. Ross, Z. King, J. Mussgnug, O. Kruse, C. Posten, M. Borowitzka, and B. Hankamer. 2010. An economic and technical evaluation of microalgal biofuels. *Nature Biotechnol.* 28: 126–128. DOI:10.1038/nbt0210-126

Sugawara, T. and T. Miyazawa. 1999. Separation and determination of glycolipids from edible plant sources by high-performance liquid chromatography and evaporative light-scattering detection. *Lipids* 34(11): 1231–1238. DOI:10.1007/s11745-999-0476-3

Teixeira, L., J. Assis, D. Mendonca, I. Santos, P. Guimarães, L. Pontes, and J. Teixeira. 2009. Comparison between conventional and ultrasonic preparation of beef tallow biodiesel. *Fuel Processing Technology* 90(9): 1164–1166. DOI: 10.1016/j.fuproc.2009.05.008

Thompson, L. and L. Doraiswamy. 1999. Sonochemistry: science and engineering. *Ind. Eng. Chem. Res.* 38(4): 1215–1249. DOI: 10.1021/ie9804172

USDA. 2012a. US Soybeans Production. National Agricultural Statistics Service. Available at: http://www.nass.usda.gov/ (accessed Nov. 20, 2012).

USDA. 2012b. US National Statistics for Canola. National Agricultural Statistics Service. Available at: http://www.nass.usda.gov/ (accessed Nov. 20, 2012).

US DOE. United States Department of Energy. 2011. US billion ton update – biomass supply for a bioenergy and bioproducts industry. Available at: http://www1.eere.energy.gov/biomass/pdfs/billion_ton_update.pdf (accessed August 30, 2012).

US EPA. 2012a. 2012 RIN Generation and Renewable Fuel Volume Production. Available at: http://www.epa.gov/otaq/fuels/rfsdata/2011emts.htm (accessed Nov. 26, 2012).

US EPA. 2012b. 2011 RIN Generation and Renewable Fuel Volume Production. Available at: http://www.epa.gov/otaq/fuels/rfsdata/2012emts.htm (accessed Nov. 26, 2012).

Van Gerpen, J. 2005. Biodiesel processing and production. *Fuel Process. Technol.* 86(10): 1097–1107. DOI: 10.1016/j.fuproc.2004.11.005

Van Gerpen, J. and B. Dvorak. 2002. The effect of phosphorus level on the total glycerol and reaction yield of biodiesel. Bioenergy 2002, the 10th Biennial Bioenergy Conference. Boise, ID. Sept. 22–26.

Van Gerpen, J., B. Shanks, R. Pruszko, D. Clements, and G. Knothe. 2004. *Biodiesel Production Technology*. Project report: NREL/SR-510-36244. Golden, CO: National Renewable Energy Laboratory, US Department of Energy.

Van Gerpen, J., B. Shanks, R. Pruszko, D. Clements, and G. Knothe. 2006. *Building a Successful Biodiesel Business*, 2nd edn. Dubuque, IA: Biodiesel Basics.

Vollmann, J., A. Damboeck, A. Eckl, H. Schrems, and P. Ruckenbauer. 1996. Improvement of *Camelina sativa*, an underexploited oilseed. In: *Progress in New Crops*, J. Janick (ed.), pp. 357–362. Alexandria, VA: ASHS Press.

Wall, J., J. Thompson, and J. Van Gerpen. 2011. Soap and glycerin removal from biodiesel using waterless processes. *Trans ASABE* 54(2): 535–541.

Wang, J., K. Chen, S. Huang, and C. Chen. 2011. Application of $Li_2SiO_3$ as a heterogeneous catalyst in the production of biodiesel from soybean oil. *Chinese Chem. Lett.* 22(11): 1363–1366. DOI: 10.1016/j.cclet.2011.05.041

Wang, L. and Yang, J., 2007. Transesterification of soybean oil with nano-MgO or not in supercritical and subcritical methanol. *Fuel* 86: 328–333. DOI: 10.1016/j.fuel.2006.07.022

Warabi, Y., D. Kusdiana, and S. Saka. 2004a. Biodiesel fuel from vegetable oil by various supercritical alcohols. *Appl. Biochem. Biotechnol.* 115: 793–801.

Warabi, Y., D. Kusdiana, and S. Saka. 2004b. Reactivity of triglycerides and fatty acids of rapeseed oil in supercritical alcohols. *Bioresour. Technol.* 91(3): 283–287. DOI: 10.1016/S0960-8524(03)00202-5

Waynick, J.A. 2005. *Characterization of biodiesel oxidation and oxidation products*. Technical literature review, TP-540-39096. Golden, CO: National Renewable Energy Laboratory, US DOE.

Wen, G., Z. Yan, M. Smith, P. Zhang, and B. Wen. 2010. Kalsilite based heterogeneous catalyst for biodiesel production. *Fuel* 89(8): 2163–2165. DOI: 10.1016/j.fuel.2010.02.016

Witting, L. 1965. Lipid peroxidation in vivo. *JAOCS* 42: 908–913. DOI: 10.1007/BF02632443

Wu, P., Y. Yang, J. Colucci, and A. Grulke. 2007. Effect of ultrasonication on droplet size in biodiesel mixtures. *JAOCS* 84(9): 877–884. DOI: 10.1007/s11746-007-1114-9

Wu, X., R. Ruan, Z. Du, and Y. Liu. 2012. Current status and prospects of biodiesel production from microalgae. *Energies* 5(8): 2667–2682. DOI: 10.3390/en508266

Xie W., Y. Liu, and H. Chun. 2012. Biodiesel preparation from soybean oil by using a heterogeneous $Ca_xMg_{2-x}O_2$ catalyst. *Catal. Lett.* 142: 352–359. DOI: 10.1007/s10562-012-0776-6

Yamaguchi, T., C. Benmore, and A. Soper. 2000. The structure of subcritical and supercritical methanol by neutron diffraction, empirical potential structure refinement, and spherical harmonic analysis. *J. Chem. Phys.* 112: 8976–8987. DOI: 10.1063/1.481530

Yang, C., Z. Fang, B. Li, and Y. Long. 2012. Review and prospects of Jatropha biodiesel industry in China. *Renewable and Sustainable Energy Reviews* 16(4): 2178–2190. DOI: 10.1016/j.rser.2012.01.043

Yang, Z. and W. Xie. 2007. Soybean oil transesterification over zinc oxide modified with alkali earth metals. *Fuel Process. Technol.* 88: 631–638. DOI: 10.1016/j.fuproc.2007.02.006

Yu, D., L. Tian, H. Wu, S. Wang, Y. Wang, D. Ma, and X. Fang. 2010. Ultrasonic irradiation with vibration for biodiesel production from soybean oil by Novozym 435. *Process Biochem.* 45(4): 519–525. DOI: 10.1016/j.procbio.2009.11.012

Zhang T., Y. Chao, N. Liu, J. Thompson, M. Garcia1-Preze, B. He, J. Van Gerpen, and S. Chen. 2011. Case study of biodiesel-diesel blends as a fuel in marine environment. *Adv. in Chem. Eng. Sci.* (1): 65–71.

# 15
# Biomethane and biohydrogen production via anaerobic digestion/fermentation

K. STAMATELATOU, Democritus University of Thrace, Greece, G. ANTONOPOULOU, Institute of Chemical Engineering Sciences, Greece and P. MICHAILIDES, Democritus University of Thrace, Greece

**DOI**: 10.1533/9780857097385.2.476

**Abstract**: Biomass is a renewable source for energy production by means of bioprocesses such as fermentation/anaerobic digestion. Biohydrogen and biogas are gaseous fuels produced by these processes. In this chapter, the basic principles and various technological aspects affecting the efficiency of fermentation and anaerobic digestion processes are presented. The hydrogen and methane yields from various types of biomass are reported, and difficulties and future trends in the area of anaerobic digestion are discussed.

**Key words**: biogas, biohydrogen, biomass, crops, wastes, food wastes.

## 15.1 Introduction

Biogas and biohydrogen are gaseous fuels produced by biological processes which occur naturally when biodegradable organic materials are found in the absence of oxygen. Biogas is a mixture of methane and carbon dioxide. It is the final product of anaerobic digestion: a process consisting of a series of reactions caused by a combination of micro-organisms having synergistic and antagonistic relationships. Hydrogen is an intermediate product of the anaerobic digestion process and is produced spontaneously but is not accumulated under normal conditions. Small-scale accumulation takes place in bio-reactors operated under controlled fermentation conditions. The sustainable production of hydrogen requires the production of biogas from the residues of the fermentation process.

Hydrogen production from biomass by means of biological processes is still in the experimental stage and has not been applied on a large scale. The biogas sector benefits from a high level of technical expertise and is attracting increasing attention. Anaerobic digestion was primarily developed as a waste treatment method but due to the developement of renewable energy resources, biogas production is now using energy crops as well as residues and wastes.

Details of the yields of biogas (methane) and hydrogen from different feedstocks (energy crops, crop residues and agricultural wastes) which may be used in a biorefinery scheme are given below. The drawbacks and potential of applying the above technologies are also discussed.

## 15.2 Basic principles of biogas and hydrogen production

### 15.2.1 Biogas

Anaerobic digestion is a complex process which involves several micro-organism groups in a consortium which converts the organic matter of the biomass into biogas (a mixture of $CO_2$ and $CH_4$). Biogas contains traces of other gases such as hydrogen, hydrogen sulphide and carbon monoxide. The final mixture is treated and enriched in methane if it is to be used through the gas grid or as a transportation fuel.

There are four separate steps comprising the anaerobic digestion process: hydrolysis, acidogenesis (fermentation), acetogenesis and methanogenesis (Fig. 15.1). *Hydrolysis* is linked to acidogenesis as micro-organisms excrete extracellular enzymes which hydrolyse the complex organic matter into smaller compounds. These can be transferred through their cellular membranes and further degraded into a mixture of acids and alcohols. Hydrolytic enzymes include cellulase, cellobiase, xylanase and amylase for converting carbohydrates into sugars, protease for hydrolysing proteins into

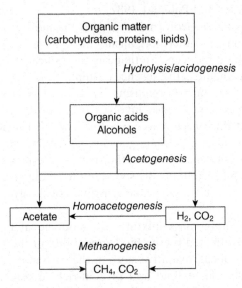

*15.1* Biogas process outline.

amino acids and lipase for degrading lipids into glycerol and long chain fatty acids (LCFA). In lignocellulosic materials, hydrolytic enzymes cannot access the polymers as these are amorphously bound with lignin. Special pretreatment methods are required for the initiation of hydrolysis (Section 15.3.4). Although hydrolysis is considered as a single step, it is actually a group of individual processes (enzyme production inside the microbial cell, diffusion through the membrane, adsorption on the polymer molecule, reaction, and enzyme deactivation). It is considered to be the slowest step of the whole process where the organic matter is in particulate form.

*Acidogenesis* involves multiple reactions for converting the products of hydrolysis into acids of low molecular weight and alcohol. The final composition of the acids and alcohol in this step depends on the quantities of sugars, amino acids and fatty acids produced from hydrolysis. In the case of sugar acidification, the microbes have the ability to shift their metabolism to more reduced organic metabolites, depending on the conditions including pH, hydrogen and partial pressure (Angelidaki et al., 2011). It is generally accepted that under conditions of low pH and high levels of hydrogen and formate, more reduced metabolites are produced, such as butyrate, lactate and ethanol.

The other two steps, **acetogenesis** and **methanogenesis**, are also linked as the hydrogen produced from acetogenesis must be scavenged by the methanogens if acetogenesis reactions are to proceed. Acetogenesis involves oxidation reactions of the acidogenesis products by micro-organisms which produce hydrogen to dispose of the electrons derived from oxidation. The production of hydrogen is thermodynamically feasible at low partial pressure ($<10^{-4}$ atm) (Harper and Pohland, 1987).

Formate is also used as an electron carrier. Hydrogen and formate are kept at low concentrations by micro-organisms such as methanogens, homoacetogens and sulphate reducers. Methane is the typical product of hydrogen and formate scavenging in the absence of sulphate. The syntrophic relationship between these micro-organisms is termed 'interspecies hydrogen transfer'. Methanogenesis also involves the transformation of acetate into methane and this process accounts for 70% of the methane produced under stable conditions. Methanogens are generally sensitive to pH and ammonia, temperature changes and other factors. Methanogenesis is considered to be the step which limits the speed of the whole process (in the case of soluble organic matter), due to the sensitivity of slowly growing methanogens.

The main constituents of biomass mixture are methane and carbon dioxide. Its composition is affected by the biodegradability of the organic matter (feedstock) and the mean oxidative state of the carbon it contains. The lower the level of carbon in lipid and protein-rich feedstocks, the richer in methane the biogas will be. The methane yields from various types of feedstocks (energy crops, residues and wastewaters) are reported in Section 15.3.

### 15.2.2 Hydrogen

Hydrogen is a colourless and odourless gas and is the simplest element present in water, biomass and fossil fuels (gasoline and natural gas). It is considered to be an alternative, environmentally friendly fuel and is described as the 'energy carrier' of the future. It is also a very useful reagent for the production of a variety of chemicals. When used as a fuel, hydrogen produces water instead of greenhouse gases and has a high energy yield (122 kJ/g, which greatly exceeds that of hydrocarbon fuels). Hydrogen does not occur naturally as a gas but is always combined with other elements (carbon, oxygen, nitrogen) through chemical bonds, the breaking of which are energy intensive. It is mainly produced from hydrocarbons through steam reforming and from water by means of electrolysis. Biomass can also be utilised for hydrogen production. Recently, biological methods have been developed which are believed to be cost effective and to have a low environmental impact when compared to the thermochemical processing of biomass.

The biological methods for hydrogen production from biomass (biohydrogen production) include photo and dark fermentation in bio-reactors and bio-electrolysis in microbial electrolysis cells (MEC). The hydrogen yield from photo fermentation is low due to a limited range of substrates (acetate, etc.), the large bio-reactor surface which is needed for efficient light penetration, and the light saturation effect. In MEC, the biodegradable material is converted into hydrogen instead of electricity as in microbial fuel cells (Call and Logan, 2008). An MEC operates under anaerobic conditions (without oxygen in the cathode) and a small external voltage is applied to the cell, so that protons and electrons produced by the anodic reaction combine at the cathode to form hydrogen. MEC is a promising technology but it is still in its infancy and many microbiological, technological and economic challenges remain to be overcome.

Dark fermentation appears more promising in terms of yield and viability when compared to alternative bio-hydrogen processes. It is directly related to the acidogenic stage of the anaerobic digestion processes described in Section 15.2.1. The cost of dark fermentation hydrogen production is estimated to be 340 times lower than that of photosynthetic processes (Morimoto, 2002). Dark fermentation is not subject to the limitations of photo fermentation in terms of feedstock suitability, complex and expensive bio-reactor engineering and poor yield. However, higher yields and process efficiency are required if it is to become a sustainable process.

Hydrogen production generally disposes of excess electrons through the enzyme hydrogenase in certain strictly anaerobic micro-organisms

(*Clostridia*, methylotrophs, rumen bacteria, methanogenic bacteria, archaea), facultative anaerobes (*Escherichia coli, Enterobacter, Citrobacter*), or even aerobes (*Alcaligenes, Bacillus*). Among the hydrogen-producing bacteria, *Clostridium* sp. and *Enterobacter*, are the most widely studied. Species of the genus *Clostridium* such as *C. butyricum, C. acetobutyricum, C. beijerinckii, C. thermolacticum, C. tyrobutyricum, C. thermocellum* and *C. paraputrificum* are examples of strict anaerobic and spore-forming micro-organisms which generate hydrogen gas during the exponential growth phase. In parallel, facultative anaerobes such as the species of genus *E. coli* and its modified strains and species of the genus *Enterobacter*, such as *E. aerogenes* and *E. cloacae* have also been used in hydrogen production. Extensive research has recently been carried out into hydrogen production at a high temperature, using thermophilic or hyperthermophilic bacteria.

The acidification of sugars under anaerobic conditions can yield high levels of hydrogen. The degradation of hexoses by mixed anaerobic microbial cultures has been extensively studied. It has been found that hydrogen and various metabolic products are produced, mainly volatile fatty acids (VFAs; acetic, propionic and butyric acid), lactic acid, and alcohols (butanol and ethanol), depending on the microbial species present and the prevailing conditions. The hydrogen yield may be correlated stoichiometrically with the final metabolic products by the principal reactions describing the individual processes of acidogenesis:

$$C_6H_{12}O_6 + 2H_2O \rightarrow 2CH_3COOH + 2CO_2 + 4H_2$$

$$C_6H_{12}O_6 \rightarrow CH_3CH_2CH_2COOH + 2CO_2 + 2H_2$$

$$C_6H_{12}O_6 + 2H_2 \rightarrow 2CH_3CH_2COOH + 2H_2O$$

Hydrogen productivity (HP) is a parameter for assessing the yield of hydrogen production through dark fermentation. It is defined as the percentage of influent substrate electrons in the hydrogen gas produced in both gaseous and aqueous phases (Kraemer and Bagley, 2005). The above reactions show that the production of acetic and butyric acids leads to the simultaneous production of hydrogen, with the fermentation of hexose to acetic acid giving the highest theoretical yield of 4 mol of $H_2$/mol of hexose (HP = 33%). The conversion to butyric acid results in 2 mol of $H_2$/mol of hexose (HP = 17%), while during the production of propionic acid, hydrogen is consumed.

Conditions prevailing during dark fermentation should shift the metabolism of the microbial consortium towards acetate and/or butyrate production in order to achieve a high hydrogen yield. *Clostridia* sp. produces a mixture of acids, with butyrate exceeding acetate during the biological degradation of glucose (Mizuno *et al.*, 2000). In practice, the production of more metabolic products (lactate or ethanol), accompanied by a negative or zero hydrogen yield, results in lower overall yields of hydrogen (HP: 10–20%). The shift of metabolism towards acetate may occur via different, non-hydrogen-yielding pathways. In mixed fermentation processes, the micro-organisms may select different pathways in converting sugars, as a response to changes in their environment (pH, sugar concentration, etc.). The absence or presence of hydrogen-consuming micro-organisms in the microbial consortium also affects the microbial metabolic balance and, consequently, the fermentation end products.

## 15.3 Biogas and biohydrogen production: technological aspects

The aspects of technology which affect biogas and hydrogen production are: the development of bio-reactors capable of operating at high organic rates, the advances in monitoring devices and control policies and the development of pretreatment methods, particularly for lignocellulosic materials which offer an abundant renewable resource. Monitoring and control of anaerobic digestion has been extensively reviewed (Pind *et al.*, 2001; Boe, 2006) and is excluded from the present section. In the sequel, the following are presented:

- a description of the basic types of anaerobic bio-reactors,
- a discussion of the key points affecting hydrogen production, and
- a review of the main pretreatment methods.

### 15.3.1 Anaerobic bio-reactors

The selection of reactor type is based mainly on the feedstock organic load and its solid matter content. Feedstocks with a high solid content are usually best converted to biogas in continuous stirred tank reactors (CSTRs) and plug flow reactors (PFR), while soluble feedstocks can be used in high-rate configurations. A brief discussion on the main bio-reactors which have been developed for biogas production follows.

- *Anaerobic lagoons (low rate)*. These are shallow covered ponds with a large surface area. The temperature is not controlled, so the ambient temperature prevails (psychrophilic level), resulting in low conversion rates. Mixing does not take place and the solids therefore accumulate at the bottom of the lagoon. The advantage of anaerobic lagoons is their low construction and operational cost. However, the low efficiency in biogas production and COD reduction cancel any economic benefit derived from the low cost infrastructure.
- *Continuous stirred tank reactors; CSTR (low rate)*. The CSTR (Fig. 15.2a) is a well-established technology in which the mixing process results in

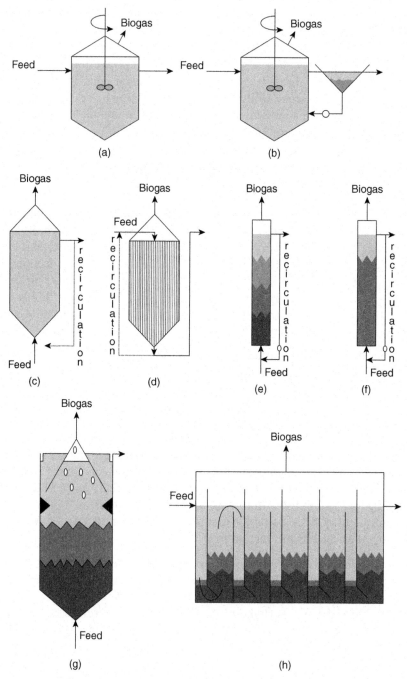

*15.2* Basic anaerobic digesters: (a) continuous stirred tank reactor (CSTR), (b) contact process, (c) downflow fixed filter, (d) upflow fixed filter, (e) expanded bed, (f) fluidised bed, (g) upflow anaerobic sludge blanket reactor (UASBR), (h) anaerobic baffled reactor (ABR).

high installation and energy costs. Homogeneous conditions prevail due to this process and result in an efficient mass transfer of nutrients, metabolites and extracellular enzymes. In feedstocks with a high solid content, dispersal of the constituents through mixing minimises the diffusion zones of the solid–liquid interface and mass transfer becomes more rapid. However, homogeneity results in the continuous removal of the solid with the liquid fraction from the reactor. This is not beneficial for two reasons. First, solid hydrolysis has been reported as a rate-limiting step and a long retention time is necessary to enhance the hydrolysis and further degradation of solids, and second, the active biomass is withdrawn with the effluent and the conversion rates which depend upon it are slow. The main configurations which have been designed to enhance solid retention in the bio-reactor are contact digesters and sequencing batch reactors. The micro-environment of the micro-organism consortium must be preserved to some extent (especially in micro-organisms with symbiotic relationships) and homogenisation should be mild.

- *Contact digester (high-rate)*. The effluent of the CSTR is allowed to remain in a separate vessel (the clarifier) and two streams are generated: a clarified liquid stream leaving the reactor system and a solid concentrated stream which is partially recirculated to the CSTR (Fig. 15.2b). This is similar to the activated sludge contact process which aims to increase solid retention time in the CSTR. The separation in the clarifier is effected by gravity and is often enhanced through inclined parallel plates (Defour *et al.*, 1994). However, biogas bubbles may hinder the settling of solids and degassing is therefore a prerequisite. Where the solid content is high, the separation process becomes slower and this type of configuration should not be used for solid concentrations above 2.5% (Burke, 2001).
- *Sequencing batch reactor; SBR (high-rate)*. In SBRs, the reactor is operated as a batch under mixing conditions. Once the desired conversion has been achieved, mixing is interrupted and the mixed liquor allowed to separate into a clarified supernatant and a sludge portion which maintains micro-organisms and unconverted solids inside the reactor. The supernatant is removed and the SBR is then refilled with a new load of waste.
- *Plug-flow digesters (low-rate)*. These are tabular reactors which provide maximal spatial distribution of the content. The feedstock enters at one end and moves towards the other while being converted to biogas. Plug-flow digesters may be in horizontal or vertical positions and a partial recirculation of the effluent takes place. Since mixing is not provided, the retention time must be sufficiently long to allow for high conversion efficiency. This type of digester is suitable for the anaerobic

digestion of feedstocks with a high solid content (>10%) and is often referred to as dry fermentation.
- *Anaerobic filters and fluidised beds (high-rate).* Anaerobic filters are elongated vessels filled with inert material (such as plastic rings). The micro-organisms are attached to the carriers and due to the high specific gravity attained, they remain in the interior of the vessel which gives high conversion rates at low hydraulic retention times (<1 day). The filter bed may be fixed or expanded (fluidised) depending on the direction; upwards (Fig. 15.2c) or downwards (Fig. 15.2d), and on the flow-rate level; low (Fig. 15.2e) or high (Fig. 15.2f) (Hall, 1992). The presence of solids in the waste stream is problematic and the danger of clogging is always present. Recirculation of the effluent is necessary in order to dilute the influent and/or fluidise the bed.
- *Upflow anaerobic sludge bed reactor; UASB (high-rate).* UASB reactors (Fig. 15.2g) are elongated vessels which contain the micro-organisms in granules and offer excellent settling properties (Lettinga et al., 1980). The active anaerobic biomass is retained in the reactor and kept in suspension by recirculating part of the effluent to the bottom of the reactor while keeping the upward velocity of the mixed liquor in the range 0.5–3 m/h (Annachhatre, 1996). High conversion efficiency rates in soluble organic compounds (preferably sugar-rich or fermented substrates) have been recorded with low hydraulic retention times (0.5–2 days). UASB has been applied in more than 900 full-scale units throughout the world in many instances of industrial wastewater containing soluble organic compounds (Garcia et al., 2008).
- *Anaerobic baffled reactor; ABR (high-rate).* ABR (Fig. 15.2h) is a horizontally elongated rectangular vessel separated into compartments which divert the horizontal flow into successive downward and upward directions (Barber and Stuckey, 1999). The waste (with low solid content) therefore comes into contact with the biomass accumulated on the lower part of the reactor.

There are several modifications of the above reactors, such as the UASB-filter reactor which combines UASB and anaerobic filter technology (Banu and Kaliappan, 2008) and the periodic anaerobic baffled reactor (PABR) which resembles a simple ABR which changes its flow pattern periodically due to the switching of influent and effluent points (Stamatelatou et al., 2009).

Apart from single-stage systems consisting of one of the main bio-reactor types described above, two or three stage systems are also widely used for the production of biogas from waste. When the waste contains a high portion of solids, the first stage serves as the hydrolytic-acidogenic step, while the second stage is the methanogenic step. The first stage requires a digester capable of handling high solid streams and operating at high retention times to enhance hydrolysis of the solids. The second stage can be a high-rate bio-reactor which further converts the first stage acidified effluent to biogas. Where the organic load of waste is mostly soluble and consists of carbohydrates, the first stage serves as the acidogenic step and takes place quickly, requiring a low hydraulic retention time. Both stages may be carried out in high-rate bio-reactors operating at different pH levels and different hydraulic retention times. The potential instability in single-stage systems, which is caused by the rapid acidification and volatile fatty acid accumulation, is therefore avoided. The different retention times required for the efficient operation of both stages determine the size of each bio-reactor. Depending on the requirements, the first bio-reactor may be larger or smaller than the second one.

Different temperature levels can be imposed on the two stages. Temperature phased systems are applied when a degree of hygienation of the effluent is required, especially in the anaerobic digestion of sewage sludge, manure, slaughterhouse wastewater and any other feedstock which may contain pathogens. The first stage is usually thermophilic, which inactivates the majority of pathogens (Song *et al.*, 2004).

The two-stage configuration concept is also applied in the case of biohydrogen production. The acidogenesis step can be regulated to yield hydrogen along with volatile fatty acids and other metabolites. In this process, elimination of hydrogen scavenging micro-organisms, pH control and regulation of the substrate and product (hydrogen) concentration in the mixed liquor are the key operating factors directing the hydrogen efficiency of the first stage, as discussed in Section 15.3.2. The second stage is a typical methanogenic bio-reactor fed on the acidified effluent of the first stage.

## 15.3.2 Factors affecting biogas and biohydrogen technology

The composite nature of an anaerobic consortium, consisting of micro-organisms grown at different rates and optimal conditions, results in shifts in the balance of the population under the effect of environmental changes. It has been widely recognised that the most important factors affecting the anaerobic digestion process which may lead to failure, are the pH, the temperature, the presence of toxic or inhibitory substances and the hydraulic retention time (HRT).

The pH determines the concentration of weak acids and bases in undissociated form. These molecules can be transferred through the cellular membrane, thus changing the intracellular pH. The pH also influences the function of the extracellular enzymes which affect the hydrolysis rate. Biogas production takes place at neutral pH, although methane production is possible at lower or higher pH values. pH also influences the activities of hydrogen-producing micro-organisms, as it directly affects hydrogenase activity as well as the metabolic pathway followed. However, a wide range of pH values has been proposed as optimum for fermentative hydrogen production from different feedstocks. The pH range of 5–7.5 (Fang et al., 2002a; Calli et al., 2008) is usually reported as optimum, even though lower or higher pH values such as pH of 4.5 (Ren et al., 2007) and 9.0 (Lee et al., 2002) have also been proposed as optimal.

The temperature influences enzyme and consequently microbial activity. Generally, the higher the temperature, the higher are the microbial reproduction and substrate conversion rates. Because of the faster rates, the inhibition effects become more severe as the inhibitory factors evolve. The micro-organisms are categorised into three types according to the temperature range in which they grow: thermophilic (optimum above 50°C), mesophilic (optimum 30–40°C) and psychrophilic (optimum below 20°C). The appropriate adaptation of micro-organisms to temperature changes may allow them to function in more than one temperature range.

Elements or compounds found to be inhibitory and even toxic to anaerobic digestion are:

- oxygen (anaerobic micro-organisms have a different tolerance to oxygen; strict anaerobes include Clostridia, methanogens, sulphate reducers and homoacetogens),
- ammonia (especially in non-ionised form, although a high degree of tolerance can be achieved through appropriate adaptation),
- long-chain fatty acids (tending to adsorb on the cellular membrane and interact with active molecules present in the membrane; however, proper acclimation of the micro-organisms and co-digestion are methods for averting the effects of toxicity) and
- metals, which are added as trace elements, but when the feedstock itself contains metals in high concentrations, toxicity results.

The hydraulic retention time (HRT), in combination with the concentration of feedstock in the influent of bio-reactors, affects the methane and hydrogen yield. In the case of biogas production, the selection of HRT and organic load are dependent upon the bio-reactor type used (Section 15.3.1). For hydrogen production from pure substrates such as glucose and sucrose, the most common values of HRT are in the range of 3–8 hours, with the lowest being 1 hour (Chang et al., 2002) and the highest 13.7 hours (Fang and Liu, 2004). For more complex substrates such as starch, an HRT of 15 or 17 hours is suggested as being necessary, due to the slow initial hydrolysis step (Hussy et al., 2003).

Hydrogen partial pressure is also important (as discussed in Section 15.2). Under stable biogas production conditions, it is sustained below $10^{-4}$ atm.

In bio-reactors designed for biohydrogen production, the hydrogen levels are higher, but at levels higher than 60–100 Pa, hydrogen production is inhibited. Several methods of keeping the hydrogen partial pressure low have been studied. Mizuno *et al.* (2000) showed that sparging with nitrogen enhanced the hydrogen yield, while Voolapalli and Stuckey (1998) used a submerged silicone membrane dissolved gas extraction system, removing hydrogen and carbon dioxide from the reactor volume. Another potentially efficient method for removing hydrogen from the gas stream based on a heated palladium-silver membrane reactor has been proposed by Nielsen *et al.* (2001).

The inoculum added to the bio-reactor for biohydrogen production is of the greatest importance. When mixed microbial cultures are used for hydrogen production by dark fermentation, hydrogen-consuming bacterial activity should be suppressed, while the activity of the hydrogen-producing bacteria must be preserved. The most common practice is to pretreat the microbial biomass used to inoculate the bio-reactors. The pretreatment method relies on the spore-forming characteristics of hydrogen-producing *Clostridium*, which is ubiquitous in anaerobic sludge and sediment (Brock *et al.*, 1994). When an anaerobic sludge is treated under harsh conditions, *Clostridium* is more likely to survive than the non-spore-forming bacteria, many of which are hydrogen consumers (Lay, 2001). Effective pretreatment processes include heating (100°C, 15 minutes), acidic or basic treatment (pH = 3, adjusted with ortho-phosphoric acid, 24 hours), aeration, chemical addition (chloroform, acetylene), and the application of an electric current (3–4.5 V). Where real biomass is used as a feedstock (wastewater, crops, etc.), it is advantageous to use the indigenous mixed microbial culture of the feedstock by applying operational conditions as proposed by Antonopoulou *et al.* (2008a, 2008b).

In general, the bio-reactors described in Section 15.3.1 can be utilised for fermentative hydrogen production in either batch or continuous mode. Batch mode operation is more suitable for research purposes, but at full scale, bio-reactors should operate on a continuous, or at least semi-continuous (fed or sequencing batch) basis. The continuous stirred tank reactor (CSTR) is the most common configuration, offering simple construction, ease of operation and effective homogeneous mixing as well as temperature and pH control. However, the solid retention time (SRT) is the same as the hydraulic retention time (HRT), resulting in a low concentration of microbial biomass and a low hydrogen production rate. The hydrogen-producing biomass in a CSTR could be self-granulated or flocculated under appropriate conditions (Fang *et al.*, 2002b; Zhang *et al.*, 2004). Another approach to increasing biomass concentration in a CSTR is to immobilise the biomass in bio-films or artificial granules made from a variety of support materials.

Another category of continuous flow reactors are those which permit the physical retention of the microbial biomass through flocculation, the formation of granules of self-immobilised microbes, microbial immobilisation on inert materials, microbial-based bio-films or retentive membranes (Hallenbeck and Ghosh, 2009). However, a potential problem posed by these types of reactors is the establishment of slow-growing methanogenic populations, due to extended retention of the biomass inside the reactor.

### 15.3.3 Pretreatment methods

Feedstocks with a high organic matter content that are difficult to degrade anaerobically generally contain lignocelluloses. Particular emphasis is therefore given to methods developed for enhancing biogas and hydrogen production from lignocellulosic feedstocks.

Lignocellulosic biomass contains complex carbohydrate polymers such as cellulose and hemicellulose, which are tightly bonded to lignin, the most recalcitrant component of plant cell walls. The higher the proportion of lignin, the higher the resistance to chemical and enzymatic/biological degradation. Softwoods usually contain more lignin than hardwoods and agricultural residues. However, regardless of the origin of the lignocellulosic biomass, some kind of pretreatment process is always necessary to break the lignin matrix and to de-polymerise the cellulose and hemicellulose. This facilitates the release of simple sugars (hexoses and pentoses) and results in a higher biofuel production yield. Pretreatment methods for lignocellulosic biomass can be divided into three main types: physical, chemical or physico-chemical and biological. It should be noted that most pretreatment methods have been limited to the experimental level as techno-economic evaluation is required to assess their cost in comparison to any resultant increase in energy gain.

Physical or mechanical pretreatment refers to milling which reduces the size of the particulate matter. The specific surface area of the solids and size of pores are therefore increased, the crystallinity and degree of polymerisation of the cellulose are decreased, and enzymes can more easily access the substrate to initiate hydrolysis. Mechanical pretreatment is always applied before any other kind of pretreatment. Delgenes *et al.* (2002) demonstrated that mechanical pretreatment by milling enhances methane production from 5 to 25%, while Menardo *et al.* (2012) showed that by reducing the size of wheat, barley, rice straw and maize stalks, the methane yields were increased by more than 80%.

During chemical or physico-chemical pretreatment, the lignocellulosic biomass is exposed to chemicals (acids, alkali or solvents) at ambient or higher temperature. The main effect is to alter the lignin structure and to dissolve the hemicelluloses. Several reviews have focused on this necessity and on the comparative study of these methods (Chandra et al., 2007; Hendriks and Zeeman, 2009).

Acid pretreatment may be performed with acids such as $H_2SO_4$, $H_3PO_4$, $HNO_3$ and HCl. During the pretreatment, the cellulose crystallinity is reduced, hemicelluloses are hydrolysed and furfural and hydoxymethyl furfural (HMF) are produced. Furfural and HMF are toxic to methanogens and hydrogen-producing bacteria (Gossett et al., 1982; Ramos, 2003; Vazquez et al., 2007). For this reason, the conditions for acid pretreatment are carefully studied. Dilute acids are preferred and used with short retention times (e.g., 5 min) at high temperature (e.g., 180°C) or relatively long retention times (e.g., 30–90 min) at lower temperatures (e.g., 120°C) (Table 15.1).

Alkali pretreatment involves the use of alkaline solutions such as NaOH, $Ca(OH)_2$ (lime) or ammonia to remove lignin and a part of the hemicelluloses by destroying the links of lignin and other polymers. Alkaline pretreatment can be generally classified into high and low concentration processes depending upon the concentration of the alkali (Mirahmadi et al., 2010). Table 15.2 lists some recent studies on alkaline pretreatment and its effect.

Thermal pretreatment takes place at high temperatures (from 150°C to 220°C). At temperatures above 160°C, the hemicelluloses are solubilised first, followed by lignin. Phenolics are released from the lignin solubilisation, some of which may be toxic to the micro-organisms (Gossett et al., 1982; Ramos, 2003). Compounds such as vanillin, furfural and HMF, which have a toxic effect on micro-organisms, may arise from the hydrolysis of hemicellulose. The degree of de-polymerisation and formation of inhibitory compounds depend significantly on conditions such as temperature. The thermal treatment technologies which have a positive effect on methane and/or hydrogen yield when applied to biomass are: steam explosion, liquid hot water and wet oxidation.

During steam explosion, lignocellulosic biomass is treated with high-pressure saturated steam for a short period of time, followed by rapid pressure relief. The temperature level of steam explosion is typically between 160 and 260°C and the duration of the process varies between a few seconds and several minutes (Sun and Cheng, 2002). The process has been thoroughly studied and applied at the laboratory and pilot scales. It is considered to be cost effective as a relatively low level of energy is required (Holtzapple et al., 1989).

Steam explosion has been studied in various biomass types. In plant biomass, increases in methane yield of 20% from wheat straw (Bauer et al., 2009) and up to 50% from salix (Estevez et al., 2012) have been recorded. The combination of steam explosion with chemical agents such as acids or

*Table 15.1* Examples of acid pretreatment of lignocellulosic materials

| Feedstock | Conditions | Effect | Reference |
|---|---|---|---|
| Olive tree biomass | 1% $H_2SO_4$, 170°C | 83% hemicellulose recovery | Cara et al. (2007) |
| Olive tree biomass | 1.4% $H_2SO_4$, 210°C | 76.5% enzymatic hydrolysis yield | |
| Barley straw and corn stalks | 1.8% w/w $H_2SO_4$ | increase in soluble sugars release but limited fermentation | Panagiotopoulos et al. (2009) |
| Corn stover | 1.69% v/v $H_2SO_4$, 121°C, 117 min | 2.24 mol $H_2$/mol sugar | Cao et al. (2009) |
| Sugarcane bagasse | 0.5% $H_2SO_4$, 121°C, 60 min | High yield in sugar | Pattra et al. (2008) |
| Wheat straw | 2% $H_2SO_4$, 120°C, 90 min | Simultaneous hydrogen production 141 mL$H_2$/gVS vs non simultaneous hydrogen production 41.9 mL$H_2$/gVS | Nasirian (2012) |
| Rapeseed and sunflower straws and meals | 2% w/w $H_2SO_4$, 121°C, 60 min | No increase in biogas yield | Antonopoulou et al. (2010) |
| Bagasse and coconut fibres | HCl | Increase in biogas by 31% and 74% respectively | Kivaisi and Eliapenda (1994) |
| Palm residues | $H_3PO_4$ | Increase in biogas by 40% | Nieves et al. (2011) |
| Hay, straw and bracken | Maleic acid, 150°C, 30 min | Increase in biogas from bracken only | Fernandes et al. (2009) |

bases has also been tested and shown positive results (Bruni et al., 2010; Teghammar et al., 2010). For example, the pretreatment of corn stover with steam explosion combined with acid resulted in high hydrogen yields of 3.0 mol mol hexose$^{-1}$, although steam explosion alone was also efficient (hydrogen yield of 2.84 mole mol hexose$^{-1}$) (Datar et al., 2007).

Liquid hot water (LHW) or hydrothermolysis involves the contact of biomass with water under high pressure and temperature (200–230°C) for around 15 minutes. It has been applied to various biomass types such as corn stover (Zeng et al., 2007), food wastes and manure (Qiao et al., 2011).

In wet oxidation, the biomass is treated with water and air (or oxygen) at temperatures above 120°C (e.g., 148–200°C) for a period of around

*Table 15.2* Examples of alkaline pretreatment of lignocellulosic materials

| Feedstock | Conditions | Effect | Reference |
| --- | --- | --- | --- |
| Soybean straws | 10% $NH_3$, 24 h, ambient temperature | Decrease of hemicellulose and lignin by 41.45% and 30.16% | Xu *et al.* (2007) |
| *Miscanthus* | NaOH or $Ca(OH)_2$, 75°C | 2.9–3.4 mol $H_2$ per mol of hexose (74–85% of the theoretical yield) | de Vrije *et al.* (2009) |
| Corn stover | 6% NaOH, 3 weeks, ambient temperature | Increase in biogas by 48.5% | Pang *et al.* (2008) |
| Corn stover | 88% moisture, 2% NaOH, 3 d, ambient temperature | Increase in biogas by 72.9% | Zheng *et al.* (2009) |
| Oil palm residues | 8% NaOH for 60 min | Increase in biogas by 100% | Nieves *et al.* (2011) |

30 min (Palonen *et al.*, 2004; Schmidt and Thomsen, 1998). Lissens *et al.* (2004) used wet oxidation to improve anaerobic biodegradability and methane yields in several raw wastes such as food waste, yard waste and digested bio-waste treated in a full-scale biogas plant. The wet oxidation process increased methane yields by approximately 35–70% when compared to untreated raw lignocellulosic wastes.

Biological pretreatment refers to the use of whole micro-organisms or purified enzymes to disrupt the lignocellulosic matrix and enhance hydrolysis. Both fungi (brown, white and soft-rot fungi) and bacteria have so far been tested for the delignification of lignocellulosic biomass. White-rot fungi such as *Phanerochaete chrysosporium*, *Trametes versicolor*, *Ceriporiopsis subvermispora*, and *Pleurotus ostreatus* are among the most effective micro-organisms for the biological pretreatment of lignocelluloses (Sun and Cheng, 2002).

Purified enzymes (ligninases such as laccase, lignin peroxidase and manganese peroxidase, cellulases or hemicellulases) have also been tested for lignin, cellulose or hemicellulose breakdown and solubilisation. Enzymatic pretreatment is usually combined with some physico-chemical method. For example, Talebnia *et al.* (2010) reported glucose yields of 98% after the enzymatic hydrolysis of wheat straw with thermal and chemical pretreatment. Other cases of biological pretreatment are reported in Table 15.3. The main advantages of biological pretreatment using whole

*Table 15.3* Examples of biological pretreatment of lignocellulosic materials

| Feedstock | Whole cell or enzymes for hydrolysis | Effect | Reference |
|---|---|---|---|
| Orange processing waste | *Sporotrichum, Aspergillus, Fusarium, Penicillium* | Increase in methane yield by 33% | Srilatha *et al.* (1995) |
| Corn straw | *Pleurotus florida* | Increase in methane yield by 120–150% | Zhong *et al.* (2011) |
| Manure | Enzymes (laccases) after treatment with steam explosion and NaOH | Increase in methane yield by 34% | Bruni *et al.* (2010) |
| Carrot pulp | Enzymes | Increase in hydrogen yield by 10% | de Vrije *et al.* (2010) |
| Rice straw | *Acinetobacter junii* F6-02 | 0.76 mol $H_2$/mol xylose | Lo *et al.* (2010) |
| Sweet sorghum bagasse | Enzymes after treatment with NaOH | 2.6 mol $H_2$/mol $C_6$ sugar | Panagiotopoulos *et al.* (2010a) |

micro-organisms are the low energy requirements, little (if any) chemical addition, and mild environmental conditions (low temperature and pressure). However, the main disadvantage is a slow conversion rate. Where purified enzymes are used in biological pretreatment, their high cost is the main disadvantage.

## 15.4 Production of biogas (methane) and biohydrogen from different feedstocks

Biomass is a versatile and abundant renewable source (plant biomass, agricultural residues, organic fraction of municipal solid wastes, agricultural wastes) and research is focused on developing its exploitation for optimum energy gains and high added value products or other useful materials (nutrients, water, etc.). In the present chapter, which considers the biorefinery concept, the potential streams for biogas and hydrogen production have been considered to be energy crops, crop residues and food production wastes (manure, food processing, wastewater). Although all these feedstocks are suitable for biogas production, hydrogen is produced in appreciable quantities from feedstocks rich in sugars.

Table 15.4 Estimation of electricity potential of various energy crops

| Energy crop | Harvest yield (t ww/ha) | Methane yield (Nm$^3$/ha) | Electricity yield (kWh/ha) |
|---|---|---|---|
| Maize silage | 50 | 4997 | 18489 |
| Sugar beet | 55 | 4673 | 17289 |
| Sudan grass | 55 | 3435 | 12711 |
| Whole plant grain silage | 40 | 3131 | 11586 |
| Grass silage | 36 | 2926 | 10826 |

Source: FNR, 2012.

### 15.4.1 Energy crops

Energy crops consist of plants cultivated for biofuel production. Some of the crops used for biogas production, their harvesting, methane and electricity yields are reported in Table 15.4 as estimated by the FNR (2012). Energy crops may also be used as a co-substrate in co-digestion systems in combinations with feedstocks with a low energy content (e.g., manure).

Other biofuels, such as hydrogen, ethanol and biodiesel, can be produced from energy crops. Methane and hydrogen yields should be related to the lignin, cellulose and hemicelluloses content of the plants as lignin is considered to be resistant to degradation. Table 15.5 summarises the yields of methane and hydrogen from various energy crops and the yields of other biofuels are cited where available.

### 15.4.2 Crop residues

Crop residues include the biomass remaining after harvesting and the extraction of sugar, cereal, oil or any other material which cannot be further utilised in the food production chain. Examples of crop residues include sugar cane and sweet sorghum bagasse, corn leaves and stover, wheat straw, and forestry residues (hardwoods) such as wood trimmings or tree residues (Table 15.6). These are primarily of a lignocellulosic nature and are considered to be an abundant renewable source.

Recent data have shown the total global sugarcane production in 2007 to be 1.59 billion metric tonnes, with an average productivity of about 67 t/ha. The major sugarcane producing countries are India, Brazil, Philippines, China, USA, Mexico, Indonesia, Australia, Colombia, Brazil and India, which together produce almost 60% of the world's sugarcane, with Brazil responsible for around 35% of total world production (McLaren, 2009).

In 2009–2010, Japan was the main contributor to total global rice production, amounting to 9.74 million tonnes (www.rice-trade.com, 2011).

Table 15.5 Biofuel yields from energy crops

| Energy Crop | Yield of biomass | Lignin | Hemicellulose | Cellulose | Used for | Biofuel yields |
| --- | --- | --- | --- | --- | --- | --- |
| | Tones (t) ha$^{-1}$ | (%) dry weight | | | | (Reference) |
| **Switch grass (*Panicum virgatum*)** Perennial grass, highly adaptable to poor soil. Chosen as the model lignocellulosic crop by the US Department of Energy in the 1990s | 13–18 (Southeastern United States) 9.63 (UK) | 12–19 | 31–37 | 29–45 | Bedding and combustion | Methane 1,200–2,600 m$^3$ha$^{-1}$ (Frigon et al., 2012) Methane (880–1,350 m$^3$ha$^{-1}$) (Massé et al., 2011) Ethanol 1,288–2,851 L ha$^{-1}$ (Propheter et al., 2010) |
| **Sweet sorghum (*Sorghum bicolor* (L) Moench)** Short growing period of 4–5 months. A wide range of growing areas | 43–150 90–140 (Greece) 40–50 (Belgium) | 7.1 | 20 | 26.3 | Sugar, alcohol, syrup, jaggery, fodder, fuel, bedding, roofing, fencing, paper and chewing | Hydrogen (946.4 m$^3$ha$^{-1}$) (Antonopoulou et al., 2008a) Ethanol (8,000 L ha$^{-1}$) (Bennett and Anex, 2009) Methane (2,639 m$^3$ha$^{-1}$) (Antonopoulou et al., 2008a) |
| **Sugarcane (*Saccharum officinarum*)** Perennial grass. Native to the warm temperate. Brazil is the world's largest sugarcane producer | 79–112 (Australia) 74.4 (Brazil) 67.7 (Japan) | 7 | 8 | 24 | Food industries (sugar), alcohol, syrup, animal feed, fertiliser | Ethanol (6,200 L ha$^{-1}$) (de Vries et al., 2010) |

| Species | Yield | | | Uses | Products |
|---|---|---|---|---|---|
| **Sugar beet (*Beta vulgaris* L.)**[a]<br>Low dry-matter content makes combustion of sugar beet pulp for heat and power production unfavourable | 47.8–57.3 (UK)<br>71.42 (Netherlands)<br>50 (Greece) | 0 | 5 | 4 | Food industries (sugar, molasses), alcohol, syrup | Ethanol (6,000 L ha$^{-1}$) (de Vries et al., 2010)<br>Ethanol (5,060 L ha$^{-1}$) (FAO, 2008)<br>Methane (95 m$^3$ t$^{-1}$ ww) (Kreuger et al., 2011)<br>Hydrogen (1,320–1,570 kg ha$^{-1}$) (Panagiotopoulos et al., 2010b) |
| **Maize (*Zea mays*)**[b]<br>The price of maize was increased due to the rising demand for maize–ethanol competing with its traditional use as food | 10.6–23.5 (USA)<br>15–30 (t dry matter ha$^{-1}$) (Germany) | | 35 | 18 | Food, animal food. 40% of the crop is used for corn ethanol. Starch from maize can also be used in plastics, fabrics, adhesives, and many other chemical products | Methane (12,390 m$^3$ ha$^{-1}$) (Amon et al., 2007)<br>Methane (300–350 m$^3$ t$^{-1}$ dry biomass) (Weiland, 2006)<br>Ethanol (386 L t$^{-1}$ maize) (Soto et al., 2005) |

*Continued*

Table 15.5 Continued

| Energy Crop | Yield of biomass | Lignin | Hemicellulose | Cellulose | Used for | Biofuel yields |
|---|---|---|---|---|---|---|
| | Tones (t) ha$^{-1}$ | (%) dry weight | | | | (Reference) |
| **Miscanthus (*Miscanthus x giganteus*)** Perennial grass. Low inputs of nutrients for cultivation, cold tolerant, provides considerable dry matter yields. Characterised as the ideal energy crop for many areas of northern Europe and the US | 8–15 (Western European Regions) 6.9–24.1 (EU, when the crop is grown on arable land) 4 (Central Germany) 44 (Northern Greece and Italy) | 10.5 | 15.9 | 57.6 | Papermaking. Energy crop | Ethanol (286 Lt$^{-1}$ dry biomass) (Deverell et al., 2009) Methane (81.48 m$^3$ t$^{-1}$ biomass) (Uellendahl et al., 2008) |
| **Poplar (*Populus trichocarpa*)** High productivity and cold/drought tolerance. Its cultivation prevents soil erosion and protect soil water | 9 (Midwestern United States) 12.4 (on non-irrigated) 22.4 (irrigated soils) | 20 | 14 | 40 | Heat and electricity production. Wood for furniture and paper. High quality timber, which can be used as sawn timber, veneer, panels and pulpwood production | Ethanol (69–244 Lt$^{-1}$ wood of poplar) (Wang et al., 2012) |

| Crop | Description | | | Use | References |
|---|---|---|---|---|---|
| **Willow (*Salix ssp*)** Harvested on a 3 year cycle | 16.9 (without fertiliser or irrigation) (Canada) | 19 | 14 | 55.4 | Manufacturing. Leaves and bark of the willow tree are used in medicine. | Methane (93.47 m³t⁻¹ biomass) (Estevez et al., 2012) |
| **Grass** Each grass has its own mowing and feeding requirements. Water use varies by species and climate | 4.2–6.4 t dry biomass | 9 | 20 | 32 | Animal food, pellets production | Methane (910 m³ha⁻¹) (Amon et al., 2007) Methane (215 m³CH₄ (tVS)⁻¹) (Cysneiros et al., 2011) Methane (306 m³CH₄ (tVS)⁻¹) (Lehtomäki et al., 2008) Methane (350 m³CH₄ (tVS)⁻¹) (Nizami et al., 2011) Methane (361 m³CH₄ (tVS)⁻¹) (Asam et al., 2011) |
| **Sunflower (*Helianthus annuus*)**[c] Annual plant | 11.02 t VSha⁻¹ (Austria) 2.47 t grain ha⁻¹ (Greece) | 9.13 | 7.83 | 23.95 | Extraction for its oil. Sunflower leaves can be used as cattle feed, and stems may be used in paper production | Methane (3,200–4,500 m³ha⁻¹) (Amon et al., 2007) Biodiesel (1,000 Lha⁻¹) (Durrett et al., 2008) |

*Continued*

Table 15.5 Continued

| Energy Crop | Yield of biomass | Lignin | Hemicellulose | Cellulose | Used for | Biofuel yields |
|---|---|---|---|---|---|---|
| | Tones (t) ha⁻¹ | (%) dry weight | | | | (Reference) |
| **Rapeseed (Brassica napus)**[b,d] 70% of all oilseeds within EU-27. | 2.01 t grain ha⁻¹ (Greece) 3–3.3 (EU) 1.8 (China) | 16 ± 5.2 | 18 ± 1.2 | 44 ± 0.5 | Extraction for its oil. Rapeseed cake is used as cattle feed | Biodiesel (1,190 L ha⁻¹) (Durrett et al., 2008) |

[a] Also contains 67% sucrose.
[b] Also contains 17% starch and 11% protein.
[c] Also contains 10.65% proteins and 26.83% lipids.
[d] The composition refers to the rapeseed cake (after oil extraction).

Table 15.6 Biofuel yields from crop residues

| Crop waste | Productivity rate or yield of biomass | Lignocellulosic composition (%) | | | Used for | Used for second generation biofuels |
|---|---|---|---|---|---|---|
| | | Lignin | Hemicellulose | Cellulose | | |
| Wheat straw | (154–185) · $10^6$ t year$^{-1}$ 2.97 t ha$^{-1}$ 5 t ha$^{-1}$ (US) | 15–21.1 | 26.1–50 | 30–39.2 | Animal feed, construction of materials such as baskets, bricks, cob | Ethanol (386 L (t dry biomass)$^{-1}$) (Naik et al., 2010) Methane (174.6 kg t$^{-1}$) (Chandra et al., 2012) Hydrogen (41.5 L (kg dry matter)$^{-1}$) (Ivanova et al., 2009) |
| Rice straw | 4.52 t ha$^{-1}$ | 20.4 | 33.5 | 44.3 | Animal feed. Straw contain silicon oxide ($SiO_2$) which could result in high quartz ashes that can cause erosion problems in the convective pass of the boiler and handling systems | Ethanol (416 L (t dry biomass)$^{-1}$) (Naik et al., 2010) Methane (167.04 kg t$^{-1}$) (Chandra et al., 2012) Hydrogen (7.4 m$^3$ (t dry matter)$^{-1}$) (Chen et al., 2012a) |
| Rice husk | (157–188) · $10^6$ t year$^{-1}$ | 18.3–19.2 | 17.4–29.3 | 34.4–38.3 | Silica production, building material, fertiliser | Hydrogen (40.38 m$^3$ (t VS)$^{-1}$) (Prakasham et al., 2009) Ethanol (210 kg t$^{-1}$) (Saha and Cotta, 2007) |
| Sugarcane bagasse | (317–380) · $10^6$ t year$^{-1}$ 25–300 kg (t sugar cane)$^{-1}$ | 30 | 20 | 45 | Burned to produce steam and electricity. Paper production | Hydrogen (17.72 L (kg dry matter)$^{-1}$) (Ivanova et al., 2009) Ethanol (3,607 kg ha$^{-1}$ or 52 kg (t dry biomass)$^{-1}$ (Kim and Day, 2011) |

*Continued*

Table 15.6 Continued

| Crop waste | Productivity rate or yield of biomass | Lignocellulosic composition (%) | | | Used for | Used for second generation biofuels |
|---|---|---|---|---|---|---|
| | | Lignin | Hemicellulose | Cellulose | | |
| Sunflower straw | $(7.5–9.0) \cdot 10^6$ t year$^{-1}$ 10 t ha$^{-1}$ (Greece) | 17 | 34.6 | 48.4 | Pectins (45%) in sunflower could be used in food and cosmetic industry. Levulinic acid | Methane (260 m$^3$ (t sunflower straw)$^{-1}$) (Antonopoulou et al., 2010) Methane (2,600–4,550 m$^3$ ha$^{-1}$) (Amon et al., 2007) |
| Sorghum bagasse | $(15–18) \cdot 10^6$ t year$^{-1}$ | 17.6 | 21.4 | 38.5 | Forage, silage, combustion energy, synthesis gas (pyrolysis) and paper | Ethanol (210 kg (t dry biomass)$^{-1}$) (Chen et al., 2012b) |
| Corn stover | $241.5 \cdot 10^6$ t year$^{-1}$(US) 609 t*$10^6$ year$^{-1}$ (world) | 8.4–10.3 | 30 | 37.5 | Forage, animal food, combustion energy | Ethanol (386 L (t dry biomass)$^{-1}$) (Naik et al., 2010) Methane (208.8 kg t$^{-1}$) (Chandra et al., 2012) Ethanol (338.12 kg t$^{-1}$) (Chandra et al., 2012) |
| Rapeseed straw | $1.4 \cdot 10^6$ t year$^{-1}$ (US) $54 \cdot 10^6$ t year$^{-1}$ (world) 5 t ha$^{-1}$ (Greece) | 18 | 19.6 | 37 | Rapeseed hulls: animal feed, heat and energy production, pyrolysis and gasification | Methane (264 m$^3$ t$^{-1}$) (Antonopoulou et al., 2010) Ethanol (140 kg t$^{-1}$) (Lu et al., 2009) |
| Sugar beet pulp | $1.6 \cdot 10^6$ t year$^{-1}$ (US) | 1.1–4.1 | 23.3–26.8 | 19.4–30 | Animal feed, microbial proteins, pectin, citric acid pectinolytic enzymes, ferulic acid and in paper making | Hydrogen (300 kg ha$^{-1}$ sugar beet) (Panagiotopoulos et al., 2010b) |

Biofuels such as hydrogen and methane generated from these feedstocks are characterised as 'second generation' as they do not compete with the production of food.

### 15.4.3 Manure

A mixture of several heterogeneous materials is implied by this designation. They include faeces, urine, hair or feathers, food, wastewater from livestock and poultry units and bedding materials (straw, sand, wood chips, etc.). This type of waste is largely produced in the initial steps of the food chain and involves meat and its products.

Manure is the waste most commonly treated by anaerobic digestion. This is because the anaerobic digestion of manure combines the energy exploitation of its organic content with its stabilisation, potential hygienation (if treated under thermophilic conditions) and odour control. Other benefits of the anaerobic digestion of manure include the reduction of greenhouse gases (which are emitted if manure is spread untreated), the conversion of organic nitrogen to ammonium nitrogen, which can be utilised in agriculture, and the production of a well-stabilised solid material possessing excellent fertilising properties. The biogas produced can be used to contribute to heating requirements in commercially available heating engines (boilers, heaters, etc.). Biogas from manure has also been used to generate electricity (Cantrell *et al.*, 2008).

There is no single practice for the anaerobic digestion of manure. Digester loading, solid content and composition of the manure will be affected by the capacity of the unit and its feeding, by the collection method (scraping or flushing) and the bedding materials. The typical composition of manure from various sources is shown in Table 15.7 and includes the main characteristics of solid and nutrient content. The presence of sand is problematic as its tendency to settle on the bottom of the digester causes an accumulation of inert material which reduces the operating volume. Sand should therefore be removed prior to anaerobic digestion via sedimentation.

The anaerobic digestion of manure requires long hydraulic retention times for the hydrolysis of solids consisting mainly of lignocellulose. The hydraulic retention time is typically in the range 15–30 days under mesophilic conditions and 10–20 days under thermophilic conditions in a CSTR (Angelidaki *et al.*, 2011).

The solid content of the influent manure also influences the process efficiency. The type of collection, scraping or flushing, affects the solid content of the final waste. In cattle livestock facilities, flushing uses 100–200 gallons (380–750 L) of water per cow per day (Burke, 2001). This causes considerable dilution of the total solids from $120 g L^{-1}$ to $8–15 g L^{-1}$ and

*Table 15.7* Solid and nutrient content of various manure types

|  | Dairy (Burke, 2001) | Pig (Prapaspongsa et al., 2010) | Poultry (Abouelenien et al., 2010; Singh et al., 2010) |
|---|---|---|---|
| Total solids (%) | 12 | 8.3 | 25 |
| Volatile solids (% TS) | 85 | 73 | 58 |
| Organic load | 1.075 g COD g$^{-1}$ TS | 1.067 g COD g$^{-1}$ TS | 0.38 g TOC g$^{-1}$ TS<br>0.289 g C g$^{-1}$ TS |
| Total Kjeldahl nitrogen (g g$^{-1}$ TS) | 0.044 | 0.048 | 0.087<br>0.084 |
| Total phosphorous (g g$^{-1}$ TS) | 0.007 | 0.016 | 0.021 |

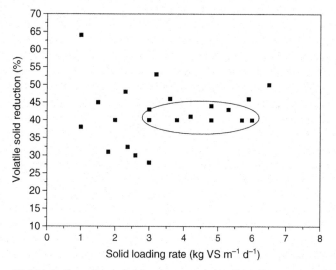

*15.3* Volatile solid reduction versus the solid loading rate at full-scale mesophilic digesters treating dairy manure (Burke, 2001).

bio-reactors such as the CSTR or plug flow may be used. The efficiency of these reactors varies between 40 and 45% reduction of the volatile solids at solid loading rates ranging between 3 and 6 kg VS m$^{-3}$ d$^{-1}$, despite the wide spread of the data (Fig. 15.3). The conversion efficiency for swine manure is slightly higher (Fig. 15.4). Poultry manure is high in solid concentration (Table 15.7). Although high solid anaerobic digestion requires small volume digesters and results in high specific methane production rates, this is not the case with poultry manure due to the high nitrogen level and inhibitory ammonia production. There are few pilot or

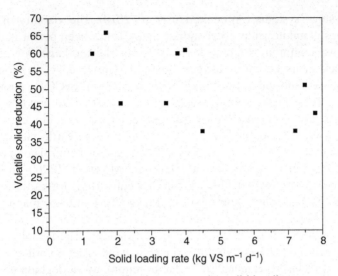

15.4 Volatile solid reduction versus the solid loading rate at mesophilic digesters treating swine manure (Chynoweth et al., 1998).

full-scale studies on poultry manure. Solares et al. (2006) reported a 49% maximum volatile solid reduction in a 40 m$^3$ thermophilic anaerobic digester operated at an HRT of 10 days with an influent solid concentration of 5.5% (the recommended influent solid concentration ranges between 5 and 6%; Singh et al., 2010).

The anaerobic digestion of manure is more stable under mesophilic than thermophilic conditions, although the conversion rate and pathogen reduction are better at high temperatures. Ammonia inhibition has been found to be more effective under thermophilic conditions (Angelidaki and Ahring, 1994) and this jeopardises the stability of the process. The degree of inhibition depends on the acclimatation of anaerobic micro-organisms and on the pH. Inhibition is caused in the range of 1.5–3 g N L$^{-1}$ at pH above 7.4, while ammonia concentrations higher than 3 g N L$^{-1}$ have been found to be toxic regardless of the pH level (Calli et al., 2005). The tolerance of methanogens to ammonia may vary as a result of micro-organism acclimatation, which is a time-consuming process. The ammonia may be reduced by diluting the manure with water or other waste (co-digestion); however, this requires high volume digesters and other wastes for co-digestion may not be available.

Physico-chemical methods such as chemical precipitation, adsorption on zeolite and clay and biological methods (nitrification-denitrification, Anammox) for ammonia removal have been studied and appear to be effective at solid concentrations of 0.5–3% (Chen et al., 2008). Ammonia may also be removed by stripping with an inert gas or biogas in two-stage or single-stage systems (Abouelenien et al., 2010). The viability of these approaches is unknown as no economic analysis has been performed.

The most common parameter for assessing the methane potential of a feedstock is its chemical oxygen demand (COD) concentration. The maximum methane generation is expected to be $0.35\,m^3\,kg\,COD^{-1}$ converted at 0°C and 1 atm. However, the breakdown of the COD content of manure into its constituents reveals that the quantities of lignin (non-biodegradable) and the slowly degradable carbohydrates are 30% in cattle manure and 20% in pig manure (Møller et al., 2004). Based on the chemical composition analysis of carbohydrates, lipids and lignin, etc., Møller et al. (2004) assessed the theoretical methane yield as ($L\,kg^{-1}$ VS converted) for cattle manure (468 ± 6 $L\,kg^{-1}$ VS), pig manure (516 ± 11 $L\,kg^{-1}$ VS) and sow manure (530 ± 6 $L\,kg^{-1}$ VS). Hill (1984) estimated a similar value for the theoretical methane yield (500 $L\,kg^{-1}$ VS) of all types of manure. The actual methane yield (L produced per kg of VS added in a batch bio-reactor until no further methane is produced) is much lower due to minimally biodegradable lignin bounded formulations (e.g., lignocelluloses) and the inhibitors present, such as the ammonia produced during protein degradation. The values determined for various manure types were 148 ± 41 $L\,kg^{-1}$ VS (cattle), 356 ± 28 $L\,kg^{-1}$ VS (pig) and 275 ± 36 $L\,kg^{-1}$ VS (sow). The difference between the various manure types is obvious; pig and sow manure have a much higher methane potential than cattle manure as they contain more proteins and lipids (yielding biogas richer in methane) and less lignocellulosic material.

The discrepancy between the theoretical and actual methane yields also holds for the particulate fraction of manure subjected to separation through centrifugation, evaporation, flocculation and de-watering processes. This is not the case for the liquid fraction (Møller et al., 2004), where the theoretical yield is close to 506 ± 25 $L\,kg^{-1}$ VS (real yield). Although the actual ultimate methane yield of the solid fractions is lower than that of the non-separated manure (a significant portion of the easily biodegradable organic matter is contained in the liquid fraction), the methane produced per volume of the particulate fractions is 2.6–3.6 higher than the pig non-separated manure and 5–6.6 higher than the sow non-separated manure. This indicates that treating the manure fractions separately will result in higher methane yields and specific production rates, as smaller digesters can be used for the particulate fraction and high-rate digesters for the liquid fraction.

A two-phase system for treating both fractions separately could therefore be beneficial. The concept of two-phase configurations could also be applied in the case of high solid manure with the aim of separating hydrolysis-acidogenesis from methanogenesis. In this case, a hydrolysis-acidogenesis bio-reactor should provide a high solid retention time for enhancing the hydrolysis, which is considered to be the rate-limiting step. Myint and Nirmalakhandan (2009) developed a leach-bed bio-reactor and increased the volatile fatty acid production while maintaining a pH between 4 and 5, which is considered to be the optimal condition for particulate hydrolysis. The enrichment of the liquid fraction in COD is achieved through recirculation of the leachate in the same leach-bed bio-reactor. The liquid fraction generated from the leach-bed can be fed to a high-rate methanogenic bioreactor.

A two-stage system was adopted by Kaparaju et al. (2009) based on CSTRs in series. Each CSTR maintains the whole anaerobic population (there being no separation between hydrolysis-acidogenesis and methanogenesis) but operates at different organic loading rates. The effluent of the first CSTR is the influent of the subsequent apparatus. A higher retention time is therefore achieved for the portion of solids that undergoes short-circuiting in the first reactor and leaves it earlier than the hydraulic retention time. It was found that the distribution of the volume of one CSTR to two CSTRs in series at ratios of 70:30 and 50:50 brought about an increase in biogas production of 16.4–17.8% when compared to the one step CSTR.

Although the anaerobic digestion of manure in developed countries is well established, biogas still remains an expensive source of renewable energy because of the high investment cost. This is particularly significant in developing countries. The payback period for a biogas plant of 500 kW of electrical energy is five years when cattle manure is used and 8.5 years when a mixture of cattle and chicken manure is used. However, small-scale biogas plants have a shorter payback period of 8–10 months. It seems that the main obstacle to widespread implementation of biogas technology can be overcome through efficient and simple digester construction and ease of operation and maintenance. State subsidies and tariffs for biogas are also significant incentives for investment (Avcioğlu and Türker, 2012).

The solid residue which remains after anaerobic digestion is another valuable product and may be used as fertiliser due to its high nitrogen and phosphorous nutrient content. However, the presence of pathogens could reduce its fertilising value. Pathogens are micro-organisms (bacteria, viruses, protozoa, etc.) which cause diseases in humans or animals and are abundant in manure. If manure is applied on land without prior treatment, it may contaminate surface water as some pathogens can survive for a long period of time. Pathogens may be adequately reduced as shown in the reports of Bendixen (1999) who studied the effect of duration and temperature on the anaerobic digestion or sanitation process in reducing the faecal streptococci which were selected as indicator pathogens. For example, mesophilic anaerobic digestion (37°C) achieves a pathogen reduction of less than 2 $\log_{10}$ units (that is, if the pathogens present in the feedstock are $10^5$/g dry weight, they are reduced to $10^3$/g dry weight during the process). Thermophilic anaerobic digestion (55°C) achieves a pathogen reduction of

3 to 5 $\log_{10}$ units. When a sanitation step (at 55°C) was included prior to mesophilic anaerobic digestion, the pathogen reduction was 3.6 $\log_{10}$ units.

### 15.4.4 Food wastes

Wastes from the food processing industry are a significant source of energy due to their high organic compound content (carbohydrates, proteins, lipids). Much research has been conducted into optimising biogas and hydrogen production from this heterogeneous category of wastes.

The key point in biogas production from food wastes is to develop simple bio-reactor configurations and to design the process to overcome problems originating from specific characteristics of the waste. For example, wastes rich in proteins, such as fish processing residues, dairy wastewater and slaughterhouse wastewater produce ammonia which can be inhibitory at high levels when subjected to anaerobic digestion, as discussed in Section 15.3.2. The presence of cations (such as sodium from sodium chloride) may also be problematic. As salts are added during food processing, the wastes may be saline or hypersaline. A concentration of sodium exceeding 10 g/L may be strongly inhibitory to methanogens (Kugelman and McCarty, 1965; Rinzema et al., 1998). As in the case of ammonia inhibition, the acclimatation of anaerobic micro-organisms to saline conditions may increase their tolerance to sodium (Omil et al., 1995; Gebauer, 2004).

Lipids are another important constituent of food wastes and yield biogas with a high methane content when converted under anaerobic conditions. Food wastes rich in lipids come from the dairy, meat, fishing and edible oil industries. The production of long-chain fatty acids which are inhibitory to anaerobic micro-organisms and are slow to biodegrade, presents a problem in the degradation of lipids. This type of inhibition has been studied and found to be due to mass transfer limitation rather than to changes in the metabolism of the micro-organisms. The long-chain fatty acids tend to adsorb on surfaces and microbial cellular membranes, so preventing the transfer of substrates and nutrient into the cell. However, when no long-chain fatty acids are present, the activity of the micro-organisms is enhanced, indicating that the inhibition effect has not altered their metabolism (Pereira et al., 2003).

The degradation of long-chain fatty acids takes place through b-oxidation by syntrophic hydrogen-producing bacteria which require hydrogen-consuming methanogenic bacteria to maintain hydrogen at low levels, thus making the conversion thermodynamically feasible. Acetogenesis is slow due to the low energy release of corresponding metabolic reactions which cause a low growth rate in long-chain fatty acid degraders.

Various physico-chemical or biological pretreatment methods have been proposed for removing or converting inhibitory compounds. Another means of decreasing toxicant concentration is to dilute wastewaters with tap water, treated effluent or wastewater devoid of these contaminants. For example, phenolic compounds present in olive mill wastewater (OMWW) can be diluted by mixing with cheese whey or manure. This last option, known as co-digestion, is the most effective, as mixing different types of wastewater results in a better balance of nutrients and alkalinity. Co-digestion may also make anaerobic digestion possible throughout the year, given the seasonal nature of the various agro-industrial wastewaters (Carrieri *et al.*, 1986, 1992; Angelidaki and Ahring, 1997; Lyberatos *et al.*, 1997; Gavala *et al.*, 1999; Angelidaki *et al.*, 2002; Marques *et al.*, 1998; Dareioti *et al.*, 2009, 2010). The main characteristics and 'niches' in the production of biogas from various food waste types are discussed in Table 15.8, along with some typical methane yields.

Hydrogen production from food wastes has recently become more widespread, especially in the case of wastes rich in carbohydrates. Table 15.9 lists some case studies and records the carbohydrate content of each feedstock. Carbohydrates form a large portion of the organic matter (as expressed by the COD content of these wastes) as seen in Table 15.9, an indicator implying the suitability of these feedstocks for hydrogen and biogas production in a two-phase process.

## 15.5 Current status and limitations

A typical use of biogas is as a fuel source to meet the partial or total thermal demand of a waste treatment facility. In cases of sufficient methane yield, there is a surplus of thermal energy which can be provided to meet the thermal demands of the manufacturing unit (e.g., cheese factory, livestock facility etc.). This is the most economically efficient way to use biogas in small- to medium-scale food production units which require thermal energy (in the form of hot water) for their processes. If the production of biogas is sufficiently high to cover the capital cost of a CHP unit producing thermal and electrical energy, additional benefits may be obtained through selling the electricity.

Biogas may also be used in the same way as natural gas (vehicle fuel, internal heating incinerators, etc.). It can be transferred through the natural gas grid if it is upgraded to meet natural gas specifications. During the upgrading process, the levels of some constituents are removed, depending upon the end use of the biogas (Table 15.10). Carbon dioxide, which is a major constituent, must be significantly decreased. The biogas is therefore enriched in methane and the end product of this process is known as biomethane. Carbon dioxide can be removed through water or polyethylene glycol absorption and by separation with carbon molecular sieves or membranes.

*Table 15.8* Methane yield from various food wastes

| Load | Methane yield | Reference |
|---|---|---|
| **Brewery** | | |

The brewery industry produces large quantities of wastewater as a result of the water consumed during brewing; 4–11 m$^3$ of water are required and 2–8 m$^3$ of wastewater are generated per m$^3$ of beer produced (Driessen and Vereijken, 2003). The organic strength of the brewery wastewaters vary from as weak as 0.6–0.9 gCOD L$^{-1}$ to as strong as 160 gL$^{-1}$, because of the different process streams. The anaerobic digesters applied are usually high-rate such as upflow anaerobic sludge blanket (UASB), expanded granular sludge bed reactor (EGSB), anaerobic filters and anaerobic sequential batch reactors (ASBR).

| Load | Methane yield | Reference |
|---|---|---|
| 5 g COD L$^{-1}$ | 256 L kg$^{-1}$ COD | Kormelinck (2003) |
| 56–62 g COD L$^{-1}$ | 370–420 L kg$^{-1}$ COD | Zupancic et al. (2007) |
| **Potato industries** | | |

Potato-based wastewater coming from the potato-starch and chip processing industry contains suspended solids (up to 7 g L$^{-1}$), easily biodegradable organics such as starch and proteins. It may also contain fats which, along with proteins, may cause foaming problems and biomass floatation. Pretreatment via coagulation decreases the solid and protein-fat content and the clarified effluent can be subjected to anaerobic digestion in high-rate bioreactors such as UASBR.

| Load | Methane yield | Reference |
|---|---|---|
| 8.5 g COD L$^{-1}$ (raw) | 263 L kg$^{-1}$ COD (37.5°C) | Kalyuzhnyi et al. (1998) |
| 9 g COD L$^{-1}$ (clarified) | 333 L kg$^{-1}$ COD (37.5°C) | |
| 2,200 g VS L$^{-1}$ | 470 L kg$^{-1}$ VS (55°C) | Fang et al. (2011) |
| Not available | 500–600 L kg$^{-1}$ VS (55°C) | Linke (2006) |
| Not available | 377 L kg$^{-1}$ VS (37.5°C) | Kryvoruchko et al. (2009) |
| **Cheese whey** | | |

Cheese whey wastewater consists of easily biodegradable compounds (lactose) in high concentrations (up to 70 g L$^{-1}$). The combination of high organic load with low alkalinity (50 meq L$^{-1}$) results in poor stability of the anaerobic process. pH control or alkalinity addition is a prerequisite to secure stability. Alternatively, use of pre-acidified cheese whey at a pH around 4 as feedstock succeeds in maintaining the pH in the methanogenic bio-reactor and secures stability. High-rate digesters including UASBR and hybrid reactors have been developed for biogas production from cheese whey (Prazeres et al., 2012).

| Load | Methane yield | Reference |
|---|---|---|
| 68.8 g COD L$^{-1}$ | 330 L kg$^{-1}$ COD (35°C) | Malaspina, et al. (1996) |
| 68.6 g COD L$^{-1}$ | 300 L kg$^{-1}$ COD (37°C) | Saddoud et al. (2007) |
| **Fish processing** | | |

Fish processing wastewater come from washing, filleting and storage and may contain organic matter in soluble, colloidal and particular form. As a result the organic load (1.3–90 g COD L$^{-1}$) and solid content (0.014–10 g TSS L$^{-1}$) vary a lot, also affected by the kind of fish being processed. As expected, nitrogen (77–1,100 mg TKN L$^{-1}$), fats, oil and grease (20–4,000 mg FOG L$^{-1}$) are present in significant concentrations too (Chowdhury et al., 2010). Anaerobic filters and UASB reactors have also been used for anaerobic digestion of the fishery wastewaters.

| Load | Methane yield | Reference |
|---|---|---|
| 1% TS | 260–280 L kg$^{-1}$ VS (37.5°C) | Eiroa et al. (2012) |

*Table 15.8* Continued

| Load | Methane yield | Reference |
|---|---|---|

**Meat processing**

During meat processing, large volumes of liquid and solid wastes are produced. The liquid stream contains on averagel 4g total solids $L^{-1}$, 2.5g COD $L^{-1}$, 0.25g nitrogen $L^{-1}$ (Marcos et al., 2010). Solid wastes such as condemned meat should be rendered first at 133°C and 3 bar pressure and the outcome of this thermal process is used for biogas production.

| Load | Methane yield | Reference |
|---|---|---|
| 1–10% TS | 351–381 L kg$^{-1}$ VS (35°C) | Wu et al. (2009) |
| 40 g VS $L^{-1}$ | 520–550 L kg$^{-1}$ VS (35°C) | Salminen and Rintala (2001) |
| 54 g VS $L^{-1}$ | 490 L kg$^{-1}$ VS (37°C) | Henjfelt and Angelidaki (2009) |

**Olive oil processing**

Extraction of olive oil is accompanied by large quantities of olive mill wastewater (OMWW). Generally, processing 100 kg of olives results in the production of 35 kg of pomace, 55–200 L OMWW (depending on the extraction process; three phase versus two phase extraction) and 5 kg of leaves. The main characteristics of OMWW are its high chemical oxygen demand (COD) concentration (45–220 mg/L), low pH (4–5), high suspended solids concentration (up to 50 g/L) and other recalcitrant organic compounds, such as water-soluble phenols and polyphenols originating from the olives (Azbar et al., 2004; Davies et al., 2004). Biogas production from OMWW is problematic for numerous reasons. The seasonal production of OMWW in large quantities at spatially scattered olive oil extraction units renders the installation and the continuous operation of a biogas plant not viable. Polyphenols, lipids and long-chain fatty acids are considered mainly responsible for the inhibitory effect on methanogenesis. In this case too, proper acclimatisation of the micro-organisms and application of anaerobic systems that allow the retention of the slow growing anaerobes in the bio-reactor result in high biogas yields. Pretreatment and/or post-treatment methods have been developed for improving methane yield and effluent quality, respectively. Both conventional (CSTR type) and high-rate digesters have been applied, although the latter require dilution of the influent to lower the incoming COD.

| Load | Methane yield | Reference |
|---|---|---|
| 54.2 g COD $L^{-1}$ | 371 L kg$^{-1}$ COD (37°C) | Sampaio et al. (2011) |
| 34–150 g COD $L^{-1}$ | 225–259 L kg$^{-1}$ COD (35°C) | Borja et al. (2003) |
| 40 g COD $L^{-1}$ (electro-Fenton pretreated) | 330 L kg$^{-1}$ COD (37°C) | Khoufi et al. (2009) |
| 19.5 g COD $L^{-1}$ | 339 L kg$^{-1}$ COD (35°C) | Stamatelatou et al. (2009) |

Table 15.9 Food wastes used for hydrogen production

| Waste/wastewater | COD (g/L) | | Carbohydrates (g/L) | | Yield of hydrogen | Reference |
|---|---|---|---|---|---|---|
| | Total | Soluble | Total | Soluble | | |
| Cheese whey | 61 | 52 | 38 | 36 | 2.49 m$^3$/m$^3$ cheese whey | Antonopoulou et al. (2008b) |
| Olive pulp | 71.5 | 20.5 | 13.4 | 6.9 | 0.8–1.9 m$^3$/t olive pulp | Koutrouli et al. (2006) |
| Soluble condensed molasses | 40 | | 5.35% w/w | | 6.794 kmol/t COD | Lay et al. (2010) |
| Apple processing wastewater | 9 | | | | 0.7–0.9 m$^3$/m$^3$ wastewater | van Ginkel et al. (2005) |
| Potato processing wastewater | 21 | | | | 2.1–2.8 m$^3$/m$^3$ wastewater | van Ginkel et al. (2005) |
| Tofu processing waste | 37.3 ± 1.8 | | 10.4 ± 2.5 | | 36.04 m$^3$/t | Kim and Lee (2010) |
| Food waste (FW), primary sludge (PS) and waste activated sludge (WAS) | FW: 19.25 ± 1.36  PS: 44.8 ± 2.16  WAS: 10.60 ± 2.89 | FW: 9.23 ± 0.30  PS: 35.9 ± 12.6  WAS: 0.24 ± 0.11 | FW: 4.48 ± 0.11  PS: 0.124 ± 0.44  WAS: 0.031 ± 0.01 | | 112 m$^3$/t VS | Zhu et al. (2008a) |
| Kitchen waste | 211.79 ± 11.13 | | | | 0.074–0.0107 m$^3$/t kitchen waste | Wang and Zhao (2009) |
| Food waste (grains 35.7% TS, vegetables 42.1% TS, meat 17.2% TS) | | | 25.0 ± 4.8 | | 120 m$^3$/t VS  20.56 m$^3$/t food waste | Shin and Youn (2005) |
| Pulverised garbage and shredded paper wastes | 111.10–17.13 | 45.4–70.7 | 31–56 | 12–30 | 56 m$^3$/t COD | Ueno et al. (2007) |
| Potato waste | 12.6 ± 0.5 | 2.22 ± 0.10 | 0.573 ± 0.152 | | 30 m$^3$/t TS  324 m$^3$/m$^3$ potato waste | Zhu et al. (2008b) |

*Table 15.10* Removal of biogas components based on the biogas utilisation

| Application | $H_2S$ | $CO_2$ | $H_2O$ |
| --- | --- | --- | --- |
| Gas heater (boiler) | <1000 ppm | No need to remove | No need to remove |
| Kitchen stove | Removal is needed | No need to remove | No need to remove |
| CHP | <1000 ppm | No need to remove | No condensation |
| Vehicle fuel | Removal is needed | Removal is recommended | Removal is needed |
| Natural gas grid | Removal is needed | Removal is needed | Removal is needed |

*Source*: IEA, Bioenergy – Biogas upgrade and utilisation, Task 24: Energy from biological conversion of organic waste.

Biogas is saturated with water because it emerges from a water-based medium. When it is stored in tanks, high pressure develops and the water is condensed and frozen. It also facilitates oxidation reactions (e.g., oxidation of sulphide to sulphate). Both sulphides and sulphates are corrosive compounds and are the most problematic constituents in biogas mixture as they have an effect on the metallic surfaces of pipelines, CHP units, incinerators, etc. The maximum permitted concentration is 5 ppm and biological and physico-chemical methods have been developed for the removal of hydrogen sulphide. The biological methods are based on the action of the sulphide oxidising micro-organisms (*Thiobacillus*) which can be activated in a micro-aerophilic environment on $CO_2$ (autotrophic). Elemental sulphur and sulphate are produced through biological sulphide transformation. The physico-chemical methods are usually based on water, polyethylene glycol or NaOH scrubbing configurations. Adsorption on activated carbon and chemical precipitation by combination with iron compounds such as iron chloride and iron oxide are also used in the biogas upgrading process.

The energy value of biogas varies between 5 and 7.5 kWh/m$^3$, depending on the methane content (average 6 kWh/m$^3$ or 21.6 MJ/m$^3$) and can be estimated if its composition is known through the energy content of the methane (9.97 kWh/m$^3$). A typical CHP unit efficiency is 30–45% (electrical energy) and 35–60% (thermal energy), while losses account for 10–15% (FNR, 2012).

The complexity of a biogas plant determines its cost but also secures its stability. The estimated cost of typical biogas units is based on their electrical capacity. The electricity and thermal requirement of a biogas plant are 7–10 and 25%, respectively, while the labour requirement amounts to 1–5 labour h/kW$_{el}$. The operating cost for biomethane upgrading depends on the volume treated: for 250 Nm$^3$ of biogas, the cost is 7.79–10.01 €cent/Nm$^3$, while for 1,000 Nm³, the cost is reduced to 5.82–6.07 €cent/Nm³. The investment costs also depend on the capacity of the biogas plant and range between 3,000 and 6,000 €/kW$_{el}$ (capacity < 500 kW$_{el}$) or lower than 3,000 €/kW$_{el}$ (capacity > 500 kW$_{el}$). Similarly, the investment cost of a CHP unit is 875 €/kW$_{el}$ (capacity = 150 kW$_{el}$), 738 €/kW$_{el}$ (capacity = 250 kW$_{el}$) or 586 €/kW$_{el}$ (capacity = 500 kW$_{el}$). The investment cost of a biogas plant which upgrades biogas to biomethane is 13–17 €/MWh (flowrate = 250 Nm³/h) or 7–13 €/MWh (flowrate = 1,000 Nm³/h).

The quantity of energy produced from biomass is expected to increase four-fold by 2035 (Angelidaki *et al.*, 2011), with a particular increase in biogas. In Germany, one of the leading countries in biogas production, the number of biogas plants in 2011 was 2,728, with a total installed electricity capacity of 7,000 MW$_{el}$, according to an estimation made by the Agency for Renewable Resources (FNR, 2012). The number of biogas plants producing biomethane reached 107 with an upgrading capacity of 68,100 Nm³ biogas/h. The feedstock of biogas plants in Germany are primarily energy crops (46%) and animal manure (45%) as recorded in 2010 (FNR, 2012). The cultivation of land for non-food crops is a controversial issue and the use of agricultural residues and waste streams as the primary source of biomass is preferred.

In contrast to Western European countries where biogas generation takes place in large-scale centralised plants, developing countries such as China and India have many small domestic biogas units as well as several medium- to large-scale plants for electricity production. Although anaerobic digestion is a mature technology, issues remain for scientific and technical consideration. These include the optimisation of process efficiency for residues and wastewaters, reactor design, the development of efficient control and monitoring and the design of small, farm-scale plants which are installed in medium-sized enterprises for cost-effective utilisation of their wastes. At present, the feasibility of electricity production from biomass through biogas is based on the subsidies and tariffs applied in each country. Nevertheless, overcoming the technical barriers will give biogas a more prominent role in energy provision.

Biohydrogen has not yet been produced on a large scale. Most research has been conducted at the laboratory scale and, less frequently, on the pilot-scale level. It is not therefore possible to perform an economic analysis. Hydrogen (as with biogas reformed to syngas), can be utilised in fuel cells, the efficiency of which is independent of the scale of the fuel cell. As a consequence, electricity production from fuel cells can be achieved at any scale (which is not the case for biogas where the size of the cogeneration unit determines the efficiency in electrical energy and its cost). The investment cost of fuel cells is still high and biohydrogen processes have not yet reached the maximum yield due to biological limitations set by the accumulation of hydrogen in the bio-reactors. The maximum hydrogen yield via dark fermentation processes is $4\,mol\,mol^{-1}$ glucose but varies within $1-2\,mol\,mol^{-1}$ glucose.

As with methane, hydrogen is produced mixed with carbon dioxide and requires enrichment. Traces of carbon monoxide cannot be tolerated at concentrations higher than 10 ppm and in order to remove trace gases (impurities) or carbon dioxide, the mixture is processed by using membrane technologies based on palladium.

## 15.6 Future trends

Biogas is a renewable source of energy which is produced by the application of a well-established, reliable and successful technology. For this reason, investments in the biogas field are supported by European governments through their subsidy policies. EU energy policy aims to meet 20% of energy demands from renewable energy sources by 2020. It has been estimated that at least 25% of total bioenergy may come from biogas produced from wet organic materials such as animal manure, crop silages, wet organic food and feed residues (Holm-Nielsen *et al.*, 2007).

Technological advances are reducing processing costs, making anaerobic digestion technology robust and stable and furthering its use. However, there is a need for further research, especially in the field of lignocellulosic biomass, which has recently emerged as a potential feedstock. Lignocellulosic biomass consists mainly of the residues of crops, forestry and gardening waste, etc., and comprises a huge category of organic material. The limited biodegradability of lignin complexes can be improved through pretreatment processes. As pretreatment costs are high, the most effective options are making use of the nutrients or other high added value materials contained in anaerobic digestion residues, together with the adoption of biorefinery concepts.

If anaerobic technology is to be disseminated worldwide, the focus must be on making the process viable at the small to medium scale and moving from economy of scale to economy of numbers. The development of simple and cheap bio-reactors and personnel training are important in spreading the technology to developing and underdeveloped countries.

Sustainable biohydrogen production will depend upon an improvement in hydrogen storage and fuel cell technology. These two factors will motivate further research on the biohydrogen process itself, thus enhancing hydrogen yield and productivity.

## 15.7 Sources of further information and advice

The European Anaerobic Digestion Network. http://www.adnett.org/index.html

Renewable Energy, Purdue University, http://www.ces.purdue.edu/bioenergy

The AD community: An independent web site: http://www.anaerobic-digestion.com/index.php

England's Official Information Portal on Anaerobic Digestion: http://www.biogas-info.co.uk/

Small-Scale Biogas Use with Biogidesters in Rural Costa Rica: http://www.ruralcostarica.com/biodigester.html

European Biomass Industry Association: http://www.eubia.org/108.0.html

EurObservER: http://www.recyclingportal.eu/artikel/25990.shtml

Biomass energy: http://www.biomassenergy.gr/en/

Hydrogen Information Network: http://www.eren.doe.gov/hydrogen/

The 'National Hydrogen Energy Roadmap'

ESF/PESC Network 'Biomass Fermentation Towards Usage in Fuel Cells': http://www.bfcnet.info/

IEA Hydrogen Program: http://www.eren.doe.gov/hydrogen/iea/

HyNet. The European Thematic Network on Hydrogen: http://www.hynet.info/

## 15.8 References

Abouelenien, F., Fujiwara, W., Namba Y., Kosseva, M., Nishio, N. and Nakashimada Y. (2010) 'Improved methane fermentation of chicken manure via ammonia removal by biogas recycle', *Bioresource Technology*, 101, 6368–6373.

Amon, T., Amon, B., Kryvoruchko, V., Machmüller, A., Hopfner-Sixt, K., Bodiroza, V., Hrbek, R., Friedel, J., Pötsch, E., Wagentristl, H., Schreiner, M. and Zollitsch, W. (2007) 'Methane production through anaerobic digestion of various energy crops grown in sustainable crop rotations', *Bioresource Technology*, 98(17), 3204–3212.

Angelidaki, I. and Ahring, B. K. (1994) 'Anaerobic thermophilic digestion of manure at different ammonia loads: effect of temperature', *Water Research*, 28(3), 727–731.

Angelidaki, I. and Ahring, B. K. (1997) 'Codigestion of olive oil mill wastewaters with manure, household waste or sewage sludge', *Biodegradation*, 8(4), 221–226.

Angelidaki, I., Ahring, B. K., Deng, H. and Schmidt, J. E. (2002) 'Anaerobic digestion of olive oil mill effluents together with swine manure in UASB reactors', *Water Science and Technology*, 45(10), 213–218.

Angelidaki, I., Karakashev, D., Batstone, D. J., Plugge, C. M. and Stams A. J. M. (2011) 'Biomethanation and its potential', *Methods in Enzymology*, 494, 327–351.

Annachhatre, A. P. (1996) 'Anaerobic treatment of industrial wastewaters', *Resources, Conservation and Recycling*, 16, 161–166.

Antonopoulou, A., Gavala, H. N., Skiadas, I. V., Angelopoulos, K. and Lyberatos, G. (2008a) 'Biofuels generation from sweet sorghum: fermentative hydrogen production and anaerobic digestion of the remaining biomass', *Bioresource Technology*, 99, 110–119.

Antonopoulou, G., Stamatelatou, K., Venetsaneas, N., Kornaros, M. and Lyberatos, G. (2008b) 'Biohydrogen and methane production from cheese whey in a two-stage anaerobic process', *Industrial and Engineering Chemistry Research*, 47(15), 5227–5233.

Antonopoulou, G., Stamatelatou, K. and Lyberatos, G. (2010) 'Exploitation of rapeseed and sunflower residues for methane generation through anaerobic digestion: the effect of pretreatment', *Chemical Engineering Transactions*, 20, 253–258.

Asam, Z., Poulsen, T. G., Nizami, A. S., Rafique, R., Kiely, G. and Murphy, J. D. (2011) 'How can we improve biomethane production per unit of feedstock in biogas plants?', *Applied Energy*, 88(6), 2013–2018.

Avcioğlu, A. O. and Türker, U. (2012) 'Status and potential of biogas energy from animal wastes in Turkey', *Renewable and Sustainable Energy Reviews*, 16(3), 1557–1561.

Azbar, N., Bayram, A., Filibeli, A., Muezzinoglu, A., Sengul, F. and Ozer, A. (2004) 'A review of waste management options in olive oil production', *Critical Reviews in Environmental Science and Technology*, 34(3), 209–247.

Banu, J. R. and Kaliappan, S. (2008) 'Treatment of tannery wastewater using hybrid upflow anaerobic sludge blanket reactor', *Journal of Environmental Engineering and Science*, 6(4), 415–421.

Barber, W. P. and Stuckey, D. C. (1999) 'The use of the anaerobic baffled reactor (ABR) for wastewater treatment: a review', *Water Research*, 33(7), 1559–1578.

Bauer, A., Bosch, P., Friedl, A. and Amon, T. (2009) 'Analysis of methane potentials of steam-exploded wheat straw and estimation of energy yields of combined ethanol and methane production', *Journal of Biotechnology*, 142, 50–55.

Bendixen, H. J. (1999) 'Hygienic safety: results of scientific investigations in Denmark (sanitation requirements in Danish biogas plants)', in Proceedings of the IEA workshop: 'Hygienic and environmental aspects of anaerobic digestion: legislation and experiences in Europe', Universität Hohenheim, Stuttgart, Germany, 27–47.

Bennett, A. S. and Anex, R. P. (2009) 'Production, transportation and milling costs of sweet sorghum as a feedstock for centralized bioethanol production in the upper Midwest', *Bioresource Technology*, 100, 1595–1607.

Boe, K. (2006) Online monitoring and control of the biogas process, PhD thesis, Technical University of Denmark.

Borja, R., Martín, A., Rincón, B. and Raposo, F. (2003) 'Kinetics for substrate utilization and methane production during the mesophilic anaerobic digestion of two phases olive pomace (TPOP)', *Journal of Agricultural and Food Chemistry*, 51(11), 3390–3395.

Brock, T. D., Madigan, M. T., Martinko, J. M. and Parker, J. (1994) *Biology of Microorganisms*. Prentice Hall, Englewood Cliffs, NJ.

Bruni, E., Jensen, A. P. and Angelidaki, I. (2010) 'Comparative study of mechanical, hydrothermal, chemical and enzymatic treatments of digested biofibers to improve biogas production', *Bioresource Technology*, 101(22), 8713–8717.

Burke, D. (2001) *Dairy Waste Anaerobic Digestion Handbook*, Environmental Energy Company, Olympia, WA.

Call, D and Logan, B. E. (2008) 'Hydrogen production in a single chamber microbial electrolysis cell lacking a membrane', *Environ Sci Technol*, 42, 3401–3406.

Calli, B., Mertoglu, B., Inanc, B. and Yenigun, O. (2005) 'Effect of high free ammonia concentrations on the performances of anaerobic bioreactors', *Process Biochemistry*, 40, 1285–1292.

Calli, B., Schoenmaekers, K., Vanbroekhoven, K. and Diels, L. (2008) 'Dark fermentative $H_2$ production from xylose and lactose – effects of on-line pH control', *International Journal of Hydrogen Energy*, 33, 522–530.

Cantrell, K. B., Ducey, T., Ro, K. S. and Hunt, P. G. (2008) 'Livestock waste-to-bioenergy generation opportunities', *Bioresource Technology*, 99, 7941–7953.

Cao, G., Ren, N., Wang, A., Lee, D. J., Guo, W., Liu, B., Feng, Y. and Zhao, Q. (2009) 'Acid hydrolysis of corn stover for biohydrogen production using *Thermoanaerobacterium thermosaccharolyticum* W16', *International Journal of Hydrogen Energy*, 34(17), 7182–7188.

Cara, C., Ruiz, E., Oliva, J. M., Saez, F. and Castro, E. (2007) 'Conversion of olive tree biomass into fermentable sugars by dilute acid pretreatment and enzymatic saccharification', *Bioresource Technology*, 99, 1869–1876.

Carrieri, C., Balice, V., Rozzi, A. and Santori, M. (1986) 'Anaerobic treatment of olive mill effluents mixed with sewage sludge. Preliminary results'. In: *Proceedings of International Symposium on Olive By-Products Valorization*, Seville, Spain, 4–7 March.

Carrieri, C., Di Pinto, A. C. and Rozzi, A. (1992) 'Anaerobic co-digestion of sewage sludge and concentrated soluble wastewaters', *Water Science Technology*, 28(2), 187–197.

Chandra, R., Bura, R., Mabee, W., Berlin, A., Pan, X. and Saddler, J. (2007) 'Substrate pretreatment: the key to effective enzymatic hydrolysis of lignocellulosics?' *Advances in Biochemical Engineering/Biotechnology*, 108, 67–93.

Chandra, R., Takeuchi, H. and Hasegawa, T. (2012) 'Methane production from lignocellulosic agricultural crop wastes: a review in context to second generation of biofuel production', *Renewable and Sustainable Energy Reviews*, 16(3), 1462–1476.

Chang, J. S., Lee, K. S. and Lin, P. J. (2002) 'Biohydrogen production with fixed-bed bioreactors', *International Journal of Hydrogen Energy*, 27, 1167–1174.

Chen, C. C., Chuang, Y. S., Lin, C. Y., Lay, C. H. and Sen, B. (2012a) 'Thermophilic dark fermentation of untreated rice straw using mixed cultures for hydrogen production', *International Journal of Hydrogen Energy*, 37(20), 15540–15546.

Chen, C., Boldor, D., Aita, G. and Walker, M. (2012b) 'Ethanol production from sorghum by a microwave-assisted dilute ammonia pretreatment', *Bioresource Technology*, 110, 190–197.

Chen, Y., Cheng, J. J. and Creamer, K. S. (2008) 'Inhibition of anaerobic digestion process: a review', *Bioresource Technology*, 99, 4044–4064.

Chowdhury, P., Viraraghavan, T. and Srinivasan, A. (2010) 'Biological treatment processes for fish processing wastewater – a review', *Bioresource Technology*, 101, 439–449.

Chynoweth, D. P., Wilkie, A. C. and Owens J. M. (1998) 'Anaerobic treatment of piggery slurry', *Proceedings of ASAE Annual International Meeting*, 28 June–4 July, Seoul, Korea.

Cysneiros, D., Thuillier, A., Villemont, R., Littlestone, A., Mahony, T. and O'Flaherty, V. (2011) 'Temperature effects on the trophic stages of perennial rye grass anaerobic digestion', *Water Science and Technology*, 64, 70–76.

Dareioti, M. A., Dokianakis, S. N., Stamatelatou, K., Zafiri, C. and Kornaros, M. (2009) 'Biogas production from anaerobic co-digestion of agroindustrial wastewaters under mesophilic conditions in a two-stage process', *Desalination*, 248, 891–906.

Dareioti, M. A., Dokianakis, S. N., Stamatelatou, K., Zafiri, C. and Kornaros, M. (2010) 'Exploitation of olive mill wastewater and liquid cow manure for biogas production'. *Waste Management*, 30, 1841–1848.

Datar, R., Huang, J., Maness, P. C., Mohagheghi, A., Czernik, S. and Chornet, E. (2007) 'Hydrogen production from the fermentation of corn stover biomass pretreated with a steam-explosion process', *International Journal of Hydrogen Energy*, 32(8), 932–939.

Davies, L. C., Novais, J. M. and Martins-Dias, S. (2004) 'Detoxification of olive mill wastewater using superabsorbent polymers', *Environmental Technology*, 25(1), 89–100.

de Vries, S. C., van de Ven, G. W. J., van Ittersum, M. K. and Giller, K. E. (2010) 'Resource use efficiency and environmental performance of nine major biofuel crops, processed by first-generation conversion techniques', *Biomass Bioenergy*, 34, 588–601.

de Vrije, T., Bakker, R. R., Budde, M. A. W., Lai, M. H., Mars, A. E. and Claassen, P. A. M. (2009) 'Efficient hydrogen production from the lignocellulosic energy crop *Miscanthus* by the extreme thermophilic bacteria *Caldicellulosiruptor saccharolyticus* and *Thermotoga neapolitana*', *Biotechnology for Biofuels*, 2, art. no. 12.

de Vrije, T., Budde, M. A. W., Lips, S. J ., Bakker, R. R., Mars, A. E. and Claassen, P. A. M. (2010) 'Hydrogen production from carrot pulp by the extreme thermophiles *Caldicellulosiruptor saccharolyticus* and *Thermotoga neapolitana*', *International Journal of Hydrogen Energy*, 35(24), 13206–13213.

Defour, D., Derycke, D., Liessens, J. and Pipyn, P. (1994) 'Field experience with different systems for biomass accumulation in anaerobic reactor technology', *Water Science and Technology*, 30, 181–191.

Delgenes, J. P., Penaud, V. and Moletta, R., (2002) 'Pretreatments for the enhancement of anaerobic digestion of solid wastes'. In: Mata-Alvarez, J. (ed.) *Biomethanization of the Organic Fraction of Municipal Solid Wastes*. IWA Publishing, London, 201–228.

Deverell, R., McDonnell, K., Ward, S. and Devlin, G. (2009) 'An economic assessment of potential ethanol production pathways in Ireland', *Energy Policy*, 37(10), 3993–4002.

Driessen, W. and Vereijken, T. (2003) 'Recent developments in biological treatment of brewery effluent', The Institute and Guild of Brewing Convention, Zambia.

Durrett, T. P., Benning, C. and Ohlrogge, J. (2008) 'Plant triacylglycerols as feedstocks for the production of biofuels', *Plant Journal*, 54(4), 593–607.

Eiroa, M., Costa, J. C., Alves, M. M., Kennes, C. and Veiga, M. C. (2012) 'Evaluation of the biomethane potential of solid fish waste', *Waste Management*, 32, 1347–1352.

Estevez, M. M., Linjordet, R. and Morken, J. (2012) 'Effects of steam explosion and co-digestion in the methane production from Salix by mesophilic batch assays', *Bioresource Technology*, 104, 749–756.

Fang, C., Boe, K. and Angelidaki, I. (2011) 'Biogas production from potato-juice, a by-product from potato-starch processing, in upflow anaerobic sludge blanket

(UASB) and expanded granular sludge bed (EGSB) reactors', *Bioresource Technology*, 102, 5734–5741.

Fang, H. H. P. and Liu, H. (2004) 'Biohydrogen production from wastewater by granular sludge', *1st International Symposium on Green Energy Revolution*, Nagaoka, Japan, 31–36.

Fang, H. H. P., Zhang, T. and Liu H. (2002a) 'Effect of pH on hydrogen production from glucose by a mixed culture', *Bioresource Technology*, 82, 87–93.

Fang, H. H. P., Liu, H. and Zhang, T. (2002b) 'Characterization of a hydrogen-producing granular sludge', *Biotechnology and Bioengineering*, 78, 44–52.

FAO (2008) *The state of food and agriculture. Biofuels: Prospects, risks and opportunities*, Rome, Italy: FAO Electronic Publishing Policy and Support Branch Communications Division, http://www.fao.org/publications/sofa-2008/en/ (accessed 10 Jan 2011).

Fernandes, T. V., Klaasse Bos, G. J., Zeeman, G., Sanders, J. P. M. and van Lier, J. B. (2009) 'Effects of thermo-chemical pre-treatment on anaerobic biodegradability and hydrolysis of lignocellulosic biomass', *Bioresource Technology*, 100(9), 2575–2579.

FNR (Fachagentur Nachwachsende Rohstoffe) (2012) *Bioenergy in Germany: Facts and Figures*. Federal Ministry of Food, Agriculture and Consumer Protection, Berlin.

Frigon, J. C., Mehta, P. and Guiot, S. R. (2012) 'Impact of mechanical, chemical and enzymatic pre-treatments on the methane yield from the anaerobic digestion of switchgrass', *Biomass and Bioenergy*, 36, 1–11.

Garcia, H., Rico, C., Garcia, P. A. and Rico, J. L. (2008) 'Flocculants effect in biomass retention in a UASB reactor treating dairy manure', *Bioresource Technology*, 99, 6028–6036.

Gavala, H. N., Skiadas, I. V. and Lyberatos, G. (1999) 'On the performance of a centralised digestion facility receiving seasonal agroindustrial wastewaters', *Water Science and Technology*, 40(1), 339–346.

Gebauer, R. (2004) 'Mesophilic anaerobic treatment of sludge from saline fish farm effluents with biogas production', *Bioresource Technology*, 93, 155–167.

Gossett, J. M., Stuckey, D. C., Owen, W. F. and McCarty, P. L. (1982) 'Heat treatment and anaerobic digestion of refuse', *Journal of the Environmental Engineering Division*, 108, 437–454.

Hall, E. R. (1992) 'Anaerobic treatment of wastewaters in suspended growth and fixed film processes', In: J. F. Malina, Jr and F. G. Pohland (eds), *Water Quality Management Library*, 7, Technomic Publishing Company, Lancaster, PA, 41–119.

Hallenbeck, P. C. and Ghosh, D. (2009) 'Review: advances in fermentative biohydrogen production: the way forward?', *Trends in Biotechnology*, 27, 287–297.

Harper, S. R. and Pohland, F. G. (1987) 'Enhancement of anaerobic treatment efficiency through process modification', *Journal of Water Pollution Control Federation*, 59, 152–161.

Henjfelt, A. and Angelidaki, I. (2009) 'Anaerobic digestion of slaughterhouse by-products', *Biomass and Bioenergy*, 33, 1046–1054.

Hendriks, A. T. W. M. and Zeeman, G. (2009) 'Pretreatments to enhance the digestibility of lignocellulosic biomass', *Bioresource Technology*, 100(1), 10–18.

Hill, D. T. (1984) 'Methane productivity of the major animal waste types', *Transactions of the ASAE*, 27(2), 530–534.

Holm-Nielsen, J. B., Oleskowicz-Popiel, P. and Al Seadi, T. (2007) 'Energy crop potentials for the future bioenergy in EU-27'. In: *Proceedings of the 15th European Biomass Conference*, 7–11 May, Berlin, Germany.

Holtzapple, M. T., Humphrey, A. E. and Taylor, J. D. (1989) 'Energy requirements for the size reduction of poplar and aspen wood', *Biotechnology and Bioengineering*, 33, 207–210.

Hussy, I., Hawkes, F. R., Dinsdale, R. and Hawkes, D. L. (2003) 'Continuous fermentative hydrogen production from a wheat starch co-product by mixed microflora', *Biotechnology and Bioengineering*, 84, 619–626.

Ivanova, G., Rákhely, G. and Kovács, K. L. (2009) 'Thermophilic biohydrogen production from energy plants by *Caldicellulosiruptor saccharolyticus* and comparison with related studies', *International Journal of Hydrogen Energy*, 34(9), 3659–3670.

Kalyuzhnyi, S., de los Santos, L. E. and Martinez J. R. (1998) 'Anaerobic treatment of raw and preclarified potato-maize wastewaters in a UASB reactor', *Bioresource Technology*, 66, 195–199.

Kaparaju, P., Ellegaard, L. and Angelidaki, I. (2009) 'Optimisation of biogas production from manure through serial digestion: lab-scale and pilot-scale studies', *Bioresource Technology*, 100, 701–709.

Khoufi, S., Aloui, F. and Sayadi S. (2009) 'Pilot scale hybrid process for olive mill wastewater treatment and reuse', *Chemical Engineering and Processing*, 48, 643–650.

Kim, M. and Day, D. F. (2011) 'Composition of sugar cane, energy cane, and sweet sorghum suitable for ethanol production at Louisiana sugar mills', *Journal of Industrial Microbiology and Biotechnology*, 38(7), 803–807.

Kim, M. S. and Lee, D. Y. (2010) 'Fermentative hydrogen production from tofu-processing waste and anaerobic digester sludge using microbial consortium' *Bioresource Technology*, 101(1 Suppl.), S48–S52.

Kivaisi, A. K. and Eliapenda, S. (1994) 'Pretreatment of bagasse and coconut fibres for enhanced anaerobic degradation by rumen micro-organisms', *Renewable Energy*, 5, 791–795.

Kormelinck, V. G. (2003) 'Optimum wastewater treatment at Paulaner Munich', *Brauwelt International*, 6, 387–390.

Koutrouli, E. C., Gavala, H. N., Skiadas, I. V. and Lyberatos, G. (2006) 'Mesophilic biohydrogen production from olive pulp', *Process Safety and Environmental Protection*, 84(4B), 285–289.

Kraemer, J. T. and Bagley, D. M. (2005) 'Continuous fermentative hydrogen production using a two-phase reactor system with recycle', *Environmental Science and Technology*, 39, 3819–3825.

Kreuger, E., Nges, I. and Björnsson, L. (2011) 'Ensiling of crops for biogas production: Effects on methane yield and total solids determination', *Biotechnology for Biofuels*, 4, art. no. 44.

Kryvoruchko, V., Machmuller, A., Bodiroza, V., Amon, B. and Amon T. (2009) 'Anaerobic digestion of by-products of sugar beet and starch potato processing', *Biomass and Bioenergy*, 33, 620–627.

Kugelman, I. J. and McCarty, P. L. (1965) 'Cation toxicity and stimulation in anaerobic waste treatment. I. Slug feed studies', *Journal-Water Pollution Control Federation*, 37, 97–116.

Lay, C. H., Wu, J. H., Hsiao, C. L., Chang, J. J., Chen, C. C. and Lin, C. Y. (2010) 'Biohydrogen production from soluble condensed molasses fermentation using anaerobic fermentation', *International Journal of Hydrogen Energy*, 35(24), 13445–13451.

Lay, J. J. (2001) 'Biohydrogen generation by mesophilic anaerobic fermentation of microcrystalline cellulose', *Biotechnology and Bioengineering*, 74, 280–287.

Lee, Y. J., Miyahara, T. and Noike, T. (2002) 'Effect of pH on microbial hydrogen fermentation', *Journal of Chemical Technology and Biotechnology*, 77, 694–698.

Lehtomäki, A., Huttunen, S., Lehtinen, T. M. and Rintala, J. A. (2008) 'Anaerobic digestion of grass silage in batch leach bed processes for methane production', *Bioresource Technology*, 99, 3267–3278.

Lettinga, G., Hobma, S. W., Klapwijk, A., Van Velsen, A. F. M. and De Zeeuw, W. J. (1980) 'Use of the upflow sludge blanket (UAS) reactor concept for biological wastewater treatment', *Biotechnology and Bioengineering*, 22, 699–734.

Linke, B. (2006) 'Kinetic study of thermophilic anaerobic digestion of solid wastes from potato processing', *Biomass and Bioenergy*, 30, 892–896.

Lissens, G., Thomsen, A. B., de Baere, L., Verstraete, W. and Ahring, B. K. (2004) 'Thermal wet oxidation improves anaerobic biodegradability of raw and digested biowaste', *Environmental Science and Technology*, 38, 3418–3424.

Lo, Y. C., Lu, W. C., Chen, C. Y. and Chang, J. S. (2010) 'Dark fermentative hydrogen production from enzymatic hydrolysate of xylan and pretreated rice straw by *Clostridium butyricum* CGS5', *Bioresource Technology*, 101(15), 5885–5891.

Lu, X., Zhang, Y. and Angelidaki, I. (2009) 'Optimization of $H_2SO_4$-catalyzed hydrothermal pretreatment of rapeseed straw for bio conversion to ethanol: focusing on pretreatment at high solids content', *Bioresource Technology*, 100, 3048–3053.

Lyberatos, G., Gavala, H. N. and Stamatelatou, A. (1997) 'An integrated approach for management of agricultural industries wastewaters', *Nonlinear Analysis, Theory, Methods and Applications*, 30(4), 2341–2351.

Malaspina, F., Cellamare, C. M., Stante, L. and Tilche, A. (1996) 'Anaerobic treatment of cheese whey with a downflow-upflow hybrid reactor', *Bioresource Technology*, 55(2), 131–139.

Marcos, A., Al-Kassir, A., Mohamad, A. A., Cuadros, F. and López-Rodríguez, F. (2010) 'Combustible gas production (methane) and biodegradation of solid and liquid mixtures of meat industry wastes', *Applied Energy*, 87, 1729–1735.

Marques, I. P., Teixeira, A., Rodrigues, L., Martins Dias, S. and Novais, J. M. (1998) 'Anaerobic treatment of olive mill wastewater with digested piggery effluent', *Water Research*, 70(5), 1056–1061.

Massé, D., Gilbert, Y., Savoie, P., Bélanger, G., Parent, G. and Babineau, D. (2011) 'Methane yield from switchgrass and reed canary grass grown in Eastern Canada', *Bioresource Technology*, 102(22), 10286–10292.

McLaren J. (2009) 'Sugarcane as a feedstock for biofuels', NCGA, Washington, DC. Available at: http://www.ncga.com/files/pdf/SugarcaneWhitePaper092810.pdf (accessed 29 March 2011).

Menardo, S., Airoldi, G. and Balsari, P. (2012) 'The effect of particle size and thermal pre-treatment on the methane yield of four agricultural by-products', *Bioresource Technology*, 104, 708–714.

Mirahmadi, K., Kabir, M. M., Jeihanipour, A., Karimi, K. and Taherzadeh, M. J. (2010) 'Alkaline pretreatment of spruce and birch to improve bioethanol and biogas production', *BioResources*, 5(2), 928–938.

Mizuno, O., Dinsdale, R., Hawkes, F. R., Hawkes, D. L. and Noike, T. (2000) 'Enhancement of hydrogen production from glucose by nitrogen gas sparging', *Bioresource Technology*, 73, 59–65.

Møller, H. B., Sommer, S. G. and Ahring, B. K. (2004) 'Methane productivity of manure, straw and solid fraction of manure', *Biomass and Bioenergy*, 26, 485–495.

Morimoto, M (2002) 'Why is the anaerobic fermentation in the production of the biohydrogen attractive?' in *The Proceedings of Conversion of Biomass into Bioenergy*. Organized by New Energy and Industrial Technology Development Organization (NEPO), Japan and Malaysian Palm Oil Board (MPOP).

Myint, M. T. and Nirmalakhandan, N. (2009) 'Enhancing anaerobic hydrolysis of cattle manure in leachbed reactors', *Bioresource Technology*, 100, 1695–1699.

Naik, S. N., Goud, V. V., Rout, P. K. and Dalai, A. K. (2010) 'Production of first and second generation biofuels: a comprehensive review', *Renewable and Sustainable Energy Reviews*, 14(2), 578–597.

Nasirian, N. (2012) 'Biological hydrogen production from acid-pretreated straw by simultaneous saccharification and fermentation', *African Journal of Agricultural Research*, 7(6), 876–882.

Nielsen, A. T., Amandusson, H., Bjorklund, R., Dannetun, H., Ejlertsson, J., Akedahl, L. G., Lumndstrom, I. and Svensson, H. H. (2001) 'Hydrogen production from organic waste', *International Journal of Hydrogen Energy*, 26, 547–550.

Nieves, D. C., Karimi, K. and Horváth, I. S. (2011) 'Improvement of biogas production from oil palm empty fruit bunches (OPEFB)', *Industrial Crops and Products*, 34(1), 1097–1101.

Nizami, A. S., Singh, A. and Murphy, J. D. (2011) 'Design, commissioning, and start-up of a sequentially fed leach bed reactor complete with an upflow anaerobic sludge blanket digesting grass silage', *Energy Fuels*, 25, 823–834.

Omil, F., Mendez, R. and Lema, J. M. (1995) 'Anaerobic treatment of saline wastewaters under high sulphide and ammonia content', *Bioresource Technology*, 54, 269–278.

Palonen, H., Thomsen, A. B.; Tenkanen, M., Schmidt, A. S. and Viikari, L. (2004) 'Evaluation of wet oxidation pretreatment for enzymatic hydrolysis of softwood', *Applied Biochemistry and Biotechnology*, 117, 1–17.

Panagiotopoulos, I. A., Bakker, R. R., Budde, M. A. W., de Vrije, T., Claassen, P. A. M. and Koukios, E. G. (2009) 'Fermentative hydrogen production from pretreated biomass: a comparative study', *Bioresource Technology*, 100(24), 6331–6338.

Panagiotopoulos, I. A., Bakker, R. R., de Vrije, T., Koukios, E. G. and Claassen, P. A. M. (2010a) 'Pretreatment of sweet sorghum bagasse for hydrogen production by *Caldicellulosiruptor Saccharolyticus*', *International Journal of Hydrogen Energy*, 35(15), 7738–7747.

Panagiotopoulos, J. A., Bakker, R. R., de Vrije, T., Urbaniec, K., Koukios, E. G. and Claassen, P. A. M. (2010b) 'Prospects of utilization of sugar beet carbohydrates for biological hydrogen production in the EU 2010', *Journal of Cleaner Production*, 18 (Suppl. 1), S9–S14.

Pang, Y., Liu, Y., Li, X., Wang, K. and Yuan, H. (2008) 'Improving biodegradability and biogas production of corn stover through sodium hydroxide solid state pretreatment', *Energy Fuels*, 22, 2761–2766.

Pattra, S., Sangyoka, S., Boonmee, M. and Reungsang, A. (2008) 'Bio-hydrogen production from the fermentation of sugarcane bagasse hydrolysate by *Clostridium butyricum*', *International Journal of Hydrogen Energy*, 33(19), 5256–5265.

Pereira, M. A., Cavaleiro, A. J., Mota, M. and Alves, M. M. (2003) 'Accumulation of long chain fatty acids onto anaerobic sludge under steady state and shock loading conditions: effect on acetogenic and methanogenic activity', *Water Science and Technology*, 48(6), 33–40.

Pind, P. F., Angelidaki, I., Ahring, B. K., Stamatelatou, K. and Lyberatos, G. (2001) 'Monitoring and control of anaerobic reactors', *Advances in Biochemical Engineering/Biotechnology*, 82, Springer, Berlin, 135–182.

Prakasham, R. S., Sathish, T., Brahmaiah, P., Subba Rao, Ch., Sreenivas Rao, R. and Hobbs, P. J. (2009) 'Biohydrogen production from renewable agri-waste blend: optimization using mixer design', *International Journal of Hydrogen Energy*, 34(15), 6143–6148.

Prapaspongsa, T., Poulsen, T. G., Hansen, J. A. and Christensen, P. (2010) 'Energy production, nutrient recovery and greenhouse gas emission potentials from integrated pig manure management systems', *Waste Management and Research*, 28, 411–422.

Prazeres, A. R., Carvalho, F. and Rivas, J. (2012) 'Cheese whey management: a review', *Journal of Environmental Management*, 110, 48–68.

Propheter. J. L., Staggenborg, S. A., Wu, X. and Wang, D. (2010) 'Performance of annual and perennial biofuel crops: yield during the first two years', *Agronomy Journal*, 102, 806–814.

Qiao, W., Yan, X., Ye, J., Sun, Y., Wang, W. and Zhang, Z. (2011) 'Evaluation of biogas production from different biomass wastes with/without hydrothermal pretreatment', *Renewable Energy*, 36, 3313–3318.

Ramos, L. P. (2003) 'The chemistry involved in the steam treatment of lignocellulosic materials', *Quimica Nova*, 26(6), 863–871.

Ren, N. Q., Chua, H., Chan, S. Y., Tsang, Y. F., Wang, Y. J. and Sin, N. (2007) 'Assessing optimal fermentation type for bio-hydrogen production in continuous-flow acidogenic reactors', *Bioresource Technology*, 98, 1774–1780.

Rinzema, A., Lier, V. V. and Lettinga, G. (1998) 'Sodium inhibition of acetoclastic methanogens in granular sludge from a UASB reactor', *Enzyme and Microbial Technology*, 10, 24–32.

Saddoud, A., Hassaïri, I. and Sayadi, S. (2007) 'Anaerobic membrane reactor with phase separation for the treatment of cheese whey', *Bioresource Technology*, 98(11), 2102–2108.

Saha, B. C. and Cotta, M. A. (2007) 'Enzymatic saccharification and fermentation of alkaline peroxide pretreated rice hulls to ethanol', *Enzyme and Microbial Technology*, 41, 528–532.

Salminen, E. and Rintala, J. (2001) 'Anaerobic digestion of organics solid poultry slaughterhouse waste – a review', *Bioresource Technology*, 83(1), 13–26.

Sampaio, M. A., Gonçalves, M. R. and Marques, I. P. (2011) 'Anaerobic digestion challenge of raw olive mill wastewater', *Bioresource Technology*, 102, 10810–10818.

Schmidt, A. and Thomsen, A. (1998) 'Optimization of wet oxidation pretreatment of wheat straw', *Bioresource Technology*, 64, 139–151.

Shin, H. S. and Youn, J. H. (2005) 'Conversion of food waste into hydrogen by thermophilic acidogenesis', *Biodegradation*, 16 (1), 33–44.

Singh, K., Lee, K., Worley, J., Risse, L. M. and Das, K. C. (2010) 'Anaerobic digestion of poultry litter: a review', *Applied Engineering in Agriculture*, 26(4), 677–688.

Solares, E. T., Bombardiere, J., Chatfield, M., Domaschko, M., Easter, M., Stafford, D. A., Angeles, S. C. and Hernadez, N. C. (2006) 'Macroscopic mass and energy balance of a pilot plant anaerobic bioreactor operated under thermophillic conditions', *Applied Biochemistry and Biotechnology*, 129-132, 959–968.

Song, C., Kwon, S. J. and Woo, J. H. (2004) 'Mesophilic and thermophilic temperature co-phase anaerobic digestion compared with single-stage mesophilic and thermophilic digestion of sewage sludge', *Water Research*, 38, 1653–1662.

Soto, R., Russell, I., Narendranath, N., Power, N. and Dawson, K. (2005) 'Estimation of ethanol yield in corn mash fermentations using mass of ash as a marker', *Journal of the Institute of Brewing*, 111(2), 137–143.

Srilatha, H. R., Nand, K., Babu, K. S. and Madhukara, K. (1995) 'Fungal pretreatment of orange processing waste by solid-state fermentation for improved production of methane', *Process Biochemistry*, 30, 327–331.

Stamatelatou, K., Kopsahelis, A., Blika, P.-S., Paraskeva, C. and Lyberatos, G. (2009) 'Anaerobic digestion of olive mill wastewater in a periodic anaerobic baffled reactor (PABR) followed by further effluent purification via membrane separation technologies', *Journal of Chemical Technology and Biotechnology*, 84, 909–917.

Sun, Y. and Cheng, J. (2002) 'Hydrolysis of lignocellulosic materials for ethanol production: a review', *Bioresource Technology*, 83, 1–11.

Talebnia, F., Karakashev, D. and Angelidaki, I. (2010) 'Production of bioethanol from wheat straw: an overview on pretreatment, hydrolysis and fermentation', *Bioresource Technology*, 101, 4744–4753.

Teghammar, A., Yngvesson, J., Lundin, M., Taherzadeh, M. J. and Horvath, S. (2010) 'Pretreatment of pater tube residuals for improved biogas production', *Bioresource Technology*, 101(4), 1206–1212.

Uellendahl, H., Wang, G., Møller, H. B., Jørgensen, U., Skiadas, I. V., Gavala, H. N. and Ahring, B. K. (2008) 'Energy balance and cost-benefit analysis of biogas production from perennial energy crops pretreated by wet oxidation', *Water Science and Technology*, 58(9), 1841–1847.

Ueno, Y., Fukui, H. and Goto, M. (2007) 'Operation of a two-stage fermentation process producing hydrogen and methane from organic waste', *Environmental Science and Technology*, 41(4), 1413–1419.

van Ginkel, S. W., Oh, S. E. and Logan, B. E. (2005) 'Biohydrogen gas production from food processing and domestic wastewaters', *International Journal of Hydrogen Energy*, 30(15), 1535–1542.

Vazquez, M., Oliva, M., Tellez-Luis, S. J. and Ramirez, J. A. (2007) 'Hydrolysis of sorghum straw using phosphoric acid: evaluation of furfural production', *Bioresource Technology*, 98, 3053–3060.

Voolapalli, R. K. and Stuckey, D. C. (1998) 'Stability enhancement of anaerobic digestion through membrane gas extraction under organic shock loads', *Journal of Chemical Technology and Biotechnology*, 73, 153–161.

Wang, X. and Zhao, Y. C. (2009) 'A bench scale study of fermentative hydrogen and methane production from food waste in integrated two-stage process', *International Journal of Hydrogen Energy*, 34(1), 245–254.

Wang, Z. J., Zhu, J. Y., Zalesny Jr., R. S. and Chen, K. F. (2012) 'Ethanol production from poplar wood through enzymatic saccharification and fermentation by dilute acid and SPORL pretreatments', *Fuel* 95, 606–614.

Weiland, P. (2006) 'Biomass digestion in agriculture: a successful pathway for the energy production and waste treatment in Germany', *Engineering in Life Sciences*, 6(3), 302–309.

Wu, G., Healy, M. G. and Zhan, X. (2009) 'Effect of the solid content on anaerobic digestion of meat and bone meal', *Bioresource Technology*, 100, 4326–4331.

Xu, Z., Wang, Q., Jiang, Z., Yang, X.-X. and Ji, Y. (2007) 'Enzymatic hydrolysis of pretreated soybean straw', *Biomass and Bioenergy*, 31, 162–167.

Zeng, M., Mosier, N. S., Huang, C. P., Sherman, D. M. and Ladisch, M. R. (2007) 'Microscopic examination of changes of plant cell structure in corn stover due to hot water pretreatment and enzymatic hydrolysis', *Biotechnology and Bioengineering*, 97, 265–278.

Zhang, J. J., Li, X. Y., Oh, S. E. and Logan, B. E. (2004) 'Physical and hydrodynamic properties of flocs produced during biological hydrogen production', *Biotechnology and Bioengineering*, 88, 854–860.

Zheng, M., Li, X., Li, L., Yang, X. and He, Y. (2009) 'Enhancing anaerobic biogasification of corn stover through wet state NaOH pretreatment', *Bioresource Technology*, 100, 5140–5145.

Zhong, W., Zhang, Z., Qiao, W., Fu, P. and Liu, M. (2011) 'Comparison of chemical and biological pretreatment of corn straw for biogas production by anaerobic digestion', *Renewable Energy*, 36, 1875–1879.

Zhu, H., Parker, W., Basnar, R., Proracki, A., Falletta, P., Béland, M. and Seto, P. (2008a) 'Biohydrogen production by anaerobic co-digestion of municipal food waste and sewage sludges', *International Journal of Hydrogen Energy*, 33(14), 3651–3659.

Zhu, H., Stadnyk, A., Béland, M. and Seto, P. (2008b) 'Co-production of hydrogen and methane from potato waste using a two-stage anaerobic digestion process', *Bioresource Technology*, 99(11), 5078–5084.

Zupancic, G. D., Straziscar, M. and Ros, M. (2007) 'Treatment of brewery slurry in thermophilic anaerobic sequencing batch reactor', *Bioresource Technology*, 98, 2714–2722.

# 16
# The production and application of biochar in soils

S. JOSEPH, University of New South Wales, Australia and
P. TAYLOR, Biochar Solutions, Australia

DOI: 10.1533/9780857097385.2.525

**Abstract**: This chapter includes a short history of biochar use over thousands of years. Biochar properties are mainly dependent on chemical and physical properties of feedstock and final heat treatment temperature. Once added to soil these properties change in complex ways that depend on environmental factors, soil properties and the types of crops grown. Different technologies are used to make biochar in the household, in rural industries, and in large scale modern industrial settings. The simplest designs made from mud and brick and/or sheet metal to more sophisticated ones being developed and commercialized in Europe, the Americas and Asia are discussed. The world market for biochar, development constraints as well as the future direction of the industry are presented.

**Key words**: biochar, pyrolysis, kilns, organomineral complexes, wood vinegar, NPK fertilizer.

## 16.1 Introduction

The International Biochar Initiative defines biochar as:

> a solid material obtained from the carbonisation of biomass. Biochar may be added to soils with the intention to improve soil functions and to reduce emissions from biomass that would otherwise naturally degrade to greenhouse gases. Biochar also has appreciable carbon sequestration value. *(IBI, 2012)*

The properties of biochars are dependent mainly on the chemical and physical properties of the feedstock and on the final heat treatment temperature (Rajkovich *et al.*, 2011). The properties are also dependent to a lesser extent on the rate of heating, the kiln pressure and atmosphere, and the type of pre- or post-treatment (Amonette and Joseph, 2009). Recent papers by Singh *et al.* (2010), Rajkovich *et al.* (2011), Kookana *et al.* (2011) and Uchimiya *et al.* (2011) have summarized a large amount of the published data on the characteristics of biochars.

Biochars can be divided into three broad categories (Joseph and Lehmann, 2009). The categories are:

1. Those that are made from a low mineral ash feedstock (<3–5%) such as wood, nut shells, bamboo and some seeds (e.g., apricots). Most of these feedstocks produce a harder biochar, that have a higher porosity, surface area and water-holding capacity than biochars in the other categories.
2. Feedstocks that have ash compositions between (3–5 and 10–13%), which includes most agricultural residues, bark and high quality greenwaste (i.e., with low contamination of plastics, soil and metals). Biochars made from agricultural residues can have a significantly higher cation exchange capacity (Van Zwieten *et al.*, 2010).
3. Feedstocks with an ash composition >13% which includes most manures, sludges, waste paper, municipal waste and rice husks. These biochars are very variable: it is hard to draw general conclusions. Some observations made in the literature are:

    - adsorption (BET) surface area is low in comparison to biochars made from lower ash feedstocks (Downie *et al.*, 2009)
    - manures and many sludge biochars have a high liming ability
    - electrical conductivity (EC) and pH are high
    - heavy metal concentrations for municipal solid waste (MSW) and sewage sludge can be much greater than from other feedstocks.

For each of these feedstocks the initial moisture content and particle size can alter the final properties of biochars.

Recent research suggests that for a given feedstock and heating rate, there is a significant difference in the physical and chemical properties of biochars depending on whether they are produced between (±50°C) 300–400°C, 400–500°C, or >500°C. There is also a considerable difference between biochars made from woody materials and those made from crop residues, manures and sludges. It should be noted that biochars made from paper sludge and chicken manure that have high contents of calcium compounds can have quite different properties from biochars made from other high mineral ash materials (Enders *et al.*, 2012).

Research undertaken by Singh *et al.* (2010), Rajkovich *et al.* (2011) and a review by Kookana *et al.* (2011) indicate that biochars made at temperatures lower than 400–450°C have a higher concentration of oxygenated functional groups, radicals, water and organic soluble compounds that can improve germination, stimulate microbial growth and induce systemic resistance and hormesis (beneficial response to low dose of toxin) (Graber *et al.*, 2010; Elad *et al.*, 2010). Compared with biochars made at above 450°C, these biochars have a lower pH, higher adsorption of ammonia (especially bamboo), but lower adsorption of most other gases (Mingjie, 2004), relatively low water-holding capacity (Krull, 2010), and little loss of metals and non-metals due to volatilization (Enders *et al.*, 2012). The electrical conductivity can be higher or lower depending on feedstock and temperature (Shinogi, 2004; Singh et al., 2010). The ability to adsorb heavy metals depends on the type of biochar.

Biochars produced from crop residues and grasses between 400 and 500°C have a high CEC and water-holding capacity. This could be partly attributed to the clay attached to the feedstock (Krull, 2010). There appears to be considerable concentration of organic compounds and acid and basic functional groups on the surfaces of wood biochars until reaction temperatures exceed 450°C (Singh et al., 2010). However the type and concentration of functional groups appear to be different between high and low mineral ash biochars and between different time and temperature regimes for producing the biochars. Considerable increase in surface area, pore volume and mesoporous area occurs for most woody biochars although not necessarily for high mineral ash biochar once the final pyrolysis temperature exceeds 450°C. Fellet et al. (2011) found that biochars produced from orchard prunings at 500°C can adsorb significant concentrations of some heavy metals. For woody biochars produced above 500°C but below 600°C, there is high adsorptivity of some organic molecules (McLaughlin et al., 2009), high surface area and volume of sub-100 nm pores (Downie et al., 2009), high volume of pores above 100 nm, high stability and degree of condensation, and high pH.

## 16.2 Effects of application of biochar to soil

Once biochars are placed in soil, their physical and chemical properties undergo complex changes. A recent review by Joseph et al. (2010a) and Lehmann et al. (2011) detail the possible reactions. Important findings are summarized in the following discussion. When biochars with a high concentration of soluble minerals and oxygenated organic molecules on their surface are added to moist soil (or are followed by a rain event) there is a change in the pH, EC, and Eh (reduction or redox potential) around the particle, probably within the first week, as the minerals dissolve and/or ions are exchanged on the surfaces of the surrounding clay particles.

Rain events can result in soil colloidal particles migrating into the pores of the biochar and reacting with the carbon surface to produce biochar-organomineral complexes (Fig. 16.1). The limited data available indicate that the type of reactions taking place and the stability of the compounds retained on the surface depend on the type of biochar and the conditions under which it was produced (Spokas and Reicosky, 2009; Nguyen et al., 2010). Any significant reaction involving the biochar, the soil minerals and dissolved organic matter could lead to a substantial change in the properties of the biochar (e.g., pH and EC).

*16.1* Wood biochar extracted from soil. Organomineral compounds evident on the surface. Source: Electron Microscopy Unit (EMU), University of NSW.

When biochars produced at low temperatures are added to moist soils, a considerable quantity of soluble organics can be released to the soil solution. As noted by Graber *et al.* (2010), Dixon (1998) and Light *et al.* (2009), some of the labile organic compounds on the surfaces of biochar can stimulate seed germination, growth of fungi, nutrient uptake and reduction in pathogens.

Enhanced $CO_2$ emissions can be observed during the first months after biochar has been added to soil. These enhanced emissions are partly attributed to biochar surface oxidation that has both a chemical and biological basis. It is possible that the mineral matter in the biochar and the water-soluble organic compounds on the external and internal surfaces provide nutrients for microorganisms to grow faster as well as catalysing the breakdown of organic matter (Amonette *et al.*, 2006). It is also possible that chemical oxidation occurs when there is a large difference in electrochemical potential between the different mineral and carbon regions in an individual particle.

Surface oxidation of biochars increases the potential for hydrophilic interactions with a range of soil organic and inorganic compounds. This is more significant in high mineral ash biochars (Lima and Marshall, 2005). Greater reactivity of biochars with mineral matter could further promote physical protection of biochar and labile organic matter, and thus long-term stability (Brodowski *et al.*, 2006; Keith *et al.*, 2011). Once roots, and in particular their root hairs, interact with the biochars, a wider range of reactions can occur, through the uptake of nutrients by the plants, and the release of root exudates. This enhances both complexation reactions and microbial activity in the rhizosphere. Over a period of time, physical disturbances, interactions with microflaura and microfauna, and complex abiotic and biotic reactions with all soil constituents, will break larger biochar particles into smaller pieces and lead to the formation of biochar organomineral aggregates (Joseph et al., 2010b).

## 16.3 Agricultural uses of biochar

The Australian Aborigines may have been the first people to produce biochar in order to increase the food available to them (Bird et al., 2008). Fire-stick farming included lighting smoky fires when grass and scrub was moist to produce high concentrations of biochar, which increased the growth of green shoots that in turn attracted the kangaroos. Aborigines likewise used fire to promote the growth of edible tubers (Jones, 1968). Aborigines in the South East of Australia also made oven mounds that comprised charcoal, minerals, organic matter and clay.

Amazonian dark earth (ADE) has been studied over the last 20 years. These soils consist of particles that have a very heterogeneous phase structure consisting of a high carbon material (probably derived from the weathering of charcoal made in open fires) surrounded by mineral phases that include clay, calcium carbonate, calcium phosphate, calcium hydroxide, potassium and sodium chloride, iron oxides and titanium dioxide. These mineral and high carbon phases (Fig. 16.2) are often bound together by organic matter, some of which is high in calcium and/or nitrogen (Chia et al., 2011). Steiner (2006, 2008) reports that Amazonian village women collect all available types of organic matter at the settlement (bones, wood, leaves, chicken manure), as well as rotten tree trunks of particular species. This mixture is burned in an open smoky fire and then it is blended with soil that had previously been burnt in a hot woody biomass fire. The final mixture is used to grow vegetables and herbs. This technique could have been used by Indians in the past, and the ageing of this biochar mineral organic mixture could have led to the formation of ADE.

Biochar has been used in agriculture in Japan and China for hundreds of years (Ogawa and Okimori, 2010). Biochar is made at different temperatures for different applications, although Ogawa and Okimori (2010) note that most biochar is made between 400°C and 600°C. Biochar has been mixed with manure and applied to rice fields and home gardens to maintain soil fertility. The smoke from the production of biochar has also been captured and refined for use as a biopesticide and a growth promotor for at least 40 years (Ogawa and Okimori, 2010).

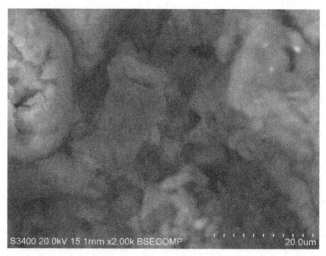

*16.2* SEM examination of the surface of a Terra Preta particle. Source: EMU, University of NSW.

In Costa Rica, a farmer has established an organic fertilizer business that uses a modified form of a technique (Bocashi) developed in Japan over 30 years ago (Joseph *et al.*, 2010b). Biochar produced at a sugar factory is mixed with manure and minerals (mainly dolomite). Microorganisms collected from the nearby mountains, as well as Japanese produced efficient micro-organisms (EM), are grown in a mixture of molasses and rice bran. The microorganisms are then mixed with the minerals, manure and biochar and composted in a pit in the absence of air before being aerobically composted.

Villagers in Northern Vietnam have developed a technique of soaking wood and bamboo in the clay and mineral-rich sludge at the bottom of ponds at the edge of rice paddies (Fig. 16.3). A significant volume of biochar and ash remains when this artificially aged biomass is burned in an open fire. The biochar and ash is then used in home gardens to grow vegetables. Ash and biochar from the burning of straw on fields is also used in conjunction with NPK fertilizer. This practice is over 40 years old. Women interviewed said that it increases the plant yield and reduces requirements for NPK fertilizer.

Large deposits of dark earths have been found in Africa (Fairhead and Leach, 2009). Rademakers (2009) notes that some tribes collect big piles of elephant grass or any other type of savannah grass, dry it, pile it so as to make long strips, cover the big rows of grass with a layer of mud, and then leave it to dry. After the mud has dried and hardened, they open one part of the strip and set fire to the grass within. The fire travels slowly through

*16.3* Wood and bamboo soaking in a pond at edge of a rice paddy field.

this 'kiln', which provides a low oxygen environment, and chars all the biomass. After this operation, they crush the mud layer, and the char beneath it. They repeat the effort several times to create layers of char and crushed mud. This then becomes their soil bed, on which they start planting crops when the rains arrive. The rains turn this soil layer into a more fertile soil.

## 16.4 Production of biochar

Many different pyrolysis kilns, retorts, ovens, and stoves have been designed and are being operated throughout the world. However, there is a lack of independent assessment of their performance especially the emissions profile ($CO$, $NOx$, $N_2O$, VOCs) during start-up, steady state operation, and shut-down.

### 16.4.1 Household devices

A range of small pyrolytic stoves for use by households or by farmers can be purchased or their designs are available on the web. Many different biochar stoves have been developed and tested (Carter and Shackley, 2011). They generally can be categorized into two distinct design types: microgasifiers operate by direct combustion of the feedstock, while retort-like stoves indirectly heat the feedstock, to convert it to biochar. The first utilizes small pieces of woody biomass, shells and pellets that are placed into a chamber where pyrolysis takes place in limited air. Pyrolysis gases are allowed to rise up from the biomass and combust in excess secondary air. The most common name used for these is a TLUD (top lit updraft) gasifier or stove. There have been many variations on this type of stove and a comprehensive overview is given by Roth (2011). Many of these newer stoves have lower fuel consumption and lower emissions than open fires (Roth, 2011).

The second type of stove has a chamber (retort) where residues such as rice husks, straw, sawdust and manure can be converted into biochar under the influence of heat from a separate combustion chamber burning scrap wood or biomass. Gases escaping from the pyrolysing residues in the retort flow into the combustion chamber and are burnt. More recently, new designs have been developed with an inner combustion chamber where large pieces of wood can be burnt. The wood combustion chamber is surrounded by another chamber where other biomass is added. This type of stove is illustrated in Fig. 16.4.

### 16.4.2 Ovens, retorts and kilns

Both the Japanese and the Chinese have developed a range of kilns that are cheap to manufacture and produce biochar over a range of temperatures.

*16.4* Clean burning biochar stove and smokey traditional stove.

Some of these kilns have been in operation for over 100 years (*Industrial and Engineering Chemistry*, 1931). Figure 16.5 illustrates two of the simple batch kilns that are being operated by small enterprises.

Japanese techniques produce high and low temperature biochars. In some of the kilns, wood is stacked vertically which allows rapid heat transfer up the internal pores (xylem and phloem). An external fire is ignited and

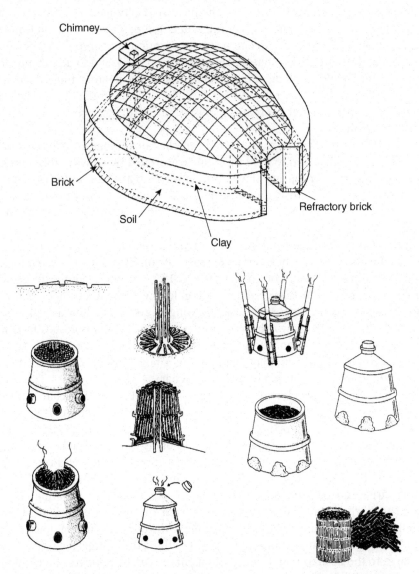

*16.5* Two different types of Japanese kilns: Iwate and the portable iron kiln.

combustion gases flow into the kiln. Steam is produced at the bottom of the wood and rises through the vertical pore channels efficiently transferring heat to the wood. The rate of heating of the biomass varies depending on the specific application. Figure 16.6 shows one practice where the biomass is slowly heated for a long period of time at temperatures below 150°C (ARECOP, 1994). This long steaming time at low temperatures slowly degrades the lignocellulosic structure of the biomass and results in a biochar that probably has quite different chemical and physical properties from those produced at higher firing rates and temperatures. The thermally treated wood may have higher water absorption, larger pore sizes and broadened pore size distribution (Hietala *et al.*, 2002), and higher diffusion coefficient of water along the tracheid axis. Long heating times lead to deacetylation, and the released acetic acid acts as a depolymerization catalyst, which further increases polysaccharide decomposition (Fengel, 1966a, 1966b).

Another firing procedure used is to bring the kiln temperature to 400°C as quickly as possible and to hold this temperature for two days before either cooling the kiln or raising the temperature to 600°C, or for very dense charcoal to 1,200°C. In these kilns there is a diffusion of moisture and low molecular weight organic compounds from the inner core of the wood to the outer char layer during the holding time at 400°C. Complex reactions take place resulting in the formation of a heterogeneous carbon mineral matrix.

Beehive kilns have been used extensively in China to make bamboo biochar and to collect wood vinegar (Fig. 16.7). Smoke coming from the kiln is captured in a funnel and then condensed in bamboo pipes before dropping into clay pots for refinement (Mingjie, 2004). The liquid is allowed to age for a month by which time three phases form. The bottom phase is tar which is used as an energy source. The top phase comprises water soluble organics. The middle phase is the vinegar which can be further refined by passing it through bamboo biochar.

A batch retort developed originally for production of fuel charcoal by Chris Adam is now being built around the world for production of biochar. Bob Wells and Peter Hirst in the USA have modified this retort (Fig. 16.8) to reduce the level of emissions and to allow the energy to be utilized for heating as well as creating the option of placing it on a mobile platform (Hirst, 2010). The wood or other feedstock is loaded from the top, the chamber is sealed, and a fire is started in the attached firebox to begin the indirect heating of the feedstock chamber. The pyrolysis gases from the heated charge are then pulled off, run through a condenser to remove the oil and other condensables, and then injected into the fire chamber to continue the indirect heating of the charge to the desired temperature. Heat energy can be removed for heating a greenhouse, hydroponics system or other structure, or can be used in a separate process. The result is a very efficient and clean running system. The

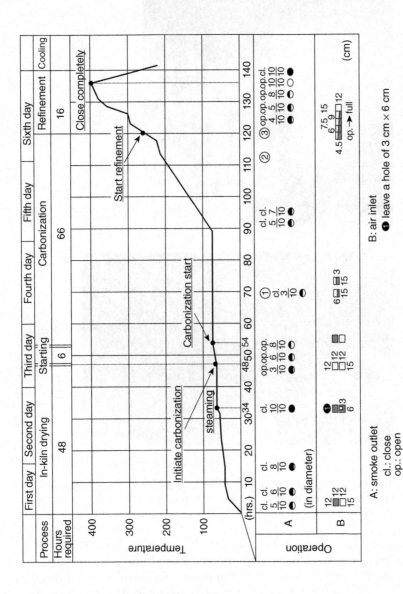

*16.6* Heating curve for Iwate kiln (ARECOP, 1994). Temperature taken in the chimney so actual temperature in the kiln is higher. Note there is a distribution of temperatures throughout the kiln.

(a)

(b)            Jar        Hollow bamboo culms

*16.7* Large-scale production of biochar and wood vinegar in China: (a) outline of mechanical furnaces; (b) saving the vinegar (Mingjie, 2004).

retort can produce about 300 kg/day of biochar as well as liquid co-products and heat energy. Unloading is fast and easy with a built-in vacuum system. The unloading system that vacuums out the finished char breaks down the particle size, and adds a mist of moisture to control dust. The cost of the kiln is approximately US$75,000 in 2012.

A ring kiln developed by Ian McChesney of Carbon Gold is being operated in Belize and the UK (see Fig. 16.9). The charge of wood or

*16.8* Modified Adam retort.

*16.9* Ring kiln design (Anon., 2011).

residues for making the biochar is placed on the outer ring of the kiln. The inner chamber has a firebox in which firewood is placed. Air is supplied through the top and the combustion gases pass through the pyrolysing wood. The syngas catches alight and is burnt in the middle chamber. Many different designs of TLUD biochar ovens manufactured from 200 litre drums are now in use throughout the world. A version has been field tested in Vietnam and has been operated by a women's group to make a mixed biochar from rice straw, rice husks, tea clippings and bamboo (Fig. 16.10).

*16.10* Drum TLUD oven being field tested in Vietnam.

To this was added clay and lime to slow the decomposition of the rice straw. The yield of biochar was approximately 37% with an average temperature of 450°C in the bed of pyrolysing material.

## 16.5 Larger-scale commercial production of biochar

Large-scale production of fuel charcoal on a continuous basis has been carried out in rotary kilns, vertical retorts and rotary hearth furnaces for over 100 years (Brown, 2009). Over the last 30 years a wide range of different pyrolysis units have been developed to produce biochar and in some cases to produce energy and bio-oils (to be used as a fuel or chemical additives). It is not possible to list all of the different designs. The following is a description of a sample of the different types of units that have been developed. There are only a small number of plants in operation throughout the world and very little data on their performance.

Kansai Corporation of Japan has developed a moving bed pyrolyser to convert rice husks and sawdust to biochar and to produce heat for drying or heating in rice mills and sawmills. Units are in operation throughout Japan, although details are not available on the numbers and the operating performance.

ICM LLC in the USA has built a screw pyrolysis system with an input of two ton/hr. In this unit the steam produced in the first section is taken to the back of the kiln where it is combined with the syngas. Air is injected along the kiln and the syngas is burnt in a thermal oxidizer. This unit has run for over 2,000 hours. The hot gas produced from this unit can be used to power an organic Rankine cycle engine or to provide heat to generate steam for a steam turbine.

Pacific Pyrolysis (previously BEST Pty Ltd) has developed a three-stage process that has as its main components a drier, a torrefier/pyrolyser and a gasifier (Fig. 16.11). The pyrolyser has a series of paddles inside that move the biomass over the hot surface of the kiln. Part of the syngas is used to heat the pyrolysis kiln. A gas clean-up system consisting of a gas cooler, particulate filter, a tar cracking unit and a scrubber can be added to produce a gas that can fuel an internal combustion engine, an organic Rankine cycle engine or a steam turbine. This system can be tailored to produce a range of different types of biochars. A pilot plant has been run extensively over 5 years and manufacture of their first demonstration plant may be operating in 2014.

BiG (www.blackisgreen.net) has built a number of rotary hearth pyrolysis kilns that can process between 250 and 500 kg/hr of biomass. Among those bigger kilns are also transportable ones which have been in operation on an irregular basis for the past two years. They cost $200,000 to $300,000. Figure 16.12(a) shows a small BiG unit used in Australia.

Pro-Natura has developed a swept drum pyrolyser (Fig. 16.12(b)) with a separate vortex burner for use in developing countries. The unit can process 250–500 kg/hr and has been tested on a range of feedstocks in Africa and France. Energy Farmers Australia Pty Ltd (Fig. 16.13) has built a continuous auger feed pyrolyser suitable for a range of feedstocks. This unit uses augers and mixers to move the biomass along a trough. An LPG burner heats the unit up and starts the feed pyrolysing. The syngas from the pyrolysing material mixes with air and ignites. The LPG burner is turned off and the radiation from the syngas keeps the pyrolysis process going.

Russell Burnett developed a screw continuous pyrolysis two-stage system (Genisis) with feed input of 200 kg/hr with moisture content of 25% (Fig. 16.14). These units use the syngas to dry and then pyrolyse the feed in two separate chambers. The unit has a scrubber and a drop-out tank to cool the wood vinegar. The wood vinegar is added to compost to increase microbial growth within the pile. Independent emissions testing has been carried out with olive seed as the input fuel. The composition of the flue gas was: CO 2.5 mg/m$^3$, NO$_x$ 100 mg/m$^3$, hydrogen sulphide 0.043 mg/m$^3$, $CO_2$ = 4.1%, and $O_2$ 17.4%. Polyaromatic hydrocarbons were not detected on the biochar produced in this machine at 550°C.

The Sanli New Energy Company has developed an open-core down-draft gasifier (Fig. 16.15) that has a throughput of approximately two tonnes/hr of mixed agricultural residue. A limited amount of air is drawn in through the top of the unit and the gas is taken out halfway down the reactor. Biochar forms in the middle of the reactor. Visual examination indicates that there is a temperature distribution between the outer walls where much of the air flows and the inner core. It is probable that the temperature of the biochar varies from 400°C in the middle to 550°C on the outer walls. The gas is cleaned and then cooled in three condensers and the different

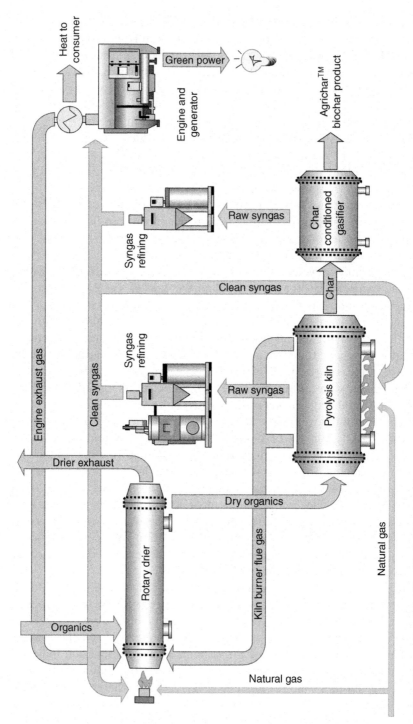

16.11 Schematic of the Pacific Pyrolysis unit.

*16.12* (a) Black is Green (big) pyrolysis kiln (courtesy Black is Green Pty Ltd); (b) Pro-Natura Kiln.

fractions of the condensate are separated. The middle fraction is further refined and sold as wood vinegar. The cool cleaned gas is then utilized to run an engine to generate electricity. Their main factory consists of three units that operate throughout the year.

## 16.6 Testing biochar properties

To fully characterize biochars involves using a range of tests some of which are complicated and expensive. Much of the research work undertaken to determine key properties of biochars uses chemical, physical, microscopic,

*16.13* Energy Farmers Australia Pty Ltd portable pyrolyser.

*16.14* Genesis continuous pyrolysis plant.

electrochemical and spectroscopic techniques. Each technique provides an important part of the puzzle and the reader is referred to Joseph and Lehmann (2009) and Joseph *et al.* (2010a). The Japanese have developed a simple standard for biochars (www.nittokusin.jp) which is presented in Table 16.1. The biochar must only be produced at a temperature above 400°C. It must meet laws related to improvement of crop fertility.

The Chinese National Standardization Technical Committee for Bamboo and Rattan (SAC/TC263/SCX) has developed a standard for bamboo biochar. The bamboo is classified by size and shape according to the finished products, which include tube charcoal, charcoal tablets, broken charcoal, granular carbon and carbon powder. There are six categories of diameter sizes: 0.18–0.5 mm, 0.5–1 mm, 1–3 mm, 3–5 mm, 5–10 mm and 10–20 mm. The

The production and application of biochar in soils 543

16.15 Sanli New Energy Company factory: (a) top of the gasifier; (b) middle of the gasifier; (c) engine producing electricity from the gas.

*Table 16.1* Japanese Standard, courtesy of Japan Special Forest Product Promotion Association

| Classification | Purpose | Woodchar | Water content | Refinement | Miscellaneous |
|---|---|---|---|---|---|
| For agriculture forestry and gardening | For materials for soil dressing (amendment) | Charcoal that is carbonized at 400°C or more (the charcoal of crop residue is included). | – | | It is subject to fertility improvement law. |

biochar is also categorized by use: 'personal use, fuel charcoal, building decorative bamboo charcoal, bamboo charcoal with environmental protection, agriculture, forestry, and horticultural charcoal'. All biochars must be odour free.

First-grade biochar must have a moisture content between 9.0 and 12.0%, ash content between 4.5 and 6.5% and a fixed carbon content between 75.0 and 85.0%. To determine the quality of the biochar, the samples are placed on white paper in a well-lit environment and then a person will smell the charcoal to ensure there is no odour, and will observe if it has a metallic lustre after it is broken, showing that it is fully pyrolysed.

After extensive consultation with stakeholders, IBI has established a protocol for testing biochars at different levels. Key measurements include pH, EC, H/Corg ratio, total and available N, P and total K, total metal content, liming capacity, and particle size distribution (See Appendix in Section 16.10). Other more complex measurements include total porosity and surface area. It is also recommended that basic toxicity measurements be done (Table 16.2).

## 16.7 Markets and uses for biochar

The most developed commercial markets for biochars exist in Japan, China, Taiwan and Korea. Biochar is an integral part of vegetable production, animal husbandry and forestry. China is now the largest producer of biochar with over 140,000 tonnes of biochar and biochar blends being produced in the last year. There also appears to be significant production of wood vinegar that is used as a biopesticide and for promoting the germination and growth of certain species of trees and plants. This increase in production

*Table 16.2* Test Category B characteristics and criteria

| Test Category B: Biochar Toxicant Reporting – Required for All Feedstocks | | |
|---|---|---|
| Requirement | Range of maximum allowed thresholds | Test method |
| Earthworm avoidance test | Pass/Fail | ISO 17512–1:2008 methodology and OECD methodology as described by Van Zwieten *et al.* (2010) |
| Germination inhibition assay | Pass/Fail | OECD methodology 3 test species, as described by Van Zwieten *et al.* (2010) |
| Polycyclic aromatic hydrocarbons (PAH) | 6–20 mg/kg TM | Method following US Environmental Protection Agency (1996) |
| Dioxin/furan (PCCD/F) | 9 ng/kg I-TEQ | Method following US Environmental Protection Agency (2007) |

has been stimulated by government grants and contracts and by the collaboration between researchers in China and Europe/US/Australia.

The principal uses of the biochar are (Pan *et al.*, 2011):

- land reclamation and remediation
- as an ingredient in an organic/mineral fertilizer
- as an ingredient in a chemical fertilizer to reduce nutrient leaching and improve plant nutrient uptake efficiency
- removal of heavy metals from waste water and contaminated land.

Essentially two approaches are being taken for developing biochar-based fertilizers. The first approach is to produce either an organic granulated fertilizer that has a high N and smaller P and K contents, utilizing fermented biomass, amino acids, extra minerals and biochar (Fig. 16.16), or a compost-biochar mixture. The fertilizer is sold (wholesale) for approximately $A300–350/tonne. The second approach mixes biochar with chemical fertilizer ingredients in various proportions. There was little detailed information provided on the formulations and the results of field trials with these different formulations. Preliminary results of field trials carried out by Nanjing Agricultural University have indicated that yield improvements of greater than 20% can be achieved when wheatstraw biochar is reacted with

*16.16* Slow release fertiliser produced from Biochar, minerals and a source of N.

clay and NPK fertilizer (Joseph *et al.*, 2013). One company (Biotechnology Co, Anhui) mixes NPK with rice husk biochar and a liquid clay binder in a vapour electro-heat roller dryer to produce a granule. Total nutrient content varies from 40–45% with $N:P_2O_5:K_2O$ ratios of 15:15:15 and 18:11:11 depending on crop and soil.

Sanli New Energy Company, situated near Shangqiu City, has been in operation for 5 years and produces more than 10,000 tonnes of biochar a year. The company has expanded and now has seven smaller plants in Henan Province, and it is selling products all over China. The factory sells biochar to other companies who then mix it with chemical and organic fertilizers, it makes its own NPK/biochar granule, and it sells wood vinegar. The factory also collects excreta from local schools, which pay a nominal amount for the collection. It then filters the waste through the biochar to increase its nutrient content. The company sells to industrial tobacco farms and small fruit and vegetable farmers who buy crop-specific versions of the biochar + fertilizer mixes. Significant purchases are made by government agencies, that purchase pure biochar for land remediation, or biochar + fertilizer for state-owned farms where they aim to reduce their inorganic fertilizer use. The selling price direct from the factory is approximately $300/tonne.

There are many smaller operations producing bamboo charcoal in areas where bamboo proliferates. The amount being produced is not known but appears to be greater than 100,000 tonnes per year (Dr Zhong, Bamboo Research Institute, pers. com.).

Japan, like China and other parts of Asia, has a long history of using biochar not only for agriculture but for health and hygiene, and for animal husbandry. Biochar is used for remediation of land and forest, for horticulture and cereal production, and for improving the quantity and quality of compost. Ogawa and Okimori (2010) have the following comments on the present industry:

> The effects of charcoal and wood vinegar were publicly recognized and authorized as a specific material for soil amendment by the Ministry of Agriculture, Forestry and Fisheries (MAFF) in 1990. At present, wood charcoal is being used mainly in agriculture, greening, tree rehabilitation, humidity control in house construction, water purification, and sewage treatment. In Japan, the total amount of non-fuel charcoal consumption has reached approximately 100,000 tonnes per year, but about half of this has been imported from South East Asia where it is produced from coconut and oil palm shells.

Bamboo, rice husk and certain types of hardwood and broadleaf bark are the preferred feedstock for most applications. Biochar is often pretreated with compost made from high-quality input materials (e.g., rice bran, chicken manure) and/or smoke water (wood vinegar) to make specific products that are used to grow high-value products such as mushrooms. Trials have been carried out with mixtures of chemical fertilizer and biochar. Ogawa and Yambe (1986) report the following:

> Bark charcoal of broad-leaved trees was mixed with 1% (w/w) of inorganic fertilizer (N-P-K, 8-8-8), urea, super lime phosphate, ammonium sulphate, and rapeseed meal, respectively. These charcoal fertilizers were stocked for one week and scattered over the soil surface at $500 g/m^2$ and $1500 g/m^2$ each before plowing. Control plots treated with $100 g/m^2$ and $200 g/m^2$ inorganic chemical fertilizer (amounts in conventional cultivation of soybeans) and one without any treatment were prepared. Finally, soy bean seedlings without root nodules were planted in each plot. Soy bean yields harvested from the plots with charcoal fertilizers of $500 g/m^2$ were mostly equal to those from the control plots of only chemical fertilizer.

Ogawa also reports that:

> Wood charcoal could improve the soil properties, but mixtures with chemical fertilizers, zeolite, wood vinegar, and organic fertilizer exhibited better effects than charcoal itself on tea plants, citrus, and vegetables (Ishigaki et al., 1990), rice and apple trees (Okutsu et al., 1990) and some leguminous plants and grasses (Sano et al., 1990).

Discussions with the Japan Biochar Association has highlighted markets where high prices are being paid (>$500/tonne) for biochar of the correct quality. These markets include the production of mushrooms, flowers, animal husbandry and ornamental trees.

The volumes of biochar being sold in Europe, Australia and North America are relatively small. Possibly the biggest use is in growing orchids. Prices range from $500 to $1500/tonne for wood charcoal and as high as $7000/tonne for high mineral ash products, e.g. Black Earth Products, www.blackearthproducts.com.au.

Carbon Gold in the UK sells horticultural products that have a mixture of biochar and other additives. Their biochar complex soil improver sells for about £7 per 1 kilo tube that contains 900 g of biochar. They also sell a seed compost with biochar and an all purpose biochar compost. The composts contain:

- biochar
- organic coir, a coconut processing by-product, which acts as a peat replacement
- a mix of organic, vegetable-based nutrients
- mycorrhizal fungi to maximize the uptake of nutrients by plant root systems
- wormcasts, to provide viable Actinomycetes bacteria that support Mycorrhizal
- kelp seaweed to promote vigour and disease resistance.

Trials in Switzerland with the use of biochar with compost in vineyards have been underway for over 3 years with reports of significant increases in amino acids and polyphenols of the grapes compared with the controls (Schmidt, 2011).

Palaterra has developed a process for mixing sludges from a biogass digester with waste wood biochar and anaerobically digesting this mixture (http://www.palaterra.eu). It is a four-stage process that takes only four weeks to complete: first the green waste and biochar are soaked in the liquid digestate and mixed. Stage two is a period of hot aerobic decomposition. The third stage is anaerobic lactic acid fermentation. Finally, it is dried and bagged as a finished product.

## 16.8 Conclusion and future trends

In most countries, except for those in North Asia, the quantities of biochar that are being produced and sold are well below 20,000 tonnes/year. A considerable number of small companies are now marketing biochar mixes mainly for use by the home gardener. Developing a viable large-scale industry in Europe, North and South America and Australia/NZ is proving to be difficult. From both producers of biochar products and technology providers the following constraints have been identified:

1. The high cost of biochar produced from existing technology. These costs are associated with the high capital cost of continuous automated plant or the high labour cost of operating batch reactors.
2. The high cost of, and long time for, permissions for the larger-scale plants (>1 tonne/hr biochar) by environmental protection agencies (EPAs) and local authorities, due to the lack of published long-term emissions data from large-scale plant utilizing residues and waste, and the lack of experience with this technology.
3. Access to capital, feedstock and markets for the final products. Most financiers require a company to show that they have relatively long-term agreements to access reliable supplies of biomass, and the long-term contracts for the final product, before they will lend the necessary capital to build a large-scale plant.
4. Reluctance of the conventional fertilizer industry to develop and market new carbon-based products.
5. Lack of published data on long-term extensive field trials carried out for a range of crop and soil combinations using a range of biochar and biochar blends.
6. Shortage of experienced and skilled engineers or scientists to develop and operate the production technology and develop biochar blends.
7. The relatively small amount of long-term R&D funding that is available for development and testing of both the production technology and the products.
8. Absence of accepted sustainability guidelines. As Leach *et al.* (2012) have pointed out, there are still people and organizations that associate biochar with the biofuels industry, and the conversion of land from production of food crops to the production of energy mainly to be used in developed countries.

To develop a viable industry, considerable technical, commercial and financial support needs to be given to the growing number of smaller companies who are either developing products or production technology. Substantial funds need to be allocated to develop biochars and biochar blends that are suitable for a range of soil types and crops. Ongoing testing and optimization of these products must be funded to ensure that enhanced soil properties and yields are maintained.

Demonstration plants in urban areas need to be funded and approvals given by local authorities and EPAs to gather long-term emissions data for a range of available residues and wastes. Larger-scale pyrolysis plants must be integrated into existing process industries that generate residues and require both heat and power. Development agencies need to allocate long-term funds to undertake biochar projects in developing countries where biochar is made both at a household and a village level from a range of available residues.

Even with the constraints noted above, the future for biochar industries appears to be very positive. Scientists, engineers and farmers are cooperating to develop designs that have significantly lower capital costs than those produced by commercial companies. Small companies are developing niche markets for blended products. Larger waste management companies are starting to collaborate with the smaller start-up ventures.

## 16.9 References

Amonette JE and Joseph S (2009) Physical properties of biochar, in J Lehmann and S Joseph (eds) *Biochar for Environmental Management: Science and Technology*, pp. 33–52 (London: Earthscan).

Amonette J, Kim J, Russell C, Hendricks M, Bashore C and Rieck B (2006) Soil charcoal – a potential humification catalyst, in *ASA-CSSA-SSSA International Annual Meetings*, 12–16 November 2006 (Madison, WI: ASA/CSSA/SSSA).

Anon. (2011) Ring Kiln design, 3rd UK Biochar Conference, Edinburgh, 25–26 May.

Asian Regional Cookstoves Programme (ARECOP) (1994) *How To Make And Operate The Iwate Black Charcoal Kiln* (Yogyakarta: ARECOP).

Bird RB, Bird DW, Codding BF, Parker CH and Jones JH (2008) The 'fire stick farming' hypothesis: Australian Aboriginal foraging strategies, biodiversity and anthropogenic fire mosaics. *Proceedings of the National Academy of Sciences* 105, 14796–14801.

Brodowski S, John B, Flessa H, Amelung W (2006) Aggregate-occluded black carbon in soil. *European Journal of Soil Science* 57(4), 539–546.

Brown R (2009) Biochar production technologies, in J Lehmann and S Joseph (eds) *Biochar for Environmental Management: Science and Technology*, pp. 127–146 (London: Earthscan).

Carter S and Shackley S (2011) Biochar Stoves: an Innovation Studies Perspective. UK Biochar Research Centre (UKBRC), School of GeoSciences, University of Edinburgh.

Chia C, Munroe P, Joseph S and Lin Y (2011) Microscopic characterisation of synthetic Terra Preta. *Soil Research* 48(7), 593–605.

Dixon K (1998) Smoke Germination of Australian Plants. RIRDC report (98/108, KPW-1A).

Downie A, Crosky A and Munroe P (2009) 'Physical properties of biochar', in J Lehmann and S Joseph (eds) *Biochar for Environmental Management: Science and Technology*, pp. 13–32 (London: Earthscan).

Elad Y, Rav David D, Meller Harel Y, Borenshtein M, Ben Kalifa H, Silber A and Graber ER (2010) Induction of systemic resistance in plants by biochar, a soil-applied carbon sequestering agent. *Phytopathology* 100, 913–921.

Enders A, Hanley K, Whitman T, Joseph S and Lehmann J (2012) Characterization of biochars to evaluate recalcitrance and agronomic performance. *Bioresource Technology* 114, 644–653.

Fairhead J and Leach M (2009) Amazonian dark earths in Africa?, in WI Woods, WG Teixeira, J Lehmann, C Steiner, AMGA WinklerPrins and L Rebellato (eds) *Amazonian Dark Earths: Wim Sombroek's Vision*, pp. 265–278 (New York: Springer).

Fellet G, Marchiol L, Delle Vedove G and Peressotti A (2011) Application of biochar on mine tailings: effects and perspectives for land reclamation. *Chemosphere* 83(9), 1262–1267.

Fengel D (1966a) On the changes of the wood and its components within the temperature range up to 200°C – Part 1. *Holz Roh-Werkst.* 24, 9–14.

Fengel D (1966b) On the changes of the wood and its components within the temperature range up to 200°C – Part 2. *Holz Roh-Werkst.* 24, 98–109.

Graber ER, Meller-Harel Y, Kolton M, Cytryn E, Silber A, Rav David D, Tsechansky L, Borenshtein M and Elad Y (2010) Biochar impact on development and productivity of pepper and tomato grown in fertigated soilless media. *Plant and Soil* 337, 481–496.

Hietala S, Maunu S, Sundholm F, Jämsä S and Viitaniemi P (2002) Structure of thermally modified wood studied by liquid state NMR measurements. *Holzforschung* 56, 522–528.

Hirst P (2010) Biochar production from colliers to retorts, in P Taylor (ed.) *The Biochar Revolution*, ch 10 (London: Global Publishing Group).

IBI (2012) What is Biochar?, International Biochar Initiative. Available at: http://www.biochar-international.org/biochar

*Industrial and Engineering Chemistry* (1931) Vol. 23, No. 6.

Ishigaki K, Fujie H and Suzuki K (1990) The effect of the soil amendment materials with charcoal and wood vinegar on the growth of citrus, tea plant and vegetables, in Technical Research Association for Multiuse of Carbonized Materials (TRA) (ed.) *The Research Report on the New Uses of Wood Charcoal and Wood Vinegar*, pp. 107–120 (Tokyo: TRA) (in Japanese).

Jones R (1968) The geographical back-ground to the arrival of man in Australia and Tasmania. *Archaeology and Physical Anthropology in Oceania* 3, 186–215.

Joseph S and Lehmann J (2009) Developing a biochar classification and test methods, in J Lehmann and S Joseph (eds) *Biochar for Environmental Management: Science and Technology*, pp. 107–126 (London: Earthscan).

Joseph S, Camps-Arbestain M, Lin Y, Munroe P, Chia CH, Hook J, van Zwieten L, Kimber S, Cowie A, Singh BP, Lehmann J, Foidl N, Smernik RJ and Amonette JE (2010a) An investigation into the reactions of biochar in soil. *Soil Research* 48(7) 501–515.

Joseph S, Major J and Taylor P (2010b) Making and using biochar mixed with organic matter, minerals and wood vinegar, in P Taylor (ed.) *The Biochar Revolution*, pp. 247–267 (London: Global Publishing Group).

Joseph S, Graber ER, Chia C, Munroe P, Donne S, Thomas T, Nielsen S, Marjo C, Rutlidge H, Pan GX, Li L, Taylor P, Rawal A and Hook J (2013) Shifting paradigms: development of high-efficiency biochar fertilizers based on nanao-structures and soluble components. *Carbon Management* 4(3), 323–343.

Keith A, Singh B and Singh BP (2011) Interactive priming of biochar and labile organic matter mineralization in a smectite-rich soil. *Environ. Sci. Technol.*, DOI: 10.1021/es202186j (accepted 28 September 2011).

Kookana RS, Sarmah AK, Van Zwieten L, Krull E and Singh B (2011) Biochar application to soil: agronomic and environmental benefits and unintended consequences. *Adv. Agron.* 112, 103–143.

Krull E (2010) Biochar from bioenergy – more than just a waste-product. Australian Bioenergy Conference.

Leach M, Fairhead J and Fraser J (2012) Green grabs and biochar: revaluing African soils and farming in the new carbon economy. *Journal of Peasant Studies* 39(2), 285–307.

Lehmann J, Rillig M, Thies J, Masiello CA, Hockaday WC and Crowley D (2011) Biochar effects on soil biota – a review. *Soil Biol Biochem* 43, 1812–1836.

Light ME, Daws MI and Van Staden J (2009) Smoke-derived butenolide: towards understanding its biological effects. *South African Journal of Botany* 75, 1–7.

Lima IM and Marshall WE (2005) Granular activated carbons from broiler manure: physical, chemical and adsorptive properties. *Bioresource Technology* 96, 699–706.

McLaughlin H, Anderson PS, Shields FE and Reed TB (2009) All biochars are not created equal, and how to tell them apart. Proceedings, North American Biochar Conference, Boulder, CO, August 2009. Available at: www.biochar-international.org/sites/default/files/All-Biochars–Version2–Oct2009.pdf.

Mingjie G (2004) *Manual for Bamboo Charcoal Production and Utilization.* (Nanjing, China: Bamboo Engineering Research Center, Nanjing Forestry University).

Nguyen B, Lehmann J, Hockaday WC, Joseph S and Masiello C (2010) Temperature sensitivity of black carbon decomposition and oxidation. *Environmental Science & Technology* 44, 3324–3331.

Ogawa M and Okimori Y (2010) Pioneering works in biochar research, Japan. *Soil Research* 48(7), 489–500.

Ogawa M and Yambe Y (1986) Effects of charcoal on VA mycorrhiza and root nodule formations of soybean: studies on nodule formation and nitrogen fixation in legume crops. Bulletin of Green Energy Program Group II 8, Ministry of Agriculture, Forestry and Fisheries, Japan, pp. 108–134.

Okutsu M, Hashimoto D, Fujiyama K, Nagayama M, Oda K, Taguchi K and Nakazutsumi K (1990) The effect of the soil amendment materials with charcoal and wood vinegar on the growth of rice plants, apple trees, and vegetables, in Technical Research Association for Multiuse of Carbonized Materials (TRA) (ed.) *The Research Report on the New Uses of Wood Charcoal and Wood Vinegar*, pp. 121–131 (Tokyo: TRA) (in Japanese).

Pan G, Lin Z, Li L, Zhang A, Zheng J and Zhang X (2011) Perspective on biomass carbon industrialization of organic 821 waste from agriculture and rural areas in China. *J. Agric. Sci. Tech.* 13, 75–82 (in Chinese).

Rademakers L (2009) African terra preta tradition? Batibo technique. Yahoo Groups – Biochar. Available from: http://tech.groups.yahoo.com/group/biochar/message/4978

Rajkovich S, Enders A, Hanley K, Hyland C, Zimmerman AR and Lehmann J (2011) Corn growth and nitrogen nutrition after additions of biochars with varying properties to a temperate soil. *Biol. Fertil. Soils* 48(3), 271–284.

Rayment GE and Higginson FR (1992) *Australian Laboratory Handbook of Soil and Water Chemical Methods* (Australia: Reed International Books/ Port Melbourne: Inkata Press.

Roth C (2011) *Micro-Gasification: Cooking with Gas from Biomass* (Berlin: GIZ HERA).

Sano H, Tatewaki E and Horio T (1990) Effects of the materials for greening with charcoal on the growth of herbaceous plants and trees (1), in Technical Research Association for Multiuse of Carbonized Materials (TRA) (ed.) *The Research Report on the New Uses of Wood Charcoal and Wood Vinegar*, pp. 155–165. (Tokyo: TRA) (in Japanese).

Schmidt H-P and Niggli C (2011) Biochar Gardening – Results 2011, *Ithaka J*. http://www.ithaka-journal.net/pflanzenkohle-in-kleingarten-resultate-2011?lang=en

Shinogi Y (2004) Nutrient leaching from carbon products of sludge. ASAE/CSAE Annual International Meeting, Paper number 044063, Ottawa, Ontario, Canada.

Singh B, Singh BP and Cowie AL (2010) Characterisation and evaluation of biochars for their application as a soil amendment. *Aust. J. Soil Res.* 48, 516–525.

Spokas K and Reicosky D (2009) Impacts of sixteen different biochars on soil greenhouse gas production. *Annals of Environmental Science* 3, 179–193.

Steiner C (2006) *Slash and Char as an Alternative to Slash and Burn*. Unpublished PhD thesis, Bayreuth University.

Steiner C, Glaser B, Teixeira WG, Lehmann J, Blum WEH and Zech W (2008) Nitrogen retention and plant uptake on a highly weathered central Amazonian Ferralsol amended with compost and charcoal. *Journal of Plant Nutrition and Soil Science – Zeitschrift für Pflanzenernahrung und Bodenkunde* 171, 893–899.

Uchimiya M, Wartelle LH, Klasson KT, Fortier CA and Lima IM (2011) Influence of pyrolysis temperature on biochar property and function as a heavy metal sorbent in soil. *J. Agric. Food Chem.* 59, 2501–2510.

US Composting Council and US Department of Agriculture (2001) Test methods for the examination of composting and compost (TMECC), Thompson W.H. (ed.) http://compostingcouncil.org/tmecc/ (accessed January 2012).

US Environmental Protection Agency (1996) Method 8275A Semivolatile organic compounds (PAHs AND PCBs) in soils/sludges and solid wastes using thermal extraction/gas chromatography/mass spectrometry (TE/GC/MS), http://www.epa.gov/osw/hazard/testmethods/sw846/pdfs/8275a.pdf (accessed September 2011).

US Environmental Protection Agency (2007) EPA Method 8290A polychlorinated dibenzo-P- dioxins (PCDDs) and polychlorinated dibenzofurans (PCDFs) by high resolution gas chromatography/high resolution mass spectrometry (HRGC/HRMS), http://www.epa.gov/osw/hazard/testmethods/sw846/pdfs/8290a.pdf (accessed September 2011).

Van Zwieten L, Kimber S, Morris S, Chan KY, Downie A, Rust J, Joseph SD and Cowie A (2010) Effect of biochar from slow pyrolysis of papermill waste on agronomic performance and soil fertility. *Plant and Soil* 327, 235–246.

## 16.10 Appendix: IBI; Standardized product definition and product testing guidelines for biochar used in soil

Test Category A: Basic Biochar Utility Properties – Required for All Biochars

| Requirement | Criteria | Unit | Test Method |
|---|---|---|---|
| Moisture | Declaration | % of total mass, dry basis | ASTM D1762-84 (specify measurement date with respect to time from production) |
| Organic Carbon | Class 1: >60%<br>Class 2: >30% and <60%<br>Class 3: >10% and <30% | % of total mass, dry basis | C, H, N analysis by dry combustion (Dumas method), before (total C) and after (organic C) HCl addition |
| $H:C_{org}$ | 0.7 (Maximum) | Molar ratio | |
| Total Ash | Declaration | % of total mass, dry basis | ASTM D1762-84 |
| Total Nitrogen | Declaration | % of total mass, dry basis | Dry combustion (Dumas method) and gas chromatography, following same procedure as for C, H, N analysis above, without HCl addition |
| pH | Declaration | pH | pH analysis procedures as outlined in section 04.11 of US Composting Council and US Department of Agriculture (2001), following dilution and sample equilibration methods from Raikovich et al. (2011) |
| Electrical Conductivity | Declaration | dS/m | EC analysis procedures as outlined in section 04.10 of US Composting Council and US Department of Agriculture (2001), following dilution and sample equilibration methods from Raikovich et al. (2011) |

| Test Category A: Basic Biochar Utility Properties – Required for All Biochars | | | |
|---|---|---|---|
| Requirement | Criteria | Unit | Test Method |
| Liming (if pH is above 7) | Declaration | % CaCO3 | Rayment & Higginson (1992) |
| Particle size distribution | Declaration | % <420 µm;<br>% 420–2,380 µm;<br>% 2,380–4,760 µm;<br>% >4,760 µm; | Progressive dry sieving with 4760 µm, 2380 µm and 420 µm sieves, as outlined in ASTM D2862-10 Method for activated carbon |

# 17
Development, properties and applications of high-performance biolubricants

D. R. KODALI, University of Minnesota, USA

DOI: 10.1533/9780857097385.2.556

**Abstract**: This chapter provides the fundamental understanding of how the inherent structural features of fatty acid ester derivatives make them suitable for lubrication applications. It provides sources of new triacylglycerol oils, their fatty acid composition, markets and performance compared to regular oils. Also discussed are the general functional requirements of lubricants and how the new functional fluids produced by various chemical modifications of fatty acid esters overcome the shortcomings of oxidative stability and low temperature fluidity. The cost effectiveness and high performance of biobased esters make them useful as base stocks and functional additives. A number of new technologies developed in the last two decades that have potential for commercial applications, their salient features and advantages along with future trends are presented.

**Key words**: biolubricants, triacylglycerols, fatty acid esters, oxidative stability, low temperature fluidity, viscosity, high oleic oils, chemical modifications.

## 17.1 Introduction

Consumers around the world are more receptive to greener products made from renewable resources and which offer societal benefits such as lower $CO_2$ emissions, reduction of waste to landfills, and reduced reliance on fossil resources. Globally, studies have shown that biobased products can reduce $CO_2$ emissions by using less fossil fuels and emitting fewer greenhouse gases than traditional petroleum-based alternatives. In the United States, approximately 10% of crude oil imports are used to produce chemicals, lubricants and plastics. Replacing petroleum-based products with products derived from renewable sources will directly reduce the dependence on crude oil. Development of biobased products of added value is an essential component to make the integrated biorefineries economically competitive. Although biobased products are much lower volume than fuels, they add disproportionate value. With new developments occurring in biotechnology and research and development in the production of fats and oils through crops, algae and fermentation and their subsequent chemical transformations, biobased products can gain market traction resulting in manufacturing economies-of-scale. Therefore, biobased products will increasingly achieve cost parity with traditional petroleum-based products in the next decade (Biotechnology Industry Organization, 2010).

Even before the invention of the wheel, early civilizations used natural fats and oils also known as triacylglycerols (TAG), for illumination, heating and lubrication. In the late nineteenth and early twentieth centuries, the availability of cheap and abundant supply of petroleum products rapidly displaced fats and oils in industrial applications. However, in the past three decades awareness and concern about the usage of petroleum-based products and their impact on the environment has created an opportunity to produce environmentally acceptable products from agricultural feedstock. The benefits of these products in comparison to petroleum derivatives include lower pollution (air, water and soil), minimal health and safety risks, and easy disposal due to their non-toxicity and facile biodegradability. Additionally, plant-derived products are sustainable, human-compatible and will not change the natural balance of our ecosystem (Kodali, 1996; Willing, 2001).

Biolubricants are broader in scope, contain base oils derived from renewable materials including plants and algae and possess desirable environmental qualities such as ready biodegradability, low toxicity and human compatibility with minimal safety and health risks. Biolubricants are also considered as environmentally friendly lubricants (EFL). A recent Environment Protection Agency (EPA) document on environmentally acceptable lubricants (EAL) describes EAL fluids as meeting standards for biodegradability, toxicity and bioaccumulation potential that minimize their likely adverse consequences in the aquatic environment, compared to conventional lubricants (EPA, 2011). The base oils used in EAL must be biodegradable, and the most common categories of base oils used in EAL fluids are vegetable oils, synthetic esters and polyalkylene glycols. However, currently there are no regulatory standards for EAL fluids.

The recent heightened interest in biolubricants derived from renewable resources such as vegetable oils is due to their ready biodegradability, low/ no toxicity, and environmentally benign nature. The realization that the petroleum-based economy cannot be sustained emphasizes the need to use renewable materials and practices to replace the petroleum products to a larger extent in the near future. The natural abundance, cost effectiveness, and inherent lubricity of vegetable oils and their ester derivatives are making inroads into various lubrication formulations such as metalworking fluids, hydraulic fluids, fuels, petroleum fuel additives and electrical insulation fluids (Hwang and Erhan, 2002; Erhan, 2005).

The commodity crops in the US produce a surplus and are partially sustained by subsidies. Global competitiveness in the production of fats and oils led to overproduction, beyond that required for food needs. Currently more than 20% of the world's vegetable oils production is being used for non-food applications (Gunstone, 2011). The total production of fats/oils over time is approximately doubling in volume every 20–25 years. For example, the total annual production of fats and oils around the world grew from about 50 million metric tons (MMT) in 1980 to more than 100 MMT in 2000 and is expected to reach close to 200 MMT by 2020. The recent prices of major types of vegetable oils are almost comparable to those of mineral oils (Gunstone, 2011). These factors strongly support the development of alternative applications for vegetable oils that provide new and value added markets.

The global demand for finished lubricants was 37.4 million tons in 2004 and the recent estimates for 2011 were very similar, with very little to no growth in the recent past. Of this, the United States accounts for 25%. The US finished lubricants are valued at about $10 billion. The major segments are about 50% for automotive lubricants and about 30% for industrial lubricants. More than 80% of the automotive fluids are engine oils, which require demanding performance. A large segment of the industrial fluids, about 40%, are hydraulic fluids, which require high performance but to a considerably lesser degree than engine oils. Other large segments of industrial lubricants include about 16% for metalworking fluids, 9% for greases and various other applications. Whatever the precise figures, approximately 50% of all lubricants used worldwide end up in the environment via total-loss applications, evaporation, spillage or accidents (Schneider, 2006). Estimates for the loss of hydraulic fluids are even higher, about 70–80% (Carnes, 2004; Miller *et al.*, 2000). Currently, over 95% of the materials used in lubricant products are based on mineral oils or synthetic oils. Mineral oils are toxic for mammals, fish and bacteria.

The important lubricant functional properties are viscosity, viscosity–temperature behavior (viscosity index, VI), reduced coefficient of friction to minimize wear of moving parts, oxidation stability, low-temperature fluidity and less volatility (low evaporation losses) and compatibility with seals (elastomers). The physical and functional properties of vegetable oils compared to mineral oils differ considerably and are related to their respective chemical structures (Kodali, 2002). The representative chemical structures of TAG oils and mineral oils are shown in Fig. 17.1. Petroleum-based mineral oils comprise straight or branched-chain paraffins, alicyclic and aromatic hydrocarbons and combinations thereof. This structural variation provides broad viscosity ranges and low temperature properties for formulation of different products from the base stocks. Also the structural heterogeneity, lack of polarity and chemically reduced state offer good low temperature fluidity and oxidative stability. However, the mineral oils by themselves are not good lubricants as they are inert and do not deliver the desired lubrication properties.

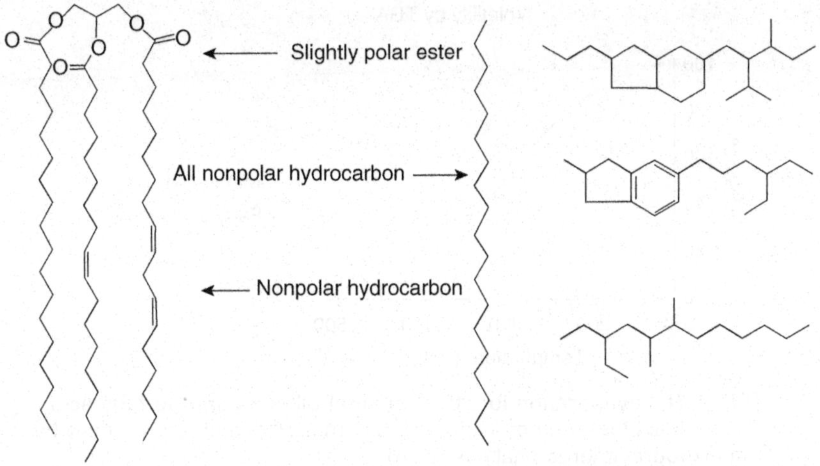

*17.1* Molecular structural features of triacylglycerol oil vs. petroleum-based mineral oil.

*Table 17.1* Coefficient of friction ($\mu$) of base oils without additives

| Base oil without additives | Coefficient of friction ($\mu$) |
|---|---|
| Mineral oil | 0.050 |
| Synthetic esters | 0.022 |
| Vegetable oil | 0.023 |

As shown in Table 17.1, mineral oils have significantly higher coefficient of friction ($\mu$) compared to ester-based fluids. The additives technologies developed since 1950 made it possible that the mineral oils can be formulated into almost all the lubrication applications. Due to lack of chemical functionality, the mineral oils lack boundary lubrication, degrade slowly, possess lower viscosity index and higher volatility. The molecular heterogeneity of mineral oil-both in structure and molecular weight contribute to higher volatility. The evaporation losses are directly proportional to the volatility of the material. The evaporation losses of mineral oil-based formulations at 250°C for 1 hr (Noack volatility) can vary from 7 to 20%, whereas the fatty acid esters are <5%, usually 1–3% under the same conditions. This point in general is illustrated in Fig. 17.2, where the weight loss is plotted against the temperature for mineral oils vs. fatty acid esters by thermogravimetric analysis (TGA).

Comparatively, vegetable oils possess the desirable characteristics of high viscosity index, high flash point, and reasonable pour points. Vegetable oils provide better boundary lubrication and load carrying capacity due to their inherent chemical structure. The polar ester region of the vegetable oil

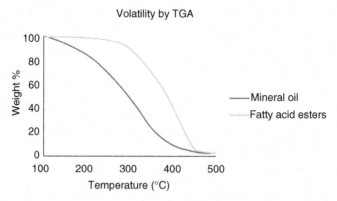

*17.2* The evaporation losses of mineral oil compared to fatty acid esters as illustrated by weight loss with increase in temperature by thermogravimetric analysis (TGA).

orients itself toward the metal surface, while the nonpolar hydrocarbon region orients away from the metal, providing a stable boundary layer that reduces wear and enhances the load and boundary lubrication properties as evidenced from Table 17.1. In addition, they are nontoxic, readily biodegradable, and a renewable resource. However, the oxidative stability and low temperature properties are a major limitation for lubrication applications. Both these properties are interdependent due to the chemical structure of triacylglycerols (TAG). Oxidative stability is related to polyunsaturated content of the oil, whereas the low temperature flow properties are related to the amount of saturation. Some conventional vegetable oils with high monounsaturated content, such as rapeseed and high oleic sunflower, are suitable for use in less severe lubrication applications. However, many high monounsaturated oils such as olive oil have limited application as lubricants mainly due to prohibitive cost and limited availability.

Many equipment and lubrication manufacturers are showing greater interest in environmentally acceptable lubricants due to regulations and the need to provide a green alternative to their customers (Battersby, 2000). The cost of vegetable oils is comparable to mineral oils, the regular feedstock used for industrial applications. Compared to synthetic esters, the high end ingredient for lubricant formulations, vegetable oils offer considerable cost advantage. The industry is looking for sustainable biobased renewable products like vegetable oils and their derivatives to replace petroleum-based products. The vegetable oil varieties with modified fatty acid profile and/or their chemically modified derivatives having good oxidative stability and low temperature fluidity can fill this new market niche.

A new class of biobased esters, derived from vegetable oils, with excellent low-temperature flow properties and oxidative stability can be produced.

One of the major advantages of these biobased esters is better performance at a lower cost compared to petroleum-based synthetic esters. This is possible owing to recent advances in the biotechnology of vegetable oils, and their chemical modifications to convert these natural esters into high-performance biolubricants.

The scope of this chapter is to provide the fundamental understanding of the structural features of TAG oils and fatty acid ester derivatives that make them (un)suitable for lubrication. Further, the structure and functional property relationships and comparison of biobased materials and petroleum products will be presented. The TAG oils production, markets, fatty acid composition and how the newer oils created by biotechnology differ from the regular oils will be provided. The generally required functional properties of lubricants will be discussed followed by how the biobased lubricants can meet these required functional properties. The TAG oils' inherent structural features, modified by chemical transformations resulting in products providing better functionality suitable for lubrication, will be described from the recent developments. A number of new technologies that have potential for commercial applications, their salient features and advantages will be discussed followed by future trends. This chapter is geared mainly towards understanding and development of high-performance biobased ester lubricants.

## 17.2 Markets for lubricants

The total lubricants consumed in 2004 amounted to 37.4 million tons worldwide of which 53% were automotive lubricants, 32% industrial lubricants, 5% marine oils, and 10% process oils. Of total industrial lubricants, 37% were hydraulic oils, 7% industrial gear oils, 31% other industrial oils, 16% metalworking fluids, and 9% greases (Mang, 2007). The total value of lubricants is about $40 billion annually. The current demand for lubricating fluids in the US is about 10 million metric tons. Engine oils and hydraulic fluids represent a substantial volume of this demand. Bremmer and Plonsker (2008) provided a comprehensive study on the biobased lubricants markets, volumes and value with emphasis on US markets.

Most of the lubricant base stocks are petroleum-derived mineral oils (Groups I, II and III), polyalphaolefins (PAO), synthetic esters and polyalkyleneglycols (PAG). The approximate price comparisons of petroleum base oils, synthetics and vegetable oils are provided in Table 17.2. Synthetic oils are high performance and value materials that are of interest in comparing biolubricants in terms of cost/performance in various applications. Synthetic oils by industry definition include the very high viscosity index (VHVI) group III oils, polyalphaolefins (PAO), polyalkyleneglycols (PAG), saturated and unsaturated synthetic esters and other modified

*Table 17.2* General price comparison of lubricant base oils and vegetable oils

| Material type | $/gal. | Average price $/lb |
|---|---|---|
| Mineral oil (Groups I and II) | 4.00–5.00 | 0.65 |
| Group III (hydrocracked oils) | 5.00–6.00 | 0.80 |
| PAO (polyalphaolefins) | 9.00–11.00 | 1.40 |
| Synthetic esters | 10.00–30.00 | 2.50 |
| PAG (polyalkyleneglycols) | 10.00–12.00 | 1.50 |
| Reg. vegetable oil | 4.00–6.00 | 0.65 |
| High oleic oil | 6.50–8.50 | 0.95 |

Price estimates are derived from recent Lube Report (www.lubereport.com) and other available public information.

materials used for lubrication. The VHVI group III oils are produced by cracking and severe hydrogenation of mineral oil and contain >90% saturates and offer good oxidative stability. They are mainly used in automotive crankcase, transmission and power steering fluids. They are a cheaper alternative to PAO in the high performance synthetic lubricants market.

Polyalphaolefins (PAO) are saturated oligomers made from long-chain alpha olefins of C8 to C12. They are relatively homogeneous materials and offer many functional advantages like higher oxidative stability, multiple viscosity grades, good thermal stability, low corrosivity and compatibility with mineral oils. The PAO markets are mainly in automotive applications with a minor fraction in industrial applications and a very small segment as aviation fluids. Most high end synthetic lubricants are formulated with PAO. However, they may require esters to compensate for or enhance the performance properties. Polyalkyleneglycols (PAG) are unique base fluids that are miscible with water. They are mainly used in fire resistant applications. The main market segments are hydraulic fluids, heat transfer fluids, and metal working fluids among others.

Synthetic esters offer many functional performance advantages. The saturated esters offer excellent thermal and oxidative stability. They offer good boundary lubrication, low wear, high viscosity index, low volatility, good additive solubility and are compatible with mineral oils and PAO. Additionally they are readily biodegradable and less toxic. There are three types of synthetic esters: diesters, polyolesters and complex esters. The polyolesters and complex esters are higher molecular weight materials and offer better oxidative stability and withstand higher operational temperatures. The excellent lubrication properties of esters combined with higher temperature performance make them the fluids of choice in jet engine lubrication. They are mainly used in industrial applications, aviation fluids and in the automotive industry. The major applications of esters are jet engine oils, refrigeration lubricants, crankcase motor oils, and compressor lubricants.

In almost all applications, the primary criterion to switch from petroleum-based fluids to biolubricants is cost/performance ratio with a rare exception of enforced regulations. The estimated base oil prices provided in Table 17.2 give a comparison of cost of base oil. The formulated product prices roughly increase 50–100% from the base oil. The high performance (operational longevity) is a very important criterion for lubricants. This is helping the synthetic lubricants to increase their market share at the expense of lower performing petroleum-based materials. The same high performance is also responsible for flat or even negative growth rate in mature markets. The new technologies in base stock production and synthetic lubricants production are increasing the performance standards of conventional products as well. For example, in the US, the GF-4 passenger car engine oil standards require higher viscosity index and lower volatility. This necessitated a switch to a higher performing Group III and PAO base stocks. Developments in this direction will also encourage use of biolubricants if the performance and cost to performance ratio of biolubricants can be improved. The biolubricants industry believes that eventually the biolubricants have the potential to replace 90% of petroleum-based products used as lubricants (Mang, 1997).

Overall public awareness, regulations and incentives for environmentally friendly fluids in the EU, North America and other developed countries (Japan, Australia, NZ, etc.) are helping to develop the biolubricants market. Biolubricants represent a very small fraction of the total lubricants market. The EU is the leading producer and consumer of biolubricants. The estimated production of biolubricants in Europe for 2006 was about 127,000 tons/year with an estimated growth of 3.7%/year (Bremmer and Plonsker, 2008). The current estimated worldwide biolubricants market is smaller than 400,000 tons/year.

In certain applications, lubrication fluids are lost into the environment due to the type of application, e.g., two-stroke oils, chain saw oils, mold release agents, drilling oils. In these instances there is a greater need to replace petroleum-based products with biorenewable materials like vegetable oils and their derivatives, especially in Europe and to a certain extent in North America. Total loss lubricant fluid applications include chain saw oils, mold release oils, two stroke engine oils and certain grease applications used for chassis and wheel flange applications. Use of lubricants in environmentally sensitive areas such as waterways, forestry, and mining also heighten the need for biocompatible, environmentally friendly, readily biodegradable and nontoxic materials. A recent study reported an analysis of marine oil losses due to vessel operational discharges and leaks of 61 million liters annually worldwide, the equivalent of one and a half to two times that of the Exxon Valdez-sized spills. The total annual estimated response and damage costs for these leaks and operational discharges are estimated to be about $322 million worldwide. About 70–80% of hydraulic fluids are lost into the environment. The recovery of lubricants used in all lubrication applications taken together is at best 50% of the total volume and the rest is lost into the environment (Bremmer and Plonsker, 2008).

In some EU countries, ready biodegradability and low toxicity are a requirement for certain environmentally sensitive applications like total loss lubricants. Consumer demand is helping the use of biolubricants due to their environmentally benign image. Regulatory issues also usually help the biolubricants. In the EU, many countries have ecofriendly labels that differentiate and support biolubricants compared to conventional products. Many eco-label standards require biodegradability, aquatic toxicity and renewability on product labels. Most of the applications that require eco-labels are total loss or high risk lubricants, e.g. chain saw lubricants, marine two stroke engine oils, forestry and mining applications. Nordic environmental label requirements for raw materials and additives were discussed with examples by Laemsae (2002). Many countries that issue eco-labels are in the EU: Blue Angel in Germany; Nordic Swan in Sweden and the Nordic Countries; and other labels in The Netherlands, Austria and France. In the United States there is some differentiation of biolubricants by some OEM manufacturers, e.g. Caterpillar's BF1 fluids standard.

Even though the annual consumption of lubricants is very large, the market share of biolubricants is a very small percentage of this volume. There are a number of factors that are responsible for this lack of market penetration. Some of the factors are related to performance and cost, whereas others are related to barriers created by historical use and the interests of the petroleum industry. For example, the established test methods and specifications are based on material properties (of petroleum derivatives) rather than application requirements. The efforts required to go through a battery of tests and protocols for new materials to prove their suitability in a given application like engine oil are very time consuming and can be prohibitively expensive (Bremmer and Plonsker, 2008).

The suitable markets for unmodified vegetable oils and their formulated products are less demanding applications like hydraulic fluids, two stroke engine oils, mold release agents, transformer fluids, cutting fluids, etc. This is partly due to performance deficiencies of TAG oils. The major advantages of using vegetable oils for lubrication applications are a renewable and environmentally friendly, nontoxic alternative at a comparable cost.

Global industry analysts forecast that the North American biolubricants market is set to reach about 250,000 tonnes by the year 2017, spurred by legislative initiatives, sustained demand from hydraulic fluids market and growth potential of product types, bio 2 cycle engine oils, greases and concrete release agents in the long term. Currently, use of biolubricants is primarily restricted to the developed markets of North America and Europe, with a lack of awareness and high pricing posing major hurdles to widespread appeal and usage in other parts of the world. In the future, increasing environmental concerns and emphasis on a shift from non-biodegradable lubricants to the environmentally safe and 'green' biolubricants may drive growth (P.R. Web, 2011).

The biolubricants market currently constitutes only a small portion of the worldwide lubricant market. Widespread usage of biolubricants is currently limited on account of its high pricing, estimated to be at least 2–3 times more expensive than the standard lubricant. Future growth in different markets worldwide will be a gradual process based on slow replacement and specific legislative norms of each country. Industrial lubricants primarily comprise non-biodegradable materials including petroleum derivatives or synthetic oils. Biodegradable lubricants, generally based on vegetable oils and ester derivatives, represent a new technological trend in the lubricants industry. These lubricants are being increasingly preferred over conventional lubricants. Although petroleum-based lubricants are less expensive, offer a high degree of stability and are universally accepted, they are non-biodegradable, and tainted with environmental hazards. Since petrochemical derivatives used as lubricants are not environmentally friendly, there is a need to develop lubricants that are high on performance, biodegradable, as well as nontoxic. The increasing number of automobiles and the growing need for oil change remain some of the main growth drivers.

According to Global industry analysts, North America and Europe represent two of the largest markets worldwide, as stated by the new research report on biolubricants. Rapid initiatives and national level legislation are in vogue in the European countries of France, Portugal, Germany, Austria, Switzerland and Sweden. Increasing activity in the manufacturing and automotive industries is a major growth impetus for the industrial lubes market, translating into higher demand for segments and applications. The hydraulic fluids segment constitutes the most important product line in both regions. Environmental regulations and fiscal incentives drive the market, with strong growth indications for the hydraulic fluids and cutting oil segments. The bio 2 cycle engine oils segment is poised to deliver robust growth of more than 12% through 2017 in the North American biolubricants market (P.R. Web, 2011).

The use of biodegradable replacements like high oleic oils and their derivatives for mineral oils in some application areas, such as chain saw oil, gearbox oils, hydraulic oils and lubricants for crude oil production, is already well established. The new developments use tailor-made fatty acid esters with specific lubricant properties. In Europe, the long-term potential for such tailor-made esters is estimated to be 10–20% of the total market

*Table 17.3* Vegetable oil production volumes and yields (estimates for 2010–11)

| Vegetable oil | Liters oil/acre | Average price ($/MT) | Total world production (MMT/yr) |
| --- | --- | --- | --- |
| Oil palm | 2700 | 1150 | 48 |
| Soybean | 200 | 1250 | 42 |
| Canola (rapeseed) | 550 | 1300 | 22 |
| Sunflower | 450 | 1600 | 11 |
| Other veg. oils (cotton, palm kernel, coconut, olive, peanut) | – | – | 23 |
| Total (2010–11) | – | – | 146 |

(500,000–1,000,000 tonnes/year). The major vegetable oils and their annual production volumes are shown in Table 17.3. Of the 146 million tons of vegetable oils produced annually, oleochemicals currently use about 20 million tons per year. Oleochemical applications are a very broad range of products where mostly the split fatty acids from TAG oils are used in surfactants, functional fluids, plastics and other specialty chemicals. In lubricants, oleic acid-based esters are used as synthetic esters. The oleic acid-rich oil fraction (triacylglycerols or TAG) are hydrolyzed to make oleic acid which is chemically modified to make key intermediates like methyl oleate, ethylhexyloleate and other polyol esters like NPG, TMP, and PE oleates.

As shown in Table 17.3, historically of the major vegetable oils, palm oil tends to be cheaper whereas sunflower oil is more expensive than other oils. This is partly due to cost of production. The palm is a perennial crop and the oil is derived from the pericarp of palm fruit, which yields 5 to 10 times higher volume per acre than other oils. A small portion of these regular vegetable oils are useful for lubrication applications where the applications are not too demanding as they do not possess high performance characteristics. In the last 20 years the fatty acid composition of the major vegetable oils has been modified through plant breeding and genetics. The high oleic oils and their derivatives thus produced have higher performance as lubricants. The fatty acid compositions of high oleic oils, their chemical modifications and their performance are discussed later.

## 17.3 Biolubricant performance requirements

The primary function of a lubricant is to reduce friction, minimize wear and dissipate heat generated by moving parts. It should also disperse deposits and sludge generated through use and contamination, inhibit corrosion and provide a seal at critical contact joints. The major constituent of a lubricating fluid is base oil (base stock) formulated with small amounts of additives. The base oil provides the primary lubrication functionality and performance. The additives enhance the performance of the base oil and also provide additional advantages and/or diminish the shortcomings of the base oil. The amount and type of additives used depend upon the severity of the application; usually the additives are from 1 to 30% of the total formulation depending on the base oil and the application requirements. Recently an excellent treatise covering all aspects of lubricants and lubrication has been edited by Mang and Dresel (2007). The important chemical and physical characteristics required for a good functional lubricant are appropriate viscosity, high viscosity index, low pour point, high stability (oxidative, hydrolytic, thermal), corrosion prevention, compatibility with additives and seals, high flash point, low volatility and good environmental acceptability, like low toxicity and high biodegradability (Odi-Owei, 1989; Kodali, 1997; Willing, 2001).

The bulk of lubrication fluids can be characterized into mineral oils and synthetics. Mineral oils are hydrocarbons derived from petroleum having different viscometrics based on molecular weight and chemical nature, having normal, branched, cyclic and aromatic structures. Mineral oil products are lower in cost as they are refined (not chemically modified) petroleum fractions, and formulated to suit an application. On the other hand, synthetics are derived from petroleum or biobased materials and offer better performance at a higher cost. Broadly the synthetic fluids encompass hydrocarbon fluids like PAO and ethers like PAG and synthetic esters. By legal definition, the hydrocracked mineral oils (Group III) having very high viscosity index (VHVI) are also classified as synthetics. Synthetic esters are condensation products of fatty acids and alcohols and usually contain two or more ester groups. A large number of carboxylate groups in the ester molecule improves the thermal properties, and provides boundary lubrication where a film can be maintained in highly loaded and high-slip contacts. Synthetic esters form a broad range of base fluids with properties varying greatly depending on the chemical structure. In principle, ester properties can be tailor-made to fit a given application. This requires understanding of structure–property relationships. For a lubricant, the molecular structure is closely related to its properties.

Viscosity and viscosity index are vital parameters for a lubricant as they determine the usefulness of a fluid for a given application. Viscosity is a measurement of internal friction of a liquid as the molecules pass each other to affect the flow. The viscosity at 40°C is a common measure of usefulness of a fluid for a specific application. The viscosity increases with molecular weight of the composition. The increase in chain length, branching, cyclic structures, and polarity (number of ester groups) will increase the viscosity of a composition. The viscosity index (VI) is a dimensionless number calculated from the variation of viscosity with temperature. The lower the variation of viscosity with temperature, the higher the viscosity index and more useful is the fluid in a given application. The viscometrics, viscosity and VI influence the ability of a fluid to form a lubricating film that reduces the friction and wear and determine its effectiveness in a given application. Both these properties are related to the molecular weight and polarity.

Vegetable oils and fatty acid esters exhibit higher viscosity index compared to mineral oils. Viscosity, VI and low temperature flow properties of various ester fluids for comparison are given in Table 17.4 (Mang and Dresel, 2007). Branching is beneficial for low-temperature properties, but decreases VI. This indicates under certain structural conditions VI and pour point have a trade-off. Recently Yao and Hammond (2006) presented the melting properties of methyl and isopropyl esters of *iso-* and *anteiso-* branched fatty acids isolated from lanolin. The branched chain fatty acid esters showed considerably lower melting temperatures and heats of fusion compared to the normal chain counterparts. Later Yao *et al.* (2008) extended this study to show the effect of structure on melting points and viscosities of oleate esters with alcohols of various chain length and branching. The cyclic structures in the acyl chain can lower the VI more than branching. The increased fatty acid chain length of TMP oleate as compared to TMP C8-C10 increases viscosity and VI, but pour point is unaffected because of the presence of double bonds in the acyl chain that prevent crystallization. The viscosity, VI and the cold flow properties are dependent upon the chemical structure, molecular weight, branching and polarity. To illustrate the point, the properties of mono- and polyolesters are given in Table 17.4.

Ready biodegradability is a major structural feature of vegetable oils and their derivatives including fatty acid esters. The biodegradability of a compound greatly depends upon the microorganisms' ability to breakdown

*Table 17.4* Dependence of lubricant properties of fatty acid esters on the molecular structure

| Chemical structure | Viscosity @ 40°C (mm$^2$/s) | Viscosity index | Pour point (°C) |
|---|---|---|---|
| Methyl oleate | 4.7 | 313 | −15 |
| 2-ethylhexyl oleate | 8.6 | 193 | −21 |
| *n*-decyl oleate | 10 | 246 | −10 |
| NPG dioleate | 30 | 192 | −33 |
| TMP trioleate | 48 | 180 | −42 |
| PE tetraoleate | 61 | 186 | −21 |
| PE tetraisostearate | 148 | 140 | −24 |

and metabolize. The compounds containing ester functionality can be broken down readily due to lipase enzymes that hydrolyze the ester function and metabolize the resulting fatty acids by β-oxidation. Lubricants can be categorized into readily biodegradable (>60% in 28 days), inherently biodegradable (30–59% in 28 days) or not biodegradable (<30% in 28 days) as measured by the ASTM D5864 biodegradability test. Mineral oils, hydrocracked oils, polyethylene glycols and polyalphaolefins biodegrade <60% in 28 days, making them fall into inherently biodegradable or not biodegradable categories, whereas almost all vegetable oils and fatty acid esters fall into the readily biodegradable category.

Lubricants are used in open and closed applications. Lubricants used in open applications are called loss lubricants (two-stroke oils, chain saw oils, mold release agents, drilling oils, etc.). They are often, if not always, used in machinery outdoors and are by definition directly emitted into the surroundings. The extent, duration and frequency of emissions depend upon the equipment, the use of the equipment and the hygiene. Loss lubrication fluids amount to 8% of all lubricants. Two-stroke engines make up the biggest part of the loss lubricant market, accounting for 30%. Other examples of loss lubricants are chain saw oils (8% of the total loss lubricant market), protective oils (10%), mold release agents (25%) and various greases (17%).

Closed systems confine the lubricant by their design. The losses in closed systems may occur due to leakages during normal operation or by accident, e.g. a broken pipe or tube rupture. Hydraulic fluids, which make up 15% of the total lubricant market in Europe, are especially susceptible to accidental spillage. The environmental damage caused by lubricants is largely associated with the approximately 50%, accounting for about 3 million tonnes/year, of predominantly mineral oil-based lubricants which are lost during use and not properly disposed of.

Most of the lubricant functional properties are provided by the base oils. However, the base oils need to be formulated with various additives to provide or enhance the required functionality. Historically, the additive formulations were based on mineral oils as they were the mainstay of the industry for a long time. The chemistry and applications of various additives differ greatly as they provide different functionality; a comprehensive account of additive chemistry is provided by Rudnick (2009). Due to structural differences, the additives requirement and the function differ from petroleum-based mineral oil and biobased lubricant formulations. Enough consideration should be taken in the formulation of biolubricants as some of the additive components (metals and chemicals) can alter the toxicity and biodegradability of the formulation. In formulating the biolubricants, the additives should not alter the ecological criteria such as toxicity, biodegradability, water pollution and waste management. A detailed account of additives for biodegradable lubricants was presented recently

(Miller, 2009). Base oils on average account for 95% of the formulation. The amount of additives in a formulation is dictated by the base oil and the application requirements, e.g., hydraulic and compressor oils require only a few percent (1–3%) additves whereas metalworking fluids and gear lubricants may require larger quantities (20–30%).

In general, various types of additives, functionality and levels of addition are presented below. Pour point depressants improve the low temperature flow properties. They modify the crystallization characteristics of the base oils by changing the kinetics, crystal size and their agglomeration, thereby preventing gelation at low temperatures. Even though pour point measurement by ASTM method is the industry standard for lubricants, in the case of biolubricants, due to their molecular size, the kinetics of gelation or solidification are slower and hence the test method usually followed to measure the low temperature flow properties is to incubate the oil at a given low temperature, say –20 or –30°C, for 24 hours, and reporting the change in flow is a more appropriate method. Increasing the heterogeneity or the addition of synthetic polyolesters with low pour point can improve the pour points of biolubricants. Polymers such as styrene esters or polymethacrylates are used as pour point depressants at 0.1–2% of the formulation.

Antifoaming agents prevent foaming. Silicon alkylates are effective antifoam agents at concentrations of 0.05–0.1%. Detergents and dispersants prevent sludge formation and keep oil-insoluble combustion products in suspension. Various surfactants, sulfonates, phosphates and phosphate esters are used at 1–5% of the formulation.

Antioxidants prevent oxidation and increase the use life of the lubricant formulation. Oxidation creates many unwanted changes such as discoloration, breakdown products, increasing polar compounds and decreasing viscosity. In the case of biolubricants where the base oil has considerable unsaturation the oxidation can lead to polymerization (thickening, increased viscosity) and to deposits forming on the surfaces. Many natural and synthetic antioxidants such as phenols like tocopherols, butylatedhydroxytoluene (BHT), alkylated phenylamines and sulfur compounds, are used at 0.1–3%. Recently, various antioxidants have been evaluated in biolubricant formulations. The natural antioxidant propyl gallate and synthetic antioxidant 4,4'-methylenebis(2,6-di-*tert*-butylphenol) were found to be most effective (Quinchia *et al.*, 2011).

Antiwear and extreme pressure (EP) additives reduce wear, extreme stress and friction. The concentration of EP additives is <5%, usually 0.20–2.0%, based on the application. Corrosion and rust inhibitors prevent metal corrosion and oxidation by acids and oxygen. Corrosion inhibitors in a formulation are in the range of 0.01–2%. Biocides are typically used in metalworking fluids and hydraulic fluids and usually in small quantities, 0.1%–0.5%.

Viscosity modifiers, are used to balance changes in viscosity of the base fluid and thickener owing to temperature changes and aging. Demulsifiers and emulsifiers prevent the formation of water-in-oil emulsions. They are all surfactants used in small concentrations depending on the application. Friction modifiers prevent stick-slip oscillations and noises by reducing frictional forces. The total additives in a formulation varies based on the application. In a hydraulic fluid they may be as low as 1%, but in engine oils they could be as high as 15–25 wt%.

## 17.4 Applications of biolubricants

The factors that determine the market penetration of a biolubricant depend upon its performance, price, safety and environmental benefits in use. The driving forces for vegetable oil use in lubricants are because the application is considered a loss lubricant and a threat to environmentally sensitive areas. The largest potential markets for vegetable oil base fluids are mining, agriculture, food machinery, outboard engine and marine applications. However, because of the performance limitations, biolubricants are used only in certain applications. Some of the promising application areas for biolubricants and the issues for each application are discussed.

Automotive applications are the biggest market for lubricants. The environmentally friendly automotive lubricants present a huge market opportunity, but tough performance requirements and the low price of petroleum alternatives make this a difficult market to enter. Automotive lubricants are not perceived as loss or high risk lubricants. The environmental issues of automotive lubricants are their impact on the fuel consumption and other issues related to the proper collection and recycling of used oil. So at best the fatty acid ester derivatives including high stability vegetable oils can be used in engine oil formulation either as an additive or as part of base fluid. This can only happen when the high stability oil or chemically modified ester derivative enhances the performance and meets the oxidative stability and low temperature flow properties while remaining cost effective.

Non-engine lubricants have lesser technical and performance demands in lubrication properties and therefore represent the best potential opportunities for vegetable oil-based ester fluids. The key applications are: two-cycle engine oils, antiwear hydraulic fluids, chain bar lubricants, gear oils, metalworking fluids, food machinery lubricants, textile lubricants, and greases.

Two-stroke oils are used in engines on boats, snowmobiles and other vehicles like two-stroke engine mopeds and personal watercrafts often used in environmentally sensitive areas. Two-cycle engines by design emit part of their fuel and lubricant unburned. An estimated 30% of the mixture of fuel and lubricant used in two-stroke engines ends up in the environment. Two-stroke oils are a complex product requiring a high amount of additives making it very difficult to formulate an environmental friendly product. Outboard motors are particularly problematic due to direct discharge into the water and use of vegetable oil-based lubricants is already mandated in parts of the world such as Europe. Vegetable oil-based systems could offer high performance and considerable savings over other ester-based lubricants. The market growth will accelerate when government legislation is enacted. The vegetable oil-based formulations need to have good low-temperature properties, good oxidative stability, and miscibility with gasoline.

Hydraulic oils make up the biggest part of industrial oils and are the second most important group of lubricants after automotive lubricants accounting for about 15% of the total lubricant consumption. Hydraulic oils are typical high risk lubricants, used in a wide range of applications both in stationery and mobile equipment in the open air (hydraulic elevators, sweepers, garage trucks, fork lifts, motor graders, front end loaders). It is important from an environmental point of view that hydraulic oils have a high degree of biodegradability and low toxicity. Hydraulic fluids represent the largest market segment growth for vegetable oils. The immediate markets are in Canada and Europe, with the US market to trail development and usage. A key aspect for growth in this area is to demonstrate higher performance and value as compared to other commodity oils. The threat of regulation is always hovering with spills and waste disposal. For this application, benefits are good biodegrability and low toxicity, anti-wear properties, protection against rust and copper corrosion, good filterability, pour point approximately −20°F, compatible with conventional hydraulic seals, miscible with mineral oils and synthetic esters. The major concerns are low pour points and poor thermal oxidative stability. Environmental awareness has already forced the conversion to more environmentally acceptable hydraulic fluids in sensitive areas such as waterways, farms and forests, and a lot of lubricant manufacturers market eco-friendly hydraulic fluids. Eco-hydraulic oils currently marketed in the EU make up the largest market of eco-lubricants and are either rapeseed-based lubricants or synthetic fluids that have been used successfully for a decade. Caterpillar, Inc. estimates this constitutes 12% of the European hydraulics market.

Chain saw oils make up 8% of the total loss lubricants market but only a very small percentage of the total lubricant market. Chain saw oils marketed as biolubricants are usually based on rapeseed oil and were launched in Europe already in the 1980s. They are low technology products, relatively low priced compared to other biolubricants. Chain saw oils of the biolubricant type are mainly used in forestry in countries with high environmental awareness: Scandinavian countries, Germany, Austria and Switzerland.

Compressor oils are used in confined systems in both industrial and mobile equipment. Compressor oils are not susceptible to accidental leakages. Gear oils may be used in open gears. For these applications, gear oils are to be considered as loss lubricants. Greases marketed as biolubricants represent a small portion of the market. Many applications of greases involve open systems like railway equipment, switch plates and others. There are products marketed today as environmentally friendly greases.

Metalworking fluids are used in the metalworking industries at the forming and cutting of metal parts. Metalworking fluids contain complex mixtures of chemicals. Metalworking fluids are no loss and high risk lubricants. Biolubricants have been introduced in this market segment in order to reduce the health and safety risks associated with the mineral oil-based products. Metalworking fluids are governed by entirely different issues than loss and high risk lubricants. These issues include occupational health issues (a potential risk for the development of allergic contact dermatitis) and environmental issues related to releases of volatile compounds in the atmosphere, generation of waste and wastewater treatment.

Concrete release agents are a typical example of loss lubricants. They make up 25% of the total loss lubricants market. Concrete release agents are usually based on rapeseed and soybean oils.

The transformer oils market in the US is close to 150,000 tons. Until recently, this market has been served by high molecular weight hydrocarbons (naphthenic oils), synthetic esters and silicone fluids. In the past decade, vegetable oil-based readily biodegradable, environmentally friendly fluids were introduced into this market. Currently in the US, there are two vegetable oil-based commercial products that are sold in this market segment: EnviroTemp FR-3 made by Cargill and Cooper and BioTemp by ABB. The EnviroTemp FR-3 fluid uses less expensive and abundant soybean oil, whereas ABB uses high oleic oils as base stock. These oils are treated with clay to remove the polar conducting materials and formulated with additives (antioxidants, metal deactivators and pour point depressants). The use of high oleic oils in transformer fluid applications has the following advantages and limitations. Advantages include renewable raw material; good biodegradability; high dielectric strength; high flash and fire point; low toxic/spill risk; and high water absorption – prolonged transformer life. Limitations include moderate oxidative stability; high viscosity; high pour point, poor low temperature properties (Biermann and Metzger, 2007).

The vegetable oil-based transformer oils gained market share over the past several years. This is mainly because of their environmentally friendly image and less the remediation risk in case of a spill. Functionally the vegetable oils are superior to mineral oils as they are more hygroscopic compared to mineral oils, thereby keeping the moisture sequestered away from the insulation paper and thus extending the transformer fluid life. The other functional properties like high fire point (necessary to get certified as fire resistant fluid) and flash point are inherent to vegetable oils. The strict controls in the refining process to remove the polar compounds like fatty acids, and maintaining the specifications for moisture after treating with clay are important steps in maintaining the quality of the transformer fluids. The oils thus prepared will meet most of the transformer fluid specifications of dielectric strength, power factor (dissipation factor), interfacial tension, acid value and moisture content. Most of the transformers used in the US are closed loop, unlike in Europe where the transformer fluid is exposed to external air (oxygen). This closed architecture of transformers made it possible to use vegetable oils like soybean oil which has marginal oxidative stability.

## 17.5 Feedstocks for biolubricants: key properties

Vegetable oils are becoming an important alternative to synthetic esters on a cost/performance ratio. There are a number of environmentally acceptable lubricant products commercially available both in Europe and North America (Bartz, 1997; Stempfel 1998). Most of these products are formulated from regular vegetable oils and perform poorly. They are used mostly in less stringent applications like hydraulic fluids and chain saw lubricants. Before vegetable oils can be considered as lubricant base stocks for severe applications like motor oils, two major limitations need to be addressed: oxidative stability and low temperature behavior (Asadauskas and Erhan, 1999; Kodali, 2002; Adhvaryu et al., 2003). The basic understanding and molecular origins of these two properties will enhance the future chemical modifications that will provide the cost-effective high performance lubricants from TAG oils.

The physical and functional properties of vegetable oils and mineral oils differ considerably and are related to their respective chemical structures. The general molecular structural features differentiating mineral oils and vegetable oils are shown in Fig. 17.1 and discussed in the introduction. In brief, mineral oils are inert hydrocarbons having different molecular weights with structural heterogeneity. The structural variation comes from normal branched, cyclic, acyclic and aromatic. The heterogeneous structural features with varying molecular weights will have different boiling points, making different cuts having different compositions and properties. Based on the functionality, the different viscosity grades are formulated to suit different applications. However, the structural features of petroleum-derived materials do not particularly offer lubrication function but give the desirable low temperature fluidity and higher oxidative stability. Compared to petroleum products, vegetable oils possess the desirable characteristics of high viscosity index, high flash point, and reasonable pour points. Vegetable oils and their ester derivatives provide better boundary lubrication and load-carrying capacity due to their inherent chemical structure. Thus the friction coefficient ($\mu$) of ester fluids is much lower than mineral oils (Table 17.1).

The polar ester region of the vegetable oil orients itself toward the metal surface and the nonpolar hydrocarbon region away from the metal, providing a stable interfacial layer thus reducing the friction coefficient. The favorable friction and wear behavior of esters compared to mineral oils reduce the need for additives and enhance the environmental acceptability. In addition, they are nontoxic, readily biodegradable, and a renewable resource. However, their oxidative stability and low temperature properties are a major limitation for lubrication applications. These properties are interdependent due to the chemical structure of triacylglycerols (TAG). Oxidative stability is related to polyunsaturated content of the oil, whereas the low temperature properties are related to the amount of saturation. In this section, an in-depth understanding of structural features of vegetable oils and fatty acid esters that are responsible for the functional shortcomings as lubricants will be discussed.

## 17.5.1 Functionality of TAG oils

The molecular structure composition of refined fats and oils is 99+% TAG, where a glycerol is esterified to three different or the same fatty acids. The TAG oil properties and their use depend upon the fatty acid composition. The industrial applications (excluding paints and coatings) require two main functional properties. They are oxidative stability and low temperature fluidity. Because of the higher importance of these properties in almost all applications, understanding their molecular origins is important. They are further explained below.

The oxidative stability of oil depends on three factors: glycerol structure, fatty acid composition and the presence of natural antioxidants like tocopherols. Almost all vegetable oils inherently contain antioxidants to protect them from oxidative deterioration. The natural antioxidants enhance the oxidative stability by reducing the rate of oxidation by scavenging the free radicals formed during the initiation, but cannot stop the oxidation process. The effectiveness of various antioxidants depends upon the type and concentration. The nature and concentration of various antioxidants present in a given oil depend upon oil type and the species' genetic disposition. With the recent advancements in molecular genetics, antioxidants are one of the traits targeted to increase the concentration and right type and ratio of antioxidants. For example, it is known that the higher total concentration of tocopherols and also higher ratio of $\gamma$-isomer compared to other isomers will provide better oxidative protection.

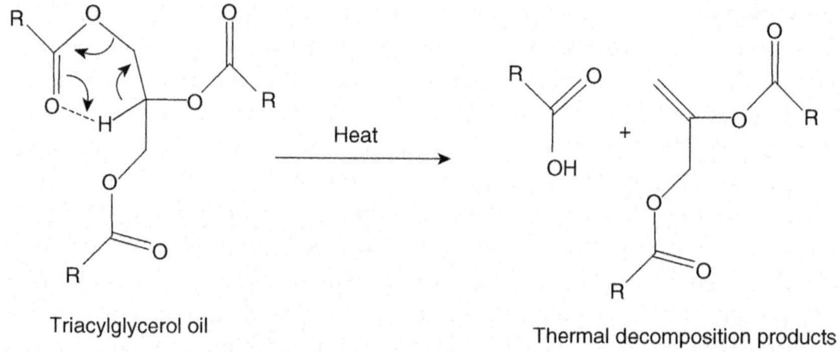

*17.3* The mechanism of glycerol β-hydrogen abstraction leading to decomposition products of vegetable oil (TAG) under thermal conditions. For simplicity, all the fatty acyl groups on TAG shown to be the same.

Glycerol, a three carbon tri-hydroxy compound, is integral to all triacylglycerol oils where the hydroxyls are esterified to fatty acids. It is known that the glycerol β-hydrogen present on the second carbon is labile, especially at high temperatures leading to degradation. The ester functionality present on the second carbon decreases the stability of secondary hydrogen and increases the thermal instability. Even though it is not proven conclusively, a possible mechanism of glycerol instability and the formation of decomposition products at high temperatures is shown in Fig. 17.3. The facile six membered ring transition formed by the carbonyl of one of the primary esters of glycerol abstracts the β-hydrogen leading to the formation of breakdown products.

The fatty acid composition of vegetable oil is the signature of a given crop and its genetic disposition. The saturated fatty acids offer the best oxidative stability; however, their concentration in any appreciable quantities leads to higher melting temperatures, making them undesirable for lubrication fluid applications. Among the unsaturated fatty acids, the oxidative stability is inversely proportional to the concentration of allylic and bis-allylic methylenes, the latter being the least desirable. The allylic methylenes are the methylenes adjacent to a double bond and the bis-allylic methylenes are the ones that are between two double bonds. For instance, as shown in Fig. 17.4, oleic acid C18:1, has two allylic methylenes and no bis-allylic methylene; linoleic acid, C18:2 has two allylic methylenes and one bis-allylic methylene; linolenic acid C18:3 has two allylic and two bis-allylic methylenes. The carbon–hydrogen bond strength of a regular methylene, allylic methylene and bis-allylic methylene are approximately 97, 88 and 78 kcal/mol, respectively. Oxidation is initiated by abstraction of hydrogen from the highly susceptible (due to lower bond

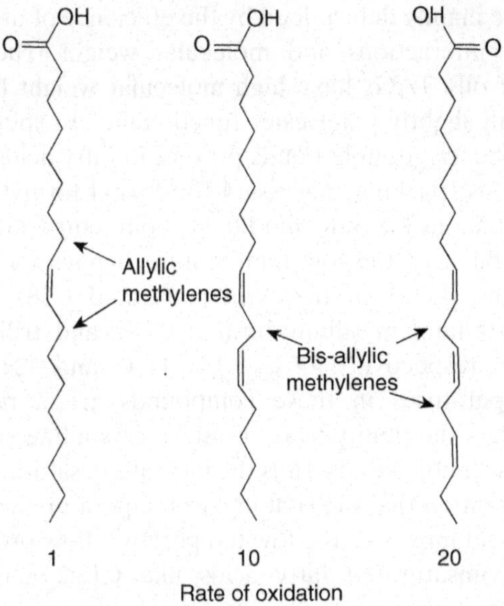

**17.4** Molecular origins of unsaturated fatty acids' susceptibility to oxidation. The monounsaturated fatty acids like oleic acid contain two allylic methylenes and polyunsaturated fatty acids like linoleic and linolenic acids additionally contain bis-allylic methylenes, that are more susceptible to oxidation than regular methylenes due to their lower carbon–hydrogen bond strength.

energy) allylic and bis-allylic methylenes of the lipid molecule. Due to the presence of allylic and bis-allylic methylenes, the relative oxidation rates of oleic (C18:1), linoleic (C18:2) and linolenic (C18:3) acids are approximately 1:10:20 respectively.

Oxidation is the major factor limiting the use of vegetable oil as lubricating fluids. Oxidation leads to polymerization and degradation. Polymerization increases the viscosity and reduces lubrication functionality. Degradation leads to breakdown products that are volatile, corrosive and detrimental to lubricant properties. The ease of oxidation depends upon the fatty acid composition of the vegetable oil. Unsaturated fatty acyl chains react with molecular oxygen to form free radicals that lead to polymerization and fragmentation products. The rate of oxidation depends upon the type and degree of unsaturation of a fatty acyl chain as shown in Fig. 17.4. The increased oxidative susceptibility of the polyunsaturated fatty acids is due mainly to the bis-allylic methylenes between the double bonds (Kodali, 2003).

The liquidity or low temperature fluidity of oils is an important characteristic for both food and industrial applications. The low temperature flow properties of oils are mainly determined by the efficiency of molecular packing, intermolecular interactions and molecular weight. The major structural component of oils, TAGs have high molecular weight between 800 and 900 Daltons with slightly polar ester functionality. In spite of the high molecular weight, the '*cis*' double bonds present in fatty acids inhibit intermolecular interactions (packing) necessary for crystal formation and growth. For this reason, the oils keep the fluidity at room temperature. The level of unsaturation influences the low temperature behavior in TAGs. For example, the melting points of tristearin (saturated C18), triolein (monounsaturated C18), tilinolein (diunsaturated C18), and trilinolenin (triunsaturated C18) are, respectively, 74°C, 5°C, –11°C, and –24°C. The decreased melting temperatures in these compounds are a result of disorganization that lowers the ability to pack into a crystalline structure due to the presence of *cis*-double bonds. Thus the increase in saturated fatty acids leads to crystal formation that melts at higher temperature, while the monounsaturated fatty acid improves the low temperature flow properties.

The presence of polyunsaturated fatty acids like C18:2 and C18:3 further improves the low temperature fluidity (better than C18:1), but at the expense of oxidative stability of the oil as discussed above. So, the monounsaturates prove to be optimal in achieving the better low temperature fluidity at a reasonable loss of oxidative stability. However, the oils having very high concentrations of oleic acid will not provide the best low temperature fluidity. This is because the higher concentration of oleic acid leads to pure oleic TAG (triolein, OOO) that phase separates and crystallizes below its melting temperature. For instance, high oleic oils having >90% oleic acid may contain >70% triolein (a pure TAG) that can separate out from the other mixed fatty acid TAG species that solidify below 5°C, the melting temperature of OOO. This demonstrates the difficulty of simultaneously achieving both low temperature fluidity and good oxidative stability in a given oil. Still, the best balance of oxidative stability and low temperature fluidity properties is achieved with vegetable oils containing high monounsaturated fatty acids. Further improvements in low temperature flow properties of high monounsaturated oils can be achieved through different chain lengths in monounsaturated fatty acids that create heterogeneous TAG (Kodali *et al.*, 2001).

Additionally, the double bonds in polyunsaturated fatty acids isomerize under thermal and catalytic conditions to form conjugated fatty acids that polymerize by Diels-Alder and radical mechanisms. Polymerization increases the molecular weight, leading to increased viscosity, gelling and loss of functionality. For this reason, the recommended operating temperatures for vegetable oil-based lubricants are lower than for mineral oils. The elimination of allylic methylenes by hydrogenation will increase oxidative stability, but the low temperature properties will be degraded. For

*Table 17.5* Fatty acid composition of high oleic vegetable oils

| Oil source | C18:1 | C18:2 | C18:3 | Total polyunsaturated | C16:0 | C18:0 | Total saturated |
|---|---|---|---|---|---|---|---|
| HO canola | 70–85 | 6–11 | 3 | 9–14 | 2 | 4 | 6–8 |
| HO sunflower | 80–92 | 3–10 | 0 | 3–10 | 3 | 4 | 7–8 |
| HO soy | 70–85 | 3–8 | 2–5 | 5–13 | 6–10 | 4 | 10–15 |
| HO safflower | 75–80 | 14–16 | 0 | 14–16 | 5 | 1 | 7–8 |
| Olive | 73–78 | 9–11 | 1 | 10–12 | 10 | 3–5 | 13–16 |
| HO palm | 59 | 16 | 0 | 16 | 18 | 7 | 25–26 |

example, hydrogenation converts unsaturated oils to *trans* and saturated fats and improves oxidation stability dramatically, but this will destroy the low temperature flow properties by producing a high melting solid fat. Altering the fatty acid profiles through advanced plant breeding and genetic engineering can create the high monounsaturated fatty acid containing oils. In high oleic oils the polyunsaturated fatty acid content has been dramatically reduced and the mononunsaturated fatty acids increased. The oxidative stability of such high monounsaturated oils is 3–10 times greater than conventional oils as determined by pressure differential scanning calorimetry (Kodali, 2005). A vegetable oil thus produced will provide both high stability and good low temperature properties. The functional properties of such ideal vegetable oils will be superior to mineral oils and comparable to the expensive synthetic oils. Some of the commercially produced high oleic vegetable oils and their fatty acid composition is provided in Table 17.5. The fatty acid composition reveals the increase in high oleic oils at the expense of polyunsaturated fatty acids and especially linolenic acid (C18:3) in certain oils is noteworthy as it provides the best stability.

### 17.5.2 High oleic oils production

Production of identity preserved (IP) crops, with specialty fatty acid composition is a long, cumbersome and expensive process. The steps involved in this process are as follows. In research and development:

- genetic analyses;
- trait stability testing;
- variety development;
- hybrid development;
- disease resistance;
- oil yield improvement;
- herbicide tolerance;
- fungal tolerance.

The R&D is followed by actual crop and oil production and marketing that involves:

- plant breeding;
- seed production;
- crop production;
- regulatory approvals;
- logistics;
- crushing/refining;
- packaging;
- marketing and sales.

In all the above steps, the quality control has to be stringent to keep identity preservation so that there is no mix up with the regular crop seeds that can ruin the product quality and specifications. For these reasons, it takes anywhere between 3–6 years to develop a new variety and millions of dollars. The new crop being developed usually results in lower yields, resulting in additional costs. To compensate for the yield losses and to incentivize farmers to grow the new variety, the companies contract the crop by paying a premium. This long process explains why the specialty oils tend to be more expensive than conventional oils.

The high oleic oils production is a target of biotechnological advancements in algal oil production and fermentation by various commercial entities around the world. As an example, a recent presentation by Solazyme Corporation on this subject matter is noteworthy (Rakitsky, 2012). Through genetic and chemical engineering capabilities, Solazyme utilizes proprietary strains of algae with standard industrial biotechnology to produce the tailor-made oils including high oleic oils. The heterotrophic algal strains grow and convert the sugars from various plant starches to produce high-value tailored oils cost effectively. The most salient feature of this new technology is that the production cycle time is reduced dramatically from a few months for a crop to grow and produce the oils to a few days to a week by this new technology using standard industrial fermentation equipment.

The advantages of using native high oleic oils (TAG) in lubricant applications either as a base oil or an additive include:

- good oxidative stability (no gumming)
- excellent boundary lubrication with reduced friction and wear
- lower cost alternative to synthetic esters
- provide high VI and solvency to hydrocarbon fluids
- natural and renewable (sustainable)
- readily biodegradable and non-toxic
- cost savings on low maintenance and disposal.

## 17.5.3 High oleic oil-based ester fluids

Due to the above advantages, high oleic oils are used as base fluids or additives in a number of applications like hydraulic fluids, transformer fluids and PAO or Group III-based synthetic oil formulations. The important feature of high oleic oils is reduction of polyunsaturates (18:2 and 18:3) and concurrent increase in monounsaturates (18:1). Another important aspect is the reduction in saturated fatty acids that are responsible for gelation. In spite of these features, high oleic oils do not meet some of the stringent lubrication requirements. For this reason, high oleic oils need to be chemically modified to make them suitable for a given application functionality such as range of viscosities, low temperature fluidity and oxidation stability. Some of these new chemical modifications reported in the literature are given in the next section.

## 17.6 Chemical modifications of biolubricant feedstocks

The use of vegetable oils in the industrial fluid applications is limited by major functional properties:

- thermal and oxidative stability,
- low temperature fluidity,
- viscometrics (lack of flexibility to create different viscosity grades), and
- hydrolytic stability.

The other factors that weigh in positively are: use of renewable materials (sustainability), ready biodegradability, nontoxic to environment (environmentally friendly), and human compatible (health and safety) materials. Above all, the products made from vegetable oils can be cost competitive compared to the commercial products. The structural features of TAG responsible for the above shortcomings arise from head group region and fatty acid chain. So the structural changes that can be achieved through chemical transformation in these two regions will be discussed in this section.

### 17.6.1 Modification of synthetic esters

As discussed earlier, the glycerol head group is not the ideal structure for optimal oxidative stability and low temperature characteristics of vegetable oils due to the presence of β-hydrogen on the glycerol carbon-2. Glycerol is a natural polyol and available in large quantities as a byproduct of biodiesel production. Recently, to overcome the shortcomings of glycerol in lubricants, 2-alkylglycerol ethers were synthesized to use as polyol to make synthetic esters (Kodali, 2011). Unlike glycerol, which is a

*Table 17.6* Thermal behavior of glycerol, 2-alkylglycerols and 2-alkylglycerol ester

| Compound | DSC crystallization | DSC melting |
|---|---|---|
| Glycerol | Crystallizes below 0°C | 18°C |
| 2-Methylglycerol | No crystallization until −45°C | No melting up to −45°C |
| 2-Ethylglycerol | No crystallization until −45°C | No melting up to −45°C |
| 2-Propylglycerol | No crystallization until −45°C | No melting up to −45°C |
| 2-Methylglycerol-1,3-di-2-ethylhexanoate | No crystallization until −45°C | No melting up to −45°C |

*Source*: Kodali (2011).

highly viscous material that melts at 18°C, the 2-alkylglcyerols and its 2-ethylhexanoate ester have low viscosity and do not crystallize or melt up to −45°C as shown in Table 17.6. The steric hindrance and reduced hydrogen bonding created by the 2-alkyl group on the glycerol is responsible for the reduced viscosity and improved low temperature properties. Also due to the ether substitution instead of ester at the glycerol 2-carbon, the thermal instability of β-hydrogen is reduced in these glycerol ether derivatives. Another advantage of 2-alkylglycerols is that the hydroxyls at carbon 1 and 3 are primary hydroxyls and can be easily esterified similar to other polyols, NPG and TMP.

Other polyols that are used frequently to make synthetic esters useful in lubrication applications are NPG, TMP and PE. These three polyols are petroleum derived and have 2, 3 and 4 primary hydroxyl groups without any β-hydrogen. The structural features of synthetic esters made from these polyols along with 2-glycerolether are shown in Fig. 17.5. The polyol esters are prepared by condensation of polyol with an appropriate fatty acid. Most of the long-chain fatty acids are derived from renewable materials by splitting the TAG. Some short-chain and branched-chain fatty acids like 2-ethylhexanoic acid are petroleum derived. The synthesis of these polyol fatty acid esters can be affected by various chemical and enzymatic processes. There are a number of condensation procedures in the literature that utilize either homogeneous or heterogeneous catalysis. The molecular weight and viscosity of these polyol esters for a given fatty acid increase with number of hydroxyls. Based on the functional requirement of an application, the right polyol and fatty acid can be determined to suit the need. In addition to these esters, complex polyol esters with higher molecular weight and viscosity are produced by using dibasic acids that crosslink the polyols, thereby increasing the molecular size.

Synthetic esters are readily biodegradable with low or no toxicity. Due to ester function they exhibit low volatility and high flash point. They are

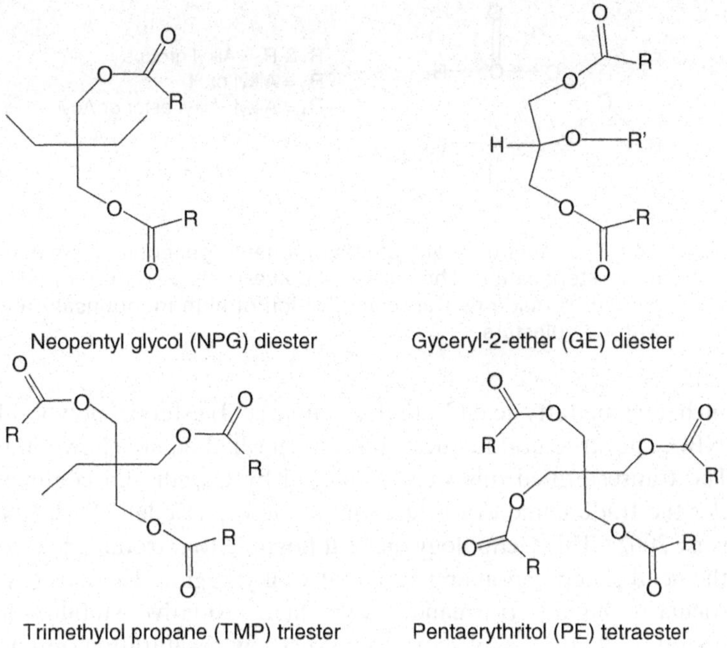

**Figure 17.5** Synthetic esters showing different polyol head group.

- Neopentyl glycol (NPG) diester
- Gyceryl-2-ether (GE) diester
- Trimethylol propane (TMP) triester
- Pentaerythritol (PE) tetraester

structurally homogeneous and usually contain saturated or monounsaturated fatty acids and hence exhibit higher oxidative and thermal stability. They also exhibit higher solvency, lubricity and hydrolytic stability. Most of the polyol esters exhibit excellent low temperature flow properties due to the presence of steric hindrance at the head group.

## 17.6.2 Heterogeneity through transesterification

An alternative approach to improve the low temperature properties is to introduce branching to disrupt the packing of the hydrocarbon chains or at the head group. Some microorganisms that live in low temperature environments maintain membrane fluidity with the help of a branched acyl chain containing phospholipids. Other methods for disrupting packing and crystallization of vegetable oils include use of different chain lengths and unsaturated fatty acids with the double bond(s) at different positions in the chain. Recently the above strategy of creating molecular asymmetry to improve the low temperature properties of oils has been successfully exploited through biotechnology and chemical modification. High-performance esters from natural oils as environmentally acceptable lubricants were created by simply transesterifying the high oleic oils with

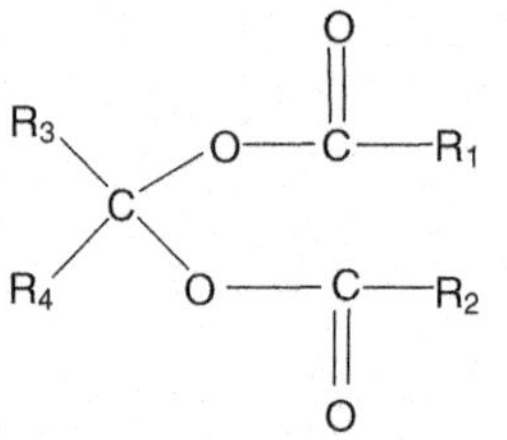

$R_1$ & $R_2$ = Alkyl groups
$R_3$ = Alkyl or H
$R_4$ = Alkyl, Alkyl ester or Acyl

*17.6* Typical structural features of a heterogeneous oil created through transesterification. The biobased polyester is a mixture of different structures made by transesterification of high monounsaturated oils with polyolesters.

branched-chain fatty acid (2-ethyl hexanoic acid) esters of polyols like TMP or NPG. The structural features of transesterified oil are shown in Fig. 17.6.

The transesterified oils were produced by Cargill and commercialized under the trade name 'Agri-Pure' and reviewed (Kodali, 2002; Kodali and Nivens, 2002). This technology made it possible to introduce heterogeneity at the head group region and at the fatty alkyl region. These esters showed particularly high performance (very high oxidative stability and low temperature fluidity) while containing very low unsaturation (iodine value). One of the major advantages of biobased synthetic esters is better performance at a lower cost compared to synthetic esters. Various polyol esters that can be used are diesters, triesters, tetraesters of various polyols with different head group structure. The transesterification process can be accomplished in a short time with a very small amount of inexpensive catalyst and an easy workup. The manufacturing cost of the product varies with the volume, with a range anywhere between 10 to 20 cents/pound above the raw material cost. The product physical properties and functionality depend upon the reactant's structure and stoichiometry.

The molecular heterogeneity in the products created by transesterification of high oleic TAG oil with TMP tri-2-ethylhexanoate is shown in Fig. 17.7. As a result of exchanging the fatty acid ester groups present in the two types of head groups, structural modification occurs both in the head group due to the incorporation of TMP along with glycerol and the ester composition on a given head group. This molecular heterogeneity creates numerous asymmetric molecular structures that show excellent low temperature flow properties. The saturated fatty acid content of the composition can be increased to >50% without sacrificing the low temperature properties, thereby producing products having excellent oxidative stability.

The salient features of transesterified polyol esters technology are that the high oleic oils can be used as is and the properties of the resulting products include:

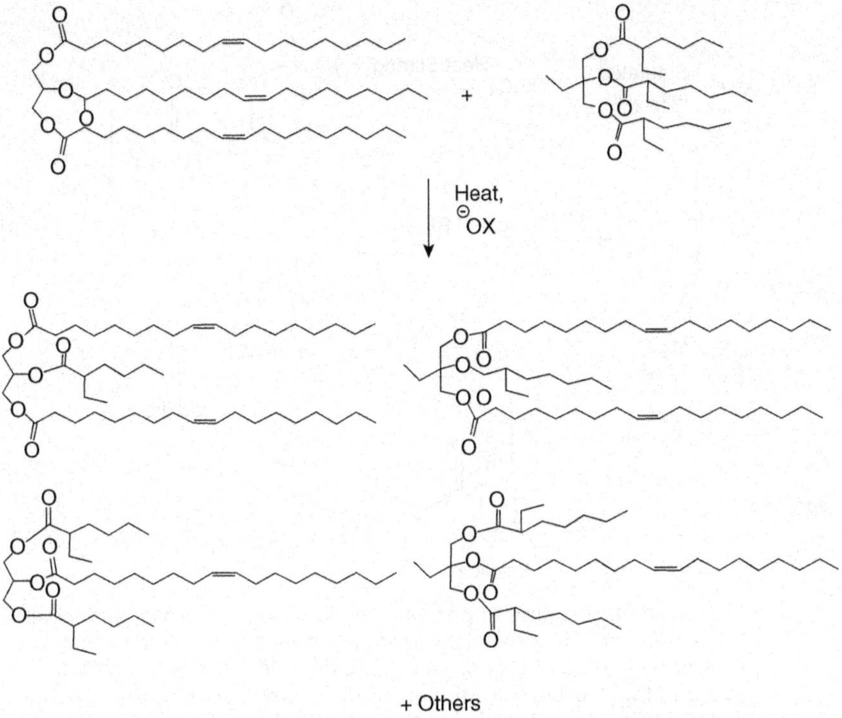

*17.7* General structural features of molecular species formed by transesterification of high oleic oil with TMP tri-2-ethylhexanoate (polyolester).

- excellent low temperature fluidity (−25°C)
- excellent oxidative stability (comparable to commercial products)
- very high viscosity index
- range of viscosities 20–50 cSt. @40°C
- low volatility
- good boundary lubrication and ready biodegradability
- most of all cost effective.

### 17.6.3 Estolide esters

Isbell and Cermak at USDA, NRR labs developed a functional fluids technology called estolides based on oleic acid, saturated fatty acid and alcohol (Isbell *et al.*, 1997, Isbell and Cermak, 2002). The synthesis and structural features of estolides are shown in Fig. 17.8. The oleic acid is reacted with saturated fatty acid in the presence of perchloric acid. The carbocation formed at the site of unsaturation undergoes nucleophilic addition of another fatty acid to form an ester linkage on the acyl chain. If the fatty acid that

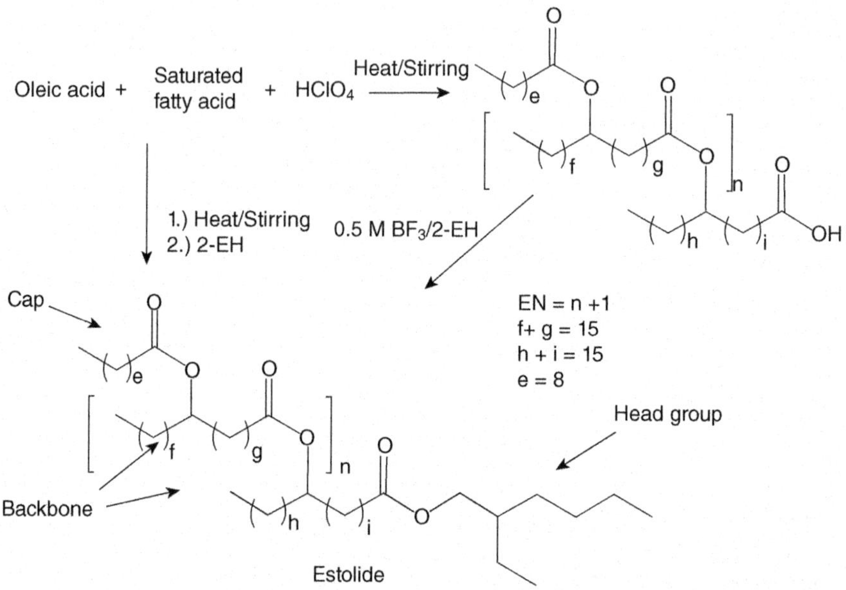

*17.8* Synthetic scheme to produce estolide esters (Cermak and Isbell, 2004). The estolide structure shows the three distinct regions: cap, back bone and head group. The cap is derived from saturated fatty acid and the backbone from oleic acid. The average molecular weight and viscosity characteristics of the estolide depend upon the estolide number, EN.

reacted with carbocation is a saturated fatty acid, it becomes a cap as it is unable to react further. If it is an oleic acid, it will continue reacting to form the backbone. The number of fatty acids in the backbone determines the estolide number and the average molecular weight. The resulting estolide acid is further esterified with a saturated alcohol, usually 2-ethylhexanol to form the head group. In short, the estolide structure as shown in Fig. 17.8, contains three structural features: backbone, head group and a cap. The backbone part is made from oleic acid derived from high oleic oils, whereas the head is a branched-chain petroleum-derived alcohol and the cap is a saturated fatty acid. In this technology, the high oleic oils are not used as is, but the oleic acid derived from high oleic oils is being used.

Recently estolide lubricants technology has been reviewed by Cermak (2011). The estolide technology addresses three characteristics: oxidative stability, low temperature flow properties and viscosity characteristics (inherent viscosity and viscosity index). The low temperature flow properties of estolides are excellent due to molecular asymmetry and heterogeniety that makes close packing (solidification) very difficult. The estolides do not have excellent oxidative stability properties but respond to commercial antioxidants very favorably, producing materials that can meet most of the stringent applications like crank-case oils. The viscosity characteristics mostly depend upon the molecular weight and nature of the estolides. The viscosity indices of the estolides are very good and comparable to synthetic lubricants. The limitations of the technology include the reaction yields and processing: the estolide yields with oleic acid (90%) as starting materials at best are 70–80%. The amount of oleic acid in the final estolide product is about 50% as it contributes mainly to the backbone of the product; the cap and head group come from other materials. The processing requires long reaction times (24 hrs) and multiple steps increasing the cost of production. Use of strong acids like perchloric acid in the reaction creates color fixation making it difficult to achieve low color materials. Purity of the end product with low acid value and hydroxyl value is required for low volatility and prolonged use-life. The distillation used to achieve the purity of the end product is expensive and throughput limiting.

### 17.6.4 Cyclopropanated oils

Another successfully employed technique was to create asymmetry and steric hindrance in the acyl chain similar to double bond geometry through cyclopropanation by inserting a carbene into the double bond. The unsaturated esters including high oleic oils are treated with methylene halide in the presence of zinc-copper reagent in a Simon-Smith type reaction. The carbine produced in the reaction was inserted into the double bond to transform it into a cyclopropane group. This procedure eliminated the unsaturation while creating asymmetry. The synthesis of cyclopropanated oils from an unsaturated TAG is shown in Fig. 17.9. The cyclopropanated oils showed excellent flow properties, being liquid at –40°C along with very high oxidative stability (Kodali and Li, 2000, 2001). However, the cyclopropanated oils were very expensive to make and were not commercialized. This technology is useful for some niche applications. The cyclopropanated oils are produced by certain plant species and it is quite possible in future to produce cyclopropanated oils cost effectively through biotechnology (Bao *et al.*, 2003).

### 17.6.5 Other esters

Epoxidized soybean oil is ring opened with acid anhydride in the presence of a catalyst to transform it into diesters (Erhan *et al.*, 2003). This technology was patented by Erhan and her coworkers from USDA, who demonstrated the usefulness of the resulting oils by utilizing them in hydraulic applications. In general, the vegetable oils having unsaturated fatty acid substituents are modified to convert sites of unsaturation to C-2–C-10 diesters. The resulting derivatives have good thermal and oxidative stability, low temperature

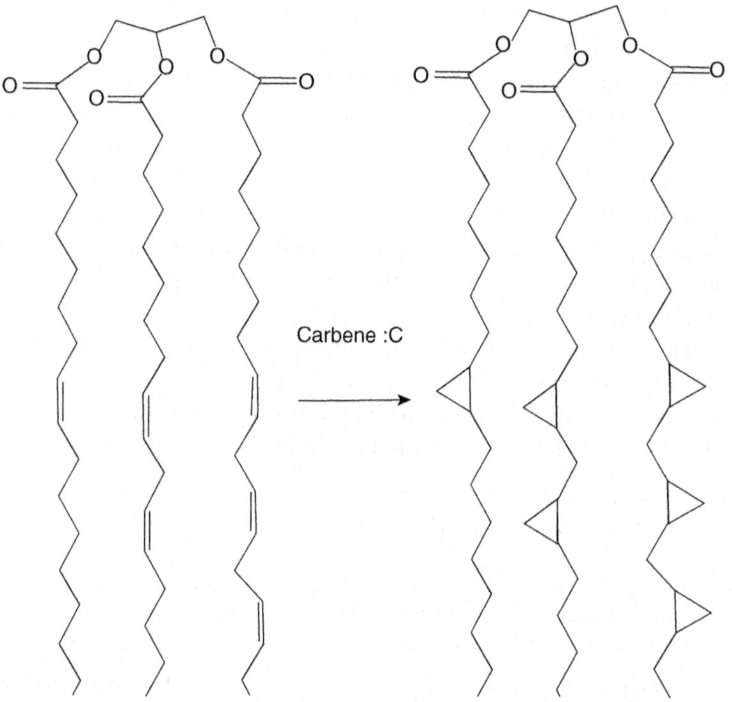

*17.9* Conversion of TAG oil into cyclopropanated TAG oil by carbene insertion into the double bond. The cyclopropanation fixes the molecular geometry similar to *cis* double bond and the cyclopropane ring creates steric hindrance.

performance properties and are environmentally friendly. They have utility as hydraulic fluids, lubricants, metalworking fluids and other industrial fluids. The TAG oils are most easily prepared via epoxidized vegetable oils which are then converted to the diesters in either a one- or two-step reaction.

Similar to the above approach, Salimon and Salih reported synthesis and evaluation of a number of potential biolubricant molecules by ring opening of epoxidized oleic acid with different fatty acids and the subsequent esterification provided di and tri-ester derivatives of C18 fatty acid having excellent low temperature fluidity and oxidative stability (Salimon and Salih, 2010). The epoxidized oils were also used to synthesize ether esters by ring opening with an alcohol followed by acylation (Lathi and Mattiasson, 2007).

Ricinoleic acid (12-hydroxy-9-*cis*-octadecenoic acid), the primary fatty acid constituent (>85%) of castor oil, even though generally more expensive than other oils, contains hydroxyl and unsaturation functionalities suitable for various chemical modifications. Recently Yao *et al.* (2010) synthesized ricinoleate and 12-hydroxystearate esters and their estolide derivatives as biolubricants. The branched ester derivatives of saturated and unsaturated fatty acid esters showed excellent low temperature flow properties and viscosity behavior suitable for lubrication applications.

The alkyl branched fatty compounds were synthesized by applying modern synthetic methods through selective functionalization of the alkyl chain (Knothe and Derksen, 1999). Carbon–carbon bond forming addition reactions afford new branched chain or elongated fatty ester compounds with interesting physical properties (Biermann *et al.*, 2000; Biermann and Metzger, 2004).

Recently, high oleic sunflower oil formulations have been tested as hydraulic fluids in agricultural applications (Mendoza *et al.*, 2011). The fatty acid composition of the base oils is not provided but the lab tests reveal good performance and the formulated product is being field tested.

Another interesting fatty acid ester chemical modification technology that is worth noting is the isomerization of unsaturated linear chain fatty acids into saturated branched-chain fatty acid isomers under different catalytic conditions. This technology has been patented by a number of researchers with different catalysts and conditions (Foglia *et al.*, 1983; Zhang *et al.*, 2005). Recently Ngo and her coworkers reported skeletal isomerization of oleic acid by a solid modified H-Ferrierite zeolite catalyst to produce predominantly saturated branched-chain fatty acids and small amounts of dimer coproducts (Ngo *et al.*, 2007, 2011). The structural features of saturated branched-chain fatty acid methyl ester derivatives are shown in Fig. 17.10. The products have excellent low temperature flow and thermal

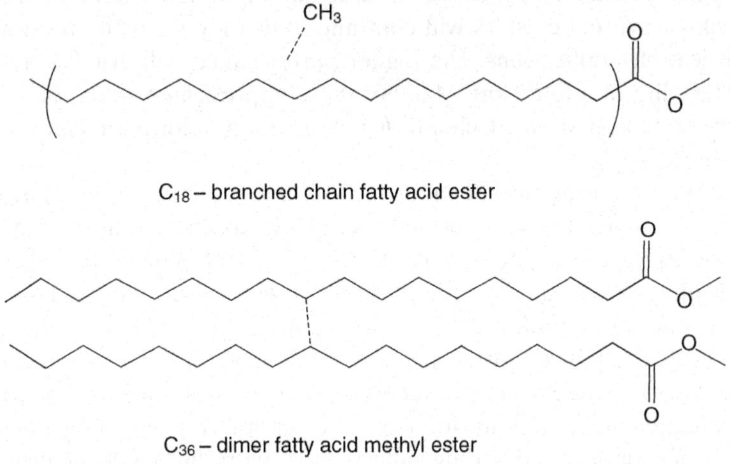

$C_{18}$ – branched chain fatty acid ester

$C_{36}$ – dimer fatty acid methyl ester

*17.10* Isomerization of oleic acid with zeolite catalyst followed by hydrogenation and esterification yield branched-chain fatty acid esters as major product and dimer acid esters as minor product. The dotted bonds in the branched-chain fatty acid ester and in the dimer denote that the bond connecting to the acyl chain(s) can be anywhere along the chain.

and oxidative stability comparable to commercial products. Very recently Manurung *et al.* (2012) reported chemical modification of sterculia oil and its fatty acid methyl esters containing cyclopropene functionality to branched ester derivatives. The chemical conversions of cyclopropene to branched-chain esters were excellent and carried out under mild conditions. The cold flow properties of the resulting branched esters are similar to the starting materials.

## 17.7 Future trends

The current and continuing future drivers for the biolubricants market segment are performance, cost, environmental acceptability, and health and safety. However, in certain applications that might affect pristine or precarious environments, regulations or legislation may force the producer and end users to develop and use environmentally acceptable biolubricants exclusively.

The use of biolubricants will continue to increase due to sustainability (derived from renewable resources) and their reduced environmental impact (ready biodegradability) in all applications as long as their cost performance ratio and availability are not a hindrance. In other words, economic and environmental factors will lead the biolubricants industry, which has a greater emphasis on saving resources, energy and reducing emissions, to gain ground along with cost and performance. The improvements in functionality in biobased natural and modified esters to match the designer synthetic esters will continue until they meet the most stringent lubrication applications. The higher performance will also include longer service life of a lubricant. Many of these improvements are going to come from a combination of chemical modification, additive technologies and biotechnology.

TAG structural modifications in the form of more stable liquids having high levels of monounsaturated oils will keep increasing in conventional crops for industrial applications and also in developing industrial crops like jatropha. Other unconventional technologies of producing TAG oils by algae and fermentation will become mainstream. As their production cycle intervals are much shorter and can be controlled, these new technologies will provide large-scale production, refining (clean-up) and restructuring through chemical modifications (e.g., ester splitting and re-synthesis). The scale of economy and finding value for the byproducts will enhance the cost effectiveness of biolubricants.

As progress in materials sciences advances, the use of various polymers as moving parts at least in ambient or low temperature applications will start increasing. This will necessitate a new type of lubricant that may use an aqueous medium. In this regard, a recent review discusses an extremely efficient synovial joint lubrication that has to last for a lifetime (Dedinaite, 2012). Lubrication of synovial joints is the most efficient and sophisticated solution to friction control in aqueous media with friction coefficients of 0.001–0.01 under high and low loads. Its efficiency is due to the complex structure of cartilage combined with the synergetic actions of self-assembled structures formed by phospholipids and biomacromolecules. Another report describes the use of polysaccharide as a superior biolubricant (Arad et al., 2006). In this study the rheological properties of a natural polymer, sulfated polysaccharide derived from red microalga indicate that it is an excellent lubricant under aqueous conditions. These two studies demonstrate that some of the future lubrication applications may rely on aqueous media and may use natural monomers and polymers that can self-assemble to interact with the moving surfaces and provide extremely efficient lubrication. The type of molecules that work in aqueous media will provide the best protection to the environment and prove to be the ultimate biolubricants.

## 17.8 Conclusion

Biolubricants are derived from renewable raw materials, are benign to the environment and possess the required functional properties for lubrication. Recent public awareness and concern about the usage of non-sustainable petroleum-based products and their impact on the environment has created an opportunity to produce environmentally acceptable products from agricultural feedstock that can lower pollution (air, water and soil), are sustainable, are human compatible with minimal health and safety risks, and are easy to dispose of due to their nontoxicity and facile biodegradability. Due to their natural abundance, cost effectiveness, inherent lubricity and the recently developed high monounsaturated oils with high stability characteristics, vegetable oils can make inroads into various lubrication applications.

Understanding the structure–functionality relationship is the key to overcoming the shortcomings of a base stock and creating high performance biolubricants. Compared to mineral oils, vegetable oils possess the desirable lubrication characteristics of high viscosity index, high flash point, low volatility, reasonable pour points, and better boundary lubrication with high load-carrying capacity and low wear characteristics. Oxidative stability and low temperature fluidity are the major limitations of vegetable oils. The high monounsaturated oils like high oleic oils created through the advancements in biotechnology or breeding address some of the limitations. Further improvements of TAG natural esters can be achieved by chemically modifying the oils. The chemical modifications include the synthesis of molecular structures that modify the head group region and the acyl chain.

The fatty acids derived from vegetable oils or petroleum can be used to make synthetic esters by condensation of polyols with the fatty acids.

Synthetic esters are versatile materials that can have uniform molecular structure with well-defined properties that can be tailored to specific applications. The knowledge of the structure–function relationship of ester fluids can be useful to design the right structure and raw materials to suit the requisite functionality and application. Ester base fluids provide both hydrodynamic and boundary lubrication. They can be used as base stocks and additives.

New technologies that are pertinent to biolubricants, that have been developed in the last two decades and have potential commercial applications are presented. These include transesterified oils, cyclopropanated oils, estolides and branched-chain esters. The structure–functionality of these new esters and their advantages and limitations are also presented.

## 17.9 Acknowledgements

The technical assistance of Lucas J. Stolp to create the figures is gratefully acknowledged.

## 17.10 References

Adhvaryu A., S.Z. Erhan and J.M. Perez. Wax appearance temperatures of vegetable oils determined by differential scanning calorimetry: effect of triacylglycerol structure and its modification. *Thermochimica Acta*, 395, 191–200, 2003.

Arad S.M., L. Rapoport, A. Moshkovich, D. van Moppes, M. Karpasas, R. Golar and Y. Golan. Superior biolubricant from a species of red microalga. *Langmuir*, 22, 7313–7317, 2006.

Asadauskas A. and S.Z. Erhan. Depression of pour points of vegetable oils by blending with diluents used for biodegradable lubricants. *J. Am. Oil Chem. Soc.*, 76, 313–316, 1999.

Bao X., J.J. Thelen, G. Bonaventure and J.B. Ohlrogge. Chracterization of cyclopropane fatty acid synthase from *Sterculia foetida*. *J. Biol. Chem.*, 278, 12846–12853, 2003.

Bartz W.J. Lubricants and the environment: new directions in tribology. Plenary invited paper. *1st World Tribology Congress Proceedings*, 103–119, 1997.

Battersby N.S. The biodegradability and microbial toxicity testing of lubricants – some recommendations. *Chemosphere*, 41, 1011–1027, 2000.

Biermann U. and J. Metzger. Catalytic C–C bond forming additions to unsaturated fatty compounds. *Topics in Catalysis*, 27, 119–130, 2004.

Biermann U. and J.O. Metzger. Applications of vegetable oil-based fluids as transformer oil. Oleochemicals under changing global conditions. Hamburg, 25–27 February 2007. Available at: http://www.abiosus.org/docs/5_Biermann _ApplicationOfVegetableOil-basedFluidsAsTransformerOil.pdf

Biermann U., W. Friedt, S. Lang, W. Lühs, G. Machmüller, J.O. Metzger, M.R. Klaas, H.J. Schäfer and M.P. Schneider. New syntheses with oils and fats as renewable

raw materials for the chemical industry. *Angew. Chem. Int. Ed.*, 39, 2206–2224, 2000.
Biotechnology Industry Organization. Biobased Chemicals and Products: A New Driver of US Economic Development and Green Jobs. 2010. Available at: www.bio.org/sites/default/files/20100310_biobased_chemicals.pdf
Bremmer B.J. and L. Plonsker. Biobased Lubricants: A Market Opportunity Study Update. Available at: http://soynewuses.org/wp-content/uploads/pdf/BioBasedLubricantsMarketStudy.pdf 2008.
Carnes K. Offroad hydraulic fluids beyond biodegradability. *Tribol. Lubr. Technol.* 60, 32–40, 2004.
Cermak S.C. Estolides: biobased lubricants. In *Surfactants in Tribology* (G. Biresaw and K.L. Mittal eds), 269–320, CRC Press, Boca Roton, FL, 2011.
Cermak S.C. and T.A. Isbell. Synthesis and physical properties of cuphea-oleic estolides and esters. *J. Am. Oil Chem. Soc.*, 81, 297–303, 2004.
Dedinaite A. Biomimetic lubrication. *Soft Matter*, 8, 273–284, 2012.
EPA 800-R-11-002. Environmentally Acceptable Lubricants, 2011. Available at: http://nepis.epa.gov/Exe/ZyPDF.cgi?Dockey=P100DCJI.PDF
Erhan S.Z. (ed.). *Industrial Uses of Vegetable Oils*, CRC Press, Boca Raton, FL, 2005.
Erhan S.Z., A. Adhvaryu and Z. Liu. Chemically modified vegetable oil based industrial fluid. U.S. Patent 6,583,302, 2003.
Foglia T.A., T. Perlstein, Y. Nakano and G. Maerker. Process for the preparation of branched chain fatty acids and esters. US Patent 4,371,469, 1983.
Gunstone F.D. Supplies of vegetable oils for non-food purposes. *Eur. J. Lipid Sci. Technol.*, 113, 3–7, 2011.
Hwang H.S. and S.Z. Erhan. Lubricant basestocks from modified soybean oil. In *Biobased Industrial Fluids and Lubricants* (S.Z. Erhan and J.M. Perez eds), 20–34, AOCS Press, Champaign IL, 2002.
Isbell T.A. and S.C. Cermak. Synthesis of triglyceride estolides from lesquerella and castor oils. *J. Am. Oil Chem. Soc.*, 79, 1227–1233, 2002.
Isbell T.A., H.B. Frykman, T.P. Abbott, J.E. Lohr and J.C. Drozd. Optimization of the sulfuric acid catalyzed estolide synthesis from oleic acid. *J. Am. Oil Chem. Soc.*, 74, 473–476, 1997.
Knothe G. and J.T.P. Derksen (eds), *Recent Developments in the Synthesis of Fatty Acid Derivatives*. AOCS Press, Champaign, IL, 1999.
Kodali D.R. Industrial applications of vegetable oils and fats: lubricants. *Proceedings of PORIM International Palm Oil Congress*, 465–474, 1996.
Kodali D.R. Industrial uses of vegetable oils and their derivatives: applications in paints, coatings and lubricants. In *Fats, Oleochemicals and Surfactants: Challenges in the 21st Century* (V.V.S. Mani and A.D. Shitole eds),119–139. Oxford &IBH Publishing Co., New Delhi, 1997.
Kodali D.R. High performance ester lubricants from natural oils. *Industrial Lubrication and Tribology*, 54, 165–170, 2002.
Kodali D.R. Biobased lubricants – chemical modification of vegetable oils. *Inform*, 14(3), 121–123, 2003.
Kodali D.R. Oxidative stability measurement of high stability oils by pressure differential scanning calorimetry (PDSC). *J. Agric. Food Chem.*, 53, 7649–7653, 2005.
Kodali D.R. Glycerol derivatives and methods of making same. US Patent 7,989,555 B2, 2011.

Kodali D.R. and K. Li. Process for modifying unsaturated triacyl-glycerol oils; resulting products and uses thereof. US Patents 6,051,539, 2000.

Kodali D.R. and K. Li. Process for modifying unsaturated triacyl-glycerol oils; resulting products and uses thereof. US Patent 6,291,409, 2001.

Kodali D.R. and S. Nivens. Oils with heterogonous chain lengths. US Patent 6,465,401 B1, 2002.

Kodali D.R., Z. Fan and L.R. DeBonte. Biodegradable high oxidative stability oils. US Patent 6,281,375 B1, 2001.

Laemsae, M. Nordic environmental label for lubricants. *Tribologie und Schmierungstechnik*, 49, 40–43, 2002.

Lathi P. and B. Mattiasson. Green approach for the preparation of biodegradable lubricant base stock from epoxidized vegetable oil. *Applied Catalysis B: Environmental*, 69, 207–212, 2007.

Mang Th. Lubricants. In *Lipid Technologies and Applications* (F.D. Gunstone and F.B. Padley, eds). Marcel Dekker, New York, 1997.

Mang Th. Lubricants and their market. In *Lubricants and Lubrication*, 2nd edn, (Th. Mang and W. Dresel, eds), Wiley-VCH, Weinheim, 2007.

Mang Th. and W. Dresel. *Lubricants and Lubrication*, 2nd edn, Wiley-VCH, Weinheim, 2007.

Manurung R., L. Daniel. H.H. van de Bovenkamp, T. Buntara, S. Maemunah, G. Kraai, I.G.B.N. Makertihartha and A.A. Broekhuis. Chemical modifications of *Sterculia foetida* L. oil branched ester derivatives. *Eur. J. Lipid Sci. Technol.*, 114, 31–48, 2012.

Mendoza G., A. Igartua, B. Fernandez-Diaz, F. Urquiola, S. Vivanco and R. Arguizoniz. Vegetable oils as hydraulic fluids for agricultural applications. *Grasas y Aceites*, 62(1), 29–38, 2011.

Miller M. Additives for bioderived and biodegradable lubricants. In *Lubricant Additives: Chemistry and Applications* (L.R. Rudnick, ed.). CRC Press, Boca Raton, FL, 445–454, 2009.

Miller S., C. Scharf and M. Miller. Utilising new crops to grow the biobased market. In *Trends in New Crops and New Uses* (J. Janick and A. Whipkey, eds), ASHS Press, Alexandria, VA, pp. 26–28, 2000.

Ngo, H.L., A. Nunez, W. Lin and T.A. Foglia. Zeolitecatalyzed isomerization of oleic acid to branched-chain isomers. *Eur. J. Lipid Sci. Technol.*, 108, 214–224, 2007.

Ngo H.L., R.O. Dunn, B. Sharma and T.A. Foglia. Synthesis and physical properties of isostearic acids and their esters. *Eur. J. Lipid Sci. Technol.*, 113, 180–188, 2011.

Odi-Owei S. Tribological properties of some vegetable oils and fats. *Lubrication Engineering*, 11, 685–690, 1989.

P.R. Web. North American Biolubricants Market to Reach 243,327 Tonnes by 2017, Global Industry Analysts, Inc. Available at: http://www.prweb.com/pdfdownload/8281822.pdf 2011.

Quinchia L.A., M.A. Delgado, C.Valencia, J.M. Franco and C. Gallegos. Natural and synthetic antioxidant additives for improving the performance of new biolubricant formulations. *J. Agric. Food Chem.*, 59, 12917–12924, 2011.

Rakitsky W. Tailored triglyceride oils for food industrial applications. Presentation given at 103rd AOCS annual meeting, 2012.

Rudnick L.R. (ed.) *Lubricant Additives: Chemistry and Applications*, 2nd edn. CRC Press, Boca Ratan, FL, 2009.

Salimon J. and N. Salih. Chemical modification of oleic acid for biolubricant industrial applications. *Australian Journal of Basic and Applied Sciences*, 4, 1999–2003, 2010.

Schneider M. Plant-oil-based lubricants and hydraulic fluids. *J. Sci. Food Agric.*, 86, 1769–1780, 2006.

Stempfel E.M. Practical experience with highly biodegradable lubricants, especially hydraulic oils and lubricating greases. *NLGI*, 62, 8–23, 1998.

Willing A. Lubricants based on renewable resources – an environmentally compatible alternative to mineral oil products. *Chemosphere*, 43, 89–98, 2001.

Yao L. and E.G. Hammond. Isolation and melting properties of branched-chain esters from lanolin. *J. Am. Oil Chem. Soc.*, 83, 547–552, 2006.

Yao L., E.G. Hammond and T. Wang. Melting points and viscosities of fatty acid esters that are potential targets for engineered oils. *J. Am. Oil Chem. Soc.*, 85, 77–82, 2008.

Yao L., E.G. Hammond, T. Wang, S. Bhuyan and S. Sundaralingam. Synthesis and physical properties of potential biolubricants based on ricinoleic acid. *J. Am. Oil Chem. Soc.*, 87, 937–945, 2010.

Zhang, S., Z. Zhang and D. Steichen. Skeletal isomerization of alkyl esters and derivatives prepared therefrom. US Patent 6,946,567, 2005.

# 18
# Bio-based nutraceuticals from biorefining

Y. LIANG and Z. WEN,
Iowa State University, USA

DOI: 10.1533/9780857097385.2.596

**Abstract**: Nutraceuticals as health promoting agents have received great attention from both the scientific community and the general public in the past decades. They have been found from diversified resources, possessing a wide range of beneficial effects in biological systems. Bio-based nutraceuticals are derived from natural renewable resources, which can be extracted as a component in the biomaterials or produced through bioprocessing. The main focus of this chapter is on bio-based nutraceuticals, specifically on the current status of the nutraceuticals market, the classification of nutraceuticals, and the properties and functions of individual classes of nutraceuticals. In addition, the future prospective of bio-based nutraceuticals is also discussed.

**Key words**: bio-based, nutraceuticals, lipid, protein, peptide, polysaccharide, phenolics, alkaloids, carotenoids.

## 18.1 Introduction

### 18.1.1 Definition and history of nutraceuticals

'Nutraceutical', a word combining 'nutrition' and 'pharmaceutical', is used to describe a special food or food ingredient with medical or health benefits such as prevention and/or treatment of disease (DeFelice, 1992). Nutraceuticals may refer to dietary supplements, plant extracts, or food designed for specific purposes (DeFelice, 1995). The earliest recognition of nutraceuticals was in the 1980s with calcium, fiber, and fish oil being recognized as health-promoting agents (DeFelice, 1995). With significant research efforts being pursued to explore functional components from plants, marine organisms, and microorganisms (actinomycetes and fungi) (Cocks *et al.*, 1995), nutraceuticals have expanded from basic dietary supplements to a vast range of compounds such as bioactive constituents extracted from plants (polyphenols and lycopene) (Carlos Espin *et al.*, 2007), food ingredients produced by microorganisms (amino acids and omega-3 fatty acids) (Hermann, 2003; Athalye *et al.*, 2009), and microorganisms themselves as an integral part of food (*Spirulina* and yogurt cultures) (Salminen and Isolauri, 2006; Peiretti and Meineri, 2011).

Nutraceutical development has been in full swing for the past decade. The global market for nutraceuticals was already worth $117.3 billion in

2007 (Ahmad et. al., 2011). With the promotion of the well-being concept and increasing health awareness, the nutraceutical market share is expected to have a steady growth rate in the coming years, and is estimated to attain $176.7 billion in 2013 (Ahmad et al., 2011).

### 18.1.2  Legislation and regulation of nutraceuticals

With the rapid expansion of the nutraceutical market, more legislation has been implemented to regulate the use and production of this group of products. Beginning in 1994, the United States Congress established the Dietary Supplement Health and Education Act (DSHEA), which provided a clear definition of a dietary supplement in order to distinguish it from medicine and food. It also created the Office of Dietary Supplements within the National Institute of Health (NIH) to institute consumer protections, increase public education, and promote research on the possible health effects of dietary supplements (Office of Legislative Policy and Analysis, n.d.).

The Food and Drug Administration (FDA) has recently established new regulations (termed the 'Final Rule') for the nutraceutical industry that are in accordance with the International Conference of Harmonisation (ICH) standards. ICH brings experts from the European Union, Japan, and the United States together to establish guidelines for safe, effective, and high quality medicine (International Conference of Harmonisation, 2012). The Final Rule, established in 2007 and phased in over three years, set standards for dietary supplement production called 'Good Manufacturing Practices' (GMPs). Specifically, it established standards for quality control, ingredient labeling, product claims, complaint records, and manufacturing procedures, among others, in an effort to insure that consumers have access to accurately labeled, high quality dietary supplements free from contamination (US Food and Drug Administration, 2009).

GMPs aim to ensure a high quality standard in dietary supplements by holding manufacturers responsible for product identity, purity, composition, and strength. This would prevent the inclusion of pesticides and other contaminants in the saleable product. However, there is no standardization of dietary supplements in the United States (Office of Dietary Supplements, n.d.).

### 18.1.3  Biorefinery and bio-based nutraceuticals

Bio-based nutraceuticals are defined as nutraceuticals derived from the biomass refinery process. In biomass refinery, where the biomass is completely utilized for sustainable fuels and chemicals production (Amidon et al., 2011), the production of bio-based nutraceuticals, which are regarded as high value co-products, can reduce the cost of the overall production process and thus enhance its economic viability. However, it should be noted that compared with fuels or bulk chemicals, the market size for bio-based nutraceuticals is rather small. If the biomass refinery industry is fully developed, the market for bio-based nutraceuticals will be saturated and thus the prices and the profit margin will inevitably be reduced.

Depending on the biorefinery processe used, production of bio-based nutraceuticals is usually achieved through two routes: physical separation/extraction from biomass materials or biological production by various microorganisms. Some biomass materials contain active compounds that can be extracted to make nutraceutical products while they are mainly used as feedstock for fuels and chemicals. For example, acetylic acids, β-carotene, vitamin B, astaxanthin, polyunsaturated fatty acids, and lutein are some valuable products found in microalgae, which can be integrated into algal-based biofuel production and potentially enhance the economic feasibility of the process (Christaki et al., 2011). Current breakthroughs in bioseparation methods such as chromatography techniques, supercritical fluid extraction (SFE), and nanotechnology separation (Walker et al., 2006) have brought in great opportunities for the economic production of bio-based nutraceuticals from biomass feedstock. For example, recent studies have shown that supercritical fluid extraction can be an economic and scalable tool in extracting functional ingredients from different natural resources, including plants, food by-products, algae, and microalgae (Herrero et al., 2006).

Many nutraceutical-based compounds can be co-produced by microorganisms during microbial fermentation for fuels and chemicals. For example, carotenoids, astaxanthins, and polyunsaturated fatty acids can be synthesized by microalgae while the algal cells are mainly used as feedstocks for fuel and energy production (Christaki et al., 2011). Currently, microbial production of nutraceuticals in general is limited by several factors such as the high cost of substrate and low production yield. Inexpensive feedstocks from biomass refineries can provide a cheap substrate for microorganisms, but may still be limited by other factors such as the lack of a robust strain with high product yield. Metabolic engineering has become an effective tool to enhance the cells' capability of synthesizing certain desired compounds for nutraceuticals (Hugenholtz and Smid, 2002). For example, fermentative production of mannitol from lactic acid bacteria has been successfully implemented by modification of the genomic information and redirection of the metabolic flux (Ferain et al., 1996; Neves et al., 2000).

In general, nutraceuticals from bio-based materials can be classified as lipid-based, protein and peptide-based, and carbohydrate-based, depending on the nature of the functional constituents. Some small molecular bioactive compounds such as secondary metabolites are also considered to be bio-based nutraceuticals. This chapter discusses the recent research progress of bio-based nutraceuticals, the advantages and limitations of bio-based nutraceutical production, and its potential opportunities and challenges.

## 18.2 Lipid-based nutraceuticals

In the past years, the production of lipid-based biofuels has been intensively studied. Lipid producers include terrestrial plant seeds (such as soybean, canola, and castor) and oleaginous microorganisms (such as microalgae, bacteria, and fungi). Compared to plant seeds, microorganisms have attracted considerable research interest due to their fast growth rate. However, the commercialization of lipid-based biofuels has been largely limited by the high production cost, even though a significant amount of research has been conducted in order to improve the microbial oil content and productivity. As a result, producing lipid-based nutraceuticals as an addition to the fuel production opens up a new opportunity for improving the process economy and reducing the overall production cost.

### 18.2.1 Types, properties, and functions of lipid-based nutraceuticals

Lipids refer to a group of naturally occurring compounds that are readily soluble in organic solvents. They are classified as neutral or polar type depending on the nature of their chemical structures. Triglyceride and its derivatives (mono-, di-glyceride, and free fatty acid), waxes, sterols, fat-soluble vitamins (A, D, E, and K), phospholipids, and sphingolipids are the major lipid types present in nature. They exist in different parts of living organisms and have various biological functions. Triglyceride and its derivatives are abundant in microorganisms as well as the fatty tissues in human bodies where they serve as energy storage molecules. Phospholipids are the major components of the cell membrane while sphingolipids work as signaling molecules in cell metabolism. Sterols are the precursors of steroid hormones and vitamins play important roles in biological systems, such as regulating cell and tissue growth and differentiation (vitamin A), promoting calcium absorption (vitamin D), and performing antioxidant function (vitamin E) (http://lipidlibrary.aocs.org/).

Lipid-based nutraceuticals encompass various products including omega-3 polyunsaturated fatty acids (PUFAs), phytosterols, lipophilic vitamins and their analogues as well as polar lipids with specific functions. The main focus of this chapter will be on PUFAs, phytosterols, and polar lipids.

*Omega-3 polyunsaturated fatty acids*

Omega-3 polyunsaturated fatty acids (PUFAs), particularly eicosapentaenoic acid (20:5 n-3, EPA) and docosahexaenoic acid (22:6 n-3, DHA) are a class of lipids with various biological functions. Clinical and epidemiological studies have shown the preventive and therapeutic effects of omega-3 PUFAs on a series of illnesses such as rheumatoid arthritis, heart disease, cancers, schizophrenia, and Alzheimer's disease (Cohen and Ratledge, 2010). The health benefits delivered by PUFAs have led to significant efforts in exploring omega-3 PUFA sources and their commercial production processes.

Fish oil has been used as a commercial source of omega-3 PUFAs. However, the peculiar taste and odor, and possible metal contamination of fish oil have limited its applications (Barclay et al., 1994). Microbial production of EPA and DHA, on the other hand, can largely avoid these problems: therefore, it has been a promising alternative for supplying high quality PUFAs (Kralovec et al., 2012).

A variety of microorganisms, including lower fungi, bacteria, and marine microalgae, are capable of synthesizing omega-3 PUFAs (Bajpai et al., 1991; Kendrick and Ratledge, 1992; Ratledge et al., 2001; Athalye et al., 2009; Johnson and Wen, 2009; Liang et al., 2011). Among these omega-3 PUFA producer candidates, bacteria are not suitable for commercial PUFA production as they cannot accumulate high levels of lipids, and the co-produced fatty acids are uncommon in other systems, which might be an issue for human consumption (Ratledge et al., 2001). In contrast, oleaginous fungi and microalgae with high PUFA content and high lipid production yield are good candidates as economical PUFA producers. For example, the marine microalga *Schizochytrium sp* (Barclay et al., 1994) and *Crypthecodinium cohnii* (Ratledge et al., 2001; de Swaaf et al., 2003) have been used for commercial DHA production. In general, microbial species belonging to the genera of *Mortierella*, *Pythium*, and *Saprolegnia* have been found to be capable of producing appreciable amounts of EPA (Cohen and Ratledge, 2010) and genetic modification has also been introduced to improve the strain performance (Cohen et al., 1992). In comparison to commercial DHA production, commercial microbial production of EPA is still rarely reported, with genetically modified yeast *Yarrowia lipolyfica* being the only strain used for commercial production by DuPont. The major reason is the low EPA yield and productivity from those microbial sources, although a considerable amount of studies have been conducted to identify EPA producers and to enhance the production yield. As a result, microbial production of EPA is still in the infancy stage. Further increase of EPA level to an economically viable point requires in-depth understanding of factors that influence EPA production.

*Phytosterols*

Phytosterols, possessing similar structures to cholesterol, are found in the fat-soluble fractions of plants. They represent a wide range of natural compounds with a structure skeleton of triterpene. Depending on the carbon side chain, the presence/absence of double bond, and attachment of sugar molecules (Moreau *et al.*, 2002), phytosterols can be classified as sterols, stanols, and conjugated glucosides. The biological synthetic pathway of triterpene skeleton starts from the reduction of 3-hydroxy-3-methylglutaryl coenzyme A (HMG-CoA) (six carbons) to mevalonate (five carbons). Six mevalonate units are then linked together to form two molecules of farnesyl diphosphate, which are connected to make squalene (30 carbons) (Hicks and Moreau, 2001). Further enzymatic reactions will promote the formation of ring structures, the cleavage of carbon–carbon bonds, and the addition of hydroxyl groups or other molecules. The end products are diversified phytosterols and triterpene alcohols (Moreau *et al.*, 2002).

Phytosterols have been reported to reduce the cholesterol level in serum through the following mechanisms:

- Phytosterols can displace cholesterol from bile salt micelles due to their higher affinity for micelles, and thus, inhibit cholesterol absorption (Ikeda and Sugano, 1998).
- Phytosterols crystallize with cholesterols in the gastrointestinal tract,which leads to the formation of poorly absorbable mixed crystals and the consequent reduction of sterol uptake in the intestinal lumen (Trautwein *et al.*, 2003).
- Phytosterols inhibit the hydrolysis of cholesterol esters, which can reduce the amount of absorbable free cholesterol, and thus interfere with cholesterol absorption (Ikeda *et al.*, 2002).

Phytosterols are present in our diet, primarily in the form of vegetable oils, cereals, fruits and vegetables. In western countries, people consume an average of 250 mg per day of phytosterols (Hicks and Moreau, 2001), which is much less than the effective dosage. Therefore, additional consumption of phytosterols is needed for an effective reduction of cholesterol. A series of phytosterol fortified foods has been developed, such as margarine, salad dressings, and milk products (Akoh, 2005). In addition to lowering serum cholesterol, phytosterols have also been used as substrates for producing high-value steroidal drugs via biological conversions and chemical reactions (Donova, 2007), which have expanded the application of phytosterol-based nutraceuticals.

With advancements in biotechnology, Microalgae have been identified as phytosterol producers in a large diversity of products. *Dunaliella salina* and *Dunaliella tertiolecta* were reported to produce 0.89% and 1.3% total sterols on a dry weight basis under defined conditions. The whole sterol extracts from *D. salina* and *D. tertiolecta* showed therapeutic effects on hypercholesterolemia, which could potentially have applications in the nutraceutical and pharmaceutical industries (Francavilla *et al.*, 2010). Although the yield of phytosterols from microalgae is relatively low, this fraction can add value to algal-based bio-fuel and bio-chemical production, thus commercialization of microalgal phytosterols might be an achievable target when a biorefinery concept is applied.

*Polar lipids*

Polar lipids with amphiphilic nature are often associated with membrane structure, and play a variety of biological functions. A majority of polar lipids (Fig. 18.1) found in cell membranes are glycerophospholipids (GPLs), which have a glycerol backbone with fatty acids attached. Some examples of GPLs include phosphatidyl choline, phosphatidyl ethanolamine, and phosphatidyl serine. When the polar lipids are derived from sphingosin, the corresponding polar lipids are classified as sphingophospholipids (SPLs). Sphingomyelin, consisting of a ceramide unit and a phosphorylcholine moiety, is the most representative SPL found in the neural system (Oshida *et al.*, 2003). By interacting with cellular membranes, dietary polar lipids can alter the membrane compositions, and thus affect a series

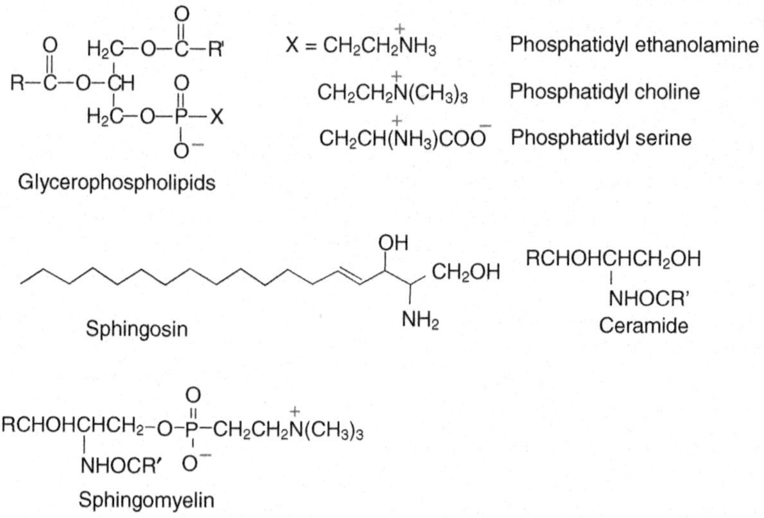

*18.1* Structure of polar lipids.

of signaling process and enzymatic functions (Küllenberg et al., 2012). Some beneficial effects exerted by polar lipids consist of anti-inflammatory, anti-cancer, cholesterol lowering, and brain development (Küllenberg et al., 2012).

Polyphosphatidyl choline (PPC), extracted from soybean, with the main active ingredient of 1,2-dilinoleoylphosphatidyl choline (DLPC) has been proven to possess beneficial effects in various liver diseases. DLPC is found to have a high content of polyunsaturated fatty acids, particularly linoleic acid. Oral administration of PPC can induce a series of changes, including restoring membrane structure, increasing membrane fluidity, enhancing membrane-associated metabolic functions, reducing peroxidative reactions, improving immune properties, and stabilizing bile compositions (Gundermann et al., 2011). These positive effects revealed from a pharmacological standpoint have also been confirmed in clinical trials. Test subjects with different types of liver disease responded positively to the consumption of PPC with rare and weak side effects. Therefore, soybean-derived PPC is a good nutraceutical supplement in preventing liver disease and promoting the restoration of physical health for patients (Gundermann et al., 2011).

Sulfoquinovosylacylglycerol (SQAG) (Fig. 18.2) is an anionicglycerolipid, associated with monogalactosyldiacylglycerol and digalactosyldiacylglycerol, and present as a structural component in the photosynthetic membranes of plants, algae, and various bacteria (Berge et al., 2002). It is one of the few lipids derived from natural sources with sulfonic acid linkages discovered to date (Sanda et al., 2001). Due to the uniqueness of the chemical nature of SQAG, studies have been conducted to extract and characterize SQAG from various marine red and blue-green algae, and a series of biological functions related to SQAG have been reported. SQAGs derived from algae have shown an inhibitory effect on both eukaryotic DNA α- and β-polymerases and HIV-reverse transcriptase. Hence, they can potentially be used as antiviral and antitumor supplements (Gustafson et al., 1989; Loya et al., 1998; Ohta et al., 1998).

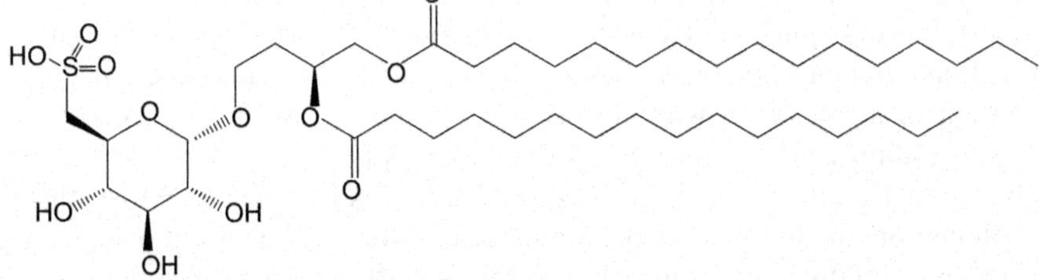

18.2 Structure of sulfoquinovosylacylglycerol.

## 18.3 Protein and peptide-based nutraceuticals

Many proteins and peptides derived from a multitude of food-based plants and animals, such as milk, soy, and fish proteins, possess specific biological functions in addition to their established nutritional roles as antioxidants and/or substrates for tissue building (Korhonen and Pihlanto, 2003; Zaloga and Siddiqui, 2004). Exogenous bioactive peptides associated with foods can be classified, based on their physiological functions, as anti-hypertensive, ACE inhibitory, antimicrobial, antithrombotic, immunomodulatory, mineral binding, opiod agonist, and opiod antagonist peptides (Owusu-Apenten, 2010).

The production of bioactive peptides can be achieved via the hydrolysis of food proteins or microbial fermentation. For example, milk protein is found to be a good substrate for both hydrolysis and fermentation for producing bioactive peptides (Pfeuffer and Schrezenmeir, 2000; Salamia et al., 2011). Many bioactive peptides can also be produced by non-food-based microorganisms, plants, and animals. For instance, cyanobacteria have been reported to produce ~40% lipopeptides, which have cytotoxic, antitumor, antiviral, and antibiotic activities (Singh et al., 2005). These bioactive peptides are diversified in chemical nature, including dipeptides, complex linear peptides, cyclic oligopeptides, or polypeptides with glycosylation, phosphorylation, and amino acid residue acylation (Schlimme and Meisel, 1995). Genetic modification is another approach for producing bioactive peptides with desirable properties through rational design and the modification of relevant genes. In this chapter, we will not explore the genetic manipulation approach for bioactive protein synthesis; instead, our main focus is on traditional ways of producing bioactive proteins and peptides, particularly from algae-derived materials.

### 18.3.1 Algae proteins

Unlike milk and egg which have been considered as good protein sources for a long time, algae-derived proteins did not receive much attention until algal biofuel research became a popular topic in recent years. Algal proteins have great potential to mitigate the high production cost of algal-based biofuel, particularly when the proteins have certain biological functions and can be sold as high value products. In general, the protein content of algae varies with species. Green and red algae have relatively high protein content (around 40%) while protein content in brown algae can be lower than 15% (Murata and Nakazoe, 2001). Most algae proteins contain all the essential amino acids with a high content of aspartic acid and glutamic acid and a low level of sulfur-containing amino acids (Fleurence, 1999).

Lectins are sugar-binding proteins naturally occurring in algae. Some bioactive lectins identified in macroalgal species such as *Ulva* sp., *Eucheuma* spp. and *Gracilaria* sp. have exhibited various biological functions including antibiotic, mitogenic, cytotoxic, anti-nociceptive, anti-inflammatory, anti-adhesion and anti-HIV activities (Bird *et al.*, 1993; Smit, 2004; Mori *et al.*, 2005). These bioactive functions are achieved mainly by binding a soluble carbohydrate or a carbohydrate moiety of glycoprotein or glycolipid, and subsequently precipitating the glycoconjugates. Lectins as agglutinins have found applications in biology, cytology, biochemistry, medicine, food science, and technology (Smit, 2004).

Phycobiliproteins are water-soluble proteins that are a part of light-harvesting protein-pigment complexes in cyanobacteria and certain microalgae. *Spirulina* has been found to be a rich source of phycobiliproteins. C-phycocyanin, a major phycobiliprotein, has been reported to possess a variety of biologically beneficial effects, including antioxidant, anti-inflammatory, and neuroprotective benefits. *Spirulina*-derived phycocyanin has been used in the food and beverage industry as a natural pigment. Phycocyanin has been commercially produced in open ponds and raceway systems at tropical and subtropical Pacific Ocean regions (Lee, 1997; Pulz, 2001; Spolaore *et al.*, 2006). In open raceway systems, the productivities of dry biomass and phycocyanin were around 14 and $0.85\,g/m^2$ a day, respectively (Jiménez *et al.*, 2003). The extraction of phycocyanin from algal biomass mainly consists of two steps: disintegration of cells and extraction of the water-soluble phycobiliproteins into aqueous media. Further purification of phycocyanin can be achieved by ammonium sulfate precipitation, chromatographic methods (Boussiba and Richmond, 1979; Zhang and Chen, 1999; Minkova *et al.*, 2003; Niu *et al.*, 2007; Abalde *et al.*, 1998; Soni *et al.*, 2006), and ultrafiltration (Herrera *et al.*, 1989). Recently, a method based on two-phase aqueous extraction followed by chromatography was successfully developed to prepare ultra-pure C-phycocyanin (Patil *et al.*, 2006).

## 18.3.2 Algae-derived bioactive peptides

Bioactive peptides from algae can be divided into two categories: endogenous peptides and enzymatically hydrolyzed peptides (Harnedy and Fitzgerald, 2011). Endogenous peptides, including linear peptides, cyclic peptides, depsipeptides, and different peptide derivatives, possess various biological activities (Harnedy and Fitzgerald, 2011). Carnosine, an antioxidant peptide due to its capability of chelating transitional metals (Brown, 1981; Fleurence 2004), has been found in macroalgae (Fleurence, 2004; Shiu and Lee, 2005). Glutathione, another antioxidant peptide found in the green macroalga *Ulva fasciata*, is produced under high oxidative conditions, in which an inherent enzymatic antioxidant system (ascorbate-glutathione) is stimulated and glutathione is produced to scavenge and detoxify oxidative free-radical molecules (Shiu and Lee, 2005). Kahalalide F, a cyclic depsipeptide, produced by the green alga *Bryopsin* sp., has been proposed as a new natural anti-cancer drug candidate (Hamann and Scheuer, 1993). It has a strong cytotoxic activity *in vitro* against different cell lines including prostate, breast, and colon carcinomas as well as neuroblastoma, chondrosarcoma, and osteosarcoma (Hamann and Scheuer, 1993; Suárez *et al.*, 2003). *In vivo* tests and some clinical trials have also been carried out, showing a great prospect for the drug application of Kahalalide F (Luber-Narod *et al.*, 2000; Provencio *et al.*, 2006; Pardo *et al.*, 2008).

Bioactive peptides from algal protein hydrolysates have shown various functionalities such as ACE-inhibitory, anti-hypertensive, antioxidant, antitumor, antityrosinase, anticoagulant, calcium-precipitation-inhibitory, antimutagenic, plasma- and hepatic-cholesterol reducing, blood-sugar-lowering, and superoxide dismutase (SOD)-like activities (Harnedy and Fitzgerald, 2011). These protein hydrolysates were mainly prepared by using food grade enzymes including protamex, trypsin, pepsin, alcalase, flavourzyme, and neutrase (Harnedy and Fitzgerald, 2011). Purification of the bioactive peptides has been achieved by ultrafiltration and nanofiltration, which render a great potential for the application of peptides as functional food ingredients (Korhonen, 2009).

## 18.4 Carbohydrate-based nutraceuticals

Polysaccharides are considered the most abundant renewable resource on earth, and have been well known as feedstocks for renewable fuels production. Unlike those cellulose- and hemicellulose-based bulk materials, in nature, there are also some specific polysaccharides possessing unique properties and biological functions that can be made into nutraceuticals. These polysaccharides will be the main emphasis in this section. Prebiotics (typically oligosaccharides) are another type of carbohydrate-based nutraceuticals, which can stimulate the growth and/or activities of gut microflora, thus increasing health benefits for humans. Research has been conducted to explore different types of prebiotics from various feedstocks. In this section, prebiotics from renewable resources will also be examined.

### 18.4.1 Bioactive polysaccharides

The first bioactive polysaccharide reported to have specific functionality is lentinan, a polysaccharide found in the mushroom *L. edodes Sing.*, with strong antitumor activity (Chichara *et al.*, 1969). Lentinan has a 1,3-β-glucopyranosidic structure with two 1,6-β-D-glucopyranosidic

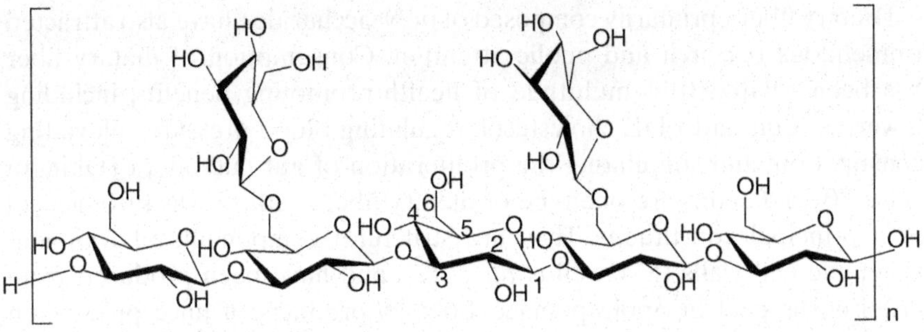

*18.3* Structure of lentinan.

branches every five glucose units in the main chain (Fig. 18.3). The strong antitumor activity of lentinan was attributed to its highly ordered structure; which can induce interleukin-12 (IL-12), a secreted protein/signaling molecule responsible for stimulating natural killer cells and T lymphocytes production (Wiederschain, 2007). Another anti-cancer polysaccharide, schizophyllan, was also isolated from the mushroom *S. commune* in the earlier days (Komatsu *et al.*, 1969). These antitumor polysaccharides possess a similar structural characteristic, i.e., the skeleton of 1,3-β-glucan with some 1,6-β-glucose branches (Yagita, 1997). Biologically active compounds can be obtained by the enzymatic hydrolysis of the hemicellulose skeleton in the cell wall of fungal mycelia, which explains the presence of a wide variety of anti-cancer polysaccharides in different fungi species.

A sulfated polysaccharide extracted from marine red alga, *Schizymenia pacifica*, was found to have anti-HIV functions. This polysaccharide was proven to be a member of λ-carrageenan with 20% sulfonate content. The molecular weight of the bioactive polysaccharide is around two million; the major sugar units are galactose and 3,6-anhydrogalactose (Nakashima *et al.*, 1987a, 1987b). This sulfated polysaccharide can selectively inhibit HIV reverse transcriptase (RT) and replication *in vitro*.

Sulfated polysaccharides are the most studied bioactive polysaccharides. In addition to their antivirus capabilities, numerous sulfated polysaccharides isolated from algae have also been reported to show antioxidant, proliferative, immuno-inflammatory, antimicrobial and antilipidemic behavior (Jiao *et al.*, 2011; Wijesinghe and Jeon, 2012). For example, a sulfated fucoidan, from commercially cultured *Cladosiphon okamuranus* TOKIDA, has demonstrated its anti-proliferative activity in human leukemia U937 cells by inducing apoptosis associated with degradation of PARP, caspase-3 and -7 activation in U937 cells (Teruya *et al.*, 2007). The producer of sulfated fucoidan, *C. okamuranus* TOKIDA has been cultivated commercially. A fucidan from *C. okamuranus* TOKIDA has been used as an additive in health foods, drinks, and cosmetics in Japan (Teruya *et al.*, 2007).

Dietary fibers, primarily composed of polysaccharides, have also attracted tremendous research and public attention. Consumption of dietary fiber has been related to a multitude of health-promoting benefits, including lowering total and LDL cholesterol, regulating blood pressure, alleviating constipation, and stimulating the proliferation of gut microflora (Dhingra et al., 2012). Pectins are one type of dietary fibers with D-galacturonic acid as a principal constituent. They are structural components of fruit and vegetable cell walls (Maxwell et al., 2012), and have been produced from dried citrus peel or apple pomace, both by-products of juice production (Sato et al., 2011; Guo et al., 2012). β-Glucans are another type of valuable dietary fibers with D-glucose monomers linked by β-glycosidic bonds. They are predominantly non-starch polysaccharides present in cereal crops, particularly in barley and oat (Ahmad et al., 2012). Both pectins and β-glucans have special physical properties such as thickening, stabilizing, emulsification, and gelation, which make them suitable to be incorporated in soups, sauces, beverages, and other food products (Burkus and Temelli, 2000; Maxwell et al., 2012).

### 18.4.2 Extraction and purification of polysaccharides

Extraction of polysaccharides from biomass sources is commonly conducted using the following procedures:

- Cell disruption and water extraction, usually achieved with the assistance of treatments like heat and ultrasonic disintegration.
- Polymer precipitation from water extract by a precipitating agent (e.g., methanol, ethanol, isopropanol or acetone).
- Drying of the precipitated polymer, namely freeze drying (laboratory scale) or drum drying (industrial scale) (Imeson, 2010; Pang et al., 2007; Yang et al., 2012).

The purification of polysaccharides can be achieved by using different chromatographic techniques, including ion exchange, gel permeation, and affinity chromatography (Srivastava and Kulshreshtha, 1989). However, these techniques are limited by their loading capacities and can only be used on a small scale. The simulated moving bed (SMB) technique has recently received more attention due to its high separation efficiency and applicability for sugar purification. It is a scalable method and suitable for industrial processing of functional carbohydrates (Walker et al., 2006).

### 18.4.3 Prebiotics

Prebiotics are defined as 'selectively fermented ingredients that allow specific changes, both in the composition and/or activity of the gastrointestinal microbiota that confers benefits upon host well-being and health' (Gibson et al., 2004). Most prebiotics are oligosaccharides, which resist digestion by pancreatic and brush border enzymes. They can be produced by extraction from plant materials, enzymatic hydrolysis of polysaccharides, and microbial/enzymatic synthesis (Figueroa-Gonzalez et al., 2011).

Inulin is one of the most established prebiotics with the structure of β(2-1)-fructans (Gibson et al., 2004; Figueroa-Gonzalez et al., 2011). It can be extracted from chicory root and *Agave tequilana* with the degree of polymerization (DP) varying from 2 to 65. Partial enzymatic hydrolysis of inulin using an endo-inulinase (EC 3.2.1.7) can lead to the production of oligofructose (DP 2-7). The food industry has developed specific products known as Synergy® (Orafti NV, Tienen, Belgium) by mixing oligofructose and long-chain inulin. Different industrial products have different DP distribution and, therefore, varying technological properties (Franck, 2002).

Galacto-oligosaccharides (GOS), which are mixtures of oligosaccharides derived from the enzymatic transglycosylation of lactose, are another type of prebiotic. β-Galactosidases are the primary enzymes responsible for the conversion of lactose to GOS. As β-galactosidases have the capability of conducting both transgalactosylation and hydrolysis reactions, the yield of GOS is low due to the presence of the hydrolysis process (Ganzle, 2011). Methods to improve the yield of GOS have been investigated including enzyme selection, protein engineering, and optimizing reaction conditions. GOS yields have been achieved from 40 to 60% in the final product (Park and Oh, 2010).

Prebiotics have also been produced through a biorefinery process as valuable products. For example, oligofructose was enzymatically synthesized when the aqueous extracts from dairy by-products were used as substrate (Smaali et al., 2012). Xylooligosaccharides with prebiotic activities have been manufactured from solid waste in malting industries by double hydrothermal processing (Gullon et al., 2011). Prebiotics, as carbohydrate-based functional compounds, have been receiving tremendous research efforts within the context of biorefinery production.

## 18.5 Other nutraceuticals

There are several types of nutraceuticals that do not fall into the aforementioned categories. These nutraceuticals are small molecules and diverse in chemical characterizations, natural presence, and biological functions. Many of them are secondary metabolites (such as phenolic compounds and alkaloids), which are found in plants and microbial cells for maintaining the physiological functions or defending the cells from external stress. Carotenoids, as pigment components, are another source of bio-based nutraceuticals.

### 18.5.1 Phenolic compounds

Phenolic compounds are the most abundant and widely represented class of plant natural products. With a growing body of evidence that bioactive phenolics exert strong health-promoting effects, phenolic compounds have a great potential to be used as nutraceutical supplements and pharmacological agents (Wildman, 2007).

Phenolic compounds are a group of structurally diversified chemicals derived from phenylalanine and tyrosine. Plant phenolic compounds comprise simple phenols, phenolic acids (both benzoic and cinnamic acid derivatives), coumarins, flavonoids, stilbenes, hydrolysable and condensed tannins, lignans, and lignins (Naczk and Shahidi, 2004). Some of those phenolic compounds have bioactive functions while others have no noticeable bioactivity, depending on their biosynthesis pathway. In general, the bioactive group of phenolic compounds are derived from shikimic acid and/or polyacetate biosynthesis pathways (Bernal et al., 2011).

Fruits and vegetables are a rich source of simple phenolics such as hydroxycinnamic acid conjugates and flavonoids. These compounds are considered to have protective effects against cancer and cardiovascular diseases due to their antioxidant activities (Boudet, 2007). Epidemiological studies have confirmed the health-promoting effects of phenolics by revealing that a high consumption of antioxidant-rich fruits and vegetables is inversely correlated with the incidence of cancer (Knekt et al., 2002). The possible mechanisms of the antioxidant activities are:

- decreasing accumulation of products from oxidant reactions (i.e., lipid peroxides),
- depressing oxidant products by inducing an external stress, and
- elevating the concentrations of endogenous antioxidants, or preventing their depletion caused by an external stress (Wildman, 2007).

Polyphenolic compounds are receiving increasing attention due to their positive roles in the treatment of neurodegenerative diseases, prevention of cardiovascular, cerebrovascular and peripheral vascular diseases, and function as phytoestrogen (Boudet, 2007). As neuroprotective agents, polyphenolics are capable of reducing or blocking neuron death induced by oxidative stress, and exert benefits through different metabolic pathways including signaling cascades, anti-apoptotic processes or the synthesis/degradation of the amyloid $\beta$ peptide (Ramassamy, 2006). Polyphenolics contained in red wine have been found to have cardioprotective function by interfering with the molecular processes related with initiation, progression, and rupture of atherosclerotic plaques (Szmitko and Verma, 2005).

Preparation of phenolic compounds from raw plant materials has been extensively studied. As plant phenolics exist in both simple and highly polymerized forms, and may also complex with carbohydrates, proteins and other plant components, the solubility of phenolics varies significantly (Naczk and Shahidi, 2004). Therefore, different solvent systems have been applied to extract specific groups of phenolic compounds. Water, ethanol, methanol, acetone, ethyl acetate, propanol, and dimethylformamide are commonly used. The combinations of those solvents have also been adopted to increase the extraction range of substances (Antolovich et al., 2000). The phenolic extracts are always a mixture of various types of phenolics and non-phenolic compounds. Therefore, additional steps are required to remove the undesired compounds. Petroleum ether, ethyl acetate, or diethyl ether extractions are usually added to concentrated phenolic extracts in order to remove lipids and unwanted polyphenols (Naczk and Shahidi, 2004). In terms of the fractionation and purification of phenolic compounds, solid phase extraction and different chromatography methods have been proven to be effective. Mateos et al. (2001) used diol-bonded phase cartridge to extract phenolics from olive oil. The oil sample was washed through the column with different solvent systems including hexane, a mixture of hexane/ethyl acetate (90:10, v/v), and methanol. The phenolics were found in the methanol eluted fraction. Mateus et al. (2001) employed a Fractogel (Toyopearl) HW-40(s) column to fractionate anthocyanin-derived pigments in red wines. The eluent was water–ethanol pigments.

In the biorefinery scenario, phenolic compounds, as valuable co-products, can be produced through either direct extraction of food or agricultural byproducts or fermentation of those byproducts as substrates. For example, olive mill wastewater has been subjected to different extraction and purification techniques to recover the polyphenols. Methods used include solid phase extraction (Bertin et al., 2011), integrated member system (Garcia-Castello et al., 2010), Azolla (aquatic fern) matrix, and active carbon adsorption (Ena et al., 2012). The major component of the total phenolics in olive mill wastewater is flavonoids with subgroups of flavanols and proanthocyanidins, which exhibited considerable antioxidant activity (El-Abbassi et al., 2012). Phenolic compounds have also been extracted from other food wastes, including potato peels (Oreopoulou and Russ, 2007), apple skins (Schieber et al., 2001), grape skins (Pinelo et al., 2006), carrot peels (Chantaro et al., 2008), raspberry waste (Laroze et al., 2010), and coffee byproducts (Murthy and Naidu, 2012).

Solid state fermentation (SSF) has been employed to increase the phenolic content in some food products as well as to produce phenolics from agro-industrial residue. Starzyńska-Janiszewska et al. (2008)

have applied *Rhizopus oligosporus* to cooked seeds of grass peas, and found that the phenolic compound content had a great increase after fermentation. Wheat grain and soybean products have also been fermented by different fungal species. The resultant products exhibited higher content of phenolics and stronger antioxidant behavior (Bhanja *et al.*, 2009; Singh *et al.*, 2010). Agricultural and forestry wastes including straw, bagasse, stover, cobs, and husks are lignocellulosic biomass with lignin as one of the major chemical components. Filamentous fungi like white-rot fungi *Phanerochaete chrysosporium*, *Trametes versicolor*, *Trametes hisuta*, and *Bjerkandera adusta* have the capability of breaking down lignin to produce phenolic compounds, which can potentially increase the nutritional value of the materials as animal feed or soil fertilizer (Nigam and Pandey, 2009).

### 18.5.2 Alkaloids

Alkaloids are low molecular weight nitrogen-containing compounds found in about 20% of plant species (Facchini, 2001). They are mostly derived from the amino acids, phenylalanine, tyrosine, tryptophan, lysine, and ornithine. Alkaloids are commonly produced in response to stressed situations as secondary metabolites with characteristic toxicity and pharmacological activity. These properties were used by human beings for hunting, execution, and warfare in old times, but now are mostly used for the treatment of disease (Mann, 1992).

Plant-derived alkaloids currently used in clinical applications consist of the analgesics morphine and codeine, the anti-cancer agents vinblastine and taxol, the gout suppressant colchicine, the muscle relaxant (C)-tubocurarine, the antiarrythmic ajmaline, the antibiotic sanguinarine, and the sedative scopolamine (Facchini, 2001).

Glucosinolates, present in cruciferous vegetables (e.g., broccoli and brussel sprouts), are a class of alkaloids with a core sulfated thiocyanate group conjugating to thioglucose, and a side chain (Clarke, 2010). They are toxic at high doses, but under subtoxic level, glucosinolates serve as activators for liver detoxification enzymes and show protective effects against carcinogenesis, mutagenesis, and other forms of toxicity (Dillard and German, 2000). Most of the beneficial effects possessed by glucosinolates are attributable to the hydrolysis product isothiocyanates produced by the plant enzyme myrosinase or intestinal microflora. One example of isothiocyanate is sulforaphane (R-1-isothiocyanato-4-methylsulfinyl butane, SF), a hydrolysis product of glucosinolate in broccoli. It has been reported that SF has the potential to reduce the risk of various types of cancers, diabetes, atherosclerosis, respiratory diseases, neurodegenerative disorders, ocular disorders, and cardiovascular diseases (Elbarbry and Elrody, 2011).

Indole-3-carbinol is another degradation product of glucosinolate that shows multiple anti-carcinogenic properties. These properties are delivered by changes in cell cycle progression, apoptosis, carcinogen bioactivation, and DNA repair (Weng *et al.*, 2008). Both SF and indole-3-carbinol are widely available as dietary supplements.

### 18.5.3 Carotenoids

Carotenoids are terpene-derived pigments found in chloroplasts and chromoplasts of plants and photosynthetic organisms. They play a vital role in mediating the electron transfer process or protecting organisms from oxidation-induced damage. In humans, carotenoids have been linked with antioxidant activities and a series of disease prevention effects (Edge *et al.*, 1997; Hughes, 2001). Therefore, carotenoids have received great commercial interest; as a result, various related products have appeared in the nutraceutical market.

Based on different chemical characteristics, carotenoids can be classified as xanthophylls and carotenes. Xanthophylls are carotenoids with oxygen present in the molecules, such as lutein, astaxanthin, and zeaxanthin. Carotenes are unoxygenated carotenoids which contain only carbon and hydrogen. Examples of carotenes include $\alpha$-carotene, $\beta$-carotene, and lycopene (Cuttriss *et al.*, 2011).

Carotenes are firstly known as vitamin A precursors and play an essential nutritional role. Recent studies have shown that $\beta$-carotene and lycopene also exhibit immune stimulatory and cancer preventive effects (Hughes, 2001). Xanthophyll typically cannot be converted to vitamin A in the human digestive tract; however, they can still offer protection for the biological system against free radicals.

Marketed carotenoids consist of $\beta$-carotene, astaxanthins, and more recently, lutein and lycopene (Walker *et al.*, 2006). The production of carotenoids can be from both natural and synthetic sources. Astaxanthin, for instance, can be produced from the microalgae *Haematococcus pluvialis*. However, nearly all commercial astaxanthin for aquaculture applications is produced synthetically due to the low cost of synthetic production as compared to natural extraction. Recent advances for the understanding and manipulation of carotenoid metabolism have shown promise for producing carotenoids from natural sources. Examples of natural carotenoid producers include the fungus *Blakeslea trispora*, microalga *Dunaliella salina*, and yeast *Phaffia rhodozyma* (Enes and Saraiva, 1996; Hejazi *et al.*, 2002; León *et al.*, 2003; Denery *et al.*, 2004). Utilization of substrates from agricultural byproducts, such as molasses and whey, also shows great potential in reducing the production cost from these microorganisms (Aksu and Eren, 2005).

## 18.6 Conclusion and future trends

Nutraceuticals, as health-enhancing and disease-preventing agents, have drawn increased attention, and thus, the market for various types of nutraceuticals has been expanding. As heart disease, cancer, osteoporosis, arthritis, and type II diabetes have become a major health concern worldwide; there has been greater awareness of the preventive effects of nutraceuticals, thus creating a bigger market for those products. The large demand for nutraceuticals is also associated with the trend of an aging population. As proportion of elderly people in the world population increases, an increased incidence of disease will occur, which can promote the use of various nutraceuticals to prevent and treat such disorders (Wildman, 2007).

By recognizing the growing nutraceutical market, food companies have dedicated hundreds of millions of dollars to discovering nutraceutical compounds and developing new products. With the advancement of biorefinery development in recent years, biotechnology companies are also jumping into this field, aiming to produce high-value nutraceuticals as part of product portfolios.

Conventional production of nutraceuticals is through extraction from natural sources, which is limited by the low content of target compounds, inefficient separation techniques, and high production cost. Current breakthroughs in biotechnology have enabled the production of high purity nutraceuticals in an economically viable way. For example, some bioactive compounds, such as DHA and EPA, naturally present in small amounts, can be produced by microorganisms through fermentation techniques. The use of agriculture byproducts or waste products shows a great potential to reduce the cost of bio-based nutraceutical production. Also, the establishment of advanced separation methods assists the enrichment and purification of bioactive compounds.

Producing bio-based nutraceuticals from renewable sources possesses tremendous potential in industrial applications. For example, Chi *et al.* (2007) used biodiesel-derived crude glycerol to replace glucose for microbial DHA production. This feedstock is less expensive and the DHA yield and productivity is similar to that from glucose culture, which indicates biodiesel-derived glycerol is a commercially competitive feedstock for DHA manufacture. Another example is ellagic acid production from pomegranate wastes. Ellagic acid is a natural phenolic antioxidant present in numerous fruits and vegetables. Recent studies have used pomegranate husks as support and nutrient sources for *Aspergillus niger* GH1 to produce ellagic acid (Aguilera-Carbo *et al.*, 2008; Robledo *et al.*, 2008). This process is economically attractive since 8 kg of ellagic acid can be produced from each ton of waste by solid state fermentation. Considering the commercial price of this compound and the low cost of feedstock, it is a quite profitable process from an industrial perspective. In addition to the aforementioned cases of producing bio-based nutraceuticals, there are still numerous examples which can illustrate the advantages of producing nutraceuticals within a biorefinery context. Therefore, bio-based nutraceuticals will no doubt promise value-added opportunities in the biorefinery process and new market opportunities for the food and pharmaceutical industry.

## 18.7 References

Abalde, J., Betancourt, L., Torres, E., Cid, A. and Barwell, C. (1998) 'Purification and characterization of phycocyanin from marine cyanobacterium *Synechococcus* sp', *Plant Sci*, 136, 109.

Aguilera-Carbo, A., Augur, C., Prado-Barragan, L. A., Favela-Torres, E. and Aguilar, C. N. (2008) 'Microbial production of ellagic acid and biodegradation of ellagitannins', *Appl Micro Biotech*, 78, 189–199.

Ahmad, M. F., Ashraf, S. A., Ahmad, F. A., Ansari, J. A. and Siddiquee, M. R. A. (2011) 'Nutraceutical market and its regulation', *Amer J Food Technol*, 6, 342–347.

Ahmad, A., Anjum, F. M., Zahoor, T., Nawaz H. and Dilshad, S. M. R. (2012) 'Beta glucan: a valuable functional ingredient in foods', *Crit Rev Food Sci Nutr*, 52, 201–212.

Akoh, C. C. (2005) *Handbook of Functional Lipids*, Boca Raton, FL, CRC Press.

Aksu, Z. and Eren, A. T. (2005) 'Carotenoids production by the yeast *Rhodotorula mucilaginosa*: use of agricultural wastes as a carbon source', *Proc Biochem*, 40, 2985–2991.

Amidon, T. E., Bujanovic, B., Liu, S. and Howard, J. R. (2011) 'Commercializing biorefinery technology: a case for the multi-product pathway to a viable biorefinery', *Forests*, 2, 929–947.

Antolovich, M., Prenzler, P., Robards, K. and Ryan, D. (2000) 'Sample preparation in the determination of phenolic compounds in fruits', *Analyst*, 125, 989–1009.

Athalye, S. K., Garcia, R. A. and Wen, Z. Y. (2009) 'Use of biodiesel-derived crude glycerol for producing eicosapentaenoic acid (EPA) by the fungus *Pythium irregulare*', *J Agri Food Chem*, 57, 2739–2744.

Bajpai, P., Bajpai, P. K. and Ward, O. P. (1991) 'Production of docosahexaenoic acid by *Thraustochytrium auerum*', *Appl Micro Biotech*, 35, 706–710.

Barclay, W. R., Meager, K. M. and Abril, J. R. (1994) 'Heterotrophic production of long chain omega-3 fatty acids utilizing algae and algae-like microorganisms', *J Appl Phycol*, 6, 123–129.

Berge, J. P., Debiton, E., Dumay, J., Durand, P. and Barthomeuf, C. (2002) '*In vitro* anti-inflammatory and anti-proliferative activity of sulfolipids from the red alga *Porphyridium cruentum*', *J Agri Food Chem*, 50, 6227–6232.

Bernal, J., Mendiola, J. A., Ibanez, E. and Cifuentes, A. (2011) 'Advanced analysis of nutraceuticals', *J Pharm Biomed Anal*, 55, 758–774.

Bertin, L., Ferri, F., Scoma, A., Marchetti, L. and Fava, F. (2011) 'Recovery of high added value natural polyphenols from actual olive mill wastewater through solid phase extraction', *Chem Eng J*, 171, 1287–1293.

Bhanja, T., Kumari, A. and Banerjee, R. (2009) 'Enrichment of phenolics and free radical scavenging property of wheat koji prepared with two filamentous fungi', *Biores Technol*, 100, 2861–2866.

Bird, K. T., Chiles, T. C., Longley, R. E., Kendrick, A. F. and Kinkema, M. D. (1993) 'Agglutinins from marine macroalgae of the southeastern United States', *J Appl Phycol*, 5, 213–218.

Boudet, A. M. (2007) 'Evolution and current status of research in phenolic compounds', *Phytochem*, 68, 2722–2735.

Boussiba, S. and Richmond, A. E. (1979) 'Isolation and characterization of phycocyanins from the blue-green alga *Spirulina platensis*', *Arch. Microbiol*, 120, 155–159.

Brown, C. E. (1981) 'Interactions among carnosine, anserine, ophidine and copper in biochemical adaptation', *J Theor Biol*, 88, 245–256.

Burkus, Z. and Temelli, F. (2000) 'Stabilization of emulsions and foams using barley beta-glucan', *Food Res Int*, 33, 27–33.

Carlos Espin, J., Teresa Garcia-Conesa, M. and Tomas-Barberan, F. A. (2007) 'Nutraceuticals: facts and fiction', *Phytochem*, 68, 2986–3008.

Chantaro, P., Devahastin, S. and Chiewchan, N. (2008) 'Production of antioxidant high dietary fiber powder from carrot peels', *LWT-Food Sci Technol*, 41, 1987–1994.

Chi, Z. Y., Pyle, D., Wen, Z. Y., Frear, C. and Chen, S. L. (2007) 'A laboratory study of producing docosahexaenoic acid from biodiesel-waste glycerol by microalgal fermentation', *Proc Biochem*, 42, 1537–1545.

Chichara, G., Maeda, Y., Hamuro, J., Sasaki, T. and Fukuoka, F. (1969) 'Inhibition of mouse sarcoma 180 by polysaccharides from *Lentinu edodes* (Berk.) Sing', *Nature*, 222, 687–688.

Christaki, E., Florou-Paneri, P. and Bonos, E. (2011) 'Microalgae: a novel ingredient in nutrition', *Int J Food Sci Nutri*, 62, 794–799.

Clarke, D. B. (2010) 'Glucosinolates, structures and analysis in food', *Anal Methods*, 2, 310–325.

Cocks, S., Wrigley, S. K., Chicarellirobinson, M. I. and Smith, R. M. (1995) 'High-performance liquid chromatography comparison of supercritical-fluid extraction and solvent-extraction of microbial fermentation products', *J Chromatogr A*, 697, 115–122.

Cohen, Z. and Ratledge, C. (2010) *Single Cell Oils: Microbial and Algal Oils*, Urbana, IL, AOCS Press.

Cohen, Z., Didi, S. and Heimer, Y. M. (1992) 'Overproduction of gamma-linolenic and eicosapentaenoic acids by algae', *Plant Physiol*, 98, 569–572.

Cuttriss, A. J., Cazzonelli, C. I., Wurtzel, E. T. and Pogson, B. J. (2011) 'Carotenoids', *Adv Bot Res*, 58, 1–36.

DeFelice, S. L. (1992) 'The nutraceutical initiative: a recommendation for US economic and regulatory reforms', *Genetic Engineering News*, 12, 13–15.

DeFelice, S. L. (1995) 'The nutraceutical revolution: its impact on food industry R&D', *Trends in Food Science & Technology*, 6, 59–61.

de Swaaf, M. E., Pronk, J. T. and Sijtsma, L. (2003) 'Fed-batch cultivation of the docosahexaenoic-acid-producing marine alga *Crypthecodinium cohnii* on ethanol', *Appl Microbiol Biotechnol*, 61, 40–43.

Denery, J. R., Dragull, K., Tang, C. S. and Li, Q. X. (2004) 'Pressurized fluid extraction of carotenoids from *Haematococcus pluvialis* and *Dunaliella salina* and kavalactones from *Piper methysticum*', *Anal Chim Acta*, 501, 175–181.

Dhingra, D, Michael, M., Rajput, H, and Patil, R. T. (2012) 'Dietary fiber in foods: a review', *J Food Sci Technol*, 49, 255–266.

Dillard, C. J. and German, J. B. (2000) 'Phytochemicals: nutraceuticals and human health', *J Sci Food Agri*, 80, 1744–1756.
Donova, M. V. (2007) 'Transformation of steroids by actinobacteria: a review', *Appl Biochem Microbiol*, 43, 1–14.
Edge, R., McGarvey, D. J. and Truscott, T. G. (1997) 'The carotenoids as antioxidants – a review', *J Photochem Photobiol B*, 41, 189–200.
El-Abbassi, A., Kiai, H. and Hafidi, A. (2012) 'Phenolic profile and antioxidant activities of olive mill wastewater', *Food Chem*, 132, 406–412.
Elbarbry, F. and Elrody, N. (2011) 'Potential health benefits of sulforaphane: a review of the experimental, clinical and epidemiological evidences and underlying mechanisms', *J Med Plant Res*, 5, 473–484.
Ena, A., Pintucci, C. and Carlozzi, P. (2012) 'The recovery of polyphenols from olive mill waste using two adsorbing vegetable matrices', *J Biotechnol*, 157, 573–577.
Enes, I. and Saraiva, P. (1996) 'Optimization of operating strategies in beta-carotene microalgae bioreactors', *Comput Chem Eng*, 20, S509–S514.
Facchini, P. J. (2001) 'Alkaloid biosynthesis in plants: biochemistry, cell biology, molecular regulation, and metabolic engineering applications', *Annu Rev Plant Physiol Plant Mol Biol*, 52, 29–66.
Ferain, T., Schanck, A. N. and Delcour, J. (1996) 'C-13 nuclear magnetic resonance analysis of glucose and citrate end products in an ldhL-ldhD double-knockout strain of *Lactobacillus plantarum*', *J Bacteriol*, 178, 7311–7315.
Figueroa-Gonzalez, I., Quijano, G., Ramirez, G. and Cruz-Guerrero, A. (2011) 'Probiotics and prebiotics–perspectives and challenges', *J Sci Food Agri*, 91, 1341–1348.
Fleurence, J. (1999) 'Seaweed proteins: biochemical, nutritional aspects and potential uses', *Trends Food Sci Technol*, 10, 25–28.
Fleurence, J. (2004) 'Seaweed proteins', in Yada, R. Y. (Ed.) *Proteins in Food Processing*. Cambridge, Woodhead Publishing Ltd, 197–213.
Francavilla, M., Trotta, P. and Luque, R. (2010) 'Phytosterols from *Dunaliella tertiolecta* and *Dunaliella salina*: a potentially novel industrial application', *Biores Technol*, 101, 4144–4150.
Franck, A. (2002) 'Technological functionality of inulin and oligofructose', *Br J Nutri*, 87, S287–S291.
Ganzle, M. G. (2011) 'Enzymatic synthesis of galacto-oligosaccharides and other lactose derivatives (hetero-oligosaccharides) from lactose', *Int Dairy J*, 22, 116–122.
Garcia-Castello, E., Cassano, A., Criscuoli, A., Conidi, C. and Drioli, E. (2010) 'Recovery and concentration of polyphenols from olive mill wastewaters by integrated membrane system', *Water Res*, 44, 3883–3892.
Gibson, G. R., Probert, H. M., Van Loo, J., Rastall, R. A. and Roberfroid, M. B. (2004) 'Dietary modulation of the human colonic microbiota: updating the concept of prebiotics', *Nutr Res Rev*, 17, 259–275.
Gullon, P., Gonzalez-Munoz, M. J. and Parajo, J. C. (2011) 'Manufacture and prebiotic potential of oligosaccharides derived from industrial solid wastes', *Biores Technol*, 102, 6112–6119.
Gundermann, K. J., Kuenker, A., Kuntz, E. and Drozdzik, M. (2011) 'Activity of essential phospholipids (EPL) from soybean in liver diseases', *Pharmacol Rep*, 63, 643–659.

Guo, X. F., Han, D. M., Xi, H. P., Rao, L., Liao, X. J., Hu, X. S. and Wu, J. H. (2012) 'Extraction of pectin from navel orange peel assisted by ultra-high pressure, microwave or traditional heating: a comparison', *Carbohydr Polym*, 88, 441–448.

Gustafson, K. R., Cardellina, J. H., Fuller, R. W., Weislow, O. S., Kiser, R. F., Snader, K. M., Patterson, G. M. L. and Boyd, M. R. (1989) 'AIDS-antiviral sulfolipids from cyanobacteria (blue-green algae)', *J Natl Cancer Inst*, 81, 1254–1258.

Hamann, M. T. and Scheuer, P. J. (1993) 'Kahalalide F: a bioactive depsipeptide from the sacoglossan mollusk *Elysia rufescens* and the green alga *Bryopsis sp*', *J Am Chem Soc*, 115, 5825–5826.

Harnedy, P. A. and Fitzgerald, R. J. (2011) 'Bioactive proteins, peptides, and amino acids from macroalgae', *J Phycol*, 47, 218–232.

Hejazi, M. A., de Lamarliere, C., Rocha, J. M. S., Vermue, M., Tramper, J. and Wijffels, R. H. (2002) 'Selective extraction of carotenoids from the microalga *Dunaliella salina* with retention of viability', *Biotechnol Bioeng*, 79, 29–36.

Hermann, T. (2003) 'Industrial production of amino acids by coryneform bacteria', *J Biotechnol*, 104, 155–172.

Herrera, A., Boussiba, S., Napoleone, V. and Hohlberg, A. (1989) 'Recovery of c-phycocyanin from the cyanobacterium *Spirulina maxima*'. *J Appl Phycol*, 1, 325–331.

Herrero, M., Cifuentes, A. and Ibanez, E. (2006) 'Sub- and supercritical fluid extraction of functional ingredients from different natural sources: plants, food-by-products, algae and microalgae: a review', *Food Chem*, 98, 136–148.

Hicks, K. B. and Moreau, R. A. (2001) 'Phytosterols and phytostanols: functional food cholesterol busters', *Food Technol Magazine*, 55, 63–67.

Hugenholtz, J. and Smid, E. J. (2002) 'Nutraceutical production with food-grade microorganisms', *Curr Opin Biotechnol*, 13, 497–507.

Hughes, D. A. (2001) 'Dietary carotenoids and human immune function', *Nutr*, 17, 823–827.

Ikeda, I. and Sugano, M. (1998) 'Inhibition of cholesterol absorption by plant sterols for mass intervention', *Curr Opin Lipidol*, 9, 527–531.

Ikeda, I., Matsuoka, R., Hamada, T., Mitsui, K., Imabayashi, S., Uchino, A., Sato, M., Kuwano, E., Itamura, T., Yamada, K., Tanaka, K. and Imaizumi, K. (2002) 'Cholesterol esterase accelerates intestinal cholesterol absorption', *Biochim Biophys Acta*, 1571, 34–44.

Imeson, A. (2010) *Food Stabilisers, Thickening and Gelling Agents*, Chichester, Wiley-Blackwell.

International Conference of Harmonisation (2012) International Conference of Harmonisation. Available from: http://www.ich.org/ (accessed 15 June 2012).

Jiao, G. L., Yu, G. L., Zhang, J. Z. and Ewart, H. S. (2011) 'Chemical structures and bioactivities of sulfated polysaccharides from marine algae', *Mar Drugs*, 9, 196–223.

Jiménez, C., Cossío, B. R., Labella, D. and Niell, F. X. (2003) 'The feasibility of industrial production of *Spirulina* (*Arthrospira*) in Southern Spain', *Aquaculture*, 217, 179–190.

Johnson, M. B. and Wen, Z. Y. (2009) 'Production of biodiesel fuel from the microalga *Schizochytrium limacinum* by direct transesterification of algal biomass', *Energy Fuels*, 23, 5179–5183.

Kendrick, A. and Ratledge, C. (1992) 'Lipids of selected molds grown for production of n-3 and n-6 polyunsaturated fatty acids', *Lipids*, 27, 15–20.
Knekt, P., Kumpulainen, J., Jarvinen, R., Rissanen, H., Heliovaara, M., Reunanen, A., Hakulinen, T. and Aromaa, A. (2002) 'Flavonoid intake and risk of chronic diseases', *Am J Clin Nutr*, 76, 560–568.
Komatsu, N., Okubo, S., Kikumoto, S., Kimura, K., Saito, G. and Sakai, S. (1969) 'Host-mediated antitumor action of schizophyllan, a glucan produced by *Schizophyllum commune*', *Gann*, 60, 137–144.
Korhonen, H. (2009) 'Milk-derived bioactive peptides: from science to applications', *J Funct Foods*, 1, 177–187.
Korhonen, H. and Pihlanto, A. (2003) 'Food-derived bioactive peptides – opportunities for designing future foods', *Curr Pharm Des*, 9, 1297–1308.
Kralovec, J. A., Zhang, S. C., Zhang, W. and Barrow, C. J. (2012) 'A review of the progress in enzymatic concentration and microencapsulation of omega-3 rich oil from fish and microbial sources', *Food Chem*, 131, 639–644.
Küllenberg, D., Taylor, L. A., Schneider, M. and Massing, U. (2012) 'Health effects of dietary phospholipids', *Lipids Health Dis*, 11, 3.
Laroze, L., Soto, C. and Zúñiga, M. E. (2010) 'Phenolic antioxidants extraction from raspberry wastes assisted by-enzymes', *Electron J Biotechnol*, 13, 6.
Lee, Y. K. (1997) 'Commercial production of microalgae in the Asia-Pacific rim', *J Appl Phycol*, 9, 403–411.
León, R., Martín, M., Vigara, J., Vilchez, C., and Vega, J. M. (2003) 'Microalgae mediated photoproduction of β-carotene in aqueous–organic two phase systems', *Biomol Eng*, 20, 177–182.
Liang, Y., Garcia, R. A., Piazza, G. J. and Wen, Z. Y. (2011) 'Nonfeed application of rendered animal proteins for microbial production of eicosapentaenoic acid by the fungus *Pythium irregulare*', *J Agric Food Chem*, 59, 11990–11996.
Loya, S., Reshef, V., Mizrachi, E., Silberstein, C., Rachamim, Y., Carmeli, S. and Hizi, A. (1998) 'The inhibition of the reverse transcriptase of HIV-1 by the natural sulfoglycolipids from cyanobacteria: contribution of different moieties to their high potency', *J Nat Prod*, 61, 891–895.
Luber-Narod, J., Smith, B., Grant, W., Jimeno, J. M., Lopez-Lazaro, L., Scotto, K., Shtil, A. *et al.* (2000) '*In vitro* safety toxicology of kahalalide F, a marine natural product with chemotherapeutic potential against selected solid tumors', *Clin Cancer Res*, 6, 4510S–4510S.
Mann, J. (1992) *Murder, Magic and Medicine*, Oxford, Oxford University Press.
Mateos, R., Espartero, J. L., Trujillo, M., Ríos, J. J., León-Camacho, M., Alcudia, F. and Cert, A. (2001) 'Determination of phenols, flavones, and lignans in virgin olive oils by solid-phase extraction and high-performance liquid chromatography with diode array ultraviolet detection', *J Agric Food Chem*, 49, 2185–2192.
Mateus, N., Silva, A. M. S., Vercauteren, J. and de Freitas, V. (2001) 'Occurrence of anthocyanin-derived pigments in red wines', *J Agric Food Chem*, 49, 4836–4840.
Maxwell, E. G., Belshaw, N. J., Waldron, K. W. and Morris, V. J. (2012) 'Pectin–an emerging new bioactive food polysaccharide', *Trends Food Sci Technol*, 24, 64–73.
Minkova, K. M., Tchernov, A. A., Tchorbadjieva, M. I., Fournadjieva, S. T., Antova, R. E. and Busheva, M. C. (2003) 'Purification of C-phycocyanin from *Spirulina* (*Arthrospira*) *fusiformis*', *J Biotechnol*, 102, 55–59.

Moreau, R. A., Whitaker, B. D. and Hicks, K. B. (2002) 'Phytosterols, phytostanols, and their conjugates in foods: structural diversity, quantitative analysis, and health-promoting uses', *Prog Lipid Res*, 41, 457–500.

Mori, T., O'Keefe, B. R., Sowder II, R. C., Bringans, S., Gardella, R., Berg, S., Cochran, P., Turpin, J. A., Buchheit Jr, R. W., McMahon, J. B. and Boyd, M. R. (2005) 'Isolation and characterization of griffithsin, a novel HIV-inactivating protein, from the red alga *Griffithsia* sp.', *J Biol Chem*, 280, 9345–9353.

Murata, M. and Nakazoe, J. (2001) 'Production and use of marine algae in Japan', *Jpn Agric Res Quart*, 35, 281–290.

Murthy, P. S. and Naidu, M. M. (2012) 'Recovery of phenolic antioxidants and functional compounds from coffee industry by-products', *Food Bioprocess Technol*, 5, 897–903.

Naczk, M. and Shahidi, F. (2004) 'Extraction and analysis of phenolics in food', *J Chromatogr A*, 1054, 95–111.

Nakashima, H., Kido, Y., Kobayashi, N., Motoki, Y., Neushul, M. and Yamamoto, N. (1987a) 'Purification and characterization of an avian myeloblastosis and human immunodeficiency virus reverse transcriptase inhibitor, sulfated polysaccharides extracted from sea algae', *Antimicrob Agents Chemother*, 31, 1524–1528.

Nakashima, H., Yoshida, O., Tochikura, T. S., Yoshida, T., Mimura, T., Kido, Y., Motoki, Y., Kaneko, Y., Uryu, T. and Yamamoto, N. (1987b) 'Sulfation of polysaccharides generates potent and selective inhibitors of human immunodeficiency virus infection and replication *in vitro*', *Jpn J Cancer Res*, 78, 1164–1168.

Neves, A. R., Ramos, A., Shearman, C., Gasson, M. J., Almeida, J. S. and Santos, H. (2000) 'Metabolic characterization of *Lactococcus lactis* deficient in lactate dehydrogenase using *in vivo* C-13-NMR', *Eur J Biochem*, 267, 3859–3868.

Nigam, P. and Pandey, A. (2009) *Biotechnology for Agro-Industrial Residues Utilisation*, Dordrecht, Springer.

Niu, J. F., Wang, G. C., Lin, X. Z. and Zhou, B. C. (2007) 'Large-scale recovery of C-phycocyanin from *Spirulina platensis* using expanded bed adsorption chromatography', *J Chromatogr B*, 850, 267–276.

Ohta, K., Mizushina, Y., Hirata, N., Takemura, M., Sugawara, F., Matsukage, A., Yoshida, S. and Sakaguchi, K. (1998) 'Sulfoquinovosyldiacylglycerol, KM043, a new potent inhibitor of eukaryotic DNA polymerases and HIV-reverse transcriptase type 1 from a marine red alga, *Gigartina tenella*', *Chem Pharm Bull*, 46, 684–686.

Office of Dietary Supplements (n.d.) Dietary Supplements. Office of Dietary Supplements. Available from: http://ods.od.nih.gov/factsheets/DietarySupplements-HealthProfessional/ (accessed 15 June 2012).

Office of Legislative Policy and Analysis (n.d.) Legislative Updates: 108th Congress. Office of Legislative Policy and Analysis. Available from: http://olpa.od.nih.gov/legislation/108/pendinglegislation/dietary.asp (accessed 15 June 2012).

Oreopoulou, V. and Russ, W. (2007) *Utilization of By-Products and Treatment of Waste in the Food Industry*, New York, Springer.

Oshida, K., Shimizu, T., Takase, M., Tamura, Y., Shimizu, T. and Yamashiro, Y. (2003) 'Effects of dietary sphingomyelin on central nervous system myelination in developing rats', *Pediatric Res*, 53, 589–593.

Owusu-Apenten, R. (2010) *Bioactive Peptides: Applications for Improving Nutrition and Health*, Boca Raton, FL, CRC Press.

Pang, X., Yao, W., Yang, X., Xie, C., Liu, D., Zhang, J. and Gao, X. (2007) 'Purification, characterization and biological activity on hepatocytes of a polysaccharide from *Flammulina velutipes* mycelium', *Carbohydr Polym*, 70, 291–297.

Pardo, B., Paz-Ares, L., Tabernero, J., Ciruelos, E., García, M., Salazar, R., López, A., Blanco, M., Nieto, A., Jimeno, J., Izquierdo, M. A. and Trigo, J. M. (2008) 'Phase I clinical and pharmacokinetic study of kahalalide F administered weekly as a 1-hour infusion to patients with advanced solid tumors', *Clin Caner Res*, 14, 1116–1123.

Park, A. R. and Oh, D. K. (2010) 'Galacto-oligosaccharide production using microbial beta-galactosidase: current state and perspectives', *Appl Microbiol Biotechnol*, 85, 1279–1286.

Patil, G., Chethana, S., Sridevi, A. S. and Raghavarao, K. S. M. S. (2006) 'Method to obtain C-phycocyanin of high purity', *J Chromatogr A*, 1127, 76–81.

Peiretti, P. G. and Meineri, G. (2011) 'Effects of diets with increasing levels of *Spirulina platensis* on the carcass characteristics, meat quality and fatty acid composition of growing rabbits', *Livestock Sci*, 140, 218–224.

Pfeuffer, M. and Schrezenmeir, J. (2000) 'Bioactive substances in milk with properties decreasing risk of cardiovascular diseases', *Br J Nutr*, 84, S155–S159.

Pinelo, M., Arnous, A. and Meyer, A. S. (2006) 'Upgrading of grape skins: significance of plant cell-wall structural components and extraction techniques for phenol release', *Trends Food Sci Technol*, 17, 579–590.

Provencio, M., Izquierdo, A., Vinolas, N., Paz-Ares, L., Feliu, J., Constenla, M., de las Heras, B., Erustes, M., Izquierdo, M. A. and Rosell, R. (2006) 'Phase II clinical trial of kahalalide F (KF) as a second line therapy in patients (pts) with advanced non-small cell lung cancer (NSCLC)', *Ann Oncol*, 17, ix 225.

Pulz, O. (2001) 'Photobioeractors: production systems for phototrophic microorganisms'. *Appl Microbiol Biotechnol*, 57, 287–293.

Ramassamy, C. (2006) 'Emerging role of polyphenolic compounds in the treatment of neurodegenerative diseases: a review of their intracellular targets', *Eur J Pharmacol*, 545, 51–64.

Ratledge, C., Kanagachandran, K., Anderson, A. J., Grantham, D. J. and Stephenson J. C. (2001) 'Production of docosahexaenoic acid by *Crypthecodinium cohnii* grown in a pH-auxostat culture with acetic acid as principal carbon source', *Lipids*, 36, 1241–1246.

Robledo, A., Aguilera-Carbó, A., Rodriguez, R., Martinez, J. L., Garza, Y. and Aguilar, C. N. (2008) 'Ellagic acid production by *Aspergillus niger* in solid state fermentation of pomegranate residues', *J Ind Microbiol Biotechnol*, 35, 507–513.

Salamia, M., Moosavi-Movahedi, A. A., Moosavi-Movahedi, F., Ehsani, M. R., Yousefi, R., Farhadi, M., Niasari-Naslaji, A., Saboury, A. A., Chobert, J. M. and Haertlé, T. (2011) 'Biological activity of camel milk casein following enzymatic digestion', *J Dairy Res*, 78, 471–478.

Salminen, S. and Isolauri, E. (2006) 'Intestinal colonization, microbiota, and probiotics', *J Pediatrics*, 149, S115–S120.

Sanda, S., Leustek, T., Theisen, M. J., Garavito, R. M. and Benning, C. (2001) 'Recombinant Arabidopsis SQD1 converts UDP-glucose and sulfite to the sulfolipid head group precursor UDP-sulfoquinovose *in vitro*', *J Biol Chem*, 276, 3941–3946.

Sato, M. D. F., Rigoni, D.C., Canteri, M. H. G., Petkowicz, C. L. D. O., Nogueira, A. and Wosiacki, G. (2011) 'Chemical and instrumental characterization of pectin from dried pomace of eleven apple cultivars', *Acta Sci Agron*, 33, 383–389.

Schieber, A., Stintzing, F. C. and Carle, R. (2001) 'By-products of plant food processing as a source of functional compounds – recent developments', *Trends Food Sci Technol*, 12, 401–413.

Schlimme, E. and Meisel, H. (1995) 'Bioactive peptides derived from milk proteins: structural, physiological and analytical aspects', *Nahrung Food*, 39, 1–20.

Shiu, C. T. and Lee, T. M. (2005) 'Ultraviolet-B-induced oxidative stress and responses of the ascorbate-glutathione cycle in a marine macroalga *Ulva fasciata*', *J Exp Bot*, 56, 2851–2865.

Singh, H. B., Singh, B. N., Singh, S. P. and Nautiyal, C. S. (2010) 'Solid-state cultivation of *Trichoderma harzianum* NBRI-1055 for modulating natural antioxidants in soybean seed matrix', *Biores Technol*, 101, 6444–6453.

Singh, S., Kate, B. N. and Banerjee, U. C. (2005) 'Bioactive compounds from cyanobacteria and microalgae: an overview', *Crit Rev Biotechnol*, 25, 73–95.

Smaali, I., Jazzar, S., Soussi, A., Muzard, M., Aubry, N. and Marzouki, M. N. (2012) 'Enzymatic synthesis of fructooligosaccharides from date by-products using an immobilized crude enzyme preparation of beta-D-fructofuranosidase from *Aspergillus awamori* NBRC 4033', *Biotechnol Bioproc Eng*, 17, 385–392.

Smit, A. J. (2004) 'Medicinal and pharmaceutical uses of seaweed natural products: a review'. *J Appl Phycol*, 16, 245–262.

Soni, B., Kalavadia, B., Trivedi, U. and Madamwar, D. (2006) 'Extraction, purification and characterization of phycocyanin from *Oscillatoria quadripunctulata* – isolated from the rocky shores of Bet-Dwarka, Gujarat, India', *Proc Biochem*, 41, 2017–2023.

Spolaore, P., Joannis-Cassan, C., Duran, E. and Isambert, A. (2006) 'Commercial applications of microalgae', *J Biosci Bioeng*, 101, 87–96.

Srivastava, R. and Kulshreshtha, D. K. (1989) 'Bioactive polysaccharides from plants', *Phytochem*, 28, 2877–2883.

Starzyńska–Janiszewska, A., Stodolak, B. and Jamróz, M. (2008) 'Antioxidant properties of extracts from fermented and cooked seeds of Polish cultivars of *Lathyrus sativus*', *Food Chem*, 109, 285–292.

Suárez, Y., González, L., Cuadrado, A., Berciano, M., Lafarga, M. and Muñoz, A. (2003) 'Kahalalide F, a new marine-derived compound, induces oncosis in human prostate and breast cancer cells', *Mol Cancer Ther*, 2, 863–872.

Szmitko, P. E. and Verma, S. (2005) 'Antiatherogenic potential of red wine: clinician update', *Am J Physiol-Heart Circ Physiol*, 288, H2023–H2030.

Teruya, T., Konishi, T., Uechi, S., Tamaki, H. and Tako, M. (2007) 'Anti-proliferative activity of oversulfated fucoidan from commercially cultured *Cladosiphon okamuranus* TOKIDA in U937 cells', *Int J Bioll Macromol*, 41, 221–226.

Trautwein, E. A., Duchateau, G. S. M. J. E., Lin, Y., Mel'nikov, S. M., Molhuizen, H. O. F. and Ntanios, F. Y. (2003) 'Proposed mechanisms of cholesterol-lowering action of plant sterols', *Eur J Lipid Sci Technol*, 105, 171–185.

US Food and Drug Administration (2009) Dietary Supplement Current Good Manufacturing Practices (CGMPs) and Interim Final Rule (IFR) Facts. Food and Drug Administration. Available from: http://www.fda.gov/Food/DietarySupplements/ GuidanceComplianceRegulatoryInformation/ RegulationsLaws/ucm110858.htm (accessed 15 June 2012).

Walker, T., Drapcho, C. and Chen, F. (2006) 'Bioprocessing technology for production of nutraceutical compounds'. In Shi J, *Functional Food Ingredients and Nutraceuticals*, New York, Taylor & Francis Group, 211–236.

Weng, J.-R., Tsai, C.-H., Kulp, S. K. and Chen, C.-S. (2008) 'Indole-3-carbinol as a chemopreventive and anti-cancer agent'. *Cancer Lett*, 262, 153–163.

Wiederschain, G. (2007) 'Polysaccharides: Structural diversity and functional versatility', *Biochem (Moscow)*, 72, 675.

Wijesinghe, W. A. J. P. and Jeon, Y.-J. (2012) 'Biological activities and potential industrial applications of fucose rich sulfated polysaccharides and fucoidans isolated from brown seaweeds: a review', *Carbohydr Polym*, 88, 13–20.

Wildman, R. C. (2007) *Handbook of Nutraceuticals and Functional Foods*. Boca Raton, FL, CRC Press.

Yagita, A. (1997) Interleukin 12 inducer and medical composition, US patent application 6238660.

Yang, W., Pei, F., Shi, Y., Zhao, L., Fang, Y. and Hu, Q. (2012) 'Purification, characterization and anti-proliferation activity of polysaccharides from *Flammulina velutipes*', *Carbohydr Polym*, 88, 474–480.

Zaloga, G. P. and Siddiqui, R. A. (2004) 'Biologically active dietary peptides', *Mini-Rev Med Chem*, 4, 815–821.

Zhang, Y. M. and Chen, F. (1999) 'A simple method for efficient separation and purification of c-phycocyanin and allophycocyanin from *Spirulina platensis*'. *Biotechnol Tech*, 13, 601–603.

# 19
Bio-based chemicals from biorefining: carbohydrate conversion and utilisation

K. WILSON, European Bioenergy Research Institute, Aston University, UK and A. F. LEE, University of Warwick, UK and Monash University, Australia

DOI: 10.1533/9780857097385.2.624

**Abstract**: The quest for sustainable sources of fuels and chemicals to meet the demands of a rapidly rising global population represents one of this century's grand challenges. Biomass offers the most readily implemented, and low cost, solution for transportation fuels, and the only non-petroleum route to organic molecules for the manufacture of bulk, fine and speciality chemicals and polymers. Chemical processing of such biomass-derived building blocks requires catalysts compatible with hydrophilic, bulky substrates to facilitate the selective deoxygenation of highly functional bio-molecules to their target products. This chapter addresses the challenges associated with carbohydrate utilisation as a sustainable feedstock, highlighting innovations in catalyst and process design that are needed to deliver high-value chemicals from biomass-derived building blocks.

**Key words**: heterogeneous catalysis, lignocellulose, platform chemicals, biofuels, porous materials.

## 19.1 Introduction

Mounting concerns over dwindling petroleum oil reserves in concert with growing governmental and public acceptance of the anthropogenic origin of rising $CO_2$ emissions and associated climate change, are driving academic and commercial routes to utilise renewable feedstocks as sustainable sources of fuel and chemicals. The quest for such sustainable resources to meet the demands of a rapidly rising global population represents one of this century's grand challenges.[1] Biomass offers the most readily implemented, and low cost, solution for transportation fuels,[2] and the only non-petroleum route to organic molecules for the manufacture of bulk, fine and speciality chemicals and polymers[3] required to meet future societal demands.[4,5]

In order to be considered truly sustainable, biomass feedstocks must be derived from sources which do not compete with agricultural land use for food production, or compromise the environment, e.g. via deforestation.[6]

Potential feedstocks include cellulosic or oil-based materials derived from plant or aquatic sources, with the so-called biorefinery concept offering the co-production of fuels, chemicals and energy,[7] analogous to today's petroleum refineries which deliver high volume/low value (e.g., fuels and commodity chemicals) and low volume/high value (e.g., fine/speciality chemicals) products, maximising biomass valorisation.[8] Unlike fossil fuels, which comprise predominantly unfunctionalised hydrocarbons, carbohydrates derived from sugars, starches or lignocellulose are highly functionalised, thus requiring new conversion technologies to yield useful chemicals.[9,10] This chapter addresses the challenges associated with carbohydrate utilisation as a sustainable feedstock,[11] highlighting recent developments in heterogeneous catalysis for the production of platform chemicals.

## 19.2 Sustainable carbohydrate sources

Feedstock selection for any bio-based process requires life-cycle and socio-economic assessments of attendant energy, resource and land use requirements to ensure sustainability. Polysaccharides sourced from starch $(C_6H_{10}O_5)_n$ derived from wheat, corn or tuberous plants, or sugar cane[12] and sorghum[13], are most easily processed to glucose via hydrolysis, but such food crops are not considered sustainable. In this respect, non-food sources of grass sugars (e.g., ryegrass)[14] and sugar beet pulp[15] are more attractive carbohydrate sources, with the latter composed of pectin (a hetero-polysaccharide comprising galactose, galacturonic acid, arabinose and xylose units) which is abundant in the cell walls of non-woody biomass (e.g., apple pomace and citrus waste).[16]

Lignocellulose derived from waste agricultural or forestry materials (e.g., logging or mill and manufacturing residues), or perennial herbaceous plants and short rotation woody crops (e.g., miscanthus, eucalyptus or willow), is considered a particularly viable option for sustainable fuels and chemicals production.[17,18] The use of waste biomass residues[19] from food processing, such as bagasse and rice husk and straw, offers a very attractive means to valorise waste materials that would otherwise be left to decompose, or in the case of rice straw, simply burned thereby releasing atmospheric $CO_2$.[20]

Lignocellulose is a biopolymer comprising cellulose (30–50% of total lignocellulosic dry mass) and hemicellulose (20–40% of total dry mass), themselves assembled from $C_6$ and $C_5$ sugars such as glucose, xylose and amylose, bound together by poly-phenolic lignin which makes up the remaining 15–25% of dry biomass and imparts rigidity to plants and trees (Fig. 19.1).

Cross-linking of the cellulose and hemicellulosic components with lignin via ester and ether linkages renders lignocellulose resistant towards

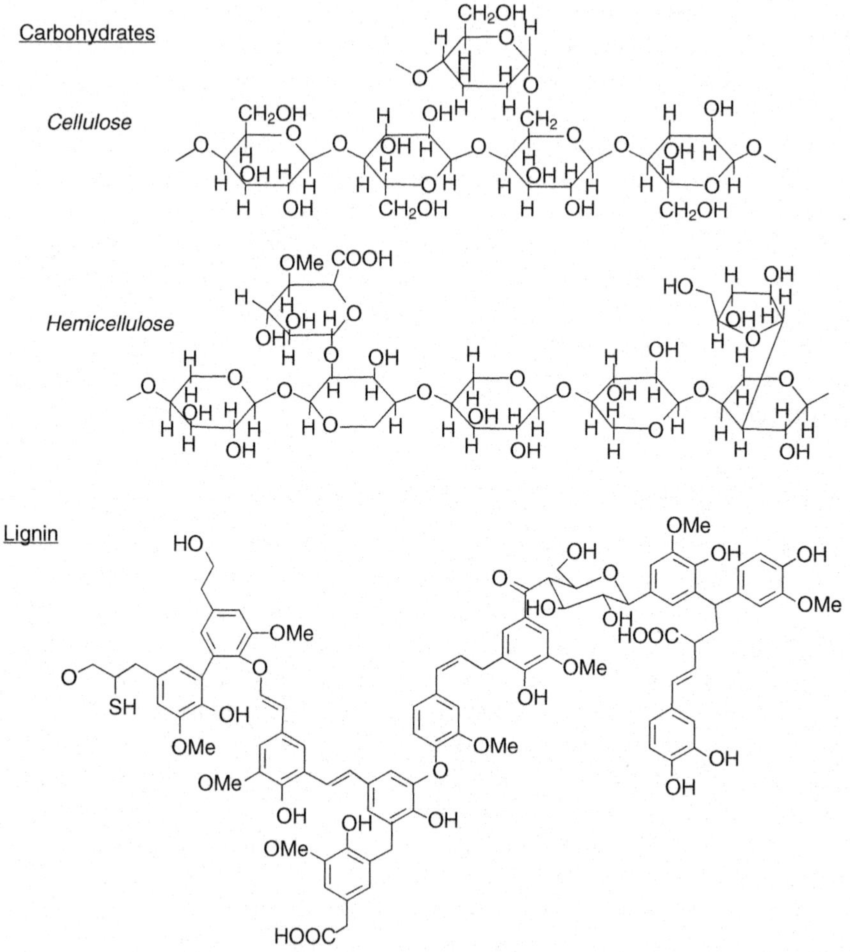

*19.1* Structure of lignin, hemicelluloses and cellulose contained within lignocellulose.

hydrolysis, hampering its chemical conversion.[19] Figure 19.2 illustrates popular approaches to lignocellulosic biomass utilisation for fuels and chemicals synthesis, encompassing sugar fermentation to ethanol, gasification to syngas ($CO/H_2$), and liquefaction or pyrolysis to bio-oils.

A limitation of lignocellulose is that it cannot be used directly in biochemical processes, thus while the conversion of sugars such as glucose to chemical feedstocks via fermentation appears an attractive prospect, it requires extensive pretreatment of the raw materials.[21] Lignin and $C_5$ sugars from hemicellulose cannot be used in fermentation processes (to produce, e.g., ethanol),[22] hence lignocellulosic biomass first requires

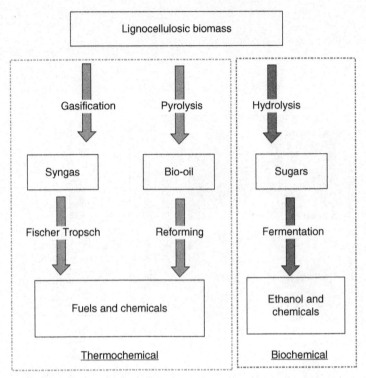

19.2 Biochemical and thermochemical routes for lignocellulose conversion to chemicals and fuels.

purification of the cellulosic component, typically through acid or base hydrolysis, steam explosion[23] or organosolv treatments[24] to separate the polysaccharide from lignin components.[25] Once separated, the cellulose fractions are typically hydrolysed to fermentable sugars for further processing into fuels and/or chemicals via enzymatic,[26] chemical (acid or base),[27,28] supercritical water[29] or more recently ionic liquid (IL)-based[30] treatments (Fig. 19.3).

Since hemicellulose $(C_5H_8O_5)_n$ is relatively amorphous, it is easier to chemically or thermally decompose than cellulose, but yields a mixture of $C_6$ and $C_5$ sugars. Current pretreatment steps are energy intensive, and generate significant quantities of waste during acid/base neutralisation steps. Hence there is great scope for improved technologies to enhance energy and atom economies. Furthermore, complete fractionation of lignocelluloses often sacrifices one or more of the components (e.g., lignin[31] or hemicellulose,[32]). For example, sulfuric acid treatments during the Kraft process result in a particularly intractable Kraft lignin material.[33] Likewise, decomposition of hemicelluloses during high temperature/pressure steam

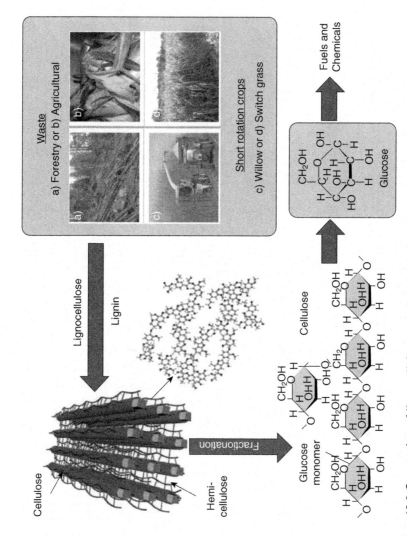

19.3 Conversion of lignocellulose to cellulose via fractionation.

explosion leads to an aqueous fraction rich in furfural and $C_1$-$C_2$ acids[34] which are problematic for subsequent fermentation since they can inhibit yeast growth.

## 19.3 Chemical hydrolysis of cellulose to sugars

### 19.3.1 Acid hydrolysis

Cellulose is composed of microcrystalline fibres, hydrogen bonded to each other by a charged water boundary layer formed from dipole–dipole interactions between aligned water molecules and the polar surface of the cellulose fibres.[35] To initiate acid hydrolysis, protons must penetrate this charged water boundary layer, and thereby catalyse hydrolysis of the β-glycosidic linkages to release glucose or cellobiose (glucose dimers) from the outermost layer of the cellulose crystallites. However, the initial hydrolysis products tend to remain in close proximity to the cellulose surface due to hydrogen bonding interactions, resulting in a viscous boundary layer that slows further hydrolysis of the underlying cellulose.[36] Efficient mixing is thus essential to displace reactively formed glucose from the surface, while the addition of $Li^+$, $Na^+$, $K^+$, $Ca^{2+}$ and $NH_4^+$ salts has also been reported to help disrupt the crystalline water matrix. While elevated reaction temperatures (via hot compressed water) can accelerate hydrolysis, they may also promote undesired side reactions if products are not continuously removed, with glucose readily undergoing subsequent reaction to by-products such as furfural and 5-HMF.[37] Without continuous removal of hydrolysis products from the 'boundary layer' at the surface of cellulose particles, conventional acid hydrolysis routes can only achieve glucose yields of around 70%, due to glucose degradation under these forcing reaction conditions.[36] Furthermore, sulfuric acid-initiated cellulose hydrolysis at acid concentrations ranging from 0.4 to 2wt%[38–40] poses additional problems due to the requirement for expensive corrosion-resistant reactors and co-production of vast quantities of gypsum waste formed during acid neutralisation via lime addition. There is thus an urgent need for low energy technologies for cellulose conversion to sugars and platform chemicals.

Catalytic aqueous-phase reforming (APR) of cellulose is one such approach which allows the direct conversion of carbohydrates into hydrogen and alkanes ($C_1$ to $C_{15}$) and could form a platform for fuels production within an integrated biorefinery.[41] APR typically involves the reaction of cellulose under acid conditions at ~200°C to yield $C_1$ to $C_6$ alkanes by aqueous-phase dehydration/hydrogenation (APD/H) of sugars, or $C_7$ to $C_{15}$ alkanes by combining aldol condensation with dehydration/hydrogenation (Fig. 19.4).

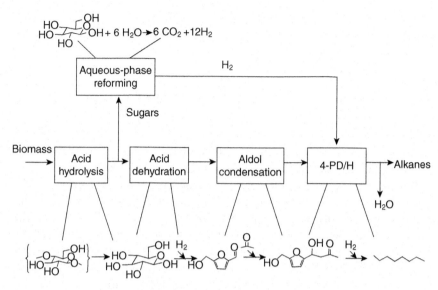

*19.4* Reaction scheme for aqueous phase reforming of cellulose to alkanes. Reproduced from Ref. 41 with permission from Elsevier.

### 19.3.2 Heterogeneous catalysts for cellulose conversion to platform chemicals

Cellulose depolymerisation to sugars through acid hydrolysis is the first step in a biorefinery,[42] hence there is much interest in developing heterogeneous catalysts to replace the conventional mineral acids (e.g., $H_2SO_4$) currently employed to achieve this.[43,44] Solid acids explored to date include sulfonic acid mesoporous silicas,[45] sulfonated porous carbons[46–48] and carbon-silicas,[49] as well as sulfonic acid polystyrene resins.[50] Sulfonated mesoporous carbons[48] are reported to produce remarkably high glucose yields (74.5%) following the 150°C hydrolysis of amorphous cellulose obtained from pretreating cellulose in a planetary ball mill at 500 rpm for 48 h. For cellulose hydrolysis by solid acids, efficient solid–solid interfacial contact is critical,[51] with the amount of water playing a key role in controlling the reaction kinetics over carbon-based catalysts. Glucose yields were optimal when the quantity of water was comparable to the weight of solid carbon catalyst; high water content is suggested to hydrate acid sites, decreasing the catalyst Brönsted acidity.

The use of magnetic sulfonated materials affords a novel and facile means to separate catalysts from aqueous media containing water-soluble sugars and solid cellulose. For example, magnetic silica nanoparticulate catalysts comprising a $CoFe_2O_4$ core with a sulfonic acid derivatised $SiO_2$ shell[52] are

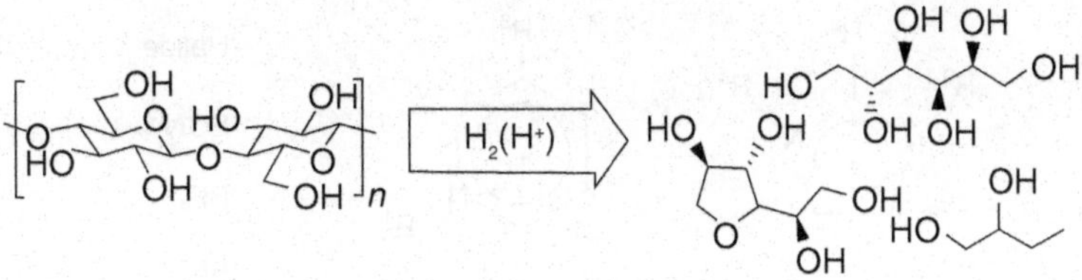

**19.5** Hydrolytic hydrogenation of cellulose to sugar polyols.

active for cellulose hydrolysis, while offering facile separation from the reaction media by a magnet, improving their reusability.

Cellulose is only soluble in water at temperatures in excess of >320°C. However, such temperatures, required to interrupt the hydrogen-bonding between the fibres,[53] result in low glucose yields due to subsequent reactions under these aggressive conditions. The hydrolytic hydrogenation of cellulose to sorbitol and sorbitan by mineral acids and supported Ru catalysts under $H_2$ pressures of 70 bar offer improved sugar yields by hydrogenating glucose as it is formed to sugar alcohols, which have higher chemical stability (Fig. 19.5).[54] Although promising, the use of mineral acids is undesirable, and the developments of processes employing only solid catalysts are preferred from safety, product recovery and catalyst recycle perspectives.[55] Ru/CMK-3[56] and Ru/ZSM-5[57] have been reported to catalyse cellulose hydrolysis via hydrolytic hydrogenation, with the active Ru species on CMK-3 proposed as a hydrated Ru oxide ($RuO_2.2H_2O$) postulated to act as a Lewis or Brönsted acid depending on the degree of hydration.

Supported metal catalysts such as $Pt/\gamma\text{-}Al_2O_3$, Pt/carbon and Ru/carbon are also reported to convert cellulose into sugar alcohols (sorbitol and mannitol) in the presence of hydrogen.[55] High selectivity to glucose was achieved by choosing appropriate reaction conditions, which include rapid heating (from 25°C to 230°C in 15 min) to hydrolyse cellulose followed by cooling to inhibit glucose degradation.

### 19.3.3 Cellulose conversion in ionic liquids

Ionic liquids (IL) have shown huge promise for applications in biomass fractionation, and attracted significant academic and industrial interest.[58] Low temperature ILs are low melting point salts, which form liquids comprising only cations and anions, having low vapour pressures and exceptional solvating properties for diverse compounds. The miscibility of ionic liquids with water or organic solvents can be controlled by varying the side chain lengths on the cation, and by careful anion choice. Some common cation and anion combinations used as ionic liquids are shown in

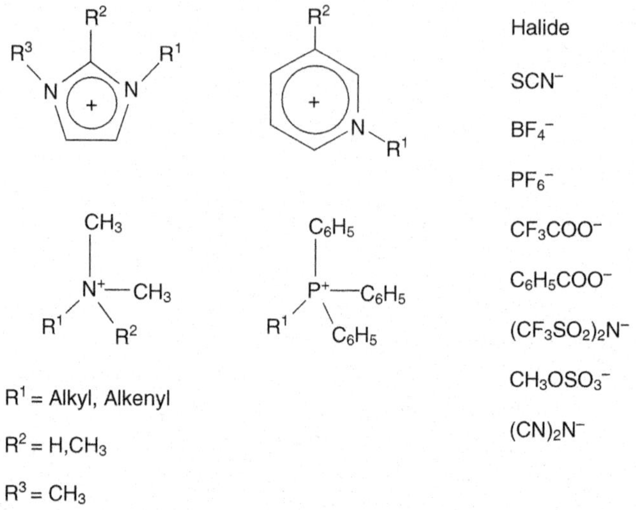

*19.6* Common cations and anions used in ionic liquids.

Fig. 19.6. Organic groups on ILs can also be functionalised to impart acid or base character. While ILs are often described as 'greener' than conventional solvents, they are not all benign. The toxicity of ILs is mainly ascribed to the alkyl chain, with the toxicity of imidazolium and pyridinium ILs increasing with cation chain length.[59] Furthermore, bmim ($BF_4$, $PF_6$, $NTf_2$ and $N(CN)_2$) ILs exhibit poor biodegradability (less than 5% after 28 days), hence large-scale use of such solvents in fuel production would require careful regulation.

As outlined above, the β-glycosidic linkages in cellulose are protected against hydrolysis by tight packing of the cellulose chains into microfibrils, hence cellulose hydrolysis requires severe conditions, such as the use of dilute sulfuric acid at high temperatures. However, cellulose[58] and wood[60,61] dissolve in alkylmethylimidazolium ILs leaving the cellulose chains accessible to chemical transformations.[62] ILs have been utilised as reaction media for the synthesis of cellulose derivatives such as carboxymethyl cellulose and cellulose acetate, while the regeneration of cellulose from ionic liquid solutions has been employed in the fabrication of films, gels and composite materials. Even though facile cellulose dissolution in IL solvents represents a significant breakthrough, the use of ILs for cellulose hydrolysis or degradation has only recently been explored.[63]

Early reports[58] demonstrated that room temperature ILs, such as 1-n-butyl-3-methylimidazolium ($C_4mim^+$) salts with $Cl^-$, $Br^-$, and $SCN^-$ anions, are capable of dissolving cellulose. High molecular weight pulp cellulose (5–10g -DP 1000) slowly dissolved in ~100g ionic liquid when heated to 100°C, yielding viscous solutions. Subsequent studies have attempted to improve upon the initial discovery of this solubility. Interactions between the carbohydrate and anion of ILs appear the dominant factor in controlling cellulose dissolution, with Cl⁻ reported to have stronger hydrogen-bonding basicity than Br⁻, SCN⁻, $PF_6^-$ and $BF_4^-$. However, the chemical structure of cations also affects carbohydrate dissolution, with cellulose solubility decreasing with increasing alkyl chain length in the imidazolium cation, e.g. use of short alkyl chains in the AMIM cation enhances solubility relative to BMIM. The use of oxygenated side chains can also enhance carbohydrate solubility via hydrogen bonding between the oxygen and the carbohydrate. High throughput screening[64] indicates that EMIM-Ac was the most efficient IL for dissolving cellulose, while AMIM-Cl (1-allyl-3methyl-imidazolium chloride) was the most effective for dissolving wood chips. A proposed mechanism by which ionic liquids break up the hydrogen bonding network in cellulose is shown in Fig. 19.7.[65] Separation of the resulting sugar from the IL was achieved using ion exclusion chromatography, enabling over 95% recovery of the ionic liquid and 94% recovery of glucose. A commercial process for industrial-scale cellulose conversion to glucose would, however, require alternative separation technologies.

In a recent study by Li et al.,[66] cellulose hydrolysis was initiated by adding catalytic amounts of $H_2SO_4$ to a cellulose–$C_4mim^+Cl^-$ solution, with a $H_2SO_4$:cellulose mass ratio of 0.92 producing total reducing sugars (TRS) and glucose in 59% and 36% yields, respectively, within 3 min. Lowering the acid:cellulose mass ratio to 0.46 produced higher yields after 42 min reaction, and for a mass ratio of only 0.11, TRS and glucose yields of 77% and 43%, respectively, were achieved after 9 h. These mild operating conditions and the use of a catalytic amount of $H_2SO_4$ offer exciting prospects for cellulose conversion.

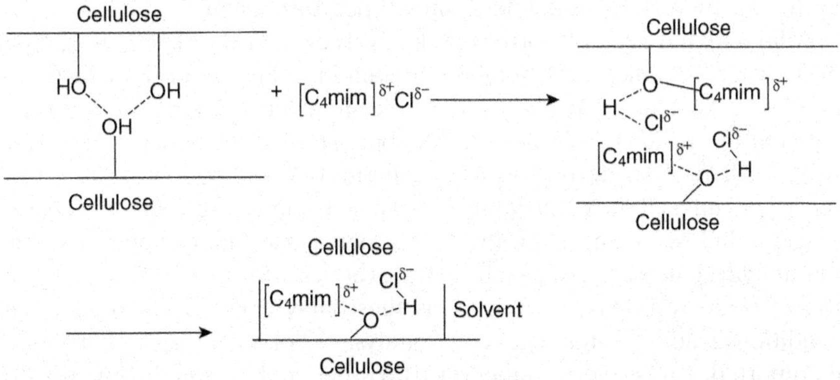

*19.7* Disruption of the hydrogen bonding network in cellulose by ILs.

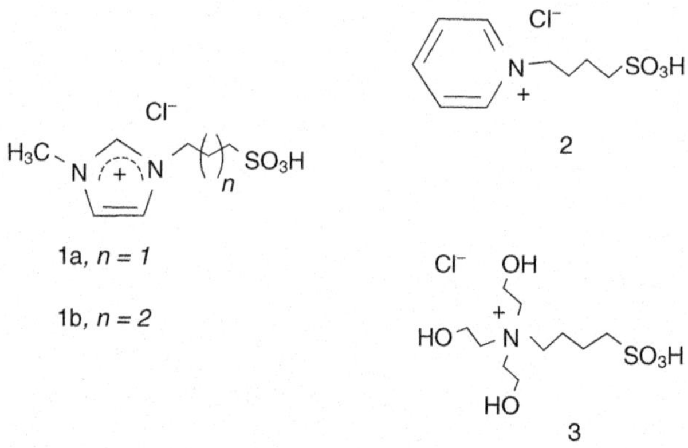

*19.8* Acidic ILs explored in cellulose hydrolysis.

Incorporation of an acidic function into the ILs has also been investigated through the synthesis of Brönsted acid ILs for the simultaneous dissolution and hydrolysis of cellulose.[67] Such Brönsted acidic ILs can act as both solvent and catalyst, eliminating waste conventionally produced during mineral acid neutralisation and separation steps. Furthermore, the high concentration of $-SO_3H$ active sites that can be introduced to ILs is expected to accelerate reactions and thus enable lower operating temperatures, facilitating cellulose dissolution and hydrolysis under moderate reaction temperatures and atmospheric pressure. Three types of Brönsted acidic ILs based on methylimidazolium (1a,b), pyridinium (2), and triethanolammonium (3) have been studied for their ability to dissolve and hydrolyse cellulose at mild reaction temperatures (Fig. 19.8). Cellulose dissolution in Brönsted acidic ionic liquids such as 1-(1-propylsulfonic)-3-methylimidazolium chloride and 1-(1-butylsulfonic)-3-methylimidazolium chloride is achievable up to 20 g/100 g IL upon gentle room temperature mixing.

Solid acids have been reported to selectively catalyse the depolymerisation of cellulose solubilised in 1-butyl-3-methylimidazolium chloride (BMIMCl) at 100 °C. Acid strength plays a key role in cellulose depolymerisation,[50] wherein the resin-based solid acid Amberlyst gave impressive results for cellulose hydrolysis, despite diffusional limitations and poor accessibility of the attendant acid sites. Reactions using solid catalysts preferentially cleave longer cellulose chains to produce oligomers consisting of approximately ten anhydroglucose units (AGU). However, care must be taken with the use of some resin-based catalysts, as these may degrade under reaction conditions leading to homogeneous catalysis.[68] For example, Amberlyst-15 is reported to be very stable in BMIMCl, but a simple change to $BMIM(CH_3COO)$, a potentially more desirable solvent because of its lower viscosity, resulted in rapid Amberlyst-15 degradation. The poor activity of traditional inorganic solid acids (e.g., Zeolite-Y and ZSM-5) has been attributed to their microporous nature and consequent poor accessibility to bulky cellulose fibres.

Cellulose has been directly converted into environmentally friendly alkyl glycoside surfactants in a one-pot transformation. Utilising BMIMCl in conjunction with an Amberlyst 15Dry (A15) catalyst, and coupling the rates of cellulose hydrolysis and glycosidation of the resultant monosaccharides with $C_4$-$C_8$ alcohols, an 82% mass yield of octyl-α-β-glucoside and octyl-α-β-xyloside was obtained.[69] The IL alone was unable to effect any cellulose conversion, possibly reflecting the low reaction temperature employed, since ILs are not known to hydrolyse cellulose below 90°C.[70]

Solid acid-promoted hydrolysis of cellulose in ionic liquids can be accelerated by microwave heating, affording similar yields of TRS and glucose to those obtained using liquid mineral acids. Protonated zeolites with a low Si/Al molar ratio and large surface area showed the highest catalytic activity, outperforming the acidic sulfated ion-exchange resin NKC-9.[71] Despite these initial successes, significant breakthroughs in catalyst and process design are essential in order to improve the overall efficiency of glucose production for subsequent bio refining.

## 19.4 Types and properties of carbohydrate-based chemicals

The US Department of Energy (DoE) has identified 12 platform chemicals obtainable through sugars via the chemical or biochemical transformation of lignocellulosic biomass (Fig. 19.9).[5] Akin to petrochemical refineries, the co-production of high value chemicals and fuels by thermo-[72] and/or biochemical[73] routes, along with heat and power in an integrated biorefinery, offers the most economically viable means to utilise biomass.[74] The idea of utilising bio-oils produced from biomass pyrolysis is an emerging concept which could potentially offer a feed stream that could be directly distilled[75] or co-refined alongside fossil fuel-derived oils in conventional petroleum refineries.[76,77] Hydrodeoxygenation (HDO) of crude bio-oils can yield hydrocarbons with similar properties to crude petroleum oil.

### 19.4.1 Biochemical conversion of carbohydrates

The use of enzymes in biochemical routes allows for selective conversion of sugars via fermentation, but faces several limitations that hinder economic biomass processing. Notably, conventional enzyme catalysed processes are unable to process pentoses, which must therefore be removed from biomass sugar feeds, requiring extra pretreatment steps in order to purify the glucose

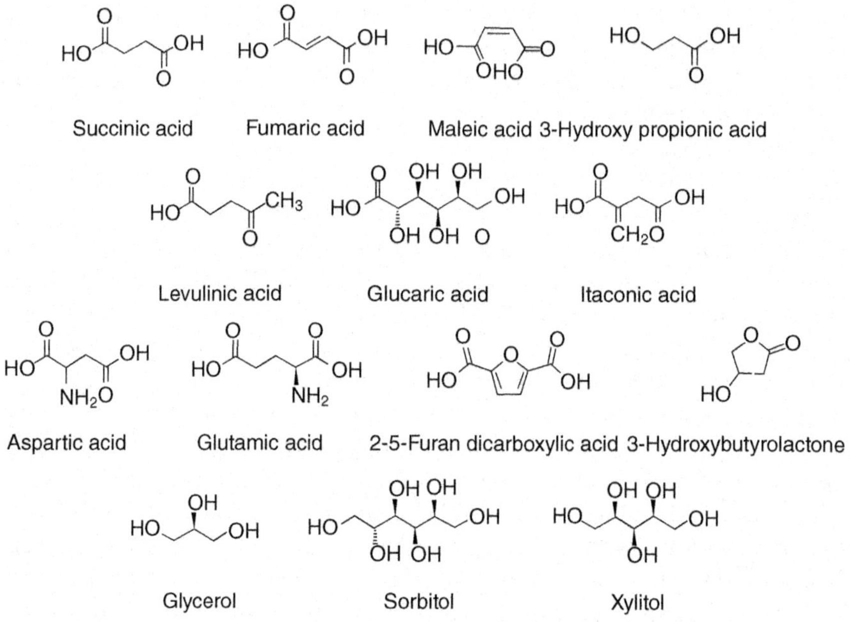

**636** Advances in Biorefineries

**19.9** Possible platform chemicals produced from biomass.[78]

feedstock. Furthermore, the platform molecules derived from fermentation are often present at low concentrations (typically <10%) in aqueous solutions, alongside other polar molecules. Purification of such fermentation broths is particularly difficult,[79,80] and energetically unfavourable, hence an ability to directly transform these aqueous solutions would be desirable.[81] Catalysts capable of driving organic chemistry in water to selectively transform platform molecules into useful chemical feedstocks,[5,9,74,82,83] and are resistant to impurities present in the fermentation broth,[84] therefore require development.

The utilisation of biomass-derived chemicals, whether from fermentation broths or sugars themselves, represents an area with extensive R&D potential for a renewable feedstock-based technology platform. Approaches to handling biomass-derived building blocks will be very different from petroleum processing, e.g. requiring reverse chemical transformations wherein highly functional bio-molecules are deoxygenated to their target product, instead of oxygenated as is usual when starting from crude oil.[85] Figure 19.10 illustrates a possible biomass synthesis of adipic acid (currently manufactured via the selective oxidation of cyclohexane) involving the selective reduction of glucose. New classes of catalyst are urgently required which are compatible with hydrophilic, bulky substrates to facilitate the move away from existing short-chain hydrocarbon supplies. Improvements

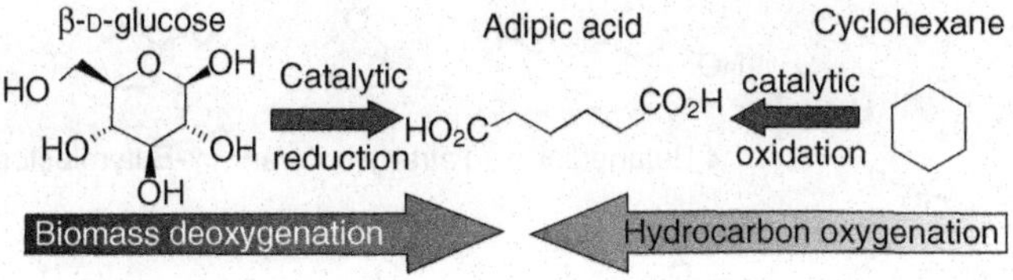

*19.10* Alternative routes to adipic acid from biomass or petroleum feedstocks.

and innovations in catalyst and processes design are needed in order to deliver high-value chemicals from biomass-derived building blocks.

These new catalysts will be hydrophilic, stable over a wide pH range and resistant to *in-situ* leaching.[86] Catalyst porosity will also be important to enable diffusion of bulky, viscous reactants to active sites; support materials with larger pores (compared to zeolites) are thus a promising starting point. Organic-inorganic hybrid catalysts may also prove interesting, as these allow catalyst hydrophobicity to be readily tuned, and in turn the adsorption strengths of polar molecules.[87] Mesoporous carbons[88] are well-suited for biomass conversion since they tend to be highly resistant to acidic and chelating media. Transforming the functional groups on platform chemicals will require catalysts capable of dehydration, hydrogenolysis and hydrogenation chemistry. Suitable catalysts will thus contain acid sites, or exhibit bifunctional character, possessing acid sites for dehydration and metal sites for (de)hydrogenation. Corma *et al.* have extensively reviewed proposed methods for transforming platform molecules into chemicals,[9] many of which employ conventional homogeneous reagents or commercial catalysts. There is thus enormous scope for rationally designing improved heterogeneously catalysed processes tailored towards biomass-derived feedstocks.

A number of reports have described pathways to important chemical intermediates from platform molecules.[9,82,83,89] Succinic acid is proposed as a valuable platform chemical, from which a range of chemical intermediates can be derived, as illustrated in Fig. 19.11 for acid catalysed esterification, or metal catalysed reductions. Carbon-based solid acid catalysts have proven effective for succinic acid esterification with ethanol.[90,91] Succinic acid may also afford new biopolymers based on polyesters, polyamides and polyesteramides.[92]

The purity of crude biorefinery feeds presents a major challenge for catalyst development, with the fermentation broth typically produced as a salty medium containing diammonium succinate rather than pure succinic acid. However, this could be exploited in the production of 2-pyrrolidone or N-methyl-pyrrolidone via hydrogenation of diammonium

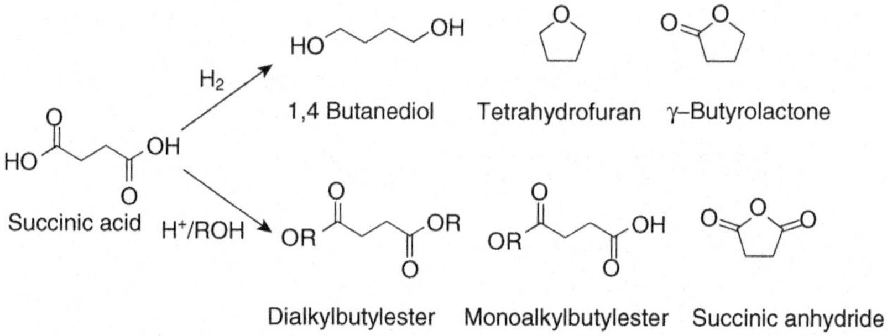

*19.11* Selected acid catalysed or hydrogenation products of succinic acid.

succinate[93] using catalysts such as Pd/ZrO$_2$/C and 2.5% Rh-2.5%Re on C, with reaction in the presence of methanol favouring *N*-methyl-pyrrolidone formation.

### 19.4.2 Thermochemical conversion of carbohydrates

Carbohydrate conversion by thermochemical routes such as fast pyrolysis, aqueous phase reforming or gasification offers alternative building blocks for chemicals or fuels production. Thermal decomposition of biomass in the absence of oxygen by pyrolysis yields a range of feedstocks depending on the reaction temperature and residence time,[77] with low temperatures and long residence times favouring charcoal formation, while moderate temperatures and short vapour residence time are optimal for liquids production. High temperatures and long residence times favour gasification,[94] with the resulting syngas available for well-established catalytic processes such as Fischer–Tropsch and methanol synthesis routes to convert CO/H$_2$ mixes to fuels and methanol. While complete gasification may be energetically costly, less initial biomass processing is required, so with efficient heat recovery modules, this approach is attractive for its compatibility with existing industrial processes. Low temperature (<200°C), thermochemical aqueous phase processing of sugars is of particular interest, as this offers a viable method to generate highly functional intermediates such as aldehydes, alcohols and esters via dehydration, hydrogenolysis and isomerisation (Fig. 19.12).[55,74,95]

Solid acids or bases are commonly employed in aqueous phase processing, although the application of bi-functional metal-doped catalysts is attracting interest for combined dehydration/hydrogenation of sugar intermediates (HDO). Some examples of key catalytic transformations of platform molecules will now be described.

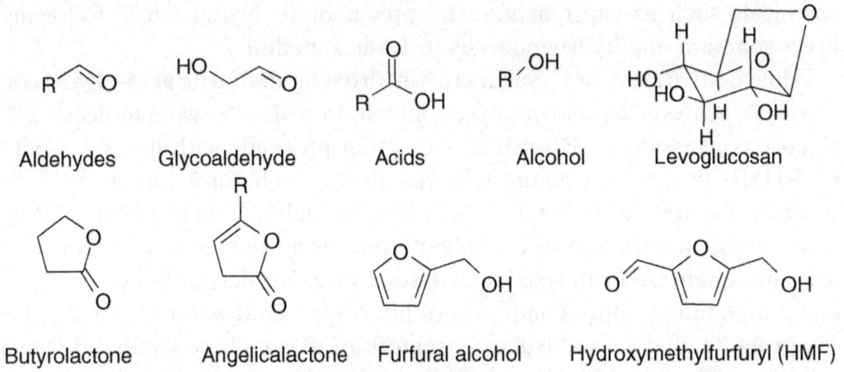

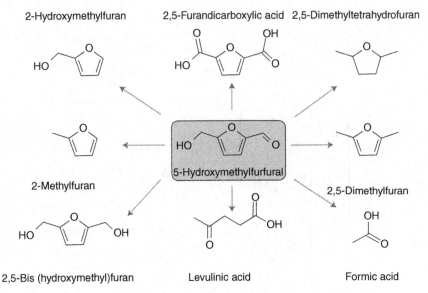

19.12 Major components formed via thermochemical processing of hemicellulose and cellulose.

19.13 Potential chemical feedstocks derived from 5-HMF.

### 19.4.3 Conversion of $C_6$ sugars to 5-HMF

5-Hydroxymethylfurfural (HMF) is a popular platform molecule with huge potential as an important bio-based commodity chemical[96,97] for the synthesis of various commercially useful acids, aldehydes, alcohols, and amines, as well as the promising fuel 2,5-dimethylfuran (DMF)[98] and renewable monomer furan dicarboxylic acid (FDCA)[99] as shown in Fig 19.13. HMF also has potential as a building block for the manufacture of commodity chemicals such as caprolactam, the precursor to Nylon 6,6,[100] following hydrogenation and hydrogenolysis to 1,6-hexanediol.

While dehydration of $C_6$ sugars to 5-hydroxymethylfurfural (5-HMF) can readily be achieved by acid-catalysed dehydration of three water molecules,[101] glucose conversion by Brönsted acids often proceeds with low selectivity to 5-HMF due to competing side reactions which form humins.[102,103] In contrast, fructose conversion proceeds with higher selectivity to 5-HMF, hence bifunctional catalysts capable of isomerising glucose to fructose prior to a subsequent acid-catalysed dehydration would be desirable. For example, under high temperature conditions of hot compressed water (200°C), $ZrO_2$ can promote glucose and fructose isomerisation via a base-catalysed route. Anatase $TiO_2$, which possesses both acid and base character, promotes glucose isomerisation and dehydration into HMF (Fig. 19.14).[104] For more selective chemistry, tailored solid catalysts capable of working at lower temperatures are sought.

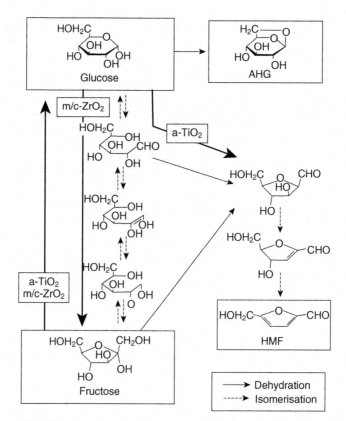

19.14 Proposed acid- and base-catalysed pathways in the dehydration of glucose to 5-HMF over $ZrO_2$ and $TiO_2$ in hot compressed water.

Initial reports of sucrose dehydration over solid acids, including acidic resins,[105] HY-zeolite,[106] aluminium-pillared montmorillonite, MCM-20 and MCM-41,[107] have sparked significant interest in 5-HMF production. Fructose dehydration to 5-HMF can be initiated by Brönsted acids in polar solvents and a range of aprotic polar solvents, such as dimethyl sulfoxide (DMSO), N,N-dimethylformamide (DMF), N,N-dimethylacetamide (DMA) and sulfolane. A variety of solid acids, including ion-exchange resins,[108] zeolites[109,110] metal oxides,[111] heteropoly acids,[112] niobic acid,[113] niobium phosphate,[114] sulfated zirconia,[115] and sulfonic acid-functionalised mesoporous silicas[116] have been explored for 5-HMF production from fructose. Detailed studies of $WO_x/ZrO_2$ catalysts reveal[111] the importance of catalyst amphoteric character in achieving high selectivity to HMF during fructose dehydration. While the overall fructose conversion correlates with acid site density, optimum selectivity towards 5-HMF of ~40% was obtained for catalysts with a base:acid site ratio of 0.3. Solid acid-catalysed conversion of fructose to 5-HMF has also been explored in ILs.[117–121]

As discussed above, the direct conversion of glucose to 5-HMF is difficult, requiring isomerisation to fructose, either via proton transfer or an intramolecular hydride shift, respectively base or Lewis acid catalysed. Base-catalysed glucose to fructose isomerisation occurs via deprotonation of the α-carbonyl carbon of glucose to form a series of enolate intermediates, and can be conducted by cation-exchanged zeolites or Mg-Al hydrotalcites.[122,123] High 5-HMF yields are obtained upon subsequent acid-catalysed dehydration of the resulting fructose.

The Lewis-Brönsted acid ratio also plays an important role in directing selectivity during glucose dehydration.[124] Davis and coworkers showed that tin-containing zeolites are highly active catalysts for glucose isomerisation in water, wherein Sn behaves as a Lewis acid catalysing an intramolecular hydride shift.[125,126] In order to minimise side reactions, biphasic systems, employing Lewis and Brönsted catalysts in conjunction with reactive extraction, have been proposed to improve 5-HMF yields (Fig. 19.15).[97,127] The combination of Lewis and Brönsted acidity is beneficial for 5-HMF formation from glucose, with Sn-β (a Lewis acid) and HCl (a homogeneous Brönsted acid) offering good 5-HMF yield from glucose in a water-tetrahydrofuransolvent mix.[128] A similar approach was adopted using homogeneous Lewis acid metal chlorides (e.g., $AlCl_3$) and HCl in a biphasic reactor comprising water and 2-sec butylphenol.[129] In this respect, amorphous niobium oxide hydrate ($Nb_2O_5 \cdot nH_2O$; niobic acid) is an interesting candidate for HMF production from glucose in water, as it possesses water-tolerant Lewis acid sites co-existing alongside Brönsted acid sites.[130]

5-HMF yields are also enhanced over mesoporous catalysts with pore diameters of 1–3 nm. Porosity was observed to have a significant effect on

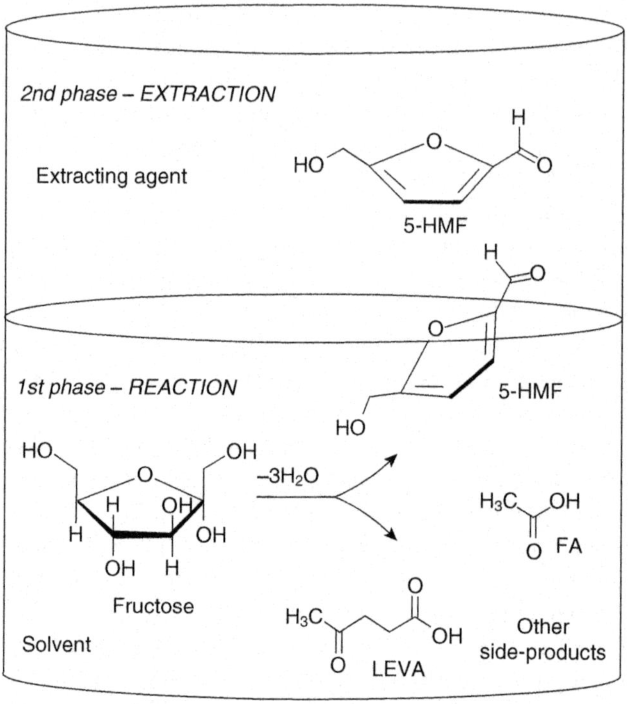

*19.15* Biphasic system for reactive extraction of 5-HMF during acid-catalysed fructose (or glucose) dehydration. Reprinted with permission from Reference 97. Copyright (2010) American Chemical Society.

5-HMF and levulinic acid yields obtained from sucrose dehydration over acid exchange resins.[105] Larger pores favour 5-HMF production, with slow 5-HMF diffusion out of smaller pores appearing to promote subsequent reaction and higher selectivity to levulinic acid.

### 19.4.4 Conversion of $C_5$ sugars to furfural

Xylose, derived from hemicelluloses, can be readily dehydrated to furfural using Brönsted acids at high temperature. A variety of Brönsted acid catalysts have been examined for furfural synthesis including heteropoly acids[131,132] and mesoporous sulfonic acid silicas,[133,134] H-type zeolites such as H-mordenite and H-Y faujasite,[109] ion-exchange resins,[135] niobium silicate,[136] $SO_4/ZrO_2$[137] and $SO_4/SnO_2$.[138]

Oxidative ring opening of furfural using $H_2O_2$ in the presence of a solid acid offers an opportunity to produce succinic acid along with maleic acid as a by-product (Fig. 19.16).[139] The yield of succinic acid was optimal using Amberlyst-15 when compared to other resin-based solid acids such as

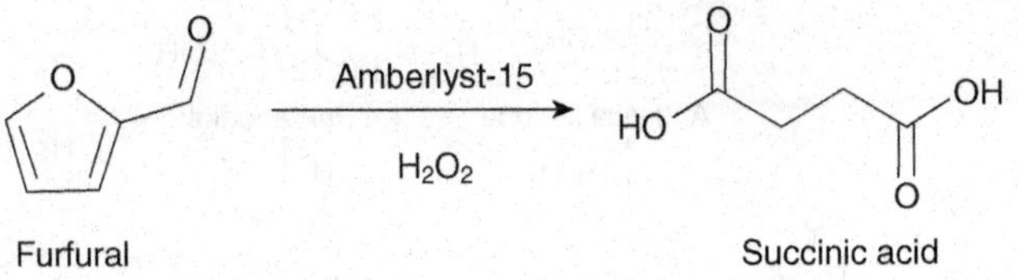

## Carbohydrate conversion and utilisation

**19.16** Oxidative ring opening of furfural to succinic acid.

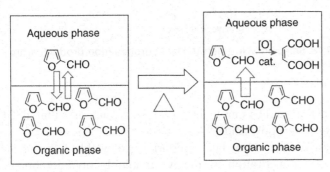

**19.17** Biphasic system for the reactive ring opening of furfural to maleic acid. Reprinted with permission from Ref. 140. Copyright (2011) American Chemical Society.

Nafion NR50 and Nafion SAC13, or other inorganic solid acid catalysts such as $Nb_2O_5$, H-ZSM-5 and $ZrO_2$.

The catalytic aerobic oxidation of furfural to maleic acid has also been explored using phosphomolybdic acid in biphasic aqueous/organic systems (Fig. 19.17), in which oxidation occurs in the aqueous phase, while the organic phase serves as the reservoir for furfural. Using this approach, maleic acid was obtained with 69% selectivity at a furfural conversion of 50%. Product separation was aided by the fact that furfural and maleic acid predominantly phase separate.[140]

### 19.4.5 Levulinic acid production

Levulinic acid is another valuable precursor to a range of chemical intermediates (Fig. 19.18), which can be used in chiral reagents, inks, coatings and batteries, and can be generated by a combination of acid-catalysed dehydration/esterification or metal-catalysed reduction.[141] While a number of studies have investigated reductions, there is surprisingly little work concerning the esterification of platform molecules using solid acid catalysts. To date, levulinic acid esterification is mostly performed using $H_2SO_4$,[142]

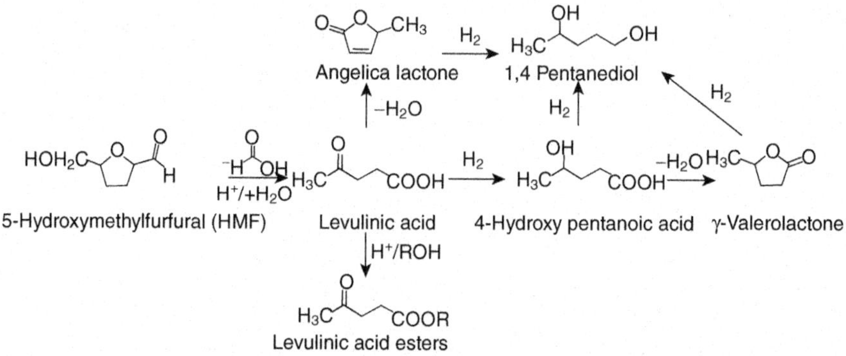

*19.18* Selected acid-catalysed or hydrogenation products from levulinic acid.

with only a few studies employing solid acids such as sulfated $TiO_2$ and $SnO_2$,[143,144] or heteropoly acids.[145,146]

Amberlyst-70 has also been reported to yield 21% of levulinic acid following dehydration of the water-soluble organics obtained via hydrothermal cellulose decomposition at 160°C.[147] When converting saccharides to levulinic acid, resin pore size was also demonstrated to have a significant effect on product selectivity.[105] There is clearly scope for the development of new catalytic systems for the conversion of biorefinery feedstocks to chemicals. However, efforts must focus on water-tolerant catalysts for the direct reaction of aqueous feeds, and the development of tailored porous-solids capable of operating under continuous flow.

### 19.4.6 Lactic acid synthesis

Sugar fermentation to lactic acid has received much attention in the context of polylactic acid (PLA) synthesis, considered the 'gold standard' for renewable polymers and applications including solvents and coatings.[148] Fermentation routes to LA proceed through homolactic or heterolactic fermentation of glucose, via glycolysis to pyruvate,[149] typically yielding >90% lactic acid as a $100\,g.l^{-1}$ aqueous solution of Ca-lactate. Subsequent $H_2SO_4$ recovery of lactic acid generates 1 kg $CaSO_4$ waste per kg lactic acid produced. Homolactic routes are less efficient, with ~44% lactate yields reported[150] at concentrations up to $55-60\,g.l^{-1}$.[151] Extraction costs account for 50% of such process economics,[152] due to the high energy consumption for separation and purification. Process improvements have been proposed employing ammonia to isolate an ammonium $L$-lactate product.[153] However, for large-scale processing, a catalytic route would be a desirable alternative.

Lactic acid synthesis has also been proposed via the retro-aldol condensation of fructose or glucose to form glyceraldehyde (GLA) and dihydroxyacetone (DHA).[154] These require Lewis acid or solid base catalysts to achieve high selectivity, with Sn-β,[155,156] Sn-SiO$_2$,[157] Sn-SiO$_2$-carbon composite,[158] and Brönsted base hydrotalcite catalysts,[159] reported as potential candidates for lactic acid or lactate synthesis from sugars. Conversion of GLA and DHA is proposed to yield pyruvaldehyde, which is subsequently hydrated to lactic acid via an internal Cannizzaro reaction, Meerwein–Ponndorf–Verley reduction or Oppenauer oxidation (Fig. 19.19).

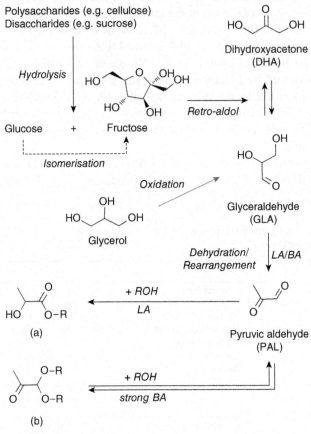

19.19 Proposed reaction scheme for converting monosaccharides such as trioses and aldo- and ketohexoses like fructose into lactic acid in aqueous medium (R = H) or into alkyl lactates (a) in alcoholic solvents (R = alkyl). The side-reaction leading to the formation of pyruvic aldehyde dialkylacetal (b) is undesired; LA = Lewis acid and BA = Brönsted acid. Reprinted with permission from Ref. 158. Copyright (2012) American Chemical Society.

## 19.5 Routes to market for bio-based feedstocks

### 19.5.1 Challenges and opportunities for introduction of bio-based products

The most significant driver for the production of renewable chemicals is the price of crude oil. When oil prices exceed US$100 per barrel, bio-based feedstocks may compete with petroleum sources.[160] However, there are other motivators for renewable feedstocks such as green marketing initiatives, individual preferences for more sustainable products, and critically their potential to mitigate pollution and $CO_2$ emissions. There are a number of potential markets for bio-derived chemicals, most notably as identical replacements for current commodity chemicals, so called 'drop-in' chemicals (e.g., bio-ethene, nylon or PET) or as completely new types of platform chemical (e.g., PLA, or PEF from FDCA).

Current and potential commercial applications of bio-based chemicals have been reviewed by Erickson *et al.*,[161] and some near-term commercial and future bio-products are summarised in Table 19.1. Succinic acid and 1,4-butanediol are identified as near-term, bio-based chemicals, while longer term targets include levulinic acid and adipic acid, the latter an important precursor to nylon.

Using polyamide synthesis as a case study, a number of building blocks have been developed around natural products using biotechnology.

*Table 19.1* (a) Near commercial and (b) future commercial bio-based chemicals

| (a) | | |
|---|---|---|
| Chemical | Companies | Applications |
| Succinic acid | Myriant BioAmber, DSM | Flavourings, dyes, perfumes, lacquers |
| 1,4-Butanediol | Genomatica | Spandex, |
| Isoprene | Genencor Amyris | Rubber, adhesives |
| Isobutanol | Gevo | Solvents, paint, biofuels |

| (b) | | |
|---|---|---|
| Chemical | Companies | Applications |
| Levulinic acid | Dupont Tate & Lyle | Plasticisers, solvents, |
| Adipic acid | Verdezyne | Fibres, plastics |
| Itaconic acid | Itaconix | Pigments, stabilisers |

*Source*: Adapted from Ref. 161.

For example, the unsaturated fatty acids sebacic or undecylenic acid derived from castor oil are employed in nylon 6,10 and 10,10 manufacture for use in automobile components (e.g., by Rhodia and Dupont),[162] while Cognis reportedly employ an enzymatic oxidative scission of oleic acids to produce muconic acid (a key precursor to adipic acid).[163] While these routes produce monomers for new materials, 'drop in' chemicals as direct replacements for those currently obtained via crude oil feedstocks are also essential for a sustainable chemicals industry. Promising biochemical routes to adipic acid via fermentation routes to muconic acid, were reported in the early 1990s. However, such approaches remain uneconomic, operating at only a 22% carbon balance, and producing dilute ~3.68 wt% aqueous solutions of muconic acid which impose excessive separation and purification costs.[164] Despite these hurdles, US biotech companies (e.g., Verdezyne) are attempting to scale-up new biocatalytic production routes for commercialisation. These efforts face stiff competition from a recent, cheaper petrochemical route to nylon 6 via butadiene carbonylation, developed by DSM, DuPont and Shell.[165] Selective catalytic routes to bio-nylon from a dipic acid or caprolactam (that avoid costly separation processes) would thus be particularly attractive.[100]

### 19.5.2 Challenges for process design

Conversion of bio-based feedstocks presents new challenges to the catalytic scientist, as the attendant reaction conditions are very different from those typical of petroleum processing, which occurs mainly through vapour phase processes >400°C. Biomass processing will be characterised by liquid phase, lower temperature[41] pathways such as hydrolysis, dehydration, isomerisation, oxidation, aldol, condensation, and hydrogenation.[166] The design of catalysts for such biomass transformations requires careful tailoring of pore structure to minimise mass transport limitations, hydrothermal stability under aqueous operation, and tunable hydrophobicity to aid product/reactant adsorption.[167] Catalyst development should thus focus on the use of tailored porous solids as high area supports to enhance reactant accessibility to active acid/basic groups. The preparation of such templated porous solids has been extensively reviewed[168–171] but generally involves the use of micellar templates to direct the growth of metal oxide frameworks as shown in Fig. 19.20. Subsequent calcination to burn out the organic template, or solvent extraction, yields materials with well-defined meso-structured pores of 2–10 nm and surface areas up to $1000\,m^2.g^{-1}$.

Macropores can also be introduced if an additional physical template, such as polystyrene microspheres, is added during templating of the mesopore network.[172] Hierarchical macroporous-mesoporous sulfonic acid SBA-15 silicas and aluminas have been synthesised via such dual-templating

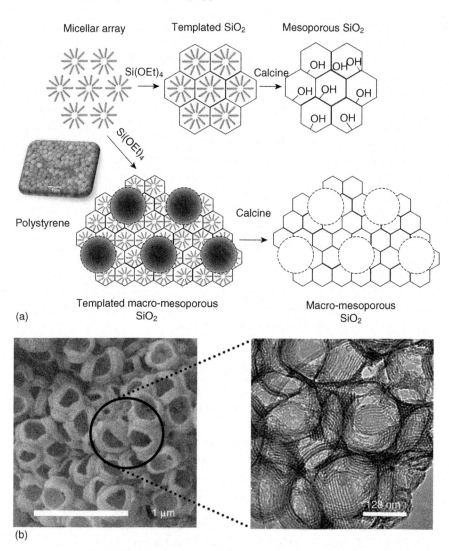

*19.20* (a) Liquid crystal templating route to form mesoporous silica and combined physical templating method using polystyrene microspheres to introduce a macropore network; (b) SEM of macroporous-mesoporous SBA-15 showing macropore network and TEM showing interconnecting mesopores.

routes, employing liquid crystalline surfactants and polystyrene beads.[173,174] These materials offer high surface areas and well-defined, interconnected macro- and mesopore networks, with respective narrow size distributions tunable over the range 100–300 nm and 3–5 nm. Such bimodal solid acid architectures offer significantly enhanced activity over mesoporous analogues. A second challenge centres around the need to better understand the selective deoxygenation and hydrogenation of polyols under acid- or base-catalysed conditions. Bifunctional catalysts need to be capable of low temperature dehydration and hydrogenation, and efficient use of $H_2$ during *in situ* hydrogenation and/or hydrogenolysis. Surface polarity is another important parameter to control and thereby permit catalyst operation in biphasic systems. As discussed in relation to Fig. 19.15, partitioning of products between aqueous and organic phase solvents via reactive extraction is a promising means to minimise side reactions.[175,176]

Hydrophobic catalysts operating in biphasic systems have been explored for 'bio-oil' obtained via biomass pyrolysis; bio-oil is a complex liquid that is only partially soluble in water or hydrocarbon solvents. In order to overcome problems of working with such mixtures, materials with tunable hydrophobicity, or even the use of micellar catalyst systems should be considered. This has recently been exploited by the group of Resasco,[176] in which a phase transfer system based upon Pd nanoparticles immobilised on carbon nanotube/silica composite support was employed. Such materials are able to catalyse the transformation of both hydrophilic and hydrophobic substrates to fuels at the oil:water interface, without the need for multiple separation steps or addition of surfactants.[177]

The development of new reactor technology will also be critical for efficient biomass processing.[178] Conventional stirred batch reactors are not suited to commodity chemical production scales, hence there is a need to devise slurry reactors that operate continuously. Intensive processing[179] could offer exciting means to improve overall process efficiency. For example, process engineering solutions available for continuous reactions include the use of fixed-bed[180] or microchannel-flow reactors,[181] pervaporation methods,[182,183] or reactive distillation.[184–186] Continuous reactors must be carefully designed to utilise the full potential of the associated heterogeneous catalyst; plug flow is a desirable reactor characteristic, as it allows tighter control of product composition and thereby reduces downstream separation processes and capital and running costs. However, conventional plug flow reactors are unsuitable for slow reactions as they necessitate very high length:diameter ratios to achieve the required mixing. Such designs are problematic due to their large footprints, pumping duties and control difficulties.

The oscillatory baffled reactor (OBR) circumvents these problems, offering good mixing and plug flow by oscillating the reaction fluid through orifice plate baffles.[187] In an OBR, mixing is decoupled from the net flow: this allows reactor designs that are very different from conventional plug flow reactors. OBR mixing is also scalable, meaning that longer reaction times can be achieved on industrial scales. Previous work[188,189] has shown that vortical mixing in the OBR is an effective, controllable method of uniformly suspending solid (e.g., catalyst) particles. In this respect, the first such demonstration of a solid acid catalyst incorporated within an OBR for the esterification of organic acids to short chain and fatty methyl esters, pertinent to fine chemicals synthesis and biofuels production, was recently reported.[190]

The development of more efficient separation methods will also be crucial[191] with advances in distillation, extraction, adsorption with molecular sieves, filtration, crystallisation and osmosis holding great promise.

## 19.6 Conclusion and future trends

This chapter has highlighted the significant progress made in recent years towards the conversion of renewable feedstocks into chemicals and fuels. Heterogeneous catalysis and process engineering hold the key to realising the potential of lignocellulosic biomass for the production of such renewable chemicals. Ultimately, catalytic chemists and engineers need to emulate the successes of heterogeneous catalysis in petroleum refining. Advances in chemistry, nanotechnology and spectroscopy will aid catalyst design, but overall process development requires an improved understanding of biomass properties and its impact on catalyst deactivation in order to accelerate biomass-to-chemicals and fuels production. Commercial heterogeneously catalysed processes will require a better understanding of individual reactant interactions with the active catalyst phase, particularly when dealing with bulky polar molecules such as those found in biorefinery feeds.

Most crucially however, widespread uptake and the development of next-generation biofuels and chemical feedstocks requires progressive government policies and incentive schemes to place biomass-derived chemicals on a comparable footing with cheaper fossil fuel-derived resources.[4,192] Biomass pretreatment to obtain sugars is one of the most wasteful steps in biorefineries, and new approaches are required to improve the processing of lignocellulose such that the initial acid hydrolysis/extraction step to form lignocellulose can be performed more efficiently. Cellulose stability itself presents a major hurdle, with environmentally friendly and energy efficient means to break up this biopolymer an ongoing challenge. In contrast to petroleum-derived oil, conventional heterogeneous catalysts cannot be employed alone in solid–solid mixtures, although some recent reports propose ball milling as an effective means to induce 'mechanocatalysis' between cellulose and clay-based catalysts with layered structures.[193] Alternative approaches are building upon the exciting discovery that ionic liquids can dissolve cellulose, and when coupled with acidic reagents can also generate selected platform chemicals.[61,194] The latter approach has been coupled with solid catalysts[50,195] to combine the ease of separation of a solid catalyst, with the dissolution strength of ILs, offering an exciting prospect for the cleaner conversion of cellulose to chemicals.

Development of new catalysts and overall process optimisation requires collaboration between catalytic chemists, chemical engineers and experts in molecular simulation to take advantage of innovative reactor designs; the future of renewable feedstock utilisation requires a concerted effort from chemists and engineers to develop catalysts and reactors in tandem. Current political concerns over the 'food versus fuel' debate also require urgent development of non-edible oil feedstocks, as well as necessary technical advances in order to ensure that catalytic routes to convert biomass to chemicals and fuels become viable processes in the renewables sector during the twenty-first century.

## 19.7 Sources of further information and advice

A number of global organisations exist performing research or acting in a consulting role concerning biomass utilisation. Useful websites include:

- Bioproducts, Sciences, and Engineering Laboratory (BSEL) at Pacific Northwest National Laboratory: http://www.pnnl.gov/biobased/bsel.stm
- NREL National Renewables Energy Laboratory: http://www.nrel.gov/
- National Non-Food Crops Centre: http://www.nnfcc.co.uk/NNFCC
- European Biomass Association: http://www.aebiom.org/

## 19.8 References

1. B. Walter, J. F. Gruson and G. Monnier, *Oil & Gas Science and Technology – Revue D Ifp Energies Nouvelles*, 2008, 63, 387–393.
2. N. Armaroli and V. Balzani, *Angew. Chem.-Int. Edit.*, 2007, 46, 52–66.
3. G.-Q. Chen and M. K. Patel, *Chemical Reviews*, 2011, 112, 2082–2099.
4. P. Azadi, O. R. Inderwildi, R. Farnood and D. A. King, *Renewable and Sustainable Energy Reviews*, 2013, 21, 506–523.
5. J. J. Bozell and G. R. Petersen, *Green Chemistry*, 2010, 12, 539–554.
6. F. Danielsen, H. Beukema, N. D. Burgess, F. Parish, C. A. Bruehl, P. F. Donald, D. Murdiyarso, B. Phalan, L. Reijnders, M. Struebig and E. B. Fitzherbert, *Conservation Biology*, 2009, 23, 348–358.
7. B. Kamm and M. Kamm, *Chemie Ingenieur Technik*, 2007, 79, 592–603.
8. B. Kamm, *Angew. Chem.-Int. Edit.*, 2007, 46, 5056–5058.
9. A. Corma, S. Iborra and A. Velty, *Chemical Reviews*, 2007, 107, 2411–2502.
10. P. Gallezot, *Catalysis Today*, 2007, 121, 76–91.
11. F. W. Lichtenthaler and S. Peters, *Comptes Rendus Chimie*, 2004, 7, 65–90.
12. M. Kim and D. F. Day, *J. Ind. Microbiol. Biotechnol.*, 2011, 38, 803–807.
13. T. L. Tew, R. M. Cobill and E. P. Richard, *BioEnergy Res.*, 2008, 1, 147–152.
14. A. Charlton, R. Elias, S. Fish, P. Fowler and J. Gallagher, *Chemical Engineering Research and Design*, 2009, 87, 1147–1161.
15. Y. Zheng, C. W. Yu, Y. S. Cheng, C. Lee, C. W. Simmons, T. M. Dooley, R. H. Zhang, B. M. Jenkins and J. S. VanderGheynst, *Appl. Energy*, 2012, 93, 168–175.

16. M. C. Edwards and J. Doran-Peterson, *Appl. Microbiol. Biotechnol.*, 2012, 95, 565–575.
17. F. Cherubini, *Energy Conversion and Management*, 2010, 51, 1412–1421.
18. V. Menon and M. Rao, *Progress in Energy and Combustion Science*, 2012, 38, 522–550.
19. M. Hoogwijk, A. Faaij, R. van den Broek, G. Berndes, D. Gielen and W. Turkenburg, *Biomass and Bioenergy*, 2003, 25, 119–133.
20. K. L. Kadam, L. H. Forrest and W. A. Jacobson, *Biomass and Bioenergy*, 2000, 18, 369–389.
21. N. Mosier, C. Wyman, B. Dale, R. Elander, Y. Y. Lee, M. Holtzapple and M. Ladisch, *Bioresource Technology*, 2005, 96, 673–686.
22. A. Garde, G. Jonsson, A. S. Schmidt and B. K. Ahring, *Bioresource Technology*, 2002, 81, 217–223.
23. W. G. Glasser and R. S. Wright, *Biomass and Bioenergy*, 1998, 14, 219–235.
24. X. Zhao, K. Cheng and D. Liu, *Appl. Microbiol. Biotechnol.*, 2009, 82, 815–827.
25. M. Fatih Demirbas, *Appl. Energy*, 2009, 86, Supplement 1, S151–S161.
26. P. Alvira, E. Tomás-Pejó, M. Ballesteros and M. J. Negro, *Bioresource Technology*, 2010, 101, 4851–4861.
27. W. S. Mok, M. J. Antal and G. Varhegyi, *Industrial & Engineering Chemistry Research*, 1992, 31, 94–100.
28. R. Rinaldi and F. Schüth, *ChemSusChem*, 2009, 2, 1096–1107.
29. M. Sasaki, B. Kabyemela, R. Malaluan, S. Hirose, N. Takeda, T. Adschiri and K. Arai, *The Journal of Supercritical Fluids*, 1998, 13, 261–268.
30. J. B. Binder and R. T. Raines, *Proceedings of the National Academy of Sciences of the United States of America*, 2010, 107, 4516–4521.
31. Suhas, P. J. M. Carrott and M. M. L. Ribeiro Carrott, *Bioresource Technology*, 2007, 98, 2301–2312.
32. F. M. Gírio, C. Fonseca, F. Carvalheiro, L. C. Duarte, S. Marques and R. Bogel-Łukasik, *Bioresource Technology*, 2010, 101, 4775–4800.
33. F. S. Chakar and A. J. Ragauskas, *Industrial Crops and Products*, 2004, 20, 131–141.
34. E. Ruiz, C. Cara, P. Manzanares, M. Ballesteros and E. Castro, *Enzyme and Microbial Technology*, 2008, 42, 160–166.
35. R. L. Dudley, C. A. Fyfe, P. J. Stephenson, Y. Deslandes, G. K. Hamer and R. H. Marchessault, *Journal of the American Chemical Society*, 1983, 105, 2469–2472.
36. R. W. Torget, J. S. Kim and Y. Y. Lee, *Industrial & Engineering Chemistry Research*, 2000, 39, 2817–2825.
37. Y. Yu, X. Lou and H. Wu, *Energy & Fuels*, 2008, 22, 46–60.
38. J. F. Saeman, *Industrial and Engineering Chemistry*, 1945, 37, 43–52.
39. A. H. Conner, B. F. Wood, C. G. Hill and J. F. Harris, *Journal of Wood Chemistry and Technology*, 1985, 5, 461–489.
40. J. Bouchard, N. Abatzoglou, E. Chornet and R. P. Overend, *Wood Science and Technology*, 1989, 23, 343–355.
41. G. W. Huber and J. A. Dumesic, *Catalysis Today*, 2006, 111, 119–132.
42. R. Rinaldi and F. Schuth, *Chemsuschem*, 2009, 2, 1096–1107.
43. S. Van de Vyver, J. Geboers, P. A. Jacobs and B. F. Sels, *Chemcatchem*, 2011, 3, 82–94.

44. H. Kobayashi, H. Ohta and A. Fukuoka, *Catalysis Science & Technology*, 2012, 2, 869–883.
45. P. L. Dhepe, M. Ohashi, S. Inagaki, M. Ichikawa and A. Fukuoka, *Catalysis Letters*, 2005, 102, 163–169.
46. S. Suganuma, K. Nakajima, M. Kitano, D. Yamaguchi, H. Kato, S. Hayashi and M. Hara, *Journal of the American Chemical Society*, 2008, 130, 12787–12793.
47. A. Onda, T. Ochi and K. Yanagisawa, *Green Chemistry*, 2008, 10, 1033–1037.
48. J. Pang, A. Wang, M. Zheng and T. Zhang, *Chemical Communications*, 2010, 46, 6935–6937.
49. S. Van de Vyver, L. Peng, J. Geboers, H. Schepers, F. de Clippel, C. J. Gommes, B. Goderis, P. A. Jacobs and B. F. Sels, *Green Chemistry*, 2010, 12, 1560–1563.
50. R. Rinaldi, R. Palkovits and F. Schuth, *Angew. Chem.-Int. Edit.*, 2008, 47, 8047–8050.
51. D. Yamaguchi, M. Kitano, S. Suganuma, K. Nakajima, H. Kato and M. Hara, *Journal of Physical Chemistry C*, 2009, 113, 3181–3188.
52. A. Takagaki, M. Nishimura, S. Nishimura and K. Ebitani, *Chemistry Letters*, 2011, 40, 1195–1197.
53. S. Deguchi, K. Tsujii and K. Horikoshi, *Chemical Communications*, 2006, 3293–3295.
54. A. A. Balandin, N. A. Vasunina, S. V. Chepigo and G. S. Barysheva, *Doklady Akademii Nauk Sssr*, 1959, 128, 941–944.
55. A. M. Ruppert, K. Weinberg and R. Palkovits, *Angew. Chem.-Int. Edit.*, 2012, 51, 2564–2601.
56. H. Kobayashi, T. Komanoya, K. Hara and A. Fukuoka, *Chemsuschem*, 2010, 3, 440–443.
57. J. Geboers, S. Van de Vyver, K. Carpentier, P. Jacobs and B. Sels, *Chemical Communications*, 2011, 47, 5590–5592.
58. R. P. Swatloski, S. K. Spear, J. D. Holbrey and R. D. Rogers, *Journal of the American Chemical Society*, 2002, 124, 4974–4975.
59. M. E. Zakrzewska, E. Bogel-Lukasik and R. Bogel-Lukasik, *Energy & Fuels*, 2010, 24, 737–745.
60. D. A. Fort, R. C. Remsing, R. P. Swatloski, P. Moyna, G. Moyna and R. D. Rogers, *Green Chemistry*, 2007, 9, 63–69.
61. A. Brandt, J. P. Hallett, D. J. Leak, R. J. Murphy and T. Welton, *Green Chemistry*, 2010, 12, 672–679.
62. O. A. El Seoud, A. Koschella, L. C. Fidale, S. Dorn and T. Heinze, *Biomacromolecules*, 2007, 8, 2629–2647.
63. A. Pinkert, K. N. Marsh, S. Pang and M. P. Staiger, *Chemical Reviews*, 2009, 109, 6712–6728.
64. M. Zavrel, D. Bross, M. Funke, J. Buchs and A. C. Spiess, *Bioresource Technology*, 2009, 100, 2580–2587.
65. L. Feng and Z. I. Chen, *Journal of Molecular Liquids*, 2008, 142, 1–5.
66. C. Li, Q. Wang and Z. K. Zhao, *Green Chemistry*, 2008, 10, 177–182.
67. A. S. Amarasekara and O. S. Owereh, *Industrial & Engineering Chemistry Research*, 2009, 48, 10152–10155.
68. R. Rinaldi, N. Meine, J. vom Stein, R. Palkovits and F. Schueth, *Chemsuschem*, 2010, 3, 266–276.

69. N. Villandier and A. Corma, *Chemical Communications*, 2010, 46, 4408–4410.
70. Y. T. Zhang, H. B. Du, X. H. Qian and E. Y. X. Chen, *Energy & Fuels*, 2010, 24, 2410–2417.
71. Z. H. Zhang and Z. B. K. Zhao, *Carbohydrate Research*, 2009, 344, 2069–2072.
72. A. V. Bridgwater, *Biomass & Bioenergy*, 2012, 38, 68–94.
73. C. A. Rabinovitch-Deere, J. W. K. Oliver, G. M. Rodriguez and S. Atsumi, *Chemical Reviews*, 2013, 113, 4611–4632.
74. S. Fernando, S. Adhikari, C. Chandrapal and N. Murali, *Energy & Fuels*, 2006, 20, 1727–1737.
75. X.-S. Zhang, G.-X. Yang, H. Jiang, W.-J. Liu and H.-S. Ding, *Sci. Rep.*, 2013, 3, 1120.
76. F. De Miguel Mercader, *Pyrolysis oil upgrading for co-processing in standard refinery units*, PhD Thesis 2010, University of Twente, The Netherlands.
77. S. Czernik and A. V. Bridgwater, *Energy & Fuels*, 2004, 18, 590–598.
78. T. Werpy and G. Petersen, *Top Value Added Chemicals From Biomass. Volume I: Results of Screening for Potential Candidates from Sugars and Synthesis Gas*, Pacific Northwest National Laboratory (PNNL) and National Renewable Energy Laboratory (NREL), August 2004.
79. K. K. Cheng, X. B. Zhao, J. Zeng, R. C. Wu, Y. Z. Xu, D. H. Liu and J. A. Zhang, *Appl. Microbiol. Biotechnol.*, 2012, 95, 841–850.
80. H. J. Huang, S. Ramaswamy, U. W. Tschirner and B. V. Ramarao, *Sep. Purif. Technol.*, 2008, 62, 1–21.
81. Y. S. Huh, Y. S. Jun, Y. K. Hong, H. Song, S. Y. Lee and W. H. Hong, *Process Biochemistry*, 2006, 41, 1461–1465.
82. A. Corma, M. Renz and M. Susarte, *Topics in Catalysis*, 2009, 52, 1182–1189.
83. M. J. Climent, A. Corma and S. Iborra, *Green Chemistry*, 2011, 13, 520–540.
84. J. H. Clark, *Journal of Chemical Technology and Biotechnology*, 2007, 82, 603–609.
85. M. Schlaf, *Dalton T*, 2006, 4645–4653.
86. R. Rinaldi and F. Schueth, *Energy & Environmental Science*, 2009, 2, 610–626.
87. J. P. Dacquin, H. E. Cross, D. R. Brown, T. Duren, J. J. Williams, A. F. Lee and K. Wilson, *Green Chemistry*, 2010, 12, 1383–1391.
88. W. C. Li, A. H. Lu and F. Schuth, *Chemistry of Materials*, 2005, 17, 3620–3626.
89. M. J. Climent, A. Corma and S. Iborra, *Chemical Reviews*, 2011, 111, 1072–1133.
90. B. Zhang, J. Ren, X. Liu, Y. Guo, Y. Guo, G. Lu and Y. Wang, *Catalysis Communications*, 2010, 11, 629–632.
91. J. H. Clark, V. Budarin, T. Dugmore, R. Luque, D. J. Macquarrie and V. Strelko, *Catalysis Communications*, 2008, 9, 1709–1714.
92. I. Bechthold, K. Bretz, S. Kabasci, R. Kopitzky and A. Springer, *Chemical Engineering & Technology*, 2008, 31, 647–654.
93. C. Delhomme, D. Weuster-Botz and F. E. Kuehn, *Green Chemistry*, 2009, 11, 13–26.
94. D. Mohan, C. U. Pittman, Jr. and P. H. Steele, *Energy & Fuels*, 2006, 20, 848–889.
95. L. Catoire, M. Yahyaoui, A. Osmont and I. Goekalp, *Energy & Fuels*, 2008, 22, 4265–4273.
96. R.-J. van Putten, J. C. van der Waal, E. de Jong, C. B. Rasrendra, H. J. Heeres and J. G. de Vries, *Chemical Reviews*, 2013, 113, 1499–1597.

97. M. E. Zakrzewska, E. Bogel-Łukasik and R. Bogel-Łukasik, *Chemical Reviews*, 2010, 111, 397–417.
98. L. Hu, G. Zhao, W. W. Hao, X. Tang, Y. Sun, L. Lin and S. J. Liu, *RSC Adv.*, 2012, 2, 11184–11206.
99. S. Dutta, S. De and B. Saha, *ChemPlusChem*, 2012, 77, 259–272.
100. T. Buntara, S. Noel, P. H. Phua, I. Melián-Cabrera, J. G. de Vries and H. J. Heeres, *Angewandte Chemie International Edition*, 2011, 50, 7083–7087.
101. Y. Roman-Leshkov, C. J. Barrett, Z. Y. Liu and J. A. Dumesic, *Nature*, 2007, 447, 982–985.
102. G. Yang, E. A. Pidko and E. J. M. Hensen, *J. Catal.*, 2012, 295, 122–132.
103. S. K. R. Patil and C. R. F. Lund, *Energy & Fuels*, 2011, 25, 4745–4755.
104. M. Watanabe, Y. Aizawa, T. Iida, R. Nishimura and H. Inomata, *Applied Catalysis A – General*, 2005, 295, 150–156.
105. R. A. Schraufnagel and H. F. Rase, *Industrial & Engineering Chemistry Product Research and Development*, 1975, 14, 40–44.
106. K. Lourvanij and G. L. Rorrer, *Industrial & Engineering Chemistry Research*, 1993, 32, 11–19.
107. K. Lourvanij and G. L. Rorrer, *Journal of Chemical Technology and Biotechnology*, 1997, 69, 35–44.
108. Y. Nakamura and S. Morikawa, *Bulletin of the Chemical Society of Japan*, 1980, 53, 3705–3706.
109. C. Moreau, R. Durand, D. Peyron, J. Duhamet and P. Rivalier, *Industrial Crops and Products*, 1998, 7, 95–99.
110. C. Moreau, R. Durand, F. R. Alies, M. Cotillon, T. Frutz and M. A. Theoleyre, *Industrial Crops and Products*, 2000, 11, 237–242.
111. R. Kourieh, V. Rakic, S. Bennici and A. Auroux, *Catalysis Communications*, 2013, 30, 5–13.
112. K.-I. Shimizu, H. Furukawa, N. Kobayashi, Y. Itaya and A. Satsuma, *Green Chemistry*, 2009, 11, 1627–1632.
113. C. Carlini, M. Giuttari, A. M. R. Galletti, G. Sbrana, T. Armaroli and G. Busca, *Applied Catalysis A – General*, 1999, 183, 295–302.
114. T. Armaroli, G. Busca, C. Carlini, M. Giuttari, A. M. R. Galletti and G. Sbrana, *Journal of Molecular Catalysis A – Chemical*, 2000, 151, 233–243.
115. X. Qi, M. Watanabe, T. M. Aida and R. L. Smith, Jr., *Catalysis Communications*, 2009, 10, 1771–1775.
116. A. J. Crisci, M. H. Tucker, M.-Y. Lee, S. G. Jang, J. A. Dumesic and S. L. Scott, *ACS Catalysis*, 2011, 1, 719–728.
117. X. Qi, M. Watanabe, T. M. Aida and R. L. Smith, Jr., *Green Chemistry*, 2009, 11, 1327–1331.
118. X. Qi, M. Watanabe, T. M. Aida and R. L. Smith, Jr., *Chemsuschem*, 2009, 2, 944–946.
119. C. Lansalot-Matras and C. Moreau, *Catalysis Communications*, 2003, 4, 517–520.
120. H. Jadhav, E. Taarning, C. M. Pedersen and M. Bols, *Tetrahedron Letters*, 2012, 53, 983–985.
121. X. Guo, Q. Cao, Y. Jiang, J. Guan, X. Wang and X. Mu, *Carbohydrate Research*, 2012, 351, 35–41.
122. A. Takagaki, M. Ohara, S. Nishimura and K. Ebitani, *Chemical Communications*, 2009, 6276–6278.

123. M. Ohara, A. Takagaki, S. Nishimura and K. Ebitani, *Applied Catalysis A – General*, 2010, 383, 149–155.
124. R. Weingarten, G. A. Tompsett, W. C. Conner and G. W. Huber, *J. Catal.*, 2011, 279, 174–182.
125. M. Moliner, Y. Roman-Leshkov and M. E. Davis, *Proc Natl Acad Sci U S A*, 2010, 107, 6164–6168.
126. Y. Roman-Leshkov, M. Moliner, J. A. Labinger and M. E. Davis, *Angewandte Chemie*, 2010, 49, 8954–8957.
127. Y. J. Pagán-Torres, T. Wang, J. M. R. Gallo, B. H. Shanks and J. A. Dumesic, *ACS Catalysis*, 2012, 2, 930–934.
128. E. Nikolla, Y. Roman-Leshkov, M. Moliner and M. E. Davis, *ACS Catalysis*, 2011, 1, 408–410.
129. Y. J. Pagán-Torres, T. F. Wang, J. M. R. Gallo, B. H. Shanks and J. A. Dumesic, *ACS Catalysis*, 2012, 2, 930–934.
130. K. Nakajima, Y. Baba, R. Noma, M. Kitano, J. N. Kondo, S. Hayashi and M. Hara, *Journal of the American Chemical Society*, 2011, 133, 4224–4227.
131. A. S. Dias, S. Lima, M. Pillinger and A. A. Valente, *Carbohydrate Research*, 2006, 341, 2946–2953.
132. A. S. Dias, M. Pillinger and A. A. Valente, *Applied Catalysis A – General*, 2005, 285, 126–131.
133. G. H. Jeong, E. G. Kim, S. B. Kim, E. D. Park and S. W. Kim, *Microporous and Mesoporous Materials*, 2011, 144, 134–139.
134. A. S. Dias, M. Pillinger and A. A. Valente, *J. Catal.*, 2005, 229, 414–423.
135. E. Lam, E. Majid, A. C. W. Leung, J. H. Chong, K. A. Mahmoud and J. H. T. Luong, *Chemsuschem*, 2011, 4, 535–541.
136. A. S. Dias, S. Lima, P. Brandao, M. Pillinger, J. Rocha and A. A. Valente, *Catalysis Letters*, 2006, 108, 179–186.
137. A. S. Dias, S. Lima, M. Pillinger and A. A. Valente, *Catalysis Letters*, 2007, 114, 151–160.
138. T. Suzuki, T. Yokoi, R. Otomo, J. N. Kondo and T. Tatsumi, *Applied Catalysis A – General*, 2011, 408, 117–124.
139. H. Choudhary, S. Nishimura and K. Ebitani, *Chemistry Letters*, 2012, 41, 409–411.
140. H. J. Guo and G. C. Yin, *Journal of Physical Chemistry C*, 2011, 115, 17516–17522.
141. R. H. Leonard, *Industrial and Engineering Chemistry*, 1956, 48, 1331–1341.
142. H. J. Bart, J. Reidetschlager, K. Schatka and A. Lehmann, *Industrial & Engineering Chemistry Research*, 1994, 33, 21–25.
143. Z. Li, R. Wnetrzak, W. Kwapinski and J. J. Leahy, *ACS Applied Materials & Interfaces*, 2012, 4, 4499–4505.
144. D. R. Fernandes, A. S. Rocha, E. F. Mai, C. J. A. Mota and V. Teixeira da Silva, *Applied Catalysis A – General*, 2012, 425, 199–204.
145. G. Pasquale, P. Vazquez, G. Romanelli and G. Baronetti, *Catalysis Communications*, 2012, 18, 115–120.
146. S. Dharne and V. V. Bokade, *Journal of Natural Gas Chemistry*, 2011, 20, 18–24.
147. R. Weingarten, W. C. Conner, Jr. and G. W. Huber, *Energy & Environmental Science*, 2012, 5, 7559–7574.
148. Y. X. Fan, C. H. Zhou and X. H. Zhu, *Catal. Rev.-Sci. Eng.*, 2009, 51, 293–324.

149. R. Auras, L.-T. Lim, S. E. M. Selke and H. Tsuji, *Polylactic Acid, Synthess Structures Properties Processing and Application*, Wiley, New York, 2010.
150. Y. D. Hang, *Biotechnology Letters*, 1989, 11, 299–300.
151. R. P. John, K. M. Nampoothiri and A. Pandley, *Journal of Basic Microbiology*, 2007, 47, 25–30.
152. K. L. Wasewar, A. A. Yawalkar, J. A. Moulijn and V. G. Pangarkar, *Industrial & Engineering Chemistry Research*, 2004, 43, 5969–5982.
153. S. Miura, L. Dwiarti, T. Arimura, M. Hoshino, L. Tiejun and M. Okabe, *Journal of Bioscience and Bioengineering*, 2004, 97, 19–23.
154. A. Onda, T. Ochi, K. Kajiyoshi and K. Yanagisawa, *Applied Catalysis A – General*, 2008, 343, 49–54.
155. M. S. Holm, S. Saravanamurugan and E. Taarning, *Science*, 2010, 328, 602–605.
156. Y. Roman-Leshkov and M. E. Davis, *ACS Catalysis*, 2011, 1, 1566–1580.
157. Z. Liu, G. Feng, C. Y. Pan, W. Li, P. Chen, H. Lou and X. M. Zheng, *Chin. J. Catal.*, 2012, 33, 1696–1705.
158. F. de Clippel, M. Dusselier, R. Van Rompaey, P. Vanelderen, J. Dijkmans, E. Makshina, L. Giebeler, S. Oswald, G. V. Baron, J. F. M. Denayer, P. P. Pescarmona, P. A. Jacobs and B. F. Sels, *Journal of the American Chemical Society*, 2012, 134, 10089–10101.
159. A. Onda, T. Ochi, K. Kajiyoshi and K. Yanagisawa, *Catalysis Communications*, 2008, 9, 1050–1053.
160. *NEXANT Inc., CHEMSYSTEMS® Prospectus Bio-Based Chemicals: Going Commercial*, 2012.
161. B. Erickson, J. E. Nelson and P. Winters, *Biotechnology Journal*, 2012, 7, 176–185.
162. F. C. Naughton, *Journal of the American Oil Chemists Society*, 1974, 51, 65–71.
163. J. J. Bozell and M. K. Patel, *Feedstocks for the Future, Renwables for the Production of Chemicals & Materials*, Washington, DC, American Chemical Society, 2004.
164. K. M. Draths and J. W. Frost, *Journal of the American Chemical Society*, 1994, 116, 399–400.
165. D. de Guzman, Green Chemicals: DSM adds adipic acid to bio-based chemicals portfolio. Available at: http://www.icis.com/Articles/2011/10/10/9498186/Green-Chemicals-DSM-adds-adipic-acid-to-bio-based-chemicals.html (accessed March 2013).
166. Y.-C. Lin and G. W. Huber, *Energy & Environmental Science*, 2009, 2, 68–80.
167. J.-P. Dacquin, A. F. Lee and K. Wilson, *Heterogeneous Catalysts for Converting Renewable Feedstocks to Fuels and Chemicals*, Springer, New York, 2012.
168. A. Davidson, *Current Opinion in Colloid & Interface Science*, 2002, 7, 92–106.
169. A. Galarneau, J. Iapichella, K. Bonhomme, F. Di Renzo, P. Kooyman, O. Terasaki and F. Fajula, *Advanced Functional Materials*, 2006, 16, 1657–1667.
170. T. Linssen, K. Cassiers, P. Cool and E. F. Vansant, *Advances in Colloid and Interface Science*, 2003, 103, 121–147.
171. J. Y. Ying, C. P. Mehnert and M. S. Wong, *Angew. Chem.-Int. Edit.*, 1999, 38, 56–77.
172. C. M. A. Parlett, K. Wilson and A. F. Lee, *Chemical Society Reviews*, 2013, 42, 3876–3893.

173. J.-P. Dacquin, J. Dhainaut, D. Duprez, S. Royer, A. F. Lee and K. Wilson, *Journal of the American Chemical Society*, 2009, 131, 12896–12897.
174. J. Dhainaut, J.-P. Dacquin, A. F. Lee and K. Wilson, *Green Chemistry*, 2010, 12, 296–303.
175. A. I. Torres, P. Daoutidis and M. Tsapatsis, *Energy & Environmental Science*, 2010, 3, 1560–1572.
176. S. Crossley, J. Faria, M. Shen and D. E. Resasco, *Science*, 2010, 327, 68–72.
177. M. P. Ruiz, J. Faria, M. Shen, S. Drexler, T. Prasomsri and D. E. Resasco, *Chemsuschem*, 2011, 4, 964–974.
178. P. Y. Dapsens, C. Mondelli and J. Perez-Ramirez, *ACS Catalysis*, 2012, 2, 1487–1499.
179. J. P. M. Sanders, J. H. Clark, G. J. Harmsen, H. J. Heeres, J. J. Heijnen, S. R. A. Kersten, W. P. M. Van Swaaij and J. A. Moulijn, *Chem. Eng. Process.*, 2012, 51, 117–136.
180. Y. Cheng, Y. Feng, Y. Ren, X. Liu, A. Gao, B. He, F. Yan and J. Li, *Bioresource Technology*, 2012, 113, 65–72.
181. A. A. Kulkarni, K.-P. Zeyer, T. Jacobs and A. Kienle, *Industrial & Engineering Chemistry Research*, 2007, 46, 5271–5277.
182. O. de la Iglesia, R. Mallada, M. Menendez and J. Coronas, *Chemical Engineering Journal*, 2007, 131, 35–39.
183. S. Assabumrungrat, W. Kiatkittipong, P. Praserthdam and S. Goto, *Catalysis Today*, 2003, 79, 249–257.
184. C. Buchaly, P. Kreis and A. Gorak, *Industrial & Engineering Chemistry Research*, 2012, 51, 896–904.
185. I. K. Lai, Y.-C. Liu, C.-C. Yu, M.-J. Lee and H.-P. Huang, *Chemical Engineering and Processing*, 2008, 47, 1831–1843.
186. A. Singh, R. Hiwale, S. M. Mahajani, R. D. Gudi, J. Gangadwala and A. Kienle, *Industrial & Engineering Chemistry Research*, 2005, 44, 3042–3052.
187. X. Ni, M. R. Mackley, A. P. Harvey, P. Stonestreet, M. H. I. Baird and N. V. R. Rao, *Chemical Engineering Research & Design*, 2003, 81, 373–383.
188. A. P. Harvey, M. R. Mackley and P. Stonestreet, *Industrial & Engineering Chemistry Research*, 2001, 40, 5371–5377.
189. M. E. Fabiyi and R. L. Skelton, *Process Safety and Environmental Protection*, 2000, 78, 399–404.
190. V. Eze, A. N. Phan, C. Pirez, A. P. Harvey, A. F. Lee and K. Wilson, *Catalysis Science & Technology*, 2013, 3, 2373–2379.
191. H.-J. Huang, S. Ramaswamy, U. W. Tschirner and B. V. Ramarao, *Sep. Purif. Technol.*, 2008, 62, 1–21.
192. S. Pinzi, I. L. Garcia, F. J. Lopez-Gimenez, M. D. Luque de Castro, G. Dorado and M. P. Dorado, *Energy & Fuels*, 2009, 23, 2325–2341.
193. S. M. Hick, C. Griebel, D. T. Restrepo, J. H. Truitt, E. J. Buker, C. Bylda and R. G. Blair, *Green Chemistry*, 2010, 12, 468–474.
194. J. B. Binder and R. T. Raines, *Journal of the American Chemical Society*, 2009, 131, 1979–1985.
195. H. Zhao, J. E. Holladay, H. Brown and Z. C. Zhang, *Science*, 2007, 316, 1597–1600.

# 20
# Bio-based chemicals from biorefining: lignin conversion and utilisation

A. L. MACFARLANE, M. MAI and
J. F. KADLA, University of
British Columbia, Canada

DOI: 10.1533/9780857097385.2.659

**Abstract**: An understanding of lignin chemical structure and properties is required before connections can be drawn between lignin sources and utilisation. Of particular interest, and one cause for the limited use of lignin, is chemical heterogeneity. The principal applications for lignin are power/fuel (short term), macromolecules (medium term) and aromatic chemicals (long term), with the preferred use ultimately directed by the global price of oil and biomass. The intention of this chapter is to compile and evaluate the properties of lignins from various sources and to identify applications that match the unique properties of each lignin.

**Key words**: lignin, lignocellulose, kraft process, sulfite process, organosolv, carbon fibres, lignosulfonate.

## 20.1 Introduction

Lignin has traditionally been derived from the kraft and sulfite processes that are ubiquitous in the pulp and paper industry. Over 50 million tonnes are produced annually worldwide (Gosselink, 2004; Dos Santos et al., 2012; Zakzeski et al., 2010). The vast majority of lignin is burned in order to recover chemicals from the pulping process and provide process heat and only 2% of all lignins are utilised for rather low-value applications (Gosselink, 2004). This led to the general perception that lignin is just a waste product of the pulping process and, as a result, this has restrained the research for value-added applications for lignin and lignin-based materials for quite some time. However, higher value products are constantly being sought.

A multitude of lignin applications have been explored at the research level so far. The current industrial products are primarily produced from lignosulfonates and sulfonated kraft lignin. Products include dispersants, emulsifiers, raw material for vanillin and dimethyl sulfoxide (DMSO) production and road-dust suppression agents. There is the opportunity to classify and refine lignin for use in many polymeric products currently being served by synthetic petroleum-derived polymers. Many biorefinery platforms produce lignin that can be incorporated into resins, foams and adhesives with improvement in properties. However, these applications still use lignin as a by-product from the pulping process. The next phase in biorefinery technology will be the design of processes with lignin as the primary product generating the greatest revenue.

## 20.2 Structure and properties of lignin

Lignin is a highly branched aromatic biopolymer only exceeded by cellulose in its abundance. Lignin is found in the cell walls of woody plants where its content reaches up to 30%. It plays a significant role in trees by giving rigidity to the cell walls and making the wood resistant to compression, impact and bending. It is also involved in the transport of water from the roots to the leaves. Through covalent bonds, lignin is interlaced with hemicellulose and forms a matrix that is reinforced by cellulose. Lignin is produced by enzyme-initiated dehydrogenative polymerisation of phenyl-propanoid groups derived from coniferyl alcohol (G), sinapyl alcohol (S), and *p*-coumaryl alcohol (H). The structures of these monomers are shown in Fig. 20.1 (Dimmel, 2010).

Softwoods, hardwoods and grasses each have different proportions of S, G and H monomers, and proportions can vary greatly between species and growing conditions (see Table 20.1). Hardwood lignins consist mainly of coniferyl and sinapyl alcohol (guaiacyl-syringyl lignin), while softwood lignins are made primarily from coniferyl alcohol. The lignin in grass and annual plants is composed mainly of *p*-coumaryl alcohol units. The difference between the three lignin precursors is the number of methoxyl groups attached to the aromatic ring which results in different amounts of certain inter-unit linkages in hardwood, softwood or grasses.

The enzymatic dehydrogenation yields several resonance-stabilised phenoxy radicals that can couple randomly with another radical and therefore form a variety of different bonds with exceptional stability (see

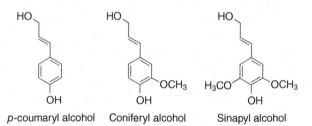

*20.1* Phenyl propane units in lignin (*p*-coumaryl alcohol, coniferyl alcohol and sinapyl alcohol).

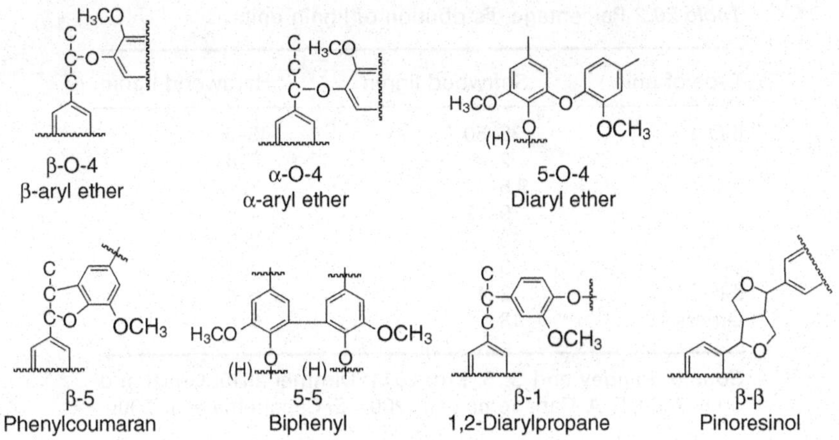

20.2 Main interunit linkages in lignin.

Table 20.1 Approximate composition of some important classes of lignins

|  | p-Coumaryl alcohol | Coniferyl alcohol | Sinapyl alcohol |
| --- | --- | --- | --- |
| Softwood lignin | 4% | 95% | 1% |
| Hardwood lignin | ~2% | ~50% | ~50% |
| Grass lignin | 5% | 70% | 25% |
| Compression wood lignin | 30% | 70% | ~0% |

Source: Dimmel (2010).

Fig. 20.2): biphenyl carbon–carbon linkages between aromatic carbons, alkyl–aryl carbon–carbon linkages between an aliphatic and aromatic carbon, and hydrolysis-resistant ether linkages. The only relatively weak and hydrolysable linkage is the α-aryl ether bond (Wool and Sun, 2005). The percentage distribution of lignin units is shown in Table 20.2.

Although it has been intensely studied for many years, the structure of native lignin still remains unclear. However, several structures have been proposed based on the analysis of degradation products and the identification of the dominant linkages between the phenylpropane units and their abundance, as well as the abundance and frequency of certain functional groups (see Figs 20.3 and 20.4). Those functional groups have a great impact on the reactivity of lignin. Lignin mostly contains methoxyl groups, phenolic and aliphatic hydroxyl groups, and a few terminal aldehyde groups. However, most of the phenolic hydroxyl groups are not available since they are occupied in interunit linkages.

Table 20.2 Percentage distribution of lignin units

| Type of unit | Softwood lignin | Hardwood lignin |
|---|---|---|
| β-O-4 | 35–50 | 45–60 |
| α-O-4 | 2–8 | 5–8 |
| 5-O-4 | 3.5–4 | 7–15 |
| β-5 | 9–12 | 4–6 |
| 5–5 | 10–23 | 2–9 |
| β-1 | 3–7 | 5–15 |
| β-β | 2–4 | 3–8 |
| Others | 13 | 5 |

Source: Pandey and C. S. Kim 2011; Dimmel 2010; Gratzl and Chen 2000; E. A. Capanema et al. 2004; E. Capanema et al. 2005.

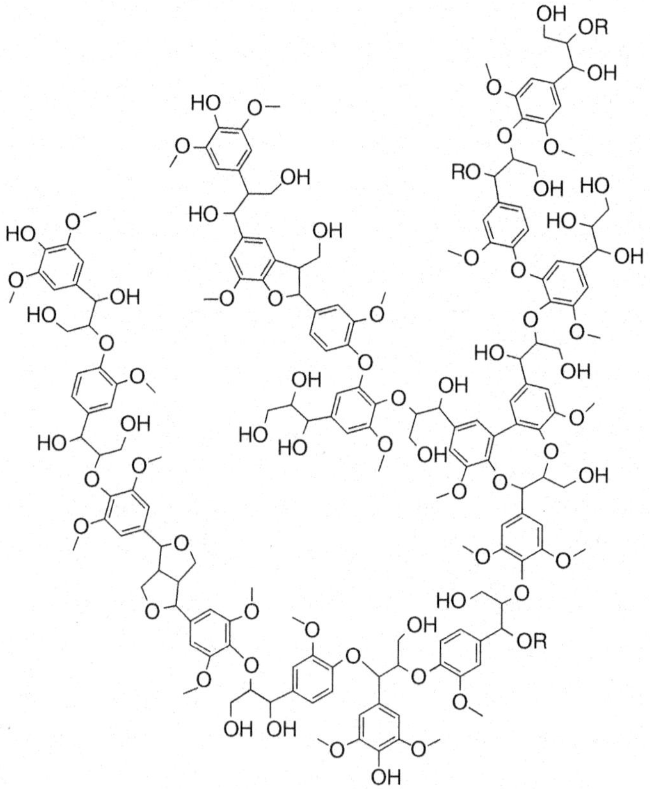

20.3 Hardwood native lignin; R = H, alkyl, aryl (from Kubo and Kadla, 2006).

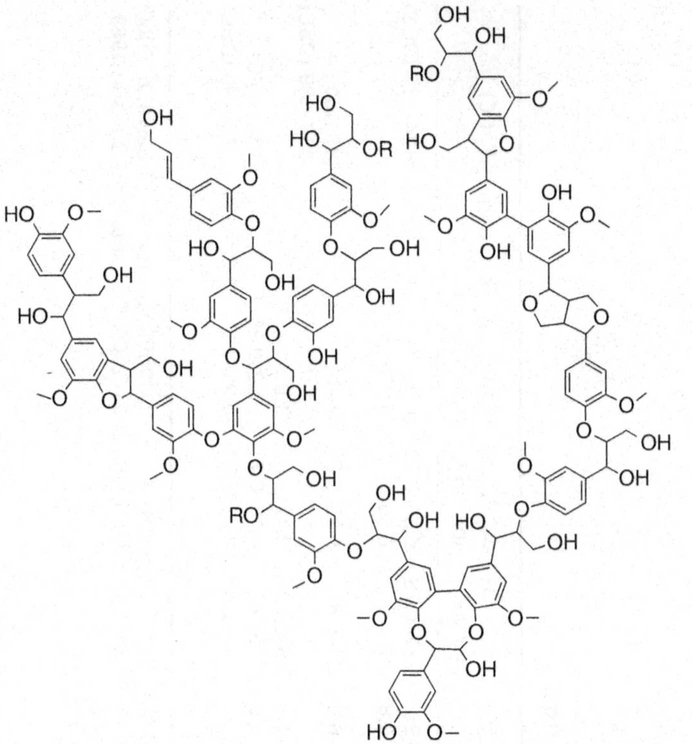

20.4 Softwood native lignin; R = H, alkyl, aryl (from Kubo and Kadla, 2006).

Depending on the delignification process employed, the amount and the nature of functional groups change (see Tables 20.3 and 20.4). Some processes hydrolyse the structure and make more phenolic hydroxyl groups available. Others introduce new functional groups to improve solubility such as for sulfonated lignin. One aim of lignin research is to investigate the structure–properties relationship of all kinds of different lignins and to find suitable applications for all of them.

## 20.3 Traditional processes for the production of lignin

The process of extracting lignin from biomass is known as delignification and disintegrates lignocellulosic material into its fibrous components. Two main categories of chemical delignification processes, namely aqueous and organic solvent-based, have been developed and include a wide range of methods. The chemical aqueous pulping processes include alkaline (or kraft pulping) and sulfite pulping, which are the two industrially most widely used methods (Wool and Sun, 2005). The primary goal of chemical pulping is to

Table 20.3 Properties of lignin from different extraction processes

| Procedure | Wood species | $M_w$ | $M_n$ | $M_w/M_n$ | Phen. OH (per C9) | Aliph. OH (per C9) | $OCH_3$ (per C9) | s:g ratio | $T_g$ (°C) |
|---|---|---|---|---|---|---|---|---|---|
| Kraft | SW | 1100–39,000 | 530–2400 | 1.4–28 | 0.57–0.78 | 0.39–0.72 | 0.77 | 1 | 124–169 (DSC) |
|  | HW | 2900–4200 | 1090–2900 | 2.2–2.8 | 0.7 | 0.35 | 1.15 |  |  |
| Organosolv |  |  |  |  |  |  |  |  |  |
| ALCELL | SW | 1750 | 700 | 2.5 |  |  |  |  |  |
| Steam explosion | HW | 1000–8000 | 600–1700 | 1.8–6.3 | 0.3–0.65 | 0.23–0.6 | 1–1.3 | 1.33 | 97–130 |
|  | SW |  |  |  |  |  |  |  | 113–139 (DSC) |
| MWL | HW | 2100–7100 | 790–1900 | 2.6–4 | 0.5–0.63 | 0.42 | 1.6 | 1–1.5 | 160 (DSC) |
|  | SW | 20,600–22,700 | 7900–9000 | 2.4–2.6 | 0.15–0.3 | 0.15–0.7 | 0.92–0.95 |  |  |
|  | HW |  |  |  | 0.19–1.58 | 0.09–0.71 | 0.21–1.6 | 1.34–1.72 | 110–130 (DSC) |
| Ligno-sulfonate | SW | 400–150,000 |  | 3.1–7.1 |  |  |  |  |  |

Source: Moerck et al. 1986; Wool and Sun, 2005; Gratzl and Chen, 2000; Gosselink et al., 2010; El Mansouri and Salvadó, 2006; Tejado et al., 2007; Lora and Glasser, 2002; Hu, 2002; Thring et al., 1996; Capanema et al., 2004, 2005; Sarkanen and Glasser, 1989.

Table 20.4 Properties of nonwood lignins

| Procedure | Wood species | $M_w$ | $M_n$ | $M_w/M_n$ | Phen. OH (per C9) | Aliph. OH (per C9) | $OCH_3$ (per C9) | $T_g$ (°C) |
|---|---|---|---|---|---|---|---|---|
| Soda | Wheat | 3300–4400 | 1200–1800 | 2.4–3.6 | 0.38–0.9 | 0.26–0.68 | 0.71–1.0 | |
| | Hemp | | 600 | | 0.49–0.6 | 0.57 | 0.86–0.9 | |
| Soda-AQ | Wheat | 3270 | 1770 | 1.8 | 0.75 | 0.87 | 1.02 | 160 |
| Organosolv | Wheat | 4430 | 800–2020 | 2.2–5.5 | 0.39–0.65 | 0.62–0.82 | 0.75–1.02 | 142 |
| | Bagasse | 2704 | 830–840 | 3.2 | 0.77 | | 0.8 | 170 |
| | Reed | | | | 0.5–0.62 | 0.52–0.56 | 0.95–1.01 | 97 (1bar) |
| | Kenaf | 1410–1480 | 650–680 | 2.2 | 0.43–0.47 | 0.58–0.76 | 1.01 | 66–70 (1bar) |
| Kraft | Bagasse | | | | 0.51 | | 0.95–1 | |

Source: Lora and Glasser, 2002; Hu, 2002, pp.268–273.

selectively remove as much lignin as possible without degradation of the carbohydrates. Therefore, the selectivity of delignification is determined by the weight ratio of lignin and carbohydrate removal. The pulping process can be divided into three stages:

1. 'initial delignification' occurring at <140°C; lignin is extracted and hemicellulose degraded
2. 'bulk delignification'; the delignification rate is accelerated at increased temperature and stays high until 90% of the lignin is removed
3. 'residual delignification' is characterized by a slow delignification rate while the degradation of carbohydrates becomes dominant.

The degree of delignification is typically indicated by the 'kappa number' of a pulp, where a high kappa number stands for high lignin content. The kappa number of unbleached softwood and hardwood pulps is 30–40 and 18–20, respectively (Kadla and Dai, 2006).

Soda pulping dates back to 1851, when it was invented in England by Burgess and Watts (Smook, 1986). It is one of the simplest pulping processes and a source of sulfur-free lignin. In soda pulping lignocellulose is treated with aqueous NaOH at a temperature between 150 and 170°C. Under these conditions both lignin and carbohydrates are degraded. Therefore the selectivity of soda pulping is low and it is currently limited to easily pulped materials like straw and some hardwoods (Kadla and Dai, 2006). To improve selectivity, additives are typically used, such as in the advanced soda-anthraquinone (AQ) process, where anthraquinone is added as a pulping additive/catalyst to decrease the carbohydrate degradation (Smook, 1986). During the soda-AQ process, the catalyst undergoes a redox cycle between AQ and the reduced form anthrahydroquinone (AHQ), which can effectively cleave β-ether bonds in lignin. The hydrolysed lignin can be further degraded by NaOH. The reducing end groups of the carbohydrates are oxidised to a carboxylic acid group during the redox cycle which is stable under alkaline conditions and prevents 'peeling' reactions. Soda lignin has been used as a biocide for wastewater treatment and as a dispersant (Gosselink, 2004). The Swiss holding company GreenValue SA (Lausanne) is a current producer of soda lignin.

The kraft process (or sulfate process) was invented in 1879 by the German chemist Dahl. It is a modified soda process in that it uses NaOH and $Na_2S$ ('white liquor') to fragment lignin until it is soluble in the alkaline medium. The pulping step is conducted at 140–180°C and lasts approximately 2 hours. The selectivity of kraft pulping is higher than soda pulping because hydrogen sulfite ions ($HS^-$) preferentially react with lignin (Gierer, 1985). As a result, a typical Kraft pulping process removes up to 80% of the lignin, 50% of the hemicellulose and less than 10% of the cellulose. Compared to soda lignin, kraft lignin is far less condensed due to the thiol groups 'blocking' the reactive benzylic centres. The thiol groups that are incorporated into the lignin structure result in a sulfur containing technical lignin. Kraft lignins are soluble in alkali (pH > 10.5) and some polar organic solvents (DMF, DMSO, methyl cellusolve, etc.). They are relatively low molecular weight and quite polydispersed.

The technological advantage of kraft pulping over other delignification processes is also one of the largest impediments to the utilisation of kraft lignin: the recovery of chemicals and energy from residual black liquor is at the core of the kraft process. It has been claimed that kraft lignin is not a viable feedstock for chemical production because of its energy value as a heat source in chemical recovery furnaces (Zakzeski et al., 2010). However, the greatest impetus for kraft lignin valorisation is that the majority of kraft mills, particularly in Canada, are recovery furnace limited and debottlenecking by lignin precipitation is considered by many pulp mills (Gosselink, 2004). Diverting lignin from heat to value-added polymers and platform chemicals would allow an increase in plant capacity and greatly improve economics. Commercial kraft lignin is produced by MeadWestvaco Corporation (Richmond, USA) and Metso Corporation (Helsinki, Finland). Up to 70% of MeadWestvaco's kraft lignin is chemically sulfonated to make it water soluble and better suitable for different applications (Zakzeski et al., 2010). The Structure of kraft pine lignin is shown in Fig. 20.5.

The sulfite process has been practised since 1874, when the first mill using the process was opened in Sweden. The active species are sulfur dioxide, hydrogen sulfite ($HSO_3^-$), and/or sulfite ions ($SO_3^{2-}$), depending on pH (Kadla and Dai, 2006). The most common industrial process is the acid sulfite process, which uses mixtures of sulfurous acid and/or its alkali salts ($Na^+$, $NH_4^+$, $Mg^{2+}$, $K^+$ or $Ca^{2+}$) to solubilise lignin through the formation of

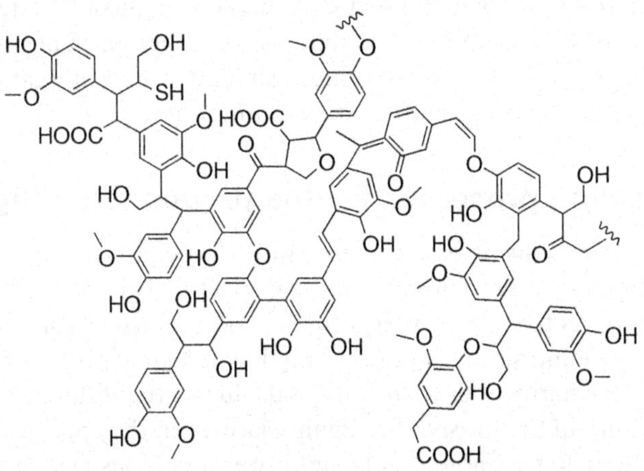

20.5 Kraft pine lignin (from Marton, 1971).

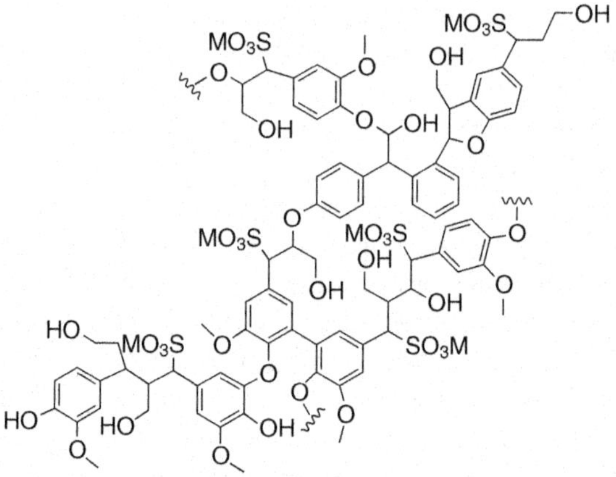

*20.6* Spruce lignosulfonate (from Gargulak and Lebo, 2000).

sulfonate functionalities and cleavage of lignin bonds (Biermann, 1996). In acid sulfite pulping, α-hydroxyl, and α-ether groups are eliminated to form carbonium ion intermediates and lignin fragmentation. The carbonium ions subsequently react with hydrated $SO_2$ and/or $HSO_3^-$, leading to sulfonation, or with other lignin fragments, leading to condensation. Sulfonation leads to increased hydrophilicity and solubility of the lignin, while condensation (which is enhanced at low pH) leads to a decrease in solubility due to a higher molecular weight (Kadla and Dai, 2006).

Worldwide, 1.06 mega tonnes of lignosulfonate are produced annually (Holladay et al., 2007). Sulfite lignin has a relatively high molecular weight ranging from 20,000 to 50,000 g/mol, a polydispersity of 6–8 and it is soluble in water from pH 0–14. One major disadvantage of lignosulfonates is the level of impurities as compared to that of kraft lignin; lignosulfonates typically contain 20–25% carbohydrate, ash and other inorganics. The structure of spruce lignosulfonate is shown in Fig. 20.6.

## 20.4 Emerging processes for the production of lignin

At the present time, solvent-based or organosolv processes are either in the developmental stage or used in small-scale production (bio-based polymers and composites). As the name implies, these processes use organic solvents to delignify lignocellulosic materials. This kind of delignification is in general more effective for hardwoods than softwoods due to the differences in the chemical structure of their respective lignins. However, they promise higher value by-products, lower capital costs and lower emissions (Pye and Lora, 1991).

The organosolv process uses an organic solvent to dissolve lignin by hydrolytic cleavage at elevated temperatures, typically between 170 and 220°C. Hemicellulose is also hydrolysed and solubilised. Various solvents have been studied including, but not limited to, acetone, methanol, ethanol, butanol, acetic acid and formic acid. Lower alcohols and specifically ethanol are the most promising of solvents due to the low cost and easier recovery by distillation (Zhao et al., 2009). The concentration of solvent is typically between 30 and 70%. An acid catalyst is often used to increase delignification rates or allow lower temperature (Aziz and Sarkanen, 1989). The lignin from organosolv using mild process conditions is reported to be virtually unmodified compared to milled wood lignin (El Hage et al., 2010) and thus allows exploitation of the natural properties of lignin such as resistance to water and biological attack. However, acidic conditions are required for effective delignification rates (Hallberg et al., 2008; Pan et al., 2006a) resulting in some etherification of lignin. The use of organosolv lignin in phenol-formaldehyde resin is identified as a possible high revenue gainer for a biorefinery (Arato et al., 2005).

The best known organosolv process is the Alcell® process (from: alcohol cellulose) using ethanol–water mixtures to delignify wood. It was developed by General Electric Corporation in the early 1970s and commercialised by Repap Enterprises Inc. (Stamford, USA) between 1978 and 1997 in a pulp mill in New Brunswick, Canada. It was later modified by Lignol Energy Corporation (Burnaby, Canada). The principal advantage of the Alcell® process is that it provides lignin, hemicelluloses and cellulose in separate streams of relatively high purity. The lignin is un-sulfonated and has a higher purity than other lignins (Zakzeski et al., 2010). The properties and chemistries of lignin can be manipulated by altering processing conditions such as temperature, solvent concentration and type (Goyal et al., 1992; Tirtowidjojo et al., 1988).

Acid hydrolysis, which entails the use of dilute acid at high temperature or concentrated acid at mild temperatures, is used to hydrolyse the cellulose and hemicelluloses portion of biomass for fermentation to ethanol. The lignin residue that remains is highly condensed, has high contamination with carbohydrates and is recovered in low yields from 50 to 70%. Nevertheless, NREL has developed a one-stage dilute acid biorefinery and acid hydrolysis lignin may be available for product development. In the 1980s, a South African company, C.G. Smith Sugar Limited, developed an acid hydrolysis lignin product with the trade name Sucrolin. It is produced by treating bagasse with superheated steam at 180°C in the presence of small quantities of acetic acid. The lignin is isolated from the lignin–cellulose residue by dissolution in sodium hydroxide, followed by neutralisation of the solution which precipitates the lignin (van der Hage et al. 1993). Sucrolin was available as a technical lignin from Sigma Aldrich (Lora and Glasser, 2002).

Acetic acid has been used as a pulping agent for both hardwoods and softwoods. The acidity increases the rate of hydrolytic cleavage of lignin but also increases rates of subsequent lignin recondensation and the production of furfural, a fermentation inhibitor (Vazquez et al. 1995, 1997; Parajo et al., 1993, 1995a, 1995b).

Formic acid pulping allows lower temperatures to be used (120°C), but results in higher cellulose losses compared to ethanol pulping (Erismann et al., 1994). For example, 99% formic acid at just 95°C results in pulps containing 8% klason lignin (Baeza et al., 1991). Formic acid and acetone in the ratio of 7:3 has been used (Baeza et al., 1999), giving high pulp yield and low lignin content (kappa ~ 25), although it is reported that high pulping pressure is necessary to achieve adequate penetration of solvent.

Lignin can also be recovered from the pyrolysis process, which heats lignocellulosic material to around 450°C in limited oxygen atmosphere to thermally degrade all components into hydrocarbons. A water insoluble product that contains lignin can be isolated from the pyrolysis oil. It is distinct from other forms of lignin because it is derived from oligomers containing up to 8 monolignols (Scholze et al., 2001). $M_w$ ranges from 600 to 1300 g/mol and $M_n$ from 300 to 600. Carbon fibre has been suggested as a high value application for pyrolysis lignin (Holladay et al., 2007; Qin and Kadla, 2012). Structures of pyrolytic lignins are shown in Fig. 20.7.

The physical delignification process of steam explosion uses saturated steam at high pressures (150–230°C) and elevated temperatures (450–500°C) for a period of time to hydrolyse lignin and hemicellulose from lignocellulosic material (Laser et al., 2002; Li et al., 2007). When $SO_2$ gas is used as catalyst, the process times are as short as 1–20 minutes (Bura et al., 2002). The wood is softened by the heat and after the explosive decompression it is widely defibrillated (Kadla and Dai, 2006), although the explosive decompression may not be necessary and refining may be used to separate fibres (Schütt et al., 2012). Steam explosion is an efficient technique to separate cellulose, hemicellulose and lignin with relatively high yield (Heitz et al., 1991; Grethlein and Converse, 1991; Kadla and Dai, 2006). Hardwoods are delignified much more easily than softwoods and, although only a small amount of lignin is dissolved during the process, the water-insoluble lignin can be extracted by 0.1 M alkaline solution or organic solvents. Stake technology uses steam explosion to produce sulfur-free lignin with a $T_g$ of 113–139°C (Heitz et al., 1991; Lora and Glasser, 2002). It has been marketed under the trade name Angiolin.

Recently, novel solvent systems such as ionic liquids (ILs) have attracted a great deal of interest. The ability to process at atmospheric pressure and dissolve any component with a well-tuned solvent has spurred significant research into ILs for lignocellulosic processing (Kim et al., 2011; Lee et al., 2009; Stark, 2011; Tan et al., 2009; Fort et al., 2007; Pu et al., 2007; Pinkert

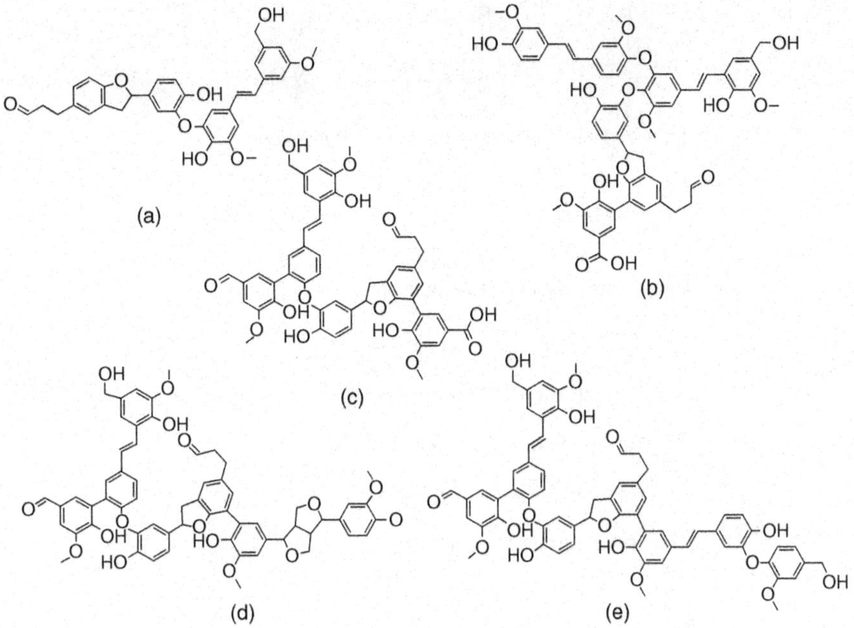

**20.7** Oligomeric structures proposed for pyrolytic lignin: (a) tetramer; (b) pentamer; (c) hexamer; (d) heptamer; (e) octamer (Bayerbach and Meier, 2009).

*et al.*, 2011). Defined as salts with melting points below 100°C (Wilkes, 2002), ILs generally feature an organic cation and inorganic anion. The major benefit of ILs for wood component solubilisation is their low vapour pressure coupled with their ability to dissolve part or all of the wood cell wall. Some of the most common IL cation and anion combinations are shown in Fig. 20.8. (Olivier-Bourbigou *et al.*, 2010).

One major benefit of IL processes includes the potential for *in-situ* derivatisation of wood components. Fort *et al.* (2007) note that very cleanly fractionated cellulose can be dissolved from wood shavings after precipitation with a suitable anti-solvent. Whole solubilisation after an extended time was not achieved for any sample and recovery of the lignin and hemicelluloses was not detailed. Relatively inexpensive ionic liquids exist that are capable of recovering 93% of the lignin from sugarcane bagasse. A mixture of 1-ethyl-3-methylimidazolium as the cation, a mixture of alkylbenzenesulfonates and xylenesulfonate as the main anion was used by Tan *et al.* (2009). Recovery of ionic liquid was between 96.1 and 99.4%, although the complexity of the recovery process was noted.

Whole wood solubilisation can be achieved but separation of the lignin from the IL is problematic. Selective dissolution of holocellulose can be

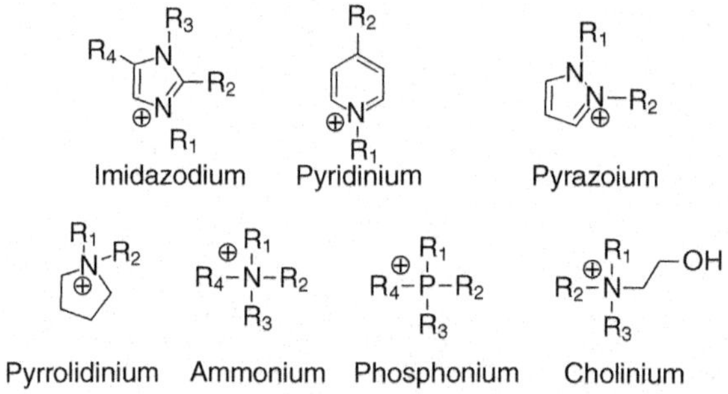

**Cations:** Imidazodium, Pyridinium, Pyrazoium, Pyrrolidinium, Ammonium, Phosphonium, Cholinium

**Anions:**

$Cl^-$, $Br^-$, $I^-$, $Al_2Cl_7^-$, $Al_3Cl_{10}^-$,

$Sb_2F_{11}^-$, $Fe_2Cl_7^-$, $Zn_2Cl_5^-$, $Zn_3Cl_7^-$

$CuCl_2^-$, $SnCl_2^-$, $NO_3^-$, $PO_4^{3-}$, $HSO_4^-$, $SO_4^{2-}$

$CF_3SO_3^-$, $ROSO_3^-$, $CF_3CO_2^-$, $C_6H_5SO_3^-$

$PF_6^-$, $SbF_5^-$, $BF_4^-$, $(CF_3SO_2)_2N^-$, $N(CN)_2^-$,

$(CF_3SO_2)_3C^-$, $BR_4^-$, $RCB_{11}H_{11}^-$

*20.8* Typical cation/anion combinations (ionic liquids) (from Olivier-Bourbigou *et al.*, 2010).

achieved allowing filtration of the solid lignin. However, efficient recovery of low molecular weight oligosaccharides has not been achieved and attention must be paid to the corrosive properties of the IL. Lignin can also be preferentially dissolved in a process analogous to the kraft process. In all cases, dissolution only occurs if biomass is thoroughly dried and substantially reduced in size. Many ionic liquids have very high viscosity, 2–3 orders of magnitude greater than solvents. This remains a challenge in ionic liquid commercial application in solubilisation processes that are mass transfer limited. The energy cost of biomass comminution may be restrictive depending on the required size (Olivier-Bourbigou *et al.*, 2010; Stark, 2011).

The cost of manufacturing ILs is limiting their use. Higher IL recovery rates or cheaper production must be achieved. Furthermore, separating solubilised wood components from the IL may require complicated processing (Olivier-Bourbigou *et al.*, 2010).

## 20.5 Applications of lignin and lignin-based products: an overview

Lignin is an excellent source of energy as it has a higher heating value than the other constituents in wood and is comparable to lignite (Raveendran and Ganesh, 1996). Pulp mills make use of this by combusting spent liquor in the recovery furnace, providing the energy for recovery of pulping chemicals. Spent liquor can also be diverted to the lime kiln or used for electricity generation for plant utilities or sale back to the grid. A great deal of research effort has been devoted to utilising lignin for applications other than heat generation, particularly specialty chemicals. These products include platform chemicals, adhesives and resins, plastics, paints, inks, soaps, coatings, cleaning compounds, lubricants and hydraulic fluids, greases, pesticides, toiletries, fragrances and cosmetics (Wool and Sun, 2005). However, very little has been realised on a commercial scale for a number of reasons. One reason is the uncertainty of lignin supply and the heterogeneity of the material (Lindberg *et al.*, 1989), as the properties of lignin depend highly on the treatment of the lignocellulosic plant and the utilisation of lignin is thus limited by the physical and chemical properties it shows after the pulping process (Wool and Sun, 2005).

Addition of lignin to improve particular products and development of novel products, rather than replace existing products and chemicals, is a promising area of research. The value of such products and additives will not be subject to competition with the oil-derived equivalent. However, in many cases, addition of lignin to a product resulted in a product with strictly worse properties, which is unacceptable. Besides large volume, low value applications, also high value, low volume products are needed to operate pulp mills economically and to make the extraction of lignin practicable. Platform chemicals that are indistinguishable from their petroleum-derived version can be made already. Although aromatics such as benzene, toluene and xylene have a large market and high value, the cost to produce these from lignin is high and spot prices are volatile.

### 20.5.1 Traditional lignin products

*Lignosulfonate*

The only lignin products obtained from a pulping process that have found a vast range of applications up to now are lignosulfonates. These are non-hazardous materials with excellent properties that are used as binders, emulsifiers and dispersants for a great variety of materials. As inexpensive components in binders, lignosulfonates are commonly used for commodities like coal briquettes, ceramics, briquetting of mineral dust, and the production of plywood or particle boards. Their ability to retain moisture and suppress dust makes them a useful tool for construction works, gravel roads, airports and sports facilities. As an anti-settling agent that also prevents lumping, lignosulfonates are used in concrete mixtures, ceramics, gypsumboard production and for leather tanning. Lignosulfonates provide flowability and plasticity to cement. This is a replacement for more expensive materials that provide set retardation such as superplasticisers, gluconates and gluconic acid. Wet-process Portland cement mills utilise lignosulfonates to increase the solids content of raw slurries. Lignin-based concrete additives are in demand and can be worth as much as $1.05–$1.32/L as an aqueous solution (Holladay et al., 2007). Sulfur-free lignin such as soda lignin has also been shown to improve flowability of mortar (Nadif et al., 2002). Lignosulfonates can also stabilise emulsions of immiscible fluids like asphalt emulsions, pesticide preparations, pigments and dyes. Due to their low toxicity, they can be used as binders in animal feed and thereby improve the feed properties of pellets. In addition, lignosulfonates show the ability to keep micronutrients in solution which is useful for micronutrient transport or as a cleaning and decontaminating agent in water and soils (Nadif et al., 2002; Calvo-Flores and Dobado, 2010).

*Copolymer/resin applications*

Copolymers and resins are a large market of medium to high value for lignin, provided lignin can improve material performance. In order to prepare formulations with lignin, it can be either used in an unmodified or chemically modified form. Due to its aromatic structure, lignin can undergo a wide range of chemical modifications that include, but are not limited to, alkylation, acylation, amination, carboxylation, halogenation, oxidation and reduction, nitration and sulfonation (Boye, 1985). Polyblending is a common industrial technique to physically mix two or more polymers in order to produce new high-performance materials from readily available materials and to reduce the production costs.

Lignin has been blended with other polymers or used as simple filler for decades in order to obtain biodegradable materials with improved mechanical properties, thermal and light stability. However, the practical results are still moderate and leave lots of room for further research (Calvo-Flores and Dobado, 2010). Already in 1983 Lyubeshkina could show that the addition of dry sulfate lignin can yield a frost-resistant polypropylene (Lyubeshkina, 1983). Numerous studies have been carried out that used lignosulfonate, kraft or organosolv lignin in blends with polyethylene, polypropylene or poly(ethylene-co-vinyl acetate) with a lignin content up to 70%. The polyblends could show good thermal stability (Lindberg et al., 1989), increased modulus (stiffness) with increased lignin content (Kubat and Stromvall, 1983), improved mechanical properties and electrical resistance (Kharade and Kale, 1998), the ability to absorb UV light (Kosikova et al., 1993), improved modulus and occasionally strength when the co-component to lignin contained polar functions (Ciemniecki and Glasser, 1989), and pronounced matrix reinforcement of blends with increased lignin content (Roesch and Muelhaupt, 1994). Gandini et al. could show that lignin acts as a stabiliser against the photo-oxidative degradation of polymers like polyolefins (Gandini and Belgacem, 2008; Gandini, 2008), which makes it possible to replace costly synthetic additives as long as the brown colour of the resulting blends does not affect the application. Besides polyolefins, also poly(vinyl chloride), poly(vinyl alcohol), hydroxypropyl cellulose and wheat starch have been blended with lignin and showed promising mechanical properties (Feldman, 2002; Rials and Glasser, 1989; Baumberger et al., 1997, 1998; Corradini et al., 1999).

Organosolv lignin has been incorporated into inks, varnishes and paints with positive effects on viscosity and misting with the addition of up to 10% w/w lignin. There were no negative side-effects of the addition apart from a brown discoloration, which was deemed irrelevant for most applications (Belgacem et al., 2003).

Stewart (2008) considers the commercial application of lignins to phenolic resin (PF resins), inevitable given the move towards sustainable practices in the chemical industry and the increasing price of phenol. The size of the phenolics market and the economic attractiveness of utilising lignin for a phenol replacement justify the large volume of research in this area. Phenol and formaldehyde react together by condensation to form a methylene bridge between phenol units at the ortho- and para- positions of the aromatic ring. As phenol can react with three molecules of formaldehyde and formaldehyde can react with two molecules of phenol, resoles are capable of forming a highly cross-linked 3-D network polymer. Because lignin consists of phenyl-propanoid units, it is capable of reacting with formaldehyde in a similar manner to phenol. The similarity of lignin and phenol is shown in Fig. 20.9.

The resin market represents a large potential sink especially for kraft and organosolv lignin, which have been incorporated into PF or epoxy resins in many studies (Cook and Sellers, 1989; Shiraishi, 1989; Ono and Sudo, 1989). In order to incorporate lignin into PF resin, lignin can be methylated or phenolated prior to addition. This ensures the chemical crosslinking of lignin into the resin. However, simple addition of lignin prior to synthesis is possible given a sufficiently reactive form of lignin and represents a significant capital cost saving over pre-synthesis reaction of lignin (Muller

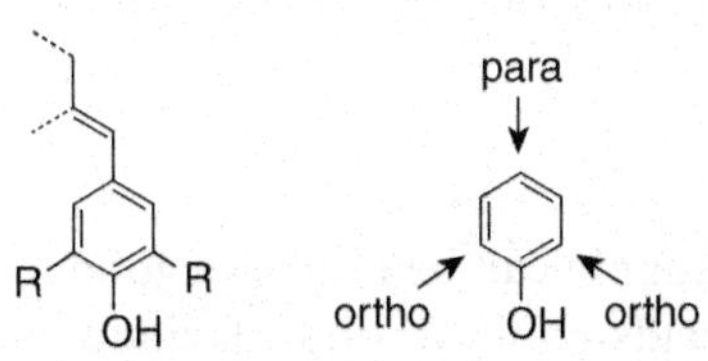

*20.9* Structure of lignin and position of formaldehyde addition onto phenol. R = H or O-CH3.

*et al.*, 1984). Lignin has been proven to reduce synthesis time (Benar *et al.*, 1999) and ammonia-modified lignin can improve thermal degradation resistance of PF resin (Alonso *et al.*, 2001).

Lignin has also successfully been used in epoxy resins (EP resins) since the 1960s and has been sold for example by Lenox Polymers Ltd (Port Huron, USA) until they liquidated in 2000. Epoxy resin containing 50% lignin can be produced with electrical and mechanical properties equivalent to non-lignin-containing epoxy. It is possible to produce large quantities of lignin-epoxy resin on the order of 500 kL/year using relatively simple purification methods (Simionescu *et al.*, 1993). Feldman (2002) could show in several studies that kraft lignin as a component in EP resins can improve the adhesive joint shear strength with a content up to 30% and the DSC data indicates a monophasic system up to 20% lignin. The production costs of the lignin-based EP resin were estimated to be 81% lower than the commercial EP resin. Studies investigating the influence of lignin type on the adhesive properties of EP–lignin polyblends showed that the addition of hardwood lignins results in a higher adhesivity than the addition of softwood lignin (Indulin). The improvement could also be related to the molecular weight of lignin. Hardwood lignin was shown to crosslink more efficiently than softwood lignin (Feldman, 2002). Methylolation of lignin in PF resins was shown to depend on the number of reactive sites (Peng *et al.*, 1993).

Another application for lignin is its use in the production of polyurethane films or foam. Polyurethane films have been produced incorporating Alcell® lignin. Flexible but weak polyurethanes were produced at low lignin content. However, relatively tough polyurethane could be produced with lignin content from 15 to 25% w/w. Above 30% w/w lignin, the polyurethanes were hard and brittle (Thring *et al.*, 1997). Precured polyurethanes have been synthesised by Gandini *et al.* from organosolv lignin, polyethylene glycol or polypropylene glycol and methylene diphenyl isocyanate. The resulting products are rigid polyurethane foams or sheet (Gandini *et al.*, 2002).

Some recent research showed that bioplastics can be produced from lignin by mixing a specific low-sulfur lignin with natural fibres and natural additives such as wax. The resulting thermoplastic material is recyclable and due to the additives can survive contact with water or saliva. The bioplastic can be injection-molded and is commercially available under the brand name ARBOFORM® (Calvo-Flores and Dobado, 2010; Nägele *et al.*, 2002).

*Metal ion absorption*

Quite a few studies have shown that lignin can be used as a good adsorbent for metal ions as well as dyes, bile acids, cholesterol, surfactants, pesticides and phenols. For kraft pine lignin, the binding strength for metals was found to be in the following order $Pb(II) > Cu(II) > Zn(II) > Cd(II) > Ca(II) >$

Sr(II). It was also found that protons or existing metals are released from the lignin during the uptake of other metal ions. At low pH, only the more tightly bound metals (Pb(II), Cu(II), Zn(II), Cd(II)) can compete with the protons for binding sites and therefore the uptake of Ca(II), Sr(II) and Li(I) only occurs at higher pH. Despite the differences between the different studies, the adsorption capacity of unmodified lignin is relatively low and inferior to activated carbon for the removal of metal ions from solutions (Suhas et al., 2007).

*High value products*

Due to the pulping process under high pressure and temperature, also low molecular weight products are obtained that can be transformed into high value products by oxidation. For instance, vanillin has been produced from lignosulfonates since 1937, and this was the dominant procedure for many years until environmental issues forced these mills to close. However, the Norwegian company Borregaard continued to produce vanillin from lignin and developed an environmentally friendly process, including an oxidation step with a copper catalyst that is economically viable and emits less $CO_2$ than other processes based on petrochemical precursors. Another valuable chemical derived from the pulping process is DMSO. It can be synthesised by oxidation of dimethyl sulfide (DMS) which occurs as a by-product of kraft pulping when lignin is treated with molten sulfur in alkaline media (Calvo-Flores and Dobado, 2010).

### 20.5.2 Emerging lignin products

*Antioxidants*

If used directly without blending with other polymers, lignin shows some interesting antioxidant behaviour and acts as a free-radical scavenger. It could be shown that free phenolic hydroxyl groups are involved in that mechanism and that ortho-methoxyl substituents help to stabilise the phenoxyl radicals by resonance (Pan *et al.* 2006b). Kraft lignin and steam-exploded lignin have shown antioxidant activity in human red blood cells, whereas certain water-soluble lignin derivatives present anti-viral activity *in vitro*. Some other derivatives have shown antibiotic and anti-carcinogenic activities which make them interesting materials for the health and pharmaceutical industry (Calvo-Flores and Dobado, 2010).

*Activated carbon/carbon black/carbon fibre*

During wartime, lignin was investigated as a substitute for carbon black and other fillers for rubbers. Since the studies could not be completed before the end of the Second World War, they were pursued afterwards (Keilen and Pollak, 1947) and as far back as 1947 lignin was added to various rubber latexes as a reinforcing agent. In 1964 the first practical test with tyres where half of the carbon black was replaced by kraft lignin was performed. Lignin is compatible with various rubber materials and does not affect the stability of those latexes even in high concentrations. Due to the low specific gravity of lignin, the resulting materials are much lighter than those filled with carbon black and the mechanical properties of lignin-reinforced rubbers can compete in terms of abrasion resistance, modulus, elongation, hardness and tensile strength (Table 20.5). There are basically two ways of introducing lignin into a rubber preparation: by co-precipitation of lignin and rubber from a lignin-rubber latex mixture, or by mixing hydrated lignin into the rubber (Feldman, 2002). However, the tyre safety and testing standards of today present high barriers to the modification of tyre composition and restrain utilisation.

Because of their highly porous structure with a large internal surface area, activated carbons have a good adsorption capacity for many substances. They are able to remove organic and inorganic pollutants from liquid as well as gaseous phases. Activated carbons are available as powdered activated carbons (PAC) or granular activated carbons (GAC) and differ in diameter and surface area. Coal, lignite, peat, wood and coconut shells are commonly used to prepare activated carbons but also lots of other materials have been investigated as a source material. Among them, lignin is a very promising material from which activated carbon with a high surface area and pore volume can be produced that is comparable to common activated carbon. The adsorption capacity of activated carbons can be controlled by the parameters of either physical or chemical activation or a post-activation treatment. The results of sorption studies of inorganic or

Table 20.5 Physical properties of lignin and carbon black reinforced styrene-butadien copolymer rubber (SBR) at 68 parts of lignin per 100 parts of rubber (oil-extended SBR)

| Property | Lignin | HAF[a] carbon black | ISAF[b] carbon black |
| --- | --- | --- | --- |
| Modulus (MPa) | 3.6 | 4.2 | 5.1 |
| Tensile strength (MPa) | 21.7 | 17.5 | 20.5 |
| Elongation (%) | 720 | 720 | 750 |
| Hardness (Shore) | 54 | 56 | 61 |
| Tear resistance (MPa) | 2.4 | 2.1 | 2.3 |
| Corrected pico abrasion | 86 | 91 | 114 |

[a]High abrasion furnace.
[b]Intermediate super abrasion furnace.

organic substances are promising, but more systematic studies are necessary because some published works are contradictory. Also the relationship between adsorption properties and molecular structure has to be investigated in more detail (Suhas et al., 2007). Although the use of lignin as precursor for activated carbon is more costly than wood dust or coconuts shells, its ability to form fibres could lead to interesting applications such as activated carbon fibres.

Carbon fibres are some of the most important and widely used advanced engineering materials: they are lightweight, fatigue resistant materials that possess high strength/modulus and high stiffness. These unique properties result from their flawless structure and the development of highly anisotropic graphic crystallites orientated along the fibre axis during the production process. Carbon fibres are manufactured by thermally treating fibres at 1,000–2,000°C in an inert atmosphere while maintaining the fibrous structure. This is aided by a stabilisation stage in which the precursor fibres are heated under tension at 200–300°C in the presence of air. This causes crosslinking on the fibre surface, among other reactions, and prevents shrinking, melting and fusing (Kadla et al., 2002b).

Advanced fibres, such as carbon fibres, are routinely used in sports equipment, marine products and the transportation industries. There are primarily two types of precursor materials of commercial significance: pitch (petroleum or coal) and polyacrylonitrile (PAN). However, increasing demand, rising energy costs and concern over the sustainability and environmental impact of fossil fuels is driving the need to find sustainable alternatives. Lignin, as a polyaromatic macromolecule, is thought to be a petroleum progenitor and therefore an ideal biopolymer precursor for the production of carbon fibres.

Although widely studied for the production of carbon fibres (Kadla et al., 2002a; Sudo and Shimizu, 1992; Otani et al., 1969; Uraki et al., 1995), lignin-based carbon fibres still suffer from poor fibre properties and limited lignin availability. As compared to pitch, lignin-based carbon fibres have an advantage of higher carbonisation yields and shorter crosslinking stage, believed to be due to the lignin structure.

Production of functional fibres by adding carbon nanotubes or magnetic particles to the lignin can open the market to applications like magnetic shielding or conductive materials. The mechanical expectations for these materials are not as high as for other applications and fibres with lower tensile strength can readily be used.

The properties of lignin-based carbon fibres depend largely upon the source of lignin. Lignosulfonates and kraft lignin typically have sodium impurities, which can lead to inclusions and microvoids arising from catalytic graphitisation and thereby decrease mechanical performance (Johnson et al., 1975). This can be eliminated via ion exchange and/or fractionation of the feed lignin (Kadla *et al.*, 2002a; Baker, 2007). By contrast, organosolv lignins have fewer inorganic impurities, have better thermal properties, and can readily be spun into fibres. However, their lower softening point can lead to problems in the subsequent thermal processing (Kadla *et al.*, 2002a). As with the utilisation of petroleum pitch, wherein thermal and/or catalytic modification produces mesogens and subsequently high performance carbon fibres (Donnet *et al.*, 1998), thermal and catalytic treatment of lignin (Sudo *et al.*, 1993) enhances fibre properties, but the underlying lignin structure seems to be a limiting factor (Dave *et al.*, 1993).

*Platform chemicals*

Several thermochemical methods have been studied in the recent past in order to depolymerise lignin and to convert it into value-added chemicals. Pyrolysis, gasification, hydrogenolysis, chemical oxidation and hydrolysis under supercritical conditions are the major methods to produce pyrolytic oil, syngas or phenols. An overview of these methods and the resulting products is shown in Fig. 20.10.

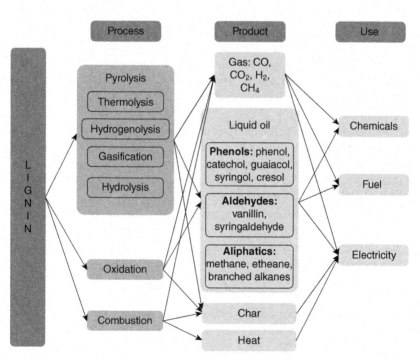

*20.10* Major thermochemical lignin conversion processes and their potential products.

Pyrolysis means the thermal breakdown of an organic substance into smaller units in the absence of air. The limited oxygen supply prevents the further combustion to $CO_2$. The pyrolysis of lignin is a complex process and is affected by the lignin type, heating rate, reaction temperature and additives. The main products of lignin pyrolysis are gaseous hydrocarbons such as $CO_2$ and CO, volatile liquids (methanol, acetone and acetaldehyde), monolignols, monophenols (phenol, guaiacol, syringol and catechol), and other monosubstituted phenols. Depending on the reaction temperature, parts of the lignin are also converted to different amounts of char (Pandey and Kim, 2011). The cleavage of weaker bonds in lignin occurs at lower temperatures while the cracking of aromatic rings happens at fairly high temperatures, thus the pyrolysis is covering a wide range of temperatures (Yang et al., 2007; Ferdous et al., 2002). Ferdous' studies of Alcell® and lignin show that the conversion at lower temperatures is higher for lower heating rates, but for temperatures above 700°C, the conversion is higher for higher heating rates. However, the fast pyrolysis has the advantage of lower char and coke formation (Windt et al., 2009). Several groups have also studied the influence of lignin source and isolation technique and it was shown that the structure of extracted lignin is (slightly) different from native lignin. Thermoanalysis of various lignin preparations showed that the highly condensed Klason lignin is more heat resistant than lignin from steam explosion, enzymatic hydrolysis or hydrochloric acid treatment, which became evident by the significantly higher solid residue and the lower yield of monophenols after pyrolysis (Gardner et al., 1985). Due to their different H/C ratio, the amount of syngas is higher after pyrolysis of kraft lignin than from Alcell lignin (Ferdous et al., 2002).

When pyrolysis is performed in the presence of hydrogen, it is known as hydrogenolysis or hydrogenation. Different solvents or catalysts can be added in order to speed up the reaction and to increase the yield. While neat pyrolysis is not the best choice for the production of liquid fuels or chemicals because it produces mainly gases and solid coke, hydrogenolysis is a very promising method with a high net conversion, high yield of monophenols and low solid residue. Shabtai et al. (1999a, 1999b) proposed a three-step process that includes base-catalysed depolymerisation, hydrodeoxygenation and hydrocracking to produce reformulated gasoline from lignin. The base-catalysed depolymerisation is usually performed with an alkali hydroxide and a supercritical alcohol such as methanol or ethanol. At temperatures of about 270°C the system generates a pressure of 140 bar. Shabtai modified this initial process by a selective hydrocracking reaction followed by an exhaustive etherification reaction and partial ring hydrogenation. This leads to a reformulated, partial oxygenated/etherified gasoline (Shabtai et al., 2001; Pandey and Kim, 2011).

A new approach for the depolymerisation of lignin has been described by Kleinert and Barth (2008). Lignin is subjected to a reductive treatment with formic acid and ethanol at temperatures of 350–400°C. The lignin is simultaneously depolymerised and deoxygenised. The product separates into a lighter organic phase which consists mainly of small substituted phenols and a heavier phase which consists of C8–C10 aliphatics. In a follow-up study, this procedure could be successfully applied to lignin isolated after enzymatic degradation of wood (Kleinert et al., 2009). Hydrogen-donating solvents such as tetralin as well as reactions under pressurised molecular hydrogen with various catalysts have been described to be advantageous for the depolymerisation of lignin (Pandey and Kim, 2011).

Just like wood-derived and vegetable bio-oils, lignin (Alcell® lignin) can be turned into gasoline range hydrocarbons by pyrolysis using the zeolite catalyst HZSM-5 (Chantal et al., 1985). Compared to typical hydrotreatment processes requiring high operating pressures and hydrogen, this process can be operated at atmospheric pressure and at moderate temperatures of at least 550–650°C. The highest yield of liquid product of 43% w/w consisting mainly of aromatic hydrocarbons such as benzene, toluene and xylene could be obtained at a temperature of 550°C and a mass flow of $5 h^{-1}$. With increasing temperature the yield of the liquid product decreases while the amount of gas increases dramatically due to extensive cleavage of the major bonds in lignin which can be concluded from the increase in $CO_2$ and CO production. Also the amount of char and coke decreases with rising temperature. At 550°C the amount of char and coke is relatively high at 38%. Those results are quite promising for the production of hydrocarbons from lignin, particularly because of the moderate reaction conditions and the high yields. In addition, the zeolite catalyst can be easily regenerated by heating in an air flow for 1 h at 600°C. The only drawback of this procedure is the utilisation of acetone as a solvent for lignin. Although it was the only feasible solvent in this process, acetone is a quite valuable chemical itself and due to the low boiling point less safe at the chosen reaction temperatures. For the future, it would be desirable to find another solvent to minimise or substitute the amount of acetone (Mullen and Boateng, 2010; Guo et al., 2008).

All these procedures seem to be promising for the depolymerisation of lignin and therefore the production of platform chemicals and gasoline. But they require quite harsh reaction conditions (temperature and pressure), expensive catalysts and the separation of a variety of different reaction products. It will be a challenging task for the future to establish an economically viable process that can compete with common petrochemical processes. Even though the platform chemical market is large, the value is not yet high enough to warrant production from isolated lignin. This is expected to change as fossil resources become more costly to extract and prices for specialty chemicals increase.

*Enzymatically treated lignins*

As a result of governmental requirements and also consumer demand for eco-friendly, biodegradable and non-toxic products, the focus of research and production worldwide has turned towards renewable and natural raw materials and cleaner industrial practices. One emerging trend is the application of biotechnological processes, which includes the use of bacteria and fungi or treatment with isolated enzymes. Also in lignin chemistry, the utilisation of enzymatic treatment is becoming increasingly popular for breakdown of lignin or the modification of the functional groups.

Up to now, lignin degradation has been studied mainly in white-rot and brown-rot fungi, which produce enzymes that can metabolise lignin (Wong, 2009). The main extracellular enzymes active in lignin degradation are the heme-containing lignin peroxidase (ligninase, LiP, EC 1.11.1.14) and manganese peroxidase (MnP, EC 21.11.1.13) and the Cu-containing laccase (benzenediol:oxygen oxidoreductase, EC1.10.3.2) (Sena-Martins *et al.*, 2008). However, these findings have not yet evolved into commercialisation of lignin breakdown, mostly because of the inherent difficulties in fungal genetics and protein expression (Ahmad *et al.*, 2011). Lignin metabolising soil bacteria has been reported (Crawford *et al.*, 1983; Zimmermann, 1990; Ramachandra *et al.*, 1988; Vicuna, 1988) but the enzymatic process of bacterial lignin degradation is not yet well characterised. Ahmad *et al.* (2011) reported two spectrophotometric assays for lignin breakdown which could identify several bacterial strains with lignin degradation activity. They were also able to identify the existence of two DyP-type peroxidases, DyP B thereof being a lignin peroxidase. On-going research in this area will help to elucidate the enzymatic cleavage of lignin linkages. The obtained degradation product could comprise small molecules as well as lower molecular weight polymers with improved properties due to new/different functionalities than present in native lignin. This could open a large area of promising applications.

The various functional groups present in lignin are highly accessible for modification by chemical, physical as well as enzymatic reactions. Compared to chemical or physical treatments, enzymatic modifications possess the advantages of high selectivity and efficiency, they are performable under mild conditions and a wide range of substrates can be used. However, enzymes are only reusable when sufficiently immobilised and they are quite sensitive to denaturating agents and several sensory or toxicological effects (Sena-Martins *et al.*, 2008). For lignin processing, the oxidative enzymes of certain ligninolytic fungi have been primarily investigated. For several reasons the industrial production of ligninolytic enzymes is still very low, making the research quite expensive and uneconomic. Nevertheless, much research is on-going in this field and the results present high potential for future use. Copolymers of straw pulp lignin with cresol and horseradish peroxidise as catalyst have been produced and can replace common phenolic resins. Also copolymers from different lignins (organosolv, Indulin AT and a synthetic hydroxypropylated lignin) with vanillic acid, diisocyante and acrylamide catalysed by laccase could be produced (Sena-Martins *et al.*, 2008). In order to reduce the amount of harmful formaldehyde in wood binders, Hüttermann *et al.* (2001) reported some approaches with binders that were derived from laccase catalysed reactions between water-insoluble lignin and common resins. The treatment of organosolv (acetosolv) lignin with polyphenoloxidase was reported to result in a higher amount of hydroxyl and carbonyl groups and therefore enhances the chelating properties of lignin. Those products can be used for the treatment of heavy metals containing wastewater (Gonçalves and Benar, 2001). In addition, environmentally friendly coatings and paintings can be produced by the enzymatic polymerisation of different lignin materials (lignosulfonates, kraft and organosolv lignin) and enzymes such as catechol oxidase, laccase or peroxidase (Bolle and Aehle, 2001).

## 20.6 Future trends

In the short term, a large market for lignin from the ubiquitous kraft process must be found. This is necessary for the de-bottlenecking of kraft mills around the world. The most promising market for kraft lignin is the resins and adhesives market, which is substantial and represents an increase in value over traditional heat and power generation. Given the push for renewables and the phenolic nature of lignin, penetration into this market is considered inevitable by some (Stewart, 2008). In addition, if lignin improves material properties, this will be reflected in its market price.

In the medium term, more lignin from unsulfonated sources will become available. This lignin, from biorefinery platforms, will be unique in properties compared to other feedstocks and have small-scale availability. Initially, the resin and adhesives market may act as a catch-all for this more reactive lignin. However, a product niche of high value and sufficient market size must be found for each lignin source. For example, hydrophobic organosolv lignins could add value to resins used for water resistant particleboard. Platform chemicals are an attractive medium-term market as there is already an existing market and no product development is needed.

In the long term, biorefineries will be tuned to produce lignin for specific materials of high value, and lignin is expected to yield greater incomes than the other components of wood.

## 20.7 Sources of further information and advice

Key books, major trade/professional bodies, research and interest groups, web sites:

- http://biomaterials.forestry.ubc.ca/
- http://www.lignol.ca/
- http://www.fpinnovations.ca/
- http://www.lignoworks.ca/

## 20.8 References

Ahmad, M., Roberts, J.N., Hardiman, E.M., Singh, R., Eltis, L.D. and Bugg, T.D.H., 2011. Identification of DypB from Rhodococcus jostii RHA1 as a lignin peroxidase. *Biochemistry*, 50(23), 5096–5107.

Alonso, M.V., Rodriguez, J.J., Oliet, M., Rodriguez, F., Garcia, J. and Gilarranz, M.A., 2001. Characterization and structural modification of ammonic lignosulfonate by methylolation. *Journal of Applied Polymer Science*, 82(11), 2661–2668.

Arato, C., Pye, E.K. and Gjennestad, G., 2005. The lignol approach to biorefining of woody biomass to produce ethanol and chemicals. *Applied Biochemistry and Biotechnology*, 121–124, 871–882.

Aziz, S. and Sarkanen, K.V., 1989. Organosolv pulping – a review. *Tappi Journal*, 72(3), 169–175.

Baeza, J., Urizar, S., Erismann, N.M., Freer, J., Schmidt, E. and Duran, N., 1991. Organosolv pulping – V: Formic acid delignification of *Eucalyptus globulus* and *Eucalyptus grandis*. *Bioresource Technology*, 37(1), 1–6.

Baeza, J., Urizar, S., Freer, J., Rodriguz, J., Peralta-Zamora, P. and Duran, N., 1999. Organosolv pulping. IX. Formic acid/acetone delignification of *Pinus radiata* and *Eucalyptus globulus*. *Cellulose Chemistry and Technology*, 33(3–4), 289–301.

Baker, F., 2007. US Patent Application 20070142225. Activated carbon fibres and engineered forms from renewable resources.

Baumberger, S., Lapierre, C., Monties, B., Lourdin, D. and Colonna, P., 1997. Preparation and properties of thermally moulded and cast lignosulfonates-starch blends. *Industrial Crops and Products*, 6(3–4), 253–258.

Baumberger, S., Lapierre, C., Monties, B. and Valle, G.D., 1998. Use of kraft lignin as filler for starch films. *Polymer Degradation and Stability*, 59(1–3), 273–277.

Bayerbach, R. and Meier, D., 2009. Characterization of the water-insoluble fraction from fast pyrolysis liquids (pyrolytic lignin). Part IV: Structure elucidation of oligomeric molecules. *Journal of Analytical and Applied Pyrolysis*, 85(1–2), 98–107.

Belgacem, M.N., Blayo, A. and Gandini, A., 2003. Organosolv lignin as a filler in inks, varnishes and paints. *Industrial Crops and Products*, 18(2), 145–153.

Benar, P., Gonclaves, A.R., Mandelli, D. and Schuchardt, U., 1999. Eucalyptus organosolv lignins: study of the hydroxymethylation and use in resols. *Bioresource Technology*, 68(1), 11–16.

Biermann, C.J., 1996. *Handbook of Pulping and Papermaking* (2nd Edn), Elsevier, San Diego, CA.

Bolle, R. and Aehle, W., 2001. US Patent 6217942, Lignin-based coating.

Boye, F., 1985. *Utilization of Lignins and Lignin Derivatives [supplement]*, Institute of Paper Chemistry, Appleton, WI.

Bura, R., Mansfield, S., Saddler, J. and Bothast, R.J., 2002. $SO_2$-catalyzed steam explosion of corn fiber for ethanol production. *Applied Biochemistry and Biotechnology*, 98–100, 59–72.

Calvo-Flores, F.G. and Dobado, J.A., 2010. Lignin as renewable raw material. *ChemSusChem*, 3(11), 1227–1235.

Capanema, E.A., Balakshin, M.Y. and Kadla, J. F., 2004. A comprehensive approach for quantitative lignin characterization by NMR spectroscopy. *Journal of Agricultural and Food Chemistry*, 52(7), 1850–1860.

Capanema, E., Balakshin, M. and Kadla, J.F., 2005. Quantitative characterization of a hardwood milled wood lignin by nuclear magnetic resonance spectroscopy. *Journal of Agricultural and Food Chemistry*, 53, 9639–9649.

Chantal, P.D., Kaliaguine, S. and Grandmaison, G.L., 1985. Reactions of phenolic compounds over HZSM-5. *Applied Catalysis*, 18(1), 133–145.

Ciemniecki, S.L. and Glasser, W.G., 1989. Polymer blends with hydroxypropyl lignin. In W.G. Glasser and S. Sarkanen, eds. *Lignin: Properties and Materials*. American Chemical Society, Washington, DC, 452–463.

Cook, P.M. and Sellers, T., 1989. Organosolv lignin-modified phenolic resins. In W.G. Glasser and S. Sarkanen, eds. *Lignin: Properties and Materials*. American Chemical Society, Washington, DC, 324–333.

Corradini, E., Pineda, E.A.G. and Hechenleitner, A.A.W., 1999. Lignin-poly (vinyl alcohol) blends studied by thermal analysis. *Polymer Degradation and Stability*, 66(2), 199–208.

Crawford, D.L., Pometto (III), A.L. and Crawford, R.L., 1983. Lignin degradation by *streptomyces viridosporus*: isolation and characterization of a new polymeric lignin degradation intermediate. *Applied and Environmental Microbiology*, 45(3), 898–904.

Dave, V., Prasad, A., Marand, H. and Glasser, W.G., 1993. Molecular organization of lignin during carbonization. *Polymer*, 34(15), 3144–3154.

Dimmel, D., 2010. Overview. In C. Heitner, D. R. Dimmel, and J. A. Schmidt, eds. *Lignin and Lignans Advances in Chemistry*. Taylor & Francis Group, Boca Raton, FL, pp. 1–10.

Donnet, J.-B., Wang, T.K., Rebouillat, S. and Peng, J.C.M., 1998. *Carbon Fibers*, 3rd edn., Marcel Dekker, New York.

Dos Santos, D.A., Rudnitskaya, A. and Evtuguin, D.V., 2012. Modified kraft lignin for bioremediation applications. *Journal of Environmental Science and Health. Part A*, 47(2), 298–307.

El Hage, R., Brosse, N., Sannigrahi, P. and Ragauskass, A., 2010. Effects of process severity on the chemical structure of *Miscanthus* ethanol organosolv lignin. *Polymer Degradation and Stability*, 95(6), 997–1003.

El Mansouri, N.-E. and Salvadó, J., 2006. Structural characterization of technical lignins for the production of adhesives: application to lignosulfonate, kraft, soda-anthraquinone, organosolv and ethanol process lignins. *Industrial Crops and Products*, 24, 8–16.

Erismann, N.M., Freer, J., Baeza, J. and Duran, N., 1994. Organosolv pulping – VII: Delignification selectivity of formic acid pulping of *Eucalyptus grandis*. *Bioresource Technology*, 47(3), 247–256.

Feldman, D., 2002. Lignin and its polyblends – a review. In T.Q. Hu, ed. *Chemical Modification, Properties and Usage of Lignin*. Springer, New York, pp. 81–99.

Ferdous, D., Dalai, A.K., Bej, S.K. and Thring, R.W., 2002. Pyrolysis of lignins: experimental and kinetics studies. *Energy & Fuels*, 16(6), 1405–1412.

Fort, D.A. Remsing, R.C., Swatloski, R.P., Moyna, P., Moyna, G. and Rogers, R.D., 2007. Can ionic liquids dissolve wood? Processing and analysis of lignocellulosic materials with 1-n-butyl-3-methylimidazolium chloride. *Green Chemistry*, 9(1), 63.

Gandini, A., 2008. Polymers from renewable resources: a challenge for the future of macromolecular materials. *Macromolecules*, 41(24), 9491–9504.

Gandini, A. and Belgacem, M.N., 2008. Lignins as components of macromolecular materials. In M.N. Belgacem and A. Gandini, eds. *Monomers, Polymers and Composites from Renewable Resources*. Elsevier Amsterdam, pp. 243–271.

Gandini, A., Belgacem, M.N., Guo, Z-X. and Montanari, S., 2002. Lignins as macromonomers for polyesters and polyurethanes. In T.Q. Hu, ed. *Chemical Modification, Properties and Usage of Lignin*. Springer, New York, pp. 57–80.

Gardner, D.J., Schultz, T.P. and McGinnis, G.D., 1985. The pyrolytic behavior of selected lignin preparations. *Journal of Wood Chemistry and Technology*, 5(1), 85–110.

Gargulak, J.D. and Lebo, S.E., 2000. Commercial use of lignin-based material. In W. Glasser, R. Northey and T. Schultz, eds. *Lignin: Historical, Biological, and Materials Perspectives*. American Chemical Society, Washington DC, pp. 304–320.

Gierer, J., 1985. Chemistry of delignification. Part 1: General concept and reactions during pulping. *Wood Science and Technology*, 19, 289–312.

Gonçalves, A R. and Benar, P., 2001. Hydroxymethylation and oxidation of Organosolv lignins and utilization of the products. *Bioresource Technology*, 79(2), 103–111.

Gosselink, R., 2004. Co-ordination network for lignin-standardisation, production and applications adapted to market requirements (EUROLIGNIN). *Industrial Crops and Products*, 20(2), 121–129.

Gosselink, R.J.A., van Dam, J., Jong, E.D., Scott, E.L., Sanders, J.P.M., Li, J. and Gellerstedt, G., 2010. Fractionation, analysis, and PCA modeling of properties of four technical lignins for prediction of their application potential in binders. *Holzforschung*, 64, 193–200.

Goyal, G., Lora, J. and Pye, E., 1992. Autocatalyzed organosolv pulping of hardwoods – effect of pulping conditions on pulp properties and characteristics of soluble and residual lignin. *Tappi Journal*, 75(2), 110–116.

Gratzl, J.S. and Chen, C.S., 2000. Chemistry of pulping: lignin reactions. In W.G. Glasser, R.A. Northey, and T.P. Schultz, eds. *Lignin: Historical, Biological, and Materials Perspectives*. American Chemical Society, Washington DC, pp. 392–421.

Grethlein, H.E. and Converse, A.O., 1991. Common aspects of acid prehydrolysis and steam explosion for pretreating wood. *Bioresource Technology*, 36, 77–82.

Guo, X., Zhang, S. and Shan, X.-Q., 2008. Adsorption of metal ions on lignin. *Journal of Hazardous Materials*, 151(1), 134–142.

Hallberg, C., O'Connor, D., Rushton, M., Pye, K.E., Gjennestad, G., Berlin, A. and MacLachlan, J.R., 2008. US Patent 7465791. Continuous counter-current organosolv processing of lignocellulosic feedstocks.

Heitz, M., Capek-Menard, E., Koeberle, P.G., Gagne, J. and Chornet, E., 1991. Fractionation of *Populus tremuloides* at the pilot plant scale: optimization of steam pretreatment conditions using the STAKE II technology. *Technology*, 35(1), 23–32.

Holladay, J.E., Bozell, J.J., White, J.F. and Johnson, D., 2007. *Top Value Added Chemicals from Biomass Vol. 2*. Pacific Northwest National Laboratory, Richland, WA.

Hu, T.Q., ed., 2002. *Chemical Modification, Properties, and Usage of Lignin*, Springer, New York.

Hüttermann, A., Mai, C. and Kharazipour, A., 2001. Modification of lignin for the production of new compounded materials. *Applied Microbiology and Biotechnology*, 55(4), 387–384.

Johnson, D., Tomizuka, I. and Watanabe, O., 1975. The fine structure of lignin-based carbon fibres. *Carbon*, 13(4), 321–325.

Kadla, J.F. and Dai, Q., 2006. Pulp. In *Kirk-Othmer Encyclopedia of Chemical Technology, 21*. Wiley, New York, pp. 1–47.

Kadla, J.F., Kubo S., Venditti, R.A., Gilbert, R.D., Compere, A.L., and Griffith, W., 2002a. Lignin-based carbon fibers for composite fiber applications. *Carbon*, 40(15), 2913–2920.

Kadla, J.F., Kubo, S., Gilbert, R.D. and Venditti, R.A., 2002b. Lignin based carbon fibers. In T.Q. Hu, ed., *Chemical Modification, Properties and Usage of Lignin*. Springer, New York, pp. 121–137.

Keilen, J.J. and Pollak, A., 1947. Lignin for reinforcing rubber. *Industrial & Engineering Chemistry*, 39(4), 480–483.

Kharade, A.Y. and Kale, D.D., 1998. Effect of lignin on phenolic novolak resins and molding powder. *Eur. Polym. J.*, 34(2), 201–205.

Kim, J.-Y., Shin, E-J., Eom, I-Y., Won, K., Kim, Y.H., Choi, D., Choi, I-G. and Choi, J.W., 2011. Structural features of lignin macromolecules extracted with ionic liquid from poplar wood. *Bioresource Technology*, 102(19), 9020–9025.

Kleinert, M. and Barth, T., 2008. Towards a lignincellulosic biorefinery: direct one-step conversion of lignin to hydrogen-enriched biofuel. *Energy & Fuels*, 22(2), 1371–1379.

Kleinert, M., Gasson, J.R. and Barth, T., 2009. Optimizing solvolysis conditions for integrated depolymerisation and hydrodeoxygenation of lignin to produce liquid biofuel. *Journal of Analytical and Applied Pyrolysis*, 85(1–2), 108–117.

Kosikova, B., Kacurakova, M. and Demianova, V., 1993. Photooxidation of the composite lignin/polypropylene films. *Chem. Papers*, 42(2), 132–136.

Kubat, J. and Stromvall, H.E., 1983. Properties of injection molded lignin-filled polyethylene and polystyrene. *Plast. Rubber Process. Appl.*, 3, 111–118.

Kubo, S. and Kadla, J. F., 2006. Effect of poly(ethylene oxide) molecular mass on miscibility and hydrogen bonding with lignin. *Holzforschung*, 60(3), 245–252.

Laser, M., Schulman, D., Allen, S.G., Lichwa, J., Antal, M.J. and Lynd, L.R., 2002. A comparison of liquid hot water and steam pretreatments of sugar cane bagasse for bioconversion to ethanol. *Bioresource Technology*, 81(1), 33–44.

Lee, S.H., Doherty, T.V., Linhardt, R.J. and Dordick, J.S., 2009. Ionic liquid-mediated selective extraction of lignin from wood leading to enhanced enzymatic cellulose hydrolysis. *Biotechnology and Bioengineering*, 102(5), 1368–1376.

Li, J., Henriksson, G. and Gellerstedt, G., 2007. Lignin depolymerization/ repolymerization and its critical role for delignification of aspen wood by steam explosion. *Bioresource Technology*, 98(16), 3061–3068.

Lindberg, J.J., Kuusela, T.A. and Levon, K., 1989. Specialty polymers from lignin. In W.G. Glasser and S. Sarkanen, eds., *Lignin: Properties and Materials*. American Chemical Society, Washington, DC, pp. 190–204.

Lora, J.H. and Glasser, W.G., 2002. Recent industrial applications of lignin – a sustainable alternative to nonrenewable materials. *Journal of Polymers and the Environment*, 10, 39–48.

Lyubeshkina, E.G., 1983. Lignins as a component of polymer composition materials. *Russ. Chem. Rev.*, 52(7), 1196–1224.

Marton, W., 1971. Lignins: occurence, formation, structure, and reactions. In K.V. Sarkanen and C.H. Ludwig, eds. *Lignins: Occurence, Formation, Structure, and Reactions*. Wiley-Interscience, New York, pp. 639–694.

Moerck, R., Yoshida, H. and Krinstad, K.P., 1986. Fractionation of kraft lignin by successive extraction with organic solvents 1. Functional groups, 13C-NMR-spectra and molecular weight distributions. *Holzforschung*, 40 (Suppl), 51–60.

Mullen, C.A. and Boateng, A.A., 2010. Catalytic pyrolysis-GC/MS of lignin from several sources. *Fuel Processing Technology*, 91(11), 1446–1458.

Muller, P.C., Kelley, S.S. and Glasser, W.G., 1984. Engineering plastics from lignin. 9. Phenolic resin synthesis and characterization. *Journal of Adhesion*, 17(3), 185–206.

Nadif, A., Hunkeler, D. and Käuper, P., 2002. Sulfur-free lignins from alkaline pulping tested in mortar for use as mortar additives. *Bioresource Technology*, 84(1), 49–55.

Nägele, H. Pfitzer, J., Nägele, E., Inone, E.R., Eisenreich, N., Eckl, W. and Eyerer, P., 2002. Arboform – a thermoplastic, processable material from lignin and natural fibers. In T.Q. Hu, ed. *Chemical Modification, Properties and Usage of Lignin*. Springer, New York, pp. 101–119.

Olivier-Bourbigou, H., Magna, L. and Morvan, D., 2010. Ionic liquids and catalysis: recent progress from knowledge to applications. *Applied Catalysis A: General*, 373(1–2), 1–56.

Ono, H.-K. and Sudo, K., 1989. Wood adhesives from phenolysis lignin. In W. Glasser and S. Sarkanen, eds. *Lignin: Properties and Materials*, American Chemical Society, Washington, DC, pp. 334–345.

Otani, S., Yoshihiko, F., Bunjiro, I. and Kesao, S., 1969. US Patent 3461082. Method for producing carbonized lignin fiber.

Pan, X., Gilkes, N., Kadla, J., Pye, K., Saka, S., Gregg, D., Ehara, K., Xie, D., Lam, D. and Saddler, J., 2006a. Bioconversion of hybrid poplar to ethanol and co-products using an organosolv fractionation process: optimization of process yields. *Biotechnology and Bioengineering*, 94(5), 851–861.

Pan, X., Kadla, J.F., Ehara, K., Gilkes, N. and Saddler, J., 2006b. Organosolv ethanol lignin from hybrid poplar as a radical scavenger: relationship between lignin structure, extraction conditions, and antioxidant activity. *Journal of Agricultural and Food Chemistry*, 54(16), 5806–5813.

Pandey, M.P. and Kim, C.S., 2011. Lignin depolymerization and conversion: a review of thermochemical methods. *Chemical Engineering & Technology*, 34(1), 29–41.

Parajo, J.C., Alonso, J.L. and Vazquez, G., 1993. On the behaviour of lignin and hemicelluloses during the acetosolv processing of wood. *Bioresource Technology*, 46(3), 233–240.

Parajo, J.C., Alonso, J. and Santos, V., 1995a. Kinetics of *Eucalyptus* wood fractionation in acetic acid-HCl-water media. *Bioresource Technology*, 51(2–3), 153–162.

Parajo, J.C., Alonso, J.L. and Santos, V., 1995b. Kinetics of catalyzed organosolv processing of pine wood. *Industrial & Engineering Chemistry Research*, 34(12), 4333–4342.

Peng, W., Riedl, B. and Barry, A.O., 1993. Study on the kinetics of lignin methylolation. *Journal of Applied Polymer Science*, 48(10), 1757–1763.

Pinkert, A., Goeke, D.F., Marsh, K.N. and Pang, S., 2011. Extracting wood lignin without dissolving or degrading cellulose: investigations on the use of food additive-derived ionic liquids. *Green Chemistry*, 13(11), 3124.

Pu, Y., Jiang, N. and Ragauskas, A.J., 2007. Ionic liquid as a green solvent for lignin. *Journal of Wood Chemistry and Technology*, 27(1), 23–33.

Pye, E. and Lora, J., 1991. The Alcell process. a proven alternative to kraft pulping. *Tappi Journal*, 74(3), 113–118.

Qin, W. and Kadla, J.F., 2012. Carbon fibers based on pyrolytic lignin. *Journal of Applied Polymer Science*, 126, E203-E212.

Ramachandra, M., Crawford, D. and Hertel, G., 1988. Characterization of an extracellular lignin peroxidase of the lignocellulolytic actinomycete *Streptomyces viridosporus*. *Appl. Environ. Microbiol.*, 54, 3057–3063.

Raveendran, K. and Ganesh, A., 1996. Heating value of biomass and biomass pyrolysis products. *Fuel*, 75(15), 1715–1720.

Rials, T.G. and Glasser, W.G., 1989. Multiphase materials with lignin. IV. Blends of hydroxypropyl cellulose with lignin. *Journal of Applied Polymer Science*, 37(8), 2399–2415.

Roesch, J. and Muelhaupt, R., 1994. Mechanical properties of organosolv-lignin-filled thermoplastics. *Polymer Bulletin*, 32, 361–365.

Sarkanen, K.V. and Glasser, W.G., 1989. *Lignin: Properties and Materials*, ACS Publications, Washington, DC.

Scholze, B., Hanser, C. and Meier, D., 2001. Characterization of the water-insoluble fraction from fast pyrolysis liquids (pyrolytic lignin) Part II. GPC, carbonyl groups, and 13 C-NMR. *Journal of Analytical and Applied Pyrolysis*, 58–59, 387–400.

Schütt, F., Westerbeng, B., Horn, S.J., Puls, J. and Saake, B., 2012. Steam refining as an alternative to steam explosion. *Bioresource Technology*, 111, 476–481.

Sena-Martins, G., Almeida-Vara, E. and Duarte, J.C., 2008. Eco-friendly new products from enzymatically modified industrial lignins. *Industrial Crops and Products*, 27(2), 189–195.

Shabtai, J.S., Zmierczak, W.W., Chornet, E. and Johnson, D.K., 1999a. Conversion of lignin. 2. Production of high-octane fuel additives. *ACS Div. Fuel Chem. Preprints of Symposia*, 44(2), 267–272.

Shabtai, J.S., Zmierczak, W.W. and Chornet, E., 1999b. US Patent 5959167. Process for conversion of lignin to reformulated hydrocarbon gasoline.

Shabtai, J.S., Zmierczak, W.W. and Chornet, E., 2001. US Patent 6172272. Process for conversion of lignin to reformulated, partially oxygenated gasoline.

Shiraishi, N., 1989. Recent progress in wood dissolution and adhesives from Kraft lignin. In W.G. Glasser and S. Sarkanen, eds. *Lignin: Properties and Materials*. American Chemical Society, Washington, DC, pp. 488–495.

Simionescu, C.I., Rusan, V., Macoveanu, M.M., Cazacu, G., Lipsa, R., Vasile, C., Stoleriu, A. and Ioanid, A., 1993. Lignin/epoxy composites. *Composites Science and Technology*, 48(1–4), 317–323.

Smook, G., 1986. *Handbook for Pulp and Paper Technologists*, 3rd ed., Angus Wilde Publications, Vancouver.

Stark, A., 2011. Ionic liquids in the biorefinery: a critical assessment of their potential. *Energy & Environmental Science*, 4(1), 19.

Stewart, D., 2008. Lignin as a base material for materials applications: chemistry, application and economics. *Industrial Crops and Products*, 27(2), 202–207.

Sudo, K. and Shimizu, K., 1992. A new carbon fiber from lignin. *Journal of Applied Polymer Science*, 44(1), 127–134.

Sudo, K., Shimizu, K., Nakashima, N. and Yokoyama, A., 1993. A new modification method of exploded lignin for the preparation of a carbon fiber precursor. *Journal of Applied Polymer Science*, 48(8), 1485–1491.

Suhas, Carrott, P.J.M. and Ribeiro Carrott, M.M.L., 2007. Lignin-from natural adsorbent to activated carbon: a review. *Bioresource Technology*, 98(12), 2301–2312.

Tan, S.S.Y., MacFarlane, D.R., Upfal, J., Edye, L.A., Doherty, W.O.S., Patti, A.F., Pringle, J.M. and Scott, J.L., 2009. Extraction of lignin from lignocellulose at atmospheric pressure using alkylbenzenesulfonate ionic liquid. *Green Chemistry*, 11(3), 339.

Tejado, A., Peria, C., Labidi, J., Echeverria, J.M. and Mondragon, I., 2007. Physico-chemical characterization of lignins from different sources for use in phenol–formaldehyde resin synthesis. *Bioresource Technology*, 98, 1655–1663.

Thring, R.W., Vanderlaan, M.N. and Griffin, S.L., 1996. Fractionation of ALCELL® lignin by sequential solvent extraction. *Journal of Wood Chemistry and Technology*, 16(2), 139–154.

Thring, R., Vanderlaan, M. and Griffin, S.L., 1997. Polyurethanes from alcell lignin. *Biomass and Bioenergy*, 13(3), 125–132.

Tirtowidjojo, S., Sarkanen, K.V., Pla, F. and McCarthy, J.L., 1988. Kinetics of organosolv delignification in batch-through and flow-through reactors. *Holzforschung*, 42(3), 177–183.

Uraki, Y., Kubo, S., Nigo, N., Sano, Y. and Sasaya, T., 1995. Preparation of carbon fibers from organosolv lignin obtained by aqueous acetic acid pulping. *Holzforschung*, 49, 343–350.

van der Hage, E.R.E., Mulder, M.M. and Boon, J.J., 1993. Structural characterization of lignin polymers by temperature-resolved in-source pyrolysis-mass spectrometry and Curie-point pyrolysis-gas chromatography/mass spectrometry. *Journal of Analytical and Applied Pyrolysis*, 25, 149–183.

Vazquez, G., Antorrena, G. and Gonzalez, J., 1995. Kinetics of acid-catalysed delignification of *Eucalyptus-globulus* wood by acetic-acid. *Wood Science and Technology*, 29(4), 267–275.

Vazquez, G., Antorrena, G., Gonzalez, J. and Freire, S., 1997. The influence of pulping conditions on the structure of acetosolv eucalyptus lignins. *Journal of Wood Chemistry and Technology*, 17(1–2), 147–162.

Vicuna, R., 1988. Bacterial degradation of lignin. *Enzyme Microb. Technol.*, 10, 646–655.

Wilkes, J.S., 2002. A short history of ionic liquids – from molten salts to neoteric solvents. *Green Chemistry*, 4(2), 73–80.

Windt, M., Meier, D., Marsman, J.H., Heeres, H.J. and de Konig, S., 2009. Micropyrolysis of technical lignins in a new modular rig and product analysis by GC–MS/FID and GC×GC–TOFMS/FID. *Journal of Analytical and Applied Pyrolysis*, 85(1–2), 38–46.

Wong, D.W.S., 2009. Structure and action mechanism of ligninolytic enzymes. *Appl. Biochem. Biotechnol.*, 157, 174–209.

Wool, R.P., and Sun, X.S., 2005. Bio-based polymers and composites. In *Bio-based Polymers and Composites*. Academic Press, San Diego, CA, pp. 551–598.

Yang, H., Yan, R., Chen, H., Lee, D.H. and Zheng, C., 2007. Characteristics of hemicellulose, cellulose and lignin pyrolysis. *Fuel*, 86(12–13), 1781–1788.

Zakzeski, J., Bruijnincx, P.C.A., Jongerius, A.L. and Weckhuysen, B.M., 2010. The catalytic valorization of lignin for the production of renewable chemicals. *Chemical Reviews*, 110(6), 3552–3599.

Zhao, X., Cheng, K. and Liu, D., 2009. Organosolv pretreatment of lignocellulosic biomass for enzymatic hydrolysis. *Applied Microbiology and Biotechnology*, 82(5), 815–27.

Zimmermann, W., 1990. Degradation of lignin by bacteria. *Journal of Biotechnology*, 13, 119–130.

# 21
# Bio-based chemicals from biorefining: lipid and wax conversion and utilization

Y. YANG and B. HU, University of Minnesota, USA

DOI: 10.1533/9780857097385.2.693

**Abstract**: This chapter reviews current bio-based lipids and wax, the chemical and biological production routes, and their applications. Lipids comprise a variety of naturally occurring compounds, such as fats/oils (triglycerides), phospholipids, diglycerides, monoglycerides, steroids and waxes. Lipids and wax are currently produced from petroleum, as well as plants and animals. Many microorganisms, like bacteria, fungi and microalgae, can also accumulate large amount of lipids and waxes in their cell biomass. Their microbial synthesis is sustainable and the microbial-derived lipids and wax are compatible with the current petroleum-based products. Lipids and wax can be applied as nutraceuticals, pharmaceuticals, fine chemicals and fuels.

**Key words**: lipids, waxes, microbial production, triglyceride, biodiesel, applications of lipids and waxes.

## 21.1 Introduction

The term 'lipid' is very general with various definitions that broadly refer to molecules that can be dissolved in organic solvent instead of water. In this case, lipids can include a wide range of compounds, including fats, waxes, sterols, fat-soluble vitamins (such as vitamins A, D, E, and K), monoglycerides, diglycerides, triglycerides, phospholipids, and others. Lipids can also be defined as a synonym for fats, which are only a subgroup of general lipids called triglycerides. Wax is an ester of a long-chain alcohol and a fatty acid, a subgroup of the broadly defined lipids. However, wax is usually listed in conjunction with lipids, whereas the term 'lipids' as used here only encompasses molecules such as fatty acids and their derivatives (including tri-, di-, monoglycerides, and phospholipids), as well as other sterol-containing metabolites, such as cholesterol. Both terms ('lipids' and 'waxes') are used in this chapter, which mainly discusses the types and properties of lipids and waxes, their sources of generation, conversion, and utilization. In addition, the current available extraction and analysis methods are discussed. Finally, future research for the lipid and waxes biorefineries is highlighted.

## 21.2 Types and properties of lipids and waxes

Lipids are a group of chemicals that comprise a variety of naturally occurring compounds. They include a diverse range of compounds such as triglycerides, monoglycerides, diglycerides, phospholipids, terpenoids, carotenoids, and steroids. Lipids mainly serve as energy storage in animals and plants and also as the structural component of the cell membrane. Waxes mainly serve as energy reserves, as well as surface protection. Some lipids are also used as hormones and vitamins for metabolism. Most lipids and waxes are insoluble in water (hydrophobic), but some types of lipids and waxes can be soluble in both water and oil (amphipathic).

### 21.2.1 Triglycerides

The mostly commonly defined lipids are triglycerides, also called fats, which are the ester of glycerol and three fatty acids. Triglycerides are the major constituents of most natural fats and oils, including vegetable oil and animal fats. A monoglyceride is the monoester of glycerol and one fatty acid; a diglyceride the diester of glycerol and two fatty acids. Triglycerides contain more than twice the energy (38 kJ/g) as carbohydrates and proteins, and they serve as fuel storage in the adipose tissue in humans and seeds of plants. Monoglycerides and diglycerides are intermediates in the degradation of triglycerides. They can be found in cell extracts and may have distinct and important biological properties.

### 21.2.2 Phospholipids

Phospholipids are mostly made from glycerides by substituting one of the three fatty acids by a phosphate group with some other molecule attached to its end. The other form of phospholipids is sphingomyelin, which is derived from sphingosine instead of glycerol. Phospholipids are soluble in both water and oil (amphiphilic) because the hydrocarbon tails of two fatty acids are still hydrophobic, but the phosphate group end is hydrophilic. Phospholipids are the major component of cell membrane to form lipid bilayers. Figure 21.1 shows the representative structure of common lipids (Gunstone *et al.*, 2007).

### 21.2.3 Wax esters

In addition to the esters of fatty acids and glycerol, the other type of fatty acid ester is wax esters. Wax esters are long chain esters of fatty acids and alcohols with chain lengths of 12 carbons or more. Wax esters have a variety of functions in organisms, such as surface protection (Gunstone *et al.*, 2007),

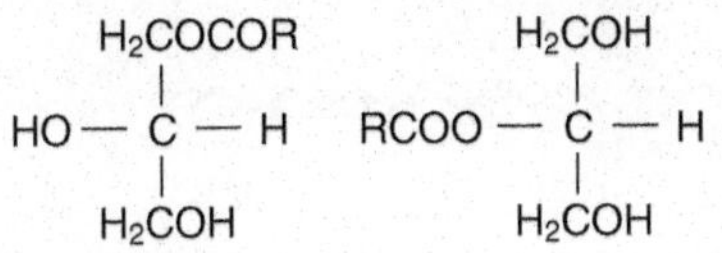

$H_2COCOR$
|
$HO-C-H$
|
$H_2COH$

$H_2COH$
|
$RCOO-C-H$
|
$H_2COH$

(a) Monoglycerides

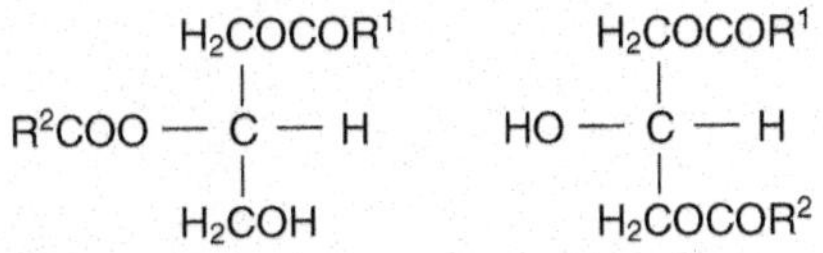

$H_2COCOR^1$
|
$R^2COO-C-H$
|
$H_2COH$

$H_2COCOR^1$
|
$HO-C-H$
|
$H_2COCOR^2$

(b) Diglycerides

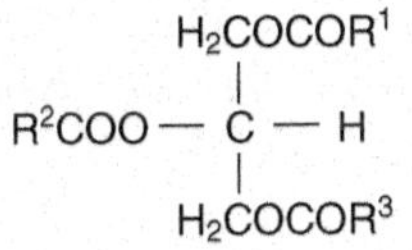

$H_2COCOR^1$
|
$R^2COO-C-H$
|
$H_2COCOR^3$

(c) Triglycerides

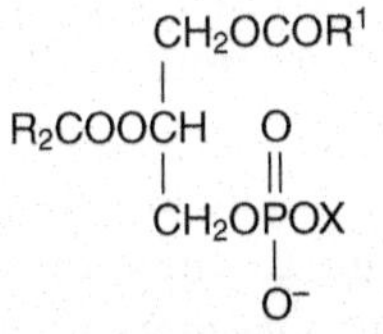

$CH_2OCOR^1$
|
$R_2COOCH$
|
$CH_2OPOX$
|
$O^-$

(d) Phospholipids derived from glycerol

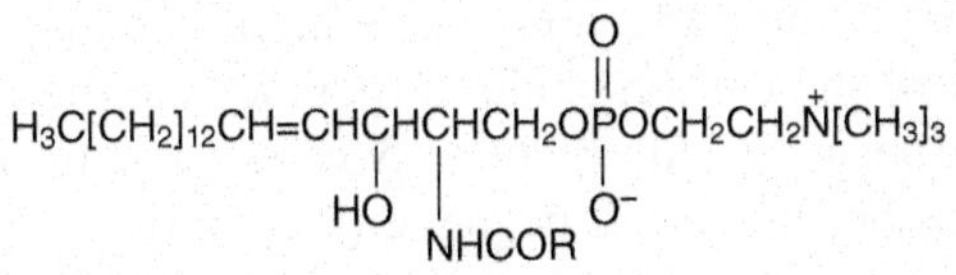

$$H_3C[CH_2]_{12}CH=CHCHCHCH_2OPOCH_2CH_2\overset{+}{N}[CH_3]_3$$
with HO, NHCOR, O⁻ substituents

(e) Example of phospholipids derived from sphingosine (Sphingomyelin)

*21.1* The representative structure of common lipids (Gunstone *et al.*, 2007): (a) Monoglycerides, (b) diglycerides, (c) triglycerides, (d) phospholipids derived from glycerol, (e) example of phospholipids derived from sphingosine (Sphingomyelin).

energy reserves (Lee, Hagen *et al.*, 2006), and constituents of the swimbladder in some myctophid fish (Phleger, 1998). High contents of wax ester can also provide thermal insulation and energy supplies under undesirable conditions (Nevenzel, 1970). Waxes are used as important ingredients in cosmetics, pharmaceuticals, lubricants, plasticizers, and polishes and the other chemical industries (Hallberg, Wang *et al.*, 1999). Figure 21.2 shows the structure of wax ester.

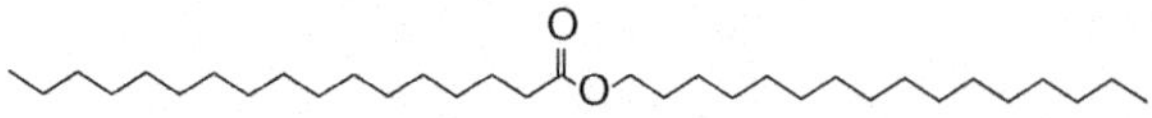

*21.2* Structure of wax ester.

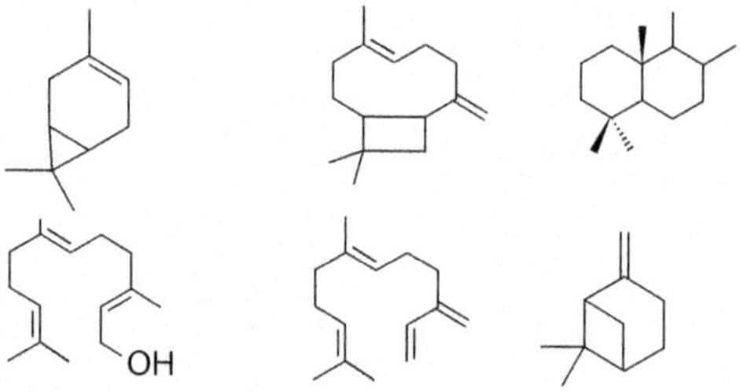

*21.3* The representative structure of examples of a monoterpene and a sesquiterpene.

### 21.2.4 Terpenoids

Terpenoids are hydrocarbon-like compounds produced by many plants, are the major constituents of the essential oils, and are universally present in small amounts in living organisms. They are a diverse class of natural products that have many functions in the plant kingdom and in human health and nutrition (Roberts, 2007). Due to their biosynthesis pathways, the carbon skeletons of terpenoids are oligomers of isoprene; however, many terpenoids undergo further modifications to obtain different structures (Abraham, 2010). Generally, based on the number of the isoprene structure and the length of the carbon chain, terpenoids can be classified as monoterpene (10 carbon), sesquiterpene (15 carbon), diterpene (20 carbon), triterpene (30 carbon), and tetraterpene (40 carbon). Figure 21.3 shows the representative molecule structures for some monoterpenes and sesquiterpenes.

### 21.2.5 Other lipid compounds

Lipids also contain other organic components that can be dissolved in the organic phase, for example, carotenoids and steroids. Carotenoids are tetraterpenoid organic pigments that are naturally occurring in chloroplasts and chromoplasts. The function of carotenoids includes absorbing blue light for photosynthesis and protecting chlorophyll from photo damage (Armstrong and Hearst, 1996). More than 600 known carotenoids have been discovered and split into two classes: xanthophylls, which contain oxygen; and carotenes, which contain no oxygen. Most carotenoids have 40 carbon atoms and belong to the tetraterpene family. Carotenoids are in the form of a polyene hydrocarbon chain and may or may not have additional oxygen atoms attached. The degree of conjugation and the isomerization state of the backbone polyene chromophore determine the absorption properties of each carotenoid (Cohen, 2011), which is directly linked to their color.

Steroids are known as precursors of certain vitamins and hormones. They are also key constituents in mammalian cell membranes to maintain a proper membrane permeability and fluidity. In animal tissues, cholesterol is by far the most abundant member of steroids. Cholesterol is the principal sterol synthesized by animals and is the precursor molecule for the synthesis of vitamin D and all the steroid hormones. Although cholesterol is important and necessary for the biological processes, high levels of cholesterol in blood is one of the major risk factors for coronary heart disease, heart attack, and stroke.

## 21.3 Sources of lipids and waxes

Lipids and waxes are currently produced from petroleum as well as plants and animals. All plants contain oils or fats, primarily in their seeds. Lipids and waxes synthesize a huge variety of fatty acids, and triglycerides are the most important form of lipids in plant seeds, followed by the phospholipids. Terpenoids and carotenoids can be found in all parts of the plants, but are relatively concentrated in certain species of plants and microalgae. Fat and steroids are predominantly found in the animal tissue. No significant amount of cholesterol is found in plant sources. Some plants contain phytosterols, which is a cholesterol-like compound and believed to have competition in cholesterol absorption and metabolism (Ostlund, Racette *et al.*, 2003). The biological-derived lipids and waxes can be a good alternative supplement to the current production from petroleum and other fossil fuel origins. However, the limited access to natural lipids and waxes results in a high cost.

### 21.3.1 Lipids and waxes extracted from plants

The most important oil plants are soybean, palm, rapeseed, and sunflower. During 2011/2012, soybeans accounted for approximately 75% of oilmeal production, while soybean and palm plants account for approximately 60% of plant oil production. Plant oil is also important feedstock for biodiesel production. The most commonly used oil plants for the production of biodiesel are soybean, sunflower, palm, rapeseed, canola, cottonseed, and jatropha.

The primary source of natural waxes produced by plants is the waxy material in the leaf and stem surface. This thin layer of wax has multiple purposes, such as limiting water diffusion and providing protection from insects. Some plants can accumulate a thick coating of wax on leaves (carnauba palm) or have the capability to produce wax ester instead of triglycerides in their seed (jojoba). The oil from the jojoba plant (*Simmondsia chinensis*) is the main biological source of wax esters. Waxes can also be generated from animal sources, with Beeswax and wool wax as the prime commodities of natural waxes from animal sources (Li, Kong et al., 2010).

One important issue for the use of plant lipids is the limited production of raw material such as soybean oil or vegetable oil, which makes the biodiesel industry suffer in terms of production capacity. Meanwhile, the most costly part of biodiesel production is the cost of feedstock, and current plant oil prices make biodiesel only profitable when it is subsidized. Waste vegetable oils, due to lower prices than original vegetable oils, are considered as a potential low-cost lipid source for biodiesel. In addition to these regular oil plants, a number of wild plant species are capable of producing high amounts of unusual fatty acids, which can be used in high-value industrial and pharmaceutical/nutraceutical applications (Dyer, Stymne et al., 2008).

### 21.3.2 Lipids extracted from microalgae

*Microalgae* are generally autotrophic eukaryotic cells that are mostly unicellular with some species that are colonial or filamentous. They can grow autotrophically and/or heterotrophically, with a wide range of tolerance to different temperatures, salinity, pH, and nutrient availabilities. The *eukaryotes microalgae* referred to include green algae, diatoms, yellow-green algae, golden algae, red algae, brown algae, dinoflagellates, and others; limited species have the capability to accumulate a high content of lipids in their cell biomass (Stoytcheva and Montero, 2011). Some algae have been found to contain more than 80% lipids, which is of great interest for a sustainable feedstock for biodiesel production. Algae oil is being seriously considered because of the large oil yields compared with that for other oilseeds. Research by the National Renewable Energy Laboratory (NREL) showed that 7.5 billion gallons of biodiesel could be produced from approximately 500,000 acres of desert land. Algae farms could also be constructed to use waste streams (either human waste or animal waste from animal farms) as a food source, to extract fertilizer high in nitrogen and phosphorous from algae production, and to provide a means for recycling nutrients. Lipid accumulation occurs within the microalgae cells and varies based on the strain and growth conditions. Table 21.1 lists several promising algae species and their growth performance. Examples include *Chlorella vulgaris*, which is a commercially important green microalgae due to its high photosynthetic efficiency. *Chlorella protothecoides* is another single-cell green microalgae, and heterotrophic growth of *C. protothecoides* supplied with acetate, glucose, or other organic compounds as carbon source, results in high biomass and high content of lipid in cells (Xu et al., 2006).

Table 21.1 Examples of microalgae cultivation for oil accumulation (Heredia-Arroyo, 2011)

| Microalgae | Cultures | | | Substrates | Growth rate (h) | Lipid content |
|---|---|---|---|---|---|---|
| | AC | MC | HC | | | |
| Chlorella protothecoides | X | | X | Glucose, acetate/$CO_2$ | 3.74 g/L in 144 h | 55.2% |
| Chlorella vulgaris | X | X | X | Glucose, acetate, lactate/$CO_2$ | 0.098 g/h | – |
| Crypthecodinium cohnii | | | X | Glucose/$CO_2$ | 40 g/L in 60–90 h | 15–30% |
| Scenedesmus obliquus | X | | X | Glucose/$CO_2$ | Double in 14 h after adaptation | 14–22% |
| Chlamydomonas reinhardtii | X | X | X | Acetate/$CO_2$ | Exponential during the first 20 h | 21% |
| Micractinium pusillum | | X | X | $CO_2$ | 0.94 g/L in 24 h | – |
| Euglena gracilis | | | X | $CO_2$ | | 14–20% |
| Schizochytrium sp | | | | Glycerol/$CO_2$ | – | 55% |
| Spirulina platensis | X | | X | Glucose/$CO_2$ | 0.008/h | – |
| Botryococcus braunii | | | X | $CO_2$ | Low growth rate | 20–86% |
| Dunaliella salina | X | X | | $CO_2$ | – | ~70% |

AC: autotrophic cultures; MC: mixotrophic cultures; HC: heterotrophic cultures.

Many nutritional and environmental factors control the cell growth and lipid contents. Nutritional factors include organic and inorganic carbon sources, nitrogen sources, and other essential macro- and micronutrients such as magnesium and copper; environmental factors include temperature, pH level, salinity, and dissolved oxygen. Many microalgae species accumulate a higher content of lipids under heterotrophic growth, using organic carbon as its source instead of carbon dioxide and sunlight. Compared with phototrophic algae, the heterotrophic growth process has the advantages of no light limitation, a high degree of process control, higher productivity, and low costs for biomass harvesting (Barclay, Meager et al., 1994). Miao and Wu (2006) reported that the oil content of heterotrophically cultured *C. protothecoides* was approximately four times greater than that in the corresponding autotrophic culture. Liu et al. (2010) demonstrated that the heterotrophically cultured cells of *C. zofingiensis* showed 411% and 900% increases in dry cell weight and lipid yield, respectively, compared with increases for autotrophically cultured cells. In addition to lipid production, high value byproducts can be obtained from heterotrophically cultured microalgae, including polyunsaturated fatty acids and carotenoids (Chen and Chen, 2006).

### 21.3.3 Lipids from yeast and fungi

Besides microalgae, many filamentous fungi species (e.g., *Mucor circinelloides* or *Mortierella isabellina*) can also accumulate a considerably high content of lipids (Xia, Zhang *et al.*, 2011; Heredia-Arroyo, Wei *et al.*, 2011). Many oleaginous yeasts were studied for lipid accumulation on different substrates, such as industrial glycerol (Meesters, Huijberts *et al.*, 1996; Papanikolaou and Aggelis, 2002), sewage sludge (Angerbauer, Siebenhofer *et al.*, 2008), whey permeate (Ykema, Verbree *et al.*, 1988; Akhtar, Gray *et al.*, 1998), sugar cane molasses (Alvarez, Rodriguez *et al.*, 1992), and rice straw hydrolysate (Huang, Zong *et al.*, 2009). Dey *et al.* (2011) screened two endophytic oleaginous fungi *Colletotrichum sp.* and *Alternaria sp.* with lipid content 30% and 58%, respectively. Furthermore, the fatty acid profile of microbial oils is quite similar to that of conventional vegetable oils, which suggests oleaginous filamentous fungi as a favorable feedstock for sustainable biodiesel industry (Peng and Chen, 2008; Zhao, Hu *et al.*, 2011). Table 21.2 shows the oil content of common fungi and yeast species (Meng, Yang *et al.*, 2009).

### 21.3.4 Waxes from bacteria

Fossil waxes are primarily generated by chemical syntheses via Fischer–Tropsch process and olefin (ethylene, propylene) polymerization. Natural waxes are mostly extracted from animals, vegetables, and minerals, and many waxes, such as jojoba oil and sperm whale oil, have important industrial use due to their excellent wetting behavior at interfaces (Hadzir, Basri

*Table 21.2* Oil content of sample fungi and yeast species (Meng, Yang *et al.*, 2009)

| Microorganisms | Oil content (% dry wt) | Microorganism | Oil content (% dry wt) |
|---|---|---|---|
| Yeast | | Fungi | |
| *Candida curvata* | 58 | *Aspergillus oryzae* | 57 |
| *Cryptococcus albidus* | 65 | *Mortierella isabellina* | 86 |
| *Lipomyces starkeyi* | 64 | *Humicola lanuginosa* | 75 |
| *Rhodotorula glutinis* | 72 | *Mortierella vinacea* | 66 |

*et al.*, 2001). However, restricted by their high price and limitation in access, industrial syntheses of wax ester by the esterification reaction of carboxylic acids with alcohols are also applied via chemical (Aracil, Martinez *et al.*, 1992) and enzymatic reactions (Trani, Ergan *et al.*, 1991).

Wax esters were found to be accumulated in some group of prokaryotes, such as bacteria (Ishige, Tani *et al.*, 2003). These microbial waxes are favorable because their composition can be manipulated by the growth condition. Some members of the genus *Acinetobacter* can store significant amounts of wax ester in cells, which can account for up to 15–25% of the cell dry matter (Kalscheuer and Steinbuchel, 2003). *Acinetobacter* can grow and synthesize wax esters on a variety of substrates, such as alkanes and aromatic hydrocarbons, as well as acetate, sugars, and acids (Waltermann, Stoveken et al., 2007). In addition, the accumulation of wax esters has also been reported in some species of the genera *Fundibacter* (Bredemeier, Hulsch *et al.*, 2003), *Micrococcus* (Russell and Volkman, 1980), and *Moraxella* (Bryn, Jantzen et al., 1977). The intracellular accumulation of wax esters was found in a few species of prokaryotes, and jojoba plant (*Simmondsia chinensis*) is the only source of eukaryotes that accumulates wax esters as intracellular storage lipid (Waltermann and Steinbuchel, 2005). Engineering bacteria with wax ester synthase has been studied for the biosynthesis of wax esters for higher production (Kalscheuer and Steinbuchel, 2003), which may substitute current wax production from crude oil or plant (jojoba) and animal oils.

In addition to plants and pure cultures of microorganisms, other sources of lipids and waxes are available. Activated sludge was discovered to have high lipid and wax content because of its high microbial biomass content, and it is technically possible to produce biofuel or biodiesel from activated sludge. A recent study found that more than 25 wt% crude lipid and waxes can be obtained from activated sludge in a food processing company in Taiwan (Huynh, Do *et al.*, 2011). Many other waste materials can also be utilized for the production of lipids and waxes, such as lanolin extracted and purified from the wool wax of wool scour wastes (Lopez-Mesas, Christoe *et al.*, 2005), and sugarcane waxes extracted from rum factory wastes (Nuissier, Bourgeois *et al.*, 2002).

## 21.3.5 Lipids and waxes generated from lignocellulosic materials and other sources

Due to the limited supply of plant-based resources, lipids and waxes generated via microbial accumulation are attracting more attention based on their availability, relatively low cost, and the development of conversion technology. The theoretical yield for converting sugars to long-chain fatty acids (lipids and waxes) is 32%; most oleaginous cell cultures can reach the yield of 20–25% because some sugars have to be diverted to support the cell growth and metabolism. Therefore, the use of non-starch biomass is critical so that lignocelluloses can be used for organic carbon supply without concerns of using food crops for fuel sources. Lignocellulosic materials are mainly composed of cellulose, hemicellulose, and lignin, whereas only cellulose and hemicellulose can be converted to fermentable sugars for microbial lipid production. Recent studies confirmed conversion of hemicellulose hydrolysate into lipids by oleaginous yeast strains and their tolerance degrees to lignocellulose degradation compounds (Chen, Li et al., 2009; Hu, Zhao et al., 2009; Huang, Zong et al., 2009).

The oleaginous cell cultivation on lignocelluloses shares many similarities with current lignocellulosic ethanol production, whereas cell fermentation is followed by pretreatment and saccharification to release monomeric sugars for microbial utilization (Wang, Wang et al., 2011). Pretreatment refers to the disruption of the naturally resistant structure of lignocellulosic biomass to make its cellulose and hemicellulose susceptible to enzymatic hydrolysis, the end goal being to generate fermentable sugars. Many pretreatment processes, including chemical, physical, physicochemical, and biological pretreatment, have been greatly developed in recent decades (Wyman, Dale et al., 2005). Sugars and phenolic compounds from cellulose, hemicellulose, and lignin, as well as their corresponding degradation products, as 5-hydroxymethyl-2-furaldehyde (HMF), 2-furaldehyde, and acetic acid, respectively, constitute the pretreatment hydrolysate. The types and concentrations of the compounds in the pretreatment hydrolysate depend on the pretreatment technology employed, specific conditions, and varieties of feedstock.

The cellulose hydrolysis/saccharification step for lignocellulosic ethanol fermentation combines enzymatic cellulose hydrolysis with yeast fermentation in a process called 'simultaneous saccharification fermentation' (SSF). SSF integrates the whole process and minimizes the substrate inhibition. It also creates some difficulties and challenges due to the higher working temperature of cellulase, compared with the lower temperatures at which most yeast strains grow. Both C5 and C6 monomeric sugars become available after hydrolysis, as well as numerous byproducts that may have inhibitive effects on cell growth, such as acetate, formic acid, furans, vanillin, levulinic acid, furfural, hydroxymethylfurfural, and hydroxybenzaldehyde (Venkatesh, 1997; Aden, Ruth et al., 2002; Yang and Wyman, 2004). Compared with lignocellulosic ethanol fermentation, microbial lipid production has its own distinctions. Many ethanol-producing yeast strains cannot utilize C5 sugar, and are severely inhibited by acetic acid and other byproducts from lignocellulosic hydrolysis, whereas oleaginous fungal strains have a wider range of utilization for both C5 and C6 sugars, less cell inhibitions, and fewer requirements for the hydrolysis.

## 21.4 Methods to extract and analyze lipids and waxes

The aim of all extraction procedures is to separate lipids and waxes from the other constituents. With the purpose to extract oil, different strategies have been developed. Considering the structure texture, sensitivities, and lipid and wax contents of plant and animal tissues and microbial cells, the methods show great diversity. The most widely used methods include solvent extraction, pressing, carbon dioxide supercritical extraction, and microwave-assisted extraction.

### 21.4.1 Solvent extraction

Lipids and waxes exist in tissues in many different physical forms. The simple lipids are often aggregates in storage tissues, while complex lipids are usually constituents of membranes and are closely associated with proteins and polysaccharides. To extract lipids and waxes from tissues, solvents must be used that not only readily dissolve the lipids but also overcome the interactions between the lipids and the tissue matrix. In addition, simultaneously perturbing both the hydrophobic and polar interactions is essential (Christie, 1993). The greatest improvement in lipid and waxes extraction was made in 1957 when Folch *et al.* (1957) described their classic extraction procedure. This procedure remains the most commonly used by researchers around the world and has become the standard against which other methods are judged. A key property of this solvent is the capacity of chloroform to associate with water molecules. With the ratio of chloroform-methanol to tissue greater than 17:1, the equivalent of 5.5% water can be solvated and remain in a single phase (Christie, 1993). At the laboratory scale, the extraction efficiency can reach approximately 90%.

### 21.4.2 Pressing extraction

Pressing may be the oldest method for oil extraction. In India, approximately 90% of the total 24 million tons of produced oilseeds are crushed using this method (Singh and Bargale, 2000). Pressing is the process of mechanically pressing liquid out of liquid-containing solids. The equipment is simple and sturdy, as well as easy to operate, with short training time. Moreover, this process can be adapted quickly for processing of different kinds of oilseeds and yields a chemical-free protein-rich cake (Pradhan, Mishra *et al.*, 2011). It is claimed that the pressing process can typically recover 86–92% of oil from oilseeds (Singh and Bargale, 2000), while the efficiency is highly dependent on the raw material used in pressing. When using hazelnut seeds to extract oil, the pressing process efficiency can reach the low limit of 6.1% (Uquiche, Jerez *et al.*, 2008). Adjusting pressing parameters, such as an increase of the internal pressure, can improve oil recovery and results in a decrease of the residual oil in the cake (Jacobsen and Backer, 1986). Suitable pretreatment of the oilseeds, such as cleaning, conditioning, decorticating, cracking, flaking, cooking, extruding, and drying to optimal moisture content, will also increase the oil recovery (Zheng, Wiesenborn *et al.*, 2003).

### 21.4.3 Carbon dioxide supercritical extraction

Carbon dioxide supercritical extraction is one application of the supercritical fluid extraction (SFE) process. SFE is a separation technology that uses supercritical fluid solvent for extraction. Carbon dioxide is the most commonly used supercritical fluid, with other choices including ethanol. Compared with traditional soxhlet extraction, SFE uses supercritical fluid to provide a broad range of useful properties. It eliminates the use of organic solvents, which reduces the problems of their storage, disposal, and environmental concerns. In the extraction process, diffusion coefficients of lipids and waxes in supercritical fluids are much higher than in liquids, therefore extraction can occur more quickly. In addition, no surface tension is present in supercritical fluids, and viscosities are much lower than in liquids, which help the supercritical fluids be able to penetrate into small pores that are inaccessible to liquid. Research has shown that the lipid extraction has the ability to reach more than 90% of theoretical value in a short time (King, 2002). Currently, carbon dioxide supercritical extraction remains at the stage of laboratory research.

Some parameters are generally considered to influence the extraction yields with carbon dioxide supercritical extraction. Temperature in supercritical conditions has no determining influence on extraction yield, while increasing of pressure will increase the extraction rate by increasing the oil solubility (Salgin, Doker *et al.*, 2006; Boutin and Badens, 2009). Moreover, many studies indicate that extraction yield increases with pressure (from 3.63 to 18.63 g $CO_2$ $kg^{-1}$ from 20 to 60 MPa) due to the solubility increase of the different compounds (mainly triglycerides) with pressure (Salgin, Doker *et al.*, 2006). Increased moisture content will reduce the extraction efficiency (King, 2002), while the moisture content itself may be affected during the extraction process (Dunford and Temelli, 1997). Pretreatment of the sample before extraction will improve mass transfer by increasing the exchange surface and seed destructuring (del Valle, Germain *et al.*, 2006).

### 21.4.4 Microwave-assisted extraction

Microwave use has only recently been applied to extraction of plant materials. Unlike conventional heating, which depends on the conduction–convection phenomenon (Mandal *et al.*, 2007), microwave energy delivered directly to materials through molecular interaction with the electromagnetic field generates heat through the sample volume to achieve uniform heating (Venkatesh and Raghavan, 2004). In this manner, the use of microwave-assisted extraction often helps to reduce processing time and save energy during the process (Uquiche, Jerez *et al.*, 2008). When combined with other extraction methods, the microwave radiation works as a pretreatment step and can highly increase the efficiency of pressing extraction from 6.1% to 45.3% (Uquiche, Jerez *et al.*, 2008). Similar results were obtained from other research in which microwave pretreatment contributed a significantly higher extraction yield than carbon dioxide supercritical extraction (Dejoye, Vian *et al.*, 2011). Under a given radiation frequency, major factors that influence dielectric properties include temperature and moisture content, chemical composition and the physical structure of the raw material, and sample density (Venkatesh and Raghavan, 2004).

The most current extraction methods in industry are solvent extraction and pressing, which can achieve decent efficiency and maintain a low cost. Carbon dioxide supercritical extraction can achieve higher efficiency, although the high cost prohibits the industrial application at this stage. The microwave pretreatment of raw material for extraction can improve efficiency, but it is currently difficult to treat large amounts of raw material.

### 21.4.5 Methods to analyze lipids and waxes

Traditional analysis of lipids and waxes focused on the nutritional aspect. Lipids are one of the major constituents of foods, providing an energy source and essential lipid nutrients. Waxes are also utilized for food purposes. Meanwhile, overconsumption of certain lipid components such as cholesterol and saturated fats can be detrimental to human health. In terms of the food science aspect, some of the most important properties of a food analyte are:

- total lipid and waxes concentration;
- type of lipids and waxes present;
- physicochemical properties of lipids and waxes, such as crystallization, melting point, smoke point, rheology, density, and color;
- structural organization of lipids and waxes (Chauhan and Varma, 2009).

The physicochemical characteristics of lipids and waxes, including solubility in organic solvents, immiscibility with water, physical characteristics, and spectroscopic properties, are used to distinguish these elements from other components in food. Analytical methods are based on these principle characteristics and can be categorized into three groups: solvent extraction, non-solvent extraction, and instrumental methods. For solvent extraction, the methods used are similar to the extraction methods described previously. However, some foods contain lipids that are complexed with proteins or polysaccharides and thus need an extra step to break the bonds to release lipid into easily extractable form (e.g., acid hydrolysis). For non-solvent extraction, the separation of lipid and waxes relies on other chemicals such as sulfuric acid and isoamyl alcohol. Standard methods include the Babcock method, the Gerber method, and the detergent method. In terms of the instrumental methods, a wide variety can be chosen. Major principles include measurement of bulk physical properties such as electrical conductivity, measurement of adsorption of radiation such as nuclear magnetic resonance (NMR) spectroscopy, and measurement of scattering of radiation like ultrasonic scattering.

As the extensive research on biofuel and biodiesel has developed recently, lipids are extracted from various sources and converted into biodiesel. The requirements of lipid analysis are then focused on the identification of components in lipids and the generated biodiesel. For this analysis purpose, gas chromatography (GC) and high-performance liquid chromatography (HPLC) are primary choices. For wax ester analysis, GC, GC-mass spectrometry (GC-MS), and HPLC are also commonly used.

GC is widely used for analysis of volatile compounds. During GC analysis, the sample is evaporated in an injection port and then transported through a column by carrier gas stream. During the transport, compounds are separated by the different retention times required to pass through the column. Detectors are connected at the end of column to quantify the amount of each compound that passes through, and normally a flame ionization detector (FID) is used for hydrocarbons and essential oils (Marriott, Shellie et al., 2001). A variety of GC columns are available to allow highly sophisticated application of samples together within a designed temperature program. When the sample contains unknown components, analysis of these hydrocarbons is often achieved by a connection with GC-MS (Koga and Morii, 2006).

Many compounds are simply not volatile enough to qualify for GC analysis; also some nonpolar compounds such as polyyne hydrocarbons and highly unsaturated fatty acids are too unstable for GC analyses (Abraham, 2010). These components are generally separated by HPLC currently. Similar to GC, the HPLC analysis also requires a column for separation and a detector for quantification. HPLC with ultraviolet or photodiode array detection is most often used (Su, Rowley et al., 2002). Similar to the GC-MS system, a mass spectrometer can also be connected with an HPLC system for the identification of unknown hydrocarbons.

## 21.5 Utilization of lipids and waxes

Lipids and waxes are mostly used as nutraceuticals, pharmaceuticals, fine chemicals, and fuels. Currently, one of the most attractive utilizations for lipids is to generate biodiesel, which is considered as a replacement for current diesel fuel.

### 21.5.1 Biofuels

As a liquid form of renewable fuel, biofuels attract great effort from research based on their capacity for application as transportation fuels. Biodiesel is an important approach for alternative biofuels. The most common type of biodiesel is esters of fatty acid, obtained by transesterification of lipid with methanol or ethanol. Biodiesel can be used in pure form (B100) or may be blended with fossil diesel at any rate. The most commonly used biodiesel is B99 because 1% of fossil fuel is applied to inhibit mold growth. Biodiesel is approximately 5% to 8% less efficient than conventional fossil diesel but is compatible with the current diesel engine and can be a total or partial (in the cold regions) replacement for the fossil diesel.

The environmental benefits to replacing fossil diesel with biodiesel are numerous because combustion of biodiesel emits far less pollutants than fossil diesel (except the $NO_x$ emission). Also, the entire process is close to carbon-neutral considering that the plant oil used to produce biodiesel is synthesized in agriculture from $CO_2$. The production and utilization of biodiesel are significant in many aspects, for example, increasing the oilseed crop market, providing domestic job opportunities to the rural community, and decreasing the dependence on imported oil; therefore biodiesel has been commercialized around the globe.

### 21.5.2 Triglycerides

The high viscosity, low volatility, and polyunsaturated character of triglycerides prohibit their direct use in an engine system as liquid biofuel. The most common industrially applied method to convert lipids to biofuel is through the transesterification reaction to convert triglycerides to fatty acid methyl ester (FAME), in which long-chain fatty acids are exchanged from triglycerides by methanol, generating FAME and glycerol (Fig. 21.4). This reaction can also be carried out by ethanol to produce fatty acid ethyl ester (FAEE) instead.

In addition to transesterification, several strategies can change the properties of the lipids to serve as biofuel. The hydrocarbon type of diesel, developed recently with direct decarboxylation of fatty acid or lipid

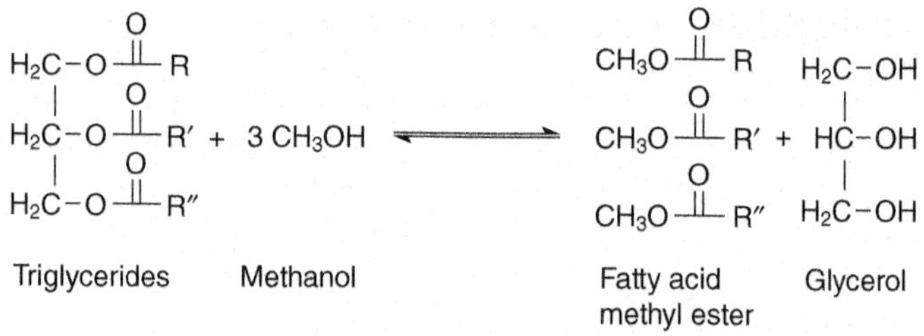

## 21.4 Representative transesterification reaction.

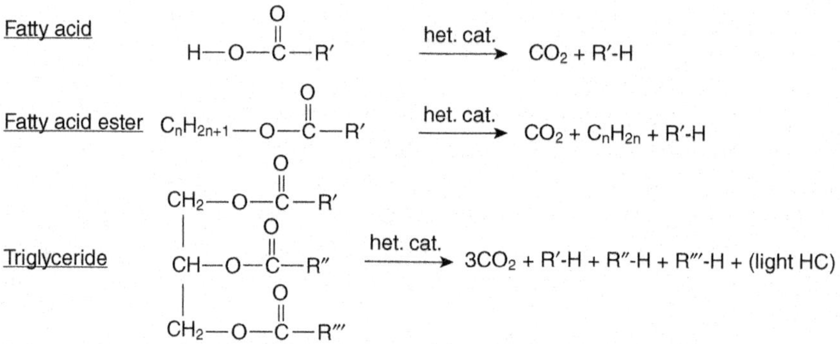

Triglycerides + 3 CH$_3$OH ⇌ Fatty acid methyl ester + Glycerol

$$H_2C-O-CO-R$$
$$|$$
$$H_2C-O-CO-R'\ +\ 3\ CH_3OH\ \rightleftharpoons\ CH_3O-CO-R\ +\ CH_3O-CO-R'\ +\ CH_3O-CO-R''\ +\ H_2C-OH,\ HC-OH,\ H_2C-OH$$
$$|$$
$$H_2C-O-CO-R''$$

## 21.5 Hydrocarbon synthesis from fatty acid or lipid.

Fatty acid:
$$H-O-\overset{O}{\underset{\|}{C}}-R' \xrightarrow{\text{het. cat.}} CO_2 + R'\text{-}H$$

Fatty acid ester:
$$C_nH_{2n+1}-O-\overset{O}{\underset{\|}{C}}-R' \xrightarrow{\text{het. cat.}} CO_2 + C_nH_{2n} + R'\text{-}H$$

Triglyceride:
$$\begin{array}{l} CH_2-O-\overset{O}{\underset{\|}{C}}-R' \\ | \\ CH-O-\overset{O}{\underset{\|}{C}}-R'' \\ | \\ CH_2-O-\overset{O}{\underset{\|}{C}}-R''' \end{array} \xrightarrow{\text{het. cat.}} 3CO_2 + R'\text{-H} + R''\text{-H} + R'''\text{-H} + \text{(light HC)}$$

(Fig. 21.5), has been shown to be superior to biodiesel in many aspects due to its resistance to water contamination; for example:

1. biodiesel is very corrosive and therefore it is not suitable for pipeline transportation;
2. biodiesel production requires oils of high quality, and algal oil has difficulty in reaching those requirements; and
3. biodiesel has shorter shelf life, and the technology to deal with biodiesel is far more immature than that for diesel.

Pyrolysis, a method of conversion of one substance into another with the aid of a catalyst in the absence of air or oxygen, is proven to produce non-oxygenated, liquid hydrocarbon mixtures from triglycerides and can be used a diesel fuel additive (Maher and Bressler, 2007).

Microemulsion, defined as clear, thermodynamically stable, isotropic liquid mixtures of oil, water, and surfactant, also shows the capability to produce a mixture of diesel fuel and plant oil with same engine performance (Singh and Singh, 2010). Other research activities in this field include hydrocarbon production from natural oils and fats over homogeneous catalysts; gas-phase selective heterogeneous catalytic decarboxylation of carboxylic acids, such as octanoic, benzoic, and salicylic acid, with palladium and nickel catalysts; unsaturated linear hydrocarbons production from saturated fatty acid and fatty acid esters with a nickel type catalyst; and unsaturated diesel-like hydrocarbons production over palladium catalysts on carbon (Snåre, Kubi ková et al., 2008). The decarboxylation process occurs at high temperature with the help of the catalyst, and many byproducts of plant or algal oils show strong toxicity to the catalyst. The chemical decarboxylation process to produce diesel still needs further development compared with the transesterification process for biodiesel production.

Additionally, the thermochemical biorefinery is an alternative method to produce lipids and waxes directly from lignocellulosic materials. Compared with the traditional lipid and wax production process of extraction and microbial accumulation, the thermochemical biorefinery is a new strategy to convert plant-based materials to chemical and energy materials. This process generates syngas and heat for energy use, with some liquid portion containing lipids and wax components. The process is mainly carried out by pyrolysis and gasification, and the feedstock can come from a variety of sources, such as wood products, agricultural residue, and municipal solid wastes. This biorefinery process includes mainly combustion, gasification, and pyrolysis, while product composition is closely related to the reaction temperature and reaction conditions. In addition to being gas generated, lipids and waxes are primarily liquid product components. This biorefinery process will be more economical and efficient when integrated with industries dealing with large amounts of biomass, such as pulp and paper mill, agricultural residue, and municipal solid waste treatment. Currently, the international corporation TRI (ThermoChem Recovery International, Inc.) has commercialized the thermochemical biorefinery integrated with the pulp and paper industry for production of liquid transportation biofuels, bio-based chemicals, substitutes for petroleum-based feedstock and products, and biomass-based heat.

## 21.5.3 Terpenoids

In addition to the biodiesel and fatty acid derived hydrocarbons, hydrocarbon-like terpenoids are other important molecules proposed for biofuel. Many terpenoid compounds are natural antibiotics, and they are under investigation for antibacterial, antineoplastic, and other pharmaceutical applications. The concept of terpenoid as biofuels is still new and not widely accepted in the scientific community. However, visionary scientific leaders such as Jay Keasling at the University of California Berkeley have promoted the concept. Many terpenoid molecules contain no oxygen, have a similar structure to hydrocarbon, and have a low boiling point to be easily separated, which makes them capable of directly serving as fuels after being purified through the refinery process.

Currently, the microbial terpenoid production systems have several major limitations to being applied for fuel production. The yield with the microbial production systems is too low (the highest yield for the amorphadiene production, including both headspace and liquid content is 22.6 and 112.2 mg/L, respectively), which is far too low for fuel production at the current stage (Martin, Pitera et al., 2003). Furthermore, the production of terpenoids with microbial systems requires the input of glucose or sucrose, which translates into high cost, lower energy conversion efficiency, and an unsustainable production system. Currently, the production of terpenoids is more suitable as high value drugs for the pharmaceutical industry.

### 21.5.4 Pharmaceuticals and nutraceuticals

Lipids and waxes are historically applied as pharmaceuticals and nutraceuticals due to their significant role in human metabolism, especially because their many components cannot be synthesized by the human body and thus need to be supplied by intake from external sources. Lipids and waxes applied in these fields are generally considered as high-value products. Their applications have already been mostly driven by the market and are currently well commercialized. The following discussion lists some applications of lipids and waxes in these areas.

*Polyunsaturated fatty acids (PUFAs)*

PUFAs are fatty acids that contain more than one double bond in their backbone. Figure 21.6 shows the structure of n-3 PUFAs (Mozaffarian and Wu, 2011). The most important PUFAs are omega-3 and omega-6. Omega-3 has the first double bond at carbon number 3, counting from the methyl end, and omega-6 has the first double bond at carbon number 6, counting from the methyl end. Major omega-3 PUFAs in the diet include alpha-linolenic acid (ALA), eicosapentaenoic acid (EPA) and docosahexaenoic acid (DHA), while major omega-6 PUFAs in the diet include linoleic acid

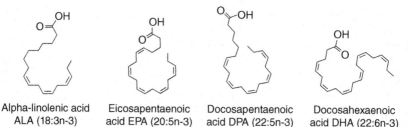

*21.6* Structure of n-3 polyunsaturated fatty acids (PUFAs).

(LA), γ-linolenic acid (GLA) and arachidonic acid (AA). The highest concentrations of the omega-3 PUFAs are found in coldwater fish such as salmon, sardines, and tuna, and omega-6 PUFAs are usually found in plants such as olive oil, sunflower seeds, and flaxseeds.

Omega-3 and omega-6 PUFAs are important structural components of the phospholipid cell membranes of the tissues, which have multiple physiological functions (Simopoulos, 1999, 2002). Omega-6 PUFAs are integral components of skin lipids; AA and DHA are the major PUFAs in the brain, nervous tissue, and retina; and DHA is essential for the visual process in the retina and for proper brain functioning. Since enzymes in the human body are not effective in the generation of precursor lipids to these complex PUFAs, directly increasing the intake of EPA, DHA, GLA, and DGLA has the highest clinical benefits.

*Monoglycerides*

Monoglycerides are fatty acid monoesters of glycerol and exist in two isomeric forms (Fig. 21.7). Their concentrations are very low in cell extracts, but they are intermediates in the degradation of triacylglycerols or diacylglycerols, and thus can be produced through a chemical process. Monoglycerides are widely used as anti-staling agents and account for approximately one third of the emulsifiers used in the baking industry. Another important use of unsaturated monoglycerides is in drug delivery systems because of their ease of processing and unique properties. When in contact with aqueous media, unsaturated monoglycerides result in a highly viscous cubic phase and provide sustained release for drugs (Chang and Bodmeier, 1997).

*Carotenoids and cholesterol*

Carotenoids constitute one of the most widespread classes of yellow, orange, and red natural pigments found in bacteria, fungi, and photosynthetic organisms, and also in eukaryotes, estimated to synthesize 108 tons every year (Cohen, 2011). Humans and animals are mostly incapable of synthesizing carotenoids; instead they obtain carotenoids from their diet for most antioxidant activity. Therefore, carotenoids are substantially

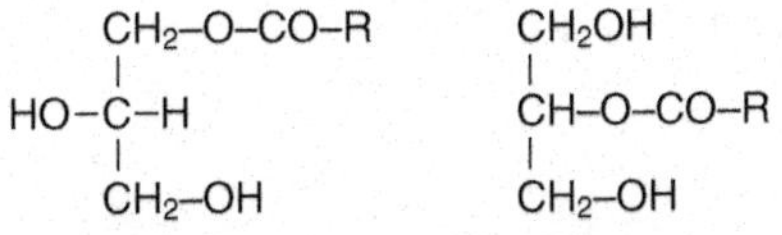

$$CH_2-O-CO-R$$
$$|$$
$$HO-C-H$$
$$|$$
$$CH_2-OH$$
sn-1-isomer

$$CH_2OH$$
$$|$$
$$CH-O-CO-R$$
$$|$$
$$CH_2-OH$$
2-isomer

*21.7* Structure of two isomeric forms of monoglycerides.

hydrophobic antioxidants, widely utilized in medicine as antioxidants or in food chemistry as colorants. Similar to carotenoids, cholesterol, as the most abundant member of steroids, is an important component in the hormonal system. Cholesterol is a building block for cell membranes and for hormones such as estrogen and testosterone. The liver produces approximately 80% of the body's cholesterol endogenously, while the remainder is from dietary sources. The major intake of dietary cholesterol is from meat sources including egg yolk, pork, beef, poultry, fish, and shrimp, while no significant amounts of cholesterol are found in plant sources. Humans need a certain level of cholesterol for proper function, but excess cholesterol can cause health problems such as the increased risk of coronary heart disease. Therefore cholesterol is one of the metabolites that people need to monitor and thus has a very limited role in serving as pharmaceuticals or nutraceuticals.

### 21.5.5 Other applications

*Cosmetics and personal care products*

Many lipid and wax products are used for cosmetics and personal care products.

- *Essential oils* – a wide category of hydrophobic liquid with volatile aroma compounds extracted from plants – are used in perfumes, cosmetics, soaps, and other products such as food and drink flavoring and household cleaning products. Among all lipid products, waxes are especially widely applied in the cosmetic industry, mainly applied as a coating agent for their excellent wetting behavior at interfaces.
- *Beeswax* – a natural wax produced in the honeybee hive – is widely used for skincare product and food additives.
- *Lanolin* – a wax ester extracted from wool scour wastes – is used mainly for cosmetics, pharmaceutical formulations, and baby care products (Lopez-Mesas, Christoe et al., 2005).
- *Octacosanol* – a straight-chain 28-carbon primary fatty alcohol that exists in plants' epicuticular waxes and rice bran wax – is used to improve exercise performance including strength, stamina, and reaction time (Chen, Cai et al., 2005).

The list is long of similar products that use lipids and waxes as main ingredients, which are well known to the public.

*Lubricants*

Lubricants, a substance (often a liquid) introduced between two moving surfaces to reduce friction, are traditionally generated from petroleum-based oil. In light of the excess amounts of lubricants that are lost into the environment, biodegradable lubricants have been considered as a suitable replacement for the petroleum-based ones. Compared to the lubricants made of petroleum, vegetable-based lubricants are much more biodegradable but inferior in many other technical characteristics. Current biodegradable lubricants account for only 2% of global consumption of lubricants; however, the market for biodegradable lubricants is expected to grow 5–10% annually (Erhan and Asadauskas, 2000). Most laws requiring the use of lubricants that are biodegradable are currently in place in several western European countries, such as Germany, Austria, and Switzerland. The increase of environmental regulations, set forth by governments, will be one of the driving factors that increase the demand for biodegradable lubricants. Meanwhile, research and development has been carried out to formulate biodegradable lubricants with technical characteristics superior or at least compatible to those based on mineral oil (petroleum).

Lubricants generally contain two major components: base stock oil and additives. Base stocks usually comprise more than 80% lubricant, determining the key performance properties of the lubricants (Erhan and Asadauskas, 2000). Natural plant-based oils are the predominant choice for the base stocks of the biodegradable lubricant, especially soybean oil, because it has excellent features such as low volatility, excellent lubricity, favorable viscosity temperature characteristics, and high solubilizing capacity for contaminants and additives. However, vegetable-based lubricants are very limited in many features, preventing them from wider applications. The most serious problem is their poor oxidative and hydrolytic stability. The oxidative instability is primarily caused by the polyunsaturation of the vegetable oil (Erhan and Asadauskas, 2000). Vegetable oils also show poor corrosion protection and the presence of ester functionality renders these oils susceptible to hydrolytic breakdown. Some of these problems can be resolved by avoiding or modifying polyunsaturation in TAG structures of vegetable oils. Another way to improve these properties of vegetable oils is chemical modification of fatty acid chains of triglycerides at sites of double bond and carboxyl groups. In particular, a high degree of branching can lower viscosity index, whereas high linearity can lead to high viscosity index, and relatively poor low-temperature characteristics. In contrast, low saturation can limit oxidation stability, whereas high saturation can result in outstanding oxidation stability (Li, Kong et al., 2010).

*Bioplastics*

Plastic materials are usually synthetic and most commonly derived from petroleum. Due to the short time of their presence in nature, no enzyme structures are capable of degrading petroleum-based plastics. They may persist for hundreds or even thousands of years, and thus environmental concerns are driving the research into biodegradable plastics. Natural biodegradable plastics are based primarily on renewable resources and can be either naturally produced or synthesized from renewable resources. The major sources of biodegradable plastics include polysaccharides, proteins, lipids, polyesters produced by plant or microorganisms, polyesters derived from bioderived monomers, and miscellaneous polymers, for example (Nampoothiri, Nair et al., 2010).

As the main components in plant oil, triglycerides provide glycerol and a mixture of fatty acids by hydrolysis, and both can be used as building blocks for the synthesis of designed monomers (Warwel, Bruse et al., 2001; Jerome, Pouilloux et al., 2008). Research developed by different groups reveals a growing interest in the reactivity of their double bonds towards olefin metathesis, which enables the straightforward synthesis of a wide variety of monomers (de Espinosa and Meier, 2011). Even though with great potential, plants oil and animal fats are generally not the primary and direct raw materials to generate bioplastics. However, from a biorefinery point of view, bioplastics manufacturing from proteins can be integrated with oilseed processing to maximize the usage of the raw materials. A typical example is the bioplastics generation from cottonseed protein, which can be extracted from the residue after organic solvent extraction for oil production. Cottonseed protein can be modified via crosslinking treatment to generate cottonseed protein bioplastics (CBPs) (Yue, Cui et al., 2012). In this scenario, the cottonseed processing will not only generate oil for biofuel generation, but also CBPs to replace plastics generated from petroleum. Similar concepts can be applied to other types of oilseed processings.

## 21.6 Conclusion and future trends

Even though lipids and waxes have been explored as fine chemicals for decades, due to the recent energy crisis, generation of liquid biofuel from lipids and wax esters has attracted attention from both academia and industry, it becoming the mainstream of research in the last few years. Numerous projects have been funded by the US Department of Energy (DOE) and the US Department of Agriculture (USDA), as well as the National Science Foundation to study the whole process for generating liquid biofuel products from non-food lipids and waxes. These projects cover a wide range of topics, including feedstock development, biological and chemical conversion technologies, harvesting and logistics, and life-cycle assessment. For example, projects funded from the Biomass Research and Development Initiative (BRDI) program under USDA and DOE are specifically focusing on this area in recent years. The DOE also funded research projects to explore possibilities for other types of drop-in biofuels, for example, hydrocarbons and terpenoids. A typical example is the project led by Professor Larry Wackett at the University of Minnesota and supported by the DOE Advanced Research Project Agency-Energy (ARPA-E) program with $2.2 million to use photosynthetic bacteria that can convert light and carbon dioxide to 'feed' a hydrocarbon-producing *Shewanella* bacteria for scaled-up production (Wackett, Gralnick *et al.*, 2010–2012).

In general, the future of the advanced biofuel production system must meet the following requirements:

1. For agriculture residues usage, especially from the average farmer, the processing facility must be compatible with the distributed biomass production system.
2. The conversion facility must be easy to operate and have a relatively simple operation if a small-size facility is chosen.
3. The conversion process must be adaptable to multiple biomass feedstock.
4. Costs related to harvest of cells and conversion of cell biomass to biofuel must be low.

Extracting lipids from cell biomass and converting them to biodiesel is not an ideal route, due to the complexity of the cell harvesting and oil extraction; however, combining the biorefinery concept to convert the biomass residue after oil extraction as animal feed may bring some extra value to offset the process cost. Direct thermal treatment of cell biomass, for example, pyrolysis, is more robust than biological conversion combined with transesterification reaction; however, the final product of the pyrolysis cannot meet the quality standard of liquid fuel, which therefore significantly affects the technical feasibility of this type of conversion. Although no process has been proven to fit all these requirements while still reaching commercial feasibility, further scientific research will advance these processes to be closer to industrial reality.

## 21.7 References

Abraham, W. R. (2010). Biosynthetic oils, fats, terpenes, sterols, waxes: analytical methods, diversity, characteristics. *Handbook of Hydrocarbon and Lipid Microbiology*. K. Timmis, Springer, Berlin, Heidelberg: 79–95.

Aden, A., M. Ruth, *et al.* (2002). Lignocellulosic biomass to ethanol process design and economics utilizing co-current dilute acid prehydrolysis and enzymatic hydrolysis for corn stover, National Renewable Energy Laboratory, Golden, CO, NREL/TP-510-32438.

Akhtar, P., J. I. Gray, *et al.* (1998). 'Synthesis of lipids by certain yeast strains grown on whey permeate.' *Journal of Food Lipids* 5(4): 283–297.

Alvarez, R. M., B. Rodriguez, *et al.* (1992). 'Lipid accumulation in *Rhodotorula glutinis* on sugar cane molasses in single-stage continuous culture.' *World Journal of Microbiology & Biotechnology* 8(2): 214–215.

Angerbauer, C., M. Siebenhofer, *et al.* (2008). 'Conversion of sewage sludge into lipids by *Lipomyces starkeyi* for biodiesel production.' *Bioresource Technology* 99(8): 3051–3056.

Aracil, J., M. Martinez, *et al.* (1992). 'Formation of a jojoba oil analog by esterification of oleic-acid using zeolites as catalyst.' *Zeolites* 12(3): 233–236.

Armstrong, G. A. and J. E. Hearst (1996). 'Carotenoids. 2. Genetics and molecular biology of carotenoid pigment biosynthesis.' *Faseb Journal* 10(2): 228–237.

Barclay, W., K. Meager, *et al.* (1994). 'Heterotrophic production of long chain omega-3 fatty acids utilizing algae and algae-like microorganisms.' *J. Appl. Phycol.* 6(2): 123–129.

Boutin, O. and E. Badens (2009). 'Extraction from oleaginous seeds using supercritical $CO_2$: experimental design and products quality.' *Journal of Food Engineering* 92(4): 396–402.

Bredemeier, R., R. Hulsch, *et al.* (2003). 'Submersed culture production of extracellular wax esters by the marine bacterium *Fundibacter jadensis*.' *Marine Biotechnology* 5(6): 579–583.

Bryn, K., E. Jantzen, *et al.* (1977). 'Occurrence and patterns of waxes in Neisseriaceae.' *Journal of General Microbiology* 102: 33–43.

Chang, C. M. and R. Bodmeier (1997). 'Swelling of and drug release from monoglyceride-based drug delivery systems.' *Journal of Pharmaceutical Sciences* 86(6): 747–752.

Chauhan A.K. and A. Varma (2009). A Textbook of Molecular Biotechnology. New Delhi, India, I K International Publishing House.

Chen, F., T. Y. Cai, *et al.* (2005). 'Optimizing conditions for the purification of crude octacosanol extract from rice bran wax by molecular distillation analyzed using response surface methodology.' *Journal of Food Engineering* 70(1): 47–53.

Chen, G.-Q. and F. Chen (2006). 'Growing Phototrophic Cells without Light.' *Biotechnology Letters* 28(9): 607–616.

Chen, X., Z. Li, *et al.* (2009). 'Screening of oleaginous yeast strains tolerant to lignocellulose degradation compounds.' *Applied Biochemistry and Biotechnology* 159(3): 591–604.

Christie, W. W. (1993). *Preparation of lipid extracts from tissues*. In *Advances in Lipid Methodology – Two*. Oily Press, Bridgwater: 195–213.

Cohen, G. N. (2011). *Microbial Biochemistry*. Springer, New York: 471–479.

de Espinosa, L. M. and M. A. R. Meier (2011). 'Plant oils: the perfect renewable resource for polymer science?!' *European Polymer Journal* 47(5): 837–852.

Dejoye, C., M. A. Vian, *et al.* (2011). 'Combined extraction processes of lipid from *chlorella vulgaris* microalgae: microwave prior to supercritical carbon dioxide extraction.' *International Journal of Molecular Sciences* 12(12): 9332–9341.

del Valle, J. M., J. C. Germain, *et al.* (2006). 'Microstructural effects on internal mass transfer of lipids in prepressed and flaked vegetable substrates.' *Journal of Supercritical Fluids* 37(2): 178–190.

Dey P., J. Banerjee and M. K. Maiti (2011). 'Comparative lipid profiling of two endophytic fungal isolates – *Colletotrichum* sp. and *Alternaria* sp. having potential utilities as biodiesel feedstock.' *Bioresource Technology* 102(10): 5815–5823.

Dunford, N. T. and F. Temelli (1997). 'Extraction conditions and moisture content of canola flakes as related to lipid composition of supercritical $CO_2$ extracts.' *Journal of Food Science* 62(1): 155–159.

Dyer, J. M., S. Stymne, et al. (2008). 'High-value oils from plants.' *Plant Journal* 54(4): 640–655.

Erhan, S. Z. and S. Asadauskas (2000). 'Lubricant basestocks from vegetable oils.' *Industrial Crops and Products* 11(2–3): 277–282.

Folch, J., M. Lees and G. H. Sloane Stanley (1957). 'A simple method for the isolation and purification of total lipides from animal tissues.' *The Journal of Biological Chemistry* 226: 497–509.

Gunstone, F. D., J. L. Harwood and A. J. Dijkstra (2007). *The Lipid Handbook*, 3rd edn. CRC Press, Boca Raton, FL.

Hadzir, N. M., M. Basri, et al. (2001). 'Enzymatic alcoholysis of triolein to produce wax ester.' *Journal of Chemical Technology and Biotechnology* 76(5): 511–515.

Hallberg, M. L., D. B. Wang, et al. (1999). 'Enzymatic synthesis of wax esters from rapeseed fatty acid methyl esters and a fatty alcohol.' *Journal of the American Oil Chemists Society* 76(2): 183–187.

Heredia-Arroyo, T. (2010). 'Oil accumulation via *Chlorella* species using carbon sources from waste.' Ann Arbor, University of Puerto Rico, Mayaguez (Puerto Rico). 1481905: 146.

Heredia-Arroyo, T., W. Wei, et al. (2011). 'Mixotrophic cultivation of *Chlorella vulgaris* and its potential application for the oil accumulation from non-sugar materials.' *Biomass & Bioenergy* 35(5): 2245–2253.

Hu, C., X. Zhao, et al. (2009). 'Effects of biomass hydrolysis by-products on oleaginous yeast *Rhodosporidium toruloides*.' *Bioresource Technology* 100(20): 4843–4847.

Huang, C., M.-H. Zong, et al. (2009). 'Microbial oil production from rice straw hydrolysate by *Trichosporon fermentans*.' *Bioresource Technology* 100(19): 4535–4538.

Huynh, L. H., Q. D. Do, et al. (2011). 'Isolation and analysis of wax esters from activated sludge.' *Bioresource Technology* 102(20): 9518–9523.

Ishige, T., A. Tani, et al. (2003). 'Wax ester production by bacteria.' *Current Opinion in Microbiology* 6(3): 244–250.

Jacobsen, L. A. and L. F. Backer (1986). 'Recovery of sunflower oil with a small screw expeller.' *Energy in Agriculture* 5(3): 199–209.

Jerome, F., Y. Pouilloux, et al. (2008). 'Rational design of solid catalysts for the selective use of glycerol as a natural organic building block.' *Chemsuschem* 1(7): 586–613.

Kalscheuer, R. and A. Steinbuchel (2003). 'A novel bifunctional wax ester synthase/acyl-CoA: diacylglycerol acyltransferase mediates wax ester and triacylglycerol biosynthesis in *Acinetobacter calcoaceticus* ADP1.' *Journal of Biological Chemistry* 278(10): 8075–8082.

King, J. W. (2002). 'Supercritical fluid extraction: present status and prospects.' *Grasas y Aceites* 53(1): 8–21.

Koga, Y. and H. Morii (2006). 'Special methods for the analysis of ether lipid structure and metabolism in archaea.' *Analytical Biochemistry* 348(1): 1–14.

Lee, R. F., W. Hagen, et al. (2006). 'Lipid storage in marine zooplankton.' *Marine Ecology – Progress Series* 307: 273–306.

Li, W., X. H. Kong, et al. (2010). 'Green waxes, adhesives and lubricants [Retracted article. See vol. 370, p. 3269, 2012].' *Philosophical Transactions of the Royal Society A – Mathematical Physical and Engineering Sciences* 368(1929): 4869–4890.

Liu, J., J. Huang, et al. (2010). 'Differential lipid and fatty acid profiles of photoautotrophic and heterotrophic *Chlorella zofingiensis*: assessment of algal oils for biodiesel production.' *Bioresource Technology* 102(1): 106–110.

Lopez-Mesas, M., J. Christoe, et al. (2005). 'Supercritical fluid extraction with cosolvents of wool wax from wool scour wastes.' *Journal of Supercritical Fluids* 35(3): 235–239.

Maher, K. D. and D. C. Bressler (2007). 'Pyrolysis of triglyceride materials for the production of renewable fuels and chemicals.' *Bioresource Technology* 98(12): 2351–2368.

Mandal, V., Y. Mohan and S. Hemalatha (2007).'Microwave assisted extraction – An innovative and promising extraction tool for medicinal plant research.' *Pharmacognosy Review* 1(1): 7–17.

Marriott, P. J., R. Shellie, et al. (2001). 'Gas chromatographic technologies for the analysis of essential oils.' *Journal of Chromatography A* 936(1–2): 1–22.

Martin, V. J. J., D. J. Pitera, et al. (2003). 'Engineering a mevalonate pathway in Escherichia coli for production of terpenoids.' *Nature Biotechnology* 21(7): 796–802.

Meesters, P. A. E. P., G. N. M. Huijberts, et al. (1996). 'High cell density cultivation of the lipid accumulating yeast *Cryptococcus curvatus* using glycerol as a carbon source.' *Applied Microbiology and Biotechnology* 45(5): 575–579.

Meng X., J. Yang, et al. (2009). 'Biodiesel production from oleaginous microorganisms.' *Renewable Energy* 34: 1–5.

Miao, X. and Q. Wu (2006). 'Biodiesel production from heterotrophic microalgal oil.' *Bioresource Technolong* 97(6): 841–846.

Mozaffarian, D. and J. H. Y. Wu (2011). 'Omega-3 fatty acids and cardiovascular disease effects on risk factors, molecular pathways, and clinical events.' *Journal of the American College of Cardiology* 58(20): 2047–2067.

Nampoothiri, K. M., N. R. Nair, et al. (2010).'An overview of the recent developments in polylactide (PLA) research.' *Bioresource Technology* 101(22): 8493–8501.

Nevenzel, J. C. (1970). 'Occurrence, function and biosynthesis of wax esters in marine organisms.' *Llpids* 5(3): 308–319.

Nuissier, G., P. Bourgeois, et al. (2002). 'Composition of sugarcane waxes in rum factory wastes.' *Phytochemistry* 61(6): 721–726.

Ostlund, R. E., S. B. Racette, et al. (2003). 'Inhibition of cholesterol absorption by phytosterol-replete wheat germ compared with phytosterol-depleted wheat germ.' *American Journal of Clinical Nutrition* 77(6): 1385–1389.

Papanikolaou, S. and G. Aggelis (2002). 'Lipid production by *Yarrowia lipolytica* growing on industrial glycerol in a single-stage continuous culture.' *Bioresource Technology* 82(1): 43–49.

Peng, X. and H. Chen (2008). 'Single cell oil production in solid-state fermentation by *Microsphaeropsis* sp. from steam-exploded wheat straw mixed with wheat bran.' *Bioresource Technology* 99(9): 3885–3889.

Phleger, C. F. (1998). 'Buoyancy in marine fishes: direct and indirect role of lipids.' *American Zoologist* 38(2): 321–330.

Pradhan, R. C., S. Mishra, et al. (2011). 'Oil expression from Jatropha seeds using a screw press expeller.' *Biosystems Engineering* 109(2): 158–166.

Roberts, S. C. (2007). 'Production and engineering of terpenoids in plant cell culture.' *Nature Chemical Biology* 3(7): 387–395.

Russell, N. J. and J. K. Volkman (1980). 'The effect of growth temperature on wax ester composition in the psychrophilic bacterium *Micrococcus cryophilus* ATCC 15174.' *Journal of General Microbiology* 118: 131–141.

Salgin, U., O. Doker, et al. (2006). 'Extraction of sunflower oil with supercritical $CO_2$: experiments and modeling.' *Journal of Supercritical Fluids* 38(3): 326–331.

Simopoulos, A. P. (1999). 'Essential fatty acids in health and chronic disease.' *American Journal of Clinical Nutrition* 70(3): 560S–569S.

Simopoulos, A. P. (2002). 'Omega-3 fatty acids in inflammation and autoimmune diseases.' *Journal of the American College of Nutrition* 21(6): 495–505.

Singh, J. and P. C. Bargale (2000). 'Development of a small capacity double stage compression screw press for oil expression.' *Journal of Food Engineering* 43(2): 75–82.

Singh, S. P. and D. Singh (2010). 'Biodiesel production through the use of different sources and characterization of oils and their esters as the substitute of diesel: a review.' *Renewable & Sustainable Energy Reviews* 14(1): 200–216.

Snåre, M., I. Kubi ková, et al. (2008). 'Catalytic deoxygenation of unsaturated renewable feedstocks for production of diesel fuel hydrocarbons.' *Fuel* 87(6): 933–945.

Stoytcheva, M. and G. Montero (2011). '*Biodiesel – Feedstocks and Processing Technologies.*' InTech, Rijeka, Croatia.

Su, Q., K. G. Rowley, et al. (2002). 'Carotenoids: separation methods applicable to biological samples.' *Journal of Chromatography B – Analytical Technologies in the Biomedical and Life Sciences* 781(1–2): 393–418.

Trani, M., F. Ergan, et al. (1991). 'Lipase-catalyzed production of wax esters.' *Journal of the American Oil Chemists Society* 68(1): 20–22.

Uquiche, E., M. Jerez, et al. (2008). 'Effect of pretreatment with microwaves on mechanical extraction yield and quality of vegetable oil from Chilean hazelnuts (*Gevuina avellana* Mol).' *Innovative Food Science & Emerging Technologies* 9(4): 495–500.

Venkatesh, K. V. (1997). 'Simultaneous saccharification and fermentation of cellulose to lactic acid.' *Bioresource Technology* 62(3): 91–98.

Venkatesh, M. S. and G. S. V. Raghavan (2004). 'An overview of microwave processing and dielectric properties of agri-food materials.' *Biosystems Engineering* 88(1): 1–18.

Wackett, L., J. Gralnick, et al. (2010–2012). *Shewanella* as an ideal platform for producing hydrocarbonfuel. University of Minnesota, Minneapolis, MN.

Waltermann, M. and A. Steinbuchel (2005). 'Neutral lipid bodies in prokaryotes: recent insights into structure, formation, and relationship to eukaryotic lipid depots.' *Journal of Bacteriology* 187(11): 3607–3619.

Waltermann, M., T. Stoveken, et al. (2007). 'Key enzymes for biosynthesis of neutral lipid storage compounds in prokaryotes: properties, function and occurrence of wax ester synthases/acyl-CoA: diacylglycerol acyltransferases.' *Biochimie* 89(2): 230–242.

Wang, M., J. Wang, et al. (2011). 'Lignocellulosic bioethanol: status and prospects.' *Energy Sources Part A – Recovery Utilization and Environmental Effects* 33(7): 612–619.

Warwel, S., F. Bruse, et al. (2001). 'Polymers and surfactants on the basis of renewable resources.' *Chemosphere* 43(1): 39–48.

Wyman, C. E., B. E. Dale, *et al.* (2005). 'Coordinated development of leading biomass pretreatment technologies.' *Bioresource Technology* 96(18): 1959–1966.

Xia, C., J. Zhang *et al.* (2011). 'A new cultivation method for bioenergy production – cell pelletization and lipid accumulation by *Mucor circinelloides*.' *Biotechnology for Biofuels* 4: 15.

Xu H., X. Miao and Q. Wu, (2006). 'High quality biodiesel production from a microalga *Chlorella protothecoides* by heterotrophic growth in fermenters.' *J Biotechnol* 126(4):499–507.

Yang, B. and C. E. Wyman (2004). 'Effect of xylan and lignin removal by batch and flowthrough pretreatment on the enzymatic digestibility of corn stover cellulose.' *Biotechnology and Bioengineering* 86(1): 88–95.

Ykema, A., E. C. Verbree, *et al.* (1988). 'Optimization of lipid production in the oleaginous yeast *Apiotrichum curvatum* in whey permeate.' *Applied Microbiology and Biotechnology* 29(2–3): 211–218.

Yue, H. B., Y. D. Cui, *et al.* (2012). 'Preparation and characterisation of bioplastics made from cottonseed protein.' *Green Chemistry* 14(7): 2009–2016.

Zhao, X., C. Hu, *et al.* (2011). 'Lipid production by *Rhodosporidium toruloides* Y4 using different substrate feeding strategies.' *Journal of Industrial Microbiology and Biotechnology* 38(5): 627–632.

Zheng, Y. L., D. P. Wiesenborn, *et al.* (2003). 'Screw pressing of whole and dehulled flaxseed for organic oil.' *Journal of the American Oil Chemists Society* 80(10): 1039–1045.

# 22
# Bio-based chemicals from biorefining: protein conversion and utilisation

E. L. SCOTT, M. E. BRUINS and J. P. M. SANDERS,
Wageningen University, The Netherlands

DOI: 10.1533/9780857097385.2.721

**Abstract**: The depletion of fossil feedstocks, fluctuating oil prices and the ecological problems associated with $CO_2$ emissions are forcing the development of alternative resources for energy, transportation fuels and chemicals: the replacement of fossil resources with biomass. The conversion of crude oil products utilises hydrocarbons and conversion to (functional) chemicals with the aid of co-reagents, such as ammonia, and various process steps. Conversely, proteins and amino acids, found in biomass, contain functionality. It is therefore attractive to exploit this to reduce the use, and preparation of, co-reagents as well as eliminating various process steps. This chapter describes how biorefineries can add value to protein containing rest streams by using amino acids as economically and ecologically interesting feedstocks and that, by taking advantage of the chemical structure in biomass rest streams, a more efficient application can be developed other than solely utilising it for the production of fuels or electricity.

**Key words**: amino acids, protein, rest streams, separation.

## 22.1 Introduction

Due to ecological concerns regarding the use of fossil resources and greenhouse gas emissions, as well as issues relating to energy security, fluctuating oil prices and the volume oil reserves, governments around the globe are looking for alternative methods to produce bulk energy (heat and electricity), transportation fuels and chemicals using alternative resources. For the production of energy there are a number of options being explored and implemented such as the use of wind, solar and hydro/geo-thermal generation. In these cases, energy can be generated without the use of carbon-containing resources. However, in the case of mass produced transportation fuels and chemicals, which rely on carbon (and other) functionality, the alternatives for production require the use of an alternative carbon source to oil. In this case, the use of biomass is the only method for their production.

Why should amino acids be considered as interesting chemical feedstocks in the first place? The petrochemical industry utilises hydrocarbon feedstocks, e.g. ethylene and propylene, together with co-reagents and conversion steps to incorporate functionality to produce a large variety of chemicals and materials. The production of ethylene and propylene from biomass has been widely reported in the literature. Generally this involves fermentation, isolation of the desired molecule and subsequent (chemical) conversion to the desired hydrocarbon. Using this approach, (partially) bio-based alternatives to fossil-derived chemicals could be generated. However, it still requires the use of reagents, conversion processes and energy (often supplied in the form of natural gas). Thus significant fossil resources are still required to produce more functionalised products. If the overall aim is to develop processes to produce industrial products from biomass which reduce fossil input, then also the process for the conversion should be considered in the whole approach and not just the feedstock.

Amino acids contain functionalities, such as $-NH_2$ and $-COOH$, which are similar to an array of functionalised industrial chemicals. In addition to this, they can be readily converted to other functionalities such as nitriles. Therefore it is attractive to exploit this in order to bypass the use and preparation of co-reagents such as ammonia, as well as eliminating various process steps. Thus potentially greater savings on the fossil energy required may be achieved (Scott *et al.*, 2007). Allied to this, it is already known that lower raw material and capital investment costs are incurred in the production of non-funtionalised chemicals compared to more functionalised compounds (Lange, 2001). Thus more efficient use of the functionality of amino acids may be beneficial in the production of functionalised chemicals.

## 22.2 Protein and amino acid sources derived from biofuel production

During the production of bioethanol and biodiesel, rest streams are generated. In the case of biodiesel, press-cake and glycerol are formed. The use of the glycerol side stream has been successfully developed to produce traditional chemicals. Solvay produces epichlorohydrin from glycerol using the Epicerol™ process (Solvay Chemicals, 2006). Other bulk chemical products that are being developed from glycerol rest streams include conversion to methanol. The consortium Bio-Methanol Chemie Nederland (BMCN) uses glycerol to produce synthesis gas which is then converted to methanol (*Chemie Magazine*, 2006). As well as this, Archer Daniels Midland (ADM) are exploring the use of carbohydrates and/or glycerol as a feedstock for the production of propylene and ethylene glycols (*Chemical Week*, 2005). The press-cake rest stream formed has received less attention in its application. In general, press-cakes are utilised as animal feed due to the significant protein content (as well as containing carbohydrates and lignin). However, not all amino acids in the protein are required for animal nutrition. It is conceivable that non-essential amino acids are removed (after isolation and hydrolysis of the protein) and used for other applications. A similar argument can be used for dried distillers grains and solubles (DDGS), the rest stream of bioethanol production. Wheat DDGS contains 40% crude protein (Nuez Ortín and Yu, 2009). It is considered that cost-effective isolation of amino acids from wheat DDGS will improve the economic value of the bioethanol by-product and will aid the economic viability of the whole bioethanol production chain (Bals et al., 2009). At present, DDGS is recognised as nutritious animal feed, but due to its protein content, it would also serve as a raw material for amino acid production. In this case, certain amino acids may become an attractive option as feedstocks for the production of chemicals. Gluten, a major protein of wheat, is high in glutamic acid that is an excellent starting material for the synthesis of bulk chemicals via γ-aminobutyric acid (GABA) (Lammens et al., 2009). Protein extraction from DDGS is not yet employed successfully at any scale.

European guidelines aim for 10% of all transportation fuels to be derived from biomass by 2020 (EU, 2009). This will therefore result in rest streams containing proteins (20–40 wt% of the dry matter). Combined with the estimated worldwide biofuel production (IEA, 2011), this will lead to a production of *ca.* 100 million tons of protein per year corresponding to *ca.* 5 wt% (*ca.* 5 million tons) of each amino acid. By 2050, the IEA predicts an even higher production of biofuels and therefore substantial generation of protein-containing rest streams. Thus considerable amounts of each individual amino acid could be available as a feedstock for the production of bulk chemicals without competing with food and feed markets.

### 22.2.1 Using rest streams derived from biofuel production

Gasoline alternatives/replacements have also been developed from starch and sugar sources. Particularly in South America, the use of sugarcane to produce ethanol has been exploited as has the use of carbohydrate from corn and grain (US and EU). The production of so-called first generation bioethanol has attracted a lot of attention arising from issues surrounding 'food versus fuel'. The reports 'World agriculture: towards 2015/2030' (FAO, 2003) and 'International Assessment of Agricultural Science and Technology for Development' (UNEP, 2008), describe sufficient food (production) for a growing global population and that new agricultural technology increases food production, but that in developing countries many will remain hungry and environmental issues attributed to agriculture still exist. The World

Bank and the International Monetary Fund (IMF) have also commented on some of the many causes of rising food prices, including bad weather, high oil prices, increased demand for meat and dairy products in some Asian countries as well as the ambitions in the West to use biofuels derived from grain to reduce oil consumption.

This has led to developments to produce second generation fuels focusing on using potentially abundant and inexpensive lignocellulosic agricultural waste streams from primary agricultural production as a feedstock which is not food competitive. The use of biorefineries, where a crop may be separated into various fractions, for food and other applications such as animal feed and raw materials for biofuels and chemicals will play a key role. As well as this, other technical challenges include obtaining fermentable sugars from (hemi-)cellulose and the conversion of other sugars, e.g. xylose to ethanol.

Rudolph Diesel demonstrated the diesel engine using peanut oil as a fuel at the World Exhibition of 1900. Today, the production of biodiesel relies on the use of crops such as soya, rape and palm which are pressed (or extracted), releasing oils (and generating a rest press-cake) which are then transesterified yielding fatty acid esters (usually methyl) and glycerol. Debate as to land use and destruction of the rainforest is a particular issue. For example, the growth of soya in South America has raised concern about the destruction of the rainforest for its production.

Rest streams from biofuel production could lead to opportunities to obtain large volumes of protein. Theoretically, the volumes would allow sufficient amino acids to be available as feedstocks for bulk functionalised chemicals. However, one should also take into account the complex issue regarding ethical, social and ecological issues surrounding biomass with respect to competition with food production and prices, land use and carbon debt. Strategies that address these suitably are maybe the most rate-determining step for approval and success in biofuel production and use of the co-produced rest streams in non-food/feed technical applications.

## 22.3 Protein isolation, hydrolysis and isolation of amino acid chemical feedstocks

Proteins need to be isolated and subsequently hydrolysed before their amino acid content can be utilised as a feedstock for chemicals. Several techniques have been developed for the isolation of protein from biomass residues, which are based on extraction using aqueous acid, base or ethanol (Fig. 22.1). It was observed that solutions combined with protease or cellulase aided solubilisation (Dale *et al.*, 2009; Bals *et al.*, 2009; Cookman and Glatz, 2009; Wolf and Lawton, 1997).

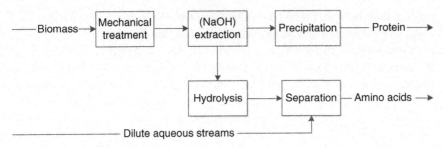

22.1 Generic scheme of amino acid or protein production from biomass.

## 22.3.1 Protein isolation from residues of vegetable oil and biofuel production

Oilseeds can be processed in two distinct ways. Traditionally, press-cakes from, for example, rapeseed and sunflower are co-produced with their respective oils in a crushing process. Oil is also often recovered from the seeds by extraction using solvents, e.g. hexane. To eliminate the use of organic solvents, other extraction media such as supercritical fluid extraction or aqueous extraction processing are also used. Thermal or enzyme treatment can be used to enhance oil availability during (aqueous) extraction processes. The method of oil production affects the quality and properties of the remaining protein fraction, which is currently used as animal feed. After oil removal, protein extraction can occur with methods, such as acid, alkaline or protease (enzyme) assisted extraction. This generally yields proteins (with acid or alkaline extraction) or a mixture primarily containing peptides and amino acids (with enzyme-assisted extraction). Alternatively, proteins can be extracted prior to oil extraction to minimise denaturation if intact protein is the main desired product.

Soya protein extraction is a well-developed process. Processing soya affects the solubility and biological activity of the proteins. The native proteins in principle are water soluble. In the process of extracting the oil and subsequent heat processing, the solubility of the protein in water is reduced. The extent of denaturation and reduction in the solubility are related to the intensity of heat treatment(s) that the bean is exposed to. Processing soya beans involves drying, cracking, dehulling and rolling into flakes. The flakes are either milled to yield full fat soya flour or extracted with hexane to remove the oil. Any residual hexane is stripped off using steam. The extracted soya bean flakes are processed into standard meal for animal feed or into special protein products. Generally, the protein content is *ca.* 45–50 wt%. To obtain a higher protein content, the removal of components other than oil, such as sugars and minerals, is required. The protein can then be dissolved and isolated by iso-electric precipitation. The isolated protein can be hydrolysed to produce amino acids.

A promising technology for protein extraction is based on the ammonia fibre expansion (AFEX) pretreatment technology (Dale, 1981). AFEX involves treating the biomass in an ammonia solution (0.65–3.5 MPa, 70–150°C, 5–15 min). This enables ammonia to permeate in the fibrous structure and solubilise the protein. With a rapid drop in pressure, the cellulosic structure is disrupted. Using this technology, protein is extracted and simultaneously cellulose is processed to be more susceptible to cellulase mediated hydrolysis to fermentable sugars. In a two-step protein extraction with AFEX, *ca.* 80% of protein present could be extracted (Dale *et al.*, 2009).

### 22.3.2 Protein hydrolysis

Once extracted, proteins need to be hydrolysed to generate amino acids. In the food industry, it is general practice to hydrolyse vegetable proteins using 6M HCl at elevated temperatures and prolonged periods of time. Major drawback of this method to produce amino acids as chemical feedstocks is the considerable amount of base required to neutralise the hydrolysate leading to high concentrations of inorganic salts. In addition to this, the harsh acid hydrolysis leads to hydrolysis of (hemi-)cellulose to sugars and furans which are dissolved in the amino acid mixture. Both salts and soluble organic molecules need to be separated from the amino acid mixture as it will have a detrimental effect on the efficiency of the downstream processing (Scott *et al.*, 2010). In addition, a number of amino acids are destroyed during the hydrolysis process leading to potential losses.

An alternative protein hydrolysis method, which does not involve formation of large quantities of salts and use elevated temperatures, is based on the use of proteases. Using protease cocktails, hydrolysis yields of up to 90% can be achieved (Hill and Schmidt, 1962). On the downside, use of proteases for protein hydrolysis incurs high enzyme costs. To enable the reuse of proteases, and reduce costs, immobilisation might prove an option. One economically attractive method could be the formation of cross-linked enzyme aggregates (CLEAs) and would also protect against autolysis (Zhu and Pittman, 2003; Sangeetha and Abraham, 2008).

A recent development is the application of supercritical water in the simultaneous extraction and hydrolysis of proteins from biomass residues (Esteban *et al.*, 2010; Kang and Chun, 2004; Sereewatthanawut *et al.*, 2008; Zhu *et al.*, 2010). Under supercritical conditions, water exhibits a lower dielectric constant and a higher ion product of water at ambient conditions. This results in a higher $H_3O^+$ and $OH^-$ ion concentration, meaning both acid and base catalysed reactions can be performed without addition of acid or base. However, reported hydrolysis yields are limited and addition of low concentrations of acid or base is still required. Major drawbacks in the use of supercritical water are its corrosiveness (which will affect the reactor investment costs) and the high energy input required for its formation.

### 22.3.3 Isolation of desired amino acids

The hydrolysis of bio-derived proteins results in a complex mixture of 20 amino acids. Some of these have the potential to be converted into chemicals, while others, such as methionine, may be more useful as animal feed (Leuchtenberger et al., 2005). To make the best use of protein, or amino acid rest streams, it is necessary to separate them in order to make them applicable for their different applications.

In the case of simple mixtures of amino acids which differ enough with respect to their isoelectric point and polarity, selective precipitation can be achieved based on pH adjustment or addition of an organic solvent, or both, in combination with cooling and concentration (Garcia, 1999). Ion-exchange chromatography is more effective in the separation of complex mixtures of amino acids. However, for large-scale applications, high and unfeasible costs are anticipated due to low throughput and the need for regeneration (Scott et al., 2010). Separation of amino acids via chromatography is still only used for analytical purposes.

A promising technology to achieve amino acid separation is electrodialysis (ED), which removes ions from a dilute solution through ion-exchange membranes to another concentrate solution by applying an electrical potential as the driving force. In principle, it could separate the protein hydrolysate into acidic, basic and neutral amino acid streams according to their charge behaviour (Hara, 1963; Sandeaux et al., 1998; Krol, 1997). Further ED separation within these streams is challenging due to the similarity of their isoelectric points. To overcome this, the charge behaviour can be specifically modified. Amino acid modification may be achieved by removing groups, such as the carboxylic acid functionality. This may be achieved using bio-catalysis. If the modification directly leads to an end product, both chemical production and separation may be achieved in a sustainable route.

An example is the two basic amino acids arginine and lysine which can be obtained by either precipitation or via ED. Both of them are interesting feedstocks for the production of diamines. Lysine decarboxylase can be used to specifically convert lysine to 1,5-pentanediamine (PDA) in the presence of arginine to produce products with different charge, thus allowing isolation of products by subsequent electrodialysis (Teng et al., 2011). In a similar approach, the acidic acids glutamic and aspartic acid may be isolated by ED and glutamic acid specifically modified to pyroglutamic acid. Based on differences in solubility, the aspartic acid can be isolated and the pyroglutamic acid converted back to glutamic acid (Teng et al., 2012). It should be noted that the presence of salt during ED separation will limit efficiency. Therefore the method of protein hydrolysis will be important for a cost-effective separation process. In general, no large-scale technique for the separation of amino acids from complex mixtures exists and this will need to be developed to enable succesful conversion from biomass to amino acids to chemicals in biorefineries.

## 22.4 (Bio)chemical conversion of amino acids to platform and speciality chemicals

A review was published which describes a number of known reactions of amino acids to a variety of chemicals which included monomers and amines. While the reactions were known, and the chemical products are of industrial significance, most reactions were only ever reported as reactions/decomposition of amino acids in a nutritional context and not as a preparative method (Scott et al., 2007). Since then, investigation into the (bio)chemical transformation of a number of amino acids has been explored.

The formation of acrylamide from asparagine and aspartic acid has been reported but only in the context of the formation of toxic compounds during cooking (Taeymans et al., 2004). Recently, Könst et al. (2009) described the conversion of aspartic acid to β-alanine using L-aspartate α-decarboxylase. Although not disclosed by the authors, routes to both acylamide and acrylonitrile were suggested.

Glutamic acid is produced by the hydrolysis of the amide functionality of glutamine. Glutamic acid can undergo enzymatic decarboxylation, resulting in the formation of γ-aminobutyric acid (GABA). Werpy and Petersen report the use of glutamic acid, produced from fermentation, as a building block for a number of chemical products such as 5-amino-1-butanol and glutaric acid (Werpy and Petersen, 2004). More recently glutamic acid as a platform for a number of chemicals has been described, as shown in Fig. 22.2. A fed batch process using glutamic acid and immobilised glutamic acid α-decarboxylase has been described (Lammens et al., 2009). The resultant GABA can then undergo ring closure and methylation using methanol in the presence of a catalytic amount of a metal halide salt to form N-methylpyrrolidone (NMP) (Lammens et al., 2010). In further articles from the same author, it was also shown that this route shows a favourable techno-economic as well as a positive ecological footprint compared to the current petrochemical route (Lammens et al., 2012b). 2-Pyrrolidone, used for the production of N-vinylpyrrolidone, can also be readily formed by the cyclisation of GABA. Further, glutamic acid can undergo

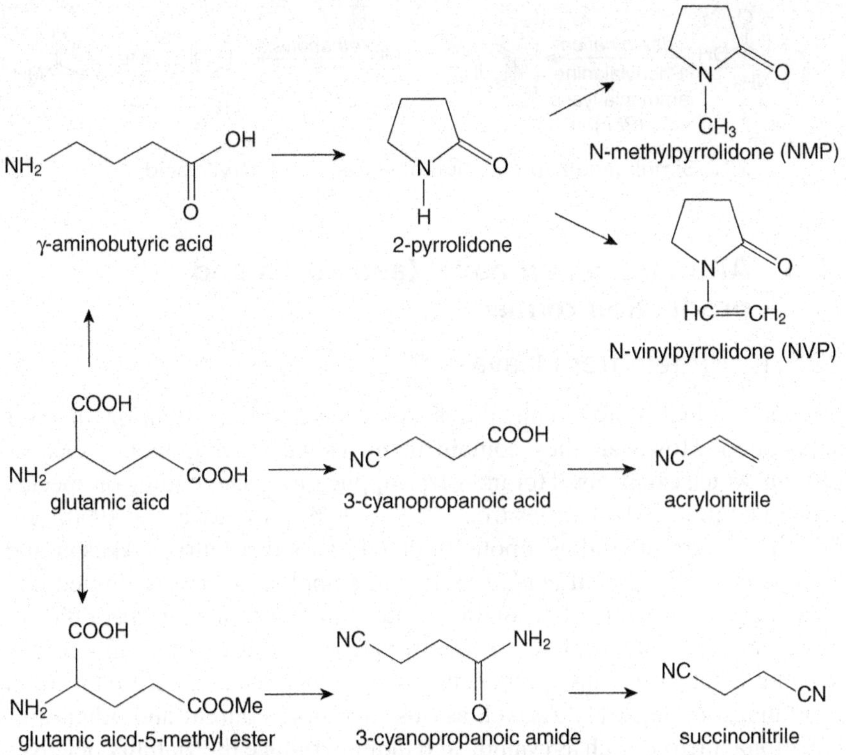

22.2 Glutamic acid as a platform for the production of chemicals.

oxidative decarboxylation to cyanopropanoic acid followed by subsequent decarbonylation elimination reaction, resulting in the formation of acrylonitrile (Le Notre et al., 2011).

The conversion of phenylalanine to cinnamic acid has been extensively described using phenylammonia lyase (Nijkamp et al., 2005; Ben-Bassat et al., 2005; Camm and Towers, 1973; Ogata et al., 1966). Subsequent decarboxylation of cinnamic acid results in the formation of styrene (Dahlig, 1955). An alternative method to produce styrene from cinnamic acid involves cross-metathesis (ethenolysis) (Sanders et al., 2010). Here, cinnamic acid undergoes a metathesis reaction with ethylene for the simulataneous production of styrene and acrylic acid, as shown in Fig. 22.3 (Spekreijse et al., 2012).

The synthesis of other aromatic compounds has been explored. Tryptophan has also been shown to undergo photocatalysis leading to intermediates that can be used in a route towards bio-based aniline (Hamdy et al., 2012; Sanders et al., 2009).

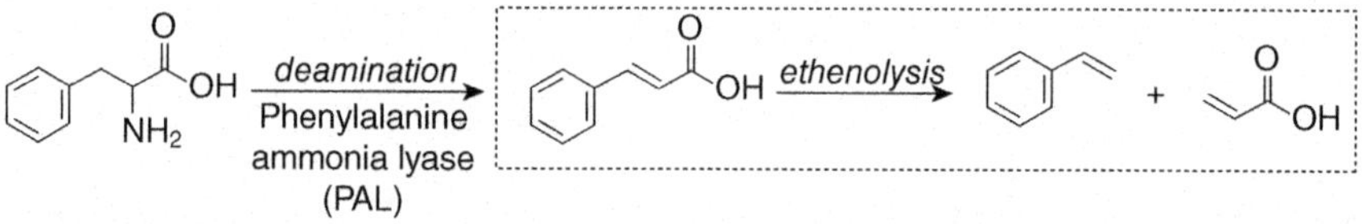

*22.3* Simultaneous production of styrene and acrylic acid.

## 22.5 Alternative and novel feedstocks and production routes

### 22.5.1 Protein from leaves

Leaves are highly abundant, but have received little attention in their application. However, they contain many valuable components, such as protein, which can be used for industrial applications. Depending on species, protein content in leaf varies from 15% to 30%. Already in the 1960s, leaf protein was considered as a potential feed or food resource (Akeson and Stahmann, 1965; Gerloff et al., 1965), but its applications were limited due to a high proportion of insoluble protein and insufficient cost-efficient processing. Soluble protein varied from 15% to 60% of total protein, depending on species and processing methods (Kammes et al., 2011; Chiesa and Gnansounou, 2011). As well as this, high fibre content and other anti-nutritional factors, such as cyanide and tannins, reduce the viability of leaves as a protein source for humans, but it is used in feed, due to the amino acid composition. Other specific leaves such as sugarbeet, cassava and grass have also been evaluated as a feedstock (Lammens et al., 2012a).

### 22.5.2 Protein from algae

Microalgae has been promoted as a future source of transportation fuels. It has been estimated that production of biodiesel from algae to replace current European diesel requirements would lead to 300 Mton of algal protein (Wijffels and Barbosa, 2010).

### 22.5.3 Protein from potato starch production residues

During the processing of potatoes for starch extraction, the main waste stream is Protamylasse™, which mainly contains sugars, organic acids, proteins and free amino acids and currently has a limited use. Allied to this, it has been shown to be a promising substrate for the production of cyanophycin, a polymer consisting of a poly(aspartic acid) backbone with arginine side chains (Elbahloul et al., 2005; Mooibroek et al., 2007). Interestingly, this polymer is insoluble under physiological conditions, therefore offering the opportunity to isolate specific amino acids from a complex mixture and aid downstream processing. Cyanophycin has been demonstrated to lead to a number of chemicals (Könst *et al.*, 2010, 2011a, 2011b).

The genes to produce cyanophycin can be incorporated into yeast for the simultaneous production of ethanol and cyanophycin in media containing sugars and amino acids. It could be conceived that such an approach incorporated into current ethanol production could lead to co-production of new value-added products.

## 22.6 Conclusion and future trends

There are a number of challenges that need to be tackled to avoid adverse effects of the use of biomass for non-food applications. Some ways to do this include:

- certification of biomass on sustainability criteria such as the influence of cultivation on the environment and social issues such as impact on food (production and market) and working conditions
- use of non-food crops on arid land which could not otherwise be cultivated for food; an example is the use of *Jatropha curcas* as a biofuel and non-food protein source
- use of (lignocellulosic) rest streams from primary agricultural production.

Improved biorefinery of agricultural materials will produce sufficient quality food and lead to suitable waste streams that can be used for other non-food technical and value-added applications. Activities and incentives to further develop the biorefinery concent are growing at a rapid pace.

Non-food application of amino acids from protein isolated from biomass required several specific technological solutions. Mild separation of proteins is vital where proteins are to be used in applications where functionality is important. This area has received some attention, but as yet cannot be carried out at low cost. The development of cost-effective techniques to isolate all single amino acids is a technology that is crucial for application of amino acids as chemical feedstocks. Some attempts to use specific conversions to aid separation have been studied and this could lead to more cost-effective *in situ* product formation and recovery. This is especially true during the isolation of specific molecules from dilute watery steams. Chromatography, while efficient, utilises significant chemicals for regeneration and the throughput for voluminous streams may be prohibitive. Concentration of product streams by distillation is both energy and cost intensive. To overcome this, an approach (using microorganisms) which concentrates molecules as insoluble materials allowing more straightforward isolation seems interesting.

In general, protein and amino acids can be important by-products in many biorefinery processes that are now aiming at producing other components like oil and sugar/ethanol. However, several other (groups) of components can be put to good use and thus create additional value for the overall chain. Examples of these components include fibres, (hemi)cellulose, phosphate, potassium (K+), organic acids, vitamins and lignin.

## 22.7 References

Akeson, W. R. and Stahmann, M. A., 1965, Leaf proteins as foodstuffs, nutritive value of leaf protein concentrate, an *in vitro* digestion study. *Journal of Agricultural and Food Chemistry*, 13, 145–148.

Bals, B., Balan, V. and Dale, B., 2009, Integrating alkaline extraction of proteins with enzymatic hydrolysis of cellulose from wet distiller's grains and solubles. *Bioresource Technology*, 100, 5876–5883.

Ben-Bassat, A., Sariaslani, F.S., Huang, L.L., Patnaik, R. and Lowe, D.J., 2005, Methods for the preparation of para-hydroxycinnamic acid and cinnamic acid at alkaline pH. US Patent 2005/0260724 A1.

Camm, E.L. and Towers, G.H.N., 1973, Phenylalanine ammonia lyase. *Phytochem*, 12, 961–973

*Chemical Week*, 2005, 'ADM plans polyols unit using renewable feedstock' 30 November/7 December, 23.

*Chemie Magazine*, 2006, 'Hoger rendement door lagereenergiekosten: Methanor verder met biomethanol', October.

Chiesa, S. and Gnansounou, E., 2011, Protein extraction from biomass in a bioethanol refinery – possible dietary applications: use as animal feed and potential extension to human consumption. *Bioresource Technology*, 102, 427–436.

Cookman, D.J. and Glatz, C.E., 2009, Extraction of protein from distiller's grain. *Bioresource Technology*, 100, 2012–2017.

Dahlig, W., 1955, Styrene from cinnamic acid. *Przemysl Chemiczny*, 11, 518–520.

Dale, B.E., 1981, Method for increasing the reactivity and digestibility of cellulose with ammonia. US Patent 4,600,590.

Dale, B.E. Allen, M.S. Laser, M. and Lynd, L.R., 2009, Protein feeds coproduction in biomass conversion to fuels and chemicals. *Biofuels, Bioproducts & Biorefining*, 3, 219–230.

Elbahloul, Y., Frey, K., Sanders, J. and Steinbüchel, A., 2005, Protamylasse, a residual compound of industrial starch production, provides a suitable medium for large-scale cyanophycin production. *Appl. Environ. Microbiol.*, 71, 7759–7767.

Esteban, M.B., García, A.J., Ramos, P. and Márquez, M.C., 2010, Sub-critical water hydrolysis of hog hair for amino acid production. *Bioresource Technology*, 101, 2472–2476.

EU Directive 2009/28/EC of the European Parliament and the Council of April the 23rd 2009, *Official Journal of the European Union.*, 2009, L140/16.

FAO, 2003, Report, 'World agriculture: towards 2015/2030'. FAO, Rome.

Garcia, A.A., 1999, Thermodynamic and transport properties. In *Bioseparation Process Science*, ed. A.A. Garcia, Wiley-Blackwell Science, Malden, MA.

Gerloff, E.D., Lima, I.H. and Stahmann, M.A., 1965, Leaf proteins as foodstuffs, amino acid composition of leaf protein concentrates. *Journal of Agricultural and Food Chemistry*, 13, 139–143.

Hamdy, M.S., Scott, E.L., Carr, R. and Sanders, J.P.M., 2012, A novel photocatalytic conversion of tryptophan to kynurine using black light as a source. *Catalysis Letters*, 142, 338–344.

Hara, Y., 1963, The separation of amino acids with an anion-exchange membrane. *Bulletin of the Chemical Society of Japan*, 36, 1372–1376.

Hill, R.L. and Schmidt, W.R., 1962, Complete enzymic hydrolysis of proteins. *The Journal of Biological Chemistry*, 237, 389–396.

IEA – www.iea.org/press/pressdetail.asp?PRESS_REL_ID=411 and related report: www.iea.org/papers/2011/biofuels_roadmap.pdf.

Kammes, K.L., Bals, B.D., Dale, B.E. and Allen, M.S., 2011, Grass leaf protein, a coproduct of cellulosic ethanol production, as a source of protein for livestock. *Animal Feed Science and Technology*, 164, 79–88.

Kang, K.-Y. and Chun, B.-S., 2004, Behavior of hydrothermal decomposition of silk fibroin to amino acids in near-critical water. *Korean Journal of Chemical Engineering*, 21, 654–659.

Könst, P.M. Franssen, M.C.R. Scott, E.L.and Sanders, J.P.M., 2009, A study on the applicability of L-aspartate alpha-decarboxylase in the biobased production of nitrogen containing chemicals. *Green Chemistry*, 11, 1646–1652.

Könst, P.M., Turras, P.M.C.C.D., Franssen, M.C.R., Scott, E.L. and Sanders, J.P.M., 2010, Stabilized and immobilized *Bacillus subtilis* arginase for the biobased production of nitrogen-containing chemicals. *Advanced Synthesis and Catalysis*, 352(9), 1493–1502.

Könst, P.M. Franssen, M.C.R. Scott, E.L. and Sanders, J.P.M., 2011a, Stabilization and immobilization of *Trypanosoma brucei* ornithine decarboxylase for the biobased production of 1,4-diaminobutane. *Green Chemistry*, 13(5), 1167–1174.

Könst, P.M., Scott, E.L., Franssen, M.C.R. and Sanders, J.P.M., 2011b, Acid and base catalyzed hydrolysis of cyanophycin for the biobased production of nitrogen containing chemicals. *Journal of Biobased Materials and Bioenergy*, 5(1), 102–108.

Krol, J.J., 1997, Monopolar and bipolar ion exchange membranes – mass transport limitations. PhD thesis, University of Twente.

Lammens, T.M., De Biase, D., Franssen, M.C.R., Scott, E.L. and Sanders, J.P.M., 2009, The application of glutamic acid alpha-decarboxylase for the valorization of glutamic acid. *Green Chemistry*, 11, 1562–1567.

Lammens, T.M., Franssen, M.C.R. ,Scott, E.L. and Sanders, J.P.M., 2010, Synthesis of biobased *N*-methylpyrrolidone by one-pot cyclization and methylation of α-aminobutyric acid. *Green Chemistry*, 12(8), 1430–1436.

Lammens, T.M., Franssen, M.C.R., Scott, E.L. and Sanders, J.P.M., 2012a, Availability of protein derived amino acids as feedstock for the production of biobased chemicals. *Biomass and Bioenergy*, 44, 168–181.

Lammens, T.M., Gangarapu, S., Franssen, M.C.R., Scott, E.L. and Sanders, J.P.M., 2012b, Techno-economic assessment of the production of bio-based chemicals from glutamic acid. *Biofuels Bioproducts and Biorefining*, 6(2), 177–187.

Lange, J.-P., 2001, Fuels and chemicals manufacturing guidelines for understanding and minimizing the production costs. *Cattech*, 5, 82–95.

Le Notre, J.E.L., Scott, E.L., Franssen, M.C.R. and Sanders, J.P.M., 2011, Biobased synthesis of acrylonitrile from glutamic acid. *Green Chemistry*, 13(4), 807–809.

Leuchtenberger, W., Huthmacher, K. and Drauz, K., 2005, Biotechnological production of amino acids and derivatives: current status and prospects. *Applied Microbiology and Biotechnology*, 69, 1–8.

Mooibroek, H., Oosterhuis, N., Giuseppin, M.L.F., Toonen, M.A.J., Franssen, H.G.J.M., Scott, E.L., Sanders, J.P.M. and Steinbuchel, A., 2007, Assessment of technological options and economic feasibility for cyanophycin biopolymer and high-value amino acid production. *Applied Microbiology and Biotechnology*, 77(2), 257–267.

Nijkamp, K., van Luijk, N. and de Bont, J.A.M., 2005, The solvent-tolerant *Pseudomonas putida* S12 as host for the production of cinnamic acid from glucose. *Appl Microbiology and Biotechnology*, 69, 170–177.

Nuez Ortín, W.G. and Yu P., 2009, Nutrient variation and availability of wheat DDGS, corn DDGS and blend DDGS from bioethanol plants. *Journal of the Science of Food and Agriculture*, 89, 1754–1761.

Ogata, K., Uchiyama, K. and Yamada, H., 1966, Microbial formation of cinnamic acid from phenylalanine. *Agr. Biol. Chem.* (Tokyo) 30(3), 311–312.

Sandeaux, J., Sandeaux, R., Gavach, C., Grib, H., Sadat, T., Belhocine, D. and Mameri, N., 1998, Extraction of amino acids from protein hydrolysates by electrodialysis. *Journal of Chemical Technology & Biotechnology*, 71, 267–273.

Sanders, J.P.M., Peter, F., Scott, E.L., Saad, M.H. and Carr, R.H., 2009, Process for converting tryptophan into kynurenine. Patent number: WO2009135843.

Sanders, J.P.M., van Haveren, J., Scott, E.L., van Es, D.S. and Le Notre, J.E.L., 2010, Olefin cross-metathesis applied to biomass. Patent number: PCT/NL2010/050286.

Sangeetha, K. and Abraham, T.E., 2008, Preparation and characterization of cross-linked enzyme aggregates (CLEA) of *Subtilisin* for controlled release applications. *International Journal of Biological Macromolecules*, 43, 314–319.

Scott, E., Peter, F. and Sanders, J., 2007, Biomass in the manufacture of industrial products – the use of proteins and amino acids. *Appl Microbiology and Biotechnology*, 75, 751–762.

Scott, E.L., Sanders, J.P.M. and Steinbüchel, A., 2010, Perspectives on chemicals from Renewable Resources. In *Sustainable Biotechnology*, eds. O.V. Singh and S.P. Harvey, Springer, Dordrecht, 195–210.

Sereewatthanawut, I., Prapintip, S., Watchiraruji, K., Goto, M., Sasaki M. and Shotipruk A., 2008, Extraction of protein and amino acids from deoiled rice bran by subcritical water hydrolysis. *Bioresource Technology*, 99, 555–561.

Solvay Chemicals, 2006, 'Solvay builds new epichlorohydrin plant to meet growing demands with innovative production process'. Press release, 31 January.

Spekreijse, J., Le Nôtre, J., van Haveren, J., Scott, E.L. and Sanders J.P.M., 2012, Ethenolysis as part of a novel route for the simultaneous production of biobased styrene and acrylates, *Green Chemistry*, 14, 2747–2751.

Taeymans, D., Wood, J., Ashby, P., Blank, I., Studer, A., Stadler, R., Gonde, P., Eijck, P., Lalljie, S., Lingnert, H., Lindblom, M., Matissek, R., Mueller, D., Tallmadge, D., O'Brien, J., Thompson, S., Silvani, D. and Whitmore, T., 2004, A review of acrylamide: an industry perspective on research, analysis, formation and control. *Food Sciences and Nutrition*, 44, 323–347.

Teng, Y., Scott, E.L., van Zeeland, A.N.T. and Sanders, J.P.M., 2011, The use of L-lysine decarboxylase as a means to separate amino acids by electrodialysis. *Green Chemistry*, 13, 624–630.

Teng, Y., Scott, E.L. and Sanders, J.P.M., 2012, Separation of L-aspartic acid and L-glutamic acid mixtures for use in the production of bio-based chemicals. *Journal of Chemical Technology and Biotechnology*, 87, 1458–1465.

UNEP, 2008, Report, 'International Assessment of Agricultural Knowledge, Science and Technology for Development'. United Notions Environment Programme, Washington DC.

Werpy, T. and Petersen, G., 2004, 'Top Value Added Chemicals from Biomass'. PNNL, NREL, EERE report 8674.

Wijffels, R.H. and Barbosa, M.J., 2010, An outlook on microalgal biofuels. *Science*, 329, 796–799.

Wolf, W.J. and Lawton, J.W., 1997, Isolation and characterization of zein from corn distillers' grains and related fractions. *Cereal Chemistry*, 74, 530–536.

Zhu, G.-Y., Zhu, X., Wan, X.-L., Fan, Q., Ma, Y.-H., Qian, J., Liu, X.-L., Shen, Y.-J. and Jiang J.-H., 2010, Hydrolysis technology and kinetics of poultry waste to produce amino acids in subcritical water. *Journal of Analytical and Applied Pyrolysis*, 88, 187–191.

Zhu, H.-J. and Pittman, C.U., 2003, Reductions of carboxylic acids and esters with $NaBH_4$ in diglyme at 162 degrees C. *Synthetic Communications*, 33, 1733–1750.

# 23
Types, processing and properties of bioadhesives for wood and fibers

A. PIZZI, University of Lorraine, France and King Abdulaziz University, Saudi Arabia

DOI: 10.1533/9780857097385.2.736

**Abstract**: The chapter describes the state of the art of biobased adhesives for wood, namely modern approaches to the already commercial tannin adhesives, as well as lignin-based adhesives, carbohydrate adhesives, protein adhesives, unsaturated oil adhesives and wood welding by mechanical friction without adhesives.

**Key words**: wood adhesives, tannin adhesives, lignin adhesives, biobased adhesives, wood welding.

## 23.1 Introduction

Wood and fiber adhesives from renewable raw materials have been a topic of considerable interest for many years. This interest, already present since the 1940s, became more intense with the world's first oil crisis in the early 1970s and subsided again as the cost of oil decreased. At the beginning of the twenty-first century this interest is becoming intense again for a number of reasons. The foreseen future scarcity of petrochemicals still appears to be reasonably far into the future. It is a contributing factor but, at this stage, it is not the main motivating force. The main impulse of today's renewed interest in bio-based adhesives is the acute sensitivity of the general public to anything that has to do with the environment and its protection. It is not even this concern *per se* that motivates such an interest. There are rather very strict, for some synthetic adhesives almost crippling, government regulations which are just starting to be put into place to allay the environmental concerns of the public.

First of all, it is necessary to define what is meant by bio-based wood adhesives, or adhesives from renewable, natural, non-oil-derived raw materials. This is necessary because in its broadest meaning the term might be considered to include urea-formaldehyde resins, urea being a non-oil derived raw material. This, of course, is not the case. The term 'bio-based adhesive' has come to be used in a very well specified and narrow sense to only include those materials of natural, non-mineral, origin which can be used as such or after small modifications to reproduce the behaviour and performance of synthetic resins. Thus, only a limited number of materials can currently be included, at a stretch, in the narrowest sense of this definition. These are tannins, lignin, carbohydrates, unsaturated oils, proteins and protein hydrolysates, dissolved wood and wood welding by self-adhesion. To this in the future will surely be added proteins, blood and collagen that are already being used to some extent based on technology from the far past. The bio-based wood adhesives approach does not mean, however, any need to go back to the technology of natural product adhesives as they existed up to the 1920s and 1930s before they were supplanted by synthetic adhesives. The bio-based adhesives we are talking about here are indeed derived from natural adhesives, but using or requiring novel technologies, formulations and methods.

Of the classes of bio-based wood adhesives mentioned above, in the case of tannins and lignins their interest has been directed primarily at substituting phenol-formaldehyde (PF) resins, because of the phenolic nature of these two classes of compounds. In some cases, some formaldehyde is still used, and in the case of lignin some other additives. It is then necessary to distinguish between bio-based adhesives in which a limited amount of synthetic additives are still used, and bio-based wood adhesives where no synthetic additives are used.

Three types of thermosetting adhesives based on bio-materials have reached commercial exploitation, although their use is not yet widespread, namely tannin-based adhesives at a level of up to 50,000 tons solids per year, lignin adhesives coupled with synthetic resins, and soy protein hydrolysate adhesives.

## 23.2 Tannin adhesives

The word tannin has been used loosely to define two different classes of chemical compounds of mainly phenolic nature: hydrolysable tannins and condensed tannins. The former are mixtures of simple phenols such as pyrogallol and ellagic acid and of esters of a sugar, mainly glucose, with gallic and digallic acids [1]. The lack of macromolecular structure in their natural state, the low level of phenol substitution they allow, their low nucleophilicity, limited worldwide production, and higher price somewhat decrease their chemical and economic importance.

Condensed tannins, on the other hand, constituting more than 90% of the total world production of commercial tannins (200,000 tons per year), are both chemically and economically more important for the preparation of adhesives and resins. Condensed tannins and their flavonoid precursors are known for their wide distribution in nature and particularly for their substantial concentration in the wood and bark of various trees. These include various *Acacia* (wattle or mimosa bark extract), *Schinopsis* (quebracho wood extract), *Tsuga* (hemlock bark extract), and *Rhus* (sumach extract) species, from which commercial tannin extracts are manufactured, and various *Pinus* bark extract species.

## 23.2.1 Condensed (polyflavonoid) tannins

Condensed tannins are polyhydroxyphenols, polyflavonoids (see Fig. 23.1), which are soluble in water, alcohols and acetone and can coagulate proteins. They are mainly obtained commercially by water extraction from wood and bark. The other main components of the extracts, called non-tannins, are simple sugars and polymeric carbohydrates. The content of non-tannins can reduce the performance and water resistance of tannin-bonded joints. The polymeric carbohydrates also increase the viscosity of the extracts.

Tannin extracts are usually sold as spray-dried powders. No purification step is usually carried out in industrial-scale production. The modification of the extracts is especially aimed at decreasing the sometimes too high viscosity to achieve better handling and application, but also a longer pot life and a better crosslinking [1,2].

As tannins contain many 'phenolic'-type subunits, one may be tempted to think that they will exhibit a similar reactive potential to that of phenol and therefore procedures used in standard PF production can be transferred

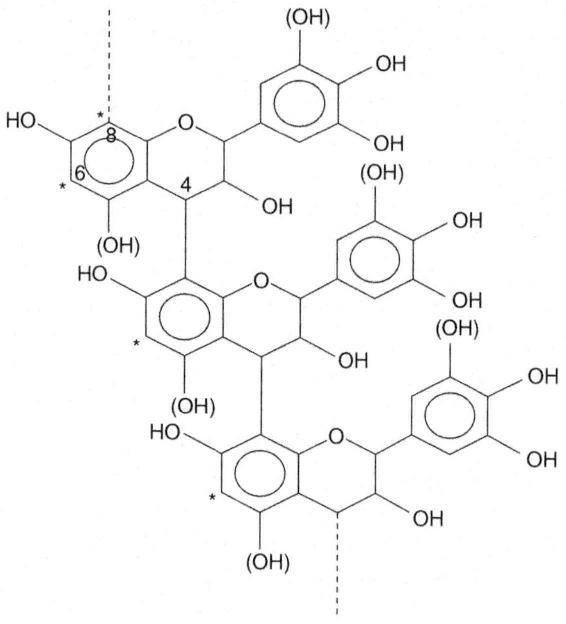

*23.1* Schematic representation of predominantly C4-C8-linked procyanidin tannin.

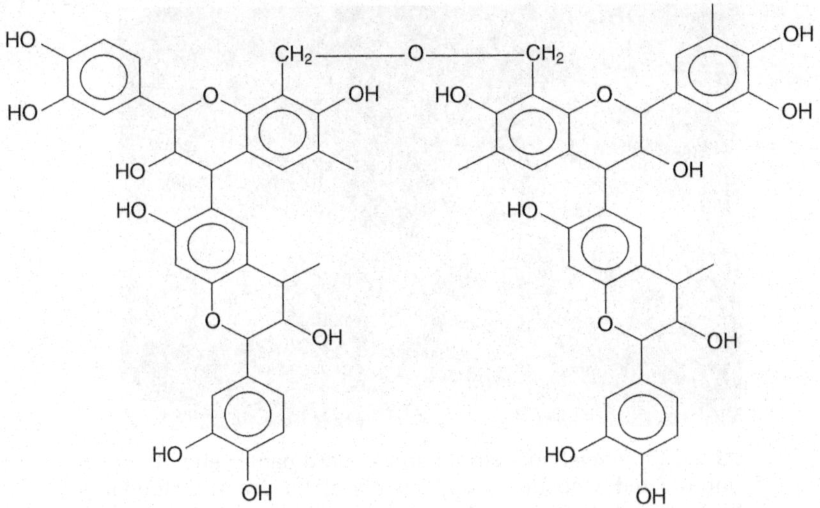

*23.2* Methylene-ether bridge between flavonoid tannin oligomers obtained by reaction with formaldehyde.

to those containing tannin. This, however, is not the case because tannins are far more reactive than phenol due to the resorcinol and phloroglucinol nuclei present in the structure of condensed tannins. This increase in hydroxyl substitution on the two aromatic rings affords in relation to phenol an increase in reactivity towards formaldehyde of 10–50 times.

Tannin crosslinking by formaldehyde via methylene or methylene ether bridges in a polycondensation reaction is the traditional way for tannins to function as exterior-grade weather-resistant wood adhesives (see Fig. 23.2). Tannin reactions are based on their phenolic character similar to those of phenol with formaldehyde. Formaldehyde reacts with tannin in an exothermic reaction forming methylene bridges. At neutral or even slightly acid pH, formaldehyde reacts rapidly with the tannin. This leads to the advantage that there is no high alkali content in tannin adhesives, contrary to synthetic phenolic resins. Thus, the neutral, hardened glue line which is obtained yields a noticeable improvement in water and weather resistance. One example that natural adhesive can present equal or even better performance of synthetic adhesives is illustrated in Fig. 23.3. Figure 23.3 shows the condition of two commercial, industrial particleboard panels after 15 years unprotected exposure to the weather at 1,500 m of altitude in a high UV radiation area. The remarkable performance of the tannin-bonded panel is easy to see in relation to a pure melamine-formaldehyde (MF) bonded panel. It must be stated clearly that in the 25 years since this picture was taken, melamine resin engineering has improved considerably to the point that similar performances to those of the tannin-bonded board

*23.3* Commercial, industrial particleboard panels after 15 years unprotected exposure to the weather at 1,500 m of altitude in a high UV radiation area. Melamine-formaldehyde on the left, tannin-paraformaldehyde on the right.

can now be obtained too with top of the range synthetic melamine-urea-formaldehyde adhesives (with at least 45:55 M:U by weight).

Due to their high reactivity, tannin adhesives, in reality natural novolaks, need addition of paraformaldehyde hardener just before use. Furthermore, their aldehyde content is much lower than for synthetic adhesives as this is exclusively used as hardener, the polymer having been built naturally already in the tree. One alternative way used industrially to add the hardener, other than just mixing it in the glue-mix, is its separate addition by, for example, dosing the paraformaldehyde via a small screw conveyor directly to the particles in the blender. Also, for a liquid crosslinker, e.g. a urea-formaldehyde prepolymer concentrate (formurea), this can be mixed with the tannin solution in a static mixer shortly before the blender. The higher viscosity of the tannin solution at higher pHs, even without addition of the hardener, can be overcome by warming up to 30–35°C or by adding water. However, the viscosity is decreased well before application by simple acid-base pretreatments to hydrolyse the polymeric carbohydrates in the tannin extract. A higher moisture content of the glued particles is no disadvantage with these adhesives, but rather necessary to guarantee a proper flow of the tannin [1–3], this being in reality a considerable industrial advantage these adhesives have over synthetic adhesives.

The hardeners used are mainly paraformaldehyde fine powder [1,2], and methylolurea mixtures such as urea-formaldehyde precondensate or formurea [4]. Tannins can also be hardened by addition of hexamethylenetetramine (hexamine) [3,5–8], whereby these boards show a very low formaldehyde emission [1–3,5,9–12]. The autocatalytic hardening of tannins without addition of formaldehyde or another aldehyde as crosslinker is possible, if small traces of alkaline $SiO_2$ is present as catalyst at a high pH, or with certain tannins just by the catalytic action induced by the wood surface [13–22]. Tannin autocondensation ensures a totally natural, environmentally-friendly adhesive, but it is only suitable for interior-grade applications.

## 23.2.2 New technologies for industrial tannin adhesives

Extensive, up-to-date and in-depth reviews of the technology of tannin adhesives based on the classical technology of tannin-formaldehyde resins, as summarized above, already exist [1,13]. These technologies are commercial now for several years and used in a number of countries. It is sufficient to state here that tannin-formaldehyde adhesives of very low emission (E0), fast pressing times and using unmodified tannin extracts are well known, are used commercially and their technology is commercially and perfectly mastered [23,24]. The new technologies are those based either on no addition of aldehydes, or on the use of hardeners which are non-emitting or manifestly non-toxic.

The quest to decrease or completely eliminate formaldehyde emissions from wood panels bonded with adhesives, although not really necessary in tannin adhesives due to their very low emission (like most phenolic adhesives), has nonetheless promoted some research to further improve formaldehyde emissions. This has centered into two lines of investigation: (i) the use of hardeners not emitting at all simply, because either no aldehyde has been added to the tannin, or because the aldehyde cannot be liberated from the system, and (ii) tannin autocondensation. Methylolated nitroparaffins and in particular the simpler and least expensive exponent of their class, namely trishydroxymethyl nitromethane [25,26], belong to the first class. They function well as hardeners of a variety of tannin-based adhesives while affording considerable side advantages to the adhesive and to the bonded wood joint. In panel products such as particleboard, medium density fiberboard and plywood, the joint performance which is obtained is of the exterior/marine grade type, while a very advantageous and very considerable lengthening in glue-mix pot-life is obtained. Furthermore, the use of this hardener is coupled with such a marked reduction in formaldehyde emission from the bonded wood panel to reduce emission exclusively to the formaldehyde emitted by heating just the wood (and slightly less, thus functioning as a mild depressant of emissions from the wood itself). Furthermore, trishydroxymethyl nitromethane can be mixed in any proportion with traditional formaldehyde-based hardeners for tannin adhesives, its proportional substitution of such hardeners inducing a proportionally marked decrease in the formaldehyde emissions of the wood panel without affecting the exterior/marine grade performance of the panel. Medium density fiberboard (MDF) industrial plant trials confirmed all the properties reported above and the trial conditions and results have been reported [25,26]. A cheaper but equally effective alternative to hydroxymethylated nitroparaffins is the use of hexamine as a tannin hardener.

*Tannin-hexamethylenetetramine (hexamine) adhesives*

These adhesives are already commercial. Under many wood adhesive application conditions, contrary to what was thought for many years, hexamine used as a hardener of a fast reacting species is not at all a formaldehyde-yielding compound, yielding extremely low formaldehyde emissions in bonded wood joints [27]. $^{13}$C NMR evidence has confirmed [28–30] that the main decomposition (and recomposition) mechanism of hexamine under such conditions is not directly to formaldehyde. It rather proceeds through reactive intermediates, hence mainly through the formation of reactive imines and iminoaminomethylene bases (Fig. 23.4). $^{13}$C NMR evidence has also confirmed [28–30] that in the presence of chemical species with very reactive nucleophilic sites, such as melamine, resorcinol and condensed flavonoid tannins, hexamine does not decompose to formaldehyde and ammonia. Instead, the very reactive but unstable intermediate fragments react with the tannin, melamine, etc., to form aminomethylene bridges before any chance to yield formaldehyde. These are also stable for 1–5 hours at temperatures as high as 120°C. The intermediate fragments of the decomposition of hexamine pass first through the formation of imines followed by their decomposition to imino-methylene bases. The latter present only one positive charge as the second methylene group is stabilized by an imine-type bond [28–30] (Fig. 23.4). Any species with a strong real or nominal negative charge under alkaline conditions, be it a tannin, resorcinol or another highly reactive phenol, be it melamine or another highly reactive amine or amide, or be it an organic or inorganic anion, is capable of reacting with the intermediate species formed by decomposition (or recomposition) of hexamine far more readily than formaldehyde [28–30]. This explains the capability of wood adhesive formulations based on hexamine to give bonded panels of extremely low formaldehyde emissions. If no highly reactive species with strong real or nominal negative charge is present, then decomposition of hexamine proceeds rapidly to formaldehyde formation as reported in the previous literature [31].

On this basis, the use of hexamine as a hardener of a tannin, hence a tannin-hexamine adhesive, is a very environmentally friendly proposition.

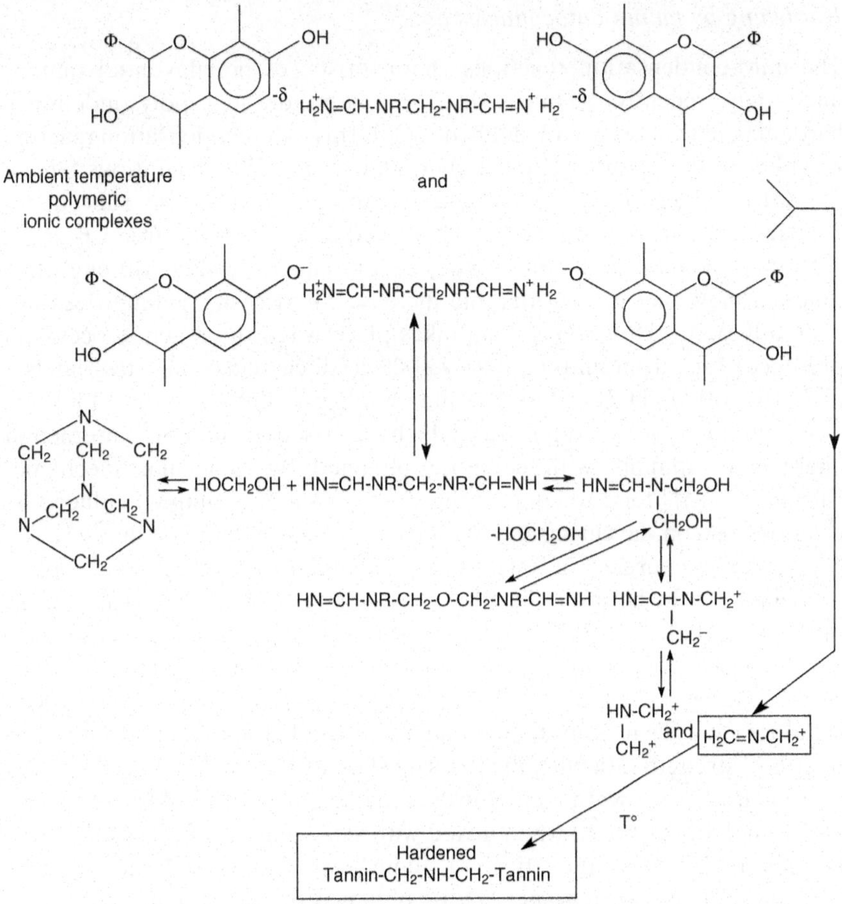

*23.4* Hexamethylenetetramine (hexamine) decomposition and reaction routes when in the presence of a reactive flavonoid tannin.

Formaldehyde emissions in a great chamber has been proved to be so low as to be limited exclusively to what is generated by the wood itself, hence truly E0 panels. The panels obtained with tannin-hexamine adhesives, according to under which conditions they are manufactured, can satisfy both interior and exterior grade standard specification requirements [32]. Steam injection presses recently have shown to be better suited to give better results for exterior grade boards using tannin-hexamine adhesives [33,34]. Comparable results are obtained with pine tannins or other procyanidins hardened with hexamine [35]. In the same reference, catalysis of the reaction in the presence of small amounts of accelerators such as a zinc salt allows even better results or faster press times.

### Hardening by tannin autocondensation

The auto-condensation reactions characteristic of polyflavonoid tannins have only recently been used to prepare adhesive polycondensates hardening in the absence of aldehydes [36]. This auto-condensation reaction is based on the opening under either alkaline or acidic conditions [36] of the O1-C2 bond of the flavonoid repeating unit and the subsequent condensation of the reactive center formed at C2 with the free C6 or C8 sites of a flavonoid unit on another tannin chain [36–40]. Although this reaction may lead to considerable increases in viscosity, gelling does not generally occur. However gelling takes place when the reaction occurs in presence of small amounts of dissolved silica (silicic acid or silicates) catalyst and some other catalysts [36–40], and on a lignocellulosic surface [36].

As in the case of other formaldehyde-based resins, the interaction energies of tannins with cellulose obtained by molecular mechanics calculations [41] tend to confirm the effect of surface catalysis induced by cellulose also on the curing and hardening reaction of tannin adhesives. The considerable energies of interactions obtained can effectively explain weakening of the heterocyclic ether bond leading to accelerated and easier opening of the pyran ring in a flavonoid unit, as well as the facility with which hardening by auto-condensation can occur. In the case of the more reactive procyanidins and prodelphinidin-type tannins, such as pine tannin, cellulose catalysis is more than enough to cause hardening and to produce boards of strength satisfying the relevant standards for interior grade panels [36]. Figure 23.5 shows that the slower reacting tannins can yield an upgraded IB strength of the board when mixed with small amounts of faster reacting tannins. In Fig. 23.5, the effect of adding pecan tannin is shown as an example, but similar upgrades can be obtained by adding pine tannin too. In the case of the less reactive tannins, however, such as mimosa and quebracho, the presence of a dissolved silica or silicate catalyst of some type

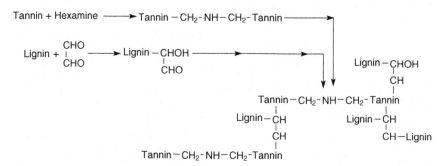

23.5 Schematic representation of the series of reactions occurring in the formation of the tannin/lignin hardened network of the formulation.

is the best manner to achieve panel strength as required by the relevant standards [13]. Auto-condensation reactions have been shown to contribute considerably to the dry strength of wood panels bonded with tannins, but to be relatively inconsequential in contributing to the bonded panels' exterior grade properties which are rather determined by polycondensation reactions with aldehydes [41,43]. Combination of tannin auto-condensation and reactions with aldehydes, and combination of radicals with ionic reactions, have been used to decrease both the proportion of aldehyde hardener as well as to decrease considerably the already low formaldehyde emissions yielded by the use of exterior tannin adhesives [41–43]. A variation on the same theme of wood adhesives by tannin autocondensation is acid-catalysed oxidative condensation [44].

## 23.3 Lignin adhesives

### 23.3.1 Lignins

Lignin is a large, amorphous, three-dimensional polymer produced by all vascular terrestrial plants. It is second only to cellulose in natural abundance and is one of the two main constituents of wood. Lignin has phenolic character. It is primarily obtained as a by-product in wood pulping processes with estimates exceeding 75 million tonnes per annum. Therefore great interest exists for possible applications. Lignins of very different chemical composition and possible applications in the wood-based panels industry (adhesives, additive to or partly replacement for adhesives, raw material for synthetic resins) have been described in a great number of scientific articles and patents. Research into lignin-based adhesives dates back over 100 years with many separate examples of resins involving lignin being cited. In reality, existing applications are very rare. No industrial use as pure adhesive for wood currently is known despite the fact that considerable research activity has been directed toward producing wood adhesives from lignins. By themselves, lignins offer no advantages in terms of chemical reactivity, product quality or color when compared to conventional wood composite adhesives. The greatest disadvantages of lignins in their application as adhesives are: (i) their low reactivity and therefore the slow hardening compared to phenol due to the lower number of reactive sites in the molecule, causing increased press curing times, and (ii) concerns about chemical variation in the feedstock. The chemical structure of lignin is very complex with the added difficulty that, unlike tannin, the individual molecules are not fixed to any particular structure, therefore no true generic molecule exists for lignin from softwood, hardwood or cereals.

Lignosulfonates can be added to synthetic glue resins as extenders (under partial replacement of resin) [45]. The partial replacement of phenol during the cooking procedure of PF-resins has no present industrial application, but the addition of between 20% and 30% pre-methylolated lignin (pre-reacted with formaldehyde) to synthetic phenol-formaldehyde resins is used industrially in some North American plywood mills [46,47]. This is so as in plywood the process and product are much less economically dependent on shorter pressing times and consequently a fairly reactive form of lignin such as pre-methylolated lignin can be useful. Methylolated lignin can react rapidly with isocyanates, and adhesive formulations for panel products succesfully using 20–30% isocyanates on methylolated lignin have been reported and have been shown to compete succesfully with synthetic adhesives also in the case of the very short press times needed for the adhesive to be of industrial significance [48,49].

### 23.3.2 Use of lignins as adhesive without adding other synthetic resins

The application of lignin as adhesive in principle is possible. The first attempt needed very long press times due to the low reactivity (Pedersen process). This process was a condensation under strong acid conditions, which lead also to considerable corrosion problems in the plant [45]. The particles were sprayed with spent sulfite liquor (raw lignosulfonate) (pH = 3–4) and pressed at 180°C. After this step the board was tempered in an autoclave under pressure at 170–200°C, and the lignosulfonate became insoluble by eliminating water and $SO_2$. Shen [50–52] modified this process by spraying the particles with spent sulfite liquor containing sulfuric acic and pressing them at temperatures well above 210°C.

Nimz et al. [45,53] described the crosslinking of lignin after an oxidation of the phenolic ring in the lignin molecule using $H_2O_2$ in the presence of a catalyst, especially $SO_2$ [53]. This leads to the formation of phenoxy radicals and with this to radical phenoxy coupling forming inter- and intramolecular C–C bonds. This reaction does not necessarily need either heat or acid conditions, but is accelerated by higher temperatures (max. 70°C) as well as lower pHs. In this way the disadvantages of the processes mentioned above (high press temperatures, long press times, use of strong acids) could be avoided [45,53].

An oxidative activation of the lignin can also be achieved biochemically by adding enzymes (phenoloxidase laccase) to the spent sulfite liquor, whereby a polymerization via a radicalic mechanism is initiated. The enzymes are obtained from nutrient solutions of white fungi. Preparing the two component adhesive is done by mixing the lignin with the enzyme solution (after filtration of the mycelium). At the beginning of the press cycle the enzyme still works, since it is stable up to a temperature of 65°C. If a higher temperature is reached, then the enzyme is deactivated. At that time, however, the level of quinone methides generated is high enough to initiate the crosslinking reaction by phenoxy coupling [54,55]. While this approach is of interest, it has a few severe drawbacks, namely: (i) the system works only at ridiculously long press curing times – for example, press times as long as 15 minutes for 3 mm thick panels have been reported when industrially viable processing times for this panel thickness are today around 20–30 seconds under the same pressing technology; (ii) this far too long a press time can be improved, and the authors have experimented with this [55], by adding to the panel a reduced amount of a very strong adhesive, namely polymeric isocyanate. The press time obtained is still about ten times longer, however, than what would be today industrially significant. The choice, then, is, to use only enzymes yielding totally unacceptable press times, or to use enzymes coupled with reduced amounts of an adhesive and still be out by a great margin on the press time. This system is not used industrially.

The temperature of pressing also has a noticeable effect [56–58] as it does influence the surface/core temperature gradient and has a direct influence on the temperature rise in the board core layer. In short, the higher the press temperature, the faster heat conduction and the faster the development of the steam gradient across the wood mat. The press temperature will influence the steam front transfer time to the core layer. The higher the initial temperature is, the faster the steam front enters the mat core. Increasing the press temperature will cause the maximum steam pressure peak to appear earlier but does not result in a higher core temperature.

### 23.3.3 Lignin adhesives technology

Much has been written about, and much research has been conducted in the use of lignins for wood panel adhesives. It can safely be said that this natural raw material has probably been the most intensely researched as regards wood adhesive applications. Lignins are phenolic materials, they are abundant and of low cost but they have lower reactivity towards formaldehyde, or other aldehydes, than even phenol. Extensive reviews on a number of proposed technologies of formulation and application do exist, and the reader is referred to these [59–68]. This field is, however, remarkable for how small has been the industrial success in using these materials. In general, lignin and lignosulfonates have been mixed in smaller proportions to synthetic resins, such as PF resins [63–66], and even UF resins [67], to decrease their cost. Their low reactivity and lower level of reactive sites, however, mean that for any percentage of lignin added, the cost advantage is abundantly lost in the lengthening of the panel pressing time this causes. The only step forward that has found industrial application in the last 20 years is to pre-react in a reactor lignin with formaldehyde to form methylolated lignin, thus to do part of the reaction with formaldehyde first, and then add this methylolated lignin to PF resins at the 20%–30% level [63,65]. These resins have been used in some North American plywood mills [63]. Particularly in plywood mills, the pressing time is not the factor determining the output rate of the factory and so one can afford to use relatively long press times with good results [63].

None of the many adhesive systems based on pure lignin resins, hence without synthetic resin addition, has succeeded commercially at an industrial level. Some were tried industrially but for one reason or another (too long a pressing time, high corrosiveness for the equipment, etc.), they did not meet with commercial success. Still notable among these is the Nimz system based on the networking of lignin in the presence of hydrogen peroxide [13,53,59]. Only one system is used successfully still today, but this only for high density hardboard, in several mills worldwide. This is the Shen system, based on the self-coagulation and crosslinking of lignin by a strong mineral acid in the presence of some aluminum salt catalysts [50,51,59]. However, attempts to extend this system to the industrial manufacture of medium density fiberboard (MDF) are known to have failed.

Of interest in the MDF field is also the system of adding laccase enzyme-activated lignin to the fibers or activating the lignin *in situ*, in the fibers also by enzyme treatment [69,70]. The results obtained, however, yielded boards that did not satisfy the relevant standards, and this at very long board pressing times. The researchers involved obviated this successfully by adding some 1% isocyanate (PMDI) to the board [69] and pressing at acceptably short press times, or by extending the pressing times to ridiculous lengths (100 s/mm board thickness while industrial press times are of the order of 3–7 s/mm board thickness) [71]. In the former case, an adhesive had to be used, with the same result obtained by pressing untreated hardboard, a 100-year-old process, hence just wasting expensive time and enzymes. The second case instead illustrates even more clearly where the problem lies and what breakthrough is necessary: enzyme mobilizing lignin works, but not fast enough. The breakthrough necessary is a new, strong catalyst of the enzymatic action capable of allowing pressing times of industrial significance. This has not been found, or even considered, as yet.

A promising new technology based on lignin use for wood adhesives is relatively recent and uses again pre-methylolated lignin in the presence of small amounts of a synthetic PF resin, and polymeric 4'4'-diphenyl methane diisocyanate (PMDI) [66,72,73]. The proportion of pre-methylolated lignin used is 65% of the total adhesive, the balance being made up of 10–15% PF resin and 20–25% PMDI. This adhesive presses at very fast pressing times, well within the fastest range used today industrially, contains a high proportion of lignin, and yields exterior grade boards [72,73]. PF resin and methylolated lignin cure accelerators such as triacetin, other esters, can also be used [72,73] notwithstanding that just the presence of the PMDI already gives a considerable acceleration to the curing rate. The system is based on crosslinking caused by the simultaneous formation of methylene bridges and of urethane bridges, overcoming with the latter the need for higher crosslinking density that has been one of the problems which has stopped lignin utilization in the past. There are movements now to test it at the industrial scale in a couple of countries. Much more interesting has been the recent development in which formaldehyde has been totally eliminated by substituting it with a non-toxic, non-volatile aldehyde, namely glyoxal [74,75]. In these formulations lignin is pre-glyoxalated in a reactor and the glyoxalated lignin obtained is mixed with tannin and with PMDI, eliminating thus the need for any formaldehyde or formaldehyde-based resins [74,75]. This technology has also brought about the total elimination of synthetic resins in the adhesive as described in Section 23.4 below, both technologies satisfying requirements for industrially significant pressing times.

A continuous flow of literature on the subject of lignin adhesives is just literature rehashing older systems all based on the substitution of some phenol in PF resins. In general, these papers do not seem to be aware of the slow pressing time problem, and they do not address it, perpetuating the myth of PF/lignin adhesives while repeating the same age-old errors. They lead new researchers in the field to believe they are doing something worthwhile with parameters that do not satisfy the requirements of press rate of the panel manufacturing industry.

Some new and rather promising technologies on lignin adhesives have, however, been developed recently. These are (a) adhesives for particleboard and other agglomerate wood panels based on a mix of tannin/hexamine with preglyoxalated lignin, and (b) similar formulations for high resin content, high performance agricultural fiber composites.

## 23.4  Mixed tannin-lignin adhesives

Mixed interior wood panel tannin adhesive formulations were developed in which lignin is in considerable proportion, 50%, of the wood panel binder and in which no 'fortification' with synthetic resins, such as the isocyanates and phenol-formaldehyde resins as used in the past, was necessary to obtain results satisfying relevant standards. A low molecular mass organosolv lignin obtained industrially by formic acid/acetic acid pulping of wheat straw was used. Environmentally-friendly, non-toxic polymeric materials of natural origin constitute up to 94% of the total panel binder. The wood panel itself is constituted of 99.5% natural materials, the 0.5% balance being composed of glyoxal, a non-toxic and non-volatile aldehyde, for the pre-glyoxalation of lignin and of hexamine already accepted as a non-formaldehyde-yielding compound when in the presence of a condensed tannin. Both particleboard and two types of plywood were shown to pass the relevant interior standards with such adhesive formulations [75–77,79].

It is of interest to understand the main mechanisms according to which this formulation works. As lignin is of much lower reactivity than a flavonoid tannin, a series of parallel reactions occur to yield co-reaction of tannin with lignin. Thus, tannin reacts very rapidly in the hot press with hexamine to form a network based on the known reactions of tannin with hexamine [5,6–8]. Lignin, much slower, is pre-reacted in a reactor with glyoxal [75,78]. Although some condensation of glyoxalated lignin occurs during the reaction, if the reaction is protracted for longer times unduly increasing the viscosity of the glyoxalated lignin, condensation needs to be minimized and addition of glyoxal on the lignin maximized. The glyoxalated lignin is then rich in methylol-type groups obtained by glyoxal addition, but the aromatic nuclei of lignin are still too slow for glyoxalated lignin alone to condense to a sufficiently hardened network in just the very brief period the board remains in the hot press. The methylol-type groups obtained by glyoxal addition to lignin are then forced to react with the much more reactive flavonoid tannin introducing, then, the lignin in the final copolymer network. A simplified scheme explaining in brief the reactions involved is shown in Fig. 23.5.

## 23.5 Protein adhesives

Intense research induced by the sponsorship of the US United Soybean Board has revised interest in protein adhesives, soya protein primarily, but also others. These technologies must not be confused with the age-old protein and bone glues used in carpentry, or blood used as an additive in plywood glue-mixes. Certainly, of the traditional technologies, one stands out head and shoulders above the others, namely casein adhesives. These are still produced, and still used industrially to very good effect in some special plywood and related products. They work as such, they are very environmentally friendly and their technology was completely mastered a long time ago. They are definitely strong candidates for future expansion, in their actual form or with some further technological improvement. What is new is some interesting work on the correlation between the bond strength obtained and the molecular architecture of the protein [80].

On the other hand, soya and even gluten adhesives are definitely new. Both addition to traditional synthetic wood adhesives, as well as their use as panels adhesives after partial hydrolysis and modifications have been reported, and with acceptable results [81–84]. These products are not widely used industrially/commercially as yet, but one industrial user has been reported in the US.

Here, too, several different technologies can be distinguished. First, technologies based on the pre-reaction of soy protein hydrolysate with formaldehyde, and this pre-formulated soy protein being mixed with a PF resin and with isocyanate (PMDI) [85,86], thus an identical technological approach to what has already presented for lignin adhesives in Section 23.3.3. Second, the evolution of this technology, again along similar lines as for lignin, in which pre-glyoxalated soy protein [87] or even soy flour, or glyoxalated gluten protein hydrolysate [88], glyoxal being a non-volatile non-toxic aldehyde, compose the glue-mix with either a PF resin or with a flavonoid tannin, the whole been added to 20–25% isocyanate (PMDI) [87,88]. Both these systems work well, but the more interesting system, and the one that has started in industrial exploitation is the third one. This is based on the pre-reaction of the soy protein hydrolysate with malei anhydride to form an adduct that is then reacted in the panel with polyethyleneimmine [82,89]. The system works well, as one company has started using it industrially in the US, but it suffers the drawback of being excessively expensive, the price quoted being considerably higher than those of isocyanates.

## 23.6 Carbohydrate adhesives

Carbohydrates in the form of polysaccharides, gums, oligomers and monomeric sugars have been employed in adhesive formulations for many years. Carbohydrates can be used as wood panel adhesives in three main ways: (i) as modifiers of existing PF and UF adhesives, (ii) by forming degradation compounds which then can be used as adhesives building blocks, and (iii) directly as wood adhesives. The second route above leads to furanic resins. Furanic resins, notwithstanding that their basic building blocks, furfuraldehyde and furfuryl alcohol, are derived from the acid treatment of the carbohydrates in waste vegetable material, are considered today as purely synthetic resins [90]. This opinion might need to change as in reality they are real natural-derived resins, and extensively used in foundry core binders. Appropriate reviews dedicated to just them do exist [90]. However, both compounds are relatively expensive, very dark-colored and furanic resins have made their industrial mark in fields where their high cost is not a disadvantage. They can be used very successfully for panel adhesives, they are used very successfully in other fields (as foundry core binders), but the relatively higher toxicity of furfuryl alcohol before it is reacted is a problem that will have to be taken into consideration if these resins are to be considered for wood products.

The use of carbohydrates directly dissolved in strong alkali as wood panel adhesives is not a new concept, but is an interesting and a topical one today. This technology has been extensively reported [91]. All sorts of agricultural cellulosic materials have been successfully adapted to this technology and the technology and its application have been extensively reported in the past [91].

Research on the first route has centered particularly on the substitution of carbohydrates for parts of PF resins. It has been reported that at laboratory level, up to 50–55% of phenol in a PF resin can be substituted with a variety of carbohydrates, from glucose to polymeric, tree-derived hemicelluloses [92–96]. Apparently reducing sugars could not be used directly as they are degraded to saccharinic acids under the acid conditions required in the formulation of the resin. Reducing sugars can be used to successfully modify PF resins if they are reduced to the corresponding alditols or converted to glycosides. Some carbohydrates appeared to be incorporated into the resin network mainly through ether bridges [92]. Generally the resin is prepared by co-reacting phenol, the carbohydrate in high proportion, a lower amount of urea and formaldehyde. Extensive and rather successful industrial trials of these resins have also been reported [96].

Carbohydrate-based adhesives in which the formulation starts with the carbohydrate itself have also been reported, but the acid system used during formulation readily degrades the original carbohydrate to furan intermediates which then polymerize. An interesting concept that was advanced early on in carbohydrate adhesives research was the conversion of the carbohydrate to furanic products *in situ*, which then homopolymerize as well as react with the lignin in wood.

Several research groups [92,97,98] have recently described the use of liquefied products from cellulosic materials, literally liquefied wood, which showed good wood adhesive properties. Lignocellulosic and cellulosic materials were liquefied in the presence of sulfuric acid under normal pressure using either phenol or ethylene glycol. The cellulosic component in wood was found to lose its pyranose ring structure when liquefied. The liquefied product contains phenolic groups when phenol is used for liquefaction. In the case of ethylene glycol liquefaction, glucosides were observed at the initial stage of liquefaction and levulinates after complete liquefaction.

## 23.7 Unsaturated oil adhesives

Saturated and unsaturated vegetable oils are now widely available as a bulk commodity for a variety of purposes and at very acceptable prices. All resin research to date has focused on oils that contain at least one double bond. These oils are predominantly a mixture of triglycerides, hence esters, with a small quantity of free fatty acid, the small proportion of free fatty acid being dictated both by the plant species and the extraction conditions. As the number of unsaturations increases, so does the overall molecule reactivity and its potential for side reactions. The majority of these technologies are not applied to wood and wood composite adhesives, but they can be translated eventually to this field. An excellent and detailed review of formulations and technology on the subject already exists [99].

Until fairly recently, only two examples could be found in the literature where seed oil derivatives were being employed as wood adhesives. Linseed oil, for example, has been used to prepare a resin that can be used as an adhesive or surface coating material [100,101]. The chemistry of this resin centers on an epoxidation of the oil double bonds followed by crosslinking with a cyclic polycarboxylic acid anhydride to build up molecular weight. The reaction is started by the addition of a small amount of polycarboxilic acid.

When the epoxidized oil resin was evaluated as a wood adhesive in composite panels, it could be tightly controlled through the appropriate selection of triglycerides and polycarboxylic anhydrides. This apparently enables a wide range of materials with quite different features to be manufactured. The use as wood adhesives is one among the many uses, the focus of the development being more on plastic materials. The literature states that this plastic is well suited for use as a formaldehyde-free binder for wood fibers and wood particles, including fibers and chips from cereal residues, such as straw and fiber mats.

The literature on this resin [100] claims that crosslinking can be varied through the addition of specialized catalysts and several samples were prepared at a range of temperatures (120–180°C) that exhibited high water tolerance even at elevated temperatures, but no actual test data were included. Since the resin of reference [100], research in a number of other countries by imitators has produced very similar epoxidized oil resins. These are suitable for a number of applications, but the author has tested one or two of them finding that for wood adhesive applications, these resins have two major defects: (i) their hot-pressing time is far too slow to be of any interest in wood panel products, with the exception perhaps of plywood (for which they have not been tried), and (ii) they are relatively expensive. Unless the slow hot-pressing problem is overcome, and at a reasonable price, these resins will remain at the stage of potential interest. There is no doubt that these resins can be of interest in other fields, but it is symptomatic that no industrial use for wood panel adhesives has been reported as yet, or is known to have occurred.

Bioresins based on soy bean and other oils have been developed also by other groups, mainly for replacement of polyester resins [102]. These liquid resins were obtained from plant and animal triglycerides by suitably functionalizing the triglyceride with chemical groups (e.g., epoxy, carboxyl, hydroxyl, vinyl, amine, etc.) that render it polymerizable. The reference claims that excellent inexpensive composites were made using natural fibers such as hemp, straw, flax and wood in fiber, particle and flake form. That soy oil-based resins have a strong affinity for natural fibers and form a good fiber-matrix interface as determined by scanning electron microscopy of fractured composites. The reference also stated that these resins can be viewed as candidate replacements for phenol-formaldehyde, urethane and other petroleum-based binders in particleboard, MDF, OSB and other panel types. However, no actual test data have been supplied, and no industrial use in wood panel adhesives has actually been reported as yet.

Cashew nut shell liquid, mainly composed of cardanol but containing also other compounds (Fig. 23.6), is an interesting candidate for wood-based resins. Its dual nature, phenolic nuclei + unsaturated fatty acid chain, makes a potential natural raw material for the synthesis of water-resistant resins and polymers. Cardanol resins are known from the past, but their use has not been very diffuse simply because the raw material itself was rather expensive. The price, however, appears to be more affordable now since the extensive cashew nut plantations in Mozambique are in production.

The phenol, often resorcinol, group and/or double bonds in the chain can be directly used to form hardened networks. Alternatively, more suitable functional groups such as aldehyde groups and others can be generated on

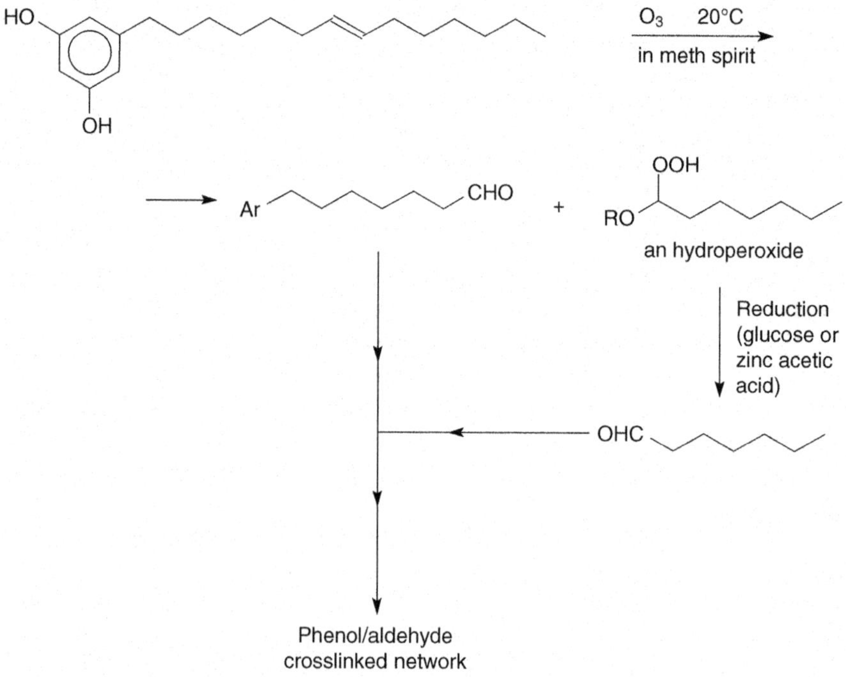

*23.6* Cardanol-based adhesive system [101].

the alkenyl chain. Generally, modifications of this kind take several reaction steps, rendering the process too expensive for commercial exploitation in wood adhesives. However, the Biocomposite Center in Wales [101] has developed a system of ozonolysis [101,103] in industrial methylated spirit [101,103] through which an aldehyde function is generated on the alkenyl chain of cardanol. The first reaction step yields as major product a cardanol hydroperoxide that following reduction by glucose or by zinc/acetic acid yields a high proportion of cardanolaldehyde groups. These crosslink with the aromatic groups of cardanol itself, thus a self-condensation of the system yielding hardened networks [101] (Fig. 23.6).

Exploratory laboratory particleboard and lap shear bonding yielded good results. Nonetheless, neither the press times used, nor other essential conditions that could help to evaluate the economical feasibility of these products were reported [101]. It remains to evaluate if the cost of the ozonolysis allows wood adhesives of a suitably low cost and, again, if the pressing times can match those of industrial resins. However, the resorcinolic structure of the cardanol phenol group would appear to indicate that the molecule should be able to achieve industrial pressing times.

More recently, alternative and very encouraging techniques involving unsaturated oils for wood and wood fiber adhesives have come to the fore [104]. Wheat straw particleboards were made using UF and acrylated epoxidized soy oil (AESO) resins with two resin content levels: 8% and 13%, and three pressing times: 8, 10 and 12 minutes. The boards' physical and mechanical properties showed that AESO-bonded particleboards have higher physical and mechanical properties than UF-bonded boards, especially in internal bonding and thickness swelling [104]. All properties of AESO-bonded boards increase by increasing both resin content and pressing time. AESO bonded boards can compete with wood particle boards according to EN standards. The good properties of these particleboards are because of high compatibility between straw particles and AESO resin due to its oil-based structure; their good properties can be maintained even at low board densities. This work was based on a number of different technologies presented and discussed in a most appropriate monograph [99].

## 23.8 Wood welding without adhesives

Assembly techniques with mechanical connectors or with adhesives are common in joining solid wood in the furniture, civil engineering and wood joinery industries. Both kinds of connections show several problems. With mechanical metal connectors, rust stains may appear on the connectors and corrosion of the connectors can and does occur. With adhesively-bonded joints working with liquid adhesives, the costs are higher than for the use of mechanical connectors as regards manufacturing equipment maintenance.

With adhesives the process is relatively longer unless high investments in adhesive materials and machinery (high-frequency or microwave systems) are made in order to speed up the hardening phase. Thermoplastic welding techniques which are widely used in the the plastic and car industries have recently been applied also to joining wood, by melting a thermoplastic polymer between the two wood surfaces to be joined. A variety of techniques such as ultrasound, mechanical friction and others have been used to melt the thermoplastic polymer *in situ*. In friction welding techniques the heat needed to melt the material is generated by pressing one of the samples to be joined against the other and to generate friction, hence heat, which increases the temperature of the weldline rapidly and considerably.

The same mechanically-induced friction welding techniques which are widely used in the plastic and car industries have recently been applied also to joining wood, without the use of any adhesive [105–111]. These work by melting some wood components and forming at the interface between the two wood surfaces to be joined a composite of entangled wood fibers drowned into a matrix of melted wood intercellular material, such as lignin and hemicelluloses [105–111] (see Figs. 23.7 and 23.8). Linear

*23.7* Welded bondline. Note the entangled wood cells (fibers) immersed in a mass of molten material. The whole superimposed on a background of elongated, undamaged, wood cells.

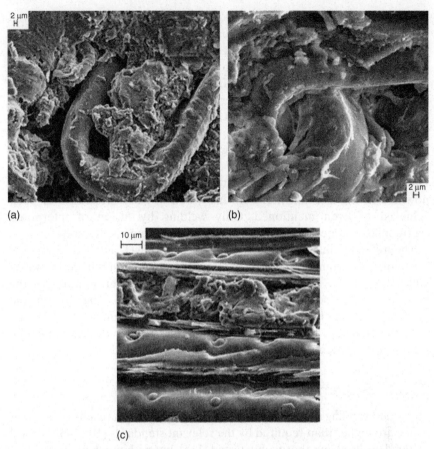

*23.8* (a) The two types of bondlines obtained in linear friction (Scot pine). Images from a scanning electron microscope with variable pressure (LEO 1450 VP). (a) and (b) Late wood tracheid entangled in the melted network of the bond. (c) Direct welding of the cell walls. Intact tracheid structure and fused intercellular material.

mechanical friction vibration has been used to yield wood joints satisfying the relevant requirements for structural applications by welding at a very rapid rate [105,107,108]. Crosslinking chemical reactions have also been shown to occur by CP-MAS $^{13}$C NMR. These reactions, however, are lesser contributors during the very short welding period proper [112,113]. They gain more, but still very limited, importance during the subsequent brief pressure holding period [105,112,113].

Also recently, high speed rotation-induced wood dowel welding, without any adhesive, has been shown to rapidly yield wood joints of considerable strength [105,110,111]. The mechanism of mechanically-induced high speed rotation wood welding is due to the temperature-induced softening and flowing of the intercellular material, mainly amorphous polymer material bonding the wood cells to each other in the structure of wood. This material is mainly composed of lignin and hemicelluloses. This flow of material induces high densification of the bonded interface [105,110,111]. Wood species, relative diameter differences between the dowel and the receiving hole, and press time were shown to be parameters yielding significant strength differences [105,110,111]. Other parameters were shown to have much lesser influence.

The insertion of dowels into solid wood, for joinery and furniture, without any wood-to-wood welding, has been used for centuries. This simple technology has also been applied to join particleboard [114]. However, nowhere in the relevant literature has wood-to-wood welding ever been achieved or even mentioned. Only welding by fusion of interposed thermoplastic materials has been mentioned in previous literature.

The relative diameter difference between dowel and substrate was the most important parameter determining joint strength performance [105,110,111]. The real determining parameter, however, is how fast the lignin/hemicelluloses melting temperature is reached. The greater the relative difference between the diameters of the dowel and of the substrate hole, the greater is the friction, hence the more rapidly the lignin melting temperature is reached and a better welding is achieved.

### 23.8.1 Systems of frictional wood welding

Two wood welding systems exist today that give strength results of the joint which are higher than required by the relevant standards [105,106]. A third system has also been thoroughly tested [115], but with much poorer results. These systems are not limiting like other welding systems such as ultrasound, microwave and/or radiofrequency heating, and high rotation or high vibration spindle welding, laser welding as well as others are all likely to afford some level of wood welding. Experiments with some of them have also been carried out, and will be reported briefly later. The material flow and melting induced by the elevated temperatures reached lead to high densification of the interface between the two profiles and interfacial loss of the cellular wood structure in the joint, hence to increased strength of the interface [105–107].

*Linear vibration welding*

The wood samples to be joined together are first brought into contact, with a pressure between 1.3 and 2 MPa, to enable the joint areas to be rubbed together with a linear reciprocating motion (Fig. 23.9). The samples are vibrated with a displacement amplitude of about 3 mm and a vibration

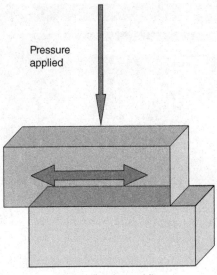

*23.9* Schematic example of frictional movement used in linear vibration wood welding.

frequency of 100 Hz in the plane of the joint. The specimens welded up to now are of a length of up to 1.0–1.8 m [116], as the capacity of the existing machines is not greater than this. The time of welding is roughly 1.5–5 sec [105,108,109] and the holding time, still under pressure, after vibration has stopped is of 5 sec [105,108,109]. The results obtained satisfy the strength requirements for structural application. It must be pointed out, however, that the parameters used for wood-to-wood welding give widely different results as regards water resistance according to the technology used. Older linear welding technologies yield joints which are not water resistant to any great extent, and which then can only be used for strictly interior applications [105]. Newer linear welding technologies instead give joints of much greater resistance to water [108,109,113], while the geometry of the joint itself in rotational dowel welding yields joints of almost exterior grade level [118,119].

Wood grain orientation differences in the two surfaces to be bonded yield bondlines of different strength in no-adhesives wood-to-wood welding. Longitudinal wood grain bonding of tangential and radial wood sections yield approximately 10% difference in strength results of the joint. Cross-grain (±90°) bonding yields instead much lower strength results, roughly half than observed for pieces bonded with the grain parallel to each other, although these results still satisfy the relevant wood-joining standards [117–119].

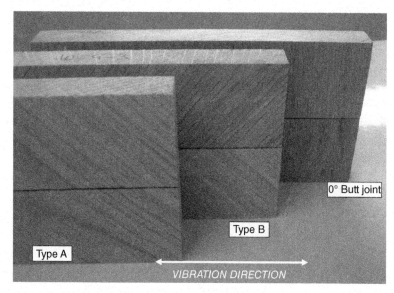

*23.10* Examples of different butt joints obtained by endgrain welding of oak wood [120].

Of particular interest are studies on wood cross-grain welding [117–120], where the principles of the anisotropy of composites rule the type of strength results that can be obtained in welding. Of particular interest is the case of wood endgrain welding [120] where the formation of very strong butt joints by this technique does open up the possibility of eliminating wood adhesive bonded fingerjoints. An example of this is shown in Fig. 23.10 for oak wood joints that have been endgrain-welded.

*High speed rotation dowel welding*

Traditionally, wooden dowels of a given diameter (most commonly 10 mm) are forced into a pre-drilled hole of lower diameter by applying pressure. Alternatively, similar dowels can be inserted into holes of the same diameter after application of an adhesive, in general PVAc. In high speed rotation welding, fluted rib beech dowels (or even smooth dowels of other woods) are inserted at high rotation speed and insertion rate within a predrilled hole of smaller diameter [106] (Fig. 23.11).

In dowel welding, generally, cylindrical beech fluted dowels of 10 mm in diameter are used. They are placed into a drill, in the place of the drill bit, and inserted within pieces of wood having pre-drilled holes of 8 mm. For best results the drill rotation rate must be higher than 1,200 revolution per minute (rpm), and if possible equal to 1500–1600 rpm. When fusion and

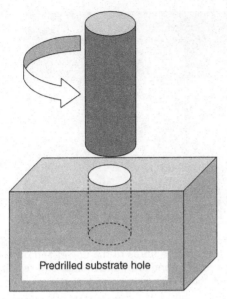

*23.11* Schematic example of frictional movement used in high speed rotational dowel welding.

*23.12* Well-welded dowel where the two pieces to be joined were maintained tight together by clamping during dowel insertion. The change in direction indicates the tightness of the interface between the two substrate pieces.

bonding are achieved, generally between 1 and 3 sec, the rotation of the dowel is stopped and the pressure may be briefly maintained [106,110,111].

Dowel welding by high speed rotation has been used to join two and even three wood block (Fig. 23.12). This is the ultimate aim of dowel welding. Strong joints were obtained [111,121]. An extremely important finding in dowel welding is that its water resistance is far superior to that obtained in linear welding [121]. This is due mainly to the geometry of the joint that allows approximately 80–90% of the dry strength to be conserved once it is wet, cold water soaked and up to 15% of the original dry strength once redried after 24 hours cold water soaking [121].

High-speed dowel rotation welding has been shown to be capable of holding together structures such as a suspended wood floor of size 4m × 4m × 0.2m without using any adhesive, any nails or any other binding system other than welded dowels [122–124]. When this technique applies

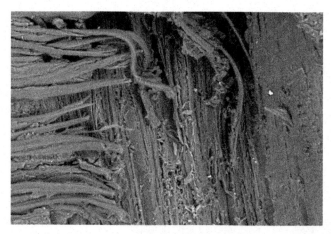

*23.13* Scanning electron microscope image of the interface of a dowel welded to the substrate. Note the fibres of the dowel and of the substrate at 90° to each other bonded by the melted and resolidified amorphous intercellular material.

to solid wood, several physical, chemical and mechanical processes occur. The rheological behavior of wood that is compressed and heated simultaneously while a rapid vibrating shift is applied changes, the moisture content at the interface is reduced, and chemical modifications of the wood structure occur during heating and the solidification of the melted joint interface.

The welding or bonding that occurs at the interface is probably mainly explained by the melting and solidification of the hemicelluloses, mainly xylans [118,121], and intercellular middle lamella's lignin and protolignin [105,118,121], and partly by the physical entanglement of the fibers interconnected between them as a result of the friction (Fig. 23.13). The direct welding of the cells is explained by the known properties of wood intercellular material: markedly thermoplastic, rigid and concentrated in the compound middle lamella of the wooden cells. Chemical phenomena occur too [105,112,113,118], mainly in the brief pressure holding phase immediately after welding, the main one of which is the formation and self-condensation of furfural [105,112]. The reactions involved are both of ionic type [105] as well as radical reactions [125].

## 23.9 Conclusion and future trends

Still many challenges confront bio-based adhesives for wood and for fibers. These can be divided into four broad classes:

1. challenges related to their performance and application in relation to synthetic adhesives,

2. challenges related to their cost in relation to the cost of synthetic adhesives,
3. challenges related to the supply of raw materials, and
4. challenges related to resistance to their introduction.

This latter is both from the psychological point of view of operators, used for decades to using the same type of synthetic adhesives, thus their natural resistance to change, as well as the lobbying by chemical companies who already control the market with synthetic products, and whose interest is often not in using alternative materials but to maximize profits with the minimal possible effort: after all why change if they are already selling something and prices are increasing?

Research and development in adhesives and resins for wood and fiber composites are driven mainly by requirements of the bonding and production process and the properties of the fiber- and wood-based composites themselves, the main topics being:

- shorter press (production) times
- better hygroscopic behavior of boards (e.g., lower thickness swelling, higher resistance against the influence of humidity and water, better outdoor performance)
- cheaper raw materials and alternative products
- modification of the wood surface
- life cycle assessment, energy and raw material balances, recycling and reuse
- reduction of emissions during the production and the use of wood-based panels.

The necessity to achieve shorter press times is omnipresent within the woodworking industry, based on a permanent and immanent pressure on costs and prices. An increased production rate is still one of the best ways to reduce production costs, as long as the market takes up the Surplus product. Shorter press times within a given production line and for a certain type of wood-based panel can be achieved, among others, by:

- highly reactive adhesive resins with quick gelling and hardening behavior and steep increase in bonding strength even at a low degree of chemical curing
- high reactive glue resin mixes, including the addition of accelerators or special hardeners, both which shall increase the gelling rate of a resin
- optimization of the pressing process, e.g. by increasing the effect of the steam shock by (i) increased press temperatures, (ii) additional steam injection, or (iii) an increased gap in the moisture content between surface and core layer.

There is quite good industrial experience on how to improve the ability of a bondline to resist against moisture and water, especially at higher temperatures. However, it is also known that, for example, in outdoors applications all adhesive systems are working at the upper limit of their performance. Hence fiber and wood-based composites with optimized properties and performance, especially for new applications in construction purposes like facades, require new bonding ideas. Cheaper raw materials are another way to reduce production costs. In this respect, bio-based adhesives are in outright competition with improvements in synthetic adhesives, and these latter can be considerable enough to deny market entrance to bio-based adhesives. It is evident that for many applications bio-based adhesives can deliver the same performance, but while they are cheaper than certain adhesives such as synthetic phenol-formaldehyde resins, isocyanates and the melamine resins, they cannot compete either in present supply quantity or in price with urea-formaldehyde adhesives. Special legislation and the public pressure for environmentally friendly products do drive this trend at present. The increase in oil costs and other cost factors are also starting to favor interest in natural or partially natural adhesives or even in synthetic adhesives totally biosourced, these being traditional adhesives obtained by synthesis but for which the reagents used are obtained from transformed raw materials of natural origin.

One particular challenge, and possibly the greater one, is the supply and availability of the correct raw materials used for bio-based adhesives. One glaring example is tannin adhesives, now a mature technology that has been in industrial use in the southern hemisphere for 40 years. These adhesives give excellent PF resin substitutes but the amount produced yearly, in the order of a few hundred thousand tons, is literally a 'drop in the ocean' if one wants seriously to substitute them for phenolic resins in many of their applications. The availability of raw material is not the problem; after all the tree barks from which tannins are extracted are a worldwide resource and can yield millions of tons of usable tannin for adhesives and other resins. The problem is their present low installed extraction capacity. Thus, two alternatives exist for the future: (1) build new tannin extraction factories to increase production, an approach that seems to have been tentatively taken by a few corporations around the world, or (2) to dilute the tannin with an alternative bio-based material to increase the amount of adhesive available, the developments in this line too, namely mixed tannin/lignin and tannin/protein adhesives, having been developed and being of interest and being pursued by industry. Equally, widely available biomaterials such as lignin from the paper industry still need systems to upgrade further their reactivity, this being at present obviated by modifying them and mixing/co-reacting them with more reactive synthetic adhesives.

The wood and fibers themselves, especially the wood and fiber surface including the interface to the bondline, also play a crucial role for the quality of bonding and hence for the quality of the fiber- and wood-based composites. Low or even no bonding strength can be caused by unfavorable properties of the substrate surface, e.g. due to low wettability. Adhesives and resins are one of the important and major raw materials of fiber- and wood-based composites. Thus, each question concerning life cycle assessment and their recycling also is a question for the adhesives and resins used. This includes, for example, the impact of the resins on various environmental topics like wastewater and effluents, emissions during the production and from the finished composite or the energetic reuse of panels. Also for several material recycling processes the type of resin has a crucial influence on feasibility and efficiency.

The emissions of gases, especially from wood-based panels during and after their production, can be caused by wood-inherent chemicals, like terpenes or free acids, as well as by volatile compounds and residual monomers of the adhesive. The emission of formaldehyde is especially a matter of concern, but also possible emissions of free phenols or other monomers. The problem of the subsequent formaldehyde emission fortunately can be regarded as more or less solved, even in the case of bio-based adhesives due to stringent regulations having been implemented in many countries, and successful long-term R&D in industry.

However, this scenario cuts further into the competitiveness of some bio-based adhesives, as the upgrading caused by developments in synthetic adhesives to solve such problems diminishes the advantages in this area which are inherent to bio-based resins.

## 23.10  References

1. A. Pizzi, Tannin based adhesives. In *Wood Adhesives – Chemistry and Technology*, Vol. I (A. Pizzi, ed.) Marcel Dekker, New York (1983), Chap. 4.
2. A. Pizzi, *Forest Prod. J.* **28**(12), 42–48 (1978).
3. F. Pichelin, C. Kamoun and A. Pizzi, *Holz Roh Werkstoff* **57**(5), 305–317 (1999).
4. A. Pizzi and P. Sorfa, *Holzforschung Holzverwertung* **31**, 5–7 (1979).
5. F. Pichelin, M. Nakatani, A. Pizzi, S. Wieland, A. Despres and S. Rigolet, *Forest Prod. J.* **56**(5), 31–36 (2006).
6. C. Kamoun, A. Pizzi and M. Zanetti, *J. Appl.Polymer Sci.* **90**(1), 203–214 (2003).
7. C. Kamoun and A. Pizzi, *Holzforschung Holzverwertung* **52**(1), 16–19 (2000).
8. C. Kamoun and A. Pizzi, *Holzforschung Holzverwertung* **52**(3), 66–67 (2000).
9. H. Heinrich, F. Pichelin and A. Pizzi, *Holz Roh Werkstoff* **54**(4), 262 (1996).
10. A. Pizzi, P. Stracke and A. Trosa, *Holz Roh Werkstoff* **55**(3), 168 (1997).

11. S. Wang and A. Pizzi, *Holz Roh Werkstoff* **55**(3), 174 (1997).
12. A. Pizzi, W. Roll and B. Dombo, Hitzehärtende Bindemittel, European patent EP-B 0 639 608 (1998); German patent DE 44 02 159 A1 (1995); US patent 5,532,330 (1996), Bakelite AG.
13. A. Pizzi, *Advanced Wood Adhesives Technology*, Marcel Dekker, New York, 1994.
14. N. Meikleham, A. Pizzi and A. Stephanou, *J. Appl. Polymer Sci.* **54**, 1827–1845 (1994).
15. A. Pizzi, N. Meikleham and A. Stephanou, *J. Appl. Polymer Sci.* **55**, 929–933 (1995).
16. A. Pizzi and N. Meikleham, *J. Appl. Polymer Sci.* **55**, 1265–1269 (1995).
17. E. Masson, A. Merlin and A. Pizzi, *J. Appl. Polymer Sci.* **60**, 263–269 (1996).
18. E. Masson, A. Pizzi and A. Merlin, *J. Appl. Polymer Sci.* **60**, 1655–1664 (1996).
19. E. Masson, A. Pizzi and A. Merlin, *J. Appl. Polymer Sci.* **64**, 243–265 (1997).
20. A. Pizzi, N. Meikleham, B. Dombo and W. Roll, *Holz Roh Werkstoff* **53**, 201–204 (1995).
21. R. Garcia and A. Pizzi, *J. Appl. Polymer Sci.* **70**(6), 1093–1109 (1998).
22. R. Garcia and A. Pizzi, *J. Appl. Polymer Sci.* **70**(6), 1111–1119 (1998).
23. A. Pizzi, *J. Macromol. Sci. Chem. Ed.* C18(2), 247–307 (1980).
24. A. Pizzi, Natural Phenolic Adhesives 1: Tannin. In *Handbook of Adhesive Technology*, 2nd edn (A. Pizzi and K.L. Mittal eds), Marcel Dekker, New York (2003).
25. A. Trosa and A. Pizzi, *Holz Roh Werkstoff* **59**(4), 266–271 (2001).
26. A. Trosa, PhD Thesis, University Henri Poincaré – Nancy 1, Nancy, France (1999).
27. A. Pizzi, *Holz Roh Werkstoff* **52**, 229 (1994).
28. F. Pichelin, C. Kamoun and A. Pizzi, *Holz Roh Werkstoff* **57**(5), 305–317 (1999).
29. C. Kamoun and A. Pizzi, *Holzforschung Holzverwertung* **52**(1), 16–19 (2000).
30. C. Kamoun, A. Pizzi and M. Zanetti, *J. Appl. Polymer Sci.* **90**(1), 203–214 (2003).
31. J.F. Walker, *Formaldehyde*, Am. Chem. Soc. Monogr. Ser. 159 (1964).
32. A. Pizzi, W. Roll and B. Dombo, European patent EP-B 0 648 807 (1998); German patent DE 44 06 825 A1 (1995); US patent 5,532,330 (1996).
33. F. Pichelin, SWOOD unpublished results (2004).
34. A. Pizzi, PhD Thesis, University of the Orange Free State, Bloemfontein, South Africa (1978).
35. A. Pizzi, J. Valenzuela and C. Westermeyer, *Holz Roh Werkstoff* **52**, 311–315 (1994).
36. N. Meikleham, A. Pizzi and A.Stephanou, *J. Appl. Polymer Sci.* **54**, 1827–1845 (1994).
37. A. Pizzi and A. Stephanou, *Holzforschung Holzverwertung* **45**(2), 30–33 (1993).
38. A. Pizzi and N. Meikleham, *J. Appl. Polymer Sci.* **55**, 1265–1269 (1995).
39. A. Pizzi, N. Meikleham and A. Stephanou, *J. Appl. Polymer Sci.* **55**, 929–933 (1995).
40. A. Pizzi, N. Meikleham, B. Dombo and W. Roll, *Holz Roh Werkstoff* **53**, 201–204 (1995).
41. A. Pizzi and A. Stephanou, *J. Appl. Polymer Sci.* **50**, 2105–2113 (1993).

42. R. Garcia and A. Pizzi, *J. Appl. Polymer Sci.* **70**(6), 1083–1091 (1998).
43. R. Garcia and A. Pizzi, *J. Appl. Polymer Sci.* **70**(6), 1093–1110 (1998).
44. H. Yamaguchi, *Japan Kokai Tokkyo Koho* (2004), Japan patent 2004-143385.
45. H.H. Nimz, Lignin-based wood adhesives. In *Wood Adhesives: Chemistry and Technology* (A. Pizzi, eds.), Marcel Dekker, New York (1983), pp. 247–288.
46. P. Md. Tahir and T. Sellers, Jr. 19th IUFRO World Congress, Montreal, Canada, August 1990.
47. L. Calvé, 19th IUFRO World Congress, Montreal, Canada, August 1990.
48. A. Stephanou and A. Pizzi, *Holzforschung* **47**(5), 439–445 (1993).
49. A. Stephanou and A. Pizzi, *Holzforschung* **47**(6), 501–506 (1993).
50. K.C. Shen, *Forest Prod. J.* **24**(2), 38–44 (1974).
51. K.C. Shen, *Forest Prod. J.* **27**(5), 32–38 (1977).
52. D.P.C. Fung, K.C. Shen and L. Calvé, Report OPX 180 E, Eastern Forest Products Laboratory, Ottawa (1977).
53. H.H. Nimz and G. Hitze, *Cell. Chem. Technol.* **14**, 371–382 (1980).
54. A. Kharazipour, A. Haars, M. Shekholeslami and A. Hütterman, *Adhäsion* **35**, 30–36 (1991).
55. A. Kharazipour and A. Hütterman, Biotechnological production of wood composites. In *Forest Products Biotechnology* (A. Bruce and J.W. Palfreyman eds), Taylor & Francis, London, (1998) pp. 141–150.
56. F. Pichelin, A. Pizzi, A. Frühwald and P. Triboulot, *Holz Roh Werkstoff* **59**(4), 256–265 (2001).
57. F. Pichelin, A. Pizzi, A. Frühwald and P. Triboulot, *Holz Roh Werkstoff* **60**(1), 9–17 (2002).
58. F. Pichelin, 1999. Herstellung von OSB mit Feuchtetoleranten Klebstoffen, Doctoral thesis, University of Hamburg, Germany.
59. H.H. Nimz, Lignin-based adhesives. In *Wood Adhesives Chemistry and Technology*, Vol. 1 (A. Pizzi, ed.), Marcel Dekker, New York (1983), pp. 247–288.
60. P. Blanchet, A. Cloutier and B. Riedl, *Wood Sci. Technol.* **34**(1), 11–19 (2000).
61. F. Lopez-Suevos and B. Riedl, *J. Adhesion Sci. Technol.* **17**(11), 1507–1522 (2003).
62. S. Kim and H.-J. Kim *J. Adhesion Sci. Technol.* **17**(10), 1369–1384 (2003).
63. L.R. Calvé, Can. Patent. 2042476 (1999).
64. K. Shimatani, Y. Sono and T. Sasaya, *Holzforschung* **48**(4), 337–342 (1994).
65. D. Gardner and T. Sellers Jr., *Forest Products J.* **36**(5), 61–67 (1986).
66. W.H. Newman and W.G. Glasser, *Holzforschung* **39**(6), 345–353 (1985).
67. V.I. Azarov, N.N. Koverniskii and G.V. Zaitseva, *Izvestjia Vysshikh Uchnykh Zavedenii, Lesnai Zhurnal* **5**, 81–83 (1985).
68. L. Viikari, A. Hase, P. Quintus-Leina, K. Kataja, S. Tuominen and L. Gadda, European Patent EP 95030 A1 (1999).
69. A. Kharazipour, A. Haars, M. Shekholeslami and A. Hüttermann, *Adhäsion* **35**(5), 30–36 (1991).
70. A. Kharazipour, C. Mai and A. Hüttermann, *Polymer Degrad. Stabil.* **59**(1–3), 237–243 (1998).
71. C. Felby, L.S. Pedersen and B.R. Nielsen, *Holzforschung* **51**(3), 281–286 (1997).
72. A. Pizzi and A. Stephanou, *Holzforschung* **47**, 439–445 (1993).
73. A. Pizzi and A. Stephanou, *Holzforschung* **47**, 501–506 (1993).

74. N.-E El Mansouri, A. Pizzi and J. Salvado, *J. Appl. Polymer Sci.* **103**(3), 1690–1699 (2007).
75. N.-E El Mansouri, A. Pizzi and J.Salvado, *Holz Roh Werkstoff* **65**(1), 65–70 (2007).
76. H.R. Mansouri, P. Navarrete, A. Pizzi, S. Tapin-Lingua, B. Benjelloun-Mlayah and S. Rigolet, *Holz Roh Werkstoff* **69**(2), 221–229 (2009).
77. P. Navarrete, H.R. Mansouri, A. Pizzi, S. Tapin-Lingua, B. Benjelloun-Mlayah and S. Rigolet, *J. Adhesion Sci.Technol.* **24**(8), 1597–1610 (2009).
78. H. Lei, A. Pizzi and G. Du, *J. Appl.Polymer Sci.* **107**(1), 203–209 (2008).
79. A. Pizzi, R. Kueny, F. Lecoanet, B. Massetau, D. Carpentier, A. Krebs, F. Loiseau, S. Molina and M. Ragoubi, *Ind.Crops & Prod.* **30**, 235–240 (2009).
80. D. Leckband, *Polymeric Materials Sci. Eng.* **90**, 266 (2004).
81. Z. Zhong, X. S. Sun, D. Wang and J.A. Ratto, *J. Polym. Environ.* **11**(4), 137–144 (2003).
82. Y. Liu and K. Li, *Macromol. Rapid Comm.* **23**(12), 739–742 (2002).
83. X. Sun and K. Bion, *J. American Oil Chemists Soc.* **76**(8), 977–980 (1999).
84. N.S. Hettiarachy, U. Kalapotly and D.J. Myers, *J. American Oil Chemists Soc.* **72**(12), 1461–1464 (1995).
85. J.M. Wescott, C.R. Frihart and L. Lorenz, Durable soy-based adhesives. Proceedings Wood Adhesives 2005, Forest Products Society, Madison, WI (2006).
86. L. Lorenz, C.R. Frihart and J.M. Wescott, Analysis of soy flour/phenol-formaldehyde adhesives for bonding wood. Proceedings Wood Adhesives 2005, Forest Products Society, Madison, WI (2006).
87. G.A. Amaral-Labat, A. A. Pizzi, R. Goncalves, A. Celzard and S. Rigolet, *J. Appl. Polymer Sci.* **108**, 624–632 (2008).
88. H. Lei, A. Pizzi, P. Navarrete, S. Rigolet, A. Redl and A. Wagner, *J. Adhesion Sci. Technol.* **24**(8), 1583–1591 (2009).
89. Y. Liu and K. Li, *Int. J. Adhesion Adhesives* **27**, 59–64 (2007).
90. M.N. Belgacem and A. Gandini, Furan-based adhesives. In *Handbook of Adhesive Technology*, 2nd edn (A. Pizzi and K.L. Mittal eds), Marcel Dekker, New York (2003), pp. 615–634.
91. C.-M. Chen, *Holzforsch. Holzverwertung* **48**(4), 58–60 (1996).
92. A.H. Conner, B.H. River and L.F. Lorenz, *J. Wood Chem. Technol.* **6**(4), 591–596 (1986).
93. A.H. Conner, L.F. Lorenz and B.H. River, Carbohydrate-modified PF resins formulated at neutral conditions, ACS Symposium series, 385 (Adhesives from Renewable Resources), pp. 355–369 (1989).
94. A. Trosa, European Patent EP 924280 (1999).
95. K.C. Shen, PCT patent WO 9837148 (1998); European patent EP 102778 (1997).
96. A. Trosa and A. Pizzi, *Holz Roh Werkstoff* **56**, 229–233 (1998).
97. M.H. Alma, M. Yoshioka, Y. Yao and N. Shiraishi, *Wood Sci. Technol.* **32**(4), 397–308 (1998).
98. M.H. Alma, M. Yoshioka, Y. Yao and N. Shiraishi, *Holzforschung* **50**(1), 85–90 (1996).
99. R.P. Wool and X.S. Sun, *Bio-based Polymers and Composites*, Academic Press, Amsterdam (2005).

100. R. Miller and U. Shonfeld, 2002. Company Literature, Preform Raumgliederungssysteme GmBH, Esbacher Weg 15, D-91555 Feuchtwangen, Germany.
101. J. Tomkinson, Adhesives based on natural resources. In *Wood Adhesion and Glued Products: Wood Adhesives* (M. Dunky, A. Pizzi and M. Van Leemput, eds), European Commission, Directorate General for Research, Brussels, pp. 46–65 (2002).
102. R.P. Wool, Proceedings of the 2nd European Panel Products Symposium, Bangor, Wales (1998).
103. P.S. Bailey, *Ozonation in Organic Chemistry*, Academic Press, New York (1978).
104. M. Tasooji, T. Tabarsa, A. Khazaeian and R.P. Wool, *J. Adhesion Sci. Technol.* **24**(8–10), 1717–1727 (2009).
105. B. Gfeller, M. Zanetti, M. Properzi, A. Pizzi, F. Pichelin, M. Lehmann and L. Delmotte, *J. Adhesion Sci. Techn.* **17**(11), 1425–1590 (2003).
106. A. Pizzi, J.-M. Leban, F. Kanazawa, M. Properzi and F. Pichelin, *J. Adhesion Sci. Technol.* **18**(11), 1263–1278 (2004).
107. J.-M. Leban, A. Pizzi, S. Wieland, M. Zanetti, M. Properzi and F. Pichelin, *J. Adhesion Sci. Technol.* **18**(6), 673–685 (2004).
108. H.R. Mansouri, P. Omrani and A. Pizzi, *J. Adhesion Sci. Technol.* **23**(1), 63–70 (2009).
109. P. Omrani, A. Pizzi, H. Mansouri, J.-M. Leban and L. Delmotte, *J. Adhesion Sci. Technol.* **23**, 827–837 (2009).
110. F. Kanazawa, A. Pizzi, M. Properzi, L. Delmotte and F. Pichelin, *J. Adhesion Sci. Technol.* **19**(12), 1025–1038 (2005).
111. C. Ganne-Chedeville, A. Pizzi, A. Thomas, J.-M. Leban, J.-F. Bocquet, A. Despres and H.R. Mansouri, *J. Adhesion Sci. Technol.* **19**(13–14), 1157–1174 (2005).
112. L. Delmotte, C. Ganne-Chedeville, J.-M. Leban, A. Pizzi and F. Pichelin, *Polymer Degrad. & Stabil.* **93**, 406–412 (2008).
113. L. Delmotte, H.R. Mansouri, P. Omrani and A. Pizzi, *J. Adhesion Sci. Technol.* **23**, 1271–1279 (2009).
114. L. Resch, A. Despres, A. Pizzi, J.-F. Bocquet and J.-M. Leban, *Holz Roh Werkstoff* **64**(5), 423–425 (2006).
115. B. Stamm, E. Windeisen, J. Natterer and G. Wegener, *Wood Science and Technology* **40**(7), 615–627 (2006).
116. B. Stamm, J. Natterer and P.Navi, *Holz Roh Werkstoff* **63**(5), 313–320 (2005).
117. J. Golé, résinés reinforcées. In *Propriétés physiques des polymères: mise en œuvre*. GFP, Groupe francais d'Etudes et d'Applications des Polymères, pp. 255–305 (1979).
118. M. Properzi, J.-M. Leban, A. Pizzi, S. Wieland, F. Pichelin and M. Lehmann, *Holzforschung* **59**(1), 23–27 (2005).
119. P. Omrani, H.R. Mansouri, A. Pizzi, E. Masson, *Eur. J. Wood Prod* **68**(1), 113–114 (2010).
120. P. Omrani, H.R. Mansouri and A. Pizzi, *J. Adhesion Sci. Technol.* **23**, 2047–2055 (2009).
121. A. Pizzi, A. Despres, H.R. Mansouri, J.-M. Leban and S. Rigolet, *J. Adhesion Sci. Technol.* **20**(5), 427–436 (2006).
122. J.F. Bocquet, A. Pizzi, A. Despres, H.R. Mansouri and L. Resch, *J. Adhesion Sci. Technol.* **21**(3–4), 301–317 (2007).

123. M. Oudjene, M. Khalifa, C. Segovia and A. Pizzi, *J. Adhesion Sci. Technol* **24**, 359–370 (2010).
124. J.-F. Bocquet, A. Pizzi and L. Resch, *J. Adhesion Sci. Technol.* **20**(15), 1727–1739 (2006).
125. S. Wieland, B. Shi, A. Pizzi, M. Properzi, M. Stampanoni, R. Abela, X. Lu and F. Pichelin, *Forest Products J.* **55**(1), 84–87 (2005).

# 24
# Types, properties and processing of bio-based animal feed

E. J. BURTON and D. V. SCHOLEY, Nottingham Trent University, UK and P. E. V. WILLIAMS, AB Vista, UK

DOI: 10.1533/9780857097385.2.771

**Abstract**: This chapter considers the history and utilisation of bio-based co-products from biorefining and their use in the animal feed sector. The path of development from traditional by-products to higher value materials is outlined with references to the influence of production scale, energy and water costs. Properties of biorefinery co-products are described together with their potential use in the animal feed sector and features that constrain their use. The chapter then expands on the impact of processing technology including potential improvements. Finally, the chapter reports on the emerging trends in bio-based co-products and discusses constraints and drivers in their development.

**Key words**: biodiesel, bioethanol, DDGS, co-products, animal feed, yeast, feed legislation, protein source.

## 24.1 Introduction

In order to consider bio-based nutrients for feed, it is necessary to make a refinement to the definition of bio-based. Beyond mineral supplements and synthetic forms of some vitamins and amino acids which are manufactured via chemical syntheses, few non-bio-based feed materials exist. Therefore in the current context, the term bio-based nutrients/feed ingredients is taken to mean feed materials that are derived directly or as co-products from fermentation and biosynthetic processes. Bio-based nutrients for feed is not a new concept but, whilst the scale has grown dramatically in recent decades (nearly 200-fold), the technology and the fundamental principles have to a great extent remained unchanged.

This chapter starts by outlining the origins of bio-based feed ingredients and how they are utilised in the global animal feed sector, before explaining in more detail the opportunities in that market. Types and properties of existing bio-based feed materials are then described with evaluation of their advantages and limitations. The chapter subsequently expands on the impact of processing technologies on the nutritional value of bio-based feed. The final sections profile emerging products with explanation of both the constraints and drivers in their development.

## 24.2 Background

One technology that has provided bio-based feed ingredients for over 200 years is that of potable alcohol production. Whilst in the early days of alcohol production, the use of the co-product as animal feed was an incidental benefit to the process, modern trends and new pressures have significantly altered the perception of how the co-products can be handled and disposed of. It was not unknown for the co-products of potable alcohol production to go into landfill or for liquid wastes to be discharged directly into the sea. There is now a considerable charge for putting waste to landfill and discharge into the sea is environmentally not permissible. In the early days, drying the co-product and marketing it as a feed for livestock was primarily a convenient means of removing the excess material from the site of production. In the meantime, a valuable business has evolved in marketing co-products as animal feed and this business is now a valuable green credential in the process of bioethanol and biodiesel production. Factors such as these, plus the need to recover as much of the available nutrients which enter the biorefining process as possible, have placed new pressures to capitalise on all available nutrients.

Therefore as a start to considering bio-based nutrients, it is worth contemplating in more detail how processes evolve to meet the needs of the animal feed industry and how the technology can be used as a blueprint for emerging novel bioprocesses. The current global approach to sustainable agriculture hinges on balancing supply of the 4Fs: feed, fuel, food and fibre. The incorporation of biorefinery co-products into animal feeds provides a major conduit for finding balance; excess fibre and feed from production of fuel may be converted into food via animal production.

The animal feed industry may be loosely divided into two sectors: one sector addressing the requirements of ruminant animals (primarily cattle and sheep) and the other sector addressing non-ruminants (primarily fish, pigs and poultry). The ability of ruminants to digest fibre as an energy source and to utilise non-protein nitrogen to meet their amino acid requirements means that the fibrous products are predominantly integrated into ruminant diets, whilst the majority of high protein biorefinery co-products are directed towards non-ruminants. The very high growth rates of commercial strains of pig and poultry render them extremely sensitive to fluctuations in the quality of feed provided and the density of protein and energy in the feed, which limits the inclusion of many co-products.

### 24.2.1 History of distilling co-products

One of the first bio-based feed materials was produced as early as 1790, when the surplus grain from the Kilbagie Distillery, Clackmannanshire, was recorded as being fed to cattle. Indeed in the early days of distilling, it was natural to feed the residues of producing the water of life 'uisage beatha' to livestock. The scale of this production, approximately 2.3 million litres alcohol per annum from approximately 8,000 tons of barley, compares with a modern bioethanol plant producing in excess of 400 million litres of ethanol from approximately one million tons of grain. Production of potable ethanol is a mature technology using enzyme liquefaction and saccharification of starch to produce glucose which is then fermented by *Saccharomyces cerevisiae* yeast to produce ethanol. The leftover mash (whole stillage) is decanted into a fibrous wet grain and a liquid component, thin stillage, which contains the majority of the yeast protein and soluble components. Thin stillage has a very high water content, with only around 60–85 g/kg dry matter (Mustafa *et al.*, 1999), so it is evaporated into a syrup, remixed with the wet grain and dried to form distillers dried grains with solubles (DDGS). The drying of the co-product is an energy-demanding, expensive necessity in order to remove all the 'waste' material not required for the production of ethanol from the distillery which, if not removed, would congest the primary process of ethanol production.

### 24.2.2 Scale of bioethanol production

The European Commission published the Biofuels Directive in 2003 (EU, 2003a) which promotes renewable fuels as a means of reducing carbon emissions. Targets were set for member states and in the UK these were incorporated into the Renewable Transport Fuel Obligation, which specifies that renewable fuels must make up 3.5% of fuel supplied on UK forecourts (RTFO, 2007), increasing up to a 5% inclusion by 2014. In 2010/11, 1,507 million litres (Ml) of biofuel were supplied in the UK, of which 41% was bioethanol (618 Ml/year), the majority of which was imported from Brazil and the USA (Department for Transport, 2012). In 2011 the EU produced 4,400 Ml, an increase of 105% on 2007 making it the third largest producer behind Brazil and the US (European Bioethanol Fuel Association, 2012). The significance of these figures for feed ingredients is based on the one-third rule: in first generation bioethanol production it is generally recognised that from the grain feedstock, one third of the grain goes into ethanol, one third into carbon dioxide and one third into animal feed. Growth in bioethanol production leads to growth in feed production.

Biodiesel production is globally some way behind bioethanol production (Fig. 24.1), but makes up the majority of the biofuel production in the EU. Feedstock accounts for 80 percent of the cost of a gallon of biodiesel, so feedstock availability is the driver for biodiesel production. This accounts for the relatively small production in the US, although this is increasing and 1.1 billion gallons were produced in 2011 (The National Biodiesel Board,

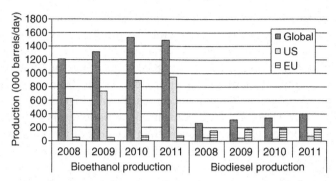

24.1 Biofuels production for 2008 to 2011 (compiled from US EIA (2013) data).

2012). The growth in renewable fuel produced from the fermentation of plant material has resulted in an equal growth in the co-product residue, which as indicated earlier has traditionally been targeted predominantly to one segment of the animal feed market, namely ruminant nutrition.

### 24.2.3 The demand for feed: attributes and trends

Increases in consumer demand for meat, especially in developing countries, will require increasing protein supplies for animal feed. Beyond the energy component of the diet, protein forms the largest dietary component for pigs, poultry and fish. The usefulness or quality of a plant protein source for animal feeds depends on three main factors: the volume of protein provided, the availability of that protein to the animal and the number of antinutritional factors (ANFs) contained with the material. Antinutritional factors are components within a plant-derived feed material that invoke a detrimental response within the animal, such as reduced growth or tumour induction. Availability of protein depends on how closely the profile of amino acids in the protein source matches the requirements of the growing animal and also how easily each amino acid may be digested.

The major protein sources in use are plant-derived meals (e.g., soya and rape meals), fishmeal and meat and bone meal, with some protein also being sourced from industry co-products and legumes. The majority of protein used in animal feed is from oilseed; Gilbert (2002) quotes an annual figure of 316 million tonnes (Mt) of oilseed protein, 14 Mt from animal by-products and 7 Mt from fishmeal. However, in the EU, legislation constrains the use of animal by-products in animal feed (TSE regulation 999/2001; ABPR 1774/2002). Similarly, although fishmeal is a high quality protein source, overfishing has led to introduction of restrictions to conserve fish stocks and Hardy (2010) stated that further reduction in use of fishmeal will be required in the future. Biorefinery co-products can go some way towards meeting the demand for sustainable animal protein. Globally, around 44 million tonnes of distillery co-products were incorporated into the production of animal feed in 2011, which is an increase of 23.7% on the previous 12-month period (DEFRA, 2011). The majority of these co-products are currently used in ruminant feed.

## 24.3 Types and properties of bio-based feed ingredients

Most commonly the starch used for bioethanol production is supplied by sugarcane (Brazil) or maize (USA), but in the UK and Canada, the feedstock is wheat. Other feedstocks include roots and tubers and molasses. For biodiesel production, the most common feedstock by a large margin is vegetable-based oils, such as soya bean oil and rapeseed oil, with some production (less than 15%) coming from Jatropha and animal fats.

Biodiesel co-products vary with feedstock; in the EU, rape meal makes up the majority, whereas soya bean meal is dominant in the smaller US market. The protein meals from soya and rape are well recognised and extensively used in the livestock feed industry. Other oilseeds supply a small but significant volume of co-product to the animal feed industry. The major biofuel co-product in terms of volume remains the traditional alcohol co-product, DDGS.

### 24.3.1 Distillers dried grains with solubles (DDGS)

Distillers dried grains with solubles (DDGS) are produced from de-alcoholised fermentation residues, after the yeast fermentation of grains to convert starch to alcohol (Weigel et al., 1997) and are a co-product of both bioethanol and potable alcohol production. DDGS is an internationally recognised protein source for cattle. An excellent review of the use of DDGS as a feed source has been written at Iowa State University (Babcock et al., 2008). Figure 24.2 shows the global DDGS production since 2002.

*Nutritional value of DDGS as a feed ingredient*

Removal of the starch from the cereal source through fermentation approximately triples the concentration of all remaining components including valuable nutrients such as protein in the DDGS product (see Table 24.1; Thacker and Widyaratyne, 2007). Unfortunately, in periods when grain may be of poor quality from moulding, for example, the concentration of the mould in the final product will also be tripled. The nutritional variability found in DDGS can affect feed manufacturing as nutrients

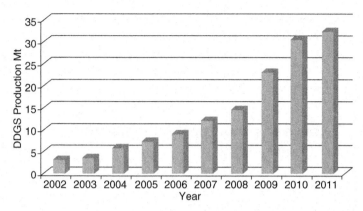

*24.2* Global production of DDGS from 2002 to 2011 (compiled using data from Renewable Fuels Agency, 2012).

*Table 24.1* Comparison of nutritional content of wheat, wheat DDGS and maize DDGS

| Variable | Wheat | Wheat DDGS | Maize DDGS |
| --- | --- | --- | --- |
| Moisture | 11.8 | 8.1 | 11.8 |
| Crude protein | 19.8 | 44.5 | 30.3 |
| Non-protein N | 4.6 | 10.2 | 5.4 |
| Fat | 1.8 | 2.9 | 12.8 |
| Ash | 2.1 | 5.3 | 4.8 |
| ADF | 2.7 | 21.1 | 14.6 |
| NDF | 9.4 | 30.3 | 31.2 |

*Source*: Thacker and Widyaratyne, 2007.

contributed by DDGS can vary so widely that the final feed can be out of tolerance for final product specification (Behnke, 2007).

The digestible amino acid content of DDGS does not necessarily correspond to its increased protein content compared to the native cereal, as some amino acids (such as proline and alanine) concentrate more rapidly than others during fermentation (e.g., histidine and leucine) (Liu, 2011). Also, the high possibility of Maillard reactions during DDGS drying leaves lysine digestibility as a major concern for the use of DDGS as a feed ingredient. Maillard browning reactions result in the formation of indigestible polymers based on the epsilon amino group of lysine and a reducing sugar.

The major feedstock for DDGS in the US is maize, whereas wheat is the major feedstock in Europe. Currently wheat DDGS is successfully used in ruminant nutrition at inclusion levels of between 25 and 35% without affecting nutrient digestibility or growth characteristics (Li *et al.*, 2011; Yang *et al.*, 2012), with maximum performance characteristics found with around

20% inclusion of DDGS (Buckner *et al.*, 2007; Klopfenstein *et al.*, 2008). The high fibre content decreases feed intake and limits nutrient utilisation in both pigs (Nyachoti *et al.*, 2005) and chicks (Thacker and Widyaratyne, 2007), as the fibre increases dietary bulk. Wheat DDGS has been used successfully in pig feed at 30% with no effect on performance traits (McDonnell *et al.*, 2011) and in broiler feeds at 15% inclusion (Thacker and Widyaratyne, 2007; Youssef *et al.*, 2008). Loar *et al.* (2010) suggested that feeding DDGS levels of 15% or higher may adversely affect young chicks less than 28 days of age, but that feed intake in older chicks is reduced if chicks are not exposed to DDGS in starter diets. Although levels up to 12% have no effect on meat quality or consumer acceptance, above these levels there may be a negative effect on thigh meat as increased fatty acids may increase oxidation (Corzo *et al.*, 2009; Schilling *et al.*, 2010).

For maize DDGS, suggested inclusion rates for broilers are 24% when fully balanced for amino acids (Shim *et al.*, 2011) and 6% when not fully supplemented (Lumpkins *et al.*, 2004). Wang *et al.* (2007) did suggest that there may be a possible loss of breast meat yield at 20% maize DDGS inclusion. Inclusion rates for layers are suggested at 10% (Lumpkins *et al.*, 2005), and Masa'deh *et al.* (2011) found that egg weight was reduced when more than 15% DDGS was included in the diet.

DDGS has also been investigated in aquaculture diets; in tilapia, Schaeffer *et al.* (2010) found that higher levels of DDGS (over 20%) reduced feed conversion ratio and bodyweight gain. These authors also hypothesised that replacing fishmeal with DDGS would require amino acid supplementation at higher inclusion levels. Again in tilapia diets, it has been shown that diets supplemented with lysine can contain up to 40% DDGS without any reduction in performance (Li *et al.*, 2011). Figure 24.3 shows the current usage of DDGS in the different animal feed sectors.

*Variability in DDGS production*

One of the major issues with DDGS as an ingredient in animal feeds is the lack of consistency, which makes accurate feed formulation difficult. Differences have been found in nutritional quality within and between production plants for both maize (Cromwell *et al.*, 1993; Spiehs *et al.*, 2002) and wheat DGGS (Lan *et al.*, 2008; Azarfar *et al.*, 2012). Some of this variation is due to feedstock differences which can lead to changes in alcohol conversion efficiency (Cottrill *et al.*, 2007) but Belyea *et al.* (2010) found that fermentation batches were the most influential source of variation. Other processing factors can affect product composition including temperature, concentration of solids and water quality (Rausch and Belyea, 2006), the mix of grains to solubles (Noll *et al.*, 2007) and the drying method (Swietkiewicz and Koreleski, 2008).

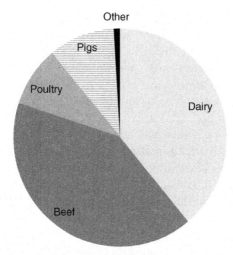

*24.3* DDGS usage globally in the animal feed sectors (compiled using data from Renewable fuels Agency, 2011).

*Processing issues with DDGS*

Much of the supplementary feed used in livestock production is pelleted to ease handling and increase bulk density for transport. DDGS has been found to deleteriously affect pellet quality; increasing DDGS content has been negatively correlated with pellet durability (Shim *et al.*, 2011) and has been shown to increase the quantity of fines (Loar *et al.*, 2010). Loar and Corzo (2011) discussed the effect of oil addition on the pelleting of DDGS, concluding that the oil addition required to add energy to DDGS-containing diets would form a more viscous diet, which required more energy to pellet. Loar *et al.* (2010) also found that higher DDGS inclusion increased energy use in the condenser due to the viscosity of the mash. The production rate was shown to decrease with 30% DDGS inclusion, which may be due to reduced supplemental rock phosphate, which has a scrubbing effect in the die (Loar *et al.*, 2010). However, the same study also showed a decrease in energy through the pellet mill to be due to the fat content, which lubricates the product through the die.

A further issue is that wheat DDGS contains varying amounts of non-starch polysaccharides (Saulnier *et al.*, 1995) which increase viscosity, causing issues for mixing and transport during processing (Smith *et al.*, 2006), and can cause uneven drying.

### 24.3.2 Oilseed meals

Oilseeds not only provide a valuable fuel source, but also a valuable feed material as the high oil content of the seed is usually coupled with high

*Table 24.2* Global production of oil bearing seeds in 2011 with the total and digestible protein content of the subsequent meals

| Protein meal source | Production (Mt million) | Total crude protein percentage (dry matter basis) | Apparent ileal digestibility coefficients of crude protein |
|---|---|---|---|
| Soya bean (*Glycine Max.*) | 251.5 | 44.8–49.9 | 0.82–0.88 |
| Rapeseed (canola) (*Brassica napus*) | 60.8 | 27.7–39.1 | 0.68–0.81 |
| Cotton seed (*Gossypium* spp.) | 46.6 | 38.3–44.9 | 0.72–0.75 |
| Peanut (*Arachishypogaea*) | 38.9 | 46.4–49.0 | 41.7–44.6 |
| Sunflower seed (*Helianthus annus*) | 35.5 | 31.0–36.6 | 0.79–0.87 |
| Palm kernel (*Elaeisguineensis*) | 13.4 | 13.6 | 0.54 |
| Copra (coconut) (*Cocos nucifera*) | 5.8 | 21.7 | 0.63 |

*Source*: Soy Stats, 2012; Bryden *et al.*, 2009; Kearl, 1982.

protein content. Post-oil extraction, the resulting meals form the backbone of protein supply to the global animal feed industry. The dominant feed protein product is soya bean meal, due to its widespread availability and balanced amino acid profile, which is generally well suited to the requirements of growing non-ruminant animals.

A number of oilseed meals other than soya also make a significant contribution to global animal feed protein. Table 24.2 shows the global production of oil-bearing seeds in 2011 and the digestible protein content of the subsequent meal.

*Soya bean meal*

Soya bean meal is used globally as a protein source in both ruminant and non-ruminant feeds due to its high protein content (around 40%) and favourable amino acid profile. It now provides the benchmark against which novel protein sources for animal feeds are measured. However, there are limits to the inclusion of soya bean meal in feed. Soya beans contain a number of anti-nutritional factors (ANFs) which cause reduced growth and feed conversion efficiency in livestock production if the soya is not appropriately heat treated (Grant, 1989; Clarke and Wiseman, 2005). The two main proteinaceous ANFs present in soya are trypsin inhibitor and lectin and it is the inactivation of these factors without reducing protein quality that forms the nexus of soya bean meal processing.

Early stages in biorefining of soya beans, such as dehulling, flaking and expanding, focus on optimising the beans for maximum oil extraction using hexane. Extraction of oil from the soya beans is achieved by washing baskets of the product with hexane in a countercurrent flow system, where the hexane is continuously recycled. Whilst the original aim of the final desolventiser-toaster stage was to maximise hexane recovery, the increasing value of the soya meal product has redirected focus to optimising protein quality by reducing trypsin inhibitor activity (TIA) levels to below 4 mg trypsin inhibition per gram of soya. However, studies into the effects of soya bean meal processing conditions on chick growth show that young broiler chicks are sensitive to variation in the quality of soya bean meal, even when samples are processed to below the recommended TIA threshold of 4 mg/g (Clarke and Wiseman, 2007). Similar sensitivity to residual TIA has been shown in pigs (Zarkadas and Wiseman, 2005).

*Rapeseed meal*

Rapeseed is usually cultivated in climates unsuited to soya production due to its high oil yield. Rapeseed oil has become the primary feedstock for biodiesel in Europe and outside Europe the dominant producers are China, India, Canada and Australia (Table 24.3). After crushing to remove oil, the resulting rapeseed cake is processed through a similar solvent extraction process to soya to produce rapeseed meal. As with many animal feed materials, its composition varies widely depending on a range of factors including origin, growing conditions, the manufacturing process and degree of oil extraction.

*Table 24.3* Top rapeseed producing countries in 2011–12

| Location | Rapeseed production (million Mt) |
|---|---|
| Canada | 14.17 |
| China | 13.00 |
| India | 6.50 |
| France | 5.36 |
| Germany | 3.87 |
| Australia | 3.19 |
| UK | 2.76 |
| Poland | 1.86 |
| Ukraine | 1.50 |
| EU 27 | 19.07 |
| World Total | 60.60 |

*Source*: United States Department of Agriculture Foreign Agricultural Service.

Rapeseed meal contains a blend of amino acids suited to non-ruminant diets but its use is limited by the presence of several ANFs. Whilst sinapine is responsible for the fishy taint of eggs from hens fed high levels of rapeseed meal, the most significant ANF with regard to feed is glucosinolate, which is highly goitrogenic and cannot be inactivated by heat processing. However, conventional (non-GM) plant breeding has addressed this through the production of 'double-zero' varieties of rapeseed containing low content of erucic acid and low content of glucosinolates. Virtually all rapeseed production in the European Union has shifted to rapeseed 00 (double zero), but unfortunately the development of low glucosinolate varieties suited to Asian countries has been less successful. It is imperative to know the source of the rapeseed meal to ensure rate of dietary inclusion does not reach levels known to impede animal performance and health. One of the first varieties of double zero rapeseed meal was developed in Canada: canola 'CANadian Oilseed, Low-Acid'. The term has now been adopted as a worldwide standard covering all double zero varieties of rapeseed meal, wherever they were produced. Table 24.4 indicates suggested inclusion levels of canola in production animal diets. Inclusion level of traditional (high glucosinolate) rapeseed meal is dictated by the concentration of glucosinolate within the meal, but halving the canola rate of inclusion has been suggested as a guide (Fenwick and Curtis, 1980).

*Other oilseed meals*

Despite the high values shown in Table 24.2 for both total protein content and proportion of digestible protein, the use of oilseed meal is often hindered by the presence of antinutritional factors or contamination with

*Table 24.4* Recommended rate of Canola inclusion in production animal diets

| Type of diet | Canola inclusion /g/kg |
|---|---|
| Cattle: Dairy compounds and blends | 250 |
| Other cattle: compounds and blends | 400 |
| Cattle: Protein concentrates | 600 |
| Sheep: Breeding compound and blends | 250 |
| Sheep: Grower/finisher compounds and blends | 150 |
| Pig grower | 30 |
| Pig finisher | 30 |
| Pig breeder | 50 |
| Poultry layer feeds | 50 |
| Poultry broiler feeds | 30 |

*Source*: Cottrill *et al.*, 2007.

toxic substances such as mycotoxins. Although mycotoxin contamination may affect almost any feed that is improperly processed or stored, peanut (sometimes referred to as groundnut) meal appears particularly susceptible to contamination by aflatoxins which are potent carcinogens. Similarly, the use of cottonseed meal in non-ruminant diets is restricted due to the presence of gossypol. In normal concentrations, gossypol has no toxic effect on cattle, but it has been shown that liveweight gain in beef cattle is reduced when the gossypol content is high. When cottonseed cake is used in poultry and pig rations, levels of up to 10% are usually recommended (Chicco and Shultz, 1977). Palm oil production is derived from two oil sources within the same plant: the majority from the mesocarp and a smaller contribution from the palm kernels. Mesocarp-sourced palm oil does not yield an animal feed co-product but the kernels are commonly used as a feed material in Africa and Asia. While there appear to be no nutritional limitations to the use of palm kernel meal, its use in ruminant feeding is restricted by its dry, unpalatable texture, in spite of its relatively high oil content. The low protein content of palm kernel oil compared to other oilseed meals also renders it unattractive for inclusion in non-ruminant feed but it has been successfully incorporated into laying hen diets at inclusions rates of up to 40% without affecting egg production (Perez *et al.*, 2000).

### 24.3.3 Corn gluten meal

The majority of global bioethanol production uses the simple dry grind process which produces DDGS as a co-product. However, there is an alternative wet grind process, which is both a more capital requiring and intensive process which separates parts of the corn kernel to make syrups, oil and other products. The final by-product of the wet grind process is corn gluten meal which is a protein product primarily used as cattle feed but which is now gaining acceptance as a protein supplement in a number of different applications including aquanutrition.

### 24.3.4 Liquid feeds

There are a number of liquid biorefinery co-products that are well suited and economical for use in feed. However, in addition to the challenges over variability in nutritional value discussed earlier for dry biorefinery feed products, transportation costs are increased by the lower nutrient density of liquid feeds.

Whilst the majority (around 70%) of bioethanol production blends fibrous and liquid fractions to create DDGS as the primary feed product, bioethanol production may also yield two feed liquid products which rising cereals costs are likely to promote as attractive feed materials. Condensed distillers solubles (CDS) from the bioethanol dry milling process are suited to liquid feeding systems used for pig production. CDS can positively affect growth performance but low palatability limits inclusions to a maximum of 20% in pig diets (de Lange et al., 2006) and presence of live yeast may lead to additional fermentation and frothing during storage. Corn steepwater (also known as condensed steepwater solubles) from the bioethanol wet milling process is viewed as an excellent source of soluble protein for beef cattle liquid feed supplements. Corn steepwater contains substantially more crude protein, ash, phosphorus and lactic acid than CDS, but the low oil levels leave it with low energy content. The similarity in consistency of CDS to molasses has raised interest in the product as a pellet quality enhancer, but the hygroscopicity of CDS is relatively high, presenting risk of mould contamination of pellets (DeFrain et al., 2003).

A final molasses-type product under consideration as an animal feed is the 5 carbon sugar product from the hemicellulose component of cellulosic bioethanol production, termed 5C molasses, which is 65% dry matter (Persson, 2009). To date, a pilot plant in Kalundborg, Denmark, is producing 11,100 tonnes per annum of 5C molasses which has been successfully trialled in pig diets (data not published) and is currently being marketed both as pig and cattle feed but may ultimately be re-directed into further bioethanol production. Another by-product of cellulosic ethanol production is lignin, which is used to produce pellet binders which increase production and enable the use of hard-to-pellet materials in animal feed, such as DDGS. Their inclusion in diets has also been shown to have a positive effect on poultry performance and caecal fermentation (Kivimäe, 1978; Moran and Conner, 1992) and may improve gastrointestinal health by providing fermentable oligosaccharides (Flickinger et al., 1998).

Biodiesel production from triglycerides yields one liquid product used in animal feed: glycerol. The 2004 boom in biodiesel production more than doubled crude glycerol production and exceeded the capacity of glycerol refineries. Existing markets required refinement of glycerol so the extra one million tonnes of glycerol produced alongside biodiesel was directed into the animal feed sector as crude glycerol. Ruminant feeds appear particularly well suited to incorporation of glycerol, as glycerol is a natural product of rumen fermentation of fat. Inclusion of up to 10% glycerol in the total mixed ration of dairy cows increased milk yield and decreased post-calving weight loss (Bodarski et al., 2005). The plummet in glycerol cost and concurrent hike in cereal costs have heightened interest in the use of glycerol as an energy source for pigs and poultry: it has been successfully incorporated into diets in poultry at levels of up to 9% (Dozier et al., 2008) and up to 10% in pig diets. These inclusion levels also improve pellet quality and production efficiency in the mill (Groesbeck et al., 2008).

## 24.4 Impact of processing technology on co-product quality

The use of wet co-products in animal nutrition is not unknown but they are not feed materials of choice. The transportation of the water is an added cost; storage over time is limited due to the need to inhibit bacterial contamination and mould formation, and systems have to be specifically constructed and adapted to supply the wet material to the animals. Perhaps the biggest example of liquid feeding is the feeding of liquid waste to pigs. For this reason, the majority of co-products are dried on the site of production. Investment in the drying capacity of the plant is a significant proportion of the overall cost of the plant and the efficient operation of the dryers is essential to the operation of the plant. A breakdown in drying can cause a shutdown of the plant. Particularly in the bioethanol process, drying of the product is therefore integral to the discussion of co-products. An additional factor coming to the fore is the fact that drying the co-product can represent approximately 40% of the overall energy requirements of the plant. Product drying is therefore a major contributor to the carbon footprint of the plant. As energy costs rise and the cost of drying increases, alternatives that obviate the need for drying and water removal by remediating water such that it can be reused in the process will become more attractive.

### 24.4.1 Drying technologies

The drying method can affect the nutritional content of the finished product, although product deterioration is usually due to the application of excess heat, rather than the moisture removal (Morris *et al.*, 2004). Drying types can be classified by either the mechanism of heating or mechanism of vapour transport. Air drying requires a high temperature air which supplies the heat and removes the water vapour, whereas vacuum drying uses a reduction in pressure to remove the vapour (Chen and Mujumder, 2009).

*Ring drying*

Ring drying is a first generation drying technology where hot air flows over an extensive area to remove water from the surface of the product. A heated air stream moves the material to be dried through a vertical column and as the particles lose moisture, they are transported to the top of the column and then further dried by moving them through one or more rings attached to the column. Adjustable splitter blades are used to convey the heavier semi-dried material back into the dryer for another pass through the system. This selective extension of residence time allows the ring dryer to process traditionally hard to dry materials. However, there is a greater possibility of burning and overheating, and consistency of product may be an issue as the material is not all dried for the same length of time. Ring drying is a relatively cost-effective method of large-scale drying producing a granular low dust product.

*Spray drying*

Spray drying is a second generation drying technology, which is defined by the formation of a spray of droplets, which are produced for optimum evaporation and contact with air. A spray dryer converts a suspension to a powder in a single processing step. A nozzle or atomiser is used to convert the liquid input into a fine spray, usually with droplet sizes of 100–200 μm (Niessen, 2002). The liquid fraction is sprayed into a hot vapour stream which vaporises the liquid. This is a very rapid method of drying which produces a consistent particle size and often a fine, free flowing, powder end product. However, spray drying is not appropriate for viscous liquids as they are difficult to atomise into a consistent, fine spray and it is a high energy demand process, which needs to be carefully managed to minimise resource use (Luna-Solano et al., 2005).

*Freeze drying*

Freeze drying is a third generation drying technology, which has four distinct stages: freezing, vacuum, sublimation and condensing. It has been shown to overcome issues of structural damage to the end product (Karel, 1975; Dalgleish, 1990) and the absence of air prevents oxidative deterioration, and there is no possibility of heat damage. Freeze drying is approximately 4–8 times more expensive than air drying (Ratti, 2001) and therefore the process is more commonly used for smaller scale applications producing very high value products. Lin Hsu et al. (2003) compared hot air, drum and freeze drying of yam flours and found antioxidants were more preserved in freeze dried samples, but otherwise little difference was observed between the drying technologies.

### 24.4.2 Maillard reactions

Maillard browning reactions are a complex series of stages, beginning with a condensation between a reducing sugar and most commonly the ε-amino group of lysine (Purlis, 2010). When pentoses and hexoses are involved, the reaction forms brown polymers (Martins et al., 2001). Maillard reactions impair the nutritional content and the bioavailability of amino acids and proteins (Moralez et al., 2007). Colour has been correlated to amino acid digestibility by a number of researchers (Batal and Dale, 2006; Fastinger *et al.*, 2006), with lysine digestibility reducing from 80% to 60% with darker DDGS (Ergul *et al.*, 2003). Very dark samples have recently been reported to give very low ileal digestibility values in pigs (Cozannet *et al.*, 2010) and in cockerels (Cozannet *et al.*, 2011a). This heat processing damage also negatively affects the digestible energy content of both maize (Fastinger *et al.*, 2006) and wheat DDGS (Cozannet *et al.*, 2011b). In wheat DDGS, amino acid digestibility was found to be lower in pigs when compared to wheat and some amino acids (including lysine) were significantly reduced by the drying process (Pederson and Lindberg, 2010).

### 24.4.3 Use of enzymes during processing

The viscosity of a co-product stream may influence the recovery efficiency, which is an effect experienced in the brewing industry. The viscosity of the stream is influenced by the composition of the dissolved solids: in particular, the presence of short chain non-starch polysaccharides (NSP) in the liquid stream will contribute to increased viscosity. Exogenous enzymes such as glucanases are available to counteract the effect of viscosity (Bamforth, 2009). Bamforth and Kanauchi (2001) hypothesised that within barley cell walls, the accessibility to β glucan is hindered by arabinoxylan content, reducing its solubility. Scheffler and Bamforth (2005) provided more evidence to support this hypothesis, and also suggested that it is β glucan rather than arabinoxylan which causes viscosity issues in mash. There is also a role for xylanase in separating the β glucan, so a mixture of enzymes is most effective for viscosity reduction in barley mashes, specifically a xylanase/glucanase mix.

## 24.5 Improving feedstocks, processes and yields

Yeast (*Saccharomyces cerevisiae*) is produced in the bioethanol process and is the most valuable component of DDGS. Spencer-Martins and Van Uden (1977) estimated that 0.071 g yeast is produced for every gram of starch fermented. Thus a 400 M litre bioethanol plant, fermenting approximately 1.1 M tonnes of wheat per annum would theoretically produce 48,000 tonnes of yeast, so pooled material from several plants of similar size has the potential to be a very valuable source of supplementary protein. Yeast contains valuable proteins, B vitamins, nucleotides and high inositol and glutamic acid levels (Silva *et al.*, 2009).

Yeast has been considered as a protein source in animal feed for many years. It has been fed successfully to chicks at up to 10% total dietary inclusion (Yalcin *et al.*, 1993; Onol and Yalcin, 1995), but higher levels have depressed performance, due to deficiencies in some amino acids (Klose and Fevold, 1945; Caballero-Cordoba and Sgarbieri, 2000) and issues with palatability and texture (Sell *et al.*, 1981; Daghir and Sell, 1982; Succi *et al.*, 1980). In fish, yeast has been used to replace 50% fishmeal, with no significant differences in growth and improved protein conversion (Oliva-Teles and Goncalves, 2001). Yeast is more commonly fed at a lower inclusion level (2% or less), and has been shown to improve performance in pigs (Spark *et al.*, 2005; Carlson *et al.*, 2005), fish (Essa *et al.*, 2011) and poultry (Miazzo *et al.*, 2005; Shareef and Al-Dabbagh, 2009).

### 24.5.1 Subfractionation

If DDGS could be modified and made more suitable for monogastric nutrition, it would enable the co-product to be more easily utilised in alternative market segments such as the large pig and poultry feed markets as opposed to the current application which is mainly for cattle. Efforts have been made to produce higher protein DDGS. The Elusieve process uses a combination of sieving and air flow to produce a higher protein enhanced DDGS (Srinivasan *et al.*, 2008, 2009). This process has been used with maize DDGS and the high protein fraction was fed to chickens resulting in a partial increase in final bodyweight compared with the non-sieved DDGS (Loar *et al.*, 2009) but no differences in digestible energy or amino acid content were observed (Kim *et al.*, 2010). Using the same process, sieved wheat DDGS improved energy digestibility in rainbow trout (Randall and Drew, 2010). The improvements to DDGS using this process to date have been small and as with DDGS, batch variability and product drying remain key issues. Whilst this product has higher protein than DDGS, its high fibre content renders it fundamentally similar to DDGS and therefore not considered by feed formulators as a yeast protein for use in non-ruminant diets.

### 24.5.2 Process improvements of bioethanol co-product

The yeast component of bioethanol co-product provides a valuable source of protein if it can be economically separated from DDGS, leaving a high fibre fraction which is nutritionally appropriate for ruminant feeding. A novel, factory scale, continuous-flow process is being developed to separate a yeast-containing, high protein fraction from the distillery stillage called yeast protein concentrate (YPC) (Williams, 2010; Williams *et al.*, 2009). Once separated, the YPC is dried to produce a powder, which may be suitable as either a protein source or a feed additive for monogastric feeds. This technology could be applied in the emerging bioethanol industry, in the existing potable alcohol industry and also in any large-scale grain fermentation facility.

The separation undertaken has three distinct stages: decanting, liquid removal and drying. Decanters are solid walled, horizontal centrifuges used to separate suspensions with a high concentration of solids. In the case of ethanol stillage, the decanter is the first stage of separation after distillation. A decanter houses a rotating horizontal bowl with a cylindrical and a conical section, and a scroll integrated in the bowl. The stillage enters the separation chamber through a centrally arranged feed pipe, and due to centrifugal forces, the solid particles are flung towards the wall of the chamber. The rotating screw in the separating bowl conveys the solids to the cone end of the bowl where they are then discharged. Regulating tubes allow the level of liquid to be altered within the bowl. The liquid phase, containing the yeast flows in the opposite direction to the solid discharge through the cylindrical part of the bowl to discharge under gravity. Further clarification of the liquid fraction is carried out using a centrifugal separator or disk stack.

The disk stack is a continuously operating nozzle centrifuge which is specially designed for liquid separation. Using a centrifuge to remove water is more cost effective than drying (V'ant Land, 1991). The disk stack comprises a rotating bowl equipped with a large number of inserts; conically arranged discs which are stacked into the bowl with small interspaces. The liquid fraction from the decanter enters the bowl through a central feed tube and the product is accelerated and conveyed into the disc portion of the separator. Solids are flung against the underside of the disk above due to their higher density and then they flow down the disc. The separated solids are continuously discharged though nozzles at the bowl periphery. The liquid in the bowl is then picked up by a centripetal pump which discharges the liquid. In the case of the alcohol stillage separation, the solid fraction from the disk stack is a yeast containing high protein cream which is dried to a powder for use as an animal feed ingredient.

However, as with DDGS, batch variability and product drying remain key issues, and any separated product may still contain some of the antinutritional factors present in wheat. Initial feeding studies have shown positive performance effects when feeding YPC to both poultry and fish. Digestible amino acid content of this yeast protein concentrate has been shown to be comparable with soya for broiler chicks, and higher than the feedstock alone (Scholey *et al.*, 2011a), although this is heavily influenced by the drying process used. In feeding studies with broiler chicks, dietary inclusion levels of 6% bioethanol YPC gave improved performance characteristics (Scholey *et al.*, 2011b). Bioethanol sourced YPC has been fed to several aquaculture species, with 20% dietary inclusion appearing optimal for performance (Omar *et al.*, 2012; Gause and Trushenski, 2011a, 2011b).

Due to the high ethanol exposure during the process, bioethanol yeast may have a thicker, toughened cell wall, which is more resistant to enzyme proteolysis (Caballero-Cordoba and Sgarbieri, 2000). Rumsey *et al.* (1991)

showed that salmonids fed disrupted yeast had an increased nitrogen absorption compared with whole yeast. This improvement in nutrient utilisation in homogenised yeast has also been shown in poultry (Vananuvat, 1977; Vananuvat and Chiraratananon, 1977) and shrimp (Coutteau *et al.*, 1990). The disk stack process exerts mechanical shear forces on the yeast, which has been shown to disrupt the cell walls, thereby increasing access to the intracellular nutrients. Yeast cell walls can be considered a prebiotic which can improve performance characteristics in poultry (Parks *et al.*, 2001; Hooge *et al.*, 2003) and pigs (Davis *et al.*, 2004) and positively affect gut morphology in broiler chicks (Santin *et al.*, 2001; Zhang *et al.*, 2005; Morales-Lopez *et al.*, 2009).

## 24.6 Regulatory issues

The feed sector of animal production is based on the principle of least cost formulation. The properties of a range of available raw materials are held in a computer matrix and linear programming is used to formulate a diet meeting the nutrient requirements of an animal in a given life stage with the cheapest possible blend of raw materials, whilst ensuring any legal or toxic boundaries are not exceeded. This allows the feed sector flexibility to use a range of raw materials based on their availability and cost. However, the matrix relies on feed materials containing known concentrations of both nutrients and anti-nutrients, so high variability renders products unattractive to feed compounders.

### 24.6.1 Registration of materials for marketing as animal feeds or feed additives

The use of materials as animal feed is tightly governed by legislation ultimately aimed at protecting human and animal health. Legislation surrounding animal feed is most stringent in the EU (EU, 2009), with severity reducing in less developed Asia-Pacific markets. A collaboration between the EU regulatory authority, the European Food Standards Agency (EFSA) and the US authority, the US Food and Drug Administration (FDA), to align EU and US standards resulted in the FDA Food Safety Modernization Act (2011). Additionally, western consumer awareness of food safety is placing pressure on countries exporting animal-based human food products to comply with western legislation.

US and EU legislation on the use and marketing of all animal feeds requires feed to fulfil stipulations relating to safety, quality, purity and traceability. Some aspects of these guidelines are of particular relevance when considering the use of biorefinery products for animal feed. The EU includes restriction on the use of medicated feed, i.e. inclusion of antibiotics

(EU, 1990), the need for specific consent for the marketing of any GMO-derived product within the EU (EU, 2003b), maximum acceptable levels of microbial contamination and specific legislation on the marketing and use of feed additives (EU, 2003c).

### 24.6.2 Antibiotics

Antibiotics are used commonly in American bioethanol production to reduce microbial overload. It is estimated that over half of plants are routinely using antibiotics (Olmstead, 2009), usually penicillin or virginiamycin (Hynes et al., 1997; Stroppa et al., 2000) and often prophylactically. There are issues relating to the presence of antibiotic residues in co-products leading to possible resistance and public health consequences (Muthaiyan and Ricke, 2010). There are other options to minimise contamination of bioethanol plants such as chlorine dioxide and natural hop-derived enzymes which are becoming more widely used. FEFAC states that antibiotic bactericides are not routinely used in the EU bioethanol industry, and not used at all in the spirit industry (FEFAC, 2008). There have been several instances in the EU where antibiotic residues have been found in DDGS from bioethanol, but these have been exclusively on Brazilian imports (Pol et al., 2009).

### 24.6.3 Mycotoxins

There are 400 recognised mycotoxins which may contaminate 25% of cereal feedstocks, but only one of these, Alflatoxin B1 has maximum permitted limit status (EU, 2002). This contamination is concentrated up to three times in DDGS, so contamination needs to be considered and treated in the feedstock. Wu and Munkvold (2008) estimated that contamination from a single mycotoxin could cost the pig industry alone up to $147 million annually, depending on level of DDGS usage.

### 24.6.4 Feed additives

Some biofuel co-products such as mannan-rich fractions of yeast cell walls may have beneficial properties beyond the simple supply of nutrients: they may improve the quality of feed or the quality of food from animal origin, or improve the animals' performance and health. Materials with such properties obviously claim a higher price than basic feed materials. However, any such claims cause the product to be classified as a feed additive rather than a feed material and therefore bound to comply with EU and US requirements for scientific verification of efficacy, health and safety of both human handlers and animal consumers of the material (EU, 2003c).

Compiled evidence is presented in a dossier that may take 3 years to produce, at an approximate cost of 1 million euros, before authorisation to market the additive within the EU is granted for a specific animal species, under specific conditions of use, for ten-year periods. Similarly in the US, a food additive petition (FAP) is required by the FDA to demonstrate that a material is safe for the proposed use in an animal feed. These requirements are likely to heavily influence decisions on whether to present novel feed products from biorefining as feed materials or feed additives.

## 24.7 Future trends

### 24.7.1 Water remediation

In a first generation bioethanol refinery, the ratio of ethanol produced to water used in the process is approximately 1:5. Approximately 40% of the energy use of the plant is used in dewatering the co-product and producing a dry product (DDGS) that can be transported off site. It is inconceivable that such a process can maintain profitability in a situation of rising energy cost, when 40% of total energy is used in dewatering to produce a feed product which has mediocre nutritional value. The stillage stream is generally a dilute solution but rich in protein and non-starch polysaccharides, which are a potential nutrient source for downstream processing. The initial challenge posed in utilising this material is the removal and disposal of the water. Developing processes that either enable the wastewater to be recycled or alternatives to heat to remove water from the co-product stream post-fermentation in a bioethanol plant offer significant benefits in terms of energy saving.

### 24.7.2 Algae

The need to address water remediation is driving bioethanol production towards adoption of technologies which either reduce reliance on high water volume or permit a higher degree of water recirculation. Algae are ubiquitous in the environment and, as such, a wide variety of organisms have developed the capacity to utilise a range of substrates as nutrients for growth under diverse environmental conditions. A characteristic of algae is the very high rate of biomass production and the fact that the organism has the ability to simultaneously produce a range of products. Large-scale algae production is recognised as a potential future energy source. The value of algae from an ethanol biorefinery is that the feedstock is already a recognised registered feedstuff, hence (barring the introduction of any extraneous components) algae produced in the system are suitable for use in feed. Algae are recognised as a valuable source of protein (Becker, 2007). It is a natural, sustainable feedstuff and for fish can be used as a part replacement of fishmeal supplying both protein and EPA (eicosapentaenoic acid) and DHA (docosahexaenoic acid) omega-3 oils. Algae may also supply a valuable source of carotenoids. Trials have already been completed with algae bioreactors co-located with bioethanol plants in the US and Scotland, demonstrating that geography and climate are not an impediment to the process. In these trials, the algae were fed to poultry and performance was comparable to high protein soya, with the algae also providing increased dietary energy content. These results are significant as they demonstrate that the co-product from a bioethanol plant can be used in the monogastric market segment rather than the traditional use of bioethanol co-products in the ruminant segment. There is the potential that algae from biofuel production could replace a third of soya in pig and poultry diets (Lei, 2012). In poultry rations, algae up to a level of 5–10% can be used safely as partial replacement for conventional proteins (Spolaore *et al.*, 2006).

The drive to discover alternative water remediation processes was also addressed by a group from Iowa State University. The same principles described above were used to grow a fungus (*Rhizopus* microspores) on stillage. They reported that the fungus removes about 60% of the organic material and most of the solids from thin stillage, allowing the water in the thin stillage to be recycled back into production. The fungus, once harvested and dried, is a protein-rich nutrient which is high in polyunsaturated fatty acids and may be suitable for use in feed for pigs and poultry.

Bioethanol plants are already adopting anaerobic digesters as a means of using the non-starch polysaccharides in the stillage to generate methane. Anaerobic digestion combines energy generation with a degree of water remediation. However, in order to efficiently operate an anaerobic digester, it has been found beneficial to reduce the nitrogen content of the stillage, the major proportion of which is contained in the spent yeast in the stillage. In an anaerobic digester excess nitrogen in the form of protein results in the generation of ammonia which can poison the digestion process. One solution to the recovery of yeast is the production of YPC. The integration of YPC and anaerobic digestion is a prospective breakthrough in novel water remediation.

### 24.7.3 The potential for genetic modification

There is potential for genetic modification to contribute to the co-product feeds that are produced from biorefining. In the long term, there is the chance that the co-products of biorefining will create more value than the current primary product of ethanol either for fuel or potable alcohol. The principle has been established with soya bean; where the original primary product was oil, now the range of protein materials produced creates more value than the oil in the bean. In biorefining, the primary feedstock is normally plant material with either single cell organisms or products from microorganisms (such as enzymes) used in the conversion process, hence there is the opportunity for genetic modification at both these points in the process.

*Genetic modification of the feedstock*

The technology of genetic modification of crops is well developed, particularly with respect to agronomic crop protection traits. However, there is a major limitation to the use of the genetic technology in developing quality traits in crops used for biorefining. The primary use of crops employed in biorefining is normally for food or feed, maize and wheat for example. Whilst the proportion of these two crops used in biorefining is increasing, the crops will always be dual purpose and interchangeably used for both food and feed. It follows, then, that any quality trait to be introduced into biorefining must be capable of being demonstrated as being totally safe when the plant is used as food. Furthermore, that the trait has sufficient value to cover the cost of identity preservation of the modified form from the point of cultivation to use and will leave a profit margin for the trait producer and the farmer. These are considerable hurdles in developing quality traits in dual-purpose crops.

*Genetic modification of microorganisms used in biorefining*

The case for modification of the microorganisms used in biorefining is more positive. These microorganisms can represent a significant proportion of the total dry mass of co-product (up to 10% in the case of yeast in first generation bioethanol) while contributing to the process by enzyme production. David (2007) patented the technology of genetic modification of the fermentation organism to create value-added products in the resulting DDGS. The technology has particular value when in conjunction with YPC yeast separation; for example, the value of the resulting yeast product would be significantly increased if the yeast is able to contain an elevated level of lysine. Other improvements could include the inclusion of high levels of carotenoids such as astaxanthin or canthaxanthin which are used as colorants in fish. Unfortunately, if the yeast cannot be separated from the residual DDGS, the value of the trait in the yeast is diluted to such an extent that it ceases to be of commercial value. Thus, whilst there appears to be considerable value (particularly for feed) in applying genetic modification to biorefining, there are major hurdles to developing a technology that encompasses food safety but maintains practical and economic viability.

## 24.8 Sources of further information and advice

Feed-related companies are not only the customer for any bio-based feed products, they are experts on the required technical and legal specification of raw feed materials. It therefore makes sense to consult or involve a feed company (or several) during development of a novel bio-based feed product, as upstream processing decisions (focused on other co-products) may profoundly affect the quality or legality of feed material produced.

Guidance on the US FDA regulations for feed registration is provided by the Association of American Feed Control Officials (AAFCO) (http://www.aafco.org/), whilst the EFSA regulations are implemented by EU member countries through individual organisations such as the UK's Food Standards Agency (http://www.food.gov.uk/).

Basic information on differing types of feed raw materials is provided on the FAO Animal Feed Resources Information System (http://www.fao.org/ag/AGA/AGAP/FRG/AFRIS/default.htm).

A more modern but narrower compendium containing more detail on processing techniques is being complied by the EU as their Feed Materials Register (http://www.feedmaterialsregister.eu/index.php?page=Register).

Emerging trends in animal feed technology and evaluations of raw materials are well summarised and promptly reported in a series of websites covering each production species published by 5M: The Pig Site (http://www.thepigsite.com/); The Poultry Site (http://www.thepoultrysite.com/); The Fish Site (http://www.thefishsite.com/). These sites also provide links to the original source of information and highlight relevant meetings and organisations.

## 24.9 References

ABPR 1774/2002: Regulation (EC) No 1774/2002 of the European Parliament and Council of 3rd October 2002. *Official Journal of the European Communities* L273/1.

Azarfar, A., Jonker, A. and Hettiarachchi-Gamage, I.K. (2012) Nutrient profile and availability of co-products from bioethanol processing. *Journal of Animal Physiology and Nutrition* **96**: 450–458.

Babcock, B.A., Hayes, D.J. and Lawrence, J.D. (2008) *Using distillers grains in the U.S. and international livestock and poultry industries*. The Midwest Agribusiness Trade Research and Information Center, Iowa State University, USA.

Bamforth, C.W. (2009) Current perspectives on the role of enzymes in brewing. *Journal of Cereal Science* **50**: 353–357.

Bamforth, C.W. and Kanauchi, M. (2001) A simple model for the cell wall of the starchy endosperm in barley. *Journal of the Institute of Brewing* **107**: 235–240.

Batal, A.B. and Dale, N.M. (2006) True metabolizable energy and amino acid digestibility of distillers dried grains with solubles. *Journal of Applied Poultry Research* **15**: 89–93.

Becker, E.W. (2007) Micro-algae as a source of protein. *Biotechnology Advances* **25**: 207–210.

Behnke, K.C. (2007) Feed manufacturing considerations for using DDGS in poultry and livestock diets. *Proceedings of the 5th Mid-Atlantic Nutrition Conference.* Zimmerman, N.G. (ed.), University of Maryland, pp. 77–81.

Belyea, R.L., Rausch, K.D., Clevenger, T.E., Singh, V., Johnston, D.B. and Tumbleson, M.E. (2010) Sources of variation in composition of DDGS. *Animal Feed Science and Technology* **159**: 122–130.

Bodarski, R., Wertelecki, T., Bommer, F. and Gosiewski, S. (2005) The changes of metabolic status and lactation performance in dairy cows under feeding TMR with glycerin (glycerol) supplement at periparturient period. *Electronic Journal of Polish Agricultural Universities*, Animal Husbandry, **8**: 1–9.

Bryden, W.L., Li, X., Ravindran, G., Hew, L.I. and Ravindran, V. (2009) *Ileal Digestible Amino Acid Values in Feedstuffs for Poultry*, Rural Industries Research and Development Corporation. Publication No. 09/071.

Buckner, C.D., Mader, T.L., Erickson, G.E., Colgan, S.L., Karges, K.K. and Gibson M.L. (2007) Optimum levels of dry distillers grains with solubles for finishing beef steers. *Nebraska Beef Cattle Report.* **MP90**: 36–38.

Caballero-Cordoba, G.M. and Sgarbieri, V.C. (2000) Nutritional and toxicological evaluation of yeast (*Saccharomyces cerevisiae*) biomass and a yeast protein concentrate. *Journal of the Science of Food and Agriculture* **80**: 341–351.

Carlson, M.S., Veum, T.L. and Turk, J.R. (2005) Effects of yeast extract versus animal plasma in weanling pig diets on growth performance and intestinal morphology. *Journal of Swine Health and Production* **13**(4): 204–209.

Chen, X.D. and Mujumder, A.S. (2009) *Drying Technologies in Food Processing.* Blackwell Publishing, Oxford.

Chicco, C.F. and Shultz, T. A. (1977) Utilization of agro-industrial by-products in Latin America. *FAO Animal Production and Health Paper* **4**: 125–146.

Clarke, E. and Wiseman, J. (2005) Effects of variability in trypsin inhibitor content of soya bean meals on true and apparent ileal digestibility of amino acids and pancreas size in broiler chicks. *Animal Feed Science and Technology* **121**: 125–138.

Clarke, E. and Wiseman, J. (2007) Effects of extrusion conditions on trypsin inhibitor activity of full fat soya beans and subsequent effects on their nutritional value for young broilers. *British Poultry Science* **48**(6): 703–712.

Corzo, A., Schilling, M.W., Loar, R.E., Jackson, V., Kin, S. and Radhakrishnan, V. (2009) Effects of feeding distillers dried grains with solubles on broiler meat quality. *Poultry Science* **88**: 432–439.

Cottrill, B., Smith, C., Berry, P., Weightman, R., Wiseman, J., White, G. and Temple, M. (2007) Opportunities and implications of using the co-products from biofuel production as feeds for livestock. *Research review No. 66*, HGCA, Kenilworth.

Coutteau, P., Lavens, P. and Sorgeloos, P. (1990) Baker's yeast as a potential substitute for live algae in aquaculture diets: *Artemia* as a case study. *Journal of the World Aquaculture Society* **21**: 1–9.

Cozannet, P., Primot, Y., Gady, C., Metayer, J.P., Lessire, M., Skiba, F. and Noblet, J. (2010) Energy value of wheat distillers grains with solubles for growing pigs and adult sows. *Journal of Animal Science* **88**: 2382–2392.

Cozannet, P., Lessire, M., Gady, C., Metayer, J.P., Primot, Y., Skiba, F. and Noblet, J. (2011a) Energy value of wheat distillers grains with solubles in roosters, broilers, layers and turkeys. *Poultry Science* **89**: 2230–2241.

Cozannet, P., Primot, Y., Gady, C., Metayer, J.P., Lessire, M., Skiba, F. and Noblet, J. (2011b) Standardised amino acid digestibility of wheat distillers dried grains with solubles in force-fed cockerels. *British Poultry Science* **52**: 72–81.

Cromwell, G.L., Herkelman, K.L. and Stahly, T.S. (1993) Physical, chemical and nutritional characteristics of distillers dried grains with solubles for chicks and pigs. *Journal of Animal Science* **71**: 679–686.

Daghir, N.J. and Sell, J.L. (1982) Amino acid limitations of yeast single-cell protein for growing chickens. *Poultry Science* **61**(2): 337–344.

Dalgleish, J.McN. (1990) *Freeze Drying for the Food Industries*. Elsevier, New York.

David, P.R. (2007) Compositions and methods for producing fermentation products and residuals. Australian Patent AU2007238228.

Davis, M.E., Maxwell, C.V., Erf, G.F., Brown, D.C. and Wistuba, T.J. (2004) Dietary supplementation with phosphorylated mannans improves growth response and modulates immune function of weanling pigs. *Journal of Animal Science* **82**: 1882–1891.

DEFRA (2011) GB animal feed statistical notice – October 2011. Available at: http://www.defra.gov.uk/statistics/files/defra-stats-foodfarm-food-animalfeed-statsnotice-111208.pdf (accessed 21 December 2011).

DeFrain, J.M, Shirley, J.E., Behnke, K.C., Titgemeyer, E.C. and Ethington, R.T. (2003) Development and evaluation of a pelleted feedstuff containing condensed corn steep liquor and raw soybean hulls for dairy cattle diets. *Animal Feed Science and Technology* **107**: 75–86.

de Lange, C.F.M., Zhu, C.H., Niven, S., Columbus, D. and Woods, D. (2006) Swine liquid feeding: nutritional considerations. *Proceedings of the Western Nutrition Conference*, Winnipeg, MB, Canada, pp. 1–13.

Department for Transport (2012) Verified 2010/11 RTFO report. Available at: http://assets.dft.gov.uk/statistics/releases/verified-rtfo-biofuel-statistics-2010-11/year-3-verified-report.pdf (accessed 28 May 2012).

Dozier, W.A., Kerr, B.J., Corzo, A., Kidd, M.T., Weber, T.E. and Bregendahl, K. (2008) Apparent metabolizable energy of glycerin for broiler chickens. *Poultry Science* **87**: 317–322.

Ergul, T., Martinez Amezcua, C., Parsons, C.M., Walters, B., Brannon, J. and Noll. S.L. (2003) Amino acid digestibility in corn distillers dried grains with solubles. *Poultry Science* **82** (Supplement 1): 70.

Essa, M.A., Mabrouk, H.A., Mohamed, R.A. and Michael, F.R. (2011) Evaluating different additive levels of yeast, *Saccharomyces cerevisiae*, on the growth and production performances of a hybrid of two populations of Egyptian African catfish, *Clarias gariepinus*. *Aquaculture* **320**: 137–141

EU (1990) Council Directive 90/167/EEC of 26 March 1990 laying down the conditions governing the preparation, placing on the market and use of medicated feeding stuffs in the Community. *Official Journal of the European Union*.

EU (2002) Directive 2002/32/EC of the European Parliament and of the Council of 7 May 2002 on undesirable substances in animal feed. *Official Journal of the European Union*.

EU (2003a) Directive 2003/30/EC of the European Parliament and of the Council of 8 May 2003 on the promotion of the use of biofuels or other renewable fuels for transport. *Official Journal of the European Union*.

EU (2003b) European Community Regulation 1829/2003 laying down the authorisation procedures for GM food and feed (the 'GM Food and Feed Regulation'). *Official Journal of the European Union*.

EU (2003c) Regulation (EC) No. 1831/2003 of the European Parliament and of the Council of 22 September 2003 on additives for use in animal nutrition. *Official Journal of the European Union*.

EU (2009) Regulation (EC) No. 767/2009 of the European Parliament and of the Council of 13 July 2009 on the placing on the market and use of feed, amending European Parliament and Council Regulation (EC) No. 1831/2003 and repealing Council Directive 79/373/EEC, Commission Directive 80/511/EEC, Council Directives 82/471/EEC, 83/228/EEC, 93/74/EEC, 93/113/EC and 96/25/EC and Commission Decision 2004/217/EC. Available at: http://eur-lex.europa.eu/LexUriServ/LexUriServ.do?uri=OJ:L:2009:229:0001:0028:EN:PDF

European Bioethanol Fuel Association (2012) http://www.epure.org/statistics/info/Productiondata (accessed 12 July 2012).

Fastinger, N.D., Latshaw, J.D. and Mahan, D.C. (2006) Amino acid availability and true metabolisable energy content of corn distillers dried grains with solubles in adult cecectomised roosters. *Poultry Science* **85**: 1212–1216.

FDA (2011) Food Safety Modernization Act. Available at: http://www.gpo.gov/fdsys/pkg/PLAW-111publ353/pdf/PLAW-111publ353.pdf

FEFAC (2008) Antibiotic residues in Brazilian feed yeasts. European Feed Manufacturers Federation Statement, 28 August 2008.

Fenwick, C.R. and Curtis, R.F. (1980) Rapeseed meal and its use in poultry diets: a review. *Animal Feed Science and Technology* **5**: 255–298.

Flickinger, E.A., Campbell, J.M., Schmitt, L.G. and Fahey, G.C. (1998) Selected lignosulphonate fractions affect growth performance, digestibility, and cecal and colonic properties in rats. *Journal of Animal Science* **76**: 1626–1635.

Gause, B. and Trushenski, J. (2011a) Replacement of fish meal with ethanol yeast in the diets of sunshine bass. *North American Journal of Aquaculture* **73**: 97–103.

Gause, B. and Trushenski, J. (2011b) Production performance and stress tolerance of sunshine bass raised on reduced fish meal feeds containing ethanol yeast. *North American Journal of Aquaculture* **73**: 168–175.

Gilbert, R. (2002) The world animal feed industry. *Protein Sources for the Animal Feed Industry, FAO Proceedings*, Bangkok, 29 April–3 May 2002.

Grant, G. (1989) Anti-nutritional effects of soya bean: a review. *Progress in Food and Nutrition Science* **13**: 317–348.

Groesbeck, C.N., McKinney, L.J., DeRouchey, J.M., Tokach, M.D., Goodband, R.D., Dritz, S.S., Nelssen, J.L., Duttlinger, A.W., Fahrenholz, A.C. and Behnke, K.C. (2008) Effect of crude glycerol on pellet mill production and nursery pig growth performance. *Journal of Animal Science* **86**: 2228–2336.

Hardy, R.W. (2010) Utilization of plant proteins in fish diets: effects of global demand and supplies of fishmeal. *Aquaculture Research* **41**: 770–776.

Hooge, D.M., Sims, M.D., Sefton, A.E., Connolly, A. and Spring, P.S. (2003) Effect of dietary mannanoligosaccharide, with and without bacitracin or virginiamycin, on live performance of broiler chickens at relatively high stocking density on new litter. *Journal of Applied Poultry Research* **12**: 461–467.

Hynes, S.H., Kjarsgaard, D.M., Thomas, K.C. and Ingledew, W.M. (1997) Use of virginiamycin to control the growth of lactic acid bacteria during alcoholic fermentation. *Journal of Industrial Microbiology and Biotechnology* **18**: 284–291.

Karel, M. (1975) Principles of food science, Part II. In: *Food Process Engineering*, (4th edn.) Prentice-Hall, Englewood Cliffs, NJ.

Kearl, L.C. (1982) 'Nutrient requirements of ruminants in developing countries'. Utah State University, Int. Feedstuffs Inst., Logan.

Kim, E.J., Parsons, C.M., Srinivasan, R. and Singh, V. (2010) Nutritional composition, nitrogen-corrected true metabolisable energy and amino acid digestibilities of new corn distillers grains with solubles produced by a new fractionation process. *Poultry Science* **89**: 44–51.

Kivimäe, A. (1978) Effects of lignosulphonates on poultry when used as a binder in compounded feed. *Arch. Geflügelk.* **42**: 238–245.

Klopfenstein, T.J., Erickson, G.E. and Bremer, V.R. (2008) The use of distillers by-products in the beef feeding industry. *Journal of Animal Science* **86**: 1223–1231.

Klose, A.A. and Fevold, H.L. (1945) Nutritional value of yeast protein to the rat and the chick. *The Journal of Nutrition* **29**(6): 421–430.

Lan, Y., Opapeju, F.O. and Nyachoti, C.M. (2008) True ileal protein and amino acid digestibilities in wheat dried distiller's grains with solubles fed to finishing pigs. *Animal Feed Science and Technology* **140**: 155–163.

Lei, X. (2012) Algae as sustainable protein alternative for animal feed. Available at:http://www.allaboutfeed.net/news/algae-as-sustainable-protein-alternative-for-animal-feed-12701.html (accessed 12 July 2012).

Li, Y.M., McAllister, T.A., Beauchemin, K.A., McKinnon, J.J. and Yang, W.Z. (2011) Substitution of wheat dried distillers grains with solubles for barley grain or barley silage in feedlot cattle diets: intake, digestibility and ruminal fermentation. *Journal of Animal Science* **89**: 2491–2501.

Lin Hsu, C., Chen, W., Weng, Y.M. and Tseng, C.Y. (2003) Chemical composition, physical properties and antioxidant activities of yam flours as affected by different drying methods. *Food Chemistry* **83**: 85–92.

Liu, K. (2011) Chemical composition of distiller's grains, a review. *Journal of Agricultural and Food Chemistry* **59**: 1508–1526.

Loar, R.E. and Corzo, A. (2011) Effects of feed formulation on feed manufacturing and pellet quality characteristics of poultry diets. *World's Poultry Science Journal* **67**: 19–28.

Loar, R.E., Srinivasan, R., Kidd, M.T., Dozier, W.A. and Corzo, A. (2009) Effects of elutriation and sieving processing (Elusieve) of distillers dried grains with solubles on the performance and carcass characteristics of male broilers. *Journal of Applied Poultry Research* **18**: 494–500.

Loar, R.E., Moritz, J.S., Donaldson, J.R. and Corzo, A. (2010) Effects of feeding distillers dried grains with solubles to broilers from 0 to 28 days post hatch on broiler performance, feed manufacturing efficiency and selected intestinal characteristics. *Poultry Science* **89**: 2242–2250.

Lumpkins, B., Batal, A. and Dale, N. (2004) Evaluation of distillers dried grains with solubles as a feed ingredient for broilers. *Poultry Science* **83**: 1891–1896.

Lumpkins, B., Batal, A. and Dale, N. (2005) Use of distillers dried grains plus solubles in laying hen diets. *Journal of Applied Poultry Research* **14**: 25–31.

Luna-Solano, G., Salgado-Cervantes, M.A., Rodriguez-Jimenes, G.C. and Garcia-Alvarado, M.A. (2005) Optimisation of brewer's yeast spray drying process. *Journal of Food Engineering* **68**: 9–18.

Martins, S.I.F.S., Jongen, W.M.F. and van Boekel, M.A.J.S. (2001) A review of Malliard reactions in food and implications to kinetic modelling. *Trends in Food Science and Technology* **11**: 364–373.

Masa'deh, M.K., Purdum, S.E. and Hanford, K.J. (2011) Dried distillers grains with solubles in laying hen diets. *Poultry Science* **90**: 1960–1966.

McDonnell, P., O'Shea, C.J., Callan, J.J. and O'Doherty, J.V. (2011) The response of growth performance, nitrogen and phosphorus excretion of growing-finisher pigs to diets containing incremental levels of maize dried distiller's grains with solubles. *Animal Feed Science and Technology* **169**: 104–112.

Miazzo, R.D., Peralta, M.F., Picco, M.L. and Nilson, A.J. (2005) Productive parameters and carcass quality of broiler chickens fed yeast (*S. cerevisiae*). *XVIIth European Symposium on the Quality of Poultry Meat*, Doorwerth, The Netherlands, 23–26 May.

Morales-Lopez, R., Auclair, E., Garcia, F., Esteve-Garcia, E. and Brufau, J. (2009) Use of yeast cell walls: beta-1,3/1,6-glucans; and mannoproteins in broiler chicken diets. *Poultry Science* **88**: 601–607.

Moralez, F.J., Acar, O.C., Serpen, A., Arribas-Lorenzo, G. and Gokmen, V. (2007) Degradation of free tryptophan in a cookie model system and its application in commercial samples. *Journal of Agricultural and Food Chemistry* **55**: 6793–6797.

Moran, E.T. and Conner, D.E. (1992) Reduction in pH of cecal contents with broiler chicks given probiotic and soluble complex carbohydrates supplemented to the starting feed. *Poultry Science* **71S**: 167 (Abstract).

Morris, A., Barnett, A. and Burrows, O.J. (2004) Effect of processing on nutrient content of foods. *Cajanus* **37**(3): 160–164.

Mustafa, A.F., McKinnon, J.J. and Christensen, D.A. (1999) Chemical characterisation and *in vitro* crude protein degradability of thin stillage derived from barley and wheat based ethanol production. *Animal Feed Science and Technology* **80**: 247–256.

Muthaiyan, A. and Ricke, S.C. (2010) Current perspectives on detection of microbial contamination in bioethanol fermenters. *Bioresource Technology* **101**: 5033–5042.

Niessen, W.R. (2002) *Combustion and Incineration Processes*. CRC Press, Boca Raton, FL.

Noll, S.L., Brannon, J. and Parsons, C. (2007) Nutritional value of corn distillers dried grains with solubles (DGS): influence of solubles addition. *Proceedings of ADSA/ASAS Joint Annual Meeting*, San Antonio, TX, Abstract M204.

Nyachoti, C.M., Haouse, J.D., Slominski, B.A. and Seddon, I.R. (2005) Energy and nutrient digestibilities in wheat dried distillers grains with solubles fed to growing pigs. *Journal of the Science of Food and Agriculture* **85**: 2581–2586.

Oliva-Teles, A. and Goncalves, P. (2001) Partial replacement of fishmeal by brewer's yeast (*Saccharomyces cerevisiae*) in diets for sea bass (*Dicentrarchus labrax*) juveniles. *Aquaculture* **202**: 269–278.

Olmstead, J. (2009) Fuelling resistance? Antibiotics in ethanol production. Institute of Agricultural and Trade Policy, Minneapolis, MN, July.

Omar, S.S., Merrifield, D.L., Kuhlwein, H., Williams, P.E.V. and Davies, S.J. (2012) Biofuel derived yeast protein concentrate (YPC) as a novel feed ingredient in carp diets. *Aquaculture* **330–333**: 54–62.

Onol, A.G. and Yalcin, S. (1995) The usage of baker's yeast in laying hen rations. *Veterinary Journal of Ankara University* **42**: 161–167.

Parks, C.W., Grimes, J.L., Ferket, P.R. and Fairchild, A.S. (2001) The effect of mannanoligosaccharides, bambermycins and virginiamycin on performance of large white male market turkeys. *Poultry Science* **80**: 718–723.

Pederson, C. and Lindberg, J.E. (2010) Ileal and total tract nutrient digestibility in wheat wet distillers solubles and wheat dried distillers grains with solubles when fed to growing pigs. *Livestock Science* **132**: 145–151.

Perez, J.F., Gernat, A.G. and Murillo, J.G. (2000) The effect of different levels of palm kernel meal in layer diets. *Poultry Science* **79**: 77–79.

Persson, M. (2009) Inbicon – production of cellulosic bioethanol on an industrial scale in sight. *International Sugar Journal* **111**: 709–712.

Pol, I., Driessen-van Lankveld, W., Van Egmond, H., Pikkemaat, M. and De Jong, J. (2009) Use of antibiotics in bioethanol production: emerging risk in feed? *Proceedings of the European Feed Microbiology Organisation*, Rostock, Germany, 21–23 September.

Purlis, E. (2010) Browning development in bakery products – a review. *Journal of Food Engineering* **99**: 239–249.

Randall, K.M. and Drew, M.D. (2010) Fractionation of wheat distillers dried grains and solubles using sieving increases digestible nutrient content in rainbow trout. *Animal Feed Science and Technology* **159**: 138–142.

Ratti, C. (2001) Hot air and freeze-drying of high value foods: a review. *Journal of Food Engineering* **49**: 311–319.

Rausch, K.D. and Belyea, R.L. (2006) The future of coproducts from corn processing. *Applied Biochemistry and Biotechnology* **128**(1): 47–85.

Renewable Fuels Agency (RFA) (2011) Renewable fuels agency statistics. http://www.ethanolrfa.org/pages/statistics

Renewable Fuels Agency (RFA) (2012) Renewable fuels agency coproducts. http://www.ethanolrfa.org/pages/industry-resources-coproducts

RTFO Renewable Transport Fuels Obligation Order (2007) Available at: www.opsi.gov.uk/si/si2007/draft/pdf/ukdsi_9780110788180_en.pdf (accessed 20 October 2013).

Rumsey, G.L., Hughes, S.G., Smith, R.R., Kinsella, J.E. and Shetty, K.J. (1991) Digestibility and energy values of intact, disrupted and extracts from brewer's dried yeast fed to rainbow trout (*Oncorhynchus mykiss*). *Animal Feed Science and Technology* **33**: 185–193.

Santin, E., Maiorka, A. and Macari, M. (2001) Performance and intestinal mucosa development of broiler chickens fed diets containing *Saccharomyces cerevisiae* cell wall. *Journal of Applied Poultry Research* **10**: 236–244.

Saulnier, L., Peneau, N. and Thibault, J-F. (1995) Variability in grain extract viscosity and water soluble arabinoxylan content in wheat. *Journal of Cereal Science* **22**: 259–264.

Schaeffer, T.W., Brown, M.L., Rosentrater, K.A. and Muthukumarappan, K. (2010) Utilization of diets containing graded levels of ethanol production co-products by Nile tipapia. *Journal of Animal Physiology and Animal Nutrition* **94**: 348–354.

Scheffler, A. and Bamforth, C.W. (2005) Exogenous β glucanases and pentosanases and their impact on mashing. *Enzyme and Microbial Technology* **36**: 813–817.

Schilling, M.W., Battula, V., Loar, R.E., Jackson, V., Kin, S. and Corzo, A. (2010) Dietary inclusion level effects of distillers dried grains with solubles on broiler meat quality. *Poultry Science* **89**: 752–760.

Scholey, D.V., Williams, P. and Burton, E.J. (2011a) Potential for alcohol co-products from potable and bioethanol sources as protein source in poultry diets. *British Poultry Abstracts* **7**(1): 23–24.

Scholey, D.V., Morgan, N., Williams, P. and Burton, E.J. (2011b) Effects of incorporating a novel bioethanol co-product in poultry diets on bird performance. *British Poultry Abstracts* **7**(1): 57.

Sell, J.L., Ashraf, M. and Bales, G.L. (1981) Yeast single-cell protein as a substitute for soybean meal in broiler diets. *Nutrition Reports International* **24**(2): 229–235.

Shareef, A.M. and Al-Dabbagh, A.S.A. (2009) Effect of probiotic (*Saccharomyces cerevisiae*) on performance of broiler chicks. *Iraqi Journal of Veterinary Sciences* **23**: 23–29.

Shim, M.Y., Pesti, G.M. and Bakalli, R.I. (2011) Evaluation of corn distillers dried grains with solubles as an alternative ingredient for broilers. *Poultry Science* **90**: 369–376.

Silva, V.K., Della Torre da Silva, J., Torres, K.A.A., de Faria Filho, D., Hirota Hada, F. and Barbosa de Moraes, V.M. (2009) Humoral immune response of broilers fed diets containing yeast extract and prebiotics in the prestarter phase and raised at different temperatures. *Journal of Applied Poultry Research* **18**: 530–540.

Smith, T.C., Kindred, D.R., Brosnan, J.M., Weightman, R.M., Shepherd, M. and Sylvester-Bradley, R. (2006) Wheat as a feedstock for alcohol production. Research Review No. 61. HGCA, Kenilworth.

Soy Stats (2012) http://soystats.com/2012-soybean-highlights/

Spark, M., Paschertz, H. and Kamphues, J. (2005) Yeast (different sources and levels) as protein source in diets of reared piglets: effects on protein digestibility and N-metabolism. *Journal of Animal Physiology and Animal Nutrition* **89**(3–6): 184–188.

Spencer-Martins, I. and Van Uden, N. (1977) Yields of yeast growth on starch. *European Journal of Applied Microbiology* **4**: 29–35.

Spiehs, M.J., Whitney, M.H. and Shurson, G.C. (2002) Nutrient database for distiller's dried grains with solubles produced from new ethanol plants in Minnesota and South Dakota. *Journal of Animal Science* **80**: 2639–2645.

Spolaore, P., Joannis-Cassan, C., Duran, E. and Isambert, A. (2006) Commercial applications of microalgae. *Journal of Bioscience and Bioengineering* **101**: 87–96.

Srinivasan, R., Moreau, R.A., Parsons, C., Lane, J.D. and Singh, V. (2008) Separation of fiber from distillers dried grains (DDG) using sieving and elutriation. *Biomass and Bioenergy* **32**: 468–472.

Srinivasan, R., To, F. and Columbus, E. (2009) Pilot scale fiber separation from distillers dried grains with solubles (DDGS) using sieving and air classification. *Bioresource Technology* **100**: 3548–3555.

Stroppa, C.T., Andrietta, M.G.S., Andrietta, S.R., Steckelberg, C. and Serra, G.E. (2000) Use of penicillin and monensin to control bacterial contamination of Brazilian alcohol fermentations. *International Sugar Journal* **102**: 78–82.

Succi, G., Pialorsi, S., Di Fiore, L. and Cardini, G. (1980) The use of methanol grown yeast LI-70 in feeds for broilers. *Poultry Science* **59**(7): 1471–1479.

Swietkiewicz, S. and Koreleski, J. (2008) The use of distillers dried grains with solubles (DDGS) in poultry nutrition. *Worlds Poultry Science Journal* **64**: 257–265.

Thacker, P.A and Widyaratyne, G.P. (2007) Nutritional value of diets containing graded levels of wheat distillers dried grains with solubles fed to broiler chicks. *Journal of the Science of Food and Agriculture* **87**: 1386–1390.

The National Biodiesel Board (2012) US Biodiesel production at 112 MMg for June, 557 MMg year-to-date. *Biodiesel Magazine*, 1 August.

TSE regulation (999/2001) http://ec.europa.eu/food/fs/bse/bse36_en.pdf (accessed 12 July 2012).

US EIA (2013) US Department of Energy, Energy Information Administration. http://www.eia.gov/biofuels

V'ant Land, C.M. (1991) *Industrial Drying Equipment*. Marcel Dekker, New York.

Vananuvat, P. (1977) Value of yeast protein for poultry feeds. *Critical Reviews in Food Science and Nutrition* **9**(4): 325–343

Vananuvat, P. and Chiraratananon, R. (1977) The use of brewery yeast in commercial type ration for poultry. *World's Poultry Science Journal* **33**(2): 88–99.

Wang, Z., Cerrate, S., Coto, C., Yan, F. and Waldroup, P.W. (2007) Utilisation of distillers dried grains with solubles (DDGS) in broiler diets using a standardised nutrient matrix. *International Journal of Poultry Science* **6**(7): 470–477.

Weigel, J.C., Loy, D. and Kilmer, L. (1997) Feed co-products of the dry corn milling process. Renewable Fuels Association and National Corn Growers Association, Washington DC and St. Louis, MO.

Williams, P.E.V. (2010) Protein recovery. UK Patent application GB2010/000577.

Williams, P., Clarke, E. and Scholey, D. (2009) The production of a high concentration yeast protein concentrate co-product from a bioethanol refinery. *Proceedings of the 17th European Symposium on Poultry Nutrition*, Edinburgh, p. 326.

Wu, F. and Munkvold, G.P. (2008) Mycotoxins in ethanol co-products: modelling economic impacts on the livestock industry and management practices. *Journal of Agriculture and Food Chemistry* **56**: 3900–3911.

Yalcin, S., Onol, A.G., Kocak, D. and Ozcan, I. (1993) The usage of baker's yeast as a protein source in broiler rations. *Doga Turkish Journal of Veterinary Animal Science* **17**: 305–309.

Yang, W.Z., McAllister, T.A., McKinnon, J.J. and Beauchemin, K.A. (2012) Wheat distillers grains with solubles in feedlot cattle diets: feeding behaviour, growth performance, carcass characteristics and blood metabolites. *Journal of Animal Science* **90**: 1301–1310.

Youssef, I.M.I., Westfahl, C., Sunder, A., Liebert, F. and Kamphues, J. (2008) Evaluation of dried distillers grains with solubles (DDGS) as a protein source for broilers. *Archives of Animal Nutrition* **62**(5): 404–414.

Zarkadas, L.N. and Wiseman, J. (2005) Influence of processing of full fat soya beans included in diets for piglets. *Animal Feed Science and Technology* **118**: 121–137.

Zhang, A.W., Lee, B.D., Lee, S.K., Lee, K.W., An, G.H., Song, K.B. and Lee C.H. (2005) Effects of yeast (*Saccharomyces cerevisiae*) cell components on growth performance, meat quality and ileal mucosa development of broiler chicks. *Poultry Science* **84**: 1015–1021.

# 25
## The use of biomass to produce bio-based composites and building materials

R. M. ROWELL, University of Wisconsin, USA

DOI: 10.1533/9780857097385.2.803

**Abstract**: There are many different types of natural fibers coming from a wide variety of plant sources. These fibers have been used by humans since the beginning of human existence for clothing, heating, shelter, weapons, tools, packaging and writing materials. This chapter will deal with the use of biomass in biocomposites. These will include waferboard, flakeboard, particleboard, fiberboard, geotextiles, filters, and sorbents as well as physical and chemical properties of natural fibers. In most cases, natural resources have been used in composites without any modification for high volume, low cost and low performance products. It is possible to improve property performance through chemical modification so that natural, renewable, recyclable, and sustainable fibers can be used in value added markets.

**Key words**: biomass, natural fibers, resources, types, isolation, properties, waferboard, flakeboard, particleboard, fiberboard, geotextiles, filters, sorbents, improved performance, value added, emerging trends.

## 25.1 Introduction

### 25.1.1 History of bio-based composites

We have used wood and other types of biomass for many applications since the beginning of the human race. The earliest humans used biomass to make shelters, cook food, construct tools, and make boats and weapons. There are human marks on a climbing pole that were made over 300,000 years ago. We have found wood in the Egyptian pyramids, Chinese temples and tombs and ancient ships that attest to the use of wood by past societies. The use of natural fibers dates back in history to about 8,000 BC. Linen and hemp fabrics are known to have existed at that time and linen textiles are known to have existed in Europe from about 4,000 BC. Reference is made to the use of textiles as reinforcements of ceramics as early as 6,500 BC. Ramie is known to have been used in mummy cloths in Egypt during the period 5,000 ± 3,300 BC. Grass and straws have been used for many generations as a reinforcing fiber in mud bricks (adobe) and in ancient Egypt 3,000 years ago, pharaoh mummies were wrapped in linen cloth impregnated with salts, resins, and honey to protect and reinforce them. The use of flax for the production of linen dates back over 5,000 years. Pictures on tombs and temple walls at Thebes depict flowering flax plants. Cotton fibers have been found in caves in Mexico that date back over 7,000 years. There are references in early Chinese history to natural fibers for papermaking. Hemp fiber implants that are over 10,000 years old have been found in pottery shards in China and Taiwan. The use of hemp fiber dates back to the Stone Age. Thomas Jefferson drafted the United States Declaration of Independence on hemp paper (Rowell, 2008).

### 25.1.2 Biomass resources

As we begin the twenty-first century, there is an increased awareness that non-renewable resources are becoming scarce, and dependence on renewable resources is growing. The twenty-first century may be the cellulosic century as we look more and more to renewable plant resources for products. It is easy to say that natural fibers are renewable and sustainable but, in fact, they are neither. Natural fibers come from plants and it is the living plants that are renewable and sustainable, not the fibers themselves. This means that we must put our emphasis on healthy forests and agricultural lands, and to manage and use our ecosystems in ways that do not put them at risk.

## 25.2 Fibrous plants

In terms of utilization, there are two general classifications of plants producing natural fibers: primary and secondary. Primary plants are those grown for their fiber content, while secondary plants are those where the fibers come as a by-product from some other primary utilization. Jute, hemp, kenaf, sisal, and cotton are examples of primary plants, while pineapple, cereal stalks, agave, oil palm, and coir are examples of secondary plants (Rowell, 2008).

Table 25.1 shows an inventory of some of the major fibers now produced in the world. While wood is the major source of fiber, plant straws and stalks combined are potentially a larger source of fiber than wood. The data for this table were extracted from several sources using estimates and extrapolations for some of the numbers. For this reason, the data should be considered as only a rough relative estimate of world fiber resources. The inventory of many agricultural resources can be found in the FAO database on its website. By using a harvest index, it is possible to determine the quantity of residue associated with a given production of a crop.

## 25.3 Fiber types and isolation

### 25.3.1 Fiber types

There are many ways to classify natural fibers. Some authors classify fibers as to industrial use, i.e. papermaking, textile, composites. Others use systems

*Table 25.1* Global inventory of biofibers

| Source | Metric tons |
|---|---|
| Wood | 1,750,000,000 |
| Straw | 1,145,000,000 |
| Stalks | 970,000,000 |
| Bagasse | 75,000,000 |
| Reeds | 30,000,000 |
| Bamboo | 30,000,000 |
| Cotton staple | 15,000,000 |
| Core fiber | 8,000,000 |
| Papyrus | 5,000,000 |
| Bast fiber | 2,900,000 |
| Cotton linters | 1,000,000 |
| Grasses | 700,000 |
| Leaf | 480,000 |
| **Total** | **4,033,080,000** |

such as hard and soft fiber, long and short fibers, cellulose content, strength, color, etc. The most common classification for natural fibers is by botanical type. Using this system, there are six basic types of natural fibers:

1. bast fibers such as jute, flax, hemp, ramie, and kenaf;
2. leaf fibers such as banana, sisal, agave, and pineapple;
3. seed fibers such as coir, cotton, and kapok;
4. core fibers such as kenaf, hemp, and jute;
5. grass and reed such as wheat, corn, and rice; and
6. all other types such as wood and roots.

Table 25.2 gives a more complete list of fiber types. Some plants yield more than one type of fiber. For example, jute, flax, hemp, and kenaf have both bast and core fibers. Agave, coconut, and oil palm have both fruit and stem fibers. Cereal grains have both stem and hull fibers (Rowell, 2008).

## 25.3.2 Isolation of bioelements

For biocomposites such as chip, flake and strand boards, wood is the major source of these elements. Green wood is debarked and either run through as chipper, flaker or strand-making equipment. The chips, flakes or strands are then dried, an adhesive is added and a mat of the elements is formed. The mat is then run into a press, the press is closed and heated for the assigned time. The panels are cooled, edge cut to size, face sanded to a standard thickness and stacked for shipment.

Table 25.2 Six general types of natural fibers

| Bast | Leaf | Seed | | | | Core | Grass/reeds | Other |
|---|---|---|---|---|---|---|---|---|
| | | Fibers | Pod | Husk | Fruit | Hulls | | |
| Hemp | Pineapple | Cotton | Kapok | | | | Kenaf | Wood |
| Ramie | Sisal | | Loofah | | | | Jute | Roots |
| Flax | Agava | | Milk weed | | | | Hemp | Galmpi |
| Kenaf | Henequen | | | | | | Flax | |
| Jute | Curaua | | | Coir | | | | |
| Mesta | Banana | | | | Oil palm | | Bamboo | |
| Urena | Abaca | | | | | Rice | Bagasse | |
| Roselle | Palm | | | | | Oat | Corn | |
| | Cabuja | | | | | Wheat | Rape | |
| | Albardine | | | | | Rye | Rye | |
| | Raphia | | | | | | Esparto | |
| | Curauà | | | | | | Sabai | |
| | | | | | | | Canary | |
| | | | | | | | grass | |

For fiberboards, many different types of biomass plants can be used. Fiber can be isolated from the plant using many different types of procedures from simple mechanical methods to chemical treatments.

It has been known since ancient times that fibers from plants such as jute, flax and hemp could be isolated by allowing the plant to rot either in the field or in a pond and then beaten on rocks to loosen the fibers from the plant. In the process, the bark is removed by enzymatic action from microorganisms (molds and bacteria) in the water. This process is also used for hemp and kenaf. The entire plant is placed in a pond and the natural decay process removes the bark and separates the long bast fiber from the core or stick. The process mainly removes pectic substances which frees the fiber bundles from the woody core. The process takes 2–3 weeks (shorter times if the water is warm). Once the bark has been removed, the fiber is removed and dried.

Jute, hemp and kenaf fibers can also be isolated using a decorticating machine. The plant is squeezed through slotted rollers that spread the plant apart and then slotted rollers with cutting edges cut and separate the fibers from the other plant tissue. The plant may have to be run through the decortication equipment several times for complete fiber isolation.

Another mechanical separation method carried out on cotton uses a cotton gin. A cotton gin separates the cotton fibers from the seed pods and the seeds. It uses a combination of a wire screen and small wire hooks that pull the cotton through the screen while brushes continuously remove the loose cotton lint to prevent jams.

Plants can be broken down into fiber bundles and single fibers by grinding or refining. In the grinding process, the wood is mechanically broken down into fibers. In the refining process, wood chips or plant parts can be broken down into fibers between one or two rotating plates in a wet environment to fibers. If the refining is done at high temperatures, the fibers tend to slip apart due to the softening of the lignin matrix between the fibers, and the fibers have a lignin rich surface. If the refining is performed at lower temperatures, the fibers tend to break apart and the surface is rich in carbohydrate polymers (Rowell, 2008).

Chemical separation can also be carried out by chemical pulping. Dilute alkali or dilute acids can be used to separate fiber bundles. Pulping is usually done to reduce the fiber bundles to separate fibers for paper making. Alkali, acid, and organo-solv pulping methods have been used to isolate fibers from plants.

## 25.4 Fiber properties

The chemical properties of some natural fibers are given in Table 25.3. The mechanical properties of some natural fibers are shown in Table 25.4. The tensile and flexural properties are shown in Table 25.5.

Table 25.3 Chemical composition of some natural fibers

| Type of fiber | Cellulose | Lignin | Pentosan | Ash | Silica |
|---|---|---|---|---|---|
| Stalk fiber: Straw | | | | | |
| Rice | 28–48 | 12–16 | 23–28 | 15–20 | 9–14 |
| Wheat | 29–51 | 16–21 | 26–32 | 4.5–9 | 3–7 |
| Barley | 31–45 | 14–15 | 24–29 | 5–7 | 3–6 |
| Oat | 31–48 | 16–19 | 27–38 | 6–8 | 4–6.5 |
| Rye | 33–50 | 16–19 | 27–30 | 2–5 | 0.5–4 |
| Cane fiber | | | | | |
| Sugar | 32–48 | 19–24 | 27–32 | 1.5–5 | 0.7–3.5 |
| Bamboo | 26–43 | 21–31 | 15–26 | 1.7–5 | 0.7 |
| Grass fiber | | | | | |
| Esparto | 33–38 | 17–19 | 27–32 | 6–8 | – |
| Sabai | – | 22 | 24 | 6 | – |
| Reed fiber | | | | | |
| Phragmites Communis | 44–46 | 22–24 | 20 | 3 | 2 |
| Bast fiber | | | | | |
| Seed flax | 43–47 | 21–23 | 24–26 | 5 | – |
| Kenaf | 44–57 | 15–19 | 22–23 | 2–5 | – |
| Jute | 45–63 | 21–26 | 18–21 | 0.5–2 | – |
| Hemp | 57–77 | 9–13 | 14–17 | 0.8 | – |
| Ramie | 87–91 | – | 5–8 | – | – |
| Core fiber | | | | | |
| Kenaf | 37–49 | 15–21 | 18–24 | 2–4 | – |
| Jute | 41–48 | 21–24 | 18–22 | 0.8 | – |
| Leaf fiber | | | | | |
| Abaca | 56–63 | 7–9 | 15–17 | 3 | – |
| Sisal | 47–62 | 7–9 | 21–24 | 0.6–1 | – |
| Seed hull fiber | | | | | |
| Cotton linter | 99–95 | 0.7–1.6 | 1–3 | 0.8–2 | – |
| Wood fiber | | | | | |
| Coniferous | 40–45 | 26–34 | 7–14 | <1 | – |
| Deciduous | 38–49 | 23–30 | 19–26 | <1 | – |

## 25.5 Types and properties of bio-based composites

Bio-based composite products started with very thick laminates for glue laminated beams, to thin veneers for plywood, to strands for strandboard, to flakes for flakeboard, to particles for particleboard, and, finally, to fibers for fiberboard. As the size of the composite element gets smaller, it is possible to either remove defects (knots, cracks, checks, etc.) or redistribute them to reduce their effect on product properties. Also, as the element size becomes smaller, the composite becomes more like a true material, i.e. consistent, uniform, continuous, predictable, and reproducible.

Composite materials can be classified by several different systems: density (e.g., medium density fiberboards, see Fig. 25.1), application (e.g., insulation

*Table 25.4* Mechanical properties of some natural fibers

| Fiber | Density (g/m³) | Length (mm) | Diameter (μm) | Elongation at break (MPa) |
|---|---|---|---|---|
| Cotton | 1.21 | 15–56 | 12–35 | 2–10 |
| Coir | 0.3–3.0 | 7–30 | 15–25 | |
| Flax | 1.38 | 10–65 | 5–38 | 1.2–3 |
| Jute | 1.23 | 0.8–6 | 5–25 | 1.5–3.1 |
| Sisal | 1.20 | 0.8–8 | 7–47 | 1.9–3 |
| Hemp | 1.35 | 5–55 | 10–51 | 1.6–4.5 |
| Henequen | 1.4 | 8–33 | 3–4.7 | |
| Ramie | 1.44 | 40–250 | 18–80 | 2–4 |
| Kenaf (bast) | 1.2 | 1.4–11 | 12–36 | 2.7–6.9 |
| Kenaf (core) | 0.31 | 0.4–1.1 | 18–37 | |
| Pineapple | 1.5 | 3–8 | 8–41 | 1–3 |
| Bagasse | 1.2 | 0.8–2.8 | 10–34 | 0.9 |
| Southern yellow pine | 0.51 | 2.7–4.6 | 32–43 | |
| Douglas fir | 0.48 | 2.7–4.6 | 32–43 | |
| Aspen | 0.39 | 0.7–1.6 | 20–30 | |

*Table 25.5* Tensile and flexural properties of some natural fibers

| Fiber | Tensile – MOR (MPa) | Tensile – MOE (GPa) | Flexural – MOR (MPa) | Flexural – MOE (GPa) |
|---|---|---|---|---|
| Wood | 30.5 | 8.2 | 55.3 | 7.5 |
| Bagasse | 27.0 | 5.4 | 47.8 | 5.1 |
| Coir | 25.9 | 3.6 | 46.9 | 3.6 |
| Curauà | 48.1 | 7.1 | 77.6 | 6.1 |
| Flax | 36.1 | 6.1 | 58.4 | 5.8 |
| Hemp | 33.6 | 6.1 | 61.5 | 6.2 |
| Jute | 34.6 | 7.2 | 57.8 | 6.9 |
| Ramie | 43.2 | 5.4 | 70.2 | 5.1 |
| Sisal | 34.3 | 7.1 | 60.0 | 6.6 |

MOR: modulus of rupture; MOE: modulus of elasticity.

board), raw material form (e.g., particleboard) and process type (e.g., dry process fiberboard). The breakdown of biomass can include large timbers, dimensional lumber, very thick laminates for glue laminated beams, to thin veneers for plywood, to strands for strandboard, to flakes for flakeboard, to chips for chipboard, to particles for particleboard, and, finally, to fibers for fiberboard. However, since the bulk of this chapter deals with the utilization of small biomass, only composites made using small bio-members, strands, chips or flakes and fiber will be covered.

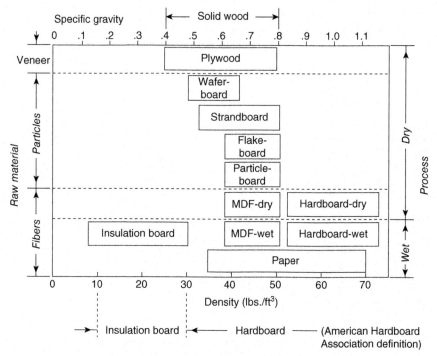

*25.1* Bio-based composites based on density.

## 25.5.1 Selection based on density

Figure 25.1 shows different types of composites based on the product density. Insulation type products have a very low density while high density hardboards have a much higher density. The density of the biomass cell wall is approximately 1.5.

## 25.5.2 Types of bio-based composites

*Waferboard and flakeboard*

Waferboard is a structural panel used in exterior applications bonded with a phenolic adhesive. Large thin wafers or smaller flakes can be produced by several methods and used to produce a composite board. Wafers are almost as wide as they are long, while flakes are much longer than they are wide. Wafers are also thicker than flakes. A waferizer slices the wood into wafers, typically 38 mm wide × 76 to 150 mm long × 7 mm thick, with a specific gravity between 0.6 and 0.8.

The first waferboard plant was opened in 1963 by MacMillian Bloedel in Saskatchewan, Canada. Aspen was the raw material and the wafers were randomly oriented. In the late 1980s, most wafers were oriented resulting in oriented waferboard (OWB) that is stronger and stiffer than the randomly oriented board. The orientation distribution may be tailored to the application. OWB and OSB compete with plywood in applications such as single layer flooring, sheathing, and underlayment in lightweight structures; however, OSB has largely replaced OWB in most places in the US. Pines, firs and spruce are usually used as well as aspen.

The flakeboard industries started in the early 1960s. These are made using an exterior grade adhesive and are used as the structural skin over wall and floor joists. The specific gravity of flakeboard is usually between 0.6 and 0.8 and flakeboard is made using a waterproof adhesive such as phenol formaldehyde or an isocyanate.

*Particleboard*

The particleboard industry started in the 1940s out of a need to use large quantities of waste products such as sawdust, planer shavings and other mill residues. Particles of various sizes are formed into an air or mechanically formed mats and glued together to produce a randomly oriented flat panel. Almost all particleboards are produced by a dry process. Particleboard is usually made in three layers with the faces made using fine particles and a core of coarser material. Most applications of particleboard are for interior use and are bonded using a urea-formaldehyde adhesive. Paraffin or microcrystalline wax emulsion is usually added to improve short-term moisture resistance. Phenol-formaldehyde, melamine-formaldehyde and isocyanates are rarely used but are when increased moisture resistance is required. The resin content ranges from 4 to 10% but is usually made using 6–9%. The resin content of the two faces is usually slightly higher than the core. Table 25.6 shows some properties of different grades of particleboard.

Particleboard is often used as a core material for veneers and laminates. These are often used in counter tops, shelving, doors, room dividers, built-ins and furniture. Particleboard generally does not warp in use. It is available in several thicknesses from 6 to 32 mm in sheets of 120 × 240 cm.

*Fiberboard*

Fiberboards can be formed using a wet-forming or a dry-forming process. In a wet-forming process, water is used to distribute the fibers into a mat and then pressed into a board. In many cases an adhesive is not used and the lignin in the fibers serves as the adhesive. In the dry process, fibers from the refiner go through a dryer and blowline where the adhesive is applied and then formed into a web which is pressed into a board.

*Table 25.6* Properties of particleboards

| Grade | Modulus of rupture (MPa) | Modulus of elasticity (MPa) | Internal bond (MPa) |
|---|---|---|---|
| H-1 | 16.5 | 2400 | 0.90 |
| H-2 | 20.5 | 2400 | 0.90 |
| H-3 | 23.5 | 2750 | 1.00 |
| M-1 | 11.0 | 1725 | 0.40 |
| M-2 | 14.5 | 2225 | 0.45 |
| M-3 | 16.5 | 2750 | 0.55 |
| LD-1 | 3.0 | 550 | 0.10 |
| LD-2 | 5.0 | 1025 | 0.15 |

H = density greater than $800 \, kg/m^3$, M = density $640–800 \, kg/m^3$, LD = density less than $640 \, kg/m^3$.
*Source*: National Plywood Association (1993).

*Table 25.7* Properties of fiberboards

| Product | Thickness (mm) | Modulus of Rupture (MPa) | Modulus of Elasticity (MPa) | Internal bond (MPa) |
|---|---|---|---|---|
| Internal MDF | | | | |
| HDF | <21 | 34.5 | 3,450 | 0.75 |
| MDF | >21 | 24.0 | 2400 | 0.60 |
| LDF | 21 | 14.0 | 1400 | 0.30 |
| Exterior | | | | |
| MDF | 21 | 34.5 | 3,450 | 0.90 |

*Source*: National Plywood Association (1994).

Low density fiberboards (LDF) have a specific gravity of between 0.15 and 0.45 and are used for insulation and for lightweight cores for furniture. They are usually produced using a dry process with a ground wood fiber. Medium density fiberboard (MDF) has a specific gravity of between 0.6 and 0.8 and is mainly used as a core for furniture and is usually made using a dry process. High density fiberboard (HDF), sometimes called hardboard, has a specific gravity of between 0.85 and 1.2 and is used as an overlay on workbenches, floors and for siding. It is usually made using a dry process but is also made using a wet fiber process. The hardboard industry started around 1950. Hardboard is produced both with wax (tempered) and without wax and sizing agents. The wax is added to give the board water resistance.

Urea-formaldehyde resin is usually used for interior applications and phenol-formaldehyde for exterior application. Table 25.7 gives the mechanical properties of LDF, MDF and HDF. Standard (without wax) and tempered (with wax) hardboards come in many different thicknesses from 2.1 to 9.5 mm and have required standard minimum average modulus of rupture of 31.0 MPa for standard and 41.4 MPa for tempered. Standard hardboard must have a required standard minimum average tensile strength parallel to the surface of 15.2 MPa and 20.7 MPa for tempered. The required standard minimum average tensile strength perpendicular to the surface is 0.62 MPa for standard and 0.90 MPa for tempered.

*Molded products*

The present bio-based composite industry mainly produces two-dimensional (flat) sheet products. In some cases, these flat sheets are cut into pieces and glued/fastened together to make shaped products such as drawers, boxes, and packaging. Flat sheet fiber composite products are made by making a gravity formed mat of fibers with an adhesive and then pressing. If the final shape can be produced during the pressing step, then the secondary manufacturing profits can be realized by the primary board producer. Instead of making low cost flat sheet type composites, it is possible to make complex shaped composites directly using the long fibers alone or combinations of long and short fibers.

Wood fibers come in two lengths: short (softwoods) and shorter (hardwoods). So wood fiber is limited to short fiber applications unless it is combined with longer agricultural fibers for applications in a wider array of products. In this technology, fiber mats are made by combining long bast or leaf fibers from such plants as kenaf, jute, cotton, sisal, agave, etc., with wood fiber and then formed into flexible fiber mats. These can be made by physical entanglement (carding), non-woven needling, or thermoplastic fiber melt matrix technologies. In carding, the fibers are combed, mixed and physically entangled into a felted mat. These are usually of high density but can be made at almost any density. A needle punched mat is produced in a machine which passes a randomly formed machine-made web through a needleboard that produces a mat in which the fibers are mechanically entangled. The density of this type of mat can be controlled by the amount of fiber going through the needleboard or by overlapping needled mats to give the desired density. In the thermoplastic fiber matrix, the bio-based fibers are held in the mat using a thermally softened thermoplastic fiber such as polypropylene or polyethylene.

During the mat formation step, an adhesive is added by dipping or spraying the fiber before mat formation or added as a powder during mat formation. The mat is then shaped and densified by a thermoforming step. The final desired shape is determined by the mold in the hot press. Within certain limits, any size, shape, thickness, and density is possible.

*Geotextiles*

Medium- to high-density fiber mats described before can be used in several ways other than for molded composites. One is for use as a geotextile. Geotextiles derive their name from the two words 'geo' and 'textile' and, therefore, mean the use of fabrics in association with the earth.

Geotextiles have a large variety of uses. They can be used for mulch around newly planted seedlings. The mats provide the benefits of natural mulch; in addition, controlled-release fertilizers, repellents, insecticides, and herbicides can be added to the mats as needed. Medium density fiber mats can also be used to replace dirt or sod for grass seeding around new home sites or along highway embankments. Grass or other types of seed can be incorporated in the fiber mat. Fiber mats promote seed germination and good moisture retention. Low and medium density fiber mats can be used for soil stabilization around new or existing construction sites, where steep slopes, without root stabilization, can lead to erosion and loss of top soil.

Medium and high density fiber mats can also be used below ground in road and other types of construction as a natural separator between different materials in the layering of the backfill. It is important to restrain slippage and mixing of the different layers by placing separators between the various layers.

It is estimated that the cost of controlling erosion in the United States is in excess of $55 million/year. This is a large potential market for forest resource composites.

*Filters and sorbents*

Filter systems are presently in use to clean our water, but new innovation in filtration technology is needed to remove contaminants from water. The development of filters to clean our water supply is big business. It is estimated that global spending on filtration (including dust collectors, air filtration, liquid cartridges, membranes and liquid macro-filtration) will increase from $17 billion in 1998 to $75 billion by 2020. The fastest-growing non-industrial application area for filter media is for the generation of clean water.

Medium to high density mats can be used as filtering aids to remove particulates out of waste and drinking water or solvents. Wood fibers have also been shown to sorb oil from water. While not as good as other agricultural fibers such as kenaf, wood fiber can remove oil from both fresh- and seawater.

### 25.5.3 Moisture, biological, ultraviolet, and thermal properties

Biomass changes dimensions with changing moisture content because the cell wall polymers contain hydroxyl and other oxygen-containing groups that attract moisture through hydrogen bonding. The hemicelluloses are mainly responsible for moisture sorption, but the accessible cellulose, non-crystalline cellulose, lignin, and surface of crystalline cellulose also play major roles. Moisture swells the cell wall, and the fiber expands until the cell wall is saturated with water (fiber saturation point, FSP). Beyond this saturation point, moisture exists as free water in the void structure and does not contribute to further expansion. This process is reversible, and the fiber shrinks as it loses moisture below the FSP.

Biomass is degraded biologically because organisms recognize the carbohydrate polymers (mainly the hemicelluloses) in the cell wall and have very specific enzyme systems capable of hydrolyzing these polymers into digestible units. Biodegradation of the cell wall matrix and the high molecular weight cellulose weakens the fiber cell. Strength is lost as the cell wall polymers and matrix undergo degradation through oxidation, hydrolysis, free radical and dehydration reactions.

Biomass exposed outdoors undergoes photochemical degradation caused by ultraviolet radiation. This degradation takes place primarily in the lignin component, which is responsible for the characteristic color changes. The lignin acts as an adhesive in the cell walls, holding the cellulose fibers together. The surface becomes richer in cellulose content as the lignin degrades. In comparison to lignin, cellulose is much less susceptible to ultraviolet light degradation. After the lignin has been degraded, the poorly bonded carbohydrate-rich fibers erode easily from the surface, which exposes new lignin to further degradative reactions. In time, this 'weathering' process causes the surface of the composite to become rough and can account for a significant loss in surface fibers.

Biomass burns because the cell wall polymers undergo pyrolysis reactions with increasing temperature to give off volatile, flammable gases. The gases are ignited by some external source and combust. The hemicelluloses and cellulose polymers are degraded by heat much before the lignin. The lignin component is the most thermally stable of the cell wall polymers and contributes to char formation. The charred layer helps insulate the composite from further thermal degradation.

## 25.6 Improving performance properties

In general, bio-based composites are used in large, low cost, low performing markets. Since we know that bio-based resources swell and shrink with changing moisture contents, decay, burn and are degraded by ultraviolet energy, we think they cannot be used in high performance applications.

For the most part, we have designed and used biomass putting up with the 'natural defects' that nature has given to us. Biomass is a hygroscopic resource that was designed to perform, in nature, in a wet environment.

Nature is programmed to recycle it in a timely way through biological, thermal, aqueous, photochemical, chemical, and mechanical degradations. In simple terms, nature builds biomass from carbon dioxide and water and has all the tools to recycle it back to the starting chemicals. We harvest a green plant and convert it into dry products, and nature, with its arsenal of degrading reactions, starts to reclaim it at its first opportunity.

The properties of any resource are, in general, a result of the chemistry of the components of that resource. In the case of wood, the cell wall polymers (cellulose, hemicelluloses, and lignin) are the components that, if modified, would change the properties of the resource. If the properties of biomass are modified, the performance of the modified biomass would be changed. This is the basis of chemical modification to change properties and improve performance.

In order to produce bio-based materials with a long service life, it is necessary to interfere with the natural degradation processes for as long as possible (Hon, 1993). This can be done in several ways. Traditional methods for decay resistance and fire retardancy, for example, are based on treating the product with toxic or corrosive chemicals which are effective in providing decay and fire resistance but can result in environmental concerns. In order to make property changes, you must first understand the chemistry of the components and the contributions each play in the properties of the resource. Following this understanding, you must then devise a way to modify what needs to be changed to get the desired change in property.

Properties of biomass, such as dimensional instability, flammability, biodegradability, and degradation caused by acids, bases, and ultraviolet radiation are all a result of chemical degradation reactions which can be prevented or, at least, slowed down if the cell wall chemistry is altered (Rowell, 2005). Two technologies to do this are already commercial: acetylation (Rowell, 2006) and heat treatments (Hill, 2006).

Almost all biomass contains some acetyl groups. Acetylation of biomass just increases that acetyl level to 15–20%. Acetylation using acetic anhydride has mainly been carried out as a liquid phase reaction. The reaction with acetic anhydride results in esterification of the accessible hydroxyl groups in the cell wall with the formation of acetic acid byproduct:

$$\text{WOOD-OH} + \text{CH}_3\text{-C(=O)-O-C(C=O)-CH}_3$$
$$\text{Acetic anhydride}$$
$$\rightarrow \text{WOOD-O-C(=O)-CH}_3 + \text{CH}_3\text{-C(=O)-OH}$$
$$\text{Acetylated wood} \qquad \text{Acetic acid}$$

Acetylation is a single-site reaction which means that one acetyl group is on one hydroxyl group with no polymerization. This means that all of the weight gain in acetyl can be directly converted into units of hydroxyl groups blocked. Fiberboards have been made from acetylated wood, jute, kenaf, bamboo and coir fiber and the boards show greatly improved dimensional stability and decay resistance to both brown- and white-rot fungi. There is not improvement in thermal properties but some level of improving resistance to weathering (UV resistance) (Rowell, 2005).

Heating biomass to improve performance dates back many thousands of years. In the early part of the twentieth century it was found that drying wood at high temperature increased dimensional stability and led to a reduction in hygroscopicity. Later, it was found that heating wood also increased resistance to microbiological attack. Along with the increase in stability and durability come increased brittleness and loss in some strength properties, including impact toughness, modulus of rupture and work to failure. The treatments usually cause a darkening of the wood and the wood has a tendency to crack and split.

## 25.7 Conclusion and future trends

If we look into the future of materials, we need to reduce our dependence on non-renewable resources and increase our use of renewable resources. Biomass is recyclable, renewable and sustainable. To replace many non-renewable materials, biomass materials will be used and they must be stable and durable. For many countries, the biomass will be wood. But for countries like India, jute and coir will play a major role, for South America, sugar cane bagasse, for China, kenaf and hemp, and for Malaysia and Singapore, oil palm fiber will play a major role.

In the United States, small diameter trees in our overgrown forests will find a major use in composites. Wood species, such as red maple, not currently used for composites will emerge. And more agricultural fibers will be brought in both alone and in combination with wood fiber. New processes for conversion of plants to composites will be developed that will use less energy and water. New fiber isolation processes will be developed with higher yields and less fiber damage.

New chemistries will be developed to improve fiber property performance that will be used for value added composites. These modified bioarchitectures will find new markets in high value, high performance products. Finally, engineered bioarchitectures will emerge where fiber component parts are taken apart and put back together with greater strength and much higher performance. Biomaterials will find 'new' markets once lost to plastics, metals and inorganics.

## 25.8 Sources of further information and advice

Fakirov, S. and D. Bhattacharyya. (2007). *Handbook of Engineering Biopolymers: Homopolymers, Blends and Composites*. Hanser Publishers, Munich.

Pickering, K.L. (2008). *Properties and Performance of Natural-Fiber Composites*. Woodhead Publishing Limited, Cambridge.

Placket, D. (2011). *Biopolymers: New Materials for Sustainable Films and Coatings*. John Wiley & Sons, Chichester.

Rowell, R.M. (ed.) (2005). *Handbook of Wood Chemistry and Wood Composites*. Taylor and Francis, Boca Raton, FL.

Rowell, R.M., F. Calderia and J.K. Rowell. (2010). *Sustainable Development in the Forest Products Industry*. Fernando Pessoa Press, Oporto, Portugal.

Thieneb, H., M. Irle and S. Milan. (2010). *Wood-based Panels: An Introduction for Specialists*. COST Action report, Brunel University Press, London, England.

## 25.9 References

Hill, C. (2006). *Wood Modification: Chemical, Thermal and Other Processes*. John Wiley & Sons, Chichester.

Hon, D.N-S. (1993). *Chemical Modification of Wood Materials*. Marcel Dekker, New York. Hon, D.N-S. (1993). *Chemical Modification of Wood Materials*. Marcel Dekker, New York.

National Plywood Association (1993) Particleboard, ANSI A208.1-1993. Gaithersburg, MD.

National Plywood Association (1994). Medium density fiberboard, ANSI A208.2-1994. Gaithersburg, MD.

Rowell, R.M. (ed.) (2005). *Handbook of Wood Chemistry and Wood Composites*. Taylor and Francis, Boca Raton, FL.

Rowell, R.M. (2006). Acetylation of wood: a journey from analytical technique to commercial reality. *Forest Products Journal*, 56(9): 4–12.

Rowell, R.M. (2008). Natural fibers: types and properties. In Pickering, K.L. (ed.) *Properties and Performance of Natural-Fiber Composites*. Woodhead Publishing Limited, Cambridge, 3–66.

# 26
## The use of biomass for packaging films and coatings

H. M. C. DE AZEREDO, Institute of Food Research, UK,
M. F. ROSA and M. DE SÁ M. SOUZA FILHO,
Embrapa Tropical Agroindustry, Brazil and
K. W. WALDRON, Institute of Food Research, UK

DOI: 10.1533/9780857097385.2.819

**Abstract**: Because of concerns involving the continual disposal of huge volumes of non-biodegradable food packaging materials, there has been an increasing tendency to replace petroleum-derived polymers with bio-based, environmentally friendly biodegradable macromolecules. There are a variety of biomass-derived structures which can be used as packaging materials, especially as films and coatings. Moreover, there are several biomass-derived compounds which can be used as additives for these materials such as plasticizers, crosslinking agents, and reinforcements, which can enhance physical properties and applicability of the materials for food packaging purposes. This review focuses on biomaterials which can be used to develop food packaging structures (especially films and coatings), their properties and interactions, and how they influence the packaging performance.

**Key words**: polysaccharides, proteins, lignocellulose, nanocomposites.

## 26.1 Introduction

People have been using natural polymeric materials such as silk, wool, cotton, wood and leather for centuries, but the advent of the petroleum-derived plastics at the beginning of the twentieth century provided the food industry with an increasingly wide variety of synthetic materials to be used as food packaging. However, fossil fuels are limited and non-renewable, and recycling is limited because of technical as well as economic difficulties. Less than 3% of the waste plastic worldwide gets recycled, compared with recycling rates of 30% for paper, 35% for metals and 18% for glass, according to Helmut Kaiser Consultancy (2012). Moreover, once discarded, petroleum-based plastics are generally non-biodegradable, in that they are resistant to microbial attack. This is due to their water insolubility, and the problem that evolutionary processes have not been sufficiently rapid to create new enzymes capable of degrading synthetic polymers during their relatively short existence in the natural environment (Mueller, 2006). In particular, the accumulated waste generated by the continuous and extensive disposal of food packaging has raised considerable concerns over their deleterious effects on wildlife and the environment. Even their incineration can produce toxic compounds such as furans and dioxins produced from burning polyvinylchloride (PVC) (Jayasekara *et al.*, 2005). In recent decades, the concerns surrounding conventional petroleum-based plastics have stimulated a focus of attention on natural macromolecules such as polysaccharides and proteins, because of their sustainable supply and biodegradability (Verbeek and van den Berg, 2010).

In 2011, the global use of biodegradable plastics was 0.85 million metric tons. BCC Research expects that the use of bioplastics will increase up to 3.7 million metric tons by 2016, a compound annual growth rate of 34.3% (BCC Research, 2012). According to studies by Helmut Kaiser Consultancy (2012), bioplastics are expected to cover approximately 25–30% of the total plastics market by 2020, and the market itself is estimated to reach over US$10 billion by 2020. Europe is one of the most important markets, partly due to an increasing environmental awareness, but also due to the limited amount of crude oil reserves.

Biodegradable materials and particularly renewable materials have been promoted as materials for use in food packaging, especially as flexible packaging, although applications for rigid packaging materials have also been mentioned (Mohareb and Mittal, 2007; Stepto, 2009).

Some biodegradable materials (actually most proteins and polysaccharides) are also edible, and can be used to develop edible packaging (films and coatings). Edible packaging materials are intended to be integral parts of foods and to be eaten with the products, thus they are also inherently biodegradable (Krochta, 2002). Edible films (or sheets) are stand-alone structures that are preformed separately from the food and then applied on the food surface or between food components, or sealed into edible pouches. Edible coatings, on the other hand, are formed directly onto the surface of the food products (Krochta and De Mulder-Johnston, 1997). Gel capsules, microcapsules and tablet coatings made from edible materials could also be considered as edible packaging (Janjarasskul and Krochta, 2010). Although edible packaging is not expected to replace conventional plastic packaging, they have an important role to play in the whole development of renewable packaging. This is because they can be used to extend food stability by reducing exchange of moisture, $O_2$, $CO_2$, lipids and flavour compounds between the food and the surrounding environment, thus increasing food stability. So, they help to improve the efficiency of food packaging and therefore reduce the amount of conventional packaging materials required for each application. Hence they have been included in this review.

The barrier requirements of an edible film or coating depend on their application and the properties of the food they are supposed to protect. For example, fresh fruits and vegetables are alive, respiring foods. Films or coatings intended for use on them should have low water vapour permeability so they reduce the desiccation rates, while the $O_2$ permeability should be low enough to reduce the respiration rates, extending the produce shelf life, but not low enough to create anaerobic conditions, leading to ethanol production and off-flavor formation. Nuts are especially susceptible to oxidation, thus the barrier against $O_2$ is the most important factor to provide an extended shelf life; barrier against UV light is also helpful to reduce oxidation rates, and moisture barrier reduces the absorption of water, which can lead to loss of the crunchy texture. Some edible coatings can be applied to food to be fried, so that the coatings act as a barrier to frying oil, reducing the oil absorption by the product and consequently its final fat content; in this case, coatings should be highly hydrophilic to have a good barrier against the hydrophobic oils. Edible films and coatings can also be applied between food components, such as between the crust and the sauce and toppings of a pizza, minimizing the moisture transfer from the sauce and toppings to the crust.

Another requirement for edible films and coatings is that they are tasteless, and do not interfere with the sensory characteristics of the food product. But there are cases where the edible films or coatings are supposed to have a characteristic desirable flavour, which should be compatible with the food to be coated. This is the case, for instance, for fruit purée-based edible films (Azeredo et al., 2009; Mild et al., 2011; Sothornvit and Pitak, 2007) and coatings (Azeredo et al., 2012b; Sothornvit and Rodsamran, 2008), fruit pomace-based edible films (Park and Zhao, 2006), and vegetable purée edible films (Wang et al., 2011).

There are a number of methods for evaluating the physical properties of the resulting film and coatings materials, such as tensile test analyses and dynamic mechanical thermal analysis (DMTA) for mechanical properties (Moates et al., 2001; Siracusa et al., 2008), determination of barrier properties (Siracusa et al., 2008), differential scanning calorimetry (DSC) for thermal properties (Abdorreza et al., 2011), FTIR spectroscopy for interactions between components (Paes et al., 2010; Yakimets et al., 2007), scanning electron microscopy for ultrastructural analyses (Bilbao-Sáinz et al., 2010), NMR spectroscopy for studying crosslinking mechanism (Zhang et al., 2010a), polymer–permeant interactions and their effects on polymer organization (Karbowiak et al., 2008, 2009), and dielectric thermal analysis (DETA) for dielectric behaviour of the films (Moates et al., 2001). However, those techniques have not been covered in this review, which is rather focused on the biomaterials which can be used for food packaging (especially films and coatings), their basic properties and interactions, and how they affect the end materials.

## 26.2 Components of packaging films and coatings from the biomass

All packaging films and coatings should have at least two components: a matrix, which usually consists of a macromolecule able to form a cohesive structure, and a plasticizer, usually required for reducing rigidity and brittleness inherent to most matrices. Additionally, some other components can be incorporated to improve the physical properties of films and coatings, such as their barrier and mechanical properties, or their resistance to moisture. Figure 26.1 presents a scheme of the basic and optional components of biomass-derived films and coatings.

## 26.3 Processes for producing bio-based films

Two basic technological approaches comprising wet and dry processes are generally used to produce biomaterials for packaging purposes.

The wet processes are based on separation of macromolecules from a solvent phase, usually by evaporation of the solvent. Usually the wet processes consist of casting a previously homogenized and vacuum degassed film forming dispersion (containing at least a biopolymer matrix, a solvent and usually a plasticizer) on a suitable base material (from which the film

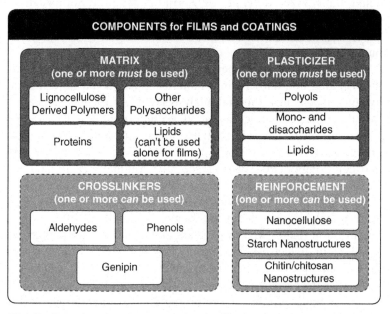

26.1 Basic and optional components for films and coatings.

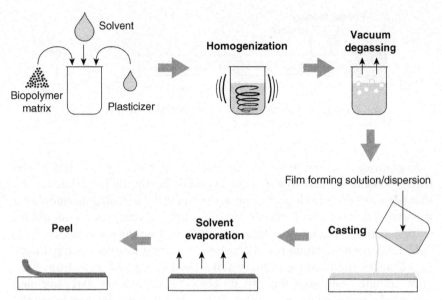

26.2 Basic scheme of the steps of a casting process to produce films.

can be easily peeled off) and later drying to a moisture content (usually 5–8%) which is optimal for peeling the film away (Lagrain et al., 2010; Tharanathan, 2003), as indicated in Fig. 26.2. Film formation generally involves inter- and intramolecular associations or crosslinking of polymer chains forming a network that entraps and immobilizes the solvent. The degree of cohesion depends on polymer structure, solvent used, temperature and the presence of other molecules such as plasticizers, reinforcements, etc. (Tharanathan, 2003).

Two continuous film casting methods are typically used to manufacture biopolymer films by wet processes: (a) casting on steel belt conveyors and (b) casting on a disposable substrate on a coating line. In (a), solutions are spread uniformly on a continuous steel belt that passes through a drying chamber. The dry film is then stripped from the steel belt and wound into mill rolls for later conversion. One of the advantages of steel belt conveyors is the ability to cast aqueous solutions directly onto the belt surface, optimizing uniformity, heat transfer and drying efficiency, while eliminating expense of a separate substrate such as polyester film or coated paper. In (b), known as web coating, solutions are spread uniformly onto a carrier web or substrate, usually a polyester film or coated paper, and the coated substrate is passed through a drying chamber. The dry film is wound into rolls while still adhering to the substrate, and is usually separated in a secondary operation (Rossman, 2009).

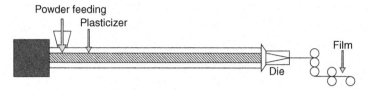

*26.3* Schematic representation of an extruder (adapted from Li *et al.*, 2011b).

In contrast, the dry processes use the thermoplastic properties of the macromolecules under low moisture conditions. The biomaterials can be shaped by existing plastic processing techniques (the so-called thermoplastic processing technologies), including thermoforming, compression moulding, extrusion, roller milling, or extrusion coating and lamination (Lagrain *et al.*, 2010). Heating amorphous polymers above their glass transition temperature ($T_g$) changes them into a rubbery state, making it possible to form films after cooling. Although the dry processes require more extensive and advanced equipment, they are efficient in large-scale production due to the low moisture contents, high temperatures, high pressures and short process times (Hernandez-Izquierdo and Krochta, 2008). Furthermore, the materials obtained are likely to exhibit more robust tensile characteristics compared with films cast with the use of plasticizers (Mangavel *et al.*, 2004) and lower water solubility, due to the creation of a highly crosslinked film network (Rhim and Ng, 2007).

Most conventional plastics, such as low density polyethylene films, are produced by extrusion. An extruder consists of a heated, fixed metal barrel containing one or two screws which convey the raw material through the heated barrel, from the feed end to the die (Fig. 26.3). The screws induce shear forces and increasing pressure along the barrel (Verbeek and van den Berg, 2010). Extrusion is one of the most important polymer processing techniques, offering several advantages over solution casting (Hernandez-Izquierdo and Krochta, 2008). However, many bio-based films are more difficult to produce by dry processes when compared to petroleum-based polymers, as they do not usually have defined melting points (due to their heterogeneous nature) and undergo decomposition upon heating (Tharanathan, 2003). There is a delicate balance required so the formulations resist the process conditions and at the same time achieve the desired film performance (Rossman, 2009). Starch (Pushpadass *et al.*, 2008; Thunwall *et al.*, 2008) and protein films (Hochstetter *et al.*, 2006; Hernandez-Izquierdo *et al.*, 2008; Kumar *et al.*, 2010) have been produced by extrusion. Some proteins which exhibit thermoplastic behaviour can be processed without further treatment, but other proteins and starch should be plasticized before processing (Rhim and Ng, 2007).

## 26.4 Processes for producing edible coatings

The processes described in the literature for producing and applying edible coatings are restricted to laboratory discontinuous and small-scale techniques. Similarly to a basic casting method, the coating dispersion is produced by homogenization and vacuum degassing of a film forming dispersion (or by melting, for lipid-based coatings). The dispersion is then applied directly on a food surface (or between food components) by dipping or spraying (Fig. 26.4). Dipping is more adequate when an irregular surface has to be coated. After dipping, excess coating material is allowed to drain from the product. Spraying is more adequate for thinner coatings materials

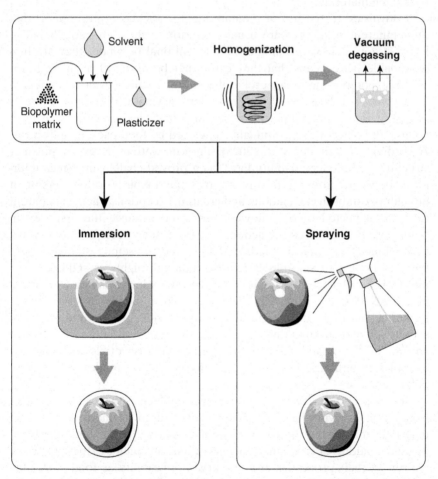

*26.4* Basic steps to prepare and apply a coating to a food surface.

and/or when only one side of the product is supposed to be coated (such as in pizza crusts, as exemplified before). Both spraying and dipping are followed by drying for polysaccharide or protein coatings, or by cooling for lipid-based coatings.

## 26.5 Products from biomass as film and/or coating matrices

There has been increasing interest in the search for new uses of biomass byproducts from the food industry. Many of these byproducts contain potential film-forming macromolecules such as polysaccharides or proteins which present opportunities for the design of bioplastics to be used as packaging materials.

A downside to the use of biomaterials for packaging purposes is that their inherent properties are usually inferior to those of petrochemical-based systems. However, unlike the conventional polymers, they are biodegradable. This means that they can either be disposed of through, for example, composting or anaerobic digestion, or might even be exploited as a source of fermentable sugars after enzymatic digestion. There is therefore increasing support for their use in order to reduce the huge volume of plastic waste continually generated by food packaging disposal. According to Van der Zee (2005), the correlations between polymer structure and biodegradability have been proved challenging, since interplays between different factors occur simultaneously, often making it difficult to establish correlations, and creating exceptions when an apparent rule was expected to be followed. For instance, since the first step in biodegradation involves the action of extracellular water-borne enzymes, hydrophilicity favours the biodegradation, and the semicrystalline nature tends to limit it to amorphous regions, although highly crystalline starch materials and bacterial polyesters are rapidly hydrolysed. Some chemical properties that are important include chemical bonds in the polymer backbone, position and chemical activity of side groups, and chemical activity of end groups. Linkages involving hetero atoms, such as ester and amide (or peptide) bonds are considered susceptible to enzymatic degradation, although there are exceptions such as polyamides and aromatic polyesters.

This chapter overviews some of the macromolecules that can be obtained from biomass byproducts which have been used as matrices for food packaging materials, as well as other biomass-derived compounds which can be used as additives, crosslinking agents or reinforcements to these matrices, improving their properties and potential applicability as food packaging materials.

## 26.5.1 Lignocellulosic biomass

*Cellulose and its derivatives*

Cellulose, the most abundant biopolymer, is formed by the repeated connection of D-glucose building blocks. Adjacent cellulose chains form a framework of aggregates (elementary fibrils) containing crystalline and amorphous regions; the crystalline regions are maintained by inter- and intramolecular hydrogen bonding. Several elementary fibrils can associate with each other to form cellulose crystallites, which are then held together by a monolayer of hemicelluloses, generating thread-like structures which are enclosed in a matrix of hemicellulose and protolignin, forming a natural composite referred to as cellulose microfibril (Ramos, 2003).

Cellulose represents about a third of the plant cell wall composition, and it is also produced by a family of sea animals called tunicates (sea squirts), by several species of algae, and by some species of bacteria and fungi (Charreau et al., 2013). Cellulose is an important structural component characterized by its hydrophilicity, chirality, biodegradability, broad capacity for chemical modification, and its formation of versatile semicrystalline fibre morphologies.

Together with starch, cellulose and its derivatives (such as ethers and esters) are the most important raw materials for elaboration of biodegradable and edible films (Peressini et al., 2003). Cellulose is an essentially linear natural polymer of $(1\rightarrow 4)$-$\beta$-D-glucopyranosyl units. Its tightly packed polymer chains and highly crystalline structure makes it insoluble in water. Water solubility can be conferred by etherification; the water-soluble cellulose ethers, including methyl cellulose (MC), hydroxypropyl cellulose (HPC), hydroxypropylmethyl cellulose (HPMC), and carboxymethyl cellulose (CMC), have good film-forming properties (Cha and Chinnan, 2004; Janjarasskul and Krochta, 2010).

Cellulose films prepared from aqueous alkali/urea solutions were reported to exhibit better oxygen barrier properties when compared to those of conventional cellophane and PVC (Yang et al., 2011). Cellulose derivatives have been used as coatings, extending the shelf life of avocados (Maftoonazad and Ramaswamy, 2005; Maftoonazad et al., 2008) and fresh eggs (Suppakul et al., 2010). Cellulose-based films have been produced by several companies such as Innovia (UK), FKuR (Germany) and Daicel Polymer (Japan).

*Hemicelluloses*

Hemicelluloses (HC) are heteropolysaccharides closely associated with cellulose in the plant cell walls (Mikkonen and Tenkanen, 2012). They are usually defined as the alkali-soluble material after the removal of pectic substances from plant cell walls (Sun *et al.*, 2004). According to Scheller and Ulvskov (2010), HC are polysaccharides in plant cell walls that have β-(1→4)-linked backbones with an equatorial configuration. They have different structures which may contain glucose, xylose, mannose, galactose, arabinose, fucose, as well as glucuronic and galacturonic acids in different proportions, depending on the source (Ebringerová and Heinze, 2000). Hemicelluloses (as well as lignin) cover cellulose microfibrils. The structural similarity between HC and cellulose generates a conformational homology leading to a hydrogen-bonded network between HC and cellulose microfibrils (O'Neill and York, 2003; Rose and Bennett, 1999).

According to their primary structure, hemicelluloses can be categorized into four main groups: xyloglycans (xylans), mannoglycans (mannans), β-glucans and xyloglucans (Ebringerová *et al.*, 2005). Xylans usually consist of a β(1→4)-D-xylopyranose backbone with side groups in position 2 or 3. Non-branched homoxylans occur in certain seaweeds, and heteroxylans include glucuronoxylans, arabinoxylans and more complex structures (Hansen and Plackett, 2008). Mannans comprise galactomannans and glucomannans. Galactomannans consist of a β(1→4)-linked mannopyranose backbone highly substituted with β(1→6)-linked galactopyranose residues (Wyman *et al.*, 2005), while glucomannans consist of alternating β-D-glucopyranosyl and β-D-mannopyranosyl units attached with (1→4) bonds (Mikkonen and Tenkanen, 2012). β-glucans have a D-glucopyranose backbone with mixed β linkages (1→3, 1→4) in different ratios (Ebringerová *et al.*, 2005). Finally, xyloglucans have a backbone of β(1→4)-linked D-glycopyranose residues with a distribution of D-xylopyranose in position 6 (Hansen and Plackett, 2008).

Some studies have been conducted on arabinoxylan-based films. Höije *et al.* (2005) obtained strong but highly hygroscopic arabinoxylan films from barley husks. β-glucan films were reported to be more compact than arabinoxylan films, with smaller nanopores, favouring the barrier properties (Ying *et al.*, 2011). Zhang *et al.* (2011) reported that the properties of arabinoxylan films were well correlated with the arabinose/xylose (Ara/Xyl) ratios. More crystalline films with lower water uptake resulted from lower arabinose contents. On the other hand, a lower chain mobility was observed in the amorphous parts for highly substituted xylans.

Galactomannan films have also been the subject of several studies. Mikkonen *et al.* (2007) reported that galactomannans with lower galactose content produced films with higher elongation at break and tensile strength, probably because galactomannans with fewer side chains can interact with other polysaccharides due to their long blocks of unsubstituted mannose units (Srivastava and Kapoor, 2005). Cerqueira *et al.* (2009) successfully used galactomannans to coat different tropical fruits, choosing the formulations taking into account parameters such as wettability, barrier to

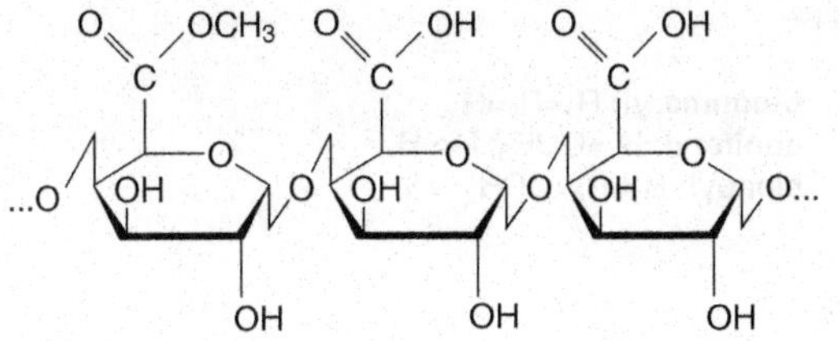

**26.5** A fragment of pectin containing esterified and non-esterified carboxyl groups of galacturonic acid.

gases and mechanical properties. Cheeses have been reported to have their shelf life extended by application of galactomannan-based coatings (Cerqueira et al., 2010; Martins et al., 2010).

*Pectins*

Pectins are water-soluble anionic heteropolysaccharides composed mainly of (1→4)-α-D-galactopyranosyluronic acid units, in which some carboxyl groups of galacturonic acid are esterified with methanol (Fig. 26.5). They are extracted from citrus peels and apple pomace by hot dilute mineral acid. Short hydrolysis times produce pectinic acids and high-methoxyl pectins (HMP), while extended acid treatment de-esterifies the methyl esters to pectic acids and generates low-methoxyl pectins (LMP). Commercial pectins are categorized according to their degree of esterification (DE), defined as the ratio of esterified to total galacturonic acid groups (Sriamornsak, 2003). HMP have a DE > 50 (usually > 69), whereas LMP have a DE < 50 (Farris et al., 2009). The ratio of esterified to non-esterified galacturonic acid determines the behaviour of pectin in food applications, since it affects solubility and gelation properties of pectin. HMP form gels with sugar and acid, whereas LMP form gels in the presence of divalent cations such as $Ca^{2+}$, which links adjacent LMP chains via ionic interactions, forming a tridimensional network (Janjarasskul and Krochta, 2010).

Pectin films, similarly to other polysaccharide films, have poor water resistance, and have been proposed for potential industrial uses where water binding either is not a problem or can provide specific advantages, such as edible bags for soup ingredients (Fishman et al., 2000). LMP films, on the other hand, when crosslinked with calcium ions, have not only improved water resistance, but also improved mechanical and barrier properties (Kang et al., 2005). Moreover, the presence of carboxyl groups carrying a negative charge at pH > $pK_a$ enables exploiting electrostatic interactions of pectin with positively charged counterparts (Farris et al., 2009).

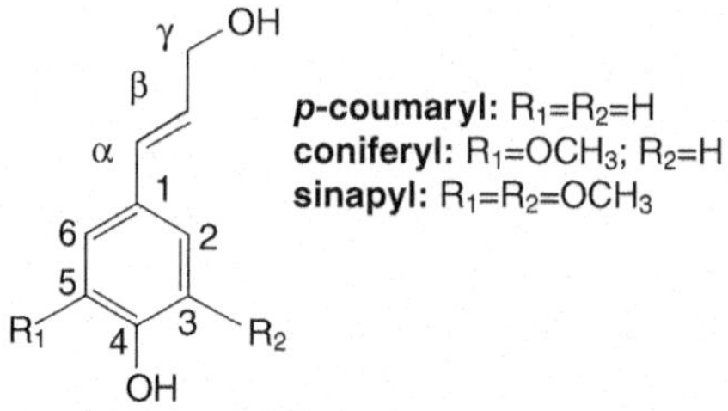

*26.6* Lignin monolignols (from Buranov and Mazza, 2008).

*Lignin*

Lignin is the second most abundant terrestrial biopolymer after cellulose, accounting for approximately 30% of the organic carbon in the biosphere (Boerjan *et al.*, 2003). It is associated with cellulose and hemicelluloses in plant cell walls, and is an abundant waste product in the pulp and paper industry.

Lignins are complex aromatic heteropolymers derived mainly from three hydroxycinnamyl alcohol monomers (monolignols) differing in their degree of methoxylation (Fig. 26.6): *p*-coumaryl, coniferyl and sinapyl alcohols (Boerjan *et al.*, 2003; Buranov and Mazza, 2008).

Lignin has some interesting properties to be used for packaging films, such as its small particle size, hydrophobicity and ability to form stable mixtures (Park *et al.*, 2008). Moreover, lignins have been shown to have efficient antibacterial and antioxidant properties (Ugartondo *et al.*, 2009). Lignin has been used as a film component in composites with gelatin (Núñez-Flores *et al.*, 2013; Ojagh *et al.*, 2011; Vengal and Srikumar, 2005), starch (Baumberger *et al.*, 1998; Vengal and Srikumar, 2005) and chitosan (Chen *et al.*, 2009). Baumberger *et al.* (1998) observed that starch films incorporated with up to 30% lignin presented higher elongation and water resistance than control starch films. On the other hand, lignin impaired the tensile strength of the films at high relative humidity (71%), reflecting the incompatibility between the hydrophilic starch and the hydrophobic lignin, which was bolstered by the water. Indeed, microscopic observations confirmed that the material consisted of two phases – a hydrophilic starch matrix filled with hydrophobic lignin aggregates. On the other hand, Chen *et al.* (2009) reported a good dispersion of lignin (up to 20%) in a chitosan matrix, evidenced by SEM, which was corroborated by FTIR results, indicating the existence of hydrogen bonding between chitosan and lignin. An FTIR spectroscopy study on gelatin-lignin films revealed strong protein conformational changes induced by lignin, producing a plasticizing effect, which was reflected in the mechanical and thermal properties (Núñez-Flores *et al.*, 2013).

A drawback from incorporating lignin in films is that they acquire a brownish colour (Mishra *et al.*, 2007; Núñez-Flores *et al.*, 2013). On the other hand, they acquire better light barrier properties (Núñez-Flores *et al.*, 2013), which could be of interest in food applications when ultraviolet-induced lipid oxidation is a problem.

## 26.5.2 Polysaccharides (other than cell wall polysaccharides)

Polysaccharides are hydrophilic polymers and therefore exhibit very low moisture barrier properties. They are of prime interest as matrices for biodegradable film formation because of their availability and rather low cost.

Most polysaccharides are neutral, although some gums are negatively charged. As a consequence of the large number of hydroxyl and other polar groups in their structure, hydrogen bonds play important roles in film formation and characteristics. Negatively charged gums, such as alginate, pectin and carboxymethyl cellulose, tend to present some different properties depending on the pH (Han and Gennadios, 2005).

Polysaccharide films are usually formed by disrupting interactions among polymer segments during coacervation and forming new intermolecular hydrophilic and hydrogen bonds upon evaporation of the solvent (Janjarasskul and Krochta, 2010). Because of their hydrophilicity, polysaccharide films provide a good barrier to $CO_2$ and $O_2$, hence they retard respiration and ripening of fruits (Cha and Chinnan, 2004). On the other hand, similarly to other hydrophilic materials, their high polarity determines their poor barrier to water vapour (Park and Chinnan, 1995) as well as their sensitivity to moisture, which may affect their functional properties (Janjarasskul and Krochta, 2010).

### Starches

Starches are polymers of D-glucopyranosyl, consisting of a mixture of the predominantly linear amylose and the highly branched amylopectin (Fig. 26.7). Native starch molecules arrange themselves in semi-crystalline granules in which amylose and amylopectin are linked by hydrogen bonding. When heat is applied to native starch in the presence of plasticizers such as water and glycerol, the granules swell and hydrate, which triggers the gelatinization process, characterized by loss of crystallinity and molecular order, followed by a dramatic increase in viscosity (Kramer, 2009). This transformation is named gelatinization and leads to the so-called thermoplastic starch (TPS) (Huneault and Li, 2007).

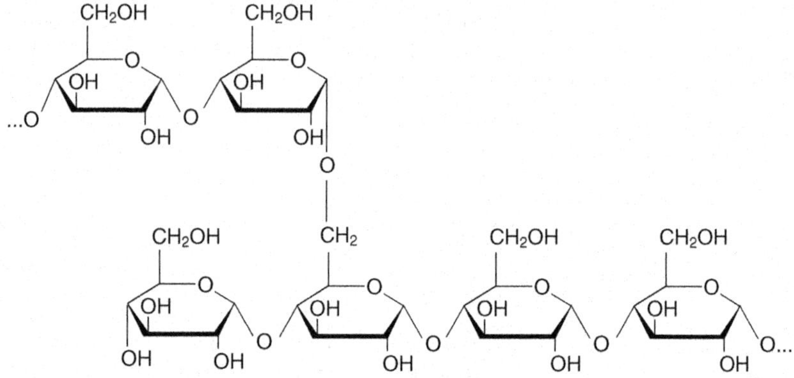

*26.7* Chemical structure of a starch fragment.

Amylose responds to the film-forming capacity of starches, since linear chains of amylose in solution tend to interact by hydrogen bonds and, consequently, amylose gels and films are stiff, cohesive and relatively strong. On the other hand, amylopectin films are brittle and non-continuous, since branched amylopectin chains in solution present little tendency to interact (Peressini *et al.*, 2003).

The first studies using starch in biodegradable food packaging were focused on substituting part of the synthetic matrix (usually polyethylene) by starch, but there were difficulties ascribed to chemical incompatibility between the polymers. Recently, studies on pure starch-based materials have predominated, usually focusing on two major drawbacks related to application of starch films. The first one is that native starch commonly exists as granules with about 15–45% crystallinity, and starch-based materials are susceptible to ageing and starch re-crystallization (retrogradation) (Forssell *et al.*, 1999; Ma *et al.*, 2006), which makes starch rigid and brittle during long-term storage, restricting its applications (Huang *et al.*, 2005). The second is their high hydrophilicity, causing its barrier properties to decrease with increasing relative humidity. Plasticizers are used to overcome the first drawback and improve material flexibility (Moates *et al.*, 2001; Peressini *et al.*, 2003; Mali *et al.*, 2004), but, since they are usually highly hydrophilic, presenting hydroxyl groups capable of interacting with water by hydrogen bonds, they tend to increase the moisture affinity of the films (Mali *et al.*, 2005).

Some companies have commercially produced starch-based packaging materials, such as Eco-Go (Thailand), Plantic (Australia) in a joint venture with Du Pont (USA), JMP (Australia), StarchTech (USA), Biome (UK), and BASF (Germany).

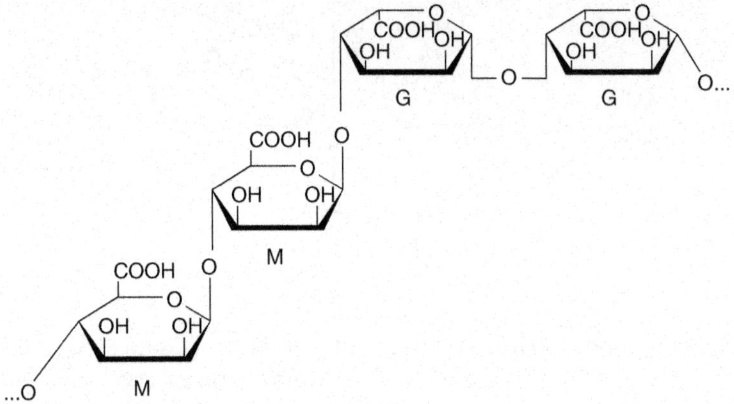

**26.8** Basic structure of alginates, containing units of mannuronic (M) and guluronic (G) acids.

## Alginates

Alginates, which are extracted from brown seaweeds, are salts of alginic acid, a linear co-polymer of D-mannuronic and L-guluronic acid monomers (Fig. 26.8), containing homogeneous poly-mannuronic and poly-guluronic acid blocks (M and G blocks, respectively) and MG blocks containing both uronic acids. The presence of carboxyl groups in each constituent residue (Ikeda *et al.*, 2000) enables sodium alginate to crosslink with di- or trivalent metal cations, especially calcium ions ($Ca^{2+}$), to produce strong gels or films (Cha and Chinnan, 2004). Calcium ions pull alginate chains together via ionic interactions, after which interchain hydrogen bonding occurs (Kester and Fennema, 1986). Films can be formed either from evaporating water from an alginate gel or by a two-step procedure involving drying of alginate solution followed by treatment with a calcium salt solution to induce crosslinking (Janjarasskul and Krochta, 2010). The strength and permeability of films may be altered by changing calcium concentration and temperature, among other factors (Kester and Fennema, 1986). Alginate films have been studied as edible coatings to be applied to a variety of foods such as fruits/vegetables (Fan *et al.*, 2009; Fayaz *et al.*, 2009) and meat products (Marcos *et al.*, 2008; Chidanandaiah *et al.*, 2009).

## Chitosan

Chitosan is a linear polysaccharide consisting of β-(1→4)-linked residues of *N*-acetyl-2-amino-2-deoxy-D-glucose (glucosamine) and 2-amino-2-deoxy-D-glucose (*N*-acetyl-glucosamine) (Fig. 26.9). It is produced from partial deacetylation of chitin, which is considered as the second most

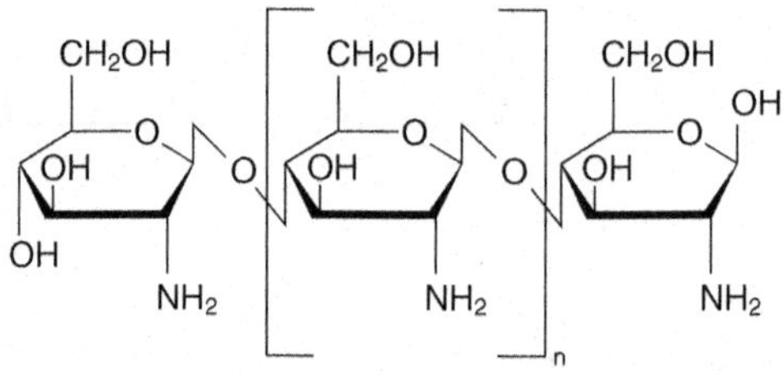

*26.9* Chemical structure of chitosan.

abundant polysaccharide in nature after cellulose (Dutta *et al.*, 2009; Aranaz *et al.*, 2010). Chitin is present in the exoskeleton of crustacea and insects, and can also be found in the cell wall of certain groups of fungi, particularly zygomycetes (Chatterjee *et al.*, 2005). It is usually extracted from crab and shrimp shells as a byproduct of the seafood industry. Since the deacetylation of chitin is usually incomplete, chitosan is a copolymer comprising D-glucosamine and *N*-acetyl-D-glucosamine with various fractions of acetylated units (Aranaz *et al.*, 2010).

Chitosan is soluble in diluted aqueous acidic solutions due to the protonation of $-NH_2$ groups at the C2 position (Aranaz *et al.*, 2010). The cationic character confers unique properties to the polymer, such as antimicrobial activity and the ability to carry and slow-release functional ingredients (Coma *et al.*, 2002). The charge density depends on the degree of deacetylation as well as the pH. Quaternization of the nitrogen atoms of amino groups has been a usual chitosan modification, whose objective is to introduce permanent positive charges along the polymer chains, providing the molecule with a cationic character independent of the aqueous medium pH (Curti *et al.*, 2003; Aranaz *et al.*, 2010).

Chitosan films have been proven effective in extending the shelf life of fruits (Hernández-Muñoz *et al.*, 2006; Chien *et al.*, 2007; Lin *et al.*, 2011; Ali *et al.*, 2011) and to have retarded microbial growth on fruit surfaces (Hernández-Muñoz *et al.*, 2006; Chien *et al.*, 2007; Campaniello *et al.*, 2008). The polycationic structure of chitosan probably interacts with the predominantly anionic components (lipopolysaccharides, proteins) of microbial cell membranes, especially Gram-negative bacteria (Helander *et al.*, 2001).

*Less conventional polysaccharides*

Several other polysaccharides are found in nature, and it is virtually impossible to mention all of them. But some examples are given in this section of less common polysaccharides which have been tested or suggested as food packaging materials.

## Carrageenans

Some red algae species (*Rhodophyta*) have a family of polysaccharides called carrageenans as cell wall polysaccharides (Van de Velde *et al.*, 2002). They are hydrophilic linear sulfated galactans consisting of alternating $(1\rightarrow 3)$-β-D-galactopyranose (G-units) and $(1\rightarrow 4)$-α-D-galactopyranose (D-units), forming a disaccharide repeating unit (Campo *et al.*, 2009). Carrageenan-based coatings have been proven to be efficient to increased stability of fresh-cut (Bico *et al.*, 2009) and fresh whole fruits (McGuire and Baldwin, 1998; Ribeiro *et al.*, 2007). Ribeiro *et al.* (2007) observed that carrageenan coatings resulted in lower weight loss and lower loss of firmness of strawberries when compared to starch coatings, probably reflecting a better moisture barrier of carrageenan coatings. Moreover, carrageenan films presented a significantly lower oxygen permeability than starch films.

## Tree exudates

Natural gums are obtained as exudates from different tree species, including gum acacia, cashew tree gum, and mesquite gum. The tree gums have been grouped into three types based on the nature of polysaccharide type, namely, arabinogalactans, substituted glucuronomannans or substituted rhamnogalacturonans (Sims and Furneaux, 2003). Cashew tree gum (CTG) is a heteropolysaccharide exudated from the cashew tree (*Anacardium occidentale*) bark (Miranda, 2009), whose composition is made up of galactose, glucose, arabinose, rhamnose and glucuronic acid (De Paula *et al.*, 1998). The greatest cultivation of cashew trees can be found in Brazil, and it is mainly focused on cashew nut production (Bezerra *et al.*, 2007). CTG films have been obtained by Carneiro-da-Cunha *et al.* (2009) and suggested to be applied as apple coatings. Azeredo *et al.* (2012a) obtained alginate–CTG blend films crosslinked with $CaCl_2$. CTG reduced tensile strength and barrier properties of the films, but favoured film extensibility. Some studies have described the use of mesquite gum as film-forming matrices (Osés *et al.*, 2009; Bosquez-Molina *et al.*, 2010). Gum acacia coatings have been proven to extend the shelf life of tomatoes (Ali *et al.*, 2010) and shiitake mushrooms (Jiang *et al.*, 2013).

## Cactus mucilages

Some hetero-polysaccharides can be obtained from cactus stems, which is waste from cactus pruning. The mucilage extracted from stems of prickly pear cactus (*Opuntia ficus-indica*), which constitutes about 14% of the cladode dry weight (Ginestra *et al.*, 2009), has been reported to contain residues of D-galactose, D-xylose, L-arabinose, L-rhamnose and D-galacturonic acid (McGarvie and Parolis, 1979). Del-Valle *et al.* (2005) studied the use of prickly pear mucilage as an edible coating to strawberries, and reported that coated strawberries presented extended physical and sensory stability when compared to uncoated ones.

Bacterial cellulose

Whilst 'cellulose' is a word originally given to the substance which constitutes a key load-bearing component of the cell wall of higher plants, bacterial cellulose (BC) is an extracellular product synthesized by bacteria belonging to some genera, its most efficient producers being the Gram-negative, acetic acid bacteria *Gluconacetobacter xylinum* (Iguchi *et al.*, 2000; Retegi *et al.*, 2010). These microfibril bundles have excellent intrinsic properties due to their high crystallinity, including a reported elastic modulus of 78 GPa (Guhados *et al.*, 2005). Compared with cellulose from plants, BC has important structural differences and it also possesses higher water-holding capacity, higher degree of polymerization (up to 8,000), and a finer web-like network (Klemm *et al.*, 2005). It is produced as a gel, and, although its solid portion is less than 1%, it is almost pure cellulose containing no lignin and other foreign substances (Iguchi *et al.*, 2000). Despite the identical chemical composition, BC is superior to plant cellulose owing to its purity, high elasticity, and nano-morphology with a large surface area (S. Chang *et al.*, 2012). It is an interesting biomaterial thanks to its fine network, excellent mechanical properties, high water-holding capacity, crystallinity and biocompatibility (Yan *et al.*, 2008; Putra *et al.*, 2008).

Retegi *et al.* (2010) obtained compression moulded bacterial cellulose (BC) films with different porosities, generated by different compression pressures. Higher pressures were found to produce films with better final mechanical properties. This behaviour was ascribed to the higher densification, reducing interfibrillar spaces, thus increasing the possibility of interfibrillar bonding zones.

### 26.5.3 Proteins

Proteins are widely available as biomass byproducts from plants (wheat gluten, maize zein, soybean proteins) and animals (collagen, gelatin, keratin, casein, whey proteins).

Proteins are linear, random copolymers built from up to 20 different monomers. The main mechanism of formation of protein films involves denaturation of the protein initiated by heat, solvents, or change in pH, followed by association of peptide chains through new intermolecular interactions (Janjarasskul and Krochta, 2010).

Proteins are distinguished from polysaccharides in that they have approximately 20 different amino acid monomers, rather than just a few or even one monomer, such as glucose in cellulose and starch. The amino acids are similar in that they contain an amino group ($-NH_2$) and a carboxyl group ($-COOH$) attached to a central carbon atom, but each amino acid has unique properties conferred by a different side group attached to the central carbon, which can be non-polar, polar uncharged, or polar (positively or negatively) charged at pH 7 (Cheftel et al., 1985; Krochta, 2002). While hydroxyl is the only reactive group in polysaccharides, proteins may be involved in several possible interactions and chemical reactions (Hernandez-Izquierdo and Krochta, 2008), such as chemical reactions through covalent (peptide and disulfide) bonds and non-covalent (ionic, hydrogen, and van der Waals) interactions. Moreover, hydrophobic interactions may occur between non-polar groups of amino acid chains (Kokini et al., 1994). In addition, protein-based films are considered to have high UV barrier properties, owing to their high content of aromatic amino acids which absorb UV light (Mu et al., 2012).

The most unique properties of proteins compared to other film-forming materials are heat denaturation, electrostatic charges, and amphiphilic character. Protein conformation can be affected by many factors, such as charge density and hydrophilic–hydrophobic balance (Han and Gennadios, 2005). Proteins have good film-forming properties and good adherence to hydrophilic surfaces. Protein-derived films provide good barriers to $O_2$ and $CO_2$ but not to water (Cha and Chinnan, 2004). Their barrier and mechanical properties are impaired by moisture owing to their inherent hydrophilic nature (Janjarasskul and Krochta, 2010).

The processing of protein films or other protein-based materials mostly requires three main steps: breaking of intermolecular bonds (non-covalent and covalent, if necessary) by chemical or physical rupturing agents; arranging and orienting polymer chains in the desired conformation; and allowing the formation of new intermolecular bonds and interactions stabilizing the film network (Jerez et al., 2007). Globular proteins are required to unfold and realign before a new three-dimensional network can be formed, and stabilized by new inter- and intra-molecular interactions (Verbeek and van den Berg, 2010).

A material to be extruded receives a considerable amount of mechanical energy, which may affect the characteristics of the final products. Extrusion requires the formation of a melt, which implies processing the protein above its softening point. Proteins have many different functional groups, and consequently a great variety of possible chain interactions that reduce molecular mobility and increase viscosity, resulting in a softening temperature which is often above decomposition temperature. Plasticizers can be useful to reduce the softening temperature, but protein extrusion is usually only possible within a limited range of processing conditions (Verbeek and van den Berg, 2010).

*Gelatin*

Not a naturally occurring protein, gelatin is produced by partial hydrolysis of collagen, which is the main constituent of animal skin, bone, and connective tissue. The insoluble collagen is treated with dilute acid or alkali, resulting in partial cleavage of the crosslinks, and the structure is broken down to such an extent that the soluble gelatin is formed (Karim and Bhat, 2009).

Gelatin is obtained mainly from pigskin and other mammalian sources, but the marine sources (fish skin and bone) have been increasingly looked upon as possible alternatives to bovine and porcine gelatin, as they are not associated with the risks of bovine spongiform encephalopathy (BSE, or 'mad cow disease') outbreaks, and because they are acceptable for religious groups which have restrictions on pig and cow derivatives (Karim and Bhat, 2009). Moreover, the fish industry sector has tried to find new outlets for their skin and bone byproducts. However, gelatins from cold water species, representing the majority of the industrial fisheries, present inferior physical properties (such as lower gelling and melting temperatures and lower modulus) when compared to mammalian gelatin (Haug *et al.*, 2004), which has been largely related to the lower contents of hydroxyproline in collagen from cold water fish species, since hydroxyproline is involved in interchain hydrogen bonding, which stabilizes the triple helical structure of collagen (Wasswa *et al.*, 2007). Gelatins from warm water fish species, like tilapia, on the other hand, have physical properties more similar to those of mammalian gelatins (Sarabia *et al.*, 2000). Moreover, fish gelatins contain more hydrophobic amino acids, so their films show significantly lower water vapour permeability when compared to films produced from mammalian gelatins (Avena-Bustillos *et al.*, 2006).

Gelatin contains a large amount of proline, hydroxyproline, lysine, and hydroxylysine, which can react in an aldol-condensation reaction to form intra- and intermolecular crosslinks among the protein chains (Dangaran *et al.*, 2009). On cooling and dehydration, gelatin films are formed with irreversible conformational changes (Badii and Howell, 2006; Dangaran *et al.*, 2009), with an increased concentration of triple helical structures. The triple helix structure is the basic unit of collagen, from which gelatin is derived. Thus, gelatin molecules partly revert back to the collagen structure during gelation (Gómez-Guillén *et al.*, 2002; Chiou *et al.*, 2006).

Several studies have described the successful use of gelatin to form films by casting (Chiou *et al.*, 2006; Jongjareonrak *et al.*, 2006; Hanani *et al.*, 2012b) as well as dry methods (Hanani *et al.*, 2012a; Krishna *et al.*, 2012).

*Milk proteins*

Total bovine milk proteins consist of about 80% casein and 20% whey proteins. Whey proteins are those which remain in milk serum after casein coagulation during cheese or casein manufacture. It is a mixture of proteins with diverse functional properties, the main ones being α-lactalbumins, β-lactoglobulins, bovine serum albumin, immunoglobulins, and proteose-peptones (Pérez-Gago and Krochta, 2002). Native whey proteins are globular complexes but become random coils upon denaturation and can form three-dimensional networks to produce biodegradable films (Zhou *et al.*, 2009). Since whey proteins have a high proportion of hydrophilic amino acid in their structure, whey protein films have low tensile strength and high water vapour permeability (McHugh *et al.*, 1994), but these properties might be improved by combining whey protein with materials with better tensile strength and hydrophobic properties, such as zein (Ghanbarzadeh and Oromiehi, 2008). The water vapour permeability of whey protein films was decreased by addition of beeswax, by both increasing the hydrophobic character of the film and decreasing the amount of hydrophilic plasticizer (glycerol) required (Talens and Krochta, 2005). Whey protein coatings have been used to increase the stability of fruits (Pérez-Gago *et al.*, 2003; Reinoso *et al.*, 2008), meat products (Shon and Chin, 2008), nuts (Min and Krochta, 2007) and eggs (Caner, 2005).

Casein is a phosphoprotein, which can be separated into various electrophoretic fractions such as α-(s1)- and α-(s2)-caseins, β-casein, and κ-casein, all of them having low solubility at low pH (pI = 4.6) (Müller-Buschbaum *et al.*, 2006). Caseins form films from aqueous solutions without further treatment due to their random-coil nature and their ability to form extensive hydrogen bonds (Lacroix and Cooksey, 2005), as well as electrostatic interactions (Gennadios *et al.*, 1994). Casein film solubility can be decreased by buffer treatments at the isoelectric point (Chen, 2002), by physical crosslinking using irradiation (Vachon *et al.*, 2000) and by chemical crosslinking using aldehydes (Ghosh *et al.*, 2009). Moreover, casein precipitated with high pressure $CO_2$ was reported to present lower water solubility than acid-precipitated casein (Dangaran *et al.*, 2006). Casein films have been used to increase stability of bread (Schou *et al.*, 2005).

*Zein*

Zein is the name given to the prolamin (alcohol-soluble) fraction of the maize proteins (Ghanbarzadeh and Oromiehi, 2008), representing about 50% of the maize endosperm proteins (Biswas *et al.*, 2005). Zein is commercially produced from maize gluten, a co-product of the maize starch production (Ghanbarzadeh and Oromiehi, 2008). Moreover, the worldwide growth of the bioethanol industry has resulted in a huge increase in zein availability, which has motivated the development of new applications for this protein (Biswas et al., 2009).

Zein films are usually prepared by dissolving zein in aqueous ethyl alcohol (Ghanbarzadeh et al., 2007). Zein is rich in non-polar amino acids, with low proportions of basic and acidic amino acids, which confers lower water vapour permeability and better water resistance to zein films when compared to other protein films (Dangaran et al., 2009; Ghanbarzadeh and Oromiehi, 2008).

*Gluten*

Wheat gluten proteins are a byproduct of starch extraction from wheat flour, which is commonly used as a functional ingredient, especially in bakery products. Wheat gluten consists of a mixture of proteins that can be classified into two types: the water insoluble glutenins, which comprise proteins with multiple peptide chains linked via interchain disulfide bonds, forming a continuous network that provides strength and elasticity; and the water soluble gliadins, which consist of single polypeptide chains associated via hydrogen bonding, hydrophobic interactions and intramolecular disulfide bonds, acting as a plasticizer to glutenin network and conferring viscosity to gluten (Balaguer et al., 2011a; Goesaert et al., 2005; Hernández-Muñoz et al., 2003).

The low quality gluten is unsuitable for flour improvement in breadmaking, but can be used for preparing plastic films, presenting adequate viscoelasticity, adhesiveness, thermoplasticity, and good film-forming properties, although the glutenins tend to aggregate upon shearing and heating (Balaguer et al., 2011a).

Gluten shows very low solubility in water, because of its low content of amino acids with ionizable side chains and high contents of non-polar amino acids and glutamine, which has a high hydrogen-bonding potential (Lagrain et al., 2010). A complex solvent system with basic or acidic conditions in the presence of alcohol and disulfide bond-reducing agents is required to prepare casting solutions (Cuq et al., 1998). Some aspects regarding gluten extrusion were approached by Lagrain et al. (2010). In contrast to what happens to thermoplastic materials, gluten viscosity does not decrease upon heating, but rather levels off or even increases due to crosslinking reactions. Therefore, gluten extrusion is only possible in a limited window of operating conditions ranging from the onset of protein flow to aggregation and eventually extensive depolymerization (Redl et al., 1999; Zhang et al., 2005), which occurs at rather low temperatures when compared to most synthetic polymers. Gluten materials are thus usually extruded between 80 and 130°C. The formation of a gluten network involves the dissociation and unraveling of the gluten proteins, which allows both glutenin and gliadin to recombine and crosslink through specific linkages in an oriented pattern (Hernandez-Izquierdo and Krochta, 2008), predominant reactions including SH oxidation and SH/SS interchange reactions leading to the formation of SS crosslinks (Lagrain et al., 2010). Overall, gluten-based materials are stable three-dimensional macromolecular networks stabilized by low-energy interactions and strengthened by covalent bonds such as SS bonds between cysteine residues (Lagrain et al., 2010).

### 26.5.4 Lipids

Unlike other macromolecules, lipids are not biopolymers, being unable to form cohesive, self-supporting films for packaging. So they are used either as coatings directly applied to food to provide a moisture barrier, or as components in stand-alone composite emulsion films, in which proteins or polysaccharides provide structural integrity and lipids respond for hydrophobic character, improving the water vapour barrier (Krochta, 2002). Moreover, the presence of lipids in composite formulations provides an appealing glassy finish to the material.

Polarity of lipids depends on the distribution of chemical groups, the length of aliphatic chains and presence and degree of unsaturation (Morillon et al., 2002). Lipids with longer chains and lower unsaturation and branching degrees have better barrier against water vapour, because of their lower polarity (Rhim and Shellhammer, 2005; Janjarasskul and Krochta, 2010). On the other hand, more polar lipids have more affinity for proteins and polysaccharides when in emulsions, producing more homogeneous films and avoiding the separation of a lipid phase (Fabra et al., 2011b).

Several types of lipids can be used for food coating or as film components, such as waxes, triglycerides, as well as di- and monoglycerides. Waxes, which are esters of long-chain aliphatic acids with long-chain aliphatic alcohols (Rhim and Shellhammer, 2005), are especially resistant to water diffusion, because of their very low content of polar groups (Kester and Fennema, 1986) and their high content in long-chain fatty alcohols and alkanes (Morillon et al., 2002). There are a variety of naturally occurring waxes, derived from vegetables (e.g., carnauba, candelilla and sugar cane waxes), minerals (e.g., paraffin and microcrystalline waxes), or animals – including insects (e.g., beeswax, lanolin and wool grease) (Rhim and Shellhammer, 2005). Some studies reported extension of shelf life of fruits resulting from application of wax-based coatings, including carnauba wax (Dang et al., 2008; Gonçalves et al., 2010) and candelilla wax (Saucedo-Pompa et al., 2009).

Most commonly, lipids have been used as hydrophobic components of emulsion films. Lipid-polysaccharide or lipid-protein emulsion films combine the complementary advantages of each component. Polysaccharides and proteins act as matrices, since they have the mechanical properties to form self-supporting films, and have good barrier against gases such as $O_2$ and $CO_2$, while lipids are useful to reduce water vapour permeability. García et al. (2000) demonstrated that lipid addition to starch films decreased their crystalline-amorphous ratio, which is expected to increase film diffusivity and consequently permeability. On the other hand, lipid addition also reduces the hydrophilic–hydrophobic ratio of films, which decreases their water solubility and therefore water vapour permeability. Emulsion films have been reported to enhance stability of fresh-cut (Chiumarelli and Hubinger, 2012) and whole fruits (Bosquez-Molina et al., 2003; Maftoonazad et al., 2007; Navarro-Tarazaga et al., 2011), vegetables (Conforti and Zinck, 2002), nuts (Mehyar et al., 2012), eggs (Wardy et al., 2011) and bakery products (Bravin et al., 2006).

## 26.6 Products from biomass as film plasticizers

Brittleness is an inherent quality attributed to most biopolymer films, due to extensive intermolecular forces such as hydrogen bonding, electrostatic forces, hydrophobic bonding, and disulfide bonding (Sothornvit and Krochta, 2005; Srinivasa et al., 2007). Plasticizers are required to break polymer–polymer interactions (such as hydrogen bonds and van der Waals forces), sometimes forming secondary bonds to polymer chains, causing the distance between adjacent chains to increase (Fig. 26.10), thus lowering the glass transition temperature $(T_g)$ and reducing film rigidity and brittleness.

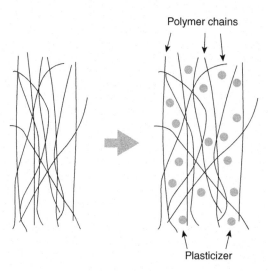

26.10 Representative scheme of the effect of plasticizers.

Moreover, plasticizers act as processing aids, since they lower the processing temperature, reduce sticking in moulds, and enhance wetting (Sothornvit and Krochta, 2005). On the other hand, plasticizers increase film permeability, and the decrease in cohesion negatively affects mechanical properties (Sothornvit and Krochta, 2005; Vieira et al., 2011).

Thanks to the low molecular size of plasticizers, they occupy spaces between polymer chains, reducing secondary forces. Most plasticizers contain hydroxyl groups which form hydrogen bonds with biopolymers, changing the three-dimensional polymer organization, reducing the energy required for mobility and the degree of hydrogen bonding between chains, resulting in increasing free volume and molecular mobility (Sothornvit and Krochta, 2005; Vieira et al., 2011).

The increased interest in bio-based packaging materials has been followed by a search for natural-based plasticizers, similarly biodegradable and of low toxicity. Commonly used plasticizers in biodegradable food packaging are polyols, mono-, di- or oligosaccharides, and fatty acids and other lipids.

### 26.6.1  Polyols

Glycerol can be produced either by microbial fermentation or synthesized chemically from petrochemical feedstock. Additionally, glycerol is a major byproduct of the increasing biodiesel production, thus creating a significant surplus resulting in a sharp decrease in glycerol prices. In this context, application of glycerol for value-added products is a necessity as well as an opportunity for the biodiesel industry (Johnson and Taconi, 2007; Yang et al., 2012). The use of glycerol as plasticizer in biopolymer-based films can be a way to help solve the existing surplus of this byproduct from biodiesel production.

Glycerol as well as other polyols, including sorbitol, propylene glycol, and polypropylene glycol, have great affinity for polysaccharide and protein films, because of their hydrophilicity. Several studies have demonstrated the effectiveness of polyols as plasticizers for polysaccharide-based and protein-based films and coatings (Bergo and Sobral, 2007; Jouki et al., 2013; Ramos et al., 2013; Tapia-Blácido et al., 2013). Qiao et al. (2011) used polyol mixtures including mixtures of glycerol and higher molecular weight polyols (HP) such as xylitol, sorbitol and maltitol. The increase of the molecular weight and the content of HP in the polyol mixture enhanced the thermal stability and mechanical strength of the resulting materials.

### 26.6.2  Mono- and disaccharides

It is not only molecular size, configuration and total number of functional hydroxyl groups that are important characteristics to be considered for an effective plasticizer, but also its compatibility with the film-forming polymer. Polymer–plasticizer compatibility is necessary to generate a homogeneous mixture without phase separation. It has been suggested that some monosaccharides work as plasticizers in polysaccharide films more effectively than polyols, because of the structure similarity between monosaccharides and polysaccharides (Zhang and Han, 2006). Indeed, Zhang and Han (2006) observed that starch films plasticized with monosaccharides (glucose, mannose and fructose) presented better overall physical properties when compared to starch films plasticized with polyols (glycerol and sorbitol). On the other hand, polyols (especially glycerol) presented better ability to lower $T_g$ of the films, indicating that polyols are more effective in terms of thermomechanical properties. Other studies have indicated monosaccharides and disaccharides as effective plasticizers in different polysaccharide-based films (Olivas and Barbosa-Cánovas, 2008; Piermaria et al., 2011). Whey protein films plasticized with sucrose presented excellent oxygen barrier properties, but sucrose tended to crystallize with time (Dangaran and Krochta, 2007).

### 26.6.3 Lipids

Generally, the purpose of adding lipids to films is to reduce their water vapour permeability and/or to provide an attractive gloss. Moreover, incorporating lipids in protein- or polysaccharide-based films may interfere with polymer chain-to-chain interactions and provide flexible domains within the film. Fabra et al. (2008) observed that oleic acid apparently interacted with the protein (sodium caseinate) matrix forming bonds through polar groups, modifiying the interaction balances in the protein network. The result can be a plasticizing effect, including reduction of film strength and increase of film flexibility, as described for whey protein films plasticized with beeswax (Talens and Krochta, 2005), caseinate films with oleic acid (Fabra et al., 2008), and gelatin films with stearic and oleic acids (Limpisophon et al., 2010).

The major drawback of using lipids as plasticizers is their low compatibility with most biopolymer matrices, because of their hydrophobicity. On the other hand, this same characteristic of lipids represents an advantage in which they reduce the water vapour permeability and moisture sensitivity of biopolymer materials.

## 26.7 Products from biomass as crosslinking agents for packaging materials

Chemical crosslinking is the process of linking polymer chains by covalent bondings, forming tridimensional networks (Fig. 26.11) which reduce the

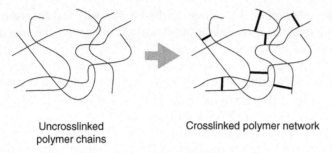

Uncrosslinked polymer chains     Crosslinked polymer network

*26.11* Schematic representation of crosslinking of polymer chains.

mobility of the structure and usually enhance its mechanical and barrier properties and its water resistance. Chemical crosslinking provides a mechanism for enhancing the performance of biopolymers. The most common crosslinking reagents are symmetrical bifunctional compounds with reactive groups with specificity for functional groups present on the matrix macromolecules (Balaguer *et al.*, 2011a). Low toxicity crosslinking agents have been explored nowadays for use in food packaging materials, such as phenolic compounds and genipin.

### 26.7.1 Phenolic compounds

Several natural phenolic compounds derived from plants have been used as crosslinkers to modify biopolymer films. Several potential interactions may be involved, such as hydrogen bonding, ionic, hydrophobic interactions, and covalent bonding, although covalent bonds are more rigid and thermally stable than other interactions (Zhang *et al.*, 2010a). The postulated chemical pathway involves oxidization of diphenol moieties of phenolic acids or other polyphenols, under alkaline conditions, producing quinone intermediates which react with nucleophiles from reactive amino acid groups such as sulfhydryl groups of cystine, amine groups of lysine and arginine, and amide groups of asparagine and glutamine, forming covalent C–N or C–S bonds between the phenolic ring and proteins. The thus regenerated hydroquinone can be reoxidized and bind a second protein chain, resulting in a crosslink (Strauss and Gibson, 2004; Zhang *et al.*, 2010a).

Caffeic and tannic acids were used as crosslinkers for gelatin (Zhang *et al.*, 2010a), and the use of high-resolution NMR technique confirmed the occurrence of chemical reactions between the phenolic groups in phenolic compounds and amino groups in gelatin, forming covalent C–N bonds; the crosslinking decreased the mobility of the gelatin matrix. A similar study by the same group (Zhang *et al.*, 2010b) indicated that the structure of a gelatin film crosslinked by tannic acid (3 wt%) was stable even under boiling, and that the crosslinking modification enhanced the mechanical properties of the protein.

Several other studies have reported beneficial effects from crosslinking protein matrices with phenolic compounds, such as tannic acid and sorghum condensed tannins (Emmambux *et al.*, 2004), procyanidin (He *et al.*, 2011) and ferulic acid (Ou *et al.*, 2005; Fabra *et al.*, 2011a).

Mathew and Abraham (2008) and Cao *et al.* (2007) reported significant increases in tensile strength and decreases in elongation at break resulting from adding ferulic acid to starch/chitosan films and gelatin-based films, respectively, which was ascribed to the crosslinking between ferulic acid and the polysaccharides or protein used. Ferulic acid is found as a naturally occurring component in cell walls, crosslinking polysaccharides (especially hemicelluloses) with each other and with other cell wall components such as lignin (Ng *et al.*, 1997; Parker *et al.*, 2005).

### 26.7.2 Genipin

Genipin (Fig. 26.12) is a hydrolysis product from genipiside, which is a component of traditional Chinese medicine, isolated from gardenia fruits (*Gardenial jasminoides* Ellis). According to studies by Sung *et al.* (1999), genipin is about 10,000 times less cytotoxic than glutaraldehyde.

Genipin reacts with nucleophilic groups such as amino groups, being an adequate crosslinking agent for protein films as well as chitosan. Mi *et al.* (2005) studied the reaction mechanism of chitosan with genipin, and found that genipin undertakes a ring-opening reaction to form an intermediate aldehyde group resulting from the nucleophilic attack by chitosan amino groups. The genipin molecules reacting with a nucleophilic reagent may further undergo polymerization (Fernandes *et al.*, 2013; Jin *et al.*, 2004; Yuan *et al.*, 2007). A dark-blue coloration appears in crosslinked materials exposed to air, which is associated with the oxygen radical-induced polymerization of genipin as well as its reaction with amino groups (Muzzarelli, 2009). Mi *et al.* (2005) observed that the colour of genipin crosslinked chitosan membranes varied from original transparent to bluish or brownish, depending on the pH value upon crosslinking. These authors ascribed the

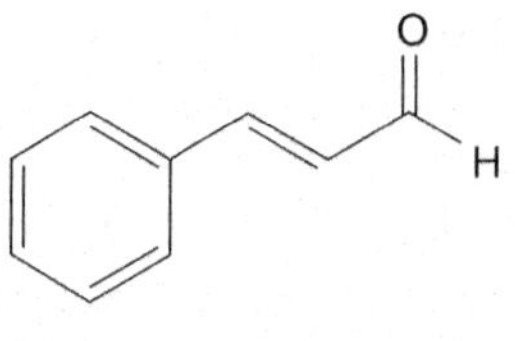

*26.12* Chemical structure of genipin (from Yuan *et al.*, 2007).

colour changes to establishment of different structures of crosslinked chitosan resulting from reaction of original genipin or polymerized genipin with primary amino groups on chitosan. The degrees of crosslinking of the genipin-crosslinked chitosan membranes depended significantly on their crosslinking pH values, being higher around pH 7.

Crosslinking with genipin has been reported to improve mechanical properties, water resistance of chitosan films, although they have turned dark bluish (Jin et al., 2004), which may be a market hindrance to many food packaging applications. Bigi et al. (2002) reported that gelatin films crosslinked with genipin presented higher Young's modulus, better thermal stability (reflected in higher denaturation temperature), and better water resistance than uncrosslinked films. After 1 month of storage in buffer solution, a small gelatin amount (about 2%) was lost from the films, but their mechanical, thermal and swelling properties were very close to those of gelatin films previously crosslinked by glutaraldehyde (Bigi et al., 2001).

Another geniposide derivative, aglycone geniposidic acid, has been used to crosslink chitosan films, improving their tensile strength and water vapour barrier, although reducing their elongation (Mi et al., 2006).

### 26.7.3 Aldehydes

Aldehydes such as glutaraldehyde, glyoxal or formaldehyde have been used as crosslinking agents to improve mechanical and barrier properties of protein films. However, due to concerns about possible toxic effects of such aldehydes, naturally occurring, less toxic aldehydes and other crosslinking agents have been explored for use in food packaging.

Cinnamaldehyde is an aromatic unsaturated aldehyde derived from cinnamon, consisting of a phenyl group attached to an unsaturated aldehyde (Fig. 26.13). It has been used as an antimicrobial agent in active packaging applications (Becerril et al., 2007), and it can also act as a crosslinking agent for proteins. Balaguer et al. (2011b) demonstrated the crosslinking effect of

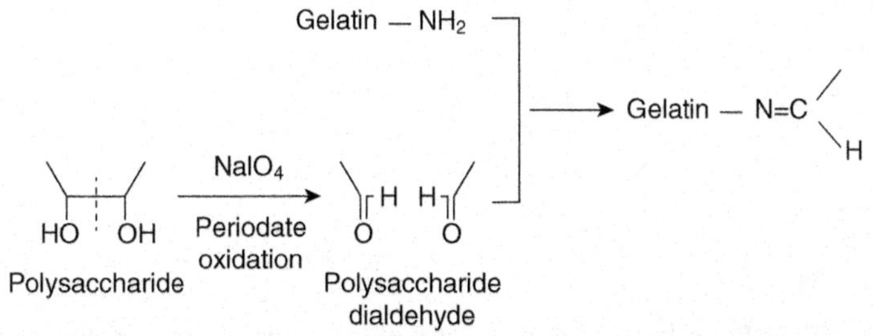

26.13 Chemical structure of cinnamaldehyde (from Balaguer et al., 2011b).

cinnamaldehyde for gliadin films, which was ascribed to the formation of intermolecular covalent bonds between polypeptide chains, polymerizing gliadins and reticulating the protein matrix. However, it is still uncertain which functional groups of proteins or other macromolecules have more potential to react with cinnamaldehyde, but Balaguer *et al.* (2011a) proposed a crosslinking mechanism involving the amino groups of proteins, although not ruling out the participation of other reactive groups.

Gliadin films presented significant improvements in their tensile strength and elastic modulus upon crosslinking with cinnamaldehyde. Such effects, as well as water resistance, were reported to be proportional to the cinnamaldehyde concentration (Balaguer *et al.*, 2011b). The crosslinked films did not disintegrate upon a 5-month immersion in water, although a weight loss was reported, indicating that part of the material was solubilized (Balaguer *et al.*, 2011a).

Dialdehyde polysaccharides have received attention as crosslinking agents of protein films. The oxidation of polysaccharides by periodate is characterized by the cleavage of the C2–C3 bond of glucose residues, resulting in the formation of two aldehyde groups per glucose unit, forming 2,3-dialdehyde polysaccharides (Li *et al.*, 2011a). The aldehyde groups can crosslink with ε-amino groups by C=N bonds, as in lysine or hydroxylysine side groups of gelatin (Fig. 26.14) to improve the properties of protein films (Dawlee *et al.*, 2005; Mu *et al.*, 2012). Dialdehyde starch (DAS) was used as a crosslinking agent for gelatin films (Martucci and Ruseckaite, 2009). Crosslinking with DAS up to 10 wt% enhanced moisture resistance and barrier properties of the films, but higher amounts of DAS conducted to phase separation, impairing transparency and tensile properties. Mu *et al.* (2012) reported that the addition of dialdehyde carboxymethylcellulose (DCMC) to gelatin films increased their tensile strength and thermal stability and reduced their water sensitivity, while keeping their transparency.

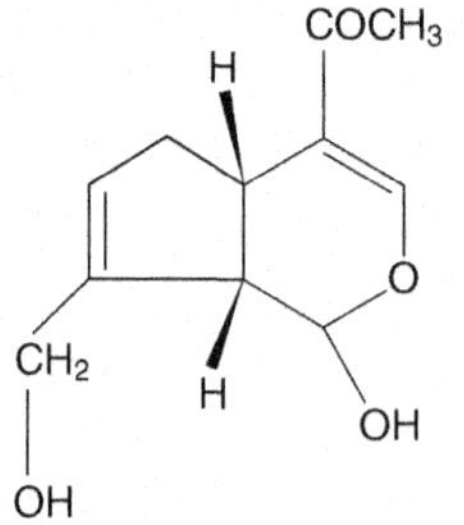

*26.14* Periodate oxidization of a glucose residue in a polysaccharide and the Schiff's reaction between gelatin and the dialdehyde polysaccharide (from Mu *et al.*, 2012).

## 26.8 Products from biomass as reinforcements for packaging materials

Polymer composites are mixtures of polymers with inorganic or organic fillers with certain geometries (fibres, flakes, spheres, particulates). When the fillers are nanoparticles, that is to say, when they have at least one nanosized dimension (up to 100 nm), the resulting material is a nanocomposite (Alexandre and Dubois, 2000). Polysaccharides are good candidates for renewable and biodegradable nanofillers, because of their partly crystalline structures, conferring good reinforcement effects (Le Corre et al., 2010).

### 26.8.1 From micro to nanoscale

The change of filler dimensions from micro to nanoscale brings about important advantages concerning the resulting composite materials. Because of their size, nanoparticles have larger surface area-to-volume ratio than their microscale counterparts. A uniform dispersion of nanoparticles leads to a very large matrix/filler interfacial area, which changes the molecular mobility, improving thermal, barrier and mechanical properties of the material. Fillers with a high ratio of the largest to the smallest dimension (i.e., aspect ratio), such as nanofibres, are particularly interesting because of their high specific surface area, providing better reinforcing effects (Azizi Samir et al., 2005; Dalmas et al., 2007). In addition, an interphase region of altered mobility surrounding each nanoparticle is induced by well-dispersed nanoparticles, resulting in a percolating interphase network playing an important role in improving the nanocomposite properties (Qiao and Brinson, 2009). According to Jordan et al. (2005), for a constant filler content, a reduction in particle size increases the number of filler particles, bringing them closer to one another, and causing the interface layers from adjacent particles to overlap, altering the bulk properties significantly.

De Moura et al. (2009) observed that the water vapour permeability of hydroxypropyl methylcellulose (HPMC) films reinforced with chitosan nanoparticles was affected by the size of the CsN – the smaller the particles, the lower the permeability. According to the authors, CsN tended to occupy the empty spaces in the pores of the HPMC matrix, thereby improving tensile and barrier properties.

### 26.8.2 Cellulosic reinforcements

Cellulose fibres are built up by smaller and mechanically stronger entities (cellulose nanoparticles) which can be extracted under proper conditions. Cellulose nanoparticles (i.e., cellulose elements having at least one dimension in the 1–100 nm range, here referred to as nanocellulose), are inherently a low cost and widely available material. Moreover, they are environmentally friendly, easy to recycle by combustion, and require low energy consumption in manufacturing (Klemm et al., 2005; Charreau et al., 2013).

A uniform dispersion of cellulose nanoparticles on a polymer matrix reduces the molecular mobility, changes the relaxation behaviour, and improves the overall thermal and mechanical properties of the material (Azizi Samir et al., 2005; Charreau et al., 2013; Qiu and Hu, 2013). In this context, nanocellulose has been presented as a promising reinforcing and barrier component for elaboration of low cost, lightweight, and high-strength bionanocomposites for food packaging purposes mainly due to its compatibility with the biopolymers (Helbert et al., 1996; Podsiadlo et al., 2005; Moon et al., 2011). Basically two different classes of nanocellulose can be obtained – whiskers and nanofibrils (Azizi Samir et al., 2005). The term 'whiskers' (or 'nanocrystals') is used to designate elongated crystalline rod-like nanoparticles, whereas the designation 'nanofibrils' is used for long flexible nanoparticles consisting of alternating crystalline and amorphous strings (Abdul Khalil et al., 2012).

Different approaches have been introduced to produce nanocellulose: top-down (nanometric structures are obtained by size reduction of bulk materials) and bottom-up approaches (nanostructures are built from individual atoms or molecules capable of self-assembling) (Yousefi et al., 2013). Any cellulosic material can be virtually considered as a potential source for top-down approaches, including crop residues and agroindustry non-food feedstock. However, variations in cellulose source and its preparation conditions lead to a broad spectrum of structures, properties and applicability, which affect performance of cellulose nanoreinforcements (Kvien and Oksman, 2007; Azeredo, 2009; Dufresne, 2012).

Acid hydrolysis has been the primary method for isolating nanocellulose, consisting basically in removing the amorphous regions present in the fibrils leaving the crystalline regions intact; the dimensions of the whiskers after hydrolysis depend on the percentage of amorphous regions in the bulk fibrils, which varies for each organism (Gardner et al., 2008). The morphology of the obtained nanowhiskers is influenced by acid-to-pulp ratio, reaction time, temperature and cellulose source. In spite of the widely varied dimensions (of 3–70 nm widths and 35–3000 nm lengths) reported from different cellulose sources and hydrolysis conditions, cellulose nanowhiskers typically consist of structures 200–400 nm in length and with an aspect ratio of about 10 (Beck-Candanedo et al., 2005; Elazzouzi-Hafraoui et al., 2008; Rosa et al., 2010). Other methods can be used to extract nanocellulose from the lignocellulosic sources, usually based on successive chemical and mechanical treatments, including high-pressure homogenization (Zimmermann *et al.*, 2010), electrospinning (Konwarh *et al.*, 2013), enzymatic hydrolysis (de Campos *et al.*, 2013), TEMPO-mediated oxidation (Isogai *et al.*, 2011), solvent-based isolation (Yousefi *et al.*, 2011), chemi-mechanical forces (Yousefi *et al.*, 2013), ultrasonication (Chen *et al.*, 2011; de Campos *et al.*, 2013), cryo crushing (Alemdar and Sain, 2008) and steam explosion (Deepa *et al.*, 2011). Such processes usually produce longer nanostructures (typically several micrometers in length), but with less uniform width (5–100 nm) (Siró and Plackett, 2010) and lower crystallinity (Iwamoto *et al.*, 2007).

Although the most important industrial source of cellulosic fibres is wood, crops such as flax, cotton, hemp, sisal and others, especially from by-products of these different plants (corn, wheat, rice, sorghum, barley, sugar cane, pineapple, bananas and coconut crops) are likely to become of increasing interest as sources of nanocellulose. These non-wood plants generally contain less lignin than wood and therefore pre-treatment processes are less demanding (Siró and Plackett, 2010; Rosa *et al.*, 2010).

In contrast to celluloses from plants, which require mechanical or chemo-mechanical processes to produce nanosized structures, the aforementioned bacterial cellulose (BC) is produced already as a nanomaterial by bacteria through cellulose biosynthesis and the building up of bundles of microfibrils (Nakagaito and Yano, 2005). In addition, BC is produced as a highly hydrated and relatively pure cellulose membrane and therefore no chemical treatments are needed to remove lignin and hemicelluloses, as is the case for plant celluloses (Siró and Plackett, 2010).

Nanocomposites for food packaging purposes have been developed by adding nanocellulose to polymers to enhance their physical and mechanical properties (Paula *et al.*, 2011; Azeredo *et al.*, 2012b; Abdollahi *et al.*, 2013; Zainuddin *et al.*, 2013). Nanocellulose has been reported to have a great effect in improving tensile strength and elastic modulus of polymers, especially at temperatures above the $T_g$ of the matrix polymer (Wu *et al.*, 2007; Azeredo *et al.*, 2012b; Zainuddin *et al.*, 2013; Abdollahi *et al.*, 2013). This effect is ascribed not only to the geometry and stiffness of the nanocellulose, but also to the formation of a fibril network within the polymer matrix, the cellulose fibres probably being linked through hydrogen bonds. Barrier properties of polymer films have also been observed to be improved by cellulose nanostructures (Sanchez-Garcia *et al.*, 2008; Paula *et al.*, 2011; Azeredo *et al.*, 2012b; Abdollahi *et al.*, 2013). The presence of crystalline fibres is thought to increase the tortuosity in the materials leading to slower diffusion processes and, hence, to lower permeability (Sanchez-Garcia *et al.*, 2008). Nanocellulose has also been reported to improve thermal properties of polymers (Petersson *et al.*, 2007; Paula *et al.*, 2011). The performance of nanocellulose has been reported to be strongly related to the content, dimensions and consequent aspect ratios of the nanostructures, as well as to the degree of matrix–cellulose interaction and percolation effects (Petersson and Oksman, 2006; Hubbe et al., 2008; Tang and Liu, 2008; Kim et al., 2009).

Since 2011, pilot plants have been opened for the production of nanocellulose in Sweden, Canada and the United States. These facilities make it possible to produce nanocellulose on a large scale for the first time, and it was an important step towards the industrialization of this technology.

### 26.8.3 Starch nanoreinforcement

Starch granules can be submitted to an extended-time hydrolysis at temperatures below that of gelatinization, when the amorphous regions are hydrolysed before the crystalline lamellae, which are more resistant to hydrolysis. The nanocrystals thus separated show platelet morphology with thicknesses of 6–8 nm (Kristo and Biliaderis, 2007). To prepare starch nanoparticles instead of nanocrystals, Ma et al. (2008) precipitated a starch solution within ethanol as the precipitant.

Similarly to cellulose nanocrystals, the reinforcing effect of starch nanocrystals (SNC) is usually ascribed to the formation of a percolating network maintained by hydrogen bonds above a given filler content (the percolation threshold) (Le Corre et al., 2010). Although not proven, this phenomenon was evidenced from experiments which indicated a changing behaviour above certain filler concentration (Angellier et al., 2005). It has been suggested that starch nanocrystals, similarly to nanoclays (which have a platelet morphology as well), create a tortuous diffusion pathway for permeant molecules through the nanocomposite materials, improving their barrier properties (Le Corre et al., 2010).

Kristo and Biliaderis (2007) reported that the addition of SNC to pullulan films resulted in improved tensile strength and modulus and decreased water vapour permeability. Moreover, the SNC promoted an increase in $T_g$, probably because of strong interactions of nanocrystals with one another and with the matrix, restricting chain mobility. Chen et al. (2008) observed that the tensile strength and elongation of polyvinyl alcohol (PVA) were only slightly improved by addition of SNC up to 10 wt% and, above this content, such properties were impaired by SNC. On the other hand, the properties of PVA nanocomposite with SNC were better than those of the composites with native starch, indicating that SNC presented a better dispersion and stronger interactions with the matrix than native starch granules.

### 26.8.4 Chitin/chitosan nanostructures

Chitin whiskers can be prepared by acid hydrolysis of chitin (Lu et al., 2004; Sriupayo et al., 2005), and have been successfully prepared from different chitin sources such as squid pens (Paillet and Dufresne, 2001), crab shells (Nair and Dufresne, 2003), and shrimp shells (Sriupayo et al., 2005). When the acid hydrolysis is followed by mechanical ultrasonication/disruption, chitin nanoparticles (nanospheres) can be formed rather than chitin whiskers (Chang et al., 2010b). Chitosan nanoparticles can be obtained by ionic gelation of chitosan, where the cationic amino groups of chitosan form electrostatic interactions with polyanionic crosslinking agents, such as tripolyphosphate (López-León et al., 2005).

Some authors reported beneficial effects from adding chitin whiskers to biopolymer films. The chitin whiskers added by Lu et al. (2004) to soy protein isolate (SPI) greatly improved the tensile properties and water resistance of the matrix. Sriupayo et al. (2005) reported that the whiskers improved the water resistance of chitosan films, and enhanced their tensile strength until a content of 2.96%, but impaired the strength when at higher contents. Similarly, Chang et al. (2010b) reported that chitin nanoparticles were uniformly dispersed and presented good interaction with a starch matrix when at low loading levels (up to 5%), improving its mechanical, thermal and barrier properties. However, aggregation of nanoparticles occurred at higher loading, impairing the performance of the matrix.

Chitosan nanoparticles have also been proven to be effective as nanoreinforcement to bio-based films. The incorporation of chitosan-tripolyphosphate (CS-TPP) nanoparticles significantly improved mechanical and barrier properties of hydroxypropyl methylcellulose (HPMC) films (De Moura et al., 2009). The authors attributed such effects to the nanoparticles filling discontinuities in the HPMC matrix. Chang et al. (2010a) added chitosan nanoparticles to a starch matrix. The tensile, barrier and thermal properties of the matrix were improved by low contents of the nanoparticles, when they were well dispersed in the matrix. Such effects were ascribed to close interactions between the nanoparticles and the matrix, due to their chemical similarities. However, higher nanoparticle loads (8% w/w) resulted in their aggregation in the nanocomposites, impairing the physical properties of the materials.

## 26.9 Future trends

### 26.9.1 Polymer surface modifications

Since most biopolymers have limitations for their applications as packaging materials, such as water sensitivity, brittleness or poor mechanical performance, chemical modification techniques can sometimes be used to generate new biomaterials with improved properties. Shi et al. (2011) grafted lauryl chloride groups onto zein molecule through an acylation reaction, and obtained seven-fold increase in elongation at break of modified zein.

Moreover, the modification increased the zein hydrophobicity, suggesting that the end material presented probably decreased moisture sensitivity and water vapour permeability.

The combination between ultraviolet light and ozone (UV-O treatment) has been suggested as an effective method to modify polymer surfaces, leading to oxidation reactions at the polymer surface, while the bulk properties, such as thermal, barrier and mechanical properties of the polymer, may not be altered (Shi et al., 2009). Shi et al. (2009) used UV-O treatment to control hydrophilicity of zein films. The treatment converted some of the surface methyl groups mainly to carbonyl groups, decreasing the water contact angles and increasing the surface hydrophilicity of zein films. Moreover, the authors suggested that, once the surface has been modified, several active compounds could have been linked to the functional groups formed, and the polymer would be not only a barrier material but a carrier to active compounds, constituting an active packaging material.

### 26.9.2 Active and bioactive biopackaging

Conventional food packaging systems are supposed to passively protect the food, acting as barriers between the food and the surrounding environment. On the other hand, an active food packaging may be defined as a system that not only acts as a barrier but also interacts with the food in some desirable way, for example by releasing desirable compounds (such as antimicrobial or antioxidant agents), or by removing a detrimental factor (such as oxygen or water vapour), usually to improve food stability, that is to say, to better maintain food quality and safety. The compounds added more frequently are antimicrobials, such as chitosan (Shen et al., 2010), acids (Guillard et al., 2009), phenolic compounds (Cerisuelo et al., 2012; Arrieta et al., 2013) and antimicrobial peptides (Sanjurjo et al., 2006; Gómez-Guillén et al., 2011).

More recently, the concept of bioactive packaging, in which a food packaging or coating has a potential to enhance food impact over consumer health, has been proposed by Lopez-Rubio et al. (2006). Enclosing bioactive compounds within packaging instead of directly incorporating them into food presents some industrial benefits, such as increasing retention of bioactives, reducing some incompatibility problems between the bioactive and the food matrix, and reducing changes to food sensory properties (Lopez-Rubio et al., 2006). Antioxidant packaging materials could be included in both active and bioactive packaging concepts, since antioxidant compounds usually have alleged benefits both for food stability and for consumer health. Some studies have described the incorporation of antioxidant compounds or extracts to biopolymer films and coatings, such as phenolic compounds to zein films (Arcan and Yemenicioglu, 2011), α-tocopherol to chitosan films (Martins *et al.*, 2012), curcumin and ascorbyl dipalmitate to cellulose-based films (Sonkaew *et al.*, 2012), and ferulic acid and α-tocopherol to sodium caseinate films (Fabra *et al.*, 2011a). Other studies have described the development of probiotic films in coatings, based on the incorporation of lactic acid bacteria, such as *Bifidobacterium lactis* incorporated to alginate and gellan films for coating of fresh fruits (Tapia *et al.*, 2007), *Lactobacillus sakei* added to caseinate films to control *Listeria monocytogenes* in fresh beef (Gialamas *et al.*, 2010), *Lactobacillus acidophilus* in starch-based coatings for breads (Altamirano-Fortoul *et al.*, 2012) and *L. acidophilus* and *Bifidobacterium bifidum* incorporated to gelatin coatings for fish (López de Lacey *et al.*, 2012).

The active and bioactive compounds are most frequently simply added to the film-forming formulation, constituting one of the formulation components, but not being chemically linked to the matrix polymer. However, some studies have suggested the covalent immobilization of active compounds onto functionalized polymer surfaces, with many possible applications for food packaging purposes. Biopolymer films may be functionalized in two basic steps, namely, the treatment of the polymer surface in order to produce reactive functional groups, and the reaction of such groups with an active compound (Kugel *et al.*, 2011). Sometimes the functionalization can be favoured by an intermediary step, such as grafting a polyfunctional agent onto the polymer surface, increasing the density of available functional groups, or by a spacer molecule to reduce steric hindrances (Goddard and Hotchkiss, 2007). An active compound which is covalently immobilized onto the packaging material is not supposed to be released, but becomes effective when in contact with the food surface (Han, 2003). The active compounds to be tethered can be enzymes, antimicrobials, biosensors, bioreactors, etc. Some studies have described the covalent attachment of active compounds onto the surface of biopolymer materials to be applied as food packaging or coatings, such as lysozyme onto chitosan coatings (Lian *et al.*, 2012) and *N*-halamine onto chitosan films (Li *et al.*, 2013).

## 26.10 Conclusion

The world biomass is rich in macromolecules with film-forming abilities which can be explored to develop materials for food packaging and coating purposes. Several studies have been conducted describing the development of food packaging materials from biomaterials, especially polysaccharides and proteins. However, to convert macromolecules from biomass into materials with both processability and performance compatible with petrochemical-based ones, research needs to address several challenges. Important gaps in knowledge remain on the structure and functionality of biomaterials. Moreover, chemical and physical changes during processing of biomaterials need to be better understood, so that the processing conditions are improved. Another major issue is to minimize the impact of environmental conditions (especially humidity) on material performance. Finally, biomaterials need to be compatible with their petrochemical counterparts in terms of price, in order to penetrate the market. One of the main challenges to make biomaterials commercially viable is their processing by conventional processing techniques used for petroleum-based polymers such as extrusion and injection moulding. The industrial application of edible coatings to food surfaces also requires developments of techniques and equipment adequate to each kind of food to be coated.

Moreover, nanotechnology has demonstrated a great potential to expand the use of biodegradable polymers, since the addition of nanofillers has led to improvements in overall performance of biopolymers, making them more competitive in a market dominated by non-biodegradable materials. However, there are still important safety concerns about nanotechnology applications to food contact materials. Considering the tiny dimensions of nanofillers, it is reasonable to assume that they might migrate from the packaging material to food. Although the properties and safety of most starting materials in their bulk form are usually well known, their nano-sized counterparts frequently exhibit different properties, because their small sizes would allow them to cross more barriers through the body, while their high surface area increases their reactivity. Hence, detailed information is still required to assess the potential toxicity and environmental effects of nanofillers to be incorporated to food packaging materials.

Finally, the biopackaging field has adopted technologies to improve the performance and/or to add functionalities to biopolymer-based materials, so their range of applications can be widened, and they can become more competitive in the polymer market.

## 26.11 Acknowledgements

The authors wish to acknowledge financial support from EMBRAPA (Brazil) and the BBSRC (UK).

## 26.12 References

Abdollahi, M, Alboofetileh, M, Rezaei, M and Behrooz, R (2013), 'Comparing alginate nanocomposite films reinforced with organic and/or inorganic nanofillers for food packaging', *Food Hydrocoll.*, 32, 416–424.

Abdorreza, M N, Cheng, L H and Karim, A A (2011), 'Effects of plasticizers on thermal properties and heat sealability of sago starch films', *Food Hydrocoll.*, 25, 56–60.

Abdul Khalil, H P S, Bhat, A H and Ireana Yusra, A F (2012), 'Green composites from sustainable cellulose nanofibrils: a review', *Carboh. Polym.*, 87, 963–979.
Alemdar, A and Sain, M (2008), 'Biocomposites from wheat straw nanofibers: morphology, thermal and mechanical properties', *Compos. Sci. Technol.*, 68, 557–565.
Alexandre, M and Dubois, P (2000), 'Polymer-layered silicate nanocomposites: preparation, properties and uses of a new class of materials', *Mater. Sci. Eng.*, 28, 1–63.
Ali, A, Maqbool, M, Ramachandran, S and Alderson, P G (2010), 'Gum arabic as a novel edible coating for enhancing shelf-life and improving postharvest quality of tomato (*Solanum lycopersicum* L.) fruit', *Postharv. Biol. Technol.*, 58, 42–47.
Ali, A, Muhammad, M T, Sijam, K and Siddiqui, Y (2011), 'Effect of chitosan coatings on the physicochemical characteristics of Eksotika II papaya (*Carica papaya* L.) fruit during cold storage', *Food Chem.*, 124, 620–626.
Altamirano-Fortoul, R, Moreno-Terrazas, R, Quezada-Gallo, A and Rosell, C M (2012), 'Viability of some probiotic coatings in bread and its effect on the crust mechanical properties', *Food Hydrocoll.*, 29, 166–174.
Angellier, H, Molina-Boisseau, S and Dufresne, A (2005), 'Mechanical properties of waxy maize starch nanocrystals reinforced natural rubber', *Macromol.*, 38, 9161–9170.
Aranaz, I, Harris, R and Heras, A (2010), 'Chitosan amphiphilic derivatives: chemistry and applications', *Curr. Org. Chem.*, 14, 308–330.
Arcan, I and Yemenicioglu, A (2011), 'Incorporating phenolic compounds opens a new perspective to use zein films as flexible bioactive packaging materials', *Food Res. Int.*, 44, 550–556.
Arrieta, M P, Peltzer, M A, Garrigós, M C and Jiménez, A (2013), 'Structure and mechanical properties of sodium and calcium caseinate edible active films with carvacrol', *J. Food Eng.*, 114, 486–494.
Avena-Bustillos, R J, Olsen, C W, Olson, D A, Chiou, B, Yee, E, Bechtel, P J and McHugh, T H (2006), 'Water vapor permeability of mammalian and fish gelatin films', *J. Food Sci.*, 71, 202–207.
Azeredo, H M C (2009), 'Nanocomposites for food packaging applications', *Food Res. Int.*, 42, 1240–1253.
Azeredo, H M C, Mattoso, L H C, Wood, D, Williams, T G, Avena-Bustillos, R J and McHugh, T H (2009), 'Nanocomposites edible films from mango puree reinforced with cellulose nanofibers', *J. Food Sci.*, 74, N31–N35.
Azeredo, H M C, Magalhães, U S, Oliveira, S A, Ribeiro, H L, Brito, E S and De Moura, M R (2012a), 'Tensile and water vapour properties of calcium-crosslinked alginate-cashew tree gum films', *Int. J. Food Sci, Technol.*, 47, 710–715.
Azeredo, H M C, Miranda, K W E, Rosa, M F, Nascimento, D M and De Moura, M R (2012b), 'Edible films from alginate-acerola puree reinforced with cellulose whiskers', *LWT – Food Sci. Technol.*, 46, 294–297.
Azizi Samir, M A S, Alloin, F and Dufresne, A (2005), 'Review of recent research into cellulosic whiskers, their properties and their application in nanocomposite field', *Biomacromol.*, 6, 612–626.
Badii, F and Howell, N K (2006), 'Fish gelatin: structure, gelling properties and interaction with egg albumen proteins', *Food Hydrocoll.*, 20, 630–640.

Balaguer, M P, Gómez-Estaca, J, Gavara, R and Hernandez-Munoz, P (2011a), 'Biochemical properties of bioplastics made from wheat gliadins cross-linked with cinnamaldehyde', *J. Agric. Food Chem.*, 59, 13212–13220.

Balaguer, M P, Gómez-Estaca, J, Gavara, R and Hernandez-Munoz, P (2011b), 'Functional properties of bioplastics made from wheat gliadins modified with cinnamaldehyde', *J. Agric. Food Chem.*, 59, 6689–6695.

Baumberger, S, Lapierre, C, Monties, B and Della Valle, G (1998), 'Use of kraft lignin as filler for starch films', *Polym. Degrad. Stabil.*, 59, 273–277.

BCC Research (2012), 'Global markets and technologies for bioplastics', Report code PLS050B, available at http://www.bccresearch.com/report/bioplastics-markets-technologies-pls050b.html.

Becerril, R, Gomez-Lus, R, Goni, P, Lopez, P and Nerin, C (2007), 'Combination of analytical and microbiological techniques to study the antimicrobial activity of a new active food packaging containing cinnamon or oregano against *E. coli* and *S. aureus*', *Anal. Bioanal. Chem.*, 388, 1003–1011.

Beck-Candanedo, S, Roman, M and Gray, D G (2005), 'Effect of reaction conditions on the properties and behavior of wood cellulose nanocrystal suspensions', *Biomacromol.*, 6, 1048–1054.

Bergo, P and Sobral, P J A (2007), 'Effects of plasticizer on physical properties of pigskin gelatin films', *Food Hydrocoll.*, 21, 1285–1289.

Bezerra, M A, Lacerda, C F, Gomes Filho, E, Abreu, C E B and Prisco, J T (2007), 'Physiology of cashew plants grown under adverse conditions', *Brazil. J. Plant Physiol.*, 19, 449–461.

Bico, S L S, Raposo, M F J, Morais, R M S C and Morais, A M M B (2009), 'Combined effects of chemical dip and/or carrageenan coating and/or controlled atmosphere on quality of fresh-cut banana', *Food Control*, 20, 508–514.

Bigi, A, Cojazzi, G, Panzavolta, S, Roveri, N and Rubini, K (2001), 'Mechanical and thermal properties of gelatin films at different degrees of glutaraldehyde crosslinking', *Biomater.*, 22, 763–768.

Bigi, A, Cojazzi, G, Panzavolta, S, Roveri, N and Rubini, K (2002), 'Stabilization of gelatin films by crosslinking with genipin', *Biomater.*, 23, 4827–4832.

Bilbao-Sáinz, C, Avena-Bustillos, R J, Wood, D F, Williams, T G and McHugh, T H (2010), 'Composite edible films based on hydroxypropyl methylcellulose reinforced with microcrystalline cellulose nanoparticles', *J. Agric. Food Chem.*, 58, 3753–3760.

Biswas, A, Sessa, D J, Lawton, J W, Gordon, S H and Willett, J L (2005), 'Microwave-assisted rapid modification of zein by octenyl succinic anhydride', *Cereal Chemistry*, 82, 1–3.

Biswas, A, Selling, G W, Woods, K K and Evans, K (2009), 'Surface modification of zein films', *Ind. Crops Prod.*, 30, 168–171.

Boerjan, W, Ralph, J and Baucher, M (2003), 'Lignin biosynthesis', *Annu. Rev. Plant Biol.*, 54, 519–546.

Bosquez-Molina, E, Guerrero-Legarreta, I and Vernon-Carter, E J (2003), 'Moisture barrier properties and morphology of mesquite gum–candelilla wax based edible emulsion coatings', *Food Res. Int.*, 36, 885–893.

Bosquez-Molina, E, Tomás, S A and Rodríguez-Huezo, M E (2010), 'Influence of $CaCl_2$ on the water vapor permeability and the surface morphology of mesquite gum based edible films', *LWT – Food Sci. Technol.*, 43, 1419–1425.

Bravin, B, Peressini, D and Sensidoni, A (2006), 'Development and application of polysaccharide–lipid edible coating to extend shelf-life of dry bakery products', *J. Food Eng.*, 76, 280–290.

Buranov, A U and Mazza, G (2008), 'Lignin in straw of herbaceous crops', *Ind. Crops Prod.*, 28, 237–259.

Campaniello, D, Bevilacqua, A, Sinigaglia, M and Corbo, M R (2008), 'Chitosan: antimicrobial activity and potential applications for preserving minimally processed strawberries', *Food Microbiol.*, 25, 992–1000.

Campo, V L, Kawano, D F, Silva Jr, D B and Carvalho, I (2009), 'Carrageenans: biological properties, chemical modifications and structural analysis – a review', *Carboh. Polym.*, 77, 167–180.

Caner, C (2005), 'Whey protein isolate coating and concentration effects on egg shelf life', *J. Sci. Food Agric.*, 85, 2143–2148.

Cao, N, Fu, Y and He, J (2007), 'Mechanical properties of gelatin films cross-linked, respectively, by ferulic acid and tannin acid', *Food Hydrocoll.*, 21, 575–584.

Carneiro-da-Cunha, M G, Cerqueira, M A, Souza, B W S, Souza, M P, Teixeira, J A and Vicente, A A (2009), 'Physical properties of edible coatings and films made with a polysaccharide from *Anacardium occidentale* L.', *J. Food Eng.*, 95, 379–385.

Cerisuelo, J P, Alonso, J, Aucejo, S, Gavara, R and Hernández-Muñoz, P (2012), 'Modifications induced by the addition of a nanoclay in the functional and active properties of an EVOH film containing carvacrol for food packaging', *J. Membr. Sci.*, 423–424, 247–256.

Cerqueira, M A, Lima, A M, Teixeira, J A, Moreira, R A and Vicente, A A (2009), 'Suitability of novel galactomannans as edible coatings for tropical fruits', *J. Food Eng.*, 94, 372–378.

Cerqueira, M A, Sousa-Gallagher, M J, Macedo, I, Rodriguez-Aguilera, R, Souza, B W S, Teixeira, J A and Vicente, A A (2010), 'Use of galactomannan edible coating application and storage temperature for prolonging shelf-life of "Regional" cheese', *J. Food Eng.*, 97, 87–94.

Cha, D S and Chinnan, M S (2004), 'Biopolymer-based antimicrobial packaging – a review', *Crit. Rev. Food Sci. Nutr.*, 44, 223–237.

Chang, P R, Jian, R, Yu, J and Ma, X (2010a), 'Fabrication and characterisation of chitosan nanoparticles/plasticized-starch composites', *Food Chem.*, 120, 736–740.

Chang, P R, Jian, R, Yu, J and Ma, X (2010b), 'Starch-based composites reinforced with novel chitin nanoparticles', *Carboh. Polym.*, 80, 420–425.

Chang, S T, Chen, L C, Lin, S B and Chen, H H (2012), 'Nano-biomaterials application: morphology and physical properties of bacterial cellulose/gelatin composites via crosslinking', *Food Hydrocoll.*, 27, 137–144.

Charreau, H, Foresti, M L and Vázquez, A (2013), 'Nanocellulose patents trends: a comprehensive review on patents on cellulose nanocrystals, microfibrillated and bacterial cellulose', *Recent Pat. Nanotechnol.*, 7, 56–80.

Chatterjee, S, Adhya, M, Guha, A K and Chatterjee, B P (2005), 'Chitosan from *Mucor rouxii*: production and physico-chemical characterization', *Proc, Biochem.*, 40, 395–400.

Cheftel, J C, Cuq, J L and Lorient, D (1985), 'Amino acids, peptides, and proteins' in *Food Chemistry*, ed. O.R. Fennema, Marcel Dekker, New York, pp. 245–369.

Chen, H (2002), 'Formation and properties of casein films and coatings' in *Protein-Based Films and Coatings*, ed. A. Gennadios, CRC Press, Boca Raton, FL, pp. 181–211.

Chen, L, Tang, C, Ning, N, Wang, C, Fu, Q and Zhang, Q (2009), 'Preparation and properties of chitosan/lignin composite films', *Chin. J. Polym. Sci.*, 27, 739–746.

Chen, W S, Yu, H P and Liu, Y X (2011), 'Preparation of millimeter-long cellulose I nanofibers with diameters of 30–80 nm from bamboo fibers', *Carboh. Polym.*, 86, 453–461.

Chen, Y, Cao, X, Chang, P R and Huneault, M A (2008), 'Comparative study on the films of poly(vinyl alcohol)/pea starch nanocrystals and poly(vinyl alcohol)/native pea starch', *Carbohy. Polym.*, 73, 8–17.

Chidanandaiah, Keshri, R C and Sanyal, M K (2009), 'Effect of sodium alginate coating with preservatives on the quality of meat patties during refrigerated ($4 \pm 1°C$) storage', *J. Muscle Foods*, 20, 275–292.

Chien, P J, Sheu, F and Yang, F H (2007), 'Effects of edible chitosan coating on quality and shelf life of sliced mango fruit', *J. Food Eng.*, 78, 225–229.

Chiou, B S, Avena-Bustillos, R J, Shey, J, Yee, E, Bechtel, P J, Iman, S H, Glenn, G M and Orts, W J (2006), 'Rheological and mechanical properties of cross-linked fish gelatins', *Polym.*, 47, 6379–6386.

Chiumarelli, M and Hubinger, M D (2012), 'Stability, solubility, mechanical and barrier properties of cassava starch – Carnauba wax edible coatings to preserve fresh-cut apples', *Food Hydrocoll.*, 28, 59–67.

Coma, V, Martial-Gros, A, Garreau, S, Copinet, A, Salin, F and Deschamps, A (2002), 'Edible antimicrobial films based on chitosan matrix', *J. Food Sci.*, 67, 1162–1169.

Conforti, F D and Zinck, J B (2002), 'Hydrocolloid-lipid coating affect on weight loss, pectin content, and textural quality of green bell peppers', *J. Food Sci.*, 67, 1360–1363.

Cuq, B, Gontard, N and Guilbert, S (1998), 'Proteins as agricultural polymers for packaging production', *Cereal Chem.*, 75, 1–9.

Curti, E, Britto, D and Campana-Filho, S P (2003), 'Methylation of chitosan with iodomethane: effect of reaction conditions on chemoselectivity and degree of substitution', *Macromol. Biosci.*, 3, 571–576.

Dalmas, F, Cavaillé, J Y, Gauthier, C, Chazeau, L and Dendievel, R (2007), 'Viscoelastic behavior and electrical properties of flexible nanofiber filled polymer nanocomposites: influence of processing conditions', *Compos. Sci. Technol.*, 67, 829–839.

Dang, K T H, Singh, Z and Swinny, E E (2008), 'Edible coatings influence fruit ripening, quality, and aroma biosynthesis in mango fruit', *J. Agric. Food Chem.*, 56, 1361–1370.

Dangaran, K L and Krochta, J M (2007), 'Preventing the loss of tensile, barrier and appearance properties caused by plasticizer crystallization in whey protein films', *Int. J. Food Sci. Technol.*, 42, 1094–1100.

Dangaran, K L, Cooke, P and Tomasula, P M (2006), 'The effect of protein particle size reduction on the physical properties of $CO_2$-precipitated casein films', *J. Food Sci.*, 71, E196–E201.

Dangaran, K, Tomasula, P M and Qi, P (2009), 'Structure and formation of protein-based edible films and coatings' in *Edible Films and Coatings for Food Applications*, ed. M.E. Embuscado and K.C. Huber, Springer, New York, pp. 25–56.

Dawlee, S, Sugandhi, A, Balakrishnan, B, Labarre, D and Jayakrishnan, A (2005), 'Oxidized chondroitin sulfate-cross-linked gelatin matrixes: a new class of hydrogels', *Biomacromol.*, 6, 2040–2048.

de Campos, A, Correa, A C, Cannella, D, Teixeira, E M, Marconcini, J M, Dufresne, A, Mattoso, L H C, Cassland, P and Sanadi, A R (2013), 'Obtaining nanofibers from curauá and sugarcane bagasse fibers using enzymatic hydrolysis followed by sonication', *Cellulose*, 20, 1491–1500.

De Moura, M R, Aouada, F A, Avena-Bustillos, R J, McHugh, T H, Krochta, J M and Mattoso, L H C (2009), 'Improved barrier and mechanical properties of novel hydroxypropyl methylcellulose edible films with chitosan/tripolyphosphate nanoparticles', *J. Food Eng.*, 92, 448–453.

De Paula, R C M, Heatley, F and Budd, P M (1998), 'Characterisation of *Anacardium occidentale* exudate polysaccharide', *Polym. Int.*, 45, 27–35.

Deepa, B, Abraham, E, Cherian, B M, Bismarck, A, Blaker, J J, Pothan, L A, Leao, A L, Souza, S F and Kottaisamy, M (2011), 'Structure, morphology and thermal characteristics of banana nano fibers obtained by steam explosion', *Bioresource Technol.*, 102, 1988–1997.

Del-Valle, V, Hernández-Muñoz, P, Guarda, A and Galotto, M J (2005), 'Development of a cactus-mucilage edible coating (*Opuntia ficus indica*) and its application to extend strawberry (*Fragaria ananassa*) shelf-life', *Food Chem.*, 91, 751–756.

Dufresne, A (2012), *Nanocellulose: From Nature to High Performance Tailored Materials*, Walter de Gruyter, Berlin.

Dutta, P K, Tripathi, S, Mehrotra, G K and Dutta, J (2009), 'Perspectives for chitosan based antimicrobial films in food applications', *Food Chem.*, 114, 1173–1182.

Ebringerová, A and Heinze, T (2000), 'Xylan and xylan derivatives – biopolymers with valuable properties, 1: Naturally occurring xylans structures, isolation procedures and properties', *Macromol. Rapid Commun.*, 21, 542–556.

Ebringerová, A, Hromadkova, Z and Heinze, T (2005), 'Hemicellulose', *Adv. Polym. Sci.*, 186: 1–67.

Elazzouzi-Hafraoui, S, Nishiyama, Y, Putaux, J L, Heux, L, Dubreuil, F and Rochas, C (2008), 'The shape and size distribution of crystalline nanoparticles prepared by acid hydrolysis of native cellulose', *Biomacromol.*, 9, 57–65.

Emmambux, M N, Stading, M and Taylor, J R N (2004), 'Sorghum kafirin film property modification with hydrolysable and condensed tannins', *J. Cereal Sci.*, 40, 127–135.

Fabra, M J, Talens, P and Chiralt, A (2008), 'Tensile properties and water vapor permeability of sodium caseinate films containing oleic acid–beeswax mixtures', *J. Food Eng.*, 85, 393–400.

Fabra, M J, Hambleton, A, Talens, P, Debeaufort, F and Chiralt, A (2011a), 'Effect of ferulic acid and α-tocopherol antioxidants on properties of sodium caseinate edible films', *Food Hydrocoll.*, 25, 1441–1447.

Fabra, M J, Pérez-Masiá, R, Talens, P and Chiralt, A (2011b), 'Influence of the homogenization conditions and lipid self-association on properties of sodium caseinate based films containing oleic and stearic acids', *Food Hydrocoll.*, 25, 1112–1121.

Fan, Y, Xu, Y, Wang, D, Zhang, L, Sun, L and Zhang, B (2009), 'Effect of alginate coating combined with yeast antagonist on strawberry (*Fragaria* × *ananassa*) preservation quality', *Postharv. Biol. Technol.*, 53, 84–90.

Farris, S, Schaich, K M, Liu, L S, Piergiovanni, L and Yam, K L (2009), 'Development of polyion-complex hydrogels as an alternative approach for the production of bio-based polymers for food packaging applications: a review', *Trends Food Sci. Technol.*, 20, 316–332.

Fayaz, A M, Balaji, K, Girilal, M, Kalaichelvan, P T and Venkatesan, R (2009), 'Mycobased synthesis of silver nanoparticles and their incorporation into sodium alginate films for vegetable and fruit preservation', *J. Agric. Food Chem.*, 57, 6246–6252.

Fernandes, S C, Santos, D M P O and Vieira, I C (2013), 'Genipin-cross-linked chitosan as a support for laccase biosensor', *Electroanal.*, 25, 557–566.

Fishman, M L, Coffin, D R, Konstance, R P and Onwulata, C I (2000), 'Extrusion of pectin/starch blends plasticized with glycerol', *Carboh. Polym.*, 41, 317–325.

Forssell, P M, Hulleman, S H D, Myllarinen, P J, Moates, G K and Parker, R (1999), 'Ageing of rubbery thermoplastic barley and oat starches', *Carboh. Polym.*, 39, 43–51.

García, M A, Martino, M N and Zaritky, N E (2000), 'Lipid addition to improve barrier properties of edible starch-based films and coatings', *J. Food Sci.*, 65, 941–947.

Gardner, D J, Oporto, G S, Mills, R and Azizi Samir, M A S (2008), 'Adhesion and surface issues in cellulose and nanocellulose', *J. Adhes Sci. Technol.*, 22, 545–567.

Gennadios, A, McHugh, T, Weller, C and Krochta, J M (1994), 'Edible coatings and films based on proteins' in *Edible Coatings and Films to Improve Food Quality*, ed. J. M. Krochta, E. A. Baldwin and M. Nisperos-Carriedo, Technomic, Lancaster, PA, pp. 231–247.

Ghanbarzadeh, B and Oromiehi, A R (2008), 'Biodegradable biocomposite films based on whey protein and zein: barrier, mechanical properties and AFM analysis', *Int. J. Biol. Macromol.*, 43, 209–215.

Ghanbarzadeh, B, Musavi, M, Oromiehie, A R, Rezayi, K, Razmi, E and Milani, J (2007), 'Effect of plasticizing sugars on water vapour permeability, surface energy and microstructure properties of zein films', *LWT – Food Sci. Technol.*, 40, 1191–1197.

Ghosh, A, Ali, M A and Dias, G J (2009), 'Effect of cross-linking on microstructure and physical performance of casein protein', *Biomacromol.*, 10, 1681–1688.

Gialamas, H, Zinoviadou, K G, Biliaderis, C G and Koutsoumanis, K P (2010), 'Development of a novel bioactive packaging based on the incorporation of *Lactobacillus sakei* into sodium-caseinate films for controlling *Listeria monocytogenes* in foods', *Food Res. Int.*, 43, 2402–2408.

Ginestra, G, Parker, M L, Bennett, R N, Robertson, J, Mandalari, G, Narbad, A, Lo Curto, R B, Bisignano, G, Faulds, C B and Waldron, K W (2009), 'Anatomical, chemical, and biochemical characterization of cladodes from prickly pear [*Opuntia ficus-indica* (L.) Mill.]', *J. Agric. Food Chem.*, 57, 10323–10330.

Goddard, J M and Hotchkiss, J H (2007), 'Polymer surface modification for the attachment of bioactive compounds', *Prog. Polym. Sci.*, 698–725.

Goesaert, H, Brijs, K, Veraverbeke, W S, Courtin, C M, Gebruers, K and Delcour, J A (2005), 'Wheat flour constituents: how they impact bread quality, and how to impact their functionality', *Trends Food Sci. Technol.*, 16, 12–30.

Gómez-Guillén, M C, Turnay, J, Fernandez-Diaz, M D, Ulmo, N, Lizarbe, M A and Montero, P (2002), 'Structural and physical properties of gelatin extracted from different marine especies: a comparative study', *Food Hydrocoll.*, 16, 25–34.

Gómez-Guillén, M C, Giménez, B, López-Caballero, M E and Montero, M P (2011), 'Functional and bioactive properties of collagen and gelatin from alternative sources: a review', *Food Hydrocoll.*, 25, 1813–1827.

Gonçalves, F P, Martins, M C, Silva Jr, G J, Lourenço, S A and Amorim, L (2010), 'Postharvest control of brown rot and *Rhizopus* rot in plums and nectarines using carnauba wax', *Postharv. Biol. Technol.*, 58, 211–217.

Guhados, G, Wan, W K and Hutter, J L (2005), 'Measurement of the elastic modulus of single bacterial cellulose fibers using atomic force microscopy', *Langmuir*, 21, 6642–6646.

Guillard, V, Issoupov, V, Redl, A and Gontard, N (2009), 'Food preservative content reduction by controlling sorbic acid release from a superficial coating', *Innov. Food Sci. Emerg. Technol.*, 10, 108–115.

Han, J H (2003), 'Antimicrobial food packaging' in *Novel Food Packaging Techniques*, ed. R. Ahvenainen, Woodhead Publishing Limited, Cambridge, pp. 50–70.

Han, J H and Gennadios, A (2005), 'Edible films and coatings: a review' in *Innovations in Food Packaging*, ed. J. H. Han, Elsevier, London, pp. 239–262.

Hanani, Z A N, Beatty, E, Roos, Y H, Morris, M A and Kerry, J P (2012a), 'Manufacture and characterization of gelatin films derived from beef, pork and fish sources using twin screw extrusion', *J. Food Eng.*, 113, 606–614.

Hanani, Z A N, Roos, Y H and Kerry, J P (2012b), 'Use of beef, pork and fish gelatin sources in the manufacture of films and assessment of their composition and mechanical properties', *Food Hydrocoll.*, 29, 144–151.

Hansen, N M L and Plackett, D (2008), 'Sustainable films and coatings from hemicelluloses: a review', *Biomacromol.*, 9, 1493–1505.

Haug, I J, Draget, K I and Smidsrod, O (2004), 'Physical and rheological properties of fish gelatin compared to mammalian gelatin', *Food Hydrocoll.*, 18, 203–213.

He, L, Mu, C, Shi, J, Zhang, Q, Shi, B and Lin, W (2011)., 'Modification of collagen with a natural cross-linker, procyanidin', *Int. J. Biol. Macromol.*, 48, 354–359.

Helander, I M, Nurmiaho-Lassila, E L, Ahvenainen, R, Rhoades, J and Roller, S (2001), 'Chitosan disrupts the barrier properties of the outer membrane of Gram-negative bacteria', *Int. J. Food Microbiol.*, 71, 235–244.

Helbert, W, Cavaillé, C Y and Dufresne, A (1996), 'Thermoplastic nanocomposites filled with wheat straw cellulose whiskers. Part I: processing and mechanical behaviour', *Polym. Compos.*, 17, 604–611.

Helmut Kaiser Consultancy (2012), 'Bioplastics market worldwide 2010/11-2015-2020-2025', available at http://www.hkc22.com/bioplastics.html.

Hernandez-Izquierdo, V M and Krochta, J M (2008), 'Thermoplastic processing of proteins for film formation – a review', *J. Food Sci.*, 73, 30–39.

Hernandez-Izquierdo, V M, Reid, D S, McHugh, T H, Berrios, J J and Krochta, J M (2008), 'Thermal transitions and extrusion of glycerol-plasticized protein mixtures', *J. Food Sci.*, 73, E169–E175.

Hernández-Muñoz, P, Kanavouras, A, Ng, P K W and Gavara, R (2003), 'Development and characterization of biodegradable films made from wheat gluten protein fractions', *J. Agric. Food Chem.*, 51, 7647–7654.

Hernández-Muñoz, P, Almenar, E, Ocio, M J and Gavara, R (2006), 'Effect of calcium dips and chitosan coatings on postharvest life of strawberries (*Fragaria ananassa*)', *Postharvest Biol. Technol.*, 39, 247–253.

Hochstetter, A, Talja, R A, Helen, H J, Hyvonen, L and Jouppila, K (2006), 'Properties of gluten-based sheet produced by twin-screw extruder', *LWT – Food Sci. Technol.*, 39, 893–901.

Höije, A, Grondahl, M, Tommeraas, K and Gatenholm, P (2005), 'Isolation and characterization of physicochemical and material properties of arabinoxylans from barley husks', *Carbohy. Polym.*, 61, 266–275.

Huang, M, Yu, J and Ma, X (2005), 'Ethanolamine as a novel plasticiser for thermoplastic starch', *Polym. Degrad. Stab.*, 90, 501–507.

Hubbe, M A, Rojas, O J, Lucia, L A and Sain, M (2008), 'Cellulosic nanocomposites: a review', *Bioresources*, 3, 929–980.

Huneault, M A and Li, H (2007), 'Morphology and properties of compatibilized polylactide/thermoplastic starch blends', *Polym.*, 48, 270–280.

Iguchi, M, Yamanaka, S and Budhiono, A (2000), 'Bacterial cellulose – a masterpiece of nature's arts', *J. Mater. Sci.*, 35, 261–270.

Ikeda, A, Takemura, A and Ono, H (2000), 'Preparation of low-molecular weight alginic acid by acid hydrolysis', *Carbohy. Polym.*, 42, 421–425.

Isogai, A, Saito, T and Fukuzumi, H (2011), 'TEMPO-oxidized cellulose nanofibers', *Nanoscale*, 3, 71–85.

Iwamoto, S, Nakagaito, A N and Yano, H (2007), 'Nano-fibrillation of pulp fibers for the processing of transparent nanocomposites', *Appl. Phys. A*, 89, 461–466.

Janjarasskul, T and Krochta, J M (2010), 'Edible packaging materials', *Annu. Rev. Food Sci. Technol.*, 1, 415–448.

Jayasekara, R, Harding, I, Bowater, I and Lornergan, G (2005), 'Biodegradability of selected range of polymers and polymer blends and standard methods for assessment of biodegradation', *J. Polym. Environ.*, 13, 231–251.

Jerez, A, Partal, P, Martinez, I, Gallegos, C and Guerrero, A (2007), 'Protein-based bioplastics: effect of thermo-mechanical processing', *Rheol. Acta*, 46, 711–720.

Jiang, T, Feng, L, Zheng, X and Li, L (2013), 'Physicochemical responses and microbial characteristics of shiitake mushroom (*Lentinus edodes*) to gum arabic coating enriched with natamycin during storage', *Food Chem.*, 138, 1992–1997.

Jin, J, Song, M and Hourston, D J (2004), 'Novel chitosan-based films cross-linked by genipin with improved physical properties', *Biomacromol.*, 5, 162–168.

Johnson, D T and Taconi, K A (2007), 'The glycerin glut: options for the value-added conversion of crude glycerol resulting from biodiesel production', *Environ. Progr.*, 26, 338–348.

Jongjareonrak, A, Benjakul, S, Visessanguan, W, Prodpran, T and Tanaka, M (2006), 'Characterization of edible films from skin gelatin of brownstripe red snapper and bigeye snapper', *Food Hydrocoll.*, 20, 492–501.

Jordan, J, Jacob, K I, Tannenbaum, R, Sharaf, M A and Jasiuk, I (2005), 'Experimental trends in polymer nanocomposites: a review', *Mat. Sci. Eng. A*, 393, 1–11.

Jouki, M, Khazaei, N, Ghasemlou, M and HadiNezhad, M (2013), 'Effect of glycerol concentration on edible film production from cress seed carbohydrate gum', *Carboh. Polym.*, 96, 39–46.

Kang, H J, Jo, C, Lee, N Y, Kwon, J H and Byun, M W (2005), 'A combination of gamma irradiation and $CaCl_2$ immersion for a pectin-based biodegradable film', *Carboh. Polym.*, 60, 547–551.

Karbowiak, T, Gougeon, R D, Rigolet, S, Delmotte, L, Debeaufort, F and Voilley, A (2008), 'Diffusion of small molecules in edible films: effect of water and interactions between diffusant and biopolymers', *Food Chem.*, 106, 1340–1349.

Karbowiak, T, Debeaufort, F, Voilley, A and Trystam, G (2009), 'From macroscopic to molecular scale investigations of mass transfer of small molecules through edible packaging applied at interfaces of multiphase food products', *Innov. Food Sci. Emerg. Technol.*, 10, 116–127.

Karim, A A and Bhat, R (2009), 'Fish gelatin: properties, challenges, and prospects as an alternative to mammalian gelatins', *Food Hydrocoll.*, 23, 563–576.

Kester, J J and Fennema, O R (1986), 'Edible films and coatings: a review', *Food Technol.*, 40, 47–59.

Kim, Y, Jung, R, Kim, H S and Jin, H J (2009), 'Transparent nanocomposites prepared by incorporating microbial nanofibrils into poly(L-lactic acid)', *Curr. Appl. Phys.*, 9, S69–S71.

Klemm, D, Heublein, B, Fink, H P and Bohn, A (2005), 'Cellulose: fascinating biopolymer and sustainable raw material', *Angew. Chem. Int. Ed.*, 44, 3358–3393.

Kokini, J L, Cocero, A M, Madeka, H and de Graaf, E (1994), 'The development of state diagrams for cereal proteins', *Trends Food Sci. Technol.*, 5, 281–288.

Konwarh, R, Karak, N and Misra, M (2013), 'Electrospun cellulose acetate nanofibers: the present status and gamut of biotechnological applications', *Biotechnol. Adv.*, 31, 421–437.

Kramer, M E (2009), 'Structure and function of starch-based edible films and coatings' in *Edible Films and Coatings for Food Applications*, ed. M. E. Embuscado and K. C. Huber, Springer, New York, pp. 113–134.

Krishna, M, Nindo, C I and Min, S C (2012), 'Development of fish gelatin edible films using extrusion and compression molding', *J. Food Eng.*, 108, 337–344.

Kristo, E and Biliaderis, C G (2007), 'Physical properites of starch nanocrystal-reinforced pullulan films', *Carboh. Polym.*, 68, 146–158.

Krochta, J M (2002), 'Proteins as raw materials for films and coatings: definitions, current status, and opportunities' in *Protein-Based Films and Coatings*, ed. A. Gennadios, CRC, Boca Raton, FL, pp. 1–41.

Krochta, J M and De Mulder-Johnston, C (1997), 'Edible and biodegradable polymer films: challenges and opportunities', *Food Technol.*, 51, 61–73.

Kugel, A, Stafslien, S and Chisholm, B J (2011), 'Antimicrobial coatings produced by "tethering" biocides to the coating matrix: a comprehensive review', *Prog. Org. Coat.*, 72, 222–252.

Kumar, P, Sandeep, K P, Alavi, S, Truong, V D and Gorga, R E (2010), 'Preparation and characterization of bio-nanocomposite films based on soy protein isolate and montmorillonite using melt extrusion', *J. Food Eng.*, 100, 480–489.

Kvien, I and Oksman, K (2007), 'Orientation of cellulose nanowhiskers in polyvinyl alcohol', *Appl. Phys. A*, 87, 641–643.

Lacroix, M and Cooksey, K (2005), 'Edible films and coatings from animal-origin proteins coatings' in *Innovations in Food Packaging*, ed. J. H. Han, Elsevier, London, pp. 301–317.

Lagrain, B, Goderis, B, Brijs, K and Delcour, J A (2010), 'Molecular basis of processing wheat gluten toward biobased materials', *Biomacromol*, 11, 533–541.

Le Corre, D, Bras, J and Dufresne, A (2010), 'Starch nanoparticles: a review', *Biomacromol.*, 11, 1139–1153.

Li, H L, Wu, B, Mu, C D and Lin, W (2011a), 'Concomitant degradation in periodate oxidation of carboxymethyl cellulose', *Carbohy. Polym.*, 84, 881–886.

Li, M, Liu, P, Zou, W, Yu, L, Xie, F, Pu, H, Liu, H and Chen, L (2011b), 'Extrusion processing and characterization of edible starch films with different amylose contents', *J. Food Eng.*, 106, 95–101.

Li, R, Hu, P, Ren, X, Worley, S D and Huang, T S (2013), 'Antimicrobial N-halamine modified chitosan films', *Carboh. Polym.*, 92, 534–539.

Lian, Z X, Ma, Z S, Wei, J and Liu, H (2012), 'Preparation and characterization of immobilized lysozyme and evaluation of its application in edible coatings', *Proc. Biochem.*, 47, 201–208.

Limpisophon, K, Tanaka, M and Osako, K (2010), 'Characterization of gelatin–fatty acid emulsion films based on blue shark (*Prionace glauca*) skin gelatin', *Food Chem.*, 122, 1095–1101.

Lin, B, Du, Y, Liang, X, Wang, X and Wang, X (2011), 'Effect of chitosan coating on respiratory behavior and quality of stored litchi under ambient temperature', *J. Food Eng.*, 102, 94–99.

López de Lacey, A M, López-Caballero, M E, Gómez-Estaca, J, Gómez-Guillén, M C and Montero, P (2012), 'Functionality of *Lactobacillus acidophilus* and *Bifidobacterium bifidum* incorporated to edible coatings and films', *Innov. Food Sci. Emerg. Technol.*, 16, 277–282.

López-León, T, Carvalho, E L S, Seijo, B, Ortega-Vinuesa, J L and Bastos-González, D (2005), 'Physicochemical characterization of chitosan nanoparticles: electrokinetic and stability behavior', *J. Colloid Interf. Sci.*, 283, 344–351.

Lopez-Rubio, A, Gavara, R and Lagaron, J M (2006), 'Bioactive packaging: turning foods into healthier foods through biomaterials', *Trends Food Sci. Technol.*, 17, 567–575.

Lu, Y, Weng, L and Zhang, L (2004), 'Morphology and properties of soy protein isolate thermoplastics reinforced with chitin whiskers', *Biomacromol.*, 5, 1046–1051.

Ma, X F, Yu, J G and Wan, J J (2006), 'Urea and ethanolamine as a mixed plasticizer for thermoplastic starch', *Carboh. Polym.*, 64, 267–273.

Ma, X, Jian, R, Chang, P R and Yu, J (2008), 'Fabrication and characterization of citric acid-modified starch nanoparticles/plasticized-starch composites', *Biomacromol.*, 9, 3314–3320.

Maftoonazad, N and Ramaswamy, H S (2005), 'Postharvest shelf-life extension of avocados using methyl cellulose-based coating', *LWT – Food Sci. Technol.*, 38, 617–624.

Maftoonazad, N, Ramaswamy, H S, Moalemiyan, M and Kushalappa, A C (2007), 'Effect of pectin-based edible emulsion coating on changes in quality of avocado exposed to *Lasiodiplodia theobromae* infection', *Carboh. Polym.*, 68, 341–349.

Maftoonazad, N, Ramaswamy, H S and Marcotte, M (2008), 'Shelf-life extension of peaches through sodium alginate and methyl cellulose edible coatings', *Int. J. Food Sci. Technol.*, 43, 951–957.

Mali, S, Grossmann, M V E, García, M A, Martino, M M and Zaritzky, N E (2004), 'Barrier, mechanical and optical properties of plasticized yam starch films', *Carboh. Polym.*, 56, 129–135.

Mali, S, Sakanaka, L S, Yamashita, F and Grossmann, M V E (2005), 'Water sorption and mechanical properties of cassava starch films and their relations to plasticizing effect', *Carboh. Polym.*, 60, 283–289.

Mangavel, C, Rossignol, N, Perronnet, A, Barbot, J, Popineau Y and Gueguen, J (2004), 'Properties and microstructure of thermopressed wheat gluten films: a comparison with cast films', *Biomacromol.*, 5, 1596–1601.

Marcos, B, Aymerich, T, Monfort, J M and Garriga, M (2008), 'High-pressure processing and antimicrobial biodegradable packaging to control *Listeria monocytogenes* during storage of cooked ham', *Food Microbiol.*, 25, 177–182.

Martins, J T, Cerqueira, M A, Souza, B W S, Avides, M C and Vicente, A A (2010), 'Shelf life extension of ricotta cheese using coatings of galactomannans from nonconventional sources incorporating nisin against *Listeria monocytogenes*', *J. Agric. Food Chem.*, 58, 1884–1891.

Martins, J T, Cerqueira, M A and Vicente, A A (2012), 'Influence of α-tocopherol on physicochemical properties of chitosan-based films', *Food Hydrocoll.*, 27, 220–227.

Martucci, J F and Ruseckaite, R A (2009), 'Tensile properties, barrier properties, and biodegradation in soil of compression-molded gelatin-dialdehyde starch films', *J. Appl. Polym. Sci.*, 112, 2166–2178.

Mathew, S and Abraham, T E (2008), 'Characterisation of ferulic acid incorporated starch-chitosan blend films', *Food Hydrocoll.*, 22, 826–835.

McGarvie, D and Parolis, H (1979), 'The mucilage of *Opuntia ficus-indica*', *Carboh. Res.*, 69, 171–179.

McGuire, R G and Baldwin, E A (1998), 'Acidic fruit coatings for maintenance of color and decay prevention on lychees postharvest', *Pr. Fl. St. Hortic. Soc.*, 111, 243–247.

McHugh, T H, Aujard, J F and Krochta, J M (1994), 'Plasticized whey protein edible films: water vapor permeability properties', *J. Food Sci.*, 59, 416–419.

Mehyar, G F, Al-Ismail, K, Han, J H and Chee, G W (2012), 'Characterization of edible coatings consisting of pea starch, whey protein isolate, and carnauba wax and their effects on oil rancidity and sensory properties of walnuts and pine nuts', *J. Food Sci.*, 77, E52–E59.

Mi, F L, Shyu, S S and Peng, C K (2005), 'Characterization of ring-opening polymerization of genipin and pH-dependent cross-linking reactions between chitosan and genipin', *J. Polym. Sci. A Polym. Chem.*, 43, 1985–2000.

Mi, F L, Huang, C T, Liang, H F, Chen, M C, Chiu, Y L, Chen, C H and Sung, H W (2006), 'Physicochemical, antimicrobial, and cytotoxic characteristics of a chitosan film cross-linked by a naturally occurring cross-linking agent, aglycone geniposidic acid', *J. Agric. Food Chem.*, 54, 3290–3296.

Mikkonen, K S and Tenkanen, M (2012), 'Sustainable food-packaging materials based on future biorefinery products: xylans and mannans', *Trends Food Sci. Technol.*, 28, 90–102.

Mikkonen, K S, Rita, H, Helén, H, Talja, R A, Hyvonen, L and Tenkanen, M (2007), 'Effect of polysaccharide structure on mechanical and thermal properties of galactomannan-based films', *Biomacromol.*, 8, 3198–3205.

Mild, R M, Joens, L A, Friedman, M, Olsen, C W, McHugh, T H, Law, B and Ravishankar, S (2011), 'Antimicrobial edible apple films inactivate antibiotic resistant and susceptible *Campylobacter jejuni* strains on chicken breast', *J. Food Sci.*, 76, M163–M168.

Min, S and Krochta, J M (2007), 'Ascorbic acid-containing whey protein film coatings for control of oxidation', *J. Agric. Food Chem.*, 55, 2964–2969.

Miranda, R L (2009), 'Cashew tree bark secretion – perspectives for its use in protein isolation strategies', *Open Glycosci*, 2, 16–19.

Mishra, S B, Mishra, A K, Kaushik, N K and Khan, M A (2007), 'Study of performance properties of lignin-based polyblends with polyvinyl chloride', *J. Mat. Proc. Technol.*, 183, 273–276.

Moates, G K, Noel, T R, Parker, R and Ring, S G (2001), 'Dynamic mechanical and dielectric characterisation of amylose–glycerol films', *Carboh. Polym.*, 44, 247–253.

Mohareb, E and Mittal, G S (2007), 'Formulation and process conditions for biodegradable/edible soy-based packaging trays', *Packag. Technol. Sci.*, 20, 1–15.

Moon, R J, Martini, A, Nairn, J, Simonsen, J and Youngblood, J (2011), 'Cellulose nanomaterials review: structure, properties and nanocomposites', *Chem. Soc. Rev.*, 40, 3941–3994.

Morillon, V, Debeaufort, F, Blond, G, Capelle, M and Voilley, A (2002), 'Factors affecting the moisture permeability of lipid-based edible films: a review', *Crit. Rev. Food Sci. Nutr.*, 42, 67–89.

Mu, C, Guo, J, Li, X, Lin, W and Li, D (2012), 'Preparation and properties of dialdehyde carboxymethyl cellulose crosslinked gelatin edible films', *Food Hydrocoll.*, 27, 22–29.

Mueller, R J (2006), 'Biological degradation of synthetic polyesters – enzymes as potential catalysts for polyester recycling', *Proc. Biochem.*, 41, 2124–2128.

Müller-Buschbaum, P, Gebhardt, R, Maurer, E, Bauer, E, Gehrke, R and Doster, W (2006), 'Thin casein films as prepared by spin-coating: influence of film thickness and of pH', *Biomacromol.*, 7, 1773–1780.

Muzzarelli, R A A (2009), 'Genipin-crosslinked chitosan hydrogels as biomedical and pharmaceutical aids', *Carboh. Polym.*, 77, 1–9.

Nair, K G and Dufresne, A (2003), 'Crab shell chitin whisker reinforced natural rubber nanocomposites. 1. Processing and swelling behavior', *Biomacromol.*, 4, 657–665.

Nakagaito, A N and Yano, H (2005), 'Novel high-strength biocomposites based on microfibrillated cellulose having nano-order-unit web-like network structure', *Appl. Phys. A Mater. Sci. Process.*, 80, 155–159.

Navarro-Tarazaga, M L, Massa, A and Pérez-Gago, M B (2011), 'Effect of beeswax content on hydroxypropyl methylcellulose-based edible film properties and postharvest quality of coated plums (Cv. Angeleno)', *LWT – Food Sci. Technol.*, 44, 2328–2334.

Ng, A, Greenshields, R N and Waldron, K W (1997), 'Oxidative cross-linking of corn bran hemicellulose: formation of ferulic acid dehydrodimers', *Carboh. Res.*, 303, 459–462.

Núñez-Flores, R, Giménez, B, Fernández-Martín, F, López-Caballero, M E, Montero, M P and Gómez-Guillén, M C (2013), 'Physical and functional characterization of active fish gelatin films incorporated with lignin', *Food Hydrocoll.*, 30, 163–172.

Ojagh, S M, Núñez-Flores, R, López-Caballero, M E, Montero, M P and Gómez-Guillén, M C (2011), 'Lessening of high-pressure-induced changes in Atlantic salmon muscle by the combined use of a fish gelatin–lignin film', *Food Chem.*, 125, 595–606.

Olivas, G I and Barbosa-Cánovas, G (2008), 'Alginate-calcium films: water vapor permeability and mechanical properties as affected by plasticizer and relative humidity', *LWT – Food Sci. Technol.*, 41, 359–366.

O'Neill, M A and York, W S (2003), 'The composition and structure of plant primary cell walls' in *The Plant Cell Wall*, ed. J. K. C. Rose, Annual Plant Reviews, vol. 8. CRC Press, Boca Raton, FL, pp. 1–54.

Osés, J, Fabregat-Vázquez, M, Pedroza-Islas, R, Tomás, S A, Cruz-Orea, A and Maté, J I (2009), 'Development and characterization of composite edible films based on whey protein isolate and mesquite gum', *J. Food Eng.*, 92, 56–62.

Ou, S, Wang, Y, Tang, S, Huang, C and Jackson, M G (2005), 'Role of ferulic acid in preparing edible films from soy protein isolate', *J Food Eng.*, 70, 205–210.

Paes, S S, Yakimets, I, Wellner, N, Hill, S E, Wilson, R H and Mitchell, J R (2010), 'Fracture mechanisms in biopolymer films using coupling of mechanical analysis and high speed visualization technique', *Eur. Polymer J.*, 46, 2300–2309.

Paillet, M and Dufresne, A (2001), 'Chitin whisker reinforced thermoplastic nanocomposites', *Macromol.*, 34, 6527–6530.

Park, H J and Chinnan, M S (1995), 'Gas and water vapor barrier properties of edible films from protein and cellulosic materials', *J. Food Eng.*, 25, 497–507.

Park, S I and Zhao, Y (2006), 'Development and characterization of edible films from cranberry pomace extracts', *J. Food Sci.*, 71, E95–E101.

Park, Y, Doherty, W O S and Halley, P J (2008), 'Developing lignin-based resin coatings and composites', *Ind. Crops Prod.*, 27, 163–167.

Parker, M L, Ng, A and Waldron, K W (2005), 'The phenolic acid and polysaccharide composition of cell walls of bran layers of mature wheat (*Triticum aestivum* L. cv. Avalon) grains', *J. Sci. Food Agric.*, 85, 2539–2547.

Paula, E L, Mano, V and Pereira, V J (2011), 'Influence of cellulose nanowhiskers on the hydrolytic degradation behavior of poly(D,L-lactide)', *Polym. Degrad. Stab.*, 96, 1631–1638.

Peressini, D, Bravin, B, Lapasin, R, Rizzotti, C and Sensidoni, A (2003), 'Starch–methylcellulose based edible films: rheological properties of film-forming dispersions', *J. Food Eng.*, 59, 25–32.

Pérez-Gago, M B and Krochta, J M (2002), 'Formation and properties of whey protein films and coatings' in *Protein-Based Films and Coatings*, ed. A. Gennadios, CRC Press, Boca Raton, FL, pp. 159–180.

Pérez-Gago, M B, Serra, M, Alonso, M, Mateos, M and Del Río, M A (2003), 'Effect of solid content and lipid content of whey protein isolate-beeswax edible coatings on color change of fresh-cut apples', *J. Food Sci*, 68, 2186–2191.

Petersson, L and Oksman, K (2006), 'Preparation and properties of biopolymer based nanocomposite films using microcrystalline cellulose' in *Cellulose Nanocomposites, Processing, Characterization and Properties*, ed. K. Oksman and M. Sain, ACS Symposium Series 938, Oxford University Press, Oxford, pp. 132–150.

Petersson, L, Kvien, I and Oksman, K (2007), 'Structure and thermal properties of poly(lactic acid)/cellulose whiskers nanocomposite materials', *Compos. Sci. Technol.*, 67, 2535–2544.

Piermaria, J, Bosch, A, Pinotti, A, Yantorno, O, Garcia, M A and Abraham, A G (2011), 'Kefiran films plasticized with sugars and polyols: water vapor barrier and mechanical properties in relation to their microstructure analyzed by ATR/FT-IR spectroscopy', *Food Hydrocoll.*, 25, 1261–1269.

Podsiadlo, P, Choi, S Y, Shim, B, Lee, J, Cuddihy, M and Kotov, N A (2005), 'Molecularly engineered nanocomposites: layer-by-layer assembly of cellulose nanocrystals', *Biomacromol.*, 6, 2914–2918.

Pushpadass, H A, Marx, D B and Hanna, M A (2008), 'Effects of extrusion temperature and plasticizers on the physical and functional properties of starch films', *Starch*, 60, 527–538.

Putra, A, Kakugo, A, Furukawa, H, Gong, J P and Osada, Y (2008), 'Tubular bacterial cellulose gel with oriented fibrils on the curved surface', *Polym.*, 49, 1885–1891.

Qiao, R. and Brinson, L C (2009), 'Simulation of interphase percolation and gradients in polymer nanocomposites', *Compos. Sci. Technol.*, 69, 491–499.

Qiao, X, Tang, Z and Sun, K (2011), 'Plasticization of corn starch by polyol mixtures', *Carboh. Polym.*, 83, 659–664.

Qiu, X and Hu, S (2013), '"Smart" materials based on cellulose: a review of the preparations, properties, and applications', *Materials*, 6, 738–781.

Ramos, L P (2003), 'The chemistry involved in the stream treatment of lignocellulosic materials', *Química Nova*, 26, 863–871.

Ramos, O L, Reinas, I, Silva, S I, Fernandes, J C, Cerqueira, M A, Pereira, R N, Vicente, A A, Poças, M F, Pintado, M E and Malcata, F X (2013), 'Effect of whey protein purity and glycerol content upon physical properties of edible films manufactured therefrom', *Food Hydrocoll.*, 30, 110–122.

Redl, A, Morel, M H, Bonicel, J, Vergnes, B and Guilbert, S (1999), 'Extrusion of wheat gluten plasticized with glycerol: influence of process conditions on flow behavior, rheological properties, and molecular size distribution', *Cereal Chem.*, 76, 361–370.

Reinoso, E, Mittal, G S and Lim, L T (2008), 'Influence of whey protein composite coatings on plum (*Prunus Domestica* L.) fruit quality', *Food Bioprocess. Technol.*, 1, 314–325.

Retegi, A, Gabilondo, N, Peña, C, Zuluaga, R, Castro, C, Gañan, P, de la Caba, K and Mondragon, I (2010), 'Bacterial cellulose films with controlled microstructure–mechanical property relationships', *Cellulose*, 17, 661–669.

Rhim, J W and Ng, P K W (2007), 'Natural biopolymer-based nanocomposite films for packaging applications', *Crit. Rev. Food Sci. Nutr.*, 47, 411–433.

Rhim, J W and Shellhammer, T H (2005), 'Lipid-based edible films and coatings' in *Innovations in Food Packaging*, ed. J. H. Han, Elsevier, London, pp. 362–383.

Ribeiro, C, Vicente, A A, Teixeira, J A and Miranda, C (2007), 'Optimization of edible coating composition to retard strawberry fruit senescence', *Postharv. Biol. Technol.*, 44, 63–70.

Rosa, M F, Medeiros, E S, Malmonge, J A, Gregorski, K S, Wood, D F, Mattoso, L H C, Glenn, G, Orts, W J and Imam, S H (2010), 'Cellulose nanowhiskers from coconut husk fibers: effect of preparation conditions on their thermal and morphological behavior', *Carboh. Polym.*, 81, 83–92.

Rose, J K C and Bennett, A B (1999), 'Cooperative disassembly of the cellulose–xyloglucan network of plant cell walls: parallels between cell expansion and fruit ripening', *Trends Plant Sci.*, 4, 176–183.

Rossman, J M (2009), 'Commercial manufacture of edible films', in *Edible Films and Coatings for Food Applications*, ed. M. E. Embuscado and K. C. Huber, Springer, Dordrecht, pp. 367–390.

Sanchez-Garcia, M D, Gimenez, E and Lagaron, J M (2008), 'Morphology and barrier properties of solvent cast composites of thermoplastic biopolymers and purified cellulose fibers', *Carboh. Polym.*, 71, 235–244.

Sanjurjo, K, Flores, S, Gerschenson, L and Jagus, R (2006), 'Study of the performance of nisin supported in edible films', *Food Res. Int.*, 39, 749–754.

Sarabia, A I, Gómez-Guillén, M C and Montero, P (2000), 'The effect of added salt on the viscoelastic properties of fish skin gelatin', *Food Chem.*, 70, 71–76.

Saucedo-Pompa, S, Rojas-Molina, R, Aguilera-Carbó, A F, Saenz-Galindo, A, de La Garza, H, Jasso-Cantú, D and Aguilar, C N (2009), 'Edible film based on candelilla wax to improve the shelf life and quality of avocado', *Food Res. Int.*, 42, 511–515.

Scheller, H V and Ulvskov, P (2010), 'Hemicelluloses', *Annu. Rev. Plant Biol.*, 61, 263–289.

Schou, M, Longares, A, Montesinos-Herrero, C, Monahan, F J, O'Riordan, D and O'Sullivan, M (2005), 'Properties of edible sodium caseinate films and their application as food wrapping', *LWT – Food Sci. Technol.*, 38, 605–610.

Shen, X L, Wu, J M, Chen, Y and Zhao, G (2010), 'Antimicrobial and physical properties of sweet potato starch films incorporated with potassium sorbate or chitosan', *Food Hydrocoll.*, 24, 285–290.

Shi, K, Kokini, J L and Huang, Q (2009), 'Engineering zein films with controlled surface morphology and hydrophilicity', *J. Agric. Food Chem.*, 57, 2186–2192.

Shi, K, Huang, Y, Yu, H, Lee, T C and Huang, Q (2011), 'Reducing the brittleness of zein films through chemical modification', *J. Agric. Food Chem.*, 59, 56–61.

Shon, J and Chin, K B (2008), 'Effect of whey protein coating on quality attributes of low-fat, aerobically packaged sausage during refrigerated storage', *J. Food Sci.*, 73, C469–C475.

Sims, I M and Furneaux, R H (2003), 'Structure of the exudates gum from *Meryta sinclairii*', *Carboh. Polym.*, 52, 423–431.

Siracusa, V, Rocculi, P, Romani, S and Dalla Rosa, M (2008), 'Biodegradable polymers for food packaging: a review', *Trends Food Sci. Technol.*, 19, 634–643.

Siró, I and Plackett, D (2010), 'Microfibrillated cellulose and new nanocomposite materials: a review', *Cellulose*, 17, 459–494.

Sonkaew, P, Sane, A and Suppakul, P (2012), 'Antioxidant activities of curcumin and ascorbyl dipalmitate nanoparticles and their activities after incorporation into cellulose-based packaging films', *J. Agric. Food Chem.*, 60, 5388–5399.

Sothornvit, R and Krochta, J M (2005), 'Plasticizers in edible films and coatings' in *Innovations in Food Packaging*, ed. J. H. Han, Elsevier, London, pp. 403–463.

Sothornvit, R and Pitak, N (2007), 'Oxygen permeability and mechanical properties of banana films', *Food Res. Int.*, 40, 365–370.

Sothornvit, R and Rodsamran, P (2008), 'Effect of a mango film on quality of whole and minimally processed mangoes', *Postharv. Biol. Technol.*, 47, 407–415.

Sriamornsak, P (2003), 'Chemistry of pectin and its pharmaceutical uses: a review', *Silpakorn Univ. Int. J.*, 3, 206–228.

Srinivasa, P C, Ramesh, M N and Tharanathan, R N (2007), 'Effect of plasticizers and fatty acids on mechanical and permeability characteristics of chitosan films', *Food Hydrocoll.*, 21, 1113–1122.

Sriupayo, J, Supaphol, P, Blackwell, J and Rujiravanit, R (2005), 'Preparation and characterization of α-chitin whisker-reinforced chitosan nanocomposite films with or without heat treatment', *Carboh. Polym.*, 62, 130–136.

Srivastava, M and Kapoor, V P (2005), 'Seed galactomannans: an overview', *Chem. Biodiv.*, 2, 295–317.

Stepto, R F T (2009), 'Thermoplastic starch', *Macromol. Symp.*, 279, 163–168.

Strauss, G and Gibson, S M (2004), 'Plant phenolics as cross-linkers of gelatin gels and gelatin-based coacervates for use as food ingredients', *Food Hydrocoll.*, 18, 81–89.

Sun, R C, Sun, X F and Tomkinson, J (2004), 'Hemicelluloses and their derivatives' in *Hemicellulose: Science and Technology*, ed. P. Gatenholm and M. Tenkanen, American Chemical Society, Washington, DC, pp. 2–22.

Sung, H W, Huang, R N, Huang, L L H and Tsai, C C (1999), '*In vitro* evaluation of cytotoxicity of a naturally occurring cross-linking reagent for biological tissue fixation', *J. Biomat. Sci. Polym. Ed.*, 10, 63–78.

Suppakul, P, Jutakorn, K and Bangchokedee, Y (2010), 'Efficacy of cellulose-based coating on enhancing the shelf life of fresh eggs', *J. Food Eng.*, 98, 207–213.

Talens, P and Krochta, J M (2005), 'Plasticizing effects of beeswax and carnauba wax on tensile and water vapor permeability properties of whey protein films', *J. Food Sci.*, 70, E239–E243.

Tang, C and Liu, H (2008), 'Cellulose nanofiber reinforced poly(vinyl alcohol) composite film with high visible light transmittance', *Compos. A Appl. Sci. Manuf.*, 39, 1638–1643.

Tapia, M S, Rojas-Graü, M A, Rodríguez, F J, Ramírez, J, Carmona, A and Martin-Belloso, O (2007), 'Alginate- and gellan-based edible films for probiotic coatings on fresh-cut fruits', *J. Food Sci.*, 72, E190–E196.

Tapia-Blácido, D R, Sobral, P J A and Menegalli, F C (2013), 'Effect of drying conditions and plasticizer type on some physical and mechanical properties of amaranth flour films', *LWT – Food Sci.Technol.*, 50, 392–400.

Tharanathan, R N (2003), 'Biodegradable films and composite coatings: past, present and future', *Trends Food Sci. Technol.*, 14, 71–78.

Thunwall, M, Kuthanová, V, Boldizar, A and Rigdahl, M (2008), 'Film blowing of thermoplastic starch', *Carboh. Polym.*, 71, 583–590.

Ugartondo, V, Mitjans, M and Vinardell, M P (2009), 'Applicability of lignins from different sources as antioxidants based on the protective effects on lipid peroxidation induced by oxygen radicals', *Ind. Crops Prod.*, 30, 184–187.

Vachon, C, Yu, H L, Yefsah, R, Alain, R, St-Gelais, D and Lacroix, M (2000), 'Mechanical and structural properties of milk protein edible films cross-linked by heating and gamma irradiation', *J. Agric. Food Chem.*, 48, 3202–3209.

Van de Velde, F, Knutsen, S H, Usov, A L, Rollema, H S and Cerezo, A S (2002), '$^1$H and $^{13}$C high resolution NMR spectroscopy of carrageenans: application in research and industry', *Trends Food Sci. Technol.*, 13, 73–92.

Van der Zee, M (2005), 'Biodegradability of polymers – mechanisms and evaluation methods' in *Handbook of Biodegradable Polymers*, ed. C. Bastioli, Rapra, Shropshire, pp. 1–31.

Vengal, J C and Srikumar, M (2005), 'Processing and study of novel lignin-starch and lignin-gelatin biodegradable polymeric films', *Trends Biomater. Artif. Organs*, 18, 238–241.

Verbeek, C J R and van den Berg, L E (2010), 'Extrusion processing and properties of protein-based thermoplastics', *Macromol. Mater. Eng.*, 295, 10–21.

Vieira, M G A, Silva, M A, Santos, L O and Beppu, M M (2011), 'Natural-based plasticizers and biopolymer films: a review', *Eur. Polym. J.*, 47, 254–263.

Wang, X, Sun, X, Liu, H, Li, M and Ma, Z (2011), 'Barrier and mechanical properties of carrot puree films', *Food Bioprod. Process.*, 89, 149–156.

Wardy, W, Torrico, D D, Jirangrat, W, No, H K, Saalia, F K and Prinyawiwatkul, W (2011), 'Chitosan-soybean oil emulsion coating affects physico-functional and sensory quality of eggs during storage', *LWT – Food Sci. Technol.*, 44, 2349–2355.

Wasswa, J, Tang, J and Gu, X (2007), 'Utilization of fish processing by-products in the gelatin industry', *Food Rev. Int.*, 23, 159–174.

Wu, Q, Henriksson, M, Liu, X and Berglund, L A (2007), 'A high strength nanocomposite based on microcrystalline cellulose and polyurethane', *Biomacromol.*, 8, 3687–3692.

Wyman, C E, Decker, S R, Himmel, M E, Brady, J W, Skopec, C E and Viikari, L (2005), 'Hydrolysis of cellulose and hemicellulose' in *Polysaccharides: Structural Diversity and Functional Versatility*, ed. S. Dumitriu, Marcel Dekker, New York, pp. 995–1033.

Yakimets, I, Paes, S S, Wellner, N, Smith, A C, Wilson, R H and Mitchell, J R (2007), 'Effect of water content on the structural reorganization and elastic properties of biopolymer films: a comparative study', *Biomacromol.*, 8, 1710–1722.

Yan, Z, Chen, S, Wang, H, Wang, B, Wang, C and Jiang, J (2008), 'Cellulose synthesized by *Acetobacter xylinum* in the presence of multi-walled carbon nanotubes', *Carboh. Res.*, 343, 73–80.

Yang, F, Hanna, M A and Sun, R (2012), 'Value-added uses for crude glycerol – a byproduct of biodiesel production', *Biotechn. Biofuels*, 5, 13.

Yang, Q, Fukuzumi, H, Saito, T, Isogai, A and Zhang, L (2011), 'Transparent cellulose films with high gas barrier properties fabricated from aqueous alkali/urea solutions', *Biomacromol.*, 12, 2766–2771.

Ying, R, Barron, C, Saulnier, L and Rondeau-Mouro, C (2011), 'Water mobility within arabinoxylan and β-glucan films studied by NMR and dynamic vapour sorption', *J. Sci. Food Agric.*, 91, 2601–2605.

Yousefi, H, Nishino, T, Faezipour, M, Ebrahimi, G and Shakeri, A (2011), 'Direct fabrication of all-cellulose nanocomposite from cellulose microfibers using ionic liquid-based nanowelding', *Biomacromol.*, 12, 4080–4085.

Yousefi, H, Faezipour, M, Hedjazi, S, Mousavi, M M, Azusa, Y and Heidari, A H (2013), 'Comparative study of paper and nanopaper properties prepared from bacterial cellulose nanofibers and fibers/ground cellulose nanofibers of canola straw', *Ind. Crop. Prod.*, 43, 732–737.

Yuan, Y, Chesnutt, B M, Utturkar, G, Haggard, W O, Yang, Y, Ong, J L and Bumgardner, J D (2007), 'The effect of cross-linking of chitosan microspheres with genipin on protein release', *Carboh. Polym.*, 68, 561–567.

Zainuddin, S Y Z, Ahmad, I and Kargarzadeh, H (2013), 'Cassava starch biocomposites reinforced with cellulose nanocrystals from kenaf fibers', *Compos. Interface*, 20, 189–199.

Zhang, X Q, Burgar, I, Do, M D and Lourbakos, E (2005), 'Intermolecular interactions and phase structures of plasticized wheat proteins materials', *Biomacromol.*, 6, 1661–1671.

Zhang, X, Do, M D, Casey, P, Sulistio, A, Qiao, G G, Lundin, L, Lillford, P and Kosaraju, S (2010a), 'Chemical cross-linking gelatin with natural phenolic compounds as studied by high-resolution NMR spectroscopy', *Biomacromol.*, 11, 1125–1132.

Zhang, X, Do, M D, Casey, P, Sulistio, A, Qiao, G G, Lundin, L, Lillford, P and Kosaraju, S (2010b), 'Chemical modification of gelatin by a natural phenolic cross-linker, tannic acid', *J. Agric. Food Chem.*, 58, 6809–6815.

Zhang, Y and Han, J H (2006), 'Mechanical and thermal characteristics of pea starch films plasticized with monosaccharides and polyols', *J. Food Sci.*, 71, E109–E118.

Zhang, Y, Pitkänen, L, Douglade, J, Tenkanen, M, Remond, C and Joly, C (2011), 'Wheat bran arabinoxylans: chemical structure and film properties of three isolated fractions', *Carboh. Polym.*, 86, 852–859.

Zhou, J J, Wang, S Y and Gunasekaran, S (2009), 'Preparation and characterization of whey protein film incorporated with $TiO_2$ nanoparticles', *J. Food Sci.*, 74, N50–N56.

Zimmermann, T, Bordeanu, N and Strub, E (2010), 'Properties of nanofibrillated cellulose from different raw materials and its reinforcement potential', *Carboh. Polym.*, 79, 1086–1093.

# Index

acetals, 369–70
acetic acid, 278, 670
acetins, 371
acetogenesis, 478
acetol, 375–6
acetylation, 816
acid-catalysed glycerol dehydration, 375–6
acid catalysts, 162
acid hydrolysis, 629–30, 669, 850–1
acid pretreatment, 489
acidogenesis, 478
*Acremonium cellulolyticus*, 212
acrolein, 375–6
acrylated epoxidised soy oil (AESO), 755
acrylic acid, 375–6
activated carbon, 677–80
activated sludge, 701
active biopackaging, 854–5
adsorption, 119–20, 365–6
adsorption capacity, 147
adsorptive beads
  removal of non-ionic impurities, 124–6
    comparative study for removal of furfural and HMF from RW-EDI-treated corn stover, 125
affinity chromatography, 608
Agency for Renewable Resources, 512
Agri-Pure, 584
Alcell process, 669
alcohol fuels, 397–405
  Brazil, 397–8
  China, 404–5
  European Union, 403–4
  spark-ignition engines, 407–17
    dedicated alcohol engines, 414–15
    efficiency, 413–14
    octane numbers, 411–12
    performance, 413
    pollutant emission, deposits and lubricant dilution, 415–17
    vapour pressure, 410–11
    volumetric energy density and stoichiometry, 408–10
  United States, 398–403
    alternative fuel consumption in the US road transport sector, 399
    evolution of ethanol production by country since 2007, 400
    renewable fuel volume requirements for RFS2, 401
alcoholysis *see* transesterification
aldehydes, 847–8
algae, 730, 791–2
  proteins, 604–5
algae-derived bioactive peptides, 605–6
alginates, 833
alkali pretreatment, 205, 489
alkaloids, 612
alkyl branched fatty compounds, 589
Alternative Motor Fuels Act (AMFA), 402–3
Amazonian dark earth (ADE), 529
Amberlyst-15, 634–5
Amberlyst 70, 171–2, 644
amino acid, 342–3
  isolation, 727–8
ammonia, 486
ammonia fibre expansion (AFEX), 132, 726
amylose, 832
anaerobic baffled reactor (ABR), 484
anaerobic bio-reactors, 481–5
  basic anaerobic digesters, 482
anaerobic digestion (AD), 108
  biomethane and biohydrogen production via fermentation, 476–513
    basic principles of biogas and hydrogen production, 477–81

875

biogas and hydrogen production and
    technological aspects, 481–92
  current status and limitations, 507,
    511–13
  future trends, 513
  methane production from different
    feedstocks, 492–507
anaerobic filters, 484
anaerobic lagoons, 481
anaerobic technology, 513
analysis methods, 705–6
Anammox, 503–4
animal fats, 455
animal feed industry, 772
antibiotics, 790
antifoaming agents, 570
antinutritional factors (ANF), 774
antioxidants, 570, 677
antisense RNA (asRNA), 251
antiwear, 570
aqueous-phase reforming (APR),
  629–30
arabinoxylan (AX), 311
arabinoxylan-based films, 828
ARBOFORM, 676
Archer Daniels Midland (ADM),
  722–3
*Arthrospira*, 97
ash, 530
Aspen Plus, 100
  models, 222
astaxanthins, 613
autocondensation reactions, 744
automotive applications, 571
auxiliary enzymes, 271

bacteria, 700–1
bacterial cellulose, 836
bagasse, 42
bakery waste, 325
bark charcoal, 547
batch retort, 534–6
battery electric vehicle (BEV), 392–3
beehive kilns, 534
beeswax, 712
bench-top fermentations, 316
benchmarking enzymes
  enzymatic conversion processes, 220–5
  state of enzyme technology,
    220–1
  techno-economic modeling, 221–5

beta-glucosidase, 214–15, 246
beta-xylosidases, 206, 217–18
bio-based animal feed
  background, 772–5
    demand for feed attributes and trends,
      774–5
    history of distilling co-products, 772–3
    scale of bioethanol production, 773–4
  types, properties and processing, 771–93
    feed ingredients, 775–83
    future trends, 791–3
    impact of process technology on
      co-product quality, 784–6
    improving feedstocks, processes and
      yields, 786–9
    regulatory issues, 789–91
bio-based chemicals
  biorefining of carbohydrate conversion
    and utilisation, 624–51
    chemical hydrolysis of cellulose to
      sugars, 629–35
    future trends, 650–1
    routes to market for bio-based
      feedstocks, 646–50
    types and properties of carbohydrate-
      based chemicals, 635–45
  biorefining of lignin conversion and
    utilisation, 659–84
    applications of lignin and lignin-based
      products, 672–84
    emerging processes for lignin
      production, 668–72
    future trends, 684
    structure and properties of lignin, 660–3
    traditional processes for lignin
      production, 663, 666–8
  biorefining of lipid and wax conversion
    and utilisation, 693–715
    future trends, 714–15
    methods of extraction and analysis,
      703–6
    sources, 697–702
    types and properties, 694–7
    utilisation, 707–14
  biorefining of protein conversion and
    utilisation, 721–32
    alternative and novel feedstocks and
      production routes, 730–1
    (bio)chemical conversion of amino
      acids to platform and specialty
      chemicals, 728–30

future trends, 731–2
protein and amino acid sources derived from biofuel production, 722–4
protein isolation, hydrolysis and isolation of amino acid and chemical feedstocks, 724–8
LCA results, 78–81
life cycle of biorefinery vs conventional products, 79
sustainable carbohydrate sources, 625–9
biochemical and thermochemical routes for lignocellulose conversion, 627
conversion of lignocellulose to cellulose via fractionation, 628
structure of lignin, hemicellulose and cellulose contained within lignocellulose, 626
bio-based composites
biomass usage to produce building materials, 803–17
fibre properties, 807–8
fibre types and isolation, 804–7
fibrous plants, 804
future trends, 817
improving performance properties, 815–17
history, 803–4
selection based on density, 810
illustration, 810
types, 808–15
fibreboard, 811–13
filters and sorbents, 814
geotextiles, 814
moulded products, 813
particleboard, 811
waferboard and flakeboard, 810–11
types and properties, 808–15
moisture, biological, ultraviolet and thermal properties, 814–15
bio-based feedstocks
challenges for process design, 647–50
liquid crystal templating route to form mesoporous silica and macromesoporous SBA-15, 648
routes to market, 646–50
challenges and opportunities for bio-based products, 646–7
bio-based films
processes for production, 822–4
basic scheme of steps of casting process to produce films, 823

schematic representation of an extruder, 824
bio-based nutraceuticals
biorefining, 596–615, 597–9
carbohydrate-based nutraceuticals, 606–9
future trends, 613–14
lipid-based nutraceuticals, 599–603
other nutraceuticals, 609–13
protein and peptide-based nutraceuticals, 604–6
bio-based products
challenges and opportunities, 646–7
near commercial and future commercial bio-based chemicals, 646
Bio-Methanol Chemie Nederland (BMCN), 722–3
bio-oil production, 144
bio-oil upgrading, 144, 172–5
deoxygenation by cracking, 173–4
hydrodeoxygenation, 174–5
bioactive biopackaging, 854–5
bioactive polysaccharides, 606–8
bioadhesives
types, processing and properties for wood and fibres, 736–65
carbohydrate adhesives, 751–2
future trends, 762–5
lignin adhesives, 745–9
mixed tanin-lignin adhesives, 749–50
protein adhesives, 750–1
tannin adhesives, 737–45
unsaturated oil adhesives, 752–5
wood welding without adhesives, 755–62
biochar
agricultural usages, 529–31
SEM examination of surface of Terra Preta particle, 530
wood and bamboo soaking in pond at edge of rice paddy field, 531
effects of application to soil, 527–9
wood biochar extracted from soil, 528
larger-scale commercial production, 538–41
Black is Green pyrolysis kiln and Pro-natural kiln, 541
Energy Farmers Australia Pty Ltd portable pyrolyser, 542
Genesis continuous pyrolysis plant, 542

Sanli New Energy Company factory, 543
schematic of Pacific Pyrolysis unit, 540
markets and usages, 544–8
  slow release fertiliser produced from minerals, 546
production, 531–8
  household devices, 531–2
  ovens, retorts and kilns, 532–8
production and application in soils, 525–50
  appendix of IBI standardised product definition and product testing guidelines, 554–5
  future trends, 548–50
testing properties, 541–4
  Japanese Standard courtesy of Japan Special Forest Product Promotion Association, 544
  test Category B characteristics and criteria, 545
biochemical biorefineries
bioproducts, 42–6
  capacity costs for lignocellulosic ethanol biorefineries employing dilute acid, 45
  capital costs for first generation ethanol biorefineries, 43
  operating costs for first generation ethanol biorefineries, 43
  operating costs for lignocellulosic ethanol biorefineries employing dilute acid, 45
biochemical catalysts, 225
biochemical conversion, 200–11
  amino acids to platform and specialty chemicals, 728–30
    glutamic acid for production of chemicals, 729
    simultaneous production of styrene and acrylic acid, 730
  basic unit process steps for conversion of biomass to product, 201
  enzymatic hydrolysis and product fermentation, 206–11
    consolidation bioprocessing (CBP), 210–11
    hybrid hydrolysis and fermentation (HHF), 209–10
    separate hydrolysis and fermentation, 208
    simultaneous saccharification and fermentation, 208–9

process integration pretreatment and hydrolysis interface, 203–6
biodegradability, 6
biodegradable food packaging, 832
biodiesel, 396–7, 405–7
  advantages and limitations, 461–5
    cold flow properties and oxidative stability, 463–5
    feedstock availability, 462–3
  production, 366
  renewable diesel production methods, 441–65
    feedstock quality issues, 458–61
    future trends, 465
    overview, 442
    renewable diesel production routes, 442–3
    traditional and emerging feedstocks, 454–8
  routes of production, 444–54
    heterogeneous catalysts, 445–51
    purification by adsorbents and resins, 453–4
    supercritical processing, 452–3
    transesterification of triacylglyceride, 444
    ultrasonic processing, 451–2
bioeconomy, 84
bioelements isolation, 805
bioenergetics, 249
bioenergy, 242
bioethanol, 355
  co-product process improvements, 787–9
  production scale, 773–4
    biofuels production for 2008 to 2011, 774
bioethanol fuel-focused biorefineries developments, 259–94
  design options for biorefining process, 279–80
  different types of ethanol biorefineries, 282–8
  ethanol biorefineries, 261–3
  process intensification and increasing dry-matter content, 280–2
  world total ethanol production since 1975 together with two major fuel producers, 260
future trends, 288–94
  future crops, 293–4
  industrialisation and process development, 288–9

lessons from LCA studies, 292–3
transition of paper industries, 289, 292
lignocellulose to ethanol process, 263
  composition of various lignocellulose feedstocks, 267–8
  fermentation, 272–3
  hydrolysis, 269, 271–2
  inhibitor tolerance, 276–8
  pentose utilisation, 273–6
  pretreatment, 263
  schematic process overview, 266
  thermo-tolerance, 278
bioethanol plants, 792
biofuel production, 725–6
  process configurations, 235, 241–2
    evolution of biomass processing strategies featuring enzymatic hydrolysis, 241
    rationale for CBP, 242
  protein and amino acid sources, 722–4
  rest streams usage, 723–4
biofuels, 225, 707
  LCA results, 74–7
    comparative environmental impacts breakdown for ethanol production, 77
    GHG emissions, 75
    GHG savings per hectare as a function of lignocellulosic crop yields, 76
  recovery, 134–41
Biofuels Directive, 773
biogas, 354–5, 477–9, 507, 511–13
  process outline, 477
biogas digester, 548
biohydrogen
  biomethane production via anaerobic digestion and fermentation, 476–513
    basic principles of biogas and hydrogen production, 477–81
    biogas and hydrogen production and technological aspects, 481–92
    current status and limitations, 507, 511–13
    future trends, 513
    methane production from different feedstocks, 492–507
  current status and limitations, 507, 511–13
    removal of biogas components based on biogas utilisation, 511
  technology, 485–8
biological attack, 203

biological pretreatment, 491
biomass, 7, 15
  components of packaging films and coatings, 822
    basic and optional components, 822
  conversion, 98–9
    application of typical conversion routes, 99
  production, 97–8
    cycles of chemicals from biomass and oil, 8
    process boundary considerations, 98
  products as crosslinking agents for packaging materials, 844–9
    aldehydes, 847–8
    genipin, 846–7
    phenolic compounds, 845–6
    schematic representation of crosslinking of polymer chains, 845
  products as film plasticisers, 842–4
    mono- and disaccharides, 843–4
    polyols, 843
    representative scheme of effect of plasticisers, 842
  products as reinforcements for packaging materials, 849–53
    cellulosic reinforcements, 849–52
    chitin and chitosan nanostructures, 852–3
    microscale to nanoscale, 849
    starch nanoreinforcements, 852
  pyrolysis, 159–60
    micrograph of SBA-15 ordered mesoporous material, 160
  resources, 804
  usage for packaging films and coatings, 819–56
    future trends, 853–5
    processes for producing bio-based films, 822–4
    processes for producing edible coatings, 825–6
    products, 826–42
  usage to produce bio-based composites and building materials, 803–17
    fibre properties, 807–8
    fibre types and isolation, 804–7
    fibrous plants, 804
    future trends, 817
    improving performance properties, 815–17
    types and properties, 808–15

biomass active enzymes, 213
biomass limit, 423–4
biomass pretreatment
　consolidated bioprocessing (CBP), 234–53
　　microorganisms, enzyme systems and bioenergetics, 245–9
　　models, 243–4
　　organism development, 249–53
　　plant biomass polymers, 236
　　process configurations for biofuel production, 235, 241–2
　　various physical pretreatment and effects on biomass structure, 237–40
biomass syngas shift, 181
biomethane
　biohydrogen production via anaerobic digestion and fermentation, 476–513
　　basic principles of biogas and hydrogen production, 477–81
　　biogas and hydrogen production and technological aspects, 481–92
　　current status and limitations, 507, 511–13
　　future trends, 513
　　methane production from different feedstocks, 492–507
bioplastics, 713–14
bioprocess, 15
biorefinery, 14–17, 34–5, 261
　defining bio-processing and, 14–15
　sustainability, 14
　economic assessment, 36–47
　　bioproducts from biochemical biorefineries, 42–6
　　bioproducts from thermochemical biorefineries, 37–42
　　capital and operating costs, 46
　　power generation, 46–7
　environmental and sustainability assessment, 67–84
　　future trends, 83–4
　　interaction between technology and environment, 71
　　life cycle assessment, 74–81
　　methodological foundations of technologies, 68–74
　　results from assessment of economic and social aspects, 81–3
　enzymatic processes and enzyme development, 199–226
　　advantages and limitations of techniques, 225

　　benchmarking enzymes and enzymatic conversion processes, 220–5
　　biochemical conversion, 200–11
　　future trends, 225–6
　　optimising enzymes, 212–20
　　technology and techniques, 211–12
　mixed feedstock source of chemicals, energy, fuels and materials, 11
　overview of concept, 90
　planning. design and development, 92–101
　　design and synthesis, 94–7
　　engineering considerations on up-scaling and implementation, 97–101
　　information flow cascades, 96
　　initial feedstock and product considerations, 92–4
　plant design, engineering and process optimisation, 89–108
　　case study, 101–4
　　future trends, 107–8
　　microalgae biomass, 91–2
　　optimising processes using process analysis, 106–7
　technological processes, 13
　types and product areas, 15–17
　　available biomass feedstocks, 16
　　thermochemical and biochemical processes, 17
　upgrading biorefinery operations, 104–6
　　biomass feedstock production and logistics, 104–5
　　biomass pretreatment and conversion, 105–6
　　process energy output and consumption, 106
biorefining
　bio-based chemicals of carbohydrate conversion and utilisation
　　chemical hydrolysis of cellulose to sugars, 629–35
　　future trends, 650–1
　　routes to market for bio-based feedstocks, 646–50
　　sustainable carbohydrate sources, 625–9
　　types and properties of carbohydrate-based chemicals, 635–45
　bio-based chemicals of lignin conversion and utilisation, 659–84
　　applications of lignin and lignin-based products, 672–84

emerging processes for lignin
production, 668–72
future trends, 684
structure and properties of lignin,
660–3
traditional processes for lignin
production, 663, 666–8
bio-based chemicals of lipid and wax
conversion and utilisation, 693–715
future trends, 714–15
methods of extraction and analysis,
703–6
sources, 697–702
types and properties, 694–7
utilisation, 707–14
bio-based chemicals of protein conversion
and utilisation, 721–32
alternative and novel feedstocks and
production routes, 730–1
biochemical conversion of amino acids
to platform ans specialty chemicals,
728–30
future trends, 731–2
protein and amino acid sources derived
from biofuel production, 722–4
protein isolation, hydrolysis and
isolation of amino acid and chemical
feedstocks, 724–8
bio-based nutraceuticals, 596–615
carbohydrate-based nutraceuticals,
606–9
future trends, 613–14
lipid-based nutraceuticals, 599–603
other nutraceuticals, 609–13
protein and peptide-based
nutraceuticals, 604–6
catalytic process and catalyst
development, 152–86
biomass products upgrading, 160–84
depolymerisation of biomass, 153–60
future trends, 184–6
current and emerging separation
technologies, 112–48
biofuels recovery by solvent extraction
in ionic liquid assisted membrane
contactor, 134–41
glycerin desalting as value added
co-product from biodiesel
production, 126–8
impurities removal from lignocellulosic
biomass hydrolysate liquor for
cellulosic sugars, 121–6

performance indices, 144–7
separation technologies, 114–21
solvent extraction and example of
recovery of value added proteins
from DSG, 130–4
succinic acid production, 128–30
trends for advanced biofuels, 141–4
biorefining process
design options, 279–80
separate hydrolysis and fermentation
(SHF) or simultaneous
saccharification and fermentation
(SSF), 279–80
bioresins, 753–4
biotechnology, 225
BioTemp, 573
black liquor, 289
bovine milk proteins, 839
brittleness, 842–3
Brönsted acids, 640
building materials
biomass usage to produce bio-based
composites, 803–17
fibre properties, 807–8
fibre types and isolation, 804–7
fibrous plants, 804
future trends, 817
improving performance properties,
815–17
types and properties, 808–15
bulk delignification, 666

$C_5$ sugars conversion
furfural, 642–3
biphasic system for reactive ring
opening of furfural to maleic acid,
643
oxidative ring opening of furfural to
succinic acid, 643
$C_6$ sugars conversion
5-HMF, 639–42
biphasic system for reactive extraction
during acid-catalysed fructose
dehydration, 642
potential chemical feedstocks derived
from 5-HMF, 639
proposed acid and base-catalysed
pathways in dehydration of glucose,
640
cactus mucilages, 835–6
caffeic, 845–6
canola oil, 454–5

capital costs, 38
carbohydrate active enzyme systems, 245–6
Carbohydrate-Active EnZYmes, 212–13
carbohydrate adhesives, 751–2
carbohydrate-based chemicals
   types and properties, 635–45
     biochemical conversion of carbohydrates, 635–8
     conversion of $C_6$ sugars to 5-HMF, 639–42
     conversion of $C_5$ sugars to furfural, 642–3
     lactic acid synthesis, 644–5
     levulinic acid production, 643–4
     possible platform chemicals produced from biomass, 636
     thermochemical conversion of carbohydrates, 638–9
carbohydrate-based nutraceuticals, 606–9
   bioactive polysaccharides, 606–8
     structure of lentinan, 607
   extraction and purification of polysaccharides, 608
   prebiotics, 608–9
carbohydrate binding module (CBM), 213
carbohydrate conversion
   bio-based chemicals from biorefining and utilisation, 624–51
   chemical hydrolysis of cellulose to sugars, 629–35
   future trends, 650–1
   routes to market for bio-based feedstocks, 646–50
   sustainable carbohydrate sources, 625–9
   types and properties of carbohydrate-based chemicals, 635–45
carbohydrate utilisation
   bio-based chemicals from biorefining and carbohydrate conversion, 624–51
   chemical hydrolysis of cellulose to sugars, 629–35
   future trends, 650–1
   routes to market for bio-based feedstocks, 646–50
   sustainable carbohydrate sources, 625–9
   types and properties of carbohydrate-based chemicals, 635–45
carbohydrates, 751–2
   biochemical conversion, 635–8
     alternative routes to adipic acid from biomass or petroleum feedstocks, 637

     selected acid catalysed or hydrogenation products of succinic acid, 638
   hydrolysing species, 245
   thermochemical conversion, 638–9
     major components formed via thermochemical processing of hemicellulose and cellulose, 639
carbon black, 677–80
carbon dioxide, 478–9
carbon dioxide supercritical extraction, 704
carbon fibre, 677–80
carbon sources, 318
carboxylation, 178
carotenes, 613
carotenoids, 93, 613, 711–12
carrageenans, 835
case studies, 61
casein, 839
cashew nut shell liquid, 754
cashew tree gum (CTG), 835
catalysis, 105
catalyst development
   catalytic process in biorefining, 152–86
     biomass products upgrading, 160–84
     depolymerisation of biomass, 153–60
     future trends, 184–6
catalysts, 444
catalytic aerobic oxidation, 643
catalytic cracking, 181–2
catalytic process
   catalyst development in biorefining, 152–86
     biomass products upgrading, 160–84
     depolymerisation of biomass, 153–60
     future trends, 184–6
cell contents, 341
cell walls, 340–1
Cellic enzymes, 223
cellobiohydrolases, 215
cellobiosephosphorylase, 247
cellulolytic bacteria, 245
cellulolytic enzymes, 253
cellulose, 340, 827
   hydrolysis, 702
   utilisation, 252
cellulose conversion
   ionic liquids, 631–5
     acidic ILs explored in cellulose hydrolysis, 634
     common cations and anions usage, 632

disruption of hydrogen bonding
  network, 633
cellulose fibres, 849–50
cellulosic biofuel producer tax credit
  (CBPTC), 49
cellulosic reinforcements, 849–52
centrifugation, 504
cereal-based biorefineries
  developments, 303–26
    fuel ethanol production from wheat,
      308–12
    future trends, 322–6
    polyhydroxyalkanoate (PHB)
      production from wheat, 316, 318–20
    succinic acid production from wheat,
      312–16
    utilisation of wheat straw, 320–2
    wheat-based biorefineries, 304–7
  future trends, 322–6
    valorisation of generic cereal-based
      waste streams, 325–6
    valorisation of industrial cereal-based
      waste and by-product streams, 323–5
chain saw oils, 572
charge-based membrane separations, 117–19
chemi-mechanical forces, 850–1
chemical crosslinking, 844–5
chemical hydrolysis
  acid hydrolysis, 629–30
    reaction scheme for aqueous phase
      reforming of cellulose to alkanes, 630
  cellulose to sugars, 629–35
    cellulose conversion in ionic liquids,
      631–5
    heterogeneous catalysts for cellulose
      conversion to
      platform chemicals, 630–1
chemical industry, 372–9
chemical oxygen demand (COD), 504
chemical precipitation, 503–4
chemical pulping, 807
chemical reaction system, 451
Chinese National Standardisation Technical
  Committee for Bamboo and Rattan,
  542–3
chitin whiskers, 852–3
chitosan, 833–4
  nanostructures, 852–3
*Chlorella* biomass, 93, 97
cholesterol, 711–12
*Chrysosporium lucknowense*, 212

cinnamaldehyde, 847–8
*Cladosiphon okamuranus*, 607
Clean Air Act Amendments, 398
clean-in-place (CIP), 116
closed systems, 569
*Clostridium thermocellum*, 212
cloud point (CP), 463
co-fermentation, 280
coatings
  biomass usage for packaging films, 819–56
    components, 822
    crosslinking agents for packaging
      materials, 844–9
    film plasticisers, 842–4
    future trends, 853–5
    processes for producing bio-based films,
      822–4
    processes for producing edible coatings,
      825–6
    products, 826–42
    reinforcements for packaging materials,
      849–53
  dispersion, 825–6
cold filter plugging point (CFPP), 463
cold flow properties, 463–5
  oxidative stability, 463–5
    examples of cloud points of biodiesel
      from different feedstocks, 464
Common Agriculture Policy (CAP), 338
complex glycoside hydrolase systems, 246–7
  structural representation of cellulosome
    as macromolecular enzyme complex,
    248
composting, 21
compressed natural gas (CNG), 400
compressor oils, 573
concrete release agents, 573
condensed distillers solubles (CDS), 782–3
condensed polyflavonoid tannins, 738–41
condensed tannins, 737–8
consolidated bioprocessing (CBP), 210–11,
  282
  biomass pretreatment, 234–53
    organism development, 249–53
    process configurations for biofuel
      production, 235, 241–2
  engineered, 244
    natural vs engineered CBP by
      differentiating unit operations, 244
  microorganisms, enzyme systems and
    bioenergetics, 245–9

bioenergetics, 249
carbohydrate active enzyme systems, 245–6
CBP microorganisms, 245
complex glycoside hydrolase systems, 246–7
mode of action, 247–9
non-complex glycoside hydrolase systems, 246
models, 243–4
ruminant or natural, 243–4
contact digester, 483
continuous stirred tank reactor (CSTRs), 481–3
conventional filtration, 115–16
conversion route divergence, 143
copolymer, 674–6
corn gluten meal, 782
corporate average fuel economy (CAFE), 420
CORTERRA, 373–4
cosmetics, 712
cost-benefit analysis (CBA), 70
cotton gin, 807
cottonseed protein bioplastics (CBP), 713–14
crops residues, 493, 499–501
biofuel yields, 499–500
crosslinking agents
aldehydes, 847–8
chemical structure of cinnamaldehyde, 847
periodate oxidation of glucose residue in polysaccharide and Schiff's reaction, 848
genipin, 846–7
chemical structure, 846
phenolic compounds, 845–6
crude glycerol phase, 366–7
crude protein, 343
cryo crushing, 850–1
crystallisation, 365–6
current efficiency, 147
cyclopropanated oils, 587
conversion of TAG oil by carbene insertion into double bond, 588

Darcy's Law, 146
dark fermentation, 479
de-watering process, 504
decanters, 787–8
decorticating machine, 807

dedicated alcohol engines, 414–15
dedicated energy crops, 47
degree of polymerisation (DP), 203
dehydration, 177–8
deoxygenation, 164–6
cracking, 173–4
deproteinated brown juice, 349
detoxification, 207
dialdehyde polysaccharides, 848
dialysis, 365–6
dietary fibres, 608
Dietary Supplement Health and Education Act (DSHEA), 597
dilute acid pretreatment, 205
dipping, 825–6
direct carbonation, 378
Directive 99/31/EC, 18
Directive 2003/108/EC, 5
Directive (EC 1907/2006), 5
disaccharides, 843–4
disk stack, 788
dissolved oxygen tension (DOT), 280
distillation, 103, 308–9
distillers dried grains with solubles (DDGS), 772–3, 775–8
global production from 2002 to 2011, 776
nutritional value as feed ingredient, 775–7
nutritional content of wheat vs wheat DDGS vs maize DDGS, 777
usage globally in animal feed sectors, 778
processing issues, 778
variability in production, 777
distiller's grains and solubles (DSG), 130–4
distinct conversion processes, 142
dolomite, 157–8
double-zero rapeseed oil, 454–5
dried distillers grains with solubles (DDGS), 308, 722–3
dry-forming process, 811–12
dry process, 824
dry washing, 444–5, 445–6
drying technologies, 784–5
freeze drying, 785
ring drying, 784–5
spray drying, 785
*Dunaliella salina*, 94–5
Dupont, 646–7

eco-hydraulic oils, 572
economic assessment, 36–47

edible coatings
  processes for production, 825–6
    basic steps to prepare and apply a coating to food surface, 825
electrodeionisation (EDI), 116, 118–19, 146–7
electrodialysis, 116, 146–7, 352, 727
electrospinning, 850–1
elucidation, 216–17
emerging feedstocks, 456–8
emerging lignin products, 677–84
  activated carbon and carbon black and carbon fibre, 677–80
    physical properties of lignin and carbon black reinforces styrene-butadien copolymer rubber, 678
  antioxidants, 677
  enzymatically treated lignins, 683–4
  platform chemicals, 680–3
    major thermochemical lignin conversion processes and potential products, 680
Emerging Sustainability Assessment Framework, 70
emission reduction units (ERU), 52
Emission Trading Scheme (ETS), 52
endo-1,4-β-glucanases, 216
endo-β-xylanases, 217–18
endoglucanases, 245
energy biorefinery, 284
energy carriers, 390–4
energy crops, 493
  biofuel yields, 494–8
  estimation of electricity potential of various energy crops, 493
Energy Policy Act (2005), 49
energy products, 354–5
  bioethanol, 355
  biogas, 354–5
  thermal combustion, 354
engine management system (EMS), 418–19
engineered consolidated bioprocessing, 244
environmental impact assessment (EIA), 70
Environmental Protection Agency (EPA), 60, 549
environmental risk assessment (ERA), 70
EnviroTemp Fr-3, 573
enzymatic catalysts, 162
enzymatic conversion processes, 220–5
enzymatic dehydrogenation, 660–1
enzymatic hydrolysis, 103, 121, 200–11, 269, 271–2, 322, 850–1

enzymatic processes
  enzyme development in biorefining, 199–226
    advantages and limitations of techniques, 225
    benchmarking enzymes and enzymatic conversion processes, 220–5
    biochemical conversion, 200–11
    future trends, 225–6
    optimising enzymes, 212–20
    technology and techniques, 211–12
enzymatically treated lignins, 683–4
enzyme development
  enzymatic processes in biorefining, 199–226
    advantages and limitations of techniques, 225
    benchmarking enzymes and enzymatic conversion processes, 220–5
    biochemical conversion, 200–11
    future trends, 225–6
    optimising enzymes, 212–20
    technology and techniques, 211–12
enzyme technology
  benchmarking, 220–1
    novozymes achieved further 1.9-fold dose reduction during second DOE-funded project, 222
    novozymes achieved sex-fold enzyme dose reduction during DOE-subcontract to NREL, 221
enzymes, 786
  catalysis, 155
epicerol, 378, 722–3
epichloridrin, 378
epoxidised soybean oil, 587–8
epoxy resins, 676
essential oils, 712
estrolide esters, 585–7
  synthetic scheme of production, 586
estrolide lubricants technology, 586–7
ethanol, 403
ethanol biorefineries, 261–3
  classification and availability of various lignocellulose sources, 264–5
  conceptual picture of biorefinery showing main technology choices, 262
  different types, 282–8
    commodity chemicals potentially derived from lignocellulose feedstocks, 286–7

conceptual figure for $C_5$-driven
    bioethanol-based biorefinery, 284
conceptual figure for energy-driven
    bioethanol-based biorefinery, 283
conceptual figure for lignin-driven
    bioethanol-based biorefinery, 283
operational demonstration plants and
    commercial plants currently under
    construction, 285
European Bioethanol Fuel Association, 773
European Food Standards Agency (EFSA),
    789
European Union (EU), 396–7
subsidy programs, 51–2
evaporation, 504
Excel-based model, 223
exhaust gas recirculation (EGR), 414–15
exoglucanases, 246
expansins, 218
extended range electric vehicles (EREV),
    392–3
extraction, 120–1, 365–6
extreme pressure (EP), 570
extrusion, 824
Exxon Valdez-sised spills, 564

Fabaceae, 336
fatty acid composition, 575–6
fatty acid esters, 568
fatty acid methyl ester (FAME), 405, 707
fed-batch mode, 319–20
feed additives, 790–1
feedstock availability, 462–3
feedstock variability, 141–2
feedstocks, 293–4
    emerging, 456–8
        fatty acid compositions of oils, 458
        quality issues, 458–61
            high content of free fatty acids, 458–9
            impurities that affect product quality, 459–61
    traditional, 454–6
        fatty acid compositions of common seed oils and animal fats, 456
fermentation, 103, 204, 272–3, 308–9, 636–7
    biomethane and biohydrogen production via anaerobic digestion, 476–513
    basic principles of biogas and hydrogen production, 477–81
    biogas and hydrogen production and technological aspects, 481–92

current status and limitations, 507, 511–13
future trends, 513
methane production from different feedstocks, 492–507
fertilisation, 344–5
fertiliser, 505–6
fibre saturation point (FSP), 814–15
fibre types, 804–5
    six general types of natural fibres, 806
fibreboard, 807, 811–13
    properties, 811–13
fibres
    bioadhesives types, processing and properties for wood, 736–65
    carbohydrate adhesives, 751–2
    future trends, 762–5
    lignin adhesives, 745–9
    mixed tanin-lignin adhesives, 749–50
    protein adhesives, 750–1
    tannin adhesives, 737–45
    unsaturated oil adhesives, 752–5
    wood welding without adhesives, 755–62
    properties, 807–8
        chemical composition of some natural fibres, 808
        mechanical properties of some natural fibres, 809
        tensile and flexural properties of some natural fibres, 809
fibrous plants, 804
    global inventory of biofibres, 805
film casting methods, 823
film plasticisers, 842–4
filter systems, 814
Fischer–Tropsch liquids, 38, 320–1
Fischer–Tropsch synthesis, 179–81, 262–3, 368, 407, 425, 638, 700–1
fish oils, 600
5-hydroxymethylfurfural (HMF), 639–40
flakeboard, 810–11
flame ionisation detector (FID), 706
flash vaporisation, 122–3
flexible-fuel vehicles (FFVs), 397–8, 402–3, 417–20
    representation of physical sensor system to detect concentration of ethanol, 420
    representation of virtual sensor system to detect concentration of ethanol, 419

flocculation, 504
fluidised beds, 484
flux, 144, 145–6
Food and Agriculture Organisation of United Nations (FAO), 325
Food and Drug Administration (FDA), 597
food packaging, 854–5
Food Safety Modernisation Act, 789
food waste, 6, 20, 325, 506–7, 611
  hydrogen production usage, 510
  methane yield, 508–9
forage-based biorefineries developments, 335–55
  field to biorefinery and impact of herbage chemical composition, 340–7
  green biorefinery products, 347–55
  overview, 336–9
forage crop harvesting, 346
formate, 478
formic acid pulping, 670
fossil waxes, 700–1
fouling, 146
Fourier transform-infrared (FTIR) spectroscopy, 830
free amino nitrogen (FAN), 310
free fatty acids (FFAs), 444
freeze drying, 785
frictional wood welding systems, 758–62
  high speed rotation dowel welding, 760–2
    schematic example of frictional movement usage, 761
    SEM image of interface of dowel welded to substrate, 762
    well-welded dowel where two pieces joined by clamping during dowel insertion, 761
  linear vibration welding, 758–60
    examples of different butt joints obtained by endgrain welding of oak wood, 760
    schematic example of frictional movement usage, 759
fuel ethanol production
  wheat, 308–12
    integrated wheat-based fuel ethanol and arabinoxylan (AX) production process, 311
    typical first generation process, 309
    worldwide fuel ethanol production from 2006 to 2011, 308
Fuel Quality Directive (FQD), 393–4, 403

fuel synthesis, 425–7
functional fibres, 679
fungal autolysis, 307
fungal enzymes, 208
fungal fermentation, 307
fungi, 700
furans, 277
  platform, 166–8
furfural, 642–3
future crops, 293–4

galacto-oligosaccharides (GOS), 609
galactomannan films, 828–9
$\gamma$-Valerolactone platform, 170–2
garapa, 42
gas chromatography (GC), 706
gas chromatography-mass spectrometry (GC-MS), 706
gas clean-up system, 539
gassification
  biomass, 156–8
    usages of syngas, 157
gel permeation, 608
gelatin, 838–9
General Algebraic Modelling System (GAMS), 310
general separations platforms, 143
generally regarded as safe (GRAS), 273
generic cereal-based waste streams, 325–6
generic fermentation feedstock, 323–4
genetic engineering, 272–3
genetic modification potential, 792–3
  feedstock, 793
  microorganisms usage in biorefining, 793
genipin, 846–7
  chemical structure, 846
*Geobacillus thermoglucosidasius,* 212
geographical information systems (GIS), 94
geotextiles, 814
GH61s proteins, 216–17
Gibbs free energy, 95
gliadin films, 848
global industry analyst forecast, 564–5
global warming potential (GWP), 75, 80
glucosinolates, 612
glutamic acid, 78, 728–9
gluten, 840–1
glyceraldehyde, 377
glyceric acid, 377
glycerin, 444–5

glycerin desalting
  value added co-product from biodiesel production, 126–8
    different pathways for production of various useful chemicals, 127
    reaction scheme of biodiesel, 126
    removal efficiency for NaCl from simulated crude glycerin solution, 128
    schematic desalting of crude glycerin stream using resin wafer-EDI, 127
glycerol, 365–7, 575, 843
  composition and purification produced from biodiesel, 365–7
    average composition of crude glycerin from Brazilian biodiesel plant, 366
  dehydration to acrolein and acrylic acid, 375–6
    oxidative dehydration, 376
  gasification, 368
  glycerol to propanediols, 372–4
    hydrogenolysis of glycerol over metal catalysts to afford 1,2 and 1,3 propanediols, 373
    production of PTT from reaction of terephthalic acid and glycerol, 373
  glycerol to propene, 374–5
    possible mechanistic pathway for hydrogenolysis over Fe-Mo supported catalyst, 375
    selective hydrogenolysis over Fe-Mo catalysts supported over activated carbon, 374
  oxidation, 377
  purification, 367
  raw material for chemical industry, 372–9
    glycerol oxidation, 377
    other transformations, 378–9
  transformation
    epicerol process, 378
    hydrogenolysis, 176–7
    main reactions involved in upgrading glycerol, 175
    oxidation, 176
    production of glycerol carbonate from carbon dioxide, 379
    three different procedures for synthesis of glycerol carbonate, 379
glycerol byproduct-based biorefineries
  applications of glycerol in fuel sector, 367–72
    biotechnological pathway of ethanol production from glycerol, 368
    complete combustion, 367
    esterification of glycerol with acetic anhydride, 371
    free radical resonance structures showing electron delocalisation, 370
    oxidation stability of soybean biodiesel with furfural/glycerol acetals, 371
    production of solketal acetate, 372
    reaction of glycerol with acetone and formaldehyde in presence of acid catalysts, 369
    reaction of glycerol with anisaldehyde, 370
    reaction of glycerol with ethanol in presence of acid catalysts, 369
  developments, 364–81
    composition and purification of glycerol produced from biodiesel, 365–7
    future trends, 379–81
    glycerol as raw material for chemical industry, 372–9
    transesterification of triglycerides to produce fatty acid methyl esters and glycerol, 365
glycerol carbonate, 178, 378
glycerol transformation, 175–8, 378–9
  carboxylation, 178
  dehydration, 177–8
  steps of glycerol to acrolein, 177
glycerophospholipids (GPLs), 602–3
Good Manufacturing Practices (GMP), 597
grass-based biorefineries
  developments, 335–55
    field to biorefinery and impact of herbage chemical composition, 340–7
    green biorefinery products, 347–55
    overview, 336–9
GRASSA green biorefinery research project, 349
grasses, 336–7, 660
grassland
  composition, 336–7
  management, 343–5
  productivity and potential availability, 339
  role and importance in Europe, 337–8
green biorefinery, 336
  press-cake fraction, 352–4
    potential applications for direct usage of separated press-cake fraction, 353

products, 347–55
  energy products, 354–5
  press-juice fraction, 347–52
green chemistry, 3–6
  changes, 7–9
  future trends, 28
  petro- to bio-refineries, 10–13
  principles, 4
  product substitution, 9–10
  product life cycle, 10
green juice, 348–51
green plant chemical composition, 340–3
  cell contents, 341
  cell walls, 340–1
    schematic representation of changes of grass with advancing maturity, 341
  lipids, 343
  non-structural carbohydrates, 342
  organic acids, 343
  proteins, 342–3
greenhouse gas (GHG) emissions, 35, 50, 75–6, 292
grinding, 58, 807

Hagen–Poiseuille's equation, 145
hardwoods, 660
harvest date, 344
heat biomass, 817
heat coagulation, 349
heat shock proteins (HSPs), 278
heat treatments, 816
hemicellulase debranching enzymes, 218
hemicellulose, 322, 340, 627–8, 827–9
herbaceous energy crops, 47
herbage proteins, 342
heterogeneity
  transesterification, 583–5
    general structure features of molecular species of high oleic oil with TMP tri-2-ethylhexanoate, 585
    typical structural features of heterogeneous oil, 584
heterogeneous alkali catalysts, 445
heterogeneous catalysts, 162
  biodiesel production, 445–51
    examples of commonly researched and their effectiveness, 447–50
  cellulose conversion to platform chemicals, 630–1
    hydrolytic hydrogenation of cellulose to sugar polyols, 631

heterologous cellulases, 252–3
heterologous gene expression, 250–1
high density fibreboards (HDF), 812
high-energy pulse electron (HEPE), 278
high oleic oil-based ester fluids, 581
high oleic oils production, 579–80
high oleic sunflower oil formulations, 589
high-performance biolubricants
  chemical modifications of feedstocks, 581–90
    cyclopropanated oils, 587
    estrolide esters, 585–7
    heterogeneity through transesterification, 583–5
    isomerisation of oleic acid with zeolite catalyst followed by hydrogenation and esterification, 589
    other esters, 587–90
    synthetic esters, 581–3
  development, properties and applications, 556–92
    applications, 571–4
    coefficient of friction of base oils without additives, 559
    evaporation losses of mineral oil vs fatty acid esters, 560
    future trends, 590–1
    markets for lubricants, 561–6
    molecular structural features of triacylglycerol oil vs petroleum-based mineral oil, 559
  feedstocks and key properties, 574–81
    functionality of TAG oils, 575–9
    high oleic oil-based ester fluids, 581
    high oleic oils production, 579–80
  performance requirements, 566–71
    dependence of lubricant properties of fatty acid esters on molecular structure, 568
high-performance liquid gas chromatography (HPLC), 706
high-pressure homogenisation, 850–1
high speed rotation dowel welding, 760–2
high value products, 677
higher heating value (HHV), 425–6
homogeneous catalysts, 162
homologous recombination, 250
household devices
  biochar, 531–2
    clean burning biochar stove and smokey traditional stove, 532

hybrid hydrolysis and fermentation
    (HHF), 209–10
  illustration, 209
hybrid processes, 281–2
hydraulic fluids, 565
hydraulic oils, 572
hydraulic retention time (HRT), 485
hydrodeoxygenation (HDO), 174–5, 635
hydrodesulfurisation (HDS), 174
hydrogen, 391–2, 479–81, 512–13
  partial pressure, 486–7
  productivity, 480
hydrogen bonds, 407–8
hydrogenated vegetable oils (HVO), 406
hydrogenation, 163–4, 442
hydrogenolysis, 155–6, 176–7, 372–3
hydrolysis, 44, 153–5, 182, 477–8, 632
  structure of lignocellulose biomass, 154
hydrolysis interface, 203–6
hydrophobic catalysts, 649
hydrothermolysis, 490
hydrotreating, 163–4

Inbicon Biorefinery
  case study, 101–4
    biomass transport and storage, 102
    biorefinery siting, 102
    enzymatic hydrolysis, 103
    fermentation and distillation, 103
    final solid/liquid separation, 103–4
    mechanical pre-treatment, 102
    thermal treatment, 102–3
industrial cereal-based by-product streams,
    323–5
industrial-scale cellulosic biofuels, 235,
    241
industrialisation
  process development, 288–9
    list of critical process development
      issues at various stages of
      biorefinery operation, 289
    recent patent applications related to
      biorefinery concept, 290–1
inhibitor tolerance, 276–8
initial delignification, 666
integrated catalytic processing (ICP), 41
integrated renewable energy system
  renewable fuels, 427–9
    integrated power, heat and transport
      system featuring large-scale
      energy storage, 428

internal combustion engines
  liquid biofuels improving usage, 389–429
    competing fuels and energy carriers,
      390–4
    future provision of renewable liquid
      fuels, 423–9
    market penetration, 394–405
    usage, 405–17
    vehicle and blending technologies for
      alcohol fuels and gasoline, 417–23
internal rate of return (IRR), 222
International Assessment of Agricultural
    Science and Technology for
    Development, 723–4
International Conference of Harmonisation
    (ICH), 597
International Energy Agency (IEA), 261
International Monetary Fund (IMF),
    723–4
International Organisation for
    Standardisation (ISO), 292
International Union of Biochemistry and
    Molecular Biology's Enzyme
    Nomenclature and Classification,
    212–13
inulin, 609
ion exchange, 365–6, 608
  purification, 367
ion exchange chromatography, 352
ion exchange resins, 366–7, 454
ionic impurities removal, 122–4
ionic liquid assisted membrane contactor,
    134–41
  concentration profile of recovered
    ethanol in ionic liquid phase, 141
  liquid-liquid extraction of ethanol, 137
  schematic of continuous ethanol
    recovery, 138
  separation factor for ethanol separation
    from water, 139
  thermodynamic calculation of theoretical
    energy consumption of membrane-
    based recovery, 140
ionic liquids, 137, 155, 631–5, 670–1
iron-based catalysts, 180–1
ISO 14040, 72, 292
ISO 14044, 72
iso-stoichiometric ternary blends, 421–3
  relationship between blend proportions of
    gasoline, ethanol and methanol, 422
isomerisation, 589, 641

Japan Biochar Association, 548
jatropha, 456–7, 590

ketals, 369–70
kilns, 532–8
  drum TLUD oven being field tested in Vietnam, 538
  heating curve for Iwate kiln, 535
  large-scale production of biochar and wood vinegar in China, 536
  modified Adam retort, 537
  ring kiln design, 537
  two different types of Japanese kilns Iwate and portable iron kiln, 533
kraft lignin, 679–80
kraft process, 666–7

lab-scale assay, 220
lactic acid, 351–2
  synthesis, 644–5
    proposed reaction scheme for converting monosaccharides, 645
land use, 77
Landfill Directive, 18
landfilling, 325
lanolin, 701, 712
leaf protein extraction, 348–9
leaves, 730
lectins, 605
legumes, 336–7
lentinan, 606–7
less conventional polysaccharides, 834–6
levulinic acid, 168, 170
  platform, 168
  production, 643–4
    selected acid-catalysed or hydrogenation products from levulinic acid, 644
Lewis–Brönsted acid ratio, 641
life cycle analysis, 60–1
life cycle assessment
  basics, 72–4
    stages, 73
  biorefineries, 74–81
    results for bio-based chemicals, 78–81
    results for biofuels, 74–7
  concept, 81–2
  studies, 292–3
life cycle costing (LCC), 81–2
life cycle impact assessment (LCIA ), 73–4, 80

life cycle inventory analysis (LCI), 73–4
life cyle thinking, 69
lignin, 202–3, 271, 340, 745–6, 830–1
  depolymerisation, 682–3
  emerging processes for production, 668–72
    oligomeric structures proposed for pyrolytic lignin, 671
    typical cation and anion combinations, 672
  reduction, 182–3
  structure and properties, 660–3
    approximate composition of some important classes, 661
    hardwood native lignin, 662
    main interunit linkages, 661
    percentage distribution of lignin units, 662
    phenyl propane units, 660
    properties from different extraction processes, 664
    properties of nonwood lignins, 665
    softwood native lignin, 663
  traditional processes for production, 663, 666–8
    kraft pine lignin, 667
    spruce lignosulfonate chemical structure, 668
  usage, 181–4
    catalytic cracking, 181–2
    hydrolysis, 182
    oxidation, 183–4
    reduction, 182–3
lignin adhesives, 745–9
  lignins, 745–6
  technology, 747–9
  usage without adding other synthetic resins, 746–7
lignin conversion
  bio-based chemicals from biorefining and lignin utilisation, 659–84
    applications of lignin and lignin-based products, 672–84
    emerging processes for lignin production, 668–72
    future trends, 684
    structure and properties of lignin, 660–3
    traditional processes for lignin production, 663, 666–8

lignin utilisation
  bio-based chemicals from biorefining and lignin conversion, 659–84
    applications of lignin and lignin-based products, 672–84
    emerging processes for lignin production, 668–72
    future trends, 684
    structure and properties of lignin, 660–3
    traditional processes for lignin production, 663, 666–8
lignocellulose, 625
lignocellulose-ethanol process, 263, 266–78
  fermentation, 272–3
  hydrolysis, 269, 271–2
  inhibitor tolerance, 276–8
    challenges faced by yeast *Saccharomyces cerevisiae* during ethanol production, 276
    schematic representation on how to deal with inhibition, 277
  pentose utilisation, 273–6
    some xylose fermenting engineered strains of *Saccharomyces cerevisiae*, 275
    xylose and arabinose pathways expressed in recombinant *Saccharomyces cerevisiae*, 274
  pretreatment, 263, 269
    summary of different methods, 270
  thermo-tolerance, 278
lignocellulosic biomass, 36, 44, 121, 488, 827–31
  cellulose and derivatives, 827
  hemicellulose, 827–9
  lignin, 830–1
    monolipids chemical structure, 830
  pectins, 829
    fragment containing esterified and non-esterified carboxyl groups of galacturonic acid, 829
lignocellulosic biomass hydrolysate liquor
  removal of impurities for cellulosic sugars production, 121–6
    adsorptive beads for removal of non-ionic impurities, 124–6
    RW-EDI for removal of ionic impurities, 122–4
    schematic of derived biofuel production route and role of intermediate separation steps, 122

lignocellulosic materials, 701–2
  depolymerisation, 153–6
  hydrogenolysis, 155–6
  hydrolysis, 153–5
lignosulfonate, 673–4, 679–80, 745–6
linear vibration welding, 758–60
lipid-based nutraceuticals, 599–603
  types, properties and functions, 599–603
    omega-3 polyunsaturated fatty acids, 600
    phytosterols, 601–2
    polar lipids, 602–3
lipid conversion
  bio-based chemicals from biorefining of wax conversion and utilisation, 693–715
    future trends, 714–15
    methods of extraction and analysis, 703–6
    sources, 697–702
    types and properties, 694–7
    utilisation, 707–14
lipid hydroprocessing, 41
lipid utilisation
  bio-based chemicals from biorefining of lipid and wax conversion, 693–715
    future trends, 714–15
    methods of extraction and analysis, 703–6
    sources, 697–702
    types and properties, 694–7
    utilisation, 707–14
lipids, 343, 506, 599, 841–2, 844
  sources, 697–702
    extracted from microalgae, 698–700
    extracted from plants, 697–8
    yeast and fungi, 700
  types and properties, 694–7
    other lipid compounds, 696–7
    phospholipids, 694
    terpenoids, 696
    triglycerides, 694
    wax esters, 694–6
liquefaction, 308–9
liquefied natural gas (LNG), 400
liquid biofuels
  competing fuels and energy carriers, 390–4
    on-board energy density and technology costs, 390–3
    environmental benefits, 393–4

well-to-wheel $CO_2$ emissions as function of fuel well-to-tank carbon intensity, 395
improving usage in internal combustion engines, 389–429
future provision of renewable liquid fuels, 423–9
usage, 405–17
vehicle and blending technologies for alcohol fuels and gasoline, 417–23
market penetration, 394, 396–405
alternative road transport fuels as fraction of global total alternative fuel supply, 396
liquid feeds, 782–3
liquid hot water (LHW), 490
liquid-liquid extraction, 134–6, 138–9, 147
liquid petroleum gas (LPG), 400
liquidity, 577–8
long-chain fatty acids, 486
loss lubricants, 569
low density fibreboards (LDF), 812
low target concentration, 142–3
lubricant dilution, 415–17
lubricants, 712–13
markets, 561–6
general price of lubricant base oils vs vegetable oils, 562
vegetable oil production volumes and yields, 566
lubrication fluids, 563–4

macropores, 647–8
magnesium silicate, 453–4
magnetic sulfonated materials, 630–1
Maillard reactions, 776, 785–6
manure, 501–6
anaerobic digestion, 503
solid and nutrient content of various manure types, 502
volatile solid reduction vs solid loading rate treating dairy manure, 502
volatile solid reduction vs solid loading rate treating swine manure, 503
mat formation, 813
material flow analysis (MFA), 70
matrix synthesis, 95
mechanically-induced friction welding techniques, 756–7
mechanocatalysis, 650–1

medium density fibreboard (MDF), 741–2, 812
membrane distillation (MD), 116
membrane separation technologies, 114–19
charge-based membrane separations, 117–19
photographs of three different sizes of commercially available ED stacks, 120
working principles of electrodeionisation, 118
working principles of electrodialysis, 117
size or solubility based-membrane separations, 116–17
metabolic engineering, 250–1
metal catalysts, 631
metal hydroxides, 446
metal ion absorption, 676–7
metal oxides, 446
metals, 486
metalworking fluids, 573
methane, 478–9
methane production
biohydrogen production from different feedstocks, 492–507
crops residues, 493, 499–501
energy crops, 493
food wastes, 506–7
manure, 501–6
methanogenesis, 478
methanol synthesis, 179, 638
methylolurea mixtures, 740–1
microalgae, 601–2, 730
biomass, 95
biorefinery, 91–2
lipids extraction, 698–700
examples of microalgae cultivation for oil accumulation, 699
oil, 457–8
microbial electrolysis cells (MEC), 479
microemulsion, 708–9
microfiltration, 115, 145–6
microwave-assisted extraction, 11, 704–5
microwave heating, 635
microwave pyrolysis, 26–7
microwave technology, 12, 27
milk proteins, 839
milling, 308–9, 488
minimum ethanol selling price (MESP), 222

894  Index

Ministry of Agriculture, Forestry and Fisheries (MAFF), 547
mixed interior wood panel tannin adhesive, 749–50
mixed tanin-lignin adhesives, 749–50
moisture, 459–60
molasses, 783
monoglycerides, 711
monosaccharides, 843–4
monovalent alkali metals, 158
Monte Carlo simulations, 62
motor octane numbers (MON), 411
moulded products, 813
moving bed pyrolyser, 538
multiple separation steps coordination, 143
mutagenesis, 250
*Myceliopthora thermophila,* 212
mycotoxins, 790

NAMASTE project, 24
nanocomposites, 851–2
nanofiltration, 115, 352
National Advanced Biofuels Consortium (NABC), 60
National Biodiesel Board, 462
National Institute of Health (NIH), 597
National Renewable Energy Laboratory (NREL), 220, 269, 271, 698–9
natural consolidated bioprocessing, 243–4
non-complex glycoside hydrolase systems, 246
non-engine lubricants, 571
non-ionic impurities removal, 124–6
non-starch polysaccharides (NSP), 786
non-structural carbohydrates, 342
nuclear magnetic resonance (NMR) spectroscopy, 705–6
nutraceuticals, 710–12
  definition and history, 596–7
  legislation and regulation, 597
Nylon 6, 639–40

octacosanol, 712
octane numbers, 411–12
  properties of 95 RON gasoline, methanol, ethanol and iso-butanol, 412
oilseed meals, 778–82
  global production of oil bearing seeds in 2011 with total and digestible protein content, 779
  other oilseed meals, 781–2
  rapeseed meal, 780–1
  recommended rate of Canola inclusion in production animal diets, 781
  top producing countries in 2011–12, 780
  soya bean meal, 779–80
oilseeds, 725
olefin polymerisation, 700–1
oleic oils, 565–6
olive mill wastewater (OMWW), 506–7, 611
olivine, 157–8
omega-3 polyunsaturated fatty acids, 600, 711
omega-6 polyunsaturated fatty acids, 711
on-board energy density
  technology costs, 390–3
    cost comparison of alternative energy vehicles, 392
    net system volumetric and gravimetric energy densities, 391
open-core down-draft gasifier, 539–40
operating conditions compatibility, 143
optimising enzymes, 212–20
  beta-glucosidase, 214–15
  cellobiohydrolases, 215
  endo-1,4-β-glucanases, 216
  endo-β-xylanases and β-xylosidases, 217–18
  fungal cellulase synergistically deconstruct cellulose microfibrils to monomeric glucose, 214
  GH61s, 216–17
  hemicellulase debranching enzymes, 218
  other activities, 218–19
  thermostabilisation and development of thermally active enzyme cocktails, 219–20
organic acids, 343
organism development, 249–53
  metabolic engineering, 250–1
  natural vs engineered GH systems, 252–3
  strategies employed in research development for CBP-enabling microorganism, 251
organosolv lignin, 675
organosolv process, 668–9
oriental waferboard (OWB), 810–11
oscillatory baffled reactor (OBR), 649–50
ovens, 532–8
oxidation, 176, 183–4, 577
oxidative activation, 746–7
oxidative dehydration, 376

oxidative stability, 463–5, 575
oxygen, 486

packaging films
　biomass usage for coatings, 819–56
　　components, 822
　　crosslinking agents for packaging materials, 844–9
　　film plasticisers, 842–4
　　future trends, 853–5
　　processes for producing bio-based films, 822–4
　　processes for producing edible coatings, 825–6
　　products, 826–42
　　reinforcements for packaging materials, 849–53
palm oil methyl ester (POME), 405
paper industries
　future transitions, 289, 292
　　integrated forest biorefinery based on Kraft pulp mill, 292
paraffins, 443
paraformaldehyde fine powder, 740–1
particleboard, 811
　properties, 812
particulate matter (PM), 417
partition coefficient, 147
pearled wheat flour, 307
pectins, 829
pennycress, 457
pentose utilisation, 273–6
peptide-based nutraceuticals, 604–6
performance indices, 144–7
　adsorption, 147
　　adsorption capacity, 147
　　electrodialysis (ED) and electrodeionisation (EDI), 146–7
　　current efficiency, 147
　　productivity, 147
　　separation efficiency, 146–7
　liquid-liquid extraction, 147
　　partition coefficient, 147
　reverse osmosis (RO), 144–5
　ultrafiltration and microfiltration, 145–6
　　flux, 145–6
　　selectivity, 146
periodic anaerobic baffled reactor (PABR), 484
personal care products, 712
pervaporation (PV), 116, 134–6

petroleum, 697
pharmaceuticals, 710–12
　carotenoids and cholesterol, 711–12
　monoglycerides, 711
　　structure of two isomeric forms, 711
　other applications, 712–14
　　bioplastics, 713–14
　　cosmetics and personal care products, 712
　　lubricants, 712–13
　polyunsaturated fatty acids (PUFA), 710–11
　　structure of n-3 PUFA, 710
phenol, 754–5
phenolic compounds, 610–12, 845–6
phenylammonia lyase, 729
phospholipids, 694
　representative structure of common lipids, 695
phosphorus, 460–1
photocatalysis, 729
photochemical degradation, 815
phycobiliproteins, 605
phytosterols, 601–2
plant cell cell polysaccharides, 201–2
plant lipids, 697–8
plant proteins, 342
platform chemicals, 78, 680–3
platform systems, 212
plug-flow digesters, 483–4
plug flow reactors (PFR), 481
plug-in hybrid electric vehicle (PHEV), 392–3
Poaceae, 336
polar lipids, 602–3
　structure of polar lipids, 602
　structure of sulfoquinovosylacylglycerol, 603
pollutant emission, 415–17
polyalkyleneglycols (PAG), 562
polyalphaolefins (PAO), 562
polyamide synthesis, 646–7
polycondensation reaction, 739–40
polyethylene (PE), 80
polyethylene terephthalate (PET), 80
polyhydroxyalkanoate production
　wheat, 316, 318–20
　　total dry weight, PHB, residual microbial biomass and PHB content in microbial cells, 319
polyhydroxyalkanoates (PHA), 80, 316, 318

polyhydroxyalkonate, 41
polyhydroxybutyrate (PHB), 316, 318
polylactic acid (PLA), 80
polymer surface modifications, 853–4
polymeric adsorptive beads, 124–6
polymeric ion exchange resin beads, 124–6
polymers, 80
polyols, 843
polyphenolic compounds, 610
polyphosphatidyl choline (PPC), 603
polysaccharide hydrolysing enzymes, 247
polysaccharides, 606, 831–6
  alginates, 833
    basic structures containing units of mannuronic and guluronic acids, 833
  chitosan, 833–4
    chemical structure, 834
  less conventional polysaccharides, 834–6
    bacterial cellulose, 836
    cactus mucilages, 835–6
    carrageenans, 835
    tree exudates, 835
  starches, 831–2
    chemical structure of starch fragment, 832
polytrimethylene (PTT), 80
polyunsaturated fatty acids (PUFA), 578, 710–11
polyurethane films, 676
potato starch production residues, 730–1
pour point (PP), 463
prebiotics, 608–9
precipitation, 727–8
preservation, 345–6
press-cake fraction, 352–4
press-juice fraction, 347–52
  green juice, 348–51
    overview of some green biorefinery activities in Europe, 350
    schematic representation of green biorefinery concept, 348
    schematic representation of Havelland green biorefinery in Germany, 351
  silage juice, 351–2
    schematic representation of Utzenaich green biorefinery in Austria, 352
pressing extraction, 703–4
pressure swing distillation (PSD), 310
pretreated slurries, 204

pretreatment, 203–6, 263, 269
  methods, 488–92
    examples of acid pretreatment of lignocellulosic materials, 490
    examples of alkaline pretreatment of lignocellulosic materials, 491
    examples of biological pretreatment of lignocellulosic materials, 492
ProAlcool, 397
process analytical technologies (PAT), 106–7
process intensification
  increasing dry-matter content, 280–2
  hybrid processes and novel concepts, 281–2
process logistics, 93
product-driven biorefineries, 16
product fermentation, 206–11
product heterogeneity, 142
product purity requirements, 142
production costs, 462–3
productivity, 147
project PIVERT, 24
propanediols, 372–4
propene, 374–5
Propylene Glycol Renewable (PGR), 373
Protamylasse, 730–1
proteases, 726
protein adhesives, 750–1
protein-based nutraceuticals, 604–6
  algae-derived bioactive peptides, 605–6
  algae proteins, 604–5
protein conversion
  alternative and novel feedstocks and production routes, 730–1
    protein from algae, 730
    protein from leaves, 730
    protein from potato starch production residues, 730–1
  bio-based chemicals from biorefining of protein utilisation, 721–32
    (bio)chemical conversion of amino acids to platform ans specialty chemicals, 728–30
    future trends, 731–2
    protein and amino acid sources derived from biofuel production, 722–4
    protein isolation, hydrolysis and isolation of amino acid and chemical feedstocks, 724–8
    generic scheme of amino acid or protein production from biomass, 725

protein hydrolysis, 726–7
protein isolation, 725–6
protein utilisation
  bio-based chemicals from biorefining of protein conversion, 721–32
    alternative and novel feedstocks and production routes, 730–1
    (bio)chemical conversion of amino acids to platform ans specialty chemicals, 728–30
    future trends, 731–2
    protein and amino acid sources derived from biofuel production, 722–4
    protein isolation, hydrolysis and isolation of amino acid and chemical feedstocks, 724–8
proteins, 342–3, 836–41
  gelatin, 838–9
  gluten, 840–1
  milk proteins, 839
  zein, 839–40
proteolysis, 346
proton exchange membrane fuel cell (PEM FC), 392–3
purification, 453–4
purified enzymes, 491–2
pyrolyser, 539
pyrolysis, 12, 41, 47, 681, 708
  process, 670

Rankine cycle, 539
rapeseed meal, 780–1
rapeseed methyl ester (RME), 405
raw glycerin phase, 366
recycling $CO_2$, 424–5
reduced order models (ROMs), 62
refinery reactor, 164
refining glycerol, 366
refining process, 807
Registration, Evaluation, Authorization and Restriction of Chemicals (REACH), 5, 7
regulatory issues, 789–91
  antibiotics, 790
  feed additives, 790–1
  mycotoxins, 790
  registration of materials for marketing as animal feeds or feed additives, 789–90
Reid vapour pressure (RVP), 410
renewable diesel, 442

biodiesel production methods, 441–65
  advantages and limitations, 461–5
  biodiesel production routes, 444–54
  feedstock quality issues, 458–61
  future trends, 465
  overview, 442
  traditional and emerging feedstocks, 454–8
production routes, 442–3
  decarboxylation and hydrogenation reactions for triolein, 443
Renewable Energy Directive (RED), 393–4, 403
renewable feedstock, 17–28, 379–80
  concept of waste biorefinery, 19
  drivers for change, 17–19
  first to second generation waste re-use, 25–8
  use of food supply chain waste, 20–4
    components of food supply chain waste, 23
    examples of by-products and volumes, 22
    generic illustration, 21
Renewable Fuel Standard (RFS), 48–9, 400–1
renewable identification number (RIN), 50–1
renewable liquid fuels, 390–1
  future provision, 423–9
    biomass limit, 423–4
    renewable fuels within an integrated renewable energy system, 427–9
    Sustainable Organic Fuels for Transport (SOFT), 424–7
renewable portfolio standards (RPS), 51
renewable power methane (RPM), 427–8
Renewable Transport Fuel Obligation (RTFO), 773
reserve osmosis, 352
residual delignification, 666
residues, 47
resin wafer electrodeionisation (RW-EDI)
  removal of ionic impurities, 122–4
    removal of sulfuric acid and acetic acid from corn stover hydrolysate liquor, 124
    schematic of different components inside RW-EDI stack, 123
resins, 674–6
resource, 7

respiratory quotient (RQ), 280
rest streams, 724
Restriction of the Use of Certain Hazardous Substances in Electrical and Electronic Equipment (ROHS), 5, 7
retorts, 532–8
retro-synthesis, 95
reverse osmosis (RO), 115–16, 144–5
reverse transcriptase (RT), 607
Rhodia, 646–7
ricinoleic acid, 588–9
ring drying, 784–5
ring kilns, 536–7
rotary hearth pyrolysis, 539
ruminant consolidated bioprocessing, 243–4

saccharification, 308–9
  *see also* enzymatic hydrolysis
*Saccharomyces cerevisiae*, 273
scanning electron microscopy (SEM), 830
*Schizymenia pacifica*, 607
screw continuous pyrolysis two-stage system, 539
screw pyrolyser system, 538
secondary air injection (SAI), 415–16
selectivity, 146
sensitivity analyses, 62
separate hydrolysis and fermentation (SHF), 208, 241, 279–80
  schematic representation of different fermentation strategies for bioethanol production, 281
  SHF vs SSF, 279
separation efficiency, 146–7
separation technologies, 114–21
  adsorption, 119–20
  current and emerging technologies in biorefining, 112–48
    biofuels recovery by solvent extraction in ionic liquid assisted membrane contactor, 134–41
    glycerin desalting as value added co-product from biodiesel production, 126–8
    impurities removal from lignocellulosic biomass hydrolysate liquor for cellulosic sugars, 121–6
    performance indices, 144–7
    solvent extraction and example of recovery of value added proteins from DSG, 130–4
    succinic acid production, 128–30
  emerging trends for advanced biofuels, 141–4
  simplified schematic of production routes of biofuels and biobased chemicals, 142
  extraction, 120–1
  membrane separation technologies, 114–19
    separation scheme of membrane process and membrane filtration spectrum, 115
sequencing batch reactor (SBR), 483
short-rotation woody crops, 47
silage juice, 351–2
simulated fermentation broth, 139–40
simulated moving bed (SMB) technique, 608
simultaneous extraction, 726–7
simultaneous saccharification and co-fermentation (SSCF), 241
simultaneous saccharification and fermentation (SSF), 208–9, 272, 279–80, 702
size based-membrane separations, 116–17
social life cycle assessment (SLCA), 81–2
soda pulping, 666
softwoods, 660
soils
  production and application of biochar, 525–50
    agricultural usages, 529–31
    appendix of IBI standardised product definition and product testing guidelines, 554–5
    effects of application, 527–9
    future trends, 548–50
    larger-scale commercial production, 538–41
    markets and usages, 544–8
    production, 531–8
    testing properties, 541–2, 544
solid acid catalysts, 153–4
solid acids, 634–5
solid-liquid separations, 207
solid retention time (SRT), 487
solid state fermentation (SSF), 611–12
solubility based-membrane separations, 116–17
solvent-based isolation, 850–1
solvent extraction, 703
  biofuels recovery in ionic liquid assisted membrane contactor, 134–41
  distillation illustration, 135

liquid-liquid extraction illustration, 136
pervaporation illustration, 135
example of recovery of value added proteins from distiller's grains and solubles (DSG), 130–4
  effect of enzyme amylase on enzymatic saccharification of cellulosic materials, 133
  effect of reaction time on enzymatic saccharification of cellulosic materials, 134
  schematic of experimental pathway of simultaneous bio-solvent based extraction, 132
sorbents, 814
soy protein isolate (SPI), 853
soya bean meal, 779–80
soya bean methyl ester (SME), 405
soya protein extraction, 725–6
soybean oil, 454
spark-ignition engines, 407–17
sphingophospholipids (SPLs), 602–3
spirulina, 605
splitting process, 208
spray drying, 785
spraying, 825–6
starch ethanol, 42
starch granules, 852
starch nanocrystals (SNC), 852
starch nanoreinforcements, 852
starches, 831–2
steam explosion (STEX), 269, 489–90, 670, 850–1
steroids, 697
sterol glucosides, 460
stoichiometry, 408–10
strategic environmental assessment (SEA), 70
subfractionation, 787
succinic acid, 637
succinic acid fermentation, 129–30
succinic acid production, 128–30
  integrated fermentation and RW-EDI separations systems, 130
  production ratio of succinic acid vs acetic acid in conventional fermentation, 129
  succinic acid, acetic acid and ethanol in recovery tank of RW-EDI, 131
  succinic acid fermentation using integrated fermentation, 131
  wheat, 312–16

alternative upstream processing strategies, 314
substrate cost vs conversion yield, 315
succinic acid concentration vs yield and productivity in batch fermentations, 317
transformations of succinic acid to added-value chemicals, 312
wheat-based process using wheat and wheat milling by-products, 313
sugar
  acidification, 480
  fermentation, 644
  intermediates, 143–4
  platform, 263
sugar modification, 166–72
  furans platform, 166–8
    reactions involved in formation of HMF, 167
    schematic upgrade of HMK to liquid fuels, 169
  γ-valerolactone platform, 170–2
    schematic illustration, 171
    upgrading to liquid fuels, 172
  levulinic acid platform, 168, 170
    production of fuels, 179
sugarcane, 42
sulfated polysaccharides, 607
sulfite process, 667–8
sulfoquinovosylacylglycerol (SQAG), 603
supercritical fluid extraction (SFE), 598, 704
supercritical processing, 452–3
supercritical water, 726–7
supported liquid membranes (SLM), 116
surface oxidation, 528–9
sustainability, 70
  assessment, 68–70, 81–3
    economic and social assessment of biorefineries, 82–3
    historical average of price of petro- and bio-based chemicals, 83
    LCA concept, 81–2
Sustainable Organic Fuels for Transport (SOFT), 424–7
  fuel synthesis, 425–7
  recycling $CO_2$, 424–5
swept drum pyrolyser, 539
symbol triangle, 95

syngas, 37–8
syngas transformations, 178–81
  Fischer–Tropsch synthesis, 179–81
    cobalt vs iron catalysts, 180
  methanol synthesis, 179
synthetic esters, 562–3
  modification, 581–3
    different polyol head group, 583
    thermal behaviour of glycerol, 2-alkylglycerols and 2-alkylglycerol ester, 582

*Talaromyces emersonii*, 212
tannic acids, 845–6
tannin adhesives, 737–45
  condensed polyflavonoid tannins, 738–41
    commercial industrial particleboard panels after 15 years unprotected exposure, 740
    methylene-ether bridge between flavonoid tannin oligomers, 739
    schematic representative of predominantly $C_4$-$C_8$-linked procyanin tannin, 738
  new technologies for industrial, 741–5
tannin autocondensation hardening, 744–5
  schematic representative of series of reaction occurring in formation of tannin, 744
tannin extracts, 738–9
tannin-hexamethylenetramine adhesives, 742–3
  decomposition and reaction routes when in presence of reactive flavonoid tannin, 743
techno-economic assessments (TEA)
  biochemical and thermochemical biorefineries, 34–63
  biorefinery economic assessment, 36–47
  future trends, 59–63
    combination of TEA and LCA, 60–1
    public support of pilot and demonstration biorefineries, 60
    risk and uncertainty quantification, 61–2
    system optimisation and statistical techniques, 62–3
  market establishment, 53–9
    biomass logistics and transport infrastructure, 54–7
    scale-up of biorefinery operations, 57–9
    US corn ethanol plant unit capital costs vs capacity, 59
    US ethanol biorefinery capacities per year, 54
    US total biomass, 56
  trade of biomass and subsidies, 47–53
    biomass cost estimates by feedstock type, 47–8
    EU member nation subsidy programs, 53
    European Union subsidy programs, 51–2
    federal subsidy programs, 48–51
    state subsidy programs, 51
techno-economic modeling, 221–5
  CTec3 performance in SHF vs SSF and HHF models, 224
  key assumptions of Novozymes techno-economic cost model, 224
technology costs, 390–3
temperate forages, 340
temperature phased systems, 485
TEMPO-mediated oxidation, 850–1
terpenoids, 696, 709–10
  representative structure of examples of monoterpene and sesquiterpene, 696
Tetrapack, 19
thermal combustion, 354
thermal pretreatment, 489
thermally active enzyme cocktails, 219–20
thermo-tolerance, 278
thermochemical biorefineries
  bioproducts, 37–42
    biomass conversion pathway intermediates, upgrading and final products, 37
    capital and operating costs for alternative biorefineries, 41
    capital costs for biomass to fuel conversion via syngas pathway, 39
    operating costs for biomass to fuel conversion via syngas pathway, 40
thermochemical biorefinery, 709
thermochemical processing, 262–3
thermoplastic starch (TPS), 831
thermostabilisation, 219–20
top lit updraft (TLUD), 531–2
torrefaction, 47, 58–9
total reducing sugars (TRS), 633
traditional feedstocks, 454–6
traditional lignin products, 673–7

copolymer and resin applications, 674–6
   structure of lignin and position of formaldehyde addition onto phenol, 675
   high value products, 677
   lignosulfonate, 673–4
   metal ion absorption, 676–7
transesterification, 160–3, 583–5
   process, 366
transformer oils, 573
tree exudates, 835
trends
   biomass depolymerisation, 185
   product upgrading, 185–6
   syngas transformations, 186
tri-flex-fuel vehicles, 420–1
triacylglyceride, 443
   hydrolysis, 453
triacylglycerols (TAG), 566
   functionality, 575–9
      fatty acid composition of high oleic vegetable oils, 579
      mechanism of glycerol β-hydrogen abstraction leading to decomposition products, 576
      molecular origins of unsaturated fatty acids susceptibility to oxidation, 577
*Trichoderma reesei*, 211–12
triglyceride, 694
triglyceride transformations, 160–6
   deoxygenation, 164–6
      reactions involved in triglycerides, 165
   hydrogenation, 163–4
      hydrogenolysis of triglycerides to produce green diesel, 163
      properties of mineral diesel and green diesel, 163
   transesterification, 160–3
      chemical reaction structure, 161
      reaction of triglyceride with methanol, 161
triglycerides, 707–9, 783
   hydrocarbon synthesis from fatty acid or lipid, 708
   representative transesterification reaction, 708
triolein, 442–3
Triton X-100, 116
trypsin inhibitor activity (TIA), 780
tryptophan, 729
turnover frequency (TOF), 164–5

two-stage configuration concept, 485
two-stroke oils, 571–2

ultrafiltration, 115, 145–6, 352
ultrasonic processing, 451–2
ultrasonic scattering, 705–6
ultrasonication, 850–1
ultrasound-assisted transesterification, 451–2
ultraviolet light and ozone (UV-O) treatment, 854
ultraviolet radiation, 815
unburned hydrocarbon emission (uHC), 415–16
United States Pharmacopoeia (USP), 365–6
unknown contaminants, 142
unsaturated oil adhesives, 752–5
   cardanol-based adhesive system, 754
upflow anaerobic sludge bed reactor (UASB), 484
upscaling, 100–1
urea-formaldehyde resin, 812–13
US Department of Energy (DOE), 211–12
US Department of Energy (DoE), 635
US Food and Drug Administration (FDA), 789
US Renewable Fuel Standard, 393–4

valorisation, 323–5
   generic cereal-based waste streams, 325–6
   industrial cereal-based waste and by-product streams, 323–5
      glucose and FAN concentrations in wheat-derived media, 324
vanadium-impregnated zeolite beta, 376–7
vapour permeation (VP), 116
vapour pressure, 410–11
   calculated variation of methanol, ethanol and iso-butanol with alcohol volume fraction, 411
vegetable oil-based transformer oils, 573–4
vegetable oil residues, 725–6
vegetable oils, 452–3, 455, 568, 574
VertecBioSolvents, 133
very high viscosity index (VHVI), 561–2
virtual sensors, 418
viscosity, 567–8
   index, 567–8
   modifiers, 571
volumetric energy density, 408–10
   stoichiometric air-fuel ratio of blends of alcohol-gasoline blends, 409

variation of alcohol-gasoline blends, 408
volumetric ethanol excise tax credit (VEETC), 49

waferboard, 810–11
waste bread, 325–6
waste valorisation, 18–19
waste vegetable oils (WVO), 455–6
water-gas shift (WGS), 180–1
  reactions, 443–4
water management, 143
water recovery, 145
water remediation, 791
water splitting phenomenon, 129–30
water washing, 445–6
wax conversion
  bio-based chemicals from biorefining of lipid conversion and utilisation, 693–715
    future trends, 714–15
    methods of extraction and analysis, 703–6
    sources, 697–702
    types and properties, 694–7
    utilisation, 707–14
wax esters, 694–6, 701
  chemical structure, 696
wax utilisation
  bio-based chemicals from biorefining of lipid and wax conversion, 693–715
    future trends, 714–15
    methods of extraction and analysis, 703–6
    sources, 697–702
    types and properties, 694–7
    utilisation, 707–14
waxes
  sources, 697–702
    bacteria, 700–1
    generated from lignocellulosic materials and other sources, 701–2
  types and properties, 694–7
wet-forming process, 811–12
wet fractionation, 346–7
wet oxidation, 490–1
wet process, 822–3
wheat
  fuel ethanol production, 308–12
  polyhydroxyalkanoate production, 316
  succinic acid production, 312–16

wheat-based biorefineries, 26, 304–7
  composition of major wheat grain functions, 305
  concept illustration, 306
wheat gluten proteins, 840
wheat straw, 25–7
  utilisation, 320–2
    scheme for second generation bioethanol production process, 321
whole-crop biorefinery, 305
whole wood solubilisation, 671–2
wood
  bioadhesives types, processing and properties for fibres, 736–65
  carbohydrate adhesives, 751–2
  future trends, 762–5
  lignin adhesives, 745–9
  mixed tanin-lignin adhesives, 749–50
  protein adhesives, 750–1
  tannin adhesives, 737–45
  unsaturated oil adhesives, 752–5
  wood welding without adhesives, 755–62
  dowel welding, 757–8
wood charcoal, 547
wood fibres, 813
wood grain, 759
wood welding
  without adhesives, 755–62
    systems of frictional wood welding, 758–62
    two types of bondlines and late wood tracheid entangled and direct welding of cell walls, 757
    welded bondline, 756
World Bank, 723–4

X-ray diffraction analysis, 374–5
xanthophylls, 613
xylooligosaccharides, 609
xylose, 642

yeast, 700, 786
  lipids from fungi, 700
    oil content of sample fungi and yeast species, 700
yeast protein concentrate (YPC), 787
yellow grease, 455–6

zein, 839–40

CPSIA information can be obtained
at www.ICGtesting.com
Printed in the USA
BVOW05s0332291216
472102BV00005B/15/P